| AMINO ACIDS AND THEIR SYMBOLS | | GENETIC CODE | | | | | |
|---|---|---|---|---|---|---|---|
| Ala | Alanine | GCA | GCC | GCG | GCU | | |
| Arg | Arginine | AGA | AGG | CGA | CGC | CGG | CGU |
| Asn | Asparagine | AAC | AAU | | | | |
| Asp | Aspartic acid | GAC | GAU | | | | |
| Cys | Cysteine | UGC | UGU | | | | |
| Glu | Glutamic acid | GAA | GAG | | | | |
| Gln | Glutamine | CAA | CAG | | | | |
| Gly | Glycine | GGA | GGC | GGG | GGU | | |
| His | Histidine | CAC | CAU | | | | |
| Ile | Isoleucine | AUA | AUC | AUU | | | |
| Leu | Leucine | UUA | UUG | CUA | CUC | CUG | CUU |
| Lys | Lysine | AAA | AAG | | | | |
| Met | Methionine | AUG | | | | | |
| Phe | Phenylalanine | UUC | UUU | | | | |
| Pro | Proline | CCA | CCC | CCG | CCU | | |
| Ser | Serine | AGC | AGU | UCA | UCC | UCG | UCU |
| Thr | Threonine | ACA | ACC | ACG | ACU | | |
| Trp | Tryptophan | UGG | | | | | |
| Tyr | Tyrosine | UAC | UAU | | | | |
| Val | Valine | GUA | GUC | GUG | GUU | | |

# CONCEPTS OF GENETICS

FOURTH EDITION

# CONCEPTS OF GENETICS

**William S. Klug**
Trenton State College at Hillwood Lakes

**Michael R. Cummings**
University of Illinois, Chicago

MACMILLAN COLLEGE PUBLISHING COMPANY
New York
MAXWELL MACMILLAN CANADA
Toronto
MAXWELL MACMILLAN INTERNATIONAL
New York   Oxford   Singapore   Sydney

Editor: Sheri Walvoord
Production Supervisor: Hudson River Studio
Production Manager: Aliza Greenblatt
Cover Designer: Leslie Baker
Cover Art: Courtesy of Cold Spring Harbor Laboratory. Sculpture by Charles Reina for Cold Spring Harbor Laboratory. Photography by Richard Megna/Fundamental Photographs.
Photo Director: Chris Migdol
Photo Researcher: Yvonne Gerin
Illustrations: Hudson River Studio
This book was set in Garamond Light by York Graphic Services and was printed and bound by R. R. Donnelley & Sons. The cover was printed by Lehigh Press.

**Macmillan Publishing Company**
**113 Sylvan Avenue, Englewood Cliffs, NJ 07632**

Library of Congress Cataloging-in-Publication Data
Klug, William S.
    Concepts of genetics/William S. Klug, Michael R.
Cummings.—4th ed.
    p.      cm.
    Includes bibliographical references and index.
    ISBN 0-02-364801-5
    1. Genetics.  I. Cummings, Michael R.   II. Title.
QH430.K574  1994
575.1—dc20
                                                  94-1547
                                                  CIP

Printing:      2  3  4  5  6  7  8  Year:  4  5  6  7  8  9  0  1  2  3

# DEDICATION

We dedicate this book to teachers of all kinds, and in all disciplines. In particular, we salute those who have influenced our own philosophy of pedagogy during our undergraduate and graduate work, and since then, our faculty colleagues whose teaching strategies and styles have rubbed off on us. Teachers have found a rewarding and honorable calling, and in our case, we feel fortunate to have the discipline of genetics as the cornerstone of our professional efforts.

W. S. K.
M. R. C.
January 1994

WILLIAM S. KLUG is currently Professor and Chairman of the Department of Biology at Trenton State College at Hillwood Lakes, New Jersey. He was previously a member of the faculty of Wabash College.

MICHAEL R. CUMMINGS, currently Associate Professor of Biological Sciences and Research Associate Professor at the Institute for the Study of Developmental Disabilities, University of Illinois, Chicago, has also served on the faculties of Northwestern University and Florida State University.

# PREFACE

*Concepts of Genetics* is now entering its second decade of successfully providing students with textual support as they study one of the most fascinating scientific disciplines. Certainly no subject area has had a more sustained impact on shaping our knowledge of the living condition. As a result of the findings in genetics over the past fifty years, we now understand with reasonable clarity the underlying mechanisms that explain how organisms develop into and then function as adults. We also better understand the basis of biological diversity and have greater insights into the evolutionary processes that generate diversity.

In the past ten years, advances in genetic technology have had a profound effect on our knowledge of human genetics. More than any other event, the launching of the Human Genome Project in 1990 symbolizes the extent of our commitment to the further pursuit of such knowledge. As we near the twenty-first century, an era awaits us where the application of our cumulative knowledge of genetics to the betterment of the human condition will be commonplace. While this era will be filled with the excitement of scientific discovery, many accompanying problems and controversies will also face us. Of these, how we utilize our knowledge of the entire nucleotide sequence of the human genome promises to be the most significant. As a consequence of this growing body of information, many legal and ethical ramifications have already arisen. In the future, the implications of genetic testing and gene therapy will become important societal issues.

As geneticists and students of genetics, the thrill of being part of this era must be balanced by a strong sense of responsibility for providing careful attention to the many related issues that will undoubtedly arise. The formulation of proper laws and policies will depend on the comprehensive knowledge of genetics and measured responses to these issues. As a result, never has there been a higher premium on the need to provide a useful and up-to-date textbook that supports the study of genetics. This has been and continues to be our goal during the preparation of the various editions of *Concepts of Genetics*.

The first edition, published in 1983, was designed around five major premises:

1. Establishing a **conceptual framework** represents the most sound approach to learning, particularly when the material at hand is extensive and complex, as is the case with genetics.

2. Emphasizing the **rich history of scientific discovery and analytical thought**, so prevalent in genetics, to provide a unique opportunity for students to hone their problem-solving abilities, explore how science works, and learn how scientists pursue the art of inquiry.

3. Covering material with a **clean, crisp organization and format**, both within each chapter and throughout the text, facilitating effective learning and use of the book.

4. **Presenting figures that teach rather than simply illustrate** the topic at hand, even for complex, analytical experiments.

5. Providing students with **clearly written, straightforward explanations** that do not oversimplify material or issues.

Creating a text based on these premises and having the opportunity to refine it over four editions has been a labor of love for us, one that has been possible because of the feedback and encouragement provided over the past decade from students, adopters, and reviewers.

# FEATURES OF THE FOURTH EDITION

## New Organization

As the field of genetics has expanded, many topics have taken on greater importance and stature within the discipline. As with each previous edition, this has necessitated a revision of the Table of Contents. This edition continues to reflect our belief that classical studies of heredity, which we refer to as Transmission Genetics, should initiate the text (Part 1). We have followed this with a group of chapters that center around DNA, culminating in the coverage of genetic technology and applications of recombinant DNA research (Part 2). With the physical basis of heredity and genetic variation established in Parts 1 and 2, Part 3 considers how genetic information is expressed and regulated. This section culminates with a chapter that discusses the genetics of cancer, where regulation of cell division has been compromised by mutated genes. Part 4 presents a series of topics in which genetic analysis serves as the cornerstone for each presentation. Current information underlying each of the latter subjects now serves as the basis for emerging subdisciplines within genetics.

While we realize that no Table of Contents in a diverse field such as Genetics can satisfy everyone, the present organization can be used in all first courses in undergraduate genetics, whether offered in a semester, quarter, or two-quarter format. The Parts and Chapters within the text may be used interchangeably, providing flexibility for the instructor.

## Improved Art Program

In the third edition, the entire art program was converted to a full-color format. In the fourth edition, we have had the opportunity to revise and refine every figure. In conjunction with the many color photographs, the text has an attractive and pedagogically superior art program. As in past editions, our emphasis has been on the creation of figures that facilitate learning. In many cases, this has involved the development of "flowchart" figures, presenting experimental approaches that illustrate major concepts.

## Modernization—Recombinant DNA and Human Genetics

One of the major trends in the field of genetics involves the application of recombinant DNA technology to the analysis of numerous biological problems. In this edition, this trend is reflected in a new chapter that considers such applications (Chapter 13), as well as applications to other research topics discussed throughout chapters in Parts 2, 3, and 4. All areas of molecular genetics have been enhanced by the ability to clone and analyze DNA. However, the discipline that has benefitted more than any other is the field of human genetics. The expansion of our knowledge of our own species represents a second major trend in the field of genetics. As a result, we have increased our coverage of human genetic examples throughout the text. In two cases, entirely new chapters ("The Genetics of Cancer" and "The Genetics of Immunity") emphasize human genetics. Other topics discussed throughout the text that are new or greatly expanded in this edition include eugenics and euphenics, genomic imprinting, genetic anticipation, trinucleotide repeat mutations, genomic instability and cancer, gene therapy, genetic disorders and their diagnosis, DNA fingerprinting and forensics, human gene mapping, and the Human Genome Project.

## Added Emphasis on Analysis

In this edition, we introduce a new feature called **Tools and Techniques**. Presented as boxed essays, these discussions provide accounts of how genetic techniques are used by researchers as tools to study biological problems. These essays appear in six chapters, and include topics such as "P Element Transposition," "YACs and Gene Transfer," "Reporter Genes," "Antisense RNA," "Enhancer Trapping," and "Knockout Mice." These are cutting edge techniques that will, in the future, have a significant impact on genetic analysis.

In addition to the Tools and Techniques feature, we have continued to emphasize genetic analysis within each chapter in the section, **Insights and Solutions**. Providing detailed solutions to genetic problems and insights into genetic analysis, this section has been expanded in most chapters and enhances the development of analytical thinking skills by students.

The **Problems and Discussion Questions** section also characterizes our emphasis on analysis, presenting problems at various levels of difficulty. In this edition, we have added even more problems considered to be of "greater difficulty." These are usually found near the end of each Problem and Discussion Question section.

## Other Pedagogic Features

Several features are provided to enhance student learning. Sections entitled **Chapter Concepts** and **Chapter Summary** precede and conclude each chapter. The former underscores our belief that within each chapter, there exists one or a few major conceptual issues that provide the framework for the chapter. The latter section provides a concise summary that may serve as a quick review of the coverage of the most important topics in each chapter. We have added a new section, **Key Terms**, to aid the student in reviewing the material. The major terms used are presented in an alphabetized list at the end of each chapter.

Finally, mention should be made of the extent and nature of the **Selected Readings** section at the end of each chapter. These sections provide comprehensive references to both historical findings and modern discoveries. In addition, many review articles are cited. Our goal is to enhance the reference value of the text and provide students with an entry into the primary and secondary literature related to topics introduced in each chapter.

## Appendices

The appendices include expanded coverage of **Experimental Methods, Solutions to Selected Even-Numbered Problems and Discussion Questions**, and an extensive **Glossary**. All three have been updated.

## Supplement Package

*Student's Handbook: A Guide to Concepts and Problem Solving*   by Harry Nickla of Creighton University reviews vocabulary, concepts, and problem solving in a chapter-by-chapter format. In addition, this handbook provides a detailed step-by-step solution or lengthy discussion for every problem and question in the text, and supplies additional problems for use by students.

*Instructor's Sourcebook*,   also by Harry Nickla, contains questions and problems an instructor can use to prepare exams. The new edition of this sourcebook also provides optional course sequences, a guide to audiovisual supplements, and several "starter references" for term papers and special research projects. The testbank portion of the sourcebook is also available in both IBM and Macintosh formats.

*Overhead Transparencies*.   A set of 125 color acetates of illustrations from the text will be available to adopters of the text.

# ACKNOWLEDGMENTS

No text in its fourth edition can be the sole work of its authors. While we assume complete responsibility for any errors herein, we gratefully acknowledge the advice, contributions, and suggestions made by reviewers of all four editions, and particularly those who were involved in this edition, listed below.

| | |
|---|---|
| Paul E. Bibbens, Jr. | *Kentucky State University* |
| Brian P. Bradley | *University of Maryland at Baltimore* |
| Rebecca V. Ferrell | *Metropolitan State College of Denver* |
| David W. Foltz | *Louisiana State University* |
| Jim Garey | *Duquesne University* |
| Sally Giordano | *Slippery Rock University of Pennsylvania* |
| Stephen L. Goldman | *University of Toledo* |
| Elliott S. Goldstein | *Arizona State University* |
| Mary Lou Guerinot | *Dartmouth College* |
| Richard Halliburton | *Western Connecticut State University* |
| Keith Hartberg | *Baylor University* |
| Richard B. Imberski | *University of Maryland* |
| Charles Jacobs | *Albion College* |
| Gustavo Maroni | *University of North Carolina–Chapel Hill* |
| C. Robertson McClung | *Dartmouth College* |
| Laurens Mets | *University of Chicago* |
| David Nash | *University of Alberta* |
| Craig L. Peebles | *University of Pittsburgh* |
| Ann E. Reynolds | *University of Washington* |
| Mark Sanders | *University of California, Davis* |
| Elizabeth Savage | *University of Alberta* |
| Henry Schaffer | *North Carolina State University* |
| Lillie L. Searles | *University of North Carolina–Chapel Hill* |
| Darrell G. Yardley | *Clemson University* |

One reviewer, Elliott Goldstein, has advised us during the preparation of all four editions. Another, Harry Nickla, has written the *Student's Handbook* and the *Instructor's Sourcebook*. We thank them both for efforts far beyond reasonable expectations. Another colleague, Elizabeth Savage, constructed the boxed essays, **Tools and Techniques**. We are very appreciative of her help and the pleasant manner she displayed throughout their preparation. We also thank David E. Sheppard at the University of Delaware for providing the initial suggestion and ideas regarding these essays.

We also thank our colleagues and our secretarial staff, who together have bolstered our efforts with encouragement, specific discussions, and endless technical support. In particular, Dr. Jim Bricker, Ms. Bette Baier, Mrs. Monica Zrada, and Anthony Maffia III have provided invaluable assistance at Trenton State College. At the University of Illinois, we wish to thank Kanchna Sri for her help in compiling the glossary.

At Macmillan Publishing, we express appreciation for the editorial guidance of Sheri Walvoord, who coordinated the entire project and to Chris Migdol and Yvonne Gerin, who carried out all the photo research. Skillful production of text and art was provided by Ed Burke and his staff at Hudson River Studio, while Dick and Beth Morel at Strong House developed the art for all figures. The latter contributions were particularly significant to both the aesthetics and pedagogy of the art program.

In spite of the intensity and associated pressures of a project of this enormity, interactions with all involved in the production of this edition has been pleasant and enjoyable. All deserve to share in any success this text enjoys. We want them to know that our gratitude goes well beyond the sentiments expressed above.

WILLIAM S. KLUG
MICHAEL R. CUMMINGS
JANUARY 1994

# Brief Contents

# CONTENTS

PART 2

# DNA: THE MOLECULAR BASIS OF HEREDITY    238

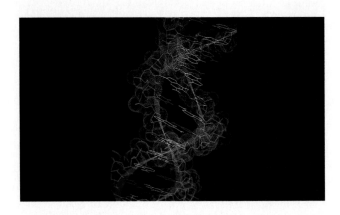

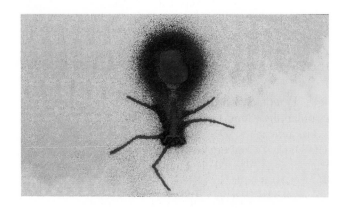
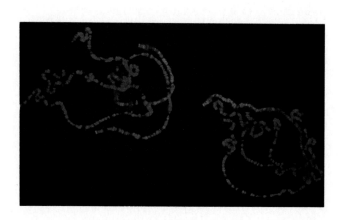

# Concepts of Genetics

# 1

# AN
# INTRODUCTION
# TO GENETICS

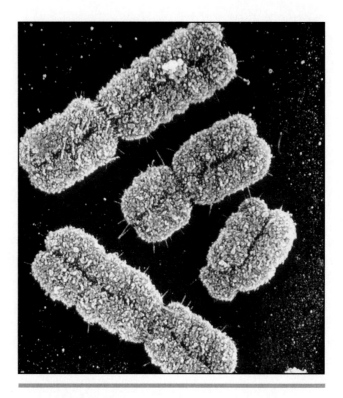

Scanning electron micrograph of human metaphase chromosomes.

*Genetics is the science of heredity. The discipline has a rich history and involves investigations of molecules, cells, organisms, and populations, using many different experimental approaches. Not only does genetic information play a significant role during evolution, but its expression influences the function of individuals at all levels. Thus, genetics unifies the study of biology and has a profound impact on human affairs.*

Welcome to the study of genetics. You are about to explore a subject that many students before you have found to be the most interesting and fascinating in the field of biology. This is not surprising because an understanding of genetic processes is fundamental to the comprehension of life. Genetic information directs cellular function, determines an organism's external appearance, and serves as the link between generations in every species. Knowing how these processes occur is important in understanding the living world. The topics studied in genetics also overlap directly with molecular biology, cell biology, physiology, evolution, ecology, systematics, and behavior. The study of each of these disciplines is incomplete without the knowledge of the genetic component underlying each of them. Genetics thus unifies biology and serves as its "core."

Fascination with this discipline further stems from the fact that, in genetics, many initially vague and abstract concepts have been so thoroughly investigated that subsequently, they have become clearly and definitively understood. As a result, genetics has a rich history that exemplifies the nature of scientific inquiry and the analytical approach used to acquire information. Scientific analysis, moving from the unknown to the known, is one of the major forces that attracts students to biology.

There is still another reason why the study of genetics is so appealing. Every year large numbers of new findings are made. Although it has been said that scientific knowledge doubles every ten years, one estimate holds that the doubling time in genetics is less than five years. Certainly, over the past five decades, no five-year period has passed without new discoveries in genetics causing us to revise our thinking or to extend our knowledge beyond a major frontier. Each advance becomes part of an ever-expanding cornerstone upon which further progress is based. It is exciting to be in the midst of these developments, whether you are studying or teaching genetics.

## THE HISTORICAL CONTEXT OF GENETICS

In the chapters that follow, we will focus on the way in which genetic information is transmitted from generation to generation as well as the way it is stored, expressed, and regulated in the individual organism. The initial basis for such information was provided by Gregor Mendel in the middle of the nineteenth century. His findings, unrecognized for about half a century, were rediscovered at a time when other significant scientific information was becoming available. By the early twentieth century, several related ideas were gaining acceptance that would become cornerstones in the understanding of biology:

1. Matter is composed of atoms;

2. Cells are fundamental units of living organisms;

3. Cells contain nuclei that house threadlike structures called chromosomes;

4. Chromosomes are constant in number and characteristic of each species.

When these ideas were integrated with Mendel's findings and with Charles Darwin's theory of evolution and natural selection, the scene was clearly set for a major breakthrough in our comprehension of the living process in both individuals and populations.

Before embarking further with our discussion of the transmission and expression of genetic information, we will backtrack well before the nineteenth century. We will briefly consider some of the ideas that preceded those of Mendel and Darwin, several of which can be traced back well over 1000 years! As we will see, their influence was still apparent in the nineteenth century!

## Prehistoric Times: Domesticated Animals and Cultivated Plants

We may never know when people first recognized the existence of heredity. However, a variety of archeological evidence (primitive art, preserved bones and skulls, dried seeds, etc.) has provided many insights. Such evidence documents the successful domestication of animals and cultivation of plants thousands of years ago. Such efforts represent artificial selection of genetic variants within populations.

Between 8000 B.C. and 1000 B.C., horses, camels, oxen, and numerous breeds of dogs (derived from the wolf family) were domesticated to serve various roles. Cultivation of plants paralleled these developments. Several plant varieties, including maize, wheat, rice, and the date palm, are thought to have been developed around 5000 B.C. Remains of maize dating to this period have been recovered in caves in the Tehucan Valley of Mexico. Assyrian art depicts artificial pollination of the date palm, which is thought to have originated in Babylonia (Figure 1.1). This deliberate selection of individual variants undoubtedly influenced the types of modern-day palms found in the region. Today, there are over 400 varieties of date palm in just four oases in the Sahara, differing from one another in traits such as fruit taste.

Prehistoric evidence of cultivated plants and domesticated animals documents our ancestors' successful attempts to manipulate the genetic composition of useful species. There is little doubt they learned that desirable and undesirable traits were passed to successive generations and that selection could be performed to obtain

**FIGURE 1.1** Relief carving depicting artificial pollination of date palms during the reign of Assyrian King Assurnasirpal II (883–859 B.C.).

more desirable varieties of animals and plants. Human awareness of heredity seems to have existed even during prehistoric times.

## The Greek Influence: Hippocrates and Aristotle

Although few, if any, ideas were put forward to explain heredity during prehistoric times, considerable attention was directed toward this subject during the Golden Age of Greek culture. This is particularly evident in the writings of the Hippocratic school of medicine (500–400 B.C.) and subsequently of the philosopher and naturalist Aristotle (384–322 B.C.), where he expressed profound interest in the origin of humans (Figure 1.2).

These ancient philosophers directed their attention toward an understanding of the source of the **physical substance** giving rise to an individual and the **generative force** that directs that substance as it materializes (develops) into a whole organism.

For example, the Hippocratic school's treatise *On the Seed* argues that male semen is formed in numerous parts of the body and is transported through blood vessels to the testicles. Active "humors" act as the bearer of hereditary traits and are drawn from various parts of the body to the semen. These humors could be healthy or diseased. Diseased humors account for the appearance of newborns exhibiting congenital disorders or deformities. Furthermore, these humors could be altered in individuals, and in their new form, be passed to their offspring. In this way, newborns could "inherit" traits that their parents had "acquired in their environment."

Aristotle, who had studied under Plato for some 20 years, was more critical and more expansive than the Hippocratic school in his analysis of human heredity. Aristotle proposed that male semen is formed from blood rather than from each organ, and that its generative power resides in a "vital heat" it contains. This vital heat has the capacity to produce offspring of the same "form" (i.e., basic structure and capacities) as the parent. He believed that it generated offspring by cooking and shaping the menstrual blood produced by the female, which was the "matter" for the offspring. The embryo would develop from the initial "setting" of the menstrual blood by the semen into a mature offspring, not because it already contained the parts in miniature (as some Hippocratics had thought), but because of the shaping power of that vital heat. These ideas constitute only one part of the Aristotelian philosophy of order in the living world.

Although to modern geneticists, the ideas of Hippocrates and Aristotle may sound naive, neither sperm nor eggs had yet been observed in any mammal, let alone vided much greater insights into the basis of life. As we shall see, although some historic ideas are now clearly recognized as incorrect, the discussions and debates surrounding them were the beginnings for the coalescence of today's explanations.

In the 1600s, the English anatomist William Harvey (1578–1657), better known for his experiments demonstrating that the blood is pumped by the heart through a circulatory system made up of the arteries and veins, also wrote a treatise on reproduction and development. In it he is credited with the earliest statement of the **theory of epigenesis**—that an organism is derived from substances present in the egg, which are assembled and differentiate during embryonic development. Patterned after Aristotle's ideas, epigenesis holds that new structures, such as body organs, are not present initially, but instead, arise *de novo* during development.

The theory of epigenesis conflicts directly with the **theory of preformationism**, first put forward in the seventeenth century. Preformationists proposed that sex cells contain a complete miniature adult called the **homunculus** (Figure 1.3). These ideas were popular well into the eighteenth century. However, work by the embryologist Casper Wolff (1733–1794) and others clearly disproved this theory, strongly favoring epigenesis. Wolff was quite convinced that several structures, such as the alimentary canal, were not present in the earliest embryos he studied, but instead, were formed later during development.

During this period, there were other significant findings in chemistry and biology. In 1808, John Dalton expounded his **atomic theory**, which states that all matter is composed of small invisible units called atoms. Improved microscopes became available, and around 1830, Matthias Schleiden and Theodor Schwann proposed the **cell theory**. This theory states that all organisms are composed of basic units called cells, which are derived from preexisting cells. By this time, the idea of **spontaneous generation**, the idea that living organisms could spontaneously arise from nonliving components, had clearly been disproved. Thus, all living organisms were considered to be derived from preexisting organisms and to consist of cells made up of atoms.

Another prevailing notion had a major influence on nineteenth-century thinking: the **fixity of species**. According to this doctrine, animal and plant groups remain unchanged from the moment of their initial appearance on earth. Embraced particularly by those also adhering to the religious belief of **special creation**, this doctrine was popularized by several people, including the Swedish physician and plant taxonomist Carolus Linnaeus (1707–1778), who is better known for devising the binomial system of nomenclature.

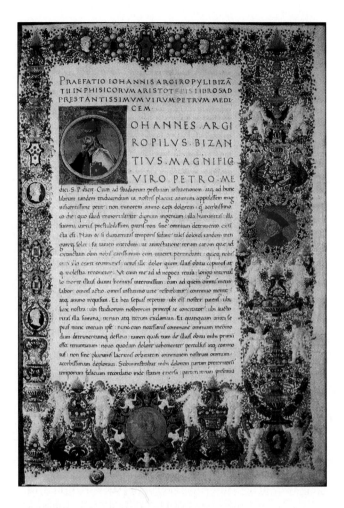

**FIGURE 1.2** Illuminated manuscript page of the preface to the Latin translation of the *Book of Aristotle's Physics* by Johannes Argyropoulos (1416–1486), a Greek scholar.

humans. In fact, eggs were not discovered until 1827! In their own right, these explanations were worthy ones in their time and for centuries to come.

## The Dawn of Modern Biology: 1600–1850

During the ensuing 1900 years (300 B.C.–A.D. 1600), the theoretical understanding of genetics was not extended by significant, new ideas, but interest in applied genetics remained strong. By the Middle Ages, naturalists, well aware of the impact of heredity on organisms they studied, were faced with reconciling their findings with current religious beliefs. The theories of Hippocrates and Aristotle still prevailed and, when applied to humans, they no doubt conflicted with some of the religious doctrines of the time.

Between 1600 and 1900, major strides were made in experimental biology. The resultant knowledge pro-

**FIGURE 1.3**    Depiction of the "homunculus," a sperm containing a miniature adult, perfect in proportion and fully formed.

The influence of these twin tenets is illustrated by considering the work of the German botanist Joseph Gottlieb Kolreuter (1733–1806), who produced findings that were potentially quite far-reaching. In work with tobacco, he crossbred two groups and derived a new hybrid form, which he then converted back to one of the parental types by repeated backcrosses. In other hybridization experiments, using carnations, he clearly observed segregation of traits, which was to become one of Mendel's principles of genetics. These results seemed to contradict the idea of species not changing with time. Because of Kolreuter's belief in both special creation and the fixity of species, he was puzzled about these outcomes, and he failed to recognize the real significance of his own work.

Like Kolreuter, Karl Friedrich Gaertner (1772–1850), experimenting with peas, obtained results similar to those Mendel would later record in 1865. While Mendel's data led him to propose the principles of dominance/recessiveness and segregation, Gaertner did not concentrate on the analysis of individual traits and failed to grasp the significance of his own work.

## Darwin: The Gap in His Theory of Evolution

In 1859, Charles Darwin published the book-length statement of his evolutionary theory, *The Origin of Species*. His many geological, geographical, and biological observations had convinced Darwin that existing species arise by descent with modification from other ancestral species. Greatly influenced by his now famous voyage on the Beagle (1831–1836), Darwin's thinking culminated in his formulation of the **theory of natural selection**, a theory that attempted to explain the causes of evolutionary change. Formulated and proposed at the same time, but independently, by Alfred Russell Wallace, natural selection is based on the observation that populations tend to consist of more offspring than the environment can support, leading to a struggle for existence among organisms in populations. In such a struggle, organisms with heritable traits that better adapt them to their environment are better able to survive and reproduce than those with less-adaptive traits. Over a long period of time, slight, but advantageous, variations will accumulate.

The primary gap in Darwin's theory was a lack of understanding of the genetic basis of variation and inheritance, leaving the theory open to reasonable criticism well into the twentieth century. Aware of this weakness in his theory of evolution, in 1868 Darwin published a second book, *Variations in Animals and Plants under Domestication*, in which he attempted to provide a more definitive explanation of how heritable variation arises gradually over time. Two of his major ideas, pangenesis and the inheritance of acquired characteristics, have their roots in the theories involving "humors," as put forward by Hippocrates and Aristotle.

In his **provisional hypothesis of pangenesis**, Darwin coined the term **gemmules** (rather than humors) to describe the physical units representing the various body parts that he thought were gathered by the blood into the semen. Darwin felt that these gemmules determine the nature or form of each body part. He further believed that gemmules could respond in an adaptive way to an individual's external environment. Once altered, such changes would be passed on to offspring, allowing for the **inheritance of acquired characteristics**. Lamarck had much earlier formalized this idea in his treatise, *Philosophie Zoologique*. Lamarck's theory, which became known as the **doctrine of use and disuse**, proposed that when organisms acquire or lose characteristics, they become heritable.

The ideas expressed in Darwin's 1868 publication were not universally embraced by his colleagues. In 1863, August Weismann, a disciple of Darwin, was to take major issue with the concept of gemmules and the inheritance of acquired characteristics. In his treatise, *The Germplasm; A Theory of Heredity*, Weismann proposed that living organisms consist of two kinds of materials, **somatoplasm** and **germplasm**. The former make up body tissues, constituting the major substance of an individual that undergoes development, growth, and ultimately, death. On the other hand, he envisioned the germplasm as constituting the immortal fragment of an organism that possesses the power of duplication of an individual. According to Weismann, the germplasm provides continuity among succeeding generations of individuals. Inconsistent with the theory of pangenesis, germplasm was not considered by Weismann to be derived from somatoplasm, nor was it formed anew with each individual; rather, it was considered to be a substance providing "a bridge of continuity" between generations.

Since offspring are not derived from somatoplasm, Weismann also rejected the idea of inheritance of acquired characteristics. Weismann's ideas were important ones that placed strong emphasis on germplasm (an hereditary material). His thinking represented a major advance leading to a more modern interpretation of inherited traits early in the twentieth century.

Even though Darwin never understood the basis for inherited variation, his ideas concerning evolution may be the most influential theory ever put forward in the history of Biology. He was able to distill his extensive observations and synthesize his ideas into a cohesive hypothesis describing the origin of the diversity of organisms populating the earth.

**FIGURE 1.4**    Gregor Johann Mendel, who in 1865 put forward the major postulates of transmission genetics as a result of experiments with the garden pea.

## Mendel: An Experimental Biologist

It is against this backdrop that the work of Gregor Johann Mendel (Figure 1.4) performed his work between 1856 and 1863, forming the basis for his classic 1865 paper. In it, Mendel demonstrated for the first time clear quantitative patterns underlying inheritance, and he developed a theory involving hereditary factors in the germ cells that explained these patterns. The strength of Mendel's work is in his straightforward experimental design and in the quantitative analysis upon which he developed his postulates. His research was, however, decades ahead of its time. It was virtually ignored until it was partially duplicated and then cited by Carl Correns, Hugo de Vries, and Eric Von Tschermak in 1900, and championed by William Bateson.

In the interval between 1865 and 1900, it gradually became clear that Weismann's "germplasm" houses the genetic material and that heredity and development are dependent on "information" contained in chromosomes, which are carried by gametes to individual offspring.

As we have seen, a rich history of scientific endeavor and thinking preceded and surrounded Mendel's work. In Chapter 3, we will return to a thorough analysis of his findings, which have served to this day as the foundation of genetics. We have, in this brief history of genetics, attempted to portray the ideas and inquiries that initiated the era of modern biological thought in the twentieth century. The groundwork was clearly in place to appreciate Mendel's work and to meaningfully explore the subject of genetics!

# BASIC CONCEPTS OF GENETICS

We turn now to a review of some of the simple but basic concepts in genetics, which you have undoubtedly already studied. By reviewing them at the outset, we can establish an initial vocabulary and proceed through the text with a common foundation. We shall approach these basic concepts by asking and answering a series of questions. You may wish to write or think through an answer before reading the explanation of each question. Throughout the text, the answers to these questions will be expanded as more detailed information is presented.

### What does "genetics" mean?

**Genetics** is the branch of biology concerned with heredity and variation. This discipline involves the study of cells, individuals, their offspring, and the populations within which organisms live. Geneticists investigate all forms of inherited variation and the nature of the underlying genetic basis of such characteristics.

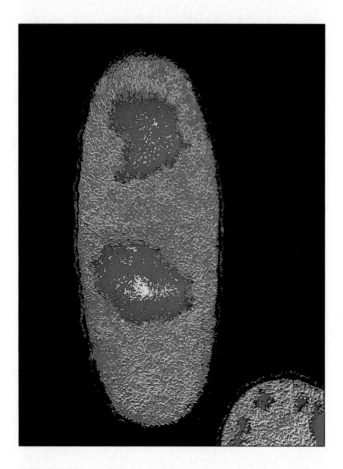

**FIGURE 1.5** Color-enhanced electron micrograph of *E. coli*, demonstrating the nucleoid regions (shown in blue). The bacterium has replicated its DNA and is about to begin cell division.

### What is the center of heredity in a cell?

In eukaryotic organisms, the **nucleus** contains the genetic material. In prokaryotes, such as bacteria, the genetic material exists in an unenclosed but recognizable area of the cell called the **nucleoid region** (Figure 1.5). In viruses, which are not true cells, the genetic material is ensheathed in the protein coat referred to as the viral head or capsid.

### What is the genetic material?

In eukaryotes and prokaryotes, **DNA** serves as the molecule storing genetic information. In viruses, either DNA or **RNA** serves this function.

### What do DNA and RNA stand for?

DNA and RNA are abbreviations for **deoxyribonucleic acid** and **ribonucleic acid**, respectively. These are the two types of nucleic acids found in organisms. Nucleic acids, along with carbohydrates, lipids, and proteins, compose the four major classes of organic biomolecules found in living things.

### How is DNA organized to serve as the genetic material?

DNA, although single stranded in a few viruses, is usually a double-stranded molecule organized as a double helix. Contained within each DNA molecule are hereditary units called **genes**, which are part of a larger element, the **chromosome**.

### What is a gene?

In simplest terms, the gene is the functional unit of heredity. In chemical terms, it is a linear array of nucleotides—the chemical building blocks of DNA and RNA. A more conceptual approach is to consider it as an informational storage unit capable of undergoing replication, mutation, and expression. As investigations have progressed, the gene has been found to be a very complex genetic element.

### What is a chromosome?

In viruses and bacteria, the chromosome (Figure 1.6) is most simply thought of as a long, usually circular DNA molecule organized into genes. In eukaryotes, the chromosome is more complex. It is composed of a linear DNA molecule associated with proteins. In addition to an abundance of genes, the chromosome contains many nongenic regions. It is not yet clear what role, if any, is played by some of these regions. Our knowledge of the chromosome, like that of the gene, is continually expanding.

### When and how can chromosomes be visualized?

If the chromosomes are released from the viral head or the bacterial cell, they can be visualized under the electron microscope (Figure 1.6). In eukaryotes, chromosomes are most easily visualized under the light microscope when they are undergoing **mitosis** or **meiosis**. In these division processes, the chromosomes are tightly coiled and condensed. Following division, they uncoil and exist as **chromatin fibers** during interphase, when they can be studied under the electron microscope.

### How many chromosomes does an organism have?

Although there are many exceptions, members of most species have a specific number of chromosomes present in each somatic cell called the **diploid number (2n)**. Upon close analysis, these chromosomes are found to occur in pairs, each member of which shares a nearly identical appearance when visible during cell division. Called **homologous chromosomes**, the members of each pair are identical in their length and in the location of the **centromere**, the point of spindle fiber attachment during division. They also contain the same sequence of gene sites, or **loci**, and pair with one another during meiosis. Thus, the number of different *types* of chromosomes in any diploid species is equal to half the diploid number and is called the **haploid (n)** number. Some organisms, such as yeast, are haploid and contain only one "set" of chromosomes. Other organisms, notably plants, are sometimes characterized by more than two sets of chromosomes and are said to be **polyploid**.

### What is accomplished during the processes of mitosis and meiosis?

**Mitosis** is the process by which the genetic material of eukaryotic cells is duplicated and distributed during cell division. **Meiosis** is the process whereby cell division produces gametes in animals and spores in most plants. While mitosis occurs in somatic tissue and yields two progeny cells with an amount of genetic material identical to the progenitor cell, meiosis creates cells with precisely one-half of the genetic material. Each gamete receives one member of each homologous pair of chromosomes, and is haploid. This accomplishment is essential if offspring arising from two gametes are to maintain a constant number of chromosomes characteristic of their parents and other members of the species.

### What are the sources of genetic variation?

Classically, there are two sources of genetic variation: **chromosomal mutations** and **gene mutations**. The former, also called **chromosomal aberrations**, includes duplication, deletion, or rearrangement of chromosome segments. Gene mutations result from a change in the stored chemical information in DNA. Such a change may include substitution, duplication, or deletion of nucleotides, which compose this chemical information. Alternative forms of the gene, which result from mutation, are called **alleles**. Genetic variation frequently, but not always, results in a change in some

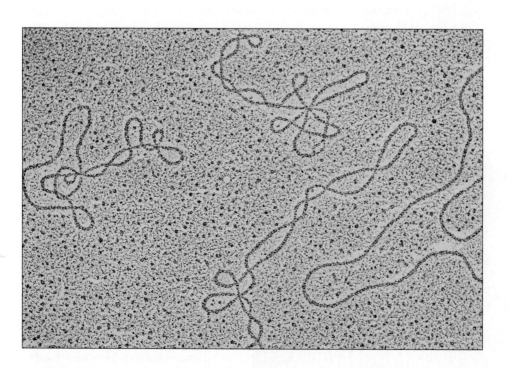

**FIGURE 1.6**    The DNA constituting the chromosome of a bacterial virus (a bacteriophage) viewed under the electron microscope.

characteristic of an organism. Once part of an organism's genetic repertoire, such variation may be dispersed into the population through various reproductive mechanisms (Figure 1.7).

### How does DNA store genetic information?

The sequence of nucleotides in a segment of DNA constituting a gene is present in the form of a **genetic code**. This code specifies the chemical nature (the amino acid composition) of proteins, which are the end product of genetic expression. Mutations are produced when the sequence of nucleotides is altered.

### How is the genetic code organized?

There are four different nucleotides in DNA, each varying in one of its components, the **nitrogenous base**. The genetic code is a triplet; therefore, each combination of three nucleotides constitutes a code word. Almost all possible triplet codes specify one of twenty **amino acids**, the chemical building blocks of proteins.

### How is the genetic code expressed?

The coded information in DNA is first transferred during a process called **transcription** into a **messenger RNA (mRNA)** molecule. The mRNA subsequently associates with the cellular organelle, the **ribosome**, where it is **translated** into a protein molecule.

### Are there exceptions where proteins are not the end product of a gene?

Yes. For example, genes coding for **ribosomal RNA (rRNA)**, which is part of the ribosome, and for **transfer RNA (tRNA)**, which is involved in the translation process, are transcribed but not translated. Therefore, RNA is sometimes the end product of stored genetic information.

### Why are proteins so important to living organisms that they serve as the end product of the vast majority of genes?

Many proteins serve as highly specific biological catalysts, or **enzymes**. In this role, these proteins control cellular metabolism, determining which carbohydrates, lipids, nucleic acids, and other proteins are present in the cell. Many other proteins perform nonenzymatic roles. For example, hemoglobin transports oxygen, collagen provides structural support and flexibility in many tissues, immunoglobulins provide the basis for the immune response, and insulin is a hormone.

### Why are enzymes necessary in living organisms?

As biological catalysts, enzymes lower the activation energy required for most biochemical reactions and speed the attainment of equilibrium. Otherwise, these reactions would proceed so slowly as to be ineffectual in organisms living under the conditions on earth. Thus, some genes control the variety of enzymes present in any cell type, which in turn dictates its overall biochemical composition.

## INVESTIGATIVE APPROACHES IN GENETICS

The scope of topics encompassed in the field of genetics is enormous. Studies have involved viruses, bacteria, and a wide variety of plants and animals and have spanned all levels of biological organization, from molecules to populations. It is helpful, before we embark on a detailed study of genetics, to know the types of investiga-

**FIGURE 1.7**    A bumblebee pollinating a flower, achieving cross-fertilization and enhancing genetic variability within the species.

tions that have been used most often in this field. Although some overlap exists, most have used one of four basic approaches.

The most classical investigative approach is the study of **transmission genetics**, in which the patterns of inheritance of traits are examined. Experiments are designed so that the transmission of traits from parents to offspring can be analyzed through several generations. Patterns of inheritance are sought that will provide insights into more general genetic principles. The first significant experimentation of this kind to have a major impact on understanding heredity was performed by Gregor Mendel in the middle of the nineteenth century. Information derived from his work serves today as the foundation of transmission genetics. In human studies, where designed matings are neither possible nor desirable, **pedigree analysis** is used. In pedigree analysis, patterns of inheritance are traced through as many generations as possible, leading to inferences concerning the mode of inheritance of the trait under investigation.

The second approach involves **cytological studies** of the genetic material. The earliest such studies used the light microscope. The initial discovery in the early twentieth century of chromosome behavior during mitosis and meiosis was a critical event in the history of genetics. In addition to playing an important role in the rediscovery and acceptance of Mendelian principles, these observations served as the basis of the **chromosomal theory of inheritance**. This theory, which viewed the chromosome as the carrier of genes and the functional unit of transmission of genetic information, was the cornerstone for further studies in genetics throughout the first half of this century.

The light microscope continues to be an important research tool. It is useful in the investigation of chromosome structure and abnormalities and is instrumental in preparing **karyotypes**, which illustrate the chromosomes characteristic of any species arranged in a standard sequence.

With the advent of electron microscopy, the repertoire of investigative approaches in genetics has grown. In high-resolution microscopy genetic molecules and their behavior during gene expression can be directly visualized.

The third general approach, **molecular and biochemical analysis**, has had the greatest impact on the recent growth of genetic knowledge. Molecular studies beginning in the early 1940s have consistently expanded our knowledge of the role of genetics in life processes. Although experimental sources were initially bacteria and the viruses that invade them, extensive information is now available concerning the nature, expression, replication, and regulation of the genetic information in eukaryotes as well. The precise nucleotide sequence has

**FIGURE 1.8**  Visualization of DNA fragments under ultraviolet light. The bands were produced using recombinant DNA technology.

been determined for genes cloned in the laboratory. **Recombinant DNA studies** (Figure 1.8), where genes from any organism are literally spliced into bacterial or viral DNA and cloned *en masse*, is the most significant and far-reaching research technology used in molecular genetic investigations. As a result, it is now possible to probe gene structure and function with a resolution heretofore impossible. Such molecular and biochemical analysis has had profound implications in medicine and agriculture.

The final approach involves the study of the **genetic structure of populations**. In these investigations scientists attempt to define how and why certain genetic variation is preserved in populations, while other variation diminishes or is lost with time. Such information is critical to the understanding of the evolutionary process. Population genetics also allows us to predict gene frequencies in future generations.

Together these approaches have transformed a subject poorly understood in 1900 into one of the most advanced disciplines in biology today. As such, the impact of genetics on society has been immense. We shall discuss many examples of the applications of genetics in the following section and throughout the text.

## GENETICS AND SOCIETY

Scientific information has the potential for wide-scale application in ways that improve society at large. The discipline of genetics has remained at the forefront of scientific discovery from the time of the reexamination of

Mendel's work early in the twentieth century. Since that time, the resultant findings have had a growing influence on society. In the 1990s, it is unusual to open a newspaper or magazine that doesn't make reference to some type of application of genetics for the improvement of human existence.

In this section, we provide a brief overview of the impact of genetics on society. As we shall see, the influence has not always been as positive as we perceive it today. We begin with two extremely negative cases.

## Eugenics: The Misguided Application of Science

There is always the danger that scientific findings will be used to formulate policies and/or actions that are unjust or even tragic. This section reviews such a case, which began near the end of the nineteenth century. At that time, Darwin's theory of natural selection provided a major influence on some people's thinking concerning the human condition. Our story recounts the initial attempt to directly apply genetic knowledge for the improvement of human existence. Championed in England by Francis Galton, the general approach is called **eugenics**, a term Galton coined in 1883.

Francis Galton, a cousin of Charles Darwin, believed that many human characteristics were inherited and subject to artificial selection if human matings could be controlled. *Positive* eugenics encouraged parents displaying favorable characteristics to have large families. Superior intelligence, intellectual achievement, and artistic talent are examples. *Negative* eugenics attempted to restrict reproduction for parents displaying unfavorable characteristics. Low intelligence, mental retardation, and criminal behavior are examples.

In the United States, the eugenics movement was a significant social force and led to state and federal laws that required sterilization of those considered "genetically inferior." Over half of the states passed such laws, commencing in 1907 with Indiana. Sterilization was mandated for "imbeciles, idiots, convicted rapists, and habitual criminals." By 1931, involuntary sterilization also applied to "sexual perverts, drug fiends, drunkards, and epileptics." Immigration to the United States from certain areas of Europe and from Asia was also restricted, to prevent the influx of what were regarded as genetically inferior people.

In addition to the violation of individual human rights, such policies were seriously flawed by an inadequate understanding of the genetic basis of various characteristics. Formulation of eugenic policies was premised on the mistaken notion that "superior" and "inferior" traits are totally under genetic control and that offspring of parents displaying these traits will also exhibit them. The potential impact of the environment was excluded as eugenic policies were developed.

In Nazi Germany in the 1930s, the concept of achieving a superior, racially pure group was an extension of the eugenics movement. Initially applied to those considered socially and physically defective, the underlying rationale of negative eugenics was soon to be applied to entire ethnic groups, including Jews and Gypsies. Fueled by various forms of racial prejudice, Adolf Hitler and the Nazi regime took eugenics to its extreme by instituting policies aimed at the extinction of these "impure" human populations. This deplorable disregard for human life was preceded by incremental policies involving forced sterilization and mercy killings. This movement, based on scientifically invalid premises, soon led to mass murder.

Even before the Nazi party came to power in 1933, English and American geneticists began separating themselves from the eugenics movement. They were concerned about the validity of the premises underlying the movement and the evidence in support of these premises. Thus, many geneticists chose not to study human genetics for fear of being grouped with those who supported eugenics.

However, since the end of World War II, tremendous strides have been made in human genetics research. Today, a new term, **euphenics**, has replaced eugenics. Euphenics refers to medical and/or genetic intervention designed to reduce the impact of defective genotypes on individuals. The use of insulin by diabetics and the dietary control of newborn phenylketonurics are longstanding examples. Today, "genetic surgery" to replace defective genes rests clearly on the horizon. Furthermore, social policies now have a solid genetic foundation upon which they may be based. Nevertheless, caution is still required to ensure that our expanded knowledge of human genetics does not obscure the role played by the environment in determining an individual's phenotype.

## Soviet Science: The Lysenko Affair

One of the most famous examples of an interaction between science and politics occurred in the former Soviet Union. This instance happened to involve genetics. In particular, it had to do with the genetic and evolutionary theory of the inheritance of acquired characteristics. As we know from our earlier discussion, the theory originated with the ancient Greeks, was formalized by Lamarck in 1809, and considered important by Darwin. However, experimental evidence early in the twentieth century argued against it. Nevertheless, in the 1930s,

Trofim Denisovich Lysenko, a young plant breeder from the Ukraine, espoused the idea that the efficiency of plant development (and thus agricultural productivity) could be improved by manipulating environmental conditions. Lysenko believed that such improvements would be incorporated into the genetic material and thus passed on to future generations of plants.

The political climate in the Soviet Union at that time provided a hospitable environment for the revival of Lamarck's ideas. First, the theory that environmental change could produce permanent genetic change was compatible with and paralleled the Marxist political thesis that the proper social conditions would induce permanent changes in human behavior. Second, Ivan Pavlov, the Soviet scientist known for his work on the learned or conditioned reflex, emphasized the importance of environmental stimuli. Pavlov claimed, although he later retreated from this statement, to have found an example of a conditioned reflex that was inherited in mice.

Perhaps most significant in the acceptance of Lysenko's ideas was the poor condition of Soviet agriculture during this period. The Soviet government was desperate to improve agricultural production, since the traditional methods of selective breeding were not perceived as producing adequate harvests.

Facing a difficult situation, Stalin was impressed with Lysenko's ideas. Lysenko proposed that germination of the winter wheat crop could be speeded up by **vernalization**, the practice of subjecting plants to an artificial cold period to shorten the dormant period of the seeds, which are usually shed in autumn. The early shedding of the seeds permitted the planting of an additional crop that could be harvested before autumn. Lysenko claimed that because the changes induced by vernalization were permanent ones, this technique need be applied only once.

In fact, this practice did not lead to a permanent increase in agricultural output, but by the time this became apparent, Lysenko had become director of the Institute of Genetics of the USSR Academy of Science. Appointed to this position in 1940, he managed to suppress throughout the Soviet Union all work in genetics that ran contrary to his own views. He was responsible for destroying research records, laboratory supplies, and experimental material of those who opposed him, and in some cases, he had his opponents arrested and sentenced to prison.

One of Russia's leading geneticists, Nikolai Ivanovich Vavilov was a favorite target of Lysenko. Director of the Lenin Academy of Agriculture, Vavilov was one of the world's experts on wheat. He was persecuted by Lysenko and eventually sentenced to prison for agricul-

tural espionage in 1941. He died from malnutrition in 1943 (see the Popovsky reference in Selected Readings at the end of the chapter).

During this time great advances were made in genetics, particularly in the United States. Scientifically based selective breeding programs had developed new varieties of hybrid grains whose yields were much greater than those of the old varieties. Yet none of this knowledge was made available to Soviet geneticists until 1964, when Lysenko finally fell from power and lost his stranglehold on Soviet genetics. One American botanist calculated that USSR corn production would have been increased by six million tons between 1947 and 1957 if only half the country's acreage had been planted with the new hybrid strains.

Unfortunately, Soviet genetics was held back for nearly a generation at a time when numerous significant advances were being made elsewhere in the world. Adherence to theories that were politically rather than scientifically acceptable was indeed costly to Soviet society. Even though the regions occupied by the former Soviet Union still experience agricultural shortfalls in the 1990s, the application of modern plant breeding information has significantly improved agricultural productivity.

**FIGURE 1.9**    Genetically improved tomatoes.

**FIGURE 1.10**    *Triticale*, a hybrid grain derived from wheat and rye, produced as a result of applied genetic research.

## Genetic Advances in Agriculture and Medicine

The major benefits to society as a result of genetic study have been in the areas of agriculture and medicine. Although cultivation of plants and domestication of animals had begun long before, the rediscovery of Mendel's work in the early twentieth century spurred scientists to apply genetic principles to these human endeavors. The Lysenko era aside, use of selective breeding and hybridization techniques has had a most significant impact in agriculture.

Plants have been improved in four major ways: (1) more efficient energy utilization during photosynthesis, resulting in more vigorous growth and increased yields (Figure 1.9); (2) increased resistance to natural predators and pests, including insects and disease-causing microorganisms; (3) production of hybrids exhibiting a combination of superior traits derived from two different strains or even two different species (Figure 1.10); and (4) selection of genetic variants with increased protein value or an increased content of limiting amino acids, which are essential in the human diet.

These improvements have resulted in a tremendous increase in yield and nutrient value in such crops as barley, beans, corn, oats, rice, rye, and wheat. It is estimated that in the United States the use of improved genetic strains has led to a threefold increase in crop yield per acre. In Mexico, where corn is the staple crop, a significant increase in protein content and yield has occurred. A substantial effort has also been made to improve the growth of Mexican wheat. Led by Norman Borlaug, a

**FIGURE 1.11**    The effects of breeding and selection, as illustrated by the production of this Vietnamese pot-bellied pig.

team of researchers was able to develop a strain of wheat that incorporated favorable genes from other strains found in various parts of the world, creating a superior variety that is now grown in many underdeveloped countries besides Mexico. Because of this effort, which led to the well-publicized "Green Revolution," Borlaug received the Nobel Peace Prize in 1970. There is little question that this application of genetics has contributed to the well-being of our own species by improving the quality of nutrition worldwide.

Applied research in genetics has also developed superior breeds of animals. Enormous increases in usable meat supplies produced per unit of food intake have occurred. For example, selective breeding has produced chickens that grow faster, produce more high-quality meat per chicken, and lay greater numbers of larger eggs. In larger animals, including pigs and cows, the use of artificial insemination has been particularly important (Figure 1.11). Sperm samples derived from a single male with superior genetic traits may now be used to fertilize thousands of females located in all parts of the world. Inherited animal disorders are also amenable to modern genetic analysis, creating the potential for their elimination through selective breeding.

Equivalent strides have been made in medicine as a result of advances in genetics, particularly since 1950. Numerous disorders in humans have been discovered to result from either a single mutation or a specific chromosomal abnormality. For example, the genetic basis of sickle-cell anemia, erythroblastosis fetalis, cystic fibrosis, hemophilia, Huntington disease, muscular dystrophy,

Tay-Sachs disease, Down syndrome, and countless metabolic disorders is now well documented and understood at the molecular level. It is estimated that more than 10 million children or adults in the United States suffer from some form of genetic affliction.

Recognition of the genetic basis of these disorders has provided direction for the development of detection and treatment. **Genetic counseling** provides parents with objective information upon which they can base rational decisions about parenting. It is estimated that every childbearing couple stands an approximately 3 percent risk of having a child with some form of genetic anomaly. There are currently thousands of documented genetic disorders.

Applied research in genetics has provided other medical benefits. Increased knowledge in **immunogenetics** has made possible compatible blood transfusions as well as organ transplants. The discovery of genetically determined, tissuebound antigens has led to the important concepts of **histocompatibility** and tissue typing. In conjunction with immunosuppressive drugs, transplant operations involving human organs, including the heart, liver, pancreas, and kidney, are increasing annually and are now considered routine surgery.

Most recent advances in human genetics have been dependent on the application of **recombinant DNA technology**. Recombinant DNA techniques play an increasing and essential role in human genetic engineering, which involves the direct manipulation of the genetic material. Human genetic engineering (or euphenics), currently in its infancy, is being used to alter the genetic constitution of individuals harboring genetic defects and to correct such defects in the developing fetus. Although such processes present ethical questions, the ability to correct serious genetic errors in members of our species ensures that such approaches will become commonplace in the near future.

This technology, built upon a foundation of gene cloning, also serves as the basis for the genetic engineering of other plants and animals. Genes cloned in bacteria may be subsequently inserted into organisms of interest. Favorable genes from any source may thus be shared by many different species. Cloned human genes that code for medically important molecules, such as insulin and interferon, serve as the source of mass production of many essential molecules.

In later chapters, these and other examples in agriculture and medicine are discussed in great detail. Although other scientific disciplines are also expanding in knowledge, none have paralleled the growth of information that is occurring annually in genetics. By the end of this course, we are confident you will agree that the present truly represents the "Age of Genetics."

## CHAPTER SUMMARY

1. The history of genetics, which emerged as a fundamental discipline of biology early in the twentieth century, dates back to prehistoric times.

2. Numerous concepts and a basic vocabulary are essential to the study of genetics.

3. Four investigative approaches are most often used in the study of genetics, including transmission genetic studies, cytogenetic analyses, molecular–biochemical experimentation, and inquiries into the genetic structure of populations.

4. Eugenics, the application of knowledge of genetics to the improvement of human existence, has a long and controversial history. Euphenics, genetic intervention designed to reduce or ameliorate the impact on individuals of defective genotypes, represents the modern approach.

5. Recombinant DNA technology has greatly expanded our research capability. It also has had a profound impact in elucidating the basis of inherited diseases and has made possible the mass production of medically important gene products.

6. Curtailment of the freedom of scientific inquiry had an extremely adverse effect in the USSR under the reign of Stalin, who supported the views and practices of Lysenko.

## KEY TERMS

allele
amino acid
atomic theory
biochemical analysis
cell theory
centromere
chromatin
chromosomal mutation
 (aberration)
chromosomal theory of
 inheritance
chromosome
cytological study
diploid number (2*n*)
DNA (deoxyribonucleic
 acid)
doctrine of use and
 disuse

enzyme
epigenesis
eugenics
euphenics
fixity of species
gemmules
gene
gene mutation
genetic code
genetic counseling
genetics
germplasm
haploid number (*n*)
histocompatibility
homologous
 chromosomes
homunculus

immunogenetics
karyotype
locus
meiosis
mitosis
molecular analysis
mRNA (messenger RNA)
natural selection
nitrogenous base
nucleoid region
nucleus
pangenesis
pedigree analysis
polyploid

population genetics
preformationism
recombinant DNA
ribosome
RNA (ribonucleic acid)
rRNA (ribosomal RNA)
somatoplasm
special creation
spontaneous generation
transcription
translation
transmission genetics
tRNA (transfer RNA)
vernalization

## PROBLEMS AND DISCUSSION QUESTIONS

1. Describe the ideas of Hippocrates and Aristotle related to the genetic basis of life.

2. Define and contrast epigenesis and preformationism.

3. Describe Darwin's and Wallace's theory of natural selection. What information was lacking from it (i.e., what gap remained in it)?

4. Describe Darwin's proposal that attempted to bridge the "gap" in his theory of natural selection.

5. Contrast chromosomes and genes and describe their role in heredity.

**6.** Describe the four major investigative approaches used in studying genetics.

**7.** Contrast eugenics with euphenics.

**8.** Norman Borlaug received the Nobel Peace Prize for his work in genetics. Why do you think he was awarded this prize?

**9.** How has genetic research been applied to the field of medicine?

**10.** Describe Lysenko's views of plant breeding. What earlier theory did he embrace?

## SELECTED READINGS

ANDERSON, W. F., and DIRCUMAKOS, E. G. 1981. Genetic engineering in mammalian cells. *Scient. Amer.* (July) 245:106–21.

BORLAUG, N. E. 1983. Contributions of conventional plant breeding to food production. *Science* 219:689–93.

BOWLER, P. J. 1989. *The Mendelian revolution: The emergence of hereditarian concepts in modern science and society.* London: Athlone.

COCKING, E. C., DAVEY, M. R., PENTAL, D., and POWER, J. B. 1981. Aspects of plant genetic manipulation. *Nature* 293:265–70.

DAY, P. R. 1977. Plant genetics: Increasing crop yield. *Science* 197:1334–39.

DUNN, L. C. 1965. *A short history of genetics.* New York: McGraw-Hill.

GARDNER, E. J. 1972. *History of biology.* 3rd ed. New York: Macmillan.

GARFIELD, E. 1981. Medical genetics: The new preventive medicine. *Current Contents—Life Sciences*, vol. 24, no. 36, pp. 5–20.

GARVER, K. L., and GARVER, B. 1991. Eugenics: Past, present, and future. *Am. J. Hum. Genet.* 49:1109–18.

HOPWOOD, D. A. 1981. The genetic programming of industrial microorganisms. *Scient. Amer.* (Sept.) 245:91–102.

HORGAN, J. 1993. Eugenics revisited. *Scient. Amer.* (June) 268:123–31.

HUTTON, R. 1978. *Bio-revolution: DNA and the ethics of man-made life.* New York: New American Library.

JORAVSKY, D. 1970. *The Lysenko affair.* Cambridge: Harvard Univ. Press.

KING, R. C., and STANSFIELD, W. D. 1990. *A dictionary of genetics.* 4th ed. New York: Oxford University Press.

MEDVEDEV, Z. A. 1969. *The rise and fall of T. D. Lysenko.* New York: Columbia Univ. Press.

MOORE, J. A. 1985. Science as a way of knowing—Genetics. *Amer. Zool.* 25:1–165.

OLBY, R. C. 1966. *Origins of Mendelism.* London: Constable.

POPOVSKY, M. 1984. *The Vavilov affair.* Hamden, CT: Archon.

SOYFER, V. N. 1989. New light on the Lysenko era. *Nature* 339:415–20.

STUBBE, H. 1972. *History of genetics: From prehistoric times to the rediscovery of Mendel.* (Translated by T. R. W. Waters.) Cambridge, MA: MIT Press.

TORREY, J. G. 1985. The development of plant biotechnology. *Amer. Scient.* 73:354–63.

WEINBERG, R. A. 1985. The molecules of life. *Scient. Amer.* (Oct.) 253:48–57.

The fruit fly, *Drosophila melanogaster.*

# PART **1**

## HEREDITY AND THE PHENOTYPE

# 2

# CELL DIVISION AND CHROMOSOMES

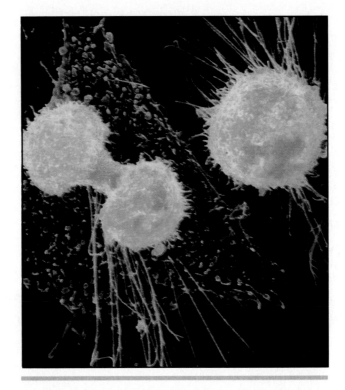

Scanning electron micrograph of human ovarian cells undergoing cell division.

*Genetic continuity between cells and organisms of sexually reproducing species is maintained by the process of mitosis and meiosis. These processes are orderly and efficient, serving to produce diploid somatic cells and haploid gametes, respectively. During these division stages, the genetic material is condensed into discrete, visible structures called chromosomes. In nondividing cells, chromosomes are unfolded and uncoiled, forming chromatin.*

In every living thing there exists a substance referred to as the **genetic material**. Except in certain viruses, this material is composed of the nucleic acid, DNA. A molecule of DNA is organized into units called **genes**, which direct all metabolic activities of cells. Genes are organized into **chromosomes**, structures that serve as the vehicle for transmission of genetic information. In this chapter, we consider the topic of genetic continuity between cells and organisms. The manner in which chromosomes are transmitted from one generation of cells to the next, and from organisms to their descendants, is exceedingly precise.

Two major processes are involved in eukaryotes: **mitosis** and **meiosis**. Although the mechanisms of the two processes are similar in many ways, the outcomes are quite different. Mitosis leads to the production of two cells with an identical number of chromosomes. Meiosis, on the other hand, reduces the amount of genetic material and the number of chromosomes by precisely half. This reduction is essential if sexual reproduction is to occur without doubling the amount of genetic material at each generation. Strictly speaking, mitosis is that portion of the cell cycle during which the hereditary components are precisely and equally divided into daughter cells. Meiosis is part of a special type of cell division leading to the production of sex cells: gametes and spores. This process is an essential step in the transmission of genetic information from an organism to its offspring.

In most cases, chromosomes are visible only when cells are actually dividing, that is, during mitosis or meiosis. When cells are not undergoing division, the genetic material making up chromosomes unfolds and uncoils into a diffuse network within the nucleus generally referred to as **chromatin**. In this chapter, we will examine this transition and also consider two specialized cases where chromosomes are extraordinarily large and amenable to investigation. The study of these examples, **polytene chromosomes** and **lampbrush chromosomes**, has greatly extended our knowledge of genetic organization and its relationship to genetic function.

## CELL STRUCTURE

Before describing mitosis and meiosis, we will briefly review the structure of cells. As we shall see, many components, such as the nucleolus, ribosome, and centriole, are involved directly or indirectly with genetic processes. Other components, the mitochondria and chloroplasts, contain their own unique genetic information. It is also useful for us to compare the structural differences of the prokaryotic bacterial cell with the eukaryotic cell. Variation in cell structure is dependent on the specific genetic expression by each cell type.

Before 1940, knowledge of cell structure was based on information obtained with the light microscope. Around 1940 the transmission electron microscope was in its early stages of development, and by 1950 many details of cell ultrastructure were unveiled. Under the electron microscope cells were seen as highly organized, precise structures. A new world of whorling membranes, miniature organelles, microtubules, granules, and filaments was revealed. These discoveries revolutionized thinking in the entire field of biology. We will be concerned with those aspects of cell structure relating to genetic study. Many of the parts of the cell are described in Figure 2.1, which depicts a typical animal cell.

### Cell Boundaries

The cell is surrounded by a **plasma membrane**, an outer covering that defines the cell boundary and delimits the cell from its immediate external environment. This membrane is not passive, but instead it controls the movement of material such as gases, nutrients, and waste products into and out of the cell. In addition to this membrane, plant cells have an outer covering called the **cell wall**. One of the major components of this rigid structure is a polysaccharide called **cellulose**. Bacterial cells also have a cell wall, but its chemical composition is quite different from that of the plant cell wall. The major component in bacteria is a complex macromolecule

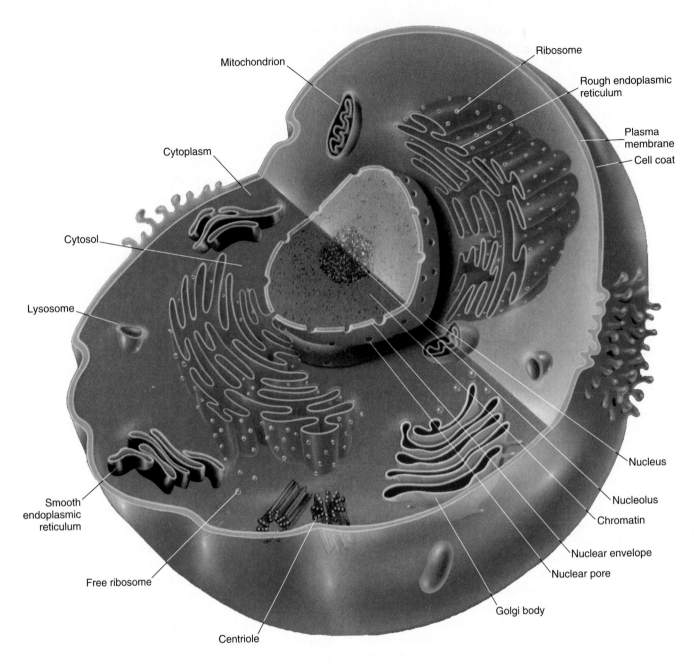

**FIGURE 2.1**    Drawing of a generalized animal cell. Emphasis has been placed on the cellular components discussed in the text.

called a **peptidoglycan**. As its name suggests, the molecule consists of peptide and sugar units. Long polysaccharide chains are cross-linked with short peptides, which impart great strength and rigidity to the bacterial cell. Some bacterial cells have still another covering, a **capsule**. It is a mucuslike polysaccharide that protects these bacteria from phagocytic activity by the host during their pathogenic invasion of eukaryotic organisms.

Activities at cell boundaries are dynamic physiological processes. Both transport in and out of cells, and com-

munication between cells are critical to normal function. Since physiological processes are biochemically based, we would expect that many genes and their related products are essential to these activities. This is indeed the case, and as such, mutations in these genes can alter or interrupt normal physiological functions, often with severe consequences. For example, the inherited disorder **Duchenne muscular dystrophy** is the result of complete loss of function of the product dystrophin, which is believed to function at the cell membrane of muscle cells. As shown in Figure 2.2, using immunofluo-rescent localization technology, dystroptin is completely absent from skeletal muscle of an afflicted individual compared to muscle from a control subject.

A quite different example is the mutation that alters the structure of the capsule surrounding the bacterium *Diplococcus pneumonia*. The result is a bacterial strain that is noninfective, failing to cause pneumonia in vertebrates such as mice. Both examples will be pursued again later in the text.

Many, if not most, animal cells have a covering over the plasma membrane called a **cell coat**. Consisting of glycoproteins and polysaccharides, the chemical composition of the cell coat differs from comparable structures in either plants or bacteria. One function served by the cell coat is to provide biochemical identity at the surface of cells. Among other forms of molecular recognition, various antigenic determinants are part of the cell coat. For example, the **AB** and **MN antigens** are found on the surface of red blood cells. In other cells, the **histocompatibility antigens**, which elicit an immune response during tissue and organ transplants, are part of the cell coat. All forms of biochemical identity at the cell surface are under genetic control, and many have been thoroughly investigated.

## The Nucleus

The presence of the **nucleus** and other membranous organelles characterizes eukaryotic cells. The nucleus houses the genetic material, DNA, which is found in association with large numbers of acidic and basic proteins. During nondivisional phases of the cell cycle this DNA/protein complex exists in an uncoiled, dispersed state called **chromatin**. As we will soon discuss, during mitosis and meiosis this material coils up and condenses into structures called **chromosomes**. Also present in the nucleus is the **nucleolus**, an amorphous component where ribosomal RNA is synthesized and where the initial stages of ribosomal assembly occur. The areas of DNA encoding rRNA are collectively referred to as the **nucleolus organizer region** or the **NOR**.

The lack of a nuclear envelope and membraneous organelles is characteristic of prokaryotes. In bacteria such as *E. coli* the genetic material is present as a long, circular DNA molecule that is compacted into an area referred to as the **nucleoid region**. Part of the DNA may be attached to the cell membrane, but in general, the nucleoid region constitutes a large area throughout the cell. Although the DNA is compacted, it does not undergo the extensive coiling characteristic of the stages of mitosis where, in eukaryotes, chromosomes become visible. Nor is the DNA in these organisms associated as extensively with proteins as is eukaryotic DNA. Figure 2.3 shows the formation of two bacteria during cell divi-

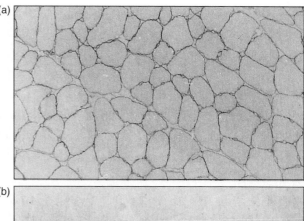

**FIGURE 2.2**    (a) Localization of dystrophin in the sarcolemmal areas of normal muscle using immunoperoxidase staining. (b) Complete lack of immunoreactive dystrophin in the skeletal muscle of a patient with Duchenne muscular dystrophy.

sion and illustrates the bacterial chromosomes in the nucleoid regions. Prokaryotic cells do not have a distinct nucleolus, but do contain genes that specify rRNA molecules.

## The Cytoplasm and Organelles

The remainder of the eukaryotic cell enclosed by the plasma membrane, and excluding the nucleus, is composed of **cytoplasm**. Cytoplasm consists of a nonparticulate, colloidal material referred to as the **cytosol**, which surrounds and encompasses the numerous types of cellular organelles. Beyond these components, an extensive system of tubules and filaments comprising the **cytoskeleton** provides a lattice of support structures within the cytoplasm. Consisting primarily of tubulin-derived microtubules and actin-derived microfilaments, this structural framework maintains cell shape, facilitates cell mobility, and anchors the various organelles. Tubulin and actin are both proteins found abundantly in eukaryotic cells.

One organelle, the membranous **endoplasmic reticulum (ER)**, compartmentalizes the cytoplasm, greatly increasing the surface area available for biochemical synthesis. ER may be smooth, in which case it serves as the site for synthesis of fatty acids and phospholipids.

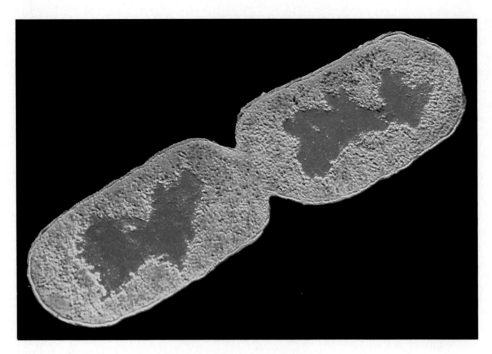

**FIGURE 2.3**    Color-enhanced electron micrograph of *E. coli* undergoing cell division. Particularly prominent are the two chromosomal areas that have been partitioned into the daughter cells.

Or, the ER may be rough because it is studded with ribosomes. Ribosomes, which will later be discussed in detail, serve as sites for the translation of genetic information contained in messenger RNA (mRNA) into proteins.

Three other cytoplasmic structures are very important in the eukaryotic cell's activities: mitochondria, chloroplasts, and centrioles. Mitochondria are found in both animal and plant cells and are the sites of the **oxidative phases of cell respiration**. These chemical reactions generate large amounts of adenosine triphosphate (ATP), an energy-rich molecule. The chloroplast is one type of plastid found in plants, algae, and some protozoans. This organelle is associated with **photosynthesis**, the major energy-trapping process on earth. Both mitochondria and chloroplasts contain a type of DNA distinct from that found in the nucleus. Furthermore, these organelles can duplicate themselves and transcribe and translate their genetic information. It is interesting to note that the genetic machinery of mitochondria and chloroplasts closely resembles that of prokaryotic cells. This and other observations have led to the proposal that these organelles were once primitive free-living structures that established a symbiotic relationship with a primitive eukaryotic cell. This theory, which describes the evolutionary origin of these organelles, is called the **endosymbiont theory**.

Animal and some plant cells also contain a pair of complex structures called the **centrioles**. These cytoplasmic bodies, contained within a specialized region called the **centrosome**, are associated with the organi-

zation of those spindle fibers that function in mitosis and meiosis. In some organisms, the centriole is derived from another structure, the **basal body**, which is associated with the formation of cilia and flagella. Over the years, there have been many reports that have suggested that centrioles and basal bodies contain DNA, which is involved in the replication of these structures. Recently, very convincing evidence has been presented that this is most certainly the case in the single-celled green alga, *Chlamydomonas*. Several genes are contained within a circular DNA structure associated with the basal body and centriole.

The organization of **spindle fibers** by the centrioles occurs during the early phases of mitosis and meiosis. Composed of arrays of microtubules, these fibers play an important role in the movement of chromosomes as they separate during cell division. The microtubules consist of polymers of alpha and beta subunits of the protein tubulin. The interaction of the chromosomes and spindle fibers will be considered later in this chapter.

## HOMOLOGOUS CHROMOSOMES, HAPLOIDY, AND DIPLOIDY

In order to discuss the processes of mitosis and meiosis, it is important to clearly understand the concept of **homologous chromosomes**. Such an understanding will also be critical to our future discussions of Mendelian genetics. Thus, before embarking further, we will address this topic and introduce other important terminology.

Chromosomes are most easily visualized during mitosis. When they are examined carefully, they are seen to take on distinctive lengths and shapes. Each contains a condensed or constricted region called the **centromere**, which establishes the general appearance of each chromosome. Figure 2.4 illustrates chromosomes with centromere placements at different points along their lengths. Extending from either side of the centromere are the arms of the chromosome. Depending on the position of the centromere, different arm ratios are produced. As Figure 2.4 illustrates, chromosomes are classified as **metacentric**, **submetacentric**, **acrocentric**, or **telocentric** on the basis of the centromere location. The shorter arm, by convention, is shown above the centromere and is called the **p arm**. The longer arm is shown below the centromere and is called the **q arm**.

When studying mitosis, we may make several other important observations. First, each somatic cell within members of the same species contains an identical number of chromosomes. This is called the **diploid number (2n)**. When the lengths and centromere placements of

all such chromosomes are examined, a second general feature is apparent. Nearly all of the chromosomes exist in pairs with regard to these two criteria. The members of each pair are called **homologous chromosomes**. For each chromosome exhibiting a specific length and centromere placement, another exists with identical features.

Figure 2.5 illustrates the nearly identical physical appearance of members of homologous chromosome pairs. There, the human mitotic chromosomes have been treated with the enzyme trypsin and then stained, leading to the banded appearance. Then, they were photographed, cut out of the print, and matched up, creating a **karyotype**. As you can see, humans have a 2n value of 46 and exhibit a diversity of sizes and centromere placements. Note also that each of the 46 chromosomes is actually a double structure. Had these chromosomes been allowed to continue dividing, the two parts of each chromosome—called **sister chromatids**—would have separated into two new cells as division continued.

Collectively, the total set of genes contained on one member of each homologous pair of chromosomes constitutes the **haploid genome** of the species. The **haploid number (n)** of chromosomes is one-half of the dip-

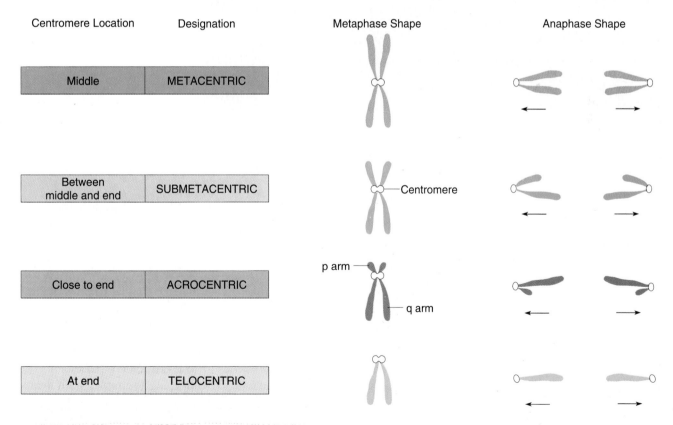

**FIGURE 2.4**    Centromere locations and designations of chromosomes based on their location. Note that the shape of the chromosome during anaphase is determined by the position of the centromere.

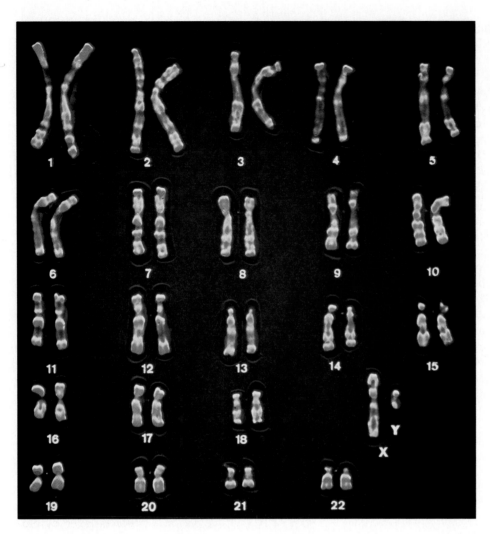

**FIGURE 2.5**    Mitotic chromosomes constituting the human male karyotype. All but the X and Y chromosomes are present in homologous pairs. Each chromosome is actually a double structure.

loid number. Table 2.1 demonstrates the wide range of *n* values found in a variety of plants and animals.

Homologous pairs of chromosomes have important genetic similarity. They contain identical gene sites along their lengths. Each gene site is called a **locus** (pl. **loci**). Thus, they have identical genetic potential. In sexually reproducing organisms, one member of each pair is derived from the maternal parent (through the ovum) and one from the paternal parent (through the sperm). Thus, each diploid organism contains two copies of each gene as a consequence of **biparental inheritance**. As will be seen in the following chapters on transmission genetics, the members of each pair of genes, while influencing the same characteristic or trait, need not be identical. Alternative forms of the same gene are called **alleles**. In a population of members of the same species, many different alleles of the same gene may exist.

The concepts of haploid number, diploid number, and homologous chromosomes may be related to the pro-

cess of meiosis. During the formation of gametes or spores, meiosis converts the diploid number of chromosomes to the haploid number. As a result, haploid gametes or spores contain precisely one member of each homologous pair of chromosomes, that is, one complete set. Following fusion of two gametes in fertilization, the diploid number is reestablished, that is, two complete sets. The constancy of genetic material from generation to generation is thus maintained.

There is one important exception to the concept of homologous pairs of chromosomes. In many species, one pair, the **sex-determining chromosomes**, may not be homologous in size, centromere placement, arm ratio, or genetic potential. For example, in humans, males contain a Y chromosome in addition to one X chromosome (Figure 2.5), while females carry two homologous X chromosomes. The X and Y chromosomes are not strictly homologous. The Y is considerably smaller and lacks most of the loci contained on the X.

**Table 2.1**  THE HAPLOID NUMBER OF CHROMOSOMES FOR REPRESENTATIVE ORGANISMS

| Common Name | Genus–Species | Haploid No. | Common Name | Genus–Species | Haploid No. |
|---|---|---|---|---|---|
| Black bread mold | *Aspergillus nidulans* | 8 | House fly | *Musca domestica* | 6 |
| Broad bean | *Vicia faba* | 6 | House mouse | *Mus musculus* | 20 |
| Cat | *Felis domesticus* | 19 | Human | *Homo sapiens* | 23 |
| Cattle | *Bos taurus* | 30 | Jimson weed | *Datura stramonium* | 12 |
| Chicken | *Gallus domesticus* | 39 | Mosquito | *Culex pipiens* | 3 |
| Chimpanzee | *Pan troglodytes* | 24 | Pink bread mold | *Neurospora crassa* | 7 |
| Corn | *Zea mays* | 10 | Potato | *Solanum tuberosum* | 24 |
| Cotton | *Gossypium hirsutum* | 26 | Rhesus monkey | *Macaca mulatta* | 21 |
| Dog | *Canis familiaris* | 39 | Roundworm | *Caenorhabditis elegans* | 6 |
| Evening primrose | *Oenothera biennis* | 7 | Silkworm | *Bombyx mori* | 28 |
| Frog | *Rana pipiens* | 13 | Slime mold | *Dictyostelium discoidium* | 7 |
| Fruit fly | *Drosophila melanogaster* | 4 | Snapdragon | *Antirrhinum majus* | 8 |
| Garden onion | *Allium cepa* | 8 | Tobacco | *Nicotiana tabacum* | 24 |
| Garden pea | *Pisum sativum* | 7 | Tomato | *Lycopersicon esculentum* | 12 |
| Grasshopper | *Melanoplus differentialis* | 12 | Water fly | *Nymphaea alba* | 80 |
| Green alga | *Chlamydomonas reinhardi* | 18 | Wheat | *Triticum aestivum* | 21 |
| Horse | *Equus caballus* | 32 | Yeast | *Saccharomyces cerevisiae* | 17 |

Nevertheless, in meiosis they behave as homologues so that gametes produced by males receive either the X or Y chromosome.

# MITOSIS AND CELL DIVISION

The process of **mitosis**, or nuclear division, is critical to all eukaryotic organisms. In single-celled organisms, such as protozoans, some fungi, and algae, mitosis, as a part of cell division, provides the basis of asexual reproduction. Multicellular diploid organisms begin life as single-celled fertilized eggs or **zygotes**. The mitotic activity of the zygote and the subsequent daughter cells is the foundation for development and growth of the organism. In adult organisms, mitotic activity associated with cell division is prominent in wound healing and other forms of cell replacement in certain tissues. For example, the epidermal skin cells of humans are continuously being sloughed off and replaced. Cell division also results in a continuous production of reticulocytes, which eventually shed their nuclei and replenish the supply of red blood cells in vertebrates. In abnormal situations, somatic cells may exhibit uncontrolled cell divisions, resulting in cancer.

It is generally observed that following cell division, the initial size of each new daughter cell is approximately one-half the size of its parent. However, the nucleus of each new cell is not appreciably smaller than the nucleus of the original cell. Quantitative measurements of DNA confirm that there are equivalent amounts of genetic material in the daughter nuclei as in the parent cell.

The process of cytoplasmic division is called **cytokinesis**. The division of cytoplasm requires a mechanism that results in a partitioning of the volume into two parts, followed by the enclosure of both new cells within a distinct plasma membrane. Cytoplasmic organelles either replicate themselves, arise from existing membrane structures, or are synthesized *de novo* (anew) in each cell. The subsequent proliferation of these structures is a reasonable and adequate mechanism for reconstituting the cytoplasm in daughter cells.

The distribution of the genetic material into daughter cells is more complex than cytokinesis and requires more precision. The chromosomes must first be exactly replicated and then accurately partitioned into daughter cells. The end result is the production of two daughter cells, each with a chromosome composition identical to the parent cell.

## Interphase and the Cell Cycle

Many cells undergo a continuous alternation between division and nondivision. The interval between each mitotic division is called **interphase**. It was once thought that the biochemical activity during interphase was devoted solely to the cell's growth and its normal function. However, we now know that another biochemical step critical to the next mitosis occurs during interphase: **the replication of the DNA of each chromosome**. Occurring midway through interphase, this period during which DNA is synthesized is called the **S phase**. The initiation and completion of synthesis can be detected by monitoring the incorporation of radioactive DNA precursors such as $^3$H-thymidine. Their incorporation can be monitored using the technique of autoradiography (Appendix A).

Investigations of this nature have demonstrated two periods during interphase, before and after S, when no DNA synthesis occurs. These are designated $G_1$ (**gap I**) and $G_2$ (**gap II**), respectively. During both of these periods, as well as during S, intensive metabolic activity, cell growth, and cell differentiation occur. By the end of $G_2$, the volume of the cell has roughly doubled, DNA has been replicated, and mitosis (M) is initiated. Continuously dividing cells then repeat this cycle ($G_1$, S, $G_2$, M) over and over. This concept of such a **cell cycle** is illustrated in Figure 2.6.

Much is known about the cell cycle based on *in vitro* (test tube) studies. When grown in culture, most cell types traverse the complete cycle in about 20 hours. The actual process of mitosis occupies only a small part of the cycle. The length of the S and $G_2$ stages of interphase are fairly consistent among different cell types. Most variation is seen in the length of time spent in the $G_1$ stage. Figure 2.7 illustrates the relative length of these periods in a typical cell.

$G_1$ is of great interest in the study of cell proliferation and its control. At a point late in $G_1$, all cells follow one of two paths. They either withdraw from the cycle and become quiescent in the $G_0$ **stage**, or they become committed to initiate DNA synthesis and complete the cycle. The time when this decision is made has been referred to as the **restriction** or **R point**. Cells that enter $G_0$ remain viable and metabolically active but are nonproliferative. Cancer cells appear to avoid entering $G_0$ or they pass through it very quickly. Other cells enter $G_0$ and never reenter the cell cycle. Still others can remain quiescent in

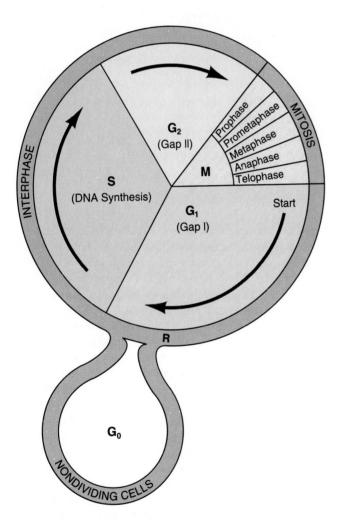

**FIGURE 2.6**    Diagrammatic representation of the stages that compose an arbitrary cell cycle. Following mitosis (M), cells initiate a new cycle ($G_1$). Cells may become nondividing ($G_0$) or continue through the restriction point (R) where they become committed to begin DNA synthesis (S) and complete the cycle ($G_2$ and M). Following mitosis, two daughter cells are produced.

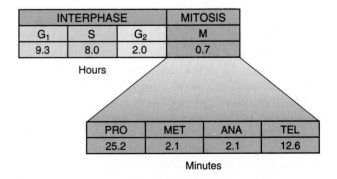

| INTERPHASE | | | MITOSIS |
|---|---|---|---|
| $G_1$ | S | $G_2$ | M |
| 9.3 | 8.0 | 2.0 | 0.7 |

Hours

| PRO | MET | ANA | TEL |
|---|---|---|---|
| 25.2 | 2.1 | 2.1 | 12.6 |

Minutes

**FIGURE 2.7**    The time spent in each phase of one complete cell cycle of a human cell in culture. Times vary according to cell types and conditions.

$G_0$, but they may be stimulated to return to $G_1$, reentering the cycle.

Cytologically, the interphase stage is characterized by the absence of the appearance of discrete chromosomes. Instead, a distinct nucleus is evident, filled with chromatin that has formed as the chromosomes have unfolded and uncoiled following the previous mitosis. This is depicted diagrammatically in part (a) of Figure 2.8(a). When viewed under the light microscope [part (a) of Figure 2.8(b)], the nucleus appears to be filled with a speckled, granular material. This is the result of having sectioned through the uncoiled chromatin fibers.

## Prophase

Once $G_1$, S, and $G_2$ are completed, mitosis is initiated. Mitosis is a dynamic period of vigorous and continual activity. For discussion purposes, the entire process is subdivided into discrete phases, and specific events are assigned to each stage. These stages, in order of occurrence, are **prophase**, **prometaphase**, **metaphase**, **anaphase**, and **telophase**. As in interphase, each of these stages is depicted in a drawing in Figure 2.8(a) and is shown as it actually occurs during plant mitosis in Figure 2.8(b).

Almost one-third of mitosis is spent in prophase; significant activities occur during this stage. One of the early events in animal and lower plant prophase involves the migration of two pairs of centrioles to opposite ends of the cell. These structures are found just outside the nuclear envelope in an area of differentiated cytoplasm called the **centrosome**. It is thought that each pair of centrioles consists of one mature unit and a smaller, newly formed centriole.

The direction of migration of the centrioles is such that two poles are established at opposite ends of the cell. This creates an axis along which chromosomal separation occurs. Following their migration, the centrioles are responsible for the organization of cytoplasmic microtubules into a series of **spindle fibers**. Although the cells of plants, fungi, and certain algae seem to lack centrioles, spindle fibers are nevertheless apparent during mitosis. If some other center that organizes microtubules into spindle fibers exists in these cells, it has yet to be discovered.

As the centrioles migrate, the nuclear envelope begins to break down and gradually disappears. In a similar fashion, the nucleolus disintegrates within the nucleus. While these events are taking place, the diffuse chromatin—the characteristic uncoiled form of the genetic material during interphase—condenses until distinct threadlike structures, or chromosomes, are visible. As condensation continues it becomes apparent near the end of prophase that each chromosome is a double structure split longitudinally except at a single point of constriction, the centromere.* The two parts of each chromosome are called **chromatids**. The DNA contained in each pair of chromatids constituting a chromosome was derived from a replication event during the S phase of the previous interphase. Thus both members of each chromosome are genetically identical and are called **sister chromatids**. In humans, with a diploid number of 46, a cytological preparation of late prophase will reveal 46 such chromosomal structures (Figure 2.5). In the cell, these are found randomly distributed in the area formerly occupied by the nucleus.

By the completion of prophase, spindle fibers have been laid down between the centrioles, which are now at opposite ends of the cell. The nucleolus and nuclear envelope are no longer visible, and sister chromatids are apparent as part of each chromosome. Part (b) of Figures 2.8(a) and 2.8(b) illustrate prophase.

## Prometaphase and Metaphase

The distinguishing event of the ensuing stages is the migration of the centromeric region of each chromosome to the equatorial plane. In some descriptions the term **metaphase** is applied strictly to the chromosome configuration following this movement. In such descriptions, **prometaphase** refers to the period of chromosome movement, as depicted in part (c) of Figures 2.8(a) and 2.8(b). The equatorial plane, also referred to as the **metaphase plate**, is the midline region of the cell, a plane that lies perpendicular to the axis established by the spindle fibers.

Migration is made possible by the binding of one or more spindle fibers to the kinetochore contained within the centromere of each chromosome. There are two major classes of spindle microtubules. Those called kinetochore microtubules have one end near the centrosome region and the other anchored in the kinetochore. The nonkinetochore microtubules grow from the centrioles of the centrosome, but have their other ends free. These sometimes interdigitate with one another, providing a framework to the spindle and maintaining the separation of the two poles during chromosome separation. At the completion of metaphase, each centromere is aligned at

---

*You may sometimes see the term *kinetochore* used synonymously with *centromere*. The kinetochore is actually a granule within the centromere that attaches to spindle fibers during mitosis. Two kinetochores form on opposite faces of the centromere, each attaching one or the other member of a pair of sister chromatids to the spindle fibers.

# MITOSIS

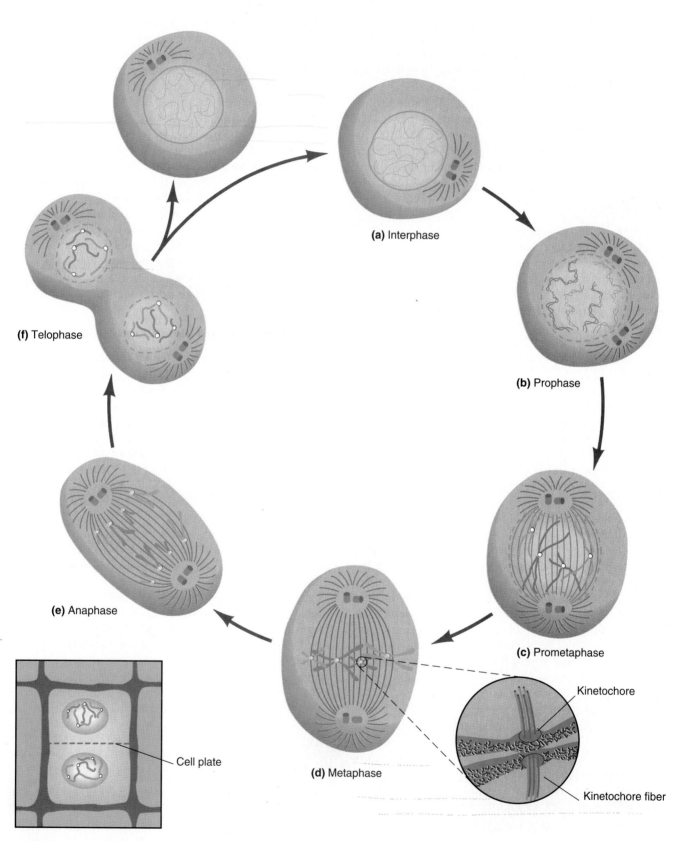

(a) Interphase

(b) Prophase

(c) Prometaphase

(d) Metaphase

Kinetochore

Kinetochore fiber

(e) Anaphase

(f) Telophase

Cell plate

(g) Plant cell telophase

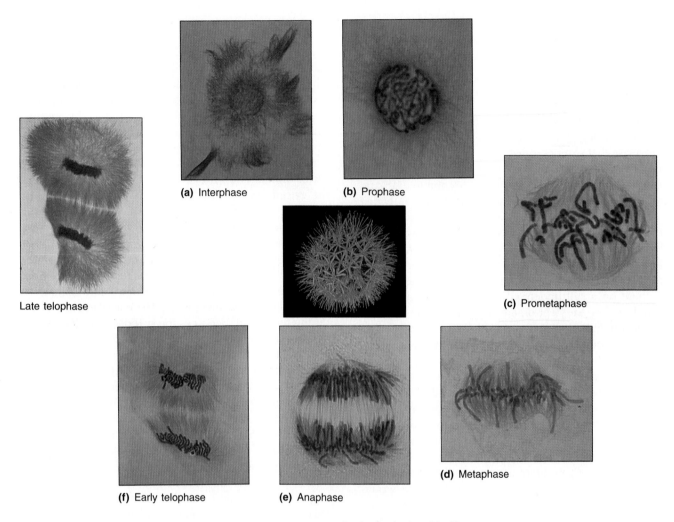

**(a)** Interphase                    **(b)** Prophase

**(c)** Prometaphase

Late telophase

**(d)** Metaphase

**(f)** Early telophase        **(e)** Anaphase

**FIGURE 2.8(b)**     Light micrographs illustrating the stages of mitosis depicted in Figure 2.8(a). These stages are derived from the flower of *Haemanthus,* shown in the center of the figure.

◀ **FIGURE 2.8(a)**     Mitosis in an animal cell with a diploid number of 4. The events occurring in each stage are described in the text. Of the two homologous pairs of chromosomes, one contains longer, metacentric members and the other shorter, submetacentric members. The maternal chromosomes and the paternal chromosomes are shown in different colors. The insert (g), showing the telophase stage in a plant cell, illustrates the formation of the cell plate and lack of centrioles.

the plate with the chromosome arms extending outward in a random array. This configuration is shown in part (d) of Figures 2.8(a) and 2.8(b).

## Anaphase

Events critical to chromosome distribution during mitosis occur during its shortest stage, **anaphase**. It is during this phase that sister chromatids of each double chromosomal structure separate from each other and migrate to opposite ends of the cell. In order for complete separation to occur, each centromeric region must be divided into two. Once this has occurred, each chromatid is now referred to as a **daughter chromosome**. Movement of the chromosomes to the opposite poles of the cell is dependent upon the centromere-spindle fiber attachment (Figure 2.9). While chromosome movement is dependent on a still undefined mechanism involving the spindle fibers, the centromeres appear to lead the way during migration, with the chromosome arms trailing behind. Depending on the location of the centromere along the chromosome, different shapes are assumed during separation (review Figure 2.4).

As the chromosomes begin their migration, the spindle elongates, extending the distance between the two centrosome regions. At the completion of anaphase, the chromosomes have migrated to the opposite poles of the cell. The steps occurring during anaphase are critical in providing each subsequent daughter cell with an identical set of chromosomes. In human cells there would now be 46 chromosomes at each pole, one from each original sister pair. Part (e) of Figures 2.8(a) and 2.8(b) illustrates anaphase prior to its completion.

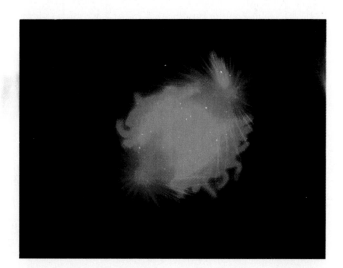

**FIGURE 2.9**    Immunofluorescent localization of spindle microtubules (yellow/green) from a dividing lung cell of a newt. Chromosomal material is stained blue.

## Telophase

Telophase is the final stage of mitosis. At its beginning, there are two complete sets of chromosomes, one at each pole. The most significant event is **cytokinesis**, the division or partitioning of the cytoplasm. Cytokinesis is essential if two new cells are to be produced from one. The mechanism differs greatly in plant and animal cells. In plant cells, a **cell plate** is synthesized and laid down across the dividing cell in the region of the metaphase plate. Animal cells, however, undergo a constriction of the cytoplasm in much the same way a loop might be tightened around the middle of a balloon. The end result is the same: Two distinct cells are formed.

It is not surprising that the process of cytokinesis varies among cells of different organisms. Plant cells, which are more regularly shaped and structurally rigid, require a mechanism for the deposition of new cell wall material around the plasma membrane. The cell plate, laid down during telophase, becomes the **middle lamella**. Subsequently, the primary and secondary layers of the cell wall are deposited between the cell membrane and middle lamella on both sides of the boundary between the two daughter cells. In animals, complete constriction of the cell membrane produces the **cell furrow** characteristic of newly divided cells.

Other events necessary for the transition from mitosis to interphase are also initiated during late telophase. They represent a general reversal of those that occurred during prophase. In each new cell, the chromosomes begin to uncoil and become diffuse chromatin once again, while the nuclear envelope reforms around them. The nucleolus gradually reforms and is completely visible in the nucleus during early interphase. The spindle fibers also disappear. Telophase in animal and plant cells is illustrated in part (f) and part (g) of Figures 2.8(a) and 2.8(b).

# MEIOSIS AND SEXUAL REPRODUCTION

The process of meiosis, unlike mitosis, reduces the amount of genetic material. While mitosis produces daughter cells with a full diploid complement, meiosis produces gametes or spores with only one haploid set of chromosomes. During sexual reproduction, gametes then combine in fertilization to reconstitute the diploid complement found in parental cells. The process itself must be very specific, because it is insufficient to produce gametes with a random array of one-half of the total number of chromosomes. Instead, each gamete must receive one member of each homologous pair of chromosomes, ensuring genetic continuity from generation to generation.

The process of sexual reproduction also ensures genetic variety among members of a species. As you study meiosis, it will become apparent that this process results in gametes with many unique combinations of maternally and paternally derived chromosomes among the haploid complement. When two gametes derived from two genetically unique individuals are combined during fertilization, a large number of chromosome combinations are possible. Furthermore, we will see that the meiotic event referred to as **crossing over** results in genetic exchange between members of each homologous pair of chromosomes. This creates intact chromosomes that are mosaics of the maternal and paternal homologues from which they are derived. This has the effect of further enhancing the potential genetic variation in gametes and the offspring derived from them.

## An Overview of Meiosis

We have already established what must be accomplished during meiosis. Before systematically considering the stages of this process (Figure 2.10) we will briefly describe how diploid cells are converted to haploid gametes or spores. Unlike mitosis, in which each paternally and maternally derived member of any given homologous pair of chromosomes behaves autonomously during division, in meiosis homologous chromosomes pair together, or **synapse**. Each synapsed structure, called a **bivalent**, gives rise to a unit, the **tetrad**, consisting of four chromatids. The presence of four chromatids demonstrates that both chromosomes have duplicated. In order to achieve haploidy, two divisions are necessary. In the first division, described as **reductional**, each tetrad separates and yields two **dyads**, each of which contains two sister chromatids joined at a common centromere. During the second division, described as **equational**, each dyad splits into two **monads** of one chromosome each. Thus, the two divisions may potentially produce four haploid cells. As in mitosis, meiosis is a continuous process. We give names to the parts of each stage of division only for the convenience of discussion.

## The First Meiotic Division: Prophase I

From a genetic standpoint, there are two critical events during prophase I. First, members of each homologous pair of chromosomes somehow find one another and undergo synapsis. Second, the exchange process referred to above as crossing over occurs between synapsed homologues. Because of the importance of these genetic events, this stage of meiosis has been subjected to continuous investigation. Most recently, the study of yeast has provided a model for our understanding of these critical events. As we discuss them, you

should be aware that replication of DNA has preceded meiosis, even though chromosomes do not immediately reveal their duplication. Recall that this also occurs in mitosis. The first meiotic prophase is complex and has been further subdivided into five substages: leptonema,* zygonema,* pachynema,* diplonema,* and diakinesis [Figure 2.10(a)].

### Leptonema
During the **leptotene stage**, the interphase chromatin material begins to condense, and the chromosomes, although still extended, become visible. Along each chromosome are **chromomeres**, localized condensations that resemble beads on a string. Recent evidence suggests that it is during leptonema that a process called **homology search** begins, one that precedes initial pairing of homologues.

### Zygonema
The chromosomes continue to shorten and thicken during the **zygotene stage**. During the process of homology search, homologous chromosomes undergo initial alignment with one another. This alignment is considered to be a **rough pairing**, and by the end of zygonema it is complete. In yeast, homologues are separated by about 300 nm, and near the end of zygonema, structures referred to as lateral elements are visible between paired homologues. As meiosis proceeds, the overall length of the lateral elements increases and a more extensive ultrastructural component, the **synaptonemal complex**, begins to form between the homologues (see Figure 2.15 on page 40).

At the completion of zygonema, the paired homologues represent structures referred to as **bivalents**. Although both members of each bivalent have already replicated their DNA, it is not yet visually apparent that each member is a double structure. The number of bivalents in each species is equal to the haploid ($n$) number.

### Pachynema
In the transition from the zygotene to the **pachytene stage**, coiling and shortening of chromosomes continues, and further development of the synaptonemal complex occurs between the two members of each bivalent. This leads to a more intimate pairing referred to as **synapsis**. Compared to the rough pairing characteristic of yeast pachynema, homologues are now separated by only 100 nm.

During pachynema, it first becomes evident that each homologue is a double structure, providing visual evidence of the earlier replication of the DNA of each chromosome. Thus, each bivalent contains four members,

---

* These are the noun forms of these stages. The adjective forms (leptotene, zygotene, pachytene, and diplotene) are also used in the text of this chapter.

# MEIOSIS I

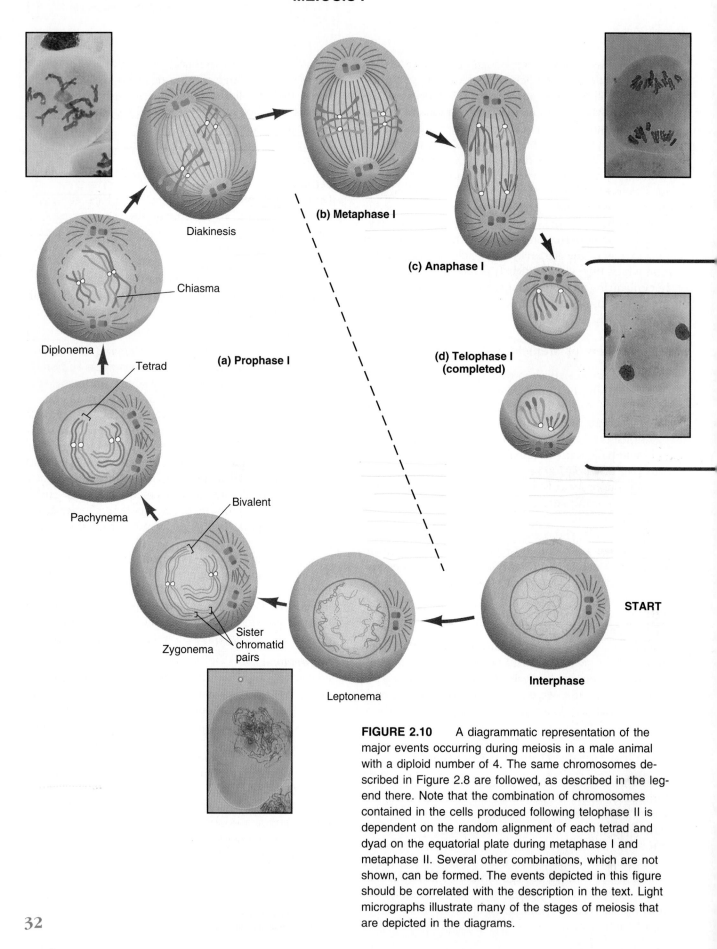

Diakinesis

**(b) Metaphase I**

**(c) Anaphase I**

Chiasma

Diplonema

**(a) Prophase I**

**(d) Telophase I (completed)**

Tetrad

Pachynema

Bivalent

START

Zygonema

Sister chromatid pairs

Interphase

Leptonema

**FIGURE 2.10** A diagrammatic representation of the major events occurring during meiosis in a male animal with a diploid number of 4. The same chromosomes described in Figure 2.8 are followed, as described in the legend there. Note that the combination of chromosomes contained in the cells produced following telophase II is dependent on the random alignment of each tetrad and dyad on the equatorial plate during metaphase I and metaphase II. Several other combinations, which are not shown, can be formed. The events depicted in this figure should be correlated with the description in the text. Light micrographs illustrate many of the stages of meiosis that are depicted in the diagrams.

# MEIOSIS II

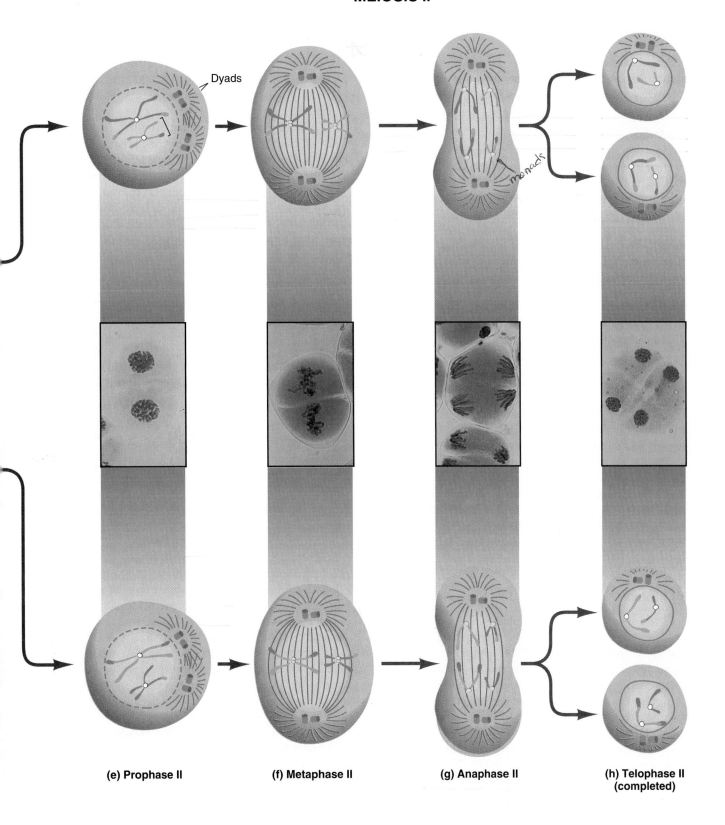

Dyads

monads

(e) Prophase II          (f) Metaphase II          (g) Anaphase II          (h) Telophase II
                                                                            (completed)

33

called **chromatids**. As in mitosis, replicates are called **sister chromatids**. The four-membered structure is also referred to as a **tetrad**, and each tetrad contains two such pairs.

### Diplonema

During observation of the ensuing **diplotene stage** it is even more apparent that each tetrad consists of two pairs of sister chromatids. Within each tetrad, each pair of sister chromatids begins to separate. However, one or more areas remain in contact where chromatids are intertwined. Each such area, called a **chiasma** (pl. **chiasmata**), is thought to represent a point where nonsister chromatids have undergone genetic exchange through the process referred to above as crossing over. Although the physical exchange between chromosome areas occurred during the previous pachytene stage, the result of crossing over is visible only when the duplicated chromosomes begin to separate. Crossing over is an important source of genetic variability. New combinations of genetic material are formed during this process.

### Diakinesis

The final stage of prophase I is **diakinesis**. The chromosomes pull farther apart, but sister chromatids on either side of the chiasmata remain loosely associated. As this separation proceeds, the chiasmata move toward the ends of the tetrad. This process, called **terminalization**, begins in late diplonema, and is completed during diakinesis. During this final period of prophase I, the nucleolus and nuclear envelope break down, and the two centromeres of each tetrad become attached to the recently formed spindle fibers.

## Metaphase, Anaphase, and Telophase I

Following the first meiotic prophase stage, steps similar to those of mitosis occur. In the **metaphase stage of the first division**, the chromosomes have maximally shortened and thickened. The terminal chiasmata of each tetrad are visible and appear to be the only factor holding the nonsister chromosomes together. Each tetrad interacts with spindle fibers, facilitating movement to the metaphase plate.

During the first division, a single centromere holds each pair of sister chromatids together. It does *not* divide. At the **first anaphase**, one-half of each tetrad (one pair of sister chromatids—called a **dyad**) is pulled toward each pole of the dividing cell. This separation event is the physical basis of the more conceptual event labeled **disjunction**. Occasionally, errors in meiosis occur and separation is not achieved. The term **nondisjunction** describes such an error, the consequences of which will soon be discussed. At the completion of the normal anaphase I, there is a series of dyads equal to the haploid number present at each pole.

If no crossing had occurred in the first meiotic prophase, each dyad at each pole would consist solely of either paternal or maternal chromatids. However, the exchanges produced by crossing over create mosaic chromatids of paternal and maternal origin. The alignment of each tetrad prior to this first anaphase stage is random. One-half of each tetrad will be pulled to one or the other pole at random, and the other half will move to the opposite pole. This random **segregation** of dyads is the basis for the Mendelian principle of **independent assortment**, which we will discuss in Chapter 3. You may wish to return to this discussion when you study this principle.

In many organisms, **telophase of the first meiotic division** reveals a nuclear membrane forming around the dyads. Then, the nucleus enters into a short interphase period. In other cases, the cells go directly from the first anaphase into the second meiotic division. If an interphase period occurs, the chromosomes do not replicate since they already consist of two chromatids. In general, meiotic telophase is much shorter than the corresponding stage in mitosis.

## The Second Meiotic Division

A second division of the sister chromatids is essential to achieve haploidy in the meiotic products. During **prophase II**, each dyad is composed of one pair of sister chromatids attached by a common centromere. During **metaphase II**, the centromeres are directed to the equatorial plate. Then, the centromeres divide, and during **anaphase II** the sister chromatids of each dyad are pulled to opposite poles. Since the number of dyads is equal to the haploid number, **telophase II** reveals one member of each pair of homologous chromosomes present at each pole. Each chromosome is referred to as a **monad**. Not only has the haploid state been achieved, but if crossing over has occurred, each monad is a combination of maternal and paternal genetic information. As a result, the offspring produced by any gamete will receive from it a mixture of genetic information originally present in his or her grandparents. Potentially, following cytokinesis in telophase II, four haploid gametes may result from a single meiotic event.

## SPERMATOGENESIS AND OOGENESIS

Although events that occur during the meiotic divisions are similar in all cells that participate in gametogenesis, there are certain differences between the production of a male gamete (spermatogenesis) and a female gamete (oogenesis) in most animal species.

**Spermatogenesis** takes place in the testes, the male

reproductive organs. The process begins with the expanded growth of an undifferentiated diploid germ cell called a **spermatogonium**. This cell enlarges to become a **primary spermatocyte**, which undergoes the first meiotic division. The products of this division are called **secondary spermatocytes**. Each secondary spermatocyte contains a haploid number of dyads. The secondary spermatocytes then undergo the second meiotic division, and each of these cells produces two haploid **spermatids**. Spermatids go through a series of developmental changes, **spermiogenesis**, and become highly specialized, motile **spermatozoa** or **sperm**. All sperm cells produced during spermatogenesis receive equal amounts of genetic material and cytoplasm. Figure 2.11 summarizes these steps.

Spermatogenesis may be continuous or occur periodically in mature male animals, with its onset determined by the nature of the species' reproductive cycle. Animals

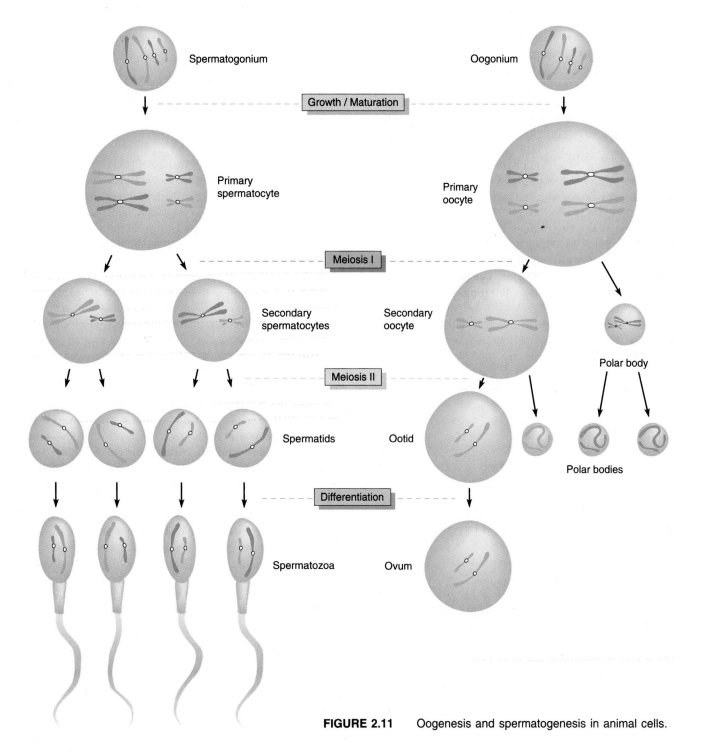

**FIGURE 2.11**    Oogenesis and spermatogenesis in animal cells.

that reproduce year-round produce sperm continuously, while those whose breeding period is confined to a particular season produce sperm only during that time.

In animal **oogenesis**, the formation of **ova** (singular: **ovum**), or eggs, occurs in the ovaries, the female reproductive organs. The daughter cells resulting from the two meiotic divisions receive equal amounts of genetic material, but they do *not* receive equal amounts of cytoplasm. Instead, during each division, almost all the cytoplasm of the **primary oocyte**, itself derived from the **oogonium**, is concentrated in one of the two daughter cells. The concentration of cytoplasm is necessary because a major function of the mature ovum is to nourish the developing embryo following fertilization.

During the first meiotic anaphase in oogenesis, the tetrads of the primary oocyte separate, and the dyads move toward opposite poles. During the first telophase, the dyads present at one pole are pinched off with very little surrounding cytoplasm to form the **first polar body**. The other daughter cell produced by this first meiotic division contains most of the cytoplasm and is called the **secondary oocyte**. The first polar body may or may not divide again to produce two small haploid cells. The mature ovum will be produced from the secondary oocyte during the second meiotic division. During this division, the cytoplasm of the secondary oocyte again divides unequally, producing an **ootid** and a **second polar body**. The ootid then differentiates into the mature ovum. Figure 2.11 illustrates the steps leading to formation of the mature ovum and polar bodies.

Unlike the divisions of spermatogenesis, the two meiotic divisions of oogenesis may not be continuous. In some animal species the two divisions may directly follow each other. In others, including the human species, the first division of all oocytes begins in the embryonic ovary, but arrests in prophase I. Many years later, the first division is reinitiated in each oocyte upon its ovulation. The second division is completed only after fertilization.

## THE SIGNIFICANCE OF MEIOSIS

The process of meiosis is critical to the successful sexual reproduction of all diploid organisms. It is the mechanism by which the diploid amount of genetic information is reduced to the haploid amount. In animals, meiosis leads to the formation of gametes, while in plants haploid spores are produced, which in turn lead to the formation of haploid gametes.

Each diploid organism contains its genetic information in the form of homologous pairs of chromosomes. Each pair consists of one member derived from the maternal parent and one from the paternal parent. Following meiosis, haploid cells contain either the paternal or maternal representative of each homologous pair of chromosomes. As a result, meiosis produces a vast amount of genetic variation. Additionally, crossing over, which occurs in the first meiotic prophase stage, further reshuffles the genetic information. Crossing over occurs between the maternal and paternal members of each homologous pair, resulting in the production of even greater amounts of genetic variation in gametes.

The two most significant points about meiosis are that the process leads to:

1. The maintenance of a constant *amount* of genetic information between generations.

2. Extensive genetic variation.

It is important that we also briefly elaborate on the important role that meiosis plays in the life cycle of fungi and plants. In many single-celled organisms in these groups, the predominant vegetative cells are haploid. They arise through meiosis and proliferate by cell division, including mitosis. In multicellular plants, the life cycle alternates between the diploid sporophyte stage and the haploid gametophyte stage. While one or the other predominates in different species during this "alternation of generations," the processes of meiosis and fertilization bridge the sporophyte and gametophyte generations (Figure 2.12).

Finally, it is important to know what happens when meiosis fails to achieve the expected outcome. As we pointed out earlier, rarely, at either the first or second division, separation or disjunction of the chromatids of a tetrad or dyad fails to occur. Instead, both members move to the same pole during anaphase. Such an event is called **nondisjunction**, because the two members fail to disjoin. If nondisjunction occurs at the first division stage, it is said to be a primary event; at the second division stage, it is called a secondary event.

The results of primary and secondary nondisjunction for just one chromosome of a diploid genome are shown in Figure 2.13. As can be seen, for the chromosome in question, some abnormal gametes may be formed containing either two members or none at all. Following fertilization of these with a normal gamete, a zygote is produced with either three members (**trisomy**) or only one member (**monosomy**) of this chromosome. While these conditions are more frequently tolerated in plants, they usually have severe or lethal effects in animals. Trisomy and monosomy will be described in greater detail in Chapter 6.

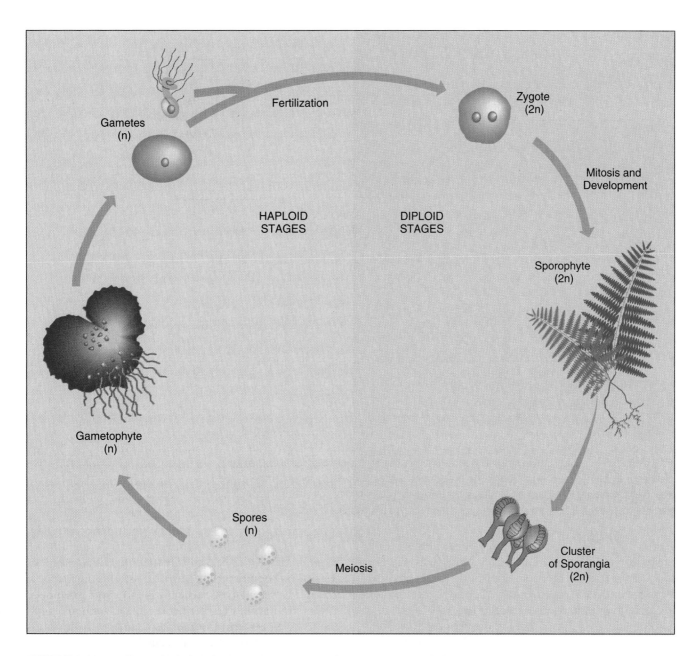

**FIGURE 2.12**    Alternation of generations between the diploid sporophyte (2*n*) and the haploid gametophyte (*n*) in a multicellular plant. The process of meiosis bridges the two phases of the life cycle.

## THE CYTOLOGICAL ORIGIN OF THE MITOTIC CHROMOSOME

Thus far in this chapter, we have focused on mitotic and meiotic chromosomes, emphasizing behavior during cell division and gamete formation. Initially, biologists knew about these chromosomes only from routine observations made with the light microscope. While chromo-somes are invisible during interphase, they appear during the prophase stage of mitosis. Geneticists were curious as to how this could happen. Based on studies using electron microscopy, we are now quite clear as to why chromosomes are visible only during mitosis.

During interphase, only dispersed chromatin fibers are present in the nucleus [Figure 2.14(a)]. **Chromatin** consists of DNA and associated proteins, particularly proteins called **histones**. It is now believed that during

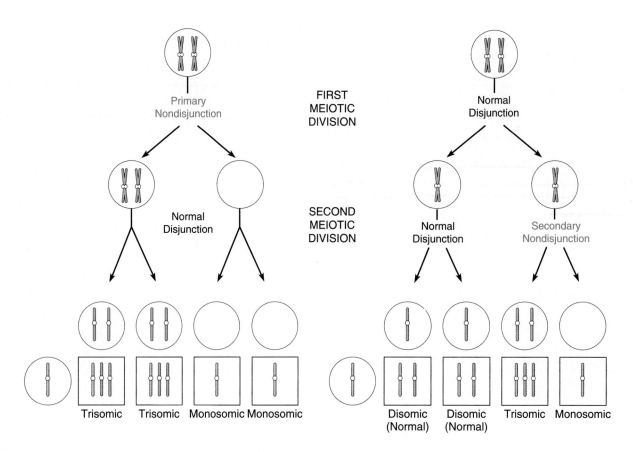

**FIGURE 2.13**    Diagram illustrating nondisjunction during the first and second meiotic divisions. In both cases, some gametes are formed either containing two members of a specific chromosome or lacking it altogether. Following fertilization by a normal haploid gamete, monosomic, disomic (normal), or trisomic zygotes result.

interphase, starting in $G_1$, mitotic chromosomes unwind to form these long chromatin fibers. It is in this physical arrangement that DNA can most efficiently function during transcription and can be replicated.

Once mitosis begins, however, the fibers coil and fold up, condensing into typical mitotic chromosomes [Figure 2.14(b)]. If the fibers making up the mitotic chromosome are loosened [Figure 2.14(c)], areas of greatest spreading reveal individual fibers similar to those seen in interphase chromatin. Very few fiber ends seem to be present. In some cases, none can be seen. Instead, individual fibers always seem to loop back into the interior. Such fibers are obviously twisted and coiled around one another, forming the regular pattern of the mitotic chromosome.

Observations of mitotic chromosomes in varying states of coiling led Ernest DuPraw to postulate the **folded-fiber model**, illustrated in Figure 2.14(d). During metaphase, each chromosome consists of two sister chromatids joined at the centromeric region. Each arm of the chromatid appears to consist of a single fiber wound up much like a skein of yarn. The fiber is composed of

double-stranded DNA and protein tightly coiled together. An orderly coiling–twisting–condensing process appears to be involved in the transition of the interphase chromatin to the more condensed, genetically inert mitotic chromosomes. During the transition from interphase to prophase, it is estimated that a 5000-fold contraction occurs in the length of DNA within the chromatin fiber!

In a later chapter, after we have provided a more thorough description of DNA structure, we will return to this topic and explore the molecular basis of the chromatin fiber (see Chapter 10).

## THE SYNAPTONEMAL COMPLEX

The electron microscope has also been used to visualize an additional ultrastructural component of the chromosome seen only in cells undergoing meiosis. This structure, first introduced during our earlier discussion of the first meiotic prophase stage, is intimately associated with synapsed homologues and is called the **synaptonemal**

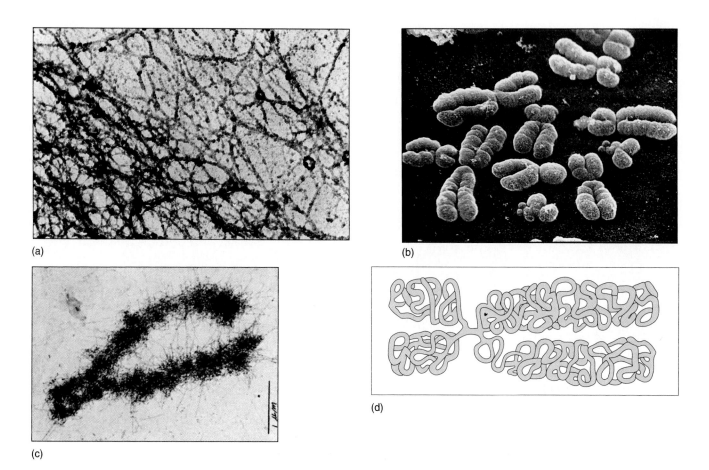

(a)

(b)

(c)

(d)

**FIGURE 2.14**    A comparison of: (a) the chromatin fibers characteristic of the interphase nucleus with (b) and (c) metaphase chromosomes that are derived from chromatin during mitosis. Part (d) depicts the folded-fiber model showing how chromatin is condensed into a metaphase chromosome. Parts (a) and (c) are transmission electron micrographs, while part (b) is a scanning electron micrograph.

**complex**.* In 1956, Montrose Moses observed this complex in spermatocytes of crayfish, and Don Fawcett saw it in pigeon and human spermatocytes. Because there was not yet any satisfactory explanation of the mechanism of synapsis or of crossing over and chiasmata formation, many researchers became interested in this structure. With few exceptions, the ensuing studies revealed the synaptonemal complex to be present in most plant and animal cells visualized during meiosis.

Figure 2.15(a) is an electron micrograph of the synaptonemal complex. It is composed primarily of a triplet set of parallel strands. The **central element** of this tripartite structure is usually less dense and thinner (100–150 Å) than the two identical outer elements (500 Å). The outer structures, called **lateral elements**, are intimately asso-

ciated with the chromatin of the synapsed homologues on either side. Selective staining has revealed that these lateral elements consist primarily of DNA and protein. This is consistent with the interpretation that the lateral elements contain chromatin. Some DNA fibrils traverse these elements, making connections with the central element, which is composed primarily of protein. Figure 2.15(b) provides a diagrammatic interpretation of the electron micrograph consistent with the above description.

The formation of the complex is initiated prior to the pachytene stage. As early as leptonema of the first meiotic prophase, lateral elements are seen in association with sister chromatids. Homologues have yet to associate with one another and are randomly dispersed in the nucleus. By the next stage, zygonema, homologous chromosomes begin to align with one another in what is called rough pairing, but remain distinctly apart by some 300 nm. Then, during pachynema, the intimate associa-

*An alternative spelling of this term is *synaptinemal complex*.

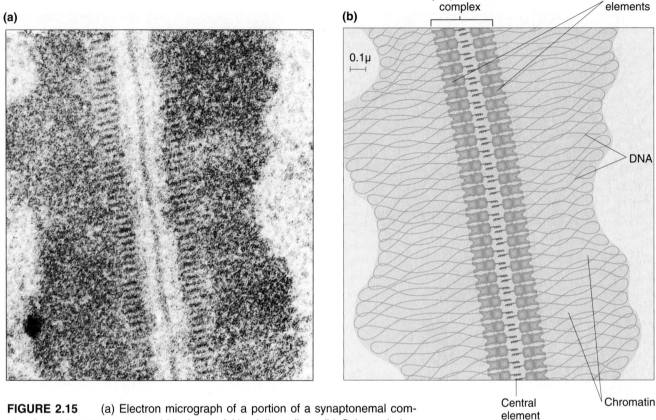

**FIGURE 2.15**    (a) Electron micrograph of a portion of a synaptonemal complex found between synapsed bivalents of *Neotiella rutilans*. (b) Schematic interpretation of the components making up the synaptonemal complex. The lateral elements, central element, and chromatin are labeled.

tion referred to as synapsis between homologues occurs as formation of the complex is completed. In some diploid organisms, this occurs in a zipperlike fashion beginning with the telomeres, which may be attached to the nuclear envelope.

It is now agreed that the synaptonemal complex is the vehicle for the pairing of homologues and their subsequent segregation during meiosis. However, some degree of synapsis can occur in certain cases where no synaptonemal complexes are formed. Thus, it is possible that the function of this structure may be more extensive than just its involvement in the formation of bivalents.

In certain instances where no synaptonemal complexes are formed during meiosis, synapsis is not so complete and crossing over is reduced or eliminated. For example, in *Drosophila melanogaster* meiotic crossing over rarely, if ever, occurs in males. Synaptonemal complexes are not usually seen during male meiosis. This observation suggests that the synaptonemal complex may be important in order for chiasmata to form and crossing over to occur.

In the yeast, *Saccharomyces cerevisiae*, study of a mutation, *ZIP1*, has provided other insights concerning chromosome pairing. Cells bearing this mutation can undergo the initial alignment stage (rough pairing) and full-length axial element formation, but fail to achieve the intimate pairing characteristics of synapsis. It has been suggested that the gene product of the *ZIP1* locus is a protein that is a component of the central element of the synaptonemal complex. In its absence, no complex is formed. This observation suggests that the synaptonemal complex functions in the transition from the initial rough alignment stage to the intimate pairing characteristic of synapsis. While synapsis and subsequent chiasmata formation are essential to the normal disjunction of chromosomes during meiosis, the available evidence does not directly link the synaptonemal complex to these latter events.

The discovery of the synaptonemal complex is significant to the field of genetics. If the complex is critical to synapsis during meiosis, then all sexual reproduction is dependent on it.

# SPECIALIZED CHROMOSOMES

We conclude this chapter by introducing two other unique types of chromosomes visible under the light microscope. It is useful to do so here in order to demonstrate several specialized forms that chromosomes may achieve besides their mitotic-like structures. Both types were studied extensively long before we understood the rationale for the appearance of chromosomes during mitosis.

## Polytene Chromosomes

Cells from a variety of organisms contain giant **polytene chromosomes**. They are found in various insect dipteran larval cells (salivary, midgut, rectal, and malpighian excretory tubules) and in several species of protozoans and plants. The large amount of information obtained from studies of these genetic structures has provided a model system for more recent investigations of chromosomes. Such structures were first observed by E. G. Balbiani in 1881 and are illustrated in Figure 2.16. What is particularly intriguing about these chromosomes is that they can be seen in the nuclei of interphase cells!

Each polytene chromosome observed under the light microscope is seen to be composed of a linear series of alternating bands and interbands (see insert in Figure 2.16). The banding pattern is distinctive for each chromosome in any given species. Individual bands are sometimes called **chromomeres**, a more generalized term describing lateral condensations of material along the axis of a chromosome. Each polytene chromosome is 200–600 $\mu$m long.

Following extensive study, it is now clear how such giant chromosomes are formed. Their large size and distinctiveness result from the many DNA strands that compose them. Many replication cycles of paired homologues occur, but without strand separation or cytoplasmic division. Thus, as replication proceeds, chromosomes are created having 1000–5000 DNA strands that remain in parallel register with one another. It is apparently the parallel register of so many DNA strands that gives rise to the distinctive band pattern along the axis of the chromosome.

It is obvious that these chromosomes bear genes that store genetic information. The presence of bands was initially interpreted as the visible manifestation of individual genes. When it became clear that the strands present in bands undergo localized uncoiling during genetic activity, this view was further strengthened. Each such uncoiling event results in what is called a **puff** because of its appearance (see Figure 2.17). That puffs are visible manifestations of gene activity (transcription that produces RNA) is evidenced by their high rate of incorporation of radioactivity labeled RNA precursors, as assayed by autoradiography. Bands that are not extended into puffs incorporate much less radioactive precursors or none at all.

The study of bands during development in insects such as *Drosophila* and the midge fly *Chironomus*, reveals differential gene activity. A characteristic pattern of band formation that is equated with gene activation is observed as development proceeds. This topic is pursued in more detail in Chapter 19. In spite of many recent genetic investigations, it is still not clear whether only a single gene is contained within each band. There

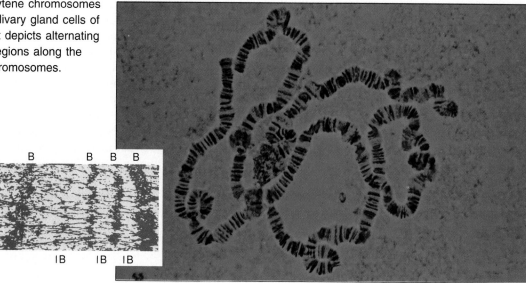

**FIGURE 2.16** Polytene chromosomes derived from larval salivary gland cells of *Drosophila*. The insert depicts alternating band and interband regions along the axis of these giant chromosomes.

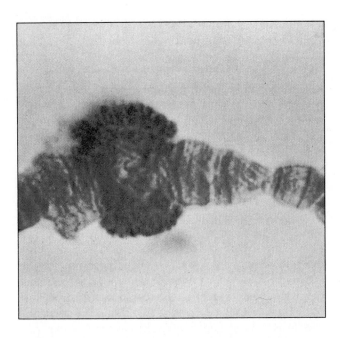

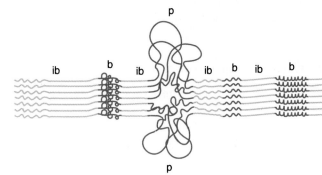

**FIGURE 2.17**    Photograph of a puff within a polytene chromosome. The schematic representation depicts the uncoiling of strands within a band (b) region to produce a puff (p) in polytene chromosomes. Interband regions (ib) are also labeled.

is much more DNA present per band than is needed to encode a single gene!

### Lampbrush Chromosomes

Another type of specialized chromosome that has provided insights into chromosomal structure is the **lampbrush chromosome**. It was given this name because it was imagined to be similar in appearance to the brushes used to clean lamp chimneys in centuries past. Lampbrush chromosomes were first discovered in 1892 in the oocytes of sharks and are now known to be characteristic of most vertebrate oocytes as well as spermatocytes of some insects. Therefore, they are meiotic chromosomes. Most experimental work has been done with material taken from amphibian oocytes.

These unique chromosomes are easily isolated from oocytes in the diplotene stage of the first prophase of meiosis, where they are active in directing the metabolic activities of the developing cell. The homologues are seen as synapsed pairs held together by chiasmata, but instead of condensing, as do most meiotic chromosomes, the lampbrush chromosomes are often extended to lengths of 500–800 $\mu$m. Later in meiosis they revert to

their normal length of 15–20 $\mu$m. Thus, lampbrush chromosomes are interpreted as uncoiled and unfolded versions of the normal meiotic chromosomes.

Figure 2.18 shows several views of these structures. Collectively, they provide significant insights into the morphology of these chromosomes. In part (a), the meiotic configuration is visualized under the light microscope. The linear axis of each structure contains large numbers of repeating condensations. As in polytene chromosomes, the more general term, chromomere, describes each condensation. Each chromomere appears to support a pair of **lateral loops**, which give the chromosome its distinctive appearance. In part (b), the scanning electron microscope (SEM) reveals an extended axis and adjacent loops. As with bands in polytene chromosomes, there is much more DNA present in each loop than is needed to encode a single gene.

Part (c) of Figure 2.18 presents another SEM, but at a much higher magnification. This micrograph provides a detailed view of one chromomere and the loop emanating from it. Each loop is thought to be composed of one DNA double helix, while the central axis is made up of two DNA helices. This hypothesis is consistent with the belief that each chromosome is composed of a pair of

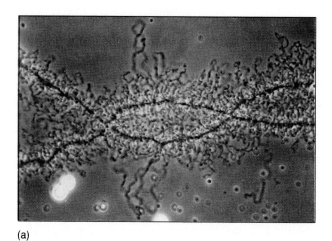

(a)

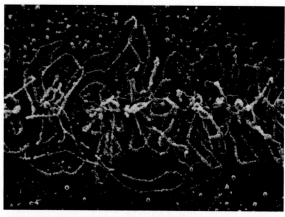

(b)

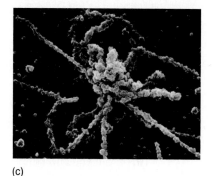

(c)

**FIGURE 2.18**    Lampbrush chromosomes derived from amphibian oocytes. Part (a) is a photomicrograph, while parts (b) and (c) are scanning electron micrographs.

sister chromatids. Studies using radioactive RNA precursors reveal that the loops are active in the synthesis of RNA. The lampbrush loops, in a way similar to puffs in polytene chromosomes, represent DNA that has been reeled out from the central chromomere axis during transcription. As with polytene chromosomes, the study of lampbrush chromosomes has provided many insights into the arrangement and function of the genetic information.

## CHAPTER SUMMARY

1.  Chromosomes, in diploid organisms, exist in homologous pairs. Each pair shares the same size, centromere placement, and gene sites. One member of each pair is derived from the maternal parent and one from the paternal parent.

2.  Mitosis and meiosis are mechanisms by which cells distribute genetic information contained in these chromosomes to their descendants in a precise, orderly fashion.

3.  Mitosis, or nuclear division, is part of the cell cycle and is the basis of cellular reproduction. Daughter cells are produced that are genetically identical to their progenitor cell.

4.  Meiosis, the underlying basis of sexual reproduction, results in the conversion of a diploid cell to a haploid gamete or spore. Each haploid cell contains one member of each homologous pair of chromosomes. Meiosis results in extensive genetic variation by virtue of the exchange between homologous chromosomes during crossing over and the random segregation of maternal and paternal chromatids.

5. Mitosis is subdivided into discrete phases: prophase, prometaphase, metaphase, anaphase, and telophase. Condensation of chromatin into chromosome structures occurs during prophase. During prometaphase, chromosomes take on the appearance of double structures, each represented by a pair of sister chromatids. In metaphase, chromosomes line up on the equatorial plane of the cell. During anaphase, sister chromatids of each chromosome are pulled apart and directed toward opposite poles. Telophase completes daughter cell formation and is characterized by cytokinesis, the division of the cytoplasm.

6. During cytokinesis in animal cells, the cytoplasm is pinched into two cells. In plants, a cell plate is synthesized and laid down, dividing the original cell into two daughter cells.

7. A major difference exists between meiosis in males and females. Spermatogenesis partitions cytoplasmic volume equally and produces four haploid sperm cells. Oogenesis, on the other hand, accumulates the cytoplasm around one egg cell and reduces the other haploid sets of genetic material to polar bodies. The extra cytoplasm contributes to zygote development following fertilization.

8. Mitotic chromosomes are produced as a result of the coiling and condensation of chromatin fibers characteristic of interphase. This transition is described by the "folded-fiber model."

9. The synaptonemal complex is an ultrastructural component of the cell present during the first meiotic prophase stage. It is important to the process of synapsis of homologues and may play a role in crossing over.

10. Polytene and lampbrush chromosomes are examples of specialized structures that have extended our knowledge of genetic organization and function.

## KEY TERMS

AB antigens
acrocentric chromosome
allele
anaphase
autoradiography
basal body
biparental inheritance
bivalent
capsule
cell coat
cell cycle
cell furrow
cell plate
cell respiration
cellulose
cell wall
centriole
centromere
centrosome
chiasma (chiasmata)
chromatid
chromatin
chromomere
chromosome
crossing over
cytokinesis
cytoplasm
cytoskeleton
cytosol
diakinesis
diploid number ($2n$)
diplotene stage
disjunction
Duchenne muscular dystrophy
dyad
dystrophin
endoplasmic reticulum (ER)
endosymbiont theory
equational division
folded-fiber model
$G_0$ stage
$G_1$ stage
$G_2$ stage
gamete
gene
genetic analysis
genetic material
genome, haploid
haploid number ($n$)
histocompatibility antigens
histones
homologous chromosomes
homology search
independent assortment
interphase
karyotype
kinetochore
lampbrush chromosome
lateral loop
leptotene stage
locus (loci)
meiosis
metacentric chromosome
metaphase
metaphase plate
middle lamella
mitosis
MN antigens
monad
monosomy
nondisjunction
nucleoid region
nucleolus
nucleolus organizer region (NOR)
nucleus
oocyte
oogenesis
oogonium
ootid
ovum (ova)
p arm
pachytene stage
peptidoglycan

| | | | |
|---|---|---|---|
| photosynthesis | q arm | spermatogenesis | synaptonemal complex |
| plasma membrane | R point | spermatogonium | telocentric chromosome |
| polar body | reductional division | spermatozoa | telophase |
| polytene chromosome | rough pairing | S phase | terminalization |
| prokaryote | segregation | spindle fibers | tetrad |
| prometaphase | sex chromosome | submetacentric | trisomy |
| prophase | sister chromatids |    chromosome | zygote |
| proto-oncogenes | spermatid | synapsis | zygotene stage |
| puff | spermatocyte | | |

# INSIGHTS AND SOLUTIONS

**W**ith this initial appearance of "Insights and Solutions," it is appropriate to provide a brief description of its value to you as a student. This section will precede the "Problems and Discussion Questions" in each chapter. One or more examples will be provided. Solutions to these problems and answers to these questions will illustrate approaches useful in **genetic analysis**. Our initial emphasis will be on insights that will help you arrive at correct solutions to ensuing problems.

1. In an organism with a haploid number of 3, how many individual chromosomal structures will align on the metaphase plate during (a) mitosis; (b) meiosis I; and (c) meiosis II? Describe each configuration.

   **SOLUTION:**   (a) In mitosis, where homologous chromosomes do not synapse, there will be 6 double structures, each consisting of a pair of sister chromatids. The number of structures is equivalent to the diploid number.

   (b) In meiosis I, the homologues have synapsed, reducing the number of structures to 3. Each is called a tetrad and consists of two pairs of sister chromatids.

   (c) In meiosis II, the same number of structures exist (3), but in this case, they are called dyads. Each consists of a pair of sister chromatids. When crossing over has occurred, each chromatid may contain part of one of its nonsister chromatids obtained during exchange in prophase I.

2. For the chromosomes illustrated in Figure 2.10 (p. 32), draw all possible alignment configurations that may occur during metaphase of meiosis I.

   **SOLUTION:**   As shown in the illustration below, there are four configurations possible when $n = 2$.

CASE I                                       CASE II

CASE III                          CASE IV

3.  As shown below, if we were to place one gene on both of the larger chromo-
    somes and use two alleles (one on each large chromosome), *A* and *a*, and place
    a second gene with two alleles on the smaller chromosomes, using *B* and *b*, cal-
    culate the probability of generating each gene combination (*AB, Ab, aB, ab*) fol-
    lowing meiosis I.

CASE I            CASE II           CASE III           CASE IV

**SOLUTION:**

Case I          *AB* and *ab*
Case II         *Ab* and *aB*
Case III        *aB* and *Ab*
Case IV         *ab* and *AB*

Total: *AB* = 2      ($p = 1/4$)
       *Ab* = 2      ($p = 1/4$)
       *aB* = 2      ($p = 1/4$)
       *ab* = 2      ($p = 1/4$)

4.  (a) How many different chromosome configurations can occur following meiosis
        I if three different pairs of chromosomes are present ($n = 3$)?

    **SOLUTION:**  If $n = 3$, then eight different configurations would be possible.
    The formula $2^n$, where $n$ equals the haploid number, will allow you to calcu-
    late the number of potential alignment patterns. As we will see in the next
    chapter, these patterns are produced as a result of the Mendelian postulate
    called segregation and they serve as the physical basis of the Mendelian pos-
    tulate of independent assortment.

    (b) Assuming that comparable chromosomes in different individuals are geneti-
        cally dissimilar because of different alleles, how many unique zygotic combi-
        nations are possible following fertilization in an organism where $n = 3$?

    **SOLUTION:**   Assuming that no crossing over occurs (and that maternal and
    paternal chromatids remain intact), then each organism can produce $2^n$ or $2^3$
    different chromosome combinations in their gametes. Following random fer-
    tilization,

    $$(2^3) \cdot (2^3) = 2^6 = 64$$

    unique combinations are possible in the offspring.

**PROBLEMS AND DISCUSSION QUESTIONS**

1. What role do the following cellular components play in the storage, expression, or transmission of genetic information: (a) chromatin, (b) nucleolus, (c) ribosome, (d) mitochondrion, (e) centriole, (f) centromere?

2. Discuss the concepts of homologous chromosomes, diploidy, and haploidy. What characteristics are shared between two chromosomes considered to be homologous?

3. If two chromosomes of a species are the same length and have similar centromere placements yet are *not* homologous, what *is* different about them?

4. Describe the events that characterize each stage of mitosis.

5. If an organism has a diploid number of 16, how many chromatids are visible at the end of mitotic prophase? How many chromosomes are moving to each pole during anaphase of mitosis?

6. How are chromosomes named on the basis of centromere placement?

7. Contrast telophase in plant and animal mitosis.

8. Outline and discuss the events of the cell cycle. What experimental technique was used to demonstrate the existence of the S phase?

9. Contrast the end results of meiosis with those of mitosis.

10. Define and discuss the following terms: (a) synapsis, (b) bivalents, (c) chiasmata, (d) crossing over, (e) chromomeres, (f) sister chromatids, (g) tetrads, (h) dyads, (i) monads, and (j) synaptonemal complex.

11. If an organism has a diploid number of 16 in an oocyte,
    (a) How many tetrads are present in the first meiotic prophase?
    (b) How many dyads are present in the second meiotic prophase?
    (c) How many monads migrate to each pole during the second meiotic anaphase?

12. Contrast spermatogenesis and oogenesis. What is the significance of the formation of polar bodies?

13. Explain why meiosis leads to significant genetic variation while mitosis does not.

14. In Figure 2.10 (p. 32), the four spermatids formed as a result of meiosis each contain a unique combination of paternally and maternally derived chromosome segments (as represented by the different colors). These four different combinations of genetic information resulted from the random alignment of both sets of chromosomes on the equatorial plate during metaphase I and metaphase II. Had they become aligned differently, many other chromosome combinations might have been produced in the spermatids. With reference to Figure 2.10, how many other combinations are possible? Draw them.

15. During oogenesis in an animal species with a haploid number of 6, one dyad undergoes secondary nondisjunction. Following the second meiotic division, the involved dyad ends up intact in the ovum. How many chromosomes are present in: (a) the mature ovum, and (b) the second polar body? (c) Following fertilization by a normal sperm, what chromosome condition is created?

16. Assume that a diploid cell contains three pairs of homologous chromosomes designated $C^1C^2$, $M^1M^2$, and $S^1S^2$ and that no crossing over occurs. What possible configurations of chromosomes will be present in: (a) two daughter cells following mitosis; (b) the first meiotic metaphase; and (c) haploid cells following meiosis II?

17. Predict the number of different haploid cells that will occur if a fourth chromosome pair ($W^1W^2$) is considered along with chromosomes $C^1C^2$, $M^1M^2$, and $S^1S^2$ from the previous problem.

18. During the first meiotic prophase, (a) when does crossing over occur? (b) When does synapsis occur? (c) During which stage are the chromosomes least condensed? (d) When are chiasmata first visible?

19. What is the role of meiosis in the life cycle of a higher plant such as an angiosperm?

20. Describe the transition of a chromatin fiber into a mitotic chromosome. What model depicts this transition?

21. When during the cell cycle (Figure 2.6) do the various transition states described in Problem 20 first occur?

22. How are giant polytene chromosomes formed?

23. What genetic process is occurring in a "puff" of a polytene chromosome?

24. During what genetic process are lampbrush chromosomes present in vertebrates?

25. What ploidy value characterizes lampbrush chromosomes?

**SELECTED READINGS**

ALBERTS, B., et al. 1989. *Molecular biology of the cell.* 2nd ed. New York: Garland.

ANGELIER, N., et al. 1984. Scanning electron microscopy of amphibian lampbrush chromosomes. *Chromosoma* 89:243–53.

BAKER, B. A., et al. 1976. The genetic control of meiosis. *Ann. Rev. Genet.* 10:53–134.

BASERGA, R., and KISIELESKI, W. 1963. Autobiographics of cells. *Scient. Amer.* (Aug.) 209:103–110.

BEERMAN, W., and CLEVER, U. 1964. Chromosome puffs. *Scient. Amer.* (April) 210:50–58.

BRACHET, J., and MIRSKY, A. E. 1961. *The cell: Meiosis and mitosis.* Vol. 3. Orlando: Academic Press.

CALLAN, H. G. 1986. *Lampbrush chromosomes.* New York: Springer-Verlag.

DUPRAW, E. J. 1970. *DNA and chromosomes.* New York: Holt, Rinehart and Winston.

GALL, J. G. 1963. Kinetics of deoxyribonuclease on chromosomes. *Nature* 198:36–38.

GOLOMB, H. M., and BAHR, G. F. 1971. Scanning electron microscopic observations of surface structures of isolated human chromosomes. *Science* 171:1024–26.

HALL, J. L., RAMANIS, Z., and LUCK, D. J. 1989. Basal body/centriolar DNA: Molecular genetic studies in *Chlamydomonas. Cell* 59:121–32.

HARTWELL, L. H. 1978. Cell division from a genetic perspective. *J. Cell Biol.* 77:627–37.

HARTWELL, L. H., et al. 1974. Genetic control of the cell division cycle in yeast. *Science* 183:46–51.

HAWLEY, R. S., and ARBEL, T. 1993. Yeast genetics and the fall of the classical view of meiosis. *Cell* 72:301–03.

HILL, R. J., and RUDKIN, G. T. 1987. Polytene chromosomes. The status of the band–interband question. *BioEssays* 7:35–40.

MAZIA, D. 1961. How cells divide. *Scient. Amer.* (Jan.) 205:101–120.

———. 1974. The cell cycle. *Scient. Amer.* (Jan.) 235:54–64.

MCINTOSH, J. R., and MCDONALD, K. L. 1989. The mitotic spindle. *Scient. Amer.* (Oct.) 261:48–56.

MOENS, P. B. 1973. Mechanisms of chromosome synapsis at meiotic prophase. *Int. Rev. Cytol.* 35:117–134.

PARDEE, A. B., et al. 1978. Animal cell cycle. *Ann. Rev. Biochem.* 47:715–50.

PRESCOTT, D. M. 1977. *Reproduction of eukaryotic cells.* Orlando: Academic Press.

PRESCOTT, D. M., and FLEXER, A. S. 1986. *Cancer, the misguided cell.* 2nd ed. Sunderland, MA: Sinauer.

SWANSON, C. P., MERZ, T., and YOUNG, W. J. 1981. *Cytogenetics, the chromosome in division, inheritance, and evolution.* 2nd ed. Englewood Cliffs, N. J.: Prentice-Hall.

WESTERGAARD, M., and vonWETTSTEIN, D. 1972. The synaptinemal complex. *Ann. Rev. Genet.* 6:71–110.

WHEATLEY, D. N. 1982. *The centriole: A central enigma of cell biology.* New York: Elsevier/ North-Holland Biomedical.

YUNIS, J. J., and CHANDLER, M. E. 1979. Cytogenetics. In *Clinical diagnosis and management by laboratory methods,* ed. J. B. Henry, vol. 1. Philadelphia: W. B. Saunders.

ZIMMERMAN, A. M., and FORER, A., eds. 1981. *Mitosis/Cytokinesis.* New York: Academic Press.

# 3

# MENDELIAN GENETICS

Open pods of the garden pea, *Pisum sativum*.

*Inherited characteristics are the result of particulate factors called genes that are transmitted from generation to generation on vehicles called chromosomes, according to rules first described by Gregor Mendel. Genetic ratios, expressed as probabilities, are subject to chance deviation and may be evaluated using statistical analysis.*

Although inheritance of biological traits has been recognized for thousands of years, the first significant insights into the mechanisms involved occurred only a little over a century ago. In 1866, Gregor Mendel published the results of a series of experiments that would lay the foundation for the formal discipline of genetics. In the ensuing years the concept of the gene as a distinct hereditary unit was established, and the ways in which genes are transmitted to offspring and control traits were clarified. Research in these areas was accelerated in the first half of the twentieth century. The resultant findings served as the foundation for the tremendous interest and research effort in genetics since about 1940. It is safe to say that studies in genetics, particularly at the molecular level, have remained continually at the forefront of biological research since that time.

In this chapter we focus on the development of the principles established by Mendel, now referred to as **Mendelian** or **transmission genetics**. These principles describe how genes are transmitted from parents to offspring and were derived directly from his experimentation.

When Mendel began his studies of inheritance using *Pisum sativum*, the garden pea, there was no knowledge of chromosomes nor of the role and mechanism of meiosis. Nevertheless, he was able to determine that distinct **units of inheritance** exist and to predict their behavior during the formation of gametes. Subsequent investigators, with access to cytological data, were able to relate their observations of chromosome behavior during meiosis to Mendel's principles of inheritance. Once this correlation was made, Mendel's postulates were accepted as the basis for the study of transmission genetics. Even today, they serve as the cornerstone of the study of inheritance.

# GREGOR JOHANN MENDEL

In 1822, Johann Mendel was born of a peasant family in the European village of Heinzendorf, now part of the Czech Republic. An excellent student in high school,

Mendel studied philosophy for several years afterward and was admitted to the Augustinian Monastery of St. Thomas in Brno in 1843. There, he took the name of Gregor and received support for his studies and research throughout the rest of his life. In 1849, he was relieved of pastoral duties and received a teaching appointment that lasted several years. From 1851 to 1853, he attended the University of Vienna, where he studied physics and botany. In 1854 he returned to Brno, where, for the next sixteen years, he taught physics and natural science.

In 1856, Mendel performed the first set of hybridization experiments with the garden pea. The research phase of his career lasted until 1868, when he was elected abbot of the monastery. Although his interest in genetics remained, his new responsibilities demanded most of his time. In 1884, Mendel died of a kidney disorder. The local newspaper paid him the following tribute: "His death deprives the poor of a benefactor, and mankind at large of a man of the noblest character, one who was a warm friend, a promoter of the natural sciences, and an exemplary priest. . . . "

## Mendel's Experimental Approach

In 1865, Mendel first reported the results of some simple genetic crosses between certain strains of the garden pea. Although, as we saw in Chapter 1, his was not the first attempt to provide experimental evidence pertaining to inheritance, Mendel's work is an elegant model of experimental design and analysis.

Mendel showed remarkable insight into the methodology necessary for good experimental biology. He chose an organism that was easy to grow and interbreed. The pea plant is self-fertilizing in nature, but is easily crossbred in designed experiments. It reproduces well and grows to maturity in a single season. Mendel worked with seven unit characters, visible features that were each represented by two contrasting forms or traits. For the character stem height, for example, he experimented with the traits *tall* and *dwarf*. He selected six other contrasting pairs of traits involving seed shape and color, pod shape and color, and pod and flower arrangement. True-breeding strains were available from local seed merchants. Each trait appeared generation after generation in self-fertilizing plants; that is, the strains exhibiting them "bred true."

Mendel's success in an area where others had failed may be attributed to several factors in addition to the choice of a suitable organism. He restricted his examination to one or very few pairs of contrasting traits in each experiment. He also kept accurate quantitative records, a necessity in genetic experiments. From the analysis of his data, Mendel derived certain postulates that have become the principles of transmission genetics.

The significance of Mendel's experiments was not realized until the early twentieth century, well after his death. Once Mendel's publications were rediscovered by geneticists investigating the function and behavior of chromosomes, the implications of his postulates were immediately apparent. He had discovered the basis for the transmission of hereditary traits!

# THE MONOHYBRID CROSS

The simplest crosses performed by Mendel involved only one pair of contrasting traits. Each such breeding experiment involves a **monohybrid cross**. A monohybrid cross is constructed by mating individuals from two parent strains, each of which exhibits one of the two contrasting forms of the character under study. Initially we will examine the first generation of offspring of such a cross, and then we will consider the offspring of **selfing** or **self-fertilizing** individuals from this first generation. The original parents are called the **$P_1$** or **parental generation**, their offspring are the **$F_1$** or **first filial generation**, and the individuals resulting from the selfing of the $F_1$ generation are called the **$F_2$** or **second filial generation**. We can, of course, continue to follow subsequent generations, if desirable.

The cross between true-breeding peas with tall stems and dwarf stems is representative of Mendel's monohybrid crosses. *Tall* and *dwarf* represent contrasting forms or traits of the character of stem height. Unless tall or dwarf plants are crossed together or with another strain, they will undergo self-fertilization and breed true, producing their respective trait generation after generation. However, when Mendel crossed tall plants with dwarf plants, the resulting $F_1$ generation consisted only of tall plants. When members of the $F_1$ generation were selfed, Mendel observed that 787 of 1064 $F_2$ plants were tall, while 277 of 1064 were dwarf. Note that in this cross (Figure 3.1) the dwarf trait disappears in the $F_1$, only to reappear in the $F_2$ generation.

Genetic data are usually expressed and analyzed as

ratios. In this particular example, many identical $P_1$ crosses were made and many $F_1$ plants—all tall—were produced. Of the 1064 $F_2$ offspring, 787 were tall and 277 were dwarf—a ratio of approximately 2.8 : 1.0, or about 3 : 1.

Mendel made similar crosses between pea plants exhibiting each of the other pairs of contrasting traits. The results of these crosses are also shown in Figure 3.1. In every case, the outcome was similar to the tall/dwarf cross just described. All $F_1$ offspring were identical to one of the parents. In the $F_2$, an approximate ratio of 3 : 1 was obtained. Three-fourths appeared like the $F_1$ plants, while one-fourth exhibited the contrasting trait, which had disappeared in the $F_1$ generation.

It is appropriate to point out one further aspect of the monohybrid crosses. In each, the $F_1$ and $F_2$ patterns of inheritance were similar regardless of which $P_1$ plant served as the source of pollen, or sperm, and which served as the source of the ovum, or egg. The crosses could be made either way—that is, pollen from the tall plant pollinating dwarf plants, or vice versa. These are called **reciprocal crosses**. Therefore, the results of Mendel's monohybrid crosses were not sex-dependent.

To explain these results, Mendel proposed the existence of particulate **unit factors** for each trait. He suggested that these factors serve as the basic units of heredity and are passed unchanged from generation to generation, determining various traits expressed by each individual plant. Using these general ideas, Mendel proceeded to hypothesize precisely how such factors could account for the results of the monohybrid crosses.

## Mendel's First Three Postulates

Using the consistent pattern of results in the monohybrid crosses, Mendel derived the following three postulates or principles of inheritance.

1. UNIT FACTORS IN PAIRS
   **Genetic characters are controlled by unit factors that exist in pairs in individual organisms**. In the monohybrid cross involving tall and dwarf stems, a specific unit factor exists for each trait. Because the factors occur in pairs, three combinations are possible: two factors for tallness, two factors for dwarfness, or one of each factor. Every individual contains one of these three combinations, which determines stem height.

2. DOMINANCE/RECESSIVENESS
   **When two unlike unit factors responsible for a single character are present in a single individual, one unit factor is dominant to the other, which is said to be recessive**.

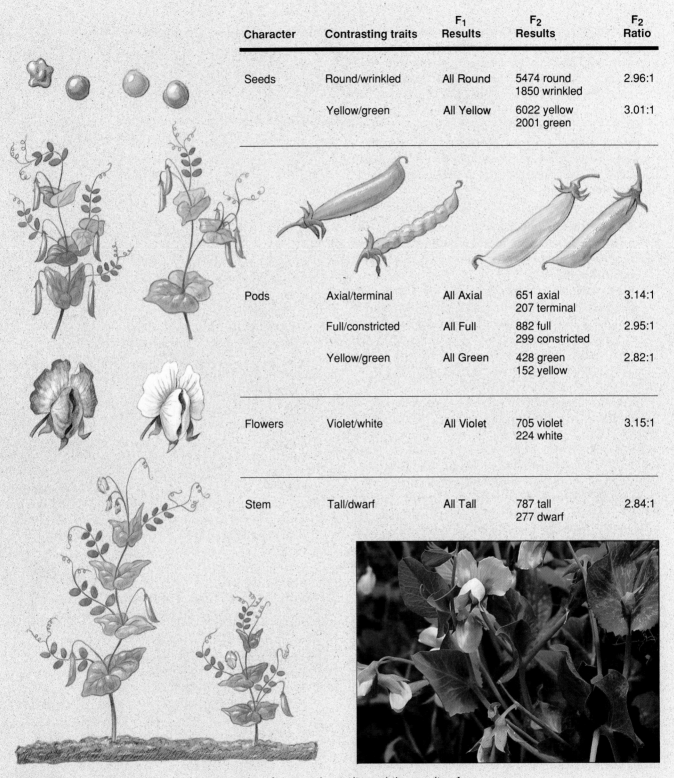

| Character | Contrasting traits | $F_1$ Results | $F_2$ Results | $F_2$ Ratio |
|---|---|---|---|---|
| Seeds | Round/wrinkled | All Round | 5474 round<br>1850 wrinkled | 2.96:1 |
| | Yellow/green | All Yellow | 6022 yellow<br>2001 green | 3.01:1 |
| Pods | Axial/terminal | All Axial | 651 axial<br>207 terminal | 3.14:1 |
| | Full/constricted | All Full | 882 full<br>299 constricted | 2.95:1 |
| | Yellow/green | All Green | 428 green<br>152 yellow | 2.82:1 |
| Flowers | Violet/white | All Violet | 705 violet<br>224 white | 3.15:1 |
| Stem | Tall/dwarf | All Tall | 787 tall<br>277 dwarf | 2.84:1 |

**FIGURE 3.1**    A summary of the seven pairs of contrasting traits and the results of Mendel's seven monohybrid crosses. In each case, pollen derived from plants exhibiting one contrasting trait was used to fertilize the ova of plants exhibiting the other contrasting trait. In the $F_1$ generation one of the two traits, referred to as dominant, was exhibited by all plants. The contrasting trait, referred to as recessive, then reappeared in approximately one-fourth of the $F_2$ plants. The garden pea (*Pisum sativum*) is shown in the photograph.

In each monohybrid cross, the trait expressed in the $F_1$ generation results from the presence of the dominant unit factor. The trait that is not expressed in the $F_1$, but which reappears in the $F_2$, is under the genetic influence of the recessive unit factor. Note that this dominance/recessiveness relationship only pertains when unlike unit factors are present together in an individual. The terms **dominant** and **recessive** are also used to designate the traits. In the above case, tall stems are said to be dominant to the recessive dwarf stems.

3. SEGREGATION
   **During the formation of gametes, the paired unit factors separate or segregate randomly so that each gamete receives one or the other with equal likelihood.**
   If an individual contains a pair of like unit factors (e.g., both specify tall), then all gametes receive one tall unit factor. If an individual contains unlike unit factors (e.g., one for tall and one for dwarf), then each gamete has a 50 percent probability of receiving either the tall *or* dwarf unit factor.

These postulates provide a suitable explanation for the results of the monohybrid crosses. The tall/dwarf cross will be used to illustrate this explanation. Mendel reasoned that $P_1$ tall plants contained identical paired unit factors, as did the $P_1$ dwarf plants. The gametes of tall plants all received one tall unit factor as a result of segregation. Likewise, the gametes of dwarf plants all received one dwarf unit factor. Following fertilization, all $F_1$ plants received one unit factor from each parent, a tall factor from one and a dwarf factor from the other, reestablishing the paired relationship. Because tall is dominant to dwarf, all $F_1$ plants were tall.

When $F_1$ plants form gametes, the postulate of segregation demands that each gamete randomly receive *either* the tall *or* dwarf unit factor. Following random fertilization events during $F_1$ selfing, four $F_2$ combinations will result in equal frequency:

        (1) tall/tall
        (2) tall/dwarf
        (3) dwarf/tall
        (4) dwarf/dwarf

Combinations (1) and (4) will clearly result in tall and dwarf plants, respectively. According to the postulate of dominance/recessiveness, combinations (2) and (3) will both yield tall plants. Therefore, the $F_2$ is predicted to consist of three-fourths tall and one-fourth dwarf, or a ratio of 3:1. This is approximately what Mendel observed in the cross between tall and dwarf plants. A similar pattern was observed in each of the other monohybrid crosses.

## Modern Genetic Terminology

In order to illustrate the monohybrid cross and Mendel's first three postulates, we must introduce several new terms as well as a set of symbols for the unit factors. Traits such as tall or dwarf are visible expressions of the information contained in unit factors. The physical appearance of a trait is called the **phenotype** of the individual.

All unit factors represent units of inheritance called **genes** by modern geneticists. For any given character, such as plant height, the phenotype is determined by the presence of different combinations of alternative forms of a single gene called **alleles**. For example, the unit factors representing tall and dwarf are alleles determining the height of the pea plant.

By one convention, the first letter of the recessive trait may be chosen to symbolize the character in question. The lowercase letter designates the allele for the recessive trait, and the uppercase letter designates the allele for the dominant trait. Therefore, we let $d$ stand for the dwarf allele and $D$ represent the tall allele. When alleles are written in pairs to represent the two unit factors present in any individual ($DD$, $Dd$, or $dd$), these symbols are referred to as the **genotype**. This term reflects the genetic makeup of an individual whether it is haploid or diploid. By reading the genotype, it is possible to know the phenotype of the individual: $DD$ and $Dd$ are tall, and $dd$ is dwarf. When identical alleles exist ($DD$ or $dd$), the individual is said to be **homozygous** or a **homozygote**; when alleles are different ($Dd$), we use the term **heterozygous** or **heterozygote**. These symbols and terms are used in Figure 3.2 to illustrate the complete monohybrid cross.

## Mendel's Analytical Approach

What led Mendel to deduce unit factors in pairs? Since there were two contrasting traits for each character, it seemed logical that two distinct factors must exist. However, why does one of the two traits or phenotypes disappear in the $F_1$ generation? Observation of the $F_2$ generation helps to answer this question. The recessive trait and its unit factor do not actually disappear in the $F_1$; they are merely hidden or masked, only to reappear in one-fourth of the $F_2$ offspring. Therefore, Mendel concluded that one unit factor for tall and one for dwarf were transmitted to each $F_1$ individual; but because the tall factor or allele is dominant to the dwarf factor or allele, all $F_1$ plants are tall. Finally, how is the 3:1 $F_2$ ratio explained? As shown in Figure 3.2, if the tall and dwarf alleles of the $F_1$ heterozygote segregate randomly into gametes, and if fertilization is random, this ratio is the natural outcome of the cross.

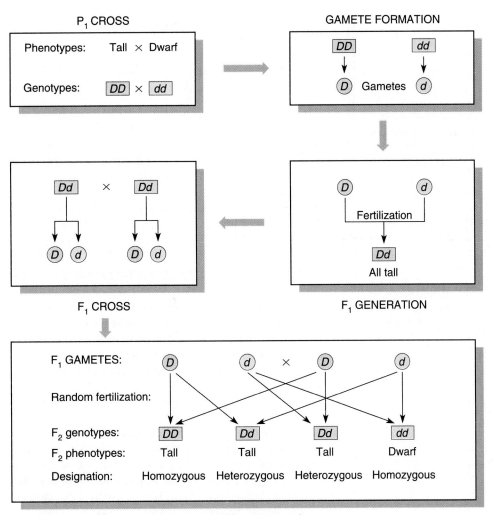

**FIGURE 3.2**   An explanation of the monohybrid cross between tall and dwarf pea plants. The symbols *D* and *d* are used to designate the tall and dwarf unit factors, respectively, in the genotypes of mature plants and gametes. All individuals are shown in rectangles. All gametes are shown in circles.

Because he operated without the hindsight that modern geneticists enjoy, Mendel's analytical reasoning must be considered a truly outstanding scientific achievement. On the basis of rather simple, but precisely executed breeding experiments, he proposed that discrete **particulate units of heredity** exist, and he explained how they are transmitted from one generation to the next!

## Punnett Squares

The genotypes and phenotypes resulting from the recombination of gametes during fertilization can be easily visualized by constructing a **Punnett square**, so named after the person who first devised this approach, Reginald C. Punnett. Figure 3.3 illustrates this method of analysis for the $F_1 \times F_1$ monohybrid cross. All possible gametes are assigned to a column or a row, with the vertical column representing those of the female parent and the horizontal row those of the male parent. After

entering the gametes in rows and columns, the new generation is predicted by combining the male and female gametic information for each combination and entering the resulting genotypes in the boxes. This process represents all possible random fertilization events. The genotypes and phenotypes of all potential offspring are ascertained by reading the entries in the boxes.

The Punnett square method is particularly useful when one is first learning about genetics and how to solve problems. In Figure 3.3, note the ease with which the 3:1 phenotypic ratio and the 1:2:1 genotypic ratio may be derived in the $F_2$ generation.

## The Test Cross: One Character

Tall plants produced in the $F_2$ generation are predicted to be of either the *DD* or *Dd* genotypes. We might ask if there is a way to distinguish the genotype. Mendel devised a rather simple method that is still used today in

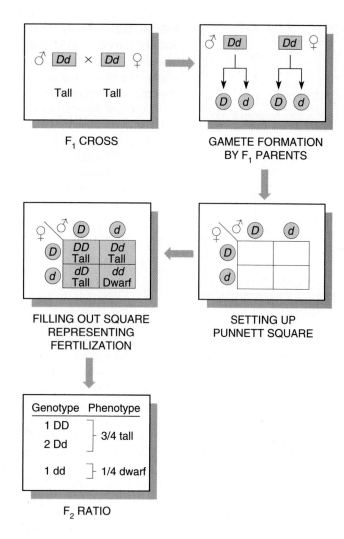

FIGURE 3.3    The use of a Punnett square in generating the F$_2$ ratio of the F$_1$ × F$_1$ cross shown in Figure 3.2.

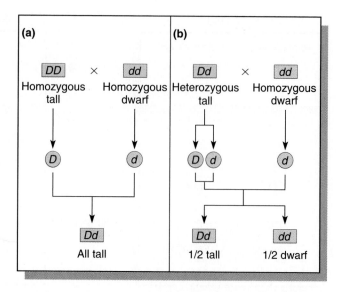

FIGURE 3.4    The test cross illustrated with a single character. In (a), the tall parent is homozygous. In (b), the tall parent is heterozygous. The genotypes of each tall parent may be determined by examining the offspring when each is crossed to the homozygous recessive dwarf plant.

breeding procedures of plants and animals: the **test cross**. The organism of the dominant phenotype but unknown genotype is crossed to a **homozygous recessive individual**. For example, if a tall plant of genotype *DD* is test-crossed to a dwarf plant, which must have the *dd* genotype, all offspring will be tall phenotypically and *Dd* genotypically. However, if a tall plant is *Dd* and is crossed to a dwarf plant (*dd*), then one-half of the offspring will be tall (*Dd*) and the other half will be dwarf (*dd*). Therefore, a 1:1 tall/dwarf ratio demonstrates the heterozygous nature of the tall plant of unknown genotype. The basis for these conclusions is illustrated in Figure 3.4. The test cross reinforced Mendel's conclusion that separate unit factors control the tall and dwarf traits.

# THE DIHYBRID CROSS

A natural extension of performing monohybrid crosses was for Mendel to design experiments where two characters were examined simultaneously. Such a cross, involving two pairs of contrasting traits, is a **dihybrid cross**. It is also called a **two-factor cross**. For example, if pea plants having yellow seeds that are also round were bred with those having green seeds that are also wrinkled, the results shown in Figure 3.5 will occur. The F$_1$ offspring are all yellow and round. It is therefore apparent that yellow is dominant to green, and that round is dominant to wrinkled. In this dihybrid cross, the F$_1$ individuals are selfed, and approximately 9/16 of the F$_2$ plants express yellow and round, 3/16 express yellow and wrinkled, 3/16 express green and round, and 1/16 express green and wrinkled.

A variation of this cross is also shown in Figure 3.5. If, instead of having one P$_1$ parent with both dominant traits (yellow, round) and one with both recessive traits (green, wrinkled), plants with yellow, wrinkled seeds can be bred with those having green, round seeds in a P$_1$ cross. In spite of the change in parental phenotypes, both the F$_1$ and F$_2$ results remain unchanged. It will become clear in the next section why this is so.

## Mendel's Fourth Postulate: Independent Assortment

We can most easily understand the results of a dihybrid cross if we consider it theoretically as consisting of two monohybrid crosses conducted separately. Think of the two sets of traits as being inherited independently of each other; that is, the chance of any plant becoming tall

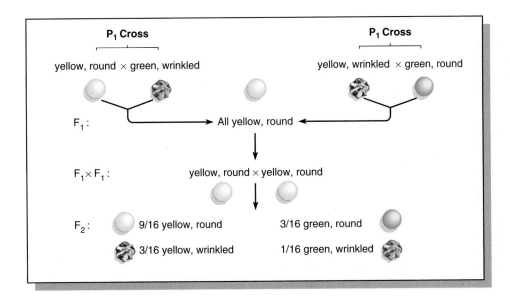

**FIGURE 3.5** The $F_1$ and $F_2$ results of Mendel's dihybrid crosses between yellow, round and green, wrinkled pea plants, and between yellow, wrinkled and green, round pea plants.

or dwarf is not at all influenced by the chance that this plant will also have round or wrinkled seeds. Thus, because yellow is dominant to green, all $F_1$ plants in the first theoretical cross would have yellow seeds. In the second theoretical cross, all $F_1$ plants would have round seeds because round is dominant to wrinkled. When Mendel examined the $F_1$ plants of the dihybrid cross, all were yellow and round, as predicted.

The predicted $F_2$ results of the first cross are 3/4 yellow and 1/4 green. Similarly, the second cross should yield 3/4 round and 1/4 wrinkled. Figure 3.5 shows that in the dihybrid cross, 12/16 of all $F_2$ plants are yellow while 4/16 are green, exhibiting the 3:1 ratio. Similarly, 12/16 of all $F_2$ plants have round seeds while 4/16 have wrinkled seeds, again revealing the 3:1 ratio.

Because the two pairs of contrasting traits are inherited independently, we can predict the frequencies of all possible $F_2$ phenotypes by applying the "product law" of probabilities: **When two independent events occur simultaneously, the combined probability of the two outcomes is equal to the product of their individual probabilities of occurrence**. For example, the probability of an $F_2$ plant having yellow *and* round seeds is (3/4)(3/4), or 9/16, because 3/4 of all $F_2$ plants should be yellow and 3/4 of all $F_2$ plants should be round.

In a like way, the probabilities of the other three $F_2$ phenotypes may be calculated: yellow (3/4) *and* wrinkled (1/4) are predicted to be present together 3/16 of the time; green (1/4) *and* round (3/4) are predicted 3/16 of the time, and green (1/4) *and* wrinkled (1/4) are predicted 1/16 of the time. These calculations are illustrated in Figure 3.6. It is now apparent why the $F_1$ and $F_2$ results are identical whether the initial cross is yellow, round

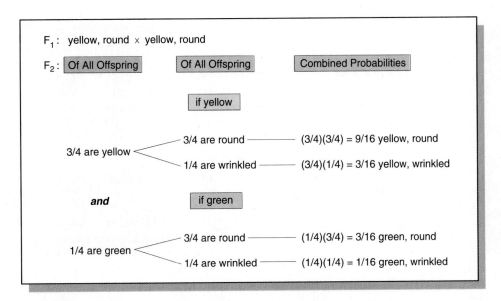

**FIGURE 3.6** The determination of the combined probabilities of each $F_2$ phenotype for two independently inherited characters. The probability of each plant being yellow or green is independent of the probability of it being round or wrinkled.

**FIGURE 3.7** Diagram of the dihybrid crosses shown in Figure 3.5. The $F_1$ heterozygous plants are self-fertilized to produce an $F_2$ generation, which is computed using a Punnett square. Both the phenotypic and genotypic $F_2$ ratios are shown. The photographs illustrate the round/wrinkled and yellow/green phenotypes of peas.

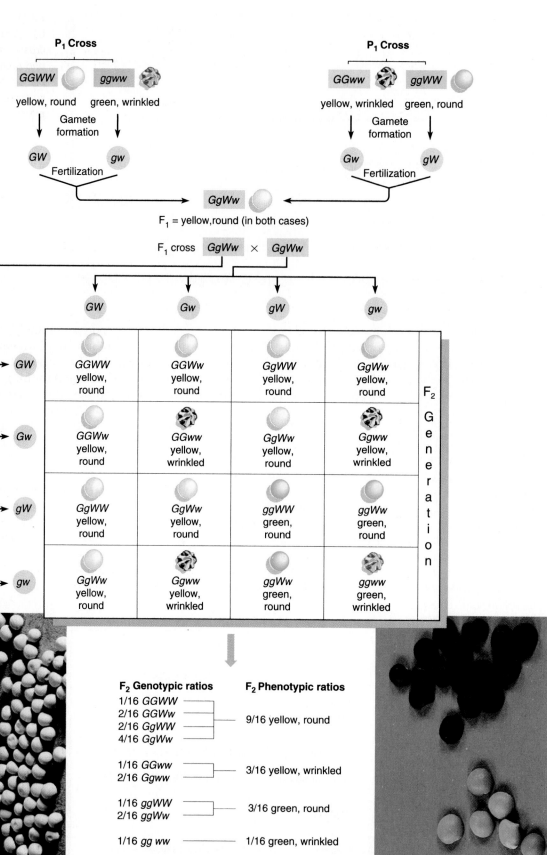

bred with green, wrinkled or if yellow, wrinkled are bred with green, round. In both crosses, the F₁ genotype of all plants is identical. Each plant is heterozygous for both gene pairs. As a result, the F₂ generation is also identical in both crosses.

On the basis of similar results in numerous dihybrid crosses, Mendel proposed a fourth postulate:

4.  INDEPENDENT ASSORTMENT
    **During gamete formation, segregating pairs of unit factors assort independently of each other**.

This postulate stipulates that segregation of any pair of unit factors occurs independently of all others. As a result of segregation, each gamete receives one member of every pair of unit factors. For one pair, whichever unit factor is received does not influence the outcome of segregation of any other pair. Thus, according to the postulate of independent assortment, all possible combinations of gametes will be formed in equal frequency.

Independent assortment is illustrated in the formation of the F₂ generation, shown in the Punnett square in Figure 3.7. Examine the formation of gametes by the F₁ plants. Segregation prescribes that every gamete receives either a *G* or *g* allele *and* a *W* or *w* allele. Independent assortment stipulates that all four combinations (*GW*, *Gw*, *gW*, and *gw*) will be formed with equal probabilities.

In every F₁ × F₁ fertilization event, each zygote has an equal probability of receiving one of the four combinations from each parent. If a large number of offspring are produced, 9/16 are yellow and round, 3/16 are yellow and wrinkled, 3/16 are green and round, and 1/16 are green and wrinkled, yielding what is designated as **Mendel's 9:3:3:1 dihybrid ratio**. This ratio is based on probability events involving segregation, independent assortment, and random fertilization. Therefore, it is an ideal ratio. Because of deviation due strictly to chance, particularly if small numbers of offspring are produced, the ideal ratio will seldom be found.

## The Test Cross: Two Characters

The test cross may also be applied to individuals that express two dominant traits, but whose genotypes are unknown. For example, the expression of the yellow, round phenotype in the F₂ generation just described may result from the *GGWW*, *GGWw*, *GgWW*, and *GgWw* genotypes. If an F₂ yellow, round plant is crossed with the homozygous recessive green, wrinkled plant (*ggww*), analysis of the offspring will indicate the correct genotype of that yellow, round plant. Each of the above genotypes will result in a different set of gametes, and in a test cross, a different set of phenotypes in the resulting offspring (Figure 3.8).

**FIGURE 3.8** The test cross illustrated with two independent characters.

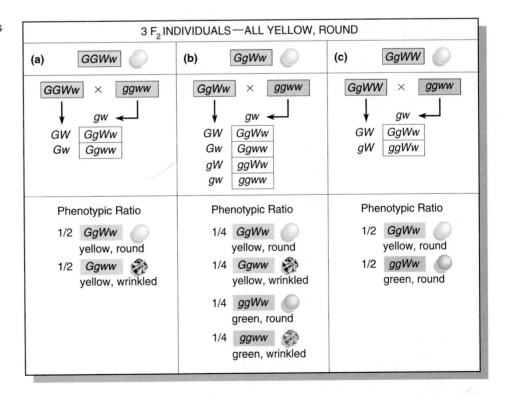

# THE TRIHYBRID CROSS

We have thus far considered inheritance by individuals of up to two pairs of contrasting traits. Mendel demonstrated that the identical processes of segregation and independent assortment apply to three pairs of contrasting traits in what is called a **trihybrid cross**, also referred to as a **three-factor cross**.

Although a trihybrid cross is somewhat more complex than a dihybrid cross, its results are easily calculated if the principles of segregation and independent assortment are followed. For example, consider the cross shown in Figure 3.9 where the gene pairs representing theoretical contrasting traits are symbolized $A/a$, $B/b$, and $C/c$. In the cross between $AABBCC$ and $aabbcc$ individuals, all $F_1$ individuals are heterozygous for all three gene pairs. Their genotype, $AaBbCc$, results in the phenotypic expression of the dominant $A$, $B$, and $C$ traits. When $F_1$ individuals are parents, they all produce 8 different gametes in equal frequencies. At this point, we could construct a Punnett square with 64 separate boxes and read out the phenotypes. Because such a method is cumbersome in a cross involving so many factors, another method has been devised to calculate the predicted ratio.

## The Forked-Line Method, or Branch Diagram

It is much less complex to consider each contrasting pair of traits separately and then to combine these results using the **forked-line method**, which was first illustrated in Figure 3.6. This method, also called a **branch diagram**, relies on the simple application of the laws of probability established for the dihybrid cross. Each gene pair is assumed to behave independently during gamete formation.

When the monohybrid cross $AA \times aa$ is made, we know that:

1. All $F_1$ individuals have the genotype $Aa$ and demonstrate the phenotype represented by the $A$ allele, which is called the $A$ phenotype in the following discussion.

2. The $F_2$ generation consists of individuals with either the $A$ phenotype or the $a$ phenotype in the ratio of 3:1, respectively.

The same generalizations may be made for the $BB \times bb$ and $CC \times cc$ crosses. Thus, in the $F_2$ generation, 3/4 of all organisms will have phenotype $A$, 3/4 will have $B$, and 3/4 will have $C$. Similarly, 1/4 of all organisms will have phenotype $a$, 1/4 will have $b$, and 1/4 will have $c$. The proportions of organisms that express each phenotypic combination may be predicted by assuming that fertilization, following the independent assortment of these three gene pairs during gamete formation, is a random process. We must simply apply once again the product law of probabilities.

The phenotypes of the $F_2$ generation calculated in this way using the forked-line method are illustrated in Figure 3.10. They fall into the trihybrid ratio of 27:9:9:9:3:3:3:1. The same method may be applied when solving crosses involving any number of gene pairs, *provided* that all gene pairs assort independently from each other. We will see later that this is not always the case. However, it appeared to be true for all of Mendel's characters.

Note in Figure 3.10 that only phenotypic ratios of the $F_2$ generation have been derived. It is possible to generate genotypic ratios as well. To do so, we again consider the $A/a$, $B/b$, and $C/c$ gene pairs separately. For example, for the $A/a$ pair the $F_1$ cross is $Aa \times Aa$. Phenotypically, an $F_2$ ratio of 3/4 $A$:1/4 $a$ is produced. Genotypically, however, the $F_2$ ratio is different; 1/4 $AA$:1/2 $Aa$:1/4 $aa$ will result. Using Figure 3.10 as a model, we would enter these genotypic frequencies on the left side of the calculation. Each would be connected by three lines to 1/4 $BB$, 1/2 $Bb$, and 1/4 $bb$, respectively. From each of these nine designations, three more lines would extend to the 1/4 $CC$, 1/2 $Cc$, and 1/4 $cc$ genotypes. Thus, on

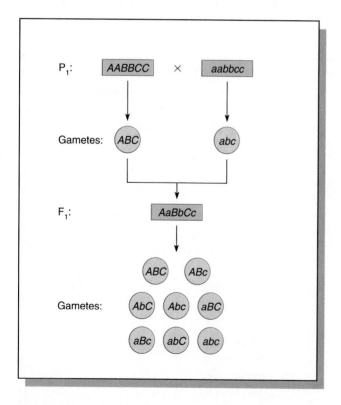

**FIGURE 3.9**    The formation of $P_1$ and $F_1$ gametes in a trihybrid cross.

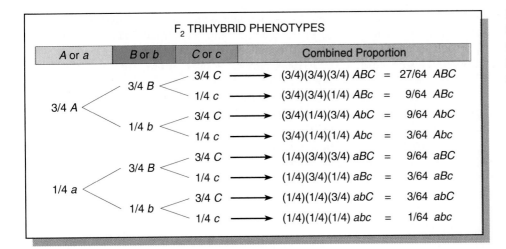

F₂ TRIHYBRID PHENOTYPES

**FIGURE 3.10**   The generation of the F₂ trihybrid phenotypic ratio using the forked-line, or branch diagram, method.

the right side of the completed diagram, 27 genotypes and their frequencies of occurrence would appear. One of the problems at the end of this chapter asks you to use the forked-line or branch diagram method to determine the genotypic ratios generated in a trihybrid cross (see Problem 16).

In crosses involving two or more gene pairs, the calculation of gametes and genotypic and phenotypic results is quite complex. There are several simple mathematical rules that will enable you to check the accuracy of various steps required in working genetic problems. First, you must determine the number of heterozygous gene pairs ($n$) involved in the cross. For example, where $AaBb \times AaBb$ represents the cross, $n = 2$; for $AaBbCc \times AaBbCc$, $n = 3$; for $AaBBCcDd \times AaBBCcDd$, $n = 3$ (because the $B$ genes are not heterozygous). Once $n$ is determined, $2^n$ is the number of different gametes that can be formed by each parent; $3^n$ is the number of

different genotypes that result following fertilization; and $2^n$ is the number of different phenotypes that are produced from these genotypes. Table 3.1 summarizes these rules, which may be applied to crosses involving any number of genes, provided that they assort independently from one another.

# THE REDISCOVERY OF MENDEL'S WORK

Mendel's work, initiated in 1856, was presented to the Brünn Society of Natural Science in 1865 and published one year later. However, his findings went largely unnoticed for about 35 years! Many reasons have been suggested to explain why the significance of his research was not immediately recognized.

**Table 3.1**   SIMPLE MATHEMATICAL RULES USEFUL IN WORKING GENETICS PROBLEMS

| Crosses between organisms heterozygous for genes exhibiting independent assortment[a] | | | |
|---|---|---|---|
| Number of Heterozygous Gene Pairs | Number of Different Types of Gametes Formed | Number of Different Genotypes Produced | Number of Different Phenotypes Produced |
| $n$ | $2^n$ | $3^n$ | $2^n$ |
| 1 | 2 | 3 | 2 |
| 2 | 4 | 9 | 4 |
| 3 | 8 | 27 | 8 |
| 4 | 16 | 81 | 16 |

[a]The fourth column assumes that dominance and recessiveness are operational for all gene pairs.

First of all, Mendel's adherence to mathematical analysis of probability events was quite an unusual approach in biological studies. Perhaps his approach seemed foreign to his contemporaries. More important, his conclusions drawn from such analyses did not fit well with the existing theories involving the cause of variation between organisms. The source of natural variation intrigued students of evolutionary theory. These individuals, stimulated by the proposal developed by Charles Darwin and Alfred Russell Wallace, believed that variation was of a **continuous nature** and that offspring were a blend of their parents' phenotypes. As we have mentioned earlier, Mendel theorized that variation was due to discrete or particulate units and was therefore of a **discontinuous nature**. For example, Mendel proposed that the $F_2$ offspring of a dihybrid cross were merely expressing traits produced by new combinations of previously existing unit factors. As a result, Mendel's theories did not fit well with the evolutionists' preconceptions about causes of variation.

Beyond this interpretation is still a further speculation as to why Mendel's contemporaries failed to grasp the significance of his findings. Perhaps they did not realize that Mendel's postulates explained *how* variation was transmitted to offspring. Instead, they may have attempted to interpret his work in a way that addressed the issue of *why* certain phenotypes survive preferentially. It was this latter question that had been addressed in the theory of natural selection, but it was not addressed by Mendel. Thus, it may well be that the collective vision of Mendel's scientific colleagues was obscured by the impact of this extraordinary theory of organic evolution.

## The Rebirth of Mendelian Genetics

Near the end of the nineteenth century, a remarkable observation set the scene for the rebirth of Mendel's work—Walter Flemming's **discovery of chromosomes** in the nuclei of salamander cells. Occurring in 1879, Flemming was able to describe the behavior of these threadlike structures during cell division. As a result of the findings of Flemming and many other cytologists, the presence of a nuclear component soon became an integral part of ideas surrounding inheritance. It was in this setting that scientists were able to reexamine Mendel's findings.

In the early twentieth century, research led to the rebirth of Mendel's work. Hybridization experiments similar to Mendel's were independently performed by three botanists, Hugo DeVries, Karl Correns, and Erich Tschermak. DeVries's work, for example, had focused on unit characters, and he demonstrated the principle of segregation in his experiments with several plant species. Apparently, he had searched the existing literature and found that Mendel's work had anticipated his own conclusions! Correns and Tschermak had independently reached conclusions similar to those of Mendel.

In 1902, two cytologists, Walter Sutton and Theodor Boveri, independently published papers linking their discoveries of the behavior of chromosomes during meiosis to the Mendelian principles of segregation and independent assortment. They pointed out that the separation of chromosomes during meiosis could serve as the cytological basis of these two postulates. Although they thought that Mendel's unit factors were probably chromosomes rather than genes on chromosomes, their findings made Mendel's work the foundation of ensuing genetic investigations.

Sutton and Boveri are credited with the initiation of the **chromosomal theory of heredity**. As we will see in Chapters 4 and 5, subsequent work by Thomas H. Morgan, Alfred H. Sturtevant, Calvin Bridges, and others using fruit flies was to establish beyond a reasonable doubt that Sutton and Boveri's theory was correct.

## Unit Factors, Genes, and Homologous Chromosomes

Because the correlation between Sutton's and Boveri's observations and Mendelian principles is the foundation for the modern interpretation of transmission genetics, we will examine this correlation before moving to the topics of the next several chapters.

As pointed out in Chapter 2, each species possesses a specific number of chromosomes in each somatic (body) cell nucleus. For diploid organisms, this number is called the **diploid number ($2n$)** and is characteristic of that species. During the formation of gametes, this number is precisely halved ($n$), and when two gametes combine during fertilization, the diploid number is reestablished. The chromosome number is not reduced in a random manner, however. It was apparent to early cytologists that the diploid number of chromosomes is composed of homologous pairs identifiable by their morphological appearance and behavior. The gametes contain one member of each pair. The chromosome complement of a gamete is thus quite specific, and the number of chromosomes in each gamete is equal to the haploid number.

With this basic information, we can see the correlation between the behavior of unit factors and chromosomes and genes. Figure 3.11 shows three of Mendel's postulates and the accepted explanation of each. Unit factors are really genes located on homologous pairs of chromosomes [Figure 3.11(a)]. Members of each pair of homologues separate, or segregate, during gamete formation [Figure 3.11(b)].

To illustrate the principle of independent assortment, it is important to distinguish between members of any

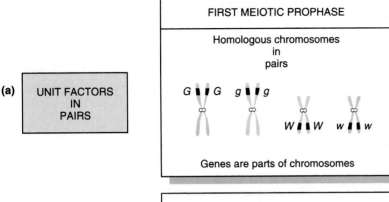

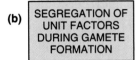

**(a)** UNIT FACTORS IN PAIRS

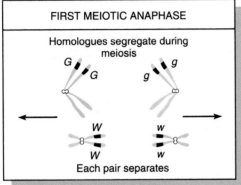

**(b)** SEGREGATION OF UNIT FACTORS DURING GAMETE FORMATION

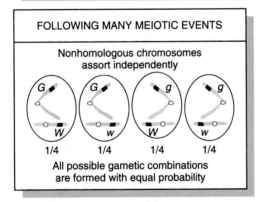

**(c)** INDEPENDENT ASSORTMENT OF SEGREGATING UNIT FACTORS

**FIGURE 3.11**    The correlation between the Mendelian postulates of (a) unit factors in pairs, (b) segregation, and (c) independent assortment, and the presence of genes located on homologous chromosomes and their behavior during meiosis.

given homologous pair of chromosomes. One member of each pair is derived from the **maternal parent**, while the other comes from the **paternal parent**. We represent their different origins by different colors. In Figure 3.11(c) the two pairs of segregating homologues behave independently during gamete formation. Each gamete always receives one homologue from each pair. All possible combinations are shown. If we add the symbols used in Mendel's dihybrid cross (*D, d* and *W, w*) to the diagram, we see why equal numbers of the four types of gametes are formed. The independent behavior of Mendel's pairs of unit factors (*D* and *W* in this example) was due to the fact that they were on separate pairs of homologous chromosomes.

From observations of the phenotypic diversity of liv-

ing organisms, we see that it is logical to assume that there are many more genes than chromosomes. Therefore, each homologue must carry genetic information for more than one trait. The currently accepted concept is that a chromosome is composed of a large number of linearly ordered, information-containing units called **genes**. Thus, Mendel's unit factors (which determine tall or dwarf stems, for example) actually constitute a pair of genes located on one pair of homologous chromosomes. The location on a given chromosome where any particular gene occurs is called its **locus** (pl. **loci**). The different forms taken by a given gene, called **alleles** (*D* or *d*), contain slightly different genetic information that determines the same character (stem length). Alleles are alternative forms of the same gene. Although we have only discussed genes with two alternative alleles, most genes have *more* than two allelic forms. We discuss the concept of **multiple alleles** in Chapter 4.

We conclude this section by reviewing the criteria necessary to classify two chromosomes as a homologous pair:

1.  During mitosis and meiosis, when chromosomes are visible as distinct figures, both members of a homologous pair are the same size and exhibit identical centromere locations.

2.  During early stages of meiosis, homologous chromosomes pair together, or synapse.

3.  Although not generally microscopically visible, homologues contain identical, linearly ordered, gene loci.

# INDEPENDENT ASSORTMENT AND GENETIC VARIATION

One of the major consequences of independent assortment is the production by individuals of genetically dissimilar gametes. Genetic variation results because the two members of any homologous pair of chromosomes are rarely, if ever, genetically identical. Because independent assortment leads to the production of all possible chromosome combinations, extensive genetic diversity results.

The number of possible gametes, each with different chromosome compositions, is $2^n$, where $n$ equals the haploid number. Thus, if a species has a haploid number of 4, then $2^4$ or 16 different gamete combinations can be formed as a result of independent assortment. Although this number is not great, consider the human species, where $n = 23$. If $2^{23}$ is calculated, we find that in excess of $8 \times 10^6$, or over 8 million, different types of gametes are represented. Because fertilization represents an event involving only one of approximately $8 \times 10^6$ possible gametes from each of two parents, each offspring represents only one of $(8 \times 10^6)^2$, or $64 \times 10^{12}$, potential genetic combinations! It is no wonder that, except for identical twins, each member of the human species demonstrates a distinctive appearance and individuality. This number of combinations of chromosomes is far greater than the number of humans who have ever lived on earth! Genetic variation resulting from independent assortment has been extremely important to the process of organic evolution in all organisms.

# PROBABILITY AND GENETIC EVENTS

Genetic ratios are most properly expressed as probabilities (e.g., 3/4 tall : 1/4 dwarf). These values predict the outcome of each fertilization event, such that the probability of each zygote having the genetic potential for becoming tall is 3/4, while the potential for becoming dwarf is 1/4. Probabilities range from 0, where an event is certain *not* to occur, to 1.0, where an event is certain *to* occur.

## The Product Law and Sum Law

When two or more events occur independently of one another, we can calculate the probability that particular outcomes of the two events will both occur. This is accomplished by applying the **product law**. As mentioned in our earlier discussion of independent assortment, it states that the probability of two or more outcomes occurring simultaneously is equal to the *product* of their individual probabilities. Two or more events are independent of one another if the outcome of each one does not affect the outcome of any of the others under consideration.

To illustrate the use of the product law, consider the possible results of an event where you toss a penny ($P$) and a nickel ($N$) at the same time and examine all combinations of heads ($H$) and tails ($T$) that can occur. There are four possible outcomes:

$$(P_H : N_H) = (1/2)(1/2) = 1/4$$
$$(P_T : N_H) = (1/2)(1/2) = 1/4$$
$$(P_H : N_T) = (1/2)(1/2) = 1/4$$
$$(P_T : N_T) = (1/2)(1/2) = 1/4$$

The probability of obtaining a head or a tail with either coin is 1/2 and is unrelated to the outcome of the other coin. All four possible combinations are predicted to occur with equal probability.

If we were interested in calculating the probability of a generalized outcome that can be accomplished in more than one way, we would apply the **sum law** to the individual mutually exclusive outcomes that accomplish this general result. For example, we can ask, what is the probability of tossing our penny and nickel and obtaining one head and one tail? In such a case, we do not care whether it is the penny or nickel that comes up heads, provided that the other coin has the alternate outcome. As we can see above, there are two ways in which the desired outcome can be accomplished [$(P_H : N_T)$ and $(P_T : N_H)$], each with a probability of 1/4. Thus, according to the sum law, the overall probability is equal to

$$(1/4) + (1/4) = 1/2$$

One-half of all such tosses are predicted to yield the desired outcome.

These simple probability laws will be useful throughout our discussions of transmission genetics, and as you solve genetics problems. In fact, we have already applied the product law earlier when the forked-line method was used to calculate the phenotypic results of Mendel's dihybrid and trihybrid crosses. When we wish to know the results of a cross, we need to only calculate the probability of each possible outcome. The results of this calculation then allow us to predict the proportion of offspring expressing each phenotype or each genotype.

There is a very important point to remember when dealing with probability. Predictions of possible outcomes are usually realized only with large sample sizes. If we predict that 9/16 of the offspring of a dihybrid cross will express both dominant traits, on the average, 9/16 will express this phenotype. However, in a small sample, variation from this expectation will be seen due to chance. The impact of deviation due strictly to chance is diminished as the sample size increases.

## Conditional Probability

Sometimes, we may wish to calculate the probability of an outcome that is dependent on a specific condition related to that outcome. Because the outcome and specific condition are not independent, we cannot apply the product law of probability. The likelihood of such an outcome is referred to as a **conditional probability**. In its simplest terms, we are asking what is the probability that one outcome will occur, given the specific condition upon which this outcome is dependent. Let us call this probability $p_c$.

For example, we might wonder what the probability is in the $F_2$ of Mendel's monohybrid cross involving tall and dwarf plants that a tall plant is heterozygous (and not homozygous). The "condition" we have set is to consider only tall $F_2$ offspring. Of any $F_2$ tall plant, what is the probability of it being heterozygous?

To solve for $p_c$, we must consider the probability of both the outcome of interest and that of the specific condition that includes the outcome. These are: (a) the probability of an $F_2$ plant being heterozygous as a result of receiving both a dominant and a recessive allele ($p_a$) and (b) the probability of the condition under which the event is being assessed, that is, being tall ($p_b$):

$p_a$ = probability of any $F_2$ plant inheriting one dominant and one recessive allele (i.e., being a heterozygote)
= 1/2

$p_b$ = probability of an $F_2$ plant of a monohybrid cross being tall
= 3/4

To calculate the conditional probability ($p_c$), we *divide* $p_a$ by $p_b$:

$$p_c = p_a/p_b$$
$$= (1/2)/(3/4)$$
$$= (1/2) \cdot (4/3)$$
$$= 4/6$$
$$p_c = 2/3$$

The conditional probability of any tall plant being heterozygous is two-thirds (2/3). Thus, on the average, two-thirds of the $F_2$ tall plants will be heterozygous. One can confirm this calculation by reexamining Figure 3.3.

Conditional probability has many applications in genetics. One application related to the case discussed above is utilized during genetic counseling. For example, the probability ($p_c$) may be calculated that an unaffected sibling of a brother or sister expressing a recessive disorder is a carrier (i.e., a heterozygote). Assuming that both parents are unaffected (and are therefore carriers), the calculation of $p_c$ is identical to the preceding example. The value of $p_c = 2/3$.

## The Binomial Theorem

The final example of probability that we shall discuss involves cases where one of two alternative outcomes is possible during each of a number of trials. By applying the **binomial theorem**, we can rather quickly calculate the probability of any specific set of outcomes among a large number of potential events. For example, in families of any size, we can calculate the probability of any combination of male and female children; for example, in a family of five, we can calculate the probability of having four children of one sex and one child of the other sex, and so on.

The expression of the binomial theorem is

$$(a + b)^n = 1$$

where $a$ and $b$ are the respective probabilities of the two alternative outcomes and $n$ equals the number of trials. For each value of $n$, the binomial must be expanded:

| $n$ | Binomial | Expanded Binomial |
|---|---|---|
| 1 | $(a + b)^1$ | $a + b$ |
| 2 | $(a + b)^2$ | $a^2 + 2ab + b^2$ |
| 3 | $(a + b)^3$ | $a^3 + 3a^2b + 3ab^2 + b^3$ |
| 4 | $(a + b)^4$ | $a^4 + 4a^3b + 6a^2b^2 + 4ab^3 + b^4$ |
| 5 | $(a + b)^5$ | $a^5 + 5a^4b + 10a^3b^2 + 10a^2b^3 + 5ab^4 + b^5$ |

To expand any binomial, the various exponents (e.g., $a^3b^2$) are determined using the pattern

$$(a + b)^n = a^n,\ a^{n-1}b,\ a^{n-2}b^2,\ a^{n-3}b^3,\ \ldots,\ b^n$$

The numerical coefficients preceding each expression may be most easily determined using Pascal's triangle:

| $n$ | | | | | | | | | | | | | | | |
|---|---|---|---|---|---|---|---|---|---|---|---|---|---|---|---|
| | | | | | | | 1 | | | | | | | | |
| 1 | | | | | | 1 | | 1 | | | | | | | |
| 2 | | | | | 1 | | 2 | | 1 | | | | | | |
| 3 | | | | 1 | | 3 | | 3 | | 1 | | | | | |
| 4 | | | 1 | | 4 | | 6 | | 4 | | 1 | | | | |
| 5 | | 1 | | 5 | | 10 | | 10 | | 5 | | 1 | | | |
| 6 | 1 | | 6 | | 15 | | 20 | | 15 | | 6 | | 1 | | |
| 7 | 1 | 7 | | 21 | | 35 | | 35 | | 21 | | 7 | | 1 | |

etc.

Notice that all numbers other than the 1s are equal to the sum of the two numbers directly above them.

Using the above methods, the initial expansion of $(a + b)^7$ is

$$a^7 + 7a^6b + 21a^5b^2 + 35a^4b^3 + \cdots + b^7$$

If we apply the binomial theorem, we might ask: What is the probability that in a family of four children, two are male and two are female?

First, assign initial probabilities to each outcome:

$$a = \text{male} = 1/2$$
$$b = \text{female} = 1/2$$

Then locate the appropriate term in the expanded binomial, where $n = 4$:

$$(a + b)^4 = a^4 + 4a^3b + 6a^2b^2 + 4ab^3 + b^4$$

In each term, the exponent of $a$ represents the number of males, and the exponent of $b$ represents the number of females. Therefore, the correct expression of $p$ is

$$p = 6a^2b^2$$
$$= 6(1/2)^2(1/2)^2$$
$$= 6(1/2)^4$$
$$= 6(1/16)$$
$$= 6/16$$
$$p = 3/8$$

Thus, the probability of families of four children having two boys and two girls is 3/8. Of all families with four children, 3 out of 8 are predicted to have two boys and two girls.

Before examining one other example, we should note that a single formula can be applied in determining the numerical coefficient for any set of exponents:

$$\frac{n!}{s!t!}$$

where $n$ = the total number of events
$s$ = the number of times outcome $a$ occurs
$t$ = the number of times outcome $b$ occurs
Therefore, $n = s + t$.

The symbol ! means "factorial." For example,

$$5! = (5) \cdot (4) \cdot (3) \cdot (2) \cdot (1) = 120$$

Note that $0! = 1$.

Using the formula, let's determine the probability, in a family of seven, that five males and two females will occur. Thus, $s = 5$, $t = 2$, and $n = 7$. We first extend our equation to include five events of outcome $a$ and two events of outcome $b$. The appropriate term is

$$p = \frac{n!}{s!\,t!}a^sb^t$$
$$= \frac{7!}{5!2!}(1/2)^5(1/2)^2$$
$$= \frac{(7) \cdot (6) \cdot (5) \cdot (4) \cdot (3) \cdot (2) \cdot (1)}{(5) \cdot (4) \cdot (3) \cdot (2) \cdot (1) \cdot (2) \cdot (1)}(1/2)^7$$
$$= \frac{(7) \cdot (6)}{(2) \cdot (1)}(1/2)^7$$
$$= \frac{42}{2}(1/2)^7$$
$$= 21(1/2)^7$$
$$= 21(1/128)$$
$$p = 21/128$$

Of families with seven children, on the average, 21/128 are predicted to have five males and two females.

Let's illustrate the use of the binomial theorem by examining one final example involving the autosomal recessive disorder albinism. Consider a family where both parents have normal pigmentation, but they have an albino child. This establishes that both of them are heterozygous carriers. If they have six more children, what is the probability that four will be normal ($a$) and two will have albinism ($b$)?

Using our formula, the appropriate expression is

$$p = \frac{6!}{4!2!}a^4b^2$$

where, based on the cross of $Aa \times Aa$,

$$a = \text{probability of a normal child} = 3/4$$
$$b = \text{probability of an albino child} = 1/4$$

Therefore,

$$p = \frac{(6) \cdot (5) \cdot (4) \cdot (3) \cdot (2) \cdot (1)}{(4) \cdot (3) \cdot (2) \cdot (1) \cdot (2) \cdot (1)} (3/4)^4 (1/4)^2$$

$$= \frac{(6) \cdot (5)}{(2) \cdot (1)} (81/256) \cdot (1/16)$$

$$= (15) \cdot (81/4096)$$

$$p = 1215/4096$$

The use of calculations involving the binomial theorem have various applications in genetics, including the analysis of polygenic traits (see Chapter 4) and in population equilibrium studies (see Chapter 24).

# EVALUATING GENETIC DATA: CHI-SQUARE ANALYSIS

Mendel's 3:1 monohybrid and 9:3:3:1 dihybrid ratios are hypothetical predictions based on the following assumptions: (1) each allele is dominant or recessive; (2) segregation is operative; (3) independent assortment occurs; and (4) fertilization is random. The last three assumptions are influenced by chance events and therefore are subject to random fluctuation. This concept, called **chance deviation**, is most easily illustrated by tossing a single coin numerous times and recording the number of heads and tails observed. In each toss, there is a probability of 1/2 that a head will occur and a probability of 1/2 that a tail will occur. Therefore, the expected ratio of many tosses is 1:1. If a coin were tossed 1000 times, usually *about* 500 heads and 500 tails would be observed. Any reasonable fluctuation from this hypothetical ratio (e.g., 486 heads and 514 tails) would be attributed to chance.

As the total number of tosses is reduced, the impact of chance deviation increases. For example, if a coin were tossed only 4 times, you wouldn't be too surprised if all 4 tosses resulted in only heads or only tails. But, for 1000 tosses, 1000 heads or 1000 tails would be most unexpected. In fact, you might believe that such a result would be impossible. Actually, all heads or all tails in 1000 tosses would be predicted to occur with a probability of only $(1/2)^{1000}$. Because $(1/2)^{20}$ is equivalent to less than 1 in 1 million times, an event occurring with a probability of only $(1/2)^{1000}$ would be virtually impossible to achieve.

Two major points are significant here:

1.  The outcomes of segregation, independent assortment, and fertilization, like coin tossing, are subject to random fluctuations from their predicted occurrences as a result of chance deviation.

2.  As the sample size increases, the average deviation from the expected decreases on a proportional basis. Therefore, a larger sample size diminishes the impact of chance deviation on the final outcome.

It is important in genetics to be able to evaluate observed deviation. When we assume that data will fit a given ratio such as 1:1, 3:1, or 9:3:3:1, we establish what is called the **null hypothesis**. It is so named because the hypothesis assumes that there is no real difference between the **measured values** (or ratio) and the **predicted values** (or ratio). Evaluation of the null hypothesis is accomplished by statistical analysis. On this basis, the null hypothesis may either: (1) be rejected, or (2) fail to be rejected. If it is rejected, any observed deviation from the expected is not attributed to chance alone. The null hypothesis and the underlying assumptions leading to it must be reexamined. If the null hypothesis fails to be rejected, any observed deviations can be attributed to chance.

Thus, statistical analysis provides a mathematical basis for examining how well observed data fit or differ from predicted or expected occurrences, testing what is called the **goodness of fit**. Assuming that the data do not "fit" exactly, just how much deviation can be allowed before the null hypothesis is rejected?

One of the simplest statistical tests devised to assess the **goodness of fit** of the null hypothesis is **chi-square analysis ($\chi^2$)**. This test takes into account the observed deviation in each component of an expected ratio as well as the sample size and reduces them to a single numerical value. This value ($\chi^2$) is then used to estimate how frequently the observed deviation can be expected to occur strictly as a result of chance. The formula used in chi-square analysis is

$$\chi^2 = \sum \frac{(o - e)^2}{e}$$

In this equation, $o$ is the observed value for a given category and $e$ is the expected value for that category. $\Sigma$ (sigma) represents the "sum of" the calculated values for each category of the ratio. Because $(o - e)$ is the deviation ($d$) in each case, the equation can be reduced to

$$\chi^2 = \sum \frac{d^2}{e}$$

Table 3.2(a) illustrates the step-by-step procedure necessary to make the $\chi^2$ calculation for the $F_2$ results of a monohybrid cross. If you were analyzing these data, you would work from left to right, calculating and entering the appropriate numbers in each column. Regardless of

## Table 3.2 CHI-SQUARE ANALYSIS

### (a) Monohybrid cross

| Expected Ratio | Observed (o) | Expected (e) | Deviation (o − e) | Deviation² (d²) | Deviation²/Expected (d²/e) |
|---|---|---|---|---|---|
| 3/4 | 740 | 3/4 (1000) = 750 | 740 − 750 = −10 | $(-10)^2 = 100$ | 100/750 = 0.13 |
| 1/4 | 260 | 1/4 (1000) = 250 | 260 − 250 = +10 | $(+10)^2 = 100$ | 100/250 = 0.40 |
| TOTAL = 1000 | | | | | $\chi^2 = 0.53$ |
| | | | | | $p = 0.48$ |

### (b) Dihybrid cross

| Expected Ratio | o | e | o − e | d² | d²/e |
|---|---|---|---|---|---|
| 9/16 | 587 | 567 | +20 | 400 | 0.71 |
| 3/16 | 197 | 189 | +8 | 64 | 0.34 |
| 3/16 | 168 | 189 | −21 | 441 | 2.33 |
| 1/16 | 56 | 63 | −7 | 49 | 0.78 |
| TOTAL = 1008 | | | | | $\chi^2 = 4.16$ |
| | | | | | $p = 0.26$ |

whether the calculated deviation ($o - e$) is initially positive or negative, it becomes positive after the number is squared. Table 3.2(b) illustrates the analysis of the $F_2$ results of a dihybrid cross. Based on your study of the calculations involved in the monohybrid cross, check to make certain that you understand how each number was calculated in the dihybrid example.

The final step in the chi-square analysis is to interpret the $\chi^2$ value. To do so, you must initially determine the value of the **degrees of freedom ($df$)**, which in these analyses, is equal to $n - 1$ where $n$ is the number of different categories into which each datum point may fall. For the 3:1 ratio, $n = 2$, so $df = 2 - 1 = 1$. For the 9:3:3:1 ratio, $df = 3$. Degrees of freedom must be taken into account because the greater the number of categories, the more deviation is expected as a result of chance.

With this accomplished, the $\chi^2$ value must now be interpreted in terms of a corresponding **probability value ($p$)**. Because this calculation is complex, the $p$ value is usually located on a table or graph. Figure 3.12 shows the wide range of $\chi^2$ and $p$ values for numerous degrees of freedom in both forms. We will use the graph to explain how to determine the $p$ value. The caption for Figure 3.12(b) explains how to use the table.

These simple steps must be followed to determine $p$:

1. Locate the $\chi^2$ value on the abscissa (the horizontal axis).

2. Draw a vertical line from this point up to the line on the graph representing the appropriate $df$.

3. Extend a horizontal line from this point to the left until it intersects the ordinate (the vertical axis).

4. Estimate, by interpolation, the corresponding $p$ value.

For our first example (the monohybrid cross) in Table 3.2, the $p$ value of 0.48 may be estimated in this way and is illustrated in Figure 3.12(a). For the dihybrid cross, use this method to see if you can determine the value. $\chi^2$ is 4.16 and $df$ equals 3. A $p$ value of 0.26 is the approximate value. Use of the table rather than the graph confirms that both $p$ values are between 0.20 and 0.50. Examine the table to confirm this.

So far, we have been concerned only with the determination of $p$. The most important aspect of $\chi^2$ analysis is understanding what the $p$ value actually means. We will use the example of the dihybrid cross ($p = 0.26$) to illustrate. In these discussions, it is simplest to think of the $p$ value as a percentage (e.g., 0.26 = 26 percent).

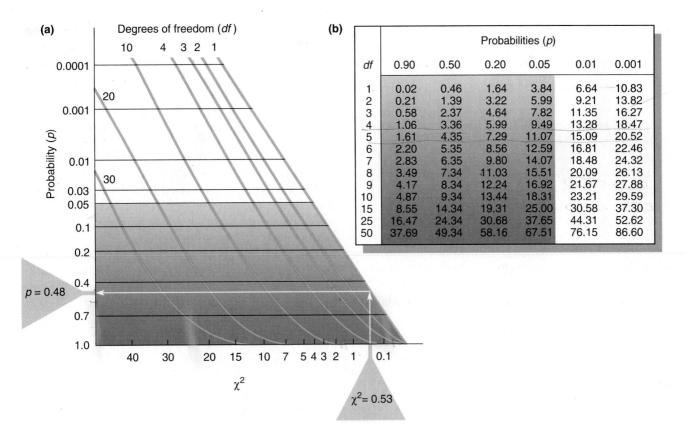

**FIGURE 3.12**   (a) A graph used to convert $\chi^2$ values to $p$ values. The interpolation of a $\chi^2$ value of 0.53 with 1 degree of freedom to an estimated probability value of 0.48 is illustrated. (b) A table showing $\chi^2$ values for a variety of combinations of $df$ and $p$. Any $\chi^2$ value greater than that shown at the $p = 0.05$ level for a particular $df$ serves as the basis to reject the null hypothesis in question. In our example, a $\chi^2$ value of 0.53 for a $df$ of 1 is converted to a probability value between 0.20 and 0.50. From our graph in (a), the more precise value ($p = 0.48$) was estimated by interpolation. All values that serve to fail to reject the null hypothesis are shaded in both the graph and the chart.

In our example, the $p$ value indicates that, were the same experiment repeated many times, 26 percent of the trials would be expected to exhibit chance deviation as great as or greater than that seen in the initial trial. Conversely, 74 percent of the repeats would show less deviation as a result of chance than initially observed.

This interpretation of the $p$ value reveals that an hypothesis (a 9:3:3:1 ratio in this case) is never proved or disproved absolutely. Instead, a relative standard must be set to serve as the basis for either rejecting or failing to reject the hypothesis. This standard is often a probability value of 0.05. In chi-square analysis, a $p$ value less than 0.05 is used to conclude that the deviation in the observed set of results was not obtained by chance alone. Instead, a $p$ value less than 0.05 indicates that the difference between the observed and predicted results is substantial and thus serves as the basis for rejecting the null hypothesis.

On the other hand, $p$ values of 0.05 or greater (1.0–0.05) fail to reject the null hypothesis. In our example where $p = 0.26$, the hypothesis of independent assortment is not rejected by the experimental data. That is, the data do not provide any reason to reject the hypothesis. Therefore, the observed deviation is attributed to chance.

## HUMAN PEDIGREES

In all crosses discussed so far, one of the two traits for each character has been dominant to the other. Based on this observation, two significant questions may be asked:

**1.**   Does the expression of all genes occur in this fashion?

**2.** Is it possible to ascertain the mode of inheritance of genes in organisms where designed crosses and the production of large numbers of offspring are impossible?

The answer to the first question is no. As we will see in Chapter 4, many modes of genetic expression exist that modify the monohybrid and dihybrid ratios observed by Mendel.

The answer to the second question is yes. Even in humans the pattern of inheritance of a specific phenotype can be studied.

The simplest way to study this pattern is to construct a family tree indicating the phenotype of the trait in question for each member. Such a family tree is called a **pedigree**. By analyzing the pedigree, we may be able to predict how the gene controlling the trait is inherited. If many similar pedigrees for the same trait are found, the prediction is strengthened.

Figure 3.13 shows the conventions used in constructing a pedigree. Circles represent females, and squares designate males. If the sex is unknown, a diamond may be used (II-2). If a pedigree traces only a single trait, as Figure 3.13 does, the circles, squares, and diamonds are shaded if the phenotype being considered is expressed. Heterozygotes, who fail to express a recessive trait, may have only the left half of their square or circle shaded (see II-3 and II-4).

The parents are connected by a horizontal line, and vertical lines lead to their offspring. All such offspring are called **sibs** and are connected by a horizontal **sibship line**. Sibs are placed from left to right according to birth order and are labeled with Arabic numerals. Each generation is indicated by a Roman numeral.

Twins are indicated by connected diagonal lines. **Monozygotic** or **identical twins** stem from a single line itself connected to the sibship line (see III-5,6 in Figure 3.13). **Dizygotic** or **fraternal twins** are connected directly to the sibship line (see III-8,9). A number within one of the symbols (see II-10–13) represents

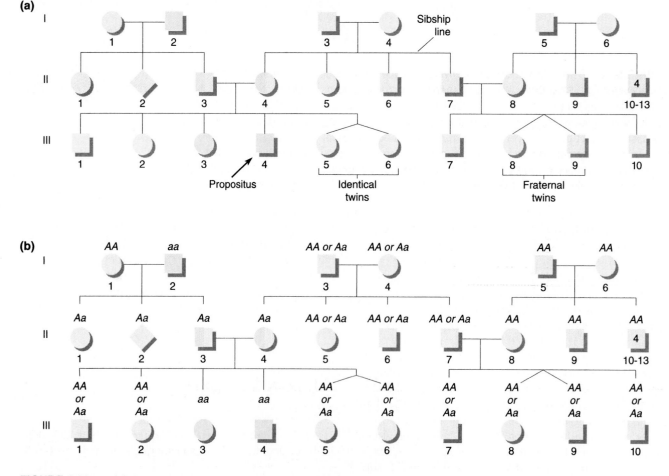

**FIGURE 3.13** (a) A representative pedigree for a single character through three generations. (b) The most probable genotypes of each individual in the pedigree.

numerous sibs of the same or unknown phenotypes. A male whose phenotype drew the attention of a physician or geneticist is called the **propositus** (a female is a **proposita**) and is indicated by an arrow (see III-4).

The pedigree shown in Figure 3.13 traces the theoretical pattern of inheritance of the human trait **albinism**. By analyzing the pedigree, we will see that albinism is inherited as a recessive trait.

One of the parents of the first generation, I-2, is affected. Because none of his offspring show the disorder, we might conclude that the unaffected female parent (I-1) was a homozygous normal individual. Had she been heterozygous, one-half of her offspring would be expected to exhibit albinism. However, such a small sample prevents any certainty in the matter.

An unaffected second generation is characteristic of a rare recessive trait. If albinism were inherited as a dominant trait, one-half of the second generation would be expected to exhibit the disorder in the crosses involving the I-1 and I-2 parents. Inspection of the offspring constituting the third generation (row III) provides further support for the hypothesis that albinism is a recessive trait. If so, parents II-3 and II-4 are both heterozygous, and ap-

proximately one-fourth of their offspring should be affected. Two of the six offspring do show albinism. This deviation from the expected ratio is characteristic of crosses with few offspring.

Based on this pedigree analysis and the conclusion that albinism is a recessive trait, the most probable genotypes of all individuals can be predicted. For both the first and second generations, this can be done with some certainty in only a few cases. In the third generation, for most normal individuals, we can only guess whether or not they are homozygous or heterozygous.

Pedigree analysis of many traits has been an extremely valuable research technique in human genetic studies. However, this approach does not usually provide the certainty in drawing conclusions that is afforded by designed crosses yielding large numbers of offspring. Nevertheless, when many independent pedigrees of the same trait or disorder are analyzed, consistent conclusions can often be drawn. Table 3.3 lists numerous human traits and classifies them according to their recessive or dominant expression. As we will see in Chapter 4, the genes controlling some of these traits are located on the sex-determining chromosomes.

**Table 3.3**  REPRESENTATIVE RECESSIVE AND DOMINANT HUMAN TRAITS

| Recessive Traits | Dominant Traits |
| --- | --- |
| Albinism | Achondroplasia |
| Alkaptonuria | Brachydactyly |
| Ataxia telangiectasia | Congenital stationary night blindness |
| Color blindness | Ehler-Danlos syndrome |
| Cystic fibrosis | Fascio-scapulo-humeral muscular dystrophy |
| Duchenne muscular dystrophy | Huntington disease |
| Galactosemia | Hypercholesterolemia |
| Hemophilia | Marfan syndrome |
| Lesch–Nyhan syndrome | Neurofibromatosis |
| Phenylketonuria | Phenylthiocarbamide tasting (PTC) |
| Sickle-cell anemia | Widow's peak |
| Tay-Sachs disease | |

**CHAPTER SUMMARY**

1. Over a century ago, Mendel studied inheritance patterns in the garden pea, establishing the principles of transmission genetics.

2. Mendel's postulates help describe the basis for the inheritance of phenotypic expression. He showed that unit factors, later called alleles, exist in pairs and exhibit a dominant/recessive relationship in determining the expression of traits.

3. Mendel postulated that unit factors must segregate during gamete formation, such that each receives only one of the two factors with equal probability.

4. Mendel's final postulate of independent assortment states that each pair of unit factors segregates independently of other such pairs. As a result, all possible combinations of gametes will be formed with equal probability.

5. The discovery of chromosomes in the late 1800s and subsequent studies of their behavior during meiosis led to the rebirth of Mendel's work, linking the behavior of his unit factors with that of chromosomes during meiosis.

6. The Punnett square and the forked-line methods are used to predict the probabilities of phenotypes (and genotypes) from crosses involving two or more gene pairs.

7. Genetic ratios are expressed as probabilities. Thus, deriving outcomes of genetic crosses relies on an understanding of the laws of probability. The sum law, the product law, conditional probability, and the use of the binomial theorem have been described.

8. Statistical analysis is used to test the validity of experimental outcomes. In genetics, variations from the expected ratios are anticipated owing to chance deviation. A chi-square analysis tests the probability of these variations being generated from chance alone. It provides the basis for assessing the null hypothesis.

9. Pedigree analysis provides a method for studying the inheritance pattern of human traits over several generations. This often provides the basis for determining the mode of inheritance of human characteristics and disorders.

## KEY TERMS

albinism
allele
binomial theorem
branch diagram
chance deviation
chi-square ($\chi^2$) analysis
chromosomal theory of heredity
conditional probability
continuous variation
degrees of freedom ($df$)
dihybrid cross
diploid number ($2n$)
discontinuous variation

dizygotic (fraternal) twins
dominance
$F_1$ (first filial) generation
$F_2$ (second filial) generation
forked-line method
gene
genotype
goodness of fit
heterozygote
homozygote
independent assortment
locus (loci)
maternal parent

Mendel's 9:3:3:1 dihybrid ratio
monohybrid cross
monozygotic (identical) twins
multiple alleles
null hypothesis
$P_1$ (parental) generation
particulate units of heredity
paternal parent
pedigree
phenotype
probability ($p$)

product law
propositus (proposita)
Punnett square
recessive
reciprocal cross
segregation
sibs
sibship line
sum law
test cross
transmission genetics
trihybrid cross
unit factors in pairs

## INSIGHTS AND SOLUTIONS

Students demonstrate their knowledge of transmission genetics by solving genetics problems. Success at this task represents not only comprehension of theory but its application to more practical genetic situations. Most students find problem solving in genetics to be challenging but rewarding. This section is designed to provide basic insights into the reasoning essential to this process.

Genetics problems are in many ways similar to algebraic word problems. The approach taken should be identical: (1) analyze the problem carefully; (2) translate words into symbols, defining each one first; and (3) choose and apply a specific technique to solve the problem. The first two steps are the most critical. The third step is largely mechanical.

The simplest problems are those that state all necessary information about the $P_1$ generation and ask you to find the expected ratios of the $F_1$ and $F_2$ genotypes and/or phenotypes. The following steps should always be followed when you encounter this type of problem:

1.  Determine the genotypes of the $P_1$ generation.

2.  Determine what gametes may be formed by the $P_1$ parents.

3.  Recombine gametes either by the Punnett square method, the forked-line method, or, if the situation is very simple, by inspection. Read the $F_1$ phenotypes directly.

4.  Repeat the process to obtain information about the $F_2$ generation.

Determining the genotypes from the given information requires an understanding of the basic theory of transmission genetics. For example, consider the following problem:

> A recessive mutant allele, *black,* causes a very dark body in *Drosophila* when homozygous. The normal, wild-type color is described as gray. What $F_1$ phenotypic ratio is predicted when a black female is crossed to a gray male whose father was black?

To work this problem, you must understand dominance and recessiveness as well as the principle of segregation. Further, you must use the information about the male parent's father. You can work out the problem as follows:

1.  Because the female parent is black, she must be homozygous for the mutant allele (*bb*).

2.  The male parent is gray; therefore, he must have at least one dominant allele (*B*). Because his father was black (*bb*) and he received one of the chromosomes bearing these alleles, the male parent must be heterozygous (*Bb*).

From here, the problem is simple:

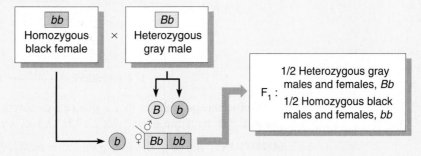

Apply this approach to the following problems.

1.  In Mendel's work, he found that full pods are dominant to constricted pods while round seeds are dominant to wrinkled seeds. One of his crosses was between full, round plants and constricted, wrinkled plants. From this cross, he obtained an $F_1$ that was all full and round. In the $F_2$, Mendel obtained his classic 9:3:3:1 ratio. Using the above information, determine the expected $F_1$ and $F_2$ results of a cross between constricted, round and full, wrinkled plants.

**SOLUTION:**   First of all, define gene symbols for each pair of contrasting traits. Select the lowercase forms of the first letter of the recessive traits to designate those phenotypes, and use the uppercase forms to designate the dominant traits. Thus, use *C* and *c* to indicate full and constricted, and use *W* and *w* to indicate the round and wrinkled phenotypes, respectively.

Now, determine the genotypes of the $P_1$ generation, form gametes, reconstitute the $F_1$ generation, and read off the phenotype(s):

$P_1$:              *ccWW*           ×          *CCww*
             constricted, round        full, wrinkled
                              ↓                         ↓

Gametes:            *cW*                       *Cw*

$F_1$:                                *CcWw*
                                 full, round

You can see immediately that the $F_1$ generation expresses both dominant phenotypes and is heterozygous for both gene pairs. Thus, we can expect that the $F_2$ generation will yield the classic Mendelian ratio of 9:3:3:1. Let's work it out anyway, just to confirm this, using the forked-line method. Because both gene pairs are heterozygous and can be expected to assort independently, we can predict the $F_2$ outcomes from each gene pair separately and then proceed with the forked-line method.

Every $F_2$ offspring is subject to the following probabilities:

$Cc \times Cc$              $Ww \times Ww$
      ↓                            ↓
   *CC*  ⎫                     *WW*  ⎫
   *Cc*  ⎬ full               *Ww*  ⎬ round
   *cC*  ⎭                     *wW*  ⎭
   *cc*     constricted     *ww*     wrinkled

The forked-line method then allows us to confirm the 9:3:3:1 phenotypic ratio. Remember that this represents proportions of 9/16:3/16:3/16:1/16. Note that we are applying the product law as we compute the final probabilities:

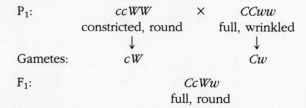

3/4 full ⟨ 3/4 round $\xrightarrow{(3/4)(3/4)}$ 9/16 full, round
              1/4 wrinkled $\xrightarrow{(3/4)(1/4)}$ 3/16 full, wrinkled

1/4 constricted ⟨ 3/4 round $\xrightarrow{(1/4)(3/4)}$ 3/16 constricted, round
              1/4 wrinkled $\xrightarrow{(1/4)(1/4)}$ 1/16 constricted, wrinkled

2.  Determine the probability that a plant of genotype *CcWw* will be produced from parental plants of the genotypes *CcWw* and *Ccww*.

**SOLUTION:**   Because the two gene pairs independently assort during gamete formation, we need only calculate the individual probabilities of the two separate events (*Cc* and *Ww*) and apply the product law to calculate the final probability:

$Cc \times Cc \rightarrow$ 1/4 *CC*:1/2 *Cc*:1/4 *cc*
$Ww \times ww \rightarrow$ 1/2 *Ww*:1/2 *ww*

$p = (1/2\ Cc)(1/2\ Ww) = 1/4\ CcWw$

3.  In another cross, involving parent plants of unknown genotype and phenotype,

the offspring shown below were obtained. Determine the genotypes and phenotypes of the parents.

Offspring: 3/8 full, round
3/8 full, wrinkled
1/8 constricted, round
1/8 constricted, wrinkled

**SOLUTION:**   This problem is more difficult and requires keener insights since you must work backwards. The best approach is to consider the outcomes of pod shape separately from those of seed texture.

Of all plants, 6/8 (3/4) are full and 2/8 (1/4) are constricted. Of the various genotypic combinations that can serve as parents, which will give rise to a ratio of 3/4:1/4? Since this ratio is identical to Mendel's monohybrid $F_2$ results, we can propose that both unknown parents share the same genetic characteristic as the monohybrid $F_1$ parents: they must both be heterozygous for the genes controlling pod color and thus are

*Cc*

Before accepting this hypothesis, let's consider the possible genotypic combinations that control seed texture. If we consider this characteristic alone, we can see that the traits are expressed in a ratio of 4/8 (1/2) round:4/8 (1/2) wrinkled. In order to generate such a ratio, the parents *cannot* both be heterozygous or their offspring would yield a 3/4:1/4 phenotypic ratio. They *cannot* both be homozygous or all offspring would express a single phenotype. Thus, we are left with testing the hypothesis that one parent is homozygous and one is heterozygous for the alleles controlling texture. The potential case of *WW* × *Ww* will not work since it also would yield only a single phenotype. This leaves us with the potential case of *Ww* × *ww*. Offspring in such a mating will yield 1/2 *Ww* (round):1/2 *ww* (wrinkled), exactly the outcome we are seeking.

Now, let's combine our hypotheses and predict the outcome of crossing. In our solution, we will use a "−" to indicate that the second allele may be either dominant or recessive, since we are only predicting phenotypes.

*CcWw* × *Ccww*

$$3/4\ C-\begin{cases}1/2\ Ww \xrightarrow{(3/4)(1/2)} & 3/8\ C-Ww\ \text{full, round}\\ 1/2\ ww \xrightarrow{(3/4)(1/2)} & 3/8\ C-ww\ \text{full, wrinkled}\end{cases}$$

$$1/4\ cc-\begin{cases}1/2\ Ww \xrightarrow{(1/4)(1/2)} & 1/8\ ccWw\ \text{constricted, round}\\ 1/2\ ww \xrightarrow{(1/4)(1/2)} & 1/8\ ccww\ \text{constricted, wrinkled}\end{cases}$$

As we can see, this cross produces offspring according to our initial information. Thus, we have solved the problem. Note that in this solution, we have used *genotypes* in the forked-line method, in contrast to the use of *phenotypes* in the earlier solution.

4.  In the laboratory, a genetics student crossed flies with normal, long wings to flies with mutant *dumpy* wings, which she believed was a recessive trait. In the $F_1$, all flies had long wings. In the $F_2$, the following results were obtained:

792 long-winged flies
208 dumpy-winged flies

The student tested the hypothesis that the *dumpy* wing is inherited as a recessive trait by performing $\chi^2$ analysis of the $F_2$ data.

(a) What ratio was hypothesized?

(b) Did the $\chi^2$ analysis support the hypothesis?

(c) What do the data suggest about the *dumpy* mutation?

**SOLUTION:**

(a) The student hypothesized that the $F_2$ data (792:208) fit Mendel's 3:1 monohybrid ratio for recessive genes.

(b) The initial step in $\chi^2$ analysis is to calculate the expected results ($e$) if the ratio is 3:1 and the deviations ($d$) between the expected and observed data:

| Ratio | $o$ | $e$ | $d$ | $d^2$ | $d^2/e$ |
|-------|-----|-----|-----|-------|---------|
| 3/4 | 792 | 750 | 42 | 1764 | 2.35 |
| 1/4 | 208 | 250 | −42 | 1764 | 7.06 |

Total = 1000

$$\chi^2 = \sum \frac{d^2}{e}$$

$$= 2.35 + 7.06$$

$$= 9.41$$

Consulting Figure 3.12 allows us to determine the probability ($p$). This value will allow us to determine whether the deviations from the null hypothesis can be attributed to chance. There are two possible outcomes ($n$), so the degrees of freedom ($df$) = $n - 1$ or 1. The table in Figure 3.12 shows that $p = 0.01$ to 0.001. The graph gives an estimate of about 0.001. That $p$ is less than 0.05 causes us to reject the null hypothesis. The data do not statistically fit a 3:1 ratio.

(c) When we accept Mendel's 3:1 ratio as a valid expression of the monohybrid cross, numerous assumptions are made. One of these may explain why the null hypothesis was rejected. We must assume *that all genotypes are equally viable.* That is, genotypes yielding long wings are equally likely to survive from fertilization through adulthood, when the data were collected, as the genotype yielding *dumpy* wings. Further study would reveal that *dumpy* flies are somewhat less viable than normal flies. As a result, we would expect *less* than 1/4 of the total offspring to express dumpy. This observation is borne out in the data, although we have not proved this.

**5.** If two parents, both heterozygous carriers of the autosomal recessive gene causing cystic fibrosis, have five children, what is the probability that three will be normal?

**SOLUTION:** First, the probability of having a normal child during each pregnancy is

$$p_a = \text{normal} = 3/4$$

while the probability of having an afflicted offspring is

$$p_b = \text{afflicted} = 1/4$$

Then apply the formula

$$\frac{n!}{s!t!}\, a^s b^t$$

where $n = 5$, $s = 3$, and $t = 2$:

$$p = \frac{(5) \cdot (4) \cdot (3) \cdot (2) \cdot (1)}{(3) \cdot (2) \cdot (1) \cdot (2) \cdot (1)} (3/4)^3 (1/4)^2$$

$$= \frac{(5) \cdot (4)}{(2) \cdot (1)} (3/4)^3 (1/4)^2$$

$$= 10(27/64) \cdot (1/16)$$

$$= 10(27/1024)$$

$$= 270/1024$$

$$p = {\sim}0.26$$

## PROBLEMS AND DISCUSSION QUESTIONS

When working genetics problems in this and succeeding chapters, always assume that members of the $P_1$ generation are homozygous, unless the information given indicates or requires otherwise.

1. In a cross between a black and a white guinea pig, all members of the $F_1$ generation are black. The $F_2$ generation is made up of approximately 3/4 black and 1/4 white guinea pigs.
   (a) Diagram this cross, showing the genotypes and phenotypes.
   (b) What will the offspring be like if two $F_2$ white guinea pigs are mated?
   (c) Two different matings were made between black members of the $F_2$ generation with the results shown below. Diagram each of the crosses.

   | Cross | Offspring |
   |---------|---------------------|
   | Cross 1 | All black |
   | Cross 2 | 3/4 black, 1/4 white |

2. Albinism in humans is inherited as a simple recessive trait. For the following families, determine the genotypes of the parents and offspring. When two alternative genotypes are possible, list both.
   (a) Two normal parents have five children, four normal and one albino.
   (b) A normal male and an albino female have six children, all normal.
   (c) A normal male and an albino female have six children, three normal and three albino.
   (d) Construct a pedigree of the families in (b) and (c). Assume that one of the normal children in (b) marries one of the albino children in (c) and that they have eight children.

3. Which of Mendel's postulates are illustrated by the pedigree in Problem 2? List and define these postulates.

4. Discuss the rationale relating Mendel's monohybrid results to his postulates.

5. What advantages were provided by Mendel's choice of the garden pea in his experiments?

6. Pigeons may exhibit a checkered or plain pattern. In a series of controlled matings, the following data were obtained:

| P₁ Cross | F₁ Progeny | |
|---|---|---|
| | Checkered | Plain |
| (a) checkered × checkered | 36 | 0 |
| (b) checkered × plain | 38 | 0 |
| (c) plain × plain | 0 | 35 |

Then, F₁ offspring were selectively mated with the following results. The P₁ cross giving rise to each F₁ pigeon is indicated in parentheses.

| F₁ × F₁ Crosses | F₁ Progeny | |
|---|---|---|
| | Checkered | Plain |
| (d) checkered (a) × plain (c) | 34 | 0 |
| (e) checkered (b) × plain (c) | 17 | 14 |
| (f) checkered (b) × checkered (b) | 28 | 9 |
| (g) checkered (a) × checkered (b) | 39 | 0 |

How are the checkered and plain patterns inherited? Select and define symbols for the genes involved and determine the genotypes of the parents and offspring in each cross.

7. Mendel crossed peas having round seeds and yellow cotyledons with peas having wrinkled seeds and green cotyledons. All the F₁ plants had round seeds with yellow cotyledons. Diagram this cross through the F₂ generation using both the Punnett square and forked-line, or branch diagram, methods.

8. Determine the genotypes of the F₁ plants given here by analyzing the phenotypes of the offspring of these crosses.

| F₂ Plants | Offspring |
|---|---|
| (a) round, yellow × round, yellow | 3/4 round, yellow<br>1/4 wrinkled, yellow |
| (b) wrinkled, yellow × round, yellow | 6/16 wrinkled, yellow<br>2/16 wrinkled, green<br>6/16 round, yellow<br>2/16 round, green |
| (c) round, yellow × round, yellow | 9/16 round, yellow<br>3/16 round, green<br>3/16 wrinkled, yellow<br>1/16 wrinkled, green |
| (d) round, yellow × wrinkled, green | 1/4 round, yellow<br>1/4 round, green<br>1/4 wrinkled, yellow<br>1/4 wrinkled, green |

9. Which of the crosses in Problem 8 is a test cross?

10. Which of Mendel's postulates can only be demonstrated in crosses involving at least two pairs of traits? Define it.

11. Correlate Mendel's four postulates with what is now known about homologous chromosomes, genes, alleles, and the process of meiosis.

12. What is the basis for homology among chromosomes?

13. Distinguish between homozygosity and heterozygosity.

14. In *Drosophila, gray* body color is dominant to *ebony* body color, while *long* wings are dominant to *vestigial* wings. Work the following crosses through the $F_2$ generation and determine the genotypic and phenotypic ratios for each generation. Assume the $P_1$ individuals are homozygous.
    (a) gray, long × ebony, vestigial
    (b) gray, vestigial × ebony, long
    (c) gray, long × gray, vestigial

15. How many different types of gametes can be formed by individuals of the following genotypes: (a) *AaBb*, (b) *AaBB*, (c) *AaBbCc*, (d) *AaBBcc*, (e) *AaBbcc*, and (f) *AaBbCcDdEe*? What are they in each case?

16. Using the forked-line, or branch diagram, method, determine the genotypic and phenotypic ratios of the trihybrid crosses (a) *AaBbCc* × *AaBBCC*, (b) *AaBBCc* × *aaBBCc*, and (c) *AaBbCc* × *AaBbCc*.

17. Mendel crossed peas with green seeds with those of yellow seeds. The $F_1$ generation produced only yellow seeds. In the $F_2$, the progeny consisted of 6022 plants with yellow seeds and 2001 plants with green seeds. Of the $F_2$ yellow-seeded plants, 519 were self-fertilized with the following results: 166 bred true for yellow and 353 produced a 3:1 ratio of yellow:green. Explain these results by diagramming the crosses.

18. In a study of black and white guinea pigs, 100 black animals were crossed individually to white animals and each cross was carried to an $F_2$ generation. In 94 of the cases, the $F_1$ individuals were all black, and an $F_2$ ratio of 3 black:1 white was obtained. In the other 6 cases, half of the $F_1$ animals were black and the other half were white. Why? Predict the results of crossing the black and white $F_2$ guinea pigs from the 6 exceptional cases.

19. Mendel crossed peas with round, green seeds to ones with wrinkled, yellow seeds. All $F_1$ plants had seeds that were round and yellow. Predict the results of test-crossing these $F_1$ plants.

20. Thalassemia is an inherited anemic disorder in humans. Individuals can be completely normal, they can exhibit a "minor" anemia, or they can exhibit a "major" anemia. Assuming that only a single gene pair and two alleles are involved in the inheritance of these conditions, which phenotype is recessive?

21. Below are shown $F_2$ results of two of Mendel's monohybrid crosses. Calculate the $\chi^2$ value and determine the $p$ value for both. Can the deviation in each case be attributed to chance or not?

    |  |  |  |
    |---|---|---|
    | (a) Full pods | | 882 |
    | Constricted pods | | 299 |
    | (b) Violet flowers | | 705 |
    | White flowers | | 224 |

22. In one of Mendel's dihybrid crosses, he observed 315 smooth, yellow, 108 smooth, green, 101 wrinkled, yellow, and 32 wrinkled, green $F_2$ plants. Analyze these data using the chi-square test to see if
    (a) they fit a 9:3:3:1 ratio.
    (b) the smooth:wrinkled data fit a 3:1 ratio.
    (c) the yellow:green data fit a 3:1 ratio.

23. A geneticist, in assessing data that fell into two phenotypic classes, observed values of 250:150. She decided to perform chi-square analysis using two different null hypotheses: (a) the data fit a 3:1 ratio; and (b) the data fit a 1:1 ratio. Calculate the $\chi^2$ values for each hypothesis. What can be concluded about each hypothesis?

**24.** The basis for rejection of any null hypothesis is arbitrary. The researcher can set more or less stringent standards by deciding to raise or lower the *p* value used to reject or fail to reject the hypothesis. In the case of chi-square analysis of genetic crosses, would the use of a standard of *p* = 0.10 be more or less stringent in failing to reject the null hypothesis? Explain.

**25.** For the following pedigree, predict the mode of inheritance and the most probable genotypes of each individual. Assume that the alleles *A* and *a* control the expression of the trait.

recessive autosomal

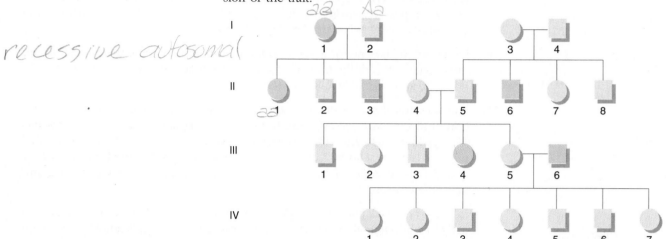

**26.** The following pedigree is for myopia in humans. Predict if the disorder is inherited as the result of a dominant or recessive allele. Determine the most probable genotype for each individual based on your prediction.

recessive
autosomal

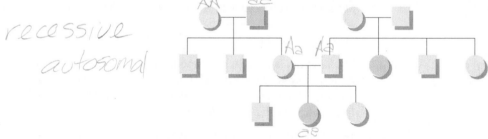

**27.** Draw all possible conclusions concerning the mode of inheritance of the trait denoted in the following limited pedigree:

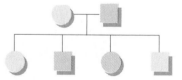

**28.** Consider three independently assorting gene pairs, *A/a*, *B/b*, and *C/c,* where each demonstrates typical dominance (*A−*, *B−*, *C−*) and recessiveness (*aa, bb, cc*). What is the probability of obtaining an offspring that is *AABbCc* from parents that are *AaBbCC* and *AABbCc*?

**29.** What is the probability of obtaining a triply recessive individual from the parents shown in Problem 28?

**30.** Of all offspring of the parents in Problem 28, what proportion will express all three dominant traits?

$$\frac{1}{1} \times \frac{3}{4} \times \frac{1}{1} = \frac{3}{4}$$

31. When a die (one of a pair of dice) is rolled, it has an equal probability of landing on any of its six sides.
    (a) What is the probability of rolling a 3 with a single throw?
    (b) When a die is rolled twice, what is the probability that the first throw will be 3 and the second will be a 6?
    (c) If two dice are rolled together, what is the probability that one will be a 3 and the other will be a 6?
    (d) If one die is rolled and it comes up as an odd number, what is the probability that it is a 5?

32. Consider the $F_2$ offspring of Mendel's dihybrid cross. Determine the conditional probability that $F_2$ plants expressing both dominant traits are heterozygous at both loci.

33. Cystic fibrosis is an autosomal recessive disorder. A male whose brother has the disease marries a female whose sister has the disease. It is not known if either the male or female is a carrier. If the male and female have one child, what is the probability that the child will have cystic fibrosis?

34. In a family of five children, what is the probability that
    (a) All are males?
    (b) Three are males and two are females?
    (c) Two are males and three are females?
    (d) All are the same sex?
    Assume that the probability of a male child is equal to the probability of a female child ($p = 1/2$).

35. In a family of eight children, where both parents are heterozygous for albinism, what mathematical expression predicts the probability that six are normal and two are albinos?

36. What do you suppose the following mathematical expression applies to? Can you think of a genetic example where it might have application?

$$\frac{n!}{s!t!u!}(a^s b^t c^u)$$

## SELECTED READINGS

CUMMINGS, M. R. 1994. *Human heredity: principles and issues.* 3rd ed. St. Paul: West.

DUNN, L. C. 1965. *A short history of genetics.* New York: McGraw-Hill.

OLBY, R. C. 1966. *Origins of Mendelism.* London: Constable.

PETERS, J., ed. 1959. *Classic papers in genetics.* Englewood Cliffs, N.J.: Prentice-Hall.

SNEDECOR, G. W., and COCHRAN, W. G. 1980. *Statistical methods.* 7th ed. Ames, Iowa: Iowa State University Press.

SOKAL, R. R., and ROHLF, F. J. 1987. *Introduction to biostatistics.* 2nd ed. New York: W. H. Freeman.

SOUDEK, D. 1984. Gregor Mendel and the people around him. *Am. J. Hum. Genet.* 36:495–98.

STERN, C., 1950. *The birth of genetics.* (Supplement to *Genetics* 35.)

STERN, C., and SHERWOOD, E. 1966. *The origins of genetics: A Mendel source book.* San Francisco: W. H. Freeman.

STUBBE, H. 1972. *History of genetics: From prehistoric times to the rediscovery of Mendel's laws.* Cambridge: MIT Press.

STURTEVANT, A. H. 1965. *A history of genetics.* New York: Harper & Row.

VOELLER, B. R., ed. 1968. *The chromosome theory of inheritance: Classical papers in development and heredity.* New York: Appleton-Century-Crofts.

# 4

# MODIFICATION OF MENDELIAN RATIOS

Yellow, black, and agouti mice.

*Specific phenotypes are often controlled by one or more gene pairs whose alleles exhibit modes of expression other than dominance and recessiveness. In all such cases, however, the Mendelian principles of segregation and independent assortment are operative during the distribution of the alleles into gametes.*

In Chapter 3, we discussed the simplest principles of transmission genetics. We saw that genes are present on homologous chromosomes and that these chromosomes **segregate** from each other and **independently assort** with other segregating chromosomes during gamete formation. These two postulates are the fundamental principles of gene transmission from parent to offspring. However, once an offspring has received the total set of genes, it is the expression of genes that determines the organism's phenotype. If gene expression does not adhere to a simple dominant/recessive mode, or if more than one pair of genes influences the expression of a single trait, the classic $3:1$ and $9:3:3:1$ $F_2$ ratios may be modified. Although in this and the next several chapters we consider more complex modes of inheritance, the fundamental principles set down by Mendel hold true in these situations as well.

In this chapter, our discussion will initially be restricted to the inheritance of traits that are under the control of only one pair of genes. In diploid organisms, where homologous pairs of chromosomes exist, two copies of each gene influence such traits. The copies need not be identical since alternative forms of genes, or **alleles**, occur within populations. How alleles act to influence a given phenotype will be our major consideration. Then we will proceed to consider how a single phenotype may be controlled by more than one gene. This general phenomenon is referred to as **gene interaction**, indicating that phenotypes are frequently under the influence of more than one gene product. Numerous examples will be presented to illustrate a variety of heritable patterns observed in such situations.

## POTENTIAL FUNCTION OF AN ALLELE

Following the rediscovery of Mendel's work in the early 1900s, research focused on the many ways in which genes can influence an individual's phenotype. This course of investigation, stemming from Mendel's findings, is called **neo-Mendelian genetics** (*neo* from the Greek word meaning since or new).

Each type of inheritance described in this chapter was investigated when observations of genetic data did not precisely conform to the expected Mendelian ratios. Hypotheses that modified and extended the Mendelian principles were proposed and tested with specifically designed crosses. The explanations for these observations were in accordance with the principle that a phenotype is under the control of one or more genes located at specific loci on one or more pairs of homologous chromosomes. If we adhere to the principles of segregation and independent assortment, we can predict accurately the transmission of any number of allele pairs.

To understand the various modes of inheritance, we must first examine the potential function of an allele. Alleles are alternative forms of the same gene. They contain modified genetic information related to the gene product specified by that gene. For example, there are many alleles of the gene that encodes the $\beta$-chain of human hemoglobin. While each allele may alter the chemical structure of the $\beta$-chain differently, all alleles are alternative forms of the same gene and affect the same gene product.

The allele that occurs most frequently in a population, or the one that is arbitrarily designated as normal, is called **wild type**. This common allele is usually dominant, such as the allele for tall plants in the garden pea, and its product is functional in the cell. Wild-type alleles are responsible, of course, for the corresponding wild-type phenotype and serve as standards for comparison against all mutations occurring at a particular locus.

The process of **mutation** is the source of new alleles. Each allele may be recognized by a change in the phenotype. A new phenotype results from a change in functional activity of the cellular product controlled by that gene. Usually, the alteration or mutation is expressed as a loss of the specific wild-type function. For example, if a gene is responsible for the synthesis of a specific enzyme, a mutation in the gene may change the conformation of this enzyme, thus eliminating its affinity for the substrate. This mutation results in a total loss of function. On the other hand, another organism may have a different mutation in this gene, which results in an enzyme with a reduced or increased affinity for binding the substrate. This mutation, representing a separate allele of this gene, may reduce or enhance rather than eliminate the functional capacity of the gene product. In either case, the phenotype may or may not be altered in a discernible way.

Although phenotypic traits may be affected by a single mutation, traits are often the result of many gene products. In the case of enzymatic reactions, most are part of complex metabolic pathways. Therefore, phenotypic traits may be under the control of more than one gene and the allelic forms of each gene involved. Diploid organisms have two copies of each gene, which may be occupied by the same allele or two different alleles.

In the initial part of this chapter, examples will be restricted to only one gene and the alleles associated with it. Thus, in these cases, we see a modification of the 3:1 monohybrid ratio. Then we will consider traits controlled by two genes and the accompanying modification of the 9:3:3:1 dihybrid ratio.

## Symbols for Alleles

In Chapter 3, we learned to symbolize alleles for very simple Mendelian traits. We used the lowercase form of the initial letter of the name of a recessive trait to denote the recessive allele and the same letter in uppercase form to refer to the dominant allele. Thus, for *tall* and *dwarf*, where *dwarf* is recessive, $D$ and $d$ represent the alleles responsible for these respective traits.

As more complex inheritance patterns were investigated, another useful system was developed to discriminate between wild-type and mutant traits. In this system, the initial letter of the name of the mutant trait is selected. If the trait is recessive, the lowercase form is used; if it is dominant, the uppercase form is used. The contrasting wild-type trait is denoted by the same letter, but with a + as a superscript.

For example, *ebony* is a recessive body color mutation in the fruit fly, *Drosophila melanogaster*. The normal wild-type body color is gray. Using the above system, *ebony* is denoted by the symbol $e$ while gray is denoted by $e^+$. If we focus on the *ebony* mutation, the responsible locus may be occupied by either the wild-type allele ($e^+$) or the mutant allele ($e$). A diploid fly may thus exhibit three possible genotypes:

$e^+/e^+$: gray homozygote (wild type)
$e^+/e$ : gray heterozygote (wild type)
$e/e$ : ebony homozygote

The slash is used to indicate that the two allele designations represent the same locus on two homologous chromosomes. If we were instead considering a dominant mutation such as *Wrinkled* (*Wr*), the three possible designations would be $Wr^+/Wr^+$, $Wr^+/Wr$, and $Wr/Wr$. The latter two genotypes express the wrinkled-wing phenotype.

One advantage of this system is that further abbreviation may be used when convenient: the wild-type allele may simply be denoted by the + symbol. Using *ebony* as an example under consideration in a cross, the designations of the three possible genotypes become:

$+/+$: gray homozygote (wild type)
$+/e$ : gray heterozygote (wild type)
$e/e$ : ebony homozygote (mutant)

As we will see in Chapter 5, this abbreviation is particularly useful when two or three genes linked together on the same chromosome are considered simultaneously.

Still other allele designations are sometimes useful. The system just described works well with alleles which are either dominant or recessive to one another. However, if no dominance exists, we may simply use uppercase letters and superscripts to denote alleles (e.g., $R^1$ and $R^2$, $L^M$ and $L^N$, $I^A$ and $I^B$). Their use will become apparent in ensuing sections of this chapter.

Finally, note that in each of the many crosses discussed in the next few chapters, only one or a few gene pairs are involved. It may be useful for you to remember that in each cross, all other genes, which are not under consideration, are assumed to be normal, or wild type.

# INCOMPLETE, OR PARTIAL, DOMINANCE

**Incomplete**, or **partial**, **dominance** in the offspring is based on the observation of intermediate phenotypes generated by a cross between parents with contrasting traits. For example, if plants such as four-o'clocks or snapdragons with red flowers are crossed with plants with white flowers, offspring may have pink flowers. It appears that neither red nor white flower color is dominant. Since some red pigment is produced in the $F_1$ intermediate-colored pink flowers, dominance appears to be incomplete or partial.

If this phenotype is under the control of a single gene and two alleles where neither is dominant, the results of the $F_1$ (pink) × $F_1$ (pink) cross can be predicted. The resulting $F_2$ generation is shown in Figure 4.1, confirming the hypothesis that only one pair of alleles determines these phenotypes. The *genotypic ratio* (1:2:1) of the $F_2$ generation is identical to that of Mendel's mono-

**FIGURE 4.1** Incomplete dominance illustrated by flower color. The photograph illustrates not only red, white, and pink snapdragons, but other colors produced as a result of the effect of other genes.

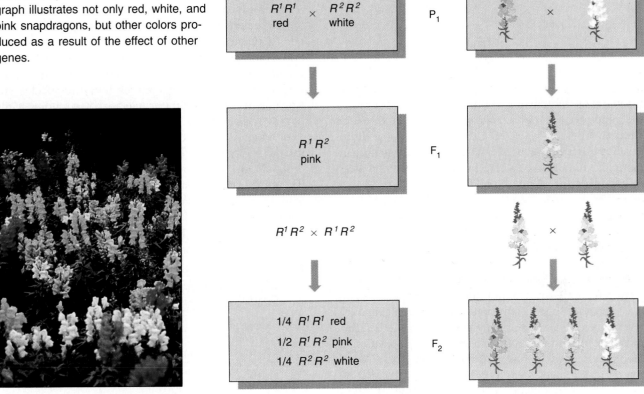

hybrid cross. Because there is no dominance, however, the *phenotypic ratio* is identical to the genotypic ratio. Note here that since neither of the alleles is recessive, we have chosen not to use upper- and lowercase letters. Instead, we have chosen $R^1$ and $R^2$ to denote the red and white alleles. We could have chosen $W^1$ and $W^2$ or still other designations such as $C^W$ and $C^R$, where $C$ indicates "color."

Clear-cut cases of incomplete dominance, which result in intermediate expression of the overt phenotype, are relatively rare. However, even when complete dominance is evident, careful examination of the level of the gene product, rather than the phenotype, often reveals intermediate gene expression. For example, in human biochemical disorders such as **Tay-Sachs disease**, homozygous recessive individuals are severely affected while heterozygotes appear phenotypically normal. In affected individuals there is almost no activity of the enzyme **hexosaminidase**. Heterozygotes, on the other hand, express only about 50 percent of the enzyme activity found in homozygous normal individuals. Fortunately, this level of enzyme activity is adequate to achieve normal biochemical function. This situation is not uncommon in enzyme disorders. It illustrates the somewhat arbitrary nature of the terms *dominance* and *recessiveness*.

## CODOMINANCE

If two alleles are responsible for the production of two distinct and detectable gene products, a situation different from the intermediate expression characteristic of incomplete dominance arises. The distinct genetic expression of both alleles in a heterozygote is called **codominance**. The **MN blood group** in humans illustrates this phenomenon and is characterized by a molecule called a **glycoprotein** found on the surface of red blood cells. Discovered by Karl Landsteiner and Philip Levine, these molecules are **native antigens** that provide biochemical and immunological identity to individuals. In the human population, two forms of this glycoprotein exist, designated M and N. An individual may exhibit either one or both of them.

The MN system is under the control of an autosomal locus found on chromosome 4 and two alleles designated $L^M$ and $L^N$. Because humans are diploid, three combinations are possible, each resulting in a distinct blood type:

| Genotype | Phenotype |
|---|---|
| $L^M L^M$ | M |
| $L^M L^N$ | MN |
| $L^N L^N$ | N |

As predicted, a mating between two MN parents may produce children of all three blood types:

$$L^M L^N \times L^M L^N$$
$$\downarrow$$
$$1/4 \quad L^M L^M$$
$$1/2 \quad L^M L^N$$
$$1/4 \quad L^N L^N$$

Codominant inheritance results in distinct evidence of the gene products of both alleles. Individual expression of each allele is apparent. This characteristic distinguishes it from other modes of inheritance, such as incomplete dominance, where heterozygotes express an intermediate, or blended, phenotype.

## MULTIPLE ALLELES

Because the information stored in any gene is extensive, mutations may modify this information in many ways. Each change has the potential for producing a different allele. Therefore, at any given locus on the chromosome, the number of alleles within a population of individuals need not be restricted to only two. When three or more alleles are found for any particular gene, the mode of inheritance is called **multiple allelism**.

*The concept of multiple alleles can only be studied in populations.* Any individual diploid organism has, at most, two homologous gene loci which may be occupied by different alleles. However, among members of a species, many alternative forms of the same gene may exist. The following examples illustrate the concept of multiple alleles. In several of the examples the relationship between genetics, immunology, and medicine is apparent.

### The ABO Blood Types

The simplest possible case of multiple alleles is that in which there are three alleles of one gene. This situation exists in the inheritance of the **ABO blood types** in humans, discovered by Landsteiner in the early 1900s. The ABO system, like the MN blood types, is characterized by the presence of native antigens on the surface of red blood cells. The ABO antigens are distinct from the MN antigens and are under the control of a different gene, located on chromosome 9. As in the MN system, one combination of alleles in the ABO system exhibits a codominant mode of inheritance.

The ABO phenotype of any individual is ascertained by mixing a blood sample with antiserum containing type A or type B antibodies. If the antigen is present on the surface of the person's red blood cells, it will react with the corresponding antibody and cause clumping or agglutination of the red blood cells.

When individuals are tested in this way, four phenotypes are revealed. Each individual has either the A antigen (A phenotype), the B antigen (B phenotype), the A and B antigens (AB phenotype), or neither antigen (O phenotype). In 1924, it was hypothesized that these phenotypes were inherited as the result of three alleles of a single gene. This hypothesis was based on studies of the blood types of many different families.

Although different designations may be used, we will use the symbols $I^A$, $I^B$, and $I^O$ for the three alleles. The $I$ designation stands for **isoagglutinogen**, another term for antigen. If we assume that the $I^A$ and $I^B$ alleles are responsible for the production of their respective A and B antigens and that $I^O$ is an allele that does not produce any detectable A or B antigens, the various genotypic possibilities can be listed and the appropriate phenotype assigned to each:

| Genotype | Antigen | Phenotype |
|---|---|---|
| $I^A I^A$ | A | |
| $I^A I^O$ | A | A |
| $I^B I^B$ | B | |
| $I^B I^O$ | B | B |
| $I^A I^B$ | A, B | AB |
| $I^O I^O$ | Neither | O |

Note that in these assignments the $I^A$ and $I^B$ alleles behave dominantly to the $I^O$ allele, but codominantly to each other.

We can test the hypothesis that three alleles control ABO blood types by examining potential offspring from many combinations of matings, as shown in Table 4.1. If we assume heterozygosity wherever possible, we can predict which phenotypes can occur. These theoretical predictions have been upheld in numerous studies examining the blood types of children of parents with all possible phenotypic combinations. The hypothesis that three alleles control ABO blood types in the human population is now universally accepted.

Our knowledge of human blood types has several practical applications. Compatible blood transfusions can be achieved and decisions about disputed parentage more accurately made. The latter cases can occur when newborns are inadvertently mixed up in hospitals, or when it is uncertain whether a specific male is the father of a child. In both cases, an examination of the ABO phenotypes as well as other inherited antigens of the possible parents and the child may help to resolve the situation. Table 4.1 demonstrates numerous cases where it is impossible for a parent of a particular ABO pheno-

**Table 4.1**  POTENTIAL PHENOTYPES IN THE OFFSPRING OF PARENTS WITH ALL POSSIBLE ABO BLOOD TYPE COMBINATIONS, ASSUMING HETEROZYGOSITY WHENEVER POSSIBLE

| P₁ Generation | | F₁ Phenotype | | | |
|---|---|---|---|---|---|
| Phenotypes | Genotypes | A | B | AB | O |
| A × A | $I^A I^O \times I^A I^O$ | 3/4 | — | — | 1/4 |
| B × B | $I^B I^O \times I^B I^O$ | — | 3/4 | — | 1/4 |
| O × O | $I^O I^O \times I^O I^O$ | — | — | — | all |
| A × B | $I^A I^O \times I^B I^O$ | 1/4 | 1/4 | 1/4 | 1/4 |
| A × AB | $I^A I^O \times I^A I^B$ | 1/2 | 1/4 | 1/4 | — |
| A × O | $I^A I^O \times I^O I^O$ | 1/2 | — | — | 1/2 |
| B × AB | $I^B I^O \times I^A I^B$ | 1/4 | 1/2 | 1/4 | — |
| B × O | $I^B I^O \times I^O I^O$ | — | 1/2 | — | 1/2 |
| AB × O | $I^A I^B \times {}^O I^O$ | 1/2 | 1/2 | — | — |
| AB × AB | $I^A I^B \times I^A I^B$ | 1/4 | 1/4 | 1/2 | — |

type to produce a child of a certain phenotype. The only mating that can result in offspring of all four phenotypes is between two heterozygous individuals, one showing the A phenotype and the other showing the B phenotype. On genetic grounds alone, a male or female may be unequivocally ruled out as the parent of a certain child. On the other hand, this type of genetic evidence *never proves* parenthood.

## The A and B Antigens

The biochemical basis of the ABO blood type system has now been carefully worked out. The A and B antigens are actually carbohydrate groups (sugars) that are bound to lipid molecules (fatty acids) protruding from the membrane of the red blood cell. The specificity of the A and B antigens is based on the terminal sugar of the carbohydrate group.

Almost all individuals possess what is called the **H substance**, to which a terminal sugar is added. As shown in Figure 4.2, the H substance itself consists of three sugar molecules, N-acetylglucosamine, galactose, and fucose, chemically linked together. The $I^A$ allele is responsible for an enzyme that can add the terminal sugar N-acetylgalactosamine to the H substance. The $I^B$ allele is responsible for a modified enzyme that cannot add N-acetylgalactosamine, but instead can add the terminal sugar galactose. Heterozygotes ($I^A I^B$) add either

one or the other entity at the many sites available. This latter phenomenon illustrates the biochemical basis of codominance in individuals of the AB blood type. Persons of type O ($I^O I^O$) cannot add either terminal sugar, possessing only the H substance protruding from the surface of their red blood cells.

The molecular basis of the mutations leading to the $I^A$, $I^B$, and $I^O$ alleles has recently been elucidated. We shall return to this topic in Chapter 11 when we discuss mutation and mutagenesis.

## The Bombay Phenotype

In 1952, a most interesting situation was observed in a woman in Bombay which provided information concerning the genetic basis of the H substance. She displayed a unique genetic history that was inconsistent with her blood type. In need of a transfusion, she was found to lack both the A and B antigens and was thus typed as O. However, as shown in the partial pedigree in Figure 4.3, one of her parents was type AB and she was the obvious donor of an $I^B$ allele to two of her offspring. Thus she was genetically type B but functionally type O!

It was subsequently shown that this woman was homozygous for a rare recessive mutation, *h*, which prevented her from synthesizing the complete H substance. The terminal portion of the carbohydrate chain protruding from the red cell membrane was shown to lack fu-

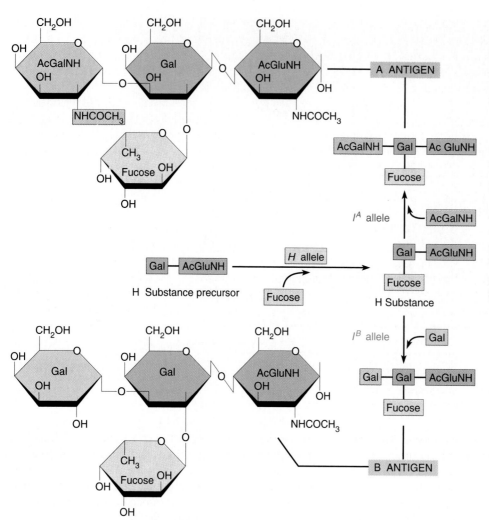

**FIGURE 4.2**   The biochemical basis of the ABO blood groups. The *H* allele, present in almost all humans, directs the conversion of a precursor molecule to the H substance by adding a molecule of fucose. Failure to do so results in the Bombay phenotype. The $I^A$ and $I^B$ alleles are then able to direct the addition of terminal sugar residues to the H substance. The $I^O$ allele is unable to direct either of these terminal additions. Gal: galactose; AcGluNH: N-acetyl-D-glucosamine; AcGalNH: N-acetylgalactosamine.

cose. In the absence of fucose, the enzymes specified by the $I^A$ and $I^B$ alleles are apparently unable to recognize the incomplete H substance as a proper substrate. Thus, neither the terminal galactose or N-acetylgalactosamine can be added, even though the enzymes capable of doing so are present and functional. As a result, the ABO system genotype cannot be expressed in individuals of genotype *hh*, and they are functionally type O. To distinguish them from the rest of the population, they are said to demonstrate the **Bombay phenotype**. The frequency of the *h* allele is exceedingly low. Thus, the vast majority of the human population is of the *HH* genotype and can synthesize the H substance.

## The Secretor Locus

Still a third gene is known to affect the expression of the ABO blood type system, found at the **secretor locus**. In about 80 percent of the human population, the A and B antigens are present in various body secretions

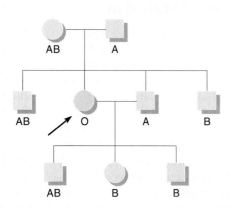

**FIGURE 4.3**   A partial pedigree of a woman displaying the Bombay phenotype. Functionally, her ABO blood group behaves as type O. Genetically, she is type B.

as well as on the membrane of red blood cells. The ability to secrete these antigens in body fluids such as saliva, gastric juice, semen, and vaginal fluids is under the influence of the dominant allele, *Se* (*Se/Se* or *Se/se*). The minority of the population who do not secrete the antigens (*se/se*) lack an enzyme that normally modifies the H substance, rendering it water soluble. Nonsecretors make the antigens as specified by the ABO loci but do not secrete them. Secretion of these antigens has great significance in forensic science (the application of scientific knowledge to law proceedings).

## The Rh Antigens

Another set of antigens thought by some geneticists to illustrate multiple allelism includes those designated Rh. Discovered by Landsteiner, Levine, and others around 1940, the Rh antigens have received a great deal of attention because of their direct involvement in the disorder **erythroblastosis fetalis**. The initial investigations led to the belief that in the human population only two alleles controlled the presence or absence of the antigen. It was thought that the *Rh*⁺ allele determined the presence of the antigen and behaved as a dominant gene. The *Rh*⁻ allele seemed to result in the absence of the antigen.

Erythroblastosis fetalis, also referred to as **hemolytic disease of the newborn (HDN)**, is a form of anemia. It occurs in an Rh-positive fetus whose mother is Rh-negative and whose father is Rh-positive, contributing that allele to the fetus. Such a genetic combination results in a potential immunological incompatibility between the mother and fetus. If fetal blood passes through the ruptured placenta at birth and enters the maternal circulation, the mother's immune system recognizes the Rh antigen as foreign and builds antibodies against it. During a second pregnancy, the antibody concentration becomes high enough that when maternal antibodies, which can pass across the placenta, enter the fetus's circulation, they begin to destroy the fetus's red blood cells. This causes the hemolytic anemia.

About 10 percent of all human pregnancies demonstrate Rh incompatibility. However, for numerous reasons, less than 0.5 percent actually result in anemia. Currently, incompatible mothers are given anti-Rh antisera immediately after giving birth to an Rh-positive baby. This destroys any Rh-positive cells that have entered the mother's circulation so that she does not produce her own anti-Rh antibodies. Before this treatment was developed, many fetuses failed to survive to term. For those that did, complete blood transfusions were often necessary. Even then, many newborns failed to survive.

With the development of more refined antisera to test for the presence of the antigen, it became apparent that

the genetic control of Rh antigens was much more complex than originally thought. For example, some presumed Rh-negative blood was found to contain the antigen, but in a different chemical form. Alexander Wiener has proposed the existence of at least eight alleles at a single Rh locus. Other workers, including Ronald A. Fisher, Robert R. Race, and Ruth Sanger, have proposed an alternative type of inheritance. These workers feel that there are three closely linked genes, each with two alleles involved in the inheritance of Rh factors. The term **linkage** is used to describe genes located on the same chromosome. We will discuss this concept in detail in Chapter 5.

The two systems of nomenclature designating the alleles are contrasted in Table 4.2. In the Wiener system, the presence of at least one of the four dominant alleles is sufficient to yield the Rh-positive blood type. Examination of the Fisher–Race system shows that the genetic locus bearing the set of *D* and *d* alleles is most critical. The presence of at least one dominant *D* allele results in the Rh-positive phenotype, while the *dd* genotype ensures the Rh-negative blood type. While the *C*, *c*, *E*, and *e* alleles specify distinguishable antigens, they are not immunologically significant. Because of the complexity of the antigenic patterns, it is difficult to favor one system over the other.

## The *white* Locus in *Drosophila*

Many other phenotypes in plants and animals are known to be controlled by multiple allelic inheritance. In *Drosophila*, for example, where the induction of mutations has been used extensively as a method of investigation, many alleles are known at practically every locus. The

**Table 4.2**  A Comparison of the Alleles Involved in the Wiener System with Those of the Fisher–Race System to Explain the Genetic Basis of the Rh Blood Groups

| Wiener Nomenclature | Fisher–Race Nomenclature | Phenotype |
|---|---|---|
| $R^1$ | CDE | |
| $R^2$ | CDe | |
| $R^0$ | cDE | Rh⁺ |
| $R^z$ | cDe | |
| $r$ | CdE | |
| $r'$ | Cde | |
| $r''$ | cdE | Rh⁻ |
| $r^y$ | cde | |

**Table 4.3**  SOME OF THE ALLELES PRESENT AT THE *WHITE* LOCUS OF *DROSOPHILA MELANOGASTER* AND THEIR EYE COLOR PHENOTYPE

| Allele | Name | Eye Color |
|--------|------|-----------|
| $w$ | white | Pure white |
| $w^a$ | white-apricot | Yellowish orange |
| $w^{bf}$ | white-buff | Light buff |
| $w^{bl}$ | white-blood | Yellowish ruby |
| $w^{cf}$ | white-coffee | Deep ruby |
| $w^e$ | white-eosin | Yellowish pink |
| $w^{mo}$ | white-mottled orange | Light mottled orange |
| $w^{sat}$ | white-satsuma | Deep ruby |
| $w^{sp}$ | white-spotted | Fine grain, yellow mottling |
| $w^t$ | white-tinged | Light pink |

recessive *white* eye mutation, discovered by Thomas H. Morgan and Calvin Bridges in 1912, is only one of over 100 alleles that may occupy this locus. In this allelic series, eye colors range from complete absence of pigment in the *white* allele, to deep ruby in the *white-satsuma* allele, to orange in the *white-apricot* allele, to a buff color in the *white-buff* allele. These alleles are designated $w$, $w^{sat}$, $w^a$, and $w^{bf}$, respectively (Table 4.3). In each of these cases, the total amount of pigment in these mutant eyes is reduced to less than 20 percent of that found in the brick red wild-type eye.

## LETHAL ALLELES

Many gene products are essential to an organism's survival. Mutations resulting in the synthesis of a gene product that is nonfunctional can sometimes be tolerated in the heterozygous state; that is, one wild-type allele may be sufficient to produce enough of the essential product to allow survival. However, such a mutation behaves as a **recessive lethal allele**, and homozygous recessive individuals will not survive. The time of death will depend upon when the product is needed during development or even adulthood.

In instances where one copy of the wild gene is not sufficient for normal development, the heterozygote will not survive. In this case, the mutation is behaving as a **dominant lethal allele** because its presence somehow overrides the expression of the wild-type product, or the amount of wild-type product is simply insufficient to support its essential function.

In some cases, the allele responsible for a lethal effect when homozygous may result in a distinctive mutant

respect to the phenotype. For example, a mutation causing a yellow coat in mice was discovered in the early part of this century. The yellow coat varied from the normal agouti coat phenotype, as illustrated in Figure 4.4. Crosses between the various combinations of the two strains yielded unusual results:

| Crosses |
|---------|
| A: agouti × agouti → all agouti |
| B: yellow × yellow → 2/3 yellow: 1/3 agouti |
| C: agouti × yellow → 1/2 yellow: 1/2 agouti |

These results are explained on the basis of a single pair of alleles. The mutant *yellow* allele $A^Y$ is dominant to the wild-type *agouti* allele $A$, so heterozygous mice will have yellow coats. However, the yellow allele also behaves as a homozygous lethal. $A^Y A^Y$ mice die before birth. As a result no homozygous yellow mice are ever recovered. The genetic basis for these three crosses is provided in Figure 4.4.

Many genes are known to behave similarly in other organisms. In *Drosophila, Curly* wing (*Cy*), *Plum* eye (*Pm*), *Dichaete* wing (*D*), *Stubble* bristle (*Sb*), and *Lyra* wing (*Ly*) behave as homozygous recessive lethals but are dominant with respect to the expression of the mutant phenotype when heterozygous.

Alleles in other organisms are known to behave as dominant lethals. In humans, a disorder called **Huntington disease** (also referred to as Huntington's chorea) is due to a dominant allele, $H$, where the onset of the disease in heterozygotes (*Hh*) is delayed, usually into adulthood. Affected individuals then undergo gradual nervous and motor degeneration until they die. This lethal disorder is particularly tragic because it has such a late onset, typically at about age 40. By that time, the affected

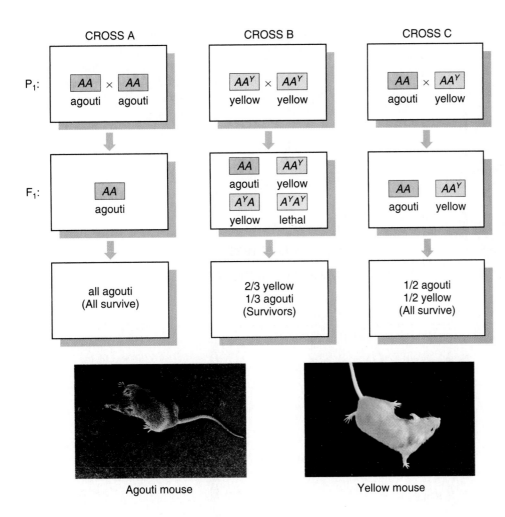

CROSS A

P₁:  $AA$ × $AA$
agouti   agouti

F₁:  $AA$
agouti

all agouti
(All survive)

Agouti mouse

CROSS B

P₁:  $AA^Y$ × $AA^Y$
yellow   yellow

F₁:
| $AA$ | $AA^Y$ |
| agouti | yellow |
| $A^YA$ | $A^YA^Y$ |
| yellow | lethal |

2/3 yellow
1/3 agouti
(Survivors)

Yellow mouse

CROSS C

P₁:  $AA$ × $AA^Y$
agouti   yellow

F₁:  $AA$   $AA^Y$
agouti   yellow

1/2 agouti
1/2 yellow
(All survive)

**FIGURE 4.4** Inheritance patterns in three crosses involving the wild type *agouti* allele (*A*) and the mutant *yellow* allele ($A^Y$) in the mouse. Note that the mutant allele behaves dominantly to the normal allele (*A*) in controlling coat color, but it also behaves as a homozygous lethal allele. The genotype $A^YA^Y$ does not survive.

and passing the lethal gene to his or her offspring. The American folk singer and composer Woody Guthrie died from this disease.

Dominant lethal alleles are rarely observed. In order for them to exist in a population, the affected individual must reproduce before the allele's lethality is expressed. If all affected individuals die during embryonic or fetal development, or before reaching the reproductive age, the mutant gene will not be passed to future generations.

## COMBINATIONS OF TWO GENE PAIRS

Each example discussed so far modifies Mendel's 3:1 F₂ monohybrid ratio. Therefore, combining any two of these modes of inheritance in a dihybrid cross will likewise modify the classical 9:3:3:1 ratio. Having established the foundation for the modes of inheritance of incomplete dominance, codominance, multiple alleles, and lethal genes, we can now deal with the situation of two modes of inheritance occurring simultaneously. Mendel's principle of independent assortment applies to

these situations when the genes controlling each character are not located on the same chromosome.

Suppose, for example, that a mating occurs between two humans who are both heterozygous for the autosomal recessive gene that causes albinism and who are both of blood type AB. What is the probability of any particular phenotypic combination occurring in each of their children? Albinism is inherited in the simple Mendelian fashion, and the blood types are determined by the series of three multiple alleles, $I^A$, $I^B$, and $I^O$. The solution to this problem is diagrammed in Figure 4.5, using the forked-line method.

Instead of this dihybrid cross yielding the classical four phenotypes in a 9:3:3:1 ratio, six phenotypes occur in a 3:6:3:1:2:1 ratio, establishing the expected probability for each phenotype. While Figure 4.5 solves the problem using the forked-line method described in Chapter 3, we could solve the problem using the more conventional Punnett square. Recall that the forked-line method simply requires that the phenotypic ratios for each trait be computed individually (which can usually be done by inspection); all possible combinations can then be calculated.

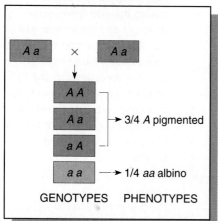

CONSIDERATION OF
PIGMENTATION ALONE

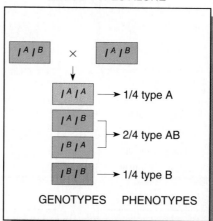

CONSIDERATION OF
BLOOD TYPES ALONE

**FIGURE 4.5**    The calculation of the probabilities in a mating involving the ABO blood type and albinism in humans using the forked-line method.

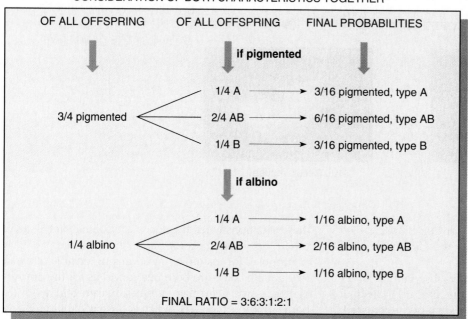

CONSIDERATION OF BOTH CHARACTERISTICS TOGETHER

FINAL RATIO = 3:6:3:1:2:1

We can deal in a similar way with any combination of two modes of inheritance. You will be asked to determine the phenotypes and their expected probabilities for many of these combinations when you solve the problems at the end of the chapter. In each case, the final phenotypic ratio is a modification of the 9:3:3:1 dihybrid ratio.

# GENE INTERACTION: DISCONTINUOUS VARIATION

Soon after the rediscovery of Mendel's work, experimentation revealed that individual characteristics displaying **discrete**, or **discontinuous**, **phenotypes**

were often under the control of more than one gene. This was a significant discovery because it revealed for the first time that genetic influence on the phenotype is much more sophisticated than envisioned by Mendel. Instead of single genes controlling the development of individual parts of the plant and animal body, it soon became clear that each character, such as eye color, hair color, or blood type, is influenced by many gene products.

The concept of **gene interaction** does not mean that two or more genes, or their products, necessarily interact directly to influence a particular phenotype. Instead, this concept implies that the cellular function of numerous gene products is related to the development of a common phenotype. To clarify this point, we will present several examples that directly illustrate gene interaction at the biochemical level.

## Epistasis

Perhaps the best examples of gene interaction leading to discontinuous variation are those illustrating the phenomenon of **epistasis**. Derived from the Greek word meaning "stoppage," epistasis can in genetics be equated with the word *masking*. The phenomenon occurs when the expression of one gene pair masks or modifies the expression of another gene pair. This masking may occur under different conditions (Figure 4.6). For example, the presence of two recessive alleles at one locus may prevent or override the expression of alleles at a second locus (or several other loci). Or, a single dominant allele at the first locus may influence the expression of the alleles at a second gene locus. In a third case, two gene pairs may complement one another such that at least one dominant allele at each locus is required to

THE CROSS

$$I^A\ I^B\ Hh \times I^A\ I^B\ Hh$$

CONSIDERATION OF BLOOD TYPES

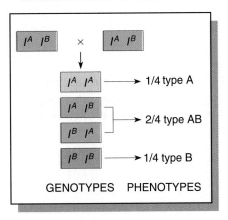

CONSIDERATION OF H SUBSTANCE

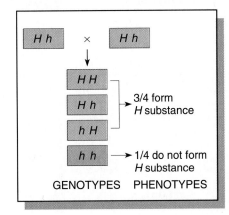

**FIGURE 4.6**   The outcome of a mating between individuals heterozygous at two genes determining their ABO blood type. Final phenotypes are calculated by considering both genes separately and then combining the results using the forked-line method.

CONSIDERATION OF BOTH GENE PAIRS TOGETHER

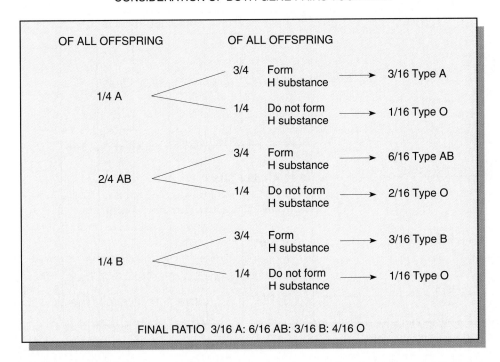

express a particular phenotype. Each of these three forms of epistasis will be examined in more detail.

An example of the homozygous recessive condition at one locus masking the expression of a second locus has been examined earlier in this chapter when we discussed the Bombay phenotype. There, the homozygous condition ($hh$) masked the expression of the $I^A$ and $I^B$ alleles. Only individuals with the $H-$ genotype can form the A or B antigens. As a result, individuals with genotypes that include the $I^A$ or $I^B$ allele who are also $hh$ express the type O phenotype.

An example of the outcome of matings between individuals heterozygous at both loci is illustrated in Figure 4.6. If many individuals of the genotype $I^A I^B Hh$ have children, the phenotypic ratio of 3 A: 6 AB: 3 B: 4 O is expected in their offspring.

Two important observations can be made when examining this cross and the predicted phenotypic ratio:

1. In illustrating gene interaction, an important distinction exists in this cross compared to the modified dihybrid cross illustrated in Figure 4.5: *only one characteristic—blood type—is being followed*. In the modified dihybrid cross in Figure 4.5, blood type *and* skin pigmentation are followed as separate phenotypic characteristics.

2. Even though only a single character was followed, the phenotypic ratio was expressed in 16s. If we knew nothing about the H substance and the genes controlling it, we could still be confident that a second gene pair, other than that controlling the A and B antigens, was involved in the phenotypic expression. *A ratio that is expressed in 16 parts (e.g., 3:6:3:4) suggests that two gene pairs are "interacting" during the expression of the phenotype under consideration.*

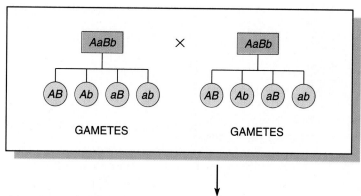

**FIGURE 4.7**   Generation of the various modified dihybrid ratios from the nine unique genotypes produced in a cross between individuals heterozygous at two genes.

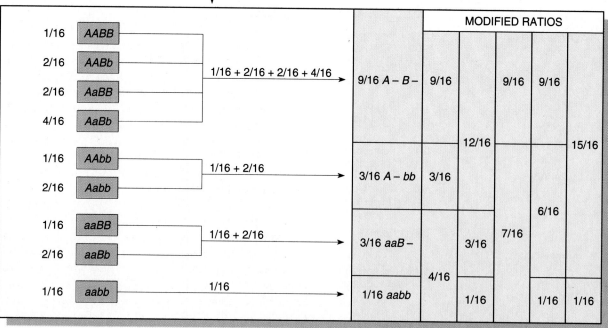

| Case | Organism | Character | AABB 1/16 | AABb 2/16 | AaBB 2/16 | AaBb 4/16 | AAbb 1/16 | Aabb 2/16 | aaBB 1/16 | aaBb 2/16 | aabb 1/16 | Final Phenotypic Ratio |
|------|----------|-----------|-----------|-----------|-----------|-----------|-----------|-----------|-----------|-----------|-----------|------------------------|
| | Pea | Mendel's dihybrid | 9/16 | | | | 3/16 | | 3/16 | | 1/16 | 9:3:3:1 |
| 1 | Mouse | Coat color | Agouti | | | | Albino | | Black | | Albino | 9:3:4 |
| 2 | Squash | Color | White | | | | | | Yellow | | Green | 12:3:1 |
| 3 | Pea | Flower color | Purple | | | | White | | | | | 9:7 |
| 4 | Squash | Fruit shape | Disc | | | | Sphere | | | | Long | 9:6:1 |
| 5 | Chicken | Color | White | | | | | | Colored | | White | 13:3 |
| 6 | Mouse | Color | White-spotted | | | | White | | Colored | | White-spotted | 10:3:3 |
| 7 | Shepherd's purse | Seed capsule | Triangular | | | | | | | | Ovoid | 15:1 |
| 8 | Flour beetle | Color | Red | Sooty | Red | Sooty | Black | | Jet | | Black | 6:3:3:4 |

**FIGURE 4.8**    The basis of modified dihybrid F₂ phenotypic ratios, resulting from crosses between doubly heterozygous F₁ individuals. The four groupings of the F₂ genotypes shown in Figure 4.7 and across the top of this figure are combined in various ways to produce these ratios.

The study of gene interaction has revealed a number of inheritance patterns that modify the classical Mendelian dihybrid $F_2$ ratio (9:3:3:1) in other ways. In several of the subsequent examples that we shall consider, epistasis has the effect of combining one or more of the four phenotypic categories in various ways. The generation of these four groups is reviewed in Figure 4.7, along with several modified ratios.

As we discuss these and other examples, we will make several assumptions and adopt certain conventions:

1. In each case, distinct phenotypic classes are produced, each clearly discernible from all others. Such traits illustrate discontinuous variation, where phenotypic categories are discrete and qualitatively different from one another.

2. The genes considered in each cross are not linked and therefore assort independently of one another during gamete formation.

3. If complete dominance exists between the alleles of any gene pair, such that $AA$ and $Aa$ or $BB$ and $Bb$ are equivalent in their genetic effects, the designations $A-$ or $B-$ will be used for both combinations. Therefore, the dash (–) indicates that either allele may be present, without consequence to the phenotype.

4. All $P_1$ crosses will involve homozygous individuals (e.g., $AABB \times aabb$, $AAbb \times aaBB$, or $aaBB \times AAbb$). Therefore, each $F_1$ generation will consist of only heterozygotes of genotype $AaBb$.

5. In each example, the $F_2$ generation produced from these heterozygous parents will be the main focus of analysis. When two genes are involved (Figure 4.7), the $F_2$ genotypes fall into four categories: 9/16 $A-B-$, 3/16 $A-bb$, 3/16 $aaB-$, and 1/16 $aabb$. Because of dominance, all genotypes in each category are equivalent in their effect on the phenotype.

Our first example is seen in the inheritance of coat color in mice (Case 1 of Figure 4.8). As we saw in our discussion of lethal alleles, wild-type coat color is agouti, a grayish pattern resulting from black hairs with a subapical band of yellow surrounding each hair just down from the tip. Agouti is dominant to black hair, caused by the homozygous expression of a recessive allele, $a$. Thus, $A-$ results in agouti, while $aa$ yields black coat color. When homozygous, a recessive mutation, $b$, at a separate locus, eliminates pigmentation altogether, yielding albino mice. Regardless of the genotype at the *a* locus, a genotype of *bb* yields albino mice. The black and agouti phenotypes are illustrated in Figure 4.9.

In a cross between agouti (*AABB*) and albino (*aabb*), members of the F₁ are all *AaBb* and have agouti coat color. In the F₂ progeny of a cross between two F₁ double heterozygotes, the following genotypes and phenotypes are observed:

**F₁**: *AaBb* × *AaBb*
↓

| F₂ Ratio | Genotype | Phenotype | Final Phenotypic Ratio |
|---|---|---|---|
| 9/16 | *A− B−* | agouti | 9/16 agouti |
| 3/16 | *A− bb* | albino | 4/16 albino |
| 3/16 | *aa B−* | black | |
| 1/16 | *aa bb* | albino | 3/16 black |

Gene interaction yielding the observed 9:3:4 F₂ ratio might be envisioned hypothetically as a two-step process:

$$\text{Precursor Molecule (colorless)} \xrightarrow[B-]{\text{Gene } B} \text{Black Pigment} \xrightarrow[A-]{\text{Gene } A} \text{Agouti Pattern}$$

In the presence of a *B* allele, black pigment can be made from a colorless substance. In the presence of an *A* allele, the black pigment is deposited during the development of hair in a pattern producing the agouti phenotype. If the *aa* genotype occurs, all of the hair remains black. If the *bb* genotype occurs, no black pigment is produced, regardless of the presence of the *A* or *a* alleles, and the mouse is albino. Therefore, the *bb* geno-

type masks or suppresses the expression of the *A* allele, illustrating epistasis.

A second type of epistasis occurs when a dominant allele at one genetic locus masks the expression of the alleles of a second locus. For instance, Case 2 of Figure 4.8 deals with the inheritance of fruit color in summer squash. Here, the dominant allele *A* results in white fruit color regardless of the genotype at a second locus, *B*. In the absence of a dominant *A* allele (the *aa* genotype), *BB* or *Bb* results in yellow color while *bb* results in green color. Therefore, if two white-colored double heterozygotes (*AaBb*) are crossed together, an interesting genetic ratio occurs because of this type of epistasis:

**F₁**: *AaBb* × *AaBb*
↓

| F₂ Ratio | Genotype | Phenotype | Final Phenotypic Ratio |
|---|---|---|---|
| 9/16 | *A− B−* | white | 12/16 white |
| 3/16 | *A− bb* | white | |
| 3/16 | *aa B−* | yellow | 3/16 yellow |
| 1/16 | *aa bb* | green | 1/16 green |

Of the offspring, 9/16 are *A− B−* and are thus white. The 3/16 bearing the genotypes *A− bb* are also white. Of the remaining squash, 3/16 are yellow (*aaB−*), while 1/16 are green (*aabb*). Thus, the modified phenotypic ratio of 12:3:1 occurs.

Our third example (Case 3 of Figure 4.8) is demonstrated in a cross between two strains of white-flowered sweet peas. Unexpectedly, the F₁ plants were all purple, and the F₂ occurred in a ratio of 9/16 purple to 7/16 white. The proposed explanation for these results suggests that the presence of at least one dominant allele of each of two gene pairs is essential in order for flowers to be purple. All other genotype combinations yield white flowers because the homozygous condition of either recessive allele masks the expression of the dominant allele at the other locus.

The cross is shown as follows:

**P₁**:   *AAbb* × *aaBB*
(white)   (white)
↓
**F₁**:   All *AaBb* (purple)
↓

| F₂ Ratio | Genotype | Phenotype | Final Phenotypic Ratio |
|---|---|---|---|
| 9/16 | *A− B−* | purple | 9/16 purple |
| 3/16 | *A− bb* | white | |
| 3/16 | *aa B−* | white | 7/16 white |
| 1/16 | *aa bb* | white | |

**FIGURE 4.9**    Mice expressing the agouti, yellow, and black phenotypes.

We can envision the way in which two gene pairs might yield such results:

Gene *A*                    Gene *B*
Precursor        ↓        Intermediate        ↓        Final
Substance   ⟶        Product        ⟶        Product
(colorless)      *A−*      (colorless)      *B−*      (purple)

At least one dominant allele from each pair of genes is necessary to ensure both biochemical conversions to the final product, yielding purple flowers. In the cross above, this will occur in 9/16 of the $F_2$ offspring. All other plants have flowers that remain white.

These three examples illustrate in a simple way how two gene products interact to influence a common phenotype. In most instances, many more than two genes interact to produce each phenotypic character. Provided that each such gene is homozygous for the wild-type alleles, these additional genes would not influence the outcome of the above crosses.

## Novel Phenotypes

Other cases of gene interaction yield novel, or new, phenotypes in the $F_2$ generation, in addition to producing modified dihybrid ratios. Case 4 in Figure 4.8 depicts the inheritance of fruit shape in the summer squash *Cucurbita pepo*. When plants with disc-shaped fruit (*AABB*) are crossed to plants with long fruit (*aabb*), the $F_1$ generation all have disc fruit. However, in the $F_2$ progeny, fruit with a novel shape—sphere—appear as well as fruit exhibiting the parental phenotypes. These phenotypes are shown in Figure 4.10.

The $F_2$ generation, with a modified 9:6:1 ratio, is as follows:

**F₁**:    *AaBb* × *AaBb*
disc         disc
↓

| F₂ Ratio | Genotype | Phenotype | Final Phenotypic Ratio |
|---|---|---|---|
| 9/16 | *A−B−* | disc | 9/16 disc |
| 3/16 | *A−bb* | sphere | 6/16 sphere |
| 3/16 | *aa B−* | sphere | |
| 1/16 | *aa bb* | long | 1/16 long |

In this example of gene interaction, both gene pairs influence fruit shape equivalently. A dominant allele at either locus ensures a sphere-shaped fruit. In the absence of dominant alleles, the fruit is long. However, if both dominant alleles (*A* and *B*) are present, the fruit is flattened into a disc shape.

Another interesting example of an unexpected phenotype arising in the $F_2$ generation is the inheritance of eye

**FIGURE 4.10**    Summer squash exhibiting various fruit-shape phenotypes.

color in *Drosophila melanogaster*. The wild-type eye color is brick red. When two autosomal recessive mutants, *brown* and *scarlet*, are crossed, the $F_1$ generation consists of flies with wild-type eye color. In the $F_2$ generation wild, scarlet, brown, and white-eyed flies are found in a 9:3:3:1 ratio. While this ratio is numerically the same as Mendel's dihybrid ratio, the *Drosophila* cross involves only one character, eye color. The diagram of this cross uses the gene symbols *A* to *B* to maintain consistency in this section. (The actual symbols for the two genes are *bw* and *st*.)

**P₁**:    *Aabb* × *aaBB*
brown        scarlet
↓
**F₁**:    *AaBb* (wild type)
↓

| F₂ Ratio | Genotype | Phenotype | Final Phenotypic Ratio |
|---|---|---|---|
| 9/16 | *A−B−* | wild type | 9/16 wild |
| 3/16 | *A−bb* | brown | 3/16 brown |
| 3/16 | *aa B−* | scarlet | 3/16 scarlet |
| 1/16 | *aa bb* | white | 1/16 white |

This cross is an excellent example of gene interaction because the biochemical basis of eye color in this organism has been determined (Figure 4.11). *Drosophila,* as a typical arthropod, has compound eyes made up of individual visual units called **ommatidia**.

The wild-type eye color is due to the deposition and mixing of two separate pigments in each ommatidium. These include the bright red pigment **drosopterin** and the brown pigment **xanthommatin**. Each pigment is

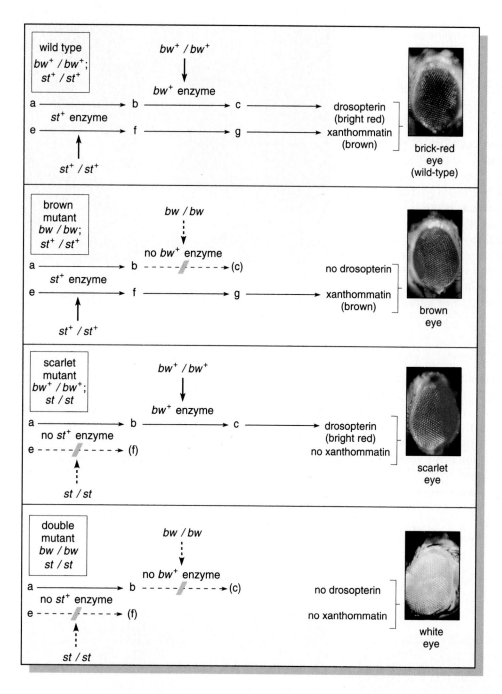

**FIGURE 4.11**    A theoretical explanation of the biochemical basis of the four eye-color phenotypes produced in a cross between flies with *brown* and *scarlet* eyes in *Drosophila melanogaster*. In the presence of at least one wild-type $bw^+$ allele, an enzyme is produced that converts substance b to c, and drosopterins are synthesized. In the presence of at least one wild-type $st^+$ allele, substance e is converted to f, and xanthommatins are synthesized. The homozygous presence of the recessive *bw* and *st* mutant alleles blocks the synthesis of these respective pigment molecules. Either none, one, or both of these pathways can be blocked, depending on the genotype.

produced by separate biosynthetic pathways. Each step of each pathway is catalyzed by a separate enzyme and is thus under the control of a separate gene. As shown in Figure 4.11, the *brown* mutation, when homozygous, interrupts the pathway leading to the synthesis of the bright red pigment. Thus, the eye contains only xanthommatin pigments and is brown. The recessive mutation *scarlet,* affecting a gene located on a separate autosome, interrupts the pathway leading to the synthesis of the brown xanthommatins and renders the eye color bright red in homozygous mutant flies. Each mutation

apparently causes the production of a nonfunctional enzyme. Flies that are double mutants and thus homozygous for both *brown* and *scarlet* lack both functional enzymes and can make neither of the pigments; they represent the novel white-eyed flies appearing in 1/16 of the $F_2$ generation.

## Other Modified Dihybrid Ratios

The remaining cases (5–8) in Figure 4.8 illustrate additional modifications of the dihybrid ratio and provide still other examples of gene interactions. All cases (1–8) have two things in common. First, in arriving at a suitable explanation of the inheritance pattern of each one, we have not violated the principles of segregation and independent assortment. Therefore, the added complexity of inheritance in these examples does not detract from the validity of Mendel's conclusions. Second, the $F_2$ phenotypic ratio in each example has been expressed in sixteenths. When a similar observation is made in crosses where the inheritance pattern is unknown, it suggests to geneticists that two gene pairs are controlling the observed phenotypes. You should make the same inference in the analysis of genetics problems. Other insights into solving genetics problems are provided in the "Insights and Solutions" section at the conclusion of this chapter.

# GENE INTERACTION: CONTINUOUS VARIATION

In the preceding section, examples in Figure 4.8 illustrated gene interaction leading to phenotypic variation that is easily classified into distinct traits. Pea plants may be tall or dwarf; squash shape may be spherical, disc-shaped, or elongated; and fruit-fly eye color may be red, brown, scarlet, or white. These phenotypes are examples of discontinuous variation where discrete phenotypic categories exist. There are many other traits in a population that demonstrate considerably more variation and are not easily categorized into distinct classes. Such phenotypes represent **continuous variation**. For example, in addition to his work with sweet peas, Mendel experimented with beans. In a cross between a purple and a white-flowered strain the $F_1$ was purple. However, the $F_2$ contained not only purple-flowered and white-flowered bean plants, but ones with numerous intermediate shades. He was unable to explain these results satisfactorily, but recognized that they were inconsistent with most of his data derived from sweet peas. It was not until more than 50 years later that the inheritance of characters exhibiting continuous variation was explained.

It is now known that traits exhibiting continuous variation are often controlled by two or more genes that make additive contributions to the phenotype. Such traits are

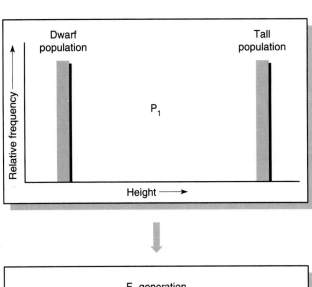

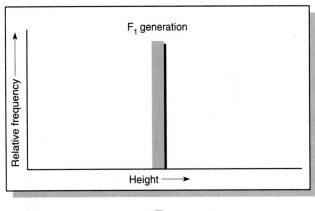

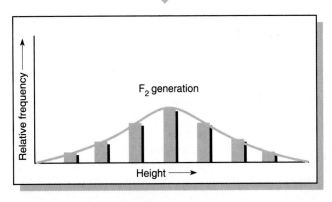

**FIGURE 4.12**     Histograms showing the relative frequency of individuals expressing various height phenotypes derived from Kolreuter's cross between dwarf and tall tobacco plants carried to the $F_2$ generation. The photograph shows a field of tobacco plants.

said to exhibit **continuous or quantitative variation** and are examples of **polygenic inheritance**. We will examine such patterns of inheritance in which a phenotypic trait is controlled by genes at two or more loci. Later in the text (Chapter 7), we will return to this general topic and outline the statistical tools used by geneticists to study traits that exhibit continuous variation.

## Quantitative Inheritance: Polygenes

In the late eighteenth century, Josef Gottlieb Kölreuter showed that when tall and dwarf tobacco plants were crossed, the $F_1$ generation was not all tall, as Mendel found with garden peas. Instead, the individual plants were all intermediate in height. When the $F_2$ generation was examined, individuals showed continuous variation in height, ranging from tall to dwarf. The majority of the $F_2$ plants were intermediate like the $F_1$, while only a few were as tall or dwarf as the $P_1$ parents. These distributions are depicted in histograms in Figure 4.12. Note that the $F_2$ data assume a normal distribution, as evidenced by the bell-shaped curve in the histogram.

At the beginning of the twentieth century, geneticists noted that many characters in different species had similar patterns of inheritance, such as height and stature in humans, seed size in the broad bean, grain color in wheat, and kernel number and ear length in corn. In each case, offspring in the succeeding generation seemed to be a blend of their parents' characteristics.

The issue of whether continuous variation could be accounted for in Mendelian terms caused considerable controversy in the early 1900s. Those such as William Bateson and Gudny Yule, who adhered to the Mendelian explanation of inheritance, suggested that a large number of factors or genes were responsible for the observed patterns. This proposal, called the **multiple-factor** or **multiple-gene hypothesis**, implied that many factors or genes contribute to the phenotype in a *cumulative* or *quantitative* way. However, some geneticists argued that Mendel's unit factors could not account for the blending of parental phenotypes characteristic of these patterns of inheritance and were thus skeptical of his ideas.

By 1920, the conclusions of several critical sets of experiments largely resolved the controversy and demonstrated that Mendelian factors could account for continuous variation. In one experiment, Edward M. East performed crosses between two strains of the tobacco plant *Nicotiana longiflora*. The fused inner petals of the flower, or corollas, of strain A were decidedly shorter than the corollas of strain B. With only minor variation, each strain was true breeding. Thus, the differences between them were clearly under genetic control.

When plants from the two strains were crossed, the $F_1$,

$F_2$, and selected $F_3$ data (Figure 4.13) demonstrated a very distinct pattern. The $F_1$ generation displayed corollas that were intermediate in length, compared with the $P_1$ varieties, and showed only minor variability among individuals. While corolla lengths of the $P_1$ plants were about 40 mm and 94 mm, the $F_1$ generation contained plants with corollas that were all about 64 mm. In the $F_2$ generation, lengths varied much more, ranging from 52 mm to 82 mm. Most individuals were similar to their $F_1$ parents, and as the deviation from this average increased, fewer and fewer plants were observed. When the data are plotted graphically (number vs. length), a bell-shaped curve results.

East further experimented with this population by selecting $F_2$ plants of various corolla lengths and allowing them to produce separate $F_3$ generations. Several are illustrated in Figure 4.13. In each case, a bell-shaped distribution was observed with most individuals similar in height to the $F_2$ parents, but with considerable variation around this value.

East's experiments thus demonstrated that although the variation in corolla length seemed continuous, experimental crosses resulted in the segregation of distinct phenotypic classes as observed in the three independent $F_3$ categories. This finding strongly suggested that the multiple-factor hypothesis could account for traits that deviate considerably in their expression.

The multiple-factor hypothesis, suggested by the observations of East and others, embodies the following major points:

1. Characters under such control can usually be quantified by measuring, weighing, counting, etc.

2. Two or more pairs of genes, located throughout the genome, account for the hereditary influence on the phenotype in an *additive way*. Because many genes may be involved, inheritance of this type is often called *polygenic*.

3. Each gene locus may be occupied by either an additive allele, which contributes a set amount to the phenotype, or by a nonadditive allele, which does not contribute quantitatively to the phenotype.

4. The total effect of each additive allele at each locus, while small, is approximately equivalent to all other additive alleles at other gene sites.

5. Together, the genes controlling a single character produce substantial phenotypic variation.

6. This genetic variation is affected by environmental factors. For example, despite their genetic basis, height in humans is influenced by individual diets, and plant characters are influenced by rainfall and soil nutrients.

**FIGURE 4.13**    The $F_1$, $F_2$, and selected $F_3$ results of East's cross between two strains of *Nicotiana* with different corolla lengths. Plants of strain A vary from 37 to 43 mm, while plants of strain B vary from 91 to 97 mm. The photograph illustrates the flower and corolla of a tobacco plant.

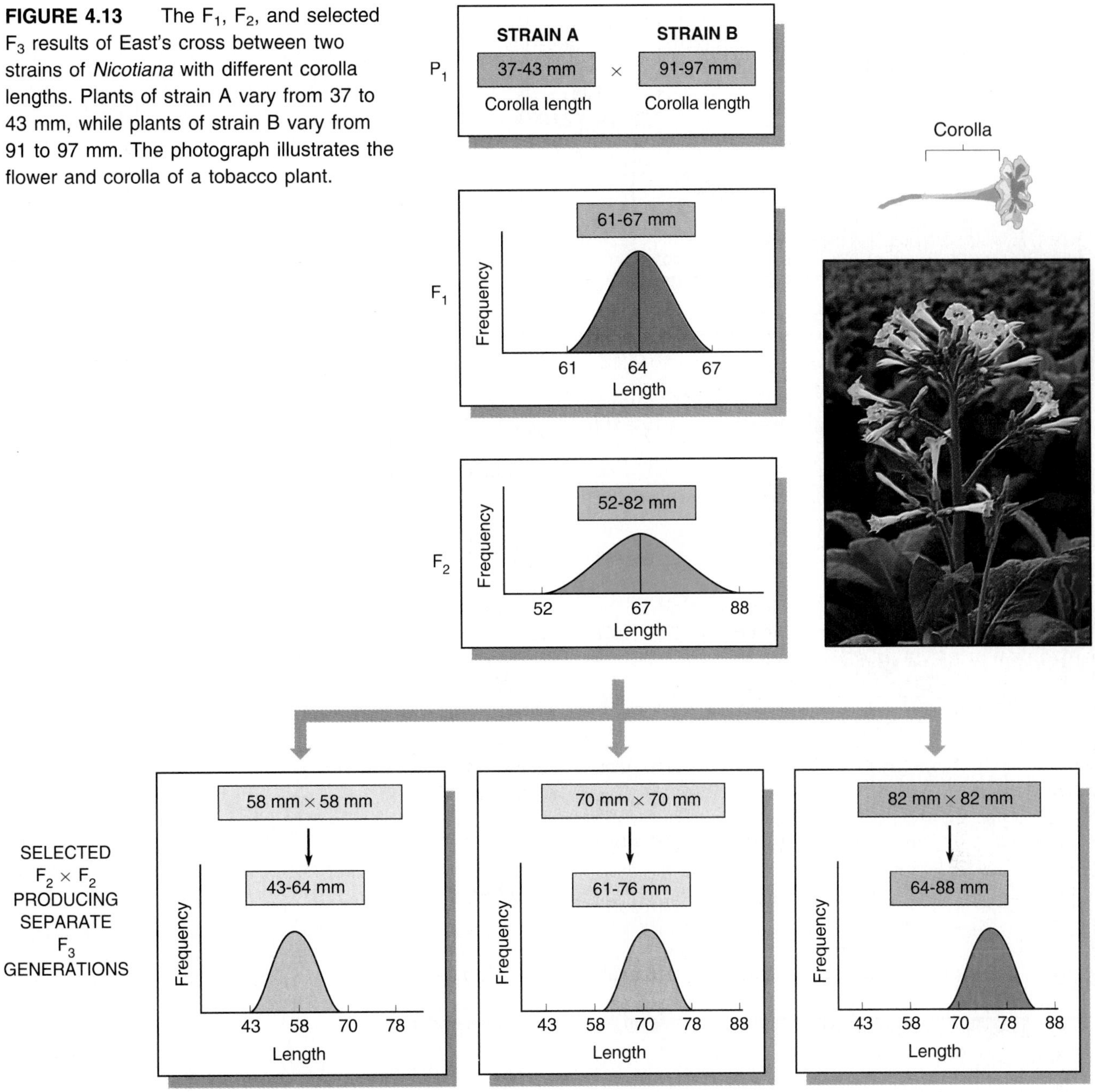

**7.** Analysis of polygenic traits requires the study of large numbers of progeny from a population of organisms.

These points, as well as an explanation of the multiple-factor hypothesis, can be illustrated by examining Herman Nilsson-Ehle's experiments involving grain color in wheat performed in the early twentieth century. In one set of experiments, wheat with red grain was crossed to wheat with white grain (Figure 4.14). The $F_1$ generation demonstrated an intermediate color. In the $F_2$, approximately 15/16 of the plants showed some de-

gree of red grain, while 1/16 of the plants showed white grain. Since the ratio occurred in sixteenths, we can hypothesize that two gene pairs control the phenotype and, if so, they segregate independently from one another in a Mendelian fashion.

Upon careful examination of the $F_2$, grain with color could be classified into four different shades of red. If two gene pairs were operating, each with one potential additive allele and one potential nonadditive allele, we can envision how the multiple-factor hypothesis could account for this variation. In the $P_1$, both parents were homozygous; the red parent contains only additive al-

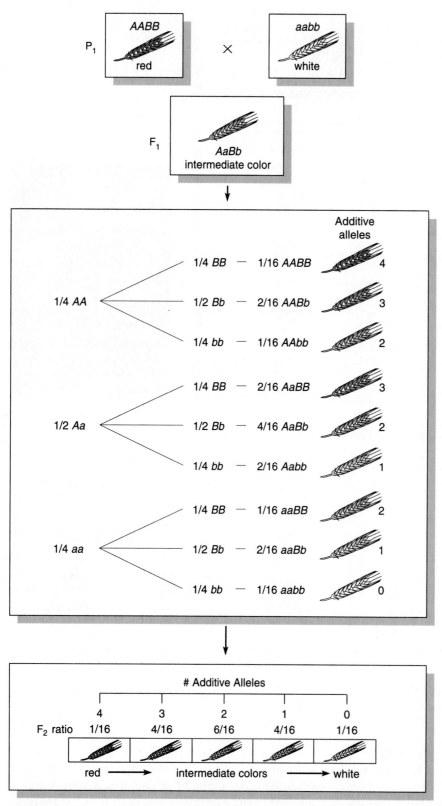

**FIGURE 4.14**    An illustration of how the multiple factor hypothesis can account for the 1:4:6:4:1 phenotypic ratio of grain color when all alleles designated by an upper-case letter are additive and contribute an equal amount of pigment to the phenotype.

leles (uppercase), while the white parent contains only nonadditive alleles (lowercase). The $F_1$, being heterozygous, contains only two additive alleles and expresses an intermediate phenotype. In the $F_2$, each offspring has either 4, 3, 2, 1, or 0 additive alleles (Figure 4.14). Wheat with no additive alleles (1/16) is white like one of the $P_1$ parents, while wheat with 4 additive alleles is red like the other $P_1$ parent. Plants with 3, 2, or 1 additive alleles constitute the other three categories of red color observed in the $F_2$, with most (6/16) having 2 additive alleles like the $F_1$ plants.

Thus, it appeared that, if examined carefully, even apparent continuous variation could be explained in a Mendelian fashion. This explanation fits nicely with the data. Multiple-factor inheritance as just described is now an accepted mechanism to account for phenotypes displaying continuous variation. This type of inheritance, in which alleles contribute additively to a phenotype, is different from any other mode of inheritance discussed thus far. Although the Nilsson-Ehle experiment involved two gene pairs, there is no reason why three, four, or more gene pairs cannot function in controlling various phenotypes. As we saw in Nilsson-Ehle's initial cross, if two gene pairs were involved, only five $F_2$ phenotypic categories, in a 1:4:6:4:1 ratio, would be expected. On the other hand, as three, four, five, or more gene pairs become involved, greater and greater numbers of classes would be expected to appear in more complex ratios. The number of phenotypes and the expected $F_2$ ratios of crosses involving up to five gene pairs are illustrated in Figure 4.15 on page 104.

### Calculating the Number of Polygenes

It is often of interest to determine the number of genes that are involved in the control of polygenic traits. If the ratio of $F_2$ individuals resembling *either* of the two most extreme phenotypes can be determined, then the number of gene pairs involved ($n$) may be calculated using the following simple formula:

$$\frac{1}{4^n} = \text{ratio of } F_2 \text{ individuals expressing either extreme phenotype}$$

In our past example, the $P_1$ phenotypes represent these two extremes. In Figure 4.14, 1/16 of the $F_2$ are either red *or* white like the $P_1$ crosses; this ratio can be substituted on the right side of the equation prior to solving for $n$:

$$\frac{1}{4^n} = \frac{1}{16}$$

$$\frac{1}{4^2} = \frac{1}{16}$$

$$n = 2$$

Table 4.4 lists the ratio and the number of $F_2$ phenotypic classes produced in crosses involving up to five gene pairs.

For low numbers of gene pairs, it is sometimes easier to use the ($2n + 1$) rule. If $n$ equals the number of gene pairs, $2n + 1$ will determine the total number of categories of possible phenotypic groups. Where $n = 2$, $2n + 1 = 5$, since each phenotypic category could have 4, 3, 2, 1, or 0 additive alleles. Where $n = 3$, $2n + 1 = 7$, since each phenotypic category could have 6, 5, 4, 3, 2, 1, or 0 additive alleles, and so on.

Polygenic control is a significant concept because it is believed to be the mode of inheritance for a vast number of traits. For example, height, weight, and stature in all plants and animals, grain yield in crops, beef and milk production in cattle, and egg production in chickens are thought to be under polygenic control. Thus, knowledge of this mode of inheritance is of prime importance in animal breeding and agriculture. In humans, the degree of skin pigmentation is thought to be under similar genetic control. In most cases of polygenic inheritance, the genotype establishes the potential range in which a particular phenotype may fall, while environmental factors determine how much of the potential will be realized. In the crosses described in this section we have assumed an optimal environment, which minimizes variation resulting from that source.

**Table 4.4** DETERMINATION OF NUMBER OF GENE PAIRS FROM POLYGENIC INHERITANCE PATTERNS

| $n$ | Ratio from Cross between Heterozygous Parents | Number of Distinct $F_2$ Phenotypic Classes |
|---|---|---|
| 1 | 1/4 | 3 |
| 2 | 1/16 | 5 |
| 3 | 1/64 | 7 |
| 4 | 1/256 | 9 |
| 5 | 1/1024 | 11 |

## GENES ON THE X CHROMOSOME: SEX LINKAGE

We near the conclusion of this chapter with a discussion of still one other mode of neo-Mendelian inheritance: **sex linkage**. This phenomenon results from the

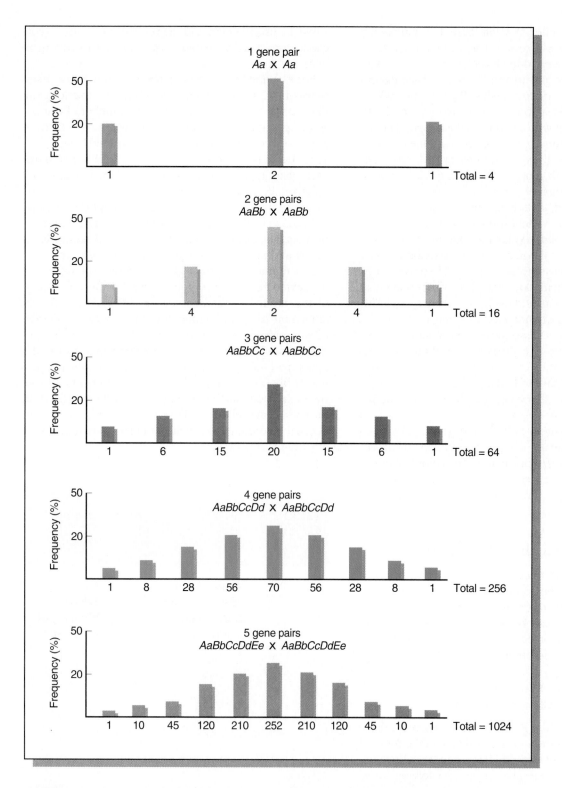

**FIGURE 4.15**    The results of crossing two heterozygotes where polygenic inheritance is in operation with one to five gene pairs. Each histogram bar indicates a distinct phenotypic class from one extreme (left end) to the other extreme (right end). Each phenotype results from a different number of additive alleles.

fact that one of the sexes in many organisms contains a pair of *unlike* chromosomes, the X and Y, which are involved in sex determination. For example, in both fruit flies (*Drosophila*) and humans, males contain an X and a Y chromosome, while females contain two X chromosomes. As we will see below, the unique pattern of inheritance stems from the fact that the Y, while behaving as a homologue to the X during meiosis, is for the most part genetically blank, or void of genes. Sex linkage, therefore, involves the transmission and expression of the normal complement of genes located on the X chromosome. To distinguish this pair of sex chromosomes, all other pairs are referred to as **autosomal chromosomes** or just **autosomes**.

## Sex Linkage in *Drosophila*

One of the first cases of sex linkage was documented in 1910 by Thomas H. Morgan during his studies of the *white* eye mutation in *Drosophila* (Figure 4.16). We will use this case to illustrate sex linkage. The normal wild-type eye color red is dominant to white.

Morgan's work established that the inheritance pattern of the white-eye trait was clearly related to the sex of the parent carrying the mutant allele. Unlike the outcome of the typical monohybrid cross, reciprocal crosses between white-eyed and red-eyed flies did not yield identical results. In contrast, in all of Mendel's monohybrid crosses, F$_1$ and F$_2$ data were very similar regardless of

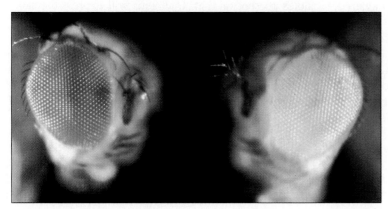

**FIGURE 4.16**    The F$_1$ and F$_2$ results of T. H. Morgan's reciprocal crosses involving the sex-linked *white* mutation in *Drosophila melanogaster*. The actual F$_2$ results are shown in parentheses. The photograph contrasts white eyes with the brick red wild-type eye color.

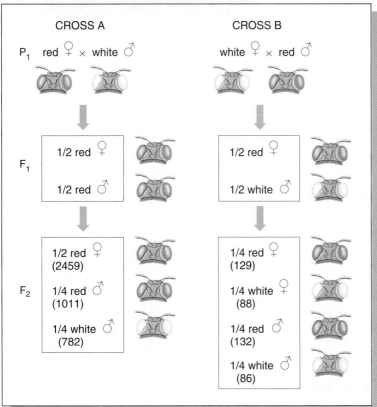

which $P_1$ parent exhibited the recessive mutant trait. Morgan's analysis led to the conclusion that the *white* locus is present on the X chromosome rather than one of the autosomes. As such, both the gene and the trait are said to be **sex-linked** or **X-linked**.

Results of reciprocal crosses between white-eyed and red-eyed flies are shown in Figure 4.16. The obvious differences in phenotypic ratios in both the $F_1$ and $F_2$ generations are dependent on whether or not the $P_1$ white-eyed parent was male or female.

Morgan was able to correlate these observations with the difference found in the sex chromosome composition between male and female *Drosophila*. He hypothesized that the recessive allele for white eye is found on the X chromosome, but its corresponding locus is absent from the Y chromosome. Females thus have two available gene sites, one on each X chromosome, while males have only one available gene site on their single X chromosome.

Morgan's interpretation of sex-linked inheritance, shown in Figure 4.17, provides a suitable theoretical explanation for his results. Since the Y chromosome lacks homology with most genes on the X chromosome, whatever alleles are present on the X chromosome of the males will be directly expressed in the phenotype. Since males cannot be either homozygous or heterozygous for sex-linked genes, this condition is referred to as being **hemizygous**. In such cases, no alternative alleles are present, and the concept of dominance and recessiveness is irrelevant.

One result of sex linkage is the **crisscross pattern of inheritance**, whereby phenotypic traits controlled by recessive sex-linked genes are passed from mothers to all sons. This pattern occurs because females exhibiting a recessive trait must contain the mutant allele on both X chromosomes. Since male offspring receive one of their mother's two X chromosomes and are hemizygous for all alleles present on that X, all sons will express the same recessive sex-linked traits as their mother.

In addition to documenting the phenomenon of sex-linkage, Morgan's work has taken on great historical significance. By 1910, the correlation between Mendel's work and the behavior of chromosomes during meiosis had provided the basis for the **chromosome theory of**

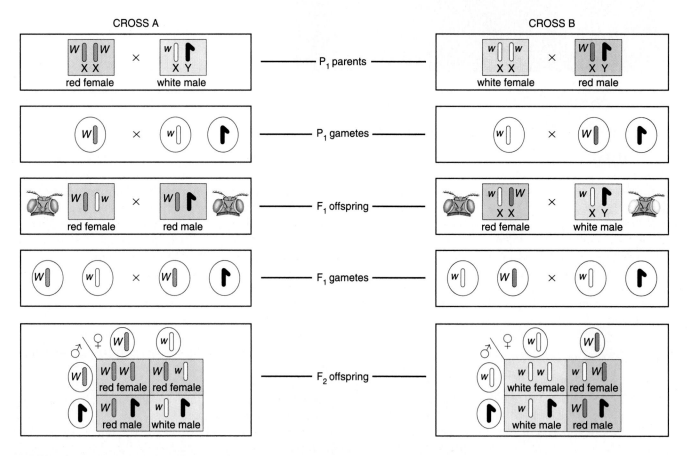

**FIGURE 4.17**    The chromosomal explanation of the results of the sex-linked crosses shown in Figure 4.16.

**(a)**

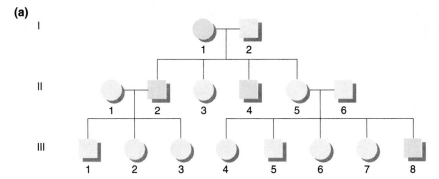

FIGURE 4.18    (a) A human pedigree of the sex-linked color blindness trait. (b) The most probable genotypes of each individual in the pedigree.

**(b)**

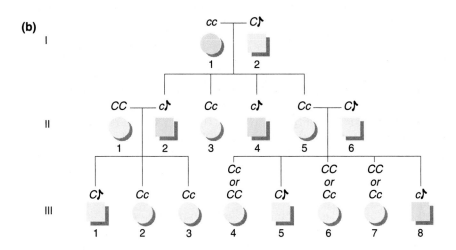

Symbols:
$c$ = color blindness
$C$ = normal vision
♪ = Y chromosome

**inheritance**, as postulated by Sutton and Boveri (see Chapter 2). The work involving sex chromosomes by Morgan, and subsequently that of his student Calvin Bridges, is considered to be the first solid experimental evidence in support of this theory. The ability of Morgan and Bridges to provide direct evidence that genes are transmitted on specific chromosomes provided the basis for support. In the ensuing two decades, these findings provided the impetus for further research, the findings of which were indisputable in support of this theory.

## Sex Linkage in Humans

In humans, many genes and the respective traits controlled by them are recognized as being linked to the X chromosome. These sex-linked traits may be easily identified in pedigrees because of the crisscross pattern of inheritance. A pedigree for one form of human color blindness is shown in Figure 4.18. The mother in generation I passes the trait to all her sons but to none of her daughters. If the offspring in generation II marry normal individuals, the color-blind sons will produce all normal male and female offspring (III-1, 2, and 3); the normal-visioned daughters will produce normal-visioned female offspring (III-4, 6, and 7), as well as color-blind (III-8) and normal-visioned (III-5) male offspring.

Many sex-linked human genes have now been identified, as shown in Table 4.5. For example, the genes controlling two forms of hemophilia and one form of muscular dystrophy are located on the X chromosome. Additionally, numerous genes whose expression yields enzymes are sex-linked. Glucose-6-phosphate dehydrogenase and hypoxanthine-guanine-phosphoribosyl transferase are two examples. In the latter case, the **Lesch–Nyhan syndrome** results from the mutant form of the X-linked gene product.

Because of the way in which sex-linked genes are transmitted, unusual circumstances may be associated with recessive sex-linked disorders in comparison to recessive autosomal disorders. For example, if a sex-linked

**Table 4.5**   HUMAN SEX-LINKED TRAITS

| Condition | Characteristics |
|---|---|
| Color blindness, deutan type | Insensitivity to green light |
| Color blindness, protan type | Insensitivity to red light |
| Fabry's disease | Deficiency of galactosidase A; heart and kidney defects, early death |
| G-6-PD deficiency | Deficiency of glucose-6-phosphate dehydrogenase; severe anemic reaction following intake of primaquines in drugs and certain foods, including fava beans |
| Hemophilia A | Classical form of clotting deficiency; lack of clotting factor VIII |
| Hemophilia B | Christmas disease; deficiency of clotting factor IX |
| Hunter syndrome | Mucopolysaccharide storage disease resulting from iduronate sulfatase enzyme deficiency; short stature, clawlike fingers, coarse facial features, slow mental deterioration, and deafness |
| Ichthyosis | Deficiency of steroid sulfatase enzyme; scaly dry skin, particularly on extremities |
| Lesch–Nyhan syndrome | Deficiency of hypoxanthine-guanine phosphoribosyl transferase enzyme (HGPRT), leading to motor and mental retardation, self-mutilation, and early death |
| Muscular dystrophy (Duchenne type) | Progressive, life-shortening disorder characterized by muscle degeneration and weakness; sometimes associated with mental retardation; deficiency of the protein dystrophin |

disorder debilitates or is lethal to the affected individual prior to reproductive maturation, the disorder occurs exclusively in males. This is the case because the only sources of the lethal allele in the population are females who carry it heterozygously but do not express the disorder. They pass the allele to one-half of their sons, who develop the disorder because they are hemizygous, but who rarely, if ever, reproduce. Heterozygous females also pass the allele to one-half of their daughters, who become carriers but do not develop the disorder. An example of such a sex-linked disorder is the Duchenne form of muscular dystrophy. The disease has an onset prior to age 6 and is often lethal prior to age 20. It normally occurs only in males.

## SEX-LIMITED AND SEX-INFLUENCED INHERITANCE

Our final topics involve inheritance affected by the sex of the organism, but not necessarily by genes on the X chromosome. There are numerous examples in different organisms where the sex of the individual plays a determining role in the expression of certain phenotypes. In some cases, the expression of a specific phenotype is absolutely limited to one sex; in others, the sex of an individual influences the expression of a phenotype that

is not limited to one sex or the other. This distinction differentiates **sex-limited inheritance** from **sex-influenced inheritance**.

In domestic fowl, tail and neck plumage is often distinctly different in males and females (Figure 4.19), demonstrating sex-limited inheritance. Cock-feathering is longer, more curved, and pointed, while hen-feathering is shorter and more rounded. The inheritance of feather type is due to a single pair of autosomal alleles whose expression is modified by the individual's sex hormones.

As shown in the following chart, hen-feathering is due to a dominant allele, $H$; but regardless of the homozygous presence of the recessive $h$ allele, all females remain hen-feathered. Only in males does the $hh$ genotype result in cock-feathering.

| Genotype | Phenotype | |
|---|---|---|
| | ♀ | ♂ |
| $HH$ | Hen-feathered | Hen-feathered |
| $Hh$ | Hen-feathered | Hen-feathered |
| $hh$ | Hen-feathered | Cock-feathered |

In the development of certain breeds of fowl, one allele or the other has become fixed in the population. In the Leghorn breed, all individuals are of the $hh$ genotype, and thus all males and females show distinctive plumage. Sebright bantams are all $HH$, showing no sexual distinction in feathering.

**FIGURE 4.19**    Hen-feathering (left) versus cock-feathering (right) in domestic fowl. The feathers in the hen are shorter and less curved.

**FIGURE 4.20**    Pattern baldness, a sex-influenced autosomal trait in humans.

Cases of sex-influenced inheritance include pattern baldness in humans, horn formation in sheep, and certain coat patterns in cattle. In such cases, autosomal genes are responsible for the contrasting phenotypes displayed by both males and females, but the expression of these genes is dependent on the hormone constitution of the individual. Thus, the heterozygous genotype may exhibit one phenotype in one sex and the contrasting one in the other. For example, **pattern baldness** in humans, where the hair is very thin on the top of the head (Figure 4.20) is inherited in the following way:

| Genotype | Phenotype | |
|----------|-----------|---|
| | ♀ | ♂ |
| *BB* | Bald | Bald |
| *Bb* | Not bald | Bald |
| *bb* | Not bald | Not bald |

Even though females may display pattern baldness, this phenotype is much more prevalent in males. When females do inherit the *BB* genotype, the phenotype is much less pronounced than in males and is expressed later in life.

## CHAPTER SUMMARY

1. Since Mendel's work was rediscovered, the study of transmission genetics has been expanded to include modes of inheritance controlling phenotypes that are often influenced by two or more genes in a variety of ways.

2. Incomplete, or partial, dominance is exhibited when intermediate phenotypic expression of a trait occurs in a heterozygote.

3. Codominance is exhibited when distinctive expression of two alleles occurs in a heterozygous organism.

4. The concept of multiple-allelism applies to populations, since a diploid organism may host only two alternative alleles at any given locus. Within a population, however, many alternative alleles of the same gene often occur.

5. Lethal mutations usually result in the inactivation or the lack of synthesis of gene products essential during development. Such mutations may be recessive or dominant. Some lethal genes, such as that causing Huntington disease, are not expressed until adulthood.

6. Mendel's classical $F_2$ ratio is often modified in instances where gene interaction results in either continuous or discontinuous variation.

7. Epistasis involves discontinuous variation, where two or more genes influence a single characteristic. Usually, the expression of one of the genes masks the expression of the other gene or genes.

8. Polygenic (or quantitative) inheritance results in continuous variation and is illustrated when products of several genes make additive contributions to phenotypes.

9. Genes located on the X chromosome display a unique mode of inheritance referred to as sex-linkage.

10. Unique sex-linked ratios result because members of the heterogametic sex (containing only one X) express all alleles present on their X chromosome, the basis of hemizygosity.

11. Sex-limited and sex-influenced inheritance occur when the sex of the organism affects the phenotype.

## KEY TERMS

ABO blood types
additive alleles
allele
antigen
autosomal chromosome
    (autosome)
Bombay phenotype
chromosome theory of
    inheritance
codominance
continuous variation
crisscross inheritance

discontinuous variation
drosopterin
epistasis
erythroblastosis fetalis
    (hemolytic disease of
    the newborn-HDN)
gene interaction
glycoprotein
H substance
hemizygous
hexosaminidase
Huntington disease

incomplete (partial)
    dominance
isoagglutinogen
Lesch–Nyhan syndrome
lethal allele
linkage
MN blood groups
multiple allelism
multiple-factor (multiple-
    gene) inheritance
mutation
neo-Mendelian genetics

ommatidium
pattern baldness
polygenic inheritance
quantitative variation
sex-influenced
    inheritance
sex-limited inheritance
sex-linked (X-linked)
    inheritance
Tay-Sachs disease
wild-type allele
xanthommatin

# INSIGHTS AND SOLUTIONS

**G**enetic problems take on an added complexity if they involve two independent characters and multiple alleles, incomplete dominance, or epistasis. The most difficult types of problems are those that were faced by pioneering geneticists in the laboratory. In these problems they had to determine the mode of inheritance by working backwards from the observations of offspring to parents of unknown genotype.

1. Consider the problem of comb shape inheritance in chickens, where walnut, rose, pea, and single are observed as distinct phenotypes. How is comb shape inherited, and what are the genotypes of the $P_1$ generation of each cross? Use the following data to answer these questions.

Cross 1:  single × single   ⟶  all single
Cross 2:  walnut × walnut  ⟶  all walnut
Cross 3:  rose × pea       ⟶  all walnut
Cross 4:  $F_1$ × $F_1$ of Cross 3
         walnut × walnut  ⟶  93 walnut
                                   28 rose
                                   32 pea
                                   10 single

Single

Walnut

Rose

Pea

**SOLUTION:**
At first glance, this problem may appear quite difficult. The approach used in solving it must involve two steps. First, carefully analyze the data for any useful information. Once you determine something concrete, follow an empirical approach; that is, formulate an hypothesis and, in a sense, test it against the given data. Look for a pattern of inheritance that is consistent with all cases.

(a) In this problem there are two immediately useful facts. First, in Cross 1, $P_1$ singles breed true. Second, while $P_1$ walnut breeds true (Cross 2), a walnut phenotype is also produced in crosses between rose and pea (Cross 3). When these $F_1$ walnuts are crossed together (Cross 4), all four comb shapes are produced in a ratio that approximates a $9:3:3:1$. This observation should immediately suggest a cross involving two gene pairs because the resulting data closely resemble the ratio of Mendel's dihybrid crosses. Since only one character is involved (comb shape), epistasis must be occurring. This may serve as a working hypothesis, and we must now propose how the two gene pairs interact to produce each phenotype.

(b) If we call the allele pairs, $A$, $a$ and $B$, $b$, we might predict that since walnut represents 9/16 in Cross 4, $A-B-$ will produce walnut. We might also hypothesize that in the case of Cross 2, the genotypes were $AABB \times AABB$ where walnut was seen to breed true. (Recall that $A-$ and $B-$ mean $AA$ or $Aa$ and $BB$ or $Bb$, respectively.)

(c) Since single is the phenotype representing 1/16 of the offspring of Cross 4, we could predict that this phenotype is the result of the $aabb$ genotype. This is consistent with Cross 1.

(d) Now we have only to determine the genotypes for rose and pea. The most logical prediction would be that at least one dominant $A$ or $B$ allele combined with the double recessive condition of the other allele pair can account for these phenotypes. For example,

$$A-bb \rightarrow \text{rose}$$
$$aa\,B- \rightarrow \text{pea}$$

If, in Cross 3, $AAbb$ (rose) were crossed with $aaBB$ (pea), all offspring would be $AaBb$ (walnut). This is consistent with the data, and we must now only look at Cross 4. We predict these walnut genotypes to be $AaBb$ (as above), and from the cross

$$\frac{AaBb}{\text{(walnut)}} \times \frac{AaBb}{\text{(walnut)}}$$

we expect

9/16 $A-B-$ (walnut)
3/16 $A-bb$ (rose)
3/16 $aa\,B-$ (pea)
1/16 $aa\,bb$ (single)

Our prediction is consistent with the information we were given. The initial hypothesis of the epistatic interaction of two gene pairs proves consistent throughout, and the problem has been solved.

This example illustrates the need to have a basic theoretical knowledge of transmission genetics. Then, you must search for the appropriate clues so that you can proceed in a stepwise fashion toward a solution. Mastering problem solving requires practice, but provides a great deal of satisfaction.

2. In radishes, flower color may be red, purple, or white. The edible portion of the radish may be long or oval. When only flower color is studied, red × white yields all purple. If these $F_1$ purples are interbred, no dominance is evident, and the $F_2$ generation consists of 1/4 red : 1/2 purple : 1/4 white. Regarding radish shape, long is dominant to oval in a normal Mendelian fashion.

(a) Determine the $F_1$ and $F_2$ phenotypes from a cross between a true-breeding red long radish and one that is white oval. Be sure to define all gene symbols initially.

**SOLUTION:** This is a modified dihybrid cross where the gene pair controlling color exhibits incomplete dominance. Shape is controlled conventionally. First, establish gene symbols:

$$RR = \text{red} \qquad Rr = \text{purple} \qquad rr = \text{white}$$

$$\text{P}_1: \quad RROO \quad \times \quad rroo$$
$$\text{(red long)} \quad \text{(white oval)}$$

$F_1$: all $RrOo$ (purple long)

$F_1 \times F_1$: $RrOo \times RrOo$

$F_2$:

| | | | |
|---|---|---|---|
| 1/4 $RR$ | 3/4 $O-$ | 3/16 $RR\ O-$ | red long |
| | 1/4 $oo$ | 1/16 $RR\ oo$ | red oval |
| 2/4 $Rr$ | 3/4 $O-$ | 6/16 $Rr\ O-$ | purple long |
| | 1/4 $oo$ | 2/16 $Rr\ oo$ | purple oval |
| 1/4 $rr$ | 3/4 $O-$ | 3/16 $rr\ O-$ | white long |
| | 1/4 $oo$ | 1/16 $rr\ oo$ | white oval |

Note above that to generate the $F_2$ results, we have used the forked-line method. First, the outcome of crossing $F_1$ parents for the color genes is considered ($Rr \times Rr$). Then the outcome of shape is considered ($Oo \times Oo$).

(b) A red oval plant was crossed with a plant of unknown genotype and phenotype, yielding the data shown below. Determine the genotype and phenotype of the unknown plant.

Offspring: 103 red long:101 red oval:98 purple long:100 purple oval

**SOLUTION:** Since the two characters are inherited independently, consider them separately. The data indicate a $1/4:1/4:1/4:1/4$ proportion. First, consider color:

$P_1$: red × ??? (unknown)

$F_1$: 204 red (1/2)

198 purple (1/2)

Since the red parent *must* be $RR$, the unknown must have a genotype of $Rr$ to produce these results. It is thus purple. Now, consider shape:

$P_1$: oval × ??? (unknown)

$F_1$: 201 long (1/2)

201 oval (1/2)

Since the oval plant *must* be $oo$, the unknown plant must have a genotype of $Oo$ to produce these results. It is thus long. The unknown plant is thus

$RrOo$ purple long

3. In a plant, height varies from 6 to 36 cm. When 6-cm and 36-cm plants are cross-fertilized, all $F_1$ plants are 21 cm. In the $F_2$ generation, about 3 of every 200 plants were as short as the 6-cm $P_1$ parent.

(a) What mode of inheritance is illustrated and how many gene pairs are involved?

**SOLUTION:** This problem illustrates polygenic inheritance, where a continuous trait (height) is studied and where alleles contribute *additively* to the phenotype. When 3/200 $F_2$ plants are as extreme as either $P_1$ parent, we first must reduce the ratio:

$$3/200 = 1/66.67$$

If two gene pairs were involved, the ratio would be about $1/16$ ($1/4^2$). If three gene pairs were involved, the ratio would be about $1/64$ ($1/4^3$). If four gene pairs were involved, then the ratio would be about $1/256$ ($1/4^4$), etc. On this basis, we conclude that three gene pairs are involved since $1/66.6$ is very close to $1/64$.

(b) How much does each additive allele contribute to height?

**SOLUTION:** The variance between the two extremes equals

$$36 - 6 = 30 \text{ cm}$$

114    **Ch. 4** / Modification of Mendelian Ratios

Since there are six potential additive alleles (*AABBCC*), then each contributes

$$30/6 = 5 \text{ cm}$$

to the base height of 6 cm, which occurs when no additive alleles are present (*aabbcc*).

(c) List all genotypes that give rise to plants that are 31 cm.

**SOLUTION:**  All combinations with five additive alleles result in a 31-cm phenotype:

> *AABBCc*
> *AABbCC*
> *AaBBCC*

(d) In a cross separate from the F$_1$ above, a plant of unknown phenotype and genotype was test-crossed with the following results:

> 1/4    11 cm
> 2/4    16 cm
> 1/4    18 cm

An astute genetics student realized that the unknown plant could be only one phenotype, but could be any of three genotypes. What were they?

**SOLUTION:**  When test-crossed (with *aabbcc*), the unknown plant must be able to contribute either one, two, or three additive alleles in its gametes in order to yield the three phenotypes in the offspring. Since no 6-cm offspring are observed, the unknown plant *never* contributes all nonadditive alleles (*abc*). Only plants that are homozygous at one locus and heterozygous at the other two loci will meet these criteria. Therefore, the unknown parent can be any of three genotypes, all of which have a phenotype of 21 cm:

> *AABbCc*
> *AaBbCC*
> *AaBBCc*

For example, in the first genotype (*AABbCc*):

$$AABbCc \times aabbcc$$

> 1/4 *AaBbCc* $\longrightarrow$ 21 cm
> 1/4 *AaBbcc* $\longrightarrow$ 16 cm
> 1/4 *AabbCc* $\longrightarrow$ 16 cm
> 1/4 *Aabbcc* $\longrightarrow$ 11 cm

which is the ratio of phenotypes observed.

4.  In humans, color blindness is inherited as a sex-linked recessive trait. A woman with normal vision, but whose father is color-blind, marries a male who has normal vision. Predict the color vision of their male and female offspring.

**SOLUTION:**  The female is heterozygous since she inherited an X chromosome with the mutant allele from her father. Her husband is normal. Therefore, the parental genotypes are

$$Cc \times C\!\!\uparrow$$

All female offspring are normal (*CC* or *Cc*). One-half of the male children will be color-blind (*c*↑), and the other half will have normal vision (*C*↑).

**PROBLEMS AND
DISCUSSION
QUESTIONS**

1. In shorthorn cattle, coat color may be red, white, or roan. Roan is an intermediate phenotype expressed as a mixture of red and white hairs. The following data were obtained from various crosses:

red × red $\longrightarrow$ all red *RR × RR*

white × white $\longrightarrow$ all white *rr × rr*

red × white $\longrightarrow$ all roan *RR × rr*

roan × roan $\longrightarrow$ 1/4 red: 1/2 roan: 1/4 white

*Rr × Rr*

How is coat color inherited? What are the genotypes of parents and offspring for each cross? *incomplete dominance*

2. Contrast incomplete dominance and codominance.

3. In foxes, two alleles of a single gene, $P$ and $p$, may result in lethality ($PP$), platinum coat ($Pp$), or silver coat ($pp$). What ratio is obtained when platinum foxes are interbred? Is the $P$ allele behaving dominantly or recessively in causing lethality? In causing platinum coat color?

4. In mice, a short-tailed mutant was discovered. When it was crossed to a normal long-tailed mouse, 4 offspring were short-tailed and 3 were long-tailed. Two short-tailed mice from the $F_1$ generation were selected and crossed. They produced 6 short-tailed and 3 long-tailed mice. These genetic experiments were repeated three times with approximately the same results. What genetic ratios are illustrated? Hypothesize the mode of inheritance and diagram the crosses.

5. List all possible genotypes for the A, B, AB, and O phenotypes. Is the mode of inheritance of the ABO blood types representative of dominance? Of recessiveness? Of codominance?

6. With regard to the ABO blood types in humans, determine the genotypes of the male parent and female parent below:

Male Parent: Blood type B; mother type O    *BO*    *OO*

Female Parent: Blood type A; father type B    *AO*    *BO*

Predict the blood types of the offspring that this couple may have and the expected proportion of each.

7. In a disputed parentage case, the child is blood type O while the mother is blood type A. What blood type would exclude a male from being the father? Would the other blood types prove that a particular male was the father?

8. The A and B antigens in humans may be found in water-soluble form in secretions, including saliva, of some individuals but not in others. The population thus contains "secretors" and "nonsecretors." The following inheritance patterns have established that this trait is inherited:

secretor × secretor $\longrightarrow$ all secretors

nonsecretor × nonsecretor $\longrightarrow$ all nonsecretors

secretor × nonsecretor $\longrightarrow$ all secretors

$F_1$ secretors $\longrightarrow$ $F_2$ = 3/4 secretors: 1/4 nonsecretors

How is the trait inherited? Determine the genotypes of all parents and offspring.

9. Distinguish between epistasis and polygenic inheritance, and between discontinuous and continuous variation.

10. In rabbits, a series of multiple alleles controls coat color in the following way: $C$ is dominant to all other alleles and causes full color. The chinchilla phenotype is due

to the $c^{ch}$ allele, which is dominant to all alleles other than $C$. The $c^b$ allele, dominant only to $c^a$ (albino), results in the Himalayan coat color. Thus, the order of dominance is $C > c^{ch} > c^b > c^a$. For each of the three cases below, the phenotypes of the $P_1$ generations of two crosses are shown, as well as the phenotype of one member of the $F_1$ generation. For each case, determine the genotypes of the $P_1$ generation and the $F_1$ offspring and predict the results of making each cross between $F_1$ individuals as shown.

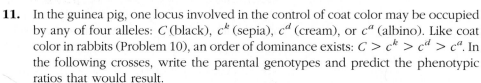

|  | $P_1$ Phenotypes |  | $F_1$ Phenotypes |
|---|---|---|---|
| (a) | Himalayan × Himalayan | ⟶ | albino |
|  |  | × ⟶ | ?? |
|  | full color × albino | ⟶ | chinchilla |
| (b) | albino × chinchilla | ⟶ | albino |
|  |  | × ⟶ | ?? |
|  | full color × albino | ⟶ | full color |
| (c) | chinchilla × albino | ⟶ | Himalayan |
|  |  | × ⟶ | ?? |
|  | full color × albino | ⟶ | Himalayan |

**11.** In the guinea pig, one locus involved in the control of coat color may be occupied by any of four alleles: $C$ (black), $c^k$ (sepia), $c^d$ (cream), or $c^a$ (albino). Like coat color in rabbits (Problem 10), an order of dominance exists: $C > c^k > c^d > c^a$. In the following crosses, write the parental genotypes and predict the phenotypic ratios that would result.

(a) sepia × cream, where both guinea pigs had an albino parent

(b) sepia × cream, where the sepia guinea pig had an albino parent and the cream guinea pig had two sepia parents

(c) sepia × cream, where the sepia guinea pig had two full color parents and the cream guinea pig had two sepia parents

(d) sepia × cream, where the sepia guinea pig had a full color parent and an albino parent and the cream guinea pig had two full color parents.

**12.** Three gene pairs located on separate autosomes determine flower color and shape as well as plant height. The first pair exhibits incomplete dominance where color can be red, pink (the heterozygote) or white. The second pair leads to personate (dominant) or peloric (recessive) flower shape, while the third gene pair produces either the dominant tall trait or the recessive dwarf trait. Homozygous plants that are red, personate, and tall are crossed to those that are white, peloric, and dwarf. Determine the $F_1$ genotype(s) and phenotype(s). If the $F_1$ plants are interbred, what proportion of the offspring will exhibit the same phenotype as the $F_1$ plants?

**13.** As in Problem 12, color may be *red, white,* or *pink,* and flower shape may be *personate* or *peloric.* For the following crosses, determine the $P_1$ and $F_1$ genotypes. What phenotypic ratios would result from crossing the $F_1$ of (a) to the $F_1$ of (b)?

(a) red peloric × white personate ⟶ $F_1$ = all pink personate

(b) red personate × white peloric ⟶ $F_1$ = all pink personate

(c) pink personate × red peloric ⟶ $F_1$ = 1/4 red personate
                                                   1/4 red peloric
                                                   1/4 pink personate
                                                   1/4 pink peloric

(d) pink personate × white peloric ⟶ $F_1$ = 1/4 white personate
                                                     1/4 white peloric
                                                     1/4 pink personate
                                                     1/4 pink peloric

**14.** Horses can be *cremello* (a light cream color), *chestnut* (a brownish color), or *palomino* (a golden color with white in the horse's tail and mane). Of these phenotypes, only palominos never breed true.

(a) From the results shown below, determine the *mode* of inheritance by assigning gene symbols and indicating which genotypes yield which phenotypes.

cremello × palomino ⟶ 1/2 cremello
1/2 palomino

chestnut × palomino ⟶ 1/2 chestnut
1/2 palomino

palomino × palomino ⟶ 1/4 chestnut
1/2 palomino
1/4 cremello

(b) Predict the $F_1$ and $F_2$ results of many initial matings between cremello and chestnut horses.

**15.** With reference to the eye color phenotypes produced by the recessive, autosomal, unlinked *brown* and *scarlet* loci in *Drosophila* (see Figure 4.11), predict the $F_1$ and $F_2$ results of the following $P_1$ crosses. Recall that when both the *brown* and *scarlet* alleles are homozygous, no pigment is produced, and the eyes are white.
(a) wild type × white
(b) wild type × scarlet
(c) brown × white

**16.** Pigment in the mouse is only produced when the *C* allele is present. Individuals of the *cc* genotype have no color. If color is present, it may be determined by the *A*, *a* alleles. *AA* or *Aa* results in agouti color, while *aa* results in black coats.
(a) What $F_1$ and $F_2$ genotypic and phenotypic ratios are obtained from a cross between *AACC* and *aacc* mice?
(b) In three crosses between gray females whose genotypes were unknown and males of the *aacc* genotype, the following phenotypic ratios were obtained:

(1) 8 agouti      (2)  9 agouti      (3)  4 agouti
    8 colorless        10 black          5 black
                                        10 colorless

What are the genotypes of these female parents?

**17.** In some plants a red pigment, cyanidin, is synthesized from a colorless precursor. The addition of a hydroxyl group ($OH^-$) to the cyanidin molecule causes it to become purple. In a cross between two randomly selected purple plants, the following results were obtained:

94 purple
31 red
43 colorless

How many genes are involved in the determination of these flower colors? Which genotypic combinations produce which phenotypes? Diagram the purple × purple cross.

**18.** In rats, the following genotypes of two independently assorting autosomal genes determine coat color:

$A-B-$  (gray)
$A-bb$  (yellow)
$aa B-$  (black)
$aa bb$  (cream)

A third gene pair on a separate autosome determines whether or not any color will be produced. The $CC$ and $Cc$ genotypes allow color according to the expression of the $A$ and $B$ alleles. However, the $cc$ genotype results in albino rats regardless of the $A$ and $B$ alleles present. Determine the $F_1$ phenotypic ratio of the following crosses:

(a) $AAbbCC \times aaBBcc$
(b) $AaBBCC \times AABbcc$
(c) $AaBbCc \times AaBbcc$
(d) $AaBBCc \times AaBBCc$
(e) $AABbCc \times AABbcc$

19. Given the inheritance pattern of coat color in rats, as described in Problem 18, predict the genotype and phenotype of the parents who produced the following $F_1$ offspring:

(a) 9/16 gray : 3/16 yellow : 3/16 black : 1/16 cream
(b) 9/16 gray : 3/16 yellow : 4/16 albino
(c) 27/64 gray : 16/64 albino : 9/64 yellow : 9/64 black : 3/64 cream
(d) 3/8 black : 3/8 cream : 2/8 albino
(e) 3/8 black : 4/8 albino : 1/8 cream

20. Consider the genes and alleles involved in the Bombay phenotype and the ABO blood groups. How would you describe the genetic interaction between these alleles? Predict the proportions of the various blood types of the offspring from parents who are both type AB and who are heterozygous ($Hh$) for the Bombay alleles.

21. In a species of the cat family, eye color can be gray, blue, green, or brown, and each trait is true-breeding. In separate crosses involving homozygous parents, the following data were obtained:

| Cross | P₁ | F₁ | F₂ |
|---|---|---|---|
| A | green × gray | all green | 3/4 green : 1/4 gray |
| B | green × brown | all green | 3/4 green : 1/4 brown |
| C | gray × brown | all green | 9/16 green : 3/16 brown : 3/16 gray : 1/16 blue |

(a) Analyze the data: How many genes are involved? Define gene symbols and indicate which genotypes yield each phenotype.
(b) In a cross between a gray-eyed cat and one of unknown genotype and phenotype, the $F_1$ generation was not observed. However, the $F_2$ resulted in the same $F_2$ ratio as in cross $C$. Determine the genotypes and phenotypes of the unknown $P_1$ and $F_1$ cats.

22. In a plant, a tall variety was crossed with a dwarf variety. All $F_1$ plants were tall. When $F_1 \times F_1$ plants were interbred, 9/16 of the $F_2$ were tall and 7/16 were dwarf.
(a) Explain the inheritance of height by indicating the number of gene pairs involved and by designating which genotypes yield tall and which yield dwarf (use dashes where appropriate).
(b) Of the $F_2$ plants, what proportion of them will be true-breeding if self-fertilized? List these genotypes.

23. In a unique species of plants, flowers may be yellow, blue, red, or mauve. All colors may be true-breeding. If plants with blue flowers are crossed to red-flowered plants, all $F_1$ plants have yellow flowers. When carried to an $F_2$ generation, the following ratio was observed:

9/16 yellow : 3/16 blue : 3/16 red : 1/16 mauve

In still another cross using true-breeding parents, yellow-flowered plants are crossed with mauve-flowered plants. Again, all $F_1$ plants had yellow flowers and the $F_2$ showed a 9:3:3:1 ratio, as shown above.

(a) Describe the inheritance of flower color by defining gene symbols and designating which genotypes give rise to each of the four phenotypes.

(b) Determine the $F_1$ and $F_2$ results of a cross between true-breeding red and true-breeding mauve-flowered plants.

**24.** Shown below are five maternal and paternal phenotypes (1–5), each designating the ABO, MN, and Rh blood group antigens. Each combination resulted in one of the five offspring shown to the right (a–e). Arrange the offspring with the correct parents such that all five cases are consistent. Is there more than one set of correct answers?

| Parental Phenotypes | Offspring |
|---|---|
| (1) A, M, $Rh^-$ × A, N, $Rh^-$ | (a) A, N, $Rh^-$ |
| (2) B, M, $Rh^-$ × B, M, $Rh^+$ | (b) O, N, $Rh^+$ |
| (3) O, N, $Rh^+$ × B, N, $Rh^+$ | (c) O, MN, $Rh^-$ |
| (4) AB, M, $Rh^-$ × O, N, $Rh^+$ | (d) B, M, $Rh^+$ |
| (5) AB, MN, $Rh^-$ × AB, MN, $Rh^-$ | (e) B, MN, $Rh^+$ |

**25.** An invading alien geneticist from a planet where genetic research was prohibited brought with him to earth two pure-breeding lines of pet frogs. One line croaked by uttering rib-it rib-it, and had purple eyes. The other line croaked more softly by muttering knee-deep knee-deep, and had green eyes. With a new-found sense of inquiry, he mated the two types of frogs. In the $F_1$, all frogs had blue eyes and uttered rib-it rib-it. He proceeded to make many $F_1 \times F_1$ crosses, and when he fully analyzed the $F_2$ data, he realized they could be reduced to the following ratio:

27/64 blue-eyed, rib-it utterer

12/64 green-eyed, rib-it utterer

9/64 blue-eyed, knee-deep mutterer

9/64 purple-eyed, rib-it utterer

4/64 green-eyed, knee-deep mutterer

3/64 purple-eyed, knee-deep mutterer

(a) How many total gene pairs are involved in the inheritance of both traits? Support your answer.

(b) Of these, how many are controlling eye color? How can you tell? How many are controlling croaking?

(c) Assign gene symbols for all phenotypes and indicate the genotypes of the $P_1$ and $F_1$ frogs.

(d) Indicate the genotypes of the six $F_2$ phenotypes.

(e) After years of experiments, the geneticist isolated pure-breeding strains of all six $F_2$ phenotypes. Indicate the $F_1$ and $F_2$ phenotypic ratios of the following cross using these pure-breeding strains:

blue-eyed, knee-deep mutterer × purple-eyed, rib-it utterer

(f) One set of crosses with his true-breeding lines initially caused the geneticist some confusion. When he crossed true-breeding purple-eyed, knee-deep mutterers with true-breeding green-eyed, knee-deep mutterers, he often got different results. In some matings, all offspring were blue-eyed, knee-deep mutterers but in other matings, all offspring were purple-eyed, knee-deep mutterers. In still a third mating, 1/2 blue-eyed, knee-deep mutterers and 1/2 purple-eyed, knee-deep mutterers were observed. Explain why the results differed.

(g) In another experiment, the geneticist crossed two purple-eyed, rib-it utterers together with the results shown below. What were the genotypes of the two parents?

> 9/16 purple-eyed, rib-it utterer
> 3/16 purple-eyed, knee-deep mutterer
> 3/16 green-eyed, rib-it utterer
> 1/16 green-eyed, knee-deep mutterer

26. Labrador retrievers may be black, brown, or golden in color. While each color may breed true, many different outcomes occur if many litters are examined from a variety of matings, where the parents are not necessarily true-breeding. Shown below are just some of the many possibilities. Propose a mode of inheritance that is consistent with these data, and indicate the corresponding genotypes of the parents in each mating. Indicate as well the genotypes of dogs that breed true for each color.

> (a) black × brown  ⟶  all black
> (b) black × brown  ⟶  1/2 black
>        1/2 brown
> (c) black × brown  ⟶  3/4 black
>        1/4 golden
> (d) black × golden  ⟶  all black
> (e) black × golden  ⟶  4/8 golden
>        3/8 black
>        1/8 brown
> (f) black × golden  ⟶  2/4 golden
>        1/4 black
>        1/4 brown
> (g) brown × brown  ⟶  3/4 brown
>        1/4 golden
> (h) black × black  ⟶  9/16 black
>        4/16 golden
>        3/16 brown

27. Assume that height in a plant is controlled by two gene pairs and that each additive allele contributes 5 cm to a base height of 20 cm (i.e., *aabb* is 20 cm).
(a) What is the height of an *AABB* plant?
(b) Predict the phenotypic ratios of $F_1$ and $F_2$ plants in a cross between *aabb* and *AABB*.
(c) List all genotypes that give rise to plants that are 25 and 35 cm in height.

28. In a cross where three gene pairs determine weight in squash, what proportion of individuals from the cross *AaBbCC* × *AABbcc* will contain only two additive alleles? Which genotype or genotypes fall into this category?

29. An inbred strain of plants has a mean height of 24 cm. A second strain of the same species from a different geographical region also has a mean height of 24 cm. When plants from the two strains are crossed together, the $F_1$ plants are the same height as the parent plants. However, the $F_2$ generation shows a wide range of heights; the majority are like the $P_1$ and $F_1$ plants, but approximately 4 of 1000 are only 12 cm high, and about 4 of 1000 are 36 cm high.
(a) What mode of inheritance is occurring here?
(b) How many gene pairs are involved?
(c) How much does each gene contribute to plant height?
(d) Indicate one possible set of genotypes for the original $P_1$ parents and the $F_1$ plants that could account for these results.

(e) Indicate three possible genotypes that could account for $F_2$ plants that are 18 cm high and $F_2$ plants that are 33 cm high.

30. In a series of crosses between plants of various heights to a 20-inch plant, the following results were obtained in the $F_1$ generations:

(a)   $4'' \times 20'' \rightarrow$ All $12''$
(b)   $8'' \times 20'' \rightarrow$ All $14''$
(c)   $12'' \times 20'' \rightarrow$ All $16''$
(d)   $16'' \times 20'' \rightarrow$ All $18''$

Propose an explanation for the inheritance of height in the above plant. Under the constraints of your explanation, predict the genotypes of plants of each height.

31. Members of a strain of inbred skunks whose stripes are uniformly 20 cm long are crossed to those of another strain whose stripes are uniformly 24 cm. The $F_1$ hybrids all have stripes that are 22 cm. When the $F_1$ skunks were crossed, the $F_2$ generation produced animals with stripes of 16, 18, 20, 22, 24, 26, and 28 cm. The 22-cm variety was most frequent and the 16- and 28-cm variety least frequent. Of 1000 $F_2$ skunks, 15 were 16 cm and 16 were 28 cm.
(a) What is the mode of inheritance illustrated above?
(b) How many gene pairs are involved?
(c) What is the contribution of each additive allele to stripe length?
(d) What are the genotypes of the $P_1$ parents and $F_1$ offspring?
(e) List four genotypes which give rise to skunks whose stripes are 20 cm long.

32. Erma and Harvey were a compatible barnyard pair, but a curious sight. Harvey's tail was only 6 cm while Erma's was 30 cm. Their $F_1$ piglet offspring all grew tails that were 18 cm. When inbred, an $F_2$ generation resulted in many piglets (Erma and Harvey's grandpigs) whose tails ranged in 4-cm intervals from 6 to 30 cm (6, 10, 14, 18, 22, 26, 30). Most had 18-cm tails, while 1/64 had 6-cm and 1/64 had 30-cm tails.
(a) Explain how tail length is inherited by describing the mode of inheritance, indicating how many gene pairs are at work, and designating the genotypes of Harvey, Erma, and their 18-cm offspring.
(b) If one of the 18-cm $F_1$ pigs were mated with the 6-cm $F_2$ pigs, what phenotypic ratio would be predicted if many offspring resulted? Diagram the cross.

33. Plants may be 10, 20, 30, 40, 50, 60, or 70 cm high where plant height is under polygenic control. A true-breeding plant that is 10 cm is crossed to another true-breeding plant that is 50 cm high. How many gene pairs are involved? What $F_1$ and $F_2$ results can be predicted?

34. A husband and wife have normal vision, although both of their fathers are red-green color-blind, which is inherited as a sex-linked recessive condition. What is the probability that their first child will be (a) A normal son? (b) A normal daughter? (c) A color-blind son? (d) A color-blind daughter?

35. In humans, the ABO blood type is under the control of autosomal multiple alleles. Color blindness is a recessive sex-linked trait. If two parents who are both type A and have normal vision produce a son who is color blind and is type O, what is the probability that their next child will be a female who is normal visioned and type O?

36. In *Drosophila,* a sex-linked recessive mutation, *scalloped (sd)*, causes irregular wing margins. Diagram the $F_1$ and $F_2$ results if: (a) A *scalloped* female is crossed with a normal male; (b) A *scalloped* male is crossed with a normal female. Compare these results to those that would be obtained if *scalloped* were not sex-linked.

37. Another recessive mutation in *Drosophila*, *ebony* (*e*), is on an autosome (chromosome 3) and causes darkening of the body compared with wild-type flies. What phenotypic $F_1$ and $F_2$ male and female ratios will result if a *scalloped*-winged female with normal body color is crossed with a normal-winged *ebony* male? Work this problem by both the Punnett square method and the forked-line method.

38. In *Drosophila*, the sex-linked recessive mutation *vermilion* (*v*) causes bright red eyes, which is in contrast to brick red eyes of wild type. A separate autosomal recessive mutation, *suppressor of vermilion* (*su-v*), causes flies homozygous or hemizygous for *v* to have wild-type eyes. In the absence of vermilion alleles, *su-v* has no effect on eye color. Determine the $F_1$ and $F_2$ phenotypic ratios from a cross between a female with wild-type alleles at the *vermilion* locus, but who is homozygous for *su-v*, with a *vermilion* male who has wild-type alleles at the *su-v* locus.

39. While *vermilion* is sex-linked and brightens the eye color, *brown* is an autosomal recessive mutation that darkens the eye. Flies carrying both mutations lose all pigmentation and are white eyed. Predict the $F_1$ and $F_2$ results of the following crosses:

   (a) vermilion females × brown males
   (b) brown females × vermilion males
   (c) white females × wild males

40. In a cross in *Drosophila* involving the X-linked recessive eye mutation *white* and the autosomally linked recessive eye mutation *sepia* (resulting in a dark eye), predict the $F_1$ and $F_2$ results of crossing true-breeding parents of the following phenotypes:

   (a) white females × sepia males
   (b) sepia females × white males

   Note that *white* is epistatic to the expression of *sepia*.

41. Consider the three pedigrees below, all involving a single human trait.
   (a) Which conditions, if any, can be *excluded*?

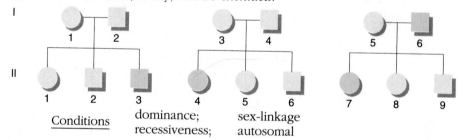

   Conditions   dominance;   sex-linkage
                recessiveness;   autosomal

   (b) For any condition that you excluded, indicate the *single individual* in generation II (e.g., II-1, II-2, etc.) that was most instrumental in your decision to exclude that condition. If none were excluded, answer "none apply."
   (c) Given your conclusions above, indicate the *genotype* of the individuals listed below. If more than one possibility applies, list all possibilities. Use the symbols *A* and *a* for the genotypes.

   II-1;   II-6;   II-9

42. In spotted cattle, the colored regions may be mahogany or red. If a red female and a mahogany male, both derived from separate true-breeding lines, are mated and the cross carried to an $F_2$ generation, the following results are obtained:

   $F_1$: 1/2 mahogany males
   1/2 red females

$F_2$: 3/8 mahogany males
1/8 red males
1/8 mahogany females
3/8 red females

When the reciprocal of the initial cross is performed (mahogany female and red male), identical results are obtained. Explain these results by postulating how the color is genetically determined. Diagram the crosses.

**43.** Predict the $F_1$ and $F_2$ results of crossing a male fowl that is cock-feathered with a true-breeding hen-feathered female fowl. Recall that these traits are sex-limited, as discussed in the text.

**SELECTED READINGS**

BRINK, R. A., ed. 1967. *Heritage from Mendel*. Madison: University of Wisconsin Press.

BULTMAN, S. J., MICHAUD, E. J., and WOYCHIK, R. P. 1992. Molecular characterization of the mouse *agouti* locus. *Cell* 71:1195–1204.

CHAPMAN, A. B. 1985. *General and quantitative genetics*. Amsterdam: Elsevier.

CLARKE, C. A. 1968. The prevention of "Rhesus" babies. *Scient. Amer.* (Nov.) 219:46–52.

CORWIN, H. O., and JENKINS, J. B. 1976. *Conceptual foundations of genetics: Selected readings*. Boston: Houghton-Mifflin.

CROW, J. F. 1983. *Genetics notes*. 8th ed. New York: Macmillan.

DRAYNA, D., and WHITE, R. 1985. The genetic linkage map of the human X chromosome. *Science* 230:753–58.

DUNN, L. C. 1966. *A short history of genetics*. New York: McGraw-Hill.

EAST, E. M. 1910. A Mendelian interpretation of variation that is apparently continuous. *Amer. Naturalist* 44:65–82.

———. 1916. Studies on size inheritance in *Nicotiana*. *Genetics* 1:164–76.

FALCONER, D. S. 1981. *Introduction to quantitative genetics*. 2nd ed. New York: Longman.

FOSTER, H. L., et al., eds. 1981. *The mouse in biomedical research. Vol. 1: History, genetics, and wild mice*. Orlando: Academic Press.

FOSTER, M. 1965. Mammalian pigment genetics. *Adv. in Genet.* 13:311–39.

GRANT, V. 1975. *Genetics of flowering plants*. New York: Columbia University Press.

LINDSLEY, D. C., and GRELL, E. H. 1967. *Genetic variations of* Drosophila melanogaster. Washington, D.C.: Carnegie Institute of Washington.

McKUSICK, V. A. 1962. On the X chromosome of man. *Quart. Rev. Biol.* 37:69–175.

MORGAN, T. H. 1910. Sex-limited inheritance in *Drosophila*. *Science* 32:120–22.

NOLTE, D. J. 1959. The eye-pigmentary system of *Drosophila*. *Heredity* 13:233–41.

PAWELEK, J. M., and KÖRNER, A. M. 1982. The biosynthesis of mammalian melanin. *Amer. Scient.* 70:136–45.

PETERS, J. A., ed. 1959. *Classic papers in genetics*. Englewood Cliffs, N.J.: Prentice-Hall.

RACE, R. R., and SANGER, R. 1975. *Blood groups in man*. 6th ed. Oxford, England: Blackwell Scientific Publishers.

STERN, C. 1973. *Principles of human genetics*. 3rd ed. New York: W. H. Freeman.

TEARLE, R. G., et al. 1989. Cloning and characterization of the *scarlet* gene of *Drosophila melanogaster*. *Genetics* 122:595–606.

VOELLER, B. R., ed. 1968. *The chromosome theory of inheritance—Classic papers in development and heredity*. New York: Appleton-Century-Crofts.

VOGEL, F., and MOTULSKY, A. G. 1986. 2nd ed. *Human genetics: Problems and approaches*. New York: Springer-Verlag.

WATKINS, M. W. 1966. Blood group substances. *Science* 152:172–81.

WIENER, A. S., ed. 1970. *Advances in blood groupings, III*. New York: Grune and Stratton.

YOSHIDA, A. 1982. Biochemical genetics of the human blood group ABO system. *Amer. J. Human Genet.* 34:1–14.

ZIEGLER, I. 1961. Genetic aspects of ommochrome and pterin pigments. *Adv. in Genet.* 10:349–403.

# 5

# LINKAGE, CROSSING OVER, AND CHROMOSOME MAPPING

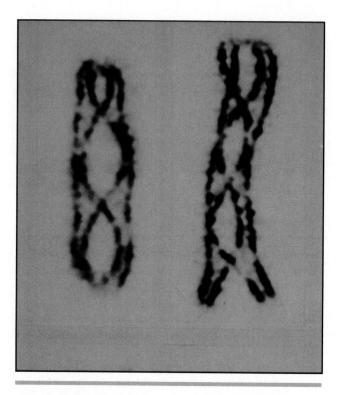

Chiasmata formation between synapsed homologues present in the tetrad stage.

*Many genes reside on a single chromosome. Unless separated by crossing over, alleles present at the many loci on each homologue segregate as a unit during gamete formation. Recombinant gametes resulting from crossing over enhance genetic variability within a species and serve as the basis for constructing chromosomal maps.*

As early as 1903, Walter Sutton, who along with Theodor Boveri united the fields of cytology and genetics, pointed out the likelihood that in organisms there are many more "unit factors" than chromosomes. Soon thereafter, genetic investigations with several organisms revealed that certain genes were not transmitted according to the law of independent assortment. When studied together in matings, these genes seemed to segregate as if they were somehow joined or linked together. Further investigations showed that such genes were part of the same chromosome, and they were indeed transmitted as a single unit.

We now know that most chromosomes consist of very large numbers of genes and, in fact, contain sufficient DNA to encode thousands of these units. Genes that are part of the same chromosome are said to be **linked** and to demonstrate **linkage** in genetic crosses.

Since the chromosome, not the gene, is the unit of transmission during meiosis, linked genes are not free to undergo independent assortment. Instead, the alleles at all loci of one chromosome should, in theory, be transmitted as a unit during gamete formation. However, in many instances this does not occur. During the first meiotic prophase, when homologues are paired or synapsed, a reciprocal exchange of chromosome segments may take place. This event, called **crossing over**, results in the reshuffling or **recombination** of the alleles between homologues.

The degree of crossing over between any two loci on a single chromosome is proportionate to the distance between them. Thus, the percentage of recombinant gametes varies, depending on which loci are being considered. This correlation serves as the basis for the construction of **chromosome maps**, which provide the relative locations of genes on the chromosomes.

Crossing over is currently viewed as an actual physical breaking and rejoining process that occurs during meiosis. This exchange of chromosome segments provides for an enormous potential variation in the gametes formed by any individual. This type of variation, in combination with that resulting from independent assort-

ment, ensures that all offspring will contain a diverse mixture of both maternal and paternal alleles. In addition to producing individual diversity, genetic variability is of paramount importance to the process of organic evolution.

In this chapter, we will discuss linkage, crossing over, and chromosome mapping. We will also consider a variety of topics involving the exchange of genetic information. We will conclude the chapter by entertaining the rather intriguing question of why Mendel, who studied seven genes, did not encounter linkage. Or did he?

## LINKAGE VERSUS INDEPENDENT ASSORTMENT

In order to provide a simplified overview of the major theme of this chapter, Figure 5.1 illustrates and contrasts the meiotic consequences of independent assortment, linkage *without* crossing over, and linkage *with* crossing over. Two homologous pairs of chromosomes are considered in the case of independent assortment and one homologous pair in the two cases of linkage.

Figure 5.1(a) illustrates the results of independent assortment of two pairs of nonhomologous chromosomes, each containing one heterozygous gene pair. Thus, no linkage is exhibited by the two genes. When a large number of meiotic events are observed, four genetically different gametes are formed in equal proportions.

We can compare these results with those that occur if the same genes are instead linked on the same chromosome. If no crossing over occurs between the two genes [Figure 5.1(b)], only two genetically different gametes are formed. Each gamete receives the alleles present on one homologue or the other, which has been transmitted intact as the result of segregation. This case illustrates **complete linkage**, which is said to result in the production of only **parental** or **noncrossover gametes**. The two parental gametes are formed in equal proportions. While complete linkage between two genes seldomly

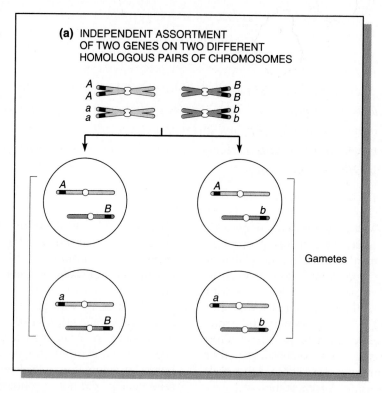

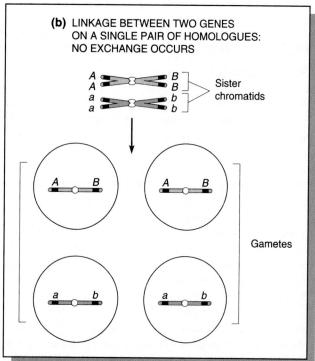

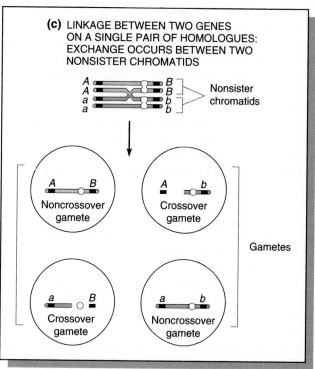

**FIGURE 5.1**     A comparison of the results of gamete formation where two heterozygous genes are (a) on two different pairs of chromosomes; (b) on the same pair of homologues, but with no exchange occurring between them; and (c) on the same pair of homologues, with an exchange between two nonsister chromatids.

occurs, consideration of the theoretical consequences of this concept is useful when studying crossing over.

Figure 5.1(c) illustrates the results when crossing over occurs between two linked genes. As you will note, this crossover involves only two nonsister chromatids of the four chromatids present in the tetrad. This exchange generates two new allele combinations, called **recombinant** or **crossover gametes**. The two chromatids not involved in the exchange result in noncrossover gametes [as those in part (b) of this figure].

The frequency with which crossing over occurs between any two linked genes is proportional to the distance separating the respective loci along the chromosome. It is theoretically possible for two randomly selected genes to be so close to each other that crossover events are too infrequent to be detected. This circumstance, complete linkage, results in the production of only parental gametes, as shown in case (b). If a distinct but small distance separates two genes, few recombinant and many parental gametes will be formed. As the distance between the two genes increases, the proportion of recombinant gametes increases and that of the parental gametes decreases. Thus, in case (c), the proportion of crossover gametes varies depending on the distance between the genes studied.

As will be explored again later in this chapter, when two linked genes whose loci are far apart are considered, the number of recombinant gametes approaches, but never exceeds 50 percent. If 50 percent recombinants occurred, a 1:1:1:1 ratio of the four types (two parental and two recombinant gametes) would result. In such a case, transmission of two linked genes would be indistinguishable from that of two unlinked, independently assorting genes. That is, the proportion of the four possible genotypes would be identical in Cases (a) and (c) of Figure 5.1.

## The Linkage Ratio

If complete linkage exists between two genes because of their close proximity, as in case (b) of Figure 5.1, and organisms with mutant alleles representing these genes are mated, a unique $F_2$ phenotypic ratio results that is characteristic of linkage. To illustrate this ratio, we will consider a cross involving the closely linked recessive mutant genes *brown* (*bw*) eye and *heavy* (*hv*) wing vein in *Drosophila melanogaster* (Figure 5.2). The normal, wild-type alleles $bw^+$ and $hv^+$ are both dominant and result in red eyes and thin wing veins, respectively.

In this cross, flies with mutant brown eyes and normal thin veins are mated to flies with normal red eyes and mutant heavy veins. In more concise terms, brown-eyed flies are crossed with heavy-veined flies. If we extend the system of using genetic symbols established in Chapter 4, linked genes may be represented by placing their allele designations above and below a single or double horizontal line. Those placed above the line are located at loci on one homologue and those placed below at the homologous loci on the other homologue. Thus, we may represent the $P_1$ generation as follows:

$$P_1: \frac{bw \quad hv^+}{bw \quad hv^+} \times \frac{bw^+ \quad hv}{bw^+ \quad hv}$$

brown, thin      red, heavy

Since the genes are located on an autosome, and not the X chromosome, no distinction for male and female is necessary.

In the $F_1$ generation each fly receives one chromosome of each pair from each parent; all flies are heterozygous for both gene pairs and exhibit the dominant traits of red eyes and thin veins:

$$F_1: \frac{bw \quad hv^+}{bw^+ \quad hv}$$

red, thin

As shown in Figure 5.2, when the $F_1$ generation is interbred because of complete linkage, each $F_1$ individual forms only parental gametes. Following fertilization, the $F_2$ generation will be produced in a 1:2:1 phenotypic and genotypic ratio. One-fourth of this generation will show brown eyes and thin veins; one-half will show both wild-type traits, red eyes and thin veins; and one-fourth will show red eyes and thick veins. In more concise terms, the ratio is 1 brown:2 wild:1 thick. Such a ratio is characteristic of complete linkage. Even though it is extremely rare for two random genes to exhibit complete linkage, theoretically, it is possible.

Figure 5.2 also demonstrates the results of a test cross with the $F_1$ flies. Such a cross produces a 1:1 ratio of brown, thin and red, thick flies. Had the genes controlling these traits been incompletely linked or located on separate autosomes, four phenotypes rather than two would have been produced.

When large numbers of mutant genes present in any given species are investigated, genes located on the same chromosome will show evidence of linkage to one another. As a result, **linkage groups** can be established, one for each chromosome. Hence, the number of linkage groups should correspond to the haploid number of chromosomes. In organisms where large numbers of mutant genes are available for genetic study, this correlation has been upheld.

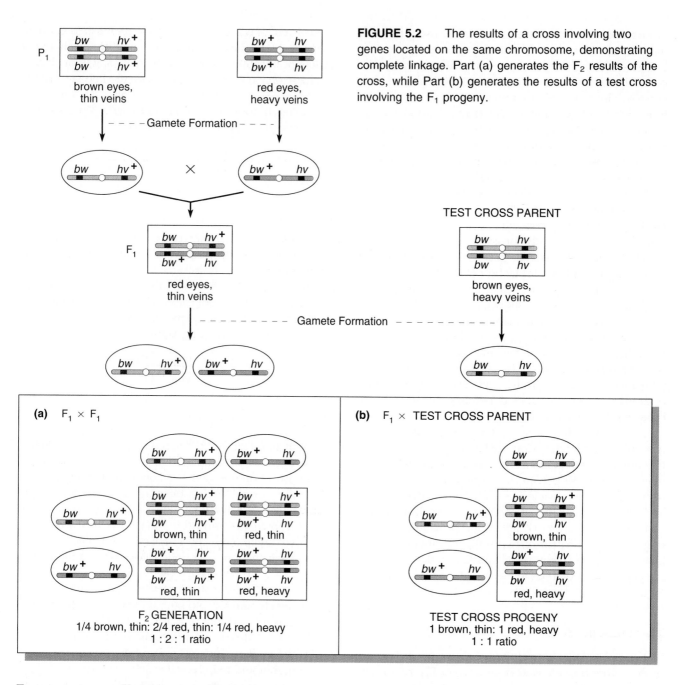

**FIGURE 5.2**    The results of a cross involving two genes located on the same chromosome, demonstrating complete linkage. Part (a) generates the $F_2$ results of the cross, while Part (b) generates the results of a test cross involving the $F_1$ progeny.

# INCOMPLETE LINKAGE, CROSSING OVER, AND CHROMOSOME MAPPING

If two genes linked on the same chromosome are selected randomly, it is unlikely that their respective loci will be contiguous to each other. Therefore, because crossing over will occur between them, complete linkage is rarely achieved. Instead, crosses involving two linked genes usually produce a percentage of offspring resulting from recombinant gametes. This percentage is variable, depending upon which two genes are involved in the cross and the distance along the chromosome that

separates them. This phenomenon was first studied and explained by two *Drosophila* geneticists, Thomas H. Morgan and his undergraduate student, Alfred H. Sturtevant.

## Morgan and Crossing Over

As you may recall from our earlier discussion of sex-linked genes in Chapter 4, Morgan first discovered the phenomenon of sex linkage. In his studies, he investigated numerous *Drosophila* mutations located on the X chromosome. When he analyzed crosses involving only one trait, he was able to deduce the mode of sex-linked inheritance. However, when he made crosses involving two sex-linked genes, his results were at first puzzling. For example, as shown in Cross A of Figure 5.3, he crossed mutant *yellow* body ($y$) and *white* eyes ($w$) fe-

males with wild-type males (gray body and red eyes). The $F_1$ females were wild type, while the $F_1$ males expressed both mutant traits. In the $F_2$, 98.7 percent of the offspring showed the parental phenotypes—

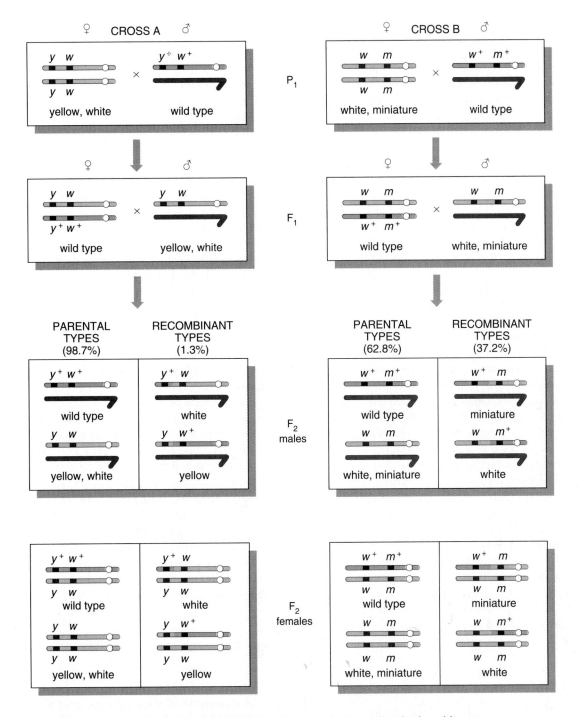

**FIGURE 5.3**    The $F_1$ and $F_2$ results of crosses involving the *yellow*-body, *white*-eye mutations and the *white*-eye, *miniature*-wing mutations. In the $F_2$ generation of Cross A, 1.3 percent of the flies demonstrate recombinant phenotypes, which express either *white* or *yellow*. In the $F_2$ generation of Cross B, 37.2 percent of the flies demonstrate recombinant phenotypes, which express either *miniature* or *white*.

yellow-bodied, white-eyed flies and wild-type flies. The remaining 1.3 percent of the flies were either yellow-bodied with red eyes or gray-bodied with white eyes. It was as if the genes had somehow separated from each other during gamete formation in the $F_1$ flies.

When he made crosses involving other sex-linked genes, the results were even more puzzling (Cross B of Figure 5.3). The same basic pattern was observed, but the proportion of $F_2$ phenotypes differed; for example, in a cross involving *white*-eye, *miniature*-wing mutants, only 62.8 percent of the $F_2$ showed the parental phenotypes, while 37.2 percent of the offspring appeared as if the mutant genes had been separated during gamete formation.

In 1911, Morgan was faced with two questions: (1) What was the source of gene separation? and (2) Why did the frequency of the apparent separation vary depending on the genes being studied? The proposed answer to the first question was based on his knowledge of earlier cytological observations made by F. Janssens and others. Janssens had observed that synapsed homologous chromosomes in meiosis wrapped around each other, creating **chiasmata** [sing., **chiasma**] where points of overlap are evident. Morgan proposed that these chiasmata could represent points of genetic exchange, the crux of the so-called **chiasmatype theory**.

In the crosses shown in Figure 5.3, Morgan postulated that if an exchange occurred between the mutant genes on the two X chromosomes of the $F_1$ females, it would lead to the observed results. He suggested that such exchanges led to 1.3 percent recombinant gametes in the *yellow-white* cross and 37.2 percent in the *white-miniature* cross. On the basis of this and other experimentation, Morgan concluded that linked genes exist in a linear order along the chromosome, and that a variable amount of exchange occurs between any two genes.

As an answer to the second question, Morgan proposed that two genes located relatively close to each other along a chromosome are less likely to have a chiasma form between them than if the two genes are farther apart on the chromosome. Thus, the closer two genes are, the less likely it is that a genetic exchange will occur between them. Morgan proposed the term **crossing over** to describe the physical exchange leading to recombination.

## Sturtevant and Mapping

Morgan's student, Alfred H. Sturtevant, was the first to realize that his mentor's proposal could be used to map the sequence of and distance between linked genes. Sturtevant argued that if the frequency of crossing over between genes is related to the distance between them, the recombination frequencies between a series of linked genes must be additive.

For example, Sturtevant compiled data on recombination between the genes represented by the *yellow, white,* and *miniature* mutants initially studied by Morgan. Frequencies of crossing over between each pair of these three genes were observed to be:

| | | |
|---|---|---|
| (1) | *yellow, white* | 0.5% |
| (2) | *white, miniature* | 34.5% |
| (3) | *yellow, miniature* | 35.4% |

Since the sum of (1) and (2) is approximately equal to (3), Sturtevant suggested that the recombination frequencies between linked genes are additive. On this basis, he predicted that the order of the genes on the X chromosome was *yellow-white-miniature*. In arriving at this conclusion, he reasoned as follows: The *yellow* and *white* genes are apparently close to each other because the recombination frequency is low. However, both of these genes are quite far apart from *miniature* because the *white, miniature* and *yellow, miniature* combinations show large recombination frequencies. Since *miniature* shows more recombination with *yellow* than with *white* (35.4 vs. 34.5), it follows that *white* is between the other two genes, not outside of them.

Sturtevant suggested that the frequency of exchange could be taken as an estimate of the relative distance between two genes or loci along the chromosome. He constructed a map of the three genes on the X chromosome, with one map unit being equated with one percent recombination between two genes.[*] In the preceding example, the distance between *yellow* and *white* would be 0.5 map unit, and between *yellow* and *miniature,* 35.4 map units. It follows that the distance between *white* and *miniature* should be (35.4 − 0.5) or 34.9. This estimate is close to the actual frequency of recombination between *white* and *miniature* (34.5). The simple map for these three genes is shown in Figure 5.4.

In addition to these three genes, Sturtevant considered three other genes on the X chromosome and produced a more extensive map including all six genes. He and a colleague, Calvin Bridges, soon began a search for autosomal linkage in *Drosophila*. By 1923, they had clearly shown that linkage and crossing over were not restricted

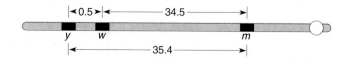

**FIGURE 5.4**    A simple map of the *yellow* (*y*), *white* (*w*), and *miniature* (*m*) genes on the X chromosome of *Drosophila melanogaster.* Each number represents the percentage of recombinant offspring produced in the three crosses, each involving two different genes.

---

[*] In honor of Morgan's work, one map unit is often referred to as a centimorgan (cM).

to sex-linked genes. Rather, they had discovered linked genes, on autosomes, between which crossing over occurred.

During this work, they made another interesting observation. In *Drosophila,* crossing over was shown to occur only in females. The fact that no crossing over occurs in males made genetic mapping much less complex to analyze in *Drosophila*. However, crossing over does occur in both sexes in most other organisms.

Although many refinements in chromosome mapping have developed since Sturtevant's initial work, his basic principles are accepted as correct. They have been used to produce detailed chromosome maps of organisms for which large numbers of linked mutant genes are known. In addition to providing the basis for chromosome mapping, Sturtevant's findings were historically significant to the field of genetics. In 1910, the **chromosomal theory of inheritance** was still being widely disputed. Even Morgan was skeptical of this theory prior to the time he conducted the bulk of his experimentation. Research has now firmly established that chromosomes contain genes in a linear order and that these genes are the equivalent of Mendel's unit factors.

## Single Crossovers

Why should the relative distance between two loci influence the amount of recombination and crossing over observed between them? The basis for this variation is explained in the following analysis.

During meiosis, a limited number of crossover events occurs in each tetrad. These recombinant events occur randomly along the length of the tetrad. Therefore, the closer two loci reside along the axis of the chromosome, the less likely it is that any single crossover event will occur between them. The same reasoning suggests that the farther apart two linked loci are, the more likely it is that a random crossover event will occur between them.

In Figure 5.5(a), a single crossover occurs between two nonsister chromatids, but not between the two loci; therefore, the crossover goes undetected because no recombinant gametes are produced. In (b), where two loci are quite far apart, the crossover occurs between them, yielding recombinant gametes.

The evidence supporting the concept that crossing over occurs in the four-strand tetrad stage will be discussed later in this chapter. However, we must point out certain consequences of this fact. When a single crossover occurs between two nonsister chromatids, the other two strands of the tetrad will not be involved in this exchange and may enter a gamete unchanged. Under these conditions, even if a single crossover occurs 100 percent of the time between two linked genes, this recombinant event will subsequently be observed in only 50 percent of the potential gametes formed. This concept is diagrammed in Figure 5.6. Theoretically, *when only single exchanges between two nonsister strands* are considered, the general rule is that the percentage of recombinant gametes is half as great as the percentage of tetrads involved in a single exchange between two genes. Therefore when only single exchanges are considered, the theoretical limit of crossing over is 50 percent.

When two linked genes are 50 map units or more apart, a crossover can theoretically be expected to occur between them in 100 percent of the tetrads. If this theoretical prediction were to be achieved, each tetrad would

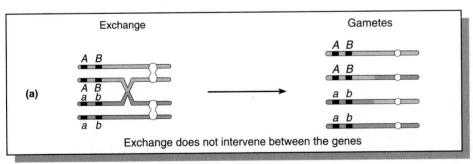

(a)

Exchange     Gametes

Exchange does not intervene between the genes

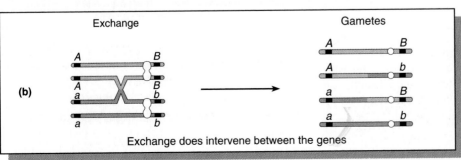

(b)

Exchange     Gametes

Exchange does intervene between the genes

**FIGURE 5.5**    Two cases of exchange between nonsister chromatids and the gametes subsequently produced. In (a) the exchange does not separate the alleles of the two genes, only parental gametes are formed, and the exchange goes undetected. In (b) the exchange separates the alleles, resulting in recombinant gametes.

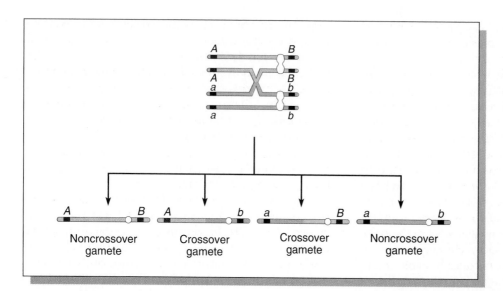

**FIGURE 5.6** The consequences of a single exchange between two nonsister chromatids occurring in the tetrad stage. Two noncrossover (parental) and two crossover (recombinant) gametes are produced.

yield equal proportions of the four gametes shown in Figure 5.6, just as if the genes were on different chromosomes and assorting independently. For a variety of reasons, this theoretical limit is never achieved.

## Multiple Crossovers

It is possible that in a single tetrad, two, three, or more exchanges will occur between nonsister chromatids as a result of several crossing over events. Double exchanges of genetic material result from **double crossovers**, as shown in Figure 5.7. For a double exchange to be studied, three gene pairs must be investigated, each heterozygous for two alleles. Before we can determine the frequency of recombination between all three loci, we must review some simple probability calculations.

The probabilities of a single exchange occurring between the *A* and *B* or the *B* and *C* genes are directly related to the physical distance between each locus. The closer *A* is to *B* and *B* is to *C*, the less likely it is that a single exchange will occur between either of the two sets of loci. In the case of a double crossover, two separate and independent events or exchanges must occur simultaneously. The mathematical probability of two independent events occurring simultaneously is equal to

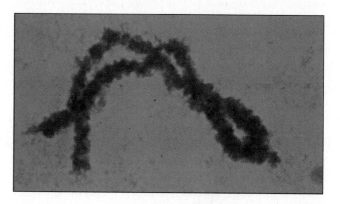

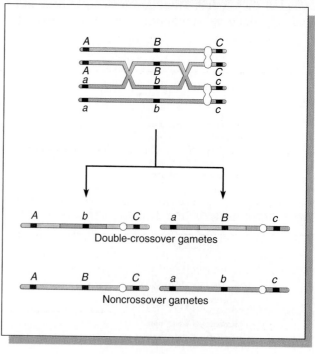

▶

**FIGURE 5.7** The results of a double exchange occurring between nonsister chromatids. Because the exchanges involve only two strands, two noncrossover gametes and two double-crossover gametes are produced. The photograph illustrates chiasmata found in a tetrad isolated during the first meiotic prophase stage.

the product of the individual probabilities. This is the product law previously introduced in Chapter 3.

Suppose that crossover gametes resulting from single exchanges between $A$ and $B$ are recovered 20 percent of the time ($p = 0.20$) and between $B$ and $C$ 30 percent of the time ($p = 0.30$). The probability of recovering a double-crossover gamete arising from two exchanges, between $A$ and $B$ and between $B$ and $C$, is predicted to be $(0.20) \cdot (0.30) = 0.06$, or 6 percent. It is apparent from this calculation that the frequency of double-crossover gametes is always expected to be much lower than that of either single-crossover class of gametes.

If three genes are relatively close together along one chromosome, the expected frequency of double-crossover gametes is extremely low. For example, consider the $A$–$B$ distance in Figure 5.7 to be 3 map units and the $B$–$C$ distance in that figure to be 2 map units. The expected double-crossover frequency would be $(0.03) \cdot (0.02) = 0.0006$, or 0.06 percent. This translates to only 6 events in 10,000. Thus, in a mapping experiment involving closely linked genes, very large numbers of offspring are required in order to detect double-crossover events. In this example, it would be unlikely that a double crossover would be observed even if 1000 offspring were examined. If these probability considerations are extended, it is evident that if four or five genes were being mapped, even fewer triple and quadruple crossovers can be expected to occur.

## Three-Point Mapping in *Drosophila*

The information presented in the previous section serves as the basis for mapping three or more linked genes in a single cross. To illustrate this, we will examine a situation involving three linked genes.

Three criteria must be met for a successful mapping cross:

1. The genotype of the organism producing the crossover gametes must be heterozygous at all loci under consideration.

2. The cross must be constructed so that genotypes of all gametes can be accurately determined by observing the phenotypes of the resulting offspring. This is necessary because the gametes and their genotypes can never be observed directly. Thus, each phenotypic class must reflect the genotype of the gametes of the parents producing it.

3. A sufficient number of offspring must be produced in the mapping experiment to recover a representative sample of all crossover classes.

These criteria are met in the three-point mapping cross from *Drosophila melanogaster* shown in Figure 5.8. In this cross, three sex-linked recessive mutant genes—*yellow* body color, *white* eye color, and *echinus* eye shape—are considered. In order to diagram the cross, we must assume some theoretical sequence, even though we do not yet know if it is correct. In Figure 5.8, we will initially assume the sequence of the three genes to be *y–w–ec*. If this is incorrect, our analysis will reveal the correct sequence.

In the $P_1$ generation, males hemizygous for all three wild-type alleles are crossed to females that are homozygous for all three recessive mutant alleles. Therefore, the $P_1$ males are wild type with respect to body color, eye color, and eye shape. They are said to have a wild-type phenotype. The females, on the other hand, exhibit the three mutant traits—*yellow* body color, *white* eyes, and *echinus* eye shape.

This cross produces an $F_1$ generation consisting of females heterozygous at all three loci and males that, because of the Y chromosome, are hemizygous for the three mutant alleles. Phenotypically, all $F_1$ females are wild type, while all $F_1$ males are *yellow, white,* and *echinus*. The genotype of the $F_1$ females fulfills the first criterion for constructing a map of the three linked genes; that is, it is heterozygous at the three loci and may serve as the source of recombinant gametes generated by crossing over. Note that because of the genotypes of the $P_1$ parents, all three mutant alleles are on one homologue and all three wild-type alleles are on the other homologue. *Other arrangements are possible.* For example, the heterozygous $F_1$ female might have the *y* and *ec* mutant alleles on one homologue and the *w* allele on the other. This would occur if, in the $P_1$ cross, one parent was *yellow* and *echinus* and the other parent was *white*.

In our cross, the second criterion is met by virtue of the gametes formed by the $F_1$ males. Every gamete will contain either an X chromosome bearing the three mutant alleles or a Y chromosome, which is genetically inert for the three loci being considered. Whichever type participates in fertilization, the genotype of the gamete produced by the $F_1$ female will be expressed phenotypically in the $F_2$ male and female offspring derived from it. As a result, all $F_1$ noncrossover and crossover gametes can be detected by observing the $F_2$ phenotypes.

With these two criteria met, we can construct a chromosome map from the crosses illustrated in Figure 5.8. First, we must determine which $F_2$ phenotypes correspond to the various noncrossovers and crossover categories. Two of these can be determined immediately.

The **noncrossover** $F_2$ phenotypes are determined by the combination of alleles present in the parental gametes formed by the $F_1$ female. Each such gamete con-

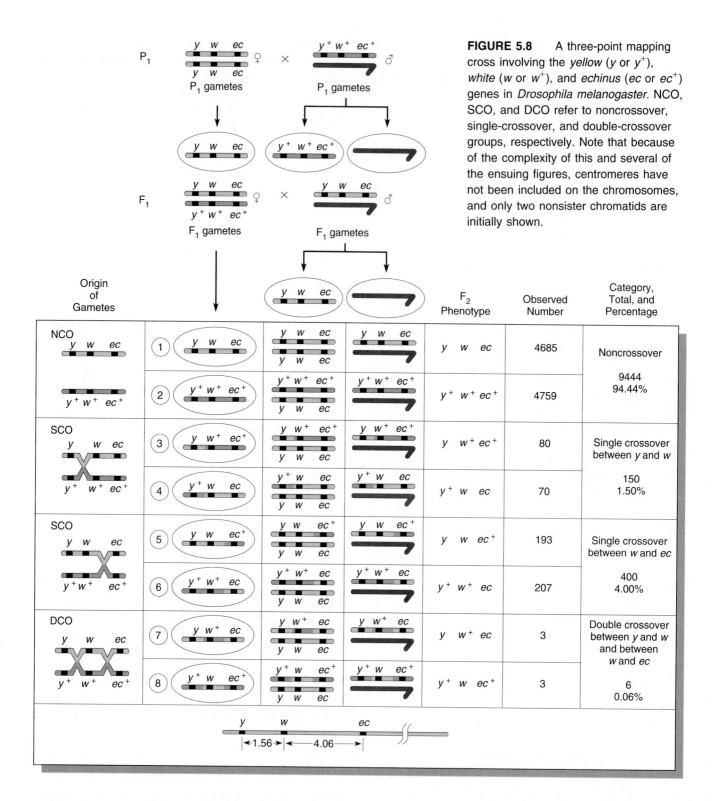

**FIGURE 5.8**   A three-point mapping cross involving the *yellow* (*y* or *y⁺*), *white* (*w* or *w⁺*), and *echinus* (*ec* or *ec⁺*) genes in *Drosophila melanogaster*. NCO, SCO, and DCO refer to noncrossover, single-crossover, and double-crossover groups, respectively. Note that because of the complexity of this and several of the ensuing figures, centromeres have not been included on the chromosomes, and only two nonsister chromatids are initially shown.

tains one or the other of the X chromosomes unaffected by crossing over. As a result of segregation, approximately equal proportions of the two types of gametes, and subsequently the F₂ phenotypes, are produced. Because they are derived from a heterozygote, the genotypes of the two parental gametes and the phenotypes of the two F₂ phenotypes complement one another. For

example, if one is wild type, the other is completely mutant. This is the case in the cross under consideration. In other situations, if one chromosome shows one mutant allele or trait, the other shows the other two mutant traits, and so on. They are therefore called **reciprocal classes** of gametes and phenotypes.

The two noncrossover phenotypes are most easily recognized because *they exist in the greatest proportion.* Figure 5.8 shows that classes (1) and (2) are present in the greatest numbers. Therefore, flies that are *yellow, white,* and *echinus* and those that are normal or wild type for all three characters constitute the noncrossover category and represent 94.44 percent of the $F_2$ offspring.

The second category that can be easily detected is represented by the double-crossover phenotypes. Because of their probability of occurrence, *they must be present in the least numbers.* Remember that this group represents two independent but simultaneous single-crossover events. Two reciprocal phenotypes can be identified: class (7), which shows the mutant traits *yellow* and *echinus* but normal eye color; and class (8), which shows the mutant trait *white* but normal body color and eye shape. Together these double-crossover phenotypes constitute only 0.06 percent of the $F_2$ offspring.

The remaining four phenotypic classes represent two categories resulting from single crossovers. Classes (3) and (4), reciprocal phenotypes produced by single-crossover events occurring between the *yellow* and *white* loci, are equal to 1.50 percent of the $F_2$ offspring. Classes (5) and (6), constituting 4.00 percent of the $F_2$ offspring, represent the reciprocal phenotypes resulting from single-crossover events occurring between the *white* and *echinus* loci.

The map distances between the three loci can now be calculated. The distance between *y* and *w,* or between *w* and *ec,* is equal to the percentage of all detectable exchanges occurring between them. For any two genes under consideration, this will include all appropriate single crossovers as well as all double crossovers. The latter are included because they represent two simultaneous single crossovers. For the *y* and *w* genes this includes classes (3), (4), (7), and (8), totaling 1.50% + 0.06%, or 1.56 map units. Similarly, the distance between *w* and *ec* is equal to the percentage of offspring resulting from an exchange between these two loci: classes (5), (6), (7), and (8), totaling 4.00% + 0.06%, or 4.06 map units. The map of these three loci on the X chromosome, based on these data, is shown at the bottom of Figure 5.8.

### Determining the Gene Sequence

In the preceding example, the order or sequence of the three genes along the chromosome was assumed to be *y–w–ec.* Our analysis shows this sequence to be consistent with the data. In most mapping experiments, the gene sequence is not known, and this constitutes another variable in the analysis. Had the gene order been unknown in this example, it could have been determined in a straightforward way. There are two methods that may be used to accomplish this. You should select one or the other method and adhere to its use in your own analysis.

**Method I**   This method is based on the fact that there are only three possible orders, each containing one of the three genes in between the other two. Note that the right-to-left direction along the chromosome is meaningless. While we assumed *y–w–ec, ec–w–y* is fully equivalent because the data provide no basis for knowing which is correct. Therefore, only three possibilities exist, and one of them must be correct:

(I) *w–y–ec* (*y* is in the middle)
(II) *y–ec–w* (*ec* is in the middle)
(III) *y–w–ec* (*w* is in the middle)

If you use the following steps during your analysis, you will be able to determine gene order:

1. Assuming any of the three orders, first determine the **arrangement of alleles** along each homologue of the heterozygous parent giving rise to noncrossover and crossover gametes (the $F_1$ female in our example).

2. Determine whether a double-crossover event occurring within that arrangement will produce the **observed double-crossover phenotypes**. Remember that these phenotypes occur least frequently and can be easily identified.

3. If this order does not produce the correct phenotypes, try each of the other two orders. One must work!

These steps will be illustrated for the cross shown in Figure 5.8. The three possible orders are labeled (I), (II), and (III), as just shown. Either *y, ec,* or *w* must be in the middle.

1. Assuming *y* is between *w* and *ec,* the arrangement of alleles along the homologues of the $F_1$ heterozygote is:

$$(\text{I})\quad \frac{w\quad y\quad ec}{w^+\quad y^+\quad ec^+}$$

We know this because of the way in which the $P_1$ generation was crossed. The $P_1$ female contributed an X chromosome bearing the *w, y,* and *ec* alleles, while the $P_1$ male contributed an X chromosome bearing the $w^+$, $y^+$, and $ec^+$ alleles.

2. A double crossover within the above arrangement would yield the following gametes:

$$\underline{w\quad y^+\quad ec}\quad \text{and}\quad \underline{w^+\quad y\quad ec^+}$$

Following fertilization, if *y* is in the middle, the $F_2$ double-crossover phenotypes will correspond to the above gametic genotypes, yielding offspring that are *white, echinus* and offspring that are *yellow*. Instead, however, determination of the actual double crossovers reveals them to be *yellow, echinus* flies and *white* flies. Therefore, our assumed order is incorrect.

3.  If we try the other orders, one with $ec/ec^+$ alleles in the middle (II) or one with the $w/w^+$ alleles in the middle (III):

$$\text{(II)} \quad \frac{y \quad ec \quad w}{y^+ \quad ec^+ \quad w^+} \quad \text{or} \quad \text{(III)} \quad \frac{y \quad w \quad ec}{y^+ \quad w^+ \quad ec^+}$$

we see that arrangement II again provides **predicted** double-crossover phenotypes that do not correspond to the **actual** double-crossover phenotypes. The predicted phenotypes are *yellow, white* flies and *echinus* flies in the $F_2$ generation. Therefore, this order is also incorrect. However, arrangement III **will** produce the **observed** phenotypes, *yellow, echinus* flies and *white* flies. Therefore, this order, where the *w* gene is in the middle, is correct.

To summarize, utilizing this method is rather straightforward. Determine the arrangement of alleles on the homologues of the heterozygote yielding the crossover gametes. Then, test each of three possible orders to determine which one yields the observed double-crossover phenotypes. Whichever of the three does so represents the correct order. Testing the three possibilities in our example is summarized in Figure 5.9.

**Method II**  This method again requires that we determine the arrangement of alleles along each homologue of the heterozygous parent. It also assumes that one specific gene of the three is in the middle. It requires one further assumption:

*Following a double-crossover event, the allele representing the middle gene will find itself present with the outside or flanking alleles present on the opposite parental homologue.*

To illustrate, assume order (I), *w–y–ec*, in the arrangement:

$$\text{(I)} \quad \frac{w \quad y \quad ec}{w^+ \quad y^+ \quad ec^+}$$

Following a double-crossover event, the *y* and $y^+$ alleles would find themselves switched to the arrangement:

$$\frac{w \quad y^+ \quad ec}{w^+ \quad y \quad ec^+}$$

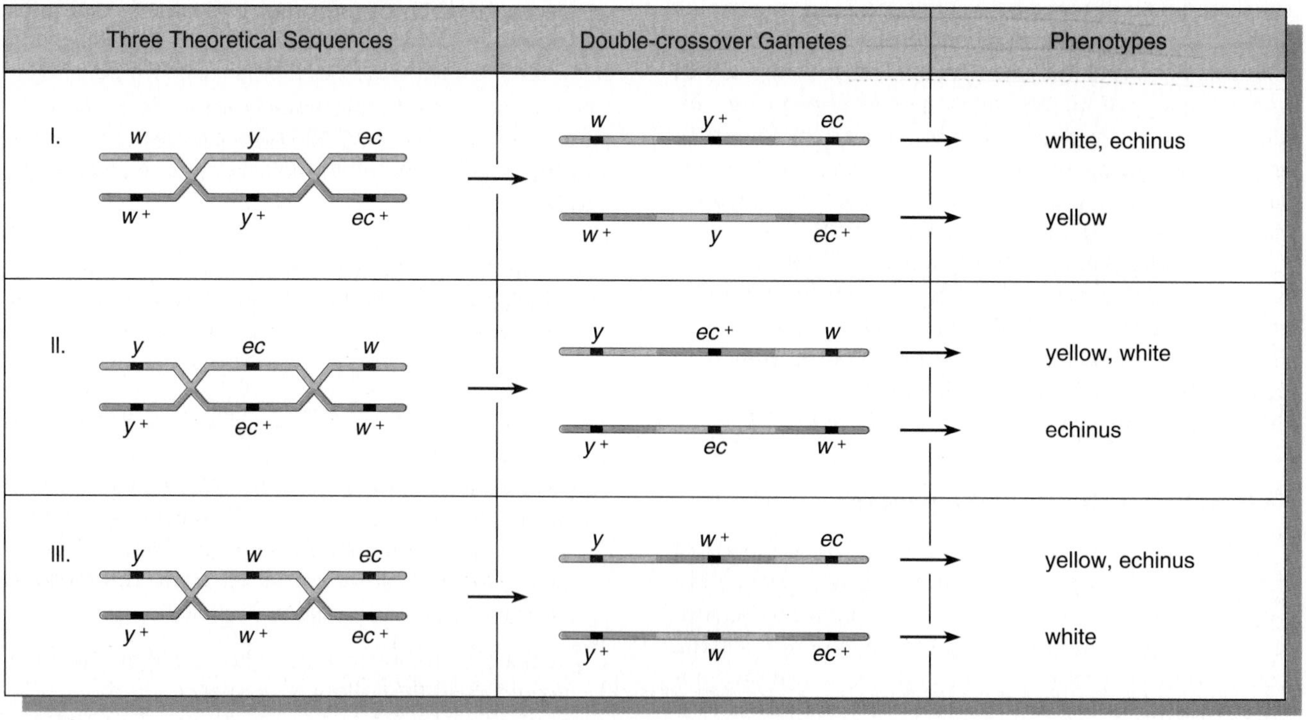

**FIGURE 5.9**  A summary of the three possible orders of the *white, yellow,* and *echinus* genes, the results of a double crossover in each case, and the resulting phenotypes produced.

Following segregation, two gametes would be formed:

$$\underline{w \quad y^+ \quad ec} \quad \text{and} \quad \underline{w^+ \quad y \quad ec^+}$$

Since the genotype of the gamete will be expressed directly in the phenotype following fertilization, the double-crossover phenotypes will be:

*white, echinus* flies   and   *yellow* flies

Note that the *yellow* allele, assumed to be in the middle, is now associated with the two outside markers of the other homologue, $w^+$ and $ec^+$. However, these predicted phenotypes do not coincide with the observed double-crossover phenotypes. Therefore, the *yellow* gene is not in the middle.

This same reasoning can be applied to the assumption that the *echinus* gene or the *white* gene is in the middle. In the former case, a negative conclusion will be reached. If we assume that the *white* gene is in the middle, the predicted and actual double crossovers coincide. Therefore, we conclude that the *white* gene is located between the *yellow* and *echinus* genes.

To summarize, this method is also straightforward. Determine the arrangement of alleles on the homologues of the heterozygote yielding crossover gametes. Then determine the actual double-crossover phenotypes. Simply select the single allele that has been switched so that it is now associated with two other alleles that have not been separated by crossing over.

In our example above, *y, ec,* and *w* are together in the $F_1$ heterozygote, as are $y^+$, $ec^+$, and $w^+$. In the $F_2$ double-crossover classes, it is *w* and $w^+$ that have been switched. The *w* allele is now associated with $y^+$ and $ec^+$, while the $w^+$ allele is now associated with the *y* and *ec* alleles. Therefore, the *white* gene is in the middle, and the *yellow* and *echinus* genes are the flanking markers.

## A Mapping Problem in Maize

Having established the basic principles of chromosome mapping, we will now consider a related problem in maize (corn) where the gene sequence and interlocus distances are unknown.

This analysis differs from the preceding discussion in several ways and therefore will expand your knowledge of mapping procedures:

1. The previous mapping cross involved sex-linked genes. Here, autosomal genes are considered.

2. In the previous discussion we initially knew the gene sequence, and we used this information to explain how to determine an unknown sequence. In this analysis of maize, the sequence is *initially* unknown.

3. In the discussion of this cross we will make a transition in the use of symbols, as first suggested in Chapter 4. Instead of using the symbols $bm^+$, $v^+$, and $pr^+$, we will simply use $+$ to denote each wild-type allele. This symbol is less complex to manipulate, but requires a better understanding of mapping procedures.

When we consider three autosomally linked genes in maize, the experimental cross must still meet the same three criteria established for the X-linked genes in *Drosophila:* (1) one parent must be heterozygous for all traits under consideration; (2) the gametic genotypes produced by the heterozygote must be apparent from observing the phenotypes of the offspring; and (3) a sufficient sample size must be available for complete analysis.

In maize, the recessive mutant genes *bm* (brown midrib), *v* (virescent seedling), and *pr* (purple aleurone) are linked on chromosome 5. Assume that a female plant is known to be heterozygous for all three traits. Nothing is known about: (1) the arrangement of the mutant alleles on the maternal and paternal homologues of this heterozygote, (2) the sequence of genes, or (3) the map distances between the genes. What genotype must the male plant have to allow successful mapping? In order to meet the second criterion, the male must be homozygous for all three recessive mutant alleles. Otherwise, offspring of this cross showing a given phenotype might represent more than one genotype, making accurate mapping impossible. Note that this is equivalent to performing a test cross.

Figure 5.10 diagrams this cross. As shown, we do not know either the arrangement of alleles or the sequence of loci in the heterozygous female. Some of the possibilities are shown, but we have yet to determine which is correct. In the heterozygous female and in the test cross male parent, *we do not know the sequence,* but must write it down initially as a random selection. Note that we have chosen to place *v* in the middle. This may or may not be correct.

The offspring have been arranged in groups of two for each pair of reciprocal phenotypic classes. The two members of each reciprocal class are derived from either no crossing over (**NCO**), one of two possible single-crossover events (**SCO**), or a double crossover (**DCO**).

To solve this problem, consider the following questions. It will be helpful to refer to Figures 5.10 and 5.11 as you consider them.

1. *What is the correct heterozygous arrangement of alleles in the female parent?*

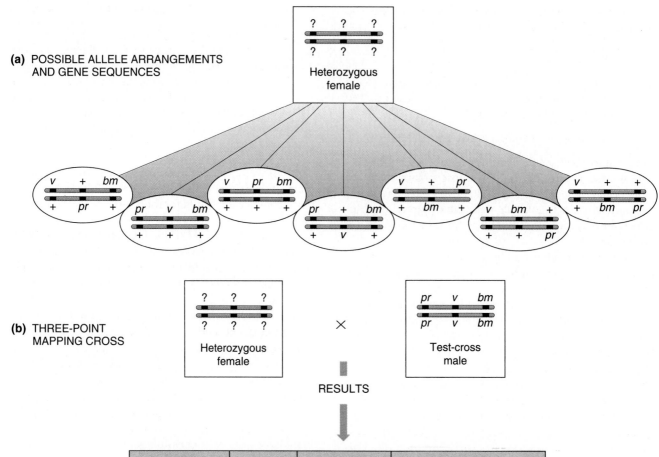

**FIGURE 5.10**    In Part (a) are shown some of the possible combinations of allele arrangements and gene sequences in a heterozygous female. The data from a three-point mapping cross, depicted in Part (b), where the female is test-crossed, provides the basis for determining which combination of arrangement and sequence is correct.

Identify the two noncrossover classes, those that occur in the highest frequency. In this case, they are $+\ v\ bm$ and $pr + +$. Therefore, the arrangement of alleles on the homologues of the female parent must be as shown in Figure 5.11(a). These homologues segregate into gametes, unaffected by any recombination event. Any other arrangement of alleles could not yield the observed noncrossover classes. (Remember that $+\ v\ bm$ is equivalent to $pr^+\ v\ bm,$ and that $pr + +$ is equivalent to $pr\ v^+\ bm^+.$)

2.  *What is the correct sequence of genes?*
    The approach described in Method I will first be used to answer this question. We know that the arrangement of alleles is

$$\dfrac{+ \quad v \quad bm}{pr \quad + \quad +}$$

But, is the assumed sequence correct? That is, will a double-crossover event yield the observed double-crossover phenotypes following fertilization? Simple observation shows that it will not [Figure 5.11(b)]. Try the other two orders [Figure 5.11(c) and (d)], keeping the same arrangement:

$$\dfrac{+ \quad bm \quad v}{pr + \quad +} \quad \text{and} \quad \dfrac{v \quad + \quad bm}{+ \quad pr \quad +}$$

Only the latter case will yield the observed double-crossover classes [Figure 5.11(d)]. Therefore, the *pr* gene is in the middle.

The same conclusion is reached if the problem is analyzed using Method II. In this case, no assumption of gene sequence is necessary. The arrangement of alleles in the heterozygous parent is

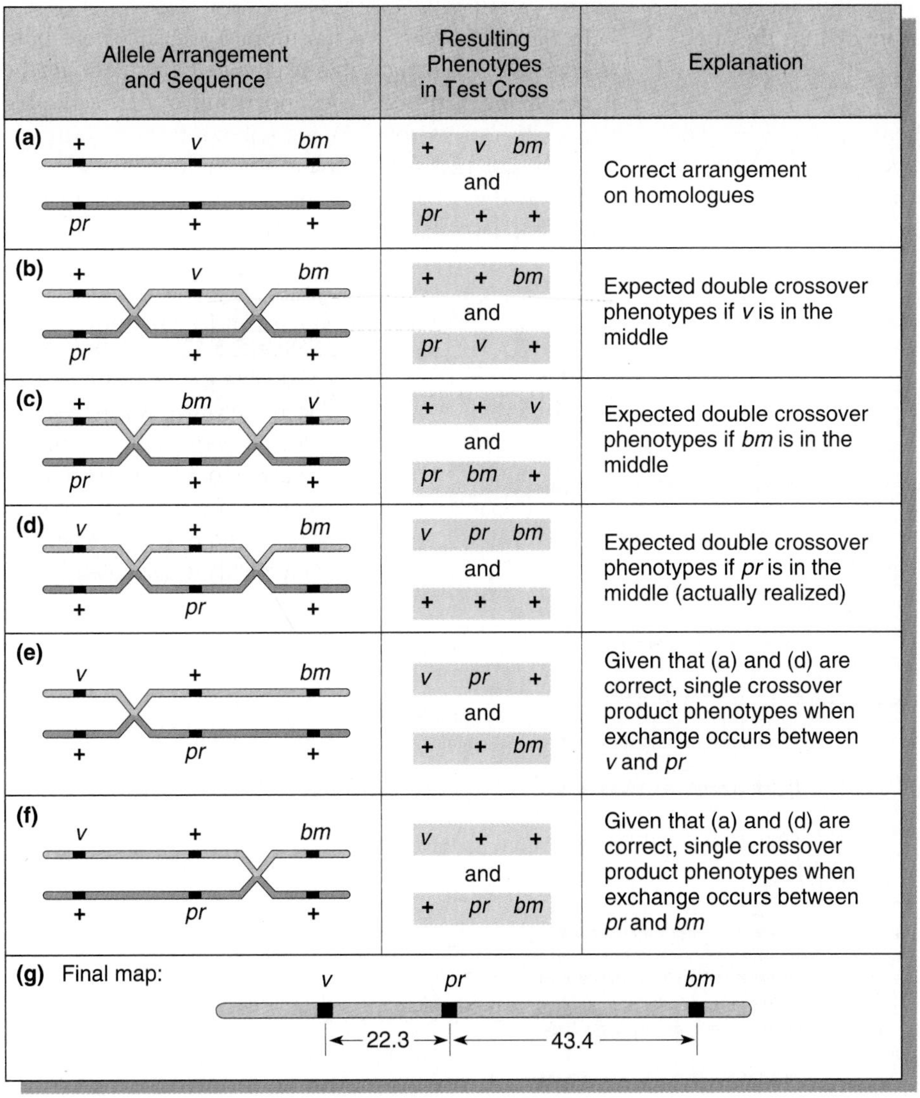

| Allele Arrangement and Sequence | Resulting Phenotypes in Test Cross | Explanation |
|---|---|---|
| **(a)** $+ \quad v \quad bm$ / $pr \quad + \quad +$ | $+ \quad v \quad bm$ and $pr \quad + \quad +$ | Correct arrangement on homologues |
| **(b)** $+ \quad v \quad bm$ / $pr \quad + \quad +$ | $+ \quad + \quad bm$ and $pr \quad v \quad +$ | Expected double crossover phenotypes if *v* is in the middle |
| **(c)** $+ \quad bm \quad v$ / $pr \quad + \quad +$ | $+ \quad + \quad v$ and $pr \quad bm \quad +$ | Expected double crossover phenotypes if *bm* is in the middle |
| **(d)** $v \quad + \quad bm$ / $+ \quad pr \quad +$ | $v \quad pr \quad bm$ and $+ \quad + \quad +$ | Expected double crossover phenotypes if *pr* is in the middle (actually realized) |
| **(e)** $v \quad + \quad bm$ / $+ \quad pr \quad +$ | $v \quad pr \quad +$ and $+ \quad + \quad bm$ | Given that (a) and (d) are correct, single crossover product phenotypes when exchange occurs between *v* and *pr* |
| **(f)** $v \quad + \quad bm$ / $+ \quad pr \quad +$ | $v \quad + \quad +$ and $+ \quad pr \quad bm$ | Given that (a) and (d) are correct, single crossover product phenotypes when exchange occurs between *pr* and *bm* |
| **(g)** Final map: $v \quad\quad pr \quad\quad bm$ $\vert\leftarrow$22.3$\rightarrow\vert\leftarrow$ 43.4 $\rightarrow\vert$ | | |

**FIGURE 5.11**    The steps used in producing a map of the three genes involved in the cross shown in Figure 5.10, where neither the arrangement of alleles nor the sequence of genes is initially known in the heterozygous female parent.

$$\begin{array}{ccc} + & v & bm \\ \hline\hline pr & + & + \end{array}$$

The double-crossover gametes are also known:

$$\underline{pr \quad v \quad bm} \quad \text{and} \quad \underline{+ \quad + \quad +}$$

We can see that the *pr* allele has shifted so as to be associated with *v* and *bm* following a double crossover. The latter two alleles were present together on one homologue, and they stayed together. Therefore, *pr* is the odd gene, so to speak, and it is in the middle.

3. *What is the distance between each pair of genes?* Having established the sequence of loci as *v–pr–bm,* we can now determine the distance between *v* and *pr* and between *pr* and *bm.* Remember that the map distance between two genes is calculated on the basis of all detectable recombinational events occurring between them. This includes both the single- and double-crossover events involving the two genes being considered.

Figure 5.11(e) shows that the phenotypes *v pr +* and *+ + bm* result from single crossovers between *v* and *pr,* accounting for 14.5 percent of the offspring. By adding the percentage of double crossovers (7.8%) to the number obtained for single crossovers, the total distance between *v* and *pr* is calculated to be 22.3 map units.

Figure 5.11(f) shows that the phenotypes *v + +* and *+ pr bm* result from single crossovers between *pr* and *bm,* totaling 35.6 percent. With the addition of the double-crossover classes (7.8%) the distance between *pr* and *bm* is calculated to be 43.4 map units.

The final map for all three genes in this example is shown in Figure 5.11(g).

## Multiple-Strand Exchanges and the Accuracy of Mapping

Thus far, for teaching purposes, we have portrayed double crossovers in the simplest way possible: *two exchanges involving two nonsister chromatids.* However, double and even triple exchanges also occur that involve more than two of the strands. These must be factored into our consideration of the theoretical limit of recombination frequency.

In Figure 5.12(a), we first depict a case where both events involved in a double exchange occur *between* the genes being mapped. The end result is that a double exchange occurs, but no recombinant chromatids are produced. Thus, recombination goes undetected.

Of greater complexity are exchanges that involve more than two strands. As illustrated in Figure 5.12(b), a

**(a)** Two-strand double exchange

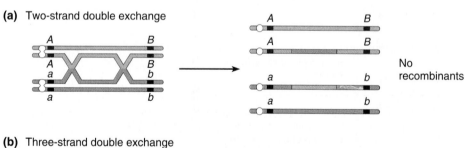

No recombinants

**(b)** Three-strand double exchange

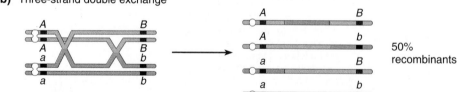

50% recombinants

**(c)** Four-strand double exchange

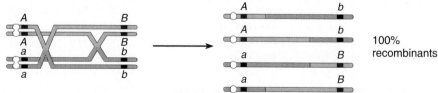

100% recombinants

**FIGURE 5.12** Three types of double exchanges that may occur between two genes. Two of them [parts (b) and (c)] involve more than two chromatids. In each case, the outcome regarding recombinant chromatids is illustrated.

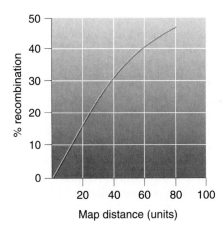

**FIGURE 5.13**    The average relationship between the actual frequency of detectable recombination and map distance, as studied in *Drosophila, Neurospora,* and *Zea mays* (corn).

three-strand double exchange yields two recombinant and two parental (nonrecombinant) gametes. A four-strand double exchange [Figure 5.12(c)] yields all recombinant gametes. These various types of exchange tend to "even out," and in aggregate, when double exchanges occur between two genes, a maximum of 50 percent of all possible chromatids will demonstrate recombination.

The next logical consideration is to ask whether the theoretical maximum is ever realized. As shown in Figure 5.13 and discussed in the next section, the phenomenon of **interference** precludes the realization of the maximum. However, when two loci are close together, the accuracy of genetic maps may be quite high.

Accurate linkage distances between widely separated genes must therefore be determined by many different experiments. These experiments should utilize numerous genes in between the two in question. For example, if genes *A* and *K* are widely separated, mapping experiments using genes *B, C, D, E, F, G, H, I,* and *J* may be necessary, assuming these to be between *A* and *K*. If each distance is determined (e.g., *A–B, B–C, C–D . . . J–K*), then the distance between *A* and *K* is additive. That is, by adding the *A–B, B–C, C–D . . . J–K* distances together, we obtain a more accurate measurement of the distance between *A* and *K*.

## Interference and the Coefficient of Coincidence

Still another factor tends to limit the accuracy of mapping data. This factor involves the actual reduction of the number of expected double crossovers when genes are reasonably close to one another along the chromosome.

This reduction, called **interference**, will be illustrated for three-point mapping.

We have already considered the probability relationships between single- and double-crossover events. In theory, the percentage of **expected double crossovers** is predicted by multiplying the percentage of the total crossovers between each pair of genes. Remember that a double crossover (DCO) represents two single-crossover events. For example, the expected double-crossover frequency of the cross illustrated in Figures 5.10 and 5.11 may be calculated in the following manner:

$$DCO_{exp} = (0.223) \times (0.434) = 0.097 = 9.7\%$$

Frequently, this predicted figure does not correspond precisely with the observed DCO frequency. Generally, there are fewer DCOs observed than predicted. In the maize cross, only 7.8 percent DCOs were observed. In some cases there are more DCOs than expected. These disparities are explained by the concept of interference, which is quantified by calculating the **coefficient of coincidence (C)**:

$$C = \frac{\text{Observed DCO}}{\text{Expected DCO}}$$

In the maize cross, we have

$$C = \frac{0.078}{0.097} = 0.804$$

Once *C* is calculated, interference (*I*) may be quantified using the simple equation

$$I = 1 - C$$

In the maize cross, we have

$$I = 1.000 - 0.804 = 0.196$$

If interference is complete and no double crossovers occur, *I* = 1.0. If fewer DCOs than expected occur, *I* is a positive number (as above) and **positive interference** has occurred. If more DCOs than expected occur, *I* is a negative number and **negative interference** has occurred.

In eukaryotic systems, positive interference is most often observed. It appears that a crossover event in one region of a chromosome inhibits a second crossover in neighboring regions of the chromosome. In general, the closer genes are to one another along the chromosome, the more positive interference is observed and the lower the *C* value. In *Drosophila*, when three genes are clustered within 10 map units, interference is often complete and no double-crossover classes are recovered. This observation suggests that interference may be explained by physical constraints that prohibit the formation of closely aligned chiasmata. Perhaps a mechanical stress is imposed on chromatids during crossing over such that one chiasma inhibits the formation of a second chiasma in the neighboring region. This interpretation is consistent with the finding that the impact of interference decreases as the genes in question are located farther apart. In the maize cross illustrated in Figures 5.10 and 5.11, the three genes are relatively far apart, and 80 percent of the expected double crossovers are observed.

### The Genetic Map of *Drosophila*

In organisms such as *Drosophila,* maize, and the

mouse, where large numbers of mutants have been discovered and experimental crosses are easy to perform, extensive maps of each chromosome have been made. Illustrated in Figure 5.14 are partial maps of the four chromosomes of *Drosophila*. Virtually every morphological feature of the fruit fly has been observed to be subject to mutation. The gene locus involved in determining an altered phenotype is first localized to one of the four chromosomes, or linkage groups, and then mapped in relation to other linked genes of that group. As can be seen, the genetic map of the X chromosome is somewhat less extensive than that of autosome 2 or 3. In compari-

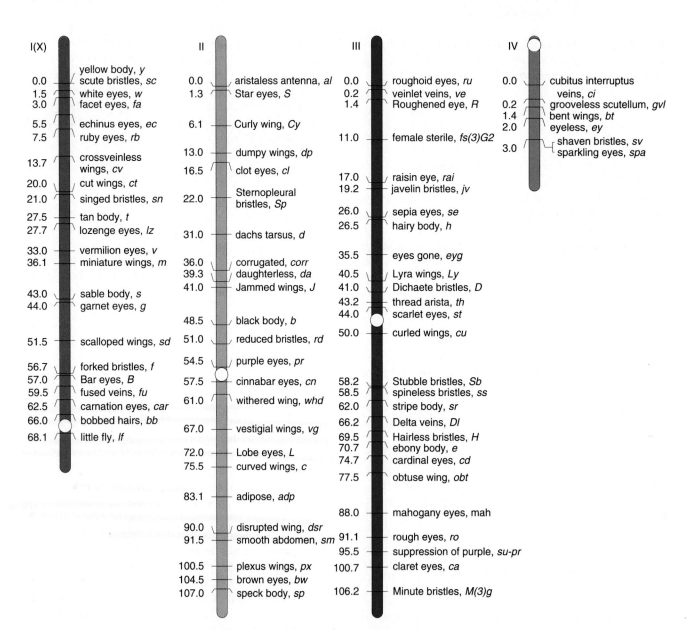

**FIGURE 5.14**    A partial genetic map of the four chromosomes of *Drosophila melanogaster*. The circle on each chromosome represents the position of the centromere.

son to these three, autosome 4 is minuscule. Based on cytological evidence, the relative lengths of the genetic maps have been found to correlate with the relative physical lengths of these chromosomes.

# OTHER ASPECTS OF GENETIC EXCHANGE

We have established that careful analysis of crossing over during gamete formation can serve as the basis for the construction of chromosome maps in both diploid and haploid organisms. We should not, however, lose sight of the real biological significance of the process, which is to generate genetic variation in gametes, and subsequently, in the offspring derived from the resultant eggs and sperm. Because of the critical role of crossing over in generating variation, the study of genetic exchange has remained an important topic for study in genetics. Many questions need to be addressed. For example, does crossing over occur in the two- or four-strand stage of meiosis? Does crossing over involve an

actual exchange of chromosome arms? Does exchange occur between paired sister chromatids during mitosis? We will briefly consider observations that attempt to answer these questions.

## Crossing Over in the Four-Strand Stage

The question of when crossing over occurs during meiosis is critical to understanding the process and consequence of genetic exchange in eukaryotes. There are two alternative times at which crossing over might occur. First, exchange could take place before the chromosomes have duplicated in the **two-strand stage**. Alternatively, exchange could occur at the **four-strand stage** after the chromosomes have duplicated.

If crossing over occurs at the two-strand stage, all four products of a single meiotic event will be recombinant gametes because each pair of sister chromatids in the tetrad is derived from one of the members of the two-strand stage. If, on the other hand, crossing over occurs between two nonsister chromatids in the four-strand stage, two parental (noncrossover) chromatids and two recombinant chromatids will be formed. These alternative possibilities are compared in Figure 5.15.

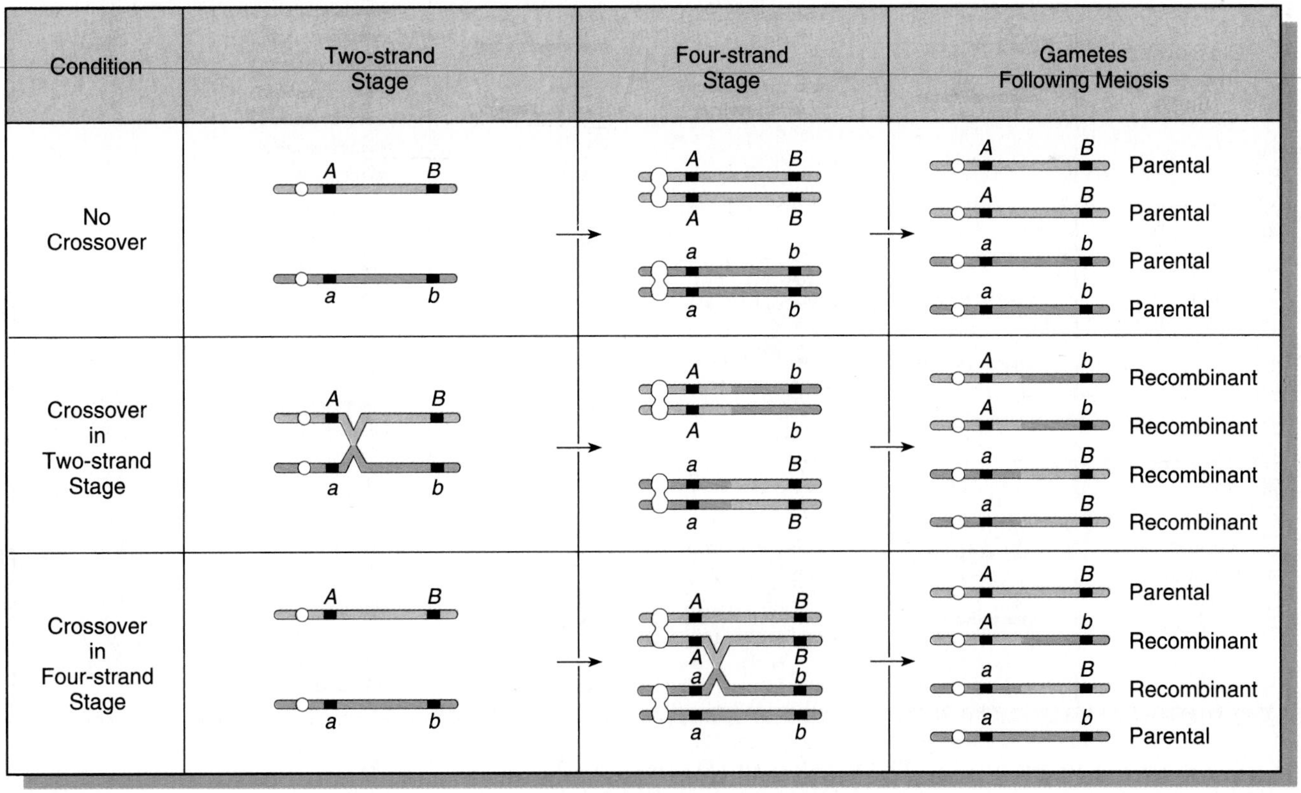

| Condition | Two-strand Stage | Four-strand Stage | Gametes Following Meiosis | |
|---|---|---|---|---|
| No Crossover | A ... B / a ... b | A B / A B / a b / a b | A B | Parental |
| | | | A B | Parental |
| | | | a b | Parental |
| | | | a b | Parental |
| Crossover in Two-strand Stage | A × B / a × b | A b / A b / a B / a B | A b | Recombinant |
| | | | A b | Recombinant |
| | | | a B | Recombinant |
| | | | a B | Recombinant |
| Crossover in Four-strand Stage | A B / a b | A B / A B / a b / a b | A B | Parental |
| | | | A b | Recombinant |
| | | | a B | Recombinant |
| | | | a b | Parental |

**FIGURE 5.15**    Comparison of the genotypes of gametes formed as a result of crossing over in the two-strand and four-strand stages with those formed when no crossing over occurs.

To decide between these two alternatives, we need to examine the results of an experiment with an organism from which all four products of single meiotic events may be recovered and observed. The organism used in this experiment is the ascomycete *Neurospora,* a haploid mold. Fertilization occurs, and meiosis results in the formation of four haploid products, all retained in a sac called the **ascus** (pl., asci). Then, a mitotic division of each product occurs, producing eight haploid cells called **ascospores**. Most important, the entire process retains the haploid ascospore products in the order in which they are formed.

Carl L. Lindegren's observations of meiotic segregation strongly support the theory that crossing over occurs in the four-strand stage. He examined the various possibilities of ascospore formation resulting from a cross between an *albino* mutant strain (*a*) and one with normal pigmentation (+). His work focused on the results of a crossover that occurred in the region between the mutant *albino* locus and the centromere. Figure 5.16 illus-

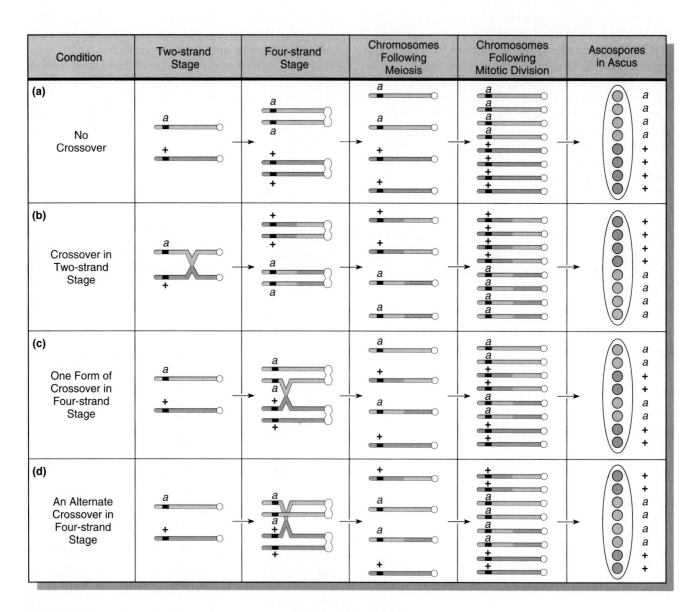

**FIGURE 5.16**    Four ways in which ascospore patterns can be generated in the asci of *Neurospora* as a result of genetic events. While the patterns produced in (a) and (b) cannot be distinguished from one another, these and the patterns produced in (c) and (d) were observed, leading to the conclusion that crossing over occurs in the four-strand stage.

trates the theoretical results for various alternatives of an exchange in this region for both the two- and four-strand stages. In case (a) no exchange occurs. In case (b) the results of crossing over in the two-strand stage are predicted. With or without a crossover event, the resulting ascus will always contain four pigmented ascospores and four unpigmented ascospores, in that sequence. The asci produced in cases (a) and (b) cannot be distinguished from each other.

If, on the other hand, a crossover occurs in this region during the four-strand stage, an alternative arrangement of ascospores in the ascus is predicted, as seen in case (c). Case (d) shows still another arrangement that occurs as a result of a slightly different exchange during the four-strand stage.

Lindegren observed asci with arrangements shown in cases (a), (b), (c), and (d). His findings are consistent with the conclusion that crossing over indeed occurs in the four-strand stage and exclude crossing over in the two-strand stage, which cannot generate arrangements (c) and (d).

Similar findings have been drawn from studies of other organisms, including *Drosophila*. To date, no experiments have been reported that seriously dispute the conclusion that crossing over occurs in the four-strand stage in eukaryotic organisms.

## Cytological Evidence for Crossing Over

Visual proof that genetic crossing over in higher or-

ganisms is accompanied by an actual physical exchange between homologous chromosomes has been demonstrated independently by Curt Stern in *Drosophila* and by Harriet Creighton and Barbara McClintock in *Zea mays* (corn). Since the experiments are similar, we will consider only the work with corn. In Creighton and McClintock's work, two linked genes on chromosome 9 were studied. At one locus, the alleles *colorless* (*c*) and *colored* (*C*) control endosperm coloration. At the other locus, the alleles *starchy* (*Wx*) and *waxy* (*wx*) control the carbohydrate characteristics of the endosperm.

A corn plant was obtained that was heterozygous at both loci and which contained two unique cytological markers on one of the homologues. The markers consisted of a densely stained knob at one end of the chromosome and a translocated piece of another chromosome (8) at the other end. The arrangement of alleles and cytological markers in this plant are shown in Figure 5.17. Creighton and McClintock crossed this plant to one homozygous for the color alleles and heterozygous for the endosperm alleles. They obtained several different phenotypes in the offspring, one of which could arise only as a result of crossing over. The chromosomes of this plant, with the colorless, waxy phenotype (Case I in Figure 5.17) were examined for the presence of the cytological markers. As expected, if genetic crossing over is accompanied by a physical exchange between homologues, the translocated chromosome is still present, but the knob is not. In a second plant (Case II), the phenotype colored, starchy is potentially the result of crossing

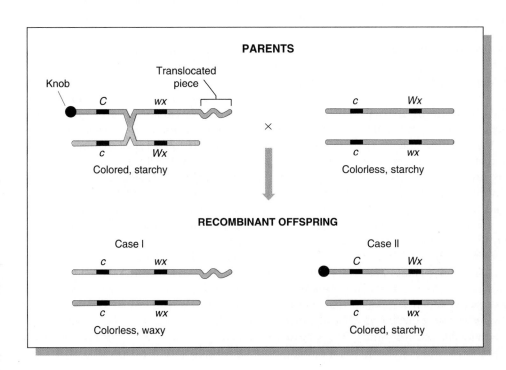

**FIGURE 5.17** The phenotypes and chromosome compositions observed in Creighton and McClintock's demonstration in maize that crossing over involves a breakage and rejoining process.

over. If so, chromosomes from it should contain the dense knob but not the translocated chromosome. This was the case, and again the findings supported the conclusion that a physical exchange took place. Similar findings by Stern using *Drosophila* leave no doubt about this conclusion involving the cytological basis of crossing over.

## The Mechanism of Crossing Over

It has long been of interest to determine how and when during meiosis crossing over occurs. If we accept the evidence just discussed, then crossing over must be considered to be the result of an actual physical exchange between DNA molecules of two homologous chromosomes. Of particular interest is the relationship between the chiasmata observed cytologically during prophase I of meiosis and the breakage and reunions presumed to occur during genetic crossing over. Are chiasmata the cytological manifestations of crossover events?

Two main theories have been proposed, both of which involve chiasmata, but in quite different ways. The **classical theory** holds that crossing over events are the result of physical strains imposed by chiasmata, the presence and location of which occur randomly. While there is not necessarily a one-to-one relationship—a chiasma may or may not induce the breakage and rejoining that leads to crossing over—*this theory holds that chiasmata are responsible for crossover events and clearly precede them*. Since we know that chiasmata are first seen rather late in meiotic prophase I, during the diplotene stage, the classical theory predicts that crossing over occurs sometime after diplonema but before the chromosomes separate at meiotic anaphase I. The order of events presumed to occur and the resultant pairing relationships created according to the classical theory are illustrated in Figure 5.18(a).

The other proposal, called the **chiasmatype theory**, was first set forth by F. A. Janssens in the early twentieth century and later modified by John Belling and C. D. Darlington around 1930. It predicts that crossing over precedes chiasma formation and occurs early in meiotic prophase I, presumably in pachytene. *As a result, chiasmata are formed at points of genetic exchange*. Thus, a chiasma present in diplotene is a cytological manifestation of one crossing over event. Chromosomes become wrapped around one another as a result of the previous exchange between homologues. These events are illustrated in Figure 5.18(b).

The chiasmatype theory has gained much support over the years, even though it does not account for all related observations. Several findings derived from the use of modern research techniques are pertinent to this

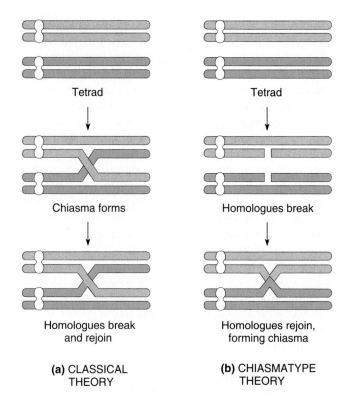

**FIGURE 5.18**   Two main theories proposed to account for the mechanism of crossing over.

subject. The first involves the discovery of DNA synthesis during meiosis. Yasuo Hotta and Herbert Stern have performed careful analysis using *Lilium* (lily) anthers, where the stages of meiotic prophase can be easily identified and studied.

They have shown that a small but measurable amount of DNA synthesis occurs during the zygotene stage of meiotic prophase I. Recall from our earlier discussion of mitosis and meiosis in Chapter 2 that DNA replication occurs during the S phase of the cell cycle, well before mitosis and meiosis begin. Thus, this finding has generated much interest. What is the role of this late-replicating DNA? It amounts to 0.3 percent of the total nuclear DNA and is distributed generally among all chromosomes. Furthermore, if this DNA synthesis is inhibited, then chromosomal synapsis is also inhibited and meiosis arrests. These latter findings suggest that the newly synthesized DNA plays an important role in chromosome alignment during meiosis.

Coincidental with this DNA synthesis is the formation of the **synaptonemal complex (SC)** during the zygotene stage. The SC is found between synapsed homologues (see Figure 2.15) and is essential to chromosome alignment. Perhaps the DNA synthesis occurring during zygonema is directly related to SC formation between homologues. Interestingly, there is a possibility that the

SC is also involved in crossing over. Several observations are particularly significant here. In male *Drosophila,* no crossing over occurs, and no SCs are found when spermatocytes are viewed under the electron microscope. In the silk moth, *Bombyx,* no crossing over occurs in females, and no synaptonemal complexes are observed in the oocytes. Because of the constant association between the presence of this structure and crossing over, the participation of the SC in this event must be considered seriously.

Many geneticists feel that these events, characteristic of the zygotene stage, are essential to genetic crossing over that occurs in the subsequent pachytene stage. Just how crossing over occurs, however, is still a matter of debate and speculation. Some interesting biochemical information derived from Stern's studies of the lily bears on this issue.

It is generally held that the physical exchange occurring during crossing over involves breakage and rejoining of the DNA strands between homologues. Stern and coworkers have demonstrated the presence of an enzyme system during zygonema and pachynema capable of breaking and resealing DNA strands. For example, certain **endonucleases** are capable of "nicking" one of the two strands of the DNA double helix. Other enzymes called **ligases** are capable of rejoining broken ends of DNA strands. Additionally, a small but discernible amount of DNA synthesis occurs during the pachytene stage in *Lilium* and other organisms, such as wheat. This synthesis, unlike that occurring during zygonema, is of the repair type. It does not result in a net increase in DNA and thus does not represent replication of DNA. Inhibitors of premeiotic S-phase and zygonema DNA synthesis are not effective in preventing the DNA synthesis that occurs during pachynema.

With good reason, these discoveries have generated much enthusiasm and the hope that we will soon come to understand the precise molecular mechanisms underlying crossing over. Several popular models have been put forth. Because we have yet to discuss the specific chemistry of DNA structure, replication, and synthesis, we will postpone their presentation. Those already versed in DNA chemistry may wish to jump ahead to Chapter 11, where a molecular model of recombination is introduced. This model provides a suitable explanation for how crossing over might occur.

## Mitotic Recombination

In 1936, Curt Stern demonstrated that exchanges similar to crossing over occur during mitosis in *Drosophila*. This finding was considered unusual because homologues do not normally pair up during mitosis in most organisms. However, such synapsis appears to be the rule in *Drosophila*. Since Stern's discovery, genetic exchange during mitosis has been shown to be a general event in certain fungi as well.

Stern observed small patches of mutant tissue in females heterozygous for the sex-linked recessive mutations *yellow* body and *singed* bristles. Under normal circumstances, a heterozygous female is completely wild type (gray-bodied with straight, long bristles). He explained these observations by postulating that, during mitosis in certain cells during development, homologous exchanges could occur between the loci for *yellow* and *singed* or between *singed* and the *centromere,* or that a double exchange could occur. These three possibilities are diagrammed in Figure 5.19. After these three types of exchanges occur, tissues derived from the progeny cells are produced with *yellow* patches, adjacent *yellow* and *singed* patches (**twin spots**), and *singed* patches, respectively. The last type of tissue, which represents the double exchange, was found in the lowest frequency, similar to the less frequent double-crossover frequency seen in meiosis. The frequency of twin spots argues strongly against the appearance of *yellow* or *singed* tissue being due to two spontaneous but independent mutational events. If the twin spots arose in such a way, their frequency would occur in only one in many million flies.

Table 5.1 compares the relative frequency of occurrence for each of the three spotting types. These frequencies are correlated with the known genetic distances between the *yellow* and *singed* loci and the centromere, which effectively serves as an additional genetic marker. The data parallel the predicted frequencies if the exchanges occur according to the rules we established earlier for meiotic crossing over. This is additional evidence that the occurrence of mutant tissue spots is due to mitotic exchange in somatic cells.

In 1958 George Pontecorvo and others described a similar phenomenon in the fungus *Aspergillus*. Although the vegetative stage is normally haploid, some cells and their nuclei fuse, producing diploid cells that divide mitotically. Occasionally, crossing over occurs between linked genes during mitosis in this diploid stage so that resulting cells are recombinant. Pontecorvo referred to these events that produce genetic variability as the **parasexual cycle**. On the basis of such exchanges, genes can be mapped by estimating the frequency of recombinant classes.

As a rule, if mitotic recombination occurs at all in an organism, it does so at a much lower frequency than meiotic crossing over. While it is assumed that there is always at least one exchange per meiotic tetrad, mitotic exchange occurs in 1 percent or less of mitotic divisions in organisms that demonstrate it. Some researchers feel that the low frequency of exchange may be explained by

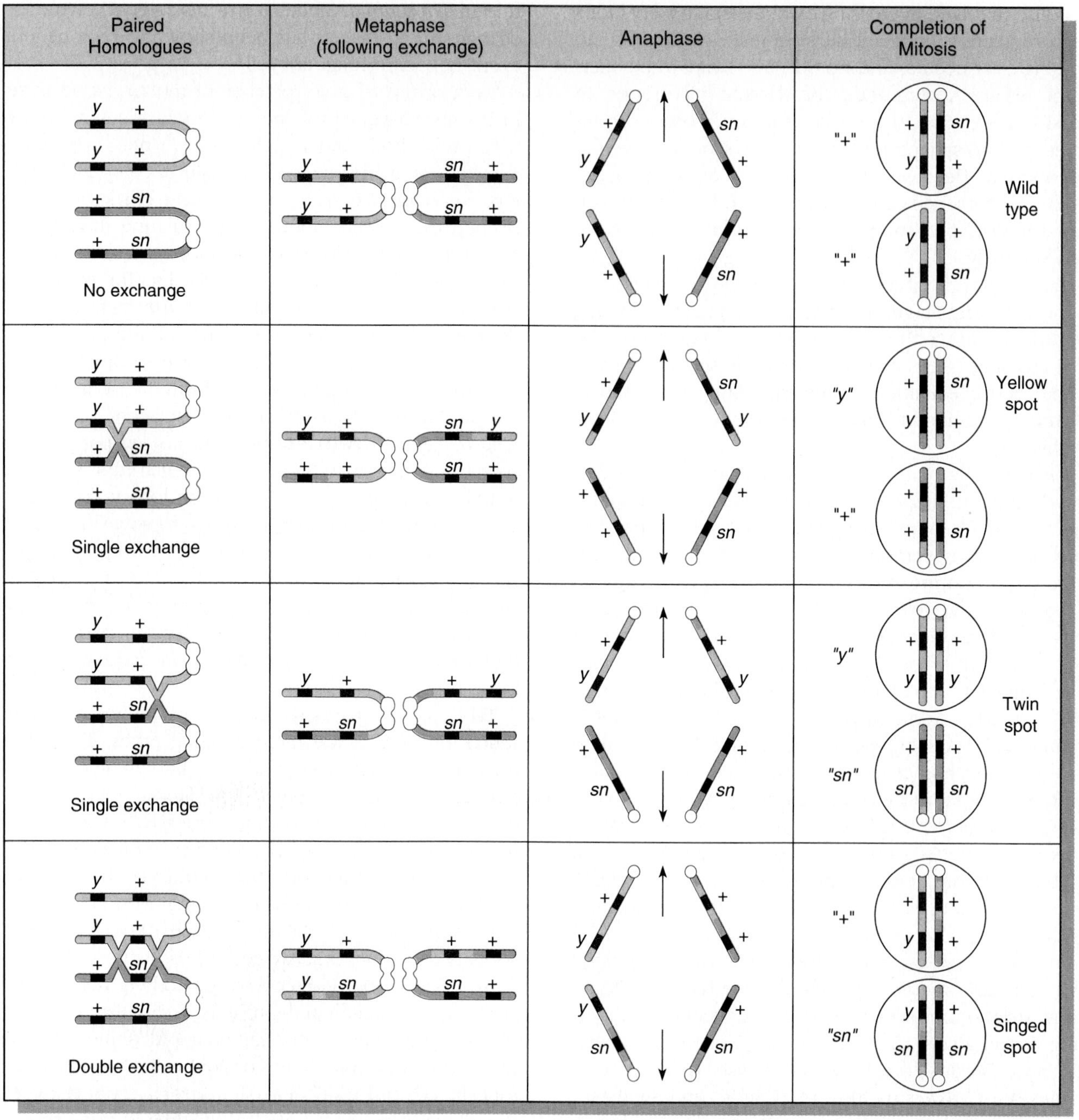

**FIGURE 5.19**   The production of mutant tissue in a female heterozygous for the recessive *yellow* and *singed* alleles as a result of mitotic recombination in *Drosophila.*

a pairing of only some portions of homologous chromosomes.

## Sister Chromatid Exchanges

Since homologous chromosomes do not usually pair up or synapse in somatic cells (*Drosophila* is an exception), each individual chromosome in prophase and metaphase of mitosis consists of two identical sister chromatids, joined at a common centromere. Surprisingly, several experimental approaches have demonstrated that reciprocal exchanges similar to crossing over occur between sister chromatids. While these **sister chromatid exchanges (SCEs)** do not produce new allelic combinations, evidence is accumulating that attaches significance to these events.

**Table 5.1**  COMPARISON OF THE VARIOUS PARAMETERS RELATED TO MITOTIC RECOMBINATION STUDIES IN *DROSOPHILA*

| Mutant Tissue Type | Region of Exchange | Relative Frequency of Exchange | Type of Exchange | Relative Distance |
|---|---|---|---|---|
| Yellow | *y-sn* | High | Single | 21 map units |
| Yellow-singed (twin spot) | *sn*-centromere | Highest | Single | 49 map units |
| Singed | *y-sn* and *sn*-centromere | Lowest | Double | ____ |

Identification and study of sister chromatid exchanges are facilitated by several modern staining techniques. In one approach, if cells are allowed to replicate for several generations in the presence of the thymidine analogue bromodeoxyuridine (BUdR), and the cells are then shifted to medium lacking the analogue, chromatids with or without BUdR are then distinguishable. Chromatids with both strands containing BUdR stain less brightly than chromatids with only one strand of the double helix containing the analogue. In Figure 5.20 numerous instances of SCE events may be detected.

While the significance of sister chromatid exchanges is still uncertain, several observations have led to great interest in this phenomenon. It is known, for example, that agents that induce chromosome damage (such as viruses, X rays, ultraviolet light, and certain chemical mutagens) also increase the frequency of sister chromatid exchanges. This frequency of SCEs is also elevated in

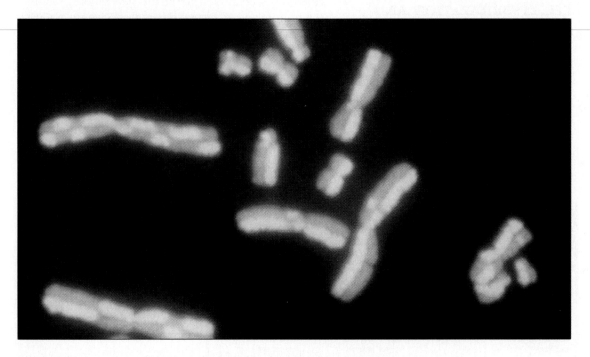

**FIGURE 5.20**    Demonstration of sister chromatid exchanges (SCEs) in mitotic chromosomes. Sometimes called **harlequin chromosomes** because of their patchlike appearance, chromatids containing the thymidine analogue BUdR in both DNA strands fluoresce *less* brightly than those with the analogue in only one strand. These chromosomes were stained with 33258-Hoechst reagent and viewed under fluorescence microscopy.

**Bloom syndrome**, an autosomal recessive disorder in humans. This rare genetic disease is characterized by retardation of growth, a great sensitivity of the facial skin to the sun, abnormal immunological function, and a predisposition to cancer. The chromosomes from cultured leucocytes, bone marrow cells, and fibroblasts derived from homozygotes (*bl/bl*) are very fragile and unstable when compared to those derived from unaffected (+/+) and heterozygous (*bl*/+) individuals. Increased breaks and rearrangements between nonhomologous chromosomes are observed in addition to excessive amounts of sister chromatid exchanges.

The mechanisms of exchange between nonhomologous chromosomes and between sister chromatids may prove to be similar because the frequency of both events increases substantially in individuals with genetic disorders. These findings suggest that further study of sister chromatid exchange may contribute to an increased understanding of recombination mechanisms and the relative stability of normal and genetically abnormal chromosomes. We shall encounter still another demonstration of SCEs in Chapter 9 when we consider replication of DNA (see Figure 9.5).

## SOMATIC CELL HYBRIDIZATION AND THE HUMAN CHROMOSOME MAPS

In humans, where neither designed matings nor large numbers of offspring are available, the earliest linkage studies were based on pedigree analysis. Attempts were made to establish whether a trait was sex-linked or autosomal. All of the traits shown to be sex-linked are linked to the X chromosome. For autosomal traits, geneticists tried to distinguish clearly whether pairs of traits demonstrated linkage or independent assortment. In this way, it was hoped that human chromosome maps could be created.

The results were discouraging because of the limitations of this approach and because of the relatively high haploid number of human chromosomes (23). By 1960, relatively few cases of either X-linkage or autosomal linkage were established, and almost no mapping information became available.

However, in the 1960s, a new technique, **somatic cell hybridization**, was developed that aided immensely in assigning human genes to their respective chromosomes. This technique, first discovered by Georges Barsky, relies on the fact that two cells in culture can be induced to fuse into a single hybrid cell. While Barsky used two mouse cell lines, it soon became evident that cells from different organisms will also fuse together. When this event occurs, an initial cell type called a **heterokaryon** is produced. The hybrid cell contains two nuclei in a common cytoplasm. Using the proper techniques, it is possible to fuse human and mouse cells, for example, and isolate the hybrids from the parental cells.

As the heterokaryons are cultured *in vitro,* two interesting changes occur. Eventually, the nuclei fuse together, creating what is termed a **synkaryon**. Then, as culturing is continued for many generations, chromosomes from one of the two parental species are gradually lost. In the case of the human–mouse hybrid, human chromosomes are lost randomly until eventually, the synkaryon has a full complement of mouse chromosomes and only a few human chromosomes. It is the result of this event that allows the assignment of human genes to the chromosomes upon which they reside.

The experimental rationale is straightforward. If a specific human gene product is synthesized in a synkaryon containing one to three human chromosomes, then the

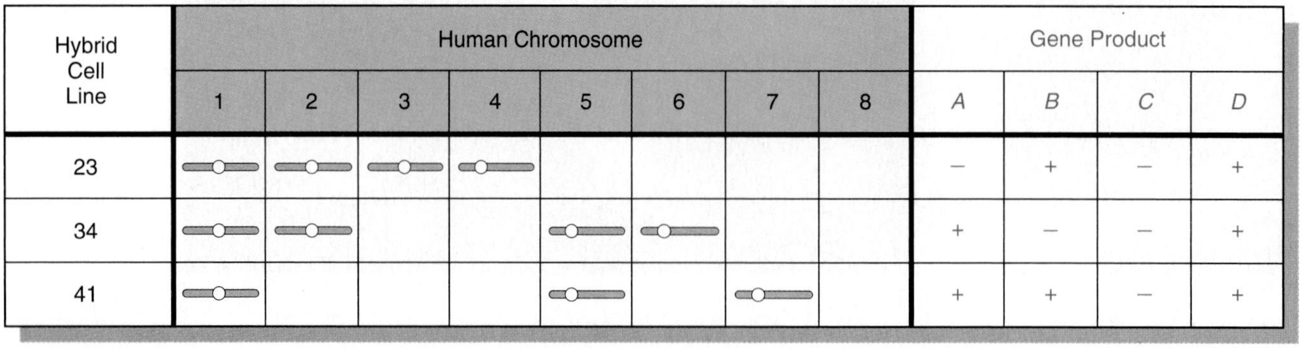

| Hybrid Cell Line | Human Chromosome | | | | | | | | Gene Product | | | |
|---|---|---|---|---|---|---|---|---|---|---|---|---|
| | 1 | 2 | 3 | 4 | 5 | 6 | 7 | 8 | *A* | *B* | *C* | *D* |
| 23 | ⊙ | ⊙ | ⊙ | ⊙ | | | | | − | + | − | + |
| 34 | ⊙ | ⊙ | | | ⊙ | ⊙ | | | + | − | − | + |
| 41 | ⊙ | | | | ⊙ | | ⊙ | | + | + | − | + |

**FIGURE 5.21**    A hypothetical grid of data used in synteny testing to assign genes to their appropriate human chromosomes. Three somatic hybrid cell lines, designated 23, 34, and 41, have each been scored for the presence or absence of human chromosomes 1 through 8, as well as for their ability to produce the hypothetical human gene products A through D.

gene responsible for that product must reside on one of the one to three human chromosomes remaining in the hybrid cell. Or, if the human gene product is absent, the responsible gene cannot be present on any of the remaining human chromosomes. Ideally, a panel of 23 hybrid cell lines, each with but one unique human chromosome, would allow the immediate assignment of any human gene for which the product could be characterized.

In practice, a panel of cell lines, each with several remaining human chromosomes is most often utilized. The correlation of the presence or absence of each chromosome with the presence or absence of each gene product is called **synteny testing**. Consider, for example, the hypothetical data provided in Figure 5.21, where four gene products (*A, B, C,* and *D*) are tested in relationship to eight human chromosomes. Let us carefully analyze the gene which produces product *A*:

1. Product *A* is not produced by cell line 23, but chromosomes 1, 2, 3, and 4 are present in cell line 23. Therefore, we can rule out the presence of gene *A* on those four chromosomes and conclude that it must be on chromosome 5, 6, 7, or 8.

2. Product *A* is produced by cell line 34, which contains chromosomes 5 and 6, but not 7 and 8. Therefore, gene *A* is on chromosome 5 or 6.

3. Product *A* is also produced by cell line 41 which contains chromosome 5 but not chromosome 6. Therefore, gene *A* is on chromosome 5, according to this analysis.

Using a similar approach, gene *B* can be assigned to chromosome 3. You should perform this analysis to demonstrate for yourself that this is correct. Gene *C* presents a unique situation. The data indicate that it is not present on any of the first seven chromosomes (1–7). While it might be on chromosome 8, no direct evidence supports this conclusion. Other panels are needed. We shall leave gene *D* for you to analyze. Upon what chromosome does it reside?

Using the approach described above, literally hundreds of human genes have been assigned to one chromosome or another. Some of the assignments shown in Figure 5.22 were derived in this way. For mapping still other genes, where the products have yet to be discovered, researchers have had to rely on other linkage techniques. For example, by combining a rather sophisticated approach using recombinant DNA technology with pedigree analysis, it has been possible to assign the genes responsible for **Huntington disease, cystic fibrosis**, and **neurofibromatosis** to their respective chromosomes, 4, 7, and 17. This approach will be discussed in Chapter 13.

We conclude this discussion by addressing how human genes can be assigned to different regions of a given chromosome. Sometimes in hybrid cell lines, fragments of a particular chromosome become transferred to another chromosome, resulting in a **translocation**. It is possible using chromosome banding techniques to identify the exact origin of the translocation and correlate its presence in hybrid cells with specific gene expression. In this way, gene maps of human chromosomes may be compiled. The partial maps of the X chromosome and chromosome 1 shown in Figure 5.22 illustrate this point. While these maps are not as specific as the genetic map of *Drosophila,* we are beginning to learn a great deal about the chromosome locations of a multitude of human genes.

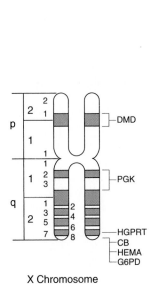

X Chromosome

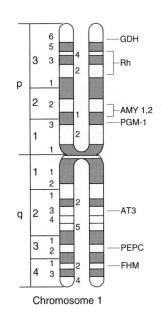

Chromosome 1

**Key:**

| | |
|---|---|
| AMY | Amylase (Salivary and Pancreatic) |
| AT-3 | Antithrombin (Clotting Factor IV) |
| CB | Colorblindness |
| DMD | Duchenne Muscular Dystrophy |
| FMH | Fumarate Hydratase (Mitochondrial) |
| GDH | Glucose Dehydrogenase |
| G6PD | Glucose-6-Phosphate Dehydrogenase |
| HEMA | Hemophilia A (Classic) |
| HGPRT | Hypoxanthine-Guanine-Phosphoribosyl Transferase (Lesch-Nyhan Syndrome) |
| PEPC | Peptidase C |
| PGK | Phosphoglycerate Kinase |
| PGM | Phosphoglucomutase |
| Rh | Rhesus Blood Group (Erythroblastosis Fetalis) |

**FIGURE 5.22** Representative regional gene assignments for human chromosome 1 and the X. Many assignments were initially derived using somatic cell hybridization techniques.

## DID MENDEL ENCOUNTER LINKAGE?

We conclude this chapter by examining a modern-day interpretation of the experiments that serve as the cornerstone of transmission genetics—the crosses with garden peas performed by Mendel.

It has been said often that Mendel had extremely good fortune in his classical experiments with the garden pea. In none of his crosses did he encounter apparent linkage relationships between any of the seven mutant charac- ters. Had Mendel obtained highly variable data characteristic of linkage and crossing over, these unorthodox observations might have hindered his successful analysis and interpretation.

The accompanying article by Stig Blixt, reprinted in its entirety, demonstrates the inadequacy of this hypothesis. As we shall see, some of Mendel's genes were indeed linked. We shall leave it to Stig Blixt to enlighten you as to why Mendel did not detect linkage.

---

## WHY DIDN'T GREGOR MENDEL FIND LINKAGE?

It is quite often said that Mendel was very fortunate not to run into the complication of linkage during his experiments. He used seven genes, and the pea has only seven chromosomes. Some have said that had he taken just one more, he would have had problems. This, however, is a gross oversimplification. The actual situation, most probably, is shown in Table 1. This shows that Mendel worked with three genes in chromosome 4, two genes in chromosome 1, and one gene in each of chromosome(s) 5 and 7. It seems at first glance that, out of the 21 dihybrid combinations Mendel theoretically could have studied, no less than four (that is, *a-i, v-fa, v-le, fa-le*) ought to have resulted in linkages. As found, however, in hundreds of crosses and shown by the genetic map of the pea[1], *a* and *i* in chromosome 1 are so distantly located on the chromosome that no linkage is normally detected. The same is true for *v* and *le* on the one hand, and *fa* on the other, in chromo-

some 4. This leaves *v-le*, which ought to have shown linkage.

Mendel, however, seems not to have published this particular combination and thus, presumably, never made the appropriate cross to obtain both genes segregating simultaneously. It is therefore not so astonishing that Mendel did not run into the complication of linkage, although he did not avoid it by choosing one gene from each chromosome.

*Weibullsholm Plant Breeding Institute, Landskrona, Sweden and Centro Energia Nucleare na Agricultura, Piraciaba, SP, Brazil*                                     STIG BLIXT

Received March 5; accepted June 4, 1975.
[1] Blixt, S. 1974, In *Handbook of genetics,* ed. R. C. King. New York: Plenum Press.

---

**Table 1**   Relationship between modern genetic terminology and character pairs used by Mendel

| Character Pair Used by Mendel | Alleles in Modern Terminology | Located in Chromosome |
|---|---|---|
| Seed color, yellow-green | *l-i* | 1 |
| Seed coat and flowers, colored-white | *A-a* | 1 |
| Mature pods, smooth expanded-wrinkled indented | *V-v* | 4 |
| Inflorescences, from leaf axiis-umbellate in top of plant | *Fa-fa* | 4 |
| Plant height, 1m-around 0.5 m | *Le-le* | 4 |
| Unripe pods, green-yellow | *Gp-gp* | 5 |
| Mature seeds, smooth-wrinkled | *R-r* | 7 |

**CHAPTER SUMMARY**

1. Genes located on the same chromosome are said to be linked. Alleles located on the same homologue, therefore, can be transmitted together during gamete formation. However, the mechanism of crossing over between homologues during meiosis results in the reshuffling of alleles, thereby contributing to genetic variability within gametes.

2. Early in this century, geneticists realized that crossing over could provide an experimental basis for mapping the location of linked genes relative to one another along the chromosome.

3. Experimental evidence using *Neurospora* has demonstrated that crossing over occurs in the four-strand tetrad stage of meiosis.

4. Cytological investigations of both corn and *Drosophila* have revealed that crossing over requires actual breaking and rejoining of nonsister chromatids.

5. An exchange of genetic material between sister chromatids may occur during mitosis as well. These events are referred to as sister chromatid exchanges (SCEs). An elevated frequency of such events is seen in the human disorder Bloom syndrome.

6. Somatic cell hybridization techniques have made possible linkage and mapping analysis of human genes.

7. Evidence now suggests that several of Mendel's seven genetic characters are linked. However, in each case, they are sufficiently far apart along the chromosome to prevent linkage from being detected.

**KEY TERMS**

allelism test
ascospore
ascus
Bloom syndrome
chiasma (chiasmata)
chiasmatype theory
chromosome maps
chromosomal theory of
    inheritance
classical theory
coefficient of coincidence
    (*C*)

complete linkage
crossing over
cystic fibrosis
double crossover
endonuclease
four-strand stage
Harlequin chromosomes
heterokaryon
Huntington disease
interference (*I*)

ligase
linkage groups
linkage ratio
mitotic recombination
neurofibromatosis
noncrossover
parasexual cycle
reciprocal classes
recombination
single crossover

sister chromatid exchange
    (SCE)
somatic cell hybridization
synaptonemal complex
synkaryon
synteny testing
tetrad analysis
translocation
twin spots
two-strand stage

# INSIGHTS AND SOLUTIONS

1. In a series of two-point map crosses involving three genes linked on chromosome 3 in *Drosophila,* the following distances were calculated.

$$cd-sr \quad 13 \text{ mu}$$
$$cd-ro \quad 16 \text{ mu}$$

(a) Determine the sequence and construct a map of these three genes.

**SOLUTION:** It is impossible to do so; there are two possibilities based on these limited data:

**Case 1:** cd ——13—— sr ——3—— ro

or

**Case 2:** ro ——16—— cd ——13—— sr

(b) What mapping data will resolve this?

**SOLUTION:** The map distance determined by crossing over between *ro* and *sr.* In fact, this distance is 29 map units (mu), demonstrating that Case 2 is correct.

(c) Can we tell which of the sequences shown below is correct?

ro ——16—— cd ——13—— sr

or

sr ——13—— cd ——16—— ro

**SOLUTION:** No; based on the mapping data, they are equivalent.

2. In rabbits, *black* (*B*) is dominant to *brown* (*b*), while *full color* (*C*) is dominant to *chinchilla* ($C^{ch}$). The genes controlling these traits are linked. Rabbits that are heterozygous for both traits and express *black, full color* were crossed to rabbits that are *brown, chinchilla,* with the following results:

31 *brown, chinchilla*
35 *black, full*
16 *brown, full*
19 *black, chinchilla*

Determine the arrangement of alleles in the heterozygous parents and the map distance between the two genes.

**SOLUTION:** This is a two-point map problem, where the two reciprocal phenotypes most prevalent are the noncrossovers. The less frequent reciprocal phenotypes arise from a single crossover. The arrangement of alleles is derived from the noncrossover phenotypes because they enter gametes intact. The cross is shown as follows:

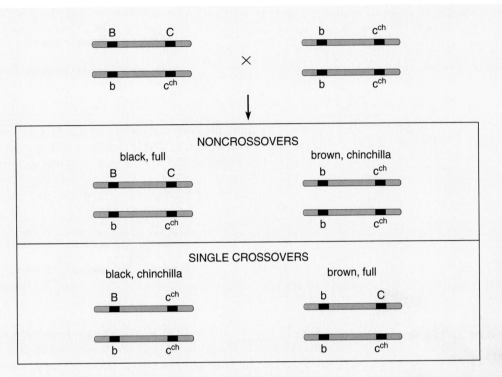

The single crossovers give rise to 35/100 offspring (35%). Therefore, the distance between the two genes is 35 map units (mu).

3. In *Drosophila, Lyra* (*Ly*) and *Stubble* (*Sb*) are dominant mutations located at locus 40 and 58, respectively, on chromosome 3. A recessive mutation with bright red eyes was discovered and shown also to be on chromosome 3. A map was obtained by crossing a female who was heterozygous for all three mutations to a male homozygous for the bright red mutation (which we will temporarily call *br*). The following data were obtained:

| | Phenotype | Number |
|---|---|---|
| (1) | *Ly Sb br* | 404 |
| (2) | + + + | 422 |
| (3) | *Ly* + + | 18 |
| (4) | + *Sb br* | 16 |
| (5) | *Ly* + *br* | 75 |
| (6) | + *Sb* + | 59 |
| (7) | *Ly Sb* + | 4 |
| (8) | + + *br* | 2 |
| | | 1000 |

Determine the location of the bright red mutation on chromosome 3. By referring to Figure 5.14, predict what mutation has been discovered. How could you be sure?

**SOLUTION:** First, determine the *arrangement* of the alleles on the homologues of the heterozygous crossover parent (the female in this case). This is done by locating the most frequent reciprocal phenotypes, which arise from the noncrossover gametes. These are phenotypes (1) and (2). Each one represents the arrangement of alleles on one of the homologues. Therefore, the arrangement is

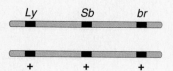

Second, determine the correct *sequence* of the three loci along the chromosome. This is done by determining which sequence will yield the *observed* double-crossover phenotypes, which are the least frequent reciprocal phenotypes (7 and 8).

If the sequence is correct as written, then a double crossover

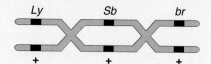

will yield $\underline{Ly + br}$ and $\underline{+ Sb +}$ as phenotypes. Inspection shows that these categories (5 and 6) are actually single crossovers, not double crossovers. Therefore, the sequence, as written, is incorrect. There are only two other possible sequences. The *br* gene is either to the left of *Ly*, or it is between *Ly* and *Sb*:

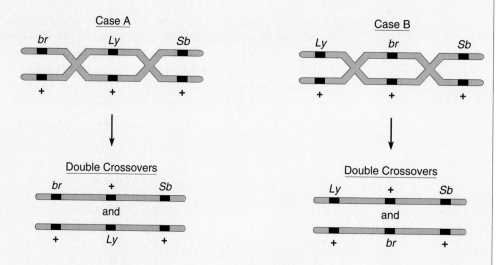

Comparison with the actual data shows that Case B is correct. The double-crossover gametes (7) and (8) yield flies that express *Ly* and *Sb*, but not *br*, or express *br*, but not *Ly* and *Sb*. Therefore, the correct arrangement *and* sequence is

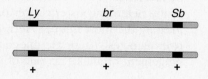

Once this has been determined, it is possible to determine the location of *br* relative to *Ly* and *Sb*. A single crossover between *Ly* and *br*

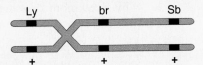

yields flies that are *Ly* + + and + *br Sb* (categories 3 and 4). Therefore, the distance between the *Ly* and *br* loci is equal to

$$\frac{18 + 16 + 4 + 2}{1000} = \frac{40}{1000} = 0.04 = 4 \text{ map units}$$

Remember that we must add in the double crossovers, since they represent two single crossovers occurring simultaneously. Because we need to know the frequency of all crossovers between *Ly* and *br*, they must be included.

Similarly, the distance between the *br* and *Sb* loci is derived mainly from single crossovers between them:

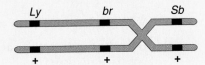

This event yields *Ly br* + and + + *Sb* phenotypes (categories 5 and 6). Therefore, the distance equals

$$\frac{75 + 59 + 4 + 2}{1000} = \frac{140}{1000} = 0.14 = 14 \text{ map units}$$

The final map shows that *br* is located at locus 44, since *Lyra* and *Stubble* are known:

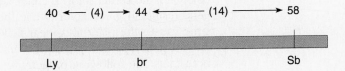

Inspection of Figure 5.14 reveals that the mutation *scarlet*, which has bright red eyes, is known to exist at locus 44, so it is reasonable to hypothesize that the bright red eye mutation is an allele of *scarlet*. To test this hypothesis, we could cross females of our bright red mutant with known *scarlet* males. If they are alleles, all progeny will reveal a bright mutant eye. If not, all progeny will show normal brick red wild-type eyes, since our mutant and scarlet are actually at slightly different loci (very near 44.0). In such a case, all progeny will be heterozygous and will not show a bright red eye. This cross represents an **allelism test**.

1.  What is the significance of genetic recombination to the process of evolution?

2.  Describe the cytological observation that suggests that crossing over occurs during the first meiotic prophase.

3.  Why does more crossing over occur between two distantly linked genes than between two genes that are very close together on the same chromosome?

4.  Why is a 50 percent recovery of single-crossover products the upper limit, even when crossing over *always* occurs between two linked genes?

5.  Why are double-crossover events expected in lower frequency than single-crossover events?

6.  What is the proposed basis for positive interference?

7.  What two essential criteria must be met in order to execute a successful mapping cross?

8.  The genes *dumpy wing* (*dp*), *clot eye* (*cl*), and *apterous wing* (*ap*) are linked on chromosome 2 of *Drosophila*. In a series of two-point mapping crosses, the following genetic distances were determined:

    $$dp–ap \quad 42$$
    $$dp–cl \quad \phantom{0}3$$
    $$ap–cl \quad 39$$

    What is the sequence of the three genes?

9.  Consider two hypothetical recessive autosomal genes *a* and *b*. Where a heterozygote is test-crossed to a double-homozygous mutant, predict the phenotypic ratios under the following conditions:
    (a) *a* and *b* are located on separate autosomes.
    (b) *a* and *b* are linked on the same autosome but are so far apart that a crossover always occurs between them.
    (c) *a* and *b* are linked on the same autosome but are so close together that a crossover almost never occurs.
    (d) *a* and *b* are linked on the same autosome about 10 map units apart.

10. In corn, colored aleurone (in the kernels) is due to the dominant allele *R*. The recessive allele *r*, when homozygous, produces colorless aleurone. The plant color (not the kernel color) is controlled by the gene pair *Y* and *y*. The dominant *Y* gene results in green color, while the homozygous presence of the recessive *y* gene causes the plant to appear yellow. In a test cross between a plant of unknown genotype and phenotype and a plant that is homozygous recessive for both traits, the following progeny were obtained:

    | | |
    |---|---|
    | Colored green | 88 |
    | Colored yellow | 12 |
    | Colorless green | 8 |
    | Colorless yellow | 92 |

    Explain how these results were obtained by determining the exact genotype and phenotype of the unknown plant, including the precise association of the two genes on the homologues (i.e., the arrangement).

11. In the cross shown below involving two linked genes, *ebony* (*e*) and *claret* (*ca*), in *Drosophila*, where crossing over does not occur in males,

$$\frac{♀}{\underset{e^+\ ca}{e\ \ ca^+}} \times \frac{♂}{\underset{e^+\ ca}{e\ \ ca^+}}$$

offspring were produced in a $2 + : 1\ ca : 1\ e$ phenotypic ratio. These genes are 30 units apart on chromosome 3. What contribution did crossing over in the female make to these phenotypes?

12. With two pairs of genes involved ($P/p$ and $Z/z$), a test cross ($ppzz$) with an organism of unknown genotype indicated that the gametes produced were in the following proportions:

*PZ,* 42.4%; *Pz,* 6.9%; *pZ,* 7.1%; and *pz,* 43.6%

Draw all possible conclusions from these data.

13. In a series of two-point map crosses involving five genes located on chromosome II in *Drosophila*, the following recombinant (single crossover) frequencies were observed:

| | |
|---|---|
| *pr–adp* | 29 |
| *pr–vg* | 13 |
| *pr–c* | 21 |
| *pr–b* | 6 |
| *adp–b* | 35 |
| *adp–c* | 8 |
| *adp–vg* | 16 |
| *vg–b* | 19 |
| *vg–c* | 8 |
| *c–b* | 27 |

(a) If *adp* gene is present near the end of the chromosome II (locus 83), construct a map of these genes.
(b) In another set of experiments, a sixth gene, *d*, was tested against *b* and *pr*:

*d–b*  17%
*d–pr*  23%

Predict the results of two-point maps between *d* and *c, d* and *vg,* and *d* and *adp.*

14. Two different female *Drosophila* were isolated, each heterozygous for the autosomally linked genes *b* (*black body*), *d* (*dachs tarsus*), and *c* (*curved wings*). These genes are in the order *d–b–c,* with *b* being closer to *d* than to *c.* Shown below is the genotypic arrangement for each female along with the various gametes formed by both. Identify which categories are noncrossovers (NCO), single crossovers (SCO), and double crossovers (DCO) in each case. Then, indicate the relative frequency in which each will be produced.

|  Female A  |  |   Female B   |  |
|---|---|---|---|
| $\dfrac{d\ b\ +}{+\ +\ c}$ | | $\dfrac{d\ +\ +}{+\ b\ c}$ | |

----Gametes----

| Female A | | Female B | |
|---|---|---|---|
| (1) *d b c* | (5) *d + +* | (1) *d b +* | (5) *d b c* |
| (2) *+ + +* | (6) *+ b c* | (2) *+ + c* | (6) *+ + +* |
| (3) *+ + c* | (7) *d + c* | (3) *d + c* | (7) *d + +* |
| (4) *d b +* | (8) *+ b +* | (4) *+ b +* | (8) *+ b c* |

**15.** In *Drosophila*, a cross was made between females expressing the three sex-linked recessive traits, *scute* (*sc*) *bristles, sable body* (*s*), and *vermilion eyes* (*v*), and wild-type males. In the $F_1$, all females were wild type, while all males expressed all three mutant traits. The cross was carried to the $F_2$ generation and 1000 offspring were counted with the results shown below. No determination of sex was made in the $F_2$ data.

| Phenotype | Offspring |
|---|---|
| *sc s v* | 314 |
| + + + | 280 |
| + *s v* | 150 |
| *sc* + + | 156 |
| *sc* + *v* | 46 |
| + *s* + | 30 |
| *sc s* + | 10 |
| + + *v* | 14 |

(a) Determine the genotypes of the $P_1$ and $F_1$ parents, using proper nomenclature.
(b) Determine the sequence of the three genes and the map distance between them.
(c) Are there more or fewer double crossovers than expected? Calculate the coefficient of coincidence. Does this represent positive or negative interference?

**16.** Another cross in *Drosophila* involved the recessive, sex-linked genes *yellow* (*y*), *white* (*w*), and *cut* (*ct*). A female that was yellow-bodied and white-eyed with normal wings was crossed to a male whose eyes and body were normal but whose wings were cut. The $F_1$ females were wild type for all three traits, while the $F_1$ males expressed the yellow-body, white-eye traits. The cross was carried to an $F_2$, and only male offspring were tallied. On the basis of the data shown below, a genetic map was constructed.

| Phenotype | Male Offspring |
|---|---|
| *y* + *ct* | 9 |
| + *w* + | 6 |
| *y w ct* | 90 |
| + + + | 95 |
| + + *ct* | 424 |
| *y w* + | 376 |
| *y* + + | 0 |
| + *w ct* | 0 |

(a) Diagram the genotypes of the $F_1$ parents.
(b) Construct a map, assuming that *white* is at locus 1.5 on the X chromosome.
(c) Were any double-crossover offspring expected?
(d) Could the $F_2$ female offspring be used to construct the map? Why or why not?

**17.** In *Drosophila, Dichaete* (*D*) is a chromosome 3 mutation with a dominant effect on wing shape. It is lethal when homozygous. The genes *ebony* (*e*) and *pink* (*p*) are chromosome 3 recessive mutations affecting the body and eye color, respectively. Flies from a *Dichaete* stock were crossed to homozygous *ebony, pink* flies, and the $F_1$ progeny, with a *Dichaete* phenotype, were back-crossed to the *ebony, pink* homozygotes. The results of this backcross were as follows:

| Phenotype | Number |
|---|---|
| Dichaete | 401 |
| ebony, pink | 389 |
| Dichaete, ebony | 84 |
| pink | 96 |
| Dichaete, pink | 2 |
| ebony | 3 |
| Dichaete, ebony, pink | 12 |
| wild type | 13 |

(a) Diagram this cross, showing the genotypes of the parents and offspring of both crosses.

(b) What is the sequence and interlocus distance between these three genes?

18. *Drosophila* females homozygous for the third chromosomal genes *pink* and *ebony* (the same genes from the previous problem) were crossed with males homozygous for the second chromosomal gene *dumpy*. Because these genes are recessive, all offspring were wild type (normal). $F_1$ females were test-crossed to triply recessive males. If we assume that the two linked genes, *pink* and *ebony*, are 20 map units apart, predict the results of this cross. If the reciprocal cross were made ($F_1$ males—where no crossing over occurs—with triply recessive females), how would the results vary, if at all?

19. In *Drosophila,* two mutations, *Stubble* (*Sb*) and *curled* (*cu*), are linked on chromosome 3. *Stubble* is a dominant gene that is lethal in a homozygous state, and *curled* is a recessive gene. If a female of the genotype

$$\frac{Sb \quad cu}{+ \quad +}$$

is to be mated to detect recombinants among her offspring, what male genotype would you choose as a mate?

20. In *Drosophila,* a heterozygous female for the sex-linked recessive traits *a, b,* and *c* was crossed to a male that was phenotypically *a b c.* The offspring occurred in the following phenotypic ratios:

| | | | |
|---|---|---|---|
| + | *b* | *c* | 460 |
| *a* | + | + | 450 |
| *a* | *b* | *c* | 32 |
| + | + | + | 38 |
| *a* | + | *c* | 11 |
| + | *b* | + | 9 |

No other phenotypes were observed.

(a) What is the genotypic arrangement of the alleles of these genes on the X chromosomes of the female?

(b) Determine the correct sequence and construct a map of these genes on the X chromosome.

(c) What progeny phenotypes are missing? Why?

**21.** *Drosophila melanogaster* has one pair of sex chromosomes (XX or XY) and three autosomes, referred to as chromosomes 2, 3, and 4. A male fly with very short legs was discovered by a genetics student. Using this male, the student was able to establish a pure breeding stock of this mutant and found that it was recessive. This mutant was then incorporated into a stock containing the recessive gene *black* (body color located on chromosome 2) and the recessive gene *pink* (eye color located on chromosome 3). A female from the homozygous *black, pink, short* stock was then mated to a wild-type male. The $F_1$ males of this cross were all wild type and were then back-crossed to the homozygous *b p sh* females. The $F_2$ results appeared as shown in the following table. No other phenotypes were observed.

| | Wild | Pink[a] | Black, Short | Black, Pink, Short |
|---|---|---|---|---|
| Females | 63 | 58 | 55 | 69 |
| Males | 59 | 65 | 51 | 60 |

[a] *Pink* indicates that the other two traits are wild type, and so on.

(a) Based on these results, the student was able to assign *short* to a linkage group (a chromosome). Which one was it? Include a step-by-step reasoning.

(b) The experiment was subsequently repeated making the reciprocal cross, $F_1$ females back-crossed to homozygous *b p sh* males. It was observed that 85 percent of the offspring fell into the above classes, but that 15 percent of the offspring were equally divided among *b + p, b + +, + sh p,* and *+ sh +* phenotypic males and females. How can these results be explained and what information can be derived from the data?

**22.** Why were there more "twin spots" observed by Stern than *singed* spots in his study of somatic crossing over? If he had been studying *tan* body color (locus 27.5) and *forked* bristles (locus 56.7) on the X chromosome of heterozygous females, what relative frequencies of tan spots, forked spots, and "twin spots" would you predict might occur?

**23.** Are mitotic recombinations and sister chromatid exchanges effective in producing genetic variability in an individual? In the offspring of individuals?

**24.** What possible conclusions can be drawn from the observations that no synaptonemal complexes are observed in male *Drosophila* and female *Bombyx* and that no crossing over occurs in these respective organisms?

**25.** A female of genotype

$$\frac{a \quad b \quad c}{+ \quad + \quad +}$$

produces 100 meiotic tetrads. Of these, 68 show no crossover events. Of the remaining 32, 20 show a crossover between *a* and *b*, 10 show a crossover between *b* and *c*, and 2 show a double crossover between *a* and *b* and between *b* and *c*. Of the 400 gametes produced, how many of each of the 8 different genotypes will be produced? Assuming the order *a–b–c* and the allele arrangement shown above, what is the map distance between these loci?

## SELECTED READINGS

ALLEN, G. E. 1978. *Thomas Hunt Morgan: The man and his science.* Princeton, NJ: Princeton University Press.

BLIXT, S. 1975. Why didn't Gregor Mendel find linkage? *Nature* 256:206.

CATCHESIDE, D. G. 1977. *The genetics of recombination.* Baltimore: University Park Press.

CHAGANTI, R., SCHONBERG, S., and GERMAN, J. 1974. A manyfold increase in sister chromatid exchange in Bloom's syndrome lymphocytes. *Proc. Natl. Acad. Sci.* 71:4508–12.

CREIGHTON, H. S., and MCCLINTOCK, B. 1931. A correlation of cytological and genetical crossing over in *Zea mays. Proc. Natl. Acad. Sci.* 17:492–97.

DOUGLAS, L., and NOVITSKI, E. 1977. What chance did Mendel's experiments give him of noticing linkage? *Heredity* 38:253–57.

EPHRUSSI, B., and WEISS, M. C. 1969. Hybrid somatic cells. *Scient. Amer.* (April) 220:26–35.

GARCIA-BELLIDO, A. 1972. Some parameters of mitotic recombination in *Drosophila melanogaster. Molec. Genet.* 115:54–72.

HOTTA, Y., TABATA, S., and STERN, H. 1984. Replication and nicking of zygotene DNA sequences: Control by a meiosis-specific protein. *Chromosoma* 90:243–53.

KING, R. C. 1970. The meiotic behavior of the *Drosophila* oocyte. *Int. Rev. Cytol.* 28:125–68.

LATT, S. A. 1981. Sister chromatid exchange formation. *Ann. Rev. Genet.* 15:11–56.

LINDSLEY, D. L., and GRELL, E. H. 1972. *Genetic variations of Drosophila melanogaster.* Washington, DC: Carnegie Institute of Washington.

MOENS, P. B. 1977. The onset of meiosis. In *Cell biology—A comprehensive treatise,* vol. 1, *Genetic mechanisms of cells,* ed. L. Goldstein and D. M. Prescott, pp. 93–109. Orlando: Academic Press.

MORGAN, T. H. 1911. An attempt to analyze the constitution of the chromosomes on the basis of sex-linked inheritance in *Drosophila. J. Exp. Zool.* 11:365–414.

NEUFFER, M. G., JONES, L., and ZOBER, M. 1968. *The mutants of maize.* Madison, Wis.: Crop Science Society of America.

PERKINS, D. 1962. Crossing-over and interference in a multiply marked chromosome arm of *Neurospora. Genetics* 47:1253–74.

RUDDLE, F. H., and KUCHERLAPATI, R. S. 1974. Hybrid cells and human genes. *Scient. Amer.* (July) 231:36–49.

STERN, C. 1936. Somatic crossing over and segregation in *Drosophila melanogaster. Genetics* 21:625–31.

STERN, H., and HOTTA, Y. 1973. Biochemical controls in meiosis. *Ann. Rev. Genet.* 7:37–66.

———. 1974. DNA metabolism during pachytene in relation to crossing over. *Genetics* 78: 227–35.

STURTEVANT, A. H. 1913. The linear arrangement of six sex-linked factors in *Drosophila,* as shown by their mode of association. *J. Exp. Zool.* 14:43–59.

———. 1965. *A history of genetics.* New York: Harper & Row.

TAYLOR, J. H., ed. 1965. *Selected papers on molecular genetics.* Orlando: Academic Press.

VOELLER, B. R., ed. 1968. *The chromosome theory of inheritance: Classical papers in development and heredity.* New York: Appleton-Century-Crofts.

VON WETTSTEIN, D., RASMUSSEN, S. W., and HOLM, P. B. 1984. The synaptonemal complex in genetic segregation. *Ann. Rev. Genet.* 18:331–414.

WOLFF, S., ed. 1982. *Sister chromatid exchange.* New York: Wiley-Interscience.

# 6

# CHROMOSOME VARIATION
# AND
# SEX DETERMINATION

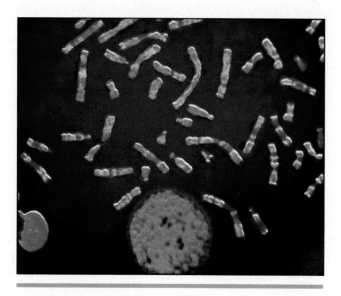

Human metaphase chromosomes from a blood cell derived
from an individual with Klinefelter syndrome.

*Genetic information of a diploid organism is delicately balanced in both content and location within the genome. As a result of studying variation in the sex chromosomes, valuable insights have been gained into the mode of sex determination in many organisms. Other studies have revealed that a change in chromosome number or in the arrangement of a chromosome region results often in phenotypic variation or disruption of development of an organism. Because the chromosome is the unit of transmission in meiosis, such variations are passed to offspring in a predictable manner, resulting in many informative genetic situations.*

U p to this point in the text, we have emphasized how mutations and the resulting alleles affect an organism's phenotype, and how traits are passed from parents to offspring according to Mendelian principles. In this chapter, we shall look at phenotypic variation occurring as a result of changes in the genetic material that are more substantial than alterations of individual genes. These involve modifications at the level of the chromosome.

Although members of diploid species normally contain precisely two haploid chromosome sets, many cases are known in which some variation from this pattern occurs. Modifications include variations in the number of individual chromosomes as well as rearrangements of the genetic material either within or among chromosomes. Taken together, such changes are called **chromosome mutations** or **chromosome aberrations**, to distinguish such genetic alterations from gene mutations. Since the chromosome is the unit transmitted according to Mendelian laws, chromosome aberrations are transmitted to offspring in a predictable manner, resulting in many interesting examples of heritable phenotypic variation.

In this chapter, we shall consider the many types of chromosomal aberrations, the consequences of them for organisms, and what we can learn from these situations. We will also discuss their role in the evolutionary process. We will see that the genetic component of a diploid organism is delicately balanced in content and location within the genome. Even minor alterations of content or location may result in some form of phenotypic variation, while more substantial changes may cause lethality, particularly in animals.

We begin this chapter with the analysis of sex chromosome variation in humans and *Drosophila*. These studies have led to an understanding of how sex is determined in these organisms. We shall also pursue several other topics related to the genetic function of sex chromosomes.

## VARIATION IN CHROMOSOME NUMBER: AN OVERVIEW

Before embarking on a discussion of variations involving the number of chromosomes in organisms and sex determination, it is useful to establish the terminology used that describes such changes. Variation in chromosome number ranges from the addition or loss of one or more chromosomes to the addition of one or more haploid sets of chromosomes. When an organism gains or loses one or more chromosomes, but not a complete set, the condition of **aneuploidy** is created. This is contrasted with the condition of **euploidy**, where complete haploid sets of chromosomes are present. If there are three or more sets, the more general term **polyploidy** is applicable. Those with three sets are specifically **triploid**; those with four sets are **tetraploid**, and so on. Table 6.1 provides a useful organizational framework for you to follow as we discuss each of these categories and the subsets within them.

## THE DIPLOID CHROMOSOME NUMBER IN HUMANS

From the time dividing cells of humans were first observed, geneticists tried to determine accurately the chromosome number of our own species. The first sig-

**Table 6.1**   TERMINOLOGY FOR VARIATION IN CHROMOSOME NUMBERS

| Term | Explanation |
| --- | --- |
| Aneuploidy | $2n \pm$ chromosomes |
| Monosomy | $2n - 1$ |
| Trisomy | $2n + 1$ |
| Tetrasomy, pentasomy, etc. | $2n + 2, 2n + 3$, etc. |
| Euploidy | Multiples of $n$ |
| Diploidy | $2n$ |
| Polyploidy | $3n, 4n, 5n, \ldots$ |
| Triploidy | $3n$ |
| Tetraploidy, pentaploidy, etc. | $4n, 5n$, etc. |
| Autopolyploidy | Multiples of the same genome |
| Allopolyploidy | Multiples of different genomes |

# CHROMOSOMES, SEX DIFFERENTIATION, AND SEX DETERMINATION IN HUMANS

In our discussion of sex linkage in Chapter 4, we pointed out that, as part of the diploid chromosome composition of both humans and *Drosophila,* females possess two X chromosomes while males possess one X and one Y chromosome. This observation might lead us to conclude that the Y chromosome causes maleness in both species; however, this is not necessarily the case. Perhaps the lack of a second X chromosome somehow causes maleness while the Y plays no part whatsoever in sex determination. Perhaps the presence of two X chromosomes causes femaleness while the Y plays no role. The evidence that clarified which explanation was correct awaited the study of variations in the sex chromosome composition of both humans and flies. As we shall see, the first explanation, where the Y determines maleness, is valid in humans but invalid in *Drosophila.*

## Klinefelter and Turner Syndromes

Around 1940 it was observed that two human abnormalities, the **Klinefelter** and **Turner syndromes**,* are characterized by aberrant sexual development. Individuals with Klinefelter syndrome have genitalia and internal ducts that are usually male, but their testes are underdeveloped and fail to produce sperm. Although masculine development occurs, feminine sexual development is not entirely suppressed. Slight enlargement of the breasts is common, for example. Intersexuality may lead to abnormal social development.

In Turner syndrome, the affected individual has female external genitalia and internal ducts, but the ovaries are rudimentary. Other characteristic abnormalities include short stature (usually under five feet); a webbed neck; and a broad, shieldlike chest.

In 1959, the karyotypes of individuals with these syndromes were independently determined to be abnormal with respect to the sex chromosomes. Individuals with Klinefelter syndrome most often have an XXY complement in addition to 44 autosomes [Figure 6.1(a)]. People with this karyotype are designated **47,XXY**. Individuals with Turner syndrome have only 45 chromosomes, including just a single X chromosome and are designated **45,X** [Figure 6.1(b)]. Note the convention used in desig-

nificant attempt was made in 1912, when H. von Winiwarter counted 47 chromosomes in a spermatogonial metaphase preparation. At that time, geneticists believed that the sex-determining mechanism in humans was based on the presence of an extra chromosome in females; that is, females were thought to have 48 chromosomes. In the 1920s, Theophilus Painter observed between 45 and 48 chromosomes in cells of testicular tissue and also discovered the small Y chromosome, which is now known to occur only in males. In his original paper, Painter favored 46 as the diploid number in humans, but he later concluded incorrectly that 48 was the chromosome number in both males and females.

For thirty years, this number was accepted. Then, in 1956, Joe Hin Tjio and Albert Levan introduced improvements in chromosome preparation techniques. These improved techniques led to a strikingly clear demonstration of metaphase stages showing that 46 was indeed the human diploid number. Later that same year, C. E. Ford and John L. Hamerton, also working with testicular tissue, confirmed this finding.

Within the normal 23 pairs of human chromosomes, one pair was shown to vary in configuration in males and females. These two chromosomes were designated the X and Y sex chromosomes. The human female has two X chromosomes, and the human male has one X and one Y chromosome. As we shall see, however, these observations are insufficient to conclude that the Y chromosome determines maleness.

---

* Although the possessive form of the names of most syndromes (eponyms) is often used (e.g., Klinefelter's), the current preference is to use the nonpossessive form for syndromes. We have adopted this convention for all human disorders and syndromes.

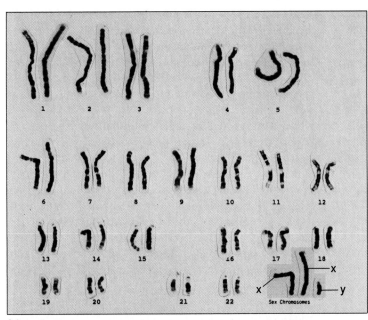

**(a)**

**FIGURE 6.1**    (a) The karyotype of an individual with Klinefelter syndrome. In the karyotype, two X chromosomes and one Y chromosome are visible, creating the 47,XXY condition. (b) The karyotype of an individual with Turner syndrome. In the karyotype, only one X chromosome is visible, creating the 45,X condition. The shading identifies the aberrant chromosome compositions.

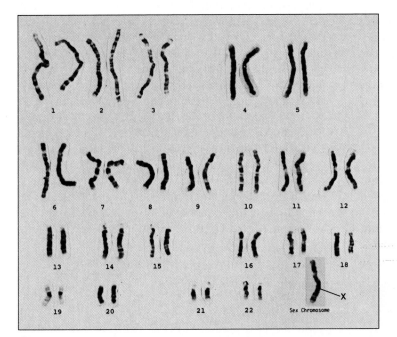

**(b)**

nating chromosome compositions. The number indicates how many chromosomes are present, and the information after the comma designates the deviation from the normal diploid content. Both conditions result from nondisjunction of the X chromosomes during meiosis (see Figure 2.13).

These karyotypes and their corresponding sexual phenotypes allow us to conclude that the Y chromosome determines maleness in humans. In its absence, the sex of the individual is female, even if only a single X chromosome is present. The presence of the Y chromosome in the individual with Klinefelter syndrome is sufficient to determine maleness, even though its expression is not complete. Similarly, in the absence of a Y chromosome, as in the case of individuals with Turner syndrome, no masculinization occurs.

Klinefelter syndrome occurs in about 2 of every 1000 male births. The karyotypes **48,XXXY**; **48,XXYY**;

**49,XXXXY**; and **49,XXXYY** are similar phenotypically to **47,XXY**, but manifestations are often more severe in individuals with a greater number of X chromosomes.

Karyotypes other than 45,X also lead to Turner syndrome. These include mosaic individuals with two apparent cell lines, the most common chromosome combinations being **45,X/46,XY** and **45,X/46,XX**. Turner syndrome is observed in only about 1 in 3000 female births, a frequency much lower than that for Klinefelter syndrome. One explanation for this difference is the observation that a substantial majority of **45,X** fetuses die *in utero* and are aborted spontaneously.

## 47,XXX Syndrome

The presence of three X chromosomes along with a normal set of autosomes **(47,XXX)** results in female differentiation. This syndrome, which is estimated to occur in about 1 of 1200 female births, is highly variable in expression. Frequently, 47,XXX women are perfectly normal. In other cases, underdeveloped secondary sex characteristics, sterility, and mental retardation may occur. In rare instances, **48,XXXX** and **49,XXXXX** karyotypes have been reported. The syndromes associated with these karyotypes are similar to but more pronounced than the 47,XXX. Thus in many cases, the presence of additional X chromosomes appears to disrupt the delicate balance of genetic information essential to normal female development.

## 47,XYY Condition

A third human trisomy involving the sex chromosomes has been discovered and intensively investigated. This case involves males with one X and two Y chromosomes in addition to the normal complement of 44 autosomes **(47,XYY)**. Many studies have attempted to clarify the effect of this chromosome composition.

In 1965, Patricia Jacobs discovered 9 of 315 males in a Scottish maximum security prison to have the 47,XYY karyotype. These males were significantly above average in height and had been involved in criminal acts of serious social consequence. Of the 9 males studied, 7 were of subnormal intelligence, and all suffered personality disorders. In several other studies, similar findings were obtained.

Because of these investigations, the phenotype and frequency of the 47,XYY condition in criminal and noncriminal populations have been examined more extensively (see Table 6.2). Above-average height and subnormal intelligence have been generally substantiated, and the frequency of males displaying this karyotype is indeed higher in penal and mental institutions compared with unincarcerated males.

The possible correlation between this chromosome composition and antisocial and criminal behavior has been of considerable interest. A particularly relevant question involves the characteristics displayed by XYY males who are not incarcerated. The only nearly constant association is that such individuals are over 6 feet tall!

A study that addressed this issue was initiated to identify 47,XYY individuals at birth and to follow their behavioral patterns during preadult and adult development. By 1974 the two primary investigators, Stanley Walzer and Park Gerald, had identified about 20 XYY newborns in 15,000 births at Boston Hospital for Women. However, they soon came under great pressure to abandon their research. Those opposed to the study argued that the investigation could not be justified and might cause great harm to those individuals who displayed this karyotype. They argued that (1) no association between the addi-

**Table 6.2**   FREQUENCY OF XYY INDIVIDUALS IN VARIOUS SETTINGS

| Setting | Restriction | Number Studied | Number XYY | Frequency XYY |
|---|---|---|---|---|
| Control population | Newborns | 28,366 | 29 | 0.10% |
| Mental–penal | No height restriction | 4,239 | 82 | 1.93 |
| Penal | No height restriction | 5,805 | 26 | 0.44 |
| Mental | No height restriction | 2,562 | 8 | 0.31 |
| Mental–penal | Height restriction | 1,048 | 48 | 4.61 |
| Penal | Height restriction | 1,683 | 31 | 1.84 |
| Mental | Height restriction | 649 | 9 | 1.38 |

SOURCE: Compiled from data presented in Hook, 1973, Tables 1–8. Copyright 1973 by the American Association for the Advancement of Science.

tional Y chromosome and abnormal behavior had been previously established in the population at large, and (2) by "labeling" these individuals, a self-fulfilling prophecy might be created. That is, as a result of participation in the study, parents, relatives, and friends might treat those identified as 47,XYY differently, perhaps leading to the type of antisocial behavior expected of them. Despite the support of a government funding agency and the faculty at Harvard Medical School, Walzer and Gerald abandoned the investigation in 1975.

Since it is now clear that many XYY males do not exhibit any form of antisocial behavior and lead normal lives, we must conclude that there is no absolute correlation between the extra Y chromosome and behavior.

## Sexual Differentiation in Humans

Once researchers had established that in humans, it is the Y chromosome that houses genetic information necessary for maleness, efforts were made to pinpoint a specific gene or genes capable of determining sex. Before pursuing a discussion of this topic, it is important to distinguish between **sex determination** and **sexual differentiation** to comprehend how adult male and female humans arise. During early development, every human embryo undergoes a period when it is potentially hermaphroditic or bisexual. Gonadal primordia arise as a pair of ridges associated with each embryonic kidney. Primordial germ cells migrate to these ridges, where an outer cortex and inner medulla form. The **cortex** is capable of developing into an ovary, while the inner **medulla** may develop into a testis. In addition, two sets of undifferentiated male (Wolffian) and female (Mullerian) ducts exist in each embryo.

The genital ridge is present after five weeks of gestation, and if these cells have the XY constitution, male development of the medullary region can be detected by the seventh week. In the absence of the Y chromosome, no male development is initiated, and the genital ridge is destined subsequently to form ovarian tissue. Parallel development of the appropriate male or female duct system then occurs, and the other duct system degenerates. There is a substantial amount of evidence that once testes differentiation is initiated, the embryonic testicular tissue secretes a hormone that supports continued sexual differentiation. For example, in rabbits, if castration occurs after testes development has begun but before the duct system has developed, embryos develop female characteristics regardless of the presence of the Y chromosome.

As the twelfth week of human female development approaches, the oogonia within the ovaries begin meiosis and primary oocytes can be detected. By the twenty-fifth week of gestation, all oocytes become arrested in meiosis, where all will remain dormant until puberty is reached some ten to fifteen years later. Primary spermatocytes, on the other hand, are not formed in males until puberty is reached.

We can conclude that genetic information on the sex chromosomes is responsible for the primary sex determination event. Under normal conditions, once testicular or ovarian development has begun, subsequent early differentiation occurs under the influence of the male or female sex hormones. These hormones support, if not determine, expression of other secondary sexual characteristics during development.

## The Y Chromosome and Male Development

Before turning to other types of chromosome variation and their effects in humans and other organisms, it is appropriate to review information available concerning *how* the Y chromosome results in male development in humans. We have previously alluded to the fact that this chromosome, unlike the X, is nearly genetically blank. Although it shares only limited homology with loci on the X chromosome, it does carry genetic information that controls sexual development.

Therefore, some region of the Y chromosome exists, presumably a gene, that is responsible for encoding a product called the **testis-determining factor (TDF)**. TDF somehow triggers the undifferentiated gonadal tissue of the embryo to form testes. In its absence, female development occurs. Research has focused on just what constitutes the region and the product.

It is now clear that a small part of the human Y chromosome contains a gene called **SRY (sex-determining region Y)**. Evidence proving that *SRY* is indeed the responsible gene has relied on the molecular geneticist's ability to identify the presence or absence of DNA sequences in rare individuals whose expected sex chromosome composition does not correspond to their sexual phenotype. For example, there are human *males* who demonstrate two X and no Y chromosomes. They have attached to one of their Xs the region of the Y containing *SRY*. There are also *females* who have only one X but also have one Y chromosome. Their Y is missing the *SRY* region. These observations argue strongly in favor of the role of *SRY* in male development.

The final proof that this gene is responsible for causing maleness involves an experiment using **transgenic mice**. Such animals arise from fertilized eggs that have foreign DNA injected into them and incorporated into the genetic composition of the developing embryo. When DNA containing the comparable gene of the mouse (*Sry*) is injected into normal XX eggs, most mice develop into males!

Thus, the relevant gene has been identified. It is present in all mammals so far examined, and thus appears to have been conserved throughout evolution. How the product of this gene triggers the embryonic gonadal tissue to develop into testes rather than ovaries is now a reasonable question to ask and one that is amenable to investigation.

## Sex Ratio in Humans

The presence of heteromorphic sex chromosomes in one sex of a species but not the other provides a potential mechanism for producing equal proportions of male and female offspring. The actual proportion of male to female offspring is called the **sex ratio**. It can be assessed in two ways. The **primary sex ratio** reflects the proportion of males and females *conceived* in a population. The **secondary sex ratio** reflects the proportion that are *born*. The secondary sex ratio is much easier to determine, but has the disadvantage of not accounting for disproportionate embryonic or fetal mortality, should it occur.

When the secondary sex ratio in the human population is determined using worldwide census data, it does not equal 1.0. For example, in the Caucasian population in the United States, the secondary ratio is 1.06, indicating that 106 males are born for each 100 females. In the black population in the United States, the ratio is 1.025. In other countries the excess of male births is even higher. In Korea, the secondary sex ratio is 1.15.

To account for this discrepancy, it has been suggested that prenatal female mortality might be greater than prenatal male mortality. If so, it is possible that the primary sex ratio is 1.0 and that it is altered before birth. However, this hypothesis has been shown to be false. In fact, just the opposite occurs. In a Carnegie Institute study, reported in 1948, the sex of approximately 6000 embryos and fetuses recovered from miscarriages and abortions was determined. On the basis of the data derived from this study, the primary sex ratio was estimated to be 1.079. More recent data have estimated that this figure is even higher—between 1.20 and 1.60! Therefore, many more males than females are conceived in the human population.

It is not clear why such a radical departure from the expected 1.0 primary sex ratio occurs. A suitable explanation can be derived only from examining the assumptions upon which the theoretical ratio is based:

1.  Because of segregation, males produce equal numbers of X- and Y-bearing sperm.

2.  Each type of sperm has equivalent viability and motility in the female reproductive tract.

3.  The egg surface is equally receptive to both X- and Y-bearing sperm.

There is no strong experimental evidence to suggest that any of these assumptions are invalid. However, it has been speculated that since the human Y chromosome is smaller than the X chromosome, Y-bearing sperm are of less mass and therefore more motile. If this is true, then the probability of a fertilization event leading to a male zygote is increased, providing a suitable explanation for the observed primary ratio.

## DOSAGE COMPENSATION IN HUMANS

The presence of two X chromosomes in normal human females and only one X in normal human males is unique compared with the equal numbers of autosomes present in the cells of both sexes. On theoretical grounds alone, it is possible to speculate that this situation should create a genetic dosage problem between males and females for all X-linked genes. Females have two copies and males only one. The additional X chromosomes in both males and females exhibiting the various syndromes discussed earlier in this chapter should compound this dosage problem. In this section, we will describe certain research findings regarding X-linked gene expression which demonstrate that a genetic mechanism underlying **dosage compensation** does indeed exist.

### Barr Bodies

Murray L. Barr and Ewart G. Bertram's experiments with female cats, and Keith Moore and Barr's subsequent study with humans, demonstrate a genetic mechanism in mammals that compensates for X chromosome dosage disparities. Barr and Bertram observed a darkly staining body in interphase nerve cells of female cats. They

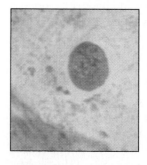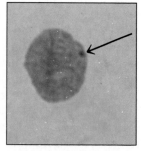

**FIGURE 6.2**    Photomicrographs comparing a cheek epithelial cell nucleus that fails to reveal a Barr body (left) with one that demonstrates a Barr body (see arrow on right). This structure, also called a sex chromatin body, represents an inactivated X chromosome.

found that this structure was absent in similar cells of males. In human females, this body can be easily demonstrated in cells derived from the buccal mucosa or in fibroblasts but not in similar male cells (Figure 6.2). This highly condensed structure, about 1 $\mu$m in diameter, lies against the nuclear envelope of interphase cells. It stains positively in the Feulgen reaction for DNA.

Current experimental evidence demonstrates that this body, called a **sex chromatin body** or simply a **Barr body**, is an inactivated X chromosome. Ohno was the first to suggest that the Barr body arises from one of the two X chromosomes. This hypothesis is attractive because it provides a mechanism for dosage compensation. If one of the two X chromosomes is inactive in the cells of females, the dosage of genetic information that may be expressed in males and females is equivalent. Convincing but indirect evidence for this hypothesis comes from the study of the sex chromosome syndromes described earlier in this chapter. Regardless of how many X chromosomes exist, all but one of them appear to be inactivated and can be seen as Barr bodies. For example, none is seen in Turner 45,X females; one is seen in Klinefelter 47,XXY males; two in 47,XXX females; three in 48,XXXX females; and so on (Figure 6.3). Therefore, the number of Barr bodies follows an $N - 1$ rule, where $N$ is the total number of X chromosomes present.

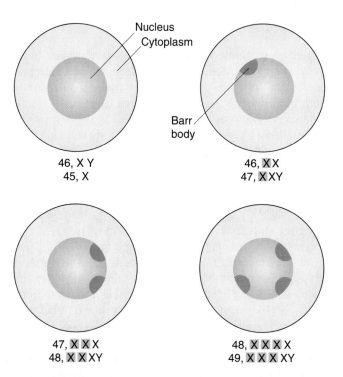

**FIGURE 6.3**     Diagrammatic representation of Barr body occurrence in various human karyotypes.

While this mechanism of inactivation of all but one X chromosome increases our understanding of dosage compensation, it further complicates our perception of other matters. Since one of the two X chromosomes is inactivated in normal human females, why then is the Turner 45,X individual not entirely normal? Why aren't females with the triplo-X and tetra-X karyotypes (47,XXX and 48,XXXX) normal? Further, in Klinefelter syndrome (47,XXY), X chromosome inactivation effectively renders such persons 46,XY. Why aren't they *unaffected* by the additional X chromosome in their nuclei?

One possible explanation is that chromosome inactivation does not normally occur in the very early stages of development of those cells destined to form gonadal tissues. Another possible explanation is that not all of each X chromosome forming a Barr body is inactivated. As a result, excessive expression of certain X-linked genes might still occur despite apparent inactivation of additional X chromosomes. Experimental support exists for both of these explanations.

## The Lyon Hypothesis

In mammalian females one X chromosome is of maternal origin, and the other is of paternal origin. Which one is inactivated? Is the inactivation random? Is the same chromosome inactive in all somatic cells? In 1961, Mary Lyon and Liane Russell independently proposed a hypothesis that answers these questions. They postulated that the inactivation of X chromosomes occurs randomly in somatic cells at a point early in embryonic development, and that once inactivation has occurred, all progeny cells have the same X chromosome inactivated.

This explanation, which has come to be called the **Lyon hypothesis**, was initially based on observations of female mice heterozygous for sex-linked coat color genes (Figure 6.4). The pigmentation of these heterozygous females was mottled, with large patches of skin expressing the color allele on one X and other patches expressing the allele on the other X. Indeed, if one or the other of the two X chromosomes was inactive in adjacent patches of cells, such a phenotypic pattern would result. Similar mottling occurs in the black and orange patches of female tortoise-shell (or calico) cats. Such sex-linked coat color patterns do not occur in male cats since all cells are hemizygous for only one sex-linked coat color allele. Two such cats are compared in Figure 6.5.

The most direct evidence in support of the Lyon hypothesis comes from studies of gene expression in clones of human fibroblast cells. Individual cells may be isolated following biopsy and cultured *in vitro*. If each culture is derived from a single cell, it is referred to as a **clone**. The synthesis of the enzyme **glucose-6-phosphate dehydrogenase (G-6-PD)** is controlled by a sex-

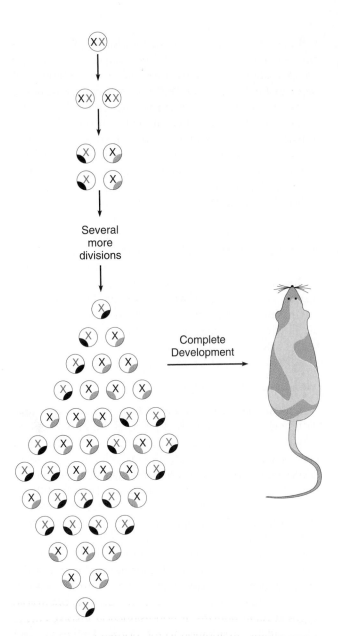

**FIGURE 6.5**   A male and female cat from the same litter, illustrating the Lyon hypothesis. The female (on the left) is a calico with orange and black patches resulting from the inactivation of one or the other X chromosome early in development. The male (on the right) is hemizygous for the orange allele. The white patches are due to still another gene.

**FIGURE 6.4**   Diagrammatic representation of the Lyon hypothesis in a representative mammal. At an early stage in development, either the maternal or paternal X chromosome (shown in two different colors) is randomly inactivated in each cell, forming a Barr body. All progeny cells then inactivate the same X chromosomes as their progenitor cell. In females heterozygous for the X-linked alleles, the adult is a mosaic for each such gene, with some groups of cells expressing one of the alleles and other groups expressing the other allele, as shown for coat color in the mouse. While we have shown initial inactivation to occur at the four-cell stage, it actually occurs later during development.

linked gene. Numerous mutant alleles of this gene have been detected, and their gene products can be differentiated from the wild-type enzyme by their migration pattern in an electrophoretic field. Each clone shows only one electrophoretic form of G-6-PD.

Fibroblasts have been taken from females heterozygous for different allelic forms of G-6-PD and studied. The Lyon hypothesis predicts the results of the examination of numerous clones derived from these female heterozygotes. If inactivation of an X chromosome occurs randomly early in development and is permanent in all progeny cells, such a female should show two types of clones in approximately equal proportion.

In 1963, Ronald Davidson and colleagues performed an experiment involving 14 clones from a single heterozygous female. Seven showed only one form of the enzyme, and seven showed only the other form. What was most important was that none of the 14 showed both forms of the enzyme. Studies of G-6-PD mutants thus provide strong support for the random permanent inactivation of either the maternal or paternal X chromosome.

Although not all of the existing evidence is as clear-cut as the preceding experiment, the Lyon hypothesis is generally accepted as valid. One extension of the hypothesis is that mammalian females are mosaics for *all heterozygous X-linked alleles*. Some areas of the body express only the maternally derived alleles, and others express only the paternally derived alleles. Two especially interesting examples involve **red-green color blindness** and **anhidrotic ectodermal dysplasia**, both X-linked recessive disorders. In the former case, hemizygous males are fully color-blind in all retinal cells. However, heterozygous females display mosaic retinas with patches of defective color perception and surrounding areas with normal color perception. In the latter disorder, hemizygous males show absence of teeth, sparse hair growth, and lack of sweat glands. The skin of heterozygous females reveals patterns of tissue with and without sweat glands. In both examples, random inactivation of one or the other X chromosome early in the development of heterozygous females has led to these occurrences.

### The Mechanism of Inactivation

The least understood aspect of the Lyon hypothesis is the mechanism of chromosome inactivation. How are most genes of an entire chromosome inactivated? Recent investigations are beginning to clarify this issue. A single region of the X chromosome, called the **X-inactivation center (XIC)**, is the major control unit. In humans, it is located on the proximal end of the p arm. Genetic expression of this region occurs *only* on the X chromosome that is inactivated. The constant association of expression and X chromosome inactivation support the conclusion that this region is an important genetic component in the process.

A gene, *XIST* (*X-inactive specific transcript*), is now believed to represent the critical locus within the XIC. A comparable region (*Xic*) and gene (*Xist*) exist in the mouse. Three interesting observations have been made regarding the RNA that is transcribed by the *XIST* and *Xist* genes. First, transcription of the genes is known to occur prior to chromosome inactivation. In the mouse, for example, the *Xist* gene becomes active about 24 hours before X-inactivation occurs in the embryo. Thus, one can argue that gene activation is more likely the cause rather than the consequence of inactivation. Second, the RNA that is made stays in the nucleus, rather than moving to cytoplasm. This is in keeping with the possible role of initiating and/or maintaining inactivation.

The third observation is a most interesting one. The nucleotide sequence of the gene and its RNA product is quite large but lacks what is called an extended **open reading frame (ORF)**. An ORF includes the information necessary for translation of the RNA product into a protein. This observation suggests that the RNA gene product serves a structural role in the nucleus, presumably in the mechanism of chromosome inactivation. This finding has led to speculation that the RNA products of *XIST* and *Xist* produce some sort of molecular "cage" that entraps an X chromosome and inactivates it.

Whatever the actual mechanism, the discovery of these genes is exciting and pushes us closer to an understanding of how dosage compensation is accomplished in mammals.

## CHROMOSOME COMPOSITION AND SEX DETERMINATION IN *DROSOPHILA*

Because males and females in *Drosophila* have the identical sex chromosome composition as humans, we might assume that the Y also causes maleness in these flies. However, the elegant work of Calvin Bridges in 1916 showed this not to be true. He studied flies with quite varied chromosome compositions leading him to the conclusion that the Y chromosome is not involved in sex determination in this organism. Instead, Bridges proposed that both the X chromosomes and autosomes together play a critical role in sex determination.

His work can be divided into two phases: (1) a study of offspring resulting from nondisjunction of the X chromosomes during meiosis in females; and (2) subsequent work with progeny of triploid ($3n$) females. As we have seen previously, nondisjunction is the failure of paired chromosomes to segregate or separate during the anaphase stage of the first or second meiotic divisions. The result is the production of two abnormal gametes, one of which contains an extra chromosome ($n + 1$) and the other that lacks a chromosome ($n - 1$). Fertilization of such gametes produces ($2n + 1$) or ($2n - 1$) aneuploid zygotes. As in humans, if nondisjunction involves the X chromosome, in addition to the normal complement of autosomes, either an XXY or an X0 sex chromosome composition results. (The zero signifies that the second chromosome is absent.) Contrary to what was subsequently found in humans, Bridges found that the XXY flies were normal females, and the X0 flies were sterile males. The presence of the Y chromosome in the XXY flies did not cause maleness, and its absence in the X0 flies did not produce femaleness. From these data he concluded that the Y chromosome in *Drosophila* lacks male-determining factors, but apparently contains genetic information essential to male fertility since the X0 males were sterile.

Bridges was able to clarify the mode of sex determination in *Drosophila* by studying the progeny of triploid females ($3n$), which have *three* copies each of the haploid complement of chromosomes. These females apparently originate from rare diploid eggs fertilized by normal haploid sperm. Triploid females have heavy-set bodies, coarse bristles, coarse eyes, and may be fertile. Because of the odd number of each chromosome (three), during meiosis, a wide range of chromosome complements is distributed into gametes that give rise to offspring with a variety of abnormal chromosome constitutions. A correlation between the sexual morphology, chromosome composition, and Bridges' interpretation is shown in Figure 6.6. *Drosophila* has a haploid number of four, thereby displaying three pairs of autosomes in addition to its pair of sex chromosomes.

Bridges realized that the critical factor in determining sex is the **ratio of X chromosomes to the number of haploid sets of autosomes present**. Normal (2X:2A) and triploid (3X:3A) females each have a ratio equal to 1.0, and both are fertile. As the ratio exceeds unity (3X:2A, or 1.5, for example), what was originally called a *superfemale* is produced. Since this female is rather weak and infertile and has lowered viability, this type is now more appropriately called a **metafemale**.

Normal (XY:2A) and sterile (X0:2A) males each have a ratio of 1:2, or 0.5. When the ratio decreases to 1:3, or 0.33, as in the case of an XY:3A male, infertile

| CHROMOSOME COMPOSITION | CHROMO-SOME FORMU-LATION | RATIO OF X CHROMOSOMES TO AUTOSOME SETS | SEX |
|---|---|---|---|
| 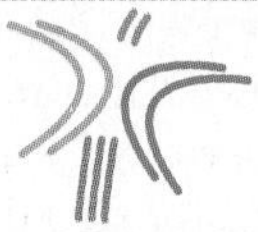 | $3X/2A$ | 1.5 | Metafemale |
|  | $3X/3A$ | 1.0 | Female |
| 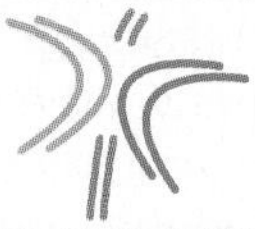 | $2X/2A$ | 1.0 | Female |
| 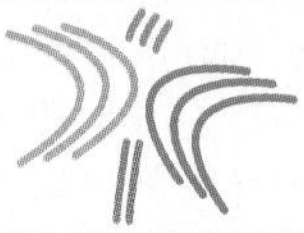 | $2X/3A$ | 0.67 | Intersex |
|  | $3X/4A$ | 0.75 | Intersex |
|  | $X/2A$ | 0.50 | Male |
| 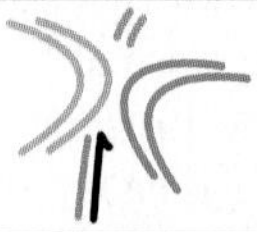 | $XY/2A$ | 0.50 | Male |
| 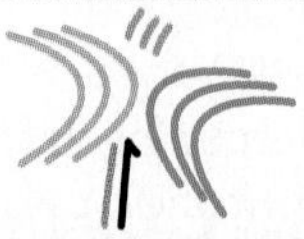 | $XY/3A$ | 0.33 | Metamale |

NORMAL DIPLOID MALE

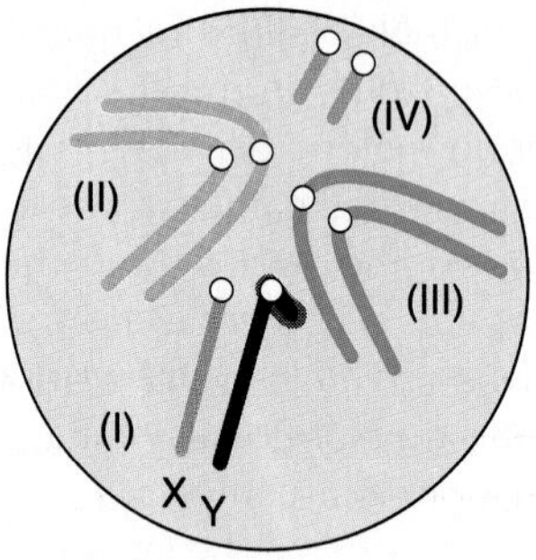

(IV)
(II)
(III)
(I)
X  Y

6 autosomes
+
X   Y

**FIGURE 6.6**    Chromosome compositions, the ratios of X chromosomes to sets of autosomes, and the resultant sexual morphology in *Drosophila melanogaster*. The normal diploid male chromosome composition is shown as a reference on the left.

**metamales** result. Other flies recovered by Bridges in these studies contained an X:A ratio intermediate between 0.5 and 1.0. These flies were generally larger, and they exhibited a variety of morphological abnormalities and rudimentary bisexual gonads and genitalia. They were invariably sterile and were designated as **intersexes**.

These results indicate that in *Drosophila*, factors that cause a fly to develop into a male are not localized on the sex chromosomes, but are instead found on the autosomes. Some female-determining factors, however, are localized on the X chromosomes. Thus, with respect to primary sex determination, male gametes containing one of each autosome plus a Y chromosome result in male offspring *not* because of the presence of the Y chromosome but because of the lack of an X chromosome. This mode of sex determination is explained by the **genic balance theory**. Bridges proposed that a threshold for maleness is reached when the X:A ratio is 1:2 (X:2A), but that the presence of an additional X (XX:2A) alters this balance and results in female differentiation.

Bridges' conclusion received further support from the experiments of Jack Schultz and Theodosius Dobzhansky. They were able to obtain males, females, and intersexes that had gained or lost variously sized fragments of the X chromosome. Some of their results are shown in Table 6.3. The effects were proportional to the extent of the addition or subtraction of the X and were consistent with Bridges' genic balance interpretation of sex determination. This finding led to the conclusion that multiple factors critical to sex determination exist on the X chromosome.

Several mutant genes have been identified that are involved in sex determination in *Drosophila*. The recessive autosomal gene *transformer* (*tra*), discovered by Alfred H. Sturtevant, is especially interesting. Homozygous mutant females are transformed into sterile males, but males are unaffected when homozygous for *tra*.

Experiments by Robert C. King and Dietrich Bodenstein showed that the transplantation of normal but immature female gonadal tissue into the abdomens of *tra/tra* sterile males results in normal ovarian development. Thus, it appears that the abdominal environment can support the production of eggs, but that the *transformer* gene has interfered with primary sex determination in these flies.

More recently, another gene, *Sex-lethal* (*Sxl*), has been shown to play a critical role in sex determination, serving as a "master switch" for the activation of a group of at least four separate regulatory genes. Activation of the *Sxl* gene, located on the X chromosome, is essential to subsequent female development, and relies on a ratio of X chromosomes to sets of autosomes that equals 1.0. In the absence of activation, resulting from an X:A ratio of 0.5, male development occurs.

While it is not yet exactly clear how this ratio influences either the group of regulatory genes or the *Sxl* locus, some insights are available. The *Sxl* locus is part of a hierarchy of gene expression and exerts control over still other genes, including the *tra* gene (discussed above) and the *dsx* (*doublesex*) gene. The wild-type allele of *tra* is activated only in females and influences the expression of *dsx*. Depending on how the initial RNA transcript of *dsx* is processed, the resultant *dsx* protein activates either male- or female-specific genes required for sexual differentiation. Although we will discuss RNA processing in more detail in Chapter 14, for the sake of the present discussion, it is important to know that alternate forms of the same RNA (following different modification) can result in different proteins. These are thought to make a major difference in the subsequent activation of other genes involved in sex determination.

Based on this information, it is evident that many genes play important roles in sex determination as well as sexual differentiation in *Drosophila*. This is an area of active investigation, so we can soon expect the complete genetic basis of sex determination to be elucidated.

**Table 6.3** THE EFFECT OF THE ADDITION OR SUBTRACTION OF FRAGMENTS OF THE *DROSOPHILA* X CHROMOSOME

| Chromosome Composition | Sex | Addition or Subtraction | Effect |
| --- | --- | --- | --- |
| 2X/3A | Intersex | + | Toward female |
| 2X/2A | Female | + | Toward metafemale |
| 2X/3A | Intersex | − | Toward male |

## Dosage Compensation in *Drosophila*

As in mammals such as humans and mice, a dosage problem exists for X-linked genes in *Drosophila*. Females contain two copies while males contain only one copy of each gene. The mechanism of dosage compensation in *Drosophila* differs considerably from mammals, since no X chromosome inactivation is observed. Instead, male X-linked genes are transcribed at twice the level of the comparable genes in females. Interestingly, if blocks of X-linked genes are moved (translocated) to autosomes, dosage compensation affects them even when they are no longer part of the X chromosome.

As in mammals, considerable gains have been made recently in understanding this phenomenon in *Drosophila*. At least four autosomal male-specific lethal genes are known to be involved in the process of dosage compensation. These are under the same master switch gene, *Sxl*, that controls sex determination. Mutations in any of the four autosomal genes reduce the increased expression of male X-linked genes and cause lethality in males, but not in females.

Evidence in support of a model for the mechanism of increased genetic activity is now available. The model predicts that one of the autosomal genes, *mle* (*maleless*), encodes a protein that binds to numerous sites along the X chromosome, causing enhancement of genetic expression. The products of the other three autosomal genes are required for *mle* binding.

The role of *Sxl* is also predicted in this model. In XX flies, *Sxl* turns on the *tra* gene, which leads ultimately to female differentiation, as we discussed above. When it is active, *Sxl* also functions to *inactivate* one or more of the male-specific autosomal genes, perhaps *mle*. In XY flies, *Sxl* is not expressed and the autosomal genes are *activated*, causing enhanced X chromosome activity. Although this model may yet be modified or refined, it is useful in guiding future research activity that will provide a more complete explanation.

The above discussion makes clear that an entirely different mechanism of dosage compensation exists in *Drosophila* (and probably many related organisms) compared to mammals. The development of elaborate mechanisms to equalize the expression of X-linked genes demonstrates the critical nature of gene expression. A delicate balance of gene products is necessary to maintain normal development of both males and females.

## *Drosophila* Mosaics

The mode of sex determination and our knowledge of sex linkage in *Drosophila* (Chapter 4) help to explain the appearance of the extremely unusual fruit fly shown in Figure 6.7. This fly was recovered from a stock where all other females were heterozygous for the sex-linked genes *white* eye (*w*) and *miniature* wing (*m*). It is a

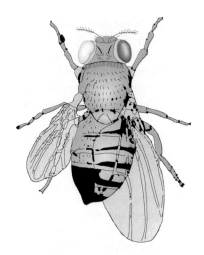

**FIGURE 6.7**    A bilateral gynandromorph of *Drosophila melanogaster* formed following the loss of one X chromosome in one of the two cells during the first mitotic division. The left side of the fly, composed of male cells containing a single X, expresses the mutant *white*-eye *miniature*-wing alleles. The right side is composed of female cells containing two X chromosomes heterozygous for the two recessive alleles.

**bilateral gynandromorph**, which means that one-half of its body has developed as a male and the other half as a female. If a zygote heterozygous for *white* eye and *miniature* wing were to lose one of the X chromosomes during the first mitotic division, the two cells would be of the XX and X0 constitution, respectively. Thus, one cell would be female and the other would be male.

In the case of the bilateral gynandromorph, each of these cells is responsible for producing all progeny cells that make up either the right half *or* the left half of the body during embryogenesis. The cell of X0 constitution apparently produced only identical progeny cells and gave rise to the left half of the fly, which was male. Since the male half demonstrated the *white, miniature* phenotype, the X chromosome bearing the $w^+$, $m^+$ alleles was lost, while the *w, m*-bearing homologue was retained. All female cells of the right side of the body remained heterozygous for both mutant genes and therefore contained normal eye–wing phenotypes. Depending on the orientation of the spindle during the first mitotic division, gynandromorphs can be produced where the "line" demarcating male versus female development occurs at almost any place along or across the fly's body.

## ANEUPLOIDY

We turn now to a consideration of variations in the number of autosomes and the genetic consequence of such changes. The most common examples of **aneuploidy**, where an organism has a chromosome number other than an exact multiple of the haploid set, are cases in which a single chromosome is either added to or lost from a normal diploid set. Such circumstances can arise as a result of primary or secondary nondisjunction (review Figure 2.13). The loss of one chromosome produces a $2n - 1$ complement and is called **monosomy**; the gain of one chromosome produces a $2n + 1$ complement and is described as **trisomy**. The $2n + 2$ and $2n + 3$ conditions are called **tetrasomy** and **pentasomy**, indicating that an individual has four or five copies of a particular chromosome in an otherwise diploid genome.

## Monosomy

Although monosomy for one of the sex chromosomes is fairly common, as we have seen in 45,X Turner syndrome in humans, monosomy for one of the autosomes is not so easily tolerated, particularly in animals. In *Drosophila*, flies monosomic for the very small chromosome 4—a condition referred to as **Haplo-IV**—develop more slowly, exhibit a reduced body size, and have impaired viability. Monosomy for the larger chromosomes 2 and 3 is apparently lethal because such flies have never been recovered.

The failure of monosomic individuals to survive in many animal species is at first quite puzzling, since at least a single copy of every gene is present in the remaining homologue. However, if just one of those genes is represented by a lethal allele, the unpaired chromosome condition leads to the death of the organism. This occurs because monosomy creates hemizygosity for the in-

volved chromosome and unmasks recessive lethals that are tolerated in heterozygotes carrying the corresponding wild-type alleles.

Another possible explanation is that the expression of genetic information during early development is very delicately regulated so that a sensitive equilibrium of gene products is required to ensure normal development. This requirement does not appear to be so stringent in the plant kingdom. Monosomy for autosomal chromosomes has been observed in maize, tobacco, the evening primrose *Oenothera*, and the Jimson weed *Datura*, among many other plants. Nevertheless, such monosomic plants are usually less viable than their diploid derivatives. Haploid pollen grains, which undergo extensive development before participating in fertilization, are particularly sensitive to the lack of one chromosome and are often sterile.

## Partial Monosomy: Cri-du-Chat Syndrome

In humans, autosomal monosomy has not been reported beyond birth. Individuals with such chromosome complements are undoubtedly conceived, but none apparently survive embryonic and fetal development. There are, however, examples of survivors with **partial monosomy**, where only part of one chromosome is lost. These cases are also referred to as **segmental deletions**. One such case was first reported by Jérôme Le-Jeune in 1963 when he described the clinical symptoms of the **cri-du-chat syndrome**. This syndrome is associated with the loss of about one-half of the short arm of chromosome 5 (Figure 6.8). Thus, the genetic constitu-

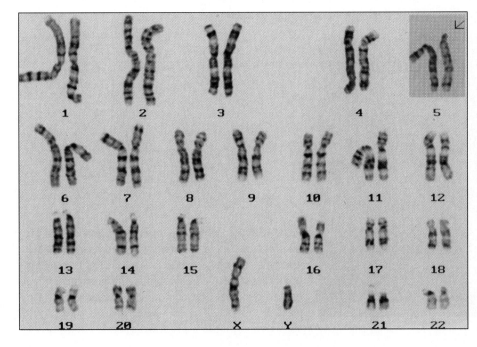

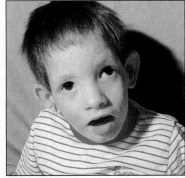

**FIGURE 6.8**    (a) A representative karyotype and (b) photograph of a child exhibiting cri-du-chat syndrome (46,5p−). In the karyotype, the arrow identifies the nearly complete absence of the short arm of one member of the chromosome 5 homologues.

tion may be designated as **46,5p−**, meaning that such an individual has all 46 chromosomes but that some of the p arm (the short arm) of chromosome 5 is missing.

Infants with this syndrome may exhibit anatomic malformations, including gastrointestinal and cardiac complications, and they are often mentally retarded. Abnormal development of the glottis and larynx is characteristic of individuals with this syndrome. As a result, the infant has a cry similar to that of the meowing of a cat, thus giving the syndrome its name.

Since 1963, over 300 cases of cri-du-chat syndrome have been reported worldwide. An incidence of 1 in 50,000 live births has been estimated. The size of the deletion appears to influence the physical, psychomotor,

and mental skill levels of these children. Although the effects of the syndrome are severe, many individuals achieve a level of social development in the trainable range. Those who receive home care and early special schooling are ambulatory, develop self-care skills, and learn to communicate verbally.

## Trisomy

In general, the effects of trisomy parallel those of monosomy. However, the addition of an extra chromosome produces somewhat more viable individuals in both animal and plant species than does the loss of a chromosome.

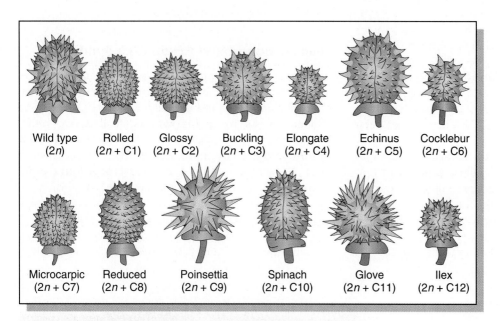

**FIGURE 6.9**   Drawings of capsule phenotypes of the fruits of *Datura stramonium.* In comparison with wild type, each phenotype is the result of trisomy of one of the twelve chromosomes characteristic of the haploid genome. The photograph illustrates the entire plant.

As in monosomy, the sex chromosome variation of the trisomic type has a less dramatic effect on the phenotype than autosomal variation. Recall from our previous discussion that *Drosophila* females with three X chromosomes and a normal complement of two sets of autosomes (3X:2A) survive and reproduce but are less viable than normal 2X:2A females. In humans, the addition of an extra X or Y chromosome to an otherwise normal male or female chromosome constitution (47,XXY, 47,XYY, and 47,XXX) leads to viable individuals exhibiting various syndromes. However, the addition of a large autosome to the diploid complement in both *Drosophila* and humans has severe effects and is usually lethal during development.

In plants, trisomic individuals are viable, but their phenotype may be altered. A classical example involves the Jimson weed *Datura*, whose diploid number is 24. Twelve different primary trisomic conditions are possible, and examples of each one have been recovered. Each trisomy alters the phenotype of the capsule of the fruit sufficiently (Figure 6.9) to produce a unique phenotype.

In plants as well as animals, trisomy may be detected during cytological observations of meiotic divisions. Since three copies of one of the chromosomes are present, pairing configurations are irregular. At any particular region along the chromosome length, only two of the three homologues may synapse (Figure 6.10). At various regions, however, different members of the trio may be paired. When three copies of a chromosome are paired, the configuration is called a **trivalent**, which may be arranged on the spindle so that during anaphase one member moves to one pole, and two go to the opposite pole. In some cases, one bivalent and one univalent (an unpaired chromosome) may be present instead of a trivalent during the first metaphase stage. Meiosis thus produces gametes that can perpetuate the trisomic condition.

## Down Syndrome

The only human autosomal trisomy in which a significant number of individuals survive longer than a year past birth was discovered in 1866 by Langdon Down. The condition is now known to result from trisomy of chromosome 21, one of the G group* (Figure 6.11), and is called **Down syndrome** or simply **trisomy 21** (designated **47,21+**). This trisomy is found in approximately 3 infants in every 2000 live births.

The external phenotype of these individuals is similar so that they bear a striking resemblance to one another. They display a prominent epicanthic fold in the corner of each eye and are characteristically short. They may have small, round heads; protruding, furrowed tongues, which cause the mouth to remain partially open; and short, broad hands with fingers showing characteristic palm and fingerprint patterns. Physical, psychomotor, and mental development is retarded, and the IQ is seldom above 70. Their life expectancy is shortened, and few individuals survive to be 50.

Down children are prone to respiratory disease and heart malformations and show an incidence of leukemia approximately 15 times higher than that of the normal population. However, careful medical scrutiny and treatment throughout their lives has extended their survival significantly.

One way in which this trisomic condition may originate is through nondisjunction of chromosome 21 during meiosis. Failure of paired homologues to disjoin during anaphase I or II can result in male or female gametes with the $n + 1$ chromosome composition. Following fertilization with a normal gamete, the trisomic condition is created. Chromosome analysis has shown that while the additional chromosome may be derived from either the mother or father, the ovum is most often the source.

Before the development of techniques that distinguish paternal from maternal homologues, this conclusion was supported by other indirect evidence derived from studies of the age of mothers giving birth to Down infants. Figure 6.12 shows the analysis of the distribution of maternal age and the incidence of Down syndrome newborns. The frequency of Down births increases dramatically as the age of the mother increases. While the

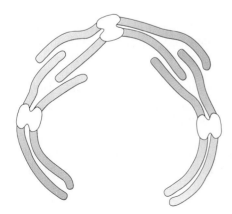

**FIGURE 6.10** Diagrammatic representation of one possible pairing arrangement of three copies of a single chromosome, forming a trivalent during meiosis I. Darker shaded areas are homologous with one another, as are lighter shaded areas.

---

* On the basis of size and centromere placement, human autosomal chromosomes are divided into seven groups: A (1–3); B (4–5); C (6–12); D (13–15); E (16–18); F (19–20); and G (21–22).

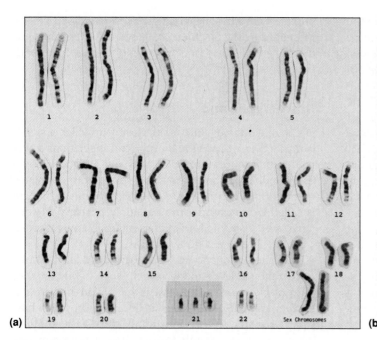

(a) (b)

**FIGURE 6.11** (a) The karyotype and (b) a photograph of a child with Down syndrome. In the karyotype, three members of the G-group chromosome 21 are present, creating the 47,21+ condition.

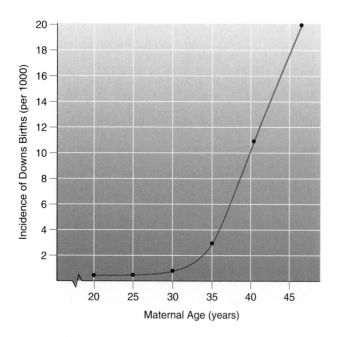

**FIGURE 6.12** Incidence of Down syndrome births contrasted with maternal age.

frequency is about 1 in 1000 at maternal age 30, a tenfold increase to a frequency of 1 in 100 is noted at age 40. The frequency increases still further to about 1 in 50 at age 45.

While the nondisjunctional event that produces Down syndrome seems more likely to occur during oogenesis in women between the ages of 35 and 45, we do not know with certainty why this is so. However, one observation may be relevant. In human females, all primary oocytes have been formed by birth. Therefore, once ovulation begins, each succeeding ovum has been arrested in meiosis for about a month longer than the one preceding it. Thus, women 30 or 40 years old produce ova that are significantly older and arrested longer than those they ovulated 10 or 20 years previously. However, it is not yet known whether ovum age is the cause of the increased incidence of nondisjunction leading to Down syndrome.

These statistics pose a serious problem for the woman who becomes pregnant late in her reproductive years. Genetic counseling early in such pregnancies serves two purposes. First, it informs the parents about the probability that their child will be affected and educates them about Down syndrome. Although some individuals with Down syndrome may be institutionalized, others benefit greatly from special education programs and may be cared for at home. Further, they are noted as being affectionate, loving children. Second, a genetic counselor may recommend a prenatal diagnostic technique such as **amniocentesis** or **chorionic villus sampling (CVS)**. These techniques require the removal and culture of fetal cells. The karyotype of the fetus may then be determined by cytogenetic analysis. If the fetus is diagnosed as having Down syndrome, a therapeutic abortion is one option the parents may consider.

Since Down syndrome appears to be caused by a random error—nondisjunction of chromosome 21 during maternal or paternal meiosis—the occurrence of the disorder is not expected to run in families. Nevertheless, in rare cases, it does so. These instances, referred to as **familial Down syndrome**, involve a **translocation** of chromosome 21, another type of chromosomal aberration, which we will discuss later in this chapter.

## Patau Syndrome

In 1960, Klaus Patau and his associates observed an infant with severe developmental malformations with a karyotype of 47 chromosomes (Figure 6.13). The additional chromosome was medium-sized, one of the acrocentric D group. It is now designated as chromosome 13. The trisomy 13 condition has since been described in many newborns and is called **Patau syndrome (47,13+)**. Affected infants are not mentally alert, are thought to be deaf, and characteristically have a harelip, cleft palate, and polydactyly. Autopsy has revealed congenital malformation of most organ systems, a condition indicative of abnormal developmental events occurring as early as five to six weeks of gestation. The average survival of these infants is less than six months.

The average maternal and paternal ages of parents of affected infants are higher than the ages of parents of normal children, but are not as high as the average maternal age in cases of Down syndrome. Both male and female parents average about 32 years of age when the affected child is born. Because the condition is so rare, occurring as infrequently as 1 in 20,000 live births, it is not known whether the origin of the extra chromosome is more often maternal or paternal, or whether it arises equally from either parent.

## Edwards Syndrome

In 1960, John H. Edwards and his colleagues reported on an infant trisomic for a chromosome in the E group, now known to be chromosome 18 (Figure 6.14). This aberration has been named **Edwards syndrome (47,18+)**. These infants are smaller than the average newborn. Their skulls are elongated in an anterior–posterior direction and their ears set low and malformed. A webbed neck, congenital dislocation of the hips, and a receding chin are often characteristic of such individuals.

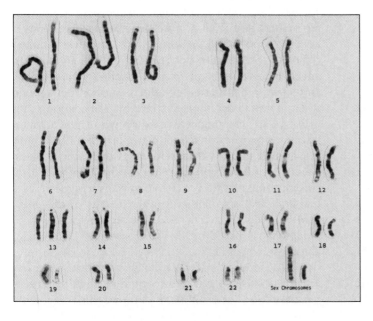

**FIGURE 6.13**    The karyotype of an infant with Patau syndrome, where three members of the D-group chromosome 13 are present, creating the 47,13+ condition.

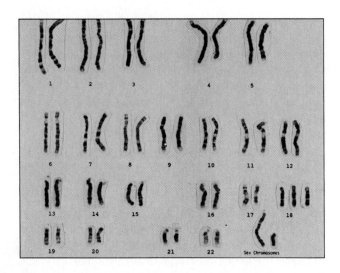

**FIGURE 6.14**    The karyotype of an infant with Edwards syndrome. Three members of the E-group chromosome 18 are present, creating the 47,18+ condition.

groups. Trisomies for every human chromosome have been recovered. Monosomies were almost never found, however, even though nondisjunction should produce $n - 1$ gametes with a frequency equal to $n + 1$ gametes. This finding suggests either that gametes lacking a single chromosome are so functionally impaired that they never participate in fertilization, or that the embryo dies so early in its development that the pregnancy goes undetected. Various forms of polyploidy (see below) and other miscellaneous chromosomal anomalies were also found in Carr's study.

These observations support the hypothesis that normal embryonic development requires a precise diploid complement of chromosomes to maintain a delicate equilibrium in the expression of genetic information. It has been estimated that as many as 5 percent of all conceptuses contain some type of chromosomal imbalance. In most cases, the prenatal mortality provides a barrier against the introduction of a variety of chromosome-based genetic anomalies into the human population.

Although the frequency of trisomy 18 is somewhat greater than that of trisomy 13, the average survival time is about the same, less than four months. Death is usually caused by pneumonia or heart failure. The phenotype of this child, like that of individuals with Down and Patau syndromes, illustrates that the presence of an extra autosome produces congenital malformations and reduced life expectancy.

Again, the average maternal age is high—34.7 years by one calculation. One unusual aspect of the syndrome is the preponderance of afflicted female infants. In one set of observations based on 143 cases, 80 percent were female. Overall, about 1 in 8000 live births exhibits this malady.

### Viability in Human Aneuploidy

The reduced viability of individuals with recognized monosomic and trisomic conditions leads us to believe that many other aneuploid conditions may arise but that the affected fetuses do not survive to term. This observation has been confirmed by karyotypic analysis of spontaneously aborted fetuses.

The majority of spontaneous abortuses demonstrating chromosomal abnormalities are aneuploids. The aneuploid with highest incidence among abortuses is the 45,X condition, which produces an infant with Turner syndrome if the fetus survives to term.

An extensive review of this subject by David H. Carr has also revealed that a significant percentage of abortuses are trisomic for one or the other chromosome

## POLYPLOIDY AND ITS ORIGINS

The term **polyploidy** describes instances where more than two multiples of the haploid chromosome set are found. The naming of polyploids is based on the number of sets of chromosomes found: a **triploid** has $3n$ chromosomes; a **tetraploid** has $4n$; a **pentaploid**, $5n$; and so forth. Several general statements may be made about polyploidy. This condition is relatively infrequent in most animal species, but is well known in lizards, amphibians, and fish. It is much more common in plant species. Odd numbers of chromosome sets are not usually maintained reliably from generation to generation because a polyploid organism with an uneven number of homologues usually does not produce genetically balanced gametes. For this reason, triploids, pentaploids, and so on, are not usually found in species that depend solely upon sexual reproduction for propagation.

Polyploidy can originate in two ways: (1) the addition of one or more extra sets of chromosomes, identical to the normal haploid complement of the same species, results in **autopolyploidy**; and (2) the combination of chromosome sets from different species may occur as a consequence of hybridization and results in **allopolyploidy**. Thus, the distinction between auto- and allopolyploidy is based on the genetic origin of the extra chromosome sets.

In our discussion of polyploidy, we will use the following symbols to clarify the origin of additional chromosome sets. For example, if $A$ represents the haploid set of chromosomes of any organism, then

$$A = a_1 + a_2 + a_3 + a_4 + \cdots + a_n$$

where $a_1$, $a_2$, and so on represent individual chromosomes, and where $n$ is the haploid number. Thus, a normal diploid organism would be represented simply as *AA*.

## Autopolyploidy

In **autopolyploidy**, each additional set of chromosomes is identical to the parent species. Therefore, **triploids** are represented as *AAA*, **tetraploids** are *AAAA*, and so forth.

**Autotriploids** may arise in several ways. A failure of all chromosomes to segregate during meiotic divisions may produce a diploid gamete. If such a gamete survives and is fertilized by a haploid gamete, a zygote with three sets of chromosomes is produced. Or, occasionally two sperm may fertilize an ovum, resulting in a triploid zygote. Triploids may also be produced under experimental conditions by crossing diploids with tetraploids. Diploid organisms produce gametes with $n$ chromosomes while tetraploids produce $2n$ gametes. Upon fertilization, the desired triploid is produced.

Because they have an even number of chromosomes, **autotetraploids** ($4n$) are theoretically more likely to be found in nature than autotriploids. Unlike triploids, which often produce genetically unbalanced gametes with odd numbers of chromosomes, tetraploids are more likely to produce balanced gametes when involved in sexual reproduction.

Tetraploid cells may be produced experimentally from diploid cells by applying cold or heat shock during mei-osis or by applying colchicine to somatic cells undergoing mitosis. **Colchicine**, an alkyloid derived from the autumn crocus, interferes with spindle formation, and thus replicated chromosomes cannot enter anaphase and migrate to the poles. When colchicine is removed, the cell may reenter interphase. When the paired sister chromatids separate and uncoil, the nucleus will contain twice the diploid number of chromosomes and is effectively $4n$. This process is illustrated in Figure 6.15.

In general, autopolyploids are larger than their diploid relatives. Often, the flower and fruit of plants are increased in size. This increase seems to be due to larger cell size rather than greater cell number. Although autopolyploids do not contain new or unique information compared with the diploid relative, such varieties may be of greater commercial value. Economically important triploid plants include several potato species of the genus *Solanum*, Winesap apples, commercial bananas, seedless watermelons, and the cultivated tiger lily *Lilium tigrinum*. These plants are propagated asexually. Diploid bananas contain hard seeds, but the commercial, triploid, "seedless" variety has edible seeds. Tetraploid alfalfa, coffee, peanuts, and McIntosh apples are also of economic value because they are either larger or grow more vigorously than their diploid or triploid counterparts. The commercial strawberry is an octoploid. These observations attest to the importance of autopolyploidy in domesticated plants.

## Allopolyploidy

Polyploidy may also result from hybridization of two closely related species. Thus, if a haploid ovum from a

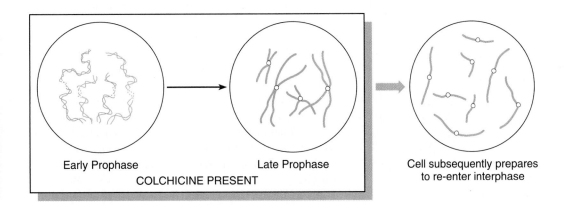

**FIGURE 6.15**    The potential involvement of colchicine in doubling the chromosome number, as occurs during the production of an autotetraploid. Two pairs of homologous chromosomes are followed. While each chromosome has replicated its DNA earlier during interphase, the chromosomes do not appear as double structures until late prophase. The impact of colchicine is in anaphase, when the two chromatids of each chromosome fail to separate and be pulled toward each pole.

species with chromosome sets *AA* is fertilized by a haploid sperm from a species with sets *BB*, the resulting hybrid is *AB*, where $A = a_1, a_2, a_3, \ldots, a_n$ and $B = b_1, b_2, b_3, \ldots, b_n$. The hybrid may be sterile because of its inability to produce viable gametes. This occurs because, most often, some or all of the *a* and *b* chromosomes cannot synapse in meiosis, and unbalanced genetic conditions result. If, however, the new *AB* genetic combination undergoes a natural or induced chromosomal doubling, a fertile *AABB* tetraploid is produced. These events are illustrated in Figure 6.16. Since this polyploid contains the equivalent of four haploid genomes and since the hybrid contains unique genetic information compared with either parent, such an organ-

ism is called an **allotetraploid** (from the Greek word *allo*, meaning other or different). An equivalent term, **amphidiploid**, is also used to describe this situation in which a hybrid organism contains two complete diploid genomes. This latter term is preferred when the original species are known.

Allopolyploidy is the most common natural form of polyploidy in plants because the chance of forming balanced gametes is much greater than in other types of polyploidy. Since two homologues of each specific chromosome are present, meiosis may occur normally (Figure 6.16), and fertilization may successfully propagate the tetraploid sexually. This discussion assumes the simplest situation, where none of the chromosomes in set *A* are homologous to those in set *B*. In hybrids of very closely related species, some homology between *a* and *b* chromosomes is likely. In this case, meiotic pairing is more complex. Multivalents, which are more complex than the trivalents shown in Figure 6.10, may be formed, resulting in the production of unbalanced gametes. In such cases, aneuploid varieties of allotetraploids may arise. Allopolyploids are rare in most animals because mating behavior is most often species-specific, and thus the initial step in hybridization is unlikely to occur.

A classical example of allotetraploidy in plants is the cultivated species of American cotton, *Gossypium* (Figure 6.17). This species has 26 pairs of chromosomes: 13 are large and 13 are much smaller. When it was discovered that Old World cotton had only 13 pairs of large chromosomes, allopolyploidy was suspected. After an examination of wild American cotton revealed 13 pairs of small chromosomes, this speculation was strengthened. J. O. Beasley was able to reconstruct the origin of cultivated cotton experimentally. He crossed the Old World strain with the wild American strain and then treated the hybrid with colchicine to double the chromosome number. The result of these treatments was a fertile allotetraploid variety of cotton. It contained 26 pairs of chromosomes and characteristics similar to the cultivated variety.

Allotetraploids often exhibit characteristics of both parental species. An interesting example, but one with no practical economic importance, is that of the hybrid formed between the radish *Raphanus sativus* and the cabbage *Brassica oleracea*. Both species have a haploid number of 9. The initial hybrid consists of 9 *Raphanus* and 9 *Brassica* chromosomes (9R + 9B). While hybrids are almost always sterile, some fertile allotetraploids (18R + 18B) have been produced. Unfortunately, the root of this allotetraploid is more like the cabbage and its shoot resembles the radish. Had the converse occurred, the hybrid might have been of economic importance.

A much more successful commercial hybridization has been performed using the grasses wheat and rye. Wheat

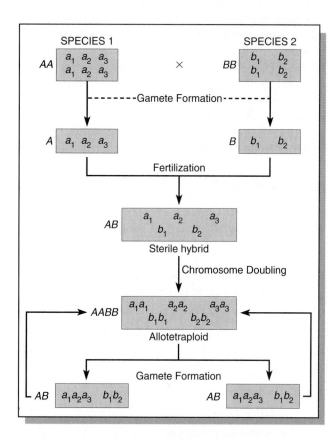

**FIGURE 6.16** The origin and propagation of an allotetraploid. Species 1 contains genome A consisting of three distinct chromosomes, $a_1$, $a_2$, and $a_3$. Species 2 contains genome B consisting of two distinct chromosomes, $b_1$ and $b_2$. Following fertilization between members of the two species and chromosome doubling, a fertile allotetraploid containing four complete haploid genomes (*AABB*) is formed.

**FIGURE 6.17**    The allotetraploid form of *Gossypium,* the cultivated cotton plant.

(genus *Triticum*) has a basic haploid genome of 7 chromosomes. The diploid species have 14 chromosomes. Cultivated autopolyploids exist, including tetraploid ($4n = 28$) and hexaploid ($6n = 42$) varieties. Rye (genus *Secale*) also has a genome consisting of 7 chromosomes. The only cultivated species is diploid ($2n = 14$).

Using the technique outlined in Figure 6.16, geneticists have produced various allopolyploid hybrids. When tetraploid wheat is crossed with diploid rye, and the $F_1$ treated with colchicine, a hexaploid variety ($6n = 42$) is derived. The hybrid, designated *Triticale* (see Figure 1.10), represents a new genus. Fertile hybrid varieties derived from various wheat and rye species may be crossed together or back-crossed. These crosses have created many variations of the genus *Triticale.*

The hybrid plants demonstrate characteristics of both wheat and rye. For example, certain hybrids combine the high protein content of wheat with the high content of the amino acid lysine of rye. The lysine content is low in wheat and thus is a limiting nutritional factor. Wheat is considered a high-yielding grain, while rye is noted for its versatility of growth in unfavorable environments. *Triticale* species, combining both traits, have the potential of significantly increasing grain production. Programs designed to improve crops through hybridization have long been underway in several underdeveloped countries of the world, as discussed in Chapter 1.

Recall that in the previous chapter, we discussed the use of **somatic cell hybrids** to map human genes. This technique has also been applied to the production of allopolyploid plants (Figure 6.18). Cells from the developing leaves of plants can be treated to remove their cell wall, resulting in **protoplasts**. These altered cells can be maintained in culture and stimulated to fuse with other protoplasts, producing somatic cell hybrids.

If cells from different plant species are fused in this way, hybrid cells can be produced with unique chromosome combinations. Since protoplasts can be induced to divide and differentiate into stems which develop leaves, the potential for producing allopolyploids is available in the research laboratory. In some cases, entire plants may be derived from cultured protoplasts. If only stems and leaves are produced, these may be grafted onto the stem of another plant. If flowers are formed, fertilization may yield mature seeds, which upon germination yield an allopolyploid plant.

There are now many examples of allopolyploids which have been created commercially using the above approach. While most of them are not true amphidiploids since one or more chromosomes are missing, precise allotetraploids are sometimes produced.

## Endopolyploidy

Certain cells in an otherwise diploid organism have been observed to be polyploid. The term **endopolyploidy** describes this general condition. In such cells, replication and separation of chromosomes occur without nuclear division. The process leading to endopolyploidy is called **endomitosis**.

Vertebrate liver cell nuclei, including human ones, often contain $4n$, $8n$, or $16n$ chromosome sets. The stem and parenchyma tissue of apical regions of flowering plants are also often endopolyploid. Cells lining the gut of mosquito larvae attain a $16n$ ploidy, but during the pupal stages such cells undergo very quick reduction divisions, giving rise to smaller diploid cells. In the water strider *Gerris*, wide variations in chromosome numbers are found in different tissues, with as many as 1024 to 2048 copies of each chromosome in the salivary gland cells. Since the diploid number in this organism is 22, the nuclei of these cells may contain over 40,000 chromosomes!

Although the role of endopolyploidy is not clear, the proliferation of chromosome copies often occurs in cells where high levels of certain gene products are required. In fact, it is well established that certain genes, whose product is in high demand in *every* cell, exist naturally in multiple copies in the genome. Ribosomal and transfer RNA genes are examples of multiple-copy genes. In certain cells of organisms, it may be necessary to replicate the entire genome, allowing an even greater rate of expression of various genes.

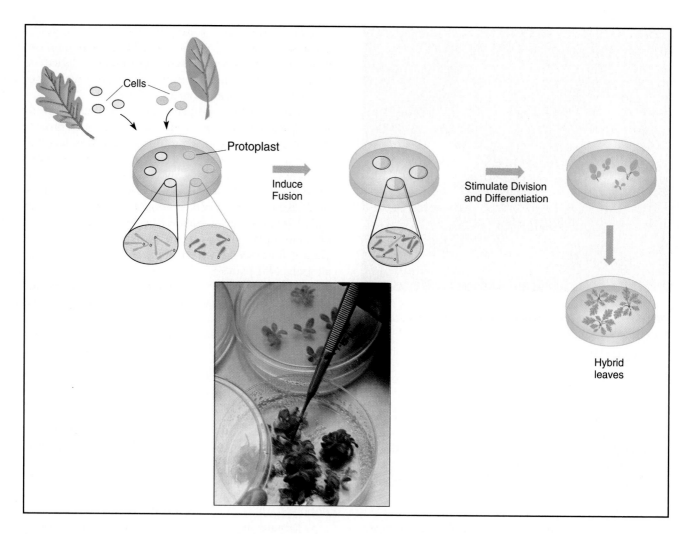

**FIGURE 6.18**    Application of the somatic cell hybridization technique in the production of an amphidiploid. Cells from the leaves of two species of plants are removed and cultured. The cell walls are digested away and the resultant protoplasts are induced to undergo cell fusion. The hybrid cell is selected and stimulated to divide and differentiate, as illustrated in the photograph. An amphidiploid has a complete set of chromosomes from each parental cell type and displays phenotypic characteristics of each. Two pairs of chromosomes from each species are depicted.

# VARIATION IN CHROMOSOME STRUCTURE AND ARRANGEMENT: AN OVERVIEW

The second general class of chromosome aberrations includes structural changes that delete, add, or rearrange substantial portions of one or more chromosomes. Included in this category are **deletions** and **duplications** of genes or part of a chromosome and **rearrangements** of genetic material in which a chromosome segment is either inverted, exchanged with a segment of nonhomologous chromosome, or merely transferred to one, leading to **translocations**. Before discussing these aberrations, we will present several general statements pertaining to them.

In most instances, these structural changes are due to one or more breaks along the axis of a chromosome, followed by either the loss or rearrangement of genetic material. Chromosomes can break spontaneously, but the rate of breakage may increase in cells exposed to chemicals or radiation. Although the actual ends of chromosomes, the **telomeres**, do not readily fuse with newly created ends of "broken" chromosomes or with other telomeres, the ends produced at points of breakage can fuse with other such ends. If breakage–rejoining does not merely reestablish the original relationship, and if the alteration occurs in germ plasm, the gametes will contain a structural rearrangement that will be heritable.

When the aberration is found in one homologue but not the other, the individual is said to be heterozygous for the aberration. In such cases, unusual but characteristic pairing configurations are formed during meiotic synapsis. These patterns are useful in identifying the type of change that has occurred. If no loss or gain of genetic material occurs, individuals bearing the aberration "heterozygously" will likely be unaffected phenotypically. However, the unusual pairing arrangements often lead to gametes that are duplicated or deficient for chromosomal regions. Thus, the offspring of "carriers" of certain aberrations often have an increased probability of demonstrating phenotypic manifestations of the aberration.

## DELETIONS

When a portion of a chromosome is lost, the missing piece is referred to as a **deletion** (or a **deficiency**). The

deletion may occur either at one end or from the interior of the chromosome. These are called **terminal** or **intercalary deletions**, respectively. Both result from one or more breaks in the chromosome. The portion retaining the centromere region will usually be maintained and that segment without the centromere will eventually be lost in progeny cells following mitosis or meiosis. In order for synapsis to occur between a chromosome with a large intercalary deficiency and a normal complete homologue, the unpaired region of the normal homologue must "buckle out." Such a configuration is called a **deficiency loop** or **compensation loop**. The origins of both types of deletion and the formation of such a loop are diagrammed in Figure 6.19.

If too much genetic information is lost as a result of a deletion, the aberration may be lethal and never becomes available for study. As seen in the cri-du-chat syndrome, where only part of the short arm of chromosome

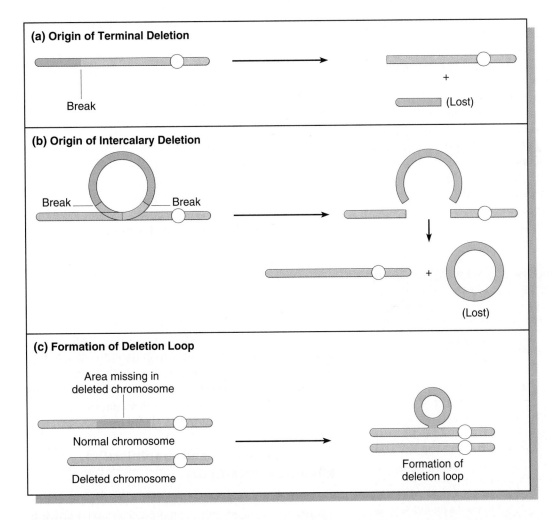

**FIGURE 6.19**    Origins of (a) a terminal and (b) intercalary deletion. In part (c), pairing occurs between a normal chromosome and one with an intercalary deletion by looping out the undeleted portion to form a deficiency loop.

5 is lost, a deficiency need not be very great before the effects become severe.

A final consequence of deletions may be noted in organisms heterozygous for a deficiency. Consider the mutant *Notch* phenotype in *Drosophila*. In these flies, the wings are notched out on the posterior and lateral margins. Data from breeding studies seem to indicate that the phenotype is controlled by a sex-linked dominant mutation since heterozygous females have notched wings and transmit this allele to one-half of their female progeny. The mutation also appears to behave as a homozygous and hemizygous lethal since such females and males are never recovered. It has also been noted that if notched-winged females are also heterozygous for the closely linked recessive mutations *white*-eye, *facet*-eye, or *split*-bristle, they express these mutant phenotypes as well as *Notch*. Because these mutations are recessive, heterozygotes should express the normal, wild-type phenotypes. These genotypes and phenotypes are summarized in Table 6.4.

These observations have been explained through a cytological examination of the polytene X chromosome of heterozygous *Notch* females. A deficiency loop was found along the X chromosome from band 3C2 through band 3C11, as shown in Figure 6.20. These bands had previously been shown to include the *white, facet,* and *split* loci, among others. On the genetic map, this region corresponds to loci 1.5 to 3.0. Thus, this region's deficiency in one of the two homologous X chromosomes has two distinct effects. First, it results in the *Notch* phenotype. Second, by deleting the loci for genes whose mutant alleles are present on the other X chromosome, the deficiency creates a partially hemizygous condition so that the recessive *white, facet,* or *split* alleles are ex-

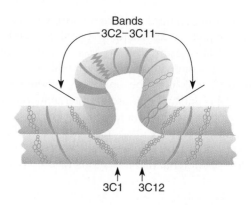

**FIGURE 6.20**    Deficiency loop formed in salivary chromosomes heterozygous for the *Notch* deletion of the X chromosome of *Drosophila melanogaster*. The deficiency encompasses bands 3C2 through 3C11.

pressed. This type of phenotypic expression of recessive genes in association with a deletion is an example of the phenomenon called **pseudodominance**.

Many independently arising *Notch* phenotypes have been investigated. The common deficient band for all *Notch* phenotypes has now been designated as 3C7. In every case that *white* was also expressed pseudodominantly, the common deficient band was 3C2. The bands that cytologically distinguish the *white* locus from the *Notch* locus have been confirmed in this manner.

## DUPLICATIONS

When any part of the genetic material—a single locus or a large piece of a chromosome—is present more than once in the genome, it is called a **duplication**. As in deletions, pairing in heterozygotes may produce a compensation loop. Duplications may arise as the result of unequal crossing over between synapsed chromosomes during meiosis (Figure 6.21) or through a replication error prior to meiosis. In the former case, both a duplication and a deficiency are produced.

Three interesting aspects of duplications will be considered. First, they may result in gene redundancy. Second, as with deletions, duplications may produce phenotypic variation. Third, according to one convincing theory, duplications have also been an important source of genetic variability during evolution.

### Gene Redundancy and Amplification: Ribosomal RNA Genes

Although many gene products are not needed in every cell of an organism, other gene products are known to be essential components of all cells. For example, ribosomal RNA must be present in abundance in order to

**Table 6.4**   *NOTCH* GENOTYPES AND PHENOTYPES

| Genotype | Phenotype |
|---|---|
| $\dfrac{N^+}{N}$ | *Notch* female |
| $\left.\dfrac{N}{N}\right\}$ ⟶ | *Lethal* |
| $\dfrac{N^+\ w}{N\ w^+}$ | *Notch, white* female |
| $\dfrac{N^+ fa\ spl}{N\ fa^+ spl^+}$ | *Notch, facet, split* female |

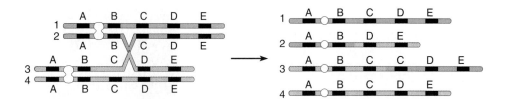

**FIGURE 6.21**    The origin of duplicated and deficient regions of chromosomes as a result of unequal crossing over. The tetrad at the left is mispaired during synapsis. A single crossover between chromatids 2 and 3 results in deficient and duplicated chromosomal regions (see chromosomes 2 and 3 respectively, on the right). The two chromosomes uninvolved in the crossover event remain normal in their gene sequence and content.

support protein synthesis. The more metabolically active a cell is, the higher is the demand for this molecule that becomes a part of each ribosome. In theory, a single copy of the gene encoding rRNA may be inadequate in many cells. Studies using the technique of molecular hybridization, which allow the determination of the percentage of the genome coding for specific RNA sequences, show that in most organisms, there are multiple copies of the genes coding for rRNA. Such DNA is called **rDNA**, and the general phenomenon is called **gene redundancy**. For example, in the common intestinal bacterium *Escherichia coli* (*E. coli*) about 0.4 percent of the haploid genome consists of rDNA. This is equivalent to 5 to 10 copies of the gene. In *Drosophila melanogaster*, 0.3 percent of the haploid genome, equivalent to 130 copies, consists of rDNA. Although the presence of multiple copies of the same gene is not restricted to those coding for rRNA, we will focus on them in this section.

Studies of *Drosophila* have documented the need for the extensive amounts of rRNA and ribosomes made possible by multiple copies of these genes. In this organism, the X-linked mutation *bobbed* has, in fact, been shown to be due to a deletion of a variable number of the 130 genes coding for rRNA. Mutant flies have low viability, are underdeveloped, and have bristles reduced in size. General development as well as bristle formation, which occurs very rapidly during normal pupal development, apparently depend on a great number of ribosomes to support protein synthesis. Many *bobbed* alleles have been studied, and each has been shown to involve a deletion, often of a unique size. The extent to which both viability and bristle length are decreased correlates well with the relative number of RNA genes deleted. Thus, we may conclude that the normal rDNA redundancy observed in wild-type *Drosophila* is near the minimum required for adequate ribosome production during normal development.

In some cells, particularly oocytes, even the normal redundancy of rDNA may be insufficient to provide adequate amounts of rRNA and ribosomes. Oocytes store abundant nutrients in the ooplasm for use by the embryo during early development. In fact, more ribosomes are included in the oocytes than in any other cell type. By considering how the amphibian *Xenopus laevis* (the South African clawed frog) acquires this abundance of ribosomes, we will see a second way in which the amount of rRNA is increased. This phenomenon is referred to as **gene amplification**.

The genes that code for rRNA are located in an area of the chromosome known as the **nucleolar organizer region (NOR)**. The NOR is intimately associated with the nucleolus, which is a processing center for ribosome production. Molecular hybridization analysis has shown that each NOR in *Xenopus* contains the equivalent of 400 redundant gene copies coding for rRNA. Even this number of genes is apparently inadequate to synthesize the vast amount of ribosomes that must accumulate in the amphibian oocyte to support development following fertilization. To further amplify the number of rRNA genes, the rDNA is selectively replicated, and each new set of genes is released from its template. Since each new copy is equivalent to the NOR, multiple small nucleoli are formed around each NOR in the oocyte. As many as 1500 of these "micronucleoli" have been observed in a single oocyte. If we multiply the number of micronucleoli (1500) by the number of gene copies in each NOR (400), we see that amplification in *Xenopus* oocytes can result in over half a million gene copies! If each copy is transcribed only 20 times during the maturation of a single oocyte, in theory, sufficient copies of rRNA may be produced to result in well over one million ribosomes.

### The *Bar* Eye Mutation in *Drosophila*

Duplications may cause phenotypic variation that at first might appear to be caused by a simple gene mutation. The *Bar* eye phenotype (Figure 6.22) in *Drosophila* is a classical example. Instead of the normal oval eye shape, *Bar*-eyed flies have narrow, slitlike eyes. This phenotype appears to be inherited as a dominant sex-linked mutation. However, since both heterozygous

**(a) Genotypes and Phenotypes**

| Genotype | Facet Number | Phenotype | Normal ⊟ = 16A segments |
|---|---|---|---|
| $B^+/B^+$ | 779 | | |
| $B/B^+$ | 358 | | |
| $B/B$ | 68 | | |
| $B^D/B^+$ | 45 | | |

**(c)**

$B^+/B^+$

$B/B^+$

$B/B$

**(b) Origin of $B^D$ allele as a result of unequal crossing over**

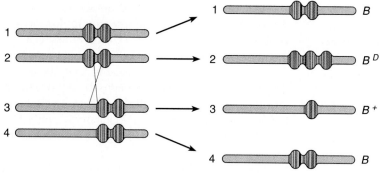

1 → 1 $B$
2 → 2 $B^D$
3 → 3 $B^+$
4 → 4 $B$

**FIGURE 6.22**    (a) Summary of the duplication genotypes and resultant *Bar* eye phenotypes. (b) The origin of the $B^D$ (double *Bar*) allele. (c) Photographs illustrate the actual *Bar* eye phenotypes compared to wild type ($B^+/B^+$).

females and hemizygous males exhibit the trait, but homozygous females show a more pronounced phenotype than either of these two cases, the inheritance is more accurately described as **semidominant**.

In the early 1920s Alfred H. Sturtevant and Thomas H. Morgan discovered and investigated this "mutation." As illustrated in Figure 6.22(a), normal wild-type females ($B^+/B^+$) have about 800 facets in each eye. Heterozygous females ($B/B^+$) have about 350 facets, while homozygous females ($B/B$) average only about 50 facets. Females are occasionally recovered with even fewer facets and are designated as *double Bar* ($B^+/B^D$).

About ten years later, Calvin Bridges and Herman J. Muller compared the polytene X chromosome banding pattern of the *Bar* fly with that of the wild-type fly. Such chromosomes contain specific banding patterns that have been well categorized into regions. Their studies

revealed that one copy of region 16A of the X chromosome was present in wild-type flies and that this region was duplicated in *Bar* flies and triplicated in *double Bar* flies. These observations provided evidence that the *Bar* phenotype is not the result of a simple chemical change in the gene, but is instead a duplication. The *double Bar* condition originates as a result of unequal crossing over, which produces the triplicated 16A region [Figure 6.22(b)].

This figure also illustrates what is referred to as a **position effect**. When the eye facet phenotypes of $B/B$ and $B^+/B^D$ flies are compared, an average of 68 and 45 facets are found, respectively. In both cases, there are two extra 16A regions. When the repeated segments are distributed on the same homologue instead of on two homologues, the phenotype is more pronounced. Thus, the same amount of genetic information produces a slightly different phenotype depending on the position of the genes.

## The Role of Gene Duplication in Evolution

One of the most intriguing aspects of the study of evolution is the consideration of the mechanisms for genetic variation. The origin of unique gene products present in phylogenetically advanced organisms but absent in less advanced, ancestral forms is a topic of particular interest. In other words, how do "new" genes arise?

In 1970, Susumo Ohno published the provocative monograph *Evolution by Gene Duplication* in which he suggests that gene duplication is essential to the origin of new genes during evolution. Ohno's thesis is based on the supposition that the gene products of essential genes, present as only a single copy in the genome, are indispensable to the survival of members of any species during evolution. Therefore, these genes are not free to accumulate mutations sufficient to alter their primary function and give rise to new genes.

However, if an essential gene were to become duplicated in a germ cell, the new copy would be inherited in all future generations. Since it is an extra copy, major mutational changes in it are tolerated because the original gene still provides the genetic information for its essential function. The duplicated copy is now free to acquire large numbers of mutational changes over long periods of time. Over short intervals, the new genetic information may be of no practical advantage. However, over long periods of evolutionary time, the duplicated gene may change sufficiently so that its product assumes a divergent role in the cell. The new function may impart an adaptability and survival advantage to organisms carrying such unique genetic information. Thus, Ohno has outlined a mechanism through which sustained genetic variability may have originated.

Ohno's thesis is supported by the discovery of genes that have a substantial amount of their DNA sequence in common, but whose gene products are distinct. For example, trypsin and chymotrypsin fit this description, as do myoglobin and hemoglobin. The DNA sequence homology is great enough in these cases to conclude that members of each gene pair arose from a common ancestral gene through duplication. During evolution, the related genes diverged sufficiently so that their products became unique.

Other support includes the presence of **gene families**, groups of contiguous genes whose products perform the same general function. Again, members of a family show DNA sequence homology sufficient to conclude that they share a common origin. The various types of globin chains that are part of hemoglobin is one example. As we will discuss in Chapter 15, various globin chains function at different times during development, but they are all a part of hemoglobin, functioning to transport oxygen. Other examples include the immunologically important **T-cell receptors** and antigens encoded by the **major histocompatibility complex (MHC)**.

## INVERSIONS

The **inversion**, another class of structural variation, is a type of chromosomal aberration in which a segment of a chromosome is turned around 180° within a chromosome. An inversion does not involve a loss of genetic information, but simply rearranges the linear gene sequence. An inversion requires two breaks along the length of the chromosome and subsequent reinsertion of the inverted segment. Figure 6.23 illustrates how one type of inversion might arise. By forming a chromosomal

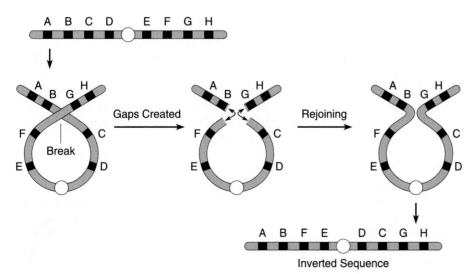

**FIGURE 6.23**    One possible origin of a pericentric inversion.

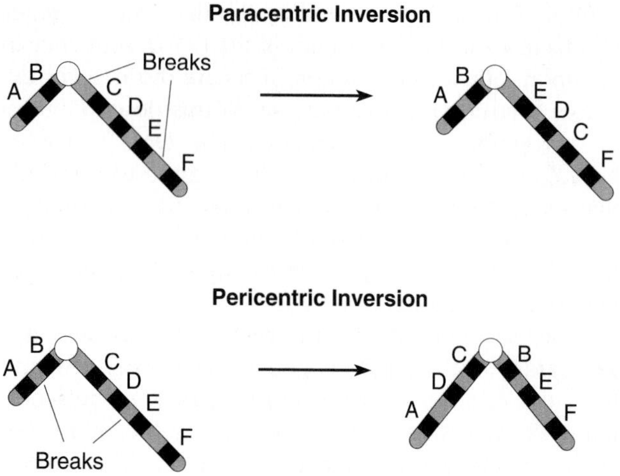

**Paracentric Inversion**

**Pericentric Inversion**

**FIGURE 6.24**    A comparison of the arm ratios before and after the occurrence of a paracentric and pericentric inversion.

loop prior to breakage, the newly created "sticky ends" are brought close together and rejoined.

The inverted segment may be short or quite long and may or may not include the centromere. If the centromere is *not* part of the rearranged chromosome segment, the inversion is said to be **paracentric**. If the centromere *is* included in the inverted segment, the term **pericentric** describes the inversion, as in Figure 6.23.

Although the gene sequence has been reversed in the paracentric inversion, the ratio of arm lengths extending from the centromere is unchanged. In contrast (Figure 6.24), some pericentric inversions create chromosomes with arms of different lengths than those of the noninverted chromosome. Thus, the **arm ratio** is often changed when a pericentric inversion is produced. The change in arm lengths may sometimes be detected during the metaphase stage of mitotic or meiotic divisions.

Although inversions may seem to have a minimal impact on the genetic material, their consequences are of interest to genticists. Organisms heterozygous for inversions may produce aberrant gametes. Inversions may also result in position effects as well as play an important role in the evolutionary process.

## Consequences of Inversions during Gamete Formation

If only one member of a homologous pair of chromosomes has an inverted segment, normal linear synapsis during meiosis is not possible. Organisms with one inverted chromosome and one noninverted homologue are called **inversion heterozygotes**. Pairing between two such chromosomes in meiosis may be accomplished

only if they form an **inversion loop**. In other cases, no loop can be formed and the homologues are seen to synapse everywhere but along the length of the inversion, where they appear separated.

If crossing over does not occur within the inverted segment of the inversion heterozygote, the homologues will segregate and result in two normal and two inverted chromatids that are distributed into gametes. However, if crossing over occurs within the inversion loop, abnormal chromatids are produced. The effect of single exchange events within a paracentric inversion is diagrammed in Figure 6.25(a).

As in any meiotic tetrad, a single crossover between nonsister chromatids produces two parental chromatids and two recombinant chromatids. One recombinant chromatid is **dicentric** (two centromeres), and one recombinant chromatid is **acentric** (lacking a centromere). Both contain duplications and deletions of chromosome segments as well. During anaphase, an acentric chromatid moves randomly to one pole or the other or may be lost, while a dicentric chromatid is pulled in two directions. This polarized movement produces **dicentric bridges** that are cytologically recognizable. A dicentric chromatid will usually break at some point so that part of the chromatid goes into one gamete and part into another gamete during the reduction divisions. Therefore, gametes containing either recombinant chromatid are deficient in genetic material. When such a gamete participates in fertilization, the zygote most often develops abnormally, if at all.

A similar chromosomal imbalance is produced as a result of a crossover event between a chromatid bearing a pericentric inversion and its noninverted homologue [Figure 6.25(b)]. The recombinant chromatids that are directly involved in the exchange have duplications and deletions. However, no acentric or dicentric chromatids are produced. Gametes receiving these chromatids also produce inviable embryos following their participation in fertilization.

Because fertilization events involving these aberrant chromosomes do not produce viable offspring, it *appears* as if the inversion suppresses crossing over since offspring bearing crossover gametes are not recovered. *Actually*, in inversion heterozygotes, the inversion has the effect of **suppressing the recovery of crossover products when chromosome exchange occurs within the inverted region**. If crossing over always occurred within a paracentric or pericentric inversion, 50 percent of the gametes would be ineffective. The viability of the resulting zygotes is therefore greatly diminished. Furthermore, up to one-half of the viable gametes have the inverted chromosome, and the inversion will be perpetuated within the species. The cycle will be repeated continuously during meiosis in future generations.

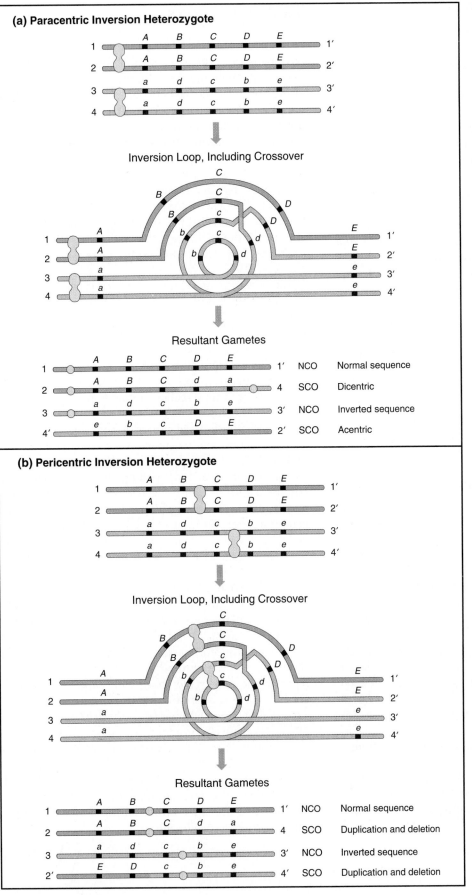

**FIGURE 6.25** The effects of a single crossover with an inversion loop in cases involving (a) a paracentric inversion and (b) a pericentric inversion. In (a), two altered chromosomes are produced, one that is acentric and one that is dicentric. Both chromosomes also contain duplicated and deficient regions. In (b), two altered chromosomes are produced, both with duplicated and deficient regions.

## Position Effects of Inversions

Another consequence of inversions involves the new positioning of genes relative to other genes and particularly to areas of the chromosome which do not contain genes, such as the centromere. If the expression of the gene is altered as a result of its relocation, a change in phenotype may result. Such a change is an example of what is called a **position effect**, as introduced in our earlier discussion of the *Bar* duplication.

In *Drosophila* females heterozygous for the sex-linked recessive mutation *white* eye ($w^+/w$), the X chromosome bearing the wild-type allele ($w^+$) may be inverted such that the *white* locus is moved to a point adjacent to the centromere. If the inversion is not present, a heterozygous female has wild-type red eyes, since the *white* allele is recessive. Females with the X chromosome inversion described above have eyes that are mottled or variegated (i.e., with red and white patches). Placement of the $w^+$ allele next to the centromere apparently inhibits wild-type gene expression, resulting in the loss of complete dominance over the *w* allele. Other genes, also located on the X chromosome, behave in a similar manner when they are relocated. Reversion to wild-type expression has sometimes been noted. When this has occurred, cytological examination has shown that the inversion has reestablished the normal gene sequence.

## Evolutionary Consequences of Inversions

One major effect of an inversion is the maintenance of a set of specific alleles at a series of adjacent loci, provided that they are contained within the inversion. Because the recovery of crossover products is suppressed in inversion heterozygotes, a particular gene sequence is preserved intact in the viable gametes. If this gene order provides a survival advantage to organisms maintaining it, the inversion is beneficial to the evolutionary survival of the species.

For example, if the set of alleles *ABcDef* is more adaptive than the sets *AbCdeF* or *abcdEF*, the favorable set will not be disrupted by crossing over if it is maintained within a heterozygous inversion. As we will see in Chapter 25, there are documented examples where inversions are adaptive in this way. Specifically, Theodosius Dobzhansky has shown that the maintenance of different inversions on chromosome 3 of *Drosophila pseudoobscura* through many generations has been highly adaptive to this species. Certain inversions seem to be characteristic of enhanced survival under specific environmental conditions.

## TRANSLOCATIONS

**Translocation**, as the name implies, involves the movement of a segment of a chromosome to a new place in the genome. Translocation may occur within a single chromosome or between nonhomologous chromosomes. The exchange of segments between two nonhomologous chromosomes is a type of structural variation called a **reciprocal translocation**. The origin of a relatively simple reciprocal exchange is illustrated in Figure 6.26(a). The least complex way for this event to occur is for two nonhomologous chromosome arms to come close to each other so that an exchange is facilitated. For this type of translocation, only two breaks are required. If the exchange includes internal chromosome segments, four breaks are required, two on each chromosome.

The genetic consequences of reciprocal translocations are, in several instances, similar to those of inversions. For example, genetic information is not lost or gained. Rather, there is only a rearrangement of genetic material. The presence of a translocation does not, therefore, directly alter viability of individuals bearing it. Like an inversion, a translocation may also produce a position effect, because it may realign certain genes in relation to other genes. This exchange may create a new genetic linkage relationship that can be detected experimentally.

Homologues heterozygous for a reciprocal translocation undergo unorthodox synapsis during meiosis. As shown in Figure 6.26(b), pairing results in a crosslike configuration. As with inversions, genetically unbalanced gametes are also produced as a result of this unusual alignment during meiosis. In the case of translocations, however, aberrant gametes are not necessarily the result of crossing over. To see how unbalanced gametes are produced, focus on the homologous centromeres in Figure 6.26(b) and (c). According to the principle of independent assortment, the chromosome containing centromere 1 will migrate randomly toward one pole of the spindle during the first meiotic anaphase; it will travel with *either* the chromosome having centromere 3 *or* centromere 4. The chromosome with centromere 2 will move to the other pole along with *either* the chromosome containing centromere 3 *or* centromere 4. This results in four potential meiotic products. The 1,4 combination contains chromosomes uninvolved in the translocation. The 2,3 combination, however, contains translocated chromosomes. These contain a complete complement of genetic information and are balanced. The other two potential products, the 1,3 and 2,4 combinations, will contain chromosomes displaying duplicated and deleted segments.

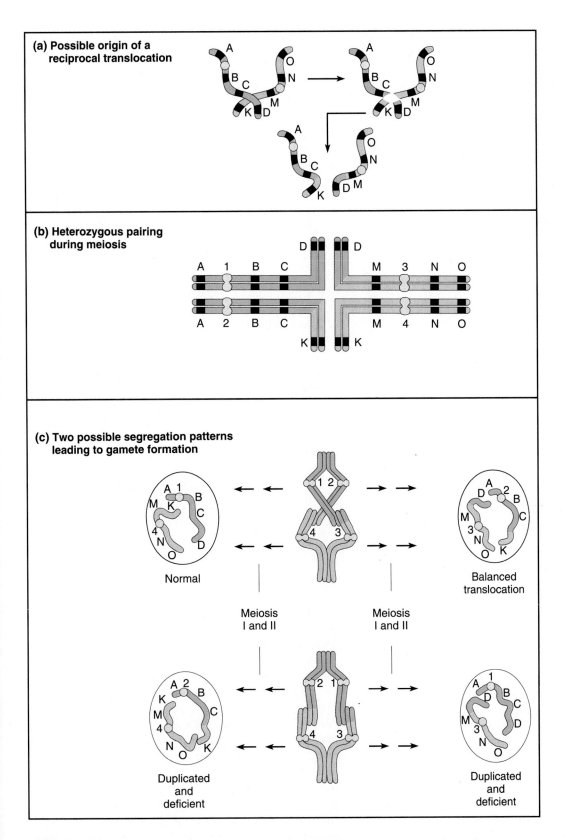

**(a) Possible origin of a reciprocal translocation**

**(b) Heterozygous pairing during meiosis**

**(c) Two possible segregation patterns leading to gamete formation**

Normal

Balanced translocation

Meiosis I and II

Meiosis I and II

Duplicated and deficient

Duplicated and deficient

**FIGURE 6.26**   Origin, synapsis, and gamete formation in a simple reciprocal translocation.

When incorporated into gametes, the resultant meiotic products are genetically unbalanced. If they participate in fertilization, lethality often results. As few as 50 percent of the progeny of parents heterozygous for a reciprocal translocation may survive. This condition, called **semisterility**, has an impact on the reproductive fitness of organisms, thus playing a role in evolution. Furthermore, in humans, such an unbalanced condition results in partial monosomy or trisomy, leading to a variety of birth defects.

## Familial Down Syndrome

Research performed since 1959 has revealed numerous translocations in members of the human population.

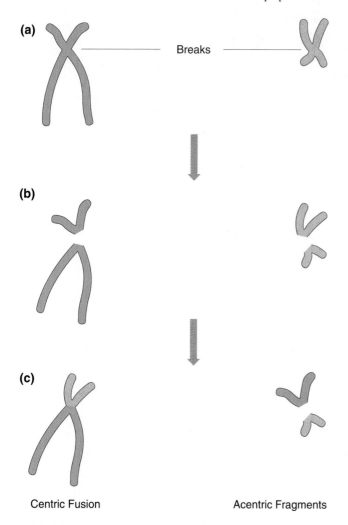

(a)

Breaks

(b)

(c)

Centric Fusion                    Acentric Fragments

**FIGURE 6.27**    The possible origin of a Robertsonian translocation. (a) Two independent breaks occur at the centromeric region on two nonhomologous chromosomes. (b) Fragments that are produced. (c) Centric fusion of the long arms of the two acrocentric chromosomes. Two acentric fragments remain.

One common type of translocation involves breaks at the extreme ends of the short arms of two nonhomologous acrocentric chromosomes. These small segments are lost, and the larger segments fuse at their centromeric region (Figure 6.27). This type of translocation produces a new, large submetacentric or metacentric chromosome and is called a **centric fusion translocation**. Such an occurrence in other organisms is called a **Robertsonian fusion**.

Just such a translocation accounts for cases in which Down syndrome is inherited or familial. Earlier in this chapter we pointed out that most instances of Down syndrome are due to trisomy 21. This chromosome composition results from nondisjunction during meiosis of one parent. Trisomy accounts for over 95 percent of all cases of Down syndrome. In such cases, the chance of the same parents producing a second afflicted child is extremely low. However, in the remaining families with a Down child, the syndrome occurs in a much higher frequency over several generations.

Cytogenetic studies of the parents and their offspring from these unusual cases explain the cause of **familial Down syndrome**. Analysis reveals that one of the parents contains a **14/21 D/G translocation** (Figure 6.28). That is, one parent has the majority of the G-group chromosome 21 translocated to one end of the D-group chromosome 14. This individual is normal even though he or she has only 45 chromosomes.

During meiosis, one-fourth of the individual's gametes will have two copies of chromosome 21: a normal chromosome and most of a second copy translocated to chromosome 14. When such a gamete is fertilized by a standard haploid gamete, the resulting zygote has 46 chromosomes but three copies of chromosome 21. These individuals exhibit Down syndrome. Other potential surviving offspring contain either the standard diploid genome (without a translocation) or the balanced translocation like the parent. Both cases result in normal individuals. Knowledge of translocations has allowed geneticists to resolve the seeming paradox of an inherited trisomic phenotype in an individual with an apparent diploid number of chromosomes.

It is of interest to note that the "carrier," who has 45 chromosomes and exhibits a normal phenotype, does not contain the *complete* diploid amount of genetic material. A small region is lost from both chromosomes 14 and 21 during the translocation event. This occurs because the ends of both chromosomes have broken off prior to their fusion. These specific regions are known to be two of many chromosomal locations housing multiple copies of the genes encoding rRNA, the major component of ribosomes. In spite of the loss of up to 10 percent of these genes, the carrier individual is unaffected.

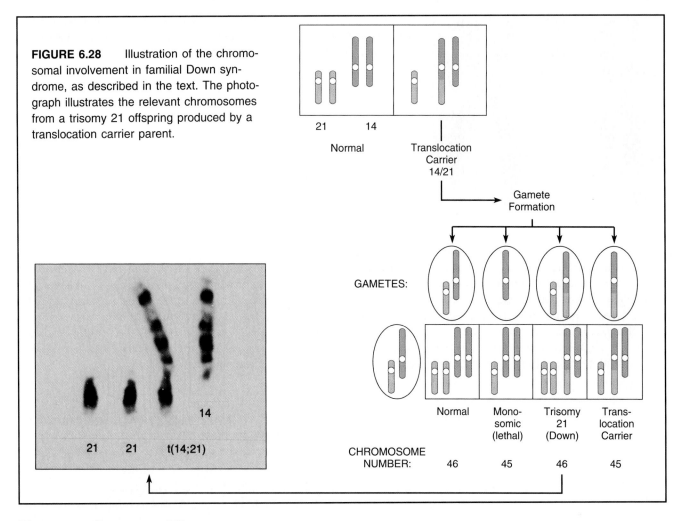

**FIGURE 6.28**    Illustration of the chromosomal involvement in familial Down syndrome, as described in the text. The photograph illustrates the relevant chromosomes from a trisomy 21 offspring produced by a translocation carrier parent.

# FRAGILE SITES IN HUMANS

We conclude this chapter by briefly discussing the results of an interesting discovery made around 1970 during observations of metaphase chromosomes prepared from human cell cultures. In cells derived from certain individuals, a specific area along one of the chromosomes failed to stain, giving the appearance of a gap. In other individuals whose chromosomes displayed or expressed such morphology, the gap appeared in different positions within the set of chromosomes. Such areas eventually became known as **fragile sites**, since they were considered susceptible to chromosome breakage when cultured in the absence of certain chemicals such as folic acid, which is normally present in the culture medium. Fragile sites were at first considered curiosities until a strong association was subsequently shown to exist between one of the sites and a form of mental retardation.

The molecular nature of fragile sites is currently being investigated. Because they appear to represent points along the chromosome susceptible to breakage, these sites may indicate regions that are not tightly coiled or compacted. It should be noted that almost all studies of fragile sites have been carried out in mitotically dividing cells, and it is unknown whether such sites are also expressed in meiotic cells.

Most fragile sites do not appear to be associated with any clinical syndrome except for a folate-sensitive site on the X chromosome (Figure 6.29). Those bearing this site exhibit **fragile X syndrome** (or **Martin-Bell syndrome**), the most common form of inherited mental retardation, affecting about 1 in 1250 males and 1 in 2500 females. Because it is a dominant trait, females carrying only one fragile X chromosome can be mentally retarded. The trait is not fully expressed, since only about 30 percent of fragile X females are retarded, while about 20 percent of fragile X-bearing males are phenotypically normal. In addition to mental retardation, affected males have characteristic long, narrow faces with protruding chins, enlarged ears, and increased testicular size.

A gene that spans the fragile site may be responsible for this syndrome. This gene, known as *FMR-1*, contains

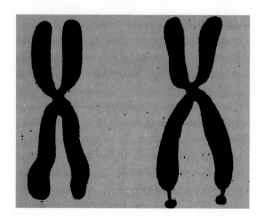

**FIGURE 6.29**    A normal human X chromosome (left) contrasted with a fragile X chromosome. The "gap" region is associated with the fragile X syndrome, including mental retardation.

a region that is increased in size in fragile X chromosomes. The fragile X breakpoint is contained in this variable length region, and full expression of the phenotype depends on both an increase in length and chemical modifications of the DNA nucleotides in this gene segment. The *FMR-1* gene is known to be expressed in the brain, but the exact nature and function of the encoded protein awaits further research. The gene may encode a DNA-binding protein based on the analysis of its nucleotide sequence. Although other genes may turn out to be involved in fragile X syndrome, the structure and function of the *FMR-1* gene is currently the leading candidate to provide insights into the molecular basis of this common form of mental retardation.

## CHAPTER SUMMARY

1. Investigations into the uniqueness of each organism's chromosomal constitution have further enhanced our understanding of genetic variation. Alterations of the precise diploid content of chromosomes are referred to as chromosomal aberrations or chromosomal mutations.

2. In humans, the study of individuals with altered sex chromosome compositions has established that the Y chromosome is responsible for male differentiation. The absence of the Y leads to female differentiation. Similar studies in *Drosophila* have excluded the Y in such a role, instead demonstrating that a balance between the number of X chromosomes and sets of autosomes is the critical factor.

3. The primary sex ratio in humans substantially favors males at conception. During embryonic and fetal development, male mortality is higher than that of females, while the secondary sex ratio at birth still favors males by a small margin.

4. Dosage compensation mechanisms limit the expression of sex-linked genes in females who have two X chromosomes, as compared to males who have only one X. In mammals, compensation is achieved as a result of the inactivation of either the maternal or paternal X early in development. This process results in the formation of Barr bodies in female somatic cells.

5. The Lyon hypothesis states that early in development, inactivation is random between the maternal and paternal X. All subsequent progeny cells inactivate the same X as their progenitor cell. Mammalian females thus develop as genetic mosaics with respect to their expression of heterozygous sex-linked alleles.

6. Deviations from the expected chromosomal number, or mutations in the structure of the chromosome, are inherited in predictable Mendelian fashion; they often result in inviable organisms or substantial changes in the phenotype.

7. Aneuploidy is the gain or loss of one or more chromosomes from the diploid content, resulting in conditions of monosomy, trisomy, tetrasomy, etc. Studies of

monosomic and trisomic disorders have increased our understanding of the delicate genetic balance that must exist in order for normal development to occur.

8. When complete sets of chromosomes are added to the diploid genome, polyploidy is created. These sets may have identical or diverse genetic origin, creating either autopolyploidy or allopolyploidy, respectively.

9. Large segments of the chromosome may be modified by deletions or duplications. Deletions may produce serious conditions such as the cri-du-chat syndrome in humans, while duplications may be particularly important as a source of redundant or new genes.

10. Inversions and translocations, while altering the gene order along chromosomes, initially cause little or no loss of genetic information or deleterious effects. However, heterozygous combinations may cause genetically abnormal gametes following meiosis, often causing lethality.

11. Fragile sites in human mitotic chromosomes have sparked research interest because one such site on the X chromosome is associated with the most common form of inherited mental retardation.

## KEY TERMS

acentric chromosome
allopolyploidy
allotetraploid
amniocentesis
amphidiploid
aneuploidy
anhidrotic ectodermal
    dysplasia
arm ratio
autopolyploidy
autotetraploid
autotriploid
Barr body
bilateral gynandromorph
centric fusion
chorionic villus sampling
    (CVS)
chromosome aberration
chromosome mutation
clone
colchicine
compensation
    (deficiency) loop
cri-du-chat syndrome
deletion (deficiency)

dicentric chromosome
dosage compensation
Down syndrome (trisomy
    21)
duplication
Edwards syndrome
    (trisomy 18)
endomitosis
endopolyploidy
euploidy
familial Down syndrome
fragile sites
fragile X syndrome
    (Martin-Bell syndrome)
gene amplification
gene families
gene redundancy
genic balance theory
intercalary deletion
intersex
inversion
inversion heterozygote
inversion loop
Klinefelter syndrome
    (47,XXY)

Lyon hypothesis
major histocompatibility
    complex (MHC)
metafemale
metamale
monosomy
nucleolar organizer
    region (NOR)
open reading frame
    (ORF)
paracentric inversion
partial monosomy
Patau syndrome (trisomy 13)
pentaploid
pentasomy
pericentric inversion
polyploidy
position effect
primary sex ratio
protoplast
pseudodominance
rDNA
reciprocal translocation
red-green color blindness
Robertsonian fusion

secondary sex ratio
segmental deletion
semidominance
semisterility
sex chromatin body
sex determination
sex-determining region
    Y (SRY)
sexual differentiation
somatic cell hybrid
T-cell receptor
telomere
terminal deletion
testis-determining factor
    (TDF)
tetraploid
tetrasomy
transgenic mice
translocation
triploid
trisomy
trivalent
Turner syndrome (45,X)
X-inactivation center
XYY condition

# INSIGHTS AND SOLUTIONS

1. In a cross using maize involving three linked genes, *a*, *b*, and *c*, a heterozygote (*abc*/+++) was test-crossed (to *abc*/*abc*). Even though the three genes were separated by at best five map units, only two phenotypes were recovered: *abc* and +++. Additionally, the cross produced significantly fewer viable plants than expected. Can you propose why no other phenotypes were recovered and why the viability was reduced?

   **SOLUTION:** One of the two chromosomes contains an inversion that overlaps all three genes, effectively precluding the recovery of any "crossover" offspring. If this is a paracentric inversion and the genes are clearly separated (assuring that a significant number of crossovers will occur between them), then numerous acentric and dicentric chromosomes will be formed, resulting in the observed reduction in viability.

2. A male *Drosophila* from a wild-type stock was discovered with only 7 chromosomes, whereas the normal 2*n* number is 8. Close examination revealed that one member of chromosome IV (the smallest chromosome) was attached to (translocated to) the distal end of chromosome II and was missing its centromere, thus accounting for the reduction in chromosome number.
   (a) Diagram all members of chromosomes II and IV during synapsis in Meiosis I.

   **SOLUTION:**

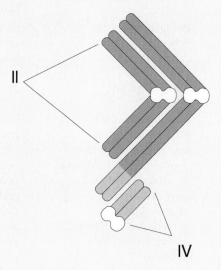

   (b) If this male mates with a female with a normal chromosome composition who is homozygous for the recessive chromosome IV mutation *eyeless* (*ey*), what chromosome compositions will occur in the offspring regarding chromosomes II and IV?

**SOLUTION:**

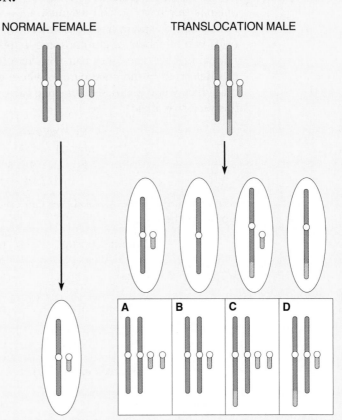

NORMAL FEMALE                  TRANSLOCATION MALE

(c) What phenotypic ratio will result regarding the presence of eyes, assuming all abnormal chromosome compositions survive?

**SOLUTION:**

A—normal (heterozygous)
B—eyeless (monosomic, contains chromosome IV from mother)
C—normal (heterozygous)
D—normal (heterozygous)

The final ratio is 3/4 normal:1/4 eyeless.

3. If a haplo-IV female *Drosophila* (containing only one chromosome 4, but an otherwise normal set of chromosomes) that has white eyes (a sex-linked trait) and normal bristles is crossed with a male with a diploid set of chromosomes and normal red eye color, but who is homozygous for the recessive chromosome 4 bristle mutation, *shaven*, (*sv*), what $F_1$ phenotypic ratio might be expected?

**SOLUTION:**   Let's first consider only the eye color phenotypes. This is a straightforward sex-linked cross. Offspring will appear as 1/2 red females:1/2 white males as shown below:

$$P_1: \quad \underset{\text{white female}}{ww} \quad \times \quad \underset{\text{red male}}{w^+//\wedge}$$

$$F_1: \quad \underset{\text{red female}}{1/2\ ww^+} \quad : \quad \underset{\text{white male}}{1/2\ w//\wedge}$$

The bristle phenotypes will be governed by the fact that the normal-bristled $P_1$

female produces gametes, one-half of which contain a chromosome 4 ($sv^+$) and one-half that have no chromosome 4 (— designates no chromosome). Following fertilization by sperm from the *shaven* male, one-half of the offspring will receive two members of chromosome 4 and be heterozygous for *sv*, expressing normal bristles. The other half will have only one copy of chromosome 4. Since its origin is from the male parent, where the chromosome bears the *sv* allele, these flies will express *shaven* since there is no wild-type allele present to mask this recessive allele.

$$P_1: \quad sv^+/— \quad \times \quad sv/sv$$
normal female      shaven male
$$F_1: \quad 1/2 \; sv^+/sv \text{ males and females}$$
$$1/2 \; sv/— \text{ males and females}$$

Using the forked-line method, we can consider both eye color and bristle phenotypes together:

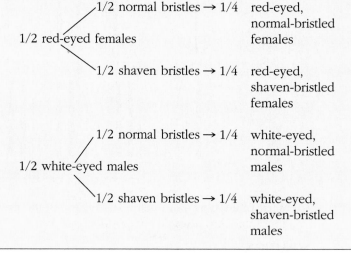

Using the forked-line method:

1/2 red-eyed females
  1/2 normal bristles → 1/4 red-eyed, normal-bristled females
  1/2 shaven bristles → 1/4 red-eyed, shaven-bristled females

1/2 white-eyed males
  1/2 normal bristles → 1/4 white-eyed, normal-bristled males
  1/2 shaven bristles → 1/4 white-eyed, shaven-bristled males

---

1. Contrast the evidence leading to the explanation of the different modes of sex determination in *Drosophila* and humans.

2. Devise a method of nondisjunction in human female gametes that would give rise to Klinefelter and Turner syndrome offspring following fertilization by a normal male gamete.

3. It has been suggested that any male-determining genes contained on the Y chromosome in humans should not be located in the limited region that synapses with the X chromosome during meiosis. What might be the outcome if such genes were located in this region?

4. What is a Barr body?

5. Indicate the expected number of Barr bodies in interphase cells of the following individuals: Klinefelter syndrome; Turner syndrome; and karyotypes 47,XYY, 47,XXX, and 48,XXXX.

6. Define the Lyon hypothesis.

7. Relate the potential effect of the Lyon hypothesis on the retina of a human female heterozygous for the X-linked red-green color blindness trait.

8. Cat breeders are aware that kittens expressing the sex-linked calico coat pattern are almost invariably females. Why?

9. What does the apparent need for dosage compensation mechanisms suggest about the expression of genetic information in normal diploid individuals?

10. A sex-linked dominant mutation in the mouse, *Testicular feminization* (*Tfm*), eliminates the normal response to the testicular hormone testosterone during sexual differentiation. An XY animal bearing the *Tfm* allele on the X chromosome develops testes, but no further male differentiation occurs. The external genitalia of such an animal are female. From this information, what might you conclude about the role of the X and Y chromosomes in sex determination and sexual differentiation in mammals?

11. The marine echiurid worm *Bonellia viridis* is an extreme example of the environment's influence on sex determination. Undifferentiated larvae either remain free-swimming and differentiate into females or they settle on the proboscis of an adult female and become males. If larvae that have been on a female proboscis for a short period are removed and placed in seawater, they develop as intersexes. If larvae are forced to develop in an aquarium where pieces of proboscises have been placed, they develop into males. Contrast this mode of sexual differentiation with that of mammals. Suggest further experimentation to elucidate the mechanism of sex determination in *Bonellia*.

12. Shown below are four graphs that plot the percentage of males that occur in various reptile groups versus the temperature fertilized eggs encounter during early development. Interpret these data as they relate to reptilian sex determination.

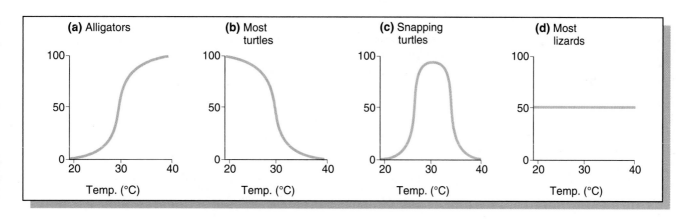

13. Discuss the possible reasons why the primary sex ratio in humans is as high as 1.40 to 1.60.

14. Define and distinguish between the following pairs of terms:

>aneuploidy/euploidy
>monosomy/trisomy
>Patau syndrome/Edwards syndrome
>autopolyploidy/allopolyploidy
>polyteny/endopolyploidy
>paracentric inversion/pericentric inversion

15. Contrast the relative survival times of individuals with Down syndrome, Patau syndrome, and Edwards syndrome. Why do you think such differences exist?

16. What evidence suggests that Down syndrome is more often the result of nondisjunction during oogenesis rather than during spermatogenesis?

17. Why do we believe that humans with aneuploid karyotypes occur, but are usually inviable?

18. What conclusions have been drawn about human aneuploidy as a result of karyotypic analyses of abortuses?

19. Contrast the fertility of an allotetraploid with an autotriploid and an autotetraploid.

20. When two plants belonging to the same genus but different species were crossed together, the $F_1$ hybrid was more viable and had more ornate flowers. Unfortunately, this hybrid was sterile and could only be propagated by vegetative cuttings. Explain the sterility of the hybrid. How might a horticulturist attempt to reverse its sterility?

21. For a species with a diploid number of 18, indicate how many chromosomes will be present in the somatic nuclei of individuals who are haploid, triploid, tetraploid, trisomic, and monosomic.

22. Discuss the origin of cultivated American cotton.

23. Predict how the synaptic configurations of homologous pairs of chromosomes might appear when one member is normal and the other member has sustained a deletion or duplication.

24. Inversions are said to "suppress crossing over." Is this terminology technically correct? If not, restate the description accurately.

25. Contrast the genetic composition of gametes derived from tetrads of inversion heterozygotes where crossing over occurs within a paracentric and pericentric inversion.

26. In a cross in *Drosophila*, a female heterozygous for the autosomally linked genes *a, b, c, d,* and *e* (*abcde*/+++++) was test-crossed to a male homozygous for all recessive alleles. Even though the distance between each of the above loci was at least 3 map units, only four phenotypes were recovered:

    | Phenotype | No. of Flies |
    |-----------|:---:|
    | + + + + + | 440 |
    | a  b  c  d  e | 460 |
    | + + + + e | 48 |
    | a  b  c  d  + | 52 |
    | Total | 1000 |

    Why are many expected crossover phenotypes missing? Can any of these loci be mapped from the data given here? If so, determine map distances.

27. Contrast the *Notch* locus with the *Bar* locus in *Drosophila*. What phenotypic ratios would be produced in a cross between *Notch* females and *Bar* males? Under what circumstances has gene duplication been essential to an organism's survival?

28. Discuss Ohno's hypothesis on the role of gene duplication in the process of evolution.

29. What roles have inversions and translocations played in the evolutionary process?

30. A human female with Turner syndrome also expresses the sex-linked trait hemophilia, as her father did. Which parent underwent nondisjunction during meiosis, giving rise to the gamete responsible for the syndrome?

31. The primrose, *Primula kewensis,* has 36 chromosomes that are similar in appearance to the chromosomes in two related species, *Primula floribunda* ($2n = 18$) and *Primula verticillata* ($2n = 18$). How could *P. kewensis* arise from these species? How would you describe *P. kewensis* in genetic terms?

32. Varieties of chrysanthemums are known that contain 18, 36, 54, 72, and 90 chromosomes; all are multiples of a basic set of 9 chromosomes. How would you describe these varieties genetically? What feature is shared by the karyotypes of each variety? A variety with 27 chromosomes was discovered. but it was sterile. Why?

33. What is the effect of a rare double crossover within a pericentric inversion present heterozygously? Within a paracentric inversion present heterozygously?

34. *Drosophila* may be monosomic or disomic for chromosome 4 and remain fertile. Contrast the $F_1$ and $F_2$ results of the following crosses involving the recessive chromosome 4 trait, *bent* bristles:
    (a) monosomic bent × disomic normal
    (b) monosomic normal × disomic bent

35. *Drosophila* may also be trisomic for chromosome 4 and remain fertile. Predict the $F_1$ and $F_2$ results of crossing:

    trisomic bent ($b/b/b$) × normal disomic ($B/B$)

36. Mendelian ratios are modified in crosses involving autotetraploids. Assume that one plant expresses the dominant trait green (seeds) and is homozygous (*WWWW*). This plant is crossed to one with white seeds that is also homozygous (*wwww*). If only one dominant allele is sufficient to produce green seeds, predict the $F_1$ and $F_2$ results of such a cross. Assume that synapsis between chromosome pairs is random during meiosis.

37. Having correctly established the $F_2$ ratio above, predict the $F_2$ ratio of a "dihybrid" cross involving two independently assorting characteristics (e.g., $P_1 =$ *WWWWAAAA* × *wwwwaaaa*).

38. A couple, in looking ahead to planning a family, were aware that through the past three generations on the male's side a substantial number of stillbirths had occurred and several malformed babies were born who died early in childhood. The female had studied genetics and urged her husband to visit a genetic counseling clinic, where a complete karyotype-banding analysis was subsequently performed. Although it was found that he had a normal complement of 46 chromosomes, banding analysis revealed that one member of the chromosome 1 pair (in Group A) contained an inversion covering 70 percent of its length. The homologue of chromosome 1 and all other chromosomes showed the normal banding sequence.
    (a) How would you explain the high incidence of past stillbirths?
    (b) What would you predict about the probability of abnormality/normality of their future children?
    (c) Would you advise the female that she would have to "wait out" each pregnancy to term in order to determine if the fetus were normal? If not, what would you suggest to her?

**SELECTED READINGS**

BARR, M. L. 1966. The significance of sex chromatin. *Int. Rev. Cytol.* 19:35–39.

BEASLEY, J. O. 1942. Meiotic chromosome behavior in species, species hybrids, haploids, and induced polyploids of *Gossypium. Genetics* 27:25–54.

BLAKESLEE, A. F. 1934. New jimson weeds from old chromosomes. *J. Hered.* 25:80–108.

BORGAONKER, D. S. 1989. *Chromosome variation in man: A catalogue of chromosomal variants and anomalies.* 5th ed. New York: Alan R. Liss.

BOUE, A. 1985. Cytogenetics of pregnancy wastage. *Adv. Hum. Genet.* 14:1–58.

BURGIO, G. R., et al., eds. 1981. *Trisomy 21.* New York: Springer-Verlag.

CARR, D. H. 1971. Genetic basis of abortion. *Ann. Rev. Genet.* 5:65–80.

COURT-BROWN, W. M. 1968. Males with an XYY sex chromosome complement. *J. Med. Genet.* 5:341–59.

CUMMINGS, M. R. 1994. *Human heredity: Principles and issues.* 3rd ed. St. Paul: West.

DAVIDSON, R., NITOWSKI, H., and CHILDS, B. 1963. Demonstration of two populations of cells in human females heterozygous for glucose-6-phosphate dehydrogenase variants. *Proc. Natl. Acad. Sci.* 50:481–85.

DEARCE, M. A., and KEARNS, A. 1984. The fragile X syndrome: The patients and their chromosomes. *J. Med. Genet.* 21:84–91.

ERICKSON, J. D. 1976. The secondary sex ratio of the United States 1969–71: Association with race, parental ages, birth order, paternal education and legitimacy. *Ann. Hum. Genet. Lond.* 40:205–12.

FELDMAN, M., and SEARS, E. R. 1981. The wild gene resources of wheat. *Scient. Amer.* (Jan.) 244:102–12.

GORDON, J. W., and RUDDLE, F. H. 1981. Mammalian gonadal determination and gametogenesis. *Science* 211:1265–78.

GORMAN, M., KURODA, M., and BAKER, B. S. 1993. Regulation of sex-specific binding of maleness dosage compensation protein to the male X chromosome in *Drosophila. Cell* 72:39–49.

GUPTA, P. K., and PRIYADARSHAN, P. M. 1982. *Triticale:* Present status and future prospects. *Adv. in Genet.* 21:256–346.

HASELTINE, F. P., and OHNO, S. 1981. Mechanisms of gonadal differentiation. *Science* 211: 1272–78.

HASSOLD, T., et al. 1980. Effect of maternal age on autosomal trisomies. *Ann. Hum. Genet. Lond.* 44:29–36.

HASSOLD, T. J., and JACOBS, P. A. 1984. Trisomy in man. *Ann. Rev. Genet.* 18:69–98.

HEAD, G., MAY, R., and PENDLETON, L. 1987. Environmental determination of sex in the reptiles. *Nature* 329:198–99.

HECHT, F. 1988. Enigmatic fragile sites on human chromosomes. *Trends in Genetics* 4:121–22.

HODGKIN, J. 1990. Sex determination compared in *Drosophila* and *Caenorhabditis. Nature* 344:721–28.

HOOK, E. B. 1973. Behavioral implications of the humans XYY genotype. *Science* 179:139–50.

HSU, T. C. 1979. *Human and mammalian cytogenetics: A historical perspective.* New York: Springer-Verlag.

HULSE, J. H., and SPURGEON, D. 1974. Triticale. *Scient. Amer.* (Aug.) 231:72–81.

JACOBS, P. A., et al. 1974. A cytogenetic survey of 11,680 newborn infants. *Ann. Hum. Genet.* 37:359–76.

KAISER, P. 1984. Pericentric inversions: Problems and significance for clinical genetics. *Hum. Genet.* 68:1–47.

KAY, G. F., et al. 1993. Expression of *Xist* during mouse development suggests a role in the initiation of X chromosome inactivation. *Cell* 72:171–82.

KHUSH, G. S. 1973. *Cytogenetics of aneuploids.* Orlando: Academic Press.

KOOPMAN, P., et al. 1991. Male development of chromosomally female mice transgenic for *Sry. Nature* 351:117–121.

LEWIS, E. B. 1950. The phenomenon of position effect. *Adv. in Genet.* 3:73–115.

LEWIS, W. H., ed. 1980. *Polyploidy: Biological relevance.* New York: Plenum Press.

LUCCHESI, J. 1983. The relationship between gene dosage, gene expression, and sex in *Drosophila. Dev. Genet.* 3:275–82.

LYON, M. F. 1961. Gene action in the X-chromosome of the mouse (*Mus musculus* L.). *Nature* 190:372–73.

LYON, M. F. 1962. Sex chromatin and gene action in the mammalian X chromosome. *Am. J. Hum. Genet.* 14:135–48.

———. 1972. X-chromosome inactivation and developmental patterns in mammals. *Biol. Rev.* 47:1–35.

———. 1988. X-chromosome inactivation and the location and expression of X-linked genes. *Am. J. Hum. Genet.* 42:8–16.

MANTELL, S. H., MATHEWS, J. A., and McKEE, R. A. 1985. *Principles of plant biotechnology: An introduction to genetic engineering in plants.* Oxford: Blackwell.

McLAREN, A. 1988. Sex determination in mammals. *TIG* 4:153–57.

McMILLEN, M. M. 1979. Differential mortality by sex in fetal and neonatal deaths. *Science* 204:89–91.

OBE, G., and BASLER, A. 1987. *Cytogenetics: Basic and applied aspects.* New York: Springer-Verlag.

OHNO, S. 1970. *Evolution by gene duplication.* New York: Springer-Verlag.

OOSTRA, B. A., and VERKERK, A. J. 1992. The fragile X syndrome: Isolation of the *FMR-1* gene and characterization of the fragile X mutation. *Chromosoma* 101:381–87.

PAGE, D. C., et al. 1987. The sex-determining region of the human Y chromosome encodes a finger protein. *Cell* 51:1091–1104.

PATTERSON, D. 1987. The causes of Down syndrome. *Sci. Amer.* (Aug.) 257:52–61.

ROONEY, D. E., and CZEPULKOWSKI, B. H., eds. 1986. *Human cytogenetics: A practical approach.* Oxford: IRL Press.

SCHIMKE, R. T., ed. 1982. *Gene amplification.* Cold Spring Harbor, NY: Cold Spring Harbor Laboratory.

SHEPARD, J. F. 1982. The regeneration of potato plants from protoplasts. *Sci. Amer.* (May) 246:154–66.

SHEPARD, J. F., et al. 1983. Genetic transfer in plants through interspecific protoplast fusion. *Science* 219:683–88.

SIMMONDS, N. W., ed. 1976. *Evolution of crop plants.* London: Longman.

SMITH, G. F., ed. 1984. *Molecular structure of the number 21 chromosome and Down syndrome.* New York: New York Academy of Sciences.

STEBBINS, G. L. 1966. Chromosome variation and evolution. *Science* 152:1463–69.

STRICKBERGER, M. W. 1990. *Evolution.* Boston: Jones and Bartlet.

SUTHERLAND, G. 1984. The fragile X chromosome. *Int. Rev. Cytol.* 81:107–43.

———. 1985. The enigma of the fragile X chromosome. *Trends in Genetics* 1:108–11.

SWANSON, C. P., MERZ, T., and YOUNG, W. J. 1981. *Cytogenetics: The chromosome in division, inheritance, and evolution.* 2nd ed. Englewood Cliffs, NJ: Prentice-Hall.

TAYLOR, A. I. 1968. Autosomal trisomy syndromes: A detailed study of 27 cases of Edwards syndrome and 27 cases of Patau syndrome. *J. Med. Genet.* 5:227–52.

THERMAN, E. 1980. *Human chromosomes.* New York: Springer-Verlag.

TJIO, J. H., and LEVAN, A. 1956. The chromosome number of man. *Hereditas* 42:1–6.

TURPIN, R., and LeJEUNE, J. 1969. *Human afflictions and chromosomal aberrations.* Oxford, England: Pergamon Press.

WESTERGAARD, M. 1958. The mechanism of sex determination in dioecious flowering plants. *Adv. in Genet.* 9:217–81.

WHARTON, K. A., et al. 1985. *opa:* A novel family of transcribed repeats shared by the *Notch* locus and other developmentally regulated loci in *D. melanogaster. Cell* 40:55–62.

WILKINS, L. E., BROWN, J. A., and WOLF, B. 1980. Psychomotor development in 65 home-reared children with cri-du-chat syndrome. *J. Pediatr.* 97:401–5.

WITKIN, H. A., et al. 1976. Criminality in XYY and XXY men. *Science* 193:547–55.

YUNIS, J. J., ed. 1977. *New chromosomal syndromes.* Orlando: Academic Press.

# 7

# ADVANCED TOPICS IN TRANSMISSION GENETICS

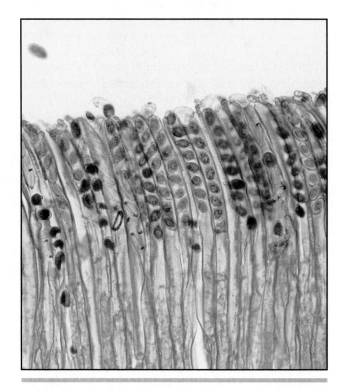

Ascospores contained in asci of the bread mold, *Neurospora crassa*.

*Of the many possibilities, our choice of relevant "advanced" topics in transmission genetics includes the consideration of phenotypic expression, analysis of quantitative inheritance, heritability, and mapping analysis in haploid eukaryotes. The expression of the phenotype is not always clear-cut and may be modified by many internal and external conditions. Quantitative traits are under the control of polygenes and are analyzed using biometric and statistical methodology. Chromosome mapping in haploid organisms utilizes a different methodology than similar studies in diploid organisms, but adheres to the same general principles.*

Thus far in the text, we have introduced a core of information that constitutes the general body of knowledge referred to as **transmission genetics**. In this chapter, we turn to several more advanced topics that are more accessible now that you have gained the foundation provided in the first six chapters. These constitute highly relevant extensions of our previous discussions.

We will begin by looking at the topic of **phenotypic expression** and reviewing numerous factors that affect it. As we shall see, expression of genetic information is not always as clear-cut as most of our previous examples might suggest. An organism bearing a mutant genotype may exhibit considerable variation in the expression of the related phenotype. In some cases, a percentage of the mutant organisms may fail to express the phenotype at all! Many factors that modify phenotypic expression will be discussed.

Next we will extend our prior discussion of polygenic inheritance from Chapter 4 with a more thorough examination of **quantitative inheritance**. As we do so, we shall consider several statistical approaches that are often used in analyzing quantitative traits and expressing variance in population data. Then we will introduce the concept of **heritability**—the assessment of the extent to which genetic factors contribute to phenotypic variation within populations. We will conclude this chapter by returning to the topics of linkage and crossing over and extending our discussion to include **chromosome mapping in haploid organisms**. We will examine this topic in eukaryotic fungus *Neurospora* and the green alga *Chlamydomonas*.

# PHENOTYPIC EXPRESSION

Gene expression is often discussed as if the genes operate in a closed, "black box" system in which the presence or absence of functional products directly determines the collective phenotype of an individual. The situation is actually much more complex. Most gene products function within the internal milieu of the cell, cells interact with one another in various ways, and the organism must survive under diverse environmental influences. Thus, gene expression and the resultant phenotype are often modified through the interaction between an individual's particular genotype and the internal and external environment.

The degree of environmental influence may vary from inconsequential to subtle to very strong. Subtle interactions are the most difficult to detect and document, and have led to unresolvable "nature–nurture" conflicts in which scientists debate the relative importance of genes versus environment. How easily such conflicts are resolved depends on the characteristic being investigated, and even then, it is sometimes impossible to provide definitive information. In this section we will deal with some of the variables known to modify gene expression.

## Penetrance and Expressivity

Some mutant genotypes are always expressed as a distinct phenotype, while others produce a proportion of individuals whose phenotypes cannot be distinguished from wild type. The variable expression of a particular trait may be quantitatively studied by determining the degree of penetrance and expressivity. The percentage of individuals that show at least some degree of expression of a mutant genotype defines the **penetrance** of the mutation. For example, the expression of many mutant alleles in *Drosophila* often overlaps with wild type. Flies homozygous for the recessive mutant gene *eyeless* yield phenotypes that range from the presence of normal eyes to the complete absence of one or both eyes (Figure 7.1). If 15 percent of mutant flies show the wild-type appearance, the mutant gene is said to have a pene-

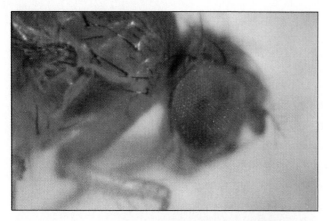

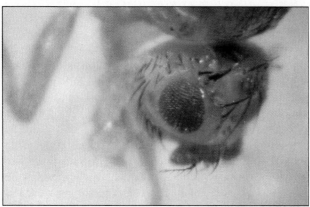

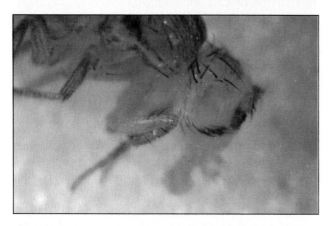

**FIGURE 7.1**    Gradations in phenotype, ranging from wild type to eyeless, associated with the *eyeless* mutation in *Drosophila*.

trance of 85 percent. On the other hand, the range of expression of the mutant genotype defines the **expressivity**. In the case of *eyeless,* although the average reduction of eye size is one-fourth to one-half, the range of expressivity is from complete loss of both eyes to completely normal eyes.

Examples such as the expression of the *eyeless* phenotype have provided the basis for experiments designed to determine the causes of phenotypic variation. If the laboratory environment is held constant and extensive variation is still observed, the genetic background may be investigated. It is possible that other genes are influencing or modifying the *eyeless* phenotype. On the other hand, if the genetic background is not the cause of the phenotypic variation, environmental factors such as temperature, humidity, and nutrition may be tested. In the case of the *eyeless* phenotype, it has been determined experimentally that both genetic background and environmental factors influence its expression.

## Genetic Background: Suppression and Position Effects

With only certain exceptions, it is difficult to assess the specific relationship between the **genetic background** and the expression of a gene responsible for determining the potential phenotype. Two of the better characterized effects of genetic background are described below.

First, the expression of other genes throughout the genome may have an effect on the phenotype produced by the gene in question. The phenomenon of **genetic suppression** is an example. Mutant genes such as *suppressor of vermilion* (*su-v*), *suppressor of forked* (*su-f*), and *suppressor of Hairy-wing* (*su-Hw*) in *Drosophila* completely or partially restore the normal phenotype to an organism that is homozygous (or hemizygous) for these recessive genes. For example, flies homozygous for both *vermilion* (a bright red eye color mutation) and *su-v* have eyes with wild-type color.

The phenomenon of suppression occurs in a wide variety of organisms. In microorganisms, where molecular studies are easier to perform, some suppressor gene products are molecules that function in the genetic translation process by correcting certain errors produced by the mutation of other genes. Transfer RNA in mutant form has been shown to have such an effect. It has also been hypothesized that a suppressor gene product might provide an alternative metabolic route to bypass a block in a biosynthetic pathway caused by the primary mutation. Thus, suppressor genes are excellent examples of the genetic background modifying primary gene effects.

Second, the physical location of a gene in relation to other genetic material may influence its expression. Such a situation is called a **position effect**. For example, if a gene is included in a **translocation** or **inversion** event (in which a region of a chromosome is relocated or rearranged), the expression of the gene may be affected. This is particularly true if the gene is relocated to or near an area of the chromosome composed of **heterochro-**

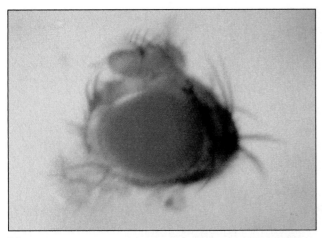

(a)

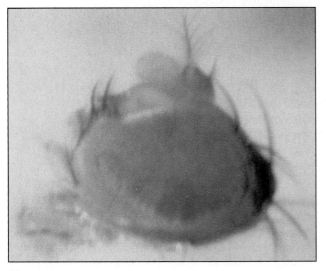

(b)

**FIGURE 7.2**    Eye phenotype in two female *Drosophila* heterozygous for the gene *white*. (a) Normal dominant phenotype showing brick red eye color. (b) Variegated color of an eye caused by translocation of the *white* gene to another location in the genome.

**matin**. These regions are thought to be genetically inert and often appear condensed in interphase.

An example of a position effect involves female *Drosophila* heterozygous for the sex-linked recessive eye-color mutant *white* (*w*). If the region of the X chromosome containing the wild-type $w^+$ allele is translocated so that it is close to heterochromatin, expression of the $w^+$ allele is modified (Figure 7.2). Before translocation, the $w^+/w$ genotype results in a wild-type brick red eye color. After translocation, the dominant effect of the $w^+$ allele is reduced. Instead of having a red color, the eyes

are variegated, or mottled with red and white patches. A similar position effect is produced if a heterochromatic region is relocated next to the *white* locus on the X chromosome. Apparently, heterochromatic regions inhibit the expression of adjacent genes. Loci in many other organisms also exhibit position effects, providing proof that alteration of the normal arrangement of genetic information can modify its expression. Recall that we saw several examples of position effects in Chapter 6.

## Temperature Effects

Because chemical activity depends on the kinetic energy of the reacting substances, which in turn depends on the surrounding temperature, we can expect temperature to influence phenotypes. For example, the evening primrose produces red flowers at 23°C and white flowers at 18°C. Siamese cats and Himalayan rabbits exhibit dark fur in regions of the nose, ears, and paws because the body temperatures of these extremities are slightly cooler (Figure 7.3). In these cases, it appears that an enzyme that is functional in pigment production at the lower temperatures present in the extremities loses its catalytic function at the slightly higher temperatures throughout the rest of the body.

Mutations such as those affecting pigment are said to be **conditional** and are called **temperature sensitive**. Examples are known in a variety of other organisms, including bacteria, viruses, fungi, and *Drosophila*. In extreme cases, an organism carrying a mutant allele may express a mutant phenotype when grown at one temperature, but express the wild-type phenotype when reared at another temperature. This type of temperature effect is useful in studying mutations that interrupt essential processes during development and are thus normally lethal to the organism. If viruses carrying a **temperature-sensitive lethal mutation** are allowed to infect bacteria cultured at 42°C, infection progresses until the essential gene product is needed and then arrests, because this is the **restrictive temperature**. If cultured at the **permissive temperature** of 25°C, infection occurs normally, new viruses are produced, and the bacteria are lysed. The use of temperature-sensitive mutations, which may be induced and isolated, has added immensely to the study of viral genetics.

Similarly, many temperature-sensitive mutations have been discovered in *Drosophila*. Not only have dominant and recessive conditional lethal mutations been recovered in great numbers, but nonlethal temperature-sensitive mutations affecting development, morphology, and behavior have also been recovered. These nonlethal mutants have proven to be valuable in the dissection of genetic control mechanisms.

(a)                                                                (b)

**FIGURE 7.3**    (a) A Himalayan rabbit. (b) A Siamese cat. Both show dark fur color at the extremities of the ears, nose, and paws. These patches are due to expression of a temperature-sensitive allele responsible for pigment production at the lower temperatures of the extremities.

## Nutritional Effects

Another example of conditional mutations involves nutrition. In microorganisms, mutations that prevent synthesis of nutrient molecules are quite common. These **nutritional mutants** arise when an enzyme essential to a biosynthetic pathway becomes inactive. In microorganisms, an organism bearing such a mutation is called an **auxotroph**. If the end product of a biochemical pathway can no longer be synthesized, and if that molecule is essential to normal growth and development, the mutation is lethal. If, for example, the bread mold *Neurospora* can no longer synthesize the amino acid leucine, proteins cannot be synthesized unless leucine is added to the growth medium. If leucine is added, the lethal effect is overcome. Nutritional mutants have been crucial to molecular genetic studies and also served as the basis for George Beadle and Edward Tatum's proposal, in the early 1940s, that one gene functions to produce one enzyme (see Chapter 15).

In humans, a slightly different set of circumstances is known. The presence or absence of certain dietary substances, which normal individuals may consume without harm, may adversely affect individuals with abnormal genetic constitutions. Often, a mutation may prevent an individual from metabolizing some substance commonly found in normal diets. For example, those afflicted with the genetic disorder **phenylketonuria** cannot metabolize the amino acid phenylalanine. Those with **galactosemia** cannot metabolize galactose. Other individuals are intolerant of the milk sugar lactose and are said to exhibit **lactose intolerance**. Those with **diabetes** or **hypoglycemia** cannot adequately metabolize glucose. In each case, excessive amounts of the molecule accumulate in the body and become toxic, and a characteristic phenotype results. However, if the dietary intake of the molecule is drastically reduced or eliminated, the associated phenotype may be reversed.

The case of lactose intolerance illustrates the general principles involved. Lactose is a disaccharide consisting of a molecule of glucose and a molecule of galactose. It is present as 7 percent of human milk and 4 percent of cow's milk. To metabolize lactose, humans require the enzyme **lactase**, which cleaves the disaccharide. Adequate amounts of lactase are produced during the first few years after birth. However, in many racial and ethnic groups, the levels of this enzyme soon drop drastically, and adults become intolerant of milk. The major phenotypic effect involves severe intestinal diarrhea, flatulence, and abdominal cramps. This condition, while not limited to, is particularly prevalent in Eskimos, Africans, Asiatics, and Americans with these heritages. In these cultures, milk is usually converted to other foods, including cheese, butter, and yogurt. In these forms, the amount of lactose is reduced significantly and adverse effects can be nearly eliminated. In the United States milk low in lactose is commercially available, and to aid in the digestion of other lactose-containing foods, lactase is now a commercial product that may be ingested.

## Onset of Genetic Expression

Not all genetic traits are expressed at the same time during an organism's life span. In most cases, the age at which a gene is expressed corresponds to the normal sequence of growth and development. In humans, the prenatal, infant, preadult, and adult phases require different kinds of genetic information. In a similar way, many genetic disorders can be expected to manifest themselves at different stages of life. Lethal genes account for many of the frequent spontaneous abortions and miscarriages occurring in the human population. These mutations are believed to alter genetic products essential to prenatal development.

In humans, many severe inherited disorders are often not manifested until after birth. For example, **Tay-Sachs disease**, inherited as an autosomal recessive, is a lethal lipid metabolism disease involving an abnormal enzyme, **hexosaminidase A**. However, newborns appear normal for the first five months. Then, progressive deterioration occurs, and the affected children die before age four.

The **Lesch–Nyhan syndrome**, inherited as an X-linked recessive, causes abnormal nucleic acid metabolism (biochemical salvage of nitrogenous purine bases), leading to hyperuricemia, mental retardation, palsy, and self-mutilation of the lips and fingers. The disorder is due to a mutation in the gene encoding **hypoxanthine-guanine phosphoribosyl transferase (HPRT)**. Newborns are normal for six to eight months prior to the onset of the first symptoms. Still another example involves **Duchenne muscular dystrophy (DMD)**, an X-linked recessive disorder associated with progressive muscular wasting. It is usually diagnosed between the ages of three and five. Even with modern medical intervention, the disease is often fatal in the early twenties.

Perhaps the most variable of all inherited human disorders regarding age of onset is the tragic **Huntington disease**. Inherited as an autosomal dominant, Huntington disease affects the frontal lobes of the cerebral cortex where progressive cell death occurs over a period of more than a decade. Brain deterioration is accompanied by spastic uncontrolled movements, intellectual and emotional deterioration, and ultimately death. While onset has been reported at all ages, it most frequently occurs between ages 30 and 50, with a mean onset of 38 years. Most often, early onset is associated with an affected male parent, although such a response is not universal to offspring of males with the disorder.

Observations such as these support the concept that the critical expression of normal genes varies throughout the life cycle of organisms, including humans. Gene products may play more essential roles at certain times. Also, it would appear that the internal physiological environment of an organism changes with age.

## Genetic Anticipation

Considerable insights have been gained concerning the variable onset of genetic expression by studying cases where clear-cut patterns of initial expression have been observed. The phenomenon of **genetic anticipation** refers to the occurrence of a genetic disorder with a progressively earlier age of onset in successive generations. As we will see, earlier onset is correlated with an increased severity of the disorder.

There are at least three human disorders that clearly demonstrate genetic anticipation: **myotonic dystrophy (DM)**, **fragile-X mental retardation**, and **spinal and bulbar muscular atrophy (Kennedy disease)**. We will focus on the information derived from studies of DM to discuss anticipation.

Myotonic dystrophy, the most common type of adult muscular dystrophy, is an autosomal dominant disorder of variable severity and age of onset. Mildly affected individuals develop cataracts as adults with little or no muscular weakness. Severely affected individuals demonstrate more severe myopathy and may be mentally retarded. In its most extreme form, the disease is fatal just after birth. Careful investigation has revealed that the increased severity is correlated with earlier onset.

A great deal of excitement has been generated recently among geneticists as a result of studies involving anticipation. In 1989, C. J. Howeler and colleagues reported on their study of 61 parent–child pairs. In 60 of 61 cases, age of onset was earlier in the child! By 1992, several research teams had discovered a possible molecular basis for this case of genetic anticipation. Located on chromosome 19 (19q1.3 contains the DM locus) is an unstable DNA insert characterized by a specific trinucleotide (CTG) that is repeated a variable number of times. The remarkable finding is the correlation between the size of the **trinucleotide repeat** and both severity and onset. In successive generations the size of the repeated segment increases. Normal individuals average about 5 copies; minimally affected individuals reveal about 50 copies, while severely affected individuals demonstrate over 1000 copies!

While it is not yet clear how an expansion of the size of this gene-associated region affects the onset and degree of phenotypic expression of this disorder, the correlation is extremely strong. Of great interest is that both fragile-X syndrome and Kennedy disease also reveal an association between a gene-associated DNA amplification and disease severity. More recently, a similar finding has been made in the gene encoding Huntington disease. We will return to this general topic in Chapter 11. For now, we can only anxiously await further information that links the molecular defect to the phenomenon of genetic anticipation.

## Genomic Imprinting

One of the major exceptions to the laws of Mendelian inheritance involves the case where genetic expression varies, based on the parental origin of the chromosome carrying a particular gene. The phenomenon is called **genomic** (or **parental**) **imprinting**. It appears that certain genes are somehow "marked" (or imprinted) for greater or lesser expression by one parent but not the other parent, leading to differential expression of the corresponding genes in the offspring. For example, in Huntington disease in humans, early onset (see the preceding section) most often occurs when the mutant gene is inherited from the father. In myotonic dystrophy, just the opposite is true. In this disorder, when early onset is observed, the affected offspring usually inherits the gene from the mother.

The imprinting step is thought to occur before or during gamete formation, leading to differentially marked genes (or whole chromosomes) in sperm-forming versus egg-forming tissues. The process is clearly different from mutation because it can be reversed every generation as genes pass from mother to son to granddaughter, and so on. Only in unusual cases, as in the diseases mentioned above, can we identify the effects of imprinting. One of the earliest examples to be observed involves the inactivation of one of the X chromosomes in mammalian females. As we discussed in Chapter 6, dosage compensation involves the random inactivation of either the paternal or maternal X chromosome during early embryonic development. However, in mice, prior to the development of the embryo proper, tissues developing from the zygote differentially inactivate the paternal X in all cells. It is only once this imprinting is "released" as development proceeds that random inactivation can occur in embryonic cells.

In 1991, three specific mouse genes were shown to be imprinted. One of them is the gene encoding insulinlike growth factor II (*Igf2*). A mouse that carries two nonmutant alleles of this gene is normal in size, whereas a mouse that carries two mutant alleles lacks a growth factor and is a dwarf. The size of a heterozygous mouse (one allele normal and one mutant—Figure 7.4) depends on the parental origin of the normal allele. The mouse is normal in size if the normal allele came from the father, but dwarf if the normal allele came from the mother.

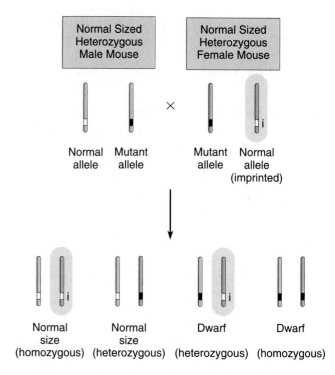

**FIGURE 7.4**    The effect of imprinting on the *Igf2* gene in the mouse, which produces dwarf mice in the homozygous condition. Heterozygous offspring that receive the normal allele from their father are normal in size. Heterozygotes that receive the normal allele from their mother, which has been imprinted (i), are dwarf.

From this it can be deduced that the normal *Igf2* gene is imprinted to function poorly during the course of egg production in females but functions normally after passing through sperm-producing tissue in males.

Imprinting continues to depend on whether the gene passes through sperm-producing or egg-forming tissue leading to the next generation. For example, a heterozygous normal-sized male (above) will donate a "normal-functioning" wild-type allele to half his offspring that will counteract a mutant allele received from the mother.

In humans, two distinct genetic disorders are thought to be caused by differential imprinting of the same region of chromosome 15 (15q1). In both cases, the disorders *appear* to be due to an identical deletion of this region in one member of the chromosome 15 pair. The first disorder, **Prader-Willi syndrome (PWS)**, results when only an undeleted maternal chromosome remains. If only an undeleted paternal chromosome remains, an entirely different disorder, **Angelman syndrome (AS)**, results.

PWS is clearly different from AS. In the former, mental retardation is noted as well as a severe eating disorder marked by an uncontrollable appetite, obesity, and diabetes. AS, on the other hand, leads to distinct clinical manifestations involving behavior as well as mental retardation. One can conclude that region 15q1 is imprinted differently in male versus female gametes and that both a maternal and paternal region are required for normal development.

The area of genomic imprinting is a relatively new field of research, and many questions remain unanswered. It is not known how many genes are subject to imprinting, nor its developmental role. While it appears that regions of chromosomes rather than specific genes are imprinted, the molecular mechanism of imprinting is still a matter for conjecture. Speculation exists that **DNA methylation** may be involved. In vertebrates, methyl groups can be added to the carbon atom at position 5 in cytosine (see Chapter 9). Methyl groups are found there when the dinucleotide CpG is present along a DNA chain. While there is some evidence that a high level of methylation is often associated with inactive genes and that active genes (or their regulatory sequences) are often undermethylated, this may or may not be involved in the mechanism of imprinting. Whatever the cause of this phenomenon, it is a fascinating topic and one that will receive significant attention in future research studies.

# CONTINUOUS VARIATION AND POLYGENES

In Chapter 4 we considered how the classical patterns of Mendelian inheritance (e.g., the 3:1 and 9:3:3:1 $F_2$ ratios) are modified because of gene interaction or the action of multiple genes on a single trait. At that time, we discussed examples of **polygenic traits** such as the inheritance of grain color in wheat, a trait that is controlled by three gene pairs. These traits were discussed to illustrate that even in complex modes of inheritance, the fundamental principles of segregation and independent assortment discovered by Mendel are operational.

Having considered the patterns of inheritance that are characteristic of polygenic traits in that earlier discussion, in this chapter we will describe the methods used by geneticists to study traits that are controlled by several genes. These methods are often statistical and involve analysis of traits using mathematical tools in addition to those of biochemistry or molecular biology.

Polygenic traits are at the heart of several disciplines in genetics, including plant breeding, livestock breeding, and wildlife management. Polygenic traits are also an important part of human genetics; traits such as intelligence, skin color, obesity, and predisposition to certain diseases are thought to be under polygenic control. In the following discussion, remember that in polygenic inheritance, as in monogenic inheritance, barring accidents, the genotype is fixed at the moment of fertilization, while the phenotype is more flexible and changes over the life span of the organism as the genotype interacts with the environment.

## Continuous versus Discontinuous Variation

Traits studied by Mendel in the garden pea could be separated into distinct qualitative categories (e.g., tall vs. dwarf plants, green vs. yellow seeds) [Figure 7.5(a)]. Such variation is described as **discontinuous**. In contrast, polygenic traits are said to exhibit **continuous variation** between the most extreme phenotypes expressed in a population. For example, Sir Francis Galton studied the diameter of sweet peas. When he crossed plants with the largest diameter to those with the smallest diameter, the $F_1$ plants had peas with an intermediate diameter. In the $F_2$, continuous variation from the largest to the smallest was exhibited, but with most intermediate in size [Figure 7.5(b)].

Analysis of continuous traits always involves some quantitation step expressed in weight, stature, volume, height, color, or other form of measurement. As you may recall from our earlier discussion of this topic in Chapter 4, we make the assumption that alleles involved in polygenic inheritance are either *additive* or *nonadditive*. We further assume that the total effect of each additive allele is small and roughly equal to all others involved in the genetic control of a specific phenotype. While these alleles are subject to the impact of genetic background and environmental influences, for simplicity we discount these in our examples.

Galton devised statistical methods to study these continuous traits, and the resulting field became known as **biometry**. While these mathematical techniques provided an important tool for the analysis of experimental data, they offered no explanation for the underlying mechanisms of inheritance. Not surprisingly, this led to many erroneous conclusions about the characteristics of continuous variation and to a subsequent debate as to whether Mendelian principles could explain this form of inheritance. As outlined in Chapter 4, this debate was settled in a series of experiments that demonstrated the role of Mendelian factors in continuous variation.

## Mapping Quantitative Trait Loci

In discussing polygenic inheritance, it is useful to know how to approximate the number of genes that control a given trait. In a given polygenic cross, the phenotype of the $F_2$ generation will show greater phenotypic variability than the $P_1$ or $F_1$ generation. The number of genes controlling a trait may be estimated by

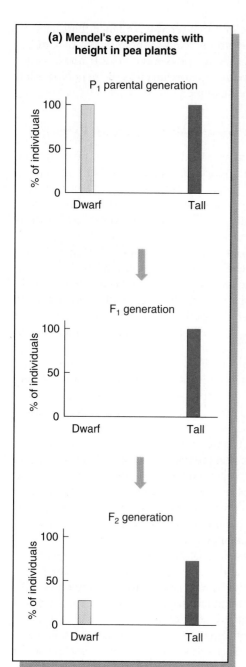

**(a) Mendel's experiments with height in pea plants**

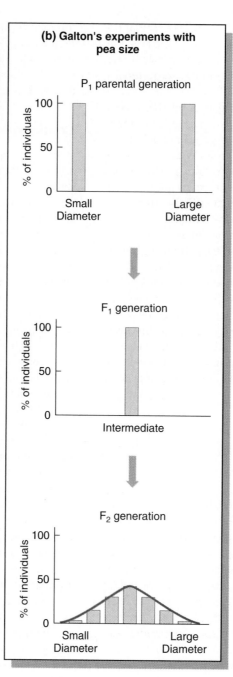

**(b) Galton's experiments with pea size**

**FIGURE 7.5**   Histograms comparing phenotypic distributions in two crosses involving pea plants. (a) The distribution of phenotypes in the $F_1$ and $F_2$ for a trait exhibiting discontinuous variation. (b) The distributions of phenotypes in the $F_1$ and $F_2$ for a trait exhibiting continuous variation.

determining the frequency of either of the most extreme (parental) phenotypes in the $F_2$ generation. Once determined, the number of involved genes ($n$) can be calculated using the term $(1/4^n)$. This calculation was discussed in Chapter 4. However, such calculations provide no information about the location of these genes, positions on chromosomes called **quantitative trait loci (QTLs)**. Because alleles occupying QTLs act on a single trait, it is not unreasonable to ask whether such genes might be physically clustered on a portion of a single chromosome, or scattered throughout the genome.

In *Drosophila,* resistance to the insecticide DDT has been shown to be a polygenic trait. To find the loci responsible, strains selected for resistance to DDT and strains selected for sensitivity were crossed to a stock carrying dominant gene markers on each chromosome (Figure 7.6). Backcrosses and $F_1 \times F_1$ matings were used to create offspring carrying combinations of marker chromosomes and chromosomes from resistant strains. These combinations were then tested for resistance by exposure to DDT. The results indicate that each of the chromosomes in the *Drosophila* karyotype contains genes that contribute to resistance. In other words, the

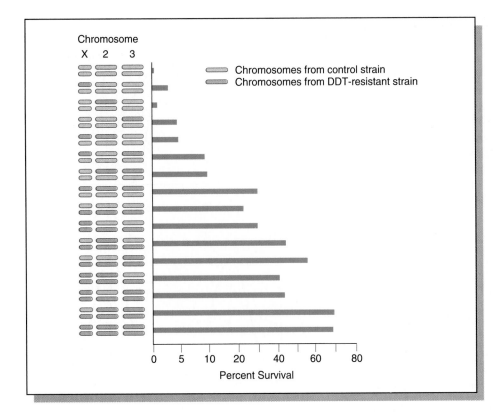

Chromosome

X   2   3

Chromosomes from control strain
Chromosomes from DDT-resistant strain

Percent Survival
0   5   10   20   40   60   80

**FIGURE 7.6**    The differential survival of *Drosophila* carrying combinations of chromosomes from DDT-resistant and DDT-sensitive strains when exposed to DDT. The results indicate that DDT resistance is polygenic, with genes on each of the major chromosomes making a major contribution. Chromosome 4 carries few genes and was omitted from this analysis.

loci that control DDT resistance are not clustered together on a single chromosome, but instead are scattered throughout the genome. This does not mean that in all cases polygenes controlling a trait are scattered throughout the genome, but it does show that polygenes need not be clustered in order to affect a single trait.

It is possible to perform more traditional mapping experiments in order to determine the map location of the QTLs responsible for DDT resistance in *Drosophila* because identifiable genetic markers are present on each chromosome. In many other organisms, especially those of agricultural importance, mutant genes, which serve as the traditional genetic markers, are not always available for each chromosome, so that systematic mapping of quantitative traits is often not feasible. However, the development of **molecular markers** (called **restriction fragment length polymorphisms** or **RFLPs**—see Chapters 12 and 13) has provided a new approach to mapping QTLs. RFLPs are polymorphisms (differences) in DNA sequence between organisms found along chromosomes that are easily assayed in individual organisms.

For example, in the tomato, the location of over 300 RFLP markers are known on the 12 chromosomes present in this plant. They are spaced about every 5 cM (one cM = one map unit) along each chromosome. Analysis is performed by crossing plants with extreme but opposite phenotypes, and through several genera-

tions, looking for *nonrandom* segregation of specific RFLPs along with the phenotypic expression of the trait controlled by the QTLs of interest. When both an RFLP molecular marker and a QTL responsible for the trait in question are closely linked on a single chromosome, they are much more likely to demonstrate an association together throughout a pedigree than if they are not closely linked. When cosegregation of the marker and the trait occur, it can be concluded that there is a QTL closely linked to the marker, thus identifying one of the QTLs.

Using the molecular map of the tomato in conjunction with some newly derived analytical methods, QTLs for fruit weight and soluble solids have been located. Genes for fruit weight were found on six chromosomes (1, 4, 6, 7, 9, and 11), and four loci for soluble solids were identified (on chromosomes 3, 4, 6, and 7). Further work using RFLP markers on each of these chromosomes has allowed localization of several loci to relatively short chromosomal segments, thus compiling a genetic map of these QTLs. The determination of the locations of QTLs for agriculturally important characteristics will open the door to their future genetic manipulation and transfer between organisms. Since establishing RFLP maps has become fairly routine in the 1990s, this approach is very feasible and will become an important tool in the repertoire of genetic engineering.

## ANALYSIS OF POLYGENIC TRAITS

Analysis of any given polygenic trait involves quantitative measurements, usually from a series of crosses. The outcome can be expressed as a frequency diagram that most often takes on a normal (bell-shaped) distribution (Figure 7.7). While it is hoped that each series of crosses is representative of the population at large, variation in small samples due strictly to chance may also influence the data that are gathered. In order to assess the experimental validity of the data, statistical techniques must be employed.

Statistical analysis serves three purposes:

1. Data can be mathematically reduced to provide a **descriptive summary** of the sample.

2. Data from a small but random sample can be utilized to infer information about groups larger than those from which the original data were obtained (**statistical inference**). The statistical values representing a typical sample are expressed using terms such as the mean, the standard error, etc. that are symbolized using italicized Roman letters (e.g. $\bar{X}$, $s$, etc.).

3. Two or more sets of experimental data may be compared to determine whether they represent significantly different populations of measurements.

Several statistical methods are useful in the analysis of traits that exhibit a normal distribution, including the

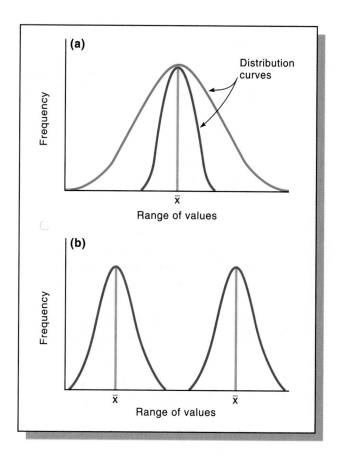

**FIGURE 7.8**    (a) Two normal frequency distributions with the same mean but different amounts of variation. (b) Two normal distributions with different means but the same amount of variation.

mean, variance, standard deviation, and standard error of the mean.

### The Mean

The distribution of phenotypic values in Figure 7.8(a) tends to cluster around a central value. This clustering is called a **central tendency**, the most common measurement of which is the mean ($\bar{X}$). The mean is simply the arithmetic average of a set of measurements or data and is calculated as

$$\bar{X} = \frac{\Sigma X_i}{n}$$

where $\bar{X}$ is the mean, $\Sigma X_i$ represents the sum of all individual values in the sample, and $n$ is the number of individual values.

Although the mean provides a descriptive summary of the sample, it is in itself of limited value. All values in the sample may be clustered near the mean, or they may be

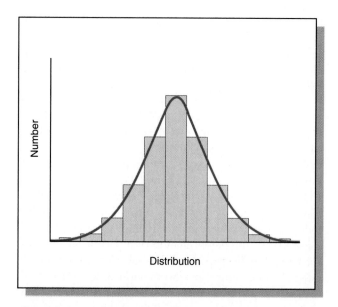

**FIGURE 7.7**    A normal frequency distribution characterized by a bell-shaped curve.

distributed widely around it. Figure 7.8(a) shows two normal (or symmetrical) distributions with identical means, but a widely different range of values. These contrasting conditions represent a different type and amount of variation within the sample. Whether due to chance or to one or more experimental variables, such variation creates the need for methods to describe variation around the mean.

## Variance

As seen in Figure 7.8(a), the range of values on either side of the mean will determine the width of the distribution curve. Measurement of this variation in a sample is called the sample variance ($s^2$ or $V$) and is used as an estimate of the variation present in an infinitely large population. The variance for a sample is calculated as

$$s^2 = V = \frac{\Sigma(X_i - \bar{X})^2}{n - 1}$$

where the sum ($\Sigma$) of the squared differences between each measured value ($X_i$) and the mean ($\bar{X}$) is divided by one less than the total sample size ($n - 1$). To avoid the numerous subtraction functions necessary in calculating $s^2$ for a large sample, we can convert the equation to its algebraic equivalent:

$$s^2 = V = \frac{\Sigma X_i^2 - n\bar{X}^2}{n - 1}$$

The variance is a valuable measure of sample variability. Two distributions may have identical means ($\bar{X}$), yet vary considerably in their concentration around the mean. The variance represents the average squared deviation of the measurements from the mean. The larger the variance, the greater the range of values on either side of the mean. The estimation of variance has been particularly valuable in determining the degree of genetic control of traits where the immediate environment also influences the phenotype.

## Standard Deviation

Since the variance is a squared value, its unit of measurement is also squared (cm$^2$, mg$^2$, etc.). To express variation around the mean in the original units of measurement, it is necessary to calculate the square root of the variance, a term called the **standard deviation ($s$)**:

$$s = \sqrt{s^2}$$

Table 7.1 shows what percentage of the individual values within a normal or bell-shaped distribution is included with different multiples of the standard deviation. The mean plus or minus one standard deviation ($\bar{X} \pm 1s$) includes 68 percent of all values in the sample. Over 95

**Table 7.1** SAMPLE INCLUSION FOR VARIOUS $s$ VALUES

| Multiples of $s$ | Percent of Sample Included |
| --- | --- |
| $\bar{X} \pm 1s$ | 68.3% |
| $\bar{X} \pm 1.96s$ | 95.0 |
| $\bar{X} \pm 2s$ | 95.5 |
| $\bar{X} \pm 3s$ | 99.7 |

percent of all values are found within two standard deviations ($\bar{X} \pm 2s$). As such, the standard deviation provides an important descriptive summary of a set of data. Furthermore, $s$ can be interpreted as a probability. The $\bar{X} \pm 1s$ indicates that there is a 68 percent probability that a measured value picked at random will fall within that range.

## Standard Error of the Mean

To estimate how much the means of other similar samples drawn from the same population might vary, we can calculate the **standard error of the mean ($S_{\bar{x}}$)**:

$$S_{\bar{x}} = \frac{s}{\sqrt{n}}$$

where $s$ is the standard deviation, and $\sqrt{n}$ is the square root of the sample size. The standard error of the mean is a measure of the accuracy of the sample mean, that is, the variation of sample means in replications of the experiment. Because it can be expected that the standard deviation of mean values will reflect less variance than the standard deviation of a set of individual measurements. Since the standard error of the mean is computed by dividing $s$ by $\sqrt{n}$, it is always a smaller value than the standard deviation.

## Analysis of a Quantitative Character

Many characteristics of importance in livestock and crop plants are controlled by polygenic systems. In Chapter 4, we discussed such phenotypes as ear length in corn and kernel color in wheat. To illustrate how biometric methods are used in the analysis of quantitative characters, we will consider a simplified example involving fruit weight in tomatoes. Let us assume that fruit weight is a quantitative character, and that one highly inbred strain produces tomatoes averaging 18 oz. in weight, and another highly inbred strain produces fruit averaging 6 oz. in weight. These two varieties are crossed and produce an $F_1$ generation with weights ranging from 10 oz. to 14 oz. (Table 7.2). The $F_2$ population contains individuals that produce fruit ranging from 6 oz. to 18 oz.

**Table 7.2**   Distribution of $F_1$ and $F_2$ Progeny

| | Weight in oz. | | | | | | | | | | | | |
|---|---|---|---|---|---|---|---|---|---|---|---|---|---|
| | **6** | **7** | **8** | **9** | **10** | **11** | **12** | **13** | **14** | **15** | **16** | **17** | **18** |
| Number of $F_1$: | | | | | 4 | 14 | 16 | 12 | 8 | | | | |
| Individuals $F_2$: | 1 | 1 | 2 | 0 | 9 | 13 | 17 | 14 | 7 | 4 | 3 | 0 | 1 |

The mean value for the fruit weight in the $F_1$ generation can be calculated as

$$\bar{X} = \frac{\Sigma X_i}{n} = \frac{626}{52} = 12.04$$

Similarly, the mean value for fruit weight in the $F_2$ generation is calculated as

$$\bar{X} = \frac{\Sigma X_i}{n} = \frac{872}{72} = 12.11$$

Average fruit weight is 12.04 oz. in the $F_1$ generation and 12.11 oz. in the $F_2$ generation. While these mean values are similar, it is apparent that there is more variation present in the $F_2$ generation, because fruit weight ranges from 6 to 18 oz. in the $F_2$ generation, but only from 10 to 14 oz. in the $F_1$ generation.

To assist in quantitating the amount of variation present in each generation, we can construct a frequency table (Table 7.3). The first column ($x$) represents the midpoint of a class interval. For example, all tomatoes weighing between 5.5 and 6.5 oz. are classified as 6 oz.

The second column (marked $f$) lists the number of tomatoes that fall into that category. The last column $f(x)$ computes the product of the first two columns ($x \times f$) and lists the total weight in oz. for each class. The sum of the $f$ column ($\Sigma f$) equals the total number of tomatoes counted, and the sum of the $f(x)$ column [$\Sigma f(x)$] equals the total weight of all the tomatoes in the sample. The mean value for fruit weight in the $F_1$ and $F_2$ generations taken from the frequency table can be calculated as

$$F_1: \frac{\Sigma f(x)}{\Sigma f} = \frac{626}{52} = 12.03$$

$$F_2: \frac{\Sigma f(x)}{\Sigma f} = \frac{842}{72} = 12.11$$

As noted above, the sample variance can be calculated as the sum of the squared differences between each value and the mean, divided by one less than the total number of observations. However, in the case where a number of observations ($f$) have been grouped into representative classes ($x$), the variance can be calculated according to the formula

**Table 7.3**   Frequency Distribution in $F_1$ and $F_2$

| | $F_1$ | | | $F_2$ | |
|---|---|---|---|---|---|
| **$x$** | **$f$** | **$f(x)$** | **$x$** | **$f$** | **$f(x)$** |
| 6 | | | 6 | 1 | 6 |
| 7 | | | 7 | 1 | 7 |
| 8 | | | 8 | 2 | 16 |
| 9 | | | 9 | | |
| 10 | 4 | 40 | 10 | 9 | 90 |
| 11 | 14 | 154 | 11 | 13 | 143 |
| 12 | 16 | 192 | 12 | 17 | 204 |
| 13 | 12 | 156 | 13 | 14 | 182 |
| 14 | 6 | 84 | 14 | 7 | 98 |
| 15 | | | 15 | 4 | 60 |
| 16 | | | 16 | 3 | 48 |
| 17 | | | 17 | | |
| 18 | | | 18 | 1 | 18 |
| | $\Sigma f = 52$ | $\Sigma f(x) = 626$ | | $\Sigma f = 72$ | $\Sigma f(x) = 872$ |

**Table 7.4**   CALCULATION OF VARIANCE

| | | **F₁** | | | | **F₂** | |
|---|---|---|---|---|---|---|---|
| $x$ | $f$ | $f(x)$ | $f(x^2)$ | $x$ | $f$ | $f(x)$ | $f(x^2)$ |
| 6 | | | | 6 | 1 | 6 | 36 |
| 7 | | | | 7 | 1 | 7 | 48 |
| 8 | | | | 8 | 2 | 16 | 128 |
| 9 | | | | 9 | | | |
| 10 | 4 | 40 | 400 | 10 | 9 | 90 | 900 |
| 11 | 14 | 154 | 1694 | 11 | 13 | 143 | 1573 |
| 12 | 16 | 192 | 2304 | 12 | 17 | 204 | 2448 |
| 13 | 12 | 156 | 2028 | 13 | 14 | 182 | 2366 |
| 14 | 6 | 84 | 1176 | 14 | 7 | 98 | 1372 |
| 15 | | | | 15 | 4 | 60 | 900 |
| 16 | | | | 16 | 3 | 48 | 768 |
| 17 | | | | 17 | | | |
| 18 | | | | 18 | 1 | 18 | 324 |
| | $52 = n$ | $626 = \Sigma f(x)$ | $7602 = \Sigma f(x^2)$ | | $72 = n$ | $872 = \Sigma f(x)$ | $10{,}864 = \Sigma f(x^2)$ |

For F₁:
$$s^2 = \frac{52 \times 7602 - (626)^2}{52(52-1)}$$
$$= \frac{395{,}304 - 391{,}876}{2652}$$
$$= 1.29$$

$$s^2 = \frac{n\Sigma f(x^2) - (\Sigma fx)^2}{n(n-1)}$$

For F₂:
$$s^2 = \frac{72 \times 10{,}864 - (872)^2}{72(72-1)}$$
$$= \frac{782{,}208 - 760{,}304}{5112}$$
$$= 4.28$$

$$s^2 = V = \frac{n\Sigma f(x^2) - (\Sigma fx)^2}{n(n-1)}$$

As shown in Table 7.4, the value for the F₁ generation is 1.29, and for the F₂ generation, it is 4.28. When converted to the standard deviation ($s = \sqrt{s^2}$), the values become 1.13 and 2.06, respectively. Thus, the distribution of tomato weight in the F₁ generation can be described as 12.04 ± 1.13, and in the F₂ as 12.11 ± 2.06. This analysis indicates that the mean fruit weight of the F₁ is identical to that of the F₂, but that the F₂ generation shows greater variability in the distribution of weights than the F₁.

The observations about the inheritance of fruit weight in crosses between these two strains of tomatoes meet the expectations for polygenic traits. For the sake of this example, if we assume that each parental strain is homozygous dominant or homozygous recessive for the genes that control fruit weight, we can calculate the number of gene pairs involved in controlling fruit weight in these two strains of tomatoes. Since 1/72 of the F₂ offspring have a phenotype that overlaps one of the parental strains (72 total F₂ offspring, one weighs 6 oz., one weighs 18 oz.; see Table 7.2), the formula $1/4^n = 1/72$ indicates that probably 3 and perhaps 4 genes control fruit weight in these tomato strains.

## HERITABILITY

Having just introduced several ways in which variation can be measured and expressed in populations, we turn to a consideration of how we can assess the extent to which genetic factors contribute to phenotypic variation within a population. Some traits demonstrate minimal variation under normal environmental circumstances, but other traits show distinct variability under the same environmental conditions. The latter case is particularly true for traits where the range of phenotypic variation is determined by polygenic inheritance. Often, polygenic traits show such extensive variation in a population that geneticists can only speculate that a large but unknown number of genes provide the genetic potential, which is then modified by the environment.

Provided that a trait can be measured quantitatively, we can approach the question of genetics versus environment analytically. Experiments on plants and animals can test the causes of variation. One approach is to use inbred strains containing individuals of a relatively homogeneous or constant genetic background. Experiments may then be designed to test the effects of environmental conditions on phenotypic variability. Variation observed *between* different highly inbred lines reared in a homogeneous environment is genetic. Variation observed *among* members of the same inbred line reared under varying environmental conditions is nongenetic.

The relative importance of genetic versus nongenetic factors may also be assessed by examining the **heritability index ($H^2$)**, which can be calculated using analysis of variance among individuals of a known genetic relationship. This may be the only practical approach in organisms with long generation times. Also called the **broad heritability**, $H^2$ measures the degree to which phenotypic variation ($V_p$) is due to genetic factors for a single population under the limits of environmental variability during the study. It is important to emphasize here that $H^2$ does not measure the proportion of the total phenotype attributed to genetic factors; rather, it measures only the observed variation in the phenotype attributed to genetic versus nongenetic factors. Phenotypic variance is due to the sum of three components: environmental variance ($V_E$), genetic variance ($V_G$), and variance resulting from the interaction of the genotype and the environment ($V_{GE}$). Therefore, phenotypic variance ($V_P$) is theoretically expressed as

$$V_P = V_E + V_G + V_{GE}$$

In most studies involving a single population in one environment, it is not practical to analyze $V_{GE}$, so it is simply included as part of $V_E$. Therefore, the simpler equation is generally used:

$$V_P = V_E + V_G$$

Broad heritability expresses that proportion of phenotypic variance due to the genetic component:

$$H^2 = \frac{V_G}{V_P}$$

A very high $H^2$ value indicates that the environmental conditions have had relatively little impact on phenotypic variance in the population studied. A very low $H^2$ value indicates that the variation in the environment has been almost solely responsible for the observed phenotypic variation.

It is not possible to obtain an *absolute* $H^2$ value for any given character. If measured in a different population under a greater or lesser degree of environmental variability, $H^2$ might well change for that character.

Practically, information regarding heritability is most applicable in animal and plant breeding as a measure of potential response to selection. A different estimate of heritability must be used in this case, based on a subcomponent $V_A$ of genetic variance ($V_G$):

$$V_G = V_A + V_D + V_I$$

$V_A$ represents the **additive variance** that results from the average effect of the additive genes themselves. $V_D$ represents **dominance variance** and accounts for variation that results when heterozygotes are not precisely intermediate between the two homozygous genotypes. $V_I$ represents **interaction variance** that occurs when two or more loci "interact" epistatically. $V_I$ reflects the variance not associated with the average effect of $V_A$. Of these subcomponents, $V_I$ is difficult to assess and is often combined with $V_D$. The sum of the two is termed "nonadditive" genetic variance.

When $V_G$ is partitioned into $V_A$ and $V_D$, a new assessment of heritability $h^2$, or **narrow heritability**, may be calculated. It is $h^2$ that is useful in assessing selection potential in animal and plant populations:

$$h^2 = \frac{V_A}{V_P}$$

Since $V_P = V_E + V_G$, and since $V_G = V_A + V_D$, then

$$h^2 = \frac{V_A}{V_E + V_A + V_D}$$

The higher $h^2$ is, the greater the impact selection may have in altering an initial population. A high $h^2$ indicates a high level of predictability in the transmission of traits from parent to offspring.

As you can imagine, measuring the components necessary to calculate $h^2$ is a complex task. Table 7.5 provides estimates of heritability for a variety of traits in different organisms. As you can see, heritability varies considerably. In general, heritability estimates are low for traits essential to an organism's survival. Egg production, litter size, and conception rate in various organisms are examples. Traits less important to survival, such as tail length and wing length, show a much higher degree of heritability. Narrow heritability estimates, while characterizing a particular population, have been very useful in agricultural programs. Based on these estimates, selection techniques have led to vast improvements in the quality and quantity of plant and animal products.

## Twin Studies in Humans

In humans traditional heritability studies are not possible. However, twins are very useful subjects for studying the heredity versus environment question. **Monozygotic** or **identical twins**, derived from the division and splitting of a single egg following fertilization, are identical in their genetic compositions. Although most identical twins are reared together and are exposed to very similar environments, some pairs are separated and raised in different settings. For any particular trait, average similarities or differences can be investigated. Such an analysis is particularly useful because characteristics that remain similar in different environments are believed to be inherited. These data can then be compared

**Table 7.5**  ESTIMATES OF HERITABILITY FOR SEVERAL TRAITS IN DIFFERENT ORGANISMS

| Trait | Heritability ($h^2$) |
|---|---|
| Mice | |
| Tail length | 60% |
| Body weight | 37 |
| Litter size | 15 |
| *Drosophila* | |
| Abdominal bristle number | 52 |
| Wing length | 45 |
| Egg production | 18 |
| Chickens | |
| Body weight | 50 |
| Egg production | 20 |
| Egg hatchability | 15 |
| Cattle | |
| Birth weight | 51 |
| Milk yield | 44 |
| Conception rate | 3 |

with a similar analysis of **dizygotic** or **fraternal twins**, who originate from two separate fertilization events. Dizygotic twins are thus no more genetically similar than any two siblings.

A form of quantitative analysis of characteristics of twins reared together may also be pursued. Twins are said to be **concordant** for a given trait if both express it *or* neither expresses it and **discordant** if one shows the trait *and* the other does not. Table 7.6 lists concordance values for various traits in both types of twins.

**Table 7.6**  A COMPARISON OF CONCORDANCE FOR VARIOUS TRAITS BETWEEN MONOZYGOTIC (MZ) AND DIZYGOTIC (DZ) TWINS

| Trait | Concordance | |
|---|---|---|
| | MZ | DZ |
| Blood types | 100% | 66% |
| Eye color | 99 | 28 |
| Mental retardation | 97 | 37 |
| Measles | 95 | 87 |
| Idiopathic epilepsy | 72 | 15 |
| Schizophrenia | 69 | 10 |
| Diabetes | 65 | 18 |
| Identical allergy | 59 | 5 |
| Tuberculosis | 57 | 23 |
| Cleft lip | 42 | 5 |
| Club foot | 32 | 3 |
| Mammary cancer | 6 | 3 |

These data must be examined very carefully before any conclusions are drawn. If the concordance value approaches 90 to 100 percent with monozygotic twins, we might be inclined to interpret this value as indicating a large genetic contribution to the expression of the trait. In some cases—blood types and eye color, for example—we know this is indeed true. In the case of measles, however, a high concordance value merely indicates that the trait is almost always induced by a factor in the environment—in this case, a virus.

Therefore, it is more meaningful to compare the *difference* between the concordance values of monozygotic and dizygotic twins. If these values are significantly higher for monozygotic twins than dizygotic twins, we suspect that there is a genetic component involved in the determination of the trait. We reach this conclusion because monozygotic twins, with identical genotypes, would be expected to show a greater concordance than genetically related, but not genetically identical, dizygotic twins. In the case of measles, where concordance is high in both types of twins, the environment is assumed to contribute significantly.

Even though a particular trait may demonstrate considerable genetically based variation, it is often difficult to formulate a precise mode of inheritance based on available data. In many cases the trait is considered to be controlled by multiple-factor inheritance. However, when the environment is also exerting a partial influence, such a conclusion is particularly difficult to prove.

# THE USE OF HAPLOID ORGANISMS IN LINKAGE AND MAPPING STUDIES

We conclude this chapter by turning to still another topic that is an extension of our study of transmission genetics: **linkage analysis and chromosome mapping in haploid eukaryotes**. As we shall see, even though analysis of the location of genes relative to one another throughout the genome of haploid organisms may *seem* a bit more complex than in diploid organisms (Chapter 4), the basic underlying principles are the same. In fact, many basic principles of inheritance were established during the study of haploid fungi.

While many single-celled eukaryotes are haploid during the vegetative stages of their life cycle, they also form reproductive cells that fuse during fertilization, forming a diploid zygote. The zygote then undergoes meiosis in order to reestablish haploidy. The haploid meiotic products are the progenitors of the subsequent members of the vegetative phase of the life cycle. Figure 7.9 illustrates this type of cycle in the green alga *Chlamydomonas*. Even though the haploid cells that fuse during fertil-

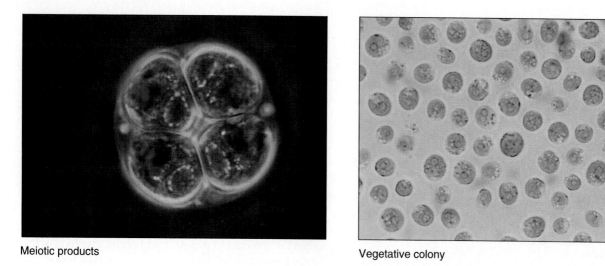

Meiotic products

Vegetative colony

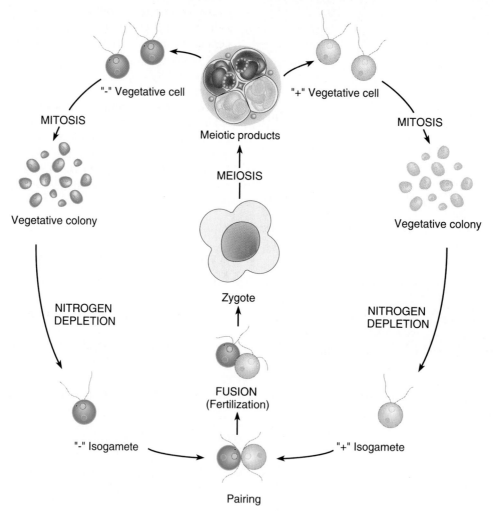

**FIGURE 7.9**    The life cycle of *Chlamydomonas.* The diploid zygote (in the center) undergoes meiosis, producing "+" or "−" haploid cells that undergo mitosis, yielding vegetative colonies. Unfavorable conditions stimulate them to form isogametes, which fuse in fertilization, producing a zygote, and repeating the cycle. Two of these stages are illustrated photographically.

ization *look* identical (and are thus called **isogametes**), a chemical identity exists on their surface. As a result, all strains are either "+" or "−" and fertilization occurs only between unlike cells.

In order to perform the genetic experiments with haploid organisms, genetic strains of different genotypes are isolated and crossed to one another. Following fertilization and meiosis, the meiotic products are retained together and may be analyzed. Such is the case in *Chlamydomonas* as well as in the fungus *Neurospora,* which we shall use as an example in the ensuing discussion. Following fertilization in *Neurospora* (Figure 7.10), meiosis occurs in a sac-like structure called the **ascus** and the initial set of haploid products, called a **tetrad**, are retained within it. This term has a quite different meaning than when it is used to describe the chromosome configuration characteristic of meiotic prophase I in diploids.

Following meiosis in *Neurospora,* each cell in the ascus divides mitotically, producing eight haploid **ascospores**. These may be dissected out and examined morphologically or tested to determine their genotypes and phenotypes. Since the eight cells reflect the sequence of their formation following meiosis, such an examination is called **tetrad analysis**, a process critical to our subsequent discussion.

## Gene to Centromere Mapping

When a single gene is analyzed in *Neurospora,* as diagrammed in Figure 7.11, the data can be used to calculate the map distance between that gene and the centromere. This process is sometimes referred to as **mapping the centromere**. It may be accomplished by experimentally determining the frequency of recombination using tetrad data.

If no crossover event occurs between the gene under study and the centromere, the pattern of ascospores in the ascus appears as shown in Figure 7.11(a) ($aaaa++++$). This pattern represents **first division segregation** because the two alleles are separated during the first meiotic division. However, a crossover event will alter this pattern, as shown in Figure 7.11(b) ($aa++aa++$) and 7.11(c) ($++aaaa++$). Actually, two recombinant patterns may occur, depending on the chromatid orientation during the second meiotic division: $++aa++aa$ and $aa++++aa$. All four of the latter patterns reflect **second division segregation** because the two alleles are not separated until the second meiotic division. Usually, the ordered tetrad data are condensed to reflect the genotypes of the identical ascospore pairs. Thus, five combinations are possible:

| First Division Segregation |
| --- |
| $aa++$ |

| Second Division Segregation |
| --- |
| $a+a+$ |
| $+a+a$ |
| $+aa+$ |
| $a++a$ |

In order to calculate the distance between the gene and the centromere, a large number of asci resulting from a controlled cross must be scored. Using these data, the distance is calculated:

$$\frac{1/2(\text{second division segregant asci})}{\text{total asci scored}}$$

The recombination percentage is only one-half the number of second division segregants because crossing over in each of them has occurred in only two of the four strands during meiosis.

To illustrate, assume that *a* represents albino and + represents wild type in *Neurospora*. In crosses between the two genetic types, suppose the following data were observed:

65 first division segregants
70 second division segregants

The distance between *a* and the centromere is thus

$$\frac{(1/2)\,(70)}{135} = 0.259$$

or about 26 map units.

As the distance increases up to 50 units, in theory, all asci should reflect second division segregation. However, numerous factors prevent this. As in diploid organisms, accuracy is greatest when the gene and the centromere are relatively close together.

## Ordered versus Unordered Tetrad Analysis

In our previous discussion, we assumed that the genotype of each ascospore and its position in the tetrad can be determined. To perform such an **ordered tetrad analysis**, individual asci must be dissected and each ascospore must be tracked as it germinates. This is a tedious process, but it is essential for two types of analysis:

1. To distinguish between first division segregation and second division segregation of alleles in meiosis.

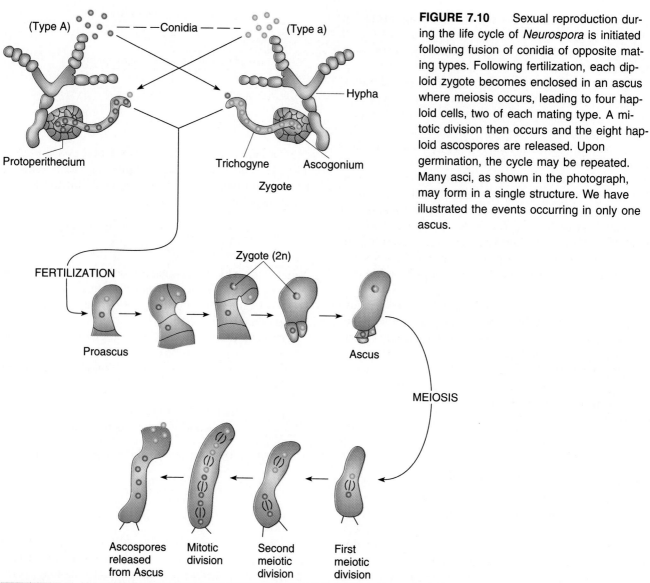

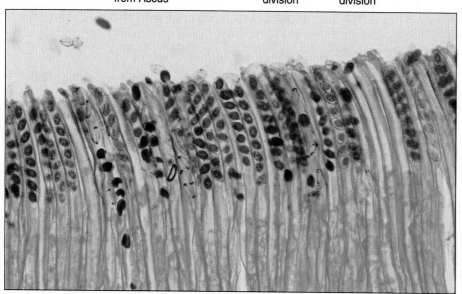

**FIGURE 7.10**   Sexual reproduction during the life cycle of *Neurospora* is initiated following fusion of conidia of opposite mating types. Following fertilization, each diploid zygote becomes enclosed in an ascus where meiosis occurs, leading to four haploid cells, two of each mating type. A mitotic division then occurs and the eight haploid ascospores are released. Upon germination, the cycle may be repeated. Many asci, as shown in the photograph, may form in a single structure. We have illustrated the events occurring in only one ascus.

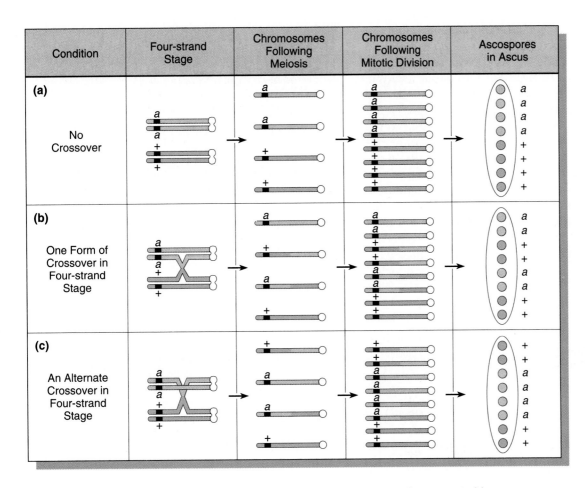

**FIGURE 7.11**    Three ways in which different ascospore patterns can be generated in *Neurospora*. Analysis of these patterns may serve as the basis of "gene to centromere" mapping, as described in the text.

2.    To determine whether recombinational events are reciprocal or not.

In the first case, such information is essential to "map the centromere" as we have just discussed. Thus, ordered tetrad analysis must be performed in order to map the distance between a gene and the centromere.

In the second case, ordered tetrad analysis has revealed that recombinational events are not always reciprocal, particularly when closely linked genes are studied in Ascomycetes. This observation has led to the investigation of the phenomenon called **gene conversion**. Since its discussion requires a background in DNA structure and analysis, we will return to this topic in Chapter 11.

It is much less tedious to isolate individual asci, allow them to mature, and then determine the genotypes of each ascospore, but not in any particular order. This approach is referred to as **unordered tetrad analysis**. As we will see in the next section, this type of analysis can be used to determine whether two genes are linked on the same chromosome or not, and if so, to determine the map distance between them.

## Linkage and Gene Mapping in Haploid Organisms

Analysis of genetic data derived from haploid organisms can be utilized in order to distinguish between linkage and independent assortment of two genes; it further allows mapping distances to be calculated between gene loci once linkage is established. In the following discussion, we will consider tetrad analysis in the alga, *Chlamydomonas*. With the exception that the four meiotic products are not ordered and *do not* undergo a mitotic division following the completion of meiosis, the general principles discussed for *Neurospora* also apply to *Chlamydomonas*.

To compare independent assortment and linkage, we will consider two theoretical mutant alleles, *a* and *b*, representing two distinct loci in *Chlamydomonas*. Suppose that 100 tetrads derived from the cross *ab* × ++ yield the data shown in Table 7.7. As you can see, all tetrads produce one of three patterns. For example, all tetrads in category I produce two ++ cells and two *ab* cells and are designated as **parental ditypes (P)**. Category II tetrads produce two *a*+ cells and two +*b* cells and are called **nonparental ditypes (NP)**. Category III tetrads produce one cell each of the four possible genotypes and are thus termed **tetratypes (T)**.

These data support the hypothesis that the genes represented by the *a* and *b* alleles are located on separate chromosomes. In order to understand why, you must refer to Figure 7.12. In parts (a) (category I) and (b) (category II) of this figure, the origin of parental and nonparental ditypes is demonstrated for two unlinked genes. According to the Mendelian principle of independent

**Table 7.7**  TETRAD ANALYSIS IN *CHLAMYDOMONAS*

| Category | I | II | III |
|---|---|---|---|
| Tetrad type | Parental (P) | Nonparental (NP) | Tetratypes (T) |
| Genotypes present | ++ <br> ++ <br> *ab* <br> *ab* | *a*+ <br> *a*+ <br> +*b* <br> +*b* | ++ <br> *a*+ <br> +*b* <br> *ab* |
| Number of tetrads | 43 | 43 | 14 |

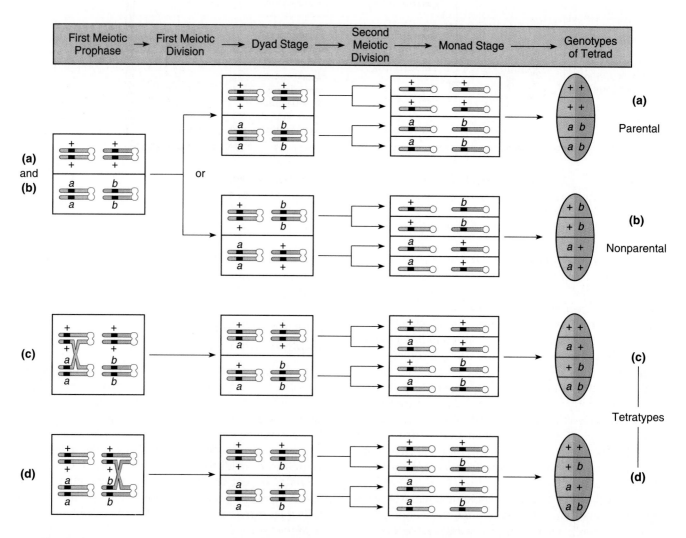

**FIGURE 7.12**    The origin of various genotypes found in tetrads in *Chlamydomonas* when two genes located on separate chromosomes are considered.

assortment of unlinked genes, approximately equal proportions of these tetrad types are predicted. Thus, when the parental ditypes are equal to the nonparental ditypes, then the two genes are not linked. The data in Table 7.7 confirm this prediction. Since independent assortment has occurred, it can be concluded that the two genes are located on separate chromosomes.

The origin of category III, the **tetratypes**, is diagrammed in Figure 7.12(c) and (d). The genotypes of tetrads in this category can be generated in two possible ways. Both involve a crossover event between one of the genes and the centromere. In Figure 7.12(c) the exchange involves one of the two chromosomes and occurs between gene $a$ and the centromere; in Figure 7.12(d), the other chromosome is involved and the exchange occurs between gene $b$ and the centromere.

The production of tetratype tetrads does not alter the final ratio of the four genotypes present in all meiotic products. If the genotypes from 100 tetrads (which yield 400 cells) are computed, 100 of each genotype are found. This 1:1:1:1 ratio is predicted according to the expectation of independent assortment.

Now consider the case where the genes $a$ and $b$ are linked (Figure 7.13). The same categories of tetrads will be produced. However, parental and nonparental ditypes will not necessarily occur in equal proportions; nor will the four genotypic combinations be found in equal numbers if the genotypes of all meiotic products are computed. For example, the following data might be encountered:

| Category I<br>P | Category II<br>NP | Category III<br>T |
|---|---|---|
| 64 | 6 | 30 |

Since the parental and nonparental categories are not produced in equal proportion, we can predict that independent assortment is not in operation. If not, we can conclude that the two genes are linked and proceed to determine the map distance between the two genes.

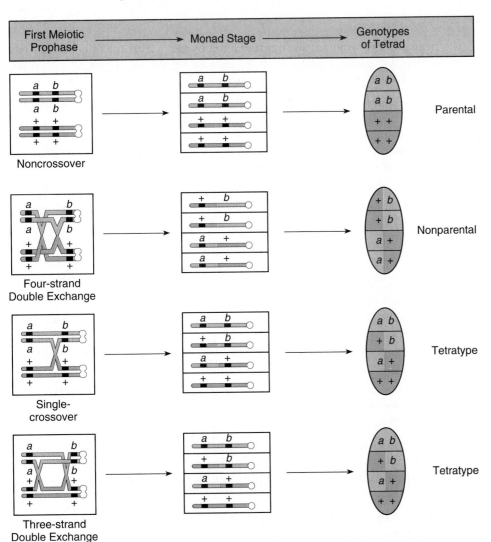

**FIGURE 7.13**   The various types of exchanges leading to the genotypes found in tetrads in *Chlamydomonas* when two genes located on the same chromosome are considered.

In the analysis of these data, we are concerned with the determination of which tetrad types represent genetic exchanges between the two genes. The **parental ditype tetrads (P)** arise only when no crossing over occurs between the two genes. The **nonparental ditype tetrads (NP)** arise only when a double exchange involving all four chromatids occurs between two genes. The **tetratype tetrads (T)** arise when either a single crossover occurs or when an alternative type of double exchange occurs between the two genes. The various types of exchanges described here are diagrammed in Figure 7.13.

When the proportion of the three tetrad types has been determined, it is possible to calculate the map distance between the two linked genes. The following formula computes the exchange frequency, which is proportional to the map distance between the two genes:

$$\text{exchange frequency (\%)} = \frac{NP + 1/2(T)}{\text{total number of tetrads}} \times 100$$

In this formula NP represents the nonparental tetrads; all meiotic products represent an exchange. The tetratype tetrads are represented by T; one-half of the meiotic products represents exchanges. The sum of the scored tetrads that fall into these categories is then divided by the total number of tetrads examined (P + NP + T). If this calculated number is multiplied by 100, it is converted to a percentage, which is directly equivalent to the map distance between the genes.

In our example, the calculation reveals that genes $a$ and $b$ are separated by 21 map units:

$$\frac{6 + 1/2(30)}{100} = \frac{6 + 15}{100} = \frac{21}{100} = 0.21 \times 100 = 21\%$$

Although we have considered linkage analysis and mapping of only two genes at a time, such studies often involve three or more genes. In these cases, the gene sequence as well as map distances can be determined.

---

**CHAPTER SUMMARY**

1. The penetrance of a mutant allele is measured by the percentage of organisms in a population that exhibits evidence of the mutant phenotype. Expressivity, on the other hand, measures the range of mutant expression within a population.

2. Phenotypic expression may be modified by factors such as genetic background, temperature, nutrition, and other environmental factors. The phenomena of genetic suppression and position effects have been used to illustrate the existence of genetic background factors. Together, all such factors constitute the total environment in which genetic information is expressed.

3. The time of onset of gene expression in all organisms varies as the need for certain gene products occurs at different periods during the processes of development, growth, and aging.

4. The degree of impact of genetic and environmental factors in the establishment of a given phenotype is difficult to ascertain. Heritability can be calculated for many characters but is especially useful in selective breeding of commercially valuable plants and animals.

5. Studies involving twins are aimed at resolving the question of heredity versus environment in human traits. The degree of concordance and discordance of a trait may be compared in monozygotic (identical) and dizygotic (fraternal) twins raised together or apart.

6. Continuous variation is exhibited in crosses involving traits under polygenic control. Such traits are quantitative in nature and are inherited as a result of the cumulative impact of additive alleles.

7. Polygenic characteristics can be analyzed using statistical methods, which include the mean, the variance, the standard deviation, and the standard error of the mean. Such statistical analysis can be descriptive, can be used to make inferences about a population, or can be used to compare sets of data.

8. Linkage analysis and chromosome mapping are possible in haploid eukaryotes. Our discussion has included gene-to-centromere and gene-to-gene mapping as well as the consideration of how to distinguish between linkage and independent assortment.

## KEY TERMS

Angelman syndrome (AS)
ascospore
ascus
auxotroph
biometry
broad heritability
central tendency
chromosome mapping in
  haploids
concordance
conditional mutation
continuous variation
discontinuous variation
discordance
dizygotic (fraternal) twins
DNA methylation
Duchenne muscular
  dystrophy (DMD)
expressivity
first division segregation
fragile-X mental
  retardation

gene conversion
genetic anticipation
genetic background
genetic suppression
genomic (parental)
  imprinting
heritability index ($H^2$)
heterochromatin
hexosaminidase A
Huntington disease
hypoxanthine-guanine
  phosphoribosyl
  transferase (HPRT)
lactase
lactose intolerance
Lesch–Nyhan syndrome
linkage analysis in
  haploids
mapping the centromere
mean

monozygotic (identical)
  twins
myotonic dystrophy (DM)
narrow heritability ($b^2$)
nonparental ditype (NP)
nutritional mutation
ordered tetrad analysis
parental ditype (P)
penetrance
permissive temperature
phenylketonuria
polygenic trait
position effect
Prader-Willi syndrome
quantitative inheritance
quantitative trait loci
  (QTL)
restriction fragment
  length polymorphism
  (RFLP)

restrictive temperature
second division
  segregation
spinal and bulbar
  muscular atrophy
  (Kennedy disease)
standard deviation
standard error of the
  mean
statistical inference
statistics
Tay–Sachs disease
temperature sensitive
  mutation
tetrad
tetrad analysis
tetratype (T)
translocation
transmission genetics
trinucleotide repeat
unordered tetrad analysis
variance

## INSIGHTS AND SOLUTIONS

1.  The following results were recorded for ear length in corn:

| | | | | | | | | | | | | | | | | | | |
|---|---|---|---|---|---|---|---|---|---|---|---|---|---|---|---|---|---|---|
| | | | | | | | Length of ear in cm | | | | | | | | | | | |
| | 5 | 6 | 7 | 8 | 9 | 10 | 11 | 12 | 13 | 14 | 15 | 16 | 17 | 18 | 19 | 20 | 21 |
| Parent A | 4 | 21 | 24 | 8 | | | | | | | | | | | | | |
| Parent B | | | | | | | | | 3 | 11 | 12 | 15 | 26 | 15 | 10 | 7 | 2 |
| $F_1$ | | | | | 1 | 12 | 12 | 14 | 17 | 9 | 4 | | | | | | | |

(a) For each of the parental strains, and the $F_1$, calculate the mean values for ear
length.

**SOLUTION:**   The mean values are calculated as follows:

$$\bar{X} = \frac{\Sigma X_i}{n} \qquad P_A: \quad \bar{X} = \frac{\Sigma X_i}{n} = \frac{378}{57} = 6.63$$

$$P_B: \quad \bar{X} = \frac{\Sigma X_i}{n} = \frac{1697}{101} = 16.80$$

$$F_1: \quad \bar{X} = \frac{\Sigma X_i}{n} = \frac{836}{69} = 12.11$$

(b) Compare the mean of the $F_1$ with that of each parental strain. What does this tell you about the type of gene action involved?

**SOLUTION:**   The $F_1$ mean (12.11) is almost midway between the parental means of 6.63 and 16.80. This indicates that the genes in question are probably additive in effect.

2. In a cross in *Neurospora* where one parent expresses the mutant allele $a$ and the other expresses a wild-type phenotype (+), the following data were obtained in the analysis of ascospores. Calculate the gene-to-centromere distance.

Asci Types

|  | 1 | 2 | 3 | 4 | 5 | 6 |
|---|---|---|---|---|---|---|
| Sequence of ascospores in ascus | + | a | a | + | a | + |
|  | + | a | a | + | a | + |
|  | + | a | + | a | + | a |
|  | + | a | + | a | + | a |
|  | a | + | a | + | + | a |
|  | a | + | a | + | + | a |
|  | a | + | + | a | a | + |
|  | a | + | + | a | a | + |
|  | 39 | 33 | 5 | 4 | 9 | 10 |

Total = 100

**SOLUTION:**   Ascus types 1 and 2 represent first division segregants (fds) where no crossing over occurred between the $a$ locus and the centromere. All others (3–6) represent second division segregation (sds). By applying the formula

$$\text{distance} = \frac{1/2\ \text{sds}}{\text{total asci}}$$

we obtain the following result:

$$d = \frac{1/2(5 + 4 + 9 + 10)}{100}$$

$$= \frac{1/2(28)}{100}$$

$$= 0.14$$

$$= 14 \text{ map units}$$

3. The mean and variance of corolla length in two highly inbred strains of *Nicotiana* and their progeny are shown below. One parent ($P_1$) has a short corolla and the other ($P_2$) has a longer corolla.

| Strain | Mean (mm) | Variance |
|---|---|---|
| $P_1$ short | 40.47 | 3.12 |
| $P_2$ long | 93.75 | 3.87 |
| $F_1(P_1 \times P_2)$ | 63.90 | 4.74 |
| $F_2(F_1 \times F_1)$ | 68.72 | 47.70 |

Calculate the heritability ($H^2$) of corolla length in this plant.

**SOLUTION:**   The formula for estimating heritability is

$H^2 = V_G/V_P$ where $V_G$ and $V_P$ are the genetic and phenotypic components of variation, respectively. The main issue in this problem is obtaining some estimate of two components of phenotypic variation: genetic and environmental factors.

$V_P$ is the combination of genetic and environmental variance. Because the two parental strains are true-breeding, they are assumed to be homozygous and the variance of 3.12 and 3.88 is considered to be the result of environmental influences. The average of these two values is 3.5. The $F_1$ is also genetically homogeneous and gives us an additional estimation of the environmental factors. By averaging with the parents

$$[(3.50 + 4.74)/2 = 4.12]$$

we obtain a relatively good idea of environmental impact on the phenotype. The phenotypic variance in the $F_2$ is the sum of the genetic ($V_G$) and environmental ($V_E$) components. We have estimated the environmental input as 4.12, so 47.71 minus 4.12 gives us an estimate of $V_G$, which is 43.59. Heritability then becomes 43.59/47.71 or 0.91. This value, when viewed in percentage form, indicates that about 91 percent of the variation in corolla length is due to genetic influences.

---

## PROBLEMS AND DISCUSSION QUESTIONS

1. Distinguish between continuous and discontinuous variation.

2. List as many human traits as you can that are likely to be under the control of a polygenic mode of inheritance.

3. Describe the difference between penetrance and expressivity.

4. Define and discuss the significance of the following terms: (a) position effect, (b) suppressor genes, (c) monozygotic and dizygotic twins, (d) concordance and discordance, and (e) heritability.

5. In the following table, average differences of height and weight between monozygotic twins (reared together and apart), dizygotic twins, and siblings are compared. Draw as many conclusions as you can concerning the effects of genetics and the environment in influencing these human traits.

| Trait | MZ Reared Together | MZ Reared Apart | DZ Reared Together | Sibs Reared Together |
|---|---|---|---|---|
| Height (cm) | 1.7 | 1.8 | 4.4 | 4.5 |
| Weight (kg) | 1.9 | 4.5 | 4.5 | 4.7 |

Source: Newman, Freeman, and Holzinger, 1937.

6. Corn plants from a test plot are measured, and the distribution of heights at 10-cm intervals is recorded below:

| Height (cm) | 100 | 110 | 120 | 130 | 140 | 150 | 160 | 170 | 180 |
|---|---|---|---|---|---|---|---|---|---|
| Plants (no.) | 20 | 60 | 90 | 130 | 180 | 120 | 70 | 50 | 40 |

Calculate (a) the mean height, (b) the variance, (c) the standard deviation, and (d) the standard error of the mean.

7. Height in humans depends on the additive action of genes. Assume that this trait is controlled by the four loci $R$, $S$, $T$, $U$ and that environmental effects are negligible. Dominant alleles contribute two units and recessive alleles contribute one unit to height.
   (a) Can two individuals of moderate height produce offspring that are much taller or shorter than either parent? If so, how?
   (b) If an individual with the minimum height specified by these genes marries an individual of intermediate or moderate height, could any of their children be taller than the tall parent? Why?

8. The mean and variance of plant height of two highly inbred parental strains ($P_1$ and $P_2$) and their $F_1$ and $F_2$ progeny are shown below. Calculate the broad heritability ($H^2$) of plant height in this species.

| Strain | Mean (cm) | Variance |
|--------|-----------|----------|
| $P_1$ | 34.2 | 4.2 |
| $P_2$ | 55.3 | 3.8 |
| $F_1$ | 44.2 | 5.6 |
| $F_2$ | 46.3 | 10.3 |

9. In a hypothetical situation, the study of Vitamin A content and cholesterol content of eggs from a large population of chickens was investigated thoroughly. The variances ($V$) were calculated as shown below:

| | Trait | |
|----------|-----------|-------------|
| Variance | Vitamin A | Cholesterol |
| $V_P$ | 123.5 | 862.0 |
| $V_E$ | 96.2 | 484.6 |
| $V_A$ | 12.0 | 192.1 |
| $V_D$ | 15.3 | 185.3 |

   (a) Calculate the narrow heritability ($h^2$) for both traits.
   (b) Which, if either, is likely to respond to selection?

10. In a cross in *Neurospora* involving two alleles $B$ and $b$, the following tetrad patterns were observed. Calculate the distance between the gene and the centromere.

| Tetrad Pattern | Number |
|----------------|--------|
| *BBbb* | 36 |
| *bbBB* | 44 |
| *BbBb* | 4 |
| *bBbB* | 6 |
| *BbbB* | 3 |
| *bBBb* | 7 |

11. In *Neurospora,* the cross $a+ \times +b$ yielded only two types of ordered tetrads in approximately equal numbers:

| | | | Spore Pair | | |
|---|---|:---:|:---:|:---:|:---:|
| | | **1–2** | **3–4** | **5–6** | **7–8** |
| Tetrad Type 1 | | $a+$ | $a+$ | $+b$ | $+b$ |
| Tetrad Type 2 | | $++$ | $++$ | $ab$ | $ab$ |

What can be concluded?

12. Below are two sets of data derived from crosses in *Chlamydomonas*, involving three genes represented by the mutant alleles *a*, *b*, and *c*. Determine as much as you can concerning genetic arrangement of these three genes relative to one another. Describe the expected results of Cross 3, assuming 100 tetrads.

| | | Tetrads | | |
|:---:|:---:|:---:|:---:|:---:|
| **Cross** | **Genes Involved** | **P** | **NP** | **T** |
| 1 | *a* & *b* | 36 | 36 | 28 |
| 2 | *b* & *c* | 79 | 3 | 18 |
| 3 | *a* & *c* | | | |

13. In *Chlamydomonas*, a cross $ab \times ++$ yielded the following unordered tetrad data where *a* and *b* are linked.

| | | | |
|:---:|:---:|:---:|:---:|
| (1) | + + <br> + + <br> a b <br> a b   38 | (4) | a b <br> a + <br> + b <br> + +   17 |
| (2) | + + <br> a b <br> + + <br> a b   5 | (5) | a b <br> + + <br> + b <br> a +   2 |
| (3) | a + <br> a + <br> + b <br> + b   6 | (6) | a b <br> + b <br> a + <br> + +   3 |

(a) Identify the categories representing parental ditypes (P), nonparental ditypes (NP), and tetratypes (T).
(b) Explain the origin of category (2).
(c) Determine the map distance between *a* and *b*.

14. The following results are ordered tetrad pairs from a cross between strain *cd* and strain $++(c^+d^+)$. They are summarized by tetrad classes.

|     | (1)   |   | (2)   |   | (3)   |   | (4)   |   | (5)   |   | (6)   |   | (7)   |   |
|-----|-------|---|-------|---|-------|---|-------|---|-------|---|-------|---|-------|---|
|     | $c$   | + | $c$   | + | $c$   | $d$ | +   | $d$ | $c$ | + | $c$   | $d$ | $c$ | + |
|     | $c$   | + | $c$   | $d$ | $c$ | $d$ | $c$ | + | +   | + | +     | + | +   | $d$ |
|     | +     | $d$ | + | + | + | + | $c$ | + | $c$ | $d$ | $c$ | $d$ | $c$ | $d$ |
|     | +     | $d$ | + | $d$ | + | + | + | $d$ | + | $d$ | + | + | +   | + |
|     | 1     |   | 17    |   | 41    |   | 1     |   | 5     |   | 3     |   | 1     |   |

(a) Name the ascus type of each class from 1 to 7 (P, NP, or T).

(b) The above data support the conclusion that the $c$ and $d$ loci are linked. State the evidence in support of this conclusion.

(c) Calculate the gene–centromere distance for each locus.

(d) Calculate the distance between the two linked loci.

(e) Draw a linkage map including the centromere and explain the discrepancy between the distances determined by the two different methods in parts (c) and (d).

(f) Describe the arrangement of crossovers needed to produce the ascus class 6 above.

**SELECTED READINGS**

BRINK, R. A., ed. 1967. *Heritage from Mendel*. Madison: University of Wisconsin Press.

CHAPMAN, A. B. 1985. *General and quantitative genetics*. Amsterdam: Elsevier.

CORWIN, H. O., and JENKINS, J. B. 1976. *Conceptual foundations of genetics: Selected readings*. Boston: Houghton-Mifflin.

CROW, J. F. 1966. *Genetics notes*. 6th ed. Minneapolis: Burgess.

DUNN, L. C. 1966. *A short history of genetics*. New York: McGraw-Hill.

FALCONER, D. 1989. *Introduction to quantitative genetics*. 3rd ed. Harlow, England: Longman Scientific and Technical Publ.

FARBER, S. L. 1980. *Identical twins reared apart*. New York: Basic Books.

FELDMAN, M. W., and LEWONTIN, R. C. 1975. The heritability hang-up. *Science* 190:1163–68.

FOSTER, M. 1965. Mammalian pigment genetics. *Adv. in Genet.* 13:311–39.

HALEY, C. 1991. Use of DNA fingerprints for the detection of major genes for quantitative traits in domestic species. *Anim. Genet.* 22:259–77.

HARPER, P. S., et al. 1992. Anticipation in myotonic dystrophy: New light on an old problem. *Am. J. Hum. Genet.* 51:10–16.

HOWLER, C. J., et al. 1989. Anticipation in myotonic dystrophy: Fact or fiction? *Brain* 112: 779–97.

LANDER, E. S., and BOTSTEIN, D. 1989. Mapping Mendelian factors underlying quantitative traits using RFLP linkage maps. *Genetics* 121:185–99.

LEWONTIN, R. C. 1974. The analysis of variance and the analysis of causes. *Amer. J. Hum. Genet.* 26:400–11.

LINDSLEY, D. C., and GRELL, E. H. 1967. *Genetic variations of* Drosophila melanogaster, Washington, D.C.: Carnegie Institute of Washington.

MAHEDEVAN, M., et al. 1992. Myotonic dystrophy mutation: An unstable CTG repeat in the 3' untranslated region of the gene. *Science* 255:1253–58.

MATHER, K. 1965. *Statistical analysis in biology*. London: Methuen.

NEWMAN, H. H., FREEMAN, F. N., and HOLZINGER, K. J. 1937. *Twins: A study of heredity and environment*. Chicago: The University of Chicago Press.

NOLTE, D. J. 1959. The eye-pigmentary system of *Drosophila*. *Heredity* 13:233–41.

PATERSON, A. H., DeVERNA, J. W., LANNI, B., and TANKSLEY, S. D. 1990. Fine mapping of quantitative trait loci using selected overlapping recombinant chromosomes in an interspecific cross of tomato. *Genetics* 124:735–42.

PATERSON, A. H., LANDER, E. S., HEWITT, J. D., PETERSON, S., LINCOLN, S. E., and TANKSLEY, S. D. 1988. Resolution of quantitative traits into Mendelian factors by using a complete linkage map of restriction fragment length polymorphisms. *Nature* 335:721–26.

PAWELEK, J. M., and KÖRNER, A. M. 1982. The biosynthesis of mammalian melanin. *Amer. Scient.* 70:136–45.

PLOMIN, R., McCLEARN, G., GORA-MASLAK, G., and NEIDERHISER, J. 1991. Use of recombinant inbred strains to detect quantitative trait loci associated with behavior. *Behav. Genet.* 21: 99–116.

PONTECORVO, G. 1958. *Trends in genetic analysis.* New York: Columbia University Press.

RANSON, R., ed. 1982. *A handbook of Drosophila development.* New York: Elsevier Biomedical Press.

REIK, W. 1992. Genomic imprinting in mammals. *Results Prob. Cell Diff.* 18:203–29.

STAHL, F. W. 1979. *Genetic recombination.* New York: W. H. Freeman.

VOELLER, B. R., ed. 1968. *The chromosome theory of inheritance—Classic papers in development and heredity.* New York: Appleton-Century-Crofts.

WILHAM, R., and WILSON, D. 1991. Genetic predictions of racing performance in quarter horses. *J. Anim. Sci.* 69:3891–94.

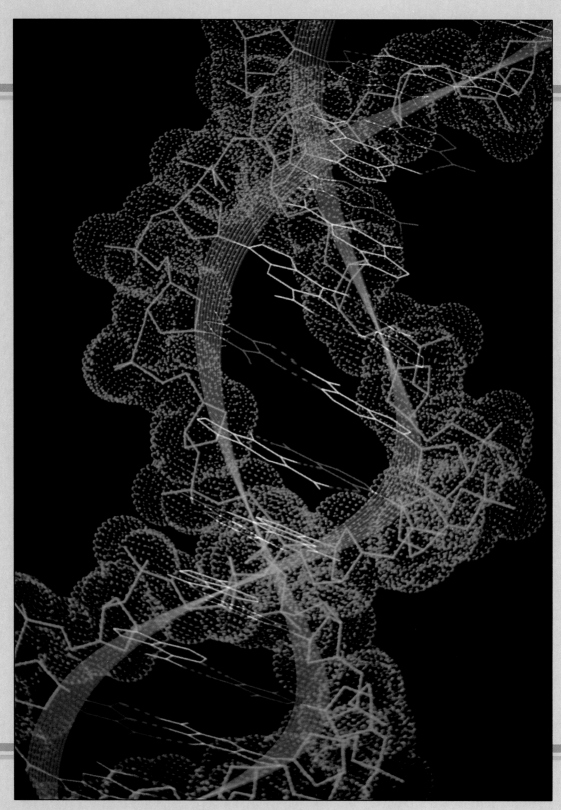

Model of the double helical configuration of DNA.

# PART 2

# DNA: THE MOLECULAR BASIS OF HEREDITY

# 8

# DNA: STRUCTURE AND ANALYSIS

James D. Watson and Francis H.C. Crick with a model of DNA.

*With few exceptions, the nucleic acid DNA serves as the genetic material in every living thing. The structure of DNA provides the chemical basis for storing and expressing genetic information within the cells, as well as transmitting it to future generations. The molecule takes the form of a double-stranded helix united by hydrogen bonds formed between complementary nucleotides. In some viruses, RNA serves as the genetic material.*

In Part 1 of the text, we discussed the presence of genes on chromosomes that control phenotypic traits and the way in which the chromosomes are transmitted through gametes to future offspring. Logically, there must be some form of information contained in genes, which, when passed to a new generation, influences the form and characteristics of the offspring; this is called the **genetic information**. We might also conclude that this same information in some way directs the many complex processes leading to the adult form.

Until 1944 it was not clear what chemical component of the chromosome makes up genes and constitutes the genetic material. Since chromosomes were known to have both a nucleic acid and a protein component, both were considered candidates. In 1944, however, there emerged direct experimental evidence that the nucleic acid, DNA, serves as the informational basis for the process of heredity.

Once the importance of DNA in genetic processes was realized, work was intensified with the hope of discerning not only the structural basis of this molecule but also the relationship of its structure to its function. Between 1944 and 1953, many scientists sought information that might answer the most significant and intriguing question in the history of biology: How does DNA serve as the genetic basis for the living process? The answer was believed to depend strongly on the chemical structure of the DNA molecule, given the complex but orderly functions ascribed to it.

These efforts were rewarded in 1953 when James Watson and Francis Crick set forth their hypothesis for the double-helical nature of DNA. The assumption that the molecule's functions would be clarified more easily once its general structure was determined proved to be correct. This chapter initially reviews the evidence that DNA is the genetic material and then discusses the elucidation of its structure.

## CHARACTERISTICS OF THE GENETIC MATERIAL

The genetic material has several characteristics: **replication**, **storage of information**, **expression of that information**, and **variation by mutation**. "Replication" of the genetic material is one facet of cell division, a fundamental property of all living organisms. Once the genetic material of cells has been replicated, it must then be partitioned equally into daughter cells. During the formation of gametes, the genetic material is also replicated but is partitioned so that each cell gets only one-half of the original amount of genetic material. This process is called meiosis. Although the products of mitosis and meiosis are different, these processes are both part of the more general phenomenon of cellular reproduction.

The characteristic of "storage" may be viewed as genetic information that is present, but that may or may not be expressed. It is clear that while most cells contain a complete complement of DNA, at any point in time they express only a part of this genetic potential. For example, bacteria turn many genes on only in response to specific environmental conditions, only to turn them off when such conditions change. In vertebrates, skin cells may display active melanin genes but never activate their hemoglobin genes; digestive cells activate many genes specific to their function, but do not activate their melanin genes.

Inherent in the concept of storage is the need for the genetic material to be able to encode the nearly infinite variety of gene products found among the countless forms of life present on our planet. The chemical language of the genetic material must be capable of this potential task as it stores information and as it is transmitted to progeny cells and organisms.

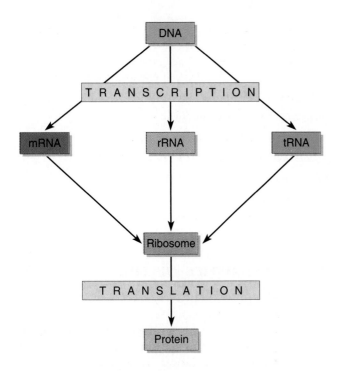

**FIGURE 8.1**    A simplified view of information flow involving DNA, RNA, and proteins within cells.

"Expression" of the stored genetic information is a complex process and is the basis for the concept of **information flow** within the cell. Figure 8.1 shows a simplified illustration of this concept. The initial event is the **transcription** of DNA, resulting in the synthesis of three types of RNA molecules: **messenger RNA (mRNA)**, **transfer RNA (tRNA)**, and **ribosomal RNA (rRNA)**. Of these, mRNAs are translated into proteins. Each type of mRNA is the product of a specific gene and leads to the synthesis of a different protein. **Translation** occurs in conjunction with rRNA-containing ribosomes and involves tRNA, which acts as an adaptor to convert the chemical information in mRNA to the amino acids that make up proteins. Collectively, these processes serve as the foundation for the **central dogma of molecular genetics**: "DNA makes RNA, which makes proteins."

The genetic material is also the source of newly arising "variability" among organisms through the process of mutation. If a change in the chemical composition of DNA occurs, the alteration will be reflected during transcription and translation, often affecting the specified protein. If a mutation is present in gametes, it will be passed to future generations and, with time, may become distributed in the population. Genetic variation, which also includes rearrangements within and between chromosomes, provides the raw material for the process of evolution.

# THE GENETIC MATERIAL: 1900–1944

The idea that genetic material is physically transmitted from parent to offspring has been accepted for as long as the concept of inheritance has existed. Beginning in the late nineteenth century, research into the structure of biomolecules progressed considerably, setting the stage for the description of the genetic material in chemical terms. Although proteins and nucleic acid were both considered major candidates for the role of the genetic material, many geneticists, until the 1940s, favored proteins. Three factors contributed to this belief.

First, proteins are abundant in cells. Although the protein content may vary considerably, these molecules compose over 50 percent of the dry weight of cells. Since cells contain such a large amount and variety of proteins, it is not surprising that early geneticists believed that some of this protein could function as the genetic material.

The second factor was the accepted proposal for the chemical structure of nucleic acids during the early to mid-1900s. DNA was first studied in 1868 by Friedrick Miescher, a Swiss chemist. He was able to separate nuclei from the cytoplasm of cells and then isolate from them an acidic substance that he called **nuclein**. Miescher showed that nuclein contained large amounts of phosphorus and no sulfur, characteristics that differentiate it from proteins.

As analytical techniques were improved, nucleic acids, including DNA, were shown to be composed of four similar molecular building blocks called nucleotides. Around 1910, Phoebus A. Levene proposed the **tetranucleotide hypothesis** to explain the chemical arrangement of these nucleotides in nucleic acids. He proposed a very simple four-nucleotide unit as shown in Figure 8.2. Levene based his proposal on studies of the composition of the four types of nucleotides. Although his actual data revealed proportions of the four that varied considerably, he assumed a 1:1:1:1 ratio. The discrepancy was ascribed to inadequate analytical technique.

Since a single covalently bound tetranucleotide structure was relatively simple, geneticists believed nucleic acids could not provide the large amount of chemical variation expected for the genetic material. Proteins, on the other hand, are made up of 20 different amino acids, thus providing the basis for substantial variation. Thus, attention was directed away from nucleic acids as important genetic biomolecules, strengthening the speculation that proteins served as the genetic material.

The third contributing factor simply concerned the

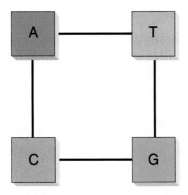

**FIGURE 8.2**    Diagrammatic depiction of Levene's proposed tetranucleotide containing one molecule each of the four nitrogenous bases: adenine (A), cytosine (C), guanine (G), and thymine (T). Each block represents a nucleotide.

areas of most active research in genetics. Before 1940, most geneticists were engaged in the study of transmission genetics and mutation. The excitement generated in these areas undoubtedly diluted the concern for finding the precise molecule that serves as the genetic material. Thus, proteins were the most promising candidate and were accepted rather passively.

Between 1910 and 1930, other proposals for the structure of nucleic acids were advanced, but they were generally overturned in favor of the tetranucleotide hypothesis. It was not until the 1940s that the work of Erwin Chargaff led to the realization that Levene's hypothesis was incorrect. Chargaff showed that, for most organisms, the 1:1:1:1 ratio was inaccurate, thus disproving Levene's hypothesis.

# EVIDENCE FAVORING DNA IN BACTERIA AND BACTERIOPHAGES

The 1944 publication by Oswald Avery, Colin MacLeod, and Maclyn McCarty concerning the chemical nature of a "transforming principle" in bacteria marked the initial event leading to the acceptance of DNA as the genetic material. Along with the subsequent findings of other research teams, this work constituted direct experimental proof that, in the organisms studied, DNA, and not protein, is the biomolecule responsible for heredity. This period marked the beginning of an era of discovery in biology that has revolutionized our understanding of life on earth. The impact of these findings parallels the work that followed the publication of Darwin's theory of evolution and that following the rediscovery of Mendel's postulates of transmission genetics. Together, these constituted three great revolutions in biology.

The initial evidence implicating DNA as the genetic material was derived from studies of prokaryotic bacteria and viruses that infect them. The reasons for the use of bacteria and bacterial viruses will become apparent as the experiments are studied. Primarily, bacteria and viruses are capable of rapid growth because they complete their life cycles in hours. They may also be experimentally manipulated, and mutations may be easily induced and selected. Thus, they are ideal for experimentation of this sort.

## Transformation Studies

The research that provided the foundation for Avery, MacLeod, and McCarty's work was initiated in 1927 by Frederick Griffith, a medical officer in the British Ministry of Health. He performed experiments with several different strains of the bacterium *Diplococcus pneumoniae.** Some were **virulent** strains, which cause pneumonia in certain vertebrates (notably humans and mice), while others were **avirulent** strains, which do not cause illness.

The difference in virulence is related to the polysaccharide capsule of the bacterium. Virulent strains have this capsule, whereas avirulent strains do not. The nonencapsulated bacteria are readily engulfed and destroyed by phagocytic cells in the animal's circulatory system. Virulent bacteria, which possess the polysaccharide coat, are not easily engulfed; they are able to multiply and cause pneumonia.

The presence or absence of the capsule is the basis for another characteristic difference between virulent and avirulent strains. Encapsulated bacteria form a **smooth**, shiny-surfaced colony (**S**) when grown on an agar culture plate; nonencapsulated strains produce **rough** colonies (**R**) (Figure 8.3). Thus, virulent and avirulent strains may be distinguished easily by standard microbiological culture techniques.

Each strain of *Diplococcus* may be one of dozens of different types called **serotypes**. The specificity of the serotype is due to the detailed chemical structure of the polysaccharide constituent of the thick, slimy capsule. Serotypes are identified by immunological techniques and are usually designated by Roman numerals. In the United States, types I and II are most common in causing pneumonia. Griffith used types II and III in the critical experiments that led to new concepts about the genetic material. Table 8.1 summarizes the characteristics of Griffith's two strains.

---

*Note that this organism is now designated *Streptococcus pneumonia*.

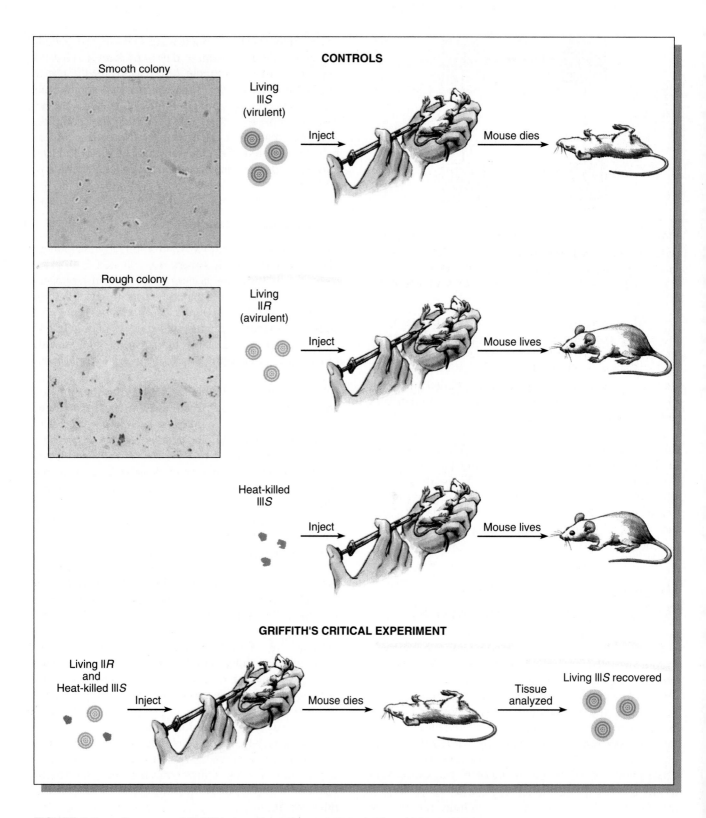

**FIGURE 8.3**    Summary of Griffith's transformation experiment. The photographs show bacterial colonies containing cells with capsules (type IIIS) and without capsules (type IIR).

**Table 8.1** STRAINS OF *DIPLOCOCCUS PNEUMONIAE* USED BY FREDERICK GRIFFITH IN HIS ORIGINAL TRANSFORMATION EXPERIMENTS

| Serotype | Colony Morphology | Capsule | Virulence |
|----------|-------------------|---------|-----------|
| II*R* | Rough | Absent | Avirulent |
| III*S* | Smooth | Present | Virulent |

Griffith knew from the work of others that only living virulent cells would produce pneumonia in mice. If heat-killed virulent bacteria are injected into mice, no pneumonia results, just as living avirulent bacteria fail to produce the disease. Griffith's critical experiment (Figure 8.3) involved an injection into mice of living II*R* (avirulent) cells combined with heat-killed III*S* (virulent) cells. Since neither cell type caused death in mice when injected alone, Griffith expected that the double injection would not kill the mice. But, after five days, all mice receiving double injections were dead. Analysis of the blood of the dead mice revealed large numbers of living type III*S* (virulent) bacteria!

As far as could be determined regarding the capsular polysaccharide, these III*S* bacteria were identical to the III*S* strain from which the heat-killed cell preparation had been made. The control mice, injected only with living avirulent II*R* bacteria for this set of experiments, did not develop pneumonia and remained healthy. This finding ruled out the possibility that the avirulent II*R* cells had simply changed (or mutated) to virulent III*S* cells in the absence of the heat-killed III*S* fraction. Instead, some type of interaction was required between living II*R* and heat-killed III*S*.

Griffith thus concluded that the heat-killed III*S* bacteria were somehow responsible for converting live avirulent II*R* cells into virulent III*S* ones. Calling the phenomenon **transformation**, he suggested that the **transforming principle** might be some part of the polysaccharide capsule *or* some compound required for capsule synthesis, although the capsule alone did not cause pneumonia. To use Griffith's term, the transforming principle from the dead III*S* cells served as a "pabulum" for the II*R* cells.

Griffith's work led other physicians and bacteriologists to research the phenomenon of transformation. By 1931, Henry Dawson, at the Rockefeller Institute, had confirmed Griffith's observations and extended his work one step further. Dawson and his coworkers showed that transformation could occur *in vitro* (in a test tube). When heat-killed III*S* cells are incubated with living II*R* cells, living III*S* cells were recovered. Thus, injection into mice was not necessary for transformation to occur. By

1933, Lionel J. Alloway had refined the *in vitro* system by using crude extracts of *S* cells and living *R* cells. The soluble filtrate from the heat-killed *S* cells was as effective in inducing transformation as were the intact cells! Alloway and others did not view transformation as a genetic event, but rather as a physiological modification of some sort. Nevertheless, the experimental evidence that a chemical substance was responsible for transformation was quite convincing.

Then, in 1944, after ten years of work, Avery, MacLeod, and McCarty published their results in what is now regarded as a classic paper in the field of molecular genetics. They reported that they had obtained the transforming principle in a purified state, and that beyond reasonable doubt, the molecule responsible for transformation was DNA. The details of their work are outlined in Figure 8.4.

These researchers began their isolation procedure with large quantities (50–75 liters) of liquid cultures of type III*S* virulent cells. The cells were centrifuged, collected, and heat-killed. Following homogenization and several extractions with the detergent deoxycholate (DOC), they obtained a soluble filtrate which, when tested, still contained the transforming principle. Protein was removed from the active filtrate by several chloroform extractions, and polysaccharides were enzymatically digested and removed. Finally, precipitation with ethanol yielded a fibrous nucleic acid that still retained the ability to induce transformation of type II*R* avirulent cells.

Further testing established beyond a reasonable doubt that the transforming principle was DNA. Treatment was performed with a protein-digesting enzyme called a **protease** and an RNA-digesting enzyme, called **ribonuclease**. Such treatment destroyed the activity of any remaining protein and RNA. Nevertheless, transforming activity still remained. Chemical testing of the final product gave strong positive reactions for DNA. The final confirmation came with experiments using crude samples of the DNA-digesting enzyme **deoxyribonuclease**, which was isolated from dog and rabbit sera. This digestion was shown to destroy transforming activity. There could be no doubt that the active transforming principle was DNA!

The great amount of work, the confirmation and reconfirmation of the conclusions drawn, and the brilliant logic involved in the research of these three scientists are truly impressive. The conclusion to the 1944 publication was, however, very simply stated: "The evidence presented supports the belief that a nucleic acid of the desoxyribose* type is the fundamental unit of the transforming principle of *Pneumococcus* type III."

---

*Desoxyribose* is now spelled *deoxyribose*, and the genus *Pneumococcus* is now referred to as *Diplococcus*.

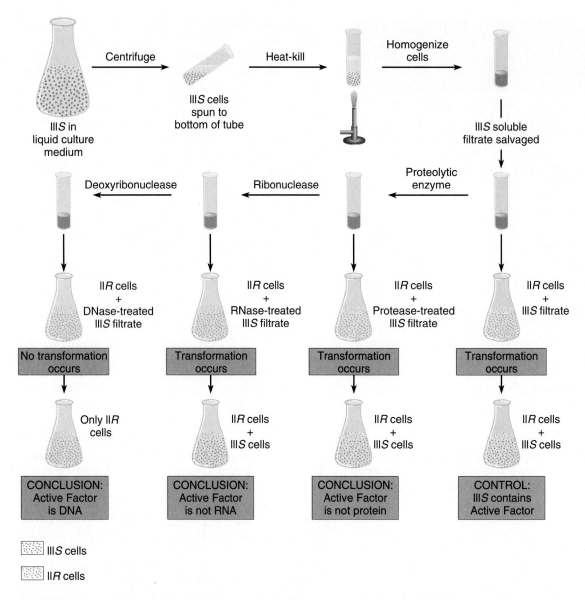

**FIGURE 8.4**    Summary of Avery, MacLeod, and McCarty's experiment demonstrating that DNA is the transforming principle.

Avery and his coworkers recognized the genetic and biochemical implications of their work. They suggested that the transforming principle interacts with the II*R* cell and gives rise to a coordinated series of enzymatic reactions that culminates in the synthesis of the type III*S* capsular polysaccharide. They emphasized that, once transformation occurs, the capsular polysaccharide is produced in successive generations. Transformation is therefore heritable, and the process affects the genetic material.

Immediately after the publication of the report, several investigators turned to or intensified their studies of transformation in order to clarify the role of DNA in genetic mechanisms. In particular, the work of Rollin Hotchkiss was instrumental in confirming that the critical factor in transformation was DNA, and not protein. In 1949, in a separate study, Harriet Taylor isolated an **extremely rough (*ER*)** mutant strain from a rough (*R*) strain. This *ER* strain produced colonies that were more irregular than *R*. DNA from *R* accomplished the transformation of *ER* to *R*. Thus, the *R* strain, which served as the recipient in the Avery experiments, was shown also to be able to serve as the DNA donor in transformation.

Transformation has now been shown to occur in *He-*

*mophilus influenzae*, *Bacillus subtilis*, *Shigella para-dysenteriae*, and *Escherichia coli*, among many other microorganisms. Transformation of numerous genetic traits other than colony morphology has also been demonstrated, including ones involving resistance to antibiotics and the ability to metabolize various nutrients. Thus, geneticists believed that most, if not all, traits could be transformed under suitable experimental conditions.

## The Hershey–Chase Experiment

The second major piece of evidence supporting DNA as the genetic material was provided by the study of the bacterial virus T2. This virus, also called a **bacteriophage** or just a **phage**, has as its host the bacterium *Escherichia coli*. The phage consists of a protein coat surrounding a core of DNA. Electron micrographs have revealed the phage's external structure to be composed of a hexagonal head plus a tail. The life cycle of bacteriophages such as T2 is shown in Figure 8.5.

In 1952, Alfred Hershey and Martha Chase published the results of experiments designed to clarify the events leading to phage reproduction. Several of the experiments clearly established the independent functions of phage protein and nucleic acid in the reproduction process associated with the bacterial cell. Hershey and Chase knew from existing data that

1. T2 phages consist of approximately 50 percent protein and 50 percent DNA.

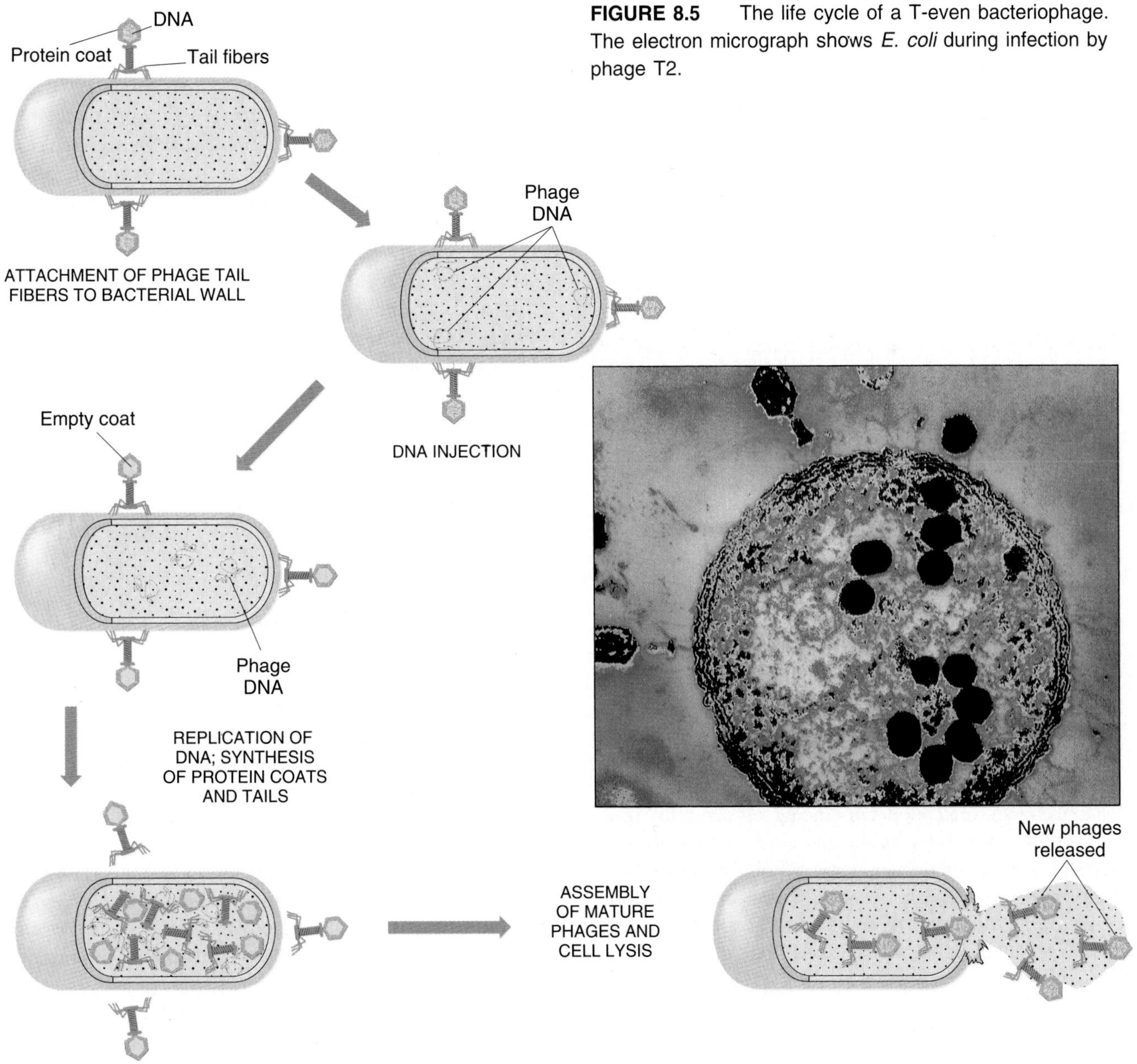

**FIGURE 8.5**    The life cycle of a T-even bacteriophage. The electron micrograph shows *E. coli* during infection by phage T2.

DNA

Protein coat    Tail fibers

ATTACHMENT OF PHAGE TAIL FIBERS TO BACTERIAL WALL

Phage DNA

DNA INJECTION

Empty coat

Phage DNA

REPLICATION OF DNA; SYNTHESIS OF PROTEIN COATS AND TAILS

ASSEMBLY OF MATURE PHAGES AND CELL LYSIS

New phages released

2. Infection is initiated by adsorption of the phage by its tail fibers to the bacterial cell.

3. The production of new viruses occurs within the bacterial cell.

It would appear that some molecular component of the phage, DNA and/or protein, enters the bacterial cell and directs viral reproduction. Which was it?

Hershey and Chase used radioisotopes to follow the molecular components of phages during infection. Both $^{32}P$ and $^{35}S$, radioactive forms of phosphorus and sulfur, were used. Since DNA contains phosphorus but not sulfur, $^{32}P$ effectively labels DNA. Since proteins contain sulfur but not phosphorus, $^{35}S$ labels protein. *This is a key point in the experiment.* If *E. coli* cells are first grown in the presence of $^{32}P$ *or* $^{35}S$ and then infected with T2 viruses, the progeny phages will have either a labeled DNA core *or* a labeled protein coat, respectively. These radioactive phages may be isolated and used to infect unlabeled bacteria (Figure 8.6).

When labeled phages and unlabeled bacteria are mixed, an adsorption complex is formed as the phages attach their tail fibers to the bacterial wall. These complexes were isolated and subjected to a high shear force by placing them in a blender. This force strips off the attached phages, which may then be analyzed separately (Figure 8.6). By tracing the radioisotopes, Hershey and Chase were able to demonstrate that most of the $^{32}P$-labeled DNA had been transferred into the bacterial cell following adsorption; on the other hand, most of the $^{35}S$-labeled protein remained outside the bacterial cell and was recovered in the phage "ghosts" (empty phage coats) after the blender treatment. Following this separation, the bacterial cells, which now contained viral DNA, were eventually lysed as new phages were produced. These progeny phages contained $^{32}P$, but not $^{35}S$.

Hershey and Chase interpreted these results to indicate that the protein of the phage coat remains outside the host cell and is not involved in directing the production of new phages. On the other hand, and most important, phage DNA enters the host cell and directs phage multiplication. Thus, they had demonstrated that in phage T2, DNA, not protein, is the genetic material.

This experimental work, along with that of Avery and his colleagues, provided convincing evidence to most geneticists that DNA was the molecule responsible for heredity. Since then, many significant findings have been based on this supposition. These many findings, constituting the field of molecular genetics, are discussed in detail in subsequent chapters.

## Transfection Experiments

During the eight years following the publication of the Hershey–Chase experiment, additional research using bacterial viruses provided even more solid proof that DNA is the genetic material. In 1957, several reports demonstrated that if *E. coli* was treated with the enzyme **lysozyme**, the outer wall of the cell could be removed without destroying the bacterium. Enzymatically treated cells are naked, so to speak, and contain only the cell membrane as the outer boundary of the cell. Such structures are called **protoplasts** (or **spheroplasts**). John Spizizen and Dean Fraser independently reported that by using protoplasts, they were able to initiate phage multiplication with disrupted T2 particles. That is, provided that the cell wall is absent, it is not necessary for a virus to be intact in order for infection to occur. This suggests that the outer protein coat structure is essential to the movement of DNA through the intact cell wall.

Similar but refined experiments were reported in 1960 by George Guthrie and Robert Sinsheimer. DNA was purified from bacteriophage $\phi$X-174, a small phage that contains a single-stranded, circular DNA molecule of some 5386 nucleotides. When added to *E. coli* protoplasts, the purified DNA resulted in the production of complete $\phi$X-174 bacteriophages. This process of infection by only the viral nucleic acid, called **transfection**, proves conclusively that $\phi$X-174 DNA alone contains all the necessary information for production of mature viruses. Thus, the evidence supporting the conclusion that DNA serves as the genetic material was further strengthened, even though all direct evidence thus far had been obtained from bacterial and viral studies.

# INDIRECT EVIDENCE FAVORING DNA IN EUKARYOTES

Most eukaryotic organisms are not amenable to the types of experiments performed to demonstrate that DNA is the genetic material in bacteria and viruses. Therefore, initial support for this conclusion in eukaryotes was based only on indirect, circumstantial evidence. Such evidence is derived from observations incidental to the hypothesis in question. While any single item of circumstantial evidence taken alone is insufficient to support the hypothesis, many different sorts of independent observations may lead to a common conclusion.

For example, since chromosomes house the genetic material, a correlation is expected between the ploidy of a cell (e.g., $n$, $2n$, etc.) and the quantitative amount of

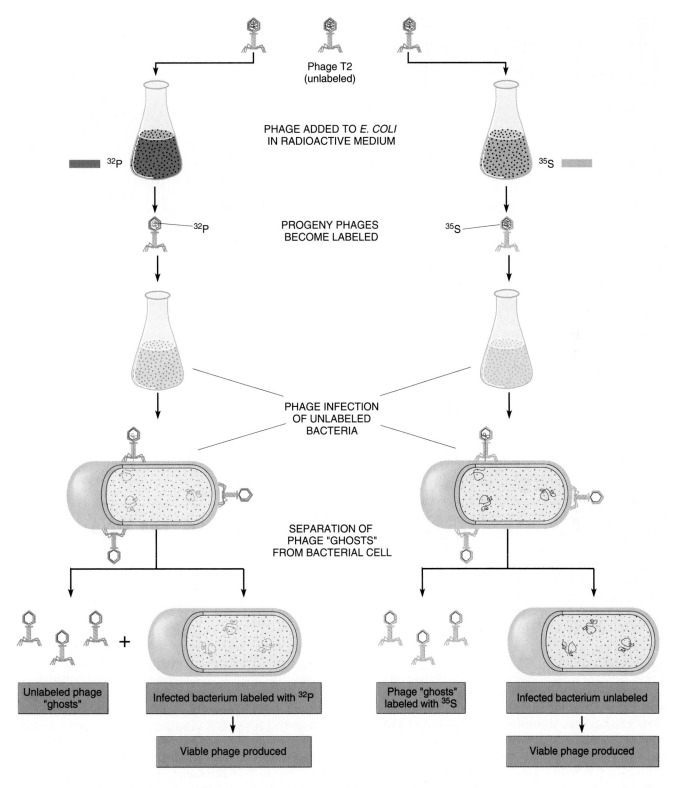

**FIGURE 8.6**    Summary of the Hershey–Chase experiment demonstrating that DNA, and not protein, is responsible for directing the reproduction of phage T2 during the infection of *E. coli.*

**Table 8.2** DNA Content of Haploid versus Diploid Cells of Various Species (in picograms)

| Organism | n | 2n |
|---|---|---|
| Human | 3.25 | 7.30 |
| Chicken | 1.26 | 2.49 |
| Trout | 2.67 | 5.79 |
| Carp | 1.65 | 3.49 |
| Shad | 0.91 | 1.97 |

NOTE: Sperm (n) and nucleated precursors to red blood cells (2n) were used to contrast DNA amounts and ploidy levels.

the molecule serving as the genetic material. Meaningful comparisons may be made between the amount of DNA and protein in gametes (sperm and eggs) and somatic (or body) cells. The latter are recognized as being **diploid** (2n) and containing twice the number of chromosomes as gametes, which are **haploid** (n).

Table 8.2 compares the amount of DNA found in haploid sperm and diploid nucleated precursors of red blood cells from a variety of organisms. There is a close correlation between the amount of DNA and the number of sets of chromosomes. No such correlation can be observed between gametes and diploid cells for proteins. These data thus provide further circumstantial evidence favoring DNA as the genetic material.

Additional evidence came from observations that mitochondria and chloroplasts, which are able to perform genetic functions and transmit hereditary information to sister organelles, contain their own DNA. With these observations and the fact that genetic material is housed in chromosomes within the nucleus, we can conclude that DNA is present in every place it needs to be. In places where it is not expected (e.g., free in the cytoplasm), it is not found.

**Mutagenesis**

**Ultraviolet (UV) light** is one of a number of agents capable of inducing mutations in the genetic material. Bacteria and other organisms may be irradiated with various wavelengths of ultraviolet light, and the effectiveness of each wavelength measured by the number of mutations it induces. When the data are plotted, an **action spectrum** of ultraviolet light as a mutagenic agent is obtained. This action spectrum may then be compared with the **absorption spectrum** of any molecule suspected to be the genetic material (Figure 8.7). The molecule serving as the genetic material is expected to absorb UV light at the wavelengths found to be mutagenic.

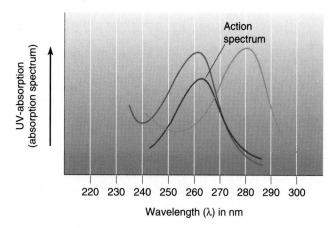

**FIGURE 8.7** The absorption spectrum of nucleic acids and proteins when subjected to ultraviolet light compared to the action spectrum (the effectiveness of inducing mutation) of ultraviolet light.

UV light is most mutagenic at the wavelength (λ) of 260 nanometers (nm). Both DNA and RNA absorb UV light most strongly at 260 nm. On the other hand, protein absorbs most strongly at 280 nm, yet no significant mutagenic effects are observed at this wavelength. Thus, this indirect evidence supports the idea that a nucleic acid is the genetic material and tends to exclude protein.

## DIRECT EVIDENCE FOR DNA: EUKARYOTIC DATA

While the circumstantial evidence just described does not constitute direct proof that DNA is the genetic material in eukaryotes, these observations spurred researchers to forge ahead, basing their work on this hypothesis. Today, there is no doubt of the validity of this conclusion; DNA *is* the genetic material in eukaryotes.

The strongest direct evidence has been provided by current experimental procedures referred to as **recombinant DNA research**. In this research, segments of eukaryotic DNA corresponding to specific genes are isolated and literally spliced into bacterial DNA. Such a complex can be inserted into a bacterial cell and its genetic expression monitored. If a eukaryotic gene is selected that is foreign to bacterial genetic information, the presence of the corresponding eukaryotic protein product demonstrates directly that this DNA is functional in expressing genetic information. This has been shown to be the case in numerous instances. For example, the human genes specifying the hormone insulin and the immunologically important molecule interferon are produced by bacteria following the insertion of human genes into the bacterial DNA using recombinant DNA procedures. As the bacterium divides, the eukaryotic DNA is replicated along with the host DNA and is distributed to the daughter cells, which also express the human genes and synthesize the corresponding proteins.

The availability of vast amounts of DNA coding for specific genes, available as a result of recombinant DNA research, has led to other direct evidence that DNA serves as the genetic material. Work in the laboratory of Beatrice Mintz has demonstrated that DNA encoding the human $\beta$-globin gene, when microinjected into a fertilized mouse egg, is later found in the adult mouse tissue and can be transmitted to that mouse's progeny! Such mice are referred to as **transgenic animals**.

More recent work has introduced *rat* DNA encoding a growth hormone into fertilized *mouse* eggs. About one-third of the resultant mice grew to twice their normal size. This indicated that the foreign DNA encoding the hormone was present *and* functional in the experimental mice. Subsequent generations received the gene and also grew larger.

We will pursue the topic of recombinant DNA again later (see Chapters 12 and 13). The point to be made here is that in eukaryotes, DNA has been shown directly to meet the requirement of expression of genetic information. Later we will see exactly how DNA is stored, replicated, expressed, and mutated.

## RNA AS THE GENETIC MATERIAL

Some viruses contain an RNA core rather than one composed of DNA. In these viruses, it would thus appear that RNA must serve as the genetic material—an exception to the general rule that DNA performs this function. In 1956, it was demonstrated that when purified RNA from **tobacco mosaic virus** (**TMV**) was spread on tobacco leaves, the characteristic lesions caused by TMV would appear later on the leaves. It was concluded that RNA is the genetic material of this virus.

Soon afterwards, another type of experiment with TMV was reported by Heinz Fraenkel-Conrat and B. Singer, as illustrated in Figure 8.8. These scientists discovered that the RNA core and the protein coat from wild-type TMV and other viral strains could be isolated separately. In their work, RNA and coat proteins were separated and isolated from TMV and a second viral strain, **Holmes ribgrass** (**HR**). Then, mixed viruses were reconstituted from the RNA of one strain and the protein of the other. When this "hybrid" virus was spread on tobacco leaves, the lesions that developed

corresponded to the type of RNA in the reconstituted virus; that is, viruses with wild-type TMV RNA and HR protein coats produced TMV lesions and vice versa. Again, it was concluded that RNA serves as the genetic material in these viruses.

In 1965 and 1966, Norman R. Pace and Sol Spiegelman further demonstrated that RNA from the phage Q$\beta$ could be isolated and replicated *in vitro*. Replication was dependent on an enzyme, **RNA replicase**, which was isolated from host *E. coli* cells following normal infection. When the RNA replicated *in vitro* was added to *E. coli* protoplasts, infection and viral multiplication occurred. Thus, RNA synthesized in a test tube can amply serve as the genetic material in these phages by directing the production of all the components necessary for viral replication.

Finally, one other group of RNA-containing viruses bears mentioning. These are the **retroviruses** which replicate in an unusual way. Their RNA serves as a template for the synthesis of the complementary DNA molecule! The process, designated as reverse transcription, occurs under the direction of an RNA-dependent DNA polymerase enzyme called **reverse transcriptase**. Because the genetic material can be represented by this DNA intermediate, it may be incorporated into the genome of the host cell. Once present, if the DNA is expressed, transcription yields retroviral RNA chromosomes. Retroviruses, including the human immunodeficiency virus (HIV) that causes AIDS, will be discussed at greater length in Chapter 22.

## NUCLEIC ACID CHEMISTRY

Having established thus far the critical importance of DNA and RNA in genetic processes, we shall now provide a brief introduction to the chemical basis of these molecules. As we shall see, the structural components of DNA and RNA are very similar. This chemical similarity is important in the coordinated functions played by these molecules during gene expression. Like the other major groups of organic biomolecules (proteins, carbohydrates, and lipids), nucleic acid chemistry is based on a variety of similar building blocks that are polymerized into chains of varying lengths.

### Nucleotides: Building Blocks of Nucleic Acids

**Nucleotides** are the building blocks of all nucleic acid molecules. These structural units consist of three essential components: a **nitrogenous base**, a **pentose sugar** (5 carbons), and a **phosphate group**. There are two kinds of nitrogenous bases: the nine-membered double-

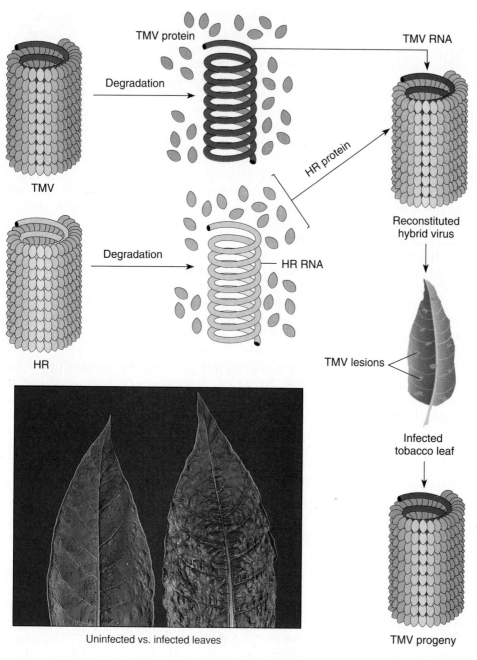

**FIGURE 8.8**    Reconstitution of hybrid tobacco mosaic viruses. In the hybrid, RNA is derived from the wild-type TMV virus, while the protein subunits are derived from the HR strain. Following infection, viruses are produced with protein subunits characteristic of the wild-type TMV strain and not those of the HR strain. The photograph shows TMV lesions on a tobacco leaf compared to an uninfected leaf.

ringed **purines** and the six-membered single-ringed **pyrimidines**. Two types of purines and three types of pyrimidines are found commonly in nucleic acids. The two purines are **adenine** and **guanine**, abbreviated **A** and **G**. The three pyrimidines are **cytosine**, **thymine**, and **uracil**, abbreviated **C**, **T**, and **U**. The chemical structures of A, G, C, T, and U are shown in Figure 8.9(a). Both DNA and RNA contain A, C, and G; only DNA contains the base T, whereas only RNA contains the base U. Each nitrogen or carbon atom of the ring structures of purines and pyrimidines is designated by an unprimed number. Note that corresponding atoms in the two rings are numbered differently in most cases.

The pentose sugars found in nucleic acids give them their names. Ribonucleic acids (RNA) contain **ribose**, while deoxyribonucleic acids (DNA) contain **deoxyribose**. Figure 8.9(b) shows the ring structures for these two pentose sugars. Each carbon atom is distinguished by a number with a prime sign (e.g., C-1', C-2'). As you can see, deoxyribose is missing one hydroxyl group at the C-2' position compared with ribose. The presence of a hydroxyl group at the C-2' position thus distinguishes RNA from DNA.

If a molecule is composed of a purine or pyrimidine base and a ribose or deoxyribose sugar, the chemical unit is called a **nucleoside**. If a phosphate group is

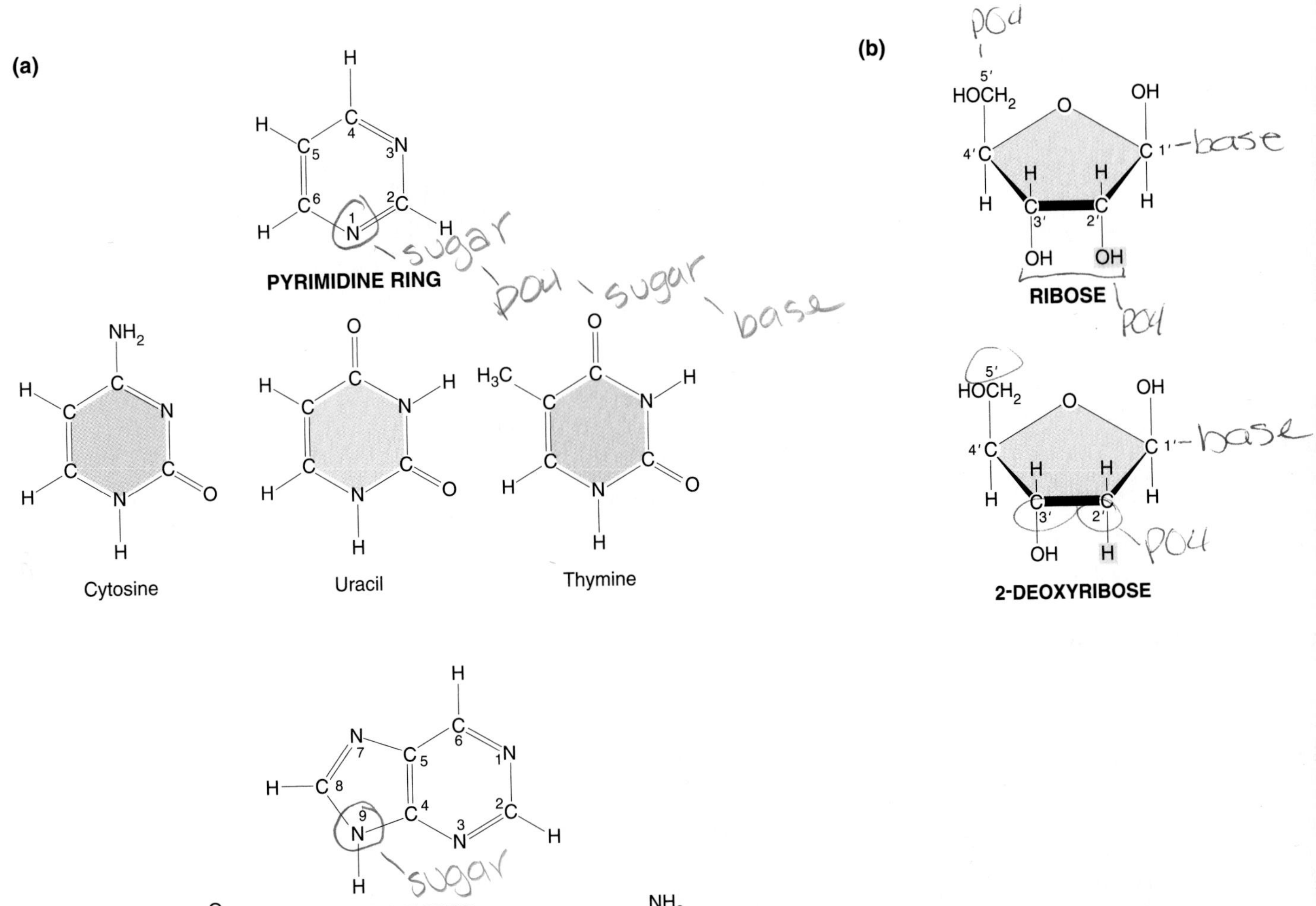

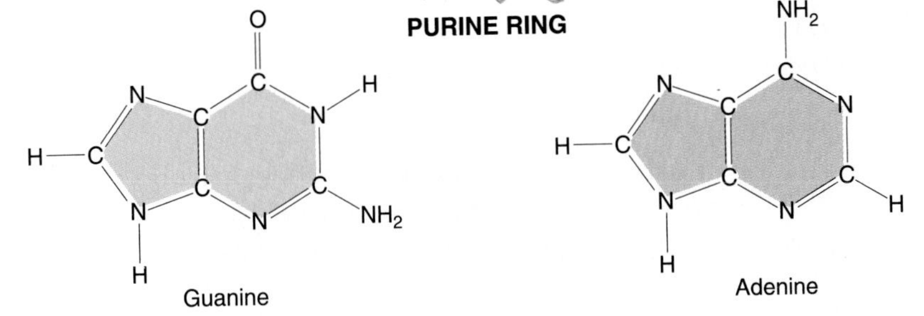

**(a)**

PYRIMIDINE RING

Cytosine      Uracil      Thymine

PURINE RING

Guanine      Adenine

**(b)**

RIBOSE

2-DEOXYRIBOSE

FIGURE 8.9    (a) Chemical structures of the pyrimidines and purines that serve as the nitrogenous bases in RNA and DNA. (b) Chemical ring structures of ribose and 2-deoxyribose, which serve as the pentose sugars in RNA and DNA, respectively.

added to the nucleoside, the molecule is now called a **nucleotide**. Nucleosides and nucleotides are named according to the specific nitrogenous base (A, T, G, C, or U) that is part of the building block. The nomenclature and general structure of the nucleosides and nucleotides are given in Figure 8.10.

The bonding between the three components of a nucleotide is highly specific. The C-1′ atom of the sugar is involved in the chemical linkage to the nitrogenous base. If the base is a purine, the N-9 atom is covalently bonded to the sugar. If the base is a pyrimidine, the bonding involves the N-1 atom. In a nucleotide, the phosphate group may be bonded to the C-2′, C-3′, or C-5′ atom of the sugar. The C-5′—phosphate configuration is shown in Figure 8.10. It is by far the most prevalent one in biological systems and that found in DNA and RNA.

## Nucleoside Diphosphates and Triphosphates

Nucleotides are also described by the term **nucleoside monophosphate** (**NMP**). The addition of one or two phosphate groups results in **nucleoside diphosphates** (**NDPs**) and **triphosphates** (**NTPs**), as illustrated in Figure 8.11. The triphosphate form is significant because it serves as the precursor molecule during nucleic acid synthesis within the cell (see Chapter 9). Additionally, the triphosphates **adenosine triphosphate** (**ATP**) and **guanosine triphosphate** (**GTP**) are important in the cell's bioenergetics because of the large

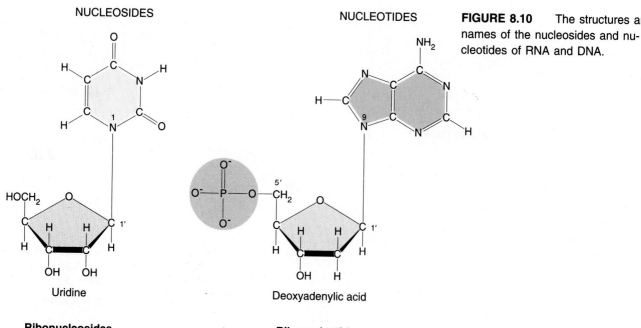

NUCLEOSIDES

NUCLEOTIDES

Uridine

Deoxyadenylic acid

**FIGURE 8.10**   The structures and names of the nucleosides and nucleotides of RNA and DNA.

**Ribonucleosides**

Adenosine
Cytidine
Guanosine
Uridine

**Ribonucleotides**

Adenylic acid
Cytidylic acid
Guanylic acid
Uridylic acid

**Deoxyribonucleosides**

Deoxyadenosine
Deoxycytidine
Deoxyguanosine
Deoxythymidine

**Deoxyribonucleotides**

Deoxyadenylic acid
Deoxycytidylic acid
Deoxyguanylic acid
Deoxythymidylic acid

amount of energy involved in the addition or removal of the terminal phosphate group. The hydrolysis of ATP or GTP to ADP or GDP and inorganic phosphate ($P_i$) is accompanied by the release of a large amount of energy in the cell. When the chemical conversion of ATP or GTP is coupled to other reactions, the energy produced may be used to drive the reactions. As a result, ATP and GTP are involved in many cellular activities, including numerous genetic events.

## Polynucleotides

The linkage between two mononucleotides consists of a phosphate group linked to two sugars. A **phosphodiester bond** is formed, because phosphoric acid has been joined to two alcohols (the hydroxyl groups on the two sugars) by an ester linkage on both sides. Figure 8.12(a) shows the resultant phosphodiester bonds in DNA and RNA. Each structure has a C-5′ end and a C-3′ end. The joining of two nucleotides forms a dinucleotide; of three nucleotides, a trinucleotide; and so forth. Short chains consisting of less than 20 nucleotides linked

together are called **oligonucleotides**. Still longer chains are referred to as **polynucleotides**.

Since drawing the structures in Figure 8.12(a) is time consuming and complex, a schematic shorthand method has been devised [Figure 8.12(b)]. The nearly vertical lines represent the pentose sugar; the nitrogenous base is attached at the top, or the C-1′ position. The diagonal line, with the (P) in the middle of it, is attached to the C-3′ position of one sugar and the C-5′ position of the neighboring sugar; it represents the phosphodiester bond. Several modifications of this shorthand method are in use, and they can be understood in terms of these guidelines.

While Levene's tetranucleotide hypothesis (described earlier in this chapter) was generally accepted before 1940, research in subsequent decades revealed it to be incorrect. It was shown that DNA does not necessarily contain equimolar quantities of the four bases. Additionally, the molecular weight of DNA molecules was determined to be in the range of $10^6$ to $10^9$ daltons, far in excess of that of a tetranucleotide. The current view of DNA is that it consists of exceedingly long polynucleotide chains.

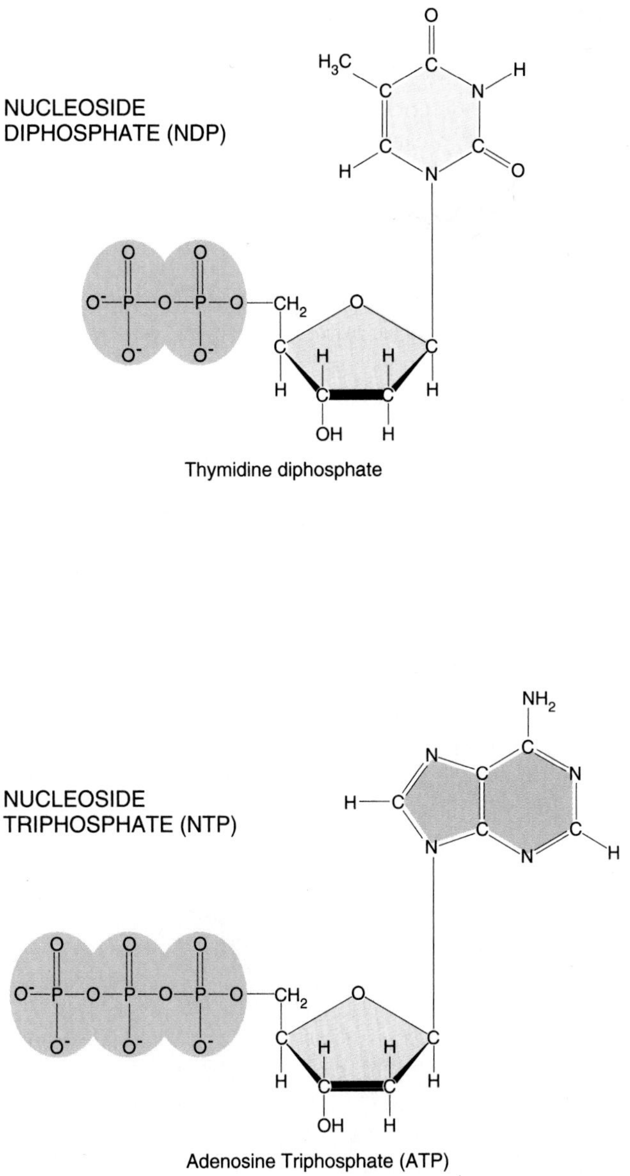

## NUCLEOSIDE DIPHOSPHATE (NDP)

Thymidine diphosphate

## NUCLEOSIDE TRIPHOSPHATE (NTP)

Adenosine Triphosphate (ATP)

**FIGURE 8.11**    The basic structures of nucleoside diphosphates and triphosphates, as illustrated by thymidine diphosphate and adenosine triphosphate.

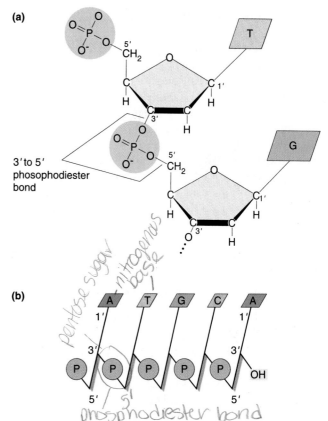

3' to 5' phosophodiester bond

*pentose sugar*

*nitrogenous base*

*phosphodiester bond*

**FIGURE 8.12**    (a) The linkage of two nucleotides by the formation of a C-3'–C-5' (3'–5') phosphodiester bond, producing a dinucleotide. (b) A shorthand notation for a polynucleotide chain.

Long polynucleotide chains would account for the observed molecular weight and provide the basis for the most important property of DNA—storage of vast quantities of diverse genetic information. If each nucleotide position in this long chain may be occupied by any one of four nucleotides, extraordinary variation is possible. For example, a polynucleotide that is only 1000 nucleotides in length may be arranged $4^{1000}$ different ways, each one different from all other possible sequences. This potential variation in molecular structure is essential if DNA is to serve the function of storing the vast amounts of chemical information necessary to direct cellular activities.

# THE STRUCTURE OF DNA

While the previous sections have established that DNA is the genetic material in all organisms (with certain viruses being the exception) and have provided details as to the basic chemical components making up nucleic acids, what remains to be deciphered is the precise structure of DNA. That is, how are polynucleotide chains organized into DNA, which serves as the genetic material? Is DNA composed of a single chain, or more than one? If the latter is the case, how do the chains relate chemically to one another? Do the chains branch? And more important, how does the structure of this molecule relate to the various genetic functions served by DNA (i.e., storage, expression, replication, and mutation)?

From 1940 to 1953, many scientists were interested in solving the structure of DNA. Among others, Erwin Chargaff, Maurice Wilkins, Rosalind Franklin, Linus Pauling, Francis Crick, and James Watson sought information that might answer the most significant and intriguing question in the history of biology: "How does DNA serve as the genetic basis for the living process?" The answer was believed to depend strongly on the chemical structure and organization of the DNA molecule, given the complex but orderly functions ascribed to it.

In 1953, two young scientists, James Watson and Francis Crick, proposed that the structure of DNA is in the form of a double helix. Their proposal was published in a short paper in *Nature*, which is reprinted in its entirety on pages 260–261. In a sense, this publication constituted the finish line in a highly competitive scientific race. This "race," as recounted in Watson's book *The Double Helix*, demonstrates the human interaction, genius, frailty, and intensity involved in the scientific effort that eventually led to the elucidation of DNA structure.

The data available to Watson and Crick, crucial to the development of their proposal, came primarily from two sources: base composition analysis of hydrolyzed samples of DNA, and X-ray diffraction studies of DNA. The analytical success of Watson and Crick may be attributed to model building that conformed to the existing data. If the structure of DNA may be analogized by a puzzle, Watson and Crick, working the Cavendish Laboratory in Cambridge, England, were the first to successfully put together all of the pieces. Given the far-reaching significance of this discovery, you may be interested to know that Watson, who entered college (the U. Chicago) at age 15, was a 24-year-old postdoctoral fellow in 1953. Crick, now considered one of the great theoretical biologists of our time, was still a graduate student.

## Base Composition Studies

Between 1949 and 1953, Erwin Chargaff and his colleagues used chromatographic methods to separate the four bases in DNA samples from various organisms. Quantitative methods were then used to determine the amounts of the four bases from each source. Table 8.3(a) provides some of Chargaff's original data. Parts (b) and (c) of this table show more recently derived base composition information from various organisms which reinforce Chargaff's findings.

On the basis of these data, which you should examine, the following conclusions may be drawn:

1. The number of adenine residues equals the number of thymine residues in the DNA of any species (columns 1, 2, and 5). Also, the number of guanine residues is equivalent to the number of cytosine residues (columns 3, 4, and 6).

2. The sum of the purines (A + G) equals the sum of the pyrimidines (C + T), as shown in column 7.

3. The ratio of (A + T)/(C + G) does not necessarily equal one; further, this ratio varies greatly among species, as shown in column 8, and as is apparent in Table 8.3(c).

These conclusions indicate definite patterns of base composition of DNA molecules. These data served as the initial clue to "the puzzle." Additionally, they directly refute the tetranucleotide hypothesis, which stated that all four bases are present in equal amounts.

## X-Ray Diffraction Analysis

When fibers of a DNA molecule are subjected to X-ray bombardment, these rays are scattered according to the molecule's atomic structure. The pattern of scatter may be captured as spots on photographic film and analyzed, particularly for the overall shape of and regularities within the molecule. This process, **X-ray diffraction** analysis, was successfully applied to the study of protein structure by Linus Pauling and other chemists. The technique had been attempted on DNA as early as 1938 by William Astbury. By 1947, he had detected a periodicity of 3.4 Å* that suggested to him that the bases were stacked like pennies on top of one another.

Between 1950 and 1953, Rosalind Franklin, working in the laboratory of Maurice Wilkins, obtained improved X-ray data from more purified samples of DNA (Figure 8.13). Her work confirmed the 3.4 Å periodicity seen by

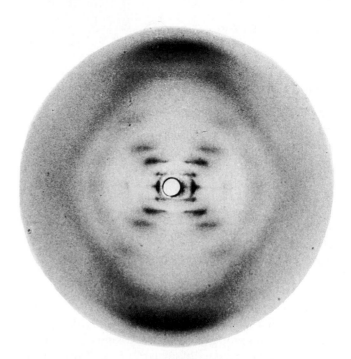

**FIGURE 8.13**    An X-ray diffraction photograph of the B form of crystallized DNA. The dark patterns at the top and bottom provide an estimate of the periodicity of nitrogenous bases, which are 3.4 Å apart. The central pattern is indicative of the molecule's helical structure.

---

*Today, measurement in nanometers (nm) is favored (1 nm = 10 Å). However, for historical purposes, we will use angstroms (Å), as reported in the original literature.

**Table 8.3**  DNA BASE COMPOSITION DATA
(a) CHARGAFF'S DATA

| Source | Approximate Percent[a] | | | |
|---|---|---|---|---|
| | 1 | 2 | 3 | 4 |
| | A | T | G | C |
| Ox thymus | 26 | 25 | 21 | 16 |
| Ox spleen | 25 | 24 | 20 | 15 |
| Yeast | 24 | 25 | 14 | 13 |
| Avian tubercle bacilli | 12 | 11 | 28 | 26 |
| Human sperm | 29 | 31 | 18 | 18 |

SOURCE: From Chargaff, 1950.
[a]Moles of nitrogenous constituent per mole of P (often, the recovery was less than 100 percent).

(c) G + C CONTENT IN SEVERAL ORGANISMS

| Organism | % G + C |
|---|---|
| Phage T2 | 36.0 |
| *Drosophila* | 45.0 |
| Maize | 49.1 |
| *Euglena* | 53.5 |
| *Neurospora* | 53.7 |

(b) BASE COMPOSITIONS OF DNAs FROM VARIOUS SOURCES

| Source | Base Composition | | | | Base Ratio | | | Asymmetry Ratio |
|---|---|---|---|---|---|---|---|---|
| | 1 | 2 | 3 | 4 | 5 | 6 | 7 | 8 |
| | A | T | G | C | A/T | G/C | (A + G)/(C + T) | (A + T)/(C + G) |
| Human | 30.9 | 29.4 | 19.9 | 19.8 | 1.05 | 1.00 | 1.04 | 1.52 |
| Sea urchin | 32.8 | 32.1 | 17.7 | 17.3 | 1.02 | 1.02 | 1.02 | 1.58 |
| *E. coli* | 24.7 | 23.6 | 26.0 | 25.7 | 1.04 | 1.01 | 1.03 | 0.93 |
| *Sarcina lutea* | 13.4 | 12.4 | 37.1 | 37.1 | 1.08 | 1.00 | 1.04 | 0.35 |
| T7 bacteriophage | 26.0 | 26.0 | 24.0 | 24.0 | 1.00 | 1.00 | 1.00 | 1.08 |

Astbury and suggested that the structure of DNA was some sort of helix. However, she did not propose a definitive model. Pauling had analyzed the work of Astbury and others and incorrectly proposed that DNA was a triple helix.

**The Watson–Crick Model**

Watson and Crick published their analysis of DNA structure in 1953 (see pp. 260–261). By building models under the constraints of the information just discussed, they proposed the double-helical form of DNA as shown in Figure 8.14(a). This model has the following major features:

1. Two right-handed helical polynucleotide chains are coiled around a central axis; the coiling is **plectonic**, meaning that the two coiled strands can only be separated by completely unwinding them.

2. The two chains are **antiparallel**; that is, their C-5′-to-C-3′ orientations run in opposite directions.

3. The bases of both chains are flat structures, lying perpendicular to the axis; they are "stacked" on one another, 3.4 Å (0.34 nm) apart and are located on the inside of the structure.

4. The nitrogenous bases of opposite chains are paired to one another as the result of the formation of **hydrogen bonds** (described in the following discussion); in DNA, only A-T and G-C pairs are allowed.

5. Each complete turn of the helix is 34 Å (3.4 nm) long; thus, 10 bases exist in each chain per turn.

6. In any segment of the molecule, alternating larger **major grooves** and smaller **minor grooves** are apparent along the axis.

7. The double helix measures 20 Å (2.0 nm) in diameter.

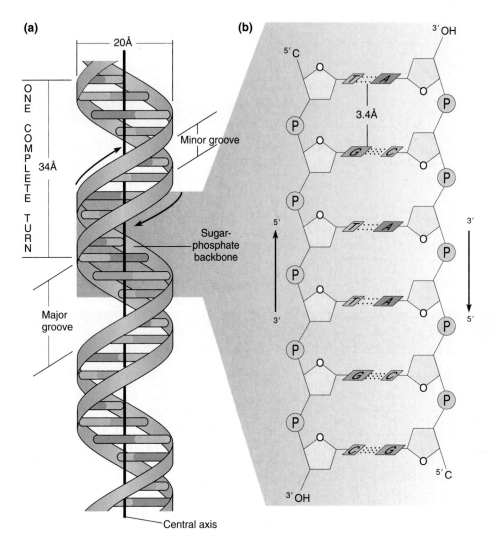

**(a)**

20Å

ONE COMPLETE TURN

34Å

Minor groove

Sugar-phosphate backbone

Major groove

Central axis

**(b)**

3′ OH

5′ C

T····A

3.4Å

G····C

5′

3′

T····A

3′

5′

T····A

G····C

C····G

5′ C

3′ OH

**FIGURE 8.14** (a) A schematic representation of the DNA double helix as proposed by Watson and Crick. The ribbon-like strands constitute the sugar-phosphate backbones, and the horizontal rungs constitute the nitrogenous base pairs, of which there are 10 per complete turn. The major and minor grooves are apparent. A solid vertical bar representing the central axis has been placed through the center of the helix. (b) A representation of the antiparallel nature of the two strands of the helix.

The antiparallel nature of the two chains is a key feature of the double-helix model. While one chain runs in the 5′-to-3′ orientation (what seems right-side-up to us), the other chain is in the 3′-to-5′ orientation (and thus appears upside-down). This is illustrated in part (b) of Figure 8.14. Given the constraints of the bond angles of the various nucleotide components, the double helix could not be constructed easily if both chains ran parallel to one another.

The right-handed nature of the helix is best appreciated by comparing such a structure to its left-handed counterpart, which is a mirror image, as shown in Figure 8.15. The conformation in space of the right-handed helix is most consistent with the data that were available to Watson and Crick. As we shall see momentarily, an alternative form of DNA (Z-DNA) does exist as a left-handed helix.

One of the most significant features of the structure proposed by Watson and Crick is the specificity of base pairing. Chargaff's data had suggested that the amounts of A equaled T and that G equaled C. Watson and Crick realized that when placed opposite each other in the model [Figure 8.14(b)], the members of each such base pair formed hydrogen bonds, providing the chemical stability necessary to hold the two chains together. Arranged in this way, both major and minor grooves become apparent along the axis. Further, with one purine (A or G) opposite one pyrimidine (T or C) as each "rung of the spiral staircase" of the proposed helix, the Watson–Crick model conformed to the 20-Å (2-nm) diameter suggested by X-ray diffraction studies.

The specific A-T and G-C base pairing is the basis for the concept of **complementarity**. This term is used to describe the chemical affinity provided by the hydrogen bonds between the bases. As we will see, this concept is very important in the processes of DNA replication and gene expression.

Two questions are particularly worthy of discussion. First, why aren't other base pairs possible? Watson and Crick discounted the A-G and C-T pairs because these

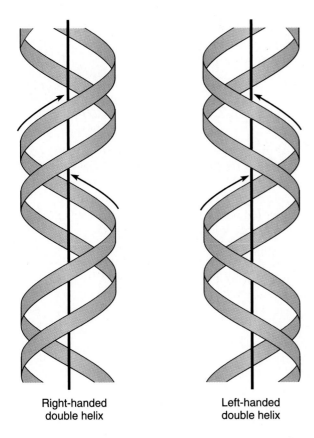

Right-handed
double helix

Left-handed
double helix

**FIGURE 8.15**    The right- and left-handed helical forms of DNA. Note that they are mirror images of one another.

represent purine-purine and pyrimidine-pyrimidine pairings, respectively. These pairings would lead to alternating diameters of more than and less than 20 Å because of the respective sizes of the purine and pyrimidine rings; additionally, the three-dimensional configurations formed by such pairings do not produce the proper alignment leading to sufficient hydrogen bond formations. It is for this latter reason that A-C and G-T pairings were also discounted, even though these pairs each consist of one purine and one pyrimidine.

The second question concerns hydrogen bonds. Just what is the nature of such a bond, and is it strong enough to stabilize the helix? A **hydrogen bond** is a very weak electrostatic attraction between a covalently bonded hydrogen atom and an atom with an unshared electron pair. The hydrogen atom assumes a partial positive charge, while the unshared electron pair—characteristic of covalently bonded oxygen and nitrogen atoms—assumes a partial negative charge. These opposite charges are responsible for the weak chemical attraction. As oriented in the double helix (Figure 8.16), adenine forms two hydrogen bonds with thymine, and guanine forms three hydrogen bonds with cytosine. Al-

though two or three hydrogen bonds taken alone are very weak, two or three thousand bonds in tandem (which would be found in two long polynucleotide chains) are capable of providing great stability to the helix.

Still another stabilizing factor is the arrangement of sugars and bases along the axis. In the Watson–Crick model, the hydrophobic or "water-fearing" nitrogenous bases are stacked almost horizontally on the interior of the axis, thus shielded from water. The hydrophilic sugar-phosphate backbone is on the outside of the axis, where both components may interact with water. These molecular arrangements provide significant chemical stabilization to the helix.

A more recent and accurate analysis of the form of DNA that served as the basis for the Watson–Crick model has revealed a minor structural difference. A precise measurement of the number of base pairs (bp) per turn has demonstrated a value of 10.4 rather than the 10.0 predicted by Watson and Crick. Where, in the classical model, each base pair is rotated around the helical axis 36°, relative to the adjacent base pair, the new finding requires a rotation of 34.6°. Thus, there are slightly more than 10 base pairs per 360° turn.

The Watson–Crick model had an immediate effect on the emerging discipline of molecular biology. Even in their initial 1953 article, the authors noted, "It has not escaped our notice that the specific pairing we have postulated immediately suggests a possible copying mechanism for the genetic material." Two months later, in a second article in *Nature*, Watson and Crick pursued this idea, suggesting a specific mode of replication of DNA—the semiconservative model. The second article also alluded to two new concepts: the storage of genetic information in the sequence of the bases, and the mutation or genetic change that would result from alteration of the bases. These ideas have received vast amounts of experimental support since 1953 and are now universally accepted. Thus, the "synthesis" of ideas by Watson and Crick was a remarkable feat and highly significant in the history of genetics and biology. For their work, they, along with Wilkins, received the Nobel Prize in Physiology and Medicine in 1962. This was to be one of many such awards bestowed for work in the field of molecular genetics.

## OTHER FORMS OF DNA

Under different conditions of isolation, several conformational forms of DNA have been recognized. At the time Watson and Crick performed their analysis, two forms—**A-DNA** and **B-DNA**—were known. Watson and

# MOLECULAR STRUCTURE OF NUCLEIC ACIDS:
## A STRUCTURE FOR DEOXYRIBOSE NUCLEIC ACID

We wish to suggest a structure for the salt of deoxyribose nucleic acid (D. N. A.). This structure has novel features which are of considerable biological interest. A structure for nucleic acid has already been proposed by Pauling and Corey.[1] They kindly made their manuscript available to us in advance of publication. Their model consists of three intertwined chains, with the phosphates near the fibre axis, and the bases on the outside. In our opinion, this structure is unsatisfactory for two reasons: (1) We believe that the material which gives the X-ray diagrams is the salt, not the free acid. Without the acidic hydrogen atoms it is not clear what forces would hold the structure together, especially as the negatively charged phosphates near the axis will repel each other. (2) Some of the van der Waals distances appear to be too small.

Another three-chain structure has also been suggested by Fraser (in the press). In his model the phosphates are on the outside and the bases on the inside, linked together by hydrogen bonds. This structure as described is rather ill-defined, and for this reason we shall not comment on it.

We wish to put forward a radically different structure for the salt of deoxyribose nucleic acid. This structure has two helical chains each coiled round the same axis. We have made the usual chemical assumptions, namely, that each chain consists of phosphate diester groups joining $\beta$-D-deoxyribofuranose residues with 3′,5′ linkages. The two chains (but not their bases) are related by a dyad perpendicular to the fibre axis. Both chains follow right-handed helices, but owing to the dyad the sequences of the atoms in the two chains run in opposite directions. Each chain loosely resembles Furberg's[2] model No. 1; that is, the bases are on the inside of the helix and the phosphates on the outside. The configuration of the sugar and the atoms near it is close to Furberg's "standard configuration," the sugar being roughly perpendicular to the attached base. There is a residue on each chain every 3.4 Å in the z-direction. We have assumed an angle of 36° between adjacent residues in the same chain, so that the structure repeats after 10 residues on each chain, that is, after 34 Å. The distance of a phosphorus atom from the fibre axis is 10 Å. As the phosphates are on the outside, cations have easy access to them.

The structure is an open one, and its water content is rather high. At lower water contents we would expect the bases to tilt so that the structure could become more compact.

The novel feature of the structure is the manner in which the two chains are held together by the purine and pyrimidine bases. The planes of the bases are perpendicular to the fibre axis. They are joined together in pairs, a single base from one chain being hydrogen-bonded to a single base from the other chain, so that the two lie side by side with identical z-co-ordinates. One of the pair must be a purine and the other a pyrimidine for bonding to occur. The hydrogen bonds are made as follows: purine position 1 to pyrimidine position 1; purine position 6 to pyrimidine position 6.

If it is assumed that the bases only occur in the structure in the most plausible tautomeric forms (that is, with the keto rather than the enol configurations) it is found that only specific pairs of bases can bond together. These pairs are: adenine (purine) with thymine (pyrimidine), and guanine (purine) with cytosine (pyrimidine).

In other words, if an adenine forms one member of a pair, on either chain, then on these assumptions the other member must be thymine; similarly for guanine and cytosine. The sequence of bases on a single chain does not appear to be restricted in any way. However, if only specific pairs of bases can be formed, it follows that if the sequence of bases on one chain is given, then the sequence on the other chain is automatically determined.

It has been found experimentally[3,4] that the ratio of the amounts of adenine to thymine, and the ratio of guanine to cytosine, are always very close to unity for deoxyribose nucleic acid.

It is probably impossible to build this structure with a ribose sugar in place of the deoxyribose, as the extra oxygen atom would make too close a van der Waals contact.

The previously published X-ray data[5,6] on deoxyribose nucleic acid are insufficient for a rigorous test of our structure. So far as we can tell, it is roughly compatible with the experimental data, but it must be regarded as unproved until it has been checked against more exact results. Some of these are given in the following communications. We were not aware of the details of the results presented there when we devised our structure, which rests mainly though not entirely on published experimental data and stereochemical arguments.

It has not escaped our notice that the specific pairing we have postulated immediately suggests a possible copying mechanism for the genetic material.

Full details of the structure, including the conditions assumed in building it, together with a set of co-ordinates for the atoms, will be published elsewhere.

We are much indebted to Dr. Jerry Donohue for constant advice and criticism, especially on interatomic distances. We have also been stimulated by a knowledge of the general nature of the unpublished experimental results and ideas of Dr. M. H. F. Wilkins, Dr. R. E. Franklin and their coworkers at King's College, London. One of us (J. D. W.) has been aided by a fellowship from the National Foundation for Infantile Paralysis.

**J. D. Watson**
**F. H. C. Crick**
*Medical Research Council Unit for the Study of the*
*Molecular Structure of Biological Systems*
*Cavendish Laboratory, Cambridge.*
*April 2.*

[1] Pauling, L., and Corey, R. B., *Nature*, 171, 346 (1953); *Proc. U.S. Nat. Acad. Sci.*, 39, 84 (1953).

[2] Furberg, S., *Acta Chem. Scand.*, 6, 634 (1952).

[3] Chargaff, E., for references see Zamenhof, S., Brawerman, G., and Chargaff, E., *Biochim. et Biophys. Acta*, 9, 402 (1952).

[4] Wyatt, G. R., *J. Gen. Physiol*, 36, 201 (1952).

[5] Astbury, W. T., Symp. Soc. Exp. Biol. 1, Nucleic Acid, 66 (Camb. Univ. Press, 1947).

[6] Wilkins, M. H. F., and Randall, J. T., *Biochim. et Biophys. Acta*, 10, 192 (1953).

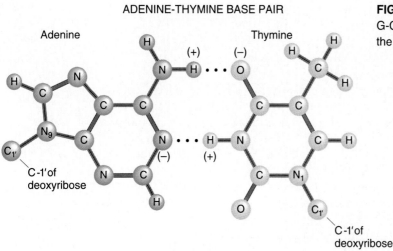

ADENINE-THYMINE BASE PAIR

GUANINE-CYTOSINE BASE PAIR

• • • Hydrogen bond

**FIGURE 8.16**    Ball-and-stick models of A-T and G-C base pairs. The rows of dots (• • •) represent the hydrogen bonds that form.

Crick's analysis was based on X-ray studies of the B form by Franklin, which is present under aqueous, low-salt conditions and is believed to be the biologically significant conformation.

While DNA studies around 1950 relied on the use of X-ray diffraction, more recent investigations have been performed using **single-crystal X-ray analysis**. The earlier studies achieved limited resolution of about 5 Å, but single crystals diffract X rays at about 1 Å, near atomic resolution. As a result, every atom is "visible," and much greater structural detail is available during analysis.

Using these modern techniques, the A form of DNA has now been scrutinized. A-DNA is prevalent under high-salt or dehydration conditions. In comparison to B-DNA (Figure 8.17), A-DNA is slightly more compact, with 11 base pairs in each complete turn of the helix, which is 23 Å in diameter. While it is also a right-handed

helix, the orientation of the bases is somewhat different. They are tilted and displaced laterally in relation to the axis of the helix. As a result of these differences, the appearance of the major and minor grooves is modified compared with those in B-DNA. It seems doubtful that A-DNA occurs under biological conditions.

Three other right-handed forms of DNA helices have been discovered. These have been designated C-, D-, and E-DNA. **C-DNA** is found to occur under even greater dehydration conditions than those observed during the isolation of A- and B-DNA. It has only 9.3 base pairs per turn and is, thus, less compact. Its helical diameter is 19 Å. Like A-DNA, C-DNA does not have its base pairs lying flat; instead they are tilted relative to the axis of the helix. Two other forms, **D-DNA** and **E-DNA**, occur in helices lacking guanine in their base composition. They have even fewer base pairs per turn: 8 and $7\frac{1}{2}$, respectively.

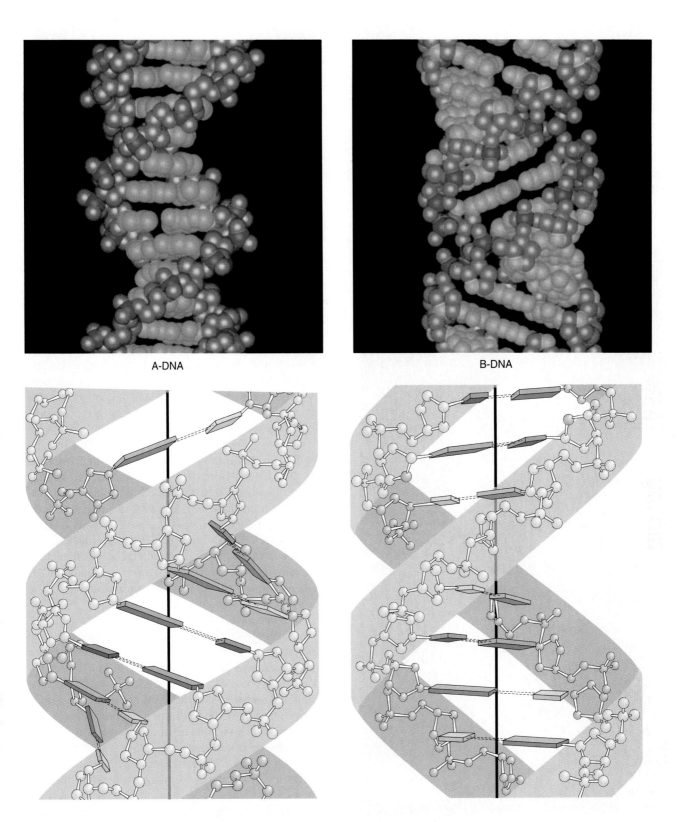

A-DNA

B-DNA

**FIGURE 8.17**    Comparison of A-DNA and B-DNA using space-filling models (top) and an artist's depiction illustrating the orientation of the base pairs (bottom). In the space-filling model of A-DNA, note the near disappearance of the grooves associated with B-DNA. In the artist's depiction, note that in A-DNA the base pairs are tilted and pulled away from the helix, while they are perpendicular to the helix in B-DNA.

Still another form of DNA, called **Z-DNA**, was discovered by Andrew Wang, Alexander Rich, and their colleagues in 1979 when they examined a small synthetic DNA fragment containing only C-G base pairs. Z-DNA takes on the rather remarkable configuration of a left-handed double helix (Figure 8.18). Like A- and B-DNA, Z-DNA consists of two antiparallel chains held together by Watson–Crick base pairs. Beyond these characteristics, Z-DNA is quite different. The left-handed helix is 18 Å (1.8 nm) in diameter, contains 12 base pairs per turn, and assumes a zigzag conformation (hence its name). The major groove present in B-DNA is nearly eliminated in Z-DNA.

Speculation has abounded over the possibility that regions of Z-DNA exist in the chromosomes of living organisms. The unique helical arrangement could provide an important recognition point for the interaction with other molecules. However, it is still not clear whether Z-DNA occurs *in vivo*.

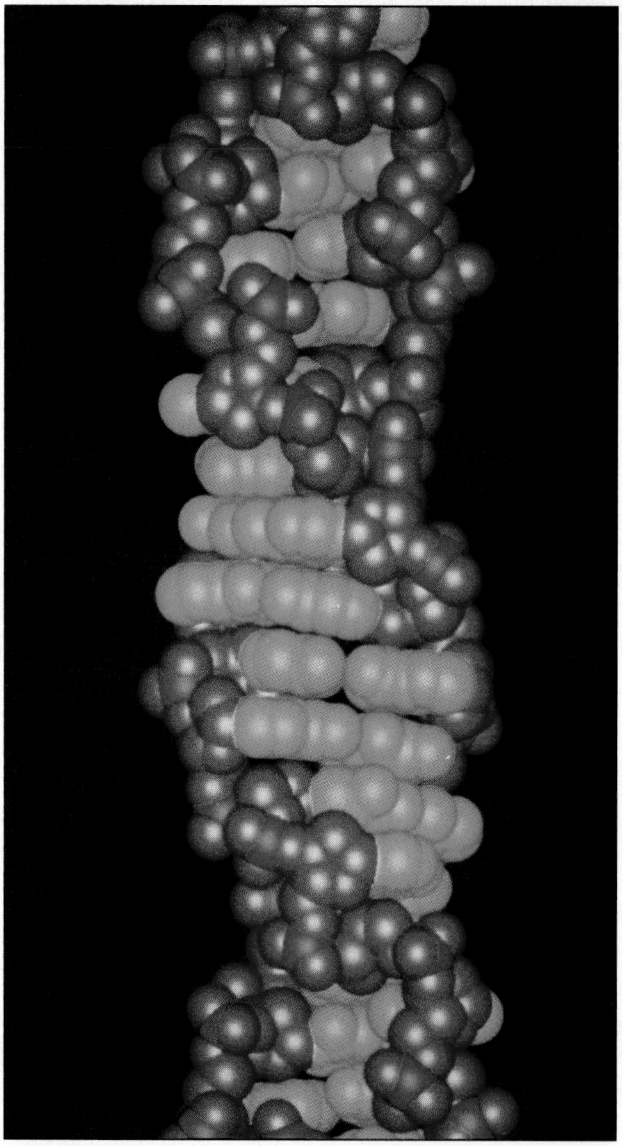

## THE STRUCTURE OF RNA

The second category of nucleic acids is the ribonucleic acids, or RNA. The structure of these molecules is similar to DNA, with several important exceptions. While RNA also has as its building blocks nucleotides linked into polynucleotide chains, ribose replaces deoxyribose and uracil replaces thymine. Another important difference is that most RNA is thought of as being single-stranded. However, RNA molecules sometimes fold back on themselves following their synthesis; such a configuration results when regions of complementarity occur in positions that allow base pairs to form. Furthermore, some animal viruses that have RNA as their genetic material contain it in the form of double-stranded helices. Thus, there are several instances where RNA does not exist strictly as a linear, single-stranded molecule.

At least three classes of cellular RNA molecules function during the expression of genetic information: **ribosomal RNA (rRNA)**, **messenger RNA (mRNA)**, and **transfer RNA (tRNA)**. These molecules all originate as complementary copies of one of the two strands of DNA segments during the process of transcription. That is, their nucleotide sequence is complementary to the deoxyribonucleotide sequence of DNA, which served as the template for their synthesis. Since uracil replaces thymine in RNA, uracil is complementary to adenine during transcription and during RNA base pairing.

Each class of RNA may be characterized by its size, sedimentation behavior in a centrifugal field, and genetic function. Sedimentation behavior depends on a molecule's density, mass, and shape, and its measure is called the **Svedberg coefficient (S)**. Table 8.4 relates the S values, molecular weights, and approximate number of nucleotides of the major forms of RNA. While higher S values almost always designate molecules of greater molecular weight, the correlation is not direct; that is, a twofold increase in molecular weight does not lead to a twofold increase in S. This is because the size and shape of the molecule impact on its rate of sedimentation (S). As you can see, there is a wide variation in size of the three classes of RNA.

**FIGURE 8.18**    A space-filling model of the Z form of DNA, which is a left-handed helix. Note the near elimination of the grooves in this molecule compared to B-DNA shown in Figure 8.17.

**Table 8.4**  Sedimentation Coefficients, Molecular Weights, and Number of Nucleotides for Various RNAs

| RNA Type | Abbreviation | Svedberg Coefficient (S) | Molecular Weight | Number of Nucleotides |
|---|---|---|---|---|
| Ribosomal RNA | rRNA | 5S | 35,000 | 120 |
| | | 5.8S | 47,000 | 160 |
| | | 18S | 700,000 | 1900 |
| | | 28S | 1,800,000 | 4800 |
| Transfer RNA | tRNA | 4S | 23,000–30,000 | 75–90 |
| Messenger RNA | mRNA | 6–50S | 25,000–1,000,000 | 100–10,000 |

Ribosomal RNA is generally the largest of these molecules and usually constitutes about 80 percent of all RNA in the cell. The various forms of rRNA found in prokaryotes and eukaryotes differ distinctly in size. The values in Table 8.4 are based on eukaryotic molecules. Ribosomal RNAs are important structural components of **ribosomes**, which function to synthesize proteins during the process of translation.

Messenger RNA molecules carry genetic information from the DNA of the gene to the ribosome, where translation occurs. They vary considerably in length, which is partially a reflection of the variation in the size of the gene serving as the template for transcription of mRNA species. Note that the values shown in Table 8.4 are only estimates, particularly at their upper limits. Precursors of many mRNAs, called **primary transcripts**, may demonstrate values considerably higher.

Transfer RNA, the smallest class of RNA molecules, carries amino acids to the ribosome during translation. Since more than one tRNA molecule interacts simultaneously with the ribosome, the molecule's smaller size facilitates these interactions.

We will discuss the functions of the three classes of RNA in much greater detail later in the text (see Chapter 14). Our purpose in this section has been to contrast the structure of DNA, which stores genetic information, with that of RNA, which functions in the expression of that information.

## ANALYSIS OF NUCLEIC ACIDS

Since 1953, the role of DNA as the genetic material and the role of RNA in transcription and translation have been clarified through detailed analysis of nucleic acids. We will consider several methods of analysis of these molecules in this chapter. Some of these, as well as other research procedures, are presented in greater detail in Appendix A.

### Absorption of Ultraviolet Light (UV)

Nucleic acids absorb ultraviolet light most strongly at wavelengths of 254 to 260 nm due to the interaction between UV light and the ring systems of the purines and pyrimidines. Thus, any molecule containing nitrogenous bases (i.e., nucleosides, nucleotides, and polynucleotides) can be analyzed using UV light. This technique is especially important in the localization, isolation, and characterization of nucleic acids.

UV analysis is used in conjunction with many standard procedures that separate molecules. As we shall see in the next section, the use of UV absorption is critical to the recovery of nucleic acids following their separation.

### Sedimentation Behavior

Nucleic acid mixtures can be separated by subjecting them to one of several possible **gradient centrifugation** procedures (Figure 8.19). The mixture may be loaded on top of a solution prepared so that a concentration gradient has been formed from top to bottom. Then the entire mixture is centrifuged at high speeds in an ultracentrifuge. The mixture of molecules will migrate downward, with each component moving at a different rate. Centrifugation is stopped and the gradient eluted from the tube. Each fraction can then be measured spectrophotometrically for absorption at 260 nm. In this way, the previous position of a nucleic acid fraction along the gradient can be predicted and the fraction isolated and studied further.

The gradient centrifugations described rely on the sedimentation behavior of molecules in solution. There are two major types of gradient centrifugation techniques employed in the analysis of nucleic acids: sedimentation equilibrium and sedimentation velocity. Both require the use of high-speed centrifugation to create large centrifugal forces upon molecules in a gradient solution.

In **sedimentation equilibrium centrifugation**

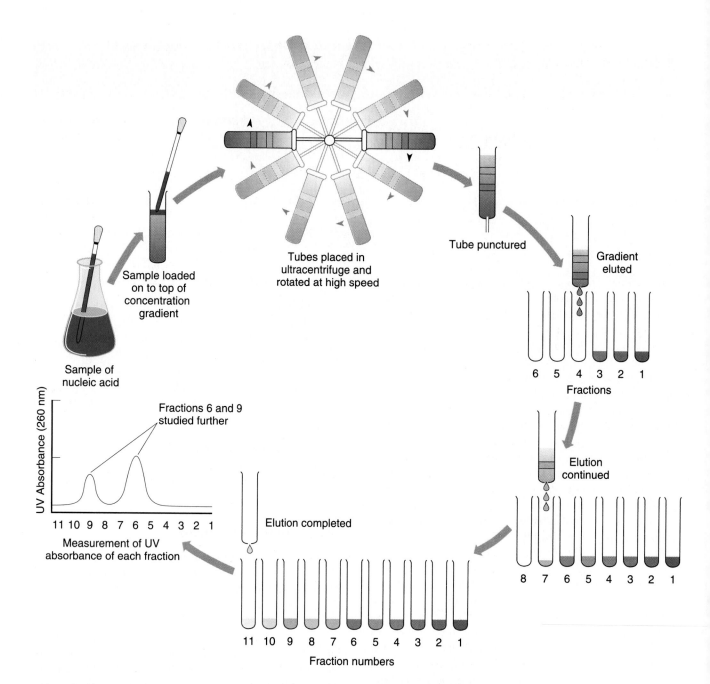

**FIGURE 8.19**    Separation of a mixture of two types of nucleic acid using gradient centrifugation. In order to fractionate the gradient, successive samples are eluted from the bottom of the tube. Each is measured for absorbance of ultraviolet light at 260 nm, producing a profile of the sample in graphic form.

(sometimes called **density gradient centrifugation**), a density gradient is created that overlaps the densities of the individual components of a mixture of molecules. Usually, the gradient is made of a heavy metal salt such as cesium chloride (CsCl). During centrifugation the molecules migrate until they reach a point of neutral buoyant density. At this point, the centrifugal force on them is equal and opposite to the upward diffusion force

and no further migration occurs. If DNAs of different densities are used, they will separate as the molecules of each density reach equilibrium with the corresponding density of CsCl. The gradient may be fractionated and the components isolated (Figure 8.19). When properly executed, this technique provides high resolution in separating mixtures of molecules varying only slightly in density.

Sedimentation equilibrium centrifugation studies may also be used to generate data on the base composition of double-stranded DNA. G-C base pairs, compared with A-T pairs, are more compact and dense. As shown in Figure 8.20, the percentage of G-C pairs in DNA is directly proportional to the molecule's buoyant density. By using this technique, we can make a useful molecular characterization of DNAs from different sources.

The second technique, **sedimentation velocity centrifugation**, employs an analytical centrifuge, which enables the migration of the molecules during centrifugation to be monitored with ultraviolet absorption optics. Thus, the "velocity of sedimentation" may be determined. This velocity has been standardized in units called Svedberg coefficients (*S*), as mentioned earlier.

In this technique, the molecules are loaded on top of the gradient, and the gravitational forces created by centrifugation drive them toward the bottom of the tube. Two forces work against this downward movement: (1) the viscosity of the solution creates a frictional resistance, and (2) part of the force of diffusion is directed upward. Under these conditions, the key variables are the mass and shape of the molecules being examined. In general, the greater the mass, the greater is the sedimentation velocity. However, the molecule's shape affects the frictional resistance. Therefore, two molecules of equal mass but different shape will sediment at different rates.

One use of the sedimentation velocity technique is the determination of **molecular weight (MW)**. If certain physical–chemical properties of a molecule under study are also known, the MW can be calculated based on the sedimentation velocity. *S* values increase with molecular weight, but are not directly proportional to it.

## Denaturation and Renaturation of Nucleic Acids

When **denaturation** of double-stranded DNA occurs, the hydrogen bonds of the duplex structure break, the duplex unwinds, and the strands separate. However, no covalent bonds break. During strand separation, which can be induced by heat or chemical treatment, the viscosity of DNA decreases, and both the UV absorption and the buoyant density increase. Denaturation as a result of heating is sometimes referred to as **melting**. The increase in UV absorption of heated DNA in solution, called the **hyperchromic effect**, is easiest to measure. This effect is illustrated in Figure 8.21.

Since G-C base pairs have one more hydrogen bond than A-T pairs, they are more stable to heat treatment. Thus, DNA with a greater proportion of G-C pairs than A-T pairs requires higher temperatures to denature completely. When absorption at 260 nm is monitored and plotted against temperature during heating, a **melting profile** of DNA is obtained. The midpoint of this profile, or curve, is called the **melting temperature ($T_m$)** and represents the point at which 50 percent of the strands are unwound or denatured (Figure 8.21). When the curve plateaus at its maximum optical density, denaturation is complete and only single strands exist. Analysis of melting profiles provides a characterization of DNA and an alternative method of estimating the base composition of DNA.

One might ask whether the denaturation process can be reversed; that is, can single strands of nucleic acids

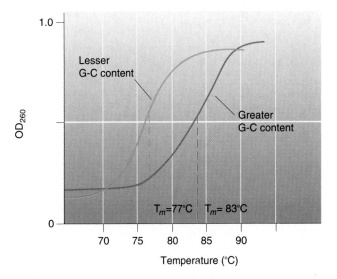

**FIGURE 8.21**    Comparison of the increase in UV absorbance with an increase in temperature for two DNA molecules with different G-C contents. The molecule with a melting point ($T_m$) of 83°C has a greater G-C content than the molecule with a $T_m$ of 77°C.

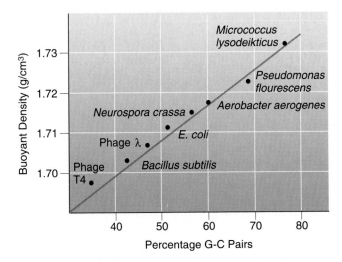

**FIGURE 8.20**    Percentage of guanine-cytosine (G-C) base pairs in DNA plotted against buoyant density for a variety of microorganisms.

reform a double helix, provided that each strand's complement is present? Not only is the answer yes, but such reassociation provides the basis for several important analytical techniques that have provided much valuable information during genetic experimentation.

If DNA that has been denatured thermally is cooled slowly, random collisions between complementary strands will result in their reassociation. At the proper temperature, hydrogen bonds will reform, securing pairs of strands into duplex structures. With time during cool-ing, more and more duplexes will form. Depending on the conditions, a complete match is not essential for duplex formation, provided that there are at least stretches of base-pairing on two reassociating strands.

## Molecular Hybridization

The property of renaturation of complementary single strands of nucleic acids is the basis for a powerful analytical technique in molecular genetics—**molecular hy-**

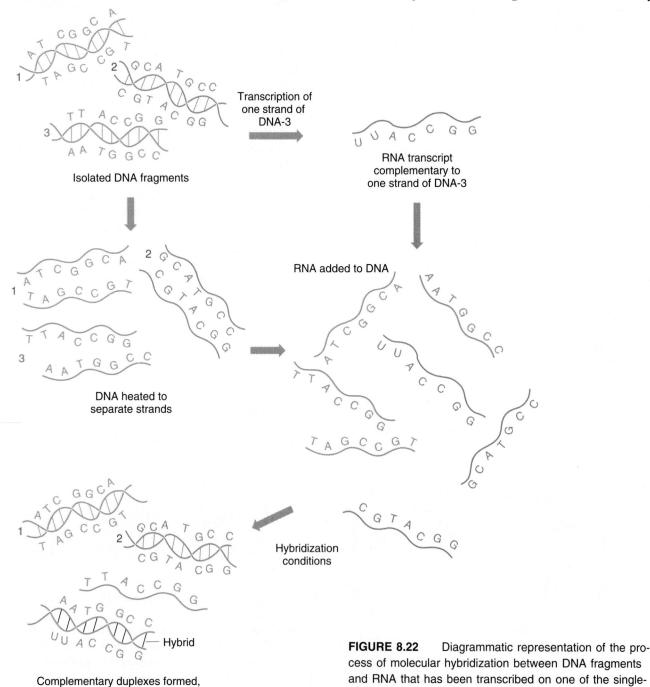

Isolated DNA fragments

Transcription of one strand of DNA-3

RNA transcript complementary to one strand of DNA-3

DNA heated to separate strands

RNA added to DNA

Hybridization conditions

Complementary duplexes formed, including a DNA/RNA hybrid

Hybrid

**FIGURE 8.22**    Diagrammatic representation of the process of molecular hybridization between DNA fragments and RNA that has been transcribed on one of the single-stranded fragments.

**bridization**. This technique derives its name based on the fact that renaturing single strands need not originate from the same nucleic acid source. For example, DNA strands may be isolated from different organisms. Provided that some degree of base complementarity exists during renaturation, **molecular hybrids** are formed. Furthermore, when a mixture of DNA and RNA strands are cooled slowly, hybridization may also occur. A case in point would be when RNA and the DNA from which it has been transcribed are present together. RNA will find its single-stranded DNA complement and renature. Since such experimental conditions are mixing *different* nucleic acid molecules, the resultant duplexes constitute still another example of molecular hybrids.

Figure 8.22 illustrates how the process of DNA-RNA hybridization occurs. In this example, the DNA strands are heated, causing strand separation, and then slowly cooled in the presence of single-stranded RNA. If the RNA has been transcribed on the DNA used in the experiment, and is therefore complementary to it, hybridization will occur. Several methods are available for monitoring the amount of double-stranded molecules produced following strand separation.

In the 1960s molecular hybridization techniques contributed to the increase in our understanding of transcriptional events occurring at the gene level. Refinements of this process have occurred continually and have been the forerunners of work in studies of molecular evolution as well as the organization of DNA in chromosomes. Hybridization can occur in solution or when DNA is bound to either a gel or a special type of filter, facilitating recovery of the newly formed hybrids.

The technique can even be performed using cytological preparations, which is called *in situ* **molecular hybridization**. In this procedure mitotic or interphase cells are fixed to slides and subjected to hybridization conditions. Radioactive single-stranded DNA or RNA is added, and hybridization is monitored. The nucleic acid that is added may be either radioactive or involve fluorescence to allow its detection. In the former case, the technique of **autoradiography** may be used (see Appendix A).

Fluorescent probes are prepared in a different way. The probe DNA is first coupled to the small organic molecule biotin (creating biotinylated DNA). Once hybridization is completed, another molecule, avidin or streptavidin, that has a high binding affinity for biotin is utilized. A fluorescent molecule such as fluorescein is linked to avidin (or streptavidin) and the complex is reacted with the cytological preparation. The end result is an extremely sensitive method for localizing the hybridized DNA. Because fluorescence is used, the technique is known by the acronym **FISH** (**fluorescent *in situ* hybridization**).

Figure 8.23 illustrates the use of FISH in identifying the DNA specific to the centromeres of human chromo-

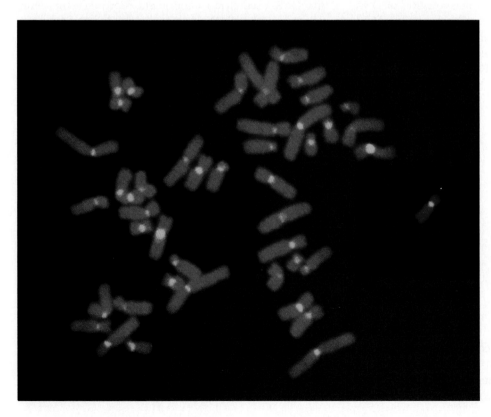

**FIGURE 8.23**    *In situ* hybridization of human metaphase chromosomes using a fluorescent technique (FISH). The probe, specific to centromeric DNA, produces a yellow fluorescence signal indicating hybridization. The red fluorescence is produced by propidium iodide counterstaining of chromosomal DNA.

somes. The resolution of FISH is great enough to detect just a single gene within an entire set of chromosomes (see Appendix A). The use of this technique in identification of chromosomal locations housing specific genetic information has been a valuable addition to the repertoire of experimental geneticists.

## Reassociation Kinetics and Repetitive DNA

One extension of molecular hybridization procedures is the technique that measures the *rate* of reassociation of complementary strands of DNA derived from a single source. This technique, called **reassociation kinetics**, was first refined and studied by Roy Britten and David Kohne.

The DNA used in such studies is first fragmented into small pieces as a result of shearing forces introduced during isolation. The resultant DNA fragments cluster around a uniform average size of several hundred base pairs. These fragments of DNA are then dissociated into single strands by heating. Next, the temperature is lowered and reassociation is monitored. During reassociation, pieces of single-stranded DNA collide randomly. If they are complementary, a stable double strand is formed; if not, they separate and are free to encounter other DNA fragments. The process continues until all matches are made.

The ideal results of such an experiment are presented in Figure 8.24. The percentage of reassociation of DNA fragments is plotted against a logarithmic scale of the product of $C_0$ (the initial concentration of DNA single strands in moles per liter of nucleotides), and $t$ (the time, usually measured in minutes). The process of renatura-

tion follows second-order rate kinetics according to the equation

$$\frac{C}{C_0} = \frac{1}{1 + kC_0t}$$

where $C$ is the single-stranded DNA concentration remaining after time $t$ has elapsed and $k$ is the second-order rate constant. Initially, $C$ equals $C_0$, and the fraction remaining single-stranded is 100 percent.

The initial shape of the curve reflects the fact that in a mixture of unique sequence fragments, each with one complement, initial matches take more time to make. Then, as many single strands are converted to duplexes, matches are made more quickly, reflecting an increase in the "slope" of the curve. Near the end of the reaction, the few remaining single strands require relatively greater time to make the final matches.

A great deal of information can be obtained from studies comparing the reassociation of DNA of different organisms. For example, we may compare the point in the reaction when one-half of the DNA is present as double-stranded fragments. This point is called the $C_0t_{1/2}$, or **half reaction time**. Provided that all pairs of single-stranded DNA complements consist of unique nucleotide sequences and all are about the same size, $C_0t_{1/2}$ varies directly with the complexity of the DNA. Designated as $X$, complexity represents the length in nucleotide pairs of all unique DNA fragments laid end to end. If the DNA used in an experiment represents the entire genome, and if all DNA sequences are different from one another, then $X$ is equal to the size of the haploid genome.

Figure 8.25 illustrates what is found when DNAs from

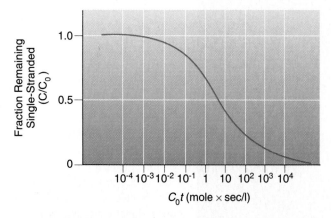

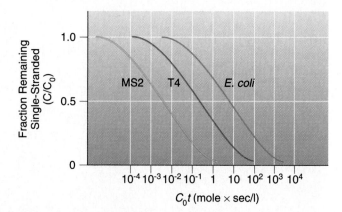

**FIGURE 8.24** The ideal time course for reassociation of DNA ($C/C_0$) when, at time zero, all DNA consists of unique fragments of single-stranded complements. Note that the abscissa ($C_0t$) is plotted logarithmically.

**FIGURE 8.25** Comparison of the reassociation rate ($C/C_0$) of DNA derived from phage MS-2, phage T4, and *E. coli*. The genome of T4 is larger than MS-2, and that of *E. coli* is larger than T4.

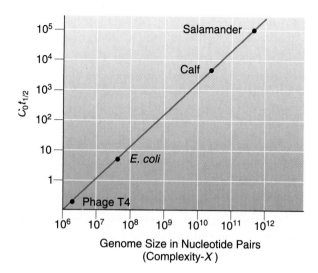

**FIGURE 8.26**   Comparison of $C_0t_{1/2}$ and genome size for phage T4, *E. coli*, calf, and salamander.

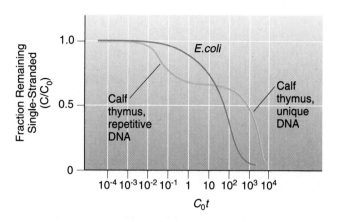

**FIGURE 8.27**   $C_0t$ curve of calf thymus DNA compared with *E. coli*. The repetitive fraction of calf DNA reassociates more quickly than that of *E. coli*, while the more complex unique calf DNA takes longer to reassociate than that of *E. coli*.

various sources are compared. As can be seen, as genome size increases, the curves obtained are shifted farther and farther to the right, indicative of an increased reassociation time.

As shown in Figure 8.26, $C_0t_{1/2}$ is directly proportional to the size of the genome. If nearly the entire genome consists of unique DNA sequences, reassociation experiments can be used to determine the genome size of organisms. This method has been useful in assessing genome size in viruses and bacteria.

However, when reassociation kinetics of DNA from eukaryotic organisms were first studied, a surprising observation was made. The data showed that some of the DNA segments reassociated even more rapidly than those derived from *E. coli*! The remainder, as expected because of its greater complexity, took longer to reassociate. For example, Britten and Kohne examined DNA derived from calf thymus tissue (Figure 8.27). Based on these observations, they hypothesized that the rapidly reassociating fraction must represent *repetitive DNA sequences* present many times in the calf genome. This interpretation would explain why these DNA segments reassociate so rapidly. On the other hand, they hypothe-

sized that the remaining DNA segments consist of unique nucleotide sequences present only once in the genome; because there are more of these unique sequences, increasing the DNA complexity in calf thymus (compared with *E. coli*), their reassociation takes longer. The *E. coli* curve has been added to Figure 8.27 for the sake of comparison.

It is now clear that repetitive DNA sequences are prevalent in the genome of eukaryotes. When analyzed carefully, it is apparent that there are various levels of repetition. Cases are known where short DNA sequences are repeated over a million times, where longer DNA sequences are repeated only a few times, and where intermediate levels of sequence redundancy are present. We will return to this topic in Chapter 10, where we will discuss the organization of DNA in genes and chromosomes. For now, we conclude this chapter by pointing out that the discovery of repetitive DNA was one of the first clues that much of the DNA in eukaryotes is not contained in genes that encode proteins! This concept will be developed and elaborated on as our coverage of the molecular basis of heredity continues.

## CHAPTER SUMMARY

1. The existence of a genetic material capable of replication, storage, expression, and mutation is deducible from the observed patterns of inheritance in organisms.

2. Both proteins and nucleic acids were initially considered as the possible candidates for genetic material. Proteins are more diverse than nucleic acids, a requirement for the genetic material, and were favored due to the advances being made in protein chemistry at the time. Additionally, Levene's tetranucleotide hypothesis had underestimated the magnitude of chemical diversity inherent in nucleic acids.

3. By 1952, transformation studies and experiments using bacteria infected with bacteriophages strongly suggested that DNA is the genetic material in bacteria and most viruses.

4. Initially, only circumstantial evidence supported the concept of DNA controlling inheritance in eukaryotes. This included the distribution of DNA in the cell, quantitative analysis of DNA, and studies of UV-induced mutagenesis. Recent recombinant DNA techniques, as well as experiments with transgenic mice, have provided direct experimental evidence that the eukaryote genetic material is DNA.

5. Numerous viruses provide an important exception to this general rule, because many of them use RNA as their genetic material. These include some bacteriophages as well as some plant and animal viruses, as well as retroviruses.

6. The establishment of DNA as the genetic material paved the way for the expansion of molecular genetics research, and has served as the cornerstone for further important studies for nearly half a century.

7. During the late 1940s and early 1950s, a considerable effort was made to integrate accumulated information on the chemical structure of nucleic acids into a model of the molecular structure of DNA. The X-ray crystallography data of Franklin and Wilkins suggested that DNA was some sort of helix. In 1953, Watson and Crick were able to assemble a model of the double helical DNA structure based on these X-ray diffraction studies as well as on Chargaff's analysis of base composition of DNA.

8. The DNA molecule exhibits antiparallel orientation and adenine-thymine and guanine-cytosine base-pairing complementarity along the polynucleotide chains. This model of DNA presents an obvious straightforward mechanism for its replication. The structure assumed by the helix appears to be a function of the nucleotide sequence and its chemical environment. Several alternative forms of the DNA helical structure exist. Watson and Crick described a B configuration, one of several right-handed helices. Wang and Rich discovered the left-handed Z-DNA currently being investigated for its physiological and genetic significance.

9. The second category of nucleic acids important in genetic function is RNA. RNA is similar to DNA with the exceptions that it is usually single-stranded, the sugar ribose replaces the deoxyribose, and the pyrimidine uracil replaces thymine. Classes of RNA—ribosomal, transfer, and messenger—facilitate the flow of information from DNA to RNA to proteins, which are the end products of most genes.

10. The structure of DNA lends itself to various forms of analysis, which have in turn led to studies of the functional aspects of the genetic machinery. Absorption of UV light, sedimentation properties, and denaturation–reassociation procedures are among the important tools for the study of nucleic acids. Reassociation kinetics analysis enabled geneticists to postulate the existence of repetitive DNA in eukaryotes, where certain nucleotide sequences are present many times in the genome.

# KEY TERMS

absorption spectrum
action spectrum
adenine
adenosine triphosphate (ATP)
A-DNA
antiparallel chains
autoradiography
bacteriophage (phage)
B-DNA
C-DNA
central dogma of molecular genetics
complementarity
cytosine
D-DNA
denaturation
deoxyribonuclease
deoxyribose
E-DNA
FISH (fluorescent *in situ* hybridization)
genetic expression
genetic information
gradient centrifugation
guanine
guanosine triphosphate (GTP)

haploid
Hershey–Chase experiment
Holmes ribgrass (HR) virus
hydrogen bond
hyperchromic effect
information flow
information storage
*in situ* molecular hybridization
lysozyme
major groove
melting profile
melting temperature ($T_m$)
messenger RNA (mRNA)
minor groove
molecular hybridization
molecular weight (MW)
nitrogenous base
nuclein
nucleoside
nucleoside diphosphate (NDP)
nucleoside monophosphate (NMP)

nucleoside triphosphate (NTP)
nucleotide
oligonucleotide
pentose sugar
phosphate group
phosphodiester bond
plectonic
polynucleotide
primary transcripts
protease
protoplast
purine
pyrimidine
reassociation kinetics
recombinant DNA technology
repetitive DNA sequences
replication
retroviruses
reverse transcriptase
ribonuclease
ribose
ribosomal RNA (rRNA)
ribosome
RNA replicase

sedimentation equilibrium centrifugation
sedimentation velocity centrifugation
serotypes
single-copy DNA
single-crystal X-ray analysis
spheroplast
Svedberg coefficient ($S$)
tetranucleotide hypothesis
thymine
tobacco mosaic virus (TMV)
transcription
transfection
transfer RNA (tRNA)
transformation
transgenic animals
translation
ultraviolet light (UV)
uracil
variation by mutation
virulent
X-ray diffraction
Z-DNA

# INSIGHTS AND SOLUTIONS

In contrast to the preceding chapters, this one does not emphasize genetic problem solving. Instead, it recounts some of the initial experimental analysis that served as the cornerstone of modern genetics. Quite fittingly, then, our Insights and Solutions section shifts its emphasis to experimental rationale and analytical thinking, an approach that will continue through the remainder of the text whenever appropriate.

1. (a) Based strictly on the transformation analysis of Avery, MacLeod, and McCarty, what objection might be made to the conclusion that DNA is the genetic material? What other conclusion might be considered?

   **SOLUTION:** Based solely on their results, it may be concluded that DNA is essential for transformation. However, DNA might have been a substance that caused capsular formation by *directly* converting nonencapsulated cells to ones with a capsule. That is, DNA may simply have played a catalytic role in capsular synthesis, leading to cells displaying smooth type III colonies.

   (b) What observations argue against this objection?

   **SOLUTION:** First, transformed cells pass the trait on to their progeny cells, thus supporting the conclusion that DNA is responsible for heredity, not for the direct production of polysaccharide coats. Second, subsequent transformation studies over the next five years showed that other traits, such as antibiotic resistance, could be transformed. Therefore, the transforming factor has a broad general effect, not one specific to polysaccharide synthesis. This observation is more in keeping with the conclusion that DNA is the genetic material.

2. If RNA was the universal genetic material, how would this have affected the Avery experiment and the Hershey–Chase experiment?

   **SOLUTION:**  In the Avery experiment, RNase rather than DNase would have eliminated transformation. Had this occurred, Avery and his colleagues would have concluded that RNA was the transforming factor. Hershey and Chase would have received identical results, since $^{32}P$ would also label RNA, but not protein. Had they been using a bacteriophage with RNA as its nucleic acid, and had they known this, they would have concluded that RNA was responsible for directing the reproduction of their bacteriophage.

3. Sea urchin DNA, which is double-stranded, was shown to contain 17.5 percent of its bases in the form of cytosine (C). What percentages of the other three bases are present in this DNA?

   **SOLUTION:**  The amount of C = G, so guanine is also present as 17.5 percent. The remaining bases, A and T, are present in equal amounts and together they represent the rest of the bases (100 − 35). Therefore, A = T = 65/2 = 32.5 percent.

4. The quest to isolate an important disease-causing organism was successful and the molecular biologists were hard at work. The organism contained as its genetic material a remarkable nucleic acid with a base composition of A = 21 percent, C = 29 percent, G = 29 percent, U = 21 percent. When heated, it showed a major hyperchromic effect, and when kinetics were studied, the nucleic acid of this organism provided the $C_0t$ curve shown below, in contrast to that of phage T4 and *E. coli*. T4 contains $10^5$ nucleotide pairs and exhibits a $C_0t_{1/2}$ of 0.5. The unknown organism produced a $C_0t_{1/2}$ of 20. Analyze this information carefully and draw *all* possible conclusions about the genetic material of this organism, based strictly on the above observations. What important, straightforward information is missing and needed to confirm your hypothesis about the nature of this molecule?

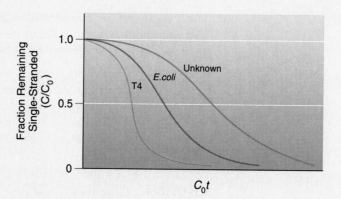

**SOLUTION:** First of all, because of the presence of uracil (U), the molecule appears to be RNA. Since $A/U = G/C = 1$, the molecule may be a double helix. The hyperchromic shift and reassociation kinetics supports this hypothesis. The kinetic study demonstrates several things. First, the shape of the $C_0t$ curve reveals that there is no repetitive sequence DNA. Further, the complexity ($X$), or total length of unique sequence DNA, is greater than that of either phage T4 or *E. coli*. $X$ can, in fact, be calculated, since a direct proportionality between $C_0t_{1/2}$ and number of base pairs exists when there are only unique sequences present:

$$\frac{0.5}{10^5} = \frac{20}{X}$$

$$0.5X = 20(10^5)$$

$$X = 40(10^5)$$

$$= 4 \times 10^6 \text{ base pairs}$$

There *may* be a greater number of genes present compared to T4 or *E. coli*, but the excessive unique sequence DNA may serve some other role, or simply play no genetic role. The missing information concerns the sugars. Our model predicts that ribose rather than d-ribose should be present. If not, the organism contains a very unusual molecule as its genetic material.

**PROBLEMS AND DISCUSSION QUESTIONS**

1. The functions ascribed to the genetic material are replication, expression, storage, and mutation. What does each of these terms mean?

2. Discuss the reasons why proteins were generally favored over DNA as the genetic material before 1940. What was the role of the tetranucleotide hypothesis in this controversy?

3. Contrast the various contributions made to an understanding of transformation by Griffith, Avery and his coworkers, and Taylor.

4. When Avery and his colleagues had obtained what was concluded to be purified DNA from the III*S* virulent cells, they treated the fraction with proteases, RNase, and DNase, followed by the assay for retention or loss of transforming ability. What were the purpose and results of these experiments? What conclusions were drawn?

5. Why were $^{32}$P and $^{35}$S chosen for use in the Hershey–Chase experiment? Discuss the rationale and conclusions of this experiment.

6. Does the design of the Hershey–Chase experiment distinguish between DNA and RNA as the molecule serving as the genetic material? Why or why not?

7. Would an experiment similar to that performed by Hershey and Chase work if the basic design were applied to the phenomenon of transformation? Explain why or why not.

8. Why is the early evidence that DNA serves as the genetic material in eukaryotes called circumstantial? List and discuss these evidences.

9. What are the exceptions to the general rule that DNA is the genetic material in all organisms? What evidence supports these exceptions?

10. Draw the chemical structure of the three components of a nucleotide and then link the three together. What atoms are removed from the structures when the linkages are formed?

11. How are the carbon and nitrogen atoms of the sugars, purines, and pyrimidines numbered?

12. Adenine may also be named 6-amino purine. How would you name the other four nitrogenous bases using this alternative system? (=O is oxy, and —$CH_3$ is methyl.)

13. Draw the chemical structure of a dinucleotide composed of A and G. Opposite this structure, draw the dinucleotide TC in an antiparallel (or upside down) fashion. Form the possible hydrogen bonds.

14. Describe the various characteristics of the Watson–Crick double-helix model for DNA.

15. What evidence did Watson and Crick have at their disposal in 1953? What was their approach in arriving at the structure of DNA?

16. Had Chargaff's data from a single source indicated the following, what might Watson and Crick have concluded?

|   | A | T | C | G |
|---|---|---|---|---|
| % | 29 | 19 | 21 | 31 |

Why would this conclusion be contradictory to Wilkins and Franklin's data?

17. How do covalent bonds differ from hydrogen bonds? Define base complementarity.

18. List three main differences between DNA and RNA.

19. What are the three types of RNA molecules? How is each related to the concept of information flow?

20. What component of the nucleotide is responsible for the absorption of ultraviolet light? How is this technique important in the analysis of nucleic acids?

21. Distinguish between sedimentation velocity and sedimentation equilibrium centrifugation.

22. What is the basis for determining base composition using density gradient centrifugation?

23. What is the physical state of DNA following denaturation?

24. Compare the following curves representing reassociation kinetics. What can be said about the DNAs represented by each set of data compared with *E. coli*?

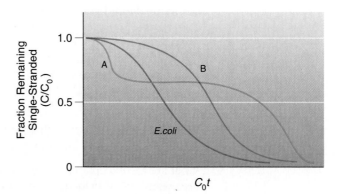

**25.** What is the hyperchromic effect? How is it measured? What does $T_m$ imply?

**26.** Why is $T_m$ related to base composition?

**27.** What is the chemical basis of molecular hybridization?

**28.** What did the Watson–Crick model suggest about the replication of DNA?

**29.** A genetics student was asked to draw the chemical structure of an adenine- and thymine-containing dinucleotide derived from DNA. His answer is shown below. The student made more than six major errors. One of them is circled, numbered ①, and explained. Find five others. Circle them, number them ②–⑥, and briefly explain each, following the example below.

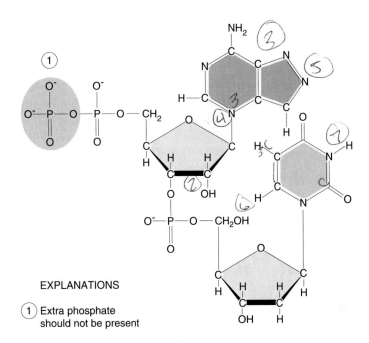

EXPLANATIONS

① Extra phosphate should not be present

**30.** The DNA of the bacterial virus T4 produces a $C_0t_{1/2}$ of about 0.5 and contains $10^5$ nucleotide pairs in its genome. How many nucleotide pairs are present in the genome of the virus MS2 and the bacterium *E. coli*, whose respective DNAs produce $C_0t_{1/2}$ values of 0.001 and 10.0?

**31.** A primitive eukaryote was discovered that displayed a unique nucleic acid as its genetic material. Analysis revealed the following observations:
   (i)   X-ray diffraction studies display a general pattern similar to DNA, but with somewhat different dimensions and more irregularity.
   (ii)  A major hyperchromic shift is evident upon heating and monitoring UV absorption at 260 nm.
   (iii) Base composition analysis reveals four bases in the following proportions.

$$
\begin{aligned}
\text{Adenine} &= \phantom{0}8\% \\
\text{Guanine} &= 37\% \\
\text{Xanthine} &= 37\% \\
\text{Hypoxanthine} &= 18\%
\end{aligned}
$$

(iv) About 75 percent of the sugars are d-ribose while 25 percent are ribose. Attempt to solve the structure of this molecule by postulating a model that is consistent with the above observations.

**32.** Considering the information in this chapter on B- and Z-DNA and right- and left-handed helices, carefully analyze the structures below and draw conclusions about the helical nature of areas (a) and (b). Which is right-handed and which is left-handed?

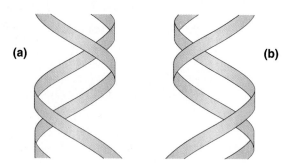

**33.** *Newsdate*: *March 1, 2005*. A unique creature has been discovered during exploration of outer space. Recently, its genetic material has been isolated and analyzed. This material is similar in some ways to DNA in its chemical makeup. It contains in abundance the 4-carbon sugar erythrose and a molar equivalent of phosphate groups. Additionally, it contains six nitrogenous bases: adenine (A), guanine (G), thymine (T), cytosine (C), hypoxanthine (H), and xanthine (X). These bases exist in the following relative proportions:

$$A = T = H \quad \text{and} \quad C = G = X$$

X-ray diffraction studies have established a regularity to the molecule and a constant diameter of about 30 Å.

Together, these data have suggested a model for the structure of this molecule.

(a) Propose a general model of this molecule. Describe it briefly.

(b) What base-pairing properties must exist for H and for X in the model?

(c) Given the contrast diameter of 30 Å, do you think that H and X are *either* (i) both purines or both pyrimidines, *or* (ii) one is a purine and one is a pyrimidine?

**SELECTED READINGS**

ALLOWAY, J. L. 1933. Further observations on the use of pneumococcus extracts in effecting transformation of type *in vitro*. *J. Exp. Med.* 57:265–78.

AVERY, O. T., MacLEOD, C. M., and McCARTY, M. 1944. Studies on the chemical nature of the substance inducing transformation of pneumococcal types. Induction of transformation by a desoxyribonucleic acid fraction isolated from pneumococcus type III. *J. Exp. Med.* 79: 137–58. (Reprinted in Taylor, J. H. 1965. *Selected papers in molecular genetics*. Orlando: Academic Press.)

BRITTEN, R. J., and KOHNE, D. E. 1968. Repeated sequences in DNA. *Science* 161:529–40.

CAIRNS, J., STENT, G. S., and WATSON, J. D. 1966. *Phage and the origins of molecular biology.* Cold Spring Harbor, NY: Cold Spring Harbor Laboratory.

CHARGAFF, E. 1950. Chemical specificity of nucleic acids and mechanism for their enzymatic degradation. *Experientia* 6:201–9.

CRICK, F. H. C., WANG, J. C., and BAUER, W. R. 1979. Is DNA really a double helix? *J. Mol. Biol.* 129:449–61.

DAVIDSON, J. N. 1976. *The biochemistry of the nucleic acids.* 8th ed. Orlando: Academic Press.

DAWSON, M. H. 1930. The transformation of pneumococcal types. I. The interconvertibility of type-specific S pneumococci. *J. Exp. Med.* 51:123–47.

DeROBERTIS, E. M., and GURDON, J. B. 1979. Gene transplantation and the analysis of development. *Scient. Amer.* (Dec.) 241:74–82.

DICKERSON, R. E. 1983. The DNA helix and how it is read. *Scient. Amer.* (June) 249:94–111.

DICKERSON, R. E. et al. 1982. The anatomy of A-, B-, and Z-DNA. *Science* 216:475–85.

DUBOS, R. J. 1976. *The professor, the Institute and DNA: Osward T. Avery, his life and scientific achievements.* New York: Rockefeller University Press.

FELSENFELD, G. 1985. DNA. *Scient. Amer.* (Oct.) 253:58–78.

FRAENKEL-CONRAT, H., and SINGER, B. 1957. Virus reconstruction. II. Combination of protein and nucleic acid from different strains. *Biochem. Biophys. Acta* 24:530–48. (Reprinted in Taylor, J. H. 1965. *Selected papers in molecular genetics.* Orlando: Academic Press.)

FRANKLIN, R. E., and GOSLING, R. G. 1953. Molecular configuration in sodium thymonucleate. *Nature* 171:740–41.

GEIS, I. 1983. Visualizing the anatomy of A, B and Z-DNAs. *J. Biomol. Struc. Dynam.* 1: 581–91.

GRIFFITH, F. 1928. The significance of pneumococcal types. *J. Hyg.* 27:113–59.

GUTHRIE, G. D., and SINSHEIMER, R. L. 1960. Infection of protoplasts of *Escherichia coli* by subviral particles. *J. Mol. Biol.* 2:297–305.

HERSHEY, A. D., and CHASE, M. 1952. Independent functions of viral protein and nucleic acid in growth of bacteriophage. *J. Gen. Phys.* 36:39–56. (Reprinted in Taylor, J. H. 1965. *Selected papers in molecular genetics.* Orlando: Academic Press.)

HOTCHKISS, R. D. 1951. Transfer of penicillin resistance in pneumococci by the desoxyribonucleate derived from resistant cultures. *Cold Spr. Harb. Symp.* 16:457–61. (Reprinted in Adelberg, E. A. 1960. *Papers on bacterial genetics.* Boston: Little, Brown.)

JUDSON, H. 1979. *The eighth day of creation: Makers of the revolution in biology.* New York: Simon & Schuster.

LEVENE, P. A., and SIMMS, H. S. 1926. Nucleic acid structure as determined by electrometric titration data. *J. Biol. Chem.* 70:327–41.

McCARTY, M. 1980. Reminiscences of the early days of transformation. *Ann. Rev. Genet.* 14: 1–16.

———. 1985. *The transforming principle: Discovering that genes are made of DNA.* New York: W. W. Norton.

McCONKEY, E. H. 1993. *Human genetics—The molecular revolution.* Boston: Jones and Bartlett.

OLBY, R. 1974. *The path to the double helix.* Seattle: University of Washington Press.

PALMITER, R. D., and BRINSTER, R. L. 1985. Transgenic mice. *Cell* 41:343–45.

PAULING, L., and COREY, R. B. 1953. A proposed structure for the nucleic acids. *Proc. Natl. Acad. Sci.* 39:84–97.

RICH, A., NORDHEIM, A., and WANG, A. H.-J. 1984. The chemistry and biology of left-handed Z-DNA. *Ann. Rev. Biochem.* 53:791–846.

SCHILDKRAUT, C. L., MARMUR, J., and DOTY, P. 1962. Determination of the base composition of deoxyribonucleic acid from its buoyant density in CsCl. *J. Mol. Biol.* 4:430–43.

SPIZIZEN, J. 1957. Infection of protoplasts by disrupted T2 viruses. *Proc. Natl. Acad. Sci.* 43: 694–701.

STENT, G. S., ed. 1981. *The double helix: Text, commentary, review, and original papers.* New York: W. W. Norton.

STEWART, T. A., WAGNER, E. F., and MINTZ, B. 1982. Human $\beta$-globin gene sequences injected into mouse eggs, retained in adults, and transmitted to progeny. *Science* 217:1046–48.

VARMUS, H. 1988. Retroviruses. *Science* 240:1427–35.

WATSON, J. D. 1968. *The double helix.* New York: Atheneum.

WATSON, J. D., and CRICK, F. C. 1953a. Molecular structure of nucleic acids. A structure for deoxyribose nucleic acids. *Nature* 171:737–38.

———. 1953b. Genetic implications of the structure of deoxyribose nucleic acid. *Nature* 171:964.

WEINBERG, R. A. 1985. The molecules of life. *Scient. Amer.* (Oct.) 253:48–57.

WILKINS, M. H. F., STOKES, A. R., and WILSON, H. R. 1953. Molecular structure of desoxypentose nucleic acids. *Nature* 171:738–40.

ZIMMERMAN, B. 1982. The three-dimensional structure of DNA. *Ann. Rev. Biochem.* 51:395–428.

# 9

# DNA: REPLICATION, SYNTHESIS, AND RECOMBINATION

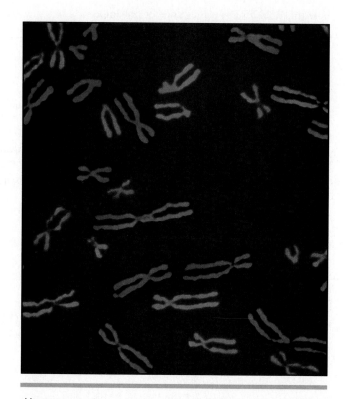

Human metaphase chromosomes that clearly demonstrate their double nature resulting from the earlier DNA replication.

*Genetic continuity between parental and progeny cells is made possible by semiconservative replication of DNA, as predicted by the Watson–Crick model. Each strand of the parent helix serves as a template for the production of its complement. Synthesis of DNA is a complex but orderly process, orchestrated by a myriad of enzymes and other molecules. Together, they function with great fidelity to polymerize nucleotides into polynucleotide chains. Genetic recombination is also a process that involves the interaction of DNA molecules with a group of enzymes.*

Following Watson and Crick's proposal for the structure of DNA, scientists focused their attention on how this molecule is replicated. This process is an essential function of the genetic material and must be executed precisely if genetic continuity between cells is to be maintained following cell division. This is an enormous, complex task. Consider for a second that some three billion base pairs exist within the 23 chromosomes of the human genome. To faithfully duplicate the DNA of just one of these chromosomes requires a mechanism of extreme precision. Even an error rate of $10^{-6}$ (one in a million) will create an excessive number of errors (3000) during each replication cycle of the genome. While not error free, an extremely accurate system of DNA replication has evolved in all organisms.

As Watson and Crick suggested in 1953, the model of the double helix gave them the initial insight into how replication could occur. This mode, called semiconservative replication, has since received strong experimental support from studies of viruses, prokaryotes, and eukaryotes.

Once the general mode of replication was made clear, research was intensified to determine the precise details of DNA synthesis. What has since been discovered is that numerous enzymes and other proteins are needed to copy a DNA helix. Because of the complexity of the chemical events during synthesis, this subject remains an extremely active area of research.

In this chapter, we will discuss the general mode of replication as well as the specific details of the synthesis of DNA. The research leading to this knowledge is still another link in our understanding of life processes at the molecular level.

## THE MODE OF DNA REPLICATION

It was apparent to Watson and Crick that because of the arrangement and nature of the nitrogenous bases, each strand of a DNA double helix could serve as a template for the synthesis of its complement (Figure 9.1).

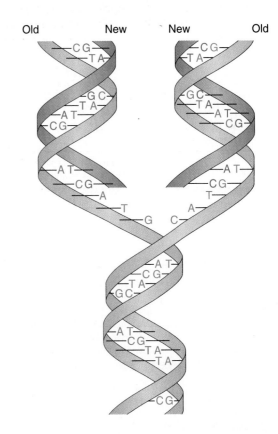

**FIGURE 9.1**   General model of semiconservative replication of DNA.

They proposed that if the helix were unwound, each nucleotide along the two parent strands would have an affinity for its complementary nucleotide. As we learned in Chapter 8, the complementarity is due to the potential hydrogen bonds that can be formed. If thymidylic acid (T) were present, it would "attract" adenylic acid (A); if guanidylic acid (G) were present, it would "attract" cytidylic acid (C); and so on. If these nucleotides were then covalently linked into polynucleotide chains along both templates, the result would be the production of two identical double strands of DNA. Each replicated DNA molecule would consist of one "old" and one "new" strand; hence the reason for the name **semiconservative replication**.

There are two other possible modes of replication (Figure 9.2) that also rely on the parental strands as a template. In **conservative replication**, synthesis of complementary polynucleotide chains would occur as described above. Following synthesis, however, the two newly created strands are brought back together, and the parental strands reassociate. The original helix is thus "conserved."

In the second alternative mode, called **dispersive replication**, the parental strands are seen to be dispersed into two new double helices following replication. Thus, each strand consists of both old and new DNA. This mode would involve cleavage of the parental strands during replication. Therefore, it is the most complex of the three possibilities. It could not, however, be ruled out as an experimental model. The theoretical results of two generations of replication by the conservative and dispersive modes are compared with those of the semiconservative mode in Figure 9.2.

## The Meselson–Stahl Experiment

In 1958, Matthew Meselson and Franklin Stahl published the results of an experiment providing strong evidence that semiconservative replication is the mode used by cells to produce new DNA molecules. *E. coli* cells were grown for many generations in a medium where $^{15}NH_4Cl$ (ammonium chloride) was the only nitrogen source. A "heavy" isotope of nitrogen, $^{15}N$ contains

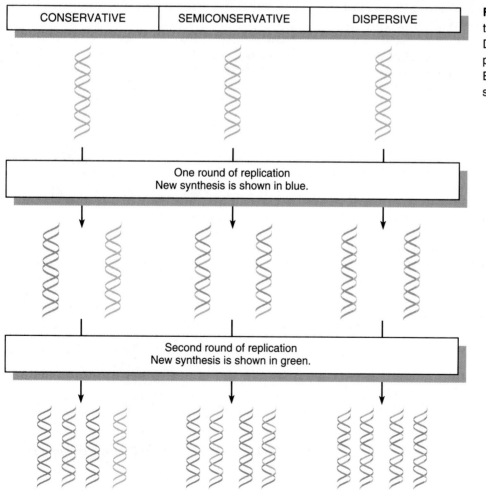

| CONSERVATIVE | SEMICONSERVATIVE | DISPERSIVE |
| --- | --- | --- |

One round of replication
New synthesis is shown in blue.

Second round of replication
New synthesis is shown in green.

**FIGURE 9.2**    The results of two rounds of replication of DNA for each of the three possible modes of replication. Each round of synthesis is shown in a different color.

one more neutron than the naturally occurring $^{14}N$ isotope. Unlike "radioactive" isotopes, $^{15}N$ is stable and is thus not radioactive. After many generations, all nitrogen-containing molecules, including the nitrogenous bases of DNA, contained the isotope in the *E. coli* cells. DNA containing $^{15}N$ may be distinguished from $^{14}N$-containing DNA by the use of **sedimentation equilibrium centrifugation**, where samples are "forced" by centrifugation through a density gradient of a heavy metal salt such as cesium chloride (see Chapter 8). The more dense $^{15}N$-DNA will reach equilibrium in the gradient at a point closer to the bottom (where the density is greater) than will $^{14}N$-DNA.

In this experiment, uniformly labeled $^{15}N$ cells were transferred to a medium containing only $^{14}NH_4Cl$. Thus, all subsequent synthesis of DNA during replication contained the "lighter" isotope of nitrogen. The time of transfer was taken as time zero ($t = 0$). The *E. coli* cells were allowed to replicate during several generations with cell samples removed at various intervals. From each sample, DNA was isolated and subjected to sedimentation equilibrium centrifugation. The results are depicted in Figure 9.3.

After one generation, the isolated DNA was all present in a single band of intermediate density—the expected result for semiconservative replication. Each replicated

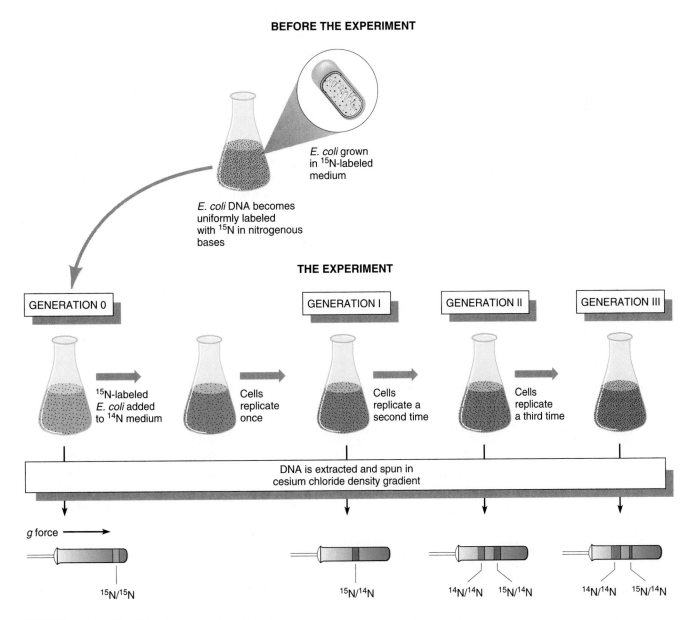

**FIGURE 9.3**    The Meselson–Stahl experiment.

molecule would be composed of one new $^{14}$N-strand and one old $^{15}$N-strand, as seen in Figure 9.4. This result was not consistent with the conservative replication mode, in which two distinct bands would have been predicted to occur.

After two cell divisions, DNA samples showed two density bands: one was intermediate and the other was lighter, corresponding to the $^{14}$N position in the gradient. Similar results occurred after a third generation, except that the proportion of the $^{14}$N-band increased. This was again consistent with the interpretation that replication is semiconservative.

Two observations served as the basis for ruling out the dispersive mode of replication. One involved analysis of DNA after the first generation of replication in an $^{14}$N-containing medium. The observation of a molecule exhibiting intermediate density is also consistent with dispersive replication. However, Meselson and Stahl isolated this hybrid molecule and analyzed it after it had been heat-denatured. Recall from Chapter 8 that heating will separate a duplex into single strands. When the densities of the single strands of the hybrid were determined, they exhibited *either* an $^{15}$N-profile *or* an $^{14}$N-profile, but *not* an intermediate density. This observation is consistent with the semiconservative mode but inconsistent with the dispersive mode.

Furthermore, if replication were dispersive, *all* generations after $t = 0$ would demonstrate DNA of an intermediate density. In each subsequent generation after the first, the ratio of $^{15}$N/$^{14}$N would decrease and the hybrid band would become lighter and lighter, eventually approaching the $^{14}$N-band. This result was not observed. Thus the Meselson–Stahl experiment provided very conclusive support in bacteria for semiconservative replication and tended to rule out the conservative and dispersive mode.

## Semiconservative Replication in Eukaryotes

In 1957, the year before the work of Meselson and his colleagues was published, evidence was also presented that supported semiconservative replication in a eukaryotic organism. J. Herbert Taylor, Philip Woods, and Walter Hughes experimented with root tips of the broad bean *Vicia faba*. Root tips are an excellent source of dividing cells. These researchers examined the chromosomes of these cells following replication of DNA. They were able to monitor the process of replication by labeling DNA with $^3$H-thymidine, a radioactive precursor of DNA, and performing autoradiography. In this experiment, labeled thymidine is found only in association with chromosomes that contain newly synthesized DNA.

The technique of autoradiography, discussed in detail in Appendix A, is a cytological procedure that allows the isotope to be localized within the cell. In this procedure, a photographic emulsion is placed over a section of cellular material (root tips in this experiment), and the preparation is stored in the dark. The slide is then developed, much as photographic film is processed. Since the radioisotope emits energy, the emulsion turns black at the approximate point of emission following development. The end result is the presence of dark spots or "grains" on the surface of the section, localizing newly synthesized DNA within the cell.

Root tips were grown for approximately one generation in the presence of the radioisotope and then placed in unlabeled medium, where cell division continued. At the conclusion of each generation, cultures were arrested at metaphase by the addition of colchicine (a chemical that "poisons" the spindle fibers), and chromosomes were examined by autoradiography. Figure 9.5 illustrates replication of a single chromosome over two division cycles as well as the distribution of grains. In this

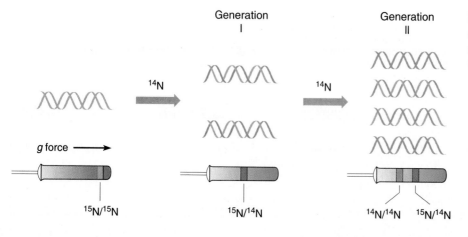

SEMICONSERVATIVE DNA REPLICATION

**FIGURE 9.4**    The expected results of two generations of semiconservative replication in the Meselson–Stahl experiment.

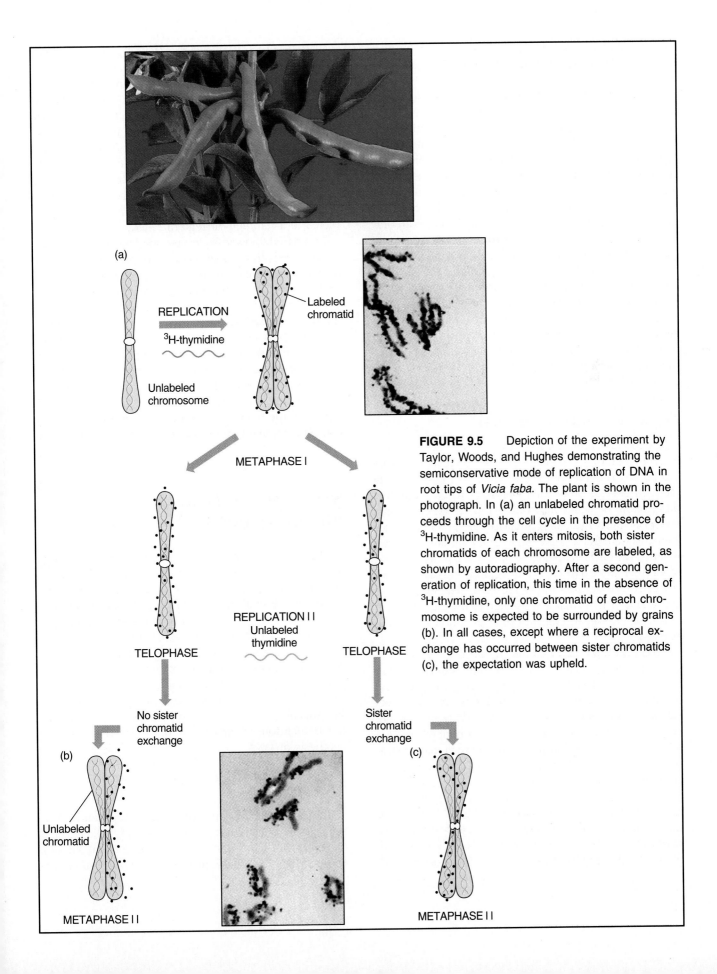

**FIGURE 9.5** Depiction of the experiment by Taylor, Woods, and Hughes demonstrating the semiconservative mode of replication of DNA in root tips of *Vicia faba*. The plant is shown in the photograph. In (a) an unlabeled chromatid proceeds through the cell cycle in the presence of $^3$H-thymidine. As it enters mitosis, both sister chromatids of each chromosome are labeled, as shown by autoradiography. After a second generation of replication, this time in the absence of $^3$H-thymidine, only one chromatid of each chromosome is expected to be surrounded by grains (b). In all cases, except where a reciprocal exchange has occurred between sister chromatids (c), the expectation was upheld.

experiment, labeled thymidine is found only in association with chromosomes that contain newly synthesized DNA.

The results are compatible with the semiconservative mode of replication. After the first replication cycle, radioactivity is detected over both sister chromatids. This finding is expected because each chromatid will contain one "new" radioactive DNA strand and one "old" unlabeled strand. After the second replication cycle, which also takes place in unlabeled medium, only one of the two new sister chromatids should be radioactive because half of the parent strands are unlabeled. With only minor exceptions of **sister chromatid exchanges** (see Chapter 5), this result was observed.

Together, the Meselson–Stahl experiment and the experiment by Taylor, Woods, and Hughes soon led to the general acceptance of the semiconservative mode of replication. The same conclusion has been reached in studies with other organisms. Since this mode is suggested by the double-helix model of DNA, these experiments also strongly supported Watson and Crick's proposal for DNA structure.

## Replication Origins and Forks

The mode of replication just established represents the general pattern by which DNA is duplicated. Before turning to the details of the biosynthesis of DNA, we will mention several other topics relevant to the complete description of semiconservative replication. The first concerns the **origin of replication** along any chromosome. Is there but a single origin or more than one point where DNA synthesis begins? And, is each point of origin random or located in a specific region along the chromosome? As we address these questions, we shall define the length of DNA that is replicated following the initiation of synthesis at a single origin as a unit called the **replicon**.

A second topic involves the direction of replication. Once it begins, does it move in a single direction or in both directions away from the origin? This consideration distinguishes between **unidirectional** and **bidirectional replication**, respectively. At each point of replication, the strands of the helix must unwind, creating what is called a **replication fork**. Bidirectional replication creates two such forks, which move apart in opposite directions away from the origin.

The evidence is reasonably clear regarding these topics. In bacteria and most bacterial viruses, which have only a single circular chromosome, there is one specific region where replication is initiated. In *E. coli* this region, called *oriC*, has been located. It consists of 245

base pairs, but only a small number are actually essential to the initiation of DNA synthesis. Since there is but a single point of origin of DNA synthesis in bacteriophages and bacteria, the entire chromosome constitutes one replicon. In *E. coli*, replication is bidirectional from *oriC* and proceeds until the entire chromosome is replicated, as illustrated in Figure 9.6. Both strands of the parent helix are copied at the two replication forks that are established. These forks move away from the origin in opposite directions and eventually merge as semiconservative replication of the entire chromosome is completed at a termination region, called *ter*.

Compared to bacteria, as described above, a major difference in replication exists in eukaryotes. While replication is bidirectional, creating two replication forks during each replication cycle, there are multiple origins along each chromosome. As a result, during the S phase of interphase, there are numerous replicating events occurring along each chromosome. Eventually, the numerous replication forks merge, completing replication of the entire chromosome. The presence of multiple replicons is undoubtedly related to the much greater length and complexity of a single eukaryotic chromosome compared to one from a bacterium. As a result, replication can be completed in a more reasonable period of time.

# SYNTHESIS OF DNA IN MICROORGANISMS

The determination that replication is semiconservative and bidirectional indicates only the pattern of DNA duplication and the association of finished strands with one another once synthesis is completed. A much more complex issue is how the *actual synthesis* of long complementary polynucleotide chains occurs on a DNA template. As in most studies of molecular biology, this question was first approached by using microorganisms. Research began about the same time as the Meselson–Stahl work, and even today this topic is an active area of investigation. What is most apparent in this research is the tremendous chemical complexity of the biological synthesis of DNA.

## DNA Polymerase I

Studies of the enzymology of DNA replication were first reported by Arthur Kornberg and colleagues in 1957. They isolated an enzyme from *E. coli* that was able to direct DNA synthesis in a cell-free (*in vitro*) system.

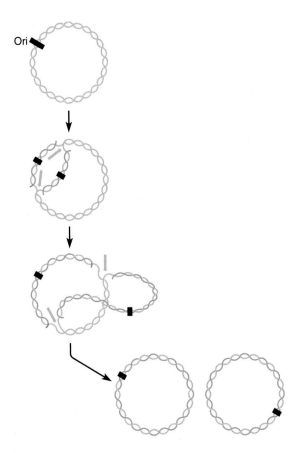

The enzyme is now called **DNA polymerase I**, since it was the first of several to be isolated. Kornberg determined that there were two major requirements for *in vitro* DNA synthesis under the direction of the enzyme:

1.  All four deoxyribonucleoside triphosphates (dATP, dCTP, dGTP, dTTP = dNTP)*

2.  Template DNA

If any one of the four deoxyribonucleoside triphosphates was omitted from the reaction, no measurable synthesis occurred. If derivatives of these precursor molecules other than the nucleoside triphosphate were used (nucleotides or nucleoside diphosphates), synthesis did not occur. If no template DNA was added, synthesis of DNA occurred but was reduced greatly. Thus, most synthesis directed by Kornberg's enzyme appeared to be exactly the type required for semiconservative replication. The reaction is summarized in Figure 9.7, where the addition of a single nucleotide is depicted. The enzyme has since been shown to consist of a single polypeptide containing 928 amino acids.

The way in which each nucleotide is added to the growing chain is a function of the specificity of DNA

**FIGURE 9.6**    Bidirectional replication of the *E. coli* chromosome.

---

*dNTP designates the deoxyribose forms of the four nucleoside triphosphates; in a similar way, dNMP refers to the monophosphate forms.

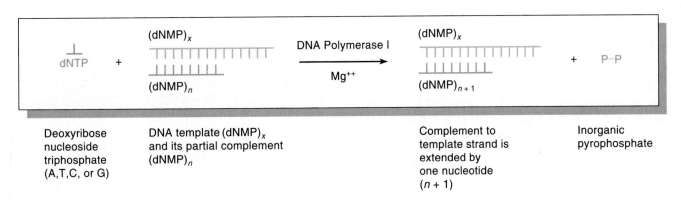

**FIGURE 9.7**    The chemical reaction catalyzed by DNA polymerase I. During each step, a single nucleotide is added to the growing complement of the DNA template, using a nucleoside triphosphate as the substrate. The release of inorganic pyrophosphate drives the reaction energetically.

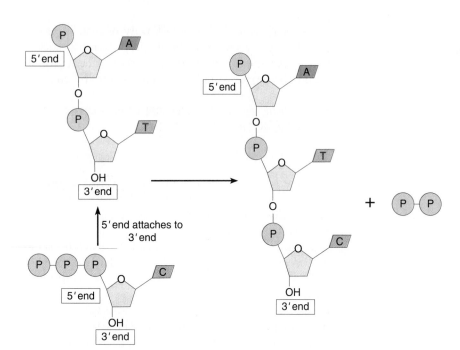

**FIGURE 9.8**    Demonstration of 5'-to-3' synthesis of DNA.

polymerase I. As shown in Figure 9.8, the precursor dNTP contains the three phosphate groups attached to the 5'-carbon of d-ribose. As the two terminal phosphates are cleaved during synthesis, the remaining phosphate attached to the 5'-carbon is covalently linked to the 3'-OH group of the d-ribose to which it is added. Each step *elongates* the growing chain by one nucleotide. Consistent with the above description and Figure 9.8, **chain elongation** occurs in the **5'-to-3' direction** by the addition of nucleotides to the growing 3' end. Each step creates a newly exposed 3'-OH group that can participate in the next addition of a nucleotide as DNA synthesis proceeds.

## Fidelity of Synthesis

Having shown how DNA was synthesized, Kornberg sought to demonstrate the accuracy, or fidelity, with which the enzyme had replicated the DNA template. Since the nucleotide sequences of the template and the product could not be determined in 1957, he had to rely initially on several indirect methods.

One of Kornberg's approaches was to compare the nitrogenous base compositions of the DNA template with those of the recovered DNA product. Table 9.1 shows Kornberg's base composition analysis of three different DNA templates. These may be compared with the DNA product synthesized in each case. Within experimental error, the base composition of each product agreed with the template DNAs used. These data, along with other types of comparisons of template and prod-

uct, suggested that the templates were replicated faithfully.

Kornberg also used the **nearest-neighbor frequency test** to test the fidelity of copying (Figure 9.9). With this technique, he determined the frequency with which any two bases occur adjacent to each other along the polynucleotide chain. This test relies on the enzyme **spleen phosphodiesterase**, which cleaves the polynucleotide chain in a way that is different from the way in which it was assembled. As we have pointed out (see Figure 9.8), during synthesis of DNA, 5'-nucleotides are inserted; that is, each nucleotide is added with the phosphate on the C-5' of deoxyribose. However, the phos-

**Table 9.1**   BASE COMPOSITION OF THE DNA TEMPLATE AND THE PRODUCT OF REPLICATION IN KORNBERG'S EARLY WORK

| Organism | Template or Product | %A | %T | %G | %C |
|---|---|---|---|---|---|
| T2 | Template | 32.7 | 33.0 | 16.8 | 17.5 |
|    | Product | 33.2 | 32.1 | 17.2 | 17.5 |
| E. coli | Template | 25.0 | 24.3 | 24.5 | 26.2 |
|    | Product | 26.1 | 25.1 | 24.3 | 24.5 |
| Calf | Template | 28.9 | 26.7 | 22.8 | 21.6 |
|    | Product | 28.7 | 27.7 | 21.8 | 21.8 |

SOURCE: From Kornberg, 1960. Copyright 1960 by the American Association for the Advancement of Science.

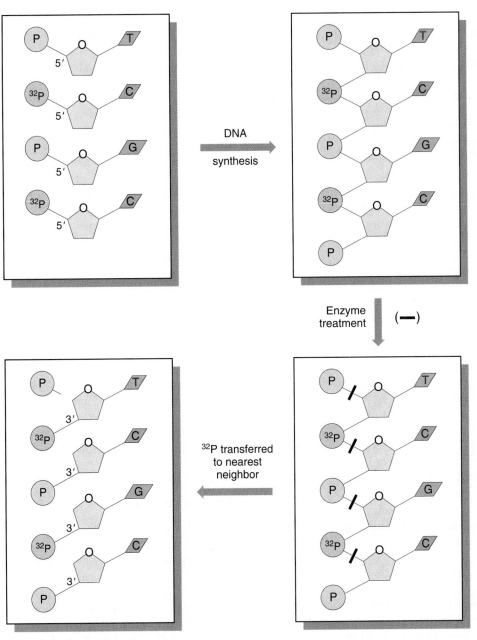

**FIGURE 9.9**    The theory of
nearest-neighbor analysis. Initially, one of the four nucleotides (C) contains $^{32}$P in the
5′-phosphate group. Following
synthesis of a polynucleotide
chain and enzymatic treatment
with phosphodiesterase, the
radioactive phosphate groups
are transferred to the nearest
neighbors (G and T). Subsequent analysis reveals the percentage of time each of the
four nucleotides (A, T, C, and
G) have become radioactive.

phodiesterase enzyme cleaves between the phosphate
and the C-5′ atom, thereby producing 3′-nucleotides. If
the phosphates on only one of the four nucleotides (cytidylic acid, for example) are radioactive during DNA synthesis, then after enzymatic cleavage a radioactive phosphate will be transferred to the base that is the "nearest
neighbor" on the 5′ side of all cytidylic acid nucleotides.
Following four separate experiments, where in each
case only one of the four nucleotide types is made radioactive, the frequency of all sixteen possible nearest
neighbors can be calculated. An example of such data is
presented in one of the latter problems at the end of this
chapter.

   When this technique was applied to the DNA template

and the resultant product from a variety of experiments,
Kornberg found general agreement between the nearest-
neighbor frequencies of the two. This type of analysis is
a more stringent measure of the fidelity of copying than
the base composition analysis. Thus, DNA polymerase I
seemed the likely candidate for the synthesis of DNA
during replication within the cell, even though Kornberg's experiments were, by necessity, performed *in
vitro.*

## Synthesis of Biologically Active DNA

   Despite Kornberg's extensive work, not all researchers
were convinced that DNA polymerase I was the enzyme

that replicates DNA within cells (*in vivo*). The primary reservations involved observations that the *in vitro* rate of synthesis was much slower than expected *in vivo*, that the enzyme was much more effective replicating single-stranded DNA than double-stranded DNA, and that the enzyme appeared to be able to *degrade* DNA as well as to *synthesize* it.

Faced with the uncertainty of the true cellular function of DNA polymerase I, Kornberg pursued another approach. He reasoned that if the enzyme could be used to synthesize **biologically active DNA** *in vitro*, then DNA polymerase I must be the major catalyzing force for DNA synthesis within the cell. The term *biological activity* means that the DNA synthesized is capable of supporting metabolic activities and directing reproduction of the organism from which it was originally duplicated.

In 1967, Mehran Goulian, Kornberg, and Robert Sinsheimer showed that the DNA of the small bacteriophage $\phi$X174 could be completely copied by DNA polymerase I *in vitro*, and that the new product could be isolated and used to successfully transfect *E. coli* protoplasts. This process produced mature phages from the synthetic DNA, thus demonstrating biological activity! The ingenious experimental design is outlined in Figure 9.10. Recall that in Chapter 8 we introduced the process of transfection, where viral DNA infects bacterial protoplasts.

The phage $\phi$X174 provided an ideal experimental system because it contains a very small (5386 nucleotides), circular, single-stranded DNA molecule as its genetic material. Since the molecule is a closed circle, the experiment depended on the isolation of a second enzyme, **DNA ligase** (also called the **polynucleotide joining enzyme**), which joins the two ends of the linear molecule following replication.

In the normal course of $\phi$X174 infection, the circular, single-stranded DNA referred to as the (+) **strand** enters the *E. coli* cell and serves as a template for the synthesis of the complementary (−) **strand**. The two strands (+ and −) remain together in a circular double helix called the **replicative form** (**RF**). The RF serves as the template for its own replication, and subsequently only (+) strands are produced. These strands are then packaged into viral coat proteins to form mature virus particles.

As shown in Figure 9.10, the experiment was carefully designed so that each newly synthesized strand could be distinguished and isolated from the template strand. Initially, the (+) strand was labeled with tritium ($^3$H). During synthesis, two "tags" were used to identify the new (−) strands. First, $^{32}$P was present in the precursor nucleotides. Second, the base analogue 5-bromouracil was used in place of thymine. In the chemical structure of this analogue, a bromine atom is substituted for a carbon atom in the methyl group at the C-5 position of the py-

rimidine ring, increasing the mass and therefore the density. As a result, newly synthesized (−) strands were radioactive and because their mass was greater, they could be isolated from the (+) strands using sedimentation equilibrium centrifugation.

Once the duplex was enzymatically "nicked" to open one strand, the BU-containing strands were isolated and represented newly synthesized DNA. The process was then repeated in the absence of any tags, using these strands as templates. The subsequent strands were then detectable based on being "lighter" and nonradioactive. Eventually, newly synthesized infectious (+) strands were isolated. The protocol of the experiment dictates that these (+) strands must have been synthesized *in vitro* under the direction of Kornberg's enzyme.

The critical test of biological activity was performed using the process of transfection. Newly synthesized (+) strands were added to bacterial protoplasts (bacterial cells minus their cell wall). Following infection by the synthetic DNA, mature phages were produced. Therefore, the synthetic DNA had successfully directed reproduction!

This demonstration of biological activity was viewed as a precise assessment of faithful copying. If even a single error had occurred to alter the coding property of even one triplet included in the 5386 nucleotides constituting the $\phi$X174 chromosome, the change might easily have caused a lethal mutation, precluding the production of viable phages.

## DNA Polymerase II and III

Although DNA synthesized under the direction of polymerase I demonstrated biological activity, a more serious reservation about the enzyme's true biological role was raised in 1969. Peter DeLucia and John Cairns reported the discovery of a mutant strain of *E. coli* that was deficient in polymerase I activity. The mutation was designated *polA1*. In the absence of the functional enzyme, this mutant strain of *E. coli* still duplicated its DNA and successfully reproduced! Other properties of the mutation led DeLucia and Cairns to conclude that in the absence of polymerase I, these cells were highly deficient in their ability to "repair" DNA. For example, the mutant strain is highly sensitive to ultraviolet light (UV) and radiation, both of which damage DNA and are mutagenic. Nonmutant bacteria are able to repair a great deal of UV-induced damage.

These observations led to two conclusions:

1. There must be at least one other enzyme present in *E. coli* cells that is capable of replicating DNA *in vivo*.

2. DNA polymerase I may serve only a secondary func-

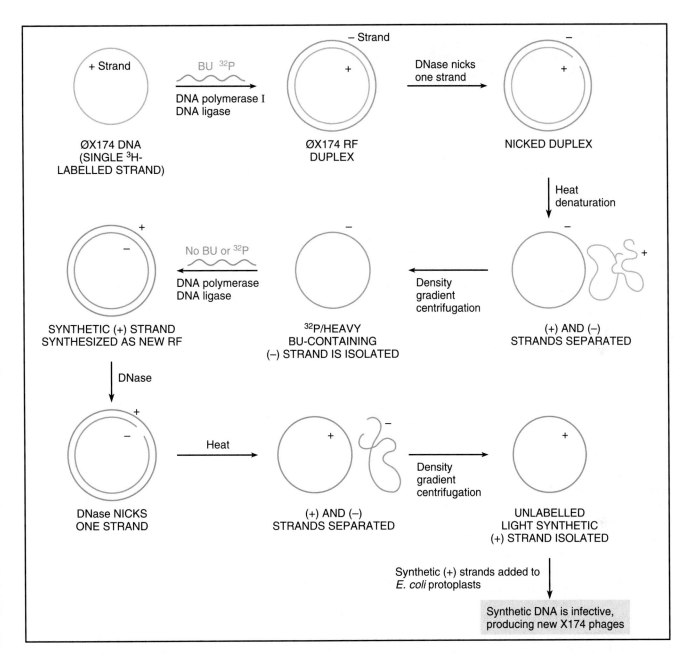

**FIGURE 9.10**   Schematic representation of Goulian, Kornberg, and Sinsheimer's experiment involving *in vitro* replication and isolation of synthetic (+) DNA strands of phage φX174. The DNA synthesized *in vitro* under the direction of DNA polymerase I successfully transfected *E. coli* protoplasts, thus demonstrating its biological activity.

tion *in vivo*. This function is now believed by Kornberg and others to be critical to the *fidelity* of DNA synthesis, but this enzyme is not the one that actually synthesizes the complementary strand.

To date, two unique DNA polymerases have been isolated from cells lacking polymerase I activity. These two enzymes have also been isolated from normal cells that also contain polymerase I.

The characteristics of the two enzymes, called **DNA polymerase II** and **III**, are contrasted with DNA polymerase I in Table 9.2. As is evident from that information, all three share several characteristics. While none can *initiate* DNA synthesis on a template, all can *elongate* an existing DNA strand, called a **primer**. As we shall see, in addition to DNA, RNA is also an adequate primer and is, in fact, what is utilized initially.

The DNA polymerase enzymes are all large, complex

**Table 9.2** COMPARATIVE PROPERTIES OF THE THREE BACTERIAL DNA POLYMERASES

| Properties | I | II | III |
|---|---|---|---|
| Initiation of chain synthesis | − | − | − |
| 5′-to-3′ polymerization | + | + | + |
| 3′-to-5′ exonuclease activity | + | + | + |
| 5′-to-3′ exonuclease activity | + | − | − |
| Molecular weight | 103,000 | 90,000 | 167,500 |
| Molecules of polymerase/cell | 400 | ? | 15 |

proteins exhibiting a molecular weight in excess of 100,000 daltons. All three possess 3′-to-5′ **exonuclease activity**. This means that they can polymerize in one direction and then reverse directions and excise nucleotides just added. As we will see, this activity provides a capacity to proofread and remove an incorrect nucleotide.

For two of the three (I and III), 5′-to-3′ exonuclease activity also is characteristic. This potentially allows the enzyme to excise nucleotides from the end where initial synthesis occurred and in the same direction. Thus, they have the potential ability to remove the primer, which is itself essential for synthesis. We will return to this topic momentarily. The final characteristics probably explain why Kornberg isolated polymerase I and not polymerase III. Polymerase I is present in much greater amounts than polymerase III and is also much more stable.

What then are the roles of the three polymerases *in vivo*? Polymerase III is considered to be the enzyme responsible for the polymerization essential to replication. Its 3′-to-5′ exonuclease activity also allows it to proofread, excise, and then correct base pairs created in error during polymerization. As we will soon see, gaps are a natural occurrence on one of the two strands during replication as RNA primers are removed. It is believed that polymerase I, originally studied by Kornberg, is responsible for removing the primer as well as for the synthesis that fills these gaps. Its exonuclease activity also allows proofreading to occur during this process. Alternatively, an RNAase has been discovered that may remove the primer prior to polymerase I-directed synthesis.

Polymerase II is suspected to be involved in repair synthesis of DNA that has been damaged by external forces such as ultraviolet light. If so, its role is also important to survival.

We end this section by emphasizing the complexity of the DNA polymerase III molecule. Its active form, called a **holoenzyme**, consists of seven separate polypeptide chains. The largest, the $\alpha$ subunit, has a molecular weight of 140,000 daltons and is responsible for the polymerization activity of the holoenzyme. The $\epsilon$ subunit possesses the 3′-to-5′ exonuclease activity. While the function of the other five subunits is not precisely clear, one binds ATP, a step essential to the initial polymerization step. It is also believed that two holoenzymes may function together as a dimer at the replication fork. Together with several other proteins at the replication fork, a complex nearly as large as a ribosome is present. It has been called a **replisome**.

## DNA SYNTHESIS: A MODEL

We have thus far established that replication is semi-conservative and bidirectional along a single replicon in bacteria and many viruses. And, we know that synthesis is in the 5′-to-3′ mode primarily under the direction of DNA polymerase III, creating two replication forks. These move in opposite directions away from the origin of synthesis. We are now ready to pursue several other aspects of DNA synthesis. We will combine this new information with that previously established into a coherent model. It takes into account the following additional points:

1. A mechanism must exist by which the helix undergoes localized unwinding and is stabilized in this "open" configuration so that synthesis may proceed along both strands.

2. As unwinding proceeds, increased coiling creates tension farther down the helix, which must be reduced.

3. A primer of some sort must be synthesized so that polymerization can commence under the direction of DNA polymerase III.

4. When DNA polymerase III commences synthesis of the complement of both strands of the parent molecule, the two strands are antiparallel to one another. As a result, continuous synthesis in the direction that the replication fork moves is possible along only one of the two strands. On the other strand, synthesis is discontinuous in the opposite direction.

5. The primers must be removed prior to completion of replication. The gaps that are temporarily created must be filled with DNA complementary to the template at each location.

6. The newly synthesized DNA strand that fills each temporary gap must be ligated to the adjacent strand of DNA.

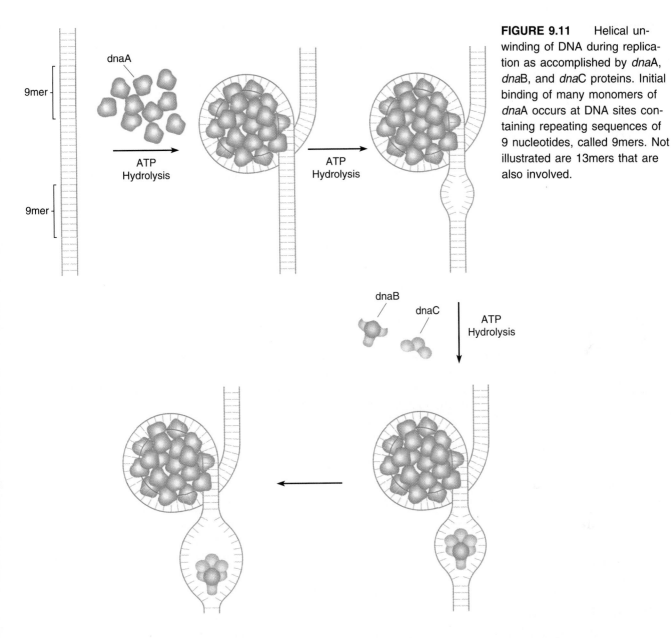

**FIGURE 9.11** Helical unwinding of DNA during replication as accomplished by *dna*A, *dna*B, and *dna*C proteins. Initial binding of many monomers of *dna*A occurs at DNA sites containing repeating sequences of 9 nucleotides, called 9mers. Not illustrated are 13mers that are also involved.

As we consider the above points, three figures (Figures 9.11, 9.12, and 9.13) will be used to illustrate how each issue is resolved.

## Unwinding the DNA Helix

As discussed earlier, there is but a single origin along the circular chromosome of most bacteria and viruses where DNA synthesis is initiated. This region of the *E. coli* chromosome has been particularly well studied. Called *oriC*, the origin of replication consists of 245 base pairs, which are characterized by the presence of repeating sequences of 9 and 13 bases (called **9mers** and **13mers**). One particular protein, called **dnaA** (because it is encoded by the gene called *dnaA*), is responsible for

the initial step in unwinding the helix. A number of subunits of the dnaA protein bind to each of several 9mers. This step is essential in facilitating the subsequent binding of **dnaB** and **dnaC** proteins that further open and destabilize the helix (Figure 9.11). Proteins such as these that require the energy normally supplied by the hydrolysis of ATP in order to break hydrogen bonds and denature the double helix are called **helicases**.

As unwinding proceeds, another group of proteins, called **single-stranded binding proteins (SSBPs)** stabilize this conformation. As unwinding proceeds, a coiling tension is created ahead of the replication fork. Oftentimes, various forms of **supercoiling** occur. These may take on the form of added twists and turns of the DNA in circular molecules, much like the coiling that

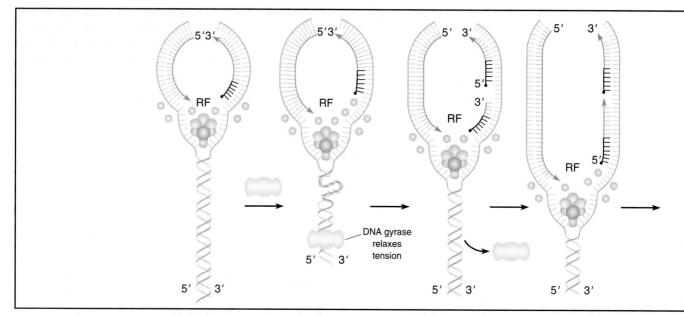

would be created in a rubber band by holding one end and twisting the other. Such supercoiling can be relaxed by the action of an enzyme, **DNA gyrase**, a member of a larger group referred to as **DNA topoisomerases** (Figure 9.12). Depending on the form of the enzyme involved, either single- or double-stranded "cuts" are made by DNA gyrase. The enzyme also catalyzes localized manipulations of DNA strands that have the effect of "undoing" the twists and knots that are created during supercoiling. The strands are then resealed. To drive these various reactions, the energy released during ATP hydrolysis is required.

In combination with the polymerase complex, these proteins comprise an array of molecules that participate in DNA synthesis and are part of what we have previously called the replisome.

### Initiation of Synthesis

Once a small portion of the helix is unwound, initiation of synthesis may occur. As previously mentioned, DNA polymerase III requires a free 3' end as a primer in order to elongate a polynucleotide chain. This prompted researchers to investigate how the first nucleotide can be added, since no free 3'-hydroxyl group is initially present. There is now evidence that RNA is involved as the primer in initiating DNA synthesis.

It is thought that a short segment of RNA, complementary to DNA, is first synthesized on the DNA template. The RNA, about 5 to 15 nucleotides long, is made under the direction of a form of RNA polymerase called **primase**. The RNA polymerase does not require a free 3' end to initiate synthesis. It is to this short segment of RNA that DNA polymerase III begins to add 5'-deoxyribonucleotides (Figure 9.12). After DNA synthesis occurs at an area adjacent to the RNA primer, the RNA segment is

clipped off and replaced with DNA. Both steps are thought to be performed by DNA polymerase I. RNA priming has been recognized in viruses, bacteria, and several eukaryotic organisms and is thought to be a universal phenomenon.

### Continuous and Discontinuous DNA Synthesis

We must now reconsider the fact that the two strands of a double helix are antiparallel to each other. One runs in the 5'-to-3' direction, while the other has the opposite 3'-to-5' polarity. Since DNA polymerase III synthesizes DNA in only the 5'-to-3' direction, simultaneous synthesis of antiparallel strands along an advancing replication fork must occur in one direction along one strand and in the opposite direction on the other.

Thus, as the strands unwind and the replication fork progresses down the helix, only one strand can serve as a template for **continuous synthesis**. This strand is called the **leading strand**. As the fork progresses, many points of initiation are necessary on the opposite, or **lagging strand** resulting in **discontinuous synthesis** (Figure 9.13).

Evidence in support of discontinuous synthesis was first provided by Reiji and Tuneko Okazaki and their colleagues. They discovered that when bacteriophage DNA is replicated in *E. coli*, some of the newly formed DNA is found as small fragments containing 1000 to 2000 nucleotides. RNA primers are part of each such fragment. These pieces, called **Okazaki fragments**, must then be enzymatically joined. As synthesis proceeds, the Okazaki fragments of low molecular weight are indeed converted into longer and longer DNA strands of higher molecular weight.

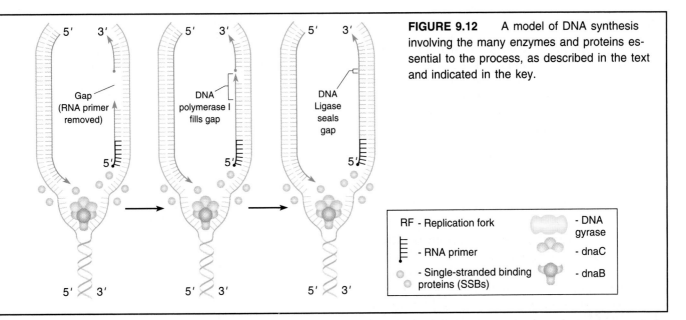

**FIGURE 9.12**    A model of DNA synthesis involving the many enzymes and proteins essential to the process, as described in the text and indicated in the key.

Discontinuous synthesis of DNA requires the removal of the RNA primer as well as an enzyme that can unite the smaller products into the longer continuous molecules that represent the lagging strand. While DNA polymerase I removes the primer and replaces the missing nucleotides, **DNA ligase** has been shown to be capable of catalyzing the formation of the phosphodiester bond, the last step in sealing the gap existing between discontinuously synthesized strands. The evidence that DNA ligase does perform this function during DNA synthesis is strengthened by the observation of a ligase-deficient mutant strain (*lig*) of *E. coli*. In this strain, Okazaki fragments accumulate in particularly large amounts. Apparently, they are not joined adequately.

Discontinuous synthesis on the lagging strand, as described above, is characteristic of bacteria as well as eukaryotic cells.

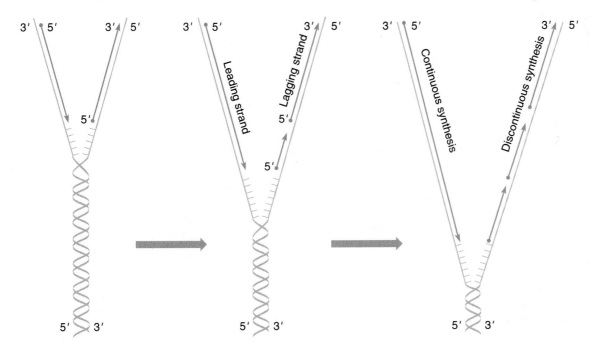

**FIGURE 9.13**    Illustration of the opposite polarity of DNA synthesis along the two strands necessitated by the requirement of 5′-to-3′ synthesis of DNA polymerase III. On the lagging strand, synthesis must be discontinuous, resulting in the production of Okazaki fragments. On the leading strand, synthesis is continuous.

## Proofreading

The underpinning of semiconservative replication is the synthesis of a new strand of DNA that is precisely complementary to the template strand at each nucleotide position. While the action of DNA polymerases is very accurate, synthesis is not perfect. Occasionally, a non-complementary nucleotide is inserted erroneously. To compensate for such inaccuracies, both polymerases I and III possess **3′-to-5′ exonuclease activity**. They are capable of detecting and removing a mismatched nucleotide by pausing, reversing their direction (3′ to 5′) and excising it. Once the mismatched nucleotide is removed, 5′-to-3′ synthesis can again proceed.

This process is called **exonuclease proofreading** or **editing** and serves to increase the fidelity of synthesis. In the case of the holoenzyme form of DNA polymerase III, the epsilon ($\epsilon$) subunit is responsible for enhancing the proofreading step. Strains of *E. coli* have been isolated where a mutation has occurred rendering the $\epsilon$ subunit nonfunctional. The error rate (the mutation rate) during DNA synthesis is increased in these strains as a result.

## Summary of DNA Synthesis

To summarize the process of DNA synthesis in bacteria and bacteriophages, the following steps provide the biochemical basis for semiconservative replication:

1. Synthesis is initiated at a specific origin within each replicon. In *E. coli*, this region is called *oriC*.

2. Unwinding proteins (called helicases) denature the helix at the origin, while other proteins (SSBPs) stabilize the denatured helix.

3. Synthesis is bidirectional, creating two replication forks which move in opposite directions away from the origin.

4. As the replication forks move away from the origin, increased coiling of the helix occurs ahead of the forks. The tension thus created is diminished by the enzymatic action of DNA gyrase. This topoisomerase cuts and reseals DNA strands after uncoiling occurs.

5. Initiation of DNA synthesis involves an RNA primer synthesized under the direction of a unique RNA polymerase called primase. The resultant RNA is complementary to its DNA template.

6. DNA polymerase III polymerizes complementary DNA strands by elongating an existing chain in the 5′-to-3′ direction.

7. As the replication forks move away from the origin, synthesis is continuous on the leading strands, but is discontinuous on the lagging strands, producing short polynucleotides called Okazaki fragments.

8. The RNA primers are removed and the resulting gaps are filled with DNA under the direction of DNA polymerase I.

9. Along the lagging strand, the Okazaki fragments are joined by DNA ligase.

10. As this process proceeds along the length of the replicon, proofreading by DNA polymerases I and III occurs, and semiconservative replication is achieved.

Since the investigation of DNA synthesis is still an extremely active area of research, this model will no doubt be extended in the future. In the meantime, it provides a summary of DNA synthesis against which genetic phenomena may be interpreted.

# THETA ($\Theta$) STRUCTURES AND ROLLING CIRCLES

In this section, we provide further details involving the replication of circular DNA molecules, characteristic of bacteria and viruses. Bidirectional DNA synthesis at a single origin along a *linear* molecule creates two replication forks and results in what is called a **replication eye** or **replication bubble**. When the replicon is a *circular* DNA molecule, the "eye" that is formed creates a characteristic $\Theta$ **structure**, as illustrated in Figure 9.14(a).

Some bacteriophages have evolved a mode of replication, called the **rolling circle model**, which is depicted in Figure 9.14(b). The initial step involves enzymatic nicking of one of the two strands, creating a free 3′-OH group. Thus, no RNA primer is required. DNA polymerase then extends this nicked chain by polymerizing the complement of the strand remaining intact. As replication proceeds, the original nicked strand is displaced at its 5′-PO$_4$ end. After one round of replication is completed, the original DNA molecule, representing a unit genome, is totally displaced, but replication may continue. In theory, the cycle can be repeated *n* number of times creating *n* unit genomes covalently bonded together. Each single-stranded DNA molecule can be converted to duplex form by the synthesis of the complementary strands. Such a structure is called a **concatemer**, which is subsequently cleaved by an endonuclease into complete chromosomes.

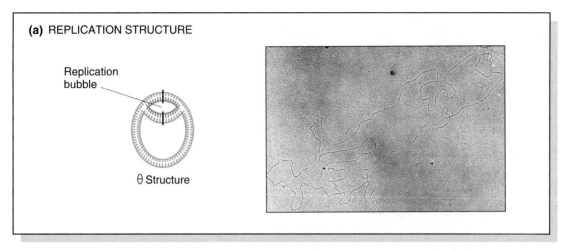

**(a)** REPLICATION STRUCTURE

Replication bubble

θ Structure

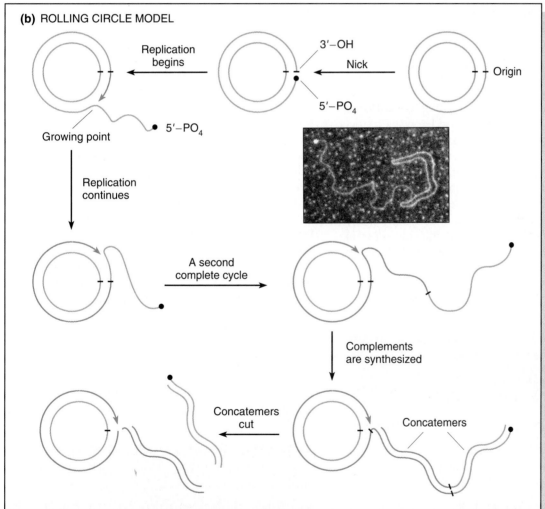

**(b)** ROLLING CIRCLE MODEL

Replication begins

3′–OH

Nick

Origin

5′–PO$_4$

Growing point

5′–PO$_4$

Replication continues

A second complete cycle

Complements are synthesized

Concatemers cut

Concatemers

**FIGURE 9.14**    (a) Semiconservative replication of a circular DNA molecule, where both strands of the parent molecule are used as a template. This creates a Θ structure, illustrated in the accompanying photograph. (b) The rolling circle model, characteristic of some bacteriophages. The newly synthesized single strands serve as templates for the synthesis of their complements. Subsequent cleavage into unit chromosomes then occurs. The micrographs illustrate these structures as visualized under the electron microscope.

This mode of replication is utilized by numerous bacterial viruses, including phage λ and φX174. In λ, the linear molecules are packaged into viral heads. Following infection of a host bacterium, the linear molecule is converted into a circular molecule to accommodate the rolling circle of replication.

## GENETIC CONTROL OF REPLICATION

Much of what we know and have outlined in the previous section concerning the details of DNA replication in viruses and bacteria has been based on the genetic analysis of the process. For example, we have already discussed the *polA1* mutation, the study of which revealed that DNA polymerase I is not the major enzyme responsible for replication. Many other mutations have been isolated that interrupt or seriously impair some aspect of replication, such as the ligase-deficient and the proofreading-deficient mutations mentioned previously. These can be studied most easily if they are **temperature-sensitive mutations** (one type of **conditional mutation**) that express the mutant condition at a restrictive temperature but function normally at a permissive temperature. Investigation of such temperature-sensitive mutants provides insights into the product and the associated function of the normal, nonmutant gene.

As shown in Table 9.3, the enzyme product or its general role in replication has been ascertained for a variety of genes in *E. coli*. For example, numerous mutations in genes specifying the subunits of polymerases I, II, and III have been isolated. Genes have also been identified that encode products involved in specification of the origin of synthesis, helix-unwinding and stabilization, initiation

and priming, relaxation of supercoiling, repair, and ligation. The discovery of such a large group of genes attests to the complexity of the process of replication, even in the relatively simple prokaryote. This complexity is not unexpected, given the enormous quantity of DNA that must be unerringly replicated in a very brief time. As we will see, the process is even more involved and therefore more difficult to investigate in eukaryotes.

## EUKARYOTIC DNA SYNTHESIS

The general scheme of DNA synthesis found in bacteria is now thought to apply also to eukaryotic systems. However, because eukaryotic cells contain about 50 times as much DNA per cell and because this DNA is complexed with proteins, eukaryotes face many problems not encountered by bacteria. As expected, these complications make the process of DNA synthesis much more complex. As a result, unraveling the mysteries of replication in eukaryotes has been more difficult.

Nevertheless, a great deal is now known about this process. As we will soon see in Chapter 10, eukaryotic DNA has complexed to it an array of histone proteins to form the chromatin fiber characteristic of interphase cells. The DNA–histone complex forms an orderly repeating structure called the **nucleosome**. The presence of these structures gives chromatin the appearance of beads on a string (Figure 9.15). Prior to the initiation of DNA synthesis at each replication fork, a process of dissociation must occur, where the histones are stripped away from the DNA. Then, immediately after DNA synthesis at the replication fork, histones reassociate with the newly formed duplexes, reestablishing the characteristic nucleosome pattern (Figure 9.15). Since, following replication of all chromosomes, the total quantity of histone proteins has doubled, DNA synthesis is tightly coupled to histone synthesis in eukaryotic cells. Both occur during the S phase of the cell cycle.

Eukaryotic cells have been found to contain four different kinds of DNA polymerases, called **α**, **β**, **γ**, and **δ**.

**Table 9.3** A LIST OF VARIOUS *E. COLI* MUTANT GENES AND THEIR PRODUCTS OR FUNCTIONS

| Mutant Gene | Enzyme or Role in Replication |
| --- | --- |
| *polA* | DNA polymerase I |
| *polB* | DNA polymerase II |
| *dnaN,Q,X,Z* | DNA polymerase III subunits |
| *dnaG* | Primase |
| *dnaA,I,P* | Initiation |
| *dnaB,C* | Helicase at *oriC* |
| *oriC* | Origin of replication |
| *gyrA,B* | Gyrase subunits |
| *lig* | Ligase |
| *rep* | Helicase |
| *ssb* | Single-stranded binding proteins |
| *rpoB* | RNA polymerase subunit |

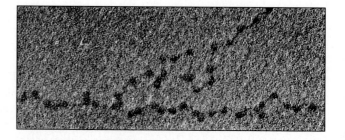

**FIGURE 9.15**    A replicating fork demonstrating the presence of histone protein-containing nucleosomes on both branches.

It appears that the α form is the major enzyme involved in the replication of nuclear DNA. The β form may be involved in DNA repair, while the γ form is the only DNA polymerase found in mitochondria. Presumably, it is unique in its replication function within that organelle. The δ form, like α, appears to function in nuclear DNA replication. DNA polymerases of eukaryotes have the same fundamental requirements for DNA synthesis as bacterial and viral systems: four deoxyribonucleoside triphosphates, a template, and a primer.

Data and observations derived from autoradiographic and electron microscopic studies have provided many other insights. As mentioned earlier, both prokaryotic and eukaryotic DNA synthesis is bidirectional, creating two replication forks from each point of origin. As would be expected because of the increased amount of DNA in eukaryotes compared to prokaryotes, many more points of origin are present. In mammals, there are about 25,000 replicons present in the genome, each consisting of an average of 100,000 to 200,000 base pairs (100–200 kb). In *Drosophila*, there are about 3500 replicons per genome of an average size of 40 kb.

In order to accommodate the increased number of replicons, many more DNA polymerase molecules are present in eukaryotic cells than found in bacteria. While *E. coli* has about 15 copies of DNA polymerase III per cell, there may be up to 50,000 copies of the α form of DNA polymerase in animal cells. In theory, this increased number makes possible the simultaneous replication of all replicons. In practice, most, but not all, are replicated at approximately the same time.

The presence of smaller replicons in eukaryotes compensates for their slower rate of DNA synthesis compared to that in prokaryotes. In *E. coli*, 100 kb are added to a growing chain per minute, while eukaryotic synthesis ranges from only 0.5 to 5 kb per minute. Nevertheless, *E. coli* requires 20 to 40 minutes to replicate its chromosome, while *Drosophila*, with 40 times more DNA, accomplishes the same task in only 3 minutes during embryonic cell divisions.

Even though DNA synthesis is slower in eukaryotes, most general aspects of chain elongation are thought to be similar. The actions of a variety of proteins—DNA helicase and SSBPs—are believed to modulate strand separation that precedes RNA priming by a primase enzyme. On the lagging strand, synthesis is discontinuous, resulting in Okazaki fragments that are linked together by DNA ligase. These fragments are about 10 times smaller (100–150 nucleotides) than in prokaryotes. While synthesis may occur continuously on the leading strand, it is not clear whether or not it can always proceed uninterrupted for the full length of a replicon. In cases where it does not occur continuously, synthesis on the leading strand is sometimes described as **semidiscontinuous**.

## DNA Synthesis at the Ends of Linear Chromosomes

One final aspect of eukaryotic versus prokaryotic DNA synthesis involves the fact that eukaryotic chromosomes are linear compared to the circular forms displayed by bacteria, as well as most bacteriophages. A special problem is encountered at the "ends" of linear molecules of DNA during replication. Such open ends are part of each telomeric region of each chromosome.

While synthesis can proceed normally to the end of the leading strand, a problem is encountered on the **lagging strand**, as illustrated in Figure 9.16. The difficulty arises as the RNA primer is removed from the lagging strand. The gap that is created is normally filled by DNA polymerase I, which begins synthesis by adding a nucleotide to the existing 3'-OH group provided during discontinuous synthesis (this would normally be present to the right of the gap in Figure 9.16). However, there is no strand present because this is the end of the chromosome! As a result, each successive round of synthesis will theoretically shorten the chromosome by the length of the RNA primer. Since this is such a significant problem, we can predict that a molecular solution would have

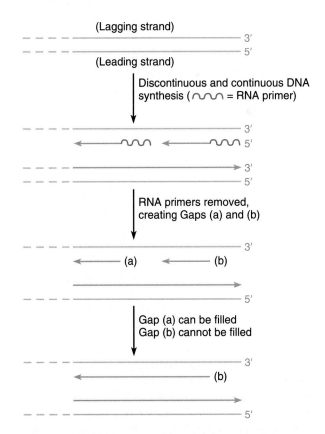

**FIGURE 9.16**    The theoretical difficulty encountered during the replication of the ends of linear chromosomes. A gap is left following synthesis on the lagging strand.

been forthcoming early in evolution, and thus be shared by all eukaryotes. Indeed, this appears to be the case.

In eukaryotes, the discovery of a unique enzyme, **telomerase**, has helped us understand how some organisms solve this problem. In the ciliated protozoan, *Tetrahymena*, the many telomeres all terminate in the sequence 5'-TTGGGG-3'. The enzyme is capable of adding repeats of TTGGGG to the ends of molecules already containing this sequence. This process, now known to occur in other organisms under the direction of a similar enzyme, prevents the telomeric ends from shortening following each replication, as described below and illustrated in Figure 9.17.

Telomerase adds several copies of the 6-nucleotide repeat to the 3' end of the lagging strand (using 5'-to-3' synthesis). These repeats appear to be capable of forming a "hairpin loop," which is stabilized by unorthodox

hydrogen bonding between opposite guanine residues (G $\approx$ G). This creates a free 3'-OH end that, following removal of the RNA primer, can serve as a substrate for DNA polymerase I to fill the gap. If the hairpin is then cleaved off, the potential loss of DNA in each subsequent replication cycle is averted. If the hairpin were simply opened, the telomere would get longer with each cycle!

Further investigation of the *Tetrahymena* telomerase enzyme, isolated and studied extensively by Elizabeth Blackburn and Carol Greider, has yielded an extraordinary finding. This enzyme adds the same TTGGGG sequence to other DNA termini. Thus, it is not using a specific terminus as a template to somehow repeat the sequence. Blackburn and Greider have now established how the enzyme works. They have discovered that the enzyme contains within its molecular structure a short piece of RNA that is essential to its catalytic activity! The functional enzyme is thus a ribonucleoprotein. The RNA component encodes the sequences used by the enzyme as a template. The RNA contains 159 bases, including the sequence 5'-AACCCC-3', which is complementary to the sequence whose synthesis it directs. Analogous enzyme functions have now been found in still other organisms, including the ciliate *Euplotes* and cultured human cells. In both cases, the RNA-containing telomerase enzyme behaves like the enzyme, reverse transcriptase (characteristic of retroviruses that synthesize DNA on an RNA template). In this case, the template is much shorter, and the enzyme supplies its own!

The analysis of telomeric DNA sequences, discussed in more detail in Chapter 10, has shown them to be highly conserved throughout evolution. Such conservation reflects not only the critical function of telomeres, but also the unusual nature of their DNA replication.

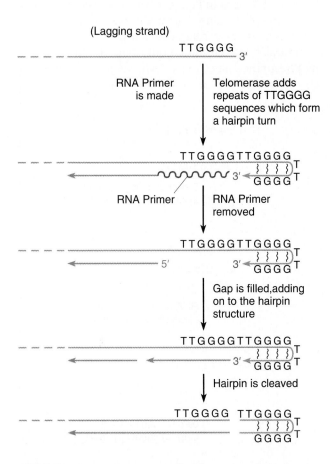

**FIGURE 9.17**    Diagram of the predicted solution to the theoretical problem posed in Figure 9.16. The enzyme telomerase directs synthesis of the TTGGGG repeated sequence resulting in the formation of a hairpin structure. As described in the text, this process averts the creation of a gap during replication of the ends of linear chromosomes.

## DNA Recombination

We conclude this chapter by returning to a topic discussed in Chapter 5—**genetic recombination**. There, it was pointed out that the process of crossing over depends on breakage and rejoining of the DNA strands between homologues. Now that we have discussed the chemistry and replication of DNA, it is appropriate to consider how recombination occurs at the molecular level. In general, the following information pertains to genetic exchange between any two homologous double-stranded DNA molecules, whether they be viral or bacterial chromosomes or eukaryotic homologues during meiosis. Genetic exchange at equivalent positions along two chromosomes with substantial DNA se-

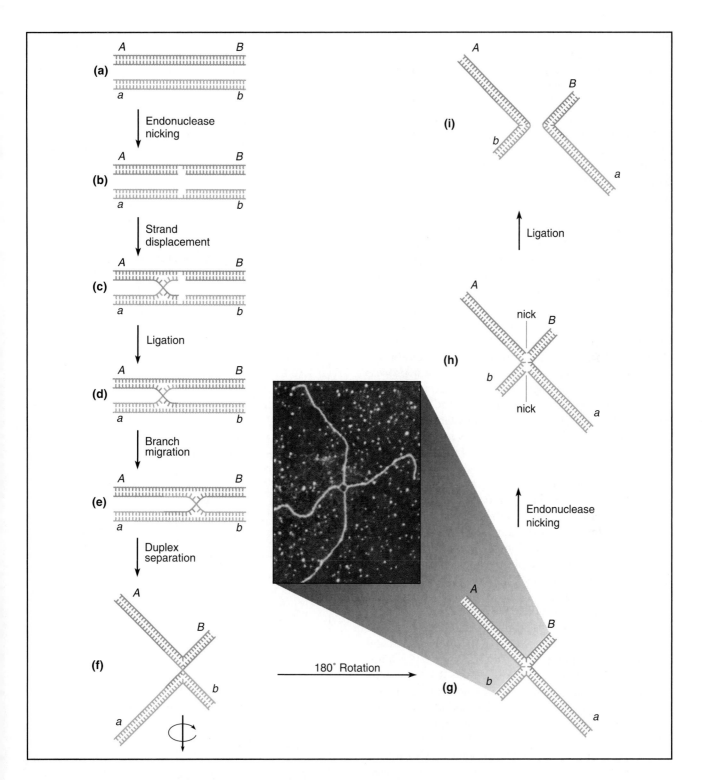

**FIGURE 9.18**    A possible molecular sequence depicting how genetic recombination occurs as a result of the breakage and rejoining of heterologous DNA strands. Each stage is described in the text. The electron micrograph shows DNA in a chi-form structure similar to the diagram in (g). The DNA is from the *Col*E1 plasmid of *E. coli*.

quence homology is referred to as **general** or **homologous recombination**.

While there are several models available to explain homologous recombination, they all share certain common features. First, all are based on the initial proposals put forth independently by Robin Holliday and Harold L. K. Whitehouse in 1964. They also depend on the complementarity between DNA strands for their precision of exchange. Finally, each model relies on a series of enzymatic processes in order to accomplish genetic recombination.

One such model is illustrated in Figure 9.18. It begins with two paired DNA duplexes or homologues (a), each of which has a single-stranded nick introduced (b) at an identical position by an endonuclease. The ends of the strands produced by these cuts are then displaced and subsequently pair with their complements on the opposite duplex (c). A ligase then seals the loose ends (d), creating hybrid duplexes called **heteroduplex DNA molecules**. The exchange creates a cross-bridged or **Holliday structure**. The position of this cross-bridge can then move down the chromosomes by a process referred to as branch migration (e). This occurs as a result of a zipperlike action as hydrogen bonds are broken and then reformed between complementary bases of the displaced strands of each duplex. This migration yields an increased length of heteroduplex DNA on both homologues.

If the duplexes now separate (f), and the bottom portions rotate 180° (g), an intermediate planar structure called a **chi form** is created. If the two strands on opposite homologues previously uninvolved in the exchange are now nicked by an endonuclease (h), and ligation occurs (i), recombinant duplexes are created. Note that the arrangement of alleles is altered as a result of the crossover that occurs.

Evidence supporting the above model includes the electron microscopic visualization of chi-form planar molecules from bacteria where four duplex arms are joined at a single point of exchange (Figure 9.18). Additionally, the discovery in *E. coli* of the **RecA protein** provides important evidence. This molecule promotes the exchange of reciprocal single-stranded DNA molecules as must occur in step (c) of the model. Further, RecA enhances the hydrogen bond formation during strand displacement, thus initiating heteroduplex formation. Finally, many other enzymes essential to the nicking and ligation process have also been discovered and investigated. The products of the *rec*B, *rec*C, and *rec*D genes are thought to be involved in the nicking and unwinding of DNA. Mutations that prevent genetic recombination have been found in a number of genes in viruses and bacteria. These are thought to represent

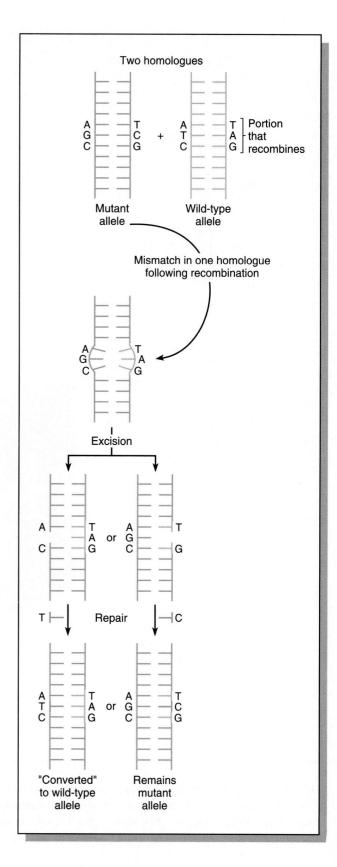

genes, the products of which play an essential role in this process.

## Gene Conversion

A modification of the above model has helped us to better understand a unique genetic phenomenon known as **gene conversion**. Initially found in yeast by Carl Lindegren and in *Neurospora* by Mary Mitchell, gene conversion is characterized by a genetic exchange ratio involving two closely linked genes that is *nonreciprocal*. If one were to cross two *Neurospora* strains each bearing a separate mutation ($a+ \times +b$), a reciprocal recombination event between the genes would yield spore pairs of the $++$ and $ab$ genotypes. However, a nonreciprocal exchange yields one pair without the other. Working with pyridoxine mutants, Mitchell observed several asci with the $++$ genotype, but not the reciprocal product ($ab$). Since the frequency of these events was higher than the predicted mutation rate, and thus could not be accounted for by that phenomenon, they were called *gene conversions*. They were so named because it appeared that one allele had somehow been "converted" to another during an event where genetic exchange also occurred. Similar findings are apparent in the study of other fungi as well.

This phenomenon is now considered to be a consequence of the process of genetic recombination, as discussed in the previous section.

One possible explanation of this genetic phenomenon interprets the conversion event as a mismatch of base pairs during heteroduplex formation involved in genetic recombination (Figure 9.19). Mismatched regions of hybrid strands may be created during recombination, but they can be repaired by excision of one of the strands and synthesis of the complement using the remaining strand as a template. Excision of either one of the strands may occur in order to accomplish the repair, yielding two possible "corrections" (Figure 9.19). One repairs the mismatched base pair and restores the original sequence. The other corrects the mismatch, but does so by copying the altered strand, creating a base pair substitution. This conversion may have the effect of creating identical alleles on the two homologues that were different initially.

In our example in Figure 9.19, suppose the $G \equiv C$ pair on one of the two homologues was responsible for the mutant allele while the $A = T$ pair was part of the wild-type gene sequence on the other homologue. Conversion of the $G \equiv C$ pair to $A = T$ would have the effect of changing the mutant allele to wild type, just as Mitchell originally observed!

Gene conversion events have helped to explain other puzzling genetic phenomena in fungi. For example, when mutant and wild-type alleles of a single gene are studied in a cross, asci should yield equal numbers of mutant and wild-type spores. However, exceptional asci with 3:1 or 1:3 ratios are sometimes observed. These ratios are more easily understood when interpreted in terms of gene conversion. The phenomenon has also been detected during mitotic events in fungi as well as during the study of unique compound chromosomes in *Drosophila*.

◀ **FIGURE 9.19** Illustration of a proposed mechanism that accounts for the phenomenon of gene conversion. A base pair mismatch occurs in one of the two homologues (bearing the mutant allele) during heteroduplex formation, which accompanies recombination in meiosis. During excision repair, one of the two mismatches is removed and the complement is synthesized. In one case, the mutant base pair is preserved. When it is subsequently included in a recombinant spore, the mutant genotype will be maintained. In the other case, the mutant base pair is converted to the wild-type sequence. When included in a recombinant spore, the wild-type genotype will be expressed, leading to a nonreciprocal exchange ratio.

**CHAPTER SUMMARY**

1. In theory, three modes of DNA replication are possible: semiconservative, conservative, and dispersive. Although all three rely on base complementarity, semiconservative replication is the most straightforward and was predicted.

2. In 1958, Meselson and Stahl resolved this problem in favor of semiconservative replication in *E. coli,* showing that newly synthesized DNA consists of one old strand and one new strand. Taylor, Woods, and Hughes used root tips of the broad bean to demonstrate semiconservative replication in eukaryotes.

3. During the same period, Kornberg isolated DNA polymerase I from *E. coli* and demonstrated it to be an enzyme capable of *in vitro* DNA synthesis, provided that a template and precursor nucleoside triphosphates were supplied.

4. The subsequent discovery of the *polA1* mutant strain of *E. coli,* capable of DNA replication in spite of its lack of polymerase I activity, cast doubt on this enzyme's *in vivo* replicative function. DNA polymerases II and III were then isolated. Polymerase III has been identified as the enzyme responsible for DNA replication *in vivo.*

5. During the process of DNA synthesis, the double helix unwinds, forming a replication fork where synthesis begins. Proteins stabilize the unwound helix and assist in relaxing the coiling tension created ahead of the replication activity.

6. Synthesis is initiated at specific sites along each template strand by RNA primase, which results in a short segment of RNA that provides a suitable 3′ end, upon which DNA polymerase III can begin polymerization.

7. Due to the antiparallel nature of the double helix, polymerase III synthesizes DNA continuously on the leading strand in a 5′-to-3′ direction. On the opposite strand, called the lagging strand, synthesis results in short Okazaki fragments that are later joined by DNA ligase.

8. DNA polymerase I removes and replaces the RNA primer with DNA, which is joined to the adjacent polynucleotide by DNA ligase.

9. The isolation of numerous phage and bacterial mutant genes affecting many of the molecules involved in the replication of DNA has helped to define the complex genetic control of the entire process.

10. DNA replication in eukaryotes is similar to but more complex than replication in prokaryotes. For example, replication at the ends (telomeres) of linear molecules poses a special problem, solved by a unique RNA-containing enzyme called telomerase.

11. Homologous recombination between genetic molecules relies on a series of enzymes that can cut, realign, and reseal DNA strands. The phenomenon of gene conversion may be best explained in terms of mismatch repair synthesis during these exchanges.

## KEY TERMS

bidirectional replication
biologically active DNA
chain elongation
chi form
concatemer
conditional mutation
conservative replication
continuous DNA
  synthesis
discontinuous DNA
  synthesis
dispersive replication
DNA gyrase
DNA helicase
DNA ligase
  (polynucleotide joining
  enzyme)

DNA polymerase (I, II,
  III, $\alpha$, $\beta$, $\gamma$, $\delta$)
DNA topoisomerase
exonuclease activity
exonuclease proofreading
  (editing)
gene conversion
genetic recombination
helicase
heteroduplex DNA
  molecules
Holliday structure
holoenzyme
homologous (general)
  recombination
lagging DNA strand
leading DNA strand

nearest-neighbor
  frequency test
9mers
nucleosome
Okazaki fragments
*oriC*
origin of replication
primer
RecA protein
replication eye (bubble)
replication fork
replicative form (RF)
replicon
replisome
RNA primase
rolling circle model
sedimentation equilibrium
  centrifugation

semiconservative
  replication
semidiscontinuous DNA
  synthesis
single-stranded binding
  proteins (SSBPs)
sister chromatid
  exchanges
spleen phosphodiesterase
supercoiling
telomerase
temperature-sensitive
  mutation
*ter*
$\Theta$ structure
13mers

## INSIGHTS AND SOLUTIONS

1.  Predict the theoretical results of conservative and dispersive models of DNA synthesis using the conditions of the Meselson–Stahl experiment. Follow the results through two generations of replication after cells have been shifted to an $^{14}$N-containing medium, using the following migration standards:

Density $\longrightarrow$

$^{14}$N/$^{14}$N          $^{15}$N/$^{14}$N          $^{15}$N/$^{15}$N

**SOLUTION:**

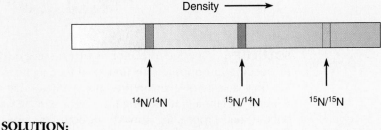

Conservative Replication

Generation I            Generation II

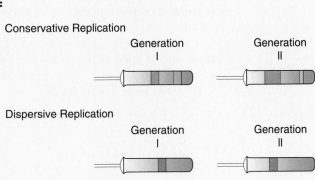

Dispersive Replication

Generation I            Generation II

2. Mutations in the *dnaA* gene of *E. coli* are lethal and can only be studied following the isolation of conditional, temperature-sensitive mutations. Such mutant strains grow nicely and replicate their DNA at the permissive temperature of 18°C, but they do not grow or replicate their DNA at the restrictive temperature of 37°C. Two observations were useful in determining the function of the *dnaA* gene product. First, *in vitro* studies using DNA templates that have been nicked (opened) do not require the dnaA protein. Second, if intact cells are grown at 18°C and then shifted to 37°C, DNA synthesis continues at this temperature until one round of replication is completed, and DNA synthesis stops. What do these observations suggest about the role of the *dnaA* gene product?

   **SOLUTION:**   These observations suggest that *in vivo* the dnaA protein is essential to the initiation of DNA synthesis. At 18°C (the permissive temperature) the mutation is not expressed and DNA synthesis begins. Following the shift to the restrictive temperature, DNA synthesis already initiated continues, but no new synthesis can begin. Since the dnaA protein is not required for synthesis of "nicked" DNA, this observation suggests that the protein functions during initiation by interacting with the intact helix and somehow facilitating the localized denaturing necessary for synthesis to proceed. In fact, both conclusions are valid.

3. DNA is allowed to replicate in moderately radioactive $^3$H-thymidine for several minutes and then switched to a highly radioactive medium for several more minutes. Synthesis is stopped and the DNA subjected to autoradiography and electron microscopy. Interpret as much as you can regarding DNA replication from the following micrograph. Can you detect in this autoradiogram an inconsistency regarding our current model for DNA synthesis?

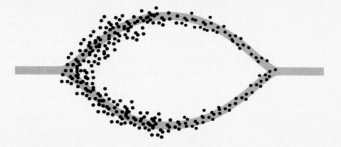

   **SOLUTION:**   One interpretation is that there are two advancing replication forks proceeding in opposite directions and that replication is therefore bidirectional. The low density of grains represents synthesis that occurred during the first few minutes, beginning at the origin (at approximately the middle of the replication bubble) and proceeding outward in both directions. The higher density of grains represents the latter minutes of synthesis when the level of radioactivity was increased.

   Since $^3$H-thymidine represents only DNA synthesis, an inconsistency exists if indeed RNA is used as the initial primer. At some point on each strand within the replication bubble, there ought to be areas devoid of grains, where RNA and not DNA has been synthesized!

1. Compare conservative, semiconservative, and dispersive modes of DNA replication.

2. Review the design, results, and conclusions of the Meselson–Stahl experiment.

3. In the Meselson–Stahl experiment, which of the three modes of replication could be ruled out after one round of replication? After two rounds?

4. Predict the results of the experiment by Taylor, Woods, and Hughes if replication were (a) conservative and (b) dispersive.

5. Reconsider Problem 33 in Chapter 8. In the model you proposed, could the molecule be replicated semiconservatively? Why? Would other modes of replication work?

6. What are the requirements for the *in vitro* synthesis of DNA under the direction of DNA polymerase I?

7. In Kornberg's initial experiments, he actually grew *E. coli* in Anheuser-Busch beer vats (he was working at Washington U. in St. Louis). Why do you think this was necessary?

8. How did Kornberg test the fidelity of copying DNA by polymerase I?

9. Which of Kornberg's tests is the more stringent assay? Why?

10. Which characteristics of DNA polymerase I led to doubts that its *in vivo* function is the synthesis of DNA leading to complete replication?

11. Explain the theory of nearest-neighbor frequency.

12. What is meant by "biologically active" DNA?

13. Why was the phage $\phi$X174 chosen for the experiment demonstrating biological activity?

14. Outline the experimental design of Kornberg's biological activity demonstration.

15. What was the significance of the *polA1* mutation?

16. Summarize the properties of polymerase I, II, and III.

17. Distinguish between (a) unidirectional and bidirectional synthesis and (b) continuous and discontinuous synthesis of DNA.

18. List the proteins that unwind DNA during *in vivo* DNA synthesis. How do they function?

19. Define and indicate the significance of (a) Okazaki fragments, (b) DNA ligase, and (c) primer RNA during DNA replication.

20. Outline the current model for DNA synthesis.

21. Why is DNA synthesis expected to be more complex in eukaryotes than in bacteria? How is DNA synthesis similar in the two types of organisms?

22. If the analysis of DNA from two different microorganisms demonstrated very similar nearest-neighbor frequencies, is the DNA of the two organisms identical in (a) amount, (b) base composition, and/or (c) nucleotide sequences?

23. Analysis of nearest-neighbor data led Josse, Kaiser, and Kornberg in 1961 (*J. Biol. Chem.* 236:864–75) to conclude that the two strands of the double helix are in opposite polarity to one another. Demonstrate your understanding of the nearest-neighbor technique by determining the outcome of such an analysis if the strands of the following molecule are (a) antiparallel versus (b) parallel:

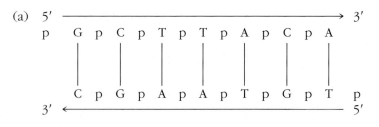

vs.

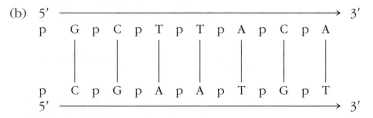

**24.** Suppose that *E. coli* synthesizes DNA at a rate of 100,000 nucleotides per minute and takes 40 minutes to replicate its chromosome.
   (a) How many base pairs are present in the entire *E. coli* chromosomes?
   (b) What is the physical length of the chromosome in its helical configuration, that is, what is the circumference of the circular chromosome?

**25.** Several temperature-sensitive mutant strains of *E. coli* are isolated that display various characteristics. Predict what enzyme or function is being affected by each mutation:
   (a) Newly synthesized DNA contains many mismatched base pairs.
   (b) Okazaki fragments accumulate, and DNA synthesis is never completed.
   (c) No initiation occurs.
   (d) Synthesis is very slow.
   (e) Supercoiled strands are found to remain and replication is never completed.

**SELECTED READINGS**

BLACKBURN, E. H. 1991. Structure and function of telomeres. *Nature* 350:569–72.

BRAMHILL, D., and KORNBERG, A. 1988. A model for initiation at origins of DNA replication. *Cell* 54:915–18.

CAMERINI-OTERO, R. D., and HSIEH, P. 1993. Parallel DNA triplexes, homologous recombination, and other homology-dependent DNA interactions. *Cell* 73:217–23.

DARNELL, J., LODISH, H., and BALTIMORE, D. 1990. *Molecular cell biology.* 2nd ed. New York: Scient. Amer. Books.

DAVIDSON, J. N. 1976. *The biochemistry of nucleic acids.* 8th ed. Orlando: Academic Press.

DELUCIA, P., and CAIRNS, J. 1969. Isolation of an *E. coli* strain with a mutation affecting DNA polymerase. *Nature* 224:1164–66.

DENHARDT, D. T., and FAUST, E. A. 1985. Eukaryotic DNA replication. *BioEssays* 2:148–53.

DRESLER, D., and POTTER, H. 1982. Molecular mechanisms in genetic recombination. *Ann. Rev. Biochem.* 51:727–61.

GELLERT, M. 1981. DNA topoisomerases. *Ann. Rev. Biochem.* 50:879–910.

GREIDER, C. W., and BLACKBURN, E. H. 1989. A telomeric sequence in the RNA of *Tetrahymena* telomerase required for telomere repeat synthesis. *Nature* 337:331–336.

HOLLIDAY, R. 1964. A mechanism for gene conversion in fungi. *Genet. Res.* 5:282–304.

HUBERMAN, J. C. 1987. Eukaryotic DNA replication: A complex picture partially clarified. *Cell* 48:7–8.

Josse, J., Kaiser, A. D., and Kornberg, A. 1961. Enzymatic synthesis of deoxyribonucleic acid. VIII. Frequencies of nearest neighbor base sequences in deoxyribonucleic acid. *J. Biol. Chem.* 236:864–75.

Kornberg, A. 1960. Biological synthesis of DNA. *Science* 131:1503–8.

———. 1969. Active center of DNA polymerase. *Science* 163:1410–18.

———. 1974. *DNA synthesis*. New York: W. H. Freeman.

———. 1979. Aspects of DNA replication. *Cold Spr. Harb. Symp.* 43:1–10.

Kornberg, A., and Baker, T. 1992. *DNA replication*. 2nd ed. New York: W. H. Freeman.

Lehman, I. R. 1974. DNA ligase: Structure, mechanism, and function. *Science* 186:790–97.

Lindegren, C. C. 1953. Gene conversion in *Saccharomyces*. *J. Genet.* 51:625–37.

Meselson, M., and Stahl, F. W. 1958. The replication of DNA in *Escherichia coli*. *Proc. Natl. Acad. Sci.* 44:671–82.

Mitchell, M. B. 1955. Aberrant recombination of pyridoxine mutants of *Neurospora*. *Proc. Natl. Acad. Sci.* 41:215–20.

Ogawa, T., and Okazaki, T. 1980. Discontinuous DNA synthesis. *Ann. Rev. Biochem.* 49:421–57.

Okazaki, T., et al. 1979. Structure and metabolism of the RNA primer in the discontinuous replication of prokaryotic DNA. *Cold Spr. Harb. Symp.* 43:203–22.

Radding, C. M. 1978. Genetic recombination: Strand transfer and mismatch repair. *Ann. Rev. Biochem.* 47:847–80.

Radman, M., and Wagner, R. 1988. The high fidelity of DNA duplication. *Scient. Amer.* (August) 259:40–46.

Schekman, R., Weiner, A., and Kornberg, A. 1974. Multienzyme systems of DNA replication. *Science* 186:987–93.

Stahl, F. W. 1979. *Genetic recombination: Thinking about it in phage and fungi*. New York: W. H. Freeman.

———. 1987. Genetic recombination. *Scient. Amer.* (Feb.) 256:90–101.

Szostak, J. W., Orr-Weaver, T. L., and Rothstein, R. J. 1983. The double-strand-break repair model for recombination. *Cell* 33:25–35.

Taylor, J. H., Woods, P. S., and Hughes, W. C. 1957. The organization and duplication of chromosomes revealed by autoradiographic studies using tritium-labeled thymidine. *Proc. Natl. Acad. Sci.* 48:122–28.

Thömmes, P., and Hübscher, U. 1992. Eukaryotic DNA helicases: Essential enzymes for DNA transactions. *Chromosoma* 101:467–73.

Tomizawa, J., and Selzer, G. 1979. Initiation of DNA synthesis in *E. coli*. *Ann. Rev. Biochem.* 48:999–1034.

Wang, J. C. 1982. DNA topoisomerases. *Scient. Amer.* (July) 247:94–108.

———. 1987. Recent studies of DNA topoisomerases. *Biochimica Biophysica Acta* 909:1–9.

Watson, J. D., et al. 1987. *Molecular biology of the gene*. Vol. 1, *General principles*. 4th ed. Menlo Park, CA: Benjamin/Cummings.

Whitehouse, H. L. K. 1982. *Genetic recombination: Understanding the mechanisms*. New York: Wiley.

Wood, W. B., Wilson, J. H., Benbow, R. M., and Hood, L. E. 1975. *Biochemistry—A problems approach*. Menlo Park, CA: Benjamin/Cummings.

Zubay, G. L., and Marmur, J., eds. 1973. *Papers in biochemical genetics*. 2nd ed. New York: Holt, Rinehart and Winston.

Zyskind, J. W., and Smith, D. W. 1986. The bacterial origin of replication, *oriC*. *Cell* 46:489–90.

# 10

# DNA:
## ORGANIZATION IN CHROMOSOMES AND GENES

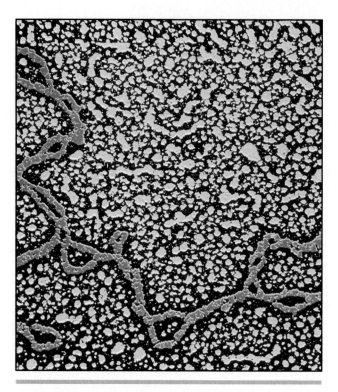

Transmission electron micrograph of supercoiled, circular DNA derived from a mitochondrion.

*DNA is organized in a manner consistent with the complexity of the host structure with which it is associated. Viruses, bacteria, mitochondria, and chloroplasts contain a shorter, often circular, DNA molecule relatively free of proteins. Eukaryotic cells contain greater amounts of DNA organized into nucleosomes and present as chromatin fibers. Studies designed to elucidate the organization of genes reveal diverse forms of organization, ranging from single copies of genes to large numbers of tandem repeats of families of related genes. A most striking feature is the degree to which noncoding DNA is associated with the organization of genes throughout the eukaryotic genome.*

Once it was understood that DNA is the genetic material, it was very important to determine the way in which DNA is organized into chromosomes and how genes are organized within these genetic structures. There has been much interest in these topics because the determination of the spatial arrangement of the genetic material and associated molecules will undoubtedly provide valuable insights into other aspects of genetics. For example, how the genetic information is stored, expressed, and regulated must be related to the organization of the genetic molecule, DNA. In eukaryotes, how the chromatin fibers characteristic of interphase are condensed into chromosome structures visible during mitosis and meiosis is also of great interest.

In this chapter, we first provide a survey of the various forms of chromosome organization, including examples from viruses, bacteria, and various eukaryotes. The genetic material has been studied using numerous approaches, including molecular analysis and direct visualization by light and electron microscopy.

We will then turn to the consideration of how genes are organized as part of chromosomes and within the entire genome of an organism. As we shall see here and in future chapters, such organization is least complex in bacteriophages and bacteria. In their genomes, a chromosome consists largely of an array of tightly packed contiguous genes. Each gene consists of an uninterrupted linear array of nucleotides, most of which either encode the amino acid sequences of proteins or are involved in gene regulation.

In comparison, a large portion of many eukaryotic genes consists of noncoding DNA sequences called introns and flanking sequences that do not become part of the final RNA transcript. Further, we now know that a large portion of the total DNA comprising the genome of eukaryotes consists of intergenic noncoding areas, some of which are occupied by repetitive DNA sequences. Beyond this information, we are learning much more about the organization of genes within the genome.

These latter topics represent some of the most exciting and unexpected discoveries made in the field of genetics. For all geneticists, but particularly those trained prior to the 1970s, the findings concerning eukaryotic gene structure serve as a source of wonderment. Perhaps we would have all been better prepared to learn of the much greater complexity of eukaryotic genes and genomes compared with those of prokaryotes and viruses if we had paid more attention to the quotation from Lewis Carroll's *Through the Looking Glass*, which was insightfully included by Richard B. Goldschmidt in his Presidential Address to the IX International Congress of Genetics in 1954:

"I can't believe that," said Alice. "Can't you?" the Queen said in a pitying tone. "Try again: draw a long breath, and shut your eyes."

Alice laughed. "There's no use trying," she said, "one can't believe impossible things." "I dare say you haven't had much practice," said the Queen. "When I was your age I did it for half-an-hour a day. Why, sometimes I've believed as many as six impossible things before breakfast."

For all of us interested in genetics, it is certain that the next "impossible thing" regarding the organization of the genetic material is just around the corner!

# VIRAL AND BACTERIAL CHROMOSOMES

In comparison with eukaryotes, the chromosomes of viruses and bacteria are much less complicated. They usually consist of a single nucleic acid molecule, largely devoid of associated proteins. Contained within the single chromosome of viruses and bacteria is much less genetic information than in the multiple chromosomes comprising the genome of higher forms. These characteristics have greatly simplified analysis, which has now provided a comprehensive view of the structure of viral and bacterial chromosomes.

The chromosomes of viruses consist of a nucleic acid molecule—either DNA or RNA—which is single- or double stranded. They may exist as circular structures (closed loops), or they may take the form of linear molecules. The single-stranded DNA of the **φX174 bacteriophage** and the double-stranded DNA of the **polyoma virus** are ring-shaped molecules housed within the protein coat of the mature virus. The **bacteriophage lambda (λ)**, on the other hand, possesses a linear double-stranded DNA molecule prior to infection, which closes to form a ring upon infection of the host cell. Still other viruses, such as the **T-even series of bacteriophages**, have linear, double-stranded chromosomes of DNA, which do not form circles inside the bacterial host.

Thus, circularity is not an absolute requirement for replication in some viruses.

Viral nucleic acid molecules have been visualized with the electron microscope. Figure 10.1 shows a mature bacteriophage lambda with its double-stranded DNA molecule in the circular configuration. One constant feature shared by viruses, bacteria, and eukaryotic cells is the ability to package an exceedingly long DNA molecule into a relatively small volume. In λ, the DNA is 17 μm long and must fit into the phage head, which is less than 0.1 μm on any side.

In Table 10.1, the measured lengths of the genetic molecules are compared with the size of the head in several viruses, including those in Figures 10.1 and 10.2. In each case, a similar packaging feat must be accomplished. Seldom does the space available in the head of a virus exceed the chromosome volume by more than a factor of two. In many cases, almost all space is filled, indicating nearly perfect packing. Once packed within the head, the genetic material is functionally inert until released into a host cell.

Bacterial chromosomes are also relatively simple in form. They always consist of a double-stranded DNA molecule, compacted into a structure sometimes referred to as the **nucleoid**. *Escherichia coli*, the most extensively studied bacterium, has a large, circular chromosome, measuring approximately 1200 μm (1.2 mm) in length. When the cell is gently lysed and the chromo-

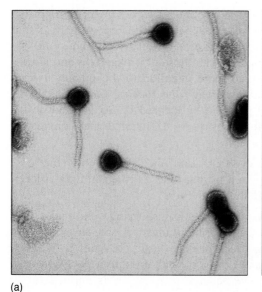

(a)

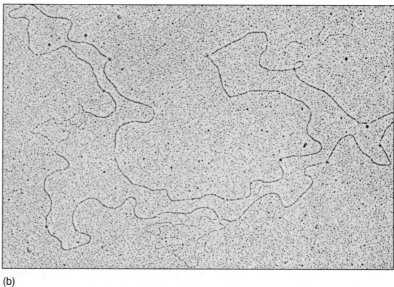

(b)

**FIGURE 10.1**    Electron micrographs of (a) phage λ and (b) the DNA isolated from it. The chromosome is 17 μm long. Note that the phages in part (a) are magnified about five times more than the DNA in part (b).

**Table 10.1**  THE GENETIC MATERIAL OF REPRESENTATIVE VIRUSES AND BACTERIA

| Organism | Nucleic Acid | | | Overall Size of Viral Head or Bacteria ($\mu$m) |
| --- | --- | --- | --- | --- |
| | Type | SS or DS | Length ($\mu$m) | |
| **VIRUSES** | | | | |
| $\phi$X174 | DNA | SS | 2.0 | $0.025 \times 0.025$ |
| Tobacco mosaic virus | RNA | SS | 3.3 | $0.30 \times 0.02$ |
| Lambda phage | DNA | DS | 17.0 | $0.07 \times 0.07$ |
| T2 phage | DNA | DS | 52.0 | $0.07 \times 0.10$ |
| **BACTERIA** | | | | |
| *Hemophilus influenzae* | DNA | DS | 832.0 | $1.00 \times 0.30$ |
| *Escherichia coli* | DNA | DS | 1200.0 | $2.00 \times 0.50$ |

SS = single-stranded; DS = double-stranded.

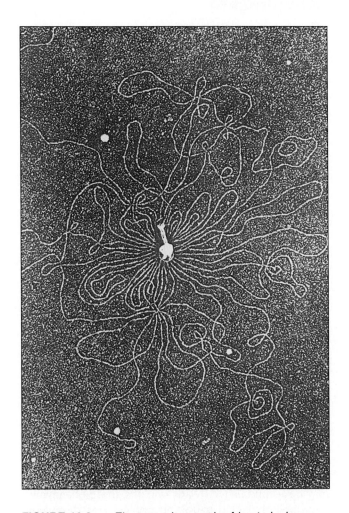

**FIGURE 10.2**    Electron micrograph of bacteriophage T2, which has had its DNA released by osmotic shock. The chromosome is 52 $\mu$m long.

some released, it can be visualized under the electron microscope (Figure 10.3).

This DNA is found to be associated with several types of **DNA-binding proteins**, including those called **HU** and **H**. They are small but abundant in the cell and contain a high percentage of positively charged amino acids that can bond ionically to the negative charges of the phosphate groups in DNA. As we will soon see, these proteins resemble structurally similar molecules called **histones** that are found associated with eukaryotic DNA. Unlike the tightly packed chromosome of a virus, the bacterial chromosome is not functionally inert. In spite of the compacted condition of the bacterial chromosome, replication and transcription readily occur.

## SUPERCOILING AND CIRCULAR DNA

One major insight into the way in which DNA is organized and packaged has come from the discovery of **supercoiled DNA**, characteristic of covalently closed circular molecules and chromosomal loops. Supercoiled DNA was first proposed as a result of a study of double-stranded DNA molecules derived from the polyoma virus, which causes tumors in mice. In 1963, it was observed that when such DNA was subjected to high-speed centrifugation, it was resolved into three distinct components, each of different density and compactness. That which was least compact, and thus least dense, demonstrated a decreased sedimentation velocity; the other two fractions each showed increasing velocity due to their greater compaction and density. All three were of identical molecular weight.

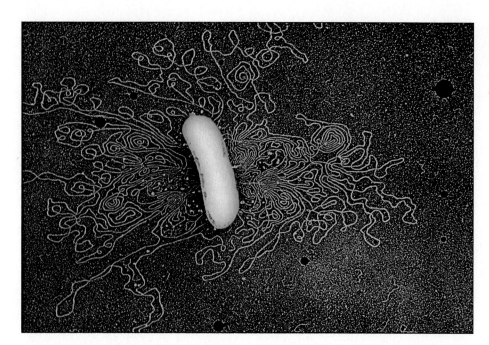

**FIGURE 10.3**   Electron micrograph of the bacterium *Escherichia coli*, which has had its DNA released by osmotic shock. The chromosome is 1200 $\mu$m long.

In 1965 Jerome Vinograd proposed an explanation for the above observations. He postulated that the two fractions of greatest sedimentation velocity both consisted of polyoma DNA molecules that are circular, while the fraction of lower sedimentation contained polyoma DNA molecules that are linear. Closed circular molecules are more compact and sediment more rapidly than linear molecules of the same length and molecular weight.

Vinograd proposed further that the more dense of the two fractions of circular molecules consisted of covalently closed DNA helices that are slightly *underwound* in comparison to the less dense circular molecules. Energetic forces stabilizing the double helix resist this underwinding, causing it to **supercoil** in order to retain normal base pairing. Vinograd proposed that it is the supercoiled shape that causes tighter packing and thus the increase in sedimentation velocity.

The transitions described above can be visualized in Figure 10.4. Consider a double-stranded linear molecule that exists in the normal Watson–Crick right-handed helix [Figure 10.4(a)]. This helix contains 20 complete turns, defining the **linking number** ($L = 20$) of this molecule. If the ends of the molecule are sealed, a closed circle is formed [Figure 10.4(b)], which is energetically *relaxed*. Suppose, however, that the circle is cut open, underwound by several full turns, and resealed [Figure 10.4(c)]. Such a structure, where $L$ is equal to 18, is energetically "strained," and as a result it will exist only temporarily in this form.

In order to assume a more energetically favorable conformation, the molecule can form supercoils in the direction opposite of the underwound helix. In our case [Figure 10.4(d)], two negative supercoils are introduced spontaneously, reestablishing the total number of original "turns" in the helix. The use of the term "negative" refers to the fact that, by definition, the supercoils are left-handed. The end result is the formation of a more compact structure with enhanced physical stability.

In most closed circular DNA molecules in bacteria and their phages, DNA is slightly underwound [as in Figure 10.4(c)]. For example, the virus **SV40** contains 5200 base pairs, where 10.4 base pairs occupy each complete turn of the helix. The linking number can be calculated as

$$L = 5200/10.4$$
$$= 500$$

When circular SV40 DNA is analyzed, it is underwound by 25 turns and $L$ is equal to only 475. Predictably, the presence of 25 negative supercoils is observed. In *E. coli*, an even larger number of supercoils is observed, greatly facilitating chromosome condensation in the nucleoid region.

When two otherwise identical molecules demonstrate a difference in their linking number, they are said to be **topoisomers** of one another. The only way in which these two topoisomers can be interconverted is by cutting one or both of the strands and winding or unwinding the helix before resealing the ends. Biologically, this may be accomplished by any one of a group of enzymes, appropriately called **topoisomerases**. First discovered by Martin Gellert and James Wang, these catalytic molecules are either type I or II, depending on whether they cleave one or both strands in the helix, respectively. In *E. coli*, topoisomerase I serves to reduce the number of

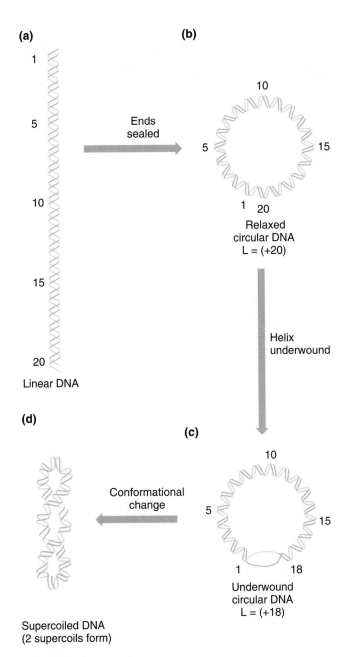

**FIGURE 10.4**    Depictions of the transformations leading to the supercoiling of circular DNA, as described in the text. *L* equals the linking number.

teristic of eukaryotes. While the chromosomes are not usually circular, supercoils are made possible when areas of DNA are embedded in a lattice of proteins associated with the chromatin fibers. This association creates "anchored" ends, which provide the stability for the maintenance of supercoils once they are introduced by topoisomerases.

As we will see, transcription of eukaryotic DNA into RNA faces several obstacles not encountered in prokaryotes. Most notable is the way in which DNA is packaged into structures called nucleosomes. In most studies to date, the formation of supercoils induced by topoisomerases has been implicated as a major factor in facilitating transcription. These enzymes may well play still other genetic roles involving DNA conformational changes.

# MITOCHONDRIAL AND CHLOROPLAST DNA

Numerous observations have demonstrated that both **mitochondria** and **chloroplasts** contain their own genetic information. Most convincing was the discovery of mutations in yeast, other fungi, and plants that were shown to alter the function of these organelles. Furthermore, transmission of these mutations did not demonstrate the biparental inheritance patterns characteristic of nuclear genes. Instead, a **maternal mode of inheritance** was observed, with traits being passed to offspring only from their mother. Since the origin of both mitochondria and chloroplasts is also maternal, these observations suggest that these organelles house their own DNA, separate from the nucleus, that influences their function.

Thus, geneticists set out to look for more direct evidence of DNA in these organelles. Electron microscopists not only documented the presence of DNA in both organelles, but they also saw DNA in a form quite unlike that seen in the nucleus of the eukaryotic cells that house these organelles. Instead, this DNA looked remarkably similar to that seen in viruses and bacteria! As we shall see, this similarity is thought to be related to the proposal that both mitochondria and chloroplasts were once primitive, free-living, bacterialike organisms.

## Molecular Organization and Function of Mitochondrial DNA

Extensive information is now available about the molecular aspects of **mitochondrial DNA (mtDNA)** and related gene function. In most eukaryotes, mtDNA (Fig-

negative supercoils in a closed-circular DNA molecule. Topoisomerase II introduces negative supercoils into DNA. This enzyme is thought to bind to DNA, twist it, cleave both strands, and then pass them through the "loop" that it has created. Once the phosphodiester bonds are reformed, the linking number is decreased and one or more supercoils form spontaneously.

Supercoiled DNA and topoisomerases are also charac-

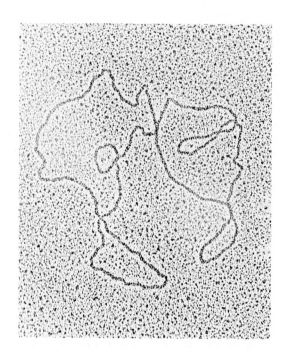

**FIGURE 10.5**    Electron micrograph of mitochondrial DNA (mtDNA) derived from *Xenopus laevis*.

**Table 10.3**  SEDIMENTATION COEFFICIENTS OF MITOCHONDRIAL RIBOSOMES

| Kingdom | Examples | Sedimentation Coefficient (S) |
|---------|----------|-------------------------------|
| Animals | Vertebrates | 55–60 |
|         | Insects | 60–71 |
| Protists | *Euglena* | 71 |
|          | *Tetrahymena* | 80 |
| Fungi | *Neurospora* | 73–80 |
|       | *Saccharomyces* | 72–80 |
| Plants | Maize | 77 |

tRNAs, and numerous products essential to the cellular respiratory functions of the organelles.

The protein-synthesizing apparatus as well as the molecular components for cellular respiration are jointly derived from nuclear and mitochondrial DNA. Ribosomes found in the organelle are different from those present in the cytoplasm. The sedimentation coefficient of these particles varies among different species. The data in Table 10.3 show that mitochondrial ribosomes vary considerably in their coefficients (55$S$ to 60$S$ in vertebrates, to 70$S$ in some algae and fungi, 80$S$ in certain protozoans and fungi).

Many nuclear-coded gene products are essential to biological activity within mitochondria: DNA and RNA polymerases, initiation and elongation factors essential for translation, ribosomal proteins, aminoacyl tRNA synthetases, and some tRNA species. As in chloroplasts, these imported components are generally regarded as distinct from their cytoplasmic counterparts, even though both sets are coded by nuclear genes. For example, the synthetases essential to charging tRNA molecules (a process essential to translation) show a distinct affinity for the mitochondrial tRNA species as compared with the cytoplasmic tRNAs. Similar affinity has been shown for the initiation and elongation factors. Furthermore, while bacterial and nuclear RNA polymerases are known to be composed of numerous subunits, the mitochondrial variety consists of only one polypeptide chain. This polymerase is generally susceptible to antibiotics that inhibit bacterial RNA synthesis but not to eukaryotic inhibitors. The relative contributions of nuclear and mitochondrial gene products are illustrated in Figure 10.6.

The above information, along with that about to be presented concerning chloroplast DNA function serve as the underlying basis for understanding modes of inheritance that are "extrachromosomal." Chapter 20 is devoted to this broad topic.

ure 10.5) is a circular duplex that replicates semiconservatively and is free of the chromosomal proteins characteristic of eukaryotic DNA. In size, mtDNA differs among organisms. This can be seen by examining the information presented in Table 10.2. In a variety of animals, mtDNA consists of about 16,000 to 18,000 base pairs (16–18 kb). In vertebrates, there are 5 to 10 DNA molecules per organelle. There is a considerably greater amount of DNA present in plant mitochondria, where 100 kb is not unusual.

Several general statements can now be made concerning mtDNA. There appear to be few or no gene repetitions, and replication is dependent upon enzymes encoded by nuclear DNA. Mitochondrial genes have been identified that code for the ribosomal RNAs, over 20

**Table 10.2**  THE SIZE OF mtDNA IN DIFFERENT ORGANISMS

| Organism | Size in Kilobases |
|----------|-------------------|
| Human | 16.6 |
| Mouse | 16.2 |
| *Xenopus* (frog) | 18.4 |
| *Drosophila* (fruit fly) | 18.4 |
| *Saccharomyces* (yeast) | 84.0 |
| *Pisum sativum* (pea) | 110.0 |

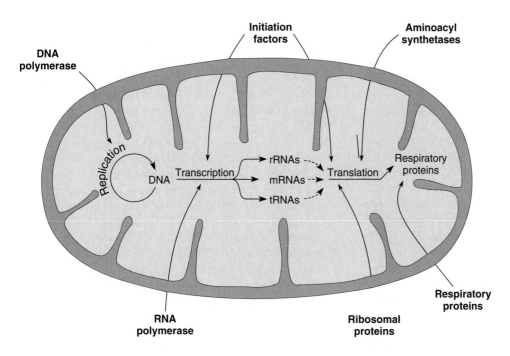

**FIGURE 10.6**    A comparison of the origin of gene products that are essential to mitochondrial function. Those shown entering the organelle are derived from the cytoplasm and encoded by the nucleus.

## Molecular Organization and Function of Chloroplast DNA

There is now available a substantial amount of molecular information about **chloroplast DNA (cpDNA)** and about the genetic function of this organelle. The chloroplast, like the mitochondrion, contains an autonomous genetic system distinct from that found in the nucleus and cytoplasm. This system includes DNA as a source of genetic information and a complete protein-synthesizing apparatus. However, the molecular components of the translation apparatus are jointly derived from both nuclear and chloroplast genetic information. As seen in Figure 10.7, chloroplast DNA is much larger than mitochondrial DNA, but it is nevertheless very similar to that found in prokaryotic cells.

DNA isolated from chloroplasts is found to be circular, double stranded, replicated semiconservatively, and free of the associated proteins characteristic of eukaryotic DNA. Compared with nuclear DNA of the same organism, it invariably shows a different buoyant density and base composition.

In *Chlamydomonas*, there are about 75 copies of the

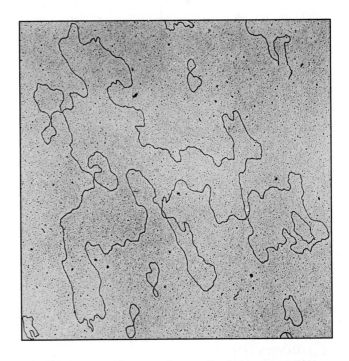

**FIGURE 10.7**    Electron micrograph of chloroplast DNA derived from corn.

chloroplast DNA molecule per organelle. Each copy consists of a length of DNA that contains 195,000 bases (195 kb). In higher plants such as the sweet pea, multiple copies of the DNA molecule are present in each organelle, but the molecule is considerably smaller than that in *Chlamydomonas*, consisting of 134 kb.

Some chloroplast gene products function to synthesize proteins. In a variety of higher plants (beans, lettuce, spinach, maize, and oats), two sets of the genes coding for the ribosomal RNAs—5$S$, 16$S$, and 23$S$ rRNA are present. Additionally, chloroplast DNA codes for at least 25 tRNA species and a number of ribosomal proteins specific to the chloroplast ribosomes. These ribosomes have a sedimentation coefficient slightly less than 70$S$, similar to that of bacteria.

Chloroplast ribosomes are sensitive to the same protein-synthesis-inhibiting antibiotics as bacterial ribosomes: chloramphenicol, erythromycin, streptomycin, and spectinomycin. Even though ribosomal proteins are derived from nuclear and chloroplast DNA, most, if not all, are distinct from the equivalent proteins of cytoplasmic ribosomes.

Still other chloroplast genes have been identified that are specific to photosynthetic function. Mutations in these genes may have the effect of inactivating photosynthesis in those chloroplasts bearing such a mutation. One of the major chloroplast gene products is the large subunit of the photosynthetic enzyme **ribulose-1-5-bisphosphate carboxylase**. Interestingly, the small subunit of this enzyme is encoded by a nuclear gene, while the large subunit, like all other chloroplast-derived gene products, is synthesized and remains within the organelle.

Based on the observation that mitochondrial and chloroplast DNA and their genetic apparatus are similar to their counterparts in bacteria, a theory of the origin of these organelles has been proposed. Championed by Lynn Margulis and others, this hypothesis, called the **endosymbiont theory**, states that mitochondria and chloroplasts may have originated as distinct bacterialike particles that became incorporated into primitive eukaryotic cells. In the evolution of this symbiotic relationship, the particles lost their ability to function independently and underwent an extensive evolution, while the eukaryotic host cell became dependent on them. Although there are many unanswered questions, the basic tenets of this theory are difficult to dispute.

# ORGANIZATION OF DNA IN CHROMATIN

The structure and organization of the genetic material in **eukaryotic cells** is much more intricate than in viruses or bacteria. This complexity is due to the greater amount of DNA per chromosome and the presence of large numbers of proteins associated with DNA in eukaryotes. For example, while DNA in the *E. coli* chromosome is 1200 $\mu$m long, the DNA in human chromosomes ranges from 14,000 to 73,000 $\mu$m in length. In a single human nucleus, all 46 chromosomes contain sufficient DNA to extend almost 2 meters! This genetic material, along with its associated proteins, is contained within a nucleus that usually measures about 5 $\mu$m in diameter.

For many reasons this intricacy is to be expected. It parallels the structural and biochemical diversity of the many types of eukaryotic cells present in a multicellular organism. Different cells assume specific functions based upon highly specific biochemical activity. While all cells carry a full genetic complement, different cells activate different sets of genes. A highly ordered regulatory system governing the readout of this information must exist if dissimilar cells performing different functions are present. Such a system must in some way be imposed on or related to the molecular structure of the genetic material.

While bacteria can reproduce themselves, they never exhibit a detailed process similar to mitosis. As described in Chapter 2, eukaryotic cells exhibit a highly organized cell cycle. During interphase, the genetic material is dispersed finely throughout the nucleus as chromatin. When mitosis begins, the chromatin condenses greatly, and during prophase it is compressed into recognizable chromosomes. This condensation represents a contraction in length of some 10,000 times for each chromatin fiber and is the basis of the **folded-fiber model** of chromosome structure (Figure 2.14). This highly regular condensation-uncoiling cycle poses special organizational problems in eukaryotic genetic material.

Early studies of the structure of eukaryotic genetic material concentrated on intact chromosomes, preferably large ones, because of the limitation of light microscopy. Discussion of two such structures—polytene and lampbrush chromosomes—was presented in Chapter 2. Subsequently, the development of techniques for biochemical analysis, as well as the examination of relatively intact eukaryotic chromatin and mitotic figures under the electron microscope, have greatly enhanced our understanding of chromosome structure.

As established earlier, the genetic material of viruses and bacteria consists of strands of DNA or RNA nearly devoid of proteins. In eukaryotes, the structure of the chromosome is much more complex. A substantial amount of protein is associated with the chromosomal DNA in all phases of the eukaryotic cell cycle. Such material is referred to as **chromatin** during interphase, when it is uncoiled. The associated proteins are divided into basic, positively charged **histones** and less positively charged **nonhistones**.

**Table 10.4** CATEGORIES AND PROPERTIES OF HISTONE PROTEINS

| Histone Type | Lysine–Arginine Content | Molecular Weight (daltons) |
|---|---|---|
| H1 | Lysine-rich | 21,000 |
| H2A | Slightly lysine-rich | 14,500 |
| H2B | Slightly lysine-rich | 13,700 |
| H3 | Arginine-rich | 15,300 |
| H4 | Arginine-rich | 11,300 |

## Nucleosome Structure

The general model for chromatin structure is based on the assumption that chromatin fibers, composed of DNA and protein, undergo extensive coiling and folding as they are condensed within the cell nucleus. Of the proteins associated with DNA, the histones clearly play the most essential structural role. Histones contain large amounts of the positively charged amino acids lysine and arginine, making it possible for them to bond electrostatically to the negatively charged phosphate groups of nucleotides. Recall that a similar interaction has been proposed for several bacterial proteins. There are five main types of histones (Table 10.4).

X-ray diffraction studies confirm that histones play an important role in chromatin structure. Chromatin produces regularly spaced diffraction rings, suggesting that repeating structural units occur along the chromatin axis. If the histone molecules are chemically removed from chromatin, the regularity of this diffraction pattern is disrupted.

The basic model for chromatin structure was worked out in the mid-1970s. Several observations are relevant to the development of this model:

1. Digestion of chromatin by certain endonucleases, such as micrococcal nuclease, yields DNA fragments that are approximately 200 base pairs in length, or multiples thereof. This demonstrates that enzymatic digestion is not random, for if it were, we would expect a wide range of fragment sizes. Thus, chromatin consists of some type of repeating unit, each of which is protected from enzymatic cleavage, except where any two units are joined. It is the area between all units that is attacked and cleaved by the nuclease.

2. Electron microscopic observations of chromatin have revealed that chromatin fibers are composed of linear arrays of spherical particles (Figure 10.8). Discovered by Ada and Donald Olins, the particles

occur regularly along the axis of a chromatin strand and resemble beads on a string. These particles are now referred to as *ν*-bodies or **nucleosomes** (*ν* is the Greek letter nu). These findings conform nicely to the earlier proposal, which suggests the existence of repeating units.

3. Results of the study of precise interactions of histone molecules and DNA in the nucleosomes constituting chromatin show that histones H2A, H2B, H3, and H4 occur as two types of tetramers: $(H2A)_2 \cdot (H2B)_2$, and $(H3)_2 \cdot (H4)_2$. First demonstrated by Roger Kornberg, he predicted that each repeating nucleosome unit consists of one of each tetramer in association with about 200 base pairs of DNA. Such a structure is consistent with previous observations and provides the basis for a model that explains the interaction of histone and DNA in chromatin.

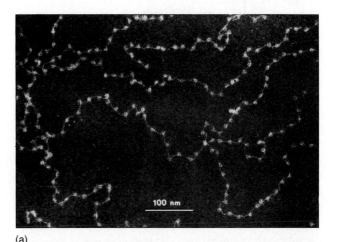

(a)

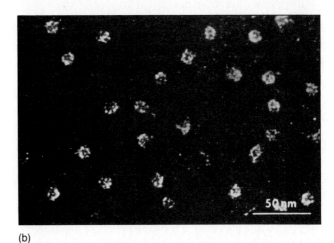

(b)

**FIGURE 10.8**   (a) Dark-field electron micrograph of nucleosomes present in chromatin derived from a chicken erythrocyte nucleus. (b) Dark-field electron micrograph of nucleosomes produced by micrococcal nuclease digestion.

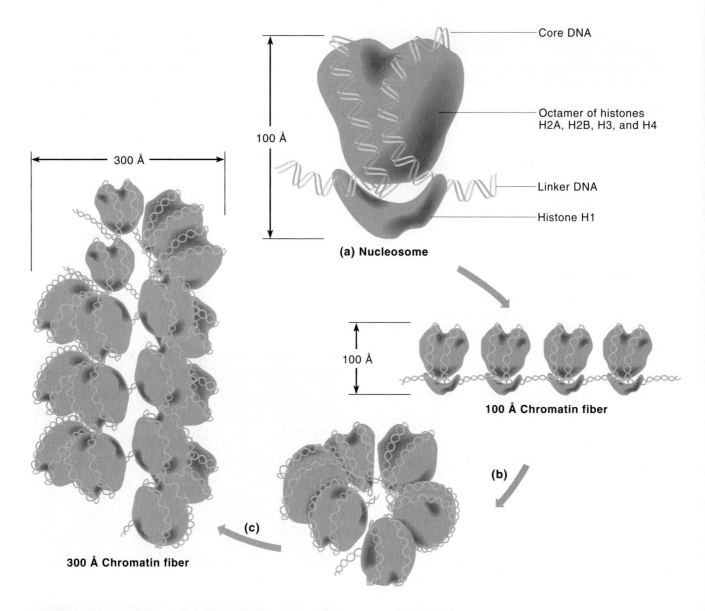

**FIGURE 10.9**    (a) General model of the association of histones and DNA in the nucleosome. (b) and (c) Schematic illustration of the way in which the chromatin fiber may be coiled into a more condensed structure.

4. When nuclease digestion time is extended, DNA is removed from both the entering and exiting strands, creating a **nucleosome core particle** consisting of 146 base pairs. This number is consistent in all organisms studied. The DNA lost in this prolonged digestion is responsible for linking nucleosomes together. This **linker DNA** is associated with histone H1.

5. On the basis of the above information as well as X-ray and neutron scattering analysis of crystallized core particles by John T. Finch, Aaron Klug, and others, a detailed model of the nucleosome was put

forward [Figure 10.9(a)]. In this model, the 146 base-pair DNA core exists as a secondary helix surrounding the octamer of histones. The coiled DNA does not quite complete two full turns. The entire nucleosome is ellipsoidal in shape, measuring about 100 Å in its greatest dimension.

The extensive investigation of nucleosomes now provides the basis for predicting how the packaging of chromatin within the nucleus occurs. In its most extended state under the electron microscope, the chromatin fiber is about 100 Å in diameter, a size consistent with the longer dimension of the ellipsoidal nucleosome. Thus, it is believed that chromatin fibers consist of long strings of repeating nucleosomes [Figure 10.9(b)].

The 100-Å chromatin fiber is apparently further folded into a structure 300 Å in diameter, as viewed under the electron microscope. This transition has been studied carefully and can be accounted for in our nucleosome model [Figure 10.9(c)]. The 300-Å fiber appears to consist of 5 or 6 nucleosomes coiled closely together.

In the overall transition from a fully extended chromatin fiber to the extremely condensed status of the mitotic chromosome, a packing ratio (the ratio of DNA length to the length of the structure containing it) of about 500 must be achieved. Our model accounts for a ratio of about 50. Obviously, the larger fiber can be further bent, coiled, and packed as even greater condensation occurs during the formation of a mitotic chromosome.

## Heterochromatin

All recent evidence supports the concept that the DNA of each eukaryotic chromosome consists of one continuous double-helical fiber along its entire length. A continuous fiber is the basis of the **unineme model of DNA** within a chromosome. This concept, along with our knowledge of nucleosomes, might lead you to suspect that the chromosome would demonstrate structural uniformity along its length. However, in the early part of this century, it was observed that some parts of the chromosome remain condensed and stain deeply during interphase, but most are uncoiled and do not stain. In 1928, the terms **heterochromatin** and **euchromatin** were coined to describe the parts of chromosomes that remain condensed and those that are uncoiled, respectively.

Subsequent investigation has revealed a number of characteristics of heterochromatin that distinguish it from euchromatin. Heterochromatic areas are genetically inactive, because they either lack genes or they contain genes that are repressed. Also, heterochromatin replicates later during the S phase of the cell cycle than does euchromatin. The discovery of heterochromatin provided the first clues that parts of eukaryotic chromosomes do not always encode proteins. Instead, some chromosome regions are thought to be involved in the maintenance of the chromosome's structural integrity in other functions such as chromosome movement during cell division.

Heterochromatin is characteristic of the genetic material of eukaryotes. Early cytological studies showed that areas of the centromeres are composed of heterochromatin. The ends of chromosomes, called telomeres, are also heterochromatic. In some cases, whole chromosomes are heterochromatic. Such is the case with the mammalian Y chromosome, which for the most part is genetically inert. And, as we discussed in Chapter 6, the inactivated X chromosome in mammalian females is condensed into an inert Barr body. In some species, such as mealy bugs, all chromosomes of one entire haploid set are heterochromatic.

When certain heterochromatic areas from one chromosome are translocated to a new site on the same or another nonhomologous chromosome, genetically active areas sometimes become genetically inert if they now lie adjacent to the translocated heterochromatin. This influence on existing euchromatin is one example of what is more generally referred to as a **position effect**. That is, the position of a gene or groups of genes relative to all other genetic material may affect their expression. We first introduced this term in Chapter 7.

## Satellite DNA and Repetitive DNA

Chapter 8 and Appendix A describe techniques useful in the analysis of DNA. Two of these—sedimentation equilibrium centrifugation and reassociation kinetics—have provided important information about the nature of DNA in chromatin. Specifically, the DNA of heterochromatin and euchromatin exhibits different molecular characteristics. The nucleotide composition of the DNA (e.g., the percentage of $G \equiv C$ versus $A = T$ pairs) of a particular species is reflected in its density, which can be measured with sedimentation equilibrium centrifugation. When the DNA of any eukaryotic species is analyzed in this way, the majority of it shows up as a single main peak or band of uniform density. However, one or more satellite peaks represent DNA that differs slightly in nucleotide composition and density from main-band DNA. This component, called **satellite DNA**, represents a variable proportion of the total DNA, depending on the species. A profile of main-band and satellite DNA from the mouse is shown in Figure 10.10. It should be noted that bacteria, as representative prokaryotes, contain only main-band DNA.

The significance of satellite DNA remained an enigma until the mid 1960s, when the techniques for measuring the reassociation kinetics of DNA became available. These techniques, developed by Roy Britten and David Kohne, allowed geneticists to determine the rate of reassociation of fragmented DNA. Britten and Kohne showed that when complementary strands of DNA fragments were separated by heating and then reassociated by cooling, certain portions were capable of reannealing more rapidly than others. They concluded that rapid reannealing was characteristic of multiple DNA fragments composed of identical or nearly identical nucleotide sequences—the basis of **repetitive DNA**.

As discussed in Chapter 8, repeating nucleotide sequences are classified as either highly or moderately re-

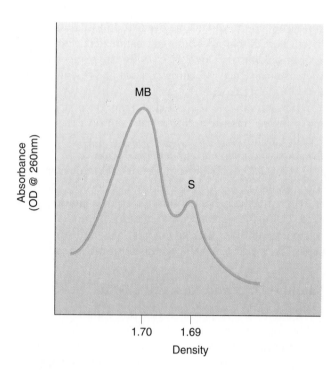

**FIGURE 10.10**    Separation of main-band (MB) and satellite (S) DNA from the mouse using ultracentrifugation in a CsCl gradient.

petitive DNA. Evidence has now accumulated linking satellite DNA to repetitive DNA. Further, this specific DNA fraction has been shown to be present in heterochromatic regions of chromosomes. The following paragraphs review information supporting and clarifying these conclusions.

When satellite DNA is subjected to analysis by reassociation kinetics, it falls into the category of highly repetitive DNA. It therefore consists of short sequences repeated a large number of times. The available evidence further suggests that these sequences are present as tandem repeats clustered at various positions in the genome.

Satellite DNA is found in very specific chromosome areas known to be heterochromatic—the **centromeric regions**. This was discovered in 1969 when several workers, including Mary Lou Pardue and Joe Gall, applied the technique of *in situ* **hybridization** to the study of satellite DNA. This technique (see Figure 8.23 and Appendix A) involves the molecular hybridization between an isolated fraction of radioactive or fluorescent DNA or RNA and the DNA contained in the chromosomes of a cytological preparation. Following the hybridization procedure, autoradiography is performed to locate the chromosome areas complementary to the fraction of DNA or RNA.

In their work, Pardue and Gall demonstrated that RNA transcribed from mouse satellite DNA hybridizes with DNA of centromeric regions of mouse mitotic chromosomes (Figure 10.11). Thus, several conclusions can be drawn. Satellite DNA differs from main-band DNA in its molecular composition, as established by buoyant density studies. It is also composed of short repetitive sequences. Finally, satellite DNA is found in the heterochromatic centromeric regions of chromosomes.

These observations establish one basis of the structural and functional differences between heterochromatin and euchromatin: the nucleotide composition is quite different. They further reflect the complexity of the genetic material in eukaryotes. Instead of viewing the chromosome as a linear series of genes, we must now see it as quite variable in its molecular organization and functional capacity. For example, there are nucleotide sequences that do not specify gene products. Are these genetically inert DNA sequences of heterochromatin related to the maintenance of the structural integrity of chromatin? Or, might they be involved in a regulatory role of transcription? Such sequences may play more than one role. There is no doubt that more extensive investigation will continually expand and modify our concept of eukaryotic genetic organization.

## Chromosome Banding

Until about 1970, mitotic chromosomes viewed under the light microscope could be distinguished only by their

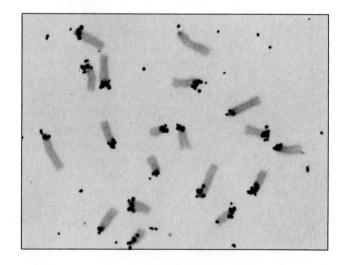

**FIGURE 10.11**    *In situ* hybridization between RNA transcribed by mouse satellite DNA and mitotic chromosomes. The grains in the autoradiograph localize the chromosome regions (the centromeres) containing satellite DNA sequences.

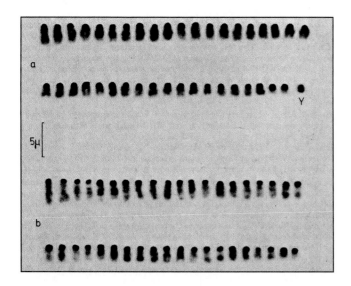

**FIGURE 10.12**    Karyotypes of a male (a) and female (b) mouse where chromosome preparations were processed to demonstrate C-bands, where only the centromeres stain.

relative sizes and the positions of their centromeres. Even in organisms with a low haploid number, two or more chromosomes are often indistinguishable from one another. However, around 1970, differential staining along the longitudinal axis of mitotic chromosomes was made possible by new cytological procedures. These methods are now called **chromosome banding techniques** because the staining patterns resemble the bands of polytene chromosomes.

One of the first chromosome banding techniques was devised by Mary Lou Pardue and Joe Gall. They found that if chromosome preparations were heat denatured and then treated with Giemsa stain, a unique staining pattern emerged. The centromeric regions of mitotic chromosomes preferentially took up the stain! Thus, this cytological technique stained a specific area of the chromosome composed of heterochromatin. A diagram of the mouse karyotype treated in this way is shown in Figure 10.12. The staining pattern is referred to as **C-banding**.

Still other chromosome banding techniques were developed about the same time. A group of Swedish researchers, led by Tobjorn Caspersson, used a technique that provided even greater staining differentiation of metaphase chromosomes. They used fluorescent dyes that bind to nucleoprotein complexes and produce unique banding patterns. When the chromosomes are treated with the fluorochrome quinacrine mustard and viewed under a fluorescent microscope, precise patterns of differential brightness are seen. Each of the 23 human

chromosome pairs can be distinguished by this technique. The bands produced by this method are called **Q-bands**.

Another banding technique produces a staining pattern nearly identical to the Q-bands. This method, producing **G-bands** (Figure 10.13), involves the digestion of the mitotic chromosomes in the cytological preparations with the proteolytic enzyme trypsin followed by Giemsa staining. Another technique results in the reverse G-band staining pattern, called an **R-band** pattern. In 1971, a meeting was held in Paris to establish the nomenclature for these various patterns (Figure 10.14). Still other banding techniques are now available.

Intense efforts are currently under way to elucidate the molecular mechanisms involved in producing these banding patterns. The variety of staining reactions under different conditions reflects the heterogeneity and complexity of chromosome composition.

## ORGANIZATION OF THE EUKARYOTIC GENOME

Having established the general view of the chromatin fiber, we now turn to a consideration of the organization of the **genome** of eukaryotic organisms. The genome

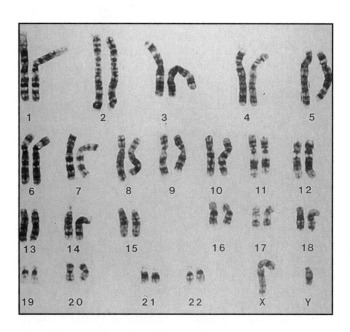

**FIGURE 10.13**    G-banded karyotype of a normal human male showing approximately 400 bands per haploid set of chromosomes. Chromosomes were derived from cells in metaphase.

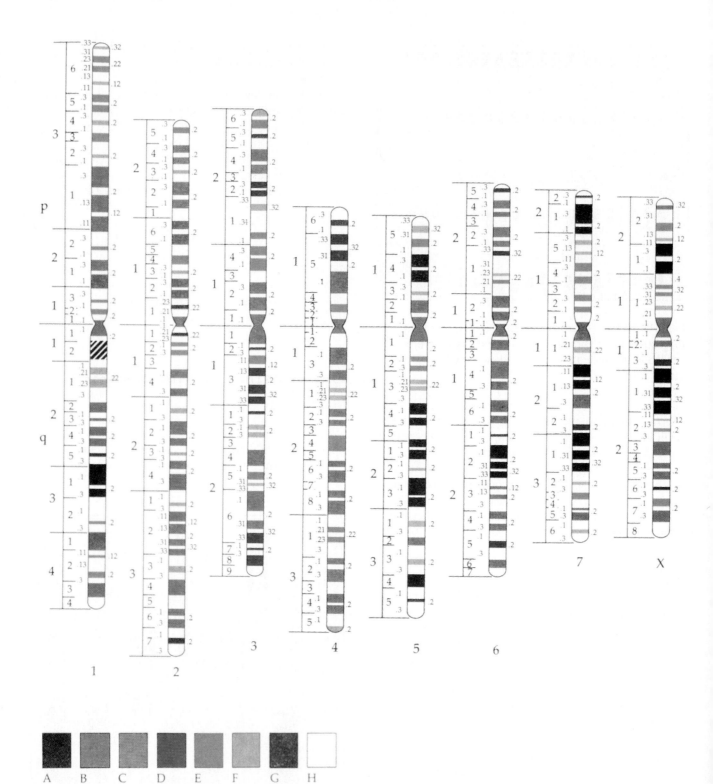

**FIGURE 10.14** Schematic representation and nomenclature of human chromosomes at the 1000-band stage using G-banding. Different shades represent varying intensities of bands, as shown in the key. The cross-hatched areas (chromosomes 1, 9, 16, and Y) represent heterochromatic regions.

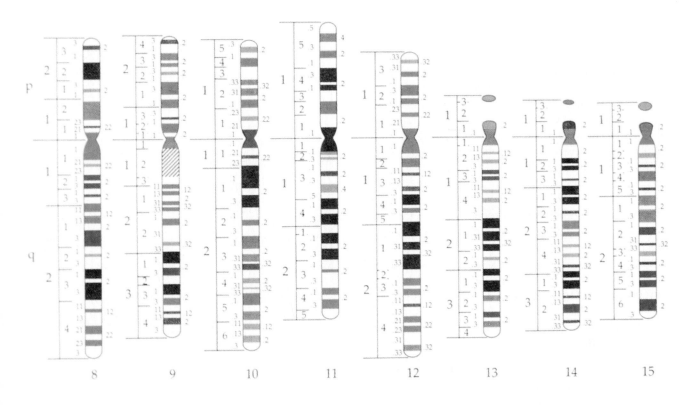

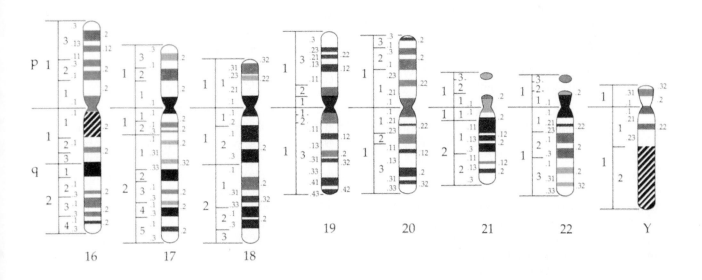

includes the entire haploid set of chromosomes constituting the genetic material of an organism.

Prior to 1970, there was little evidence at the molecular level involving eukaryotic gene structure and organization. Nevertheless, many geneticists believed that when the structure of genes and the organization of the genome of eukaryotes was revealed, fundamental differences from viruses and bacteria might be evident. Multicellular eukaryotes, it was reasoned, have several unique requirements distinguishing them from phages and bacteria. For example, cell differentiation, tissue organization, and coordinated development and function depend on genetic expression and its regulation. Do these requirements also depend on different modes of gene structure and organization?

When the technologies of DNA analysis were developed (Chapter 12), the answer to this question began to emerge. The eukaryotic gene is far more complex than we could ever have imagined. In this section, we will review many findings related to the organization of the eukaryotic genome and the structure of genes.

## Eukaryotic Genomes and the *C* Value Paradox

The amount of DNA contained in the haploid genome of a species is called the *C* **value**. When such values were determined for a large variety of eukaryotic organisms (Figure 10.15) several trends were apparent. The most notable trends are that:

1. Eukaryotes contain substantially more DNA in their genomes than viruses or prokaryotes and exhibit a wide variation among different groups.

2. Evolutionary progression has been accompanied by increased amounts of DNA. In many major phylogenetic groups, more evolutionarily advanced forms generally contain more DNA than most less advanced forms.

It might be argued that increasing *C* values are simply the result of a greater need for increased amounts and varieties of gene products in more complex organisms. However, several observations make this explanation unacceptable. First, the increases are very dramatic. While viruses and bacteria contain $10^4$ to $10^7$ nucleotide pairs in their single chromosomes, eukaryotes contain $10^7$ to $10^{11}$ nucleotide pairs in each set of chromosomes. Does a eukaryote require $10^4$ (10,000) times as many genes than the most complex bacterium, as these figures might suggest? Second, closely related organisms with the same degree of complexity in body form and tissue and organ types often vary tenfold or more in DNA con-

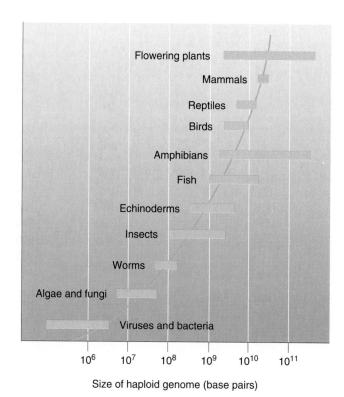

**FIGURE 10.15**    The range of DNA content of the haploid genome of many representative groups of organisms.

tent. Third, the DNA contents of amphibians and flowering plants, which vary as much as 100-fold within their taxonomic classes, are often greater than those of other more recently evolved eukaryotes. In spite of this observation, there is no correlation between increased genome size and the morphological complexity of these groups.

Therefore, it is doubtful that the development of greater complexity during evolution can account for the amount of DNA found in eukaryotic genomes. This conclusion is the basis of what has been called the *C* **value paradox**: *Excess DNA is present that does not seem to be essential to the development or evolutionary divergence of eukaryotes.* Is such DNA vital to the organisms carrying it in their genomes, or is it simply DNA that has somehow accumulated during evolution and has no function?

## Repetitive DNA: Centromeres and Telomeres

Careful analysis of the eukaryotic genome is beginning to partially unravel the mystery surrounding the *C* value paradox. For example, in Chapter 8 we established that noncoding sequences are prevalent in the form of various types of repetitive DNA. Earlier in this chapter, we demonstrated that satellite DNA of mammals such as the mouse is highly repetitive and is found to be associated with the centromere. As we will see below, an understanding of the molecular organization of this and a second heterochromatic region, the telomere, provides important insights into the molecular organization of the genome.

We shall begin our discussion with a consideration of the **centromere**, the region of the chromosome that attaches to one or more spindle fibers and moves to one of the poles during the anaphase stages of mitosis and meiosis (see Chapter 2).

Separation of chromatids is essential to the fidelity of chromosome distribution during mitosis and meiosis. Most estimates of infidelity during the mitotic transmission of chromosomes are exceedingly low: $1 \times 10^{-5}$ to $1 \times 10^{-6}$, or 1 error per 100,000 to 1,000,000 cell divisions. As a result, it has been generally assumed that analysis of the DNA sequence of centromeric regions will provide insights into the rather remarkable features of this chromosomal region. This DNA region is designated the **CEN**.

Analysis of the CEN regions of yeast *Saccharomyces cerevisiae* chromosomes provided the basis for a model system first described by John Carbon and Louis Clarke. Since each centromere serves an identical function, it is not surprising that all CENs were found to be remarkably similar in their organization. The CEN region of yeast chromosomes consists of about 225 base pairs, which can be divided into three regions (Figure 10.16). The first and third regions (I and III) are relatively short and highly conserved, consisting of only 8 and 26 base pairs,

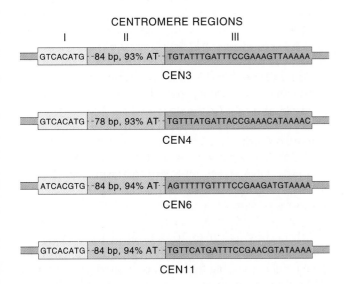

**FIGURE 10.16**     Nucleotide sequence information derived from DNA of the three major centromere regions of chromosomes 3, 4, 6, and 11 of yeast.

respectively. Region II, which is larger (80–85 base pairs) and extremely A-T rich (up to 95%), varies in sequence between different chromosomes.

Mutational analysis suggests that regions I and II are less critical to centromere function than region III. Genetic alteration of the former regions is often tolerated, but mutations in region III are usually disruptive to centromere function. The DNA sequence of this region may be essential to binding to spindle fibers.

The amount of DNA associated with centromeres of multicellular eukaryotes is much more extensive than in yeast. Recall from an earlier discussion in this chapter that highly repetitive "satellite" DNA is localized in the centromere regions of mice. Such sequences, absent from yeast but characteristic of most multicellular organisms, vary considerably in size. For example, the 10 base-pair sequence AATAACATAG is tandemly repeated many times in the centromeres of all four chromosomes of *Drosophila*. In humans, one of the most recognized satellite DNA sequences is the **alphoid family**. Found mainly in the centromere regions, alphoid sequences, each about 170 base pairs in length, are present in tandem arrays of up to 1 million base pairs.

The role of this highly repetitive DNA in centromere function remains unclear; it is known that the sequences are not transcribed. There is some sequence variation among members of the alphoid family; the number of repeats is specific to each human chromosome. Perhaps there is a shorter sequence common to the alphoid family that, like yeast CEN DNA, is absolutely critical to centromere function.

The second important structure of chromosomes is that portion of the linear chromosome constituting each end, called the **telomere**. One of its major functions is to provide stability, aiding in the maintenance of individual integrity of the chromosome. This role is accomplished by rendering chromosome ends generally inert in interactions with other chromosome ends. In contrast to broken chromosomes, whose ends may rejoin other such ends, telomere regions do not fuse with one another or with broken ends. Thus, it has been thought that some aspect of the DNA sequence of chromosome ends must be unique compared with most other chromosome regions.

As with centromeres, the problem was first approached by investigating the smaller chromosomes of simple eukaryotes, such as protozoans and yeast. The idea that all telomeres in a given species might share a common nucleotide sequence has now been borne out.

Two types of telomere sequences have been discovered. The first type, called simply **telomeric DNA sequences**, consists of short tandem repeats. It is this group that contributes to the stability and integrity of the chromosome. In the ciliate *Tetrahymena*, over fifty tandem repeats of the hexanucleotide sequence GGGGTT occur. In another ciliate, *Oxytricha*, GGGGTTTT is repeated many times. In humans, the sequence GGGATT is repeated numerous times. The analysis of telomeric DNA sequences has shown them to be highly conserved throughout evolution, reflecting the critical role they play in maintaining the integrity of chromosomes.

The second type is called **telomere-associated sequences**. These sequences are also repetitive and are found both adjacent to and within the telomere. These sequences vary from one another in different organisms, and their significance remains unknown.

While it is not yet clear how telomeric DNA sequences render the ends of chromosomes inert, the role of these sequences in the process of DNA replication at the ends of chromosomes is more apparent (see Chapter 9). At chromosome ends, replication of the lagging strand poses difficulty for initiation of synthesis. As a result, the chromosome could theoretically be shortened during each replication cycle. A unique enzyme, **telomerase**, whose function is closely linked to the telomeric DNA sequences, overcomes this problem. Thus, these sequences play an important role during DNA replication at the ends of chromosomes.

## Repetitive DNA: SINEs, LINEs, and VNTRs

A brief review of other prominent categories of repetitive DNA sheds further light on our understanding of the eukaryotic genome. In this current discussion, we will also account for more of the genomic DNA that fails to encode proteins.

In addition to highly repetitive DNA, which constitutes about 5 percent of the human genome (and 10% of the mouse genome), a second category, **moderately** (or **middle**) **repetitive DNA**, is fairly well characterized. Since a great deal is being learned about the human genome, we will use our own species to illustrate the part played by this category of DNA in genome organization.

Moderately repetitive DNA consists of either interspersed or tandemly repeated sequences. That which is interspersed is either short or long. **Short interspersed elements**, called **SINEs**, are less than 500 base pairs long and are present as much as 500,000 times.

The best characterized human SINE is a set of closely related sequences called the ***Alu* family** (the name is based on the presence of DNA recognition sequences by the restriction endonuclease *Alu*). Members of this DNA family, also found in other mammals, are 200 to 300 base pairs long and are present 700,000 to 900,000 times throughout the genome, both between and within genes. In humans, this family encompasses almost 10 percent of the entire genome.

*Alu* sequences are particularly interesting, even though their function, if any, is yet undefined. Members of the *Alu* family are sometimes transcribed. The role of this RNA is unclear, but in some cases it may be related to the observation that *Alu* elements are potentially transposable within the genome. *Alu* sequences are thought to have arisen from an RNA element whose DNA complement was dispersed throughout the genome as a result of the activity of reverse transcriptase (an enzyme that synthesizes DNA on an RNA template).

**Long interspersed elements (LINEs)** represent still another category of moderately repetitive DNA. In humans, the most prominent example is a family of designated **L1**. Members of this sequence family are around 6400 base pairs long and are present between 3000 and 40,000 times, according to different estimates. Their 5′ end is highly variable, while their role within the genome has yet to be defined.

Moderately repetitive DNA may also be clustered as tandem repeats rather than interspersed throughout the genome. In some cases, this category includes functional genes present in multiple copies. For example, many copies of genes encoding 18*S* and 28*S* rRNA in humans are clustered on the p arm of the acrocentric chromosomes 13, 14, 15, 21, and 22. The genes encoding 5.0*S* rRNA are clustered together on the terminal portion of the p arm of chromosome 1. We will return momentarily to discuss this arrangement of tandemly repeated genes within the genome.

A second type of tandem repeats includes those called **variable number tandem repeats (VNTRs)**. The repeating DNA sequence or VNTRs may be 15 to 100 base pairs long. The number of tandem copies of any specific sequence varies in individuals, creating localized regions of 1000 to 5000 base pairs in length. As will see in Chapter 13, such regions in humans are the basis for the forensic technique referred to as **DNA fingerprinting**. VNTRs may be found within and between genes.

Together, the various forms of moderately repetitive DNA comprise up to 30 percent of the human genome. However, we are left with over 60 percent of the genome still unaccounted for. Such observations are not uncommon. While the proportion of the genome consisting of repetitive DNA varies between organisms, one feature seems to be shared: *only a small part of the genome codes for proteins*. For example, sea urchins contain an estimated 20,000 to 30,000 genes coding for proteins, which occupy less than 10 percent of the genome. In *Drosophila*, only 5 to 10 percent of the genome is occupied by genes coding for proteins. In humans, it appears that only 1 to 2 percent of the genome encodes proteins! We must conclude that while there is sufficient DNA to code for over 1 million genes in many eukaryotic organ-

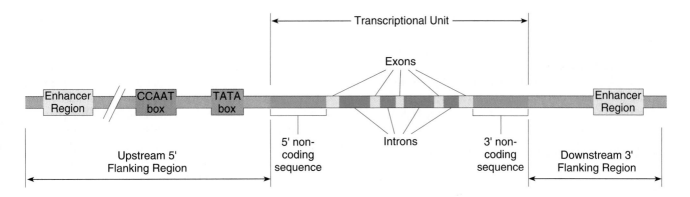

**FIGURE 10.17**    Modern concept of the eukaryotic gene.

isms, *the majority of DNA does not represent genes encoding proteins.*

## Eukaryotic Gene Structure

We turn now to a discussion of the structure and organization of the eukaryotic gene. This information serves as the foundation of our understanding of how genes function, a topic to be pursued in depth in Chapter 14. As our discussion ensues, refer to Figure 10.17, which presents a generalized model of the eukaryotic gene.

An important insight has been derived from a comparison of mRNA molecules and the DNA of the genes from which they are derived. Such studies involve direct sequencing of these nucleic acids and molecular hybridization studies. It has been found that many internal base sequences of genes referred to as intervening sequences, or just **introns**, never appear in the mature mRNAs that are translated. The remaining areas of each gene that are ultimately translated into the amino acid sequence of the encoded protein are called **exons**. While the entire gene is transcribed into a large precursor mRNA, called **heterogeneous nuclear RNA (hnRNA)**, the introns are excised and the exons are spliced back together prior to translation. We will pursue the topic of introns and splicing in greater detail in Chapter 14.

Given this information about internal gene sequences, we can ask whether the nucleotide sequences found in introns solve the *C* value paradox. That is, do introns account for the remainder of the noncoding sequences beyond those of repetitive DNA? The answer is emphatically no! As shown in Table 10.5, the presence of introns may increase the average gene by four or five times our estimate of the nucleotide length occupied by single-copy genes. However, since this estimate is seldom above 10 percent of the genome and usually much less, a great deal of noncoding DNA is still unaccounted for.

Aside from introns, we shall review briefly the pres-

ence and nature of the noncoding flanking regions to complete our discussion of the eukaryotic gene. It is now clear that the 5′ region upstream from the sequence encoding a protein is critical to efficient RNA transcription. Closest to the coding sequence is the **promoter** with its **TATA box**, where RNA polymerase binds prior to initiation of transcription. The TATA box is so named because of a sequence of that nature has been conserved in most eukaryotic genes. Farther upstream is the **CCAAT box**, which is somehow involved in the regulation of transcription. In numerous cases, an **enhancer** region is also present. Enhancers may also be found within the coding sequence or as part of the 3′ downstream region, beyond the coding sequence. These regions play a role in modulating transcription. Note also that in Figure 10.17 there are noncoding sequences that are part of the transcriptional unit. These are part of the

**Table 10.5**    A Comparison of mRNAs and the Initial Transcripts from Which They Are Derived

| Gene | mRNA Size | Minimum Transcript Length | Ratio |
|---|---|---|---|
| Rabbit β-globin | 589 | 1295 | 2.2 |
| Mouse $\beta^{maj}$-globin | 620 | 1382 | 2.2 |
| Mouse $\beta^{min}$-globin | 575 | 1275 | 2.2 |
| Mouse α-globin | 585 | 850 | 1.5 |
| Rat insulin I | 443 | 562 | 1.3 |
| Rat insulin II | 443 | 1061 | 2.4 |
| Chick ovalbumin | 1859 | 7500 | 4.0 |
| Chick ovomucoid | 883 | 5600 | 6.3 |
| Chick lysozyme | 620 | 3700 | 6.0 |

primary transcript of a gene, but they are trimmed off as the mRNA matures.

Since flanking regions are critical to specific gene function, they are considered part of each gene. Their discovery has extended our knowledge of gene structure considerably. In Chapter 14, we shall return to these topics as we discuss the process of transcription in detail.

## Exon Shuffling and Protein Domains

At first glance, the discovery of introns and exons is most puzzling. Why, during the evolution of eukaryotes, would their genes acquire noncoding introns between the coding portions of genes whose presence requires highly precise splicing following transcription and whose replication requires a substantial expenditure of energy? While we do not yet have an absolute answer, important clues are emerging primarily as a result of our ability to determine the nucleotide sequence of many genes. With the knowledge of the sequence of the exons of genes, the amino acid sequences of the corresponding proteins can be predicted accurately. When composite data bases are established and compared by computer analysis, the uniqueness of genes can be examined. As a result of this approach, some very interesting findings have been forthcoming, as first recognized and proposed by Walter Gilbert in 1977. He suggested that genes in higher organisms may consist of collections of exons originally present in ancestral genes that are brought together through recombination during the course of evolution. Referring to this process as **exon shuffling**, he proposed that exons are modular in the sense that each might encode a functional domain in the structure of a protein. For example, a particular amino acid sequence encoded by a single exon might always create a specific type of fold in a protein. This **protein domain** may impart a unique but characteristic type of molecular interaction to all molecules containing such a fold. Serving as the basis of useful parts of a protein, many similar domains might be functional in a variety of proteins. In Gilbert's proposal, during evolution, many exons could be mixed and matched to form unique genes in eukaryotes.

Several observations lend support to this proposal. Most exons are fairly small, averaging about 150 nucleotides and encoding about 50 amino acids. This size is consistent with the production of functional domains in proteins. Second, recombinational events that lead to exon shuffling would be expected to occur within areas of genes represented by introns. Since introns are free to accumulate mutations without harm to the organism, recombinational events would tend to further randomize their nucleotide sequences. Over extended evolutionary periods, diverse sequences would tend to accumulate. This is, in fact, what is observed. Introns range from 50 to 20,000 bases and exhibit fairly random base sequences.

Since 1977, a vast research effort has been directed toward the analysis of split genes. In 1985, more direct evidence in favor of Gilbert's proposal of exon modules was presented. For example, analysis was made of the human gene encoding the membrane receptor for low density lipoproteins (LDL). This **LDL receptor protein** is essential to the transport of plasma cholesterol into the cell. It mediates **endocytosis** and is expected to have numerous functional domains. These include the capability of this protein to bind specifically to the LDL substrates and to interact with other proteins at different levels of the membrane during transport across it. In addition, this receptor molecule is modified posttranslationally by the addition of a carbohydrate; a domain must exist that links to this carbohydrate.

Detailed analysis of the gene encoding this protein supports the concept of exon modules and their shuffling during evolution. The gene is quite large—45,000 nucleotides—and contains 18 exons. These represent only slightly less than 2600 nucleotides. These exons are related to the functional domains of the protein *and* appear to have been recruited from other genes during evolution.

Figure 10.18 illustrates these relationships. The first exon encodes a signal sequence that is removed from the protein before the LDL receptor becomes part of the membrane. The next five exons represent the domain specifying the binding site for cholesterol. This domain is made up of a 40-amino-acid sequence repeated seven times. The next domain consists of a sequence of 400 amino acids bearing a striking homology to the mouse peptide hormone **epidermal growth factor (EGF)**. This region is encoded by eight exons and contains three repetitive sequences of 40 amino acids. A similar sequence is also found in three blood-clotting proteins. The fifteenth exon specifies the domain for the posttranslational addition of the carbohydrate, while the remaining two exons specify regions of the protein that are part of the membrane, anchoring the receptor to specific sites called coated pits on the cell surface.

These observations are fairly compelling in favor of the theory of exon shuffling during evolution. Certainly, there is no disagreement concerning the concept of protein domains being responsible for specific molecular interactions.

What does remain controversial and evocative in the exon shuffling theory is the question of when introns first appeared on the evolutionary scene. In 1978, W. Ford Doolittle proposed that these intervening sequences were part of the genome of the most primitive ancestors of modern-day eukaryotes. In support of this idea, Gilbert has argued that if intron similarities in DNA sequence are found in identical positions within genes

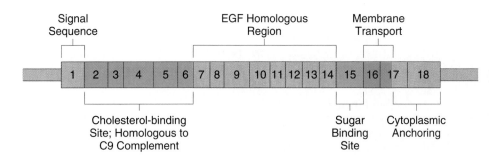

**FIGURE 10.18** A comparison of the 18 exons making up the LDL receptor protein gene and their organization into five functional domains and one signal sequence.

shared by totally unrelated eukaryotes (such as humans, chickens, and corn), they must have also been present in primitive ancestral genomes.

If Doolittle's proposal is correct, why are introns absent in most prokaryotes? Gilbert argues that they were present at one point during evolution, but as the genome of these primitive organisms evolved, they were lost. This occurred as a result of strong selection pressure to streamline their chromosomes to minimize energy expenditure supporting replication and gene expression. Further, streamlining leads to more error-free mRNA production. However, supporters of the opposing "intron-late" school, including Jeffrey Palmer, argue against this speculation and believe that much later during evolution, introns became a part of a single group of eukaryotes that are ancestral to modern-day members but not prokaryotes. Palmer has also argued that intron patterns are quite different in animals as compared to plants. While it may be difficult to resolve, it seems certain that this most interesting controversy will persist for many years to come.

## Multigene Families: The α- and β-Globin Genes

Still another aspect of the organization of the eukaryotic genome includes the distribution of related genes. Called **multigene families**, members share DNA sequence homology, and their gene products are functionally related. Often, but not always, they are found together in a single location along a chromosome. The examination of multigene families further provides valuable insights into the *C* value paradox.

We will first examine cases where groups of genes encode very similar, but not identical, polypeptide chains that become part of proteins that are very closely related in function. The globin gene families, responsible for encoding the various polypeptides that are part of hemoglobin molecules, are examples that provide many insights into the organization of the genome. Then, we will consider the case of identical, tandemly repeated genes, as represented by histone and ribosomal RNA genes.

The human **alpha-** and **beta-globin gene families** are two of the most extensive and best characterized groups. The alpha family resides on the short arm of chromosome 16 and contains five genes, while the beta family resides on the short arm of chromosome 11 and contains six genes. Members of both families show homology to one another, but not nearly to the extent members within the same family exhibit homology to one another. Members of both families encode globin polypeptides that combine into a single tetrameric molecule, which interacts with a heme group that can reversibly bind to oxygen (see Figure 15.13). And, within each group of genes, members are coordinately turned on and off during the embryonic, fetal, and adult stages of development in the precise order in which they occur along the chromosome within each family.

The alpha family [Figure 10.19(a)] spans more than 30,000 nucleotide pairs (30 kb) and contains five genes: the zeta gene ($\zeta$), expressed only in the embryonic stage, two nonfunctional **pseudogenes**, and two copies of the alpha ($\alpha$) gene, expressed during the fetal and adult stages. Pseudogenes, found in both families, are similar in sequence to the other genes in their family, but contain significant nucleotide substitutions, deletions, and duplications that prevent their transcription. The first pseudogene in the alpha family shows greater homology to the zeta gene, while the second shows greater homology to the alpha gene. Pseudogenes are designated by the prefix $\psi$ (the Greek letter psi), followed by the symbol of the gene they most resemble. Thus, the designation $\psi\alpha1$ indicates a pseudogene of the first alpha gene.

An examination of the organization of the members of the alpha family as well as their intragenic distribution of introns and exons [Figure 10.19(a)] reveals several interesting features. First, only a small portion of the chromosomal region housing the entire family is occupied by the three functional genes. The region consists almost entirely of intergenic regions. Second, the functional genes in this family contain two introns at precisely the same positions within the genes. The nucleotide sequences within corresponding exons are nearly identical when the zeta and alpha genes are compared. Both encode polypeptide chains that are 141 amino acids long.

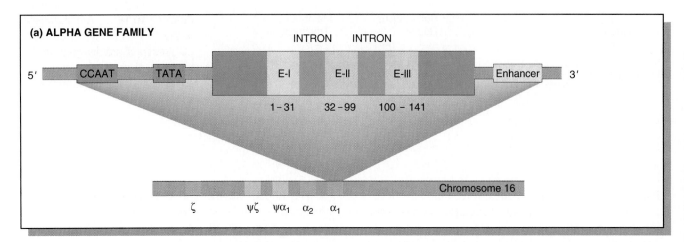

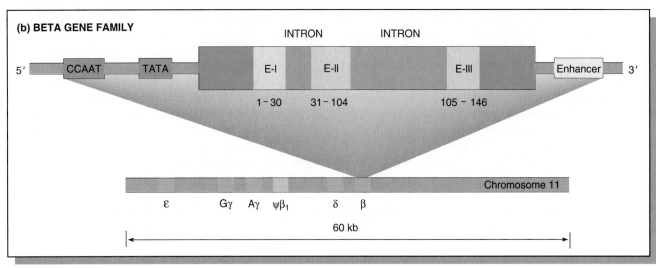

**FIGURE 10.19**    The comparative arrangement of the human alpha and beta gene families, depicted in parts (a) and (b), respectively. The internal organization of the functional genes from each family are also shown. All genes contain three exons (E–I, II, and III) and two introns, as well as 5' and 3' noncoding regions. The arabic numbers designate the amino acid positions encoded by each exon.

However, the intron sequences are extremely divergent, even though they are about the same size. Significantly, a high percentage of the nucleotide sequence within each gene is contained in these noncoding introns. The above information bears on both the structural organization of a gene family as well as on our earlier discussion of the *C* value paradox.

The beta-globin family in humans is even more extensive, containing six genes and occupying a 60-kb sequence of DNA [Figure 10.19(b)]. As with the alpha family, the sequence of genes along the chromosome parallels the order of expression during development. A pseudogene is also present.

Of the six genes, three are expressed prior to birth: the epsilon ($\epsilon$) gene is embryonic, while two nearly identical gamma genes ($^G\gamma$ and $^A\gamma$) are fetal. The two gamma gene products vary by only a single amino acid. The two remaining functional genes, delta ($\delta$) and beta ($\beta$), are expressed following birth. Finally, a single pseudogene $\psi\beta1$ is present within the family.

All five functional genes encode products that are 146 amino acids long and contain two similarly sized introns at exactly the same positions. The second intron is significantly larger than its counterpart in the functional alpha genes. These introns and their locations within the gene are compared in Figure 10.19.

The flanking regions of each gene have also been studied. In all functional genes, both a TATA box (29 or 30 base pairs upstream) and a CCAAT region (70–78 base pairs upstream from the initial codon) have been observed. Downstream, globin enhancers have been identified.

Several observations of the β-globin genes pertain to the *C* value paradox. Only about 5 percent of the 60-kb region consists of coding sequences. The remaining 95 percent includes the introns, flanking regions, and spacer DNA found between genes. Of this percentage, only about 11 percent consists of introns. The remainder of the DNA serves no known function and consists of sequences not found elsewhere in the genome. The majority of DNA in the cluster is therefore either excess nonfunctional DNA *or* DNA whose function has yet to be discovered.

## The Histone Gene Family

A variation on the theme established in the globin gene families is illustrated by the **histone genes**. Here, a cluster of five related but nonidentical genes, separated from each other by highly divergent nontranscribed spacer sequences, is tandemly repeated many times. As you recall, histones are positively charged (basic) proteins that interact with the negatively charged phosphate groups of DNA to form the nucleosomes characteristic of the chromatin fiber. The need for histone proteins is extremely great because of the large amount of DNA in eukaryotic chromosomes. In rapidly dividing cells of some organisms, not only is the need great, but the necessary quantity must appear rapidly as one cell cycle runs into the next, time and time again, without pause to catch up with the cell's biochemical needs. To accommodate this need, evolution appears to have seized upon the duplication of the entire cluster in order to increase the number of copies of each histone gene.

Many interesting correlations exist that support this contention. Yeast cells, with much less DNA than other eukaryotes, have only two copies of each of four histone genes (they lack histone H1). Contrast this with the number of times the entire cluster is tandemly repeated in many evolutionarily advanced organisms, as displayed in Table 10.6. From 10 to 600 repeats occur. Sea urchins, which display the largest number of duplications of the cluster, are noted for their extremely rapid rate of cell division during development.

Beyond the repeat nature of the histone gene family, there are several other differences in comparison to the globin families. First, almost all histone genes lack introns. The entire cluster of genes is considerably smaller

| Organism | Repeats |
|---|---|
| Chickens | 10 |
| Mammals | 22 |
| *Xenopus* | 40 |
| *Drosophila* | 100 |
| Sea urchins | 300–600 |

**Table 10.6** THE NUMBER OF TANDEMLY REPEATED HISTONE GENE CLUSTERS FOUND IN A VARIETY OF EUKARYOTIC ORGANISMS

than either globin cluster. While some variation exists between organisms, due primarily to the different sizes of the noncoding spacer sequences, most clusters are less than 10 kb in size. The other interesting difference is the observation that the individual genes within the cluster are often oriented in opposite directions.

As shown in Figure 10.20, the arrangement of genes and their polarity of transcription vary in the sea urchin and *Drosophila*, two organisms where this cluster has been well studied. Polarity differences exist in other organisms studied, as well. Despite polarity differences, the amino acid sequences of the various histones (reflecting the DNA encoding each one) are remarkably similar in highly divergent organisms. Homology of histone genes is one of the best examples known of sequence conservation throughout evolution.

In mice and humans, members of this gene family appear to be clustered but not present as a tandem array of repeat units. For example, a region of the distal tip of the long arm of the human chromosome 7 contains all members of the histone gene family, but they may be interspersed with other nonhistone genes. Therefore, no single pattern of organization applies to all organisms.

## Tandem Repeat Families: rRNA Genes

We conclude our discussion of the organization of multigene families by considering the example of **tandem repeats** of the gene family encoding rRNA molecules present in ribosomes. This gene family encodes three molecules in the following order along the chro-

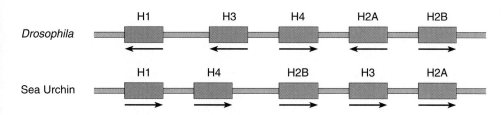

**FIGURE 10.20** The histone gene clusters in *Drosophila* and the sea urchin. The arrows indicate the polarity of transcription.

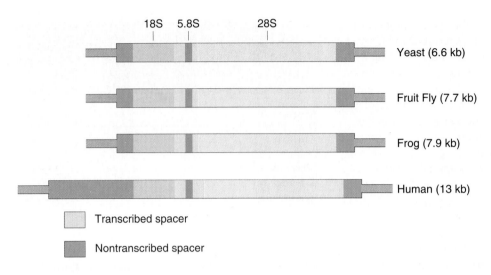

**FIGURE 10.21**    Comparison of the repeating rRNA gene unit in various eukaryotic organisms.

mosome: 18*S*, 5.8*S*, and 28*S* rRNAs (Figure 10.21). These RNAs are transcribed as a single unit and are not translated. Each repeated unit contains substantial regions that are part of the initial transcript, but that are not found in the final rRNA gene products. Thus, the initial transcript must be processed in a manner *analogous* to the removal of introns from the primary transcript of most mRNAs. In addition, there is a nontranscribed spacer region between each unit. Note that one other RNA component—5.0*S* rRNA—is encoded elsewhere in the genome.

The basic repeating unit that is transcribed varies in length in different organisms from almost 7 kb in yeast to 13 kb in humans (Figure 10.21). Together, each gene family unit and its accompanying spacer regions occupy a substantial portion of DNA. In the human rRNA gene family, a length of about 43 kb of DNA is utilized in each tandem repetition. This amount of DNA is as large as the smaller globin gene cluster.

Multiple copies of a single transcriptional unit encoding rRNAs also exist in bacteria. In *E. coli*, seven dispersed copies are present, presumably to accommodate the need for the rapid production of the approximately 15,000 ribosomes found in each cell. It is not surprising, therefore, to find many more than seven copies in the larger, more complex eukaryotic cells. Yeast contains over 100, while *Xenopus, Drosophila*, and humans have up to 400 copies per genome. A single eukaryotic cell may have a million or more ribosomes if it is actively producing proteins.

If all of these genes were present in a single cluster in an organism such as humans, a substantial part of one chromosome would be occupied by them (400 × 45 kb = 18,000 kb or 18 million base pairs). This amount of DNA is almost five times greater than the entire *E. coli* chromosome ($4 \times 10^6$ base pairs)! As it turns out, they are present in clusters dispersed throughout the genome. In humans, for example, the clusters are found near the ends of chromosomes 13, 14, 15, 21, and 22. In *Drosophila*, they are present on both the X and Y chromosomes. Within the nucleus, these various gene regions are collectively referred to as the **nucleolus organizer regions (NORs)**, because of their association with the production of ribosomes in the nucleolus.

This information extends our knowledge and overall picture of how some genes are organized within the genome. It also provides insights into the vast amount of DNA that may be utilized in order to provide cells with one set of gene products. Examination of the same gene family in different organisms also provides information valuable to the study of evolution at the molecular level.

**CHAPTER SUMMARY**

1. Knowledge of the organization of the molecular components forming chromosomes is essential to the understanding of the function of the genetic material. Largely devoid of associated proteins, bacteriophage and bacterial chromosomes contain DNA molecules in a form equivalent to the Watson–Crick model.

2. Mitochondria and chloroplasts contain DNA that encodes products essential to their biological function. This DNA is remarkably similar in form and appearance to some bacterial and phage DNA, lending support to the endosymbiont theory, which suggests that these organelles were once free-living organisms.

3. The eukaryotic chromatin fiber is a nucleoprotein organized into repeating units called nucleosomes. Composed of about 200 base pairs of DNA and an octamer of four types of histones, the nucleosome is important in facilitating the conversion of the extensive chromatin fiber characteristic of interphase into the highly condensed chromosome seen in mitosis.

4. The structural heterogeneity of the chromosome axis has been established as a result of both biochemical and cytological investigation. Heterochromatin, prematurely condensed in interphase, is genetically inert. The centromeric and telomeric regions, the Y chromosome, and the Barr body are examples.

5. DNA analysis has unique nucleotide sequences in both the heterochromatic centromere and telomere regions. These sequences distinguish these structures from other parts of the chromosome and provide the basis for the critical features imparted to the chromosome by these regions.

6. The *C* value paradox, made apparent during the study of the organization of eukaryotic DNA, suggests that eukaryotic organisms contain much more DNA than necessary to encode only gene products essential to the development and normal functions of an organism.

7. Detailed analysis of eukaryotic gene structure has revealed numerous categories of noncoding DNA sequences, much of which include various forms of repetitive DNA. Introns, internal noncoding sequences, are represented in the initial mRNA molecules or heterogeneous RNA, but never appear in the mature mRNAs from which proteins are synthesized.

8. The flanking regions of genes, particularly those giving rise to the 5′ initiation point of mRNA, have also been analyzed. Three upstream regions that appear to be essential to efficient transcription of mRNA are the TATA box, the CCAAT box, and the enhancer element.

9. Multigene families are composed of structurally related genes that are often clustered together and encode functionally similar products. The human alpha- and beta-globin families are two examples.

10. Histone and ribosomal RNA genes provide an example of tandem repeat families, where the same gene is duplicated many times.

---

## KEY TERMS

alpha- and beta-globin
  gene families
alphoid family
*Alu* family
bacteriophage lambda
CCAAT box
C-banding
CEN region
centromere
chloroplast
chloroplast DNA
  (cpDNA)
chromatin
chromosome banding
*C* value paradox
DNA-binding proteins
DNA fingerprinting
endocytosis
endosymbiont theory
enhancer

epidermal growth factor
  (EGF)
euchromatin
eukaryotic cells
exon
exon shuffling
folded-fiber chromosome
  model
G-banding
genome
heterochromatin
heterogeneous nuclear
  RNA (hnRNA)
histone
H and HU proteins
*in situ* hybridization
intron
linking number (*L*)
long interspersed element
  (LINE)
L1 family

maternal mode of
  inheritance
mitochondrial DNA
  (mtDNA)
mitochondrion
moderately (middle)
  repetitive DNA
multigene families
nonhistone proteins
nucleoid
nucleolus organizer
  region (NOR)
nucleosome (nu (ν-)
  body)
nucleosome core particle
polyoma virus
position effect
promoter
protein domain
pseudogenes
Q-banding

R-banding
repetitive DNA
ribulose-1-5-bisphosphate
  carboxylase (RuBP)
satellite DNA
short interspersed
  element (SINE)
supercoiled DNA
SV40 virus
tandem repeats
TATA box
telomerase
telomere
telomeric DNA sequences
T-even series of
  bacteriophages
topoisomer
topoisomerase
unineme model of DNA
variable number tandem
  repeat (VNTR)

---

## INSIGHTS AND SOLUTIONS

**A** previously undiscovered single-celled organism was found living at a great depth on the ocean floor. Its nucleus contained only a single linear chromosome containing $7 \times 10^6$ nucleotide pairs of DNA coalesced with three types of histonelike proteins.

1. A short micrococcal nuclease digestion yielded DNA fractions consisting of 700, 1400, and 2100 base pairs. What do these fractions represent? What conclusions can be drawn?

   **SOLUTION:** The chromatin fiber may consist of a variation of nucleosomes containing 700 base pairs of DNA. The 1400- and 2100-bp fractions represent two and three nucleosomes, respectively, linked together. Enzymatic digestion may have been incomplete, leading to the latter two fractions.

2. Analysis of individual nucleosomes revealed that each unit contained one copy of each protein, and that the short linker DNA contained no protein bound to it. If the entire chromosome consists of nucleosomes (discounting any linker DNA), how many are there, and how many total proteins are needed to form them?

   **SOLUTION:** Since the chromosome contains $7 \times 10^6$ base pairs of DNA, the number of nucleosomes, each containing $7 \times 10^2$ base pairs, is equal to

   $$\frac{7 \times 10^6}{7 \times 10^2} = 10^4 \text{ nucleosomes}$$

The chromosome thus contains $10^4$ copies of each of the three proteins, for a total of $3 \times 10^4$ molecules.

**3.** Analysis then revealed DNA to be a double helix similar to the Watson–Crick model, but containing 20 base pairs per complete turn of the right-handed helix. The physical size of the nucleosome was exactly double the volume occupied by that found in all other known eukaryotes, by virtue of increasing the distance along the fiber axis by a factor of two. Compare the degree of compaction of this organism's nucleosome to that found in other eukaryotes.

**SOLUTION:**   The unique organism compacts a length of DNA consisting of 35 complete turns of the helix (700 base pairs per nucleosome/20 base pairs per turn) into each nucleosome. The normal eukaryote compacts a length of DNA consisting of 20 complete turns of the helix (200 base pairs per nucleosome/10 base pairs per turn) into a nucleosome one-half the volume of that in the unique organism. The degree of compaction is therefore less in the unique organism.

**4.** No further coiling and compaction of this unique chromosome occur in the unique organism. Compare this to a eukaryotic chromosome. Do you think an interphase human chromosome $7 \times 10^6$ base pairs in length would be a shorter or longer chromatin fiber?

**SOLUTION:**   The eukaryotic chromosome contains still another level of condensation in the form of "solenoids," which are dependent on the H1 histone molecule associated with linker DNA. Solenoids condense the eukaryotic fiber by still another factor of five.

The length of the unique chromosome is compacted into $10^4$ nucleosomes, each containing an axis length twice that of the eukaryotic fiber. The eukaryotic fiber consists of $7 \times 10^6/2 \times 10^2 = 3.5 \times 10^4$ nucleosomes, 3.5 more than the unique chromosome organism. However, they are compacted by the factor of 5 in each solenoid. Therefore, the chromosome of the unique organism is a longer chromatin fiber.

**PROBLEMS AND DISCUSSION QUESTIONS**

**1.** Compare and contrast the chemical nature, size, and form assumed by the genetic material of viruses and bacteria.

**2.** Contrast the DNA associated with mitochondria and chloroplasts.

**3.** Why might it be predicted that the organization of eukaryotic genetic material would be more complex than that of viruses or bacteria?

**4.** Describe the sequence of research findings leading to the model of chromatin structure. What is the molecular composition and arrangement of the nucleosome? What is the solenoid (see Insights and Solutions)?

**5.** When chloroplasts and mitochondria are isolated, they are found to contain ribosomes that are similar to prokaryotic cells. How does this observation relate to the endosymbiont theory?

**6.** Provide a comprehensive definition of heterochromatin and list as many examples as you can.

**7.** Mammals contain a diploid genome consisting of at least $10^9$ base pairs. If this amount of DNA is present as chromatin fibers where each group of 200 base pairs of DNA is combined with 9 histones into a nucleosome, and each group of 5 nucleosomes is combined into a solenoid, achieving a final packing ratio of 50, determine:

(a) The total number of nucleosomes in all fibers.
(b) The total number of solenoids in all fibers.
(c) The total number of histone molecules combined with DNA in the diploid genome.
(d) The combined length of all fibers.

**8.** Assume that a viral DNA molecule is in the form of a 50-μm-long circular strand of a uniform 20-Å diameter. If this molecule is contained within a viral head that is a sphere with a diameter of 0.08 μm, will the DNA molecule fit into the viral head, assuming complete flexibility of the molecule? Justify your answer mathematically.

**9.** How many base pairs are in a molecule of phage T2 DNA, which is 52 μm long?

**10.** Describe what is meant by exon shuffling.

**11.** The β-globin gene family consists of 60 kb of DNA yet only 5 percent of the DNA encodes β-globin gene products. Account for the remaining 95 percent of the DNA.

**12.** Compare the histone gene family and the rRNA gene family with the globin gene families.

**13.** Describe the rRNA gene family in eukaryotes. What accounts for the major difference in the size of tandem repeats in different organisms?

**14.** "In the 1990s, the *C* value paradox is not so paradoxical." Agree or disagree with this statement, and support your position.

**SELECTED READINGS**

BAUER, W. R., CRICK, F. H. C., and WHITE, J. H. 1980. Supercoiled DNA. *Scient. Amer.* (July) 243:118–33.

BLACKBURN, E. H. 1991. Structure and function of telomeres. *Nature* 350:569–73.

BLACKBURN, E. H., and SZOSTAK, J. W. 1984. The molecular structure of centromeres and telomeres. *Ann. Rev. Biochem.* 53:163–94.

BLOOM, K., HILL, A., and YEH, E. 1986. Structural analysis of a yeast centromere. *BioEssays* 4:100–05.

CARBON, J. 1984. Yeast centromeres: Structure and function. *Cell* 37:352–53.

CHEN, T. R., and RUDDLE, F. H. 1971. Karyotype analysis utilizing differential stained constitutive heterochromatin of human and murine chromosomes. *Chromosoma* 34:51–72.

CORNEO, G., et. al. 1968. Isolation and characterization of mouse and guinea pig satellite DNA. *Biochemistry* 7:4373–79.

CRICK, F. H. C. 1979. Split genes and RNA splicing. *Science* 204:264–71.

DORIT, R. L., and GILBERT, W. 1991. The limited universe of exons. *Curr. Opin. Genet. Dev.* 1:464–69.

FRITSCHE, E. F., LAWN, R. M., and MANIATIS, T. 1980. Molecular cloning and characterization of the human β-like globin gene cluster. *Cell* 19:959–72.

GILBERT, W., MARCHIONNA, M., and MCKNIGHT, G. 1986. On the antiquity of introns. *Cell* 46:151–53.

GREEN, B. R., and BURTON, H. 1970. *Acetabularia* chloroplast DNA: Electron microscopic visualization. *Science* 168:981–82.

HENTSCHEL, C. C., and BIRNSTIEL, M. L. 1981. The organization and expression of histone gene families. *Cell* 25:301–13.

HEWISH, D. R., and BURGOYNE, L. 1973. Chromatin sub-structure. The digestion of chromatin DNA at regularly spaced sites by a nuclear deoxyribonuclease. *Biochem. Biophys. Res. Comm.* 52:504–10.

HSU, T. C. 1973. Longitudinal differentiation of chromosomes. *Ann. Rev. Genet.* 7:153–77.

KORENBERG, J. R., and RYKOWSKI, M. C. 1988. Human genome organization: *Alu*, LINES and the molecular structure of metaphase chromosome bands. *Cell* 53:391–400.

KORNBERG, R. D. 1975. Chromatin structure: A repeating unit of histones and DNA. *Science* 184:868–71.

KORNBERG, R. D., and KLUG, A. 1981. The nucleosome. *Scient. Amer.* (Feb.) 244:52–64.

MANIATIS, T., et al. 1980. The molecular genetics of human hemoglobins. *Ann. Rev. Genet.* 14:145–78.

MOYZIS, R. K. 1991. The human telomere. *Scient. Amer.* (Aug.) 265:48–55.

OLINS, A. L., and OLINS, D. E. 1974. Spheroid chromatin units (*v* bodies). *Science* 183:330–32.

———. 1978. Nucleosomes: The structural quantum in chromosomes. *Amer. Scient.* 66: 704–11.

PALMER, J. D., and LOGSDON, J. 1991. The recent origins of introns. *Curr. Opin. Genet. Dev.* 1:470.

SCHMID, C. W., and DEININGER, P. L. 1975. Sequence organization of the human genome. *Cell* 6:345–58.

SCHMID, C. W., and JELINEK, W. R. 1982. The *Alu* family of dispersed repetitive sequences. *Science* 216:1065–70.

SINGER, M. F. 1982. SINEs and LINEs: Highly repeated short and long interspersed sequences in mammalian genomes. *Cell* 28:433–34.

SUDHOF, T. C., et al. 1985. The LDL receptor gene: A mosaic of exons shared with different proteins. *Science* 228:815–22.

VAN HOLDE, K. E. 1989. *Chromatin*. New York: Springer-Verlag.

VERMA, R. S., ed. 1988. *Heterochromatin: Molecular and structural aspects*. Cambridge, England: Cambridge University Press.

WILLARD, H. F. 1990. Centromeres of mammalian chromosomes. *Trends in Genet.* 6:410–416.

WILLARD, H. F., and WAYE, J. S. 1987. Hierarchical order in chromosome-specific human alpha satellite DNA. *Trends in Genet.* 3:192–198.

YUNIS, J. J. 1981. Chromosomes and cancer: New nomenclature and future directions. *Hum. Pathol.* 12:494–503.

# 11

# DNA: MUTATION, REPAIR, AND TRANSPOSABLE ELEMENTS

A mutant erythrocyte derived from an individual with sickle-cell anemia.

*Aside from chromosomal mutations, the major basis of genetic diversity among organisms, even those that are closely related, is variation at the level of the gene. The origin of such variation is the process of mutation, whereby groups of nucleotide sequences are altered as a result of base substitution or the addition or deletion of one or more bases within those sequences. Relocation of mobile genetic elements within the genome also disrupts normal expression of genes, creating mutations. Fortunately, elaborate mechanisms have evolved to repair various errors and lesions within DNA. Gene mutations are the working tools of the geneticists and the raw material upon which evolution relies.*

In Chapter 8, we defined the four characteristics or functions ascribed to the genetic information: replication, storage, expression, and variation by mutation. In a sense, *mutation* is a failure to store the genetic information faithfully. If a change occurs in the stored information, it may be reflected in the expression of that information and will be propagated by replication. Historically, the term mutation includes both chromosomal changes and changes within single genes. We have discussed the former alterations in Chapter 6, referring to them collectively as chromosomal aberrations. In this chapter, we will be concerned with the so-called **gene mutations**. As we will see, a change may be a simple substitution of a single nucleotide or involve the addition or deletion of one or more nucleotides within the normal sequence of DNA.

The term mutation was coined in 1901 by Hugo DeVries to explain the variation he observed in crosses involving the evening primrose, *Oenothera lamarckiana*. Most of the variation was actually due to multiple translocations, but two cases were subsequently shown to be caused by gene mutations. As the studies of mutations progressed, it soon became clear that gene mutations serve as the source of most alleles and thus are the origin of much of the existing genetic variability within populations. As new alleles arise, they constitute the raw material to be tested by the evolutionary process of natural selection, which will determine whether they are detrimental, neutral, or beneficial.

Mutations provide the basis for genetic studies. The resulting phenotypic variability allows the geneticist to study the genes that control the traits that have been modified. In this regard, mutations serve as identifying "markers" of genes so that they may be followed during their transmission from parents to offspring. Without the phenotypic variability that mutations provide, all genetic crosses would be meaningless. For example, if all pea plants displayed a uniform phenotype, Mendel would have had no basis for his experimentation. Because of the importance of mutations, great attention has been given to their origin, induction, and classification.

Certain organisms lend themselves to induction of mutations that can be easily detected and studied throughout reasonably short life cycles. Viruses, bacteria, fungi, fruit flies, certain plants, and mice fit these criteria to various degrees. Thus, these organisms have often been used in studying mutation and mutagenesis and through other studies have also contributed to more general aspects of genetic knowledge.

Once we have completed our presentation of mutation, we will turn our attention to two related topics: DNA repair and transposable genetic elements. These topics are logical extensions of our consideration of gene mutation. Repair processes serve to counteract mutation. Transposable elements often disrupt the normal structure of the gene and therefore create mutations.

## CLASSIFICATION OF MUTATIONS

There are various schemes by which mutations are classified. They are not mutually exclusive, but instead depend simply on which aspects of mutation are being investigated or discussed. In this section, we will describe three sets of distinctions concerning mutations.

## Spontaneous versus Induced Mutations

All mutations are described as either **spontaneous** or **induced**. While these two categories overlap to some degree, **spontaneous mutations** are those that arise randomly in nature. No specific agent—other than natural forces—is associated with their occurrence. Both spontaneous and induced mutations result from changes in the nucleotide sequences of genes.

What causes spontaneous mutations? Although their origin is not fully understood, many such mutations can be linked to normal chemical processes or phenomena that result in rare errors. Such errors may cause alterations in the chemical structure of nitrogenous bases that are part of existing genes. They occur more frequently, however, during the enzymatic process of DNA replication, which we will explore later in this chapter. It is generally agreed that any natural phenomenon that heightens chemical reactivity in cells will lead to more errors. For example, background radiation from cosmic and mineral sources and ultraviolet radiation from the sun are energy sources to which most organisms are exposed. As such, they may be factors leading to spontaneous mutations. Once an error is present in the genetic code, it may be reflected in the amino acid composition of the specified protein. If the changed amino acid is present in a part of the molecule critical to its biochemical activity, a functional alteration may result.

In contrast to such spontaneous events, those that arise as a result of the influence of any artificial factor are considered to be **induced mutations**. The earliest demonstration of the induction of mutation occurred in 1927, when Herman J. Muller reported that X rays could cause mutations in *Drosophila*. In 1928, Lewis J. Stadler reported the same finding in barley. In addition to various forms of radiation, a wide spectrum of chemical agents is also known to be mutagenic, the aspects of which we will examine later in this chapter.

## Gametic versus Somatic Mutations

When considering the effects of mutation in eukaryotic organisms, it is important to distinguish whether the change occurs in somatic cells or gametes. Mutations arising in somatic cells are not transmitted to future generations. In tissues of an adult organism, thousands upon thousands of cells may be performing a similar function. Thus, a mutation in a single cell may not impair the organism, even if the mutation is detrimental. First, a random mutation might occur in a gene that is not active or essential to the function of that cell. Second, even if an active gene is affected, there are still thousands of unaffected cells to perform the function of the tissue in question.

Somatic mutations will have a greater impact if they are dominant or if they are sex-linked, since such mutations are most likely to be immediately expressed. Similarly, the impact will be more noticeable if a somatic mutation occurs early in development, where an undifferentiated cell will give rise to several differentiated tissues or organs.

There is a very important exception to the above description. If a mutation occurs that disrupts the control of cell proliferation, such a mutation may cause cells to divide uncontrollably, causing a cancerous condition. This topic will be pursued extensively in Chapter 18.

Mutations in gametes or gamete-forming tissue are part of the germline and are of greater concern because they have the potential of being expressed in all cells of an offspring. Such a mutation may also be transmitted to future generations, gradually increasing in frequency within the population. **Dominant autosomal mutations** will be expressed phenotypically in the first generation. **Sex-linked recessive mutations** arising in the gametes of a heterogametic female may be expressed in hemizygous male offspring. This will occur provided the male offspring receives the affected X chromosome. Because of heterozygosity, the occurrence of an **autosomal recessive mutation** in the gametes of either males or females (even one resulting in a lethal allele) may go unnoticed for many generations until it has become widespread in the population. The new allele will become evident only when a chance mating brings two copies of it together in the homozygous condition.

## Other Categories of Mutation

Various types of mutations are classified on the basis of their effect on the organism. Even though some of these categories were introduced earlier, we will briefly review several of them again. A single mutation may well fall into more than one category.

The most obvious mutations are those affecting a **morphological trait**. For example, all of Mendel's pea characters and most genetic variations encountered in the study of *Drosophila* fit this designation, since they cause obvious changes in the morphology of the organism during development. All such variations deviate from the normal or wild-type phenotype.

A second broad category includes mutations that are **nutritional** or **biochemical variations** from the norm. In bacteria and fungi, the inability to synthesize an amino acid or vitamin is an example of a typical nutritional mutation. In humans, hemophilia is an example of a biochemical mutation. While such mutations in these organisms are not visible and do not usually affect specific morphological characters, they can have a more general effect on the well-being and survival of the affected individual.

A third category consists of mutations that affect behavior patterns of an organism. For example, mating behavior or circadian rhythms of animals may be altered. The primary effect of **behavior mutations** is often difficult to discern. For example, the mating behavior of a fruit fly may be impaired if it cannot beat its wings. However, the defect may be in: (1) the flight muscles; (2) the nerves leading to them; or (3) the brain, where the nerve impulses that initiate wing movements originate. The study of behavior and the genetic factors influencing it has benefited immensely from investigations of behavior mutations.

Still another type of mutation may affect the regulation of genes. A regulatory gene may produce a product that controls the transcription of another gene. In other instances, a region of DNA either close to or far away from a gene may modulate its activity. In either case, **regulatory mutations** may disrupt this process and permanently activate or inactivate a gene. Our knowledge of genetic regulation has been dependent on the study of mutations that disrupt this process.

Another group consists of **lethal mutations**. Nutritional and biochemical mutations may also fall into this category. A mutant bacterium that cannot synthesize a specific amino acid it needs will cease to grow if plated on a medium lacking that amino acid. Various human biochemical disorders, such as Tay-Sachs disease and Huntington disease, are lethal at different points in the life cycle of humans.

Finally, any of the above groups can exist as **conditional mutations**. Even though a mutation is present in the genome of an organism, it may not be evident under certain conditions. The best examples are the **temperature-sensitive mutations**, found in a variety of organisms. At certain permissive temperatures, a mutant gene product functions normally, only to lose its functional capability at a different restrictive temperature. When shifted to this temperature, the impact of the mutation becomes apparent, even lethal, and is amenable to investigation. The study of conditional mutations has been extremely important in experimental genetics, particularly in understanding the function of genes essential to the viability of organisms.

# DETECTION OF MUTATION

Before geneticists can study directly the mutational process or obtain mutant organisms for genetic investigations, they must be able to detect mutations. The ease and efficiency of detecting mutations in a particular organism has by and large determined the organism's usefulness in genetic studies. In this section, we will use several examples to illustrate how mutations are detected.

## Detection in Bacteria and Fungi

Detection of mutations is most efficient in haploid microorganisms such as bacteria and fungi. Detection depends on a selection system where mutant cells are isolated easily from nonmutant cells. The general principles are similar in bacteria and fungi. To illustrate, we will describe how nutritional mutations in the fungus *Neurospora crassa* are detected. The life cycle of this organism was diagrammed in Figure 7.10.

*Neurospora* is a pink mold that normally grows on bread. It may also be cultured in the laboratory. This eukaryotic mold is haploid in the vegetative phase of its life cycle. Thus, mutations may be detected without the complications generated by heterozygosity in diploid organisms. Pioneering biochemical genetic studies with *Neurospora* were performed by George Beadle and Edward Tatum around 1940.

Visible mutants such as *albino* have been well studied, but the full potential of *Neurospora* genetics was attained during the investigation of nutritional mutants. Wild-type *Neurospora* grows on a **minimal culture medium** of glucose, a few inorganic acids and salts, a nitrogen source such as ammonium nitrate, and the vitamin biotin. Induced nutritional mutants will not grow on minimal medium, but will grow on a supplemented or **complete medium** that also contains numerous amino acids, vitamins, nucleic acid derivatives, and so forth. Microorganisms that are nutritional wild types (requiring only minimal medium) are called **prototrophs**, while those mutants that require a specific supplement to the minimal medium are called **auxotrophs**.

The procedural details for the detection of nutritional mutants are illustrated in Figure 11.1. Haploid conidia bearing such a mutation may be detected and isolated by their failure to grow on minimal medium and their ability to grow on complete medium. The mutant cells can no longer synthesize some essential compound absent in minimal medium but present in complete medium [Figure 11.1(a)]. Once a nutritional mutant is detected and isolated, the missing compound may be determined by attempts to grow the mutant strain in a series of tubes, each containing minimal medium supplemented with a single compound [Figure 11.1(b) and (c)]. In this example, mutants can grow if tyrosine is supplied. Thus, the mutation is a tyrosine auxotroph ($tyr^-$). Further details of this procedure will be presented in Chapter 15.

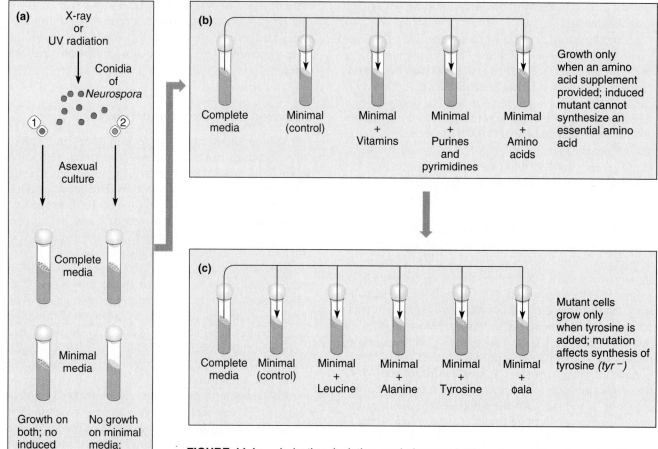

**FIGURE 11.1** Induction, isolation, and characterization of a nutritional auxotrophic mutation in *Neurospora*. In (a), conidia 1 is not affected, but conidia 2 contains such a mutation. In (b) and (c), the precise nature of the mutation is determined to involve the biosynthesis of tyrosine.

In Chapter 19, where the topics of bacterial and viral genetics are introduced, we will return to the topic of mutation. There, we will see that it was not until 1943 that spontaneous mutations were believed to occur in bacteria. Techniques for the detection and study of such mutations will be discussed. Analysis of mutations in microorganisms has been particularly important to the study of molecular genetics.

## Detection in *Drosophila*

Muller, in his studies demonstrating that X rays are mutagenic, developed a number of detection systems in *Drosophila melanogaster*. These systems allow the estimation of the spontaneous and induced rates of sex-linked and autosomal recessive lethal mutations. We will consider two techniques: the **ClB** and the **attached-X procedures**, which Muller devised. The *ClB* system (Figure 11.2) detects the rate of induction of sex-linked recessive lethal mutations. The *ClB* stock involves *C*, an inversion that suppresses the recovery of crossover

products; *l*, a recessive lethal allele; and *B*, the dominant gene duplication causing *Bar* eye, described in Chapter 6. All of these traits are located on the X chromosome.

In this technique, wild-type males are treated with a mutagenic agent—in this case, X rays—and mated to untreated, heterozygous *ClB* females. In the $F_1$ generation, females expressing *Bar* are selected. They received one X from their mother (the *ClB* chromosome) and one X from their father. Of the paternal X chromosomes, some will bear an induced X-linked lethal mutation (shown in color). When a female that becomes heterozygous for such a lethal mutation is backcrossed to a wild-type male, no male progeny will result. One-half of them die because they are hemizygous for the *ClB* chromosome (remember, *l* is lethal) and the other half die because they are hemizygous for the newly induced lethal allele. Following many single-pair matings such as described, the percentage of vials bearing *only* females represents the frequency of induction of X-linked lethal mutations.

For example, if 500 such $F_2$ culture bottles were in-

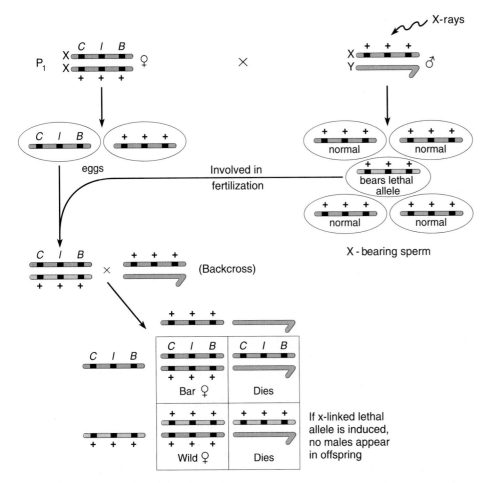

**FIGURE 11.2** Muller's classical *ClB* technique for the detection of induced, sex-linked recessive lethal mutations in *Drosophila*.

spected, and 25 contained only females, the induced rate of such mutations would be 5 percent. This method also permits the detection of sex-linked, morphological mutations. If a mutation of this type has been induced, all surviving $F_2$ males will show the trait.

The second technique, using females with attached-X chromosomes, is even simpler to use in detecting recessive morphological mutations because it requires only one generation. These females have two X chromosomes attached to a single centromere and one Y chromosome, in addition to the normal diploid complement of autosomes. When attached-X females are mated to males with normal sex chromosomes (XY), four types of progeny result: triplo-X females that die; viable attached-X females (XXY); YY males that also die; and viable XY males (who received their X from their father and their Y from their mother). Figure 11.3 illustrates how $P_1$ males that have been treated with a mutagenic agent produce $F_1$ male offspring that express any induced sex-linked recessive mutation. In contrast to the *ClB* technique, which tests only a single X chromosome, the attached-X method tests numerous X chromosomes at one time in a single cross.

Detection techniques have also been devised for recessive autosomal lethals in *Drosophila*. In these techniques, dominant marker mutations are followed through a series of three generations. While more cumbersome to perform, these techniques are also fairly efficient.

## Detection in Plants

Genetic variation in plants is extensive. Mendel's peas, for example, were the basis for the fundamental postulates of transmission genetics. Studies of plants have also enhanced our understanding of gene interaction, polygenic inheritance, linkage, sex determination, chromosome rearrangements, and polyploidy. Most variations are detected simply by visual observation. However, there are also techniques for the detection of biochemical mutations in plants. The first is the analysis of the biochemical composition of plants. For example, the isolation of proteins from maize endosperm, hydrolysis of the proteins, and determination of the amino acid composition have revealed that the **opaque-2 mutant strain** contains significantly more lysine than do other,

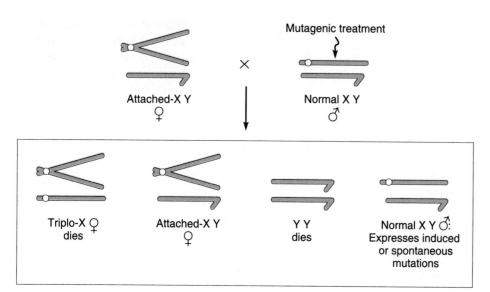

**FIGURE 11.3**    The attached-X method for detection of induced morphological mutations in *Drosophila*.

nonmutant lines. Plant geneticists now routinely analyze the amino acid compositions of new strains of grain crops, including corn, rice, wheat, barley, and millet. The results of these analyses are useful in combating malnutrition diseases resulting from inadequate protein or the lack of essential amino acids in the diet.

The second detection technique involves tissue culture of plant cell lines in defined medium. The plant cells are handled as microorganisms, and biochemical requirements may be determined by adding or deleting nutrients in the culture medium. There are other advantages to this method. Techniques associated with conditional lethal mutants can be used on plant cells in tissue culture and then applied to the genetics of higher plants. Also, this method provides a detection system that is generally not useful with the intact plant. Temperature-sensitive mutations in plants are beginning to be explored, particularly in tobacco. These studies may add significantly to our understanding of plant growth, metabolism, and genetics.

## Detection in Humans

Since humans are obviously not suitable experimental organisms, the techniques that have been developed for the detection of mutations in organisms such as *Drosophila* are not available to human geneticists. To determine the mutational basis for any human characteristic or disorder, geneticists must first analyze a pedigree that traces the family history as far back as possible. Once any trait has been shown to be inherited, it is possible to predict whether the mutant allele is behaving as a dominant or a recessive and whether it is sex-linked or autosomal.

Dominant mutations are the simplest to detect. If they

are present on the X chromosome, affected fathers pass the phenotypic trait to all their daughters. If dominant mutations are autosomal, approximately 50 percent of the offspring of an affected heterozygous individual are expected to show the trait. Figure 11.4 shows a hypothetical pedigree illustrating the initial occurrence of an autosomal dominant allele for cataracts of the eye. The parents in generation I were unaffected, but one of their three offspring (Generation II) developed cataracts. Presumably, the original mutation occurred in a gamete of one of the parents. The affected female, the proband, produced two children, of which the male child was affected (Generation III). Of his six offspring, four of six were also affected (Generation IV). These observations are consistent with, but do not prove, an autosomal dominant mode of inheritance. However, the high percentage of affected offspring in generation IV favors this conclusion. Also, the presence of an unaffected daughter in this generation argues against sex-linkage because she received her X chromosome from her affected father. Provided that the mutant allele is completely penetrant, such a conclusion is soundly based.

Sex-linked recessive mutations may also be detected by pedigree analysis, as discussed in Chapter 5. The most famous case of a sex-linked mutation in humans is that of **hemophilia**,\* which was found in the descendants of Queen Victoria. The recessive mutation for hemophilia has occurred many times in human populations, but the political consequences of the mutation that occurred in the royal family have been sweeping. Inspection of the pedigree in Figure 11.5 leaves little doubt

---

\* Robert Massie's *Nicholas and Alexandra* and Robert and Suzanne Massie's *Journey* (see the list of selected readings at the end of this chapter) provide fascinating reading on the topic of hemophilia.

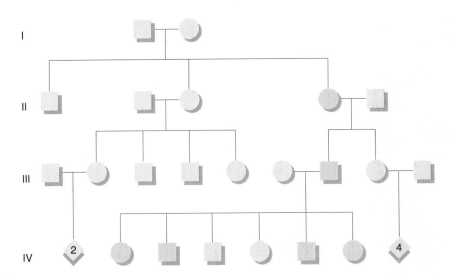

**FIGURE 11.4**    A hypothetical pedigree of inherited cataract of the eye in humans.

that Victoria was heterozygous (*Hh*) for the trait. Her father was not affected, and there is no reason to believe that her mother was a carrier, as was Victoria.

In a similar manner, it is possible to detect autosomal recessive alleles. Because this type of mutation is "hidden" when heterozygous, it is not unusual for the trait to appear only intermittently through a number of generations. An affected individual and a homozygous normal individual will produce unaffected carrier children. Matings between two carriers will produce, on the average, one-fourth affected offspring.

In addition to pedigree analysis, human cells may now be routinely cultured *in vitro*. This procedure has allowed the detection of many more mutations than any other form of analysis. Analysis of enzyme activity, protein migration in electrophoretic fields, and direct sequencing of DNA and proteins are among the techniques that have identified mutations and demonstrated wide genetic variation between individuals in human populations.

## SPONTANEOUS MUTATION RATE

The types of detection systems just described allow geneticists to estimate mutation rates. It is of considerable interest to determine the rate of spontaneous mutation. Such information provides insights into evolution and provides the baseline for measuring the rate of experimentally induced mutation. Induction of mutation can only be ascertained when the induced rate clearly exceeds the spontaneous rate for the organism under study.

Examination of this rate in a variety of organisms reveals many interesting points (see Table 11.1). First, the rate is exceedingly low for all organisms studied. Second, the rate is seen to vary considerably in different organisms. Third, even within the same species, the spontaneous mutation rate varies from gene to gene.

Viral and bacterial genes undergo spontaneous mutation on an average of about 1 in 100 million ($10^{-8}$) cell divisions. While *Neurospora* exhibits a similar rate, maize, *Drosophila*, and humans demonstrate a rate several orders of magnitude higher. The genes studied in these groups average between 1/1,000,000 and 1/100,000 ($10^{-6}$ and $10^{-5}$) mutations per gamete formed. Mouse genes are still another order of magnitude higher in their spontaneous mutation rate, 1/100,000 to 1/10,000 ($10^{-5}$ to $10^{-4}$). It is not clear why such a large variation occurs in mutation rate. The variation might reflect the relative efficiency of enzyme systems whose function is to repair errors created during replication. Repair systems will be discussed later in this chapter.

## MOLECULAR BASIS OF MUTATION

Since we have yet to describe in detail the genetic code and the processes of transcription and translation, we will use a *simplified* definition of a gene in the description of the molecular basis of mutation. In this context, it is easiest to consider a gene as a linear sequence of nucleotide pairs representing stored chemical information. Since the genetic code is a triplet, each sequence of three nucleotides specifies a single amino acid in the corresponding polypeptide. Beyond the coding sequences, genes also contain various types of regulatory sequences that are critical to their transcription. Any change that disrupts these sequences or the coded infor-

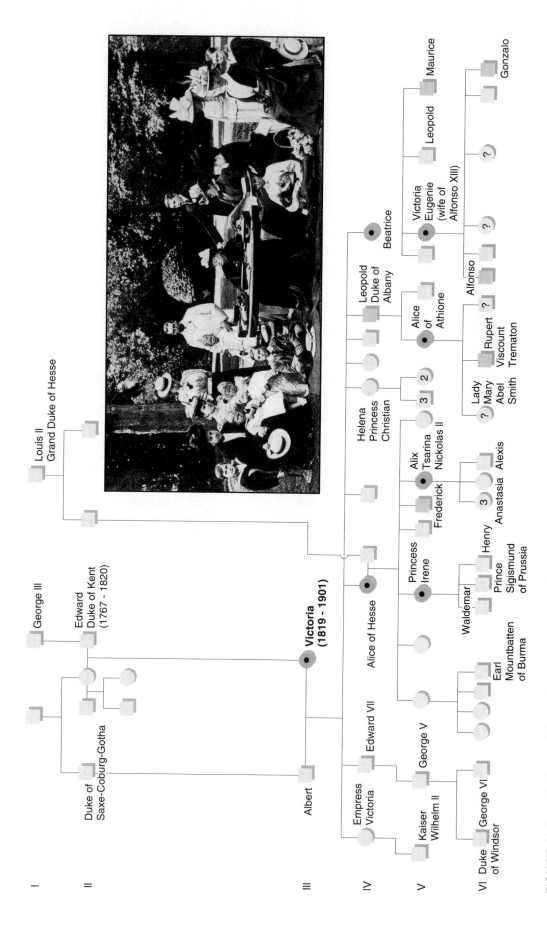

**FIGURE 11.5** Pedigree of hemophilia in the royal family descended from Queen Victoria. The pedigree is typical of the transmission of sex-linked recessive traits. Circles with a dot in them indicate presumed female carriers heterozygous for the trait. Circles with a question mark indicate females whose status is uncertain. The photograph shows Queen Victoria (seated front center) and some of her immediate family.

**Table 11.1** Rates of Spontaneous Mutations at Various Loci in Different Organisms

| Organism | Character | Gene | Rate | Units |
|---|---|---|---|---|
| Bacteriophage T2 | Lysis inhibition | $r \rightarrow r^+$ | $1 \times 10^{-8}$ | Per gene replication |
| | Host range | $b^+ \rightarrow b$ | $3 \times 10^{-9}$ | |
| E. coli | Lactose fermentation | $lac^- \rightarrow lac^+$ | $2 \times 10^{-7}$ | Per cell division |
| | Lactose fermentation | $lac^+ \rightarrow lac^-$ | $2 \times 10^{-6}$ | |
| | Phage T1 resistance | $T1\text{-}s \rightarrow T1\text{-}r$ | $2 \times 10^{-8}$ | |
| | Histidine requirement | $his^+ \rightarrow his^-$ | $2 \times 10^{-6}$ | |
| | Histidine independence | $his^- \rightarrow his^+$ | $4 \times 10^{-8}$ | |
| | Streptomycin dependence | $str\text{-}s \rightarrow str\text{-}d$ | $1 \times 10^{-9}$ | |
| | Streptomycin sensitivity | $str\text{-}d \rightarrow str\text{-}s$ | $1 \times 10^{-8}$ | |
| | Radiation resistance | $rad\text{-}s \rightarrow rad\text{-}r$ | $1 \times 10^{-5}$ | |
| | Leucine independence | $leu^- \rightarrow leu^+$ | $7 \times 10^{-10}$ | |
| | Arginine independence | $arg^- \rightarrow arg^+$ | $4 \times 10^{-9}$ | |
| | Tryptophan independence | $try^- \rightarrow try^+$ | $6 \times 10^{-8}$ | |
| Salmonella typhimurium | Trytophan independence | $try^- \rightarrow try^+$ | $5 \times 10^{-8}$ | Per cell division |
| Diplococcus pneumoniae | Penicillin resistance | $pen^s \rightarrow pen^r$ | $1 \times 10^{-7}$ | Per cell division |
| Chlamydomonas reinhardi | Streptomycin sensitivity | $str^r \rightarrow str^s$ | $1 \times 10^{-6}$ | Per cell division |
| Neurospora crassa | Inositol requirement | $inos^- \rightarrow inos^+$ | $8 \times 10^{-8}$ | Mutant frequency among asexual spores |
| | Adenine independence | $ade^- \rightarrow ade^+$ | $4 \times 10^{-8}$ | |
| Zea mays | Shrunken seeds | $sh^+ \rightarrow sh^-$ | $1 \times 10^{-6}$ | Per gamete per generation |
| | Purple | $pr^+ \rightarrow pr^-$ | $1 \times 10^{-5}$ | |
| | Colorless | $c^+ \rightarrow c$ | $2 \times 10^{-6}$ | |
| | Sugary | $su^+ \rightarrow su$ | $2 \times 10^{-6}$ | |
| Drosophila melanogaster | Yellow body | $y^+ \rightarrow y$ | $1.2 \times 10^{-6}$ | Per gamete per generation |
| | White eye | $w^+ \rightarrow w$ | $4 \times 10^{-5}$ | |
| | Brown eye | $bw^+ \rightarrow bw$ | $3 \times 10^{-5}$ | |
| | Ebony body | $e^+ \rightarrow e$ | $2 \times 10^{-5}$ | |
| | Eyeless | $ey^+ \rightarrow ey$ | $6 \times 10^{-5}$ | |
| Mus musculus | Piebald coat | $s^+ \rightarrow s$ | $3 \times 10^{-5}$ | Per gamete per generation |
| | Dilute coat color | $d^+ \rightarrow d$ | $3 \times 10^{-5}$ | |
| | Brown coat | $b^+ \rightarrow b$ | $8.5 \times 10^{-4}$ | |
| | Pink eye | $p^+ \rightarrow p$ | $8.5 \times 10^{-4}$ | |
| Homo sapiens | Hemophilia | $b^+ \rightarrow b$ | $2 \times 10^{-5}$ | Per gamete per generation |
| | Huntington disease | $Hu^+ \rightarrow Hu$ | $5 \times 10^{-6}$ | |
| | Retinoblastoma | $R^+ \rightarrow R$ | $2 \times 10^{-5}$ | |
| | Epiloia | $Ep^+ \rightarrow Ep$ | $1 \times 10^{-5}$ | |
| | Aniridia | $An^+ \rightarrow An$ | $5 \times 10^{-6}$ | |
| | Achondroplasia | $A^+ \rightarrow A$ | $5 \times 10^{-5}$ | |

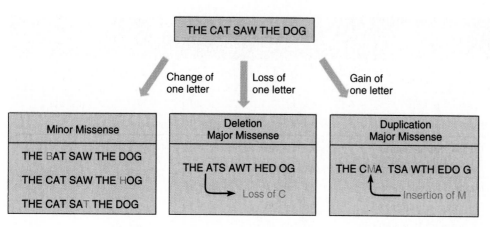

**FIGURE 11.6**    The impact of the substitution, addition, or deletion of one letter in a sentence composed of three-letter words, creating various levels of missense.

mation provides sufficient basis for a mutation. The simplest change resulting in a mutation is the substitution of a single nucleotide. In Figure 11.6 such a change is compared with our own written language, using three-letter words to be consistent with the genetic code. A single change can obviously alter the meaning of the sentence "THE CAT SAW THE DOG," creating what is called *missense*. These are analogies to what are most appropriately referred to as **base substitutions** or **point mutations**. The mutation has turned information that makes sense into various forms of missense.

Two terms are often used to describe nucleotide substitutions. If a pyrimidine replaces a pyrimidine or a purine replaces a purine, a **transition** has occurred. If a purine and a pyrimidine are interchanged, a **transversion** has occurred.

A second type of change that could occur is the insertion or deletion of one or more nucleotides at any point along the gene. As also illustrated in Figure 11.6 (provided that the insertion or deletion is not a multiple of three), the remainder of the words becomes garbled, creating even more extensive missense. These examples are called **frameshift mutations** because the frame of reading has become altered. We will return momentarily to a discussion of this type of mutation as we introduce acridine dyes (see below).

## Tautomeric Shifts

In 1953, immediately after they had proposed a molecular structure of DNA, Watson and Crick published a paper in which they discussed the genetic implications of this structure. They recognized that the purines and pyrimidines found in DNA could exist in **tautomeric forms**; that is, each can exist in several chemical forms, differing by only a single proton shift in the molecule.

Watson and Crick suggested that **tautomeric shifts** could result in base pair changes or mutations. The biologically important tautomers involve keto-enol pairs for thymine and guanine, and amino-imino pairs for cytosine and adenine (Figure 11.7).

The *infrequent* tautomer is capable of hydrogen bonding with a normally noncomplementary base. However, the pairing is always between a pyrimidine and a purine. Figure 11.8 compares the normal base-pairing relationships with the rare unorthodox pairings.

The effect occurs at a time of DNA replication when a rare tautomer in the template strand matches with a noncomplementary base. In the next round of replication, the "mismatched" members of the base pair are separated, and each specifies its normal complementary base. The end result is a mutation (see Figure 11.9).

## Base Analogues

**Base analogues**, which are mutagenic chemicals, are molecules that may substitute for purines or pyrimidines during nucleic acid biosynthesis. The halogenated derivative of uracil in the number-5 position of the pyrimidine ring **5-bromouracil (5-BU)*** is a good example. Figure 11.10 compares the structure of this thymine analogue with the structure of thymine. The presence of the bromine atom in place of the methyl group increases the probability that a tautomeric shift will occur. If 5-BU is incorporated into DNA in place of thymine and a tautomeric shift occurs, the result is an A = T to G ≡ C transition (see Figure 11.9). If the tautomeric shift to the enol form occurs before the analogue is incorporated into

---

*If 5-BU is chemically linked to d-ribose, the nucleoside analogue bromodeoxyuridine (BUdR) is formed.

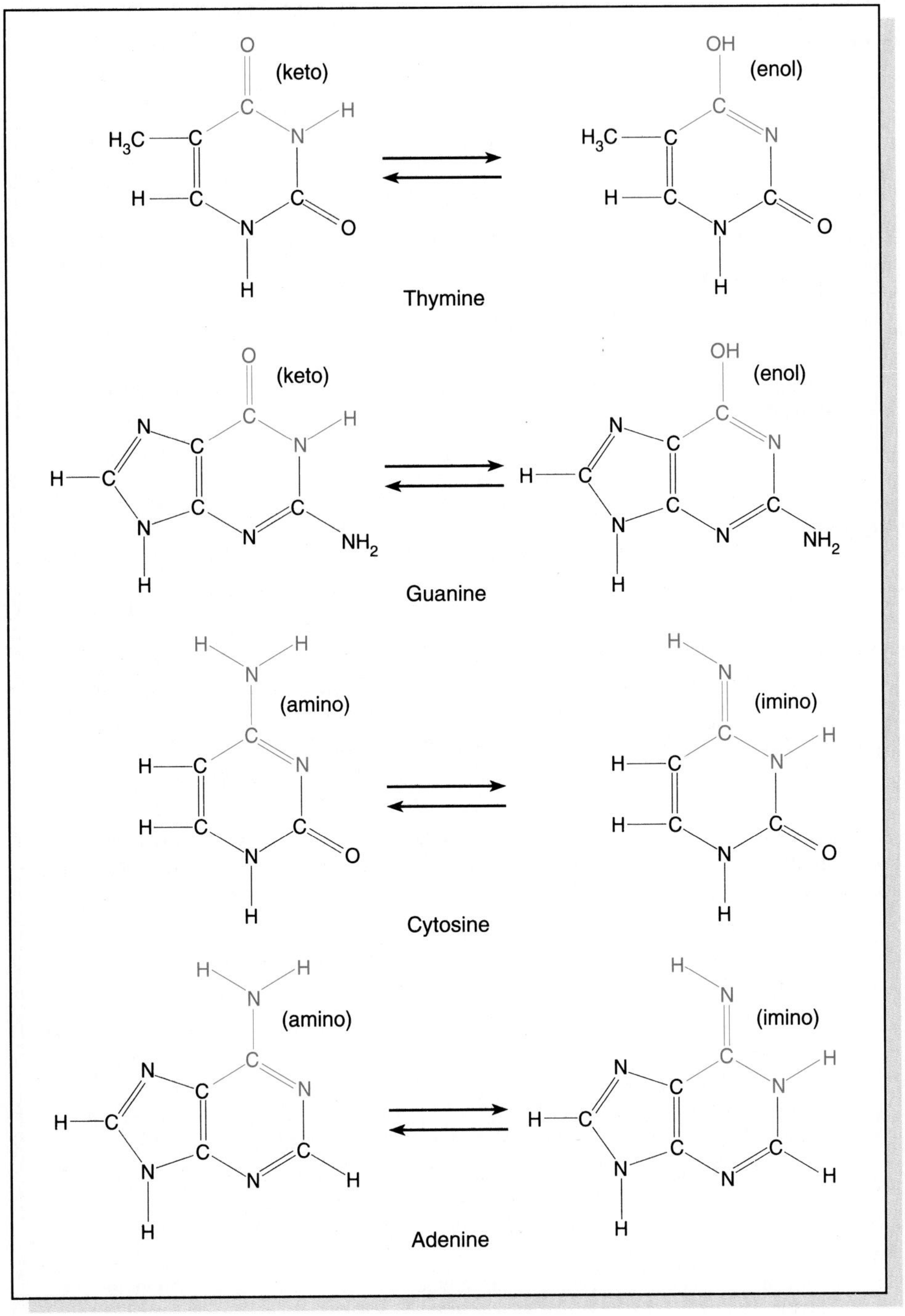

**FIGURE 11.7**    Rate tautomeric shifts that occur in the chemical structure of the four nitrogenous bases of DNA.

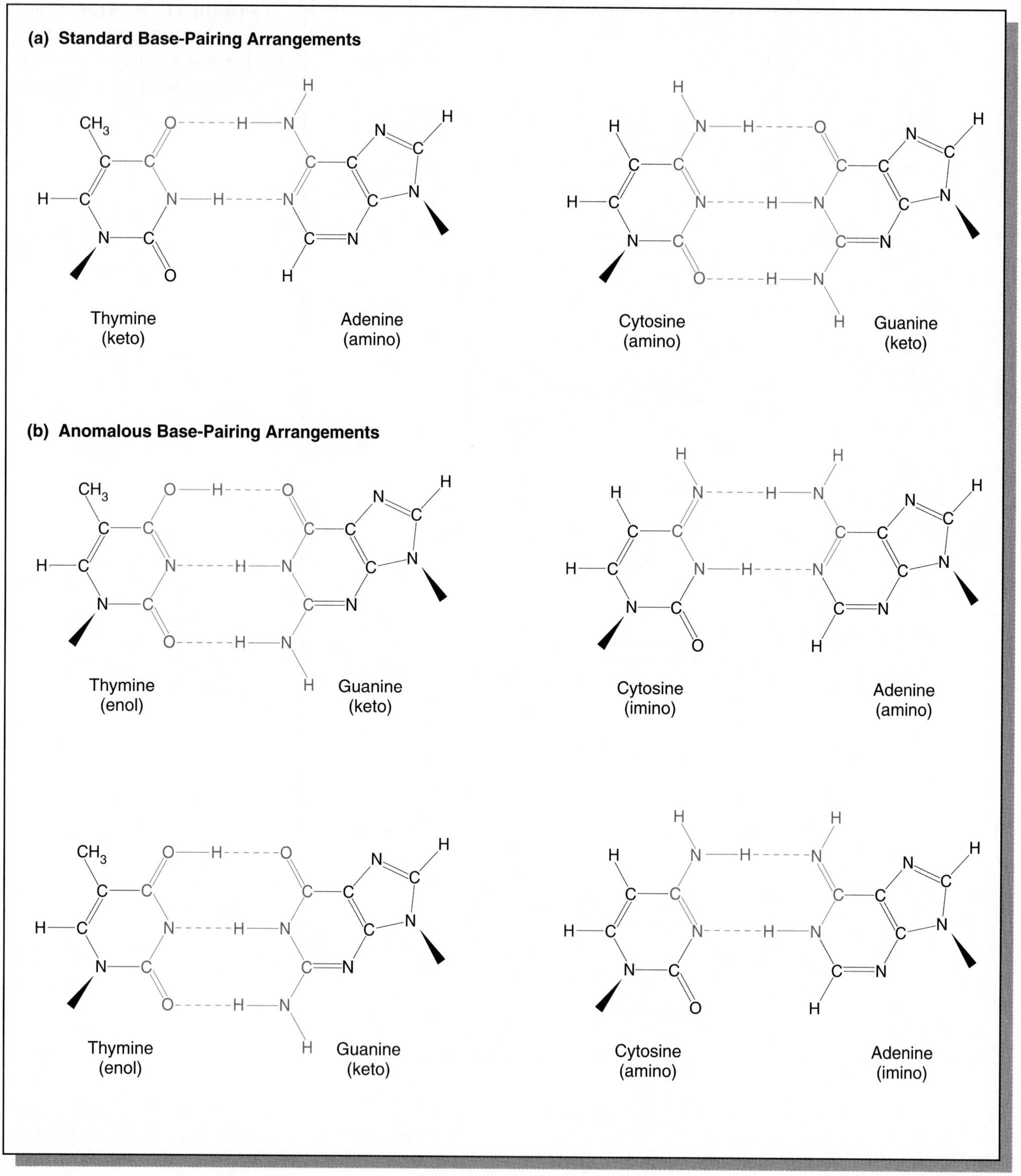

**FIGURE 11.8**    The standard base-pairing relationships compared with anomalous arrangements occurring as a result of tautomeric shifts. The dense arrow indicates the point of bonding to the pentose sugar.

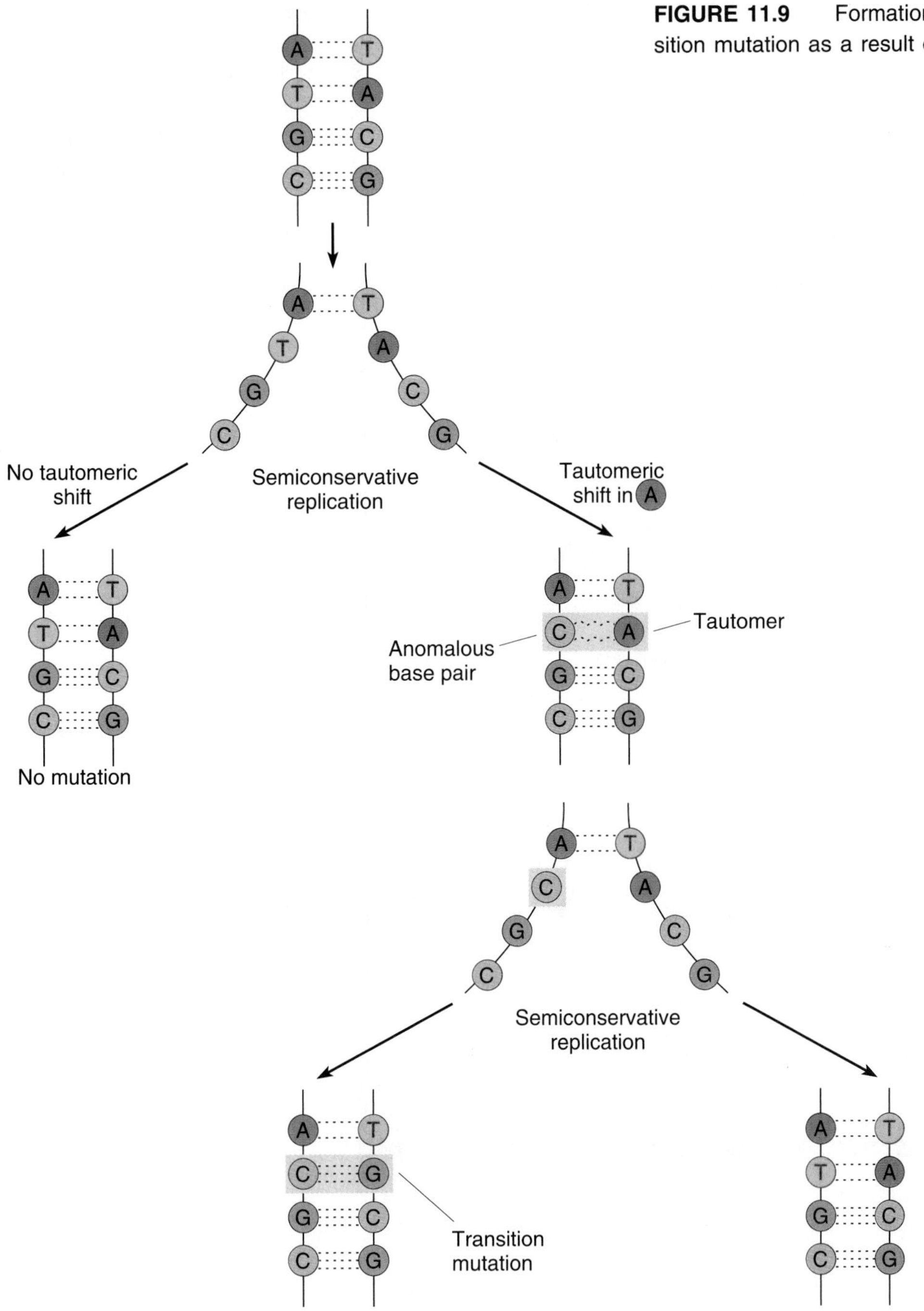

**FIGURE 11.9**    Formation of an A = T to a G ≡ C transition mutation as a result of a tautomeric shift in adenine.

DNA, and 5-BU is mistaken for C, the transition is from G ≡ C to A = T.

There are other base analogues that are mutagenic. One, **2-amino purine** (**2-AP**), can serve successfully as an analogue of adenine. In addition to its base-pairing affinity with thymine, 2-AP can also base pair with cytosine. As such, transitions from A = T to G ≡ C may result following replication.

Because of the specificity by which base analogues such as 2-AP induce transition mutations, base ana-

logues may also be used to induce reversion to the wild-type nucleotide sequence. This alteration is called **reverse mutation**. The process can also occur spontaneously, but at a much lower rate.

## Alkylating Agents

The sulfur-containing **mustard gases** were one of the first groups of chemical mutagens discovered. This discovery was made as a result of studies concerned with

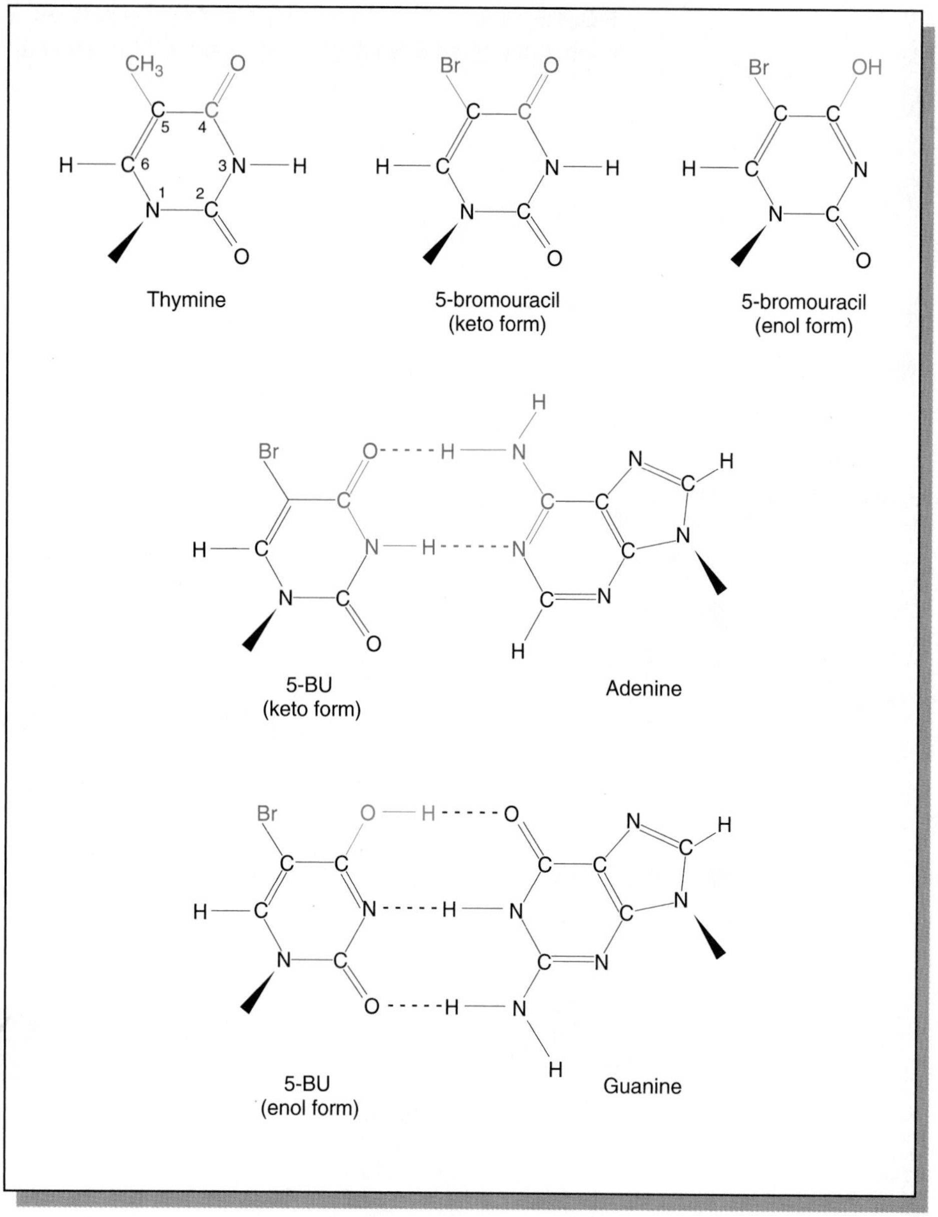

**FIGURE 11.10**    Similarity of 5-bromouracil (5-BU) structure to thymine structure. In the common keto form, 5-BU pairs normally with adenine, behaving as an analogue. In the rare enol form, it pairs anomalously with guanine.

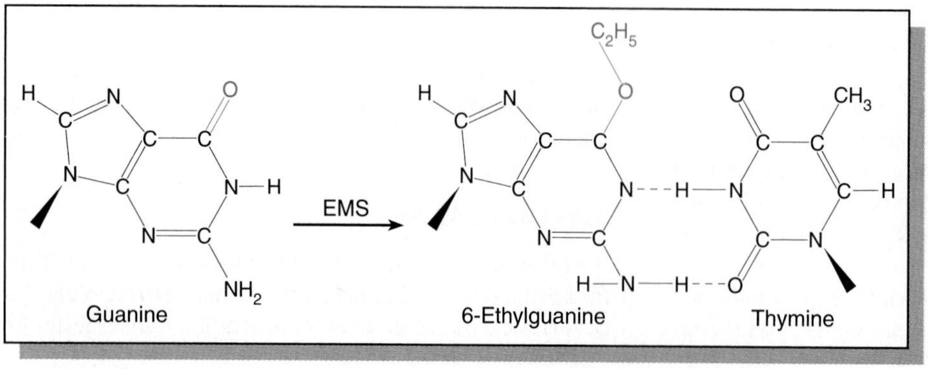

**FIGURE 11.11**    Conversion of guanine to 6-ethylguanine by the alkylating agent ethylmethane sulfonate (EMS). 6-ethylguanine base pairs with thymine.

**Table 11.2**   ALKYLATING AGENTS

| Common Name or Symbol | Chemical Name | Chemical Structure |
|---|---|---|
| Mustard gas (sulfur) | Di-(2-chloroethyl)sulfide | $Cl-CH_2-CH_2-S-CH_2-CH_2-Cl$ |
| EMS | Ethylmethane sulfonate | $CH_3-CH_2-O-\overset{\displaystyle O}{\underset{\displaystyle O}{S}}-CH_3$ |
| EES | Ethylethane sulfonate | $CH_3-CH_2-O-\overset{\displaystyle O}{\underset{\displaystyle O}{S}}-CH_2-CH_3$ |

chemical warfare during World War II. Mustard gases are **alkylating agents**; that is, they donate an alkyl group such as $CH_3-$ or $CH_3-CH_2-$ to amino or keto groups in nucleotides. **Ethylmethane sulfonate (EMS)**, for example, alkylates the keto groups in the number-6 position of guanine and in the number-4 position of thymine (Figure 11.11). As with base analogues, base-pairing affinities are altered and transition mutations result. In the case of **6-ethyl guanine**, this molecule acts like a base analogue of adenine, causing it to pair with thymine. Table 11.2 lists the chemical names and structures of several frequently used alkylating agents known to be mutagenic.

## Acridine Dyes and Frameshift Mutations

Other chemical mutagens cause **frameshift mutations**. These result from the addition or removal of one or more base pairs in the polynucleotide sequence of the gene. Inductions of frameshift mutations have been studied in detail with a group of aromatic molecules known as **acridine dyes**. The structures of **proflavin**, the most widely studied acridine mutagen, and **acridine orange** are shown in Figure 11.12. Acridine dyes are of about the same dimension as a nitrogenous base pair and are known to intercalate or wedge between purines and pyrimidines of intact DNA. Intercalation of acridine dyes induces contortions in the DNA helix, causing deletions and additions.

One model suggests that the resultant frameshift mutations are generated at gaps produced in DNA during replication, repair, or recombination. During these events, there is the possibility of slippage and improper base pairing of one strand with the other. The model suggests that intercalation of the acridine into an improperly base-paired region can extend the existence of these slippage structures. If so, the probability increases that the mispaired configuration will exist when synthesis and re-

joining occurs, thereby resulting in an addition or deletion of one or more bases from one of the strands.

The impact of a frameshift mutation is usually severe. The genetic code is read in three-letter groups, one after the other during translation. As pointed out earlier, the addition or deletion of any group of nucleotides other than a multiple of three shifts the frame of reading from that point on. Initially, this creates missense triplets (see Figure 11.6). Eventually, one of the new three-letter groups may be one of the three stop codons, resulting in premature termination of translation. This will result in an incomplete polypeptide chain. Loss of function usually results.

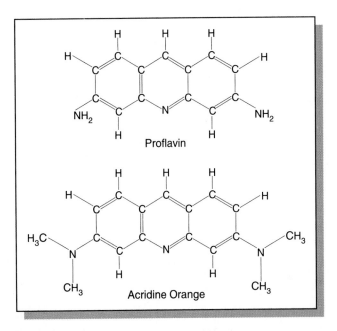

**FIGURE 11.12**    Chemical structures of proflavin and acridine orange, which intercalate with DNA and cause frameshift mutations.

## Apurinic Sites and Other Lesions

Still another type of mutation involves the spontaneous loss of one of the nitrogenous bases in an intact double-helical DNA molecule. Most frequently, such an event involves either guanine or adenine. These sites, created by "breaking" of the glycosidic bond linking the 1'-C of d-ribose and the 9-N of the purine ring, are called **apurinic sites (AP sites)**. It has been estimated that thousands of such spontaneous lesions are formed daily in the DNA of mammalian cells in culture.

The absence of a nitrogenous base at an AP site will alter the genetic code if the involved strand is transcribed and translated. If replication occurs, the AP site is an inadequate template and may cause replication to stall. If a nucleotide is inserted, it is frequently incorrect, causing still another mutation! Fortunately, as we will soon see, cells contain repair systems that often counteract and correct this type of lesion.

Several other types of lesions are known to be the source of some mutations. In the process of **deamination**, an amino group is converted to a keto group in cytosine and adenine (Figure 11.13). In these two cases, cytosine is converted to uracil and adenine is changed to hypoxanthine.

The major effect of these changes is to alter the base-pairing specificities of these two molecules during DNA replication. For example, cytosine normally pairs with guanine. Following its conversion to uracil, which pairs with adenine, the original G ≡ C pair is converted to an A = U pair and, following an additional replication, to an A = T pair. When adenine is deaminated, an original A = T pair is converted to a G ≡ C pair because hypoxanthine pairs naturally with cytosine. Nitrous acid is a known mutagen capable of inducing deamination of bases in DNA.

The final type of mutational lesion that we shall mention is the group caused by oxidation reactions. Active forms of oxygen radicals such as hydrogen peroxide $(H_2O_2)$ and superoxides $(O_2^-)$ can cause damage to DNA bases and result in mispairing during replication.

## Ultraviolet Radiation, Thymine Dimers, and the SOS Response

In Chapter 8 we emphasized the fact that purines and pyrimidines absorb **ultraviolet (UV) radiation** most intensely at a wavelength around 260 nm. This property has been used extensively in the detection and analysis of nucleic acids. In 1934, as a result of studies involving *Drosophila* eggs, it was discovered that ultraviolet radiation is mutagenic. By 1960, several studies concerning the *in vitro* effect on the components of nucleic acids had been completed with the following conclusions. The major effect of UV radiation is on pyrimidines,

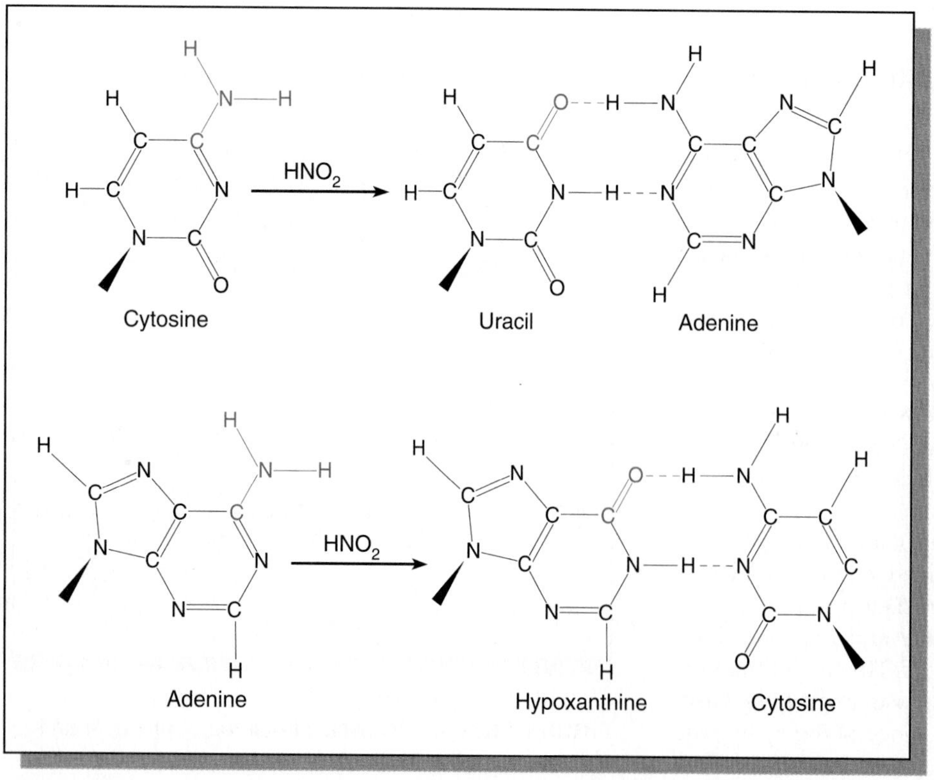

**FIGURE 11.13**    Deaminations caused by nitrous acid $(HNO_2)$ leading to new base-pairing arrangements and mutations. G ≡ C to A = T and A = T to G ≡ C transition mutations result.

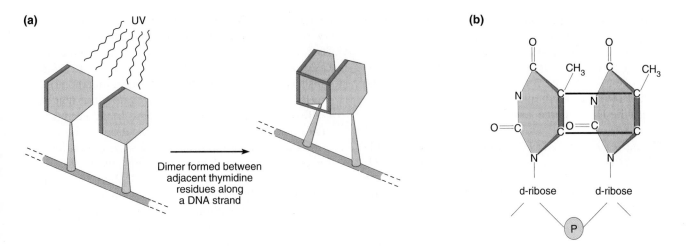

**FIGURE 11.14**     Formation of a thymine dimer induced by ultraviolet radiation. Part (a) depicts the distortion that results, while part (b) shows the cyclobutane ring that forms the basis for the photodimer.

where dimers are formed, particularly between two thymine residues (Figure 11.14). While cytosine–cytosine and thymine–cytosine dimers may also be formed, they are less prevalent. It is believed that the dimers distort the DNA conformation and inhibit normal replication, which seems to be responsible for the killing effects of UV radiation on microorganisms.

In order for this UV-induced lesion to be mutagenic and not lethal, cells must somehow overcome the inhibition of replication, even if it means inserting an incorrect nucleotide during synthesis. A system in bacteria has been discovered that, when activated, appears to allow the "block" to be bypassed. The products of several genes in *E. coli*, including *rec*A, somehow allow the strict adherence of the insertion of complementary bases to be relaxed. While this decreases the fidelity during replication, the process allows the survival of otherwise lethal effects of UV radiation. As a result, this error-prone system is called the **SOS response**.

This survival response is likely to be also activated by other types of lesions that block replication. For example, the AP sites just discussed undoubtedly block replication in a manner similar to pyrimidine dimers, and also induce the SOS response, allowing error-prone replication to occur.

# CASE STUDIES OF MUTATIONS IN HUMANS

The preceding section summarizes the molecular basis of mutation. The information presented relies almost exclusively on the study of nucleic acid chemistry and

has described numerous ways in which one nucleotide pair can be converted to another, either spontaneously or as a result of an induced change. Ways in which frameshift mutations occur have also been discussed. We know that these various types of mutations actually occur, primarily because analysis of amino acid sequences of many proteins within populations of a species show substantial diversity. This diversity has arisen during evolution and is a reflection of changes in the triplet codons following substitution, duplication, or deletion of one or more nucleotides in the DNA sequences constituting genes.

As our ability to analyze DNA more directly increases, we are better able to look specifically at the actual nucleotide sequence of genes and gain greater insights into mutation. Several techniques capable of accurate, rapid sequencing of DNA are available (see Chapter 12 and Appendix A). These and other approaches involving the analysis of DNA have greatly extended our knowledge in molecular genetics.

## ABO Blood Types versus Muscular Dystrophy

In this section, we examine the results of two studies that have investigated the actual gene sequence of various mutations affecting humans. The first provides an interesting insight into the molecular basis of the **ABO antigens**, originally presented in Chapter 4 as an example of multiple alleles. The second case involves the nature of the mutations that have led to the devastating sex-linked disorder **muscular dystrophy**.

The ABO system is based on a series of antigenic determinants found on erythrocytes and other cells, particularly epithelial types. As previously discussed, three alleles of a single gene exist, the product of which is designed to modify the H substance. The modification involves glycosyltransferase activity, converting the H substance to either the A or B antigen, as a result of the product of the $I^A$ or $I^B$ allele, respectively, or failing to modify the H substance, as a result of the $I^O$ allele (see Figure 4.2).

Using recombinant DNA technology, the responsible gene has been examined in 14 cases of varying ABO status. Four consistent nucleotide substitutions were found when the DNAs of the $I^A$ and $I^B$ alleles were compared. It is assumed that the changes in the amino acid sequence of the glycosyltransferase gene product resulting from these substitutions lead to the different modifications of the H substance.

The situation with the $I^O$ allele is unique and interesting. Individuals homozygous for this allele are type O, lack glycosyltransferase activity, and fail to modify the H substance. Analysis of the DNA of this allele shows one consistent change compared to that of the other alleles: the deletion of a single nucleotide early in the coding sequence, causing a frameshift mutation. A complete messenger RNA is transcribed, but upon its translation, the frame of reading shifts at the point of deletion and continues out of frame for about 100 nucleotides before a "stop" codon is encountered. At this point, premature termination of the resulting polypeptide chain occurs, producing a nonfunctional product.

These results provide a direct molecular explanation of the ABO allele system and the basis for the biosynthesis of the corresponding antigens. The molecular basis for the antigenic phenotypes is clearly the result of structural alterations, or mutations, of the nucleotide sequence of the gene encoding the glycosyltransferase enzyme.

The second case of mutational analysis involves the severe disorder muscular dystrophy. It is characterized by progressive muscular degeneration, or myopathy, resulting in the death of affected individuals by early adulthood. Because the condition is recessive and sex linked, and since affected males do not reproduce, females are rarely affected by the disorder. The incidence of 1/3500 live male births makes muscular dystrophy one of the most common life-shortening hereditary disorders known. Two related forms exist: **Duchenne muscular dystrophy (DMD)** is more common and more severe than the allelic form called **Becker muscular dystrophy (BMD)**.

The region containing the gene has been analyzed extensively and consists of over two million base pairs. In unaffected individuals, transcription results in a messenger RNA containing about 14,000 bases (14 kb) that is translated into the protein **dystrophin**, consisting of 3685 amino acids. This protein can be detected in most cases of the less severe BMD, but is rarely found in DMD. This has led to the hypothesis that most mutations causing BMD do not alter the reading frame, but that most DMD mutations change the reading frame early in the gene, resulting in premature termination of dystrophin translation. This hypothesis is consistent with the observed differences in severity of these two forms.

In an extensive analysis of the DNA of 194 patients (160 DMD and 34 BMD), J. T. Den Dunnen and associates found that 128 of these mutations (65 percent) consisted of substantial deletions or duplications. Of 115 deletions, 17 occurred in BMD, and of 13 duplications, 1 was in BMD, with the remainder being found in DMD cases. In most cases, the above results were consistent with the "reading frame" hypothesis. With few exceptions, DMD mutations changed the frame of reading. BMD mutations usually did not alter the reading frame.

Perhaps the most noteworthy finding is the high percentage of deleterious mutations studied that represent the deletion or duplication of nucleotides within the gene. This observation reflects the fact that a mutation caused by a random single-nucleotide substitution within a gene is more likely to be tolerated without the devastating effect of muscular dystrophy than the addition or loss of numerous nucleotides that may alter the frame of reading. There are three reasons for this:

**1.** A nucleotide substitution may not change the encoded amino acid since the code is degenerate.

**2.** If an amino acid substitution does result, the change may not be present at a location within the protein that is critical to its function.

**3.** Even if the altered amino acid is present at a critical region, it may still have little or no effect on the function of the protein. For example, an amino acid might be changed to another with nearly identical chemical properties or to one with very similar recognition properties, such as shape.

As a result, single-base substitutions may have little or no effect on protein function, or may simply reduce the efficiency but not eliminate the functional capacity of the gene product. It is clear that we cannot look at mutations with the oversimplified expectation that most of them are single-base substitutions. As more of them are analyzed directly, our picture of mutation will become increasingly clear.

## Trinucleotide Repeats in Fragile-X Syndrome, Myotonic Dystrophy, and Huntington Disease

Between 1991 and 1993, molecular analysis of the DNA representing the genes responsible for various human disorders provided a remarkable set of observations. In the cases of three disparate genes, responsible for the X-linked **fragile-X syndrome** and the autosomal disorders **myotonic dystrophy** and **Huntington disease**, an intriguing similarity was discovered: Each gene was found to contain a unique trinucleotide DNA sequence repeated many times. While each repeated sequence is also present in the nonmutant (normal) allele of each gene, what characterizes each mutation is a significant, variable increase in the number of times the trinucleotide is repeated.

You may recall that we introduced this topic earlier in Chapter 7 when we discussed the onset of gene expression. In several cases, a correlation has been found between the number of repeats and the onset of mutant gene expression. The greater the number, the earlier disease onset occurs.

Of great interest and significance is the fact that the number of repeats may increase in each subsequent generation. This general phenomenon, which we called **anticipation** in Chapter 7, represents a unique form of mutation related to an instability of specific regions of the three different genes. While other normal genes in humans, as well as many other species, are known to contain trinucleotide repeats, these repeats do not balloon in size, and the genes maintain normal function.

The gene responsible for fragile-X syndrome, *FMR-1*, may have several hundred to several thousand copies of the trinucleotide sequence CGG. Individuals with up to 50 copies are normal and do not display mental retardation associated with the syndrome. Individuals with 50 to 200 copies are considered "carriers." While they are normal, their offspring may contain even more copies and express the syndrome.

Myotonic dystrophy, the most common form of adult muscular dystrophy, contains multiple copies of the sequence CTG. Fewer than 35 copies results in normal gene expression. Above this number, symptoms range from mild myopathy and cataracts to severe dystrophy, retardation, and even death in early childhood. Both severity and onset are directly correlated with the size of the repeated sequence.

Most recently, the gene responsible for Huntington disease has been found to demonstrate a similar pattern. The trinucleotide CAG, present 10 to 35 times in normal individuals, increases significantly in diseased individuals. Recall that the onset of this disease varies tremendously, most often being expressed in the mid- to late thirties. Much earlier onset occurs when very large numbers of copies are present. Interestingly, in still another disorder, **spinobulbar muscular atrophy** (**Kennedy disease**), the gene contains repeated copies of the same triplet. However, diseased individuals usually contain only 30 to 60 copies of the CAG sequence.

The role of such repeated sequences in normal and mutant genes remains a mystery. Their locations within the gene vary in each case. While Huntington and Kennedy diseases contain the repeat within the coding portion of the gene, this is not the case in the other two disorders. In the gene responsible for fragile-X syndrome, the repeat is upstream (5′) in an area that is most often involved in regulating gene expression. In the case of myotonic dystrophy, the repeat is downstream (the 3′ end).

In addition to the role of the repeat in normal and mutant genes, the mechanism by which the sequence expands from generation to generation is of great interest. How such an instability during DNA replication affects only specific areas of certain genes is currently an important research topic. This instability seems to be more prevalent in humans than in many other organisms.

A final note of interest is to point out that these findings are very recent. Thus, it appears likely that other mutant genes will be found to parallel those discussed above. Within the next several years, other examples of this phenomenon are likely to be uncovered.

## DETECTION OF MUTAGENICITY: THE AMES TEST

There is particular concern about the possible mutagenic properties of any chemical that enters the human body, whether through the skin, digestive tract, or respiratory tract. For example, great attention has been given to residual materials of air and water pollution, food preservatives and additives, artifical sweeteners, herbicides, pesticides, and pharmaceutical products. While mutagenicity may be tested in various organisms, including *Drosophila*, mice, and cultured mammalian cells, the most common test involves bacteria and was devised by Bruce Ames.

The **Ames test** utilizes four tester strains of the bacterium *Salmonella typhimurium* that were selected for sensitivity and specificity for mutagenesis. One strain is used to detect base-pair substitutions, and the other three detect various frameshift mutations. Each mutant strain is unable to synthesize histidine, and therefore requires histidine for growth ($his^-$). The assay measures the frequency of reverse mutation, which yields wild-type ($his^+$) bacteria. Greater sensitivity to mutagens occurs because these strains bear other mutations that eliminate the DNA excision repair system (discussed later in this chapter) and the lipopolysaccharide barrier that coats and protects the surface of the bacteria.

It is very interesting to note that many substances that enter the human body are relatively innocuous until activated metabolically to a more chemically reactive electrophilic product. This usually occurs in the liver. Thus, the Ames test, which is performed *in vitro*, includes a step in which the test compound is incubated in the presence of a mammalian liver extract. Or, test compounds are actually injected into the mouse, which is later sacrificed and the liver removed. Extracts are then tested.

In the initial use of Ames testing in the 1970s, a large number of known carcinogens were examined. Over 80 percent of these were shown to be strong mutagens! This is not surprising, since transformation of cells to the malignant state undoubtedly occurs as a result of some alteration of DNA. Although a positive response as a mutagen does not prove the carcinogenic nature of a test compound, the Ames test is useful as a preliminary screening device. It is used extensively in conjunction with the industrial and pharmaceutical development of chemical compounds.

# REPAIR OF DNA

In previous sections of this chapter we have established that replicating and nonreplicating DNA is vulnerable to various forms of errors and lesions that constitute or lead to gene mutations. Living systems have evolved a variety of very elaborate repair systems that are able to counteract many of the forms of DNA damage that lead to mutation. As we will see, such repair systems are essential to the survival of organisms on earth. For example, in humans, the loss of just one type of repair system leads to a devastating, life-shortening genetic disorder.

## UV Radiation, Thymine Dimers, and Photoreactivation Repair

The study of mutagenicity of UV radiation paved the way for the discovery of many forms of natural repair of DNA damage. The first relevant discovery concerning UV repair in bacteria was made in 1949 when Albert Kelner observed the phenomenon of **photoreactiva-**

**tion repair**. He showed that the UV-induced damage to *E. coli* DNA could be partially reversed if, following irradiation, the cells were exposed briefly to light in the blue range of the visible spectrum. The photoreactivation repair process has subsequently been shown to be temperature dependent, suggesting that the light-induced mechanisms involve an enzymatically controlled chemical reaction. Visible light appears to induce the repair process of the DNA damaged by UV radiation.

Further studies of photoreactivation have revealed that the process is due to a protein called the **photoreactivation enzyme (PRE)**. This molecule can be isolated from extracts of *E. coli* cells. The enzyme's mode of action is to cleave the bonds between thymine dimers, thus reversing the effect of UV radiation on DNA [Figure 11.15(a)]. While the enzyme will associate with a dimer in the dark, it must absorb a photon of light to cleave the dimer.

The gene(s) encoding PRE have been preserved throughout evolution. Activity of this repair system has been detected in human cells in culture, as well as in other eukaryotes. The conservation of the genetic components over millions of years suggests that this repair system is an important one to all organisms.

## Excision Repair

Investigations in the early 1960s suggested that a repair system or systems that do not require light also exist in *E. coli*. Paul Howard-Flanders and coworkers isolated several independent mutants demonstrating increased sensitivity to ultraviolet radiation. One group was designated *uvr* (ultraviolet repair) and included the *uvr*A, *uvr*B, and *uvr*C mutations. These genes and their protein products were subsequently shown to be involved in a process called **excision repair**. During this process, three steps have been shown to occur, as illustrated in Figure 11.15(b):

1. The distortion of the strand caused by the UV-induced dimer is recognized and enzymatically clipped out by a nuclease that cleaves the phosphodiester bonds. This "excision," which may include several nucleotides adjacent to the dimer as well, leaves a gap in the helix. The *uvr* gene products operate at this step.

2. **DNA polymerase I** fills this gap by inserting d-ribonucleotides complementary to those on the intact strand. The enzyme adds these bases to the 3'-OH end of the clipped DNA.

3. The joining enzyme **DNA ligase** seals the final "nick" that remains at the 3'-OH end of the last base inserted, closing the gap.

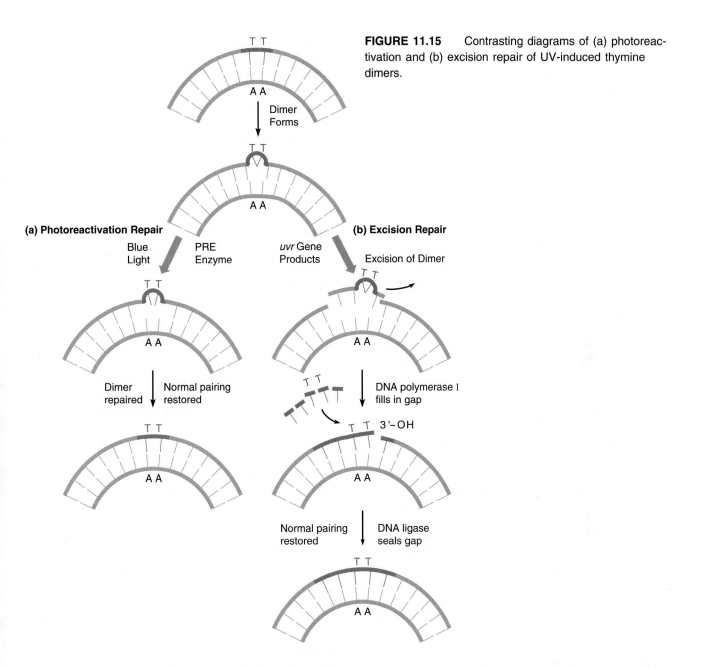

**FIGURE 11.15** Contrasting diagrams of (a) photoreactivation and (b) excision repair of UV-induced thymine dimers.

DNA polymerase I, the enzyme discovered by Kornberg, was once assumed to be the universal DNA-replication enzyme (see Chapter 8). The discovery of the *polA1* mutation demonstrated that this is not the case. *E. coli* cells carrying the *polA1* mutation lack functional polymerase I. However, replication of DNA takes place normally. Cells with this mutation are unusually sensitive to UV radiation. Apparently, such cells are unable to fill the gap created by the excision of the thymine dimers. This finding demonstrates the importance of excision repair mechanisms in counteracting the effects of UV radiation.

It has been shown subsequently that the process of excision repair can be activated in response to any damage to DNA that distorts the helix. For example, as we

discussed earlier, the loss of a purine from d-ribose of one strand creates what is called an **apurinic site** (**AP site**). The complementary pyrimidine on the opposite strand has nothing with which to form hydrogen bonds. Such a sugar with a missing base is recognized by an enzyme called **AP endonuclease**. The endonuclease makes a cut in the polynucleotide chain at the AP site. This creates the distortion that is recognized by the excision repair system that is then activated, leading ultimately to the correction of the error (the AP site).

Other enzymes recognize incorrect bases and stimulate repair. For example, one specific member of a group of enzymes called **DNA glycosylases** recognizes the presence of uracil when it is part of DNA. It cuts the glycosidic bond between the base and sugar, creating an AP site, which is then repaired as discussed above. Glycosylases are important repair components because when created by deamination of cytosine, uracil will lead to a $C \equiv G$ to $T = A$ transition mutation after replication if it is not repaired.

In theory, excision repair may serve as the final step in a variety of repair processes, provided that the lesion or distortion in DNA may be recognized.

### Proofreading and Mismatch Repair

As we pointed out in our discussion of DNA synthesis in Chapter 9, DNA polymerases I and III exhibit 3'-to-5' exonuclease activity. This enzymatic action provides the initial step of error correction when a mismatched base pair has been inserted during synthesis. We have referred to this process as **proofreading**. In bacterial systems, proofreading is thought to increase fidelity during synthesis by two orders of magnitude. If mismatches occur in $1/10^5$ nucleotide pairs (a rate of $10^{-5}$), proofreading decreases final mismatches to $1/10^7$ (a rate of $10^{-7}$).

Once proofreading has occurred, still another mechanism, called **mismatch repair**, may be activated. Proposed over 20 years ago by Robin Holliday, the molecular basis of this process is now well established. Like other DNA lesions, (1) the alteration or mismatch must be detected, (2) the incorrect nucleotide must be removed, and (3) replacement with the correct base must occur. But a special problem exists with correction of a mismatch. It is not clear which strand is correct (the parental strand) and which contains the mismatch (the newly synthesized strand). How the repair system discriminates and recognizes the nonparental strand puzzled geneticists for decades. If the mismatch is recognized, but no discrimination occurred and excision was random, half the time the strand bearing the correct base would be clipped out. The concept of strand discrimination by a repair enzyme is thus the critical step.

At least in some bacteria, including *E. coli,* this process has been elucidated and is based on the process of **DNA methylation**. These bacteria contain an enzyme, **adenine methylase**, that recognizes the DNA sequence

$$5' \ldots GATC \ldots 3'$$
$$3' \ldots CTAG \ldots 5'$$

as a substrate. Upon recognition, a methyl group is added to each of the adenine residues. This modification is stable throughout the cell cycle.

Following a further round of replication, the newly synthesized strands remain temporarily unmethylated. It is at this point that the repair enzyme recognizes the mismatch and preferentially binds to the unmethylated

strand. It is excised and the correct complement is inserted! Interestingly, the GATC sequence need only be within several thousand base pairs of the mismatch. A series of *E. coli* gene products, mutH, L, S, and U are involved in the discrimination step. Mutations in each result in strains quite deficient in mismatch repair.

While it is agreed that mismatch repair undoubtedly also occurs in the DNA of higher organisms, the question of strand discrimination remains speculative in the absence of GATC methylation.

### Recombinational Repair

The final mode of repair to be discussed was discovered in an excision-defective strain of *E. coli.* Called **recombinational repair**, this system is thought to respond when damaged DNA has escaped repair and the damage disrupts the process of replication. The cells that show this phenomenon are dependent on the product of a gene, *recA,* which is involved in several types of recombination in *E. coli.*

When such DNA is being replicated, DNA polymerase at first stalls at and then skips over the distortion, creating a gap on one of the newly synthesized strands. To counteract this, the RecA protein directs a recombinational exchange process whereby this gap is filled as a result of the insertion of a segment initially present on the intact homologous strand. The resulting gap now present on the "donor" strand is filled by repair synthesis as replication proceeds.

This mode of repair is a "rescue operation" that allows replication to create accurate and complete DNA copies from a template that is damaged. Phil Hanawalt and Paul Howard-Flanders, among others, have established that many different gene products are involved. Of particular interest is the LexA product, which serves to repress the function of the *recA* and *uvr* genes. However, when a RecA protein binds to DNA in the area of a gap, this binding somehow activates a second function of the RecA protein—the ability to cleave the LexA repressor molecule. The absence of a functional repressor molecule allows the activation of the *recA* and *uvr* genes, leading to an increased production of the proteins for which they code.

## UV RADIATION AND HUMAN SKIN CANCER: XERODERMA PIGMENTOSUM

The essential nature of any biological system can be assessed by examining the effects when the system fails. Regrettably, we can determine just how essential DNA repair is in humans by examining individuals who exhibit an inherited loss of function of one of the major repair systems. **Xeroderma pigmentosum (XP)**, a rare autosomal recessive disorder in humans, predisposes individuals to epidermal pigment abnormalities. Exposure to ultraviolet radiation present in sunlight results in malignant growth of the skin. Figure 11.16 contrasts two XP individuals, one of whom has been detected early and protected from sunlight.

The condition is very severe and may be lethal. Since sunlight contains UV radiation, a causal relationship has been predicted between thymine dimer production and XP. It has also been of great interest to determine which of the three forms of repair processes counteracting the effects of UV-induced damage to DNA (if any) are operating in humans. Furthermore, it was suspected that XP individuals might lack one or more repair systems, which may cause them to be susceptible to UV-induced skin damage.

The various modes of repair of UV-induced lesions have been investigated in human fibroblast cultures derived from XP and normal individuals. Fibroblasts are undifferentiated connective tissue cells. The results are varied and suggest that the XP phenotype may be caused by more than one mutant gene.

In 1968, James Cleaver showed that cells from XP patients were deficient in the **unscheduled DNA synthesis** elicited in normal cells by UV radiation. This assay is thought to represent activity of the excision repair system, suggesting that XP cells are deficient in this form of repair. In 1974, the presence of a **photoreactivation enzyme (PRE)** was established in normal human cells.

Betsy Sutherland identified the enzyme first in leukocytes and subsequently in fibroblast cells. Sutherland demonstrated further that some XP cultures contain a lower PRE activity than control cultures. The activity in various XP cell strains ranges from 0 to 50 percent of normal in cultures established from different patients.

Cultured cells from any two XP variants may be induced to fuse together, forming a **heterokaryon** where the two nuclei share a common cytoplasm. This process is one aspect of **somatic cell genetics**, which we first introduced in Chapter 5. When fusion is performed with cells of different XP variants, excision repair, as assayed by unscheduled DNA synthesis, is sometimes reestablished. When this occurs, the two variants are said to undergo **complementation**. Alone, neither cell type demonstrates excision repair, but together in a heterokaryon the process occurs. In genetic terms, this is strong evidence that different genes are involved in the defect of each variant. Complementation occurs because the heterokaryon has at least one normal copy of each gene. In many studies, variants have been divided into seven complementation groups, which suggests that as many as seven different genes may be involved in excision repair. Cells from an eighth form of XP does not undergo complementation with any other group. It is thought to be caused by a defect in postreplication repair.

These findings establish that normal individuals are probably susceptible to UV-induced damage of DNA by exposure to sunlight. However, this damage activates the repair systems that counteract it. We can expect that future work will clarify the precise role and mechanism of repair of UV-induced damage in humans.

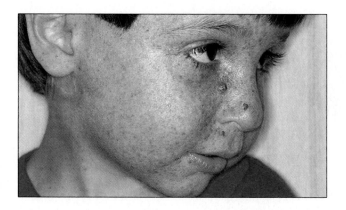 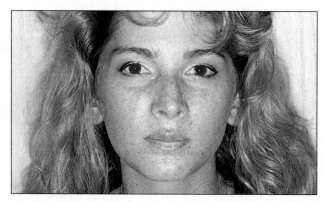

**FIGURE 11.16** Xeroderma pigmentosum. The four-year-old boy on the left shows marked skin lesions induced by sunlight. Mottled redness (erythema) and irregular pigment changes that are a response to cellular injury are apparent. Two nodular cancers are present on his nose. The eighteen-year-old girl on the right has been carefully protected from sunlight since the diagnosis of xeroderma pigmentosum made in infancy. Several cancers have been removed and she works as a successful model.

# HIGH-ENERGY RADIATION

Within the electromagnetic spectrum, energy varies inversely with wavelength. Figure 11.17 compares the relative wavelengths of the various portions of the electromagnetic spectrum. **X rays**, **gamma rays**, and **cosmic rays** have even shorter wavelengths than ultraviolet radiation and are therefore more energetic. As a result, they are strong enough to penetrate deeply into tissues, causing ionization of the molecules encountered along the way. These sources of **ionizing radiation** can be predicted to be mutagenic. Indeed, Herman Muller detected such effects in 1927 while working with *Drosophila*. Since that time, the effects of ionizing radiation, particularly X rays, have been studied intensely.

As X rays penetrate cells, electrons are ejected from the atoms of molecules encountered by the radiation. Thus, stable molecules and atoms are transformed into free radicals and reactive ions. Along the path of a high-energy ray, a trail of ions is left that can initiate a variety of chemical reactions. These reactions can affect directly or indirectly the genetic material, altering the purines and pyrimidines in DNA and resulting in point mutations. Such ionizing radiation is also capable of breaking phosphodiester bonds, thus disrupting the integrity of chromosomes. This results in a variety of aberrations.

Figure 11.18 shows a plot of induced sex-linked recessive lethal mutations versus the dose of X rays administered. The graph shows a straight line that, if extrapo-lated, intersects near the zero axis. A linear relationship is evident between X-ray dose and the induction of mutation. For each doubling of dose, twice as many mutations are induced. Because the line intersects near the zero axis, this graph suggests that even very small doses of irradiation are mutagenic.

These observations may be interpreted in the form of the **target theory**, first proposed in 1924 by J. A. Crowther and F. Dessauer. The theory proposes that there are one or more sites, or targets, within cells and that a single event of irradiation at one site will bring about a damaging effect, or mutation. In a simple form, the target theory says that one "hit" of irradiation will cause one "event" or mutation, suggesting that the X rays interact directly with the genetic material.

Two other observations concerning irradiation effects are of particular interest. First, in some organisms studied, the *intensity of the dose* (dose-rate) administered seems to make little difference in mutagenic effect. That is, a 100-roentgen exposure (a **roentgen** is a measure of energy dose), whether occurring in a single acute dose or cumulatively in many smaller chronic doses, seems to produce the same mutagenic effect. *Drosophila*, for example, shows this response. In mammals such as mice and humans, however, this is not the case. Repair of the damage seems to occur during the intervals between irradiation. Thus, several smaller doses are not as potent as a single large dose.

The second observation is that certain portions of the

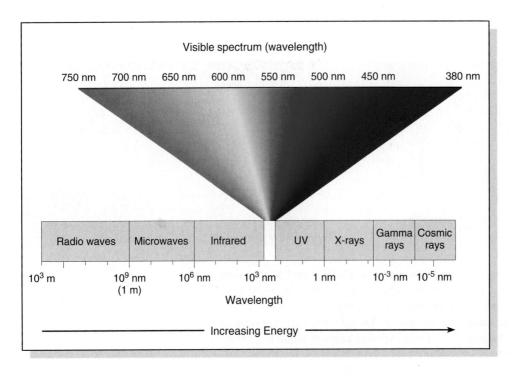

**FIGURE 11.17**    The components of the electromagnetic spectrum and their associated wavelengths.

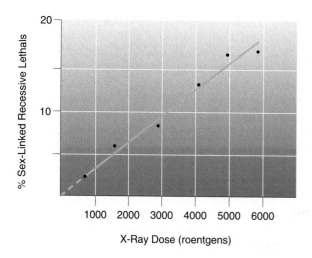

**FIGURE 11.18** Plot of sex-linked recessive mutations induced by increasing doses of X rays. If extrapolated, the graph intersects the zero axis.

cell cycle are more susceptible to radiation effects. As mentioned previously, in addition to a mutagenic effect, X rays can also break chromosomes, resulting in terminal or intercalary deletions, translocations, and general chromosome fragmentation. Damage occurs most readily when chromosomes are greatly condensed in mitosis. This property constitutes one of the reasons why radiation is used to treat human malignancy. Since tumorous cells are more often undergoing division than their nonmalignant counterparts, they are more susceptible to the immobilizing effect of radiation.

## SITE-DIRECTED MUTAGENESIS

This section introduces a useful experimental technique that allows tailor-made changes within genes: **site-directed mutagenesis**. The technique relies on a number of manipulations involving recombinant DNA technology, which will be introduced in Chapter 12. However, the underlying principles are based on previous information. This technique is of great interest here because it illustrates the application of basic knowledge of a topic such as mutation to current research investigations.

The goal of the technique is to specifically alter one or more nucleotides within a gene in order to change a specific triplet codon. Upon transcription and translation, this change will cause the insertion of a "mutant" amino acid in the protein encoded by the original gene. Such designed mutations are particularly useful in studying the effects of specific mutations on protein function.

The first step is to determine the nucleotide sequence of the gene being studied. This can be accomplished by DNA sequencing techniques, or it can be predicted if the amino acid sequence of the protein is known by utilizing our knowledge of the genetic code.

The next step is to isolate the DNA of known sequence and obtain from it one of the two complementary strands. A decision is then made as to which nucleotides are to be changed. Then, a small piece of DNA complementary to that region is chemically synthesized. It is complementary at all points, except in the triplet sequence (or sequences) that is to be altered. This triplet encodes the amino acid that will change in the protein.

As illustrated in Figure 11.19, this short piece of DNA hybridizes with the original parent strand, forming a partial duplex because of its complementarity along most of its length. If DNA polymerase and DNA ligase are then added to this hybrid complex, the short sequence is extended so that a duplex of the entire gene is formed. The two strands are perfectly complementary except at the point of alteration.

When this DNA undergoes semiconservative replication, two types of duplexes are formed: one is like the original, unaltered gene and the other contains the newly designed sequence. Using recombinant DNA technology, it is possible not only to complete the above manipulations, but also to allow the altered gene to be expressed so that large amounts of the desired protein are available for study. This approach is useful in investigating the role of specific amino acids during protein function.

## TRANSPOSABLE GENETIC ELEMENTS

We conclude this chapter by introducing the phenomenon of **transposable genetic elements**. As the name suggests, this topic encompasses a group of genetic units that are mobile. They can move, or be transposed, within the genome. The discussion of them in this chapter on mutation is appropriate because the movement of genetic units from one place in the genome to another often disrupts genetic function and results in phenotypic variation. As such, the impact of transposition of genetic units often fits into a broad definition of mutation.

As we shall see, transposable elements were first discovered almost 50 years ago in maize. However, even in the 1950s and 1960s the idea that genetic information was *not* fixed within the genome of an organism was slow to find acceptance. Such a notion was quite alien to the classical interpretation of genes on chromosomes. It was not until other transposable elements were discov-

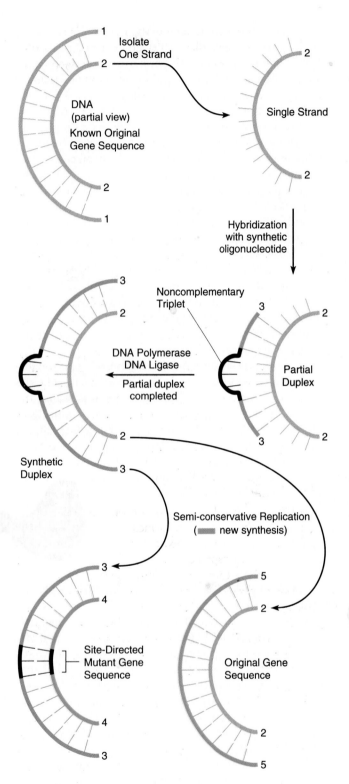

**FIGURE 11.19** Site-directed mutagenesis, achieved by first obtaining a single strand of DNA from a gene of interest. This is hybridized with a synthetic oligonucleotide containing a triplet altered so as to encode an amino acid of choice. The partial duplex is completed under the direction of DNA polymerase and DNA ligase. Following semiconservative replication, a different complementary base pair is present in one of the new duplexes. Upon transcription and translation, a mutant protein, "designed" in the laboratory, will be produced.

ered, and their molecular basis revealed, that the phenomenon was found to be nearly a universal phenomenon.

## Insertion Sequences

Even though the presence of transposable elements in maize had been predicted earlier by Barbara McClintock, the first observation at the molecular level involved **insertion sequences (ISs)** in *E. coli*. Discovered in the early 1970s by a number of independent researchers, including Peter Starlinger and James Shapiro, insertion sequences were first visualized as a unique class of mutations affecting different genes in various bacterial strains. For example, the expression of a cluster of related genes involving galactose metabolism was repressed as a result of one such mutation.

This phenotypic effect was heritable, but was found not to be caused by a base pair change characteristic of conventional gene mutations. Instead, it was shown that a short, specific DNA segment had been inserted into the bacterial chromosome at the beginning of the galactose gene cluster. When this segment was spontaneously excised from the bacterial chromosome, wild-type function was restored!

It was subsequently revealed that several other distinct DNA segments could behave in a similar fashion, inserting into the chromosome and affecting gene function. These DNA segments are relatively short, not exceeding 2000 base pairs [2 kilobases (kb)]. For example, the first insertion sequence to be characterized in *E. coli* was IS1. It is about 800 base pairs long, while IS2, 3, 4, and 5 are about 1400 base pairs in length.

Analysis of the DNA sequences of most IS units reveals a most interesting feature. At each end of any given double-stranded unit, the nucleotide sequence consists of a **perfect inverted repeat** of the other end. While Figure 11.20 shows this terminal repeating unit to consist of only a few nucleotides, many more are actually involved. For example, in *E. coli*, the IS1 termini contain

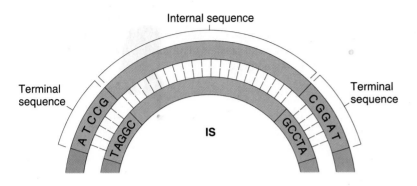

Internal sequence

Terminal
sequence

Terminal
sequence

ATCCG  TAGGC  GCCTA  CGGAT

IS

**FIGURE 11.20**   Diagrammatic representation of an insertion sequence (IS). The terminal sequences are perfect inverted repeats of one another.

about 25 nucleotide pairs, IS2 about 40 pairs, and IS4 about 18 pairs. It seems likely that these terminal sequences are an integral part of the mechanism of insertion of IS units into DNA. That insertion of IS units is more likely to occur at certain DNA regions than others suggests that IS termini can recognize certain target sequences in the DNA during the process of insertion.

Careful investigation has revealed that IS units are present in the wild-type *E. coli* chromosome as well as in other autonomous segments of bacterial DNA called plasmids. Thus, their presence does not always result in mutation. In the *E. coli* chromosome, five or more copies of IS1, IS2, and IS3 are known to be present. The exact number of each varies, depending on the strain examined.

## Bacterial Transposons

In addition to their potential mutational effects, IS units play an even more significant role in the formation and movement of the larger **transposon (Tn) elements**. Transposons in bacteria consist of IS units that contain within their internal DNA sequence genes whose functions are unrelated to the insertion process. Like IS units, Tn elements are mobile in bacterial and viral chromosomes as well as plasmids. Tn elements provide a mechanism for movement of genetic information from place to place both within and between organisms.

Transposons were first discovered to move between DNA molecules as a result of observations of antibiotic-resistant bacteria. In the mid-1960s, Susumu Mitsuhashi first suggested that genes responsible for resistance to several antibiotics were mobile and could move between bacterial plasmids and chromosomes. Electron microscopic studies have confirmed that the two terminal-inverted repeat sequences are indeed present within plasmids harboring transposons. When double-stranded DNA from such a plasmid is separated into single strands and each is allowed to reanneal separately, the inverted repeat units are complementary. As might

be predicted, all areas other than the terminal repeat units remain single stranded and form loops on either end of the double-stranded stem (Figure 11.21).

Transposons have become the focus of increased interest, particularly because they have been found in organisms other than bacteria. Bacteriophages that demonstrate the ability to insert their genetic material into the host chromosome behave in a similar fashion. The **bacteriophage mu**, consisting of over 35,000 nucleotides, can insert its DNA at various places in the *E. coli* chromosome. Like IS units, if insertion occurs within a gene, mutant behavior at that locus results. Transposons have also been discovered in higher organisms, including yeast, corn, *Drosophila*, and humans.

## The *Ac–Ds* System in Maize

With our knowledge of insertion sequences and transposons in bacteria, it comes as no surprise that mobile genetic units also exist in eukaryotes. They are, in fact, more widespread, and they often have the effect of altering the expression of genetic information.

About 20 years before the discovery of transposons in prokaryotic organisms, Barbara McClintock analyzed the genetic behavior of two mutations, ***Dissociation (Ds)*** and ***Activator (Ac)***, in corn plants (maize). Often, analysis involves an examination of the phenotypes of the kernels of maize resulting from genes expressed in either the endosperm or aleurone layers (Figure 11.22). Initially, she determined that *Ds* is located on chromosome 9. Provided that *Ac* is also present in the genome, *Ds* has the effect of inducing breakage at a point on the chromosome adjacent to its location. If breakage occurs in somatic cells during their development, progeny cells often lose part of chromosome 9, causing a variety of phenotypic effects.

Subsequent analysis suggested to McClintock that both the *Ds* and *Ac* genes are sometimes transposed to different chromosomal locations. While *Ds* moves only if *Ac* is also present, *Ac* is capable of autonomous move-

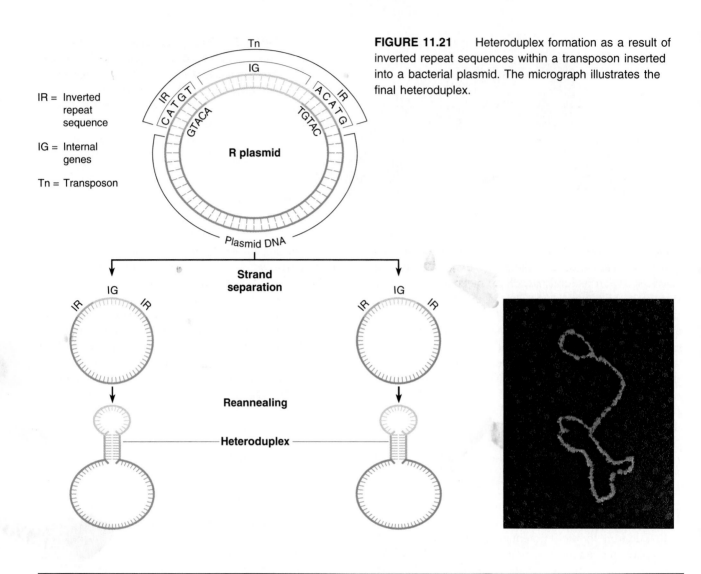

IR = Inverted
     repeat
     sequence

IG = Internal
     genes

Tn = Transposon

**FIGURE 11.21**    Heteroduplex formation as a result of inverted repeat sequences within a transposon inserted into a bacterial plasmid. The micrograph illustrates the final heteroduplex.

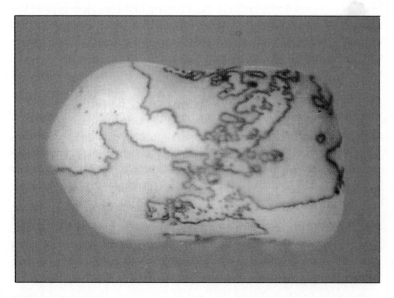

**FIGURE 11.22**    Corn kernel showing spots of colored aleurone produced by genetic transposition involving the *Ac–Ds* system.

ment. Where *Ds* comes to reside determines its genetic effect. Although it might cause chromosome breakage, it might instead inhibit gene expression. In cells where expression is inhibited, *Ds* might move again, releasing this inhibition. In these cases, the *Ds* element is believed to insert into a gene and subsequently to depart from it, causing changes in gene expression. Figure 11.23 illustrates the sort of movements and effects of the *Ds* and *Ac* elements described above. McClintock concluded that the *Ds* and *Ac* genes are **transposable controlling elements**.

It was not until many years later that anything comparable to the controlling elements in maize was recognized in other organisms. When insertion sequences and transposons were discovered, many parallels were evident. Transposons and insertion sequences were seen to move into and out of chromosomes, to insert at different positions, and to affect gene expression at the point of insertion.

Several *Ac* and *Ds* elements have now been isolated and carefully analyzed (Figure 11.24). As a result of this information, the relationship between the two elements

has been clarified. The first *Ac* element sequence is 4563 bases long and strikingly similar to the bacterial transposon **Tn3**. This sequence contains two 11-base-pair imperfect terminal-inverted repeats, two **open reading frames** (**ORFs**), and three noncoding regions. Open reading frames contain initiation and termination sequences and are considered to encode genetic products. The first *Ds* element studied (*Ds-a*) is nearly identical in structure to *Ac* except for a 194-base segment that has been deleted from the largest open reading frame (Figure 11.24). There is some evidence that this gene encodes a **transposase enzyme**, essential to transposition of both *Ac* and *Ds* elements. The deletion of part of this gene in the *Ds-a* element explains its dependence on the *Ac* element for transposition. Several other *Ds* elements have also been sequenced, and each reveals an even larger deletion in the same region. In each case, however, the terminal repeats are retained and seem to be essential for transposition, provided that a functional transposase enzyme is supplied by the gene in the *Ac* element.

**(a)**

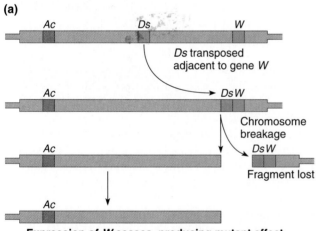

**Expression of *W* ceases, producing mutant effect**

**(b)**

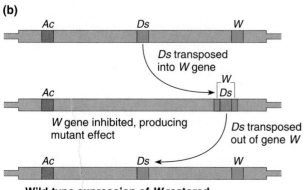

**Wild-type expression of *W* restored**

**FIGURE 11.23**    Two consequences of the influence of the *Activator* (*Ac*) element on the *Dissociation* (*Ds*) element. In (a), *Ds* is transposed to a region adjacent to a theoretical gene *W*. Subsequent chromosomal breakage is induced, the *W*-bearing segment is lost, and mutant gene expression occurs. In (b), *Ds* is transposed to a region within the *W* gene, causing immediate mutant expression. *Ds* may also "jump" out of the *W* gene with the acc nying restoration of *W* gene activity and its w'' pression.

# P Element Transposition

The analysis of mutations in individual genes is paramount in the study of gene expression and function. Because mutation represents one of the most critical tools for geneticists, methods for the induction of mutations have received much attention. Traditionally, mutations have been induced by chemicals and radiation, but it has been discovered in some organisms, including the fruit fly *Drosophila,* that parts of the genome itself in the form of transposable DNA segments are important sources of mutation.

Transposable elements (transposons) cause mutation by disrupting genes located at sites at which insertion occurs. In *Drosophila* this process is understood well enough so that transposable elements can be used as a means of experimentally inducing mutations. In fact, transposons called P elements are now used for many studies on mutagenesis in *Drosophila.* P elements were discovered through their ability to cause mutations when females from laboratory stocks are mated with males isolated from natural populations. The progeny from such crosses show a high mutation rate, sterility, and chromosome aberrations. It was soon discovered that this phenomenon was associated with the insertion of mobile DNA segments (now called P elements) into genes, causing their inactivation.

The movement of P elements to new sites in the genome occurs only in germ cells and depends on expression of two genes encoded by the mobile element itself. The first is a transposase that catalyzes the movement of the element from one chromosomal site to another, and a second gene encoding a repressor that inhibits transposase production, and thus prevents movement of the P element. Because transposition is repressed in strains already carrying P elements, transposition is restricted to offspring of crosses between females without P elements (true of most laboratory strains which were collected early in this century) and males with P elements. These properties make it possible to use P elements to locate genes, transfer genes into the germ cells of a recipient, and as a mutagen in *Drosophila.*

P element mutagenesis can be achieved by crossing appropriate strains of flies; stable mutations can be achieved with this method by transposing single P elements into strains lacking any P elements. Microinjection of P element DNA into *Drosophila* embryos can also be used to create mutants. In this process, P elements carrying two different defects are transferred. One P element does not encode transposase, but can insert into a chromosome; the other carries a transposase gene, but cannot insert into a chromosome. In place of a transposase gene, the first P element carries a gene that is to be mobilized for transposition and insertion into the *Drosophila* genome. This gene can be from *Drosophila* or any other organism. Once transferred, the newly acquired gene is inherited in a Mendelian fashion, and its expression at various stages of the *Drosophila* life cycle can be studied.

The process of P element transposition can be simplified by using *Drosophila* strains constructed to initiate transposon jumping in appropriate genetic crosses. One strain, referred to as a "jumpstarter," carries a P element that produces transposase, but cannot transpose. The other strain carries a defective P element that can transpose but cannot produce transposase. Crosses between these strains produce some flies that carry both types of P element and initiate transposition in their germline cells. Offspring from these flies will carry new P element insertions, and show new mutant phenotypes. Subsequent crosses can be used to stabilize these new mutations by eliminating the chromosome carrying the integrated P element producing transposase.

These methods have been used to generate a large number of P element insertion mutants that can be characterized and used in many areas of research, including studies of gene regulation, behavior, and development. Insertional mutants are particularly useful since their chromosomal position can be determined by *in situ* hybridization of labeled transposon probes to polytene chromosomes. Transposons are large enough (0.5 to 3.0 kb) to alter the pattern of DNA fragments produced using enzymes in recombinant DNA analyses. In addition, they provide convenient tags that simplify isolation and cloning of the gene into which they are inserted.

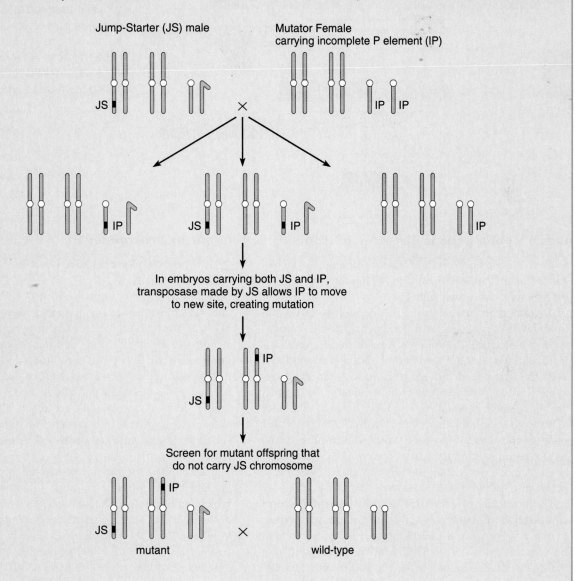

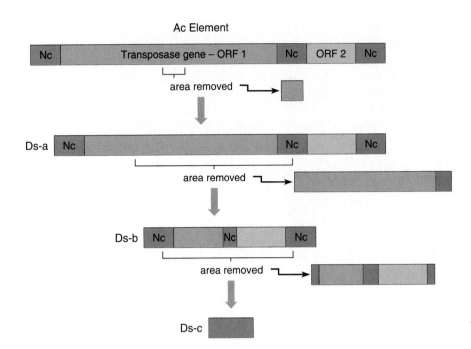

**FIGURE 11.24** A comparison of the structure of an *Ac* element with three *Ds* elements, all of which have been isolated and sequenced. The imperfect and inverted repeats are at the ends of the *Ac* element. The transposase gene is in an open reading frame (ORF 1). No function has yet been assigned to ORF 2. Noncoding regions are designated by Nc. As this scheme shows, *Ds-a* appears to be simply an *Ac* element containing a small deletion in the gene encoding the transposase enzyme.

## Other Mobile Genetic Elements in Plants

More recent work on transposable elements in plants has led us full circle to the union of Mendel's own observations with molecular genetics. Recall that one of the first phenotypes investigated by Mendel involved the inheritance of round and wrinkled peas. The two phenotypes are produced by alleles of a single gene, *rugosus*. It is now known that the wrinkled phenotype is associated with the absence of an enzyme, starch-branching enzyme (SBEI), that controls the formation of branch points in starch molecules. The lack of starch synthesis leads to the accumulation of sucrose and a higher water content and osmotic pressure in the developing seeds. As the seeds mature, those that are wrinkled (genotype *rr*) lose more water than the smooth seeds (*RR* or *Rr*), producing the wrinkled phenotype (Figure 11.25).

The structural gene for SBEI has now been cloned and characterized in both wild-type and mutant genotypes. In the *rr* genotype, the SBEI protein is nonfunctional, presumably because the SBEI gene is interrupted by a 0.8-kb insertion, resulting in the production of an abnormal RNA transcript. The inserted DNA has 12-bp inverted repeats at each end that are highly homologous to the terminal sequences in the transposable element *Ac* from maize, and other *Ac*-like elements from snapdragons and parsley. Terminal repeated sequences and the genetic information encoding a transposase enzyme appear to be universal components of transposons in all organisms studied.

## Copia in *Drosophila*

Transposable elements have been discovered in other eukaryotic organisms, notably in yeast, *Drosophila*, and primates, including humans. In 1975, David Hogness and his colleagues David Finnigan, Gerald Rubin, and Michael Young identified a class of genes in *Drosophila melanogaster*, which they designated as *copia*. These genes transcribe "copious" amounts of RNA (thus, their name), which can be isolated in the mRNA fraction. Present up to 30 times in the genome of cells, *copia* genes are nearly identical in nucleotide sequence. Mapping studies show that they are transposable to different chromosomal locations and are dispersed throughout the genome.

*Copia* appears to be only one of the approximately 30 families of transposable elements in *Drosophila*, each of which is present from a few copies up to 50 to 100 times in the genome. Since the discovery of *copia*, many other families of transposable elements have been recognized. Some are referred to as *copia*-like. Together, the many families constitute about 5 percent of the *Drosophila* genome and over half of the middle repetitive DNA of this organism.

Despite the variability in DNA sequence between the members of different families, they share a common structural organization thought to be related to the insertion and excision processes of transposition. Each *copia* gene consists of approximately 5000 to 8000 base pairs of DNA, including a long family-specific **direct termi-**

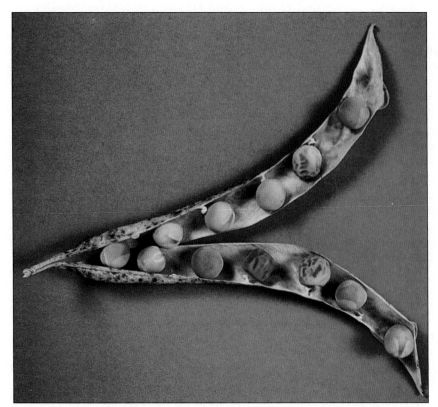

**FIGURE 11.25**    The *wrinkled* trait in garden peas, studied by Gregor Mendel, is caused by the insertion of a transposable element into the structural gene for starch-branching enzyme.

nal repeat (**DTR**) sequence of 276 base pairs at each end. Within each repeat is a short **inverted terminal repeat (ITR)** of 17 base pairs. These features are illustrated in Figure 11.26. The DTR sequences are found in other transposons in other organisms but are not universal. However, the shorter ITR sequences are considered universal in *copia* elements.

Insertion of *copia*, as with other transposable elements, appears to be dependent on ITR sequences and appears to occur at specific target sites in the genome. In general, eukaryotic transposons are strikingly similar to one another and share many features with those in bacteria.

The *copia*-like elements demonstrate regulatory effects at the point of their insertion in the chromosome. Certain mutations, including ones affecting eye color and segment formation, have been found to be due to insertions within genes. For example, the eye color mutation *white-apricot* ($w^a$), an allele of the *white* ($w$) mutation, contains a *copia* insertion element within the gene. Removing the transposable element sometimes restores the wild-type allele.

## P Elements in *Drosophila*

Still another interesting category of transposable elements in *Drosophila* is the family called **P elements**. These were discovered while studying the phenomenon of **hybrid dysgenesis**, a condition causing sterility, elevated mutation rate, and chromosome rearrangement in

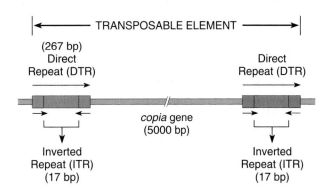

**FIGURE 11.26**    Structural organization of a *copia* transposable element in *Drosophila melanogaster*.

the offspring of crosses between certain strains of fruit flies. Such "dysgenic" offspring occur when certain females reared in the laboratory (M cell type) are crossed to males derived from a natural population (P cell type). The reciprocal cross fails to produce affected offspring.

Many of the mutations that result are unstable and frequently revert to wild type, suggesting that a transposable element might be involved. Subsequently, P elements were discovered to be associated with P cell types (the males in our original cross mentioned above). Present in up to 50 copies per genome, various elements range from 500 to 3000 base pairs in length. Regardless of the size, a 31-base-pair perfect inverted repeat is found at both ends of each element.

The internal portions of P elements may encode up to three proteins. One is thought to function as a **transposase enzyme** that is responsible for transposition. A second is thought to encode a molecule that represses expression of the transposase gene. In P strains, repressor products effectively block transposition in this way. However, when sperm from P males fertilize eggs from M-type females (who are lacking P elements and therefore lacking the repressor), hybrid dysgenesis results as the P elements are mobilized and transposition ensues. Note that the reciprocal cross (P♀ × M♂) does not produce dysgenesis since the ooplasm of the eggs of the P females contain ample amounts of the repressor gene product.

## Transposable Elements in Humans

In Chapter 10 we introduced the *Alu* family of short interspersed elements (SINEs), which are characteristic of moderately repetitive DNA in mammals. In humans, some 300,000 copies of this 200 to 300 base-pair sequence are present in the genome.

These sequences are considered transposable based on several lines of evidence. Most important is the fact that they contain a 300-bp sequence flanked on either side by direct repeat sequences consisting of 7 to 20 bp. This observation is similar to that of bacterial insertion sequences. These flanking regions are related to the insertion process during transposition. The second line of evidence involves the observation that clustered regions of *Alu* sequences vary in the DNA of normal and diseased individuals and in different tissues of the same individual. Additionally, the sequences have been found extrachromosomally.

The potential mobility and mutagenic effect of these and other elements have far-reaching effects. A recent example involves a situation where a transposon has been "caught in the act." The case involves a male child with **hemophilia**, as investigated by Haig Kazazian and his colleagues. One cause of hemophilia is a defect in blood-clotting factor VIII, the product of an X-linked gene. Kazazian found two cases where inserted within this gene, there was a transposable element much longer than *Alu* sequences. Also present elsewhere in the genome, this sequence was shown to be an example of a long interspersed element (LINE).

There has been great interest in determining if one of the mother's X chromosomes also contains this specific LINE. If so, the unaffected mother would be heterozygous and pass the LINE-containing chromosome to her son. The startling finding is that the LINE sequence is *not* present on either of her *X chromosomes*, but *was* detected on *chromosome 22* of both parents. This suggests that this mobile element may have "moved" from one chromosome to another in the gamete-forming cells of the mother, prior to being transmitted to the son.

Many questions remain concerning this and other transposable elements. What is their origin? Were they once some sort of retrovirus (see Chapter 19)? Exactly how do they move, and what has been their role during evolution? These and other questions will intrigue researchers for many years to come.

**CHAPTER SUMMARY**

1. The phenomenon of mutation not only provides the basis for most of the inherent variation present in living organisms, but also serves as the working tool of the geneticist in studying and understanding the nature of genes and the mechanisms governing genetic processes.

2. Mutations are distinguished by the tissues affected. Somatic mutations are those that may affect the individual, but are not heritable. Mutations arising in the gametic tissues may produce new alleles that can be passed on to offspring and enter the gene pool.

3. Another classification of mutations relies on their effect. Morphological mutations, for example, may be detected visibly. Other types include biochemical, lethal, conditional, and regulatory mutations, groups that are not mutually exclusive.

4. Organisms in which mutations can be easily induced and detected are most often used in genetic studies. Viruses, bacteria, fungi, and *Drosophila* are frequently used because of these properties, as well as the fact that they have short life cycles.

5. Spontaneous mutations may arise naturally as a result of rare chemical rearrangements of atoms, or tautomeric shifts, and as the result of errors occurring during DNA replication. While spontaneous mutations are very rare, the rate of mutagenesis may be increased experimentally by a variety of mutagenic agents.

6. Mutagenic agents such as nitrous acid, hydroxylamine, alkylating agents, and base analogues cause chemical changes in nucleotides that alter their base-pairing affinities. As a result, base substitution mutations arise following DNA replication.

7. Frameshift mutations arise when the addition or deletion of one or more nucleotides (but not multiples of three) occurs. Acridine dyes, which intercalate with DNA, are potent inducers of frameshift mutations.

8. Direct analysis of DNA from individuals carrying specific alleles that have arisen through mutations, including the ABO blood types and muscular dystrophy, has been informative. When complete loss of function occurs, as is the case in blood type O and the Duchenne form of muscular dystrophy, deletions or duplications of nucleotides have resulted in a shift in reading frame, eventually causing premature termination of translation of the resulting messenger RNA.

9. Another form of mutation, discovered in several human disorders, includes unstable trinucleotide units that balloon in size during DNA replication, and are inherited in this form through successive generations. Increasing numbers of repeats often correlate with early onset and severity of disease.

10. Ultraviolet radiation and high-energy radiation from gamma, cosmic, or X-ray sources are another form of potent mutagenic agents. UV radiation induces the formation of pyrimidine dimers in DNA, while high-energy radiation is able to penetrate deeper into tissues, causing the ionization of molecules in its path, including DNA.

11. Various forms of repair of DNA lesions, such as pyrimidine dimers, have been discovered, including photoreactivation, excision, mismatch, and recombinational repair. Loss of repair function by mutation in humans results in the severe disorder xeroderma pigmentosum.

12. Site-directed mutagenesis is a technique allowing researchers to create specific alterations in the nucleotide sequence of the DNA of genes.

13. Insertion sequences in bacteria and other transposable elements in eukaryotes have a profound effect on genetic expression, thus serving as a distinct category of mutagenic agents. A recent example involves a mutation causing hemophilia in humans.

## KEY TERMS

ABO antigens
acridine dyes
Activator (*Ac*) element
alkylating agents
Ames test
anticipation
AP endonuclease
apurinic sites (AP sites)
attached-X procedure
autosomal mutation
auxotroph
bacteriophage mu
base analogue
base substitution
Becker muscular
    dystrophy (BMD)
behavior mutation
biochemical mutation
*ClB* procedure
complementation
complete medium
conditional mutation
cosmic rays
direct terminal repeat
    (DTR)
Dissociation (*Ds*) element

DNA glycosylase
DNA ligase
DNA methylation
DNA polymerase I
dominant mutation
Duchenne muscular
    dystrophy (DMD)
dystrophin
ethylmethane sulfonate
    (EMS)
excision repair
5-bromouracil (5-BU)
fragile-X syndrome
frameshift mutation
gamma rays
gene mutation
hemophilia
heterokaryon
Huntington disease
hybrid dysgenesis
induced mutation
insertion sequence (IS)
inverted terminal repeat
    (ITR)
ionizing radiation
lethal mutation

*lex*A repressor
minimal culture medium
mismatch repair
morphological trait
muscular dystrophy
mustard gases
myotonic dystrophy
nutritional mutation
opaque-2 mutant strain
open reading frame
    (ORF)
P elements
perfect inverted repeat
photoreactivation enzyme
    (PRE)
photoreactivation repair
point mutation
proflavin
proofreading
prototroph
*rec*A gene
recombinational repair
regulatory mutation
reverse mutation
roentgen
sex-linked mutation

site-directed mutagenesis
6-ethyl guanine
somatic cell genetics
SOS response
spinobulbar muscular
    atrophy (Kennedy
    disease)
spontaneous mutation
target theory
tautomeric forms
tautomeric shift
temperature-sensitive
    mutation
transition mutation
transposable genetic
    elements
transposase enzyme
transposon (Tn) elements
transversion mutation
2-amino purine (2-AP)
ultraviolet (UV) radiation
unscheduled DNA
    synthesis
xeroderma pigmentosum
    (XP)
X rays

# INSIGHTS AND SOLUTIONS

1.  How could you isolate a mutant strain of bacteria that is resistant to penicillin, an antibiotic that inhibits cell wall synthesis?

    **SOLUTION:**  Grow a culture of bacterial cells in liquid medium and plate the cells on agar medium to which penicillin has been added. Only penicillin-resistant cells will reproduce and form colonies. Each colony will, in all likelihood, represent a cloned group of cells with the identical mutation. Isolate members of each colony. To enhance the chance of such a mutation arising, you might want to add a mutagen to the liquid culture.

2.  The base analogue 2-amino purine (2-AP) substitutes for adenine during DNA replication, but it may base pair with cytosine. The base analogue 5-bromouracil (5-BU) substitutes for thymidine, but it may base pair with guanine. Follow the double-stranded trinucleotide sequence shown below through three rounds of replication, assuming that in the first round, both analogues are present and become incorporated wherever possible. In the second and third round of replication, they are removed. What final sequences occur?

    **SOLUTION:**  The solution is illustrated on the next page.

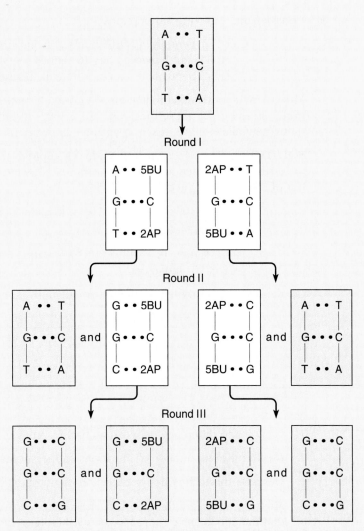

3. A rare dominant mutation was studied in humans that was expressed at birth. Records showed that six cases were discovered in 40,000 live births. Family histories revealed that in two cases, the mutation was already present in one of the parents. Calculate the spontaneous mutation rate for this mutation and discuss the validity of this estimate.

**SOLUTION:** Only four cases represent a new mutation. Since each live birth represents two gametes, the sample size is from 80,000 meiotic events. The rate is equal to

$$\frac{4}{80,000} = \frac{1}{20,000} = 0.5 \times 10^{-4}$$

There are various reasons why this mutation rate may be inaccurate. We have assumed that the mutant gene is fully penetrant and is expressed in each individual bearing it. If it is not fully penetrant, our calculation may be an underestimate because one or more mutations may have gone undetected. We have also assumed that the screening was 100 percent accurate. One or more mutant individuals may have been "missed," again leading to an underestimate. Finally, we assume that the viability of mutant and nonmutant individuals is equivalent and that they survive equally *in utero*. Therefore, our assumption is that the number of mutant individuals at birth is equal to the number at conception. If this were not true, our calculation would again be inaccurate.

4. Consider the following assumptions:

   (i)   There are $4 \times 10^9$ humans living on this planet.
   (ii)  Each individual has $10^5$ genes.
   (iii) The average mutation rate at the $10^5$ loci is $10^{-5}$.

How many spontaneous mutations are currently present in the human population? Assuming that these mutations are equally distributed among all genes, how many new alleles have arisen at each locus in the human population?

**SOLUTION:**   First, since each individual is diploid, there are two copies of each gene per person, each arising from a separate gamete. Therefore, the total number of spontaneous mutations is

$$\left(\frac{2 \times 10^5 \text{ genes}}{\text{individual}}\right)\left(4 \times 10^9 \text{ individuals}\right)\left(\frac{10^{-5} \text{ mutations}}{\text{gene}}\right)$$

$$= (2 \times 10^5) \cdot (4 \times 10^9) \cdot (10^{-5}) \text{ mutations}$$
$$= 8 \times 10^9 \text{ mutations}$$

On the average, in the entire human population there are

$$\frac{8 \times 10^9 \text{ mutations}}{10^5 \text{ genes}} = \frac{8 \times 10^4 \text{ mutations}}{\text{gene}}$$

---

**PROBLEMS AND DISCUSSION QUESTIONS**

1. What is the difference between a chromosomal aberration and a gene mutation?

2. Discuss the importance of mutations to the successful study of genetics.

3. Describe the technique for detection of nutritional mutants in *Neurospora*.

4. Most mutations are thought to be deleterious. Why, then, is it reasonable to state that mutations are essential to the evolutionary process?

5. Why do you suppose that a random mutation is more likely to be deleterious than beneficial?

6. Most mutations in a diploid organism are recessive. Why?

7. What is meant by a conditional lethal mutation?

8. Contrast the concerns about mutation in somatic and gametic tissue.

9. In the *ClB* technique for detection of mutation in *Drosophila*:
   (a) What type of mutation may be detected?
   (b) What is the importance of the *l* gene?
   (c) Why is it necessary to have the crossover suppressor (*C*) in the genome?
   (d) What is *C*?

10. In an experiment with the *ClB* system, a student irradiated wild-type males and subsequently scored 500 $F_2$ cultures, finding all of them to have both males and females present. What conclusions can you draw about the induced mutation rate?

11. In *Drosophila*, induced mutations on chromosome 2 that are recessive lethals may be detected using a second chromosomal stock, *Curly, Lobe/Plum* (*Cy L/Pm*). These alleles are all dominant and lethal in the homozygous condition. Detection

is performed by crossing *Cy L/Pm* females to wild-type males that have been subjected to a mutagen. Three generations are required. In the $F_1$, *Cy L* males are selected and individually backcrossed to *Cy L/Pm* females. In the $F_2$ of each cross, flies expressing *Cy* and *L* are mated to produce a series of $F_3$ generations. Diagram these crosses and predict how an $F_3$ culture will vary if a recessive lethal was induced in the original test male compared to the case where no lethal mutation resulted. In which $F_3$ flies would a recessive morphological mutation be expressed?

12. Describe tautomerism and the way in which this chemical event may lead to mutation.

13. Contrast and compare the mutagenic effects of deaminating agents, alkylating agents, and base analogues.

14. Acridine dyes induce frameshift mutations. Is such a mutation likely to be more detrimental than a point mutation where a single pyrimidine or purine has been substituted?

15. Why is X-radiation a more potent mutagen than UV radiation?

16. Contrast the induction of mutations by UV radiation and X-radiation.

17. Contrast the various types of DNA repair mechanisms known to counteract the effects of UV radiation and other DNA damage. What is the role of visible light in photoreactivation?

18. Mammography is an accurate screening technique for the early detection of breast cancer in humans. Because this technique uses X rays diagnostically, it has been highly controversial. Can you explain why?

19. Describe the Ames assay for screening potential environmental mutagens. Why is it thought that a compound that tests positively in the Ames assay may also be carcinogenic?

20. Describe the methodology used in site-directed mutagenesis.

21. Describe the disorder xeroderma pigmentosum in humans.

22. Presented below are theoretical findings from studies of heterokaryons formed from seven human xeroderma pigmentosum cell strains:

|  | XP1 | XP2 | XP3 | XP4 | XP5 | XP6 | XP7 |
|---|---|---|---|---|---|---|---|
| XP1 | 0 | | | | | | |
| XP2 | 0 | 0 | | | | | |
| XP3 | 0 | 0 | 0 | | | | |
| XP4 | + | + | + | 0 | | | |
| XP5 | + | + | + | + | 0 | | |
| XP6 | + | + | + | + | 0 | 0 | |
| XP7 | + | + | + | + | 0 | 0 | 0 |

NOTE: + = complementing; 0 = noncomplementing.

These data represent the occurrence of unscheduled DNA synthesis in the fused heterokaryon when neither of the strains alone showed synthesis. What does unscheduled DNA synthesis represent? Which strains fall into the same complementation groups? How many different groups are revealed based on these limited data? How does one interpret the presence of these complementation groups?

**23.** In a bacterial culture where all cells were unable to synthesize leucine ($leu^-$) a potent mutagen was added, and the cells were allowed to undergo one round of replication. At that point, samples were taken and a series of dilutions was made prior to plating the cells on either minimal medium or minimal medium to which leucine was added. The first culture condition (minimal medium) allows the detection of mutations from $leu^-$ to $leu^+$, while the second culture condition allows the determination of total cells, since all bacteria can grow. From the following results, determine the frequency of mutant cells. What is the rate of mutation at the locus involved with leucine biosynthesis?

| Culture Condition | Dilution | Colonies |
|---|---|---|
| minimal medium | $10^{-1}$ | 18 |
| minimal + leucine | $10^{-7}$ | 6 |

**24.** Contrast the various transposable genetic elements in bacteria, maize, *Drosophila*, and humans. What properties do they share?

**25.** The initial discovery of IS units in bacteria involved their presence upstream (5′) to three genes controlling galactose metabolism. All three genes were affected in spite of no direct physical interaction with the IS unit. Speculate as to why this might occur.

**26.** Ty, a transposable element in yeast, has been found to contain an open reading frame (ORF) that encodes the enzyme reverse transcriptase. This enzyme synthesizes DNA from an RNA template. Speculate as to the role of this gene product in the transposition of Ty within the yeast genome.

**SELECTED READINGS**

AMES, B. N., McCANN, J., and YAMASAKI, E. 1975. Method for detecting carcinogens and mutagens with the *Salmonella*/mammalian microsome mutagenicity test. *Mut. Res.* 31:347–64.

AUERBACH, C. 1978. Forty years of mutation research: A pilgrim's progress. *Heredity* 40:177–87.

AUERBACH, C., and KILBEY, B. J. 1971. Mutations in eukaryotes. *Ann. Rev. Genet.* 5:163–218.

BALTIMORE, D. 1981. Somatic mutation gains its place among the generators of diversity. *Cell* 26:295–96.

BEADLE, G. W., and TATUM, E. L. 1945. *Neurospora* II. Methods of producing and detecting mutations concerned with nutritional requirements. *Amer. J. Bot.* 32:678–86.

BECKER, M. M., and WANG, Z. 1989. Origin of ultraviolet damage in DNA. *J. Mol. Biol.* 210:429–38.

BERG, D., and HOWE, M., eds. 1989. *Mobile DNA*. Washington, D.C.: Amer. Soc. Microbiol.

BRENNER, S., BARNETT, L., CRICK, F. H. C., and ORGEL, L. 1961. The theory of mutagenesis. *J. Mol. Biol.* 3:121–24.

CARTER, P. 1986. Site-directed mutagenesis. *Biochem. J.* 237:1–7.

CLEAVER, J. E. 1968. Defective repair of replication of DNA in xeroderma pigmentosum. *Nature* 218:652–56.

———. 1990. Do we know the cause of xeroderma pigmentosum? *Carcinogenesis* 11:875–82.

CLEAVER, J. E., and KARENTZ, D. 1986. DNA repair in man: Regulation by a multiple gene family and its association with human disease. *Bioessays* 6:122–27.

COHEN, S. N., and SHAPIRO, J. A. 1980. Transposable genetic elements. *Scient. Amer.* (Feb.) 242:40–49.

CROW, J. F., and DENNISTON, C. 1985. Mutation in human populations. *Adv. Hum. Genet.* 14:59–216.

DEERING, R. A. 1962. Ultraviolet radiation and nucleic acids. *Scient. Amer.* (Dec.) 207:135–44.

DEN DUNNEN, J. T., et al. 1989. Topography of the Duchenne muscular dystrophy (DMD) gene. *Am. J. Hum. Genet.* 45:835–47.

DEVORET, R. 1979. Bacterial tests for potential carcinogens. *Scient. Amer.* (Aug.) 241:40–49.

DORING, H. P., and STARLINGER, P. 1984. Barbara McClintock's controlling elements: Now at the DNA level. *Cell* 39:253–59.

DRAKE, J. W. 1970. *Molecular basis of mutation.* San Francisco: Holden-Day.

———. 1991. A constant rate of spontaneous mutation in DNA-based microbes. *Proc. Nat. Acad. Sci.* 88:7160–64.

DRAKE, J. W., et al. 1975. Environmental mutagenic hazards. *Science* 187:505–14.

DRAKE, J. W., GLICKMAN, B. W., and RIPLEY, L. S. 1983. Updating the theory of mutation. *Amer. Scient.* 71:621–30.

FEDOROFF, N. V. 1984. Transposable genetic elements in maize. *Scient. Amer.* (June) 250:85–98.

FINNEGAN, D. J. 1985. Transposable elements in eukaryotes. *Int. Rev. Cytol.* 93:281–326.

HANAWALT, P. C., and HAYNES, R. H. 1967. The repair of DNA. *Scient. Amer.* (Feb.) 216:36–43.

HASELTINE, W. A. 1983. Ultraviolet light repair and mutagenesis revisited. *Cell* 33:13–17.

HOWARD-FLANDERS, P. 1981. Inducible repair of DNA. *Scient. Amer.* (Nov.) 245:72–80.

KELNER, A. 1951. Revival by light. *Scient. Amer.* (May) 184:22–25.

KNUDSON, A. G. 1979. Our load of mutations and its burden of disease. *Am. J. Hum. Genet.* 31:401–13.

KRAEMER, F. H., et al. 1975. Genetic heterogeneity in xeroderma pigmentosum: Complementation groups and their relationship to DNA repair rates. *Proc. Natl. Acad. Sci.* 72:59–63.

LITTLE, J. W., and MOUNT, D. W. 1982. The SOS regulatory system of *E. coli. Cell* 29:11–22.

MASSIE, R. 1967. *Nicholas and Alexandra.* New York: Atheneum.

MASSIE, R., and MASSIE, S. 1975. *Journey.* New York: Knopf.

McCANN, J., CHOI, E., YAMASAKI, E., and AMES, B. 1975. Detection of carcinogens as mutagens in the *Salmonella*/microsome test: Assay of 300 chemicals. *Proc. Natl. Sci.* 72:5135–39.

McCLINTOCK, B. 1956. Controlling elements and the gene. *Cold Spr. Harb. Symp.* 21:197–216.

MACDONALD, M. E., et al. 1993. A novel gene containing a trinucleotide repeat that is expanded and unstable in Huntington's disease chromosome. *Cell* 72:971–80.

McKUSICK, V. A. 1965. The royal hemophilia. *Scient. Amer.* (Aug.) 213:88–95.

MULLER, H. J. 1927. Artificial transmutation of the gene. *Science* 66:84–87.

———. 1955. Radiation and human mutation. *Scient. Amer.* (Nov.) 193:58–68.

NEWCOMBE, H. B. 1971. The genetic effects of ionizing radiation. *Adv. in Genet.* 16:239–303.

O'HARE, K. 1985. The mechanism and control of P element transposition in *Drosophila. Trends in Genet.* 1:250–54.

OSSANA, N., PETERSON, K. R., and MOUNT, D. W. 1986. Genetics of DNA repair in bacteria. *Trends in Genet.* 2:55–58.

RADMAN, M., and WAGNER, R. 1988. The high fidelity of DNA duplication. *Scient. Amer.* (August) 259(2):40–46.

SHORTLE, D., DiMARIO, D., and NATHANS, D. 1981. Directed mutagenesis. *Ann. Rev. Genet.* 15:265–94.

SIGURBJORHSSON, B. 1971. Induced mutations in plants. *Scient. Amer.* (Jan.) 224:86–95.

STADLER, L. J. 1928. Mutations in barley induced by X-rays and radium. *Science* 66:84–87.

SUTHERLAND, B. M. 1981. Photoreactivation. *Bioscience* 31:439–44.

TOMLIN, N. V., and APRELIKOVA, O. N. 1989. Uracil DNA glycosylases and DNA uracil repair. *Intern. Rev. Cytol.* 114:81–124.

TOPAL, M. D., and FRESCO, J. R. 1976. Complementary base pairing and the origin of substitution mutations. *Nature* 263:285–89.

VOGEL, F. 1992. Risk calculations for hereditary effects of ionizing radiation in humans. *Hum. Genet.* 89:127–46.

WILLS, C. 1970. Genetic load. *Scient. Amer.* (March) 222:98–107.

WOLFF, S. 1967. Radiation genetics. *Ann. Rev. Genet.* 1:221–44.

YAMAMOTO, F., et al. 1990. Molecular genetic basis of the histo-blood group ABO system. *Nature* 345:229–33.

# 12

# DNA: CLONING AND MANIPULATION

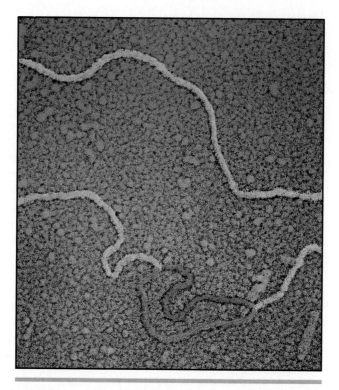

Transmission electron micrograph of a bacterial plasmid DNA molecule into which has been inserted a single gene (blue).

*Recombinant DNA technology depends in part on the ability to cleave and rejoin DNA segments at specific base sequences. Using this methodology, individual DNA segments can be transferred to viruses or bacteria and amplified, isolated, and identified. The use of this technology has brought about significant advances in gene mapping, disease diagnosis, the commercial production of human gene products, and the trans-specific transfer of genes in plants and animals.*

The independent rediscovery of Mendel's work by de Vries, Correns, and von Tschermak in 1900 marks the beginning of genetics as an organized discipline. During the course of its growth and development, several key discoveries have served as landmarks or turning points, each accelerating the rate at which our knowledge of genetics has grown, and, in turn, opening new fields of investigation.

One of the first turning points was the concept that chromosomes are the repository for genes. This concept, proposed by Sutton and Boveri in 1902, was developed by Morgan and his colleagues using the fruit fly, *Drosophila*. From these studies came our understanding of transmission genetics, sex determination, linkage, and the use of polytene chromosomes to map genes to specific cytological loci.

A second landmark was the discovery by Avery, MacLeod, and McCarty that DNA is the macromolecular carrier of genetic information. This work stimulated the use of viruses and bacteria as organisms for genetic research and led to the Watson–Crick model for the structure of DNA. From this model has come knowledge of the molecular basis for genetic coding, transcription, translation, and gene regulation.

We are now in the beginning stages of another and perhaps the most profound transition in the history of genetics—the development and application of **recombinant DNA technology**, also known as **gene splicing** or **genetic engineering**. This technology is being used to generate new knowledge and to develop new commercial products such as vaccines and drugs. It has also raised fears about epidemics or widespread ecological changes that might result from the release of genetically engineered organisms into the environment. In this chapter we will review some of the methods used in recombinant DNA technology to isolate, replicate, and analyze genes, and in the following chapter we will discuss some of the applications of this technology to agriculture, medicine, and industry.

## RECOMBINANT DNA TECHNOLOGY: AN OVERVIEW

The term **recombinant DNA** refers to the creation of a new association between DNA molecules or segments of DNA molecules that are not found together naturally. Although genetic mechanisms such as crossing over technically produce recombinant DNA, the term is generally reserved for DNA molecules produced by joining segments derived from different biological sources.

Recombinant DNA technology uses techniques derived from the biochemistry of nucleic acids coupled with genetic methodology originally developed for the study of bacteria and viruses. The basic procedures involve a series of steps:

1. DNA fragments are generated by using enzymes called **restriction endonucleases** that recognize and cut DNA molecules at specific nucleotide sequences.

2. These segments are joined to other DNA molecules that serve as **vectors**. Vectors can replicate autonomously and thus facilitate the manipulation and identification of the newly created recombinant DNA molecule.

3. The vector, carrying an inserted DNA segment, is transferred to a host cell. Within this cell, the recombinant DNA molecule composed of the vector and the inserted DNA segment is replicated, producing dozens of identical copies known as **clones**.

4. The cloned DNA segments can be recovered from the host cell, purified, and analyzed.

5. Host cells containing recombinant DNA pass this on to all progeny cells, creating a population of identical cells, all carrying the cloned sequence.

6. Potentially, the cloned DNA can be transcribed, its mRNA translated, and the gene product isolated and studied.

The development of genetic engineering techniques affords new opportunities for research, making it easier to obtain large amounts of DNA encoding specific genes and facilitating studies of gene organization, structure, and expression. This methodology has also given impetus to the development of a burgeoning biotechnology industry that has delivered a growing number of products to the marketplace.

## Restriction Enzymes

The cornerstone of recombinant DNA technology is a class of enzymes called **restriction endonucleases**. These enzymes, isolated from bacteria, received their name because they restrict or prevent viral infection by degrading the invading nucleic acid. Restriction enzymes recognize a specific nucleotide sequence and produce a double-stranded break within or near that sequence. Two classes of restriction endonucleases have been discovered, and more than 80 enzymes of both types are now recognized. Type I restriction enzymes recognize a specific nucleotide sequence and cut both strands of the DNA at a random location at some distance from the recognition site. Type II enzymes recognize a specific sequence and precisely cut both strands within the sequence. In Type II enzymes, the recognition site is **palindromic**, that is, it reads the same in the 5'-to-3' direction on complementary strands. The DNA of bacteria producing a specific restriction enzyme is protected by **modification enzymes** that methylate the nucleotides contained in recognition sites, thereby protecting them from the action of restriction endonucleases.

Of the two classes, Type II enzymes are most often used in genetic engineering. One of the first such enzymes discovered was from *E. coli* and was designated *Eco*RI. Its palindromic recognition site and points of cleavage are shown in Figure 12.1(a).

The complementary tails produced by cutting with *Eco*RI are said to be "sticky" because under hybridization conditions, they can reanneal with each other. If DNAs from different sources share the same palindromic recognition sites, then both will contain complementary single-stranded tails when treated with a restriction endonuclease (Figure 12.2). If placed together under proper conditions, the DNA fragments from these two sources can form recombinant molecules by annealing of their sticky ends. The enzyme DNA ligase is used to chemically join these fragments (Figure 12.2). DNA ligase is the enzyme that ligates Okazaki fragments during DNA replication. It is commercially available as a purified preparation.

Other Type II restriction enzymes such as *Sma*I, however, cleave DNA in a manner that produces blunt-end fragments [Figure 12.1(b)]. Ligation of DNA fragments with blunt ends can also be used to create recombinant

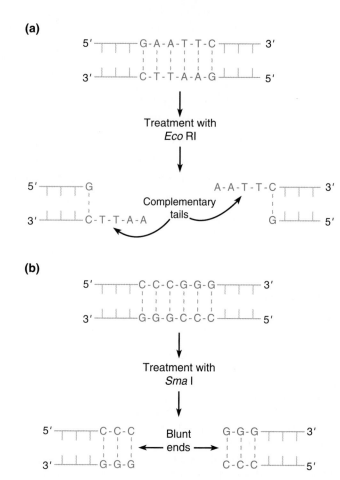

**FIGURE 12.1**    (a) The restriction enzyme *Eco*RI recognizes and binds to the palindromic nucleotide sequence GAATTC. Cleavage with this enzyme produces complementary single-stranded tails at the cutting site. These single-stranded tails can anneal with similar ends from other DNA fragments to form recombinant DNA molecules. (b) The restriction enzyme *Sma*I recognizes and binds to the palindromic nucleotide sequence CCCGGG. Cleavage with this enzyme produces blunt-end fragments. Such fragments must be modified before use in forming recombinant DNA molecules.

molecules, but this is not as efficient as with sticky ends, unless the DNA is modified (Figure 12.3). Commonly, the enzyme **terminal deoxynucleotidyl transferase** is used to enhance annealing. This enzyme extends single-stranded ends by the addition of nucleotide "tails." If a poly dA tail is added to DNA fragments from one source (such as plasmid DNA), and a poly dT tail is added to eukaryotic DNA, complementary tails are created, and the fragments can anneal. Recombinant molecules can then be created by ligation (Figure 12.3). Table 12.1 lists some common restriction enzymes and their recognition sequences, many of which yield cohesive or "sticky" ends, while others generate blunt ends. After its

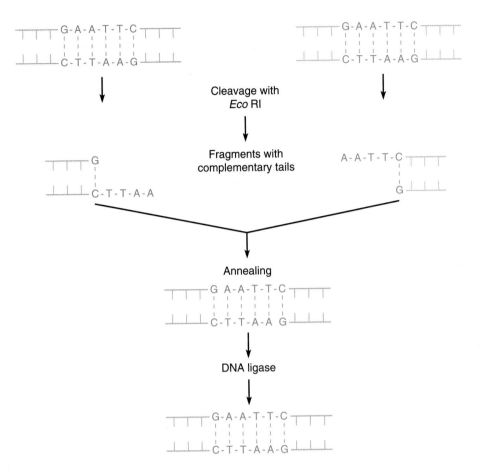

**FIGURE 12.2**    In forming recombinant DNA molecules, DNA from different sources is cleaved with *Eco*RI and mixed together under proper conditions to allow annealing into recombinant molecules. The enzyme DNA ligase is then used to chemically bond these annealed fragments into an intact recombinant DNA molecule.

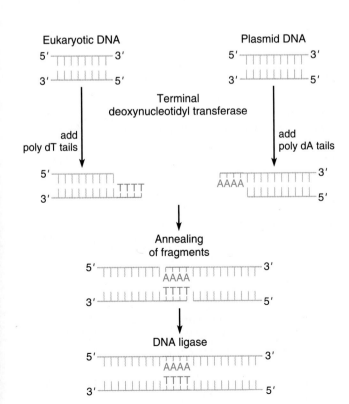

**FIGURE 12.3**    Recombinant DNA molecules can also be formed from DNA cut with enzymes that leave blunt ends. In this method, the enzyme terminal deoxynucleotidyl transferase is used to create complementary tails by the addition of poly A and poly T to the cut fragments. These tails serve to anneal DNA from different sources and create recombinant DNA molecules that are chemically bonded by treatment with DNA ligase.

**Table 12.1** SOME RESTRICTION ENZYMES, CLEAVAGE PATTERN AND SOURCE

| Enzyme | Recognition, Cleavage Sequence | Source | Enzyme | Recognition, Cleavage Sequence | Source |
|---|---|---|---|---|---|
| EcoRI | ↓<br>GAATTC<br>CTTAAG<br>    ↑ | E. coli | AluI | ↓<br>AGCT<br>TCGA<br> ↑ | Arthrobacter luteus |
| HindIII | ↓<br>AAGCTT<br>TTCGAA<br>    ↑ | Hemophilus influenza | HaeIII | ↓<br>GGCC<br>CCGG<br> ↑ | Hemophilus aegypticus |
| BamHI | GGATCC<br>CCTAGG<br>    ↑ | Bacillus amyloliquefaciens | BalI | TGGCCA<br>ACCGGT<br>   ↑ | Brevibacterium albidum |
| TaqI | ↓<br>TCGA<br>AGCT<br> ↑ | Thermus aquaticus | Sau3A | ↓<br>GATC<br>CTAG<br> ↑ | Staphylococcus aureus |

formation, the recombinant molecule must be inserted into a host organism for replication.

## Plasmid Vectors

After insertion into a vector or cloning vehicle, a DNA segment can gain entry into a host cell and be replicated or cloned. Vectors are, in essence, carrier DNA molecules. To serve as a vector, a DNA molecule must have several properties:

1. It must be able to replicate itself and inserted DNA segments independently of the chromosome(s) of the host cell by carrying an origin of replication ($ori^+$).

2. It should contain a number of unique restriction enzyme cleavage sites for insertion of DNA segments.

3. It should carry a selectable marker (usually in the form of antibiotic resistance genes or genes for enzymes missing in the host cell) to allow identification of host cells that contain it.

4. It should be easy to recover from the host cell.

There are a number of vectors currently in use, including **plasmids**, **bacteriophages**, and **cosmids**.

Plasmids are naturally occurring, extrachromosomal double-stranded DNA molecules that replicate autonomously within bacterial cells (Figure 12.4). The genetics of plasmids and their host bacterial cells will be discussed in Chapter 16. In this section, emphasis will be on the role of plasmids as vectors. For use in genetic engineering, many plasmids have been modified or engineered to contain a limited number of restriction sites and specific antibiotic resistance genes.

One of the first genetically engineered plasmids to be used in recombinant DNA work was pBR322 (Figure 12.5). This plasmid carries an origin of replication ($ori^+$), genes for resistance to the antibiotics ampicillin and tetracycline, and a number of unique sites for restriction cleavage. Sites for the enzymes BamHI, SpHI, Sal I, XmaIII, and NruI are within the tetracycline gene, and unique sites for RruI, PvuI, and Pst I are within the ampicillin resistance gene. Introduction of pBR322 into a plasmid-free, antibiotic-sensitive bacterial cell will result in a cell resistant to both tetracycline and ampicillin. If a DNA fragment is inserted into the RruI, PvuI, or Pst I sites, it will inactivate the ampicillin gene, but not affect the tetracycline gene. If this recombinant plasmid is transferred into a plasmid-free bacterial cell, such cells can be identified by their resistance to tetracycline and sensitivity to ampicillin.

Over the years, more sophisticated plasmid vectors have been developed, offering a number of useful features. One such plasmid, derived from pBR322, is pUC18 (Figure 12.6), which has the following properties:

1. The plasmid is about half the size of pBR322, allowing larger DNA fragments to be inserted.

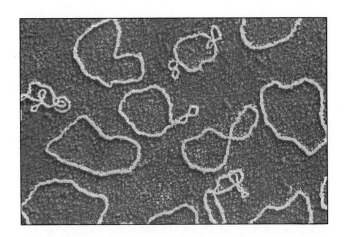

**FIGURE 12.4**  Circular plasmids isolated from the bacterium *E. coli*. Genetically engineered plasmids are used as vectors for cloning DNA segments.

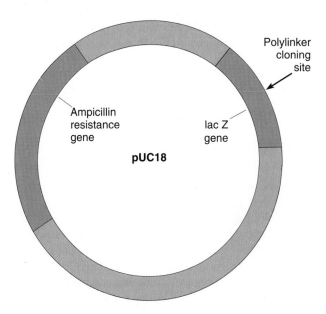

pUC18

Polylinker cloning site

Ampicillin resistance gene

lac Z gene

Restriction sites in polylinker

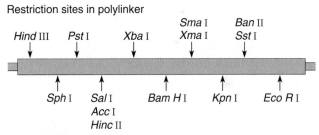

*Hind* III   *Pst* I   *Xba* I   *Sma* I / *Xma* I   *Ban* II / *Sst* I

*Sph* I   *Sal* I / *Acc* I / *Hinc* II   *Bam* H I   *Kpn* I   *Eco* R I

**FIGURE 12.6**  The plasmid pUC18 offers several advantages as a vector for cloning. It can accept large DNA fragments for cloning, replicates to a high copy number, and has a large number of restriction sites within the polylinker, located within a *lac* Z gene. Bacteria carrying pUC18 produce blue colonies when grown on a medium containing X-gal. DNA inserted into the polylinker site disrupts the *lac* Z gene, resulting in white colonies, allowing identification of bacterial colonies carrying cloned DNA sequences.

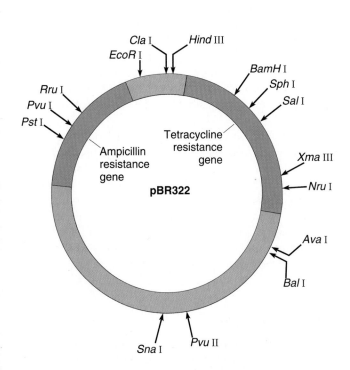

*Cla* I   *Hind* III   *EcoR* I   *BamH* I   *Sph* I   *Sal* I   *Rru* I   *Pvu* I   *Pst* I   Ampicillin resistance gene   Tetracycline resistance gene   pBR322   *Xma* III   *Nru* I   *Ava* I   *Bal* I   *Sna* I   *Pvu* II

**FIGURE 12.5**  Restriction map of the plasmid pBR322, showing the locations of the restriction enzyme sites that cleave the plasmid only once. Also shown are the locations of the genes for antibiotic resistance.

2. pUC18 replicates to produce about 500 copies per cell, 5 to 10 times the number of copies produced by pBR322.

3. A larger number of unique restriction sites are present in pUC18, and these are clustered in one region, known as a **polylinker** region.

4. The polylinker is situated within the *E. coli lacZ* gene. Bacterial cells carrying this plasmid will produce blue colonies when grown on medium containing a compound called *X-gal.* DNA inserted into the polylinker disrupts the *lacZ* gene, resulting in white colonies, making it easy to select cells carrying foreign DNA sequences.

### Bacteriophage Vectors

A bacteriophage widely used in recombinant DNA work is the phage lambda (Figure 12.7). Its genes have all been identified and mapped, and the DNA sequence of the entire genome is known. The middle third of the lambda chromosome contains a cluster of genes necessary for the lysogenic phase of its life cycle, but not for the lytic phase (review the life cycle of phages in Chapter 16). As a result, this region can be replaced with foreign DNA without affecting the ability of the phage to infect cells and form plaques (Figure 12.8). Over 100 vectors based on lambda phages have been derived by removing various portions of the central gene cluster. Lambda vectors carrying inserted foreign DNA can be introduced into bacterial host cells in a process called transfection,

in which the host cells have been made permeable by a chemical treatment. Each phage vector can be amplified by growth on plated bacterial cells to form plaques or by infecting cells in liquid medium and harvesting the lysed cells.

Other bacteriophages are also used as vectors, including the single-stranded filamentous phage known as M13 (Figure 12.9). When M13 infects a bacterial cell, the single strand (+ strand) replicates to produce a double-stranded molecule known as the **replicative form (RF)**. RF molecules can be regarded as similar to plasmids, and foreign DNA can be inserted into single restriction enzyme cleavage sites present in the genome. When reinserted into bacterial cells, the progeny of RF molecules consists of a single (+) strand, containing one strand of the inserted DNA segment. The cloned single-stranded DNA produced by M13 can be used for DNA sequencing or as a template for mutational alteration of the cloned sequence.

### Cosmid Vectors and Shuttle Vectors

Cosmids are vectors constructed in the laboratory using the *cos* sequence of phage lambda, necessary for the assembly of chromosomes into phage heads, and plasmid (the *mid* in cosmid) sequences for antibiotic resistance and replication (Figure 12.10). Cosmid DNA containing inserted foreign DNA is packaged into lambda protein heads. These constructs will infect bacterial cells, and replicate as plasmids within host cells. Because almost the entire lambda genome has been deleted, cosmids allow the cloning of large segments of DNA, up to 50 kb in length. Phage vectors, on the other hand, can accommodate DNA inserts of about 20 kb, while plasmids are usually limited to inserts of 10 kb of foreign DNA.

Other hybrid vectors constructed with origins of replication derived from different sources (e.g., plasmids and animal viruses such as SV40) are able to function in more than one type of host cell and are known as **shuttle vectors**. These vectors usually contain genetic markers that are selectable in both host systems, and can be used to shuttle foreign DNA sequences back and forth between hosts. Often, such vectors are employed in studying gene expression. Other vectors, used in specific applications, will be described in following sections.

### Cloning in *E. coli*

A variety of cell types can be used as hosts for replication of DNA segments inserted into vectors. One of the most commonly used hosts is the laboratory *E. coli* strain K12. This and other *E. coli* strains are used as hosts because they are genetically well characterized and can

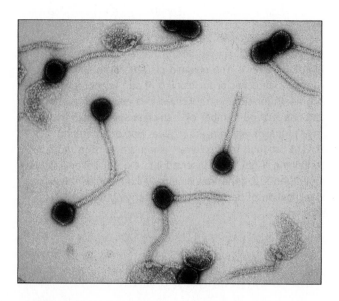

**FIGURE 12.7**    An electron micrograph showing a cluster of the bacteriophage lambda, widely used as a vector in recombinant DNA work.

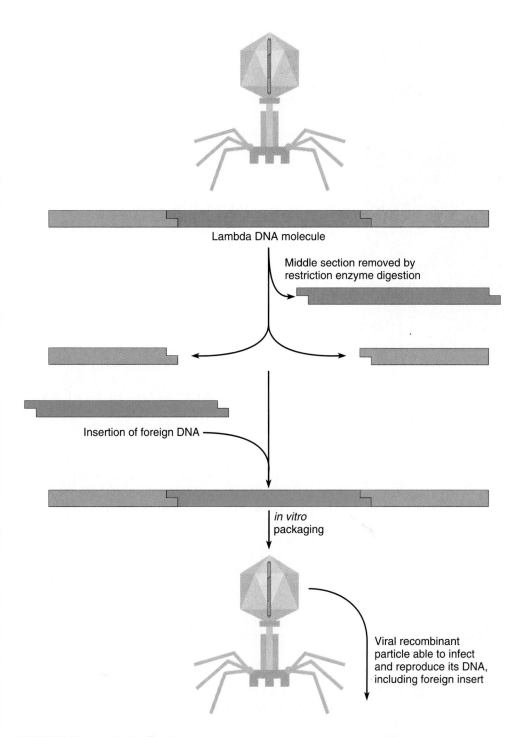

**FIGURE 12.8**     A diagrammatic representation of the lambda chromosome, showing how the middle section can be removed and replaced with foreign DNA. The recombinant chromosome is then packaged into phage proteins to form a recombinant virus. This virus is able to infect bacterial cells and replicate its chromosome, including the foreign DNA insert.

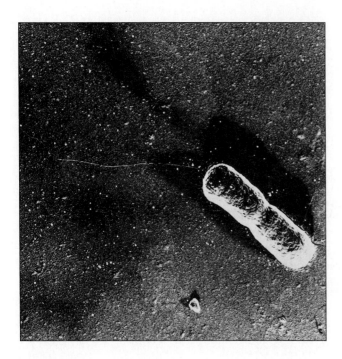

**FIGURE 12.9**    An electron micrograph of the single-stranded, filamentous phage M13. Foreign DNA inserted into the M13 chromosome is replicated as a single strand within bacterial cells and extruded into the medium. The recovered inserts are used for DNA sequencing or as templates for mutagenic experiments.

serve as hosts for a wide range of vectors, including plasmids, phages, and cosmids.

Several steps are required to create recombinant DNA molecules (Figure 12.11). Using a plasmid vector as an example, the following steps are carried out:

1. Isolate the DNA to be cloned and treat it with a restriction enzyme to create fragments ending in specific sequences.

2. Ligate these fragments to plasmid molecules cut with the same restriction enzyme.

3. Introduce the recombinant vector into bacterial host cells by transformation, a process in which DNA molecules are transferred across the cell membrane and into the host cell.

4. Select for cells that have taken up recombinant plasmids.

To select for cells that contain recombinant plasmids, the transformed bacterial cells are plated on selective medium. Cells that have taken up a plasmid will grow and form discrete colonies (Figure 12.12). Since each colony consists of cells derived from a single ancestor, all cells in the colony, and the plasmids they contain, are genetically identical, or **clones**.

In the case of pBR322, if the DNA to be cloned is inserted into the gene for tetracycline resistance, this gene will be inactivated. Colonies formed from a cell carrying this vector will not grow in the presence of tetracycline, but will grow on plates containing ampicillin. If the host cell had taken up a pBR322 vector with no insert, it would grow on a medium containing both tetracycline and ampicillin.

Other vectors such as pUC18 have been engineered to produce blue colonies when in bacterial cells plated on medium containing a compound called X-gal. The pUC18 polylinker restriction sites are in the *lacZ* gene

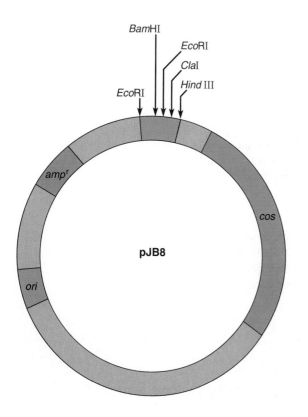

◄ **FIGURE 12.10**    The cosmid pJB8 contains a bacterial origin of replication (*ori*), a single *cos* site (*cos*), an ampicillin resistance gene (*amp*), and a region containing four restriction sites for cloning (*Bam*HI, *Eco*RI, *Cla*I, and *Hind*III). Because the vector is only 5.4 kb long, it can accept foreign DNA segments between 33 and 46 kb in length. The *cos* site allows cosmids carrying large inserts to be packaged into lambda viral coat proteins as if they were viral chromosomes. The viral coats carrying the cosmid can be used to infect a suitable bacterial host, and the vector, carrying a foreign DNA insert, will be transferred into the host cell. Once inside, the *ori* sequence allows the cosmid to replicate as a bacterial plasmid.

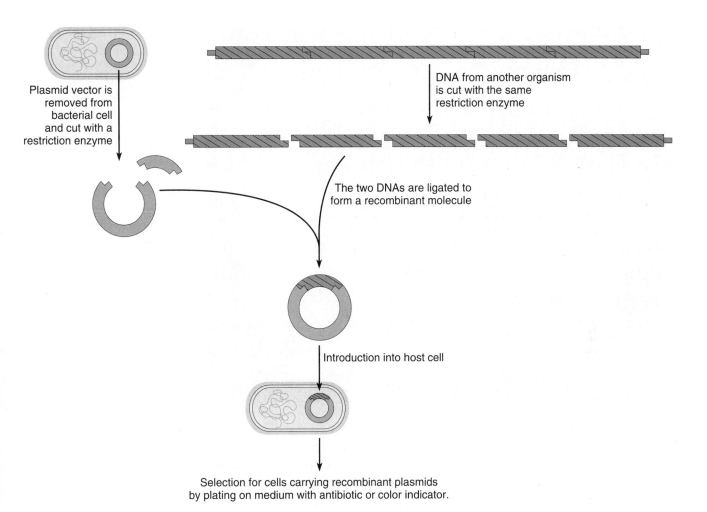

Plasmid vector is removed from bacterial cell and cut with a restriction enzyme

DNA from another organism is cut with the same restriction enzyme

The two DNAs are ligated to form a recombinant molecule

Introduction into host cell

Selection for cells carrying recombinant plasmids by plating on medium with antibiotic or color indicator.

**FIGURE 12.11**   Summary of steps for cloning in a bacterial host. Plasmid vectors are isolated and cut with a restriction enzyme. DNA from another organism is cut with the same restriction enzyme, resulting in a collection of fragments. The DNA fragments are spliced into the vector and transferred to a bacterial host for replication. Those bacterial cells carrying plasmids with foreign DNA can be identified and isolated. The cloned DNA can be recovered from the bacterial host for further analysis.

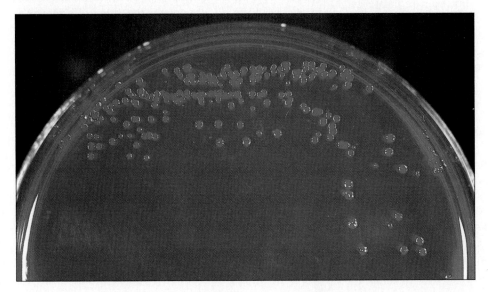

**FIGURE 12.12**   A petri plate containing bacterial colonies. Each colony is composed of millions of cells, all derived from a single ancestor; each colony, therefore, can be regarded as a clone.

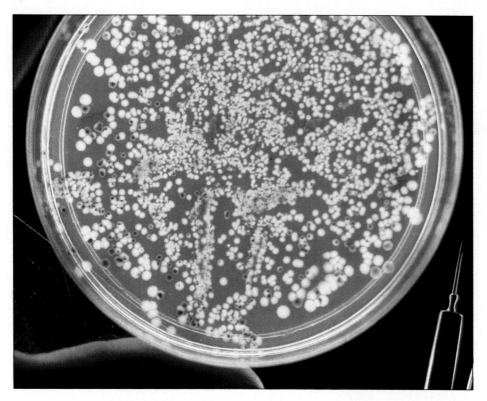

**FIGURE 12.13** A petri plate showing colonies derived from a cloning experiment. The medium on the plate contains a compound called X-gal. The vector has been engineered so that foreign DNA inserts disrupt the gene responsible for the formation of blue colonies. As a result, blue colonies do not carry any cloned foreign DNA, while clear colonies contain cloned vectors carrying foreign DNA inserts.

responsible for the ability to form blue colonies, and insertion of DNA into the polylinker destroys this ability. (For a discussion of the *lac* gene, see Chapter 16.) As a result, plasmids carrying inserted segments of DNA produce white colonies (Figure 12.13), whereas those without inserts produce blue colonies. Similarly, phages containing foreign DNA can be used to transfect *E. coli*, and when plated on solid medium, the resulting plaques each represent a cloned descendant of a single ancestral phage, and are regarded as clones. Cosmids behave as phages and infect bacterial cells, but then replicate as plasmids inside the host cell, and can be identified and recovered in the same way as plasmids. In a subsequent section, we will consider the use of eukaryotic vectors and hosts in cloning DNA segments.

## LIBRARY CONSTRUCTION

Since each cloned DNA segment is relatively small, many separate clones must be constructed in order to include even a small portion of the total genome of an organism. A set of cloned DNA segments derived from a single individual is called a **library**. Cloned libraries can represent the entire genome of an individual, the DNA from a single chromosome, or the set of genes that are actively transcribed in a single cell type.

### Genomic Libraries

A genomic library is usually constructed to provide a reference source for clones from a specific organism. Ideally, a genomic library contains at least one copy of all sequences represented in the genome. Each vector molecule can contain only a relatively few kilobases of foreign DNA, so one of the primary tasks in preparing a genomic library is the selection of the proper vector to carry all sequences in the smallest number of clones. The number of clones required to contain all the sequences in a genome is dependent on the average size of the cloned inserts and the size of the genome to be cloned. This number can be represented by the following formula:

$$N = \frac{\ln(1 - P)}{\ln(1 - f)}$$

where $N$ is the number of required clones, $P$ is the probability of recovering a given sequence, and $f$ represents the fraction of the genome present in each clone. Suppose that we wish to prepare a library of the human genome using a lambda phage vector. The human genome is $3.0 \times 10^6$ kb, and if the average size of cloned inserts in the vector is 17 kb, a library of about $8.1 \times 10^5$ phages would be required (where $f = 1.7 \times 10^4/3.0 \times 10^9$) to have a 99 percent ($P = 0.99$) probability that any human gene is present in at least one copy. If we had

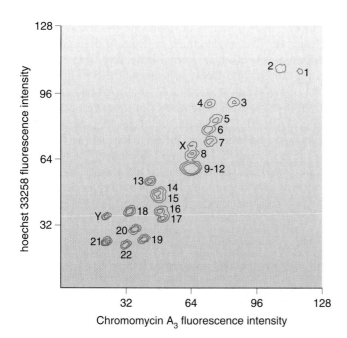

**FIGURE 12.14**    Human chromosomes after separation by flow sorting.

selected pBR322 as the vector, with an average insert of about 5 kb, several million clones would be needed to contain the library.

## Chromosome-Specific Libraries

A library made from a subgenomic fraction such as a single chromosome can be of great value in the selection of specific genes and the study of chromosome organiza-tion. In an early attempt to prepare a library from a specific region of the genome, DNA from a small segment of the X chromosome of *Drosophila* corresponding to a region of about 50 polytene bands was isolated by microdissection. DNA was extracted from this chromosome, cut with a restriction endonuclease and cloned into a lambda vector. This region of the chromosome contains the genes *white*, *zeste*, and *Notch*, as well as the original site of a transposing element that can translocate a chromosomal segment to more than a hundred chromosomal sites. Although technically difficult, this procedure produces a library that contains only the genes of interest and their adjacent sequences.

Libraries derived from individual human chromosomes have been prepared using a technique known as **flow cytometry**. In this procedure, chromosomes from mitotic cells are stained with two fluorescent dyes, one that binds to A-T pairs, the other to G-C pairs. The stained chromosomes flow past a laser beam that stimulates them to fluoresce, and a photometer sorts and fractionates the chromosomes by differences in dye binding and light scattering (Figure 12.14). Using this technique, the National Laboratories at Los Alamos and Lawrence-Berkley have prepared cloned libraries for each of the human chromosomes and made them available for distribution to research scientists all over the world.

Recently, using a modification of electrophoresis known as **pulse field gel electrophoresis**, yeast chromosomes have been separated from each other and used to construct chromosome-specific libraries (Figure 12.15). Libraries constructed using these techniques can be used to gain access to genetic loci for which there are

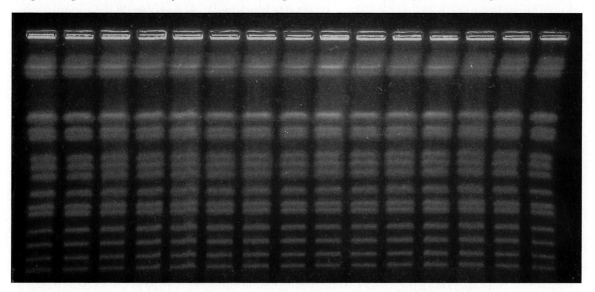

**FIGURE 12.15**    Intact yeast chromosomes separated using a method of electrophoresis employing contour-clamped homogeneous electric field (CHEF). In each lane, 15 of the 16 yeast chromosomes are visible, separated by size, with the largest chromosomes at the top.

no convenient selection systems using DNA, mRNA, or proteins, or where the gene product is unknown. In addition, chromosome-specific libraries provide a means for studying the molecular organization and even the nucleotide sequence in a defined region of the genome.

## cDNA Libraries

A library representing the structural genes active in a specific cell type at a specific time can be constructed by first synthesizing **cDNA (complementary DNA)** molecules. If we begin with a population of mRNA molecules containing 3′ poly A tails, poly dT (deoxythymidine) can be used as a primer to pair with the poly A residues (Figure 12.16). This poly dT serves as the starting point for the synthesis of a DNA strand using the enzyme **reverse transcriptase**. This enzyme is an RNA-dependent DNA polymerase that copies a single-stranded RNA template into a single-stranded DNA. The result is an RNA–DNA double-stranded duplex. The RNA strand can be

removed by treatment with alkali or the enzyme **ribonuclease H**. The single strand of DNA is then used as a template to synthesize a complementary strand of DNA using the enzyme **DNA polymerase I**. Because the 3′ end of the single strand of DNA often loops back upon itself, it can serve as the primer for the synthesis of the second strand. The result is a DNA duplex with the strands joined together at one end. The hairpin loop can be opened using the enzyme **$S_1$ nuclease**. The result is a double-stranded DNA molecule that can be cloned by attaching sequences called linkers to both ends to allow it to be inserted into a plasmid or phage vector.

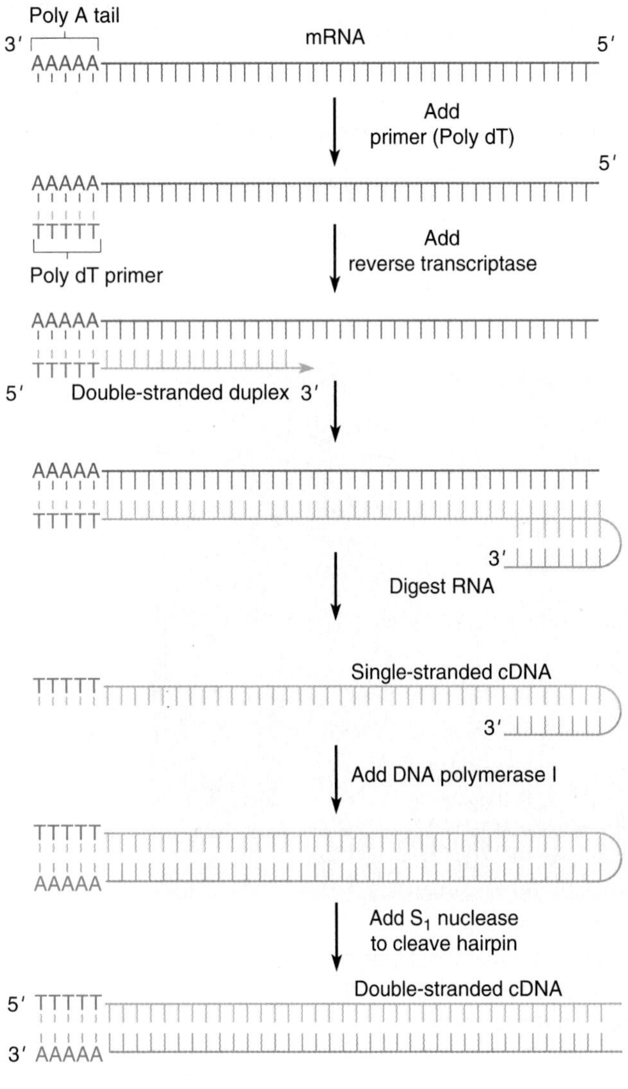

# SELECTION OF RECOMBINANT CLONES

As we calculated above, a library can contain several hundred thousand clones. The problem is to identify and select only the clone or clones that contain a DNA sequence or gene of interest. Several techniques can be employed to accomplish this task, and the choice often depends on circumstances and available information. The methods described below utilize somewhat different approaches to this problem.

## Probes to Identify Specific Clones

Most procedures to select specific clones from a library employ a radioactive polynucleotide that contains a base sequence complementary to all or part of the gene of interest. This radioactively labeled nucleic acid is commonly referred to as a **probe**. Probes can be derived from a variety of sources; even related genes isolated from other species can serve as probes if enough of the DNA sequence has been conserved. For example, extra-

◀ **FIGURE 12.16**    The production of cDNA from mRNA. Since many eukaryotic mRNAs have a polyadenylated tail of variable length ($A_n$) at their 3′ end, a short oligonucleotide sequence composed of thymidine nucleotides can be annealed to this tail. This oligonucleotide acts as a primer for the enzyme reverse transcriptase, which uses the mRNA as a template to synthesize a complementary DNA strand. A characteristic hairpin loop is often formed as synthesis is terminated on the template. The mRNA is removed by alkaline treatment of the complex, and DNA polymerase is used to synthesize the second DNA strand. $S_1$ nuclease is used to open the hairpin loop, and the result is a double-stranded cDNA molecule that can be cloned into a suitable vector, or used as a probe in library screening.

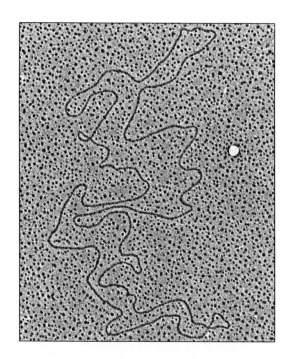

**FIGURE 12.17**    Electron micrograph showing extrachromosomal circles of *Xenopus* ribosomal DNA recovered from oocytes by centrifugation. These circles can be treated with restriction enzymes, and the resulting fragments cloned into plasmid vectors.

chromosomal copies of the ribosomal genes of the clawed frog *Xenopus laevis* were isolated by centrifugation (Figure 12.17) and cloned into plasmid vectors. Because ribosomal gene sequences have been highly conserved during eukaryotic evolution, the *Xenopus* clones were used as probes to isolate the human ribosomal genes from a cloned library.

If the gene to be selected from a genomic library is transcriptionally active in certain cell types, a cDNA probe can be prepared. This technique is particularly useful when purified mRNA can be obtained. For example, hemoglobin mRNA is the predominant messenger RNA produced at a certain stage of red blood cell development. Similarly, ovalbumin mRNA can be obtained from hormonally stimulated avian uterine tissue. Purified mRNA is copied by reverse transcriptase into a cDNA molecule that can be used as a probe to recover the structural gene from a cloned genomic library.

Still another approach to obtaining a probe is the use of a technique dubbed **reverse translation**. If the protein product of the gene is available, and at least part of its amino acid sequence is known, a polynucleotide probe can be constructed with this information. Beginning with the amino acid sequence, the nucleotide sequence encoding these amino acids can be deduced (Figure 12.18). Note that because of degeneracy in the

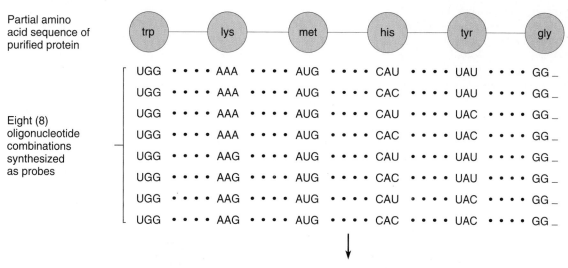

**FIGURE 12.18**    The process of reverse translation. From the partial amino acid sequence of a purified protein, the coding nucleotide sequence can be deduced. In this case, two amino acids (*trp* and *met*) have only one codon each. Three others (*lys*, *his*, and *tyr*) are encoded by two codons. The sixth amino acid (*gly*) is encoded by a degenerate series of codons that encompasses all base combinations at the third base position. Consequently, the eight coding sequences shown encompass all possible combinations. The exact coding sequence of the gene must be one of the eight. Using a radioactive mixture of these sequences as probes, a cloned library can be screened to isolate the structural gene for the complete protein.

genetic code, several combinations of nucleotide sequences are possible. In the example given, eight different oligonucleotides could encode the amino acid sequence. Each of these oligonucleotides is chemically synthesized using radioactive nucleotides and are used as probes to identify the clone encoding the gene of interest. Note that it is not necessary to know the entire amino acid sequence of the protein or to synthesize the nucleotide sequence of the entire gene. A fragment of 18–21 nucleotides corresponding to 6–7 amino acids is usually sufficient to produce a probe.

## Colony and Plaque Hybridization

The **colony hybridization** method was developed by Grunstein and Hogness for screening a plasmid library (Figure 12.19). Bacterial colonies to be screened are grown on plates, forming hundreds or thousands of colonies. A replica of the colonies on the plate is made by gently pressing a filter made of nitrocellulose or other DNA-binding material onto the surface in order to transfer bacterial cells from the colonies to the filter. The colonies on the filter are lysed, and the double-stranded DNA

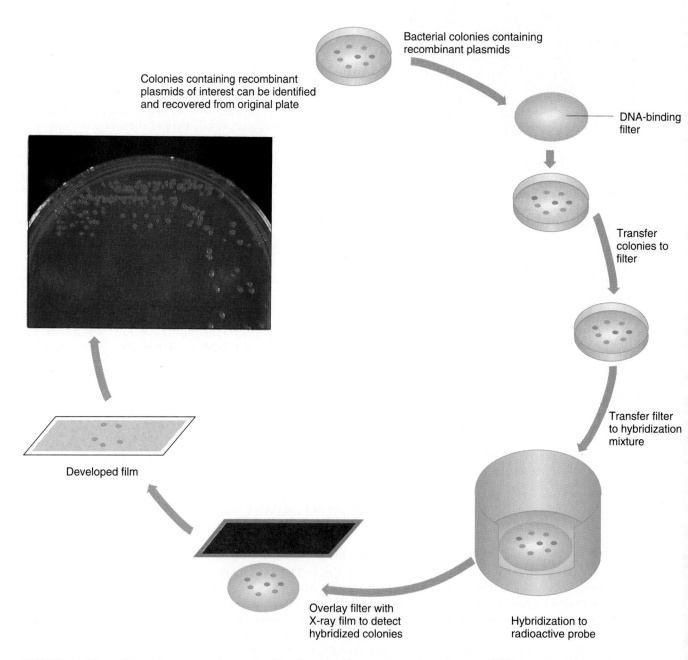

**FIGURE 12.19**    The colony screening method for the detection and recovery of cloned DNA inserts of interest.

from the bacterial cells is denatured and converted to single strands.

The colonies on this filter are screened by incubation with a nucleic acid probe. This method requires some knowledge about the gene or DNA sequence being screened for. In this form of screening, the filter is placed in a solution with a radioactive probe to allow hybridization between the probe and complementary sequences present on the filter (review the discussion of hybridization in Chapter 10). After excess probe is washed away, the filter is overlaid with a piece of photographic film. If the probe is complementary to the DNA of one of the lysed colonies, the hybridized colony will be identified as a dark spot on the developed film. That colony can be recovered from the initial plate, and the cloned DNA it carries can be used in further experiments.

A similar technique known as **plaque hybridization** has been developed to screen phage vectors carrying DNA inserts. Because phage plaques are much smaller than bacterial colonies, many more plaques can be grown on a plate, and consequently, more plaques can be screened on a single filter, making this method more efficient for screening large genomic libraries.

## METHODS FOR ANALYSIS OF CLONED SEQUENCES

### Physical Mapping Using Restriction Enzymes

A restriction map is a compilation of the number, order, and distance between restriction enzyme cutting sites along a cloned segment of DNA. The map units are expressed in **base pairs (bp)** or for longer distances, **kilobase pairs (kb)**. The fragments generated by cutting DNA with restriction enzymes can be separated by **gel electrophoresis**, a method that separates fragments by size, with the smallest pieces moving farthest (see Appendix A for a detailed description of electrophoresis). The fragments appear as a series of bands that can be visualized by staining the DNA with ethidium bromide and viewing under ultraviolet illumination (Figure 12.20). The size of individual fragments can be determined by running a set of fragments of known size on the same gel in another lane. Figure 12.21 shows the construction of a restriction map from a cloned DNA segment that contains cutting sites for two restriction enzymes. For the construction of this map, let us begin with a cloned DNA segment 7.0 kb in length (Figure 12.21). Three samples of the DNA are digested with restriction enzymes: one with *Hind*III, one with *Sal*I, and one with both *Hind*III and *Sal*I. The fragments generated by digestion with the restriction enzymes are separated by electrophoresis (Figure 12.21). The sizes of the separated

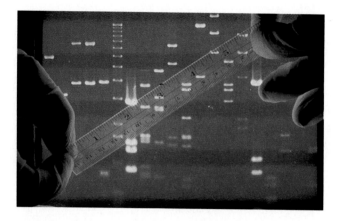

**FIGURE 12.20**    An agarose gel containing separated DNA fragments, stained with a dye (ethidium bromide) and visualized under ultraviolet light. Smaller fragments migrate faster and farther than larger fragments, resulting in the distribution shown.

fragments can be estimated by comparison to a set of standards run in adjacent lanes. The results indicate that when the DNA is cut with *Hind*III, two fragments are produced, indicating that there is only one cutting site for this enzyme (located 0.8 kb from one end). Similarly, when the DNA is cut with *Sal*I, two fragments result, with the single cutting site located 1.2 kb from one end. Taken together, the results show that there is one restriction site for each enzyme, but the relationship between the two sites is unknown. From the information available to this point, two different maps are possible. In one model, the *Hind*III site is located 0.8 kb from one end, and the *Sal*I site is located 1.2 kb from the same end. In the alternative model, the *Hind*III site is located 0.8 kb from one end, and the *Sal*I site is located 1.2 kb from the other end.

The correct model can be selected by considering the results from the digestion with both *Hind*III and *Sal*I. The first model predicts that digestion with both enzymes will generate three fragments of 0.8, 0.4, and 6.2 kb; the second model predicts that digestion with both enzymes will generate three fragments of 0.8, 1.2, and 5.0 kb. The actual fragment pattern observed after cutting with both enzymes indicates that model 1 is correct (Figure 12.21). In most cases, the situation is more complex, involving a number of sites for each enzyme, and a greater number of enzymes. In addition, the DNA to be analyzed is often contained in a plasmid (and is a circular molecule), but the methodology and logic employed are similar to the example above. Additional sites on this segment can be mapped through the use of other restriction enzymes.

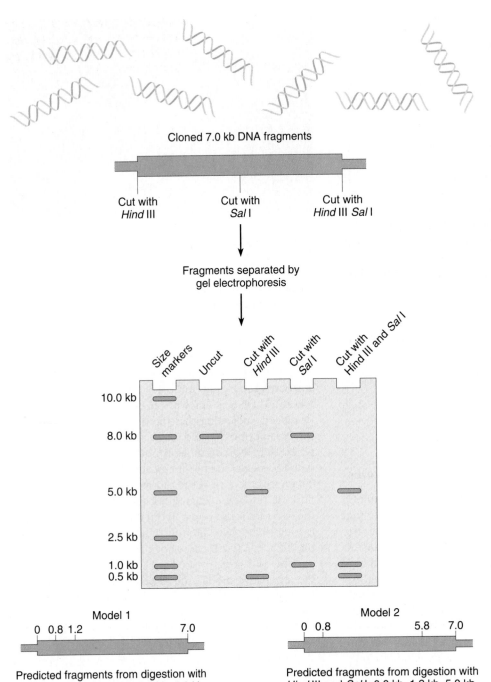

FIGURE 12.21    Method of restriction map construction. Three samples of the 7.0-kb DNA fragments are digested with restriction enzymes: one with *Hind*III, one with *Sal*I, and one with both *Hind*III and *Sal*I. The fragments generated by restriction enzyme digestion are separated by gel electrophoresis. The sizes of the separated fragments can be estimated by comparison with molecular weight standards in an adjacent lane. *Hind*III generates two fragments: 0.8 kb and 6.2 kb. *Sal*I produces two fragments: 1.2 kb and 6.8 kb. Digestion with *Hind*III and *Sal*I results in three fragments: 0.4 kb, 0.8 kb, and 5.8 kb. The results are interpreted and a map is constructed in two steps. First, two models are constructed, based on the data from the cuts with *Hind*III and with *Sal*I. These models can be used to predict the fragment sizes generated by digestion with both *Hind*III and *Sal*I. Model 1: 0.4 kb, 0.8 kb, 5.8 kb; model 2: 0.8 kb, 1.2 kb, 5.0 kb. Comparison of the predicted fragments with those observed on the gel indicate that model 1 is the correct restriction map.

Cloned 7.0 kb DNA fragments

Cut with *Hind* III        Cut with *Sal* I        Cut with *Hind* III *Sal* I

Fragments separated by gel electrophoresis

Size markers    Uncut    Cut with *Hind* III    Cut with *Sal* I    Cut with *Hind* III and *Sal* I

10.0 kb
8.0 kb
5.0 kb
2.5 kb
1.0 kb
0.5 kb

Model 1
0  0.8 1.2                7.0

Predicted fragments from digestion with *Hind* III and *Sal* I: 0.4 kb, 0.8 kb, 5.8 kb

Model 2
0  0.8            5.8   7.0

Predicted fragments from digestion with *Hind* III and *Sal* I: 0.8 kb, 1.2 kb, 5.0 kb

Fragments generated by cutting with *Hind* III and *Sal* I are 0.4, 0.8 and 5.8 kb in length, indicating that model 1 is correct.

Restriction maps provide an important way of characterizing a DNA segment, and can be constructed in the absence of any information about the genetic content or function of the mapped DNA. In conjunction with other techniques, restriction mapping can be used to define the boundaries of a gene and provides a way of dissecting the molecular organization within a gene and its flanking regions (see Chapter 11). Mapping can also serve as a starting point for the isolation of an intact gene from cloned segments of DNA, and provides a means for locating mutational sites within genes.

Restriction maps, also called physical maps, can also be used to refine genetic maps. To a large extent, the accuracy of maps constructed by genetic analysis is based on the frequency of recombination between genetic markers, and the number of genes or genetic markers used in construction of the map. If there is a large distance between markers and/or variation in recombination frequency, the genetic map may not correspond to the physical map of the chromosome region. In humans, for example, the genome size is large ($3.2 \times 10^9$ bp), and the number of mapped genes is small (a few thousand), meaning that each map unit is composed of millions of base pairs of DNA. The result is a poor correlation between the genetic and physical maps of chromosomes. Restriction enzyme cutting sites can themselves be used as genetic markers, reducing the distance between sites on a map, increasing the accuracy of maps, and providing reference points for the correlation of genetic and physical maps.

Restriction sites have played an important role in mapping genes to specific human chromosomes and to defined regions of individual chromosomes. In addition, if a restriction site maps close to a mutant gene, it can be used as a marker in a diagnostic test. This has proven especially useful because the mutant genes underlying many human genetic diseases are poorly characterized at the molecular level, and closely linked restriction sites have been successfully used in the detection of afflicted individuals and heterozygotes at risk for having affected children.

## Nucleic Acid Blotting

DNA fragments cloned into vectors can be used in hybridization reactions to characterize the identity of specific genes, to locate coding regions or flanking regulatory regions within cloned sequences, and to study the molecular organization of genomic sequences.

The transfer of DNA to nitrocellulose or nylon membranes for use in hybridization experiments is a technique developed by Edward Southern. Known as a **Southern blot**, this procedure has many applications in genetics. In this procedure, the DNA to be probed is first cut into fragments with one or more restriction enzymes, and the fragments separated by gel electrophoresis (Figure 12.22). The DNA in the gel is denatured into single-stranded fragments by treatment with alkali. These fragments are transferred to a sheet of DNA-binding material, usually nitrocellulose or a nylon derivative. The transfer is effected by placing the transfer membrane on top of the gel and causing buffer to flow through the gel and the nitrocellulose or nylon sheet by capillary action. The buffer flows through both, pulling the DNA out of the gel, and immobilizing it in the membrane. The denatured DNA is then hybridized with a radioactive probe. Only those single-stranded DNA fragments imbedded in the membrane that are complementary to the base sequence of the probe will form hybrids. The unbound probe is washed away, and the hybridized fragments are visualized by autoradiography. (See Appendix A for a detailed discussion of autoradiography.)

A related blotting technique can be used to determine whether a cloned gene is transcriptionally active in a given cell or tissue type by probing for the presence of RNA that is complementary to a cloned segment of DNA. This is accomplished by extracting RNA from one or several cell and/or tissue types. The RNA is fractionated by gel electrophoresis and the pattern of RNA bands is transferred to a sheet of RNA-binding membrane as in the Southern blot. The sheet is then hybridized to a radioactive single-stranded DNA probe, derived from cloned genomic DNA or cDNA. If RNA complementary to the DNA probe is present, it will be detected by autoradiography as a band on the film. Because the original procedure using DNA bound to a filter became known as a Southern blot, this procedure using RNA bound to a filter was called a **northern blot**. (Following this somewhat perverse logic, another procedure involving proteins bound to a filter is known as a **western blot**.)

Northern blots provide information about whether RNA transcripts complementary to cloned DNA segments are present in a cell or tissue, and also provide two additional bits of information. If marker RNAs of known size are run in an adjacent lane, the size and molecular weight of the transcribed RNA can be calculated. Furthermore, the amount of transcribed RNA present in the cell or tissue is reflected by the density of the RNA band on the autoradiograph. This can be quantified by measuring the density of the band, providing a measurement of the degree of transcriptional activity. Thus, northern blotting can be used to characterize and quantify the transcriptional activity of a cloned DNA segment in different cells and tissues.

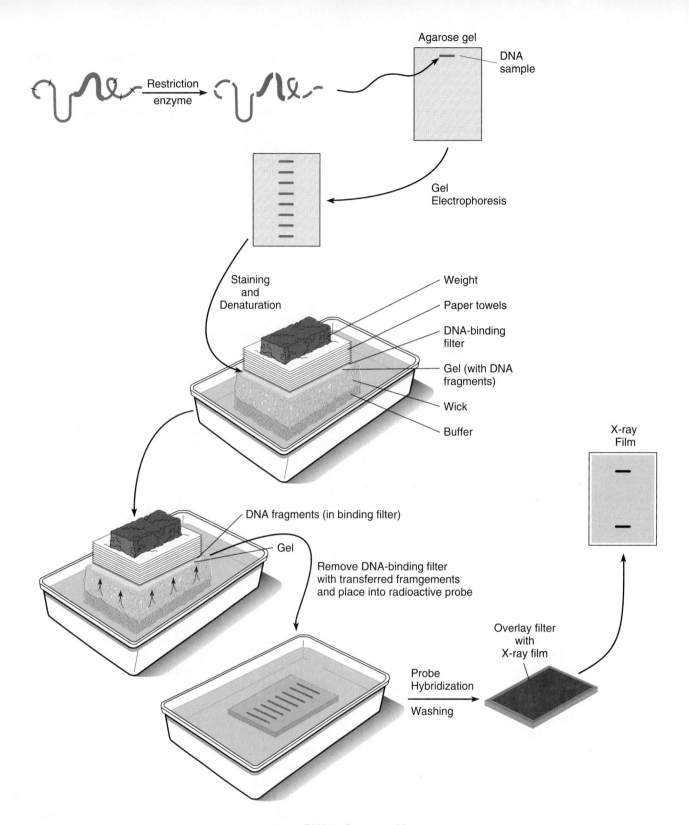

**FIGURE 12.22**    The Southern blotting technique. DNA is first cut with one or more restriction enzymes and the fragments separated by gel electrophoresis. The pattern of fragments can be visualized and photographed under ultraviolet illumination by staining the gel with ethidium bromide. The gel is then placed on a filter paper wick in contact with a buffer solution and covered with a sheet of a DNA-binding filter. Layers of paper towels or blotting paper are placed on top of the DNA-binding filter and held in place with a weight. Capillary action draws the buffer through the gel, transferring the pattern of DNA fragments from the gel to the DNA-binding filter. The DNA fragments on the binding filter are then denatured and hybridized with a radioactive probe, washed, and overlaid with a piece of X-ray film for autoradiography. The hybridized fragments show up as bands on the X-ray film.

## DNA Sequencing

In a sense, the ultimate characterization of a cloned DNA segment is the determination of its nucleotide sequence. The ability to sequence cloned DNA has added immensely to our understanding of gene structure and the mechanisms of gene regulation. Although techniques were available in the 1940s to determine the base composition of DNA (see Chapter 8), it was not until the 1960s that methods for the determination of nucleotide sequences were developed. In 1965 Robert Holley determined the sequence of a tRNA molecule consisting of 74 nucleotides. This work required about one year of concentrated effort. In the 1970s, more efficient and direct methods were developed, and it is now possible for a molecular biology laboratory to sequence more than a thousand bases in a week. In the near future, with widespread automation of this process, it will be commonplace to sequence thousands of nucleotides in a single day.

DNA sequencing depends on the ability of cloning methods to provide large amounts of specific DNA fragments. The use of vectors such as M13 or modifications of the PCR technique are also important in providing single-stranded DNA segments that can be directly sequenced.

Two methods of DNA sequencing are in general use. One is a chemical method devised by Alan Maxam and Walter Gilbert. The Maxam–Gilbert method was used to determine the base sequence of the 4362 nucleotides in the plasmid pBR322. The second method, developed by Fredrick Sanger and his colleagues, is an enzymatic method based on the 5'-to-3' elongation of DNA molecules by DNA polymerase. Both sequencing strategies generate a series of single-stranded DNA molecules of the same sequence, but of differing lengths. Each method uses a series of four reactions, each in a separate tube, to determine the sequence of the four bases in a DNA segment. These sequences, differing in length by as little as one base, are separated from each other by gel electrophoresis in four adjacent lanes. The result is a series of bands that form a ladderlike pattern. The sequence can be read directly from the band pattern in the four lanes (Figure 12.23). Details of each of the methods of DNA sequencing are presented in Appendix A.

DNA sequencing has provided information about the organization of genes and the nature of mutational events that alter both genes and gene products, confirming the conclusion from genetic analysis that genes and proteins are colinear molecules. Sequencing has also been used to study the organization of regulatory regions that flank prokaryotic and eukaryotic genes, and to derive the amino acid sequence of proteins. The gene for cystic fibrosis, an autosomal recessive human genetic disorder, was first identified by the fact that it was a DNA

**FIGURE 12.23**   Photograph of a DNA sequencing gel, showing the separation of bases in the four sequencing reactions (one per lane). The gel is analyzed to reveal the base sequence of the DNA fragment. To do this, the gel is read from the bottom, beginning with the lowest band in any lane, then the next lowest, then the next, etc. The sequence of this gel begins with TCGCATAATATATTT, etc.

sequence with an open reading frame, since the nature of the protein product was unknown. The DNA sequence was then used to infer that the gene encodes a protein of 1480 amino acids with properties similar to membrane proteins that function in ion transport. This discovery serves to illustrate how the use of DNA sequencing now plays an important role in genetic analysis.

## PCR Analysis

Recombinant DNA techniques were developed in the early 1970s, and in the subsequent years they revolutionized the way geneticists and molecular biologists conduct research. In 1986, a technique, called the **polymerase chain reaction (PCR)**, was developed, and has extended the power of recombinant DNA research. It has even replaced some of the earlier methods. PCR analysis has found applications in a wide range of disciplines, including molecular biology, human genetics, evolution, development, and even forensics.

One of the prerequisites for many recombinant DNA techniques is the availability of large quantities of a specific DNA segment. Often such quantities are obtained through tedious and labor-intensive efforts in cloning and recloning. PCR allows a more direct amplification of specific DNA segments, and can be used on fragments of DNA that are initially present in infinitesimally small quantities. The PCR method is based on the amplification of a DNA segment using DNA polymerase and oligonucleotide primers that hybridize to opposite strands of the sequence to be amplified (Figure 12.24). There are three basic steps in the PCR reaction, and the amount of amplified DNA produced is limited theoretically only by the number of times this series of steps is repeated.

The first step is the denaturation of the DNA to be amplified. The DNA to be amplified does not have to be purified, and can come from any number of sources, including genomic DNA, forensic samples such as dried blood or semen, dried samples stored as part of medical records, single hairs, mummified remains, and fossils. The double-stranded DNA is denatured by heating (at about 90°C) until it dissociates into single strands (usually about 5 minutes).

The second step is the annealing of primer sequences to the single strands of DNA. These primers are synthetic oligonucleotides that anneal to sequences flanking the segment to be amplified. Two different primers are most often used in PCR. Each primer has a sequence complementary to only one of the two strands of DNA. The primers align themselves with their 3′ ends facing each other since they anneal to opposite strands (Figure 12.24).

The third step in the PCR is the actual amplification

**FIGURE 12.24**    PCR amplification. In the polymerase chain reaction (PCR) method, the DNA to be amplified is first denatured into single strands, and then each strand is annealed to a different primer. The primers are synthetic oligonucleotides complementary to sequences flanking the region to be amplified. DNA polymerase is added along with nucleotides to extend the primers in the 3′ direction, resulting in a double-stranded DNA molecule with the primers incorporated into the newly synthesized strand. In a second PCR cycle, the products of the first cycle are denatured into single strands, primers are added, and they are annealed and extended by DNA polymerase. Repeated cycles can amplify the original DNA sequence by more than a million times.

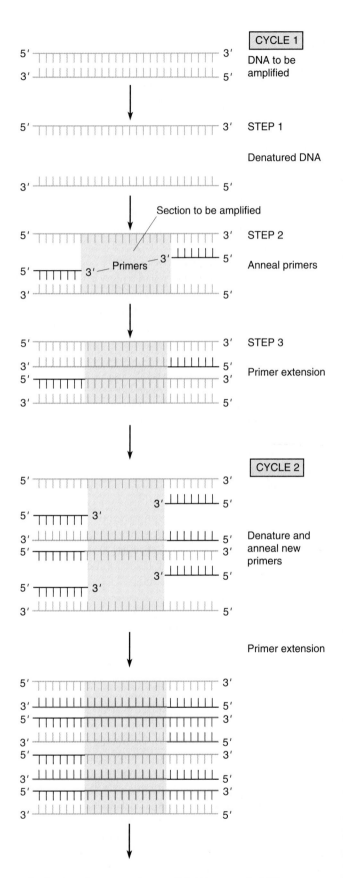

Cycles 3-25 for greater than a $10^6$-fold increase in DNA

event, carried out by extension of the primers using the single-stranded DNA as a template (Figure 12.24). This is carried out by adding DNA polymerase to the reaction mixture. The polymerase extends the oligonucleotide primers in the 5′-to-3′ direction, using the single-stranded DNA bound to the primer as a template. The product is a double-stranded DNA molecule with the primers incorporated into the final product (Figure 12.24).

Each set of three steps—denaturation of the double-stranded product, annealing of primers, and extension of polymerase—is referred to as a cycle. Generally, each step of the cycle is performed at a different temperature. The cycle can be repeated by carrying out each of the steps again. Twenty-five cycles of amplification result in a several-million-fold increase in the amount of target DNA. The process has been automated through the development of an instrument called a thermocycler that can be programmed to carry out a predetermined number of cycles, yielding large amounts of amplified DNA segments that can be used in other procedures such as cloning, sequencing, preparation of probes, diagnosis of infectious disease, and genetic screening.

# TRANSFERRING GENES INTO EUKARYOTIC ORGANISMS

We have described in some detail the use of *E. coli* as a host for cloning. Other species of bacteria can also be used as hosts for cloning, including *B. subtilis* and *Streptomyces*. These bacterial host systems use methodologies similar to those we have already described for *E. coli*. To study expression and regulation, however, it is convenient and even necessary to use eukaryotic hosts for some cloned eukaryotic genes. Several such systems have been developed, and in this section we will describe three gene cloning systems that employ eukaryotic cells as hosts.

## Transferring Genes into Yeast

Although yeast is a eukaryotic organism, it can be manipulated and grown in much the same way as bacterial cells. Further, the genetics of yeast has been intensively studied, providing a large catalog of mutations and highly developed genetic maps of the yeast chromosomes. Finally, a naturally occurring plasmid named the **2-micron plasmid** is present in yeast, and forms the basis for a number of vector systems. By combining bacterial plasmid sequences with those of the yeast 2-micron plasmid, shuttle vectors with many useful properties can be produced.

A second type of yeast vector is the **yeast artificial chromosome (YAC)**. This vector has two sets of yeast telomeric sequences and contains a yeast origin of replication, which confers the ability to replicate autonomously. It has a yeast centromere to allow distribution of replicated YACs at cell division, a yeast selectable marker to allow identification of cells carrying YACs, and restriction sties for the integration of foreign DNA (Figure 12.25).

These vectors have been used to perform a wide range of studies, including the expression of mammalian

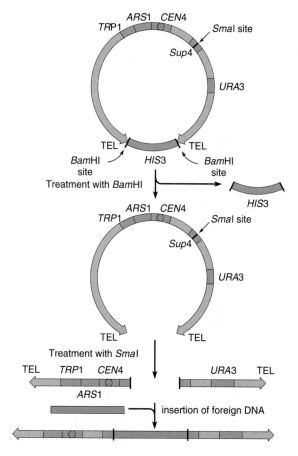

**FIGURE 12.25**   A summary of the yeast artificial chromosome (YAC) cloning system. YAC2 is constructed from pBR322 sequences and the *SUP4*, *TRP1*, *HIS3*, and *URA3* genes from yeast. The *ARS1* and *CEN4* sequences represent a chromosome segment adjacent to the *TRP1* gene on yeast chromosome 4. The *TEL* sequences act as telomeres and are derived from *Tetrahymena* ribosomal gene clusters. The centromere (*CEN4*), telomere, and *ARS* (autonomously replicating sequence, in effect an origin of replication) confer the basic properties of yeast chromosomes on the construct. Cleavage with *Bam*HI and *Sma*I results in the production of two arms that can be ligated to foreign DNA, followed by transformation into yeast host cells.

# TOOLS AND TECHNIQUES

## Using YACs for Gene Transfer of Eukaryotic DNA

Yeast artificial chromosomes (YACs) are cloning vectors that can replicate in yeast and are passed on to daughter cells during cell division. Segments of foreign DNA more than 1 megabase long (1 Mb = $10^6$ base pairs) can be inserted and successfully cloned in these vectors. The ability to clone large pieces of DNA in YACs has made them an important tool for constructing physical maps of eukaryotic genomes. Physical maps are created by ordering overlapping clones, and the use of YACs significantly reduces the number of clones required to span a particular chromosome region. Physical maps for the human Y chromosome and the long arm of human chromosome 21 have been constructed using human DNA cloned in YACs. These first physical maps represent a major step forward in the Human Genome Project.

Recently, it has been demonstrated that YACs can be used to greatly increase the efficiency of transferring genes into the germline of mice. Gene transfer is dependent on the introduction of DNA sequences into the nucleus of an appropriate mouse cell such as a fertilized egg or an embryonic stem cell, followed by the integration of the DNA into a chromosomal site. Mice that carry such integrated DNA sequences are called transgenic mice, and the new genes they carry are called transgenes.

Earlier experiments with transgenes in mice were hampered by the relatively short length of DNA that could be introduced using plasmid cloning vectors. In many cases, mammalian genes to be transferred were longer than the cloning capacity of the vectors, and in other cases, where the gene could be transferred, its expression was inhibited or inappropriate because adjacent or distant regulatory sequences were not transferred. The large cloning capacity of YACs overcomes this problem and researchers are now using YACs to develop strains of transgenic mice for a number of purposes.

YACs are transferred to mice in several ways. The first involves the microinjection of purified YAC DNA into the pronucleus of a mouse zygote. The zygotes are then transferred to the uteri of foster mothers for normal development. Other techniques use mouse embryonic stem (ES) cells as hosts. One of these is called lipofection. In this method, YAC DNA carrying a gene to be inserted is coated with lipids to form a DNA–lipid complex. The lipid coat facilitates uptake of the DNA into the stem cell by a process called endocytosis. An alternative approach fuses a yeast cell carrying a YAC with a stem cell, transferring the YAC and all or a portion of the yeast genome to the stem cell. The transgenic ES cells are then transferred into blastocyst stage mouse embryos, where they participate in the formation of adult tissues, including those that form germ cells.

The ability to transfer large DNA segments into mice has applications in many areas of research. It permits the precise localization of a gene on a large DNA fragment, something that is often a laborious task. Introducing a series of mouse genomic fragments into mutant mice and looking for correction of the mutant phenotype is one way to identify a gene. For example, YACs that carry a normal copy of the tyrosinase gene rescue the mutant phenotype of albino mice, providing them with pigmented coats. Using YACs, transgenic mice can be constructed carrying mutant human genes to produce mouse models of human genetic disorders. Large DNA transfers may also be used to produce mouse models for partial human trisomies, such as Down syndrome.

Transgenic mice also offer other important possibilities for the study and treatment of human disorders. For example, researchers are developing transgenic mice that carry human immunoglobulin (antibody) genes. These mice have the potential to produce human antibodies that can be used to provide therapeutic treatment for infectious disease. The development of this technique of DNA transfer represents an important development in molecular biology and genetics, and offers the prospect of improvements in understanding and eventually treating human genetic disorders.

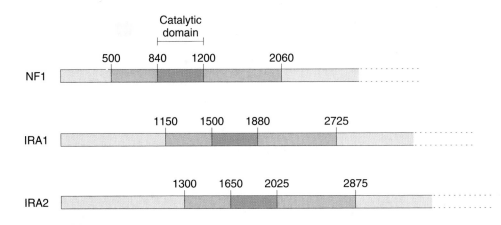

**FIGURE 12.26** Structural similarities among human NF1 protein and yeast *IRA*1 and *IRA*2 proteins. The darkly shaded regions correspond to the catalytic domain of the proteins. Hatched regions represent other regions of shared homology. Numbers represent amino acid positions. All three are large proteins: NF1 has at least 2485 amino acids; *IRA*1 has 2938 and *IRA*2 has 3079 amino acids, respectively.

genes in yeast. Yeasts and mammals both contain genes belonging to the *ras* family, a group of genes that encodes cytoplasmic proteins involved in the transfer of signals from receptors in the plasma membrane to signal pathways in the cytoplasm. In transferring these signals, *ras* proteins interact with a cytoplasmic protein called GAP. In humans, an autosomal dominant disorder called neurofibromatosis (NF1) is caused by mutation in a GAP-related protein. To determine whether the function of this protein can be studied in yeast, a segment of the normal human NF1 gene carrying the GAP-related region was cloned into a yeast vector and transferred to yeast cells from a mutant strain (Figure 12.26). This mutant strain, called *ira*-1⁻, has a nonfunctional GAP protein. The presence of the human NF1 gene segment in the yeast converts the phenotype of the mutant yeast strain to wild type, indicating that it can replace the mutant yeast *ira*-1 gene. Further studies are now under way using the GAP-related region from a mutant NF1 gene, allowing geneticists to dissect the molecular basis of a human genetic disorder by using yeast cells.

YACs have also proven to be very useful in the cloning of human genes, where the size range of inserted DNA exceeds the cloning capacity of lambda vectors and cosmids. A human genomic library has been cloned using YAC vectors and yeast as the host cells. YACs have also been used to construct a complete library of human chromosome 21 sequences. If the average size of the insert in each YAC is 440 kb, the entire chromosome would be represented with some redundancy in only 330 YACs.

Although cloning in yeast is at present the most advanced eukaryotic host system, other fungal hosts are being developed, and other eukaryotes such as plant cells and mammalian tissue culture cells show great promise for use as host systems.

## Transferring Genes into Higher Plant Cells

As strange as it might seem, cloning using higher plants as hosts has been achieved using a bacterial plasmid. The infectious soil bacterium *Agrobacterium tumifaciens* produces tumors or galls in many species of dicotyledonous plants. The ability to cause tumor formation is associated with the presence of a tumor-inducing (Ti) plasmid in the bacteria (Figure 12.27). When bacteria infect the plant cells, a segment of the Ti plasmid, known as T-DNA, is transferred into the chromosomal DNA of the host plant cell. The T-DNA controls tumor formation and the synthesis of small molecules known as opines that are required for growth of the infecting bacteria. Foreign genes can be inserted into the T-DNA segment of Ti, and the altered Ti can be transferred into a plant cell by the infecting bacteria. The foreign DNA is inserted into the plant genome when the T-DNA is integrated into a host cell chromosome. Such genetically altered single cells can be induced to grow in tissue culture to form a cell mass known as a **callus**. By manipulating the culture medium, the callus can be induced to form roots and shoots, and eventually develop into mature plants carrying a foreign gene. Plants (or animals) carrying a foreign gene are said to be **transgenic**. In the next chapter, we will see how gene transfer in plants has been used to genetically alter agriculturally important crop plants.

## Transferring Genes into Mammalian Cells

DNA can be taken up into mammalian cells by several methods, including coprecipitation with calcium phosphate and endocytosis, direct microinjection, exposure of cells to short pulses of high voltage (electroporation), and by encapsulation of DNA into artificial membranes

Cloning site for several enzymes

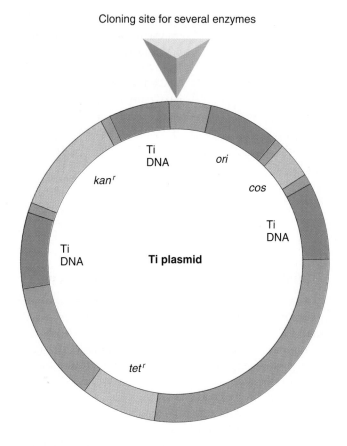

**FIGURE 12.27**    A Ti plasmid designed for cloning in plants. Segments of Ti DNA, including those necessary for opine synthesis and integration, are combined with bacterial segments that incorporate cloning sites and antibiotic resistance genes (*kan^r* and *tet^r*). The vector also contains an origin of replication (*ori*) and a lambda *cos* site that permits recovery of cloned inserts from the host plant cell.

scriptase into a double-stranded DNA. This DNA is integrated into the host genome and passed on to daughter cells as part of the host chromosome. The retroviral genome can be engineered to remove some viral genes, creating vectors that accept foreign DNA, including human structural genes (Figure 12.28). These vectors are used in the treatment of genetic disorders by human gene therapy, a topic that will be discussed in the following chapter.

The development of techniques for producing millions of copies of DNA segments by cloning opened the way for structural and functional studies of DNA from plants and animals. In turn, it led to the isolation of specific genes from organisms including humans, the generation of the array of research and commercial applications that has revolutionized the life sciences, medicine, and biotechnology, and the international effort to analyze the human genome at the nucleotide level known as the Human Genome Project. These applications of recombinant DNA technology will be discussed in the next chapter.

(liposomes) followed by fusion with cell membranes. DNA can also be transferred using a variety of vectors based on viruses. DNA introduced into a mammalian cell by any of these methods usually results in the stable integration of the foreign DNA into the host genome. This range of DNA transfer methods has been used successfully to transfer genes to fertilized eggs, producing transgenic animals; to study molecular aspects of gene expression; and to replicate cloned genes using mammalian cells as hosts.

Vectors for the delivery of functional genes into mammalian cells have been developed using avian and mouse retroviruses. The genome of such retroviruses is a single-stranded RNA that is transcribed by reverse tran-

Moloney MLV
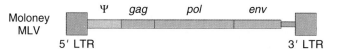

SAX
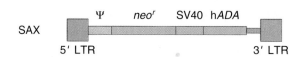

**FIGURE 12.28**    Retroviral vectors constructed from the Moloney murine leukemia virus (Moloney MLV). The native MLV genome contains a *psi* sequence required for encapsulation and genes that encode viral coat proteins (*gag*), an RNA-dependent DNA polymerase (*pol*), and surface glycoproteins (*env*). At each end, the genome is flanked by long terminal repeat (LTR) sequences that control transcription and integration into the host genome. The SAX vector retains the LTR and *psi* sequences, and includes a bacterial neomycin resistance (*neo^r*) gene that can be used as a selective marker. As shown, the vector carries a cloned human adenosine deaminase (*hADA*) gene, fused to an SV40 early region promoter/enhancer. The SAX construct is typical of retroviral vectors that will be used in human gene therapy.

**CHAPTER SUMMARY**

1. Recombinant DNA refers to the creation of a new association between DNA molecules or segments not usually found together. The cornerstone of recombinant DNA technology is a class of enzymes called restriction endonucleases that cut DNA at specific recognition sites. The fragments produced are joined with DNA vector segments using the enzyme DNA ligase to form recombinant DNA molecules. Vectors have been constructed from many sources, including bacterial plasmids, viruses, and artificial vectors such as cosmids and synthetic yeast chromosomes.

2. Recombinant DNA molecules are transferred into a host, and cloned copies are produced during host cell replication. A variety of host cells may be used for replication, including yeast, bacteria, cells of higher plants, and mammalian cells in tissue culture. However, the most common host is *E. coli*. Cloned copies of foreign DNA sequences can be recovered, purified, and analyzed.

3. Cloned libraries have been constructed containing DNA sequences from entire genomes, single chromosomes, or chromosome segments. Researchers use libraries to select probes complementary to all or part of a gene being surveyed.

4. Once cloned, DNA sequences can be analyzed using a variety of methods, including restriction mapping and DNA sequencing. Other methods such as blotting and hybridization can be used to identify genes and flanking regulatory regions within the cloned sequences.

5. For studies of gene regulation in cloned eukaryotic genes, eukaryotic host systems have been developed, including yeast and the cells of higher plants and animals.

## KEY TERMS

bacteriophage
cDNA (complementary DNA) molecules
clone
colony hybridization
cosmid
DNA polymerase I
*Eco* RI
flow cytometry
gel electrophoresis
gene splicing (genetic engineering)

kilobase pairs (kb)
library
modification enzyme
northern blot
nucleotidyl transferase
palindrome
plaque hybridization
plasmid
polylinker
polymerase chain reaction (PCR)

probe
pulse field gel electrophoresis
recombinant DNA
replicative form (RF)
restriction endonuclease
reverse transcriptase
reverse translation
ribonuclease H
shuttle vector
S$_1$ nuclease

Southern blot
terminal deoxynucleotidyl transferase
transgenic
2-micron plasmid
vector
western blot
X-gal
yeast artificial chromosome (YAC)

# INSIGHTS AND SOLUTIONS

The recognition site for the restriction enzyme *Sau*3A is GATC (Table 12.1). The recognition site for the enzyme *Bam*HI is GGATCC, where the four internal bases are identical to the *Sau*3A site. This means that the single-stranded ends produced by the two enzymes are identical. Suppose that you have a cloning vector containing a *Bam*HI site, and foreign DNA that you have cut with *Sau*3A.

1. Can this DNA be ligated into the *Bam*HI site of the vector? Why?

   **SOLUTION:** DNA cut with *Sau*3A can be ligated into the *Bam*HI site of the vector because the single-stranded ends generated by the two enzymes are compatible.

2. Can the DNA segment cloned into this site be cut from the vector with *Sau*3A? With *Bam*HI? What potential problems do you see with the use of *Bam*HI?

   **SOLUTION:** The DNA can be cut from the vector with *Sau*3A since the recognition site for this enzyme is maintained. The DNA may be recovered from the vector by *Bam*HI, but there is a potential problem with the use of this enzyme. When the *Sau*3A fragment is ligated into the *Bam*HI site, a base at each end of the *Bam*HI site must be added (arrows). Even though a template is present on the opposite strand, insertion of the wrong base will occur in a fraction of cases, destroying the *Bam*HI site, making it impossible to remove the insert with *Bam*HI.

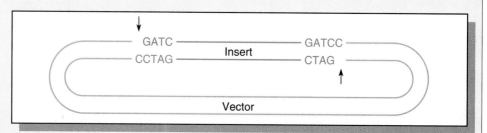

3. Calculate the average size of the fragments generated when cutting DNA with *Sau*3A and with *Bam*HI.

   **SOLUTION:** According to the Watson–Crick model of DNA, the sequence of bases in a DNA molecule can be at random. Therefore, if an A is present in a strand of DNA, the next base in that strand can be any base, even another A. Thus, the chance that any given base in DNA can be an A is 1/4, and the chance that the next base is a T is also 1/4. As a result, the chance that the four-base sequence GATC will occur is $4^4$, or 1 in every 256 bases. The chance of encountering the six-base sequence GGATCC is $4^6$, or 1 in every 4096 bases. Accordingly, cuts with *Sau*3A should generate fragments that average 256 bp in length, while cuts with *Bam*HI should generate fragments that average 4096 base pairs in length. This, of course, is only an approximation, and assumes that each base is present about 25 percent of the time. In reality, however, many genomes depart from this ideal, and may have only 40 or 42 percent G + C content, instead of the 50 percent ideal.

1. In recombinant DNA studies, what is the role of each of the following: restriction endonucleases, terminal transfersase, vectors, calcium chloride, and host cells?

2. Why is poly dT an effective primer for reverse transcriptase?

3. An ampicillin-resistant, tetracycline-resistant plasmid is cleaved with *Eco* RI, which cuts within the ampicillin gene. The cut plasmid is ligated with *Eco* RI-digested *Drosophila* DNA to prepare a genomic library. The mixture is used to transform *E. coli* K12.
   (a) Which antibiotic should be added to the medium to select cells that have incorporated a plasmid?
   (b) What antibiotic resistance pattern should be selected to obtain plasmids containing *Drosophila* inserts?
   (c) How can you explain the presence of colonies that are resistant to both antibiotics?

4. Clones from the preceding question are found to have an average length of 5 kb. Given that the *Drosophila* genome is $1.5 \times 10^5$ kb long, how many clones would be necessary to give a 99 percent probability that this library contains all genomic sequences?

5. Type II restriction enzymes recognize palindromic sequences in intact DNA molecules and cleave the double-stranded helix at these sites. Inasmuch as the bases are internal in a DNA double helix, how is this recognition accomplished?

6. In a control experiment, a plasmid containing a *Hind*III site within a kanamycin-resistant gene is cut with *Hind*III, religated, and used to transform *E. coli* K12 cells. Kanamycin-resistant colonies are selected, and plasmid DNA from these colonies subjected to electrophoresis. Most of the colonies contain plasmids that produce single bands that migrate at the same rate as the original, intact plasmid. A few colonies, however, produce two bands, one of original size, and one that migrates much higher in the gel. Diagram the origin of this slow band during the relegation process.

7. When making cDNA, the single-stranded DNA produced by reverse transcriptase can be made double stranded by treatment with DNA polymerase I. However, no primer is required with the DNA polymerase. Why is this?

8. What facts should you consider in deciding which vector to use in constructing a genomic library of eukaryotic DNA?

9. Using DNA sequencing on a cloned DNA segment, you recover the following nucleotide sequence: CAGTATCCTAGGCAT. Does this segment contain a palindromic recognition site for a restriction enzyme? What is the double-stranded sequence of the palindrome? What enzyme would cut at this site? (Consult Table 12.1 for a list of restriction enzyme recognition sites.)

10. Table 12.1 lists restriction enzymes that recognize sequences of four bases and six bases. How frequently should four- and six-base recognition sequences occur in a genome? If the recognition sequence consists of eight bases, how frequently would such sequences be encountered? Under what circumstances would you select such an enzyme for use?

11. List the steps involved in cloning a DNA insert into the *Sal*I site of the plasmid pBR322, up to and including the identification of host cells carrying recombinant plasmids.

**12.** You have recovered a cloned DNA segment of interest, and determined that the insert is 1300 bp in length. To characterize this cloned segment, you have isolated the insert and decide to construct a restriction map. Using enzyme I and enzyme II, followed by gel electrophoresis, you determine the number and size of the fragments produced by enzymes I and II alone and in combination as follows:

| Enzymes | Restriction Fragment Sizes (bp) |
|---------|--------------------------------|
| I | 350 bp, 950 bp |
| II | 200 bp, 1100 pb |
| I and II | 150 bp, 200 bp, 950 bp |

Construct a restriction map from these data, showing the positions of the restriction sites relative to one another, and the distance between them in base pairs.

**13.** Although the potential benefits of cloning in higher plants are obvious, the development of this field has lagged behind cloning in bacteria, yeast, and mammalian cells. Can you think of any reason for this?

**14.** cDNA can be cloned into vectors to create a cDNA library. In analyzing cDNA clones, it is often difficult to find clones that are full length, that is, extend to the 5′ end of the mRNA. Why is this so?

**15.** List the steps involved in screening a genomic library by reverse translation. What needs to be known before starting such a procedure? What are the potential problems with such a procedure? How can they be overcome or minimized?

**SELECTED READINGS**

ALWINE, J. C., KEMP, D. J., and STARK, G. R. 1977. Method for detection of specific RNAs in agarose gels by transfer to diazobenzyloxymethyl paper and hybridization with DNA probes. *Proc. Natl. Acad. Sci.* 74:5350–54.

ANDERSON, W. F., and DIACUMAKOS, E. G. 1981. Genetic engineering in mammalian cells. *Scient. Amer.* (July) 245:106–21.

BERG, P. 1981. Dissections and reconstructions of genes and chromosomes. *Science* 213: 296–303.

BOYER, H. W. 1971. DNA restriction and modification mechanisms in bacteria. *Ann. Rev. Microbiol.* 25:153–76.

COHEN, S., et al. 1973. Construction of biologically functional bacterial plasmids *in vitro*. *Proc. Natl. Acad. Sci.* 70:3240–44.

COLLINS, J., and HOHN, B. 1978. Cosmids: A type of plasmid gene cloning vector that is packageable *in vitro* in bacteriophage heads. *Proc. Natl. Acad. Sci.* 75:4242–46.

DAVIES, K. E., et al. 1981. Cloning of a representative genomic library of the human X chromosome after sorting by flow cytometry. *Nature* 293:374–76.

GLOVER, D. M. 1984. *Gene cloning: The mechanism of DNA manipulation.* London: Chapman and Hall.

GRAF, L. H. 1982. Gene transformation. *Amer. Scient.* 70:496–505.

GRIESBACH, R. J., KOIVUNIEMI, P. J., and CARLSON, P. S. 1981. Extending the range of plant genetic manipulation. *Bioscience* 31:754–56.

GUBLER, U. and HOFFMAN, B. J. 1983. A simple and very effective method for generating cDNA libraries. Gene 25:263–269.

KRUMLAUFF, R., JEANPIERRE, M., and YOUNG, B. D. 1982. Construction and characterization of genomic libraries from specific human chromosomes. *Proc. Natl. Acad. Sci.* 79:2971–75.

MESSING, J., and VIERIA, J. 1982. The pUC plasmids, an M13mp7-derived system for insertion mutagenesis and sequencing with synthetic universal primers. *Gene* 19:259–268.

MOLLIS, K. B. 1990. The unusual origin of the polymerase chain reaction. *Scient. Amer.* (April) 262:56–65.

OLD, R. W., and PRIMROSE, S. B. 1985. *Principles of genetic manipulation: An introduction to genetic engineering.* Palo Alto, CA: Blackwell Scientific.

OSTE, C. 1988. Polymerase chain reaction. *BioTechniques* 6:162–67.

SAMBROOK, J., FRITSCH, E., and MANIATIS, T. 1992. *Molecular Cloning: A Laboratory Manual.* 2nd. ed. Cold Springs Harbor: Cold Springs Harbor Laboratory Press.

SOUTHERN, E. M. 1975. Detection of specific sequences among DNA fragments separated by gel electrophoresis. *J. Mol. Biol.* 98:503–17.

THOMAS, T. L., and HALL T. C. 1985. Gene transfer and expression in plants: Implications and potential. *BioEssays* 3:149–53.

TORREY, J. G. 1985. The development of plant biotechnology. *Amer. Scient.* 73:354–63.

WATSON, J., GILMAN, M., WITKOWSKI, J., and ZOLLER, M. 1992. *Recombinant DNA.* 2nd. ed. New York: Scientific American Books.

WHITE, R. 1985. DNA sequence polymorphisms revitalize linkage approaches in human genetics. *Trends in Genetics.* 1:177–80.

# 13

# DNA:
## APPLICATIONS OF
## RECOMBINANT TECHNOLOGY

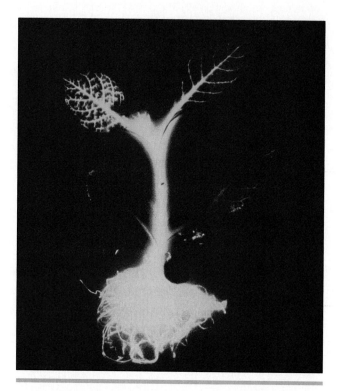

Tobacco plant expressing the firefly gene encoding the enzyme luciferase, which catalyzes the reaction releasing light energy.

*The use of recombinant DNA technology has brought about significant advances in experimental genetics, gene mapping, and the diagnosis and treatment of disease. These techniques have also formed the basis for the biotechnology industry and the commercial production of human gene products for therapeutic uses, and the transfer of genes across species in agricultural plants and animals.*

In 1971, a paper published by Hamilton Smith, Daniel Nathans, and Walter Arber marked the beginning of the recombinant DNA era. This paper described the isolation of an enzyme from a strain of bacteria and the use of this enzyme to cleave viral DNA. It contains the first published photograph of DNA cut with a restriction enzyme. From this modest beginning, recombinant DNA technology has revolutionized all fields of experimental biology and spread beyond the research laboratory to many other fields. In the intervening years, this technology has expanded to become part of everyday life; recombinant DNA techniques are now being used to help produce medicines, milk, and soon, even the food at the supermarket.

In the last chapter we discussed the methods used to create and analyze recombinant DNA molecules. In this chapter we will describe how these techniques are being used in research and commercial applications. We will consider a cross section of applications that illustrate the power of recombinant DNA technology to map and identify human genes, diagnose and treat disease, and to generate new plants and animals. We will begin by considering how recombinant DNA has changed some of the basic methods of research in genetics, allowing the focus to shift from individual mutant genes to entire genomes. Next, we will examine the impact of recombinant DNA on human genetics and medicine, from the prenatal analysis of genotypes to the treatment of genetic disorders by repairing or replacing defective genes. Finally, we will discuss some of the products and methods being used in the multibillion dollar biotechnology industry generated by recombinant DNA technology.

# RECOMBINANT DNA AND GENETIC ANALYSIS

Conventional genetic analysis depends on the isolation and mapping of mutant genes to provide information about the number, location, and function of genes in an organism's genome. Over the past 90 years, geneticists have developed effective methods for generating and mapping mutants, but one drawback to this method is that genes can be characterized only if mutants of those genes are identified and isolated. To exhaustively characterize a genome by this method, at least one mutation for each of the genes in the genome is required, and this is often a difficult, labor-intensive task.

Geneticists are now using recombinant DNA techniques in a different, more direct approach to genetic analysis. Instead of screening for mutants and constructing genetic maps, this approach involves creating a library of clones encompassing the whole genome, mapping overlapping clones to establish complete genetic and physical maps, and determining the nucleotide sequence of the entire genome. The product is a physical map, with all genes in the genome identified by their nucleotide sequence, without the need to screen for mutants, and often, without any initial knowledge of the function or phenotype of the mapped genes. With this information in hand, further work can then proceed from the nucleotide level to the phenotypic level.

Genome projects involving several species are under way (Table 13.1). Some species, such as *E. coli*, yeast, *Drosophila*, and the mouse, were selected because detailed genetic maps are already available. *C. elegans* is being used because of the small number of cells in the organism, and *Arabadopsis* was selected as a model for higher plants. Humans are being studied in order to learn about genetic diseases and about ourselves.

Two basic approaches are used in genome projects: (1) the *bottom-up* approach, which starts with small, overlapping clones and assembles them into larger segments that eventually cover the entire genome; and (2) the *top-down* approach, which isolates very large, randomly cloned DNA segments that encompass the entire genome and breaks these down into smaller units for mapping and sequence analysis. A discussion of each of these approaches will serve as an introduction to the

**Table 13.1** ORGANISMS STUDIED BY GENOME PROJECTS

| Organism | Genome Size (bp) |
|---|---|
| *E. coli* (bacterium) | $4 \times 10^6$ |
| *Saccharomyces cerevisciae* (yeast) | $4 \times 10^6$ |
| *Arabadopsis thalia* (plant) | $1 \times 10^8$ |
| *Caenorhabditis elegans* (nematode) | $1 \times 10^8$ |
| *Drosophila melanogaster* (insect) | $1.65 \times 10^8$ |
| *Mus musculus* (mouse) | $3 \times 10^9$ |
| *Homo sapiens* (human) | $3.2 \times 10^9$ |

Human Genome Project and its goal of isolating and identifying the 50,000 to 100,000 genes in the human genome.

## *E. coli* Genome Project

The bottom-up method is being used to analyze the *E. coli* genome. A cloned genomic library of the *E. coli* K12 strain was prepared in lambda vectors by Yuji Kohara and his colleagues. Clones from this library were characterized by their pattern of restriction fragments, and these data were entered into a computer. Then overlapping clones were identified by the presence of shared restriction sites (Figure 13.1). The resulting contiguous segment of DNA covered by the overlapping clones is

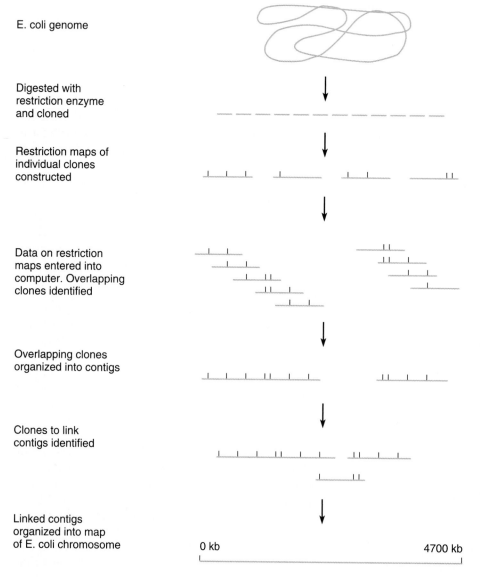

E. coli genome

Digested with restriction enzyme and cloned

Restriction maps of individual clones constructed

Data on restriction maps entered into computer. Overlapping clones identified

Overlapping clones organized into contigs

Clones to link contigs identified

Linked contigs organized into map of E. coli chromosome

0 kb                    4700 kb

**FIGURE 13.1**    Bottom-up strategy for *E. coli* genome project. The *E. coli* chromosome is digested with restriction enzymes into small fragments and cloned into a vector. Individual clones are characterized by restriction mapping. Information about these restriction maps is analyzed by computer to identify overlapping clones. These overlapping clones are organized into contiguous segments of DNA called contigs. Next, more clones are screened to identify those spanning contigs. These are used to reduce gaps, bringing the number of contigs from 70 to 7 and finally to 1, which corresponds to the entire *E. coli* chromosome. Individual clones are being sequenced, and eventually the nucleotide sequence of the entire chromosome will be known.

called a **contig**. In the initial analysis, some 70 contigs were identified, covering 94 percent of the genome. Finding clones that spanned the small gaps between contigs reduced the number of contigs to 7, and finally, gaps between these 7 contigs were closed. The contigs were then arranged in a physical map that covers the entire 4700 kilobases (kb) of DNA in the *E. coli* genome.

The second stage of the project involves determining the nucleotide sequence of the clones that make up the contig library. To date, about 30 percent of the genome has been sequenced, covering about 1.5 million base pairs of DNA. At this stage, it is fair to ask what has been found in this analysis. In a recent study, Fred Blattner and colleagues reported the nucleotide sequence of 91.4 kb of *E coli* DNA. They found that genes (identified by the presence of open reading frames within their nucleotide sequence) covered 84 percent of the bases in this region. About half of these corresponded to genes already identified by conventional mutant screening and mapping, but the rest were previously unknown, encoding as yet unidentified gene products. This result indicates that years of intensive efforts to map bacterial genes by conventional genetic analysis have identified only about half the genes in the *E. coli* genome, and that many more genes remain to be identified. In addition, this work demonstrates that a recombinant DNA-based approach using physical maps at the nucleotide level is effective in finding and mapping all the genes in a genome.

### *Drosophila* Genome Project

For the top-down approach being used to analyze the *Drosophila* genome, a cloned genomic library has been constructed using a YAC vector. The average size of the *Drosophila* DNA fragment cloned into each YAC is about 200 kb, a span that encompasses many genes. The library contains the equivalent of 1.7 genomes and is carried in 965 YACs. For mapping, a given YAC is separated from other yeast chromosomes by pulse field gel electrophoresis, and its physical location in the genome is identified by *in situ* hybridization to the banded polytene chromosomes of larval salivary glands (Figure 13.2). Because of their large size, the average YAC covers six to eight chromosome bands; contigs are constructed by identifying YACs that have two or more chromosome bands in common. The 965 YACs examined to date cover about 80 percent of the euchromatic region of the genome in 161 contigs, with just over 150 gaps to be closed.

At higher resolution, many YACs and derived subclones can be assigned to individual bands. To study individual genes or regions, YACs are broken down by restriction digestion and subcloned into other vectors.

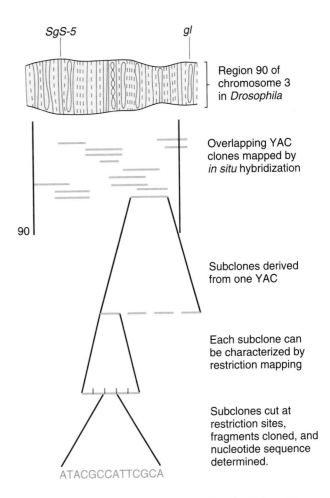

**FIGURE 13.2**    Top-down strategy for the *Drosophila* genome project. A genomic library is constructed using large (~200 kb) fragments in YACs. The physical location of YACs are mapped to polytene chromosomes by *in situ* hybridization. The region shown is region 90, located on chromosome 3. This region contains about 890 kb of DNA and approximately 34 genes. About 14 percent of the region is transcribed as RNA. Each YAC covers several polytene bands and presumably includes several genes. Individual YACs are digested with restriction enzymes and the individual fragments are subcloned. The subclones are characterized by restriction mapping, and restriction fragments from these subclones are used for nucleotide sequence analysis.

Placing genes on this physical map is more difficult than in *E. coli*, in part because the relatively large size of the genome (see Table 13.1 to compare the genome sizes) makes sequencing a less efficient method of mapping. To map genes at the level of nucleotide sequence, each gene is defined as a stretch of a unique DNA sequence. Therefore, any unique short DNA segment within a gene can be used as a marker or **sequence tagged site (STS)** to place a gene on the physical map. Note that this can be accomplished without knowing the nature of the gene, the gene product, or its phenotype.

Three sources of STS markers are used in this approach. First, a large number (over 3800) of mutant genes have been identified and mapped in *Drosophila*, and many of these have been completely or partially sequenced. Using STS markers derived from DNA sequence information, these genes can be placed on the physical map by *in situ* hybridization, where they serve as important reference points. As a second source of STS markers, genes expressed at high levels in certain tissues are used to produce cDNA molecules. When used as probes, these cDNA molecules identify additional genes that can be placed on the physical map. This method generates map positions for some genes that have not

been recovered as mutants. Lastly, the genome of *Drosophila* strains from the wild contains transposable DNA segments called **P elements**. Introduced into laboratory strains, the sites where these elements insert into the genome can be used as STS markers. Other mapping efforts in *Drosophila* use cosmid clones made from microdissected chromosome regions to construct contigs, and other projects are utilizing phage vectors. A computerized database of information about the *Drosophila* mapping project that includes genes, gene products, assigned functions, map positions, and chromosome aberrations is available by electronic access to investigators and students worldwide. At its completion, the *Drosophila* genome project will provide a molecular map of the *Drosophila* genome at the nucleotide level for research and teaching purposes.

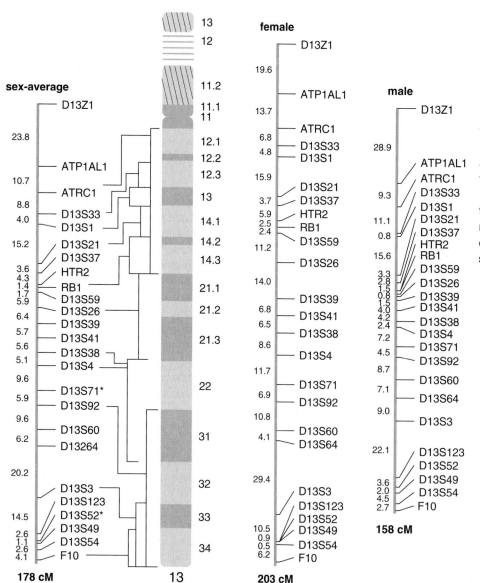

**FIGURE 13.3** A genetic and a physical map of human chromosome 13. The genetic map for females is 203 cM, and that for males is 158 cM. The discrepancy reflects differences in the frequencies of recombination between females and males. When the two maps are averaged together, the result is the sex-averaged map of 178 cM shown at left. The location of markers on the physical map is indicated by the brackets adjacent to the chromosome.

## Human Genome Project

The **Human Genome Project** includes scientists from laboratories around the world engaged in a coordinated effort to construct a physical map of the entire 3 billion base pairs in the haploid human genome. In the United States, consideration of a genome project began in 1986, and in 1988 the National Institutes of Health and the Department of Energy created a joint committee to develop a five-year plan. The plan, called the Human Genome Project, got under way in 1990. Other countries, notably France, Britain, and Japan, began similar projects, which are now coordinated by an international organization, the Human Genome Organization (HUGO).

This complex task is proceeding in a series of well-defined stages. The first stage involves the construction of high-resolution genetic maps for each of the 22 autosomes and the sex chromosomes (Figure 13.3). In these maps, identified genes, RFLP markers, or STS sites are mapped to loci on each chromosome. In the first step, this will require somewhere between 600 to 1500 markers; it is hoped that this map will be available by 1995 or 1996. This genetic map, with an average distance of 2 million base pairs of DNA between markers, will be used to order the contigs being generated by physical mapping. The second task is the construction of physical maps, using either the bottom-up approach of assembling contigs from DNA segments randomly cloned in cosmids or YACs, or the top-down approach, beginning with chromosomes isolated in somatic cell hybrids. The ultimate goal of the project is sequencing the 3.2 billion nucleotides in the human genome, a task that in itself may take a decade or more and will require the development of new technology for sequencing DNA, as well as new technology for information storage, analysis, and retrieval. Knowing the entire nucleotide sequence of the human genome will not tell us everything there is to know about being human, but will provide the tools needed to explore the inner workings of our cells and ourselves.

## RFLP Markers and Linkage

Although an increasing number of the 50,000 to 100,000 human genes have been localized to chromosomal sites, in the majority of human genetic disorders the primary defect is unknown, and without a marker the gene cannot be mapped. A defective gene product has been identified in just under 400 single-gene disorders, and the nature of the mutation is known in only about 40 of these. In the majority of cases, the mutant gene product responsible for a genetic disease has not been identified, and as a result, classical genetic methods cannot be used to identify the chromosomal locus of the disorder.

The development of recombinant DNA technology has provided a new method for mapping allelic variants without the need to identify the gene product. This method makes use of the fact that single-nucleotide changes in the recognition sequence of restriction enzymes can alter the pattern of cuts made in DNA. If one or more nucleotides in the restriction recognition site are altered, the enzyme will fail to cut at that site. If a restriction site is present on the DNA from one chromosome, but absent on its homologue, the two chromosomes can be distinguished from one another by their pattern of restriction fragments (Figure 13.4). These detectable var-

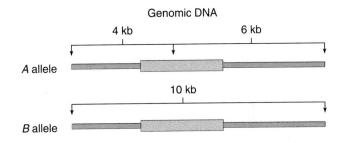

Genomic DNA

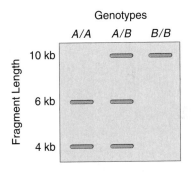

Genotypes

**FIGURE 13.4**    Restriction fragment length polymorphisms (RFLPs). The *A* and *B* alleles represent DNA segments from homologous chromosomes. The thick region represents a segment that can be detected with a DNA probe. Arrows indicate the pattern of restriction enzyme cutting sites that define the alleles. In allele *A*, three cutting sites generate fragments of 6 and 4 kb. In allele *B*, only two cutting sites are present, generating a fragment 10 kb in length. The absence of the cutting site in *B* could be the result of a single base mutation within the enzyme recognition/cutting site. Because the alleles are codominant, three genotypes (*AA*, *AB*, and *BB*) can be generated. The allele combination carried by any individual can be detected by restriction digestion of genomic DNA (obtained from leukocytes or skin fibroblasts), followed by gel electrophoresis, transfer to a DNA binding filter, and hybridization to the appropriate probe. The fragment patterns for the three possible genotypes are shown as they would appear on a Southern blot.

**(a)**

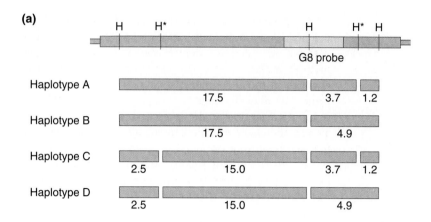

**(b)**

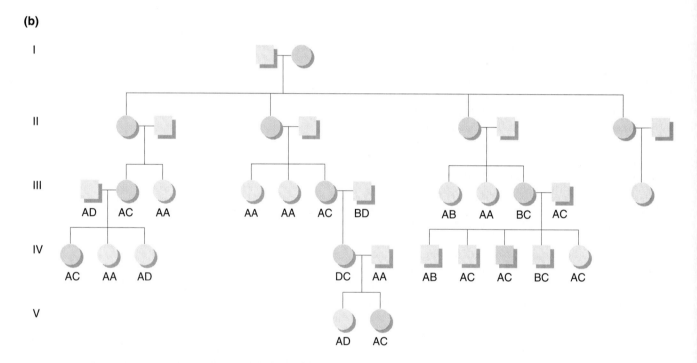

**FIGURE 13.5**    RFLP analysis and Huntington disease (HD). (a) The G8 locus was used to map the HD locus to human chromosome 4. The *Hind*III sites in the G8 region are shown. The two sites that are polymorphic are indicated with an asterisk; the remaining sites are invariant. Below are shown the patterns of fragments generated when either or both of the polymorphic sites are present. The region detected by the G8 probe is indicated by the hatched region. Note that the probe does not detect all fragments. These include: the 1.2-kb fragment in haplotype A, the 2.5-kb fragment and the 1.2-kb fragment in haplotype C, and the 2.5-kb fragment in haplotype D. (b) A pedigree showing the segregation of HD haplotypes (filled symbols) detected by the G8 probe. All affected individuals carry the C allele of the G8 locus. Note that having a C allele is not diagnostic for HD, since unrelated spouses (III-12) can carry the C allele without HD and introduce this allele into the family pedigree. In this family, the C allele detected by RFLP analysis helps track HD through the pedigree.

iations in DNA fragment length are called **restriction fragment length polymorphisms (RFLPs)**. These cutting-site variations are inherited as codominant alleles and can be used to follow the inheritance of individual chromosomes from generation to generation.

RFLPs are distributed randomly throughout the genome, and in 1980 it was proposed that these markers, in combination with known genes, could be used to construct a linkage map of the human genome. This proposal, made by David Botstein, Mark Skolnick, and Ray White, accelerated the mapping of human genes, and a partial linkage map of the human genome was produced in the next few years. This may not seem like rapid progress, but consider that from the 1930s to about 1970, only five cases of autosomal linkage were known in humans. From 1970 to 1980, additional examples of linkage were discovered through the use of somatic cell hybrids, but by 1980 only a relative handful of linked genes had been identified. Presently, RFLP analysis has made it possible to obtain high-resolution linkage maps for some human chromosomes, such as chromosome 21, and linkage maps using hundreds of RFLP markers are now available for all 23 human chromosomes.

To construct a linkage map using RFLP markers requires a multigenerational family in which an RFLP marker and the gene to be mapped are inherited. In this case, the RFLP acts as a codominantly inherited genetic marker (Figure 13.5), and the inheritance of the RFLP and the gene are followed through the family. In linkage analysis, the probability that the observed pattern of inheritance (of the gene and the RFLP) could occur by chance alone is calculated. This analysis begins with the assumption that the gene and the RFLP marker are unlinked. Then the analysis is repeated, calculating the probability of the observed pattern of inheritance if the gene and the marker have a certain degree of linkage. The ratio of the two probabilities (no linkage/certain degree of linkage) is computed and expressed as the odds for that degree of linkage. This calculation is known as the **logarithm of the odds** or **lod score**. As a standard, a lod score of 3 or greater is taken as evidence of linkage between the gene and the RFLP marker. A lod score of 3 means that the odds are 1000:1 that the gene and the marker are linked.

In genetics, the unit of linkage is the **centimorgan (cM)**, named after the geneticist T. H. Morgan. One centimorgan is equal to a recombination frequency of 1 percent between two markers. Genetic maps indicate the distance between markers in centimorgans. The genetic distance between two genes expressed in centimorgans is not directly correlated with the physical distance expressed in nucleotides, but on average, there are about $1 \times 10^6$ bp of DNA for every centimorgan. Thus, in mapping human genes, a distance of 2 cM corresponds to

roughly 2 million nucleotides. By compiling many family studies using RFLP markers and genetic traits, a genetic map for a human chromosome can be constructed (Figure 13.3).

## Mapping Human Genetic Disorders

The value of RFLP analysis in human genetics can be illustrated by considering the search for the chromosomal locus of type 1 neurofibromatosis (NF1). This disorder is an autosomal dominant condition with an incidence of about 1 in 3000. NF1 is associated with a range of nervous system disorders, including tumors and an increased incidence of learning disorders. The spontaneous mutation rate at this locus is very high, and it is estimated that almost 50 percent of all cases represent new mutations. To map this autosomal gene by conventional methods in a systematic fashion would be an almost impossible task that would require the identification of large families carrying genes mapped to each of the 22 autosomes, and also carrying neurofibromatosis.

Mapping of the NF1 gene was accomplished in several steps. First, a number of laboratories compared the inheritance of NF1 in multigenerational families to that of dozens of RFLP markers, each representing a specific human chromosome or chromosome region. This produced an **exclusion map**, indicating the chromosomes and chromosome regions where NF1 was not located, and pointed to chromosomes 5, 10, and 17 as likely candidates. A second step used a collection of RFLP markers to determine that the disorder segregated with an RFLP near the centromere of chromosome 17 (Figure 13.6). Further efforts using over 30 RFLP markers from the pericentromeric region of chromosome 17 and analysis of 13,000 genotypes refined the location of this gene to a small region on the long arm designated as 17q11.2. Finally, using a collection of cloned DNA sequences that spanned a subregion of 17q11.2, the gene for NF1 was identified in 1990. This feat was accomplished in a little over three years, beginning with no direct knowledge of either the nature of the gene product or the mutational events that result in the NF1 phenotype.

Subsequently, the DNA sequence of the NF1 gene was determined, and probes have been constructed to study the patterns of expression in specific tissues and developmental stages. It has also been possible to reconstruct the amino acid sequence of the gene product from the nucleotide sequence and deduce its functional properties. The NF1 gene product has been identified as a cytoplasmic protein that may control cell growth by interacting with other proteins in pathways that regulate proliferation. This information has, in turn, allowed the construction of antibodies to study its pattern of distribution in the cell. It is hoped that the sum of this knowl-

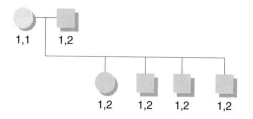

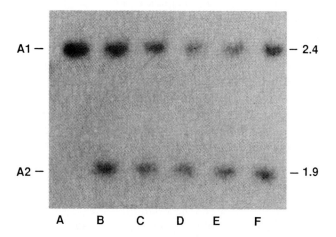

**FIGURE 13.6**   The segregation of RFLP allele *A2* with type 1 neurofibromatosis (NF1) in each of four affected offspring and their father. This RFLP is detected by probe pA10-41, which is known to be a DNA segment near the centromere of human chromosome 17. On the basis of this and results from the use of other probes, the locus for NF1 was assigned to chromosome 17.

edge will allow the development of therapeutic strategies to effectively treat NF1.

The use of RFLPs in gene mapping is a significant breakthrough in human genetics and has been used to map an increasing number of human genes associated with genetic disorders.

## Cloning Genes Responsible for Genetic Disorders

The mapping and cloning of the gene for NF1 described above is an example of **positional cloning**. The use of this recombinant DNA-based method is a departure from previous approaches that work back from the identified gene product to the gene locus. In positional cloning, the gene can be mapped, isolated, and cloned with no knowledge of the gene product. Using this strategy, an ever-increasing number of human genes are being mapped and isolated. Where RFLP markers do not show close linkage to a gene, other methods, or a com-

bination of methods, have been used to map and isolate genes associated with genetic disorders.

In the case of Duchenne muscular dystrophy (DMD), an X-linked disorder associated with progressive muscle wasting and premature death, a combination of RFLP analysis and cytogenetics was used in cloning the gene. RFLP analysis was used to map the DMD gene to the p21 region of the X chromosome. However, the nearest RFLP was estimated to be 15 cM or at least 15 million base pairs away from the DMD locus, making it difficult to isolate the gene. In rare cases DMD affects women, and some of these cases involved reciprocal translocations between the X chromosome and an autosome. In all cases, the translocation occurred in Xp21, reinforcing the conclusion that the DMD gene locus was in this chromosome region. In one woman, the translocation involved the X chromosome and chromosome 21 t(X;21) and placed a block of ribosomal genes from chromosome 21 adjacent to the region containing the DMD gene (Figure 13.7). Since the ribosomal genes (rDNA) had already been cloned, rDNA probes were used to clone X chromosome DNA sequences adjacent to the breakpoint. Analysis of these clones indicated that they contain a portion of the DMD gene, establishing the locus of the gene. The complete gene was later cloned by Louis Kunkel and his colleagues using DNA isolated from a translocation that covered a larger portion of Xp21. The DMD gene turns out to be one of the largest human genes known, covering more than 2.5 million base pairs, with 65 exons; one intron is more than 250 kb in length. DNA sequence analysis was used to deduce the amino acid sequence and putative function of the predicted protein. In the case of DMD, comparisons with the amino acid sequence of well-characterized proteins indicated that the DMD protein, called **dystrophin**, is related to cytoskeletal proteins, providing a clue to its function. Further work was revealed that dystrophin is involved in anchoring the cytoskeletal system to the plasma membrane and that in DMD this missing connection causes the muscle cell plasma membrane to tear under the mechanical stress of contraction, leading to the death of the muscle fiber.

An alternative approach to cloning genes for genetic disorders is the selection and investigation of candidate genes. This approach involves having some information about the molecular basis of a genetic defect and having some detailed knowledge of a gene that might carry out the function missing in the mutant condition. One such gene cloned by this approach is that for Marfan syndrome, an autosomal dominant disorder of connective tissue that affects about 1 in 10,000 individuals. It has been speculated that Abraham Lincoln, the 16th president of the United States, had this disorder. Given the nature of the defect, attention was focused on connective tissue proteins as the primary candidates. However,

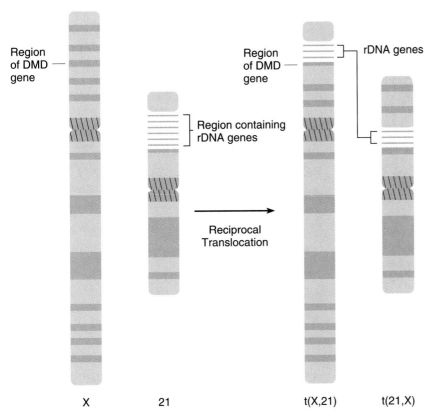

Region
of DMD
gene

Region
of DMD
gene

Region containing
rDNA genes

rDNA genes

Reciprocal
Translocation

X          21          t(X,21)          t(21,X)

**FIGURE 13.7**    Translocation associated with muscular dystrophy. The locus for the X-linked disorder Duchenne muscular dystrophy (DMD) is located in Xp21. To isolate the gene, a translocation between the X chromosome and chromosome 21 was used. The translocation on the X was known to be in the region of the DMD gene, and since ribosomal DNA (rDNA) genes were moved from chromosome 21 to this region, rDNA probes were used to clone sequences that spanned the translocation breakpoint on the X chromosome. A portion of the DMD gene was recovered, pin-pointing the location of the DMD gene.

part of the problem in identifying the molecular basis of Marfan syndrome is that many connective tissue proteins are posttranslationally modified, often making it difficult to identify the primary gene product.

Fibrillin is a connective tissue protein, first identified in 1986 as a component of the eye, aorta, and other elastic tissues, all of which are affected in Marfan syndrome. Using RFLP analysis, the gene for Marfan syndrome was mapped to a large region on the long arm of chromosome 15. *In situ* hybridization confirmed that the gene for fibrillin mapped to a locus on the long arm of chromosome 15. These circumstances made fibrillin a candidate gene for Marfan syndrome. The relationship between fibrillin and Marfan syndrome was defined by cloning and sequencing the fibrillin gene. Next, a mutation in the fibrillin gene was identified in three families with Marfan syndrome. In all three families, the mutation was a missense mutation leading to the substitution of a proline for arginine at position 239 in fibrillin, establishing the link between the candidate gene and the disorder.

# MOLECULAR DIAGNOSIS OF HUMAN DISEASE

The most widely used methods for prenatal detection of genetic diseases are **amniocentesis** and **chorionic villus sampling (CVS)**. In amniocentesis, a needle is used to withdraw amniotic fluid (Figure 13.8). The fluid and the cells it contains can be analyzed for chromosomal or single-gene disorders. In CVS, a catheter is inserted into the uterus and used to retrieve a small tissue sample of the fetal chorion. This tissue is used for cytogenetic and biochemical prenatal diagnosis.

Coupled with these sampling methods, recombinant DNA technology has proven to be a highly sensitive and accurate tool for the detection of genetic disorders. The use of cloned DNA sequences has the advantage of allowing direct examination of the genotype rather than a potentially unknown or unexpressed gene product. The prenatal detection of two different defects in the beta globin gene serves to illustrate the application of recombinant DNA technology to disease detection. Disorders of beta globin are difficult to detect prenatally by traditional means because the beta globin gene is not active until a few days after birth.

## Deletions in Thalassemia

Beta thalassemia is an autosomal recessive condition associated with lowered or absent production of beta globin. At the molecular level, this condition can be caused by a variety of deletions or point mutations. One such deletion covers about 600 nucleotides, and includes the third exon of the gene (Figure 13.9). Restric-

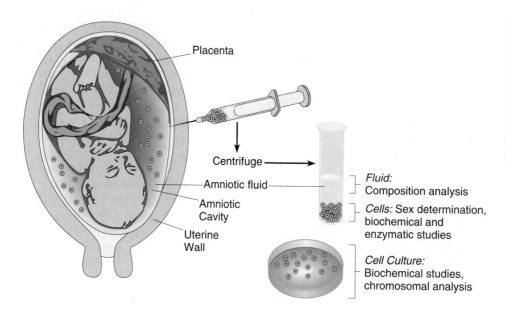

FIGURE 13.8    The technique of amniocentesis. The position of the fetus is first determined by ultrasound, and then a needle is inserted through the abdominal and uterine wall to recover fluid and fetal cells for cytogenetic and/or biochemical analysis.

tion enzyme digestion and Southern blot hybridization to a cloned beta globin probe can be used to diagnose the genetic status of a fetus and other members of a family (Figure 13.9). Because the Southern blot detects fragments separated on the basis of size, this method is useful in studying mutations that involve deletions, insertions, and rearrangements.

Deletions, however, are relatively rare mutational events, and account for only a small fraction (about 5%) of all mutations. More commonly, mutations are associated with changes in one or a small number of nucleotides (point mutations). Even changes in a single nucleotide can have devastating clinical and phenotypic effects, as discussed below.

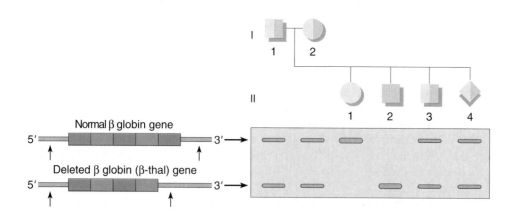

FIGURE 13.9    Diagnosis of beta thalassemia caused by a partial deletion of the beta globin gene. The family pedigree (□ = male, ○ = female, ◇ = fetus) is shown positioned above the individuals' respective genotypes on a Southern blot. The normal and deleted beta globin genes that produce the genotypic band patterns are shown at the right. Arrows indicate the cutting sites for restriction enzymes used in this analysis. The normal gene produces larger fragments shown as the top row of fragments on the gel; the smaller fragments produced by the deleted gene are represented at the bottom of the gel. Thin lines on the gel represent one copy of an allele, thick lines represent two copies.

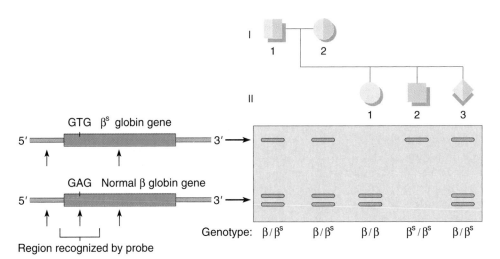

**FIGURE 13.10**    Southern blot diagnosis of sickle-cell anemia. Arrows below the genes represent restriction enzyme cutting sites. In the mutant ($\beta^s$) globin gene, a point mutation (GAG → GTG) has destroyed a restriction enzyme cutting site, resulting in an altered pattern of DNA fragments. In the pedigree, the family has one unaffected homozygous normal daughter (II-1), an affected son (II-2), and an unaffected fetus (II-3).

## Sickle-Cell Anemia and Prenatal Genotyping

Sickle-cell anemia is an autosomal recessive condition common in those with family origins in areas of West Africa and the Mediterranean basin. Sickle-cell anemia is caused by a single nucleotide substitution. While the mutation does not change the length of the beta globin gene, it does destroy a cutting site for several restriction enzymes, including *Mst*II and *Cvn*I. This results in changes in the length of restriction fragments, allowing diagnosis of genotypes by Southern blot analysis. One method of detecting the mutant gene is by DNA digestion with a restriction enzyme that recognizes the normal nucleotide sequence (such as *Mst*II), followed by gel electrophoresis to separate fragments by size, and Southern blot hybridization with a radioactive cloned probe. With this procedure, it is possible to determine fetal genotypes as well as those of the parents (Figure 13.10).

Naturally, not every point mutation involving a single-base change alters a restriction site. In fact, it is estimated that only about 5 to 10 percent of all such mutations can be detected by restriction analysis. Alternatively, if the mutant gene is well characterized, it is possible to use synthetic oligonucleotides as probes.

## Allele-Specific Nucleotides and Detection of Mutations

Under the proper conditions, a probe known as an **allele-specific oligonucleotide (ASO)** will hybridize only with its complementary sequence, and not with other sequences that might vary by only a single nucleotide. When coupled with PCR amplification, the method can be used to detect situations in which one or a small number of point mutants are responsible for causing a disorder. A newer method based on ASOs and PCR analysis has been developed to screen for genotypes related to sickle-cell anemia. In this method, white blood cells are lysed and heated to denature the DNA, and the beta globin gene in the crude cell lysate is amplified by PCR.

An aliquot of the amplified DNA is spotted on nitrocellulose filters, and each filter is hybridized to an ASO (Figure 13.11). The genotype can be read directly from the

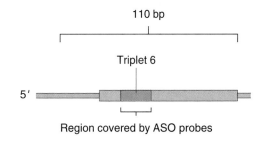

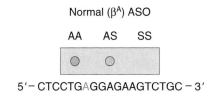

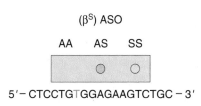

**FIGURE 13.11**    Genotype determination using allele-specific oligonucleotides (ASOs). In this technique, DNA extracted from leukocytes is denatured and amplified by PCR. Aliquots of the amplified DNA are spotted onto strips of DNA binding filters. Each strip is hybridized to a specific ASO. The patterns expected for each genotype can be read directly from the filter. *AA*—homozygous normal hemoglobin; *AS*—heterozygous sickle-cell anemia; and *SS*—homozygous for sickle-cell anemia.

filters. This rapid, inexpensive, and highly accurate technique will probably be the method of choice for diagnosis of a wide range of genetic disorders caused by point mutations.

ASOs can also be used to conduct screening for genetic disorders, where the nucleotide sequence of the normal gene is known and the molecular nature of the mutant gene is known. For example, about 70 percent of cystic fibrosis mutations are caused by a three-nucleotide deletion called delta 508, causing the deletion of phenylalanine at position 508 in the gene product, a protein called the **cystic fibrosis transmembrane conductance regulator (CFTR)**. To detect heterozygote carriers for this CF mutation, nucleotide primers are used to make allele-specific oligonucleotides by PCR from the normal allele and the mutant allele. DNA samples prepared from white blood cells of individuals to be tested are applied to a nylon or nitrocellulose filter and hybridized to each of the ASOs (Figure 13.12). In affected individuals, only the ASO made from the mutant allele will hybridize. In heterozygotes, both ASOs will hybridize, and in normal homozygotes, only the ASO from the normal allele will hybridize.

Using ASOs from the 508 deletion and two other mutations, heterozygotes for 85 to 95 percent of all CF mutations can be detected. Because CF affects approximately 1 in 2000 individuals of European descent, the availability of this technique has led to calls for population screening to detect and advise heterozygote carriers of their status for CF. However, the cost-effectiveness of undertaking such screening at the present time is in question. A negative result does not eliminate possible carrier status, since not all mutations can be assayed, and it is possible that not all mutations of the CF gene are known. Such screening will probably be commonplace once screening tests can cover 98 to 99 percent of all possible CF mutations.

# GENE THERAPY

The ability to isolate and clone specific human genes, originally developed as a research tool, is now being used in medicine to treat inherited disorders by replacing defective genes with copies of normal genes. This process, known as *gene therapy*, is at the heart of a new set of medical treatments for an ever-increasing range of diseases and disorders. Several methods have been developed that allow the delivery of genes into human cells. These include the use of viruses as vectors to deliver genes to cells, chemically assisted transfer that promotes the transport of genes across cell membranes, and the fusion of cells with artificially constructed vesicles

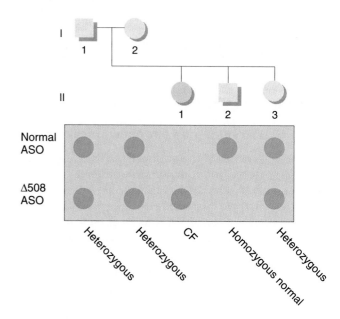

**FIGURE 13.12**    Screening for cystic fibrosis (CF) by allele-specific oligonucleotides (ASOs). ASOs for the region spanning the most common mutation in CF, a three-nucleotide deletion (D508), are prepared from normal CF genes and D508 CF genes. In screening, DNA samples from a family with an affected member are amplified by PCR and spotted on a DNA-binding membrane. The membrane is hybridized to each of the ASOs, and the genotypes can be read from the filter. DNA from individual I-1 hybridizes to both ASOs, indicating that he carries a normal allele and a mutant allele and is heterozygous. The DNA from individual II-1 hybridizes only to the D508 ASO, indicating that she is homozygous for the mutation. The DNA from II-2 hybridizes only to the normal ASO, indicating that he carries two normal alleles.

containing cloned DNA sequences. Gene therapy relies on the use of cloned DNA to treat specific diseases.

The number of gene therapy trials is growing rapidly. Since 1992, more than a dozen different gene therapy protocols have been proposed. This experimental form of treatment is not undertaken without extensive review. At the national level, there are several levels of review at the National Institutes of Health (NIH). Proposals are reviewed by panels composed of scientists, lawyers, ethicists, and others who review and approve the trials. Finally, the director of NIH must approve the procedure. At the local level, a review board at the medical center or hospital where the gene therapy will take place must review and approve the proposal and monitor the trial to protect the interests of the patients.

The guidelines for gene therapy are well established and include several requirements. First, the gene must be isolated and available for transfer, usually through cloning. Second, there must be an effective means of transferring the gene. At present, this involves the use of retroviral vectors, although other methods, including adenovirus vectors and physical and chemical techniques, are under investigation. Third, the target tissue must be accessible for gene transfer. The first gene therapy trials have used white blood cells or their precursors as an available target tissue, although newer trials are using liver cells as targets. Fourth, there must be no other form of effective therapy available, and the gene therapy must not harm the patient. For retroviral vectors, one concern is achieving an appropriate level of gene expression. For recessive disorders, 1–50 percent of normal expression may be enough to produce a therapeutic effect, but in some cases, overexpression of the transferred gene could cause problems.

Several disorders are currently being treated with gene therapy, including severe combined immunodeficiency and familial hypercholesterolemia, while others such as cystic fibrosis and muscular dystrophy will soon be treated by gene therapy.

## Retroviral Vectors and ADA Deficiency

**Severe combined immunodeficiency (SCID)** is a rare autosomal recessive disorder in which affected individuals have no functioning immune system and usually die from otherwise minor infections (see the "boy in the bubble" described in Chapter 22). One form of SCID is caused by a defect in a gene that encodes the enzyme **adenosine deaminase (ADA)**. A young girl affected with this form of SCID was the first patient to ever receive gene therapy. This treatment, begun in 1990, starts with the isolation of specialized white blood cells called **T cells** (Figure 13.13). These cells are mixed with a genetically modified virus that carries a normal copy of the human ADA gene. The virus infects the T cells, inserting a functional copy of the ADA gene into the T cell's genome (Figure 13.13). The modified T cells are grown in the laboratory to ensure that the gene is active, and the patient is treated by periodically re-implanting a billion or so genetically altered T cells into the body. Results on a small number of SCID patients have been encouraging, and most are able to attend school and have been ex-

posed to common childhood diseases, such as chicken pox, which otherwise might be fatal.

Another genetic disorder being treated by gene therapy is familial hypercholesterolemia. This relatively common autosomal dominant condition is characterized by the inability to properly metabolize dietary fats and results in elevated levels of blood cholesterol, increased deposition of cholesterol in arterial plaques, and premature death from coronary heart disease.

For gene therapy, a section of liver is removed from a person with hypercholesterolemia and separated into individual cells. A genetically modified mouse virus, carrying a human gene for cell-surface cholesterol receptors, is used to infect the liver cells. The treated cells are then injected into the patient, using a vein that supplies the liver. The injected cells are carried through the vein into the liver where they take up residence and begin to metabolize cholesterol, reducing blood levels of cholesterol to the normal range.

## New Vectors and Target Cell Strategies

The first round of gene therapy trials used genetically altered retroviruses as vectors for gene transfer. These vectors were constructed to allow the virus to retain its ability to infect cells, but not replicate after introduction into the body. Although retroviral vectors are effective, they have several drawbacks that limit their widespread use in gene therapy. First, integration of the retroviral genome (including the cloned human gene) into the host cell genome only occurs if the cells are replicating their DNA. Under *in vivo* conditions, there is little DNA synthesis in many highly differentiated cell types that might be target tissues. Second, insertion of viral genomes into the host chromosome can inactivate or mutate an indispensable gene. Third, retroviruses have a low cloning capacity and cannot carry inserted sequences much larger than 8 kb. Many human genes, even without introns, exceed this size. Finally, there is the possibility of producing an infectious virus by recombination between the vector and retroviral sequences already present in the host genome. For these reasons, other viral vectors and strategies for targeting cells are being developed. Some of the second generation vectors are listed in Table 13.2.

Vectors made from adenoviruses have the advantage of being able to transform cells in the absence of cell division. This is important since most differentiated cells are nondividing. In addition, adenovirus vectors have specific targets such as lung cells and are being used in trial therapies for cystic fibrosis and emphysema caused by alpha-1-antitrypsin deficiency. At the moment, the central issue concerning the use of adenovirus vectors is the lack of information about safety. Among the dangers

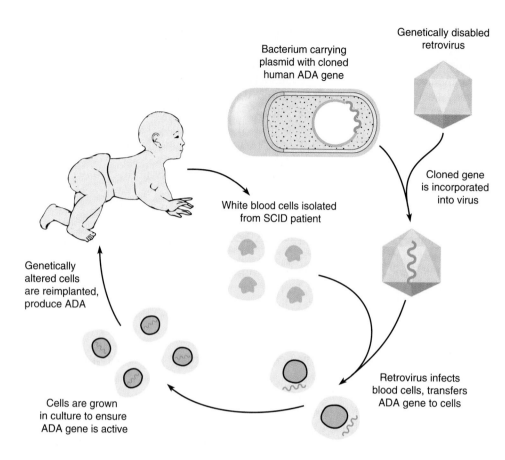

Bacterium carrying
plasmid with cloned
human ADA gene

Genetically disabled
retrovirus

White blood cells isolated
from SCID patient

Cloned gene
is incorporated
into virus

Genetically
altered cells
are reimplanted,
produce ADA

Retrovirus infects
blood cells, transfers
ADA gene to cells

Cells are grown
in culture to ensure
ADA gene is active

**FIGURE 13.13** Gene therapy for treatment of severe combined immunodeficiency (SCID), a fatal disorder of the immune system, caused by lack of the enzyme adenosine deaminase (ADA). The cloned human ADA gene is transferred into a viral vector, which is used to infect white blood cells removed from the patient. The transferred ADA gene is incorporated into a chromosome, and becomes active. After growth to enhance their numbers, the cells are reimplanted into the patient, where they produce ADA, allowing the development of an immune response.

is that co-infection of a genetically engineered adenovirus along with an intact adenovirus might spread the transferred gene to cells in nontarget tissues.

Adeno-associated viruses (AAV) offer the advantage of integrating into the genome at highly selective sites, but can transfer only small amounts of human DNA. In addition, it is difficult to purify AAV to eliminate contaminating adenoviruses. Further development of these vectors will be needed before they are used in clinical trials.

Vectors based on herpes viruses are being developed in order to selectively target cells of the nervous system. Since these cells do not divide, it is not necessary for this

vector to integrate into the host genome in order to be retained in cells. This will prevent the danger of insertional mutagenesis associated with the use of retroviruses.

In addition to the development of new viral vectors, other approaches are also being tested for use in gene therapy. One of the problems in designing gene therapy for the X-linked disorder muscular dystrophy is that cloning an intact gene into a vector for delivery to cells is very difficult because the gene is so large. One alternative approach to gene therapy by vector delivery systems is now being tested for the treatment of muscular dystro-

| **Table 13.2** NEW VIRAL VECTORS FOR GENE THERAPY | | | |
|---|---|---|---|
| | **Adenovirus** | **Adeno-Associated Virus** | **Herpes Virus** |
| CELL TARGETS | Lung cells, cells of respiratory tract | Fibroblasts, T cells | Nerve cells, glial cells |
| CLONING CAPACITY | 7–35 kb | 3 kb | Up to 150 kb |
| INTEGRATION INTO CHROMOSOMAL DNA | No | Yes | No, but retained in nucleus |

phy. Muscle fibers are multinucleate synctia formed by the fusion of embryonic precursor cells known as myoblasts. Experiments using a mouse model for muscular dystrophy have shown that normal myoblasts injected into muscles of mutant mice result in the production of dystrophin-positive muscle fibers.

Although these results are encouraging, it seems unlikely that muscle cell therapy will become the primary method for treatment of muscular dystrophy. Aside from problems of scaling the procedure up from the mouse to humans, myoblast fusion does not correct the lack of dystrophin in cardiac muscle. Progressive loss of heart muscle fibers is a leading cause of death in muscular dystrophy. Consequently, to be effective, therapy must treat both skeletal and cardiac muscle. In addition, there is the problem of an unfavorable immune response to injected myoblasts. In spite of these difficulties, such work holds the possibility that direct gene transfer to muscle cells may be a viable method of gene therapy for muscle-related genetic disorders.

Gene therapy is also being used or contemplated as a treatment for skin cancer, breast cancer, brain cancer, and AIDS. In the last few years, at least a dozen biotech companies have been founded specifically to develop products for gene therapy, and a number of such products should reach the marketplace beginning in 1995. In the twenty-first century, gene therapy will be a commonplace method of treatment for a large number of disorders.

# DNA FINGERPRINTS

Restriction fragment length polymorphisms have been used to distinguish normal from mutant copies of single genes and can be used as genetic markers, since they are inherited in a codominant fashion. The phenotype of these markers is an array of DNA fragments of different sizes present on a Southern blot after the DNA has been cut with a restriction enzyme. A second type of restriction length polymorphism arises from variations in the number of tandemly repeated DNA sequences present between two restriction enzyme sites. These sequences are derived from **minisatellites**, repeat sequences from 2 to 20 nucleotides in length. For example, the base sequence

GGAAGGGAAGGGAAGGGAAG

consists of four tandem repeats of a 5-nucleotide sequence (GGAAG). Such groups of sequences are widely dispersed in the human genome, and the number of alleles at these loci vary from 2 to more than 20. These loci are known as **variable-number tandem repeats** (VNTRs).

The patterns of bands produced when VNTR sequences are cut with restriction enzymes and visualized by Southern blotting are known as **DNA fingerprints** (Figure 13.14). These RFLPs are the equivalent of fingerprints because the pattern of bands generated varies from one individual to another, but is always the same

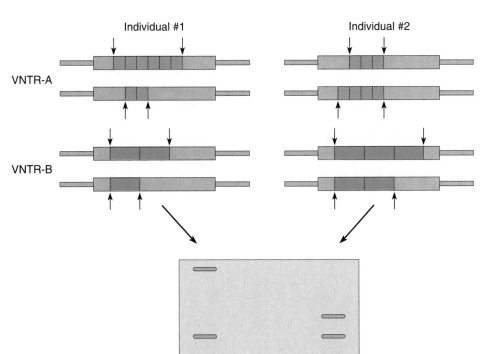

**FIGURE 13.14** VNTR loci and DNA fingerprints. VNTRs at two loci are shown for each individual. Arrows mark restriction cutting sites flanking the VNTRs. Restriction digestion produces a series of fragments that can be detected as bands on a Southern blot (below). Because of differences in the number of repeats to each locus, the overall pattern of bands is distinct for each individual, even though one band is shared. Such a pattern is known as a DNA fingerprint.

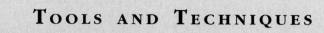

# TOOLS AND TECHNIQUES

## Reporter Genes

Transcriptional regulation of gene expression depends on the action and interaction of DNA-binding proteins with specific DNA sequences called promoters and enhancers. Promoters are binding sites adjacent to the protein coding sequences of a gene and mark the beginning of transcription; enhancers are usually found some distance away on the same DNA strand and regulate the rate of transcription. To understand the molecular basis of this regulation, especially tissue-specific regulation, researchers have developed recombinant DNA techniques that enable them to isolate and analyze the action of promoters and enhancers. One of the more informative approaches is the use of recombinant DNA constructs known as "reporter genes."

Reporter genes code for proteins that are not normally found in the cell under study and can be quantitatively detected in very sensitive and reliable assays. In constructing a reporter gene, DNA carrying the regulatory sequence (promoter or enhancer) to be studied is ligated to a reporter gene, and the expression of the reporter gene under various experimental conditions is monitored.

Reporter gene vectors consist of plasmids carrying reporter genes, into which potential regulatory DNA sequences are inserted. These vector constructs are then introduced into a host such as cells in culture or intact organisms (such as *Drosophila*). If the inserted DNA contains a regulatory element, the reporter gene will be expressed, and its protein product can be detected. When a regulatory element, say an enhancer, has been identified, its location can be further refined by linking smaller and smaller pieces of DNA to the reporter gene. Once its position has been established, its function can be determined by introducing nucleotide substitutions or deletions into the regulatory sequence and testing for altered patterns of reporter gene expression. Although this is an indirect way to measure transcriptional regulation of a particular gene, important conclusions can be inferred from these tests.

A variety of reporter gene vectors have been developed that can be adapted for different types of studies, and several of these are now commercially available. The following section describes a few of the most commonly used reporter genes.

The chloramphenicol acetyltransferase (CAT) gene of *E. coli* codes for an enzyme not normally found in mammalian cells. When expressed as a reporter gene in mammalian cells, the enzyme can be extracted and analyzed in a sensitive assay. A second *E. coli* gene, beta galactosidase, is also used as a reporter gene. The expression of this gene can be determined by a staining reaction in which cells producing the enzyme turn blue. This reporter gene can be used to detect tissue-specific regulation and it is used as the reporter gene in enhancer detection vectors (see Tools and Techniques in Chapter 21). More recently, the luciferase gene from the firefly has been used as a reporter gene. This gene encodes an enzyme that acts on the substrate luciferin to produce a flash of light. The light emission is proportional to the amount of luciferase present and can be measured with an instrument called a luminometer. Because the substrate luciferin can enter intact cells, expression of this reporter gene can be accomplished without preparing cell extracts.

The use of reporter gene technology has enabled researchers to identify and characterize the regulatory sequences adjacent to many different genes in a variety of organisms. These studies show that transcriptional regulation is an exceedingly complex process involving several different sequences within a regulatory element. For example, a promoter region can contain sequences that determine a baseline level of expression, sequences that are necessary for induced expression, and sequences that can repress or switch off expression. In addition to studies on transcriptional regulation, reporter genes can be used in studies of *in vivo* mutagenesis, protein export, and cellular differentiation.

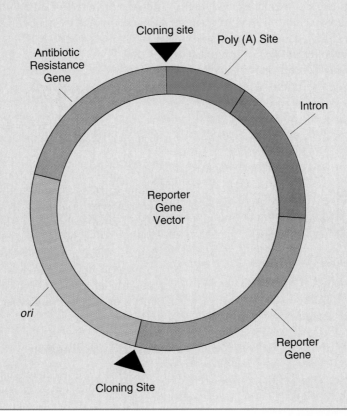

for a given individual, no matter what tissue is used as the source of DNA. In addition, there is so much variation from individual to individual in the band pattern that theoretically each person's pattern is unique, like the individualized pattern in a fingerprint (Figure 13.14). DNA fingerprint analysis can be performed on very small samples of material [less than 60 microliters ($\mu$l) of blood] and on samples that are old (RFLP analysis has been performed on Egyptian mummies over 2400 years old), increasing its usefulness in legal cases.

## Forensic Applications

In the United States, DNA fingerprints have been used as evidence in criminal trials since 1988 and have also been used to settle immigration cases, disputes involving purebred dogs, paternity cases (Figure 13.15), and in animal conservation studies. The use of DNA fingerprints in criminal trials has been the subject of some dispute, however. The controversy centers on two facets of the method, one technical, and the other scientific. On the technical side, there has been little standardization of the methods used to perform DNA fingerprinting, and the quality control in some commercial laboratories has been called into question. Scientifically, the method depends on showing that the pattern of any individual is unique enough that he or she is the sole source of such a pattern among individuals in a population. Such statements can be made only when information is available about the frequency of specific RFLPs in various populations. A 1992 report by the National Academy of Sciences addressed these issues and generally supported the use of DNA fingerprints in criminal trials. At this time, there is still some lingering controversy about the use of DNA fingerprints in criminal trials. However, in the meantime, several states are establishing DNA fingerprint files on those convicted of certain felonies and are using DNA fingerprints as evidence in trials.

## BIOTECHNOLOGY

Although recombinant DNA techniques were originally developed to facilitate basic research into gene organization and regulation of expression, scientists have not been blind to the commercial possibilities of this technology. As a result, scientists have participated in the formation of biotechnology companies to exploit recombinant DNA technology. The commercial production of human gene products, such as hormones and clotting factors, from sequences cloned into bacterial cells are examples of the commercial use of recombinant DNA. In the last decade, the biotechnology industry has grown into a multibillion dollar segment of the economy.

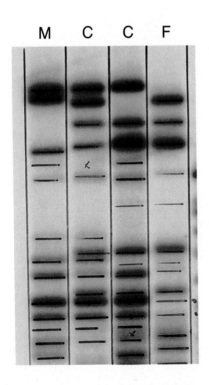

M    C    C    F

**FIGURE 13.15**    DNA fingerprinting in a family consisting of mother (M), child (C), child (C), and father (F). The bands represent DNA fragments separated by gel electrophoresis. Note that the pattern of banding in the children is made up of bands from one or the other or both parents.

## Insulin Production

The first recombinant human gene product licensed for therapeutic use was human insulin, which became available in 1982. Insulin is a protein hormone that regulates sugar metabolism, and an inability to produce insulin results in diabetes, a disease that in its more severe form affects more than 2 million individuals in the United States. Beginning in the 1930s, this protein was extracted from pigs and cows and used for the treatment of diabetes.

In humans, insulin is synthesized in cells embedded in the pancreas as a precursor peptide known as **preproinsulin.** During the process of secretion, extra amino acids are cleaved from the end and from the middle of the chain to produce the mature insulin molecule, consisting of two polypeptide chains (the A and B chains), joined by disulfide bonds. The initial method for producing insulin by recombinant DNA methods is instructive, as it shows both the promises and difficulties of this technology.

First, oligonucleotides were used to construct synthetic genes for the A and B subunits (the A subunit has 21 amino acids, and the B subunit has 30). Each was inserted into a vector at a position adjacent to a bacterial gene encoding beta galactosidase. Thus, the bacterial promoter for beta galactosidase transcribes a fusion gene product consisting of beta galactosidase and one of the insulin subunits (Figure 13.16). The fusion proteins are purified from bacterial extracts and treated with cyanogen bromide, which cleaves proteins adjacent to methionine residues. The fusion gene was engineered to insert

a methionine at the junction between beta galactosidase and the insulin subunit, so that an intact insulin subunit is released by this chemical treatment. The insulin subunits are then isolated, purified, and mixed together to form an active insulin molecule.

A number of genetically engineered proteins for therapeutic use have been produced by similar methods or are in clinical trials (Table 13.3). In most cases, these human proteins are produced by cloning a human gene into a plasmid vector and inserting the construct into a bacterial host. After ensuring that the transferred gene is active, large quantities of the transformed bacteria are produced, and the human protein is recovered and purified.

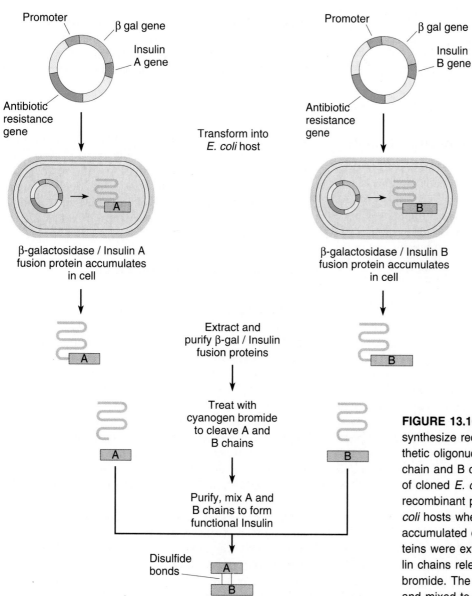

**FIGURE 13.16**    Original method used to synthesize recombinant human insulin. Synthetic oligonucleotides encoding the insulin A chain and B chain were inserted at the tail end of cloned *E. coli* β-galactosidase genes. These recombinant plasmids were transferred to *E. coli* hosts where the β-gal/insulin fusion protein accumulated during growth. The fusion proteins were extracted and purified and the insulin chains released by treatment with cyanogen bromide. The insulin subunits were purified and mixed to produce a functional insulin molecule.

**Table 13.3** GENETICALLY ENGINEERED PHARMACEUTICAL PRODUCTS AVAILABLE OR IN CLINICAL TESTING

| Gene Product | Condition Being Treated |
| --- | --- |
| Atrial natriuretic factor | Heart failure, hypertension |
| Epidermal growth factor | Burns, skin transplants |
| Erythropoietin | Anemia |
| Factor VIII | Hemophilia |
| Gamma interferon | Cancer |
| Granulocyte colony-stimulating factor | Cancer |
| Hepatitis B vaccine | Hepatitis |
| Human growth hormone | Dwarfism |
| Insulin | Diabetes |
| Interleukin-2 | Cancer |
| Superoxide dismutase | Transplants |
| Tissue plasminogen activator | Heart attacks |

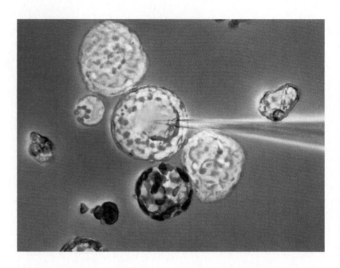

**FIGURE 13.17**   Injection of genetic material. A micropipette is used to transfer genetic material into the nucleus of a mammalian zygote. The injected zygote will then be transferred to the uterus of a surrogate mother for development.

## Pharmaceutical Production in Animal Hosts

Although the use of bacterial hosts has served to produce the first generation of recombinant proteins, there are some disadvantages in using prokaryotic hosts to synthesize eukaryotic proteins. For example, bacterial cells are unable to process and modify eukaryotic proteins, and cannot add the sugars and phosphate groups needed for full biological activity. In addition, eukaryotic proteins produced in prokaryotic cells often do not fold into the proper three-dimensional configuration, and as a result, are inactive. To overcome these difficulties and to increase yields, the next generation of human proteins are being produced in eukaryotic hosts. As an alternative to bacterial hosts or even mammalian cell cultures, human proteins such as alpha-1-antitrypsin are being produced in the milk of livestock, as described below.

A deficiency in alpha-1-antitrypsin is associated with the heritable form of emphysema, a progressive and fatal respiratory disorder common among those of European ancestry. To produce alpha-1-antitrypsin by genetic engineering, the human gene was cloned into a vector at a site adjacent to a sheep DNA sequence that regulates expression of milk-associated proteins. This sequence, called a promoter, limits expression of its adjacent gene to mammary tissue. This fusion gene was then microinjected into *in vitro* fertilized sheep oocytes (Figure 13.17), which in turn were implanted into foster mothers. The resulting transgenic sheep developed normally and, after mating, produced milk that contained high concentrations of functional alpha-1-antitrypsin. This

human protein is present in concentrations up to 35 grams/liter of milk, or about $\frac{1}{3}$ pound of human protein per gallon. While this method is not yet used in commercial production, it is easy to envision that a small herd of lactating sheep could easily provide an adequate supply of this protein and that herds of other transgenic animals, acting as biofactories, might become part of the pharmaceutical industry.

## Herbicide-Resistant Crop Plants

In agriculture, gene transfer vectors have been used to transfer traits such as herbicide resistance to crop plants. One herbicide, glyphosate, is widely used for weed control, but cannot be used on fields containing crop plants because it kills them along with the weeds. Weed growth is a serious agricultural problem and damage from weeds reduces the yield on many crops by more than 10 percent. As a herbicide, glyphosate is effective at very low concentrations, is not toxic to humans, and is rapidly degraded by soil microorganisms. At the molecular level, glyphosate works by inhibiting the action of a chloroplast enzyme called EPSP synthase, active in amino acid synthesis. Without these vital amino acids, plants wither and die.

One way of generating glyphosate resistance is by increasing the synthesis of EPSP synthase. To do this, a fusion gene was created by cloning the EPSP synthase gene into a vector under control of a promoter sequence from a plant virus. This fusion gene was put into a Ti vector, which was in turn transferred into the bacterium *A. tumifaciens* (Figure 13.18). Plasmid-carrying bacteria

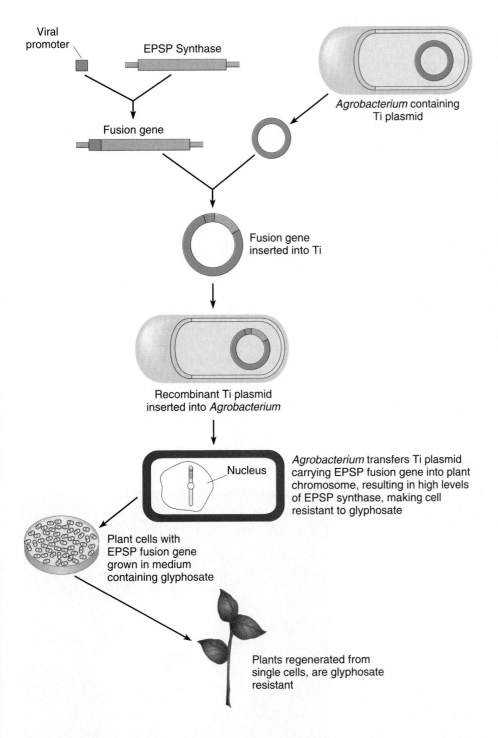

Viral promoter

EPSP Synthase

Fusion gene

*Agrobacterium* containing Ti plasmid

Fusion gene inserted into Ti

Recombinant Ti plasmid inserted into *Agrobacterium*

Nucleus

*Agrobacterium* transfers Ti plasmid carrying EPSP fusion gene into plant chromosome, resulting in high levels of EPSP synthase, making cell resistant to glyphosate

Plant cells with EPSP fusion gene grown in medium containing glyphosate

Plants regenerated from single cells, are glyphosate resistant

**FIGURE 13.18** Gene transfer of glyphosate resistance. The EPSP gene is fused to a promoter from cauliflower mosaic virus. This chimeric or fusion gene is then transferred to a Ti plasmid vector, and the recombinant vector is inserted into an *Agrobacterium* host. Agrobacterium infection of cultured plant cells transfers the EPSP fusion gene into a plant cell chromosome. Cells that acquire the gene are able to synthesize large quantities of EPSP synthase, making it resistant to the herbicide glyphosate. Resistant cells are selected by growth in herbicide-containing medium. Plants regenerated from these cells are herbicide-resistant.

were used to infect cells in discs cut from plant leaves. Calluses formed from these discs were then selected for their ability to grow on glyphosate. Transgenic plants generated from glyphosate-resistant calluses were grown and sprayed with glyphosate at concentrations four times higher than needed to kill wild-type plants. The transgenic plants overproducing the EPSP synthase grew and developed, while the control plants withered and died. Similar methods have been used to transfer resistance to virus infection, insects, and drought to crop plants. Other work has been directed at improving the nutritional value of crops such as soybeans and corn. Many of these projects are in the developmental stage, but one of the first genetically engineered plants that will soon reach the marketplace is a transgenic tomato that has improved flavor and ripening characteristics. Other transgenic products will reach the marketplace over the next few years.

## Transgenic Plants and Vaccines

One of the potentially most valuable applications of recombinant DNA technology is the production of vaccines. Vaccines act as antigens to stimulate the immune system to produce antibodies against a disease-causing organism and thereby confer immunity against the disease. Two types of vaccines are commonly used: **inactivated vaccines**, which are prepared from killed samples of the infectious virus or bacteria, and **attenuated vaccines**, which are live viruses or bacteria but are modified so that they no longer can reproduce and cause disease when present in the body.

Using recombinant DNA technology, a new type of vaccine called a **subunit vaccine** is being produced. These vaccines consist of one or more surface proteins of the virus or bacterium that acts as antigens to stimulate the immune system. At present, the only licensed subunit vaccine is one for hepatitis B, a virus that causes liver damage and cancer (see Chapter 18 for a discussion of hepatitis B). The gene for the hepatitis B surface protein was cloned into a yeast expression vector and is grown in commercial quantities using yeast as a host.

In research efforts currently under way, recombinant DNA technology is being used to produce the hepatitis B surface protein in plants, to develop plant leaves or fruits as the source of oral vaccines that can be eaten instead of injected. Such plant-produced vaccines would be inexpensive, since extensive purification and industrial investment would be unnecessary. Further, such vaccines would not require injection and might be most useful in developing countries where medical services and delivery of health care is not highly developed. Recently, the antigenic subunit of hepatitis B vaccine has been transferred to tobacco plants and expressed in the leaves (Figure 13.19). The tobacco plant is being used as a host system for the *Agrobacterium*-based vectors described in Chapter 12. For use as a source of vaccine, the gene would be inserted into food plants such as grain or vegetable plants.

The hepatitis B surface protein produced in the plant forms aggregates with phospholipids and is similar if not identical to the antigens present in the serum of people infected with hepatitis B, indicating that the plant-produced antigen may stimulate the immune system and confer immunity. Other experiments are directed at producing an edible vaccine against cholera toxin in alfalfa plants. In these experiments, the antigenic B subunit (Figure 13.20) of the bacillus *Vibrio* has been cloned and inserted into alfalfa plants. Experiments are now under way to determine whether the antigens produced by genetically engineered plants will induce antibody production when ingested, using mice as a test system. If these tests are successful, there are still formidable hur-

**FIGURE 13.19**    Tobacco leaves expressing the hepatitis B antigen. Transgenic tobacco plants carrying an antigenic subunit of hepatitis B virus were generated. Leaves from transgenic plants were treated with antibodies against hepatitis B antigen, showing that the plants produced the antigen. The unstained central region of the photograph contains a leaf from a normal tobacco plant, which is unstained.

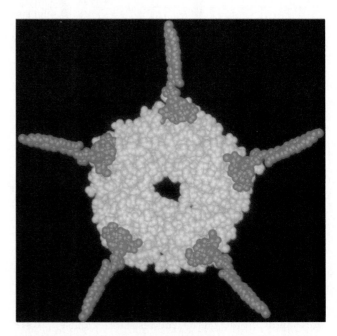

**FIGURE 13.20**    A molecular model of the cholera toxin B subunit. The gene encoding this subunit has been transferred to alfalfa plants to see if the cholera B protein will induce antibody formation when eaten, making possible an edible vaccine against cholera.

dles to overcome before we can be vaccinated against disease by eating our vegetables, but there is widespread enthusiasm for this approach.

As discussed in Chapter 1, plants and animals were domesticated some 8000 to 10,000 years ago, and by selective breeding we have been modifying these organisms ever since, producing the diversity of domesticated plants and animals present today. The development of recombinant DNA technology has changed the rate at which new plants and animals can be developed, and to some degree, altered the kind of changes that can be made. Unlike selective breeding programs, biotechnology has generated concerns about the release of genetically modified organisms into the environment, and about the safety of eating such products. If biotechnology is to achieve a new green revolution, these concerns need to be addressed through prudent research and education.

## CHAPTER SUMMARY

1. Recombinant DNA technology offers the geneticist a new approach to the problem of genetic analysis. Instead of relying on isolation and mapping of mutant genes, it is now possible to begin by cloning the entire genome and by manipulating large cloned segments of the genome to establish genetic and physical maps using molecular markers instead of phenotypes visible at the level of the organism.

2. Genome projects are under way for several organisms, including *E. coli*, yeast, *Drosophila*, the mouse, and humans. The goals of these genome projects are to identify and map all genes in the genome and to determine the complete nucleotide sequence of the genome.

3. Cloned DNA is being used in a wide variety of applications, including gene mapping and the identification and isolation of genes responsible for genetic disorders. The method known as positional cloning, based on recombinant DNA methodology, allows the mapping and identification of a gene without any knowledge of the nature or function of the gene product.

4. Recombinant DNA techniques are being used in the prenatal diagnosis of human genetic disorders. This method allows direct examination of the genotype, whereas previous methods relied on gene expression and the identification of the gene product. These methods can also be used to identify carriers of genetic disorders and are the basis of proposals to screen the population for a number of genetic disorders, including sickle-cell anemia and cystic fibrosis.

5. The availability of cloned human genes has led to their use in replacement of mutant genes in somatic tissues. This somatic gene therapy is carried out by transferring a cloned normal copy of a gene into a vector and using the vector as a means of transferring the gene to a target tissue that takes up and expresses the cloned copy of the gene, altering the mutant phenotype. The development of new and more effective vector systems probably means that gene therapy will become a standard method for the treatment of genetic disorders in the near future.

6. The use of recombinant DNA techniques in detecting allelic variants of variable tandem nucleotide repeats (DNA fingerprints) has found applications in forensics, although there are issues yet to be resolved before this method is universally accepted as evidence in criminal cases.

7. The biotechnology industry is using recombinant DNA methods to produce human gene products in a variety of hosts, ranging from bacteria to farm animals. In addition, gene transfer techniques are being used to improve crop plants by transfer of herbicide resistance and to improve produce such as tomatoes. In the near future, it may be possible to offer vaccination against infectious disease through food plants, making resistance to infectious agents almost universal.

## KEY TERMS

adenosine deaminase
  (ADA)
allele-specific
  oligonucleotide (ASO)
amniocentesis
attenuated vaccine
centimorgan (cM)
chorionic villus sampling
  (CVS)

contig
cystic fibrosis
  transmembrane
  conductance regulator
  (CFTR)
DNA fingerprints
dystrophin
exclusion map
Human Genome Project

inactivated vaccine
logarithm of the odds
  (lod) score
minisatellite
P element
preproinsulin
restriction fragment
  length polymorphism
  (RFLP)

sequence tagged site
  (STS)
severe combined
  immunodeficiency
  (SCID)
subunit vaccine
T cell
variable-number tandem
  repeat (VNTR)

## INSIGHTS AND SOLUTIONS

1.  DNA fingerprints have been used to test forensic specimens, to identify criminal suspects, and to settle immigration cases and paternity disputes. Probes for DNA fingerprinting can be derived from a single locus or multiple loci. Two multiple-loci probes have been widely used in both criminal and civil cases and derive from minisatellite loci on chromosome 1 (1cen-q24) and chromosome 7 (7q31.3). These probes are widely used because they produce a highly individual fingerprint. These probes have been used to determine paternity in thousands of cases over the last few years.

    (a)  Shown below are the results of DNA fingerprinting of a mother (M), putative father (F), and child (C) using these probes.

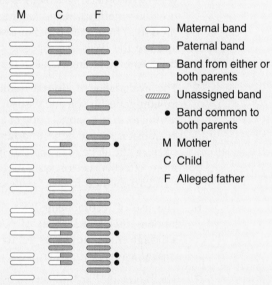

As shown, the child has 6 maternal bands, 11 paternal bands, and 5 bands shared between the mother and the alleged father. Based on this fingerprint, can you conclude that the man tested is the father of the child?

**SOLUTION:** All bands present in the child can be assigned as coming from either the mother or the father. In other words, all the bands in the child's DNA fingerprint that are not maternal are present in the father. Because the father and the child share 11 bands in common and the child has no unassigned bands, paternity can be assigned with confidence. In fact, the chance that the man tested is not the father is on the order of $10^{-13}$.

(b) In a second case, the fingerprints of the mother, the alleged father, and the child are shown below.

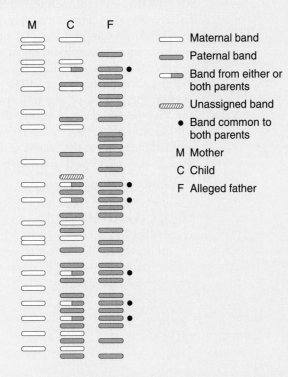

In this case, the child has 8 maternal bands, 15 paternal bands, 6 bands that are common to both the mother and the alleged father, and 1 band that is not present in either the mother or the alleged father. What are the possible explanations for the presence of this band? Based on your analysis of the band pattern, which explanation is most likely?

**SOLUTION:** In this case, one band in the child cannot be assigned to either parent. Two possible explanations for this finding are that the child is mutant for one band, or that the man tested is not the father. In estimating probability of paternity, the mean number of resolved bands ($n$) is determined, and the mean probability ($x$) that a band in individual A matches that in a second unrelated individual, B, is calculated. Since in this case the child and the father share 15 bands in common, the probability that the man tested is not the father is very low (probably $10^{-7}$ or lower). As a result, the most likely explanation is that the child is a mutant for a single band. In fact, in 1419 cases of genuine paternity resolved by the minisatellite probes on chromosomes 1 and 7, single-mutant bands in the children were recorded in 399 cases, accounting for 28 percent of all cases.

2. Infection by HIV-1 (human immunodeficiency virus) is responsible for destruction of cells in the immune system and results in the symptoms of AIDS (acquired immunodeficiency syndrome). HIV infects and kills cells of the immune system that carry a cell surface receptor known as CD4. An HIV surface protein known as gp120 binds to the CD4 receptor and allows the virus entry into the cell. The gene encoding the CD4 protein has been cloned. How can this clone be used along with recombinant DNA techniques to combat HIV infection?

**SOLUTION:** Several methods utilizing the CD4 gene are being explored to combat HIV infection. First, since infection depends on an interaction between the viral gp120 protein and the CD4 protein, the cloned CD4 gene has been modified to produce a soluble form of the protein (sCD4). The idea is that HIV can be prevented from infecting cells if the gp120 protein of the virus is bound up with the soluble form of the CD4 protein, and thus unable to bind to CD4 proteins on the surface of immune system cells. Studies in cell culture systems indicate that the presence of sCD4 effectively prevents HIV infection of tissue culture cells. However, studies in HIV-positive humans has been somewhat disappointing, mainly because the strains of HIV used in the laboratory are different from those found in infected individuals.

HIV-infected cells carry the viral gp120 protein on their surface. To kill such cells, the CD4 gene has been fused with those encoding bacterial toxins. The resulting fusion protein contains the CD4 regions that bind to gp120 and the toxin regions that kill the infected cell. In tissue culture experiments, cells infected with HIV are killed by the fusion protein, while noninfected cells survive. It is hoped that the targeted delivery of drugs and toxins can be used in therapeutic applications to treat HIV infection.

**PROBLEMS AND
DISCUSSION
QUESTIONS**

1. In attempting to vaccinate against diseases by eating antigens, the antigen (such as the cholera toxin) must be presented to the cells of the small intestine. What are some potential problems in this method? Why don't absorbed food molecules stimulate the immune system and make you allergic to the food you eat?

2. Outline the steps involved in transferring glyphosate resistance to a crop plant. Do you envision that this trait can escape from the crop plant and make weeds glyphosate resistant? Why or why not?

3. Although not yet completed, what lessons from the *E. coli* genome project can be applied to the Human Genome Project?

4. Gene therapy for human genetic disorders involves transferring a copy of the normal human gene into a vector and using the vector to transfer the cloned human gene into target tissues. Presumably, the gene enters the target tissue, becomes active, and the gene product relieves the symptoms. Although now being used to treat several disorders, there are some unresolved problems with this method.
   (a) Why are disorders such as muscular dystrophy difficult to treat by gene therapy?
   (b) What are the potential problems in using retroviruses as vectors?
   (c) In the long run, should gene therapy involve germ tissue instead of somatic tissue? What are some of the potential ethical problems associated with this approach?

5. In producing physical maps of markers and cloned sequences, what advantage does *Drosophila* offer that other organisms including humans cannot?

6. Outline the steps in identifying a gene by positional cloning. What steps may cause difficulty in this process? Once a region on a chromosome has been identified as containing a given gene, what kind of mutations would speed the process of identifying the locus?

7. The phenotype of many behavior traits such as manic depression or schizophrenia may be controlled by several genes, each at a different locus. Can positional cloning be used to map and isolate such genes? What if a trait is controlled by six genes, each contributing equally in an additive way to the phenotype? Can positional cloning be used in this case? Why or why not?

8. Suppose that you develop a screening method for cystic fibrosis that allows you to identify the predominant mutation (delta 508) and the next six most prevalent mutations. What do you need to consider before using this method in screening the population at large for this disorder?

9. The DNA sequence surrounding the site of the sickle-cell mutation in the beta globin is shown for normal and mutant genes:

5' G A C T C C T GAG GAG A A G T  3'          5' G A C T C C T G T G GAG A A G T - 3'

3' C T G A G G A C T C C T C T T C A  5'          3' C T G A G G A C A C C T C T T C A - 5'

Normal DNA                                           Sickle-cell DNA

Each type of DNA is denatured into single strands and applied to a filter. The paper containing the two spots is hybridized to an ASO of the following sequence: 5' GACTCCTGAGGAGAAGT 3'. Which (if either) spot will hybridize to this probe? Why?

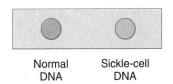

Normal          Sickle-cell
DNA               DNA

10. You are asked to help in prenatal genetic testing of a couple who have been found to be carriers for a deletion in the beta globin gene that produces beta thalassemia when homozygous. The couple already has one child who is unaffected and is not a carrier. The woman is pregnant, and they wish to know the status of the fetus. You receive DNA samples obtained from the fetus by amniocentesis and from the rest of the family by extraction from white blood cells. Using a probe that detects the deletion, the following blot is obtained (see Figure 13.11).

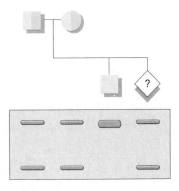

Is the fetus affected? What is its genotype for the beta globin gene?

**11.** One form of hemophilia, an X-linked disorder of blood clotting, is caused by mutation in clotting factor VIII. Many single-nucleotide mutations of this gene have been described, making detection of mutant genes by Southern blots inefficient. There is, however, an RFLP for the enzyme *Hind*III contained in an intron of the factor VIII gene that can often be used in screening, as shown below.

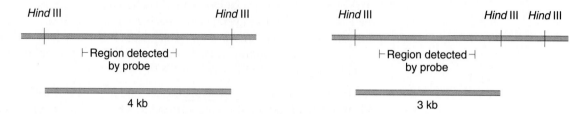

A female whose brother has hemophilia is at 50 percent risk of being a carrier of this disorder. To test her status, DNA is obtained from white blood cells from family members, cut with *Hind*III, and the fragments probed and visualized by Southern blotting. The results are shown below.

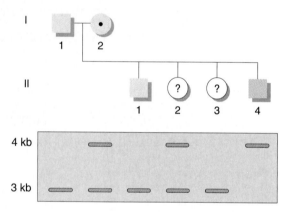

Determine whether any of the females in generation II are carriers for hemophilia.

**12.** The human insulin gene contains a number of introns. In spite of the fact that bacterial cells will not excise introns from mRNA, how can a gene like this be cloned into a bacterial cell and produce insulin?

**13.** In mice transfected with the rabbit beta globin gene, the rabbit gene is active in a number of tissues including spleen, brain, and kidney. In addition, some mice suffer from thalassemia, caused by an imbalance in the coordinate production of alpha and beta globins. What problems associated with gene therapy are illustrated by these findings?

**SELECTED READINGS**

Amos, B., and Pemberton, J. 1993. DNA fingerprinting in non-human populations. *Curr. Opin. Genet. Dev.* 2:857–860.

Anderson, W. F., and Diacumakos, E. G. 1981. Genetic engineering in mammalian cells. *Scient. Amer.* (July) 245:106–21.

Antonarakis, S. 1989. Diagnosis of genetic disorders at the DNA level. *New Engl. J. Med.* 320:153–163.

Antonarakis, S. E., et al. 1985. Hemophilia A: Detection of molecular defects and carriers by DNA analysis. *New Engl. J. Med.* 313:842–48.

Barker, D., et al. 1987. Gene for von Recklinghausen neurofibromatosis is in the pericentromeric region of chromosome 17. *Science* 236:1001–1102.

CAWTHON, R. M., et al. 1990. A major segment of the neurofibromatosis type 1 gene: cDNA sequence, genomic structure, and point mutations. *Cell* 62:193–201.

CHANG, J. C., and KAN, Y. W. 1981. Antenatal diagnosis of sickle-cell anemia by direct analysis of the sickle mutation. *Lancet*, 2, pp. 1127–29.

DANNA, K., and NATHANS, D. 1971. Specific cleavage of simian virus 40 DNA by restriction endonuclease of *Hemophilus influenzae. Proc. Nat. Acad. Sci.* 68:2913–2917.

DAVIDSON, B., ALLEN, E., KOZARSKY, K., WILSON, J., and ROESSLER, B. 1993. A model system for *in vivo* gene transfer into the central nervous system using an adenoviral vector. *Nature Genet.* 3:219–223.

ELLIS, K. P., and DAVIES, K. E. 1985. An appraisal of the application of recombinant DNA techniques to chromosomal defects. *Biochem. J.* 226:1–11.

FLOTTE, T. 1993. Prospects for virus-based gene therapy for cystic fibrosis. *J. Bioenerg. Biomembr.* 25:37–42.

FRIEDMANN, T. 1989. Progress toward human gene therapy. *Science* 244:1275–81.

GLOVER, D. M. 1984. *Gene cloning: The mechanism of DNA manipulation.* London: Chapman and Hall.

HARTL, D., and LOZOVSKAYA, E. 1992. The *Drosophila* genome project: Current status of the physical map. *Comp. Biochem. Physiol.* [B] 103:1–8.

HOPWOOD, D. A. 1981. Genetic programming of industrial microorganisms. *Scient. Amer.* (Sept.) 245:91–102.

INBAL, A., ENGLANDER, T., KORNBROT, N., RANDI, A., CASTAMAN, G., MANNUCCI, P., and SADLER, J. 1993. Identification of three candidate mutations causing type IIA von Willebrand disease using a rapid, nonradioactive, allele-specific hybridization method. *Blood* 82:830–836.

KNORR, D., and SINSKEY, A. J. 1985. Biotechnology in food production and processing. *Science* 229:1224–29.

KRUMLAUFF, R., JEANPIERRE, M., and YOUNG, B. D. 1982. Construction and characterization of genomic libraries from specific human chromosomes. *Proc. Natl. Acad. Sci.* 79:2971–75.

LEWONTIN, R., and HARTL, D. 1991. Population genetics in forensic DNA typing. *Science* 254:1745–1750.

MANIATIS, T., FRITSCH, E. F., and SAMBROOK, J. 1988. *Molecular cloning: A laboratory manual.* 2nd ed. Cold Spring Harbor, NY: Cold Spring Harbor Laboratory.

MASON, H., LAM, D., and ARNTZEN, C. 1992. Expression of hepatitis B surface antigen in transgenic plants. *Proc. Nat. Acad. Sci.* 89:11745–11749.

MOLDOVEANU, Z., NOVAK, M., HUANG, W., GILLEY, R., STAAS, J., SCHAFER, D., et al. 1993. Oral immunization with influenza virus in biodegradable microspheres. *J. Infect. Dis.* 167:84–90.

MOLLIS, K. B. 1990. The unusual origin of the polymerase chain reaction. *Scient. Amer.* (April) 262:56–65.

NAHREINI, P., WOODY, M., ZHOU, S., and SRIVASTAVA, A. 1993. Versatile adeno-associated virus 2-based vectors for constructing recombinant virions. *Gene* 124:257–262.

OLD, R. W., and PRIMROSE, S. B. 1985. *Principles of genetic manipulation: An introduction to genetic engineering.* Palo Alto, CA: Blackwell Scientific.

OSTE, C. 1988. Polymerase chain reaction. *BioTechniques* 6:162–67.

REISS, J., and COOPER, D. N. 1990. Application of the polymerase chain reaction to the diagnosis of human genetic disease. *Human Genetics* 85:1–8.

SAIKI, R. K., et al. 1986. Analysis of enzymatically amplified beta-globin and HLA-DQ alpha DNA with allele-specific oligonucleotide probes. *Nature* 324:163–66.

SOUTHERN, E. M. 1975. Detection of specific sequences among DNA fragments separated by gel electrophoresis. *J. Mol. Biol.* 98:503–17.

TAYLOR, R. 1993. Food for thought: Seropositive plants may yield cheap oral vaccines. *J. N.I.H. Research* 5:49–53.

THOMAS, T. L., and HALL, T. C. 1985. Gene transfer and expression in plants: Implications and potential. *BioEssays* 3:149–53.

VILLA-KOMAROFF, L., et al. 1978. A bacterial clone synthesizing proinsulin. *Proc. Natl. Acad. Sci.* 75:3727–31.

WAINWRIGHT, B. J. 1993. The isolation of disease genes by positional cloning. *Med. J. Austral.* 159:170–174.

ZUO, J., ROBBINS, C., BAHARLOO, S., COX, D., and MYERS, R. 1993. Construction of cosmid contigs and high-resolution restriction mapping of the Huntington disease region of human chromosome 4. *Hum. Mol. Genet.* 2:889–899.

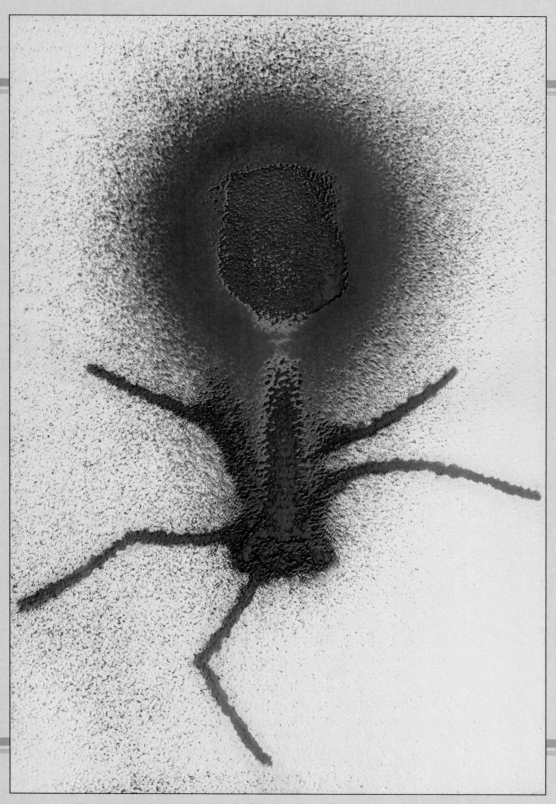

Transmission electron micrograph of bacteriophage T$_4$, a virus that infects *E. coli*.

# PART 3

## GENE EXPRESSION AND REGULATION

# 14

# STORAGE AND EXPRESSION OF GENETIC INFORMATION

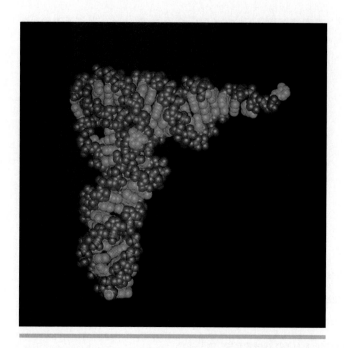

Space-filling model of transfer RNA, the adaptor molecule
that functions during protein synthesis.

*Genetic information, stored in DNA and transferred to RNA during the process of transcription, is present as three-letter codes. Using four different letters, corresponding to the four ribonucleotides in RNA, the 64 triplet codons specify the 20 different amino acids found in proteins, as well as providing signals that initiate and terminate protein synthesis. This unique language is the basis of life as we know it.*

In previous chapters, we have established that DNA is the genetic material and explored the structure and organization of DNA as it serves as the basis for heredity in all living things. In this chapter, we want to explore two additional aspects of molecular genetics. First, we will discuss how genetic information is encoded. Second, we will examine how this information is decoded during expression of genetic information. As we shall see, the elaborate system that provides the physical and chemical basis for the storage and expression of genetic information has the potential to produce a nearly endless variety of protein molecules. Although we have yet to discuss the details of the processes, we know that the DNA of an active gene is first transcribed into an RNA complement that is then translated into a protein. Thus, the sequence of deoxyribonucleotides in DNA is first copied into the complementary sequence of ribonucleotides in RNA, which then specifies the order of insertion of amino acids during protein synthesis. The end products of various genes are responsible for the normal and mutant phenotypes of organisms.

A fundamental question, then, is how an RNA molecule consisting of only four different types of ribonucleotides (A, U, C, and G) can specify 20 different amino acids. This question poses an intriguing theoretical problem. When ingenious analytical research was applied to this problem, the genetic code was deciphered. It was established that the code is triplet in nature. Code words, or **codons**, consisting of three ribonucleotides direct the insertion of amino acids into a polypeptide chain during its synthesis.

The process by which RNA molecules are synthesized on a DNA template is called **transcription**. The ribonucleotide sequence of RNA is then capable of directing the process of **translation**. During translation, polypeptide chains—which mature into proteins—are synthesized. Protein synthesis is dependent on a series of **transfer RNA (tRNA)** molecules, which serve as adaptors between the codons of mRNA and the amino acids specified by them. In addition, the process of translation occurs only in conjunction with an intricate cellular component, the **ribosome**.

The processes of transcription and translation are complex molecular events. Like the replication of DNA, both rely heavily on base-pairing affinities between complementary nucleotides. The initial transfer from DNA to mRNA produces a molecule complementary to the gene sequence of one of the two strands of the double helix. Then, each triplet codon is complementary to the anticodon region of tRNA as the corresponding amino acid is correctly inserted into the polypeptide chain during translation.

In this chapter, we will describe in detail how the code was deciphered and how gene expression occurs. The work leading to these discoveries occurred most intensively in the late 1950s and early 1960s, one of the most exciting periods in the study of molecular genetics. This research revealed the intricacies of the specific chemical language that serves as the basis of all life on earth.

# AN OVERVIEW OF THE GENETIC CODE

Before considering the various analytical approaches used in arriving at our current understanding of the genetic code, we shall provide a summary of the general features that characterize it:

1. The code is written in linear form using the ribonucleotide bases that compose mRNA molecules as the letters. The ribonucleotide sequence is, of course, derived from its complement in DNA.

2. Each word within the mRNA contains three letters. Thus, the code is a **triplet**. Called a **codon**, each group of *three* ribonucleotides specifies *one* amino acid.

3. The code is **unambiguous**, meaning that each triplet specifies only a single amino acid.

4. The code is **degenerate**, meaning that more than one triplet specifies a given amino acid. This is the case for 18 of the 20 amino acids.

5. The code is **ordered**. Degenerate codons for a given amino acid are grouped together, most often varying by only the third base.

6. The code contains "start" and "stop" punctuation signals. Certain triplets are necessary to **initiate** and to **terminate** translation.

7. No commas (or internal punctuation) are used in the code. Thus, the code is said to be **commaless**. Once translation of mRNA begins, each three ribonucleotides are read in turn, one after the other.

8. The code is **nonoverlapping**. Once translation commences, any single ribonucleotide at a specific location within the mRNA is part of only one triplet.

9. The code is almost **universal**. With only minor exceptions, a single coding dictionary is used by almost all viruses, prokaryotes, and eukaryotes.

# EARLY THINKING ABOUT THE CODE

Before it became clear that mRNA serves as an intermediate in transferring genetic information from DNA to proteins, it was thought that DNA itself might directly encode proteins during their synthesis. The central question, whether DNA or RNA houses the code, was how only four letters—the four nucleotides—could specify 20 words—the amino acids. Once mRNA was discovered, it was clear that even though genetic information is stored in DNA, the code that is translated into proteins resides in RNA.

As to the size of the code, Sidney Brenner argued on theoretical grounds that it must be a triplet since three-letter words represent the minimal use of four letters to specify 20 amino acids. For example, four nucleotides, taken two at a time provide only 16 unique code words ($4^2$). While a triplet code provides 64 words ($4^3$)—clearly more than the 20 needed—it is much simpler than a four-letter code, where 256 words ($4^4$) would be specified.

Brenner also argued that the code was nonoverlapping. Assuming a triplet code, Brenner considered restrictions that might be placed on it if it were overlapping. For example, he considered sequences within a protein consisting of three consecutive amino acids. In the nucleotide sequence GTACA, parts of the central triplet, TAC, are shared by the outer triplets, GTA and ACA. He reasoned that if this were the case, then only certain amino acids should be found adjacent to the one encoded by the central triplet, as shown below:

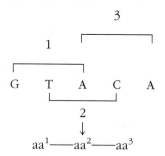

In fact, for any given central amino acid, only 16 combinations ($2^4$) of adjacent amino acids are theoretically possible.

Therefore, Brenner concluded that if the code were overlapping, tripeptide sequences of proteins should be somewhat limited. Looking at the available amino acid sequences of proteins that had been studied, he found no such restriction in tripeptide sequences. For any central amino acid, many more than 16 different tripeptides were found.

A second major argument against an overlapping code involved the effect of a single nucleotide change characteristic of a point mutation. With an overlapping code, two adjacent amino acids would be affected. But, mutations in the genes coding for the protein coat of tobacco mosaic virus (TMV), human hemoglobin, and the bacterial enzyme tryptophan synthetase invariably revealed only single amino acid changes.

The third argument against an overlapping code was presented by Francis Crick in 1957, when he predicted that DNA does not serve as a direct template for the formation of proteins. Crick reasoned that any affinity between nucleotides and an amino acid would require hydrogen bonding, Chemically, however, such affinities seemed unlikely. Instead, he proposed that there must be an **adaptor molecule** that could covalently bind to the amino acid, yet be capable of hydrogen bonding to a nucleotide sequence. Because various adaptors would somehow have to overlap one another at nucleotide sites, Crick reasoned that there simply would not be room for this to occur during translation. As we will see later in this chapter, Crick's prediction was correct; transfer RNA (tRNA) serves as the adaptor in protein synthesis.

Crick's and Brenner's arguments, taken together, strongly suggested that during translation, the genetic code is **nonoverlapping**. Without exception, this concept has been upheld.

# THE CODE: FURTHER DEVELOPMENTS

Between 1958 and 1960, information related to the genetic code continued to accumulate. In addition to his adaptor proposal, Crick hypothesized, on the basis of genetic evidence, that the code is **comma-free**; that is, he believed no internal punctuation occurs along the reading frame. He also speculated that only 20 of the 64 possible triplets specify an amino acid and that the remaining 44 carry no coding assignment.

At the time, however, there was as yet no experimental evidence that the code was indeed a triplet; nor had the concept of messenger RNA, as an intermediate between DNA and protein, been established. Thus, since ribosomes had already been identified, the current thinking was that information in DNA is transferred in the nucleus to the RNA of the ribosome, which serves as the template for protein synthesis in the cytoplasm.

This concept soon became untenable as accumulating evidence demonstrated that the template intermediate was unstable. The RNA of ribosomes, on the other hand, was found to be extremely stable. As a result, in 1961 François Jacob and Jacques Monod postulated the existence of **messenger RNA (mRNA)**. The scene was thus set to demonstrate the triplet nature of the code, as contained in the intermediate mRNA, and to decipher the specific codon assignments.

Many questions beyond the triplet assignments also remained. Are there start–stop **punctuation signals**? Is the code **ambiguous**, with one triplet specifying more than one amino acid? Was Crick wrong with respect to the 44 "blank" codes? That is, is the code **degenerate**, with more than one triplet assignment for each amino acid? Is the code **universal**? As we shall see, these and other questions were answered in the next decade.

## The Study of Frameshift Mutations

Before discussing the experimentation in which specific codon assignments were deciphered, we will consider the ingenious experimental work of Crick, Leslie Barnett, Brenner, and R. J. Watts-Tobin. Their work represented the first solid evidence for the triplet nature of the code.

In their work, these researchers induced insertion and deletion mutations in the B cistron of the *rII* locus of phage T4. The B cistron is one of two functional sites in this locus which, in mutant form, causes rapid lysis and distinctive plaques. Mutants in the *rII* locus will successfully infect strain B of *E. coli* but cannot reproduce on a separate strain of *E. coli*, designated K12. Crick and his colleagues used the acridine dye **proflavin** to induce

mutations (see Figure 11.12, pp. 355). This mutagenic agent intercalates within the double helix of DNA, often causing the insertion (or the deletion) of an extra nucleotide during replication. As shown in Figure 14.1(a), the frame of reading will therefore shift, changing all subsequent triplets to the right of the insertion. Upon translation, the protein subsequently synthesized will be garbled, so to speak, from the point of change. These mutations are called **frameshifts**, representing a concept first introduced in Chapter 11. When they are present at the *rII* locus, T4 will not reproduce on *E. coli* K12.

They reasoned that if phages with these induced mutations were treated again with proflavin, still other insertions or deletions would occur. This second change might result in a revertant phage, which would behave like wild type and successfully infect *E. coli* K12. For example, if the original mutant contained an insertion (+), a second event causing a deletion (−) close to the insertion would restore the original reading frame. In the same way, an event resulting in an insertion (+) might correct an original deletion (−).

In studying many mutations of this type, these researchers were able to compare various mutant combinations together on the same DNA molecule. They

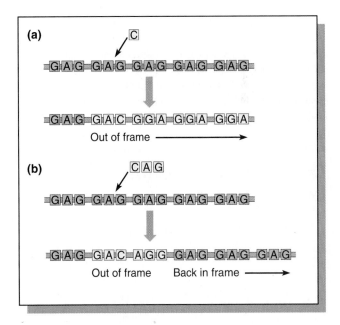

**FIGURE 14.1** Schematic diagram of the effect of frameshift mutations on a DNA sequence repeating the triplet sequence GAG. In part (a) the insertion of a single nucleotide has shifted all subsequent reading frames. In part (b) the insertion of three nucleotides changes only two reading frames, but all remaining frames are retained in their original sequence.

found that various combinations of one (+) and one (−) indeed caused reversion to wild-type behavior. Still other observations shed light on the number of nucleotides constituting the genetic code. When two (+)s were together or when two (−)s were together, the correct reading frame *was not* reestablished. This argued against a doublet (two-letter) code. However, when three (+)s [Figure 14.1(b)] or three (−)s were present together, the original frame *was* reestablished. These observations strongly supported the triplet nature of the code.

These data further suggested that the code is degenerate, which was contrary to Crick's earlier proposal. A degenerate code is one in which more than one codon specifies the same amino acid. The reasoning leading to this conclusion is as follows. In the cases where wild-type function is restored— (+) and (−), (+++), and (−−−) —the original frame of reading is also restored. However, there may be numerous triplets between the various additions and deletions that would still be out of frame. If 44 of the 64 possible triplets were blank and did not specify an amino acid, one of these so-called **nonsense triplets** would very likely occur in the length of nucleotides still out of frame. If a nonsense code were encountered during protein synthesis, it was reasoned that the process would stop or be terminated at that point. If so, the product of the *rII-B* locus would not be made, and restoration would not occur. Since the various mutant combinations were able to reproduce on *E. coli* K12, Crick and his colleagues concluded that, in all likelihood, most if not all of the remaining 44 codes were not blank. It follows that the genetic code is **degenerate**. As we shall see, this reasoning proved to be correct.

# DECIPHERING THE CODE: INITIAL STUDIES

In 1961 Marshall Nirenberg and J. Heinrich Matthaei

published results characterizing the first specific coding sequences. These results served as a cornerstone for the complete analysis of the code. Their success was dependent on the use of two experimental tools: **a cell-free protein-synthesizing system**, and an enzyme, **polynucleotide phosphorylase**, which allowed the production of synthetic mRNAs. These mRNAs served as templates for polypeptide synthesis in the cell-free system.

In the cell-free (*in vitro*) system, amino acids can be incorporated into polypeptide chains. This *in vitro* mixture, as might be expected, must contain the essential factors for protein synthesis in the cell: ribosomes, tRNAs, amino acids, and other molecules essential to translation. In order to follow (or trace) protein synthesis, one or more of the amino acids must be radioactive. Finally, an mRNA must be added, which serves as the template to be translated.

In 1961 mRNA had yet to be isolated. However, the use of the enzyme polynucleotide phosphorylase allowed artificial synthesis of RNA templates, which could be added to the cell-free system. This enzyme, isolated from bacteria, catalyzes the reaction shown in Figure 14.2. Discovered in 1955 by Marianne Grunberg-Manago and Severo Ochoa, the enzyme functions metabolically in bacterial cells to degrade RNA. However, *in vitro*, with high concentrations of ribonucleoside diphosphates, the reaction can be "forced" in the opposite direction to synthesize RNA, as illustrated.

In contrast to RNA polymerase, polynucleotide phosphorylase requires no DNA template. As a result, ribonucleotides are assembled at random, according to the relative concentration of the four ribonucleoside diphosphates added to the reaction mixtures. *This point is absolutely critical to understanding the work of Nirenberg and others in the ensuing discussion.*

Taken together, the cell-free system for protein synthesis and the availability of synthetic mRNAs provided a means of deciphering the ribonucleotide composition of various triplets encoding specific amino acids.

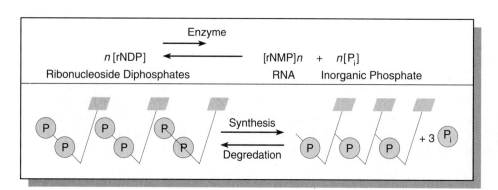

**FIGURE 14.2** The reaction catalyzed by the enzyme polynucleotide phosphorylase. Note that the equilibrium of the reaction favors the degradation of RNA but can be "pushed" in favor of synthesis.

## Nirenberg and Matthaei's Homopolymer Codes

In their initial experiments, Nirenberg and Matthaei synthesized **RNA homopolymers**, each consisting of only one type of ribonucleotide. Therefore, the mRNA added to the *in vitro* system was either UUUUUU . . . , AAAAAA . . . , CCCCCC . . . , or GGGGGG. . . . In testing each mRNA, they were able to determine which, if any, amino acids were incorporated into newly synthesized proteins. They determined this by labeling one of the 20 amino acids added to the *in vitro* system and conducting a series of experiments, each with a different amino acid made radioactive.

For example, consider one of the initial experiments, where $^{14}$C-phenylalanine was used (Table 14.1). From these and related experiments, Nirenberg and Matthaei concluded that the message poly U (polyuridylic acid) directs the incorporation of only phenylalanine into the homopolymer polyphenylalanine. Assuming a triplet code, they had determined the first specific codon assignment! UUU codes for phenylalanine.

In the same way, they quickly found that AAA codes for lysine and CCC codes for proline. Poly G did not serve as an adequate template, probably because the molecule folds back on itself. Thus, the assignment for GGG had to await other approaches. Note that the specific triplet codon assignments were possible only because of the use of homopolymers. This method yields only the composition of triplets, not the sequence. However, three U's, C's, or A's can have only one possible sequence (i.e., UUU, CCC, and AAA).

## The Use of Mixed Copolymers

With these techniques in hand, Nirenberg and Matthaei, and Ochoa and coworkers turned to the use of **RNA heteropolymers**. In this technique, two or more different ribonucleoside diphosphates are added in combination to form the message. These researchers reasoned that if the relative proportion of each type of ribonucleoside diphosphate is known, the frequency of any particular triplet codon occurring in the synthetic mRNA can be predicted. If the mRNA is then added to the cell-free system and the percentage of any particular amino acid present in the new protein is ascertained, correlations may be made and composition assignments predicted.

This concept is illustrated in Figure 14.3. Suppose that A and C are added in a ratio of 1A:5C. Now, the insertion of a ribonucleotide at any position along the RNA molecule during its synthesis is determined by the ratio of A:C. Therefore, there is a 1/6 possibility for an A and a 5/6 chance for a C to occupy each position. On this basis, we can calculate the frequency of any given triplet appearing in the message.

For AAA, the frequency is $(1/6)^3$, or about 0.4 percent. For AAC, ACA, and CAA, the frequencies are identical— that is, $(1/6)^2(5/6)$, or about 2.3 percent for each. Together, all three 2A:1C triplets account for 6.9 percent of the total three-letter sequences. In the same way, each of three 1A:2C triplets accounts for $(1/6)(5/6)^2$, or 11.6 percent (or a total of 34.8 percent). CCC is represented by $(5/6)^3$, or 57.9 percent of the triplets.

By examining the percentages of any given amino acid incorporated into the protein synthesized under the direction of this message, we may make tentative composition assignments (Figure 14.3). Since proline appears 69 percent of the time and since 69 percent is close to 57.9 percent + 11.6 percent, we can deduce that proline is coded by CCC and by one triplet of the 2C:1A variety. Histidine, at 14 percent, is probably coded by one 2C:1A (11 percent) and one 1C:2A (2 percent). Threonine, at 12 percent, is likely coded by only one 2C:1A. Asparagine and glutamine appear each to be coded by one of the 1A:2C triplets, and lysine appears to be coded by AAA.

In similar experiments, some using as many as all four ribonucleotides to construct the mRNA, many possible combinations were tested. Although the determination of the *composition* of triplet code words corresponding to all 20 amino acids represented a very significant breakthrough, *specific sequences* of triplets were still unknown. Their determination awaited still other approaches.

## The Triplet Binding Technique

It was not long before more advanced techniques were developed. In 1964 Nirenberg and Philip Leder developed the **triplet binding assay**, which led to specific assignments of triplets. The technique took advantage of the observation that ribosomes, when presented with an RNA sequence as short as three ribonucleotides,

**Table 14.1** INCORPORATION OF $^{14}$C-PHENYLALANINE INTO PROTEIN

| Artificial mRNA | Radioactivity (counts/min) |
|---|---|
| None | 44 |
| Poly U | 39,800 |
| Poly A | 50 |
| Poly C | 38 |

SOURCE: After Nirenberg and Matthaei, 1961, p. 1595.

| Possible Compositions | Probability of Occurrence of Any Triplet | Possible Triplets | Final % |
|---|---|---|---|
| 3A | $(1/6)^3 = 1/216 = 0.4\%$ | AAA | 0.4 |
| 2A:1C | $(1/6)^2(5/6) = 5/216 = 2.3\%$ | AAC ACA CAA | $3 \times 2.3 = 6.9$ |
| 1A:2C | $(1/6)(5/6)^2 = 25/216 = 11.6\%$ | ACC CAC CCA | $3 \times 11.6 = 34.8$ |
| 3C | $(5/6)^3 = 128/216 = 57.9\%$ | CCC | 57.9 |
| | | | 100.0% |

**FIGURE 14.3**    Results and interpretation of a mixed copolymer experiment where a ratio of 1A:5C is used (1/6A:5/6C).

Transcription of
Synthetic Message

CCCCCCCCCACCCCCCAACCACCCCCACCCCCACCCAAACCCCCACCCCCC    RNA

Translation of
Message

| Percentage of Amino Acids in Protein | | Probable Base Composition Assignments |
|---|---|---|
| Proline | 69 | CCC,2C:1A |
| Histidine | 14 | 2C:1A, 1C:2A |
| Threonine | 12 | 2C:1A |
| Asparagine | 2 | 1C:2A |
| Glutamine | 2 | 1C:2A |
| Lysine | 1 | AAA |

will bind to it and attract the charged tRNA corresponding to the triplet codon. For example, if ribosomes are presented with an RNA triplet UUU and tRNA$^{phe}$, a complex will form that is similar to what actually occurs *in vivo*. The triplet acts like a codon in mRNA and it attracts the complementary sequence called the **anticodon** found within tRNA (Figure 14.4). Anticodons are those triplet sequences in tRNAs that are complementary to the codons of mRNA. Although it was not yet feasible to chemically synthesize long stretches of RNA, triplets of

known sequence could be constructed in the laboratory to serve as templates.

All that was needed now was a method to determine which tRNA-amino acid was bound to the triplet RNA-ribosome complex. The test system devised was quite simple. The amino acid to be tested is made radioactive, and a charged tRNA (tRNA bonded to its amino acid) is produced. Since code compositions were known, it was possible to narrow the decision as to which amino acids should be tested for each specific triplet.

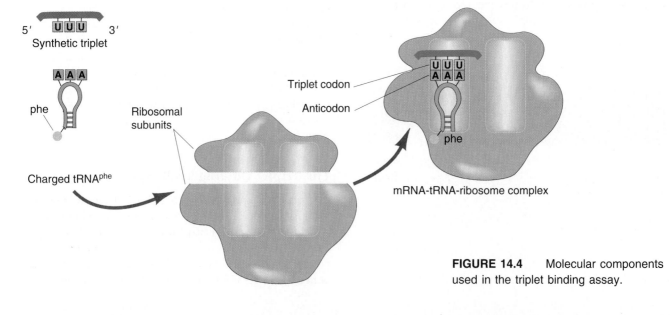

5′  UUU  3′
Synthetic triplet

AAA

phe

Charged tRNA$^{phe}$

Ribosomal
subunits

Triplet codon

Anticodon

UUU
AAA

phe

mRNA-tRNA-ribosome complex

**FIGURE 14.4**    Molecular components used in the triplet binding assay.

**Table 14.2** Amino Acid Assignments to Specific Trinucleotides Derived from the Triplet Binding Assay

| Trinucleotide | Amino Acid |
|---|---|
| UGU UGC | Cysteine |
| GAA GAG | Glutamic acid |
| AUU AUC AUA | Isoleucine |
| UUA UUG CUU | Leucine |
| CUC CUA CUG | Leucine |
| AAA AAG | Lysine |
| AUG | Methionine |
| UUU UUC | Phenylalanine |
| CCU CCC | Proline |
| CCG CCA | Proline |
| UCU UCC | Serine |
| UCA UCG | Serine |

The radioactive charged tRNA, the RNA triplet, and ribosomes are incubated together on a nitrocellulose filter, which will retain the larger ribosomes but not the other smaller components, such as charged tRNA. If radioactivity is not retained on the filter, an incorrect amino acid has been tested. If radioactivity remains on the filter, it is retained because the charged tRNA has bound to the triplet associated with the ribosome. In such a case, a specific codon assignment may be made.

Work proceeded in several laboratories, and in many cases clearcut, unambiguous results were obtained.

Table 14.2, for example, shows 26 triplets assigned to 9 amino acids. However, in some cases the degree of binding was insufficient, and unambiguous assignments were not possible. Eventually, about 50 of the 64 triplets were assigned. The binding technique was a major innovation in deciphering the code. Based on these specific assignments of triplets to amino acids, two major conclusions were drawn. The genetic code is **degenerate**; that is, one amino acid may be specified by more than one triplet. The code is also **unambiguous**; that is, a single triplet specifies only one amino acid. As we shall see later in this chapter, these conclusions have been upheld with only minor exceptions.

## The Use of Repeating Copolymers

Still another innovative technique used to decipher the genetic code was developed by Gobind Khorana. He was able to chemically synthesize long RNA molecules consisting of short sequences repeated many times, which could be used in the cell-free protein synthesizing system. First, he created shorter sequences (e.g., di-, tri-, and tetranucleotides), which were then replicated many times and finally joined enzymatically to form the long polynucleotides.

As illustrated in Figure 14.5, a dinucleotide made in this way is converted to a message with two repeating triplets. A trinucleotide is converted to one with three potential triplets, depending on the point at which initiation occurs. Similarly, a tetranucleotide creates four repeating triplets.

When these synthetic mRNAs were added to a cell-free system, the predicted number of different amino acids incorporated was upheld. Several examples are

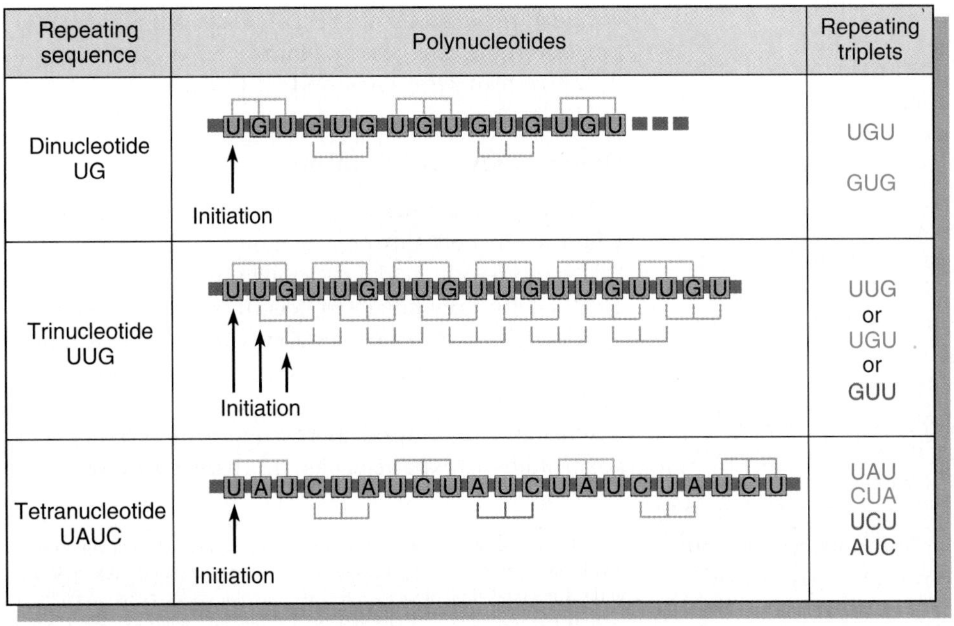

**FIGURE 14.5** The conversion of di-, tri-, and tetranucleotides into repeating copolymers. The triplet codons produced in each case are shown.

**Table 14.3**  AMINO ACIDS INCORPORATED USING REPEATING SYNTHETIC COPOLYMERS OF RNA

| Repeating Copolymer | Codons Produced | Amino Acids in Polypeptide |
|---|---|---|
| UG | UGU | Cysteine |
|  | GUG | Valine |
| AC | ACA | Threonine |
|  | CAC | Histidine |
| UUC | UUC | Phenylalanine |
|  | UCU | Serine |
|  | CUU | Leucine |
| AUC | AUC | Isoleucine |
|  | UCA | Serine |
|  | CAU | Histidine |
| UAUC | UAU | Tyrosine |
|  | CAU | Leucine |
|  | UCU | Serine |
|  | AUC | Isoleucine |
| GAUA | GAU |  |
|  | AGA |  |
|  | UAG | None |
|  | AUA |  |

shown in Table 14.3. When such data were combined with conclusions drawn from other approaches (composition assignment, triplet binding), specific assignments were possible.

One example of specific assignments made from such data will illustrate the value of Khorana's approach. Consider the following three experiments in concert with one another. The repeating trinucleotide sequence UUCUUCUUC . . . produces three possible triplets: UUC, UCU, and CUU, depending on the first nucleotide to initiate reading. When placed in a cell-free translation system, the polypeptides containing phenylalanine (phe), serine (ser), and leucine (leu) are produced. On the other hand, the repeating dinucleotide sequence UCUCUCUC . . . produces the triplets UCU and CUC with the incorporation of leucine and serine. Therefore, the triplets UCU and CUC specify leucine and serine, but it cannot be determined which is which. One can further conclude that *either* the CUU *or* the UUC triplet also encodes leucine *or* serine, while the other encodes phenylalanine.

In order to derive more specific information, we can examine the results of using the repeating tetranucleotide sequence UUAC, which produces the triplets UUA, UAC, ACU, and CUU. The CUU triplet is one of the two

in which we are interested. Three amino acids are incorporated: leucine, threonine, and tyrosine. Because CUU must specify only serine or leucine, and because, of these two, only leucine appears, we may conclude that CUU specifies leucine.

Once this is established, we can logically determine all other specific assignments. Of the two triplet pairs remaining (UUC and UCU from the first experiment *and* UCU and CUC from the second experiment) whichever triplet is common to both must encode serine. This is UCU. By elimination, UUC is determined to encode phenylalanine and CUC is determined to encode leucine. While the logic must be carefully followed, four specific triplets encoding three different amino acids have been assigned from these experiments.

From such interpretations, Khorana reaffirmed triplets already deciphered and filled in gaps left from other approaches. For example, the use of two tetranucleotide sequences, GAUA and GUAA, suggested that at least two triplets were termination signals. This conclusion was drawn because neither of these sequences directed the incorporation of any amino acids into a polypeptide. Since there are no triplets common to both messages, it was predicted that each repeating sequence contains at least one triplet that terminates protein synthesis. The possible triplets for the poly-(GAUA) sequence are shown in Table 14.3, one of which, UAG, is a termination codon.

## THE CODING DICTIONARY

The various techniques applied to decipher the genetic code have yielded a dictionary of 61 triplet codon–amino acid assignments. The remaining three triplets are termination signals, not specifying any amino acid. Figure 14.6 designates the assignments in a particularly illustrative form first suggested by Crick.

### Degeneracy and Wobble

The degenerate nature of the code becomes apparent when we inspect this presentation of the genetic code. While the amino acids tryptophan and methionine are encoded by only single triplets, most amino acids are specified by two, three, or four triplets. Three amino acids (serine, arginine, and leucine) are coded by six triplets each.

What also becomes evident is the pattern of degeneracy, leading to what is called an **ordered code**.* Most

---

*Another aspect of "order" in the code is that similar types of amino acids may be grouped by the middle base of triplets encoding them; that is, U or G in the second position often specifies hydrophobic amino acids, and A in the second position specifies charged or hydrophilic amino acids. Chemical properties of amino acids are discussed in Chapter 15.

Second Position

| First Position (5′ end) | U | C | A | G | Third Position (3′ end) |
|---|---|---|---|---|---|
| U | UUU<br>UUC phe<br>UUA<br>UUG leu | UCU<br>UCC<br>UCA ser<br>UCG | UAU<br>UAC tyr<br>UAA Stop<br>UAG Stop | UGU<br>UGC cys<br>UGA Stop<br>UGG trp | U<br>C<br>A<br>G |
| C | CUU<br>CUC<br>CUA leu<br>CUG | CCU<br>CCC<br>CCA pro<br>CCG | CAU<br>CAC his<br>CAA<br>CAG gin | CGU<br>CGC<br>CGA arg<br>CGG | U<br>C<br>A<br>G |
| A | AUU<br>AUC ile<br>AUA<br>AUG met | ACU<br>ACC<br>ACA thr<br>ACG | AAU<br>AAC asn<br>AAA<br>AAG lys | AGU<br>AGC ser<br>AGA<br>AGG arg | U<br>C<br>A<br>G |
| G | GUU<br>GUC<br>GUA val<br>GUG | GCU<br>GCC<br>GCA ala<br>GCG | GAU<br>GAC asp<br>GAA<br>GAG glu | GGU<br>GGC<br>GGA gly<br>GGG | U<br>C<br>A<br>G |

Initiation Termination

**FIGURE 14.6** The coding dictionary. AUG encodes methionine, which initiates most polypeptide chains. All other amino acids except tryptophan, which is encoded by UGG, are represented by two to six triplets. The triplets UAA, UAG, and UGA are termination signals and do not encode any amino acids.

often in a set of codons specifying the same amino acid, the first two letters are the same, with only the third differing. This interesting pattern prompted Crick in 1966 to postulate the **wobble hypothesis**.

Crick concluded that the first two ribonucleotides of triplet codes are more critical than the third member in attracting the correct tRNA. Thus, Crick proposed that hydrogen bonding at the third position of the codon–anticodon interactions need not adhere as specifically as the first two members to the established base-pairing rules. After examining the coding dictionary carefully, he proposed a new set of base-pairing rules at the third position of the codon.

This relaxed base-pairing requirement, or "wobble," allows the anticodon of a single tRNA species to pair with more than one triplet in mRNA. The code's degeneracy would often allow this to occur without changing the amino acid. Consistent with the coding assignments, it appears that U at the third position of the anticodon of tRNA may pair with A or G at the third position of the triplet in mRNA, and that G may likewise pair with U or C. Inosine, one of the modified bases found in tRNA, may pair with C, U, or A. Applying these wobble rules, a minimum of about 30 different tRNA species is necessary to accommodate the 61 triplets specifying an amino acid. If nothing more, wobble can be considered an economy measure, provided that fidelity of translation is not compromised.

## Initiation, Termination, and Suppression

Initiation of protein synthesis is a highly specific process. In bacteria, the initial amino acid inserted into all polypeptide chains is a modified form of methionine—**N-formylmethionine (fmet)**. Only one codon, AUG, codes for methionine, and it is sometimes called the **initiator codon**. However, when AUG appears internally in mRNA, unformylated methionine is inserted into the polypeptide chain. Rarely, still another triplet, GUG, specifies methionine during initiation. It is not clear why this occurs, since GUG normally encodes valine.

In bacteria, either the formyl group is removed from the initial methionine upon completion of synthesis of a protein, or the entire formylmethionine residue is removed. In eukaryotes, methionine is also the initial amino acid during polypeptide synthesis. However, it is not formylated.

As mentioned in the preceding section, three other triplets (UAG, UAA, and UGA) serve as punctuation signals and do not code for any amino acid. They are not recognized by a tRNA molecule, and termination of translation occurs when they are encountered. Mutations occurring internally in a gene that produce any of the three triplets also result in termination. As a result, only a partial polypeptide is synthesized, since it is prematurely released from the ribosome. When such a change occurs, it is called a **nonsense mutation**. The terms **amber** (UAG), **ochre** (UAA), and **opal** (UGA) have been used to distinguish the three possibilities.

Interestingly, a distinct mutation in a second gene may cause **suppression** of premature termination. These mutations cause the chain-termination signal to be read as a sense codon. The "correction" usually inserts an amino acid other than that found in the wild-type protein. However, if the protein's structure is not altered drastically, it may function almost normally. Therefore, this second mutation has "suppressed" the mutant character resulting from the initial change to the termination codon.

Some suppressor mutations occur in genes specifying tRNAs. If the mutation results in a change in the anticodon such that it becomes complementary to a termination code, there is the potential for insertion of an amino acid and suppression.

Other types of suppression involve mutations in genes coding for aminoacyl synthetases, which are responsible for attaching the amino acid to tRNA (the process called *charging*), and in genes coding for ribosomal proteins. In both cases, suppression occurs as a result of misreading the mutant codon. That a ribosomal protein can cause ambiguity of translation demonstrates the intimate relationship between the ribosome, mRNA, and tRNA during translation.

# CONFIRMATION OF CODE STUDIES: PHAGE MS2

All aspects of the genetic code discussed so far yield a fairly complete picture. The code is triplet in nature, degenerate, unambiguous, and commaless, but contains punctuation with respect to start and stop signals. These individual principles have been confirmed by the detailed analysis of the RNA-containing bacteriophage MS2 by Walter Fiers and his coworkers.

**MS2** is a bacteriophage that infects *E. coli*. Its nucleic acid (RNA) contains only about 3500 ribonucleotides, making up only three genes. These genes specify a coat protein, an RNA-directed replicase, and a maturation protein (the A protein). This simple system of a small genome and few gene products allowed Fiers and his colleagues to sequence the genes and their products. The amino acid sequence of the coat protein was completed in 1970, and the nucleotide sequence of the gene and a number of nucleotides on each end of it were reported in 1972.

The coat protein contains 129 amino acids, and the gene contains 387 nucleotides, as expected for a triplet code. Each amino acid and triplet corresponds in linear sequence to the correct codon in the RNA code word dictionary, confirming the earlier experimental work that served as the basis for deciphering the genetic code. These findings further provided direct proof of the **colinear relationship** between nucleotide sequence and amino acid sequence. The codon for the first amino acid is preceded by AUG, the common initiator codon; and the codon for the last amino acid is succeeded by two consecutive termination codons, UAA and UAG.

By 1976, the other two genes and their protein products were sequenced, providing similar confirmation. The analysis clearly shows that the genetic code as established in bacterial systems is identical in this virus. We shall now briefly consider other evidence suggesting that the code is also identical in eukaryotes.

# UNIVERSALITY OF THE CODE

Between 1960 and 1978, it was generally assumed that the genetic code would be found to be universal, applying equally to viruses, bacteria, and eukaryotes. Certainly, the nature of mRNA and the translation machinery seemed to be very similar in these organisms. For example, cell-free systems derived from bacteria could translate eukaryotic mRNAs. Poly U was shown to stimulate translation of polyphenylalanine in cell-free systems when the components were derived from eukaryotes.

Many recent studies involving recombinant DNA technology (see Chapters 12 and 13) have revealed that eukaryotic genes can be inserted into bacterial cells and transcribed and translated. Within eukaryotes, mRNAs from mice and rabbits have been injected into amphibian eggs and efficiently translated. For the many eukaryotic genes that have been sequenced, notably those for hemoglobin molecules, the amino acid sequence of the encoded proteins adheres to the coding dictionary established from bacterial studies.

However, several 1979 reports on the coding properties of DNA derived from yeast and human mitochondria (mtDNA) altered the principle of universality of the genetic language. Since then, mtDNA has been examined in many other organisms.

Not only do mitochondria contain DNA, but transcription and translation occur within these organelles. Cloned mtDNA fragments were sequenced and compared with the amino acid sequences of various mitochondrial proteins, revealing several exceptions to the coding dictionary (Table 14.4). Most surprising was that the codon UGA, normally causing termination, specifies the insertion of tryptophan during translation in yeast and human mitochondria. In human mitochondria, AUA, which normally specifies isoleucine, directs the internal insertion of methionine. In yeast mitochondria, threonine is inserted instead of leucine when CUA is encountered in mRNA.

More recently, in 1985, several other exceptions to the

**Table 14.4** EXCEPTIONS TO THE UNIVERSAL CODE

| Triplet | Normal Code Word | Altered Code Word | Source |
|---|---|---|---|
| UGA | termination | trp | Human and yeast mitchondria Mycoplasma |
| CUA | leu | thr | Yeast mitochondria |
| AUA | ile | met | Human mitchondria |
| AGA AGG | arg | termination | Human mitchondria |
| UAA | termination | gln | *Paramecium Tetrahymena Stylonychia* |
| UAG | termination | gln | *Paramecium* |

standard coding dictionary have been discovered. These and prior aberrant codes are also summarized in Table 14.4. Such changes have been observed in the bacterium *Mycoplasma capricolum*, and in the protozoan ciliates *Paramecium, Tetrahymena*, and *Stylonychia*. As shown, each change converts one of the termination codons (UGA) to tryptophan. These changes are significant because both a prokaryote and several eukaryotes are involved, representing distinct species that have evolved over a long period of time.

Note the apparent pattern in several of the altered codon assignments. The change in coding capacity involves only a shift in recognition of the third, or wobble, position. For example, AUA specifies isoleucine during translation in the cytoplasm and methionine in the mitochondrion. In cytoplasmic translation, methionine is specified by AUG. In a similar way, UGA calls for termination in the cytoplasmic system but tryptophan in the mitochondrion. In the cytoplasm, tryptophan is specified by UGG. Although it has been suggested that such changes in codon recognition may represent an evolutionary trend toward reducing the number of tRNAs needed in mitochondria, the significance of these findings is not yet clear, It is known that only 22 tRNA species are encoded in human mitochondria. However, until still other examples are revealed, the differences must be considered as exceptions to the previously established general coding rules.

## READING THE CODE: THE CASE OF OVERLAPPING GENES

In this chapter we established that the genetic code is nonoverlapping. This means that each ribonucleotide of an mRNA which specifies a polypeptide chain is part of only one triplet. However, this characteristic of the code does not rule out the possibility that a single mRNA may have multiple initiation points for translation. If so, these points could theoretically create several different frames of reading within the same mRNA, thus specifying more than one polypeptide. This concept, which would create **overlapping genes**, is illustrated in Figure 14.7(a).

That this might actually occur in some viruses was suspected when phage $\phi$X174 was carefully investigated. The circular DNA chromosome consists of 5386 nucleotides, which should encode a maximum of 1795 amino acids, sufficient for five or six proteins. However, it was realized that this small virus in fact synthesizes 11 proteins consisting of more than 2300 amino acids! Comparison of the nucleotide sequence of the DNA and the amino acid sequences of the polypeptides synthesized has clarified this paradox. At least four cases of multiple initiation have been discovered, creating overlapping genes [Figure 14.7(b)].

The sequences specifying the K and B polypeptides are initiated with separate reading frames within the se-

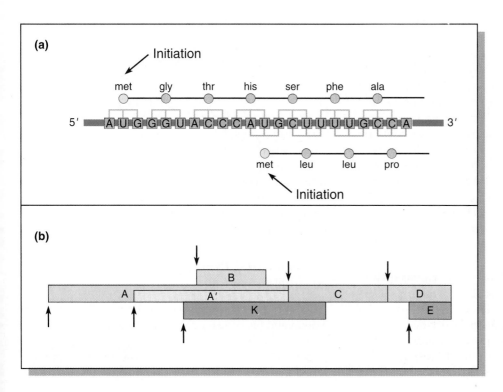

**FIGURE 14.7** Illustration of the concept of overlapping genes. (a) An mRNA sequence initiated at two different AUG positions out of frame with one another will give rise to two distinct amino acid sequences. (b) The relative positions of the sequences encoding seven polypeptides of the phage $\phi$X174. Those encoding B, K, and E are out of frame.

quence specifying the A polypeptide. The K sequence overlaps into the adjacent sequence specifying the C polypeptide. The E sequence is out of frame with, but initiated in, that of the D polypeptide. Finally, the A′ sequence, while in frame, begins in the middle of the A sequence. They both terminate at the identical point. In all, seven different polypeptides are created from a DNA sequence that might otherwise have specified only three (A, C, and D).

A similar situation has been observed in other viruses, including phage G4 and the animal virus SV40. Like ϕX174, phage G4 contains a circular, single-stranded DNA molecule.The use of overlapping reading frames optimizes the use of a limited amount of DNA present in these small viruses. However, such an approach to storing information has the distinct disadvantage that a single mutation may affect more than one protein and thus increase the chances that the change will be deleterious or lethal. In the case discussed above, a single mutation at the junction of genes *A* and *C* could affect three proteins (the A, C, and K proteins). It may be for this reason that such an approach has not become common in other organisms.

# EXPRESSION OF GENETIC INFORMATION: AN OVERVIEW

Even while the genetic code was being studied, it was quite clear that proteins were the end products of many genes. Thus, while some geneticists were attempting to elucidate the code, other research efforts were directed toward the nature of genetic expression. The central question was how DNA, a nucleic acid, is able to specify a protein composed of amino acids. Put still another way, How is information transferred between DNA and protein? We shall return in the next chapter to examine the evidence supporting the conclusion that proteins are specified by genes, as well as to discuss protein structure and function. Here we shall emphasize the concept of **information flow** as it occurs between DNA and protein. This topic was extremely interesting and exciting in the early 1960s, and it remains no less so in the 1990s!

Genetic information, stored in DNA, was shown to be transferred to RNA during the initial stage of gene expression. The process by which RNA molecules are synthesized on a DNA template is called **transcription**. The ribonucleotide sequence of RNA, written in a genetic code, is then capable of directing the process of **translation**. During translation, polypeptide chains— the precursors of proteins—are synthesized. Protein synthesis is dependent on a series of **transfer RNA (tRNA)** molecules, which serve as adaptors between the

codons of mRNA and the amino acids specified by them. In addition, the process occurs only in conjunction with a large, intricate cellular organelle, the **ribosome**.

The processes of transcription and translation are complex molecular events. Like the replication of DNA, both rely heavily on base-pairing affinities between complementary nucleotides. The initial transfer from DNA to mRNA produces a molecule complementary to the gene sequence of one of the two strands of the double helix. Then, each triplet codon in mRNA is complementary to the anticodon region of tRNA as the corresponding amino acid is correctly inserted into the polypeptide chain during translation. In the following sections, we will describe in detail how these processes were discovered and how they are executed.

# TRANSCRIPTION: RNA SYNTHESIS

The idea that RNA is involved as an intermediate molecule in the process of information flow between DNA and protein is suggested by the following observations:

1. DNA is, for the most part, associated with chromosomes in the nucleus of the eukaryotic cell. However, protein synthesis occurs in association with ribosomes located outside the nucleus in the cytoplasm. Therefore, DNA does not appear to participate directly in protein synthesis.

2. RNA is synthesized in the nucleus of eukaryotic cells, where DNA is found, and is chemically similar to DNA.

3. Following its synthesis, most RNA migrates to the cytoplasm, where protein synthesis occurs.

4. The amount of RNA is generally proportional to the amount of protein in a cell.

Collectively, these observations suggested that genetic information, stored in DNA, is transferred to an RNA intermediate, which directs the synthesis of proteins. As with most new ideas in molecular genetics, the initial supporting experimental evidence was based on studies of bacteria and their phages.

## Experimental Evidence for the Existence of mRNA

In two papers published in 1956 and 1958, Elliot Volkin and his colleagues reported their analysis of RNA produced immediately after bacteriophage infection of *E. coli*. Using the isotope $^{32}$P to follow newly synthesized RNA, they found that its base composition closely resembled that of the phage DNA but was different from that of bacterial RNA (Table 14.5). Although this newly synthesized RNA was unstable, or short-lived, its production was shown to precede the synthesis of new phage proteins. Thus, they considered the possibility that synthesis of RNA is a preliminary step in the process of protein synthesis.

Although ribosomes were known to participate in protein synthesis, their role in this process was not clear. One possible role was that each ribosome is specific for the protein synthesized in association with it. That is, perhaps genetic information in DNA is transferred to the RNA of a ribosome during its synthesis so that each class of ribosome specifies a particular protein. The alternative hypothesis was that ribosomes are nonspecific "workbenches" for protein synthesis and that specific genetic information rests with a "messenger" RNA.

In an elegant experiment using the *E. coli*–phage system, the results of which were reported in 1961, Sidney Brenner, François Jacob, and Matthew Meselson clarified this question. They labeled uninfected *E. coli* ribosomes with "heavy" isotopes (see Appendix A) and then allowed phage infection to occur in the presence of radioactive RNA precursors. They demonstrated that phage proteins were synthesized on bacterial ribosomes that were present prior to infection. Therefore, the ribosomes appeared to be nonspecific, strengthening the case that another type of RNA serves as an intermediary in the process of protein synthesis.

That same year, Sol Spiegelman and his colleagues isolated $^{32}$P-labeled phage RNA following infection of bacteria and used it in molecular hybridization studies. They tried hybridizing this RNA to the DNA of both phages and bacteria in separate experiments. The RNA hybridized only with the phage DNA, showing that it was complementary in base sequence to the viral genetic information.

The results of all these experiments agree with the concept of a **messenger RNA (mRNA)** being made on a DNA template and then directing the synthesis of specific proteins in association with ribosomes. This concept was formally proposed by François Jacob and Jacques Monod in 1961 as part of a model for gene regulation in bacteria. Since then, mRNA has been isolated and thoroughly studied. There is no longer any question about its role in genetic processes.

## RNA Polymerase

In order to prove that RNA may be synthesized on a DNA template, it was necessary to demonstrate that there is an enzyme capable of directing this synthesis. By 1959, several investigators, including Samuel Weiss, had independently discovered such a molecule from rat liver. Called **RNA polymerase**, it has the same general substrate requirements as DNA polymerase, the major exception being that the nucleotides that are substrates contain the ribose rather than the deoxyribose form of the sugar. Unlike DNA polymerase, no primer is required to initiate synthesis. The initial base remains as NTP. The overall reaction that summarizes the synthesis of RNA on a DNA template may be expressed as:

$$n(\text{NTP}) \xrightarrow[\text{enzyme}]{\text{DNA}} (\text{NMP})_n + n(\text{PP}_i)$$

As the equation reveals, nucleoside triphosphates

**Table 14.5**  BASE COMPOSITIONS (IN MOLE PERCENTS) OF RNA PRODUCED IMMEDIATELY FOLLOWING INFECTION OF *E. COLI* BY THE BACTERIOPHAGES T2 AND T7 IN CONTRAST TO THE COMPOSITION OF RNA OF UNINFECTED *E. COLI*

|  | Adenine | Thymine | Uracil | Cytosine | Guanine |
|---|---|---|---|---|---|
| Postinfection RNA in T2-infected cells | 33 | — | 32 | 18 | 18 |
| T2 DNA | 32 | 32 | — | 17[a] | 18 |
| Postinfection RNA in T7-infected cells | 27 | — | 28 | 24 | 22 |
| T7 DNA | 26 | 26 | — | 24 | 22 |
| *E. coli* RNA | 23 | — | 22 | 18 | 17 |

[a] 5-hydroxymethyl cytosine.

SOURCE: From Volkin and Astrachan, 1956; and Volkin, Astrachan, and Countryman, 1958.

(NTPs) serve as substrates for the enzyme. It catalyzes the polymerization of nucleoside monophosphates (NMPs), or nucleotides, into a polynucleotide chain $(NMP)_n$. Nucleotides are linked during synthesis by 5'-to-3' phosphodiester bonds (see Figure 8.12 on pp. 255). The energy created by cleaving the triphosphate precursor into the monophosphate form drives the reaction and inorganic phosphates $(PP_i)$ are produced.

A second equation summarizes the sequential addition of each ribonucleotide as the process of transcription progresses:

$$(NMP)_n + NTP \xrightarrow[\text{enzyme}]{\text{DNA}} (NMP)_{n+1} + PP_i$$

As this equation shows, each step of transcription involves the addition of one ribonucleotide (NMP) to the growing polyribonucleotide chain $(NMP)_{n+1}$, using a nucleoside triphosphate (NTP) as the precursor.

RNA polymerase from *E. coli* has been extensively characterized and shown to consist of subunits designated $\alpha$, $\beta$, $\beta'$, and $\sigma$. The active form of the enzyme (the holoenzyme) $\alpha_2\beta\beta'\sigma$ has a molecular weight of almost 500,000 daltons. Of these various subunits, it is the $\boldsymbol{\beta}$ and $\boldsymbol{\beta'}$ **polypeptides** that provide the catalytic basis and active site for transcription. As we will see, the $\boldsymbol{\sigma}$ **(sigma) subunit** plays a regulatory function involving the initiation of RNA transcription.

While there is but a single form of the enzyme in *E. coli*, separate forms of RNA polymerase are involved in the transcription of the three types of RNA in eukaryotes. The nomenclature used in describing these is summarized in Table 14.6. The three eukaryotic polymerases all consist of a greater number of polypeptide subunits than the bacterial form of the enzyme.

One way in which the three eukaryotic forms may be distinguished is by their differential sensitivity to the mushroom poison, $\boldsymbol{\alpha}$**-amanitin**. Polymerase I, which catalyzes the transcription leading to most mRNAs, is insensitive to the poison; polymerase II, which catalyzes the transcription producing mRNAs, is extremely sensitive (inhibited); polymerase III, which catalyzes transcription yielding tRNA and a small 5$S$ rRNA component, is of intermediate sensitivity, being inhibited only at much higher concentrations. It is not immediately clear what functional differences exist among the three forms of this complex enzyme, but it is clear that they recognize different promoter sequences, thus providing the specificity of their transcriptive activities.

## The Sigma Subunit, Promoters, and Template Binding

Transcription results in the synthesis of a single-stranded RNA molecule complementary to a region along one of the two strands of the DNA double helix. For the purpose of future discussion, we will call the DNA strand that is transcribed the **template strand**. Its complement will be referred to as the **partner strand**. The initial step is referred to as **template binding**, where RNA polymerase interacts physically with DNA [Figure 14.8(a)]. The accuracy of this initial binding in bacteria is achieved as a result of the recognition of specific DNA sequences called **promoters** by the sigma subunit ($\sigma$) of the holoenzyme.

These regions are located upstream (to the left in the illustration) from the point of initial transcription of a gene. Most likely, the enzyme "explores" a length of DNA until the promoter region is recognized, and a tightly bound complex results. Once this occurs, the helix is denatured or unwound locally, making the DNA template accessible to the action of the enzyme.

A great deal is now known about promoters and template binding. The enzyme is a complex molecule, large enough to bind about 60 nucleotide pairs of the helix, 40 of which are upstream from the point of initial transcription.

The importance of promoter sequences cannot be overemphasized. They govern the efficiency of initiation of transcription. Both strong promoters and weak promoters are recognized, leading to a variation of initiation from once every 1 to 2 seconds to only once every 10 to 20 minutes. In fact, mutations in promoter sequences may have the effect of severely reducing the initiation of gene expression. And, as we shall see later in this chapter, a consensus sequence rich in adenine and thymine residues (called the TATA box) is present in the promoter region of almost all eukaryotic genes.

A **consensus sequence** is one that is found to be present in similar positions in the genes of evolutionarily diverse organisms. The high degree of conservation through evolution attests to the critical nature of consen-

**Table 14.6** RNA Polymerases in Eukaryotes

| Type | Product | Location |
|------|---------|----------|
| I | rRNA | Nucleolus |
| II | mRNAs | Nucleoplasm |
| III | 5$S$ RNA | Nucleoplasm |
| | tRNAs | Nucleoplasm |

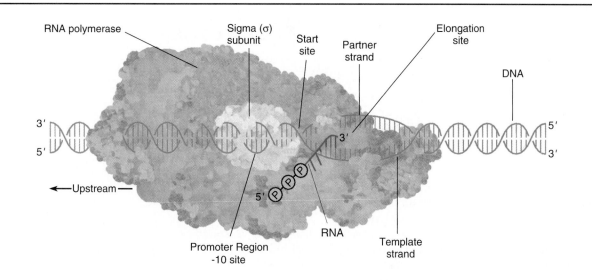

3′
5′

⟵ Upstream ⟶

5′  Ⓟ Ⓟ Ⓟ

**(a)** Template binding and initiation
of transcription

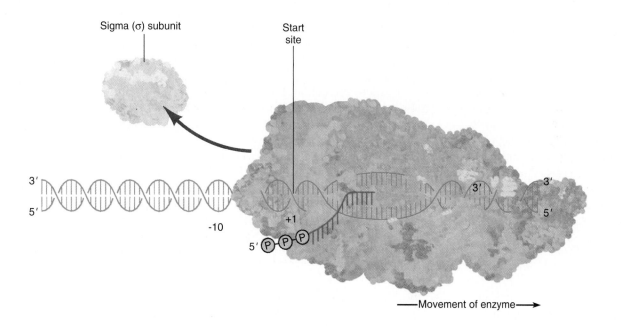

3′
5′

-10        +1

5′ Ⓟ Ⓟ Ⓟ

⟶ Movement of enzyme ⟶

**(b)** Chain elongation

**FIGURE 14.8**    Schematic representation of the early stages of transcription, including: (a) template binding and initiation involving the sigma subunit of RNA polymerase;
and (b) chain elongation, after sigma has dissociated from the transcription complex.

# TOOLS AND TECHNIQUES

## Antisense RNA

Gene expression is a two-step process: first, a single-stranded messenger RNA (mRNA) is copied (transcribed) from the strand of a duplex DNA molecule that encodes genetic information. In the second step, the mRNA moves to the cytoplasm, is complexed to ribosomes, and its genetic information is translated into the amino acid sequence of a polypeptide.

In the early 1980s it was shown that sometimes the *other* DNA strand of a gene can be copied into a single-stranded RNA. The RNA produced by transcription from the "wrong" strand is called antisense RNA. The base sequence of the antisense RNA is complementary to that of the gene's normal mRNA. Just as with complementary strands of a DNA molecule, complementary strands of RNA can form duplex, double-stranded molecules. The formation of duplex structures between complementary antisense and messenger RNA prevents the translation of the mRNA into a polypeptide. The precise mode of inactivation has not yet been fully determined. It may render the whole duplex subject to action by nucleolytic enzymes that recognize and destroy such structures, or in eukaryotes it may prevent processing of the mRNA, transport to the cytoplasm, or the binding of mRNA to ribosomes.

The formation of antisense RNA was first discovered as a naturally occurring phenomenon in bacteria, where it may play a role in gene regulation. This discovery led researchers to explore the use of antisense RNA as a means of inhibiting specific genes. Production of an effective antisense RNA potentially eliminates activity in the complementary strand of mRNA; hence, introduction of an antisense RNA could be equivalent to introducing a mutation into a normal gene, allowing study of altered cell function in the absence of a specific gene. Besides use as a research tool to study gene function, antisense technology provides a powerful means of suppressing unwanted gene function, for example, in viral infections.

RNA is not the only nucleic acid that can bind to and inhibit mRNA; short complementary strands of DNA (antisense oligodeoxynucleotides or ODNs) can also be synthesized and used. However, practical use of this technique requires that the antisense ODN be able to enter the cell. Although cells take up nucleic acids and ODNs, they do so only at low levels, and once inside, the nucleic acids and ODNs are subject to rapid degradation by nucleolytic enzymes. To circumvent this, a DNA sequence complementary to the antisense RNA is cloned into a vector and transferred into a host cell or organism. Once inside, the DNA is transcribed into the antisense RNA. Alternatively, the ODNs can be chemically modified so that they pass into the cell more efficiently and are more resistant to enzymatic degradation.

sus sequences and, in the above case, the important role they play in transcription.

### The Synthesis of RNA

Once the promoter has been recognized and bound by the enzyme complex, RNA polymerase catalyzes the insertion of the first 5'-ribonucleoside triphosphate, which is complementary to the first nucleotide at the start site of the DNA template strand. Unlike DNA synthesis, no primer is required. Subsequent ribonucleotide complements are inserted and linked together by phosphodiester bonds as RNA polymerization proceeds. This process, called **chain elongation** [Figure 14.8(a)], continues in the 5'-to-3' direction, creating a temporary DNA/RNA duplex whose chains run antiparallel to one another.

After a few ribonucleotides have been added to the growing RNA chain, the $\sigma$ subunit is dissociated from the holoenzyme, and elongation proceeds under the direction of the core enzyme [Figure 14.8(b)]. In *E. coli* this process proceeds at the rate of about 50 nucleotides/second at 37°C.

Eventually, the enzyme traverses the entire gene and encounters a termination signal, a specific nucleotide sequence. Synthesis is completed usually in conjunction with the **termination factor, rho ($\rho$)**. At that point, the transcribed RNA molecule is released from the DNA tem-

Although the use of antisense technology is still quite new, it has been used to inhibit replication of human immunodeficiency virus (HIV). In another application, plants genetically engineered to produce antisense RNA against a viral gene of the tomato golden mosaic virus (TGMV) develop significant resistance to infection.

More recently, antisense technology has been tested as a therapeutic method for cancer treatment. Mice, transformed to carry human leukemia cells, were given antisense oligodeoxynucleotides against a human gene called *c-myb*. This gene, known to be a cancer-causing gene when mutated, encodes a protein that stimulates replication of white blood cells, a process that is uncontrolled in leukemia. Treatment with *c-myb* antisense ODN significantly prolonged survival of treated mice up to 3.5 times longer than control animals. These encouraging results indicate that antisense therapy may prove to be effective in the treatment of human leukemias and other forms of cancer.

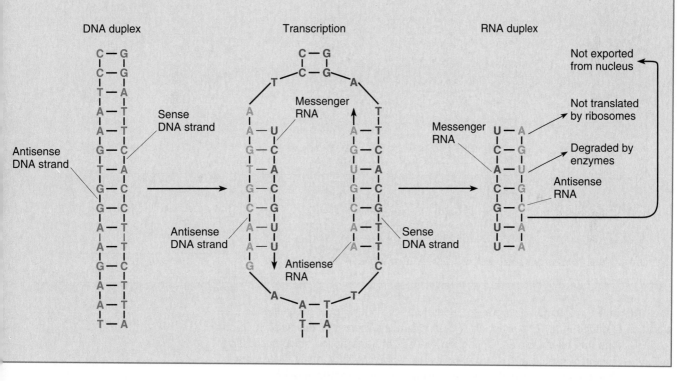

plate, and the core enzyme dissociates. Under the direction of RNA polymerase, an RNA molecule is synthesized that is precisely complementary to a DNA sequence representing the template strand of a gene. Wherever an A, T, C, or G residue existed, a corresponding U, A, G, or C residue has been incorporated into the RNA molecule, respectively. The significance of this synthesis is enormous, for it is the initial step in the process of information flow within the cell.

## Visualization of Transcription

Electron microscope studies by Oscar Miller, Jr., Barbara Hamkalo, and Charles Thomas have provided strik-

ing visual demonstrations of the transcription process. Figure 14.9 shows micrographs and interpretive drawings from two organisms, the bacterium *E. coli* and the newt *Notophthalmus viridescens*. In both cases multiple strands of RNA are seen to emanate from different points along the more central DNA template. Many RNA strands result because numerous transcription events are occurring simultaneously along each gene. Progressively longer RNA strands are found farther downstream from the point of initiation of transcription, while the shortest strands are closest to the point of initiation.

An interesting picture emerges from the study of *E. coli* [Figure 14.9(a)]. Because prokaryotes lack nuclei, cytoplasmic ribosomes are not separated physically from

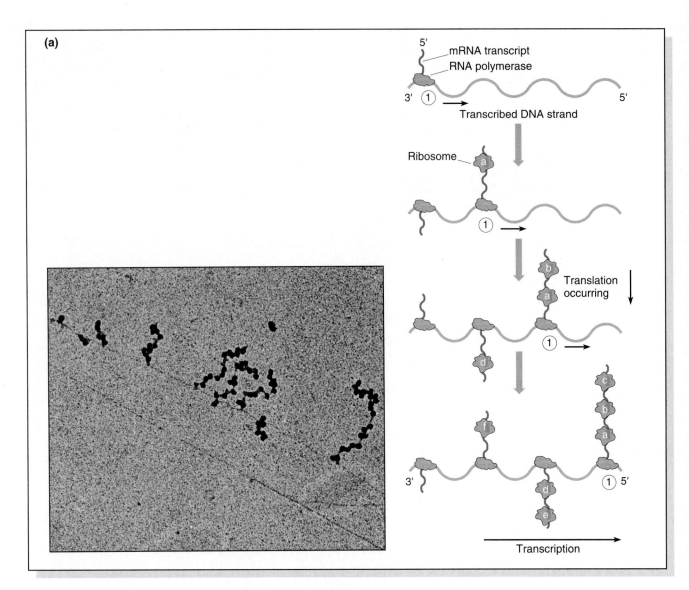

**FIGURE 14.9**   Electron micrographs and interpretive drawings of simultaneous transcription of genes in (a) *E. coli* and (b) *Notophthalmus* (*Triturus*) *viridescens*. In *E. coli*, a single gene has been isolated. During the process of transcription, each successive mRNA is engaged in translation by ribosomes. In *Notophthalmus*, many rRNA genes are present and many rRNA molecules are being transcribed on each gene. The diagrams illustrate the sequence of genetic events leading to what is visualized under the electron microscope.

the chromosome. As a result, ribosomes are free to attach to *partially* transcribed mRNA molecules and initiate translation. The longer RNA strands demonstrate the greatest number of ribosomes (see the later discussion of polyribosomes). In the case of the newt, the segment of DNA is derived from oocytes that are known to produce an enormous amount of rRNA. To accomplish this synthesis, the genes specific for RNA (**rDNA**) are replicated

many times in the oocyte. This process is called **gene amplification**. The micrograph in Figure 14.9(b) shows many of these genes in tandem, each being transcribed simultaneously numerous times. For each rRNA gene, longer and longer strands of incomplete rRNA molecules are produced as the enzymes move along the DNA strand. Visualization of transcription confirms our expectations based on the biochemical analysis of this process.

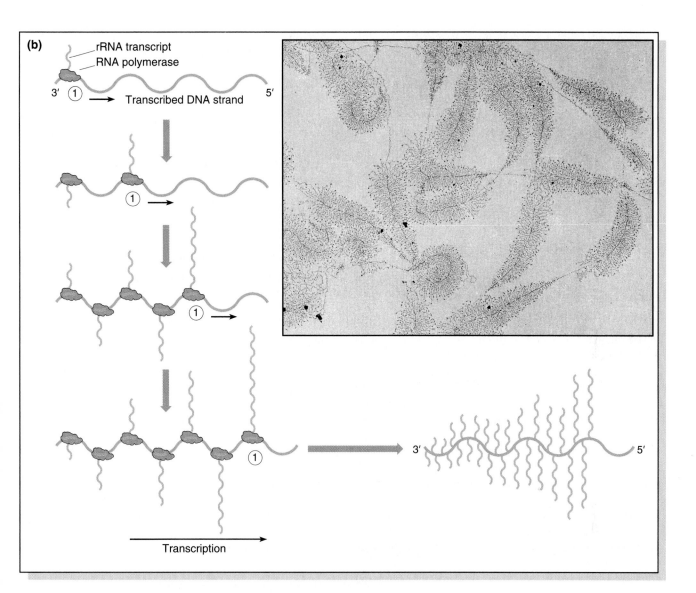

# TRANSCRIPTION IN EUKARYOTES

Much of our knowledge of transcription has been derived from studies of prokaryotes. Many general aspects of the mechanics of these processes are similar in eukaryotes. There are, however, numerous notable differences, several of which will be discussed below. We will first summarize some of the major differences and then expand on several of these in greater detail.

1. Transcription in eukaryotes occurs within the nucleus under the direction of three separate forms of RNA polymerase. (See Table 14.6 on page 456.) Unlike prokaryotes, the RNA transcript is not free to associate with ribosomes prior to the completion of transcription. For the mRNA to be translated, it must move out of the nucleus into the cytoplasm.

2. The initiation and regulation of transcription involve more extensive nucleotide sequences found in DNA upstream from the point of initial transcription. There are, in addition to promoters, other control units called enhancers.

3. Maturation of eukaryotic mRNA from the primary transcript involves many complex stages referred to generally as "processing." An initial step involves the addition of a 5' cap and a 3' tail to most transcripts destined to become mRNAs.

4. Most notably, extensive modifications occur to the internal sequence of nucleotides of eukaryotic RNA

transcripts that eventually serve as mRNAs. The initial transcripts are most often much larger than those that are eventually translated. Thus, they are called **pre-mRNAs** and are thought to constitute a group of molecules found only in the nucleus—a group referred to generally as **heterogeneous nuclear RNA (hnRNA)**. Only about 25 percent of hnRNA molecules are converted to mRNA. Those that are converted have substantial amounts of their ribonucleotide sequence excised, while the remaining segments are spliced back together prior to translation. This phenomenon has given rise to the concepts of **split genes** and **splicing** in eukaryotes.

In the final sections of this chapter, we will elaborate on each of these differences.

## Eukaryotic Promoters, Enhancers, and Transcription Factors

The recognition of certain highly specific DNA regions by RNA polymerase is at the heart of orderly genetic function in all cells. Both the polymerases and promoters leading to template binding have been found to be more complex in eukaryotes. RNA polymerase, as discussed earlier, exists in three forms and each is larger and more intricate than the prokaryotic counterpart of the enzyme. In regard to the initial template binding step and promoter regions, most is known about polymerase II, which transcribes all mRNAs in eukaryotes.

There are at least three areas of a gene that are essential to the efficient initiation of transcription by polymerase II. The first is called the **Goldberg–Hogness** or **TATA box**, which is found about 25 nucleotide pairs upstream ($-25$) from the start point of transcription. The consensus sequence is a heptanucleotide consisting solely of A and T residues. The sequence and function are analogous to that found in the $-10$ region of prokaryotic genes. While mutations in this sequence reduce or eliminate transcription of a variety of genes, the TATA box may be fairly nonspecific and simply have the responsibility for fixing the site of initiation of transcription by facilitating the denaturation of the helix. Such a conclusion is supported by the fact that A = T base pairs are less stable than G ≡ C pairs.

The second region of interest is found farther upstream from the TATA box. In different genes studied, positions anywhere from 5 to 500 nucleotides upstream appear also to modulate transcription. These regions of DNA have been located on the basis of the effects of their deletion on transcription. Their loss appears to drastically reduce *in vivo* transcription. Some regions, such as those associated with globin and the viral SV40 genes, are about 50 to 100 nucleotides from the TATA sequence. Others, such as those associated with the sea

urchin H2A gene and the *Drosophila* glue protein gene, are 200 to 500 nucleotides upstream. Because GC-rich sequences as well as the sequence CCAAT are frequently part of these regions, they are sometimes called **CCAAT boxes**.

The third region is represented by elements called **enhancers**. While their location may vary, enhancers often may be found even farther upstream than the regions discussed above. They have the effect of modulating transcription from a distance. Thus, they may not participate directly in template binding, even though they are essential to highly efficient initiation of transcription. We will return to a discussion of these elements in Chapter 17.

Aside from the more extensive regulatory sequences present in eukaryotes, a second major difference exists in the facilitation of template binding: the presence of protein transcription factors. It is generally believed that RNA polymerase II cannot bind directly to promoter sites and initiate transcription without the presence of such factors.

It appears that this enzyme recognizes a molecular complex consisting of DNA bound by several specific proteins. For example, there is a **TATA-factor** (also called **TFIID**), which binds to the promoter sequence and is critical to transcription. Still other factors specific to the CCAAT box and other DNA regions have been discovered and studied. It may be that these transcription factors supplant the role of the sigma factor in the prokaryotic enzyme, thus providing a much greater degree of specific control of transcription. We will return in Chapter 17 to a consideration of the role of some of these factors in eukaryotic gene regulation.

## Heterogeneous RNA and Its Processing

Still other insights into regions of DNA that do not directly encode proteins have come from the study of RNA. This research has provided detailed knowledge of eukaryotic gene structure. The genetic code is written in the ribonucleotide sequence of mRNA. This information originated, of course, in the template strand of DNA, where complementary sequences of deoxyribonucleotides exist. In bacteria, the relationship between DNA and RNA appears to be quite direct. The DNA base sequence is transcribed into an mRNA sequence, which is then translated into an amino acid sequence according to the genetic code.

However, in eukaryotes the situation is much more complex than in bacteria. It has been found that many internal base sequences of a gene may never appear in the mature mRNA that is translated. Other modifications occur at the beginning and the end of the mRNA prior to translation. These findings have made it clear that in eukaryotes, complex processing of mRNA occurs before it participates in translation.

By 1970, accumulating evidence showed that eukaryotic mRNA is transcribed initially as a much larger precursor molecule than that which is translated. This notion was based on the observation by James Darnell and coworkers of **heterogeneous nuclear RNA (hnRNA)** in mammalian cells. Heterogeneous RNA is of large but variable size (up to $10^7$ daltons), is found only in the nucleus, and is rapidly degraded. Nevertheless, hnRNA was found to contain nucleotide sequences common to the smaller mRNA molecules of the cytoplasm. Thus, it was proposed that the initial transcript of a gene results in a large RNA molecule that must first be processed in the nucleus before it appears in the cytoplasm as a mature mRNA molecule.

A subsequent discovery provided further evidence for this proposal. Both hnRNAs and mRNAs were found to contain at the 3' end a stretch of up to 22 adenylic acid residues. Such **poly A** sequences are added **posttran-** **scriptionally** to the initial gene transcript. In higher eukaryotes, for example, transcription is terminated, and the 3' end of the initial transcript is reduced in length close to an AAUAAA sequence and then polyadenylated. Subsequent investigation has shown poly A at the 3' end of almost all mRNAs studied in a variety of eukaryotic organisms. The exceptions seem to be the products of histone genes and some yeast genes.

Poly A is added to the initial RNA transcript, which is then processed before its transport to the cytoplasm. Processing reduces the size of the initial transcript. A further posttranscriptional modification of eukaryotic mRNA involves the 5' end of these molecules, where a **7-methyl guanosine (7mG) cap** is added. The cap appears to be important to the successful translation of the message, perhaps by serving as a recognition point for ribosome binding. Methylation of the two initial nucleotides is part of the capping process and may provide added recognition. These various modifications are summarized in Figure 14.10.

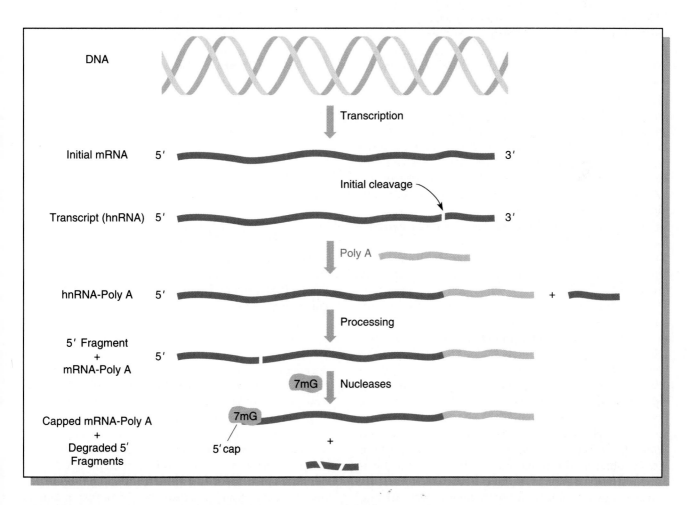

**FIGURE 14.10** The conversion in eukaryotes of heterogeneous nuclear RNA (hnRNA) to messenger (mRNA), which contains a 5' cap and a 3' poly A tail.

## Intervening Sequences and Split Genes

One of the most exciting discoveries in the history of molecular genetics occurred in 1977. At this time, direct evidence was provided by Susan Berget, Philip Sharp, and Richard Roberts that the genes of animal viruses contain internal nucleotide sequences that are not expressed in the amino acid sequence of the proteins that they encode. That is, certain internal sequences in DNA do not always appear in the mature mRNA that is translated into a protein.

Such nucleotide segments have been called **intervening sequences**, contained within **split genes**. Those DNA sequences not present in the final mRNA product are also called **introns** ("int" for intervening), and those retained and expressed are called **exons** ("ex" for expressed). Removal of introns occurs as a result of an excision and rejoining process referred to as **splicing**.

Similar discoveries were soon to be made in eukaryotes. Two approaches have been most fruitful. The first involves molecular hybridization of purified, functionally mature mRNAs with DNA containing the genes specifying that message. When hybridization occurs between nucleic acids that are not perfectly complementary, **heteroduplexes** are formed that may be visualized with the electron microscope. As illustrated in Figure 14.11, introns present in DNA but absent in mRNA must loop out and remain unpaired. This figure shows an electron micrograph and an interpretive drawing derived from hybridization between the template strand of the gene encoding chicken ovalbumin and the mature RNA prior to its translation into that protein. There are seven introns (A–G) whose sequences are present in DNA but not in the final mRNA.

The second approach provides more specific information. It involves a comparison of nucleotide sequences of DNA with those of mRNA and amino acid sequences. Such an approach allows the precise identification of all intervening sequences.

So far, a large number of genes from diverse eukaryotes have been shown to contain introns. One of the first so identified was the **beta-globin gene** in mice and rabbits, as studied independently by Philip Leder and Richard Flavell. The mouse gene contains an intron 550 nucleotides long, beginning immediately after the codon specifying the 104th amino acid. In the rabbit (Figure 14.12), there is an intron of 580 base pairs near the codon for the 110th amino acid. Additionally, a second intron of about 120 nucleotides exists earlier in both genes. Similar introns have been found in the beta-globin gene in all mammals examined.

Several genes, notably those coding for histones and interferon, appear to contain no introns. However, intervening sequences have been identified in most other eukaryotic genes that have been examined.

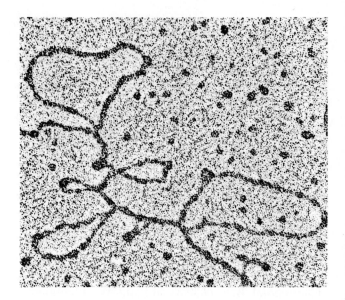

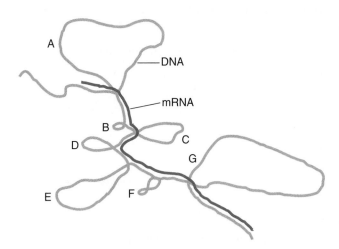

**FIGURE 14.11**    An electron micrograph and an interpretive drawing of the hybrid molecule formed between the template DNA strand of the ovalbumin gene and the mature ovalbumin mRNA. Seven DNA introns, A–G, produce unpaired loops.

As pointed out above, a more extensive set of introns has been located in the **ovalbumin gene** of chickens. The gene has been extensively characterized by Bert O'Malley in the United States and Pierre Chambon in France. As shown in Figure 14.12, the gene contains seven introns. Notice that the majority of the gene's DNA sequence is "silent," being composed of introns. The initial RNA transcript is four times the length of the mature mRNA. You should compare the information on the ovalbumin gene presented in Figures 14.11 and 14.12. Can you match the unpaired loops in Figure 14.11 with the sequence of introns specified in Figure 14.12?

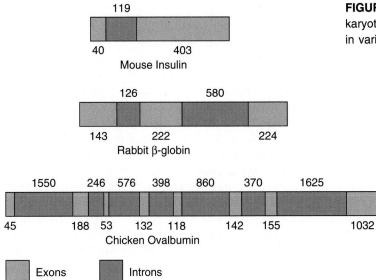

**FIGURE 14.12**    Intervening sequences in various eukaryotic genes. The numbers indicate nucleotides present in various intron and exon regions.

The list of genes containing intervening sequences is growing rapidly. In fact, few eukaryote genes seem to be without introns. An extreme example of the number of introns in a single gene is that found in one of the chicken genes, *pro-α-2(1) collagen*. One of several genes coding for a subunit of this connective tissue protein, *pro-α-2(1) collagen* contains about 50 introns. The precision with which cutting and splicing occur must be extraordinary if errors are not to be introduced into the mature mRNA.

## Splicing Mechanisms

The discovery of split genes represents one of the most exciting genetic findings in recent years. As a result, intensive investigation is now in progress to elucidate the mechanism by which introns of RNA are excised and exons are spliced back together. A great deal of progress has already been made. Interestingly, it appears that somewhat different mechanisms are utilized for each of the three types of RNA as well as for RNAs produced in mitochondria and chloroplasts.

Based on splicing mechanisms, there are four groups of introns (I–IV). Group I includes those that are part of the primary transcript of rRNAs derived from nuclei, mitochondria, and chloroplasts. Contrary to what was anticipated, there are no high-energy transformations or enzymatic reactions involved in splicing the primary transcript that leads to mature rRNAs. Instead, a **self-excision process** occurs (Figure 14.13). This amazing discovery, made in 1982 by Thomas Cech and his colleagues in their studies of the ciliated protozoan *Tetrahymena*, revealed that RNA can demonstrate autonomous catalytic properties. As a result, the RNAs that undergo splicing are sometimes called **ribozymes**.

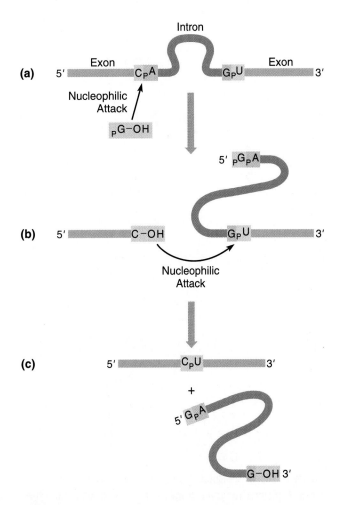

**FIGURE 14.13**    Splicing mechanism involved with group I introns removed from the primary transcript leading to rRNA. The process is one of self-excision.

Chemically, two nucleophilic reactions occur, the first involving guanosine and the primary transcript [Figure 14.13(a)] and then involving an intermediate structure [Figure 14.13(b)], leading to the spliced RNA [Figure 14.13(c)]. Cech demonstrated that proteolytic enzymes were *not* involved when he transcribed rDNA *in vitro*. The primary transcript of rRNA then underwent self-splicing in the absence of any other exogenous proteins.

Group II introns are those removed from primary transcripts leading to mRNAs in mitochondria and chloroplasts. Like group I molecules, these are also autocatalytic but involve an adenosine residue found within the intron [Figure 14.14(a)]. The intermediate forms a characteristic lariat-like structure [Figure 14.14(b)] that is then excised [Figure 14.14(c)]. Thus, a slightly different mechanism has evolved for self-splicing by these transcripts.

Group III introns are the largest group, being found in nuclear-derived transcripts representing mRNAs. Introns in mRNA, in comparison to other RNAs thus far dis-

cussed, can be much larger— up to 20,000 nucleotides— and they are more plentiful. Thus, their removal appears to require a much more complex mechanism, which until recently has been very difficult to define.

Many clues are now emerging. First, the nucleotide sequence around different introns is often similar. Most begin at the 5′ end with a GU dinucleotide sequence and terminate at the 3′ end with an AG dinucleotide sequence. These, as well as other consensus sequences shared by introns, may attract a molecular complex essential to ligation and splicing. Such a complex has been identified in extracts of yeast as well as mammalian cells. Called a **spliceosome**, it is very large, being $40S$ in yeast and $60S$ in mammals. One group of components of these complexes consists of a unique set of small nuclear RNAs (snRNAs) that are usually 250 nucleotides or less and complexed with proteins. Found only in the nucleus, these complexes are called **small nuclear ribonucleoproteins** (snRNPs—or snurps). The snRNAs have been arbitrarily designated U1, U2, U3, . . . , U6, and the list continues to grow. The U1 snRNA bears a nucleotide sequence that is homologous to that of the 5′ end of the intron. Base pairing resulting from this homology promotes binding that represents the initial step in the formation of the spliceosome. Following the addition of the other snurps (U2, U4, U5, U6), splicing commences and introns are removed.

The simplest splicing mechanism appears to be used to process group IV introns removed from **tRNA molecules**. As you may recall, these are small RNAs consisting of approximately 80 nucleotides. Some of them contain a short internal intron of about 15 nucleotides, generally located one nucleotide from the anticodon (on the 3′ side of it). In such cases, initial folding of the molecule creates a **pre-tRNA** with an extra loop representing the intron. This is enzymatically removed. John Abelson has demonstrated this enzymatic process *in vitro*. A nuclease makes appropriate cuts in the RNA, leaving the ends that are to be joined in juxtaposition to one another. A ligase-like reaction then links them together.

Even given the above knowledge, the complexity of the splicing complex has made analysis most difficult. While we have discussed the removal of one intron in a single RNA transcript, large numbers often must be cleaved out and the many exons spliced back together with absolute precision. To ensure such precision, an extremely efficient mechanism must be at work.

The finding of "genes in pieces," as split genes have been described, raises many interesting questions and has provided great insights into the organization of eukaryotic genes. In addition, the processing involved in splicing represents a potential regulatory step during gene expression. Thus, we will return to this topic again in Chapter 17.

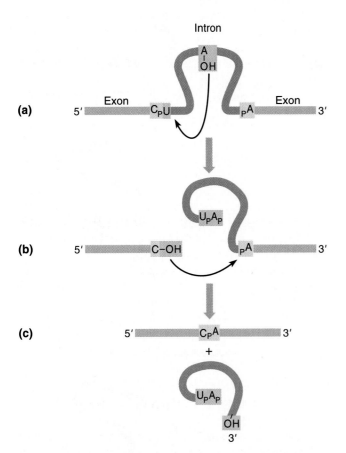

**FIGURE 14.14**   Splicing mechanism involved with group II introns removed from mRNAs. The lariat structure in the intermediate stage is characteristic of this mechanism.

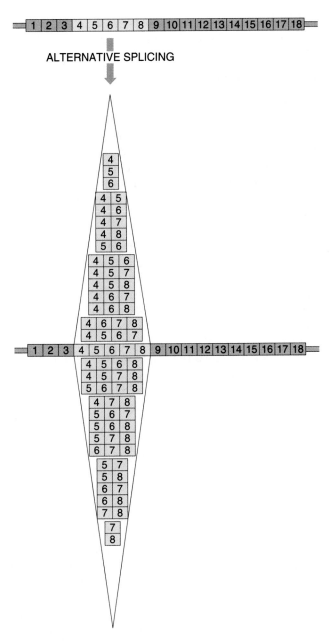

Alternative Exon Combinations

**FIGURE 14.15**    Illustration of the 32 combinations of exons that may result from alternative splicing of the troponin T gene. All combinations of exons 4–8 are generated in addition to the constant presence of exons 1–3 and 9–18. *Not shown* is the potential alternative splicing involving exons 16 and 17 that may generate two versions ($\alpha$ and $\beta$ forms) of each of the 32 combinations shown.

## Alternative Splicing

We conclude our coverage of splicing by considering several cases where group III introns present in pre-mRNAs *derived from the same gene* are spliced in more than one way, thereby yielding different collections of exons in the mature mRNA. This process, referred to as **alternative splicing**, yields a group of mRNAs, that upon translation, result in a series of related proteins called **isoforms**. A growing number of examples are now found in organisms ranging from viruses to *Drosophila* to humans.

Our first example includes the myosin light chain (MLC) from vertebrates. In this case two promoters exist, and depending on which is utilized by RNA polymerase II, initial RNA transcripts of slightly different lengths are formed. Alternative splicing yields related but distinct mRNAs, differing in their 5′ exons. The shorter of the two lacks one of the exons present in the longer RNA transcript.

A second type of alternative splicing results when the pre-mRNA transcript is polyadenylated at different locations. In mammals, the calcitonin gene transcript may yield two separate mRNAs, one with four exons (1–4) and one with five exons (1–3 and 5–6). The specificity of tissues determines which splice sites are utilized based on the poly A location.

The third example involves the case where a common group of exons are present in all pre-mRNA transcripts and a "mix-and-match" approach to splicing yields a diverse group of exon combinations. For example, in the muscle protein troponin T, all combinations of the presence or absence of five exons along with the remaining 14 fixed exons yields 32 different mRNAs (Figure 14.15). In another example, RNA present in the viral genome of retroviruses is commonly alternatively spliced during infection, producing different mRNAs, which upon translation yield different protein components.

Alternative splicing of pre-mRNAs provides the basis for producing related proteins from a single gene. We shall return to this topic in our discussion of the regulation of gene expression in eukaryotes (Chapter 17).

## TRANSLATION: COMPONENTS NECESSARY FOR PROTEIN SYNTHESIS

Translation of mRNA is the biological polymerization of amino acids into polypeptide chains. The process, alluded to in our earlier discussion of the genetic code, occurs only in association with ribosomes. The central question in translation is how triplet ribonucleotides of mRNA direct specific amino acids into their correct position in the polypeptide. This question was answered once transfer RNA (tRNA) was discovered. This class of molecules adapts specific triplet codons in mRNA to their correct amino acids. This adaptor role of tRNA was postulated by Crick in 1957.

In association with a ribosome, mRNA presents a triplet codon that calls for a specific amino acid. Because a specific tRNA molecule contains within its composition three consecutive ribonucleotides complementary to the **codon**, they are called the **anticodon** and can base pair with the codon. Another region of this tRNA is covalently bonded to the amino acid called for. Inside the ribosome, hydrogen bonding of tRNAs to mRNA holds the amino acid in proximity so that a peptide bond can be formed. As this process occurs over and over as mRNA runs through the ribosome, amino acids are polymerized into a polypeptide.

In our discussion of translation, we will first consider the structure of the ribosome and transfer RNA, two of the major components essential for protein synthesis.

## Ribosomal Structure

Because of its essential role in the expression of genetic information, the ribosome has been extensively analyzed. One bacterial cell contains about 10,000 of these structures, while a eukaryotic cell contains many times more. Electron microscopy has revealed that the bacterial ribosome is about 250 Å in its largest diameter and consists of a larger and a smaller subunit. Both subunits consist of one or more molecules of rRNA and an array of **ribosomal proteins**.

The specific differences between prokaryotic and eukaryotic ribosomes are summarized in Figure 14.16. The subunit and rRNA components are most easily isolated and characterized on the basis of their sedimentation

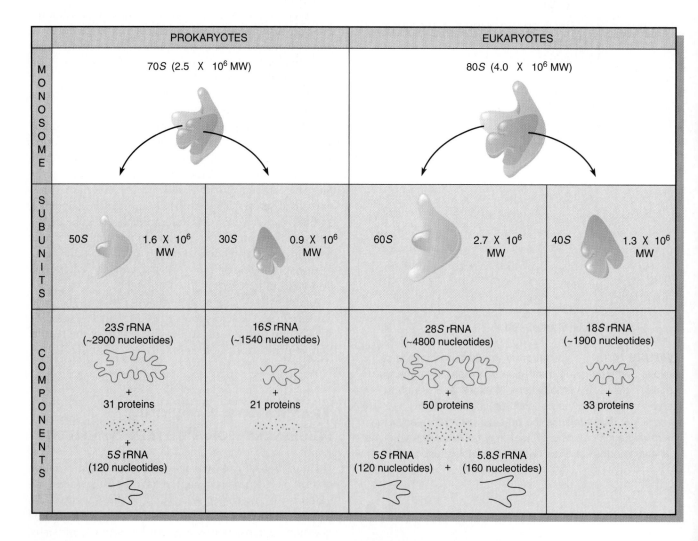

**FIGURE 14.16**    A comparison of the components in the prokaryotic and eukaryotic ribosomes.

behavior (their rate of migration) in sucrose gradients (see Chapter 8). The two subunits associated with each other constitute a **monosome**. In prokaryotes the monosome is a $70S$ particle, and in eukaryotes it is approximately $80S$. Sedimentation coefficients, which reflect the variable rate of migration of different-sized particles and molecules, are not additive. For example, the $70S$ monosome consists of a $50S$ and a $30S$ subunit, and the $80S$ monosome consists of a $60S$ and a $40S$ subunit.

The larger subunit in prokaryotes consists of a $23S$ RNA molecule, a $5S$ rRNA molecule, and 31 ribosomal proteins. In the eukaryotic equivalent, a $28S$ rRNA molecule is accompanied by a $5.8S$ and $5S$ rRNA molecule and about 50 proteins. In the smaller prokaryotic subunits, a $16S$ rRNA component and 21 proteins are found. In the eukaryotic equivalent, an $18S$ rRNA component and about 33 proteins are found. The approximate molecular weights and number of nucleotides of these components are shown in Figure 14.16.

Molecular hybridization studies have established the degree of redundancy of the genes coding for the rRNA components. The *E. coli* genome contains seven copies of a single sequence that codes for all three components—$23S$, $16S$, and $5S$. The initial transcript of these genes produces a $30S$ RNA molecule that is enzymatically cleaved into these smaller components. Having the genetic information encoding these three rRNA components coupled together ensures that, following multiple transcription events, equal quantities of all three will be present as ribosomes are assembled.

In eukaryotes, many more copies of a sequence encoding the $28S$ and $18S$ components are present. In *Drosophila*, approximately 120 copies per haploid genome are each transcribed into a molecule of about $34S$. This is processed to the $28S$, $18S$, and $5.8S$ rRNA species. These are homologous to the three rRNA components of *E. coli*. In *X. laevis*, over 500 copies per haploid genome are present. In mammalian cells, the initial transcript is $45S$. The rRNA genes are part of the moderately repetitive DNA fraction and are present in clusters at various chromosomal sites. Each cluster consists of **tandem repeats**, with each unit separated by a noncoding **spacer DNA** sequence [Figure 14.9(b)]. In humans, these gene clusters have been localized on the ends of chromosomes 13, 14, 15, 21, and 22.

The unique $5S$ rRNA component of eukaryotes is not part of the larger transcript. Instead, genes coding for this ribosomal component are distinct and located separately. In humans, a gene cluster encoding them has been located on chromosome 1.

In spite of the detailed knowledge available on the structure and genetic origin of the ribosomal components, a complete understanding of the function of these components has eluded geneticists. This is not surprising, because the ribosome is perhaps the most intricate of all cellular organelles. In bacteria, the monosome has a combined molecular weight of 2.5 million daltons!

## tRNA Structure

Because of their small size and stability in the cell, tRNAs have been investigated extensively. In fact, tRNAs are the best characterized RNA molecules. They are composed of only 75 to 90 nucleotides, displaying a nearly identical structure in bacteria and eukaryotes. In both types of organisms, tRNAs are transcribed as larger precursors, which are cleaved into mature $4S$ tRNA molecules. In *E. coli*, for example, tRNA[tyr] (the superscript identifies the specific tRNA and the amino acid that binds to it) is composed of 77 nucleotides, yet its precursor contains 126 nucleotides.

In 1965, Robert Holley and his coworkers reported the complete sequence of tRNA[ala] isolated from yeast. Of great interest was the finding that there are a number of nucleotides unique to tRNA. As shown in Figure 14.17, each is a modification of one of the four nitrogenous bases expected in RNA (G, C, A, and U). These include **inosinic acid**, which contains the purine **hypoxanthine**, **ribothymidylic acid**, and **pseudouridine** among others. These modified structures, sometimes referred to as *unusual*, *rare*, or *odd bases*, are created **posttranscriptionally**. That is, the unmodified base is inserted during transcription, and then enzymatic reactions catalyze the modifications.

Holley's sequence analysis led him to propose the two-dimensional **cloverleaf model** of transfer RNA. It had been known that tRNA demonstrates secondary structure due to base pairing. Holley discovered that he could arrange the linear model in such a way that several stretches of base pairing would result. This arrangement created a series of paired stems and unpaired loops resembling the shape of a cloverleaf. Loops consistently contained modified bases, which do not generally form base pairs. Holley's model is shown in Figure 14.18.

Since the triplets GCU, GCC, and GCA specify alanine, Holley looked for the theoretical anticodon of his tRNA[ala] molecule. He found it in the form of CGI, at the bottom loop of the cloverleaf. Recall from Crick's wobble hypothesis that I (inosinic acid) is predicted to pair with U, C, or A. Thus, the anticodon loop was established.

As other tRNA species were examined, numerous constant features were observed. First, at the 3′ end, all tRNAs contain the sequence **. . . pCpCpA-3′**. It is to the terminal adenosine residue that the amino acid is joined covalently during charging. At the 5′ terminus, all tRNAs contain **. . . pG-5′**.

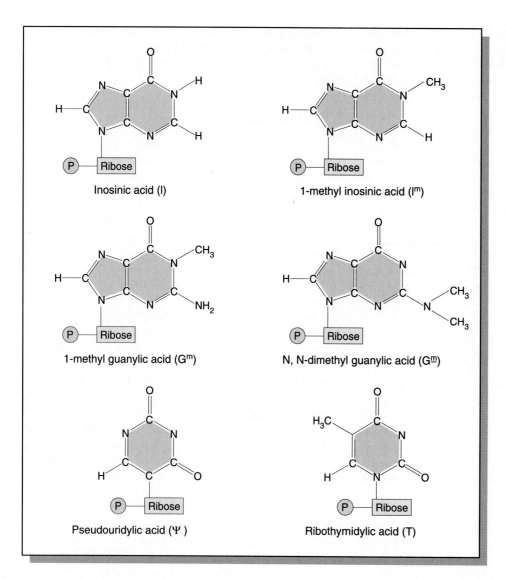

FIGURE 14.17    Unusual nitrogenous bases found in transfer RNA.

Inosinic acid (I)

1-methyl inosinic acid (Iᵐ)

1-methyl guanylic acid (Gᵐ)

N, N-dimethyl guanylic acid (G$\underline{m}$)

Pseudouridylic acid (Ψ)

Ribothymidylic acid (T)

Additionally, the lengths of various stems and loops are very similar. Each tRNA examined also contains an anticodon complementary to the known amino acid code for which it is specific. All anticodon loops are present in the same position of the cloverleaf.

The cloverleaf model was predicted strictly on the basis of nucleotide sequence. Thus, there was great interest in X-ray crystallographic examination of tRNA, which reveals three-dimensional structure. By 1974, Alexander Rich and his coworkers in the United States, and J. Roberts, B. Clark, Aaron Klug, and their colleagues in England had been successful in crystallizing tRNA and

performing X-ray crystallography at a resolution of 3 Å. At such resolution, the pattern formed by individual nucleotides is discernible.

As a result of these studies, a complete three-dimensional model is now available (Figure 14.19). The model reveals tRNA to be L-shaped. At one end of the L is the anticodon loop and stem, and at the other end is the 3'-acceptor region where the amino acid is bound. It has been speculated that the shapes of the intervening loops may be recognized by specific enzymes responsible for adding the amino acid to tRNA, the subject to which we now turn our attention.

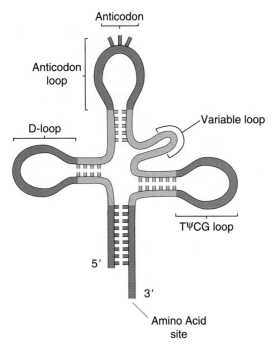

Anticodon

Anticodon loop

D-loop

Variable loop

TΨCG loop

5′

3′

Amino Acid site

**FIGURE 14.18**    The two-dimensional cloverleaf model of transfer RNA.

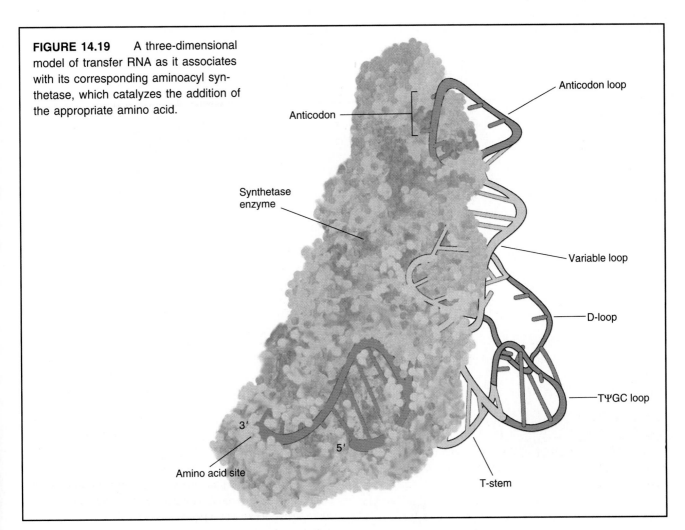

**FIGURE 14.19**    A three-dimensional model of transfer RNA as it associates with its corresponding aminoacyl synthetase, which catalyzes the addition of the appropriate amino acid.

Anticodon

Anticodon loop

Synthetase enzyme

Variable loop

D-loop

TΨGC loop

3′

5′

Amino acid site

T-stem

## Charging tRNA

Before translation can proceed, the tRNA molecules must be chemically linked to their respective amino acids. This activation process, called **charging**, occurs under the direction of enzymes called **aminoacyl tRNA synthetases**. Because there are 20 different amino acids, there must be at least 20 different tRNA molecules and as many different enzymes. In theory, since there are 61 triplet codes, there could be the same number of specific tRNAs and enzymes. Because of the ability of the third member of a triplet code to "wobble," however, it is now thought that there are about 30 different tRNAs; it is also believed that there are only 20 synthetases, one for each amino acid, regardless of a greater number of corresponding tRNAs.

The charging process is outlined in Figure 14.20. In the initial step, the amino acid is converted to an activated form, reacting with ATP to form an **aminoacyladenylic acid**. A covalent linkage is formed between the 5' phosphate group of ATP and the carboxyl end of the amino acid. This molecule remains associated with the enzyme, forming a complex that then reacts with a specific tRNA molecule. In this second step, the amino acid is transferred to the appropriate tRNA and bonded covalently to the adenine residue at the 3' end. The charged tRNA may participate directly in protein synthesis. Aminoacyl-tRNA synthetases are highly specific enzymes because they recognize only one amino acid and only a subset of corresponding tRNAs called isoaccepting tRNAs. This is a crucial point if fidelity of translation is to be maintained. The basis for this recognition and subsequent binding has sometimes been referred to as the **second genetic code**, although this is not a particularly apt descriptor.

## TRANSLATION: THE PROCESS

In a way similar to transcription, the process of translation can be best described by breaking it into discrete steps. Be aware, however, that translation is also a dynamic, ongoing process. Correlate the following discussion with the step-by-step characterization of the process in Figure 14.21. Many of the protein factors and their roles in translation are summarized in Table 14.7.

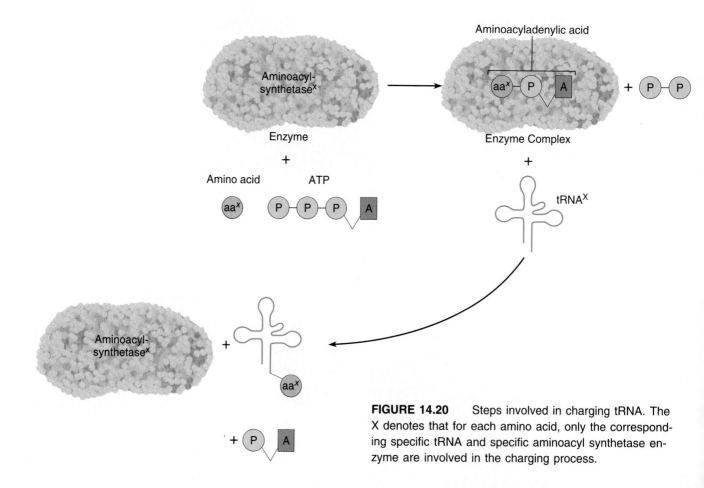

**FIGURE 14.20**    Steps involved in charging tRNA. The X denotes that for each amino acid, only the corresponding specific tRNA and specific aminoacyl synthetase enzyme are involved in the charging process.

**Table 14.7**    VARIOUS PROTEIN FACTORS INVOLVED DURING TRANSLATION IN *E. COLI*

| Process | Factor | Role |
|---|---|---|
| Initiation | IF1 | Stabilizes 30*S* subunit |
| | IF2 | Binds fmet-tRNA to 30*S*–mRNA complex; Binds to and stimulates GTP hydrolysis |
| | IF3 | Binds 30*S* subunit to mRNA; dissociates monosomes into subunits following termination |
| Elongation | EF-Tu | Binds GTP; brings aminoacyl-tRNA to the A site |
| | EF-Ts | Generates active EF-Tu |
| | EF-G | Stimulates translocation; GTP-dependent |
| Termination | RF1 | Catalyzes release of the polypeptide chain from tRNA and dissociation of the translocation complex; specific for UAA and UAG termination codons. |
| | RF2 | Behaves like RF1; specific for UGA and UAA codons |
| | RF3 | Stimulates RF1 and RF2 |

## Initiation (Steps 1–3)

Recall that the ribosome serves as a nonspecific workbench for the translation process. Most ribosomes, while not involved in translation, are dissociated into their large and small subunits. Initiation of translation in *E. coli* involves these subunits: an mRNA molecule, a specific charged initiator tRNA, GTP, $Mg^{++}$, and at least three proteinaceous **initiation factors (IFs)**. The initiator factors are not part of the ribosome, but are required to enhance the binding affinity of the various translational components, as described in Table 14.7. In prokaryotes, the initiation codon of mRNA, AUG, calls for the modified amino acid **formylmethionine**.

The small ribosomal subunit binds to several initiation proteins, and this complex in turn binds to mRNA. In bacteria, this binding involves a sequence of up to six ribonucleotides, which precedes the initial AUG codon of mRNA. This sequence (which is purine-rich and called the **Shine-Dalgarno sequence**) base pairs with a region of the 16*S* rRNA of the small ribosomal subunit.

Another initiation protein then facilitates the binding of charged formylmethionyl-tRNA to the small subunit in response to the AUG triplet. This step "sets" the reading frame so that all subsequent groups of three ribonucleotides are translated accurately. This aggregate represents the **initiation complex**, which then combines with the large ribosomal subunit. In this process, a molecule of GTP is hydrolyzed, providing the required energy, and the initiation factors are released.

## Elongation (Steps 4–9)

Once both subunits of the ribosome are assembled with the mRNA, binding sites for two charged tRNA molecules are formed. These are designated as the **P**, or **peptidyl**, and the **A**, or **aminoacyl**, **sites**. The initiation tRNA binds to the P site, provided that the AUG triplet of mRNA is in the corresponding position of the small subunit. The sequence of the second triplet in mRNA dictates which charged tRNA molecule will become positioned at the A site. Once it is present, **peptidyl transferase** catalyzes the formation of the peptide bond, which links the two amino acids together. This enzyme is part of the large subunit of the ribosome. At the same time, the covalent bond between the amino acid and the tRNA occupying the P site is hydrolyzed (broken). The product of this reaction is dipeptide (see Chapter 15), which is attached to the 3′ end of tRNA at the A site. The step in which the growing polypeptide chain has increased in length by one amino acid is called **elongation**.

Before elongation can be repeated, the tRNA attached to the P site, which is now uncharged, must be released from the large subunit. The entire **mRNA–tRNA–aa₂–aa₁** complex now shifts in the direction of the P site by a distance of three nucleotides. This event requires several protein **elongation factors (EFs)** as well as the energy derived from hydrolysis of GTP (Table 14.7). The result is that the third triplet of mRNA is now in a position to direct another specific charged tRNA into the A site. One

**FIGURE 14.21** Schematic representation of the process of translation, depicting the steps involved in protein synthesis.

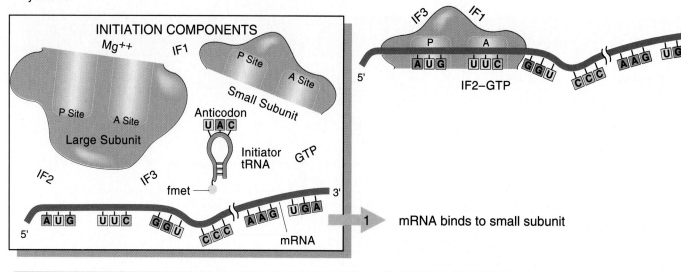

**INITIATION COMPONENTS**

1  mRNA binds to small subunit

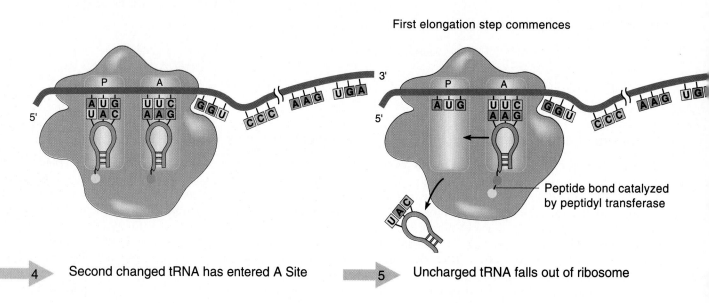

First elongation step commences

Peptide bond catalyzed by peptidyl transferase

4  Second changed tRNA has entered A Site

5  Uncharged tRNA falls out of ribosome

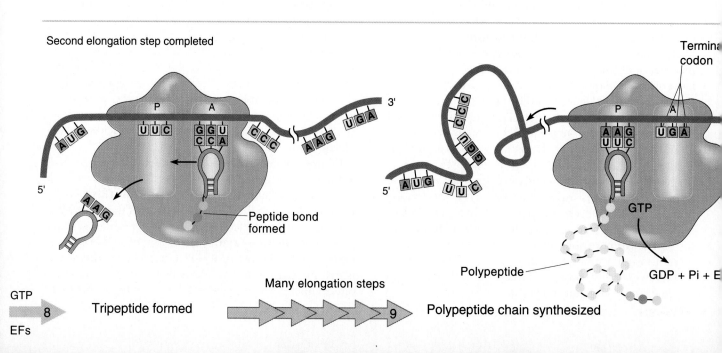

Second elongation step completed

Terminal codon

Peptide bond formed

Polypeptide

GDP + Pi + E

GTP
EFs

8  Tripeptide formed

Many elongation steps

9  Polypeptide chain synthesized

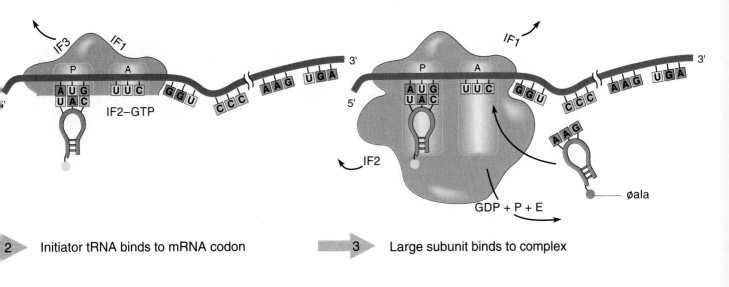

2 ▷ Initiator tRNA binds to mRNA codon

3 ▷ Large subunit binds to complex

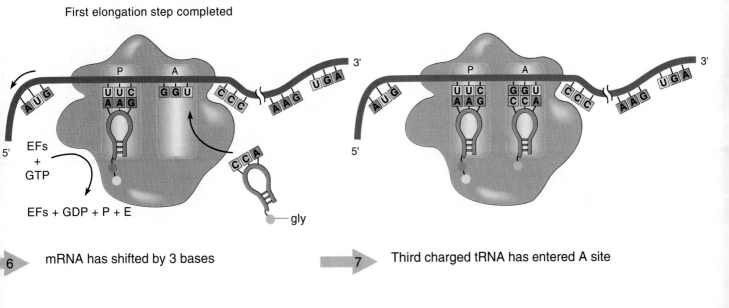

First elongation step completed

6 ▷ mRNA has shifted by 3 bases

7 ▷ Third charged tRNA has entered A site

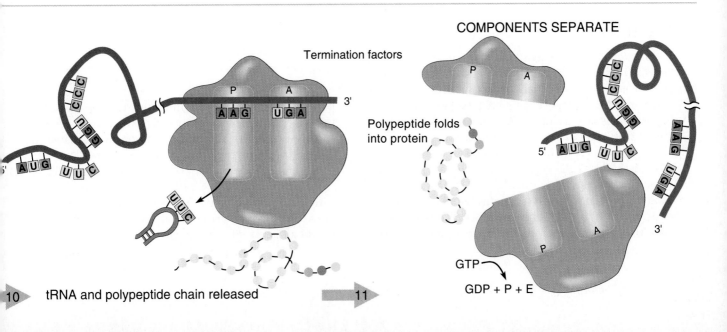

COMPONENTS SEPARATE

Termination factors

Polypeptide folds into protein

10 ▷ tRNA and polypeptide chain released

11 ▷

simple way to distinguish the two sites in your mind is to remember that, *following the shift*, the P site contains a tRNA attached to a peptide chain (*P* for peptide), while the A site contains a tRNA with an amino acid attached (*A* for amino acid).

The sequence of elongation is repeated over and over. An additional amino acid is added to the growing polypeptide chain each time the mRNA advances through the ribosome. The efficiency of the process is remarkably high; the observed error rate is only about $10^{-4}$. An incorrect amino acid will thus occur in every 20 polypeptides of an average length of 500 amino acids! In *E. coli*, elongation occurs at a rate of about 15 amino acids per second at 37°C. The process can be likened to a tape moving through a tape recorder. As the tape moves, sequential sound is emitted from the recorder. Likewise, as mRNA moves, a growing polypeptide is produced by the ribosome.

## Termination (Steps 10–11)

The termination of protein synthesis is signaled by one or more of three triplet codes: UAG, UAA, or UGA. These codons do not specify an amino acid, nor do they direct tRNA into the A site. These codons are called **stop codons**, **termination codons**, or **nonsense codons**. The finished polypeptide is therefore still attached to the terminal tRNA at the P site, and the A site is empty. The termination codon signals the action of **GTP-dependent release factors** (Table 14.7), which cleave the polypeptide chain from the terminal tRNA, releasing it from the translation complex. Once this cleavage occurs, the tRNA is released from the ribosome, which then dissociates into its subunits. If a termination codon should appear in the middle of an mRNA molecule as a result of mutation, the same process occurs, and the polypeptide chain is prematurely terminated.

## Polyribosomes

As elongation proceeds and the initial portion of mRNA has passed through the ribosome, the message is free to associate with another small subunit to form another initiation complex. This process can be repeated several times with a single mRNA and results in what are called **polyribosomes** or just **polysomes**.

Polyribosomes can be isolated and analyzed following a gentle lysis of cells. Figure 14.22(a) and (b) illustrates these complexes as seen under the electron microscope; note the presence in part (a) of this figure of mRNA between the individual ribosomes. The micrograph in part (b) is even more remarkable since it shows the poly-

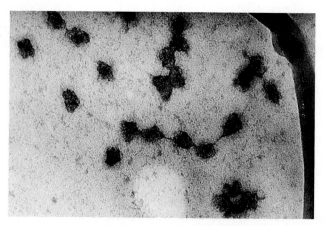

(a)

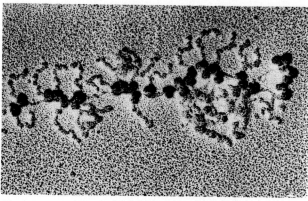

(b)

**FIGURE 14.22** Polyribosomes visualized under the electron microscope. Those viewed by electron microscopy were derived in part (a) from rabbit reticulocytes engaged in the translation of hemoglobin mRNA and in part (b) from giant salivary gland cells of the midgefly, *Chironomus thummi*. In part (b), the nascent polypeptide chain is apparent as it emerges from each ribosome. Its length increases as translation proceeds from left (5′) to right (3′) along the mRNA.

peptide chains emerging from the ribosomes during translation. The formation of polysome complexes represents an efficient use of the components available for protein synthesis during a unit time.

To complete the analogy with tapes (mRNA) and tape recorders (ribosomes), in polysome complexes one tape would be played simultaneously by several recorders, but at any given moment, the transcripts (polypeptides) would all be at different stages of completion.

# TRANSLATION IN EUKARYOTES

The general features of the model just presented were initially derived from investigations of the translation process in bacteria. During that discussion, we pointed out one of the main differences between translation in prokaryotes and eukaryotes. In the latter group, translation occurs on ribosomes that are larger and whose rRNA and protein components are more complex than prokaryotes (see Figure 14.16).

In addition, several other notable differences need to be mentioned. Eukaryotic mRNAs are much longer lived than their prokaryotic counterparts. Most exist for hours rather than minutes prior to their degradation by nucleases in the cell. Thus eukaryotic mRNAs are available much longer to orchestrate protein synthesis.

Two aspects that involve the initiation of translation are different in eukaryotes. First, as we discussed during our consideration of mRNA maturation, the 5′ end is "capped" with a 7-methylguanosine residue. The presence of this cap, absent in prokaryotes, is essential to efficient translation. RNAs lacking the cap are translated poorly. The cap seems to play a similar role to the Shine-Dalgarno sequence of prokaryotic mRNA in the initial binding of mRNA to the small subunit of the ribosome.

The second factor related to initiation of translation involves the insertion of the first amino acid. Initiation of eukaryotic translation does not require the amino acid formylmethionine. However, as in prokaryotes, the AUG triplet is essential to the formation of the translational complex, and a unique transfer RNA ($tRNA_i^{met}$) is used during initiation. Thus, the first amino acid in eukaryotic polypeptides is methionine.

Finally, protein factors similar to those in prokaryotes guide initiation, elongation, and termination of translation in eukaryotes. Many of these eukaryotic factors are clearly homologous to their counterparts in prokaryotes. However, there may be a greater number of factors required during each of these steps, and they may be somewhat more complex in prokaryotes.

## CHAPTER SUMMARY

1. The genetic code, stored in DNA, is copied to RNA, where it is used to direct the synthesis of polypeptide chains. It is degenerate, unambiguous, nonoverlapping, and comma-free.

2. The complete coding dictionary, determined using various experimental approaches, reveals that of the 64 possible codons, 61 encode the 20 amino acids found in proteins, while three triplets terminate translation. One of these 61 is the initiation codon and specifies methionine.

3. The observed pattern of degeneracy often involves only the third letter of a triplet series and led Crick to propose the wobble hypothesis.

4. Confirmation for the coding dictionary, including codons for initiation and termination, was obtained by comparing the complete nucleotide sequence of phage MS2 with the amino acid sequence of the corresponding proteins. Other findings support the belief that, with only minor exceptions, the code is universal for all organisms.

5. In some bacteriophages multiple initiation points may occur during the transcription of RNA, resulting in multiple reading frames and overlapping genes.

6. Transcription and translation—RNA and protein biosynthesis, respectively—are the fundamental processes essential to the expression of genetic information.

7. The processes of transcription and translation, like DNA replication, can be subdivided into the stages of initiation, elongation, and termination. Also like DNA replication, both processes rely on base-pairing affinities between complementary nucleotides.

8. Transcription describes the synthesis, under the direction of RNA polymerase, of a strand of RNA complementary to a DNA template.

9. The complex energy-requiring process of translation involves charged tRNA molecules, numerous proteins, ribosomes, and mRNA. Transfer RNA (tRNA) serves as the adaptor molecule between an mRNA triplet and the appropriate amino acid. The ribosome serves as the workbench for translation.

10. The processes of transcription and translation are more complex in eukaryotes than in prokaryotes. The primary transcript representing mRNA in eukaryotes must be modified in various ways, including the addition of a cap and poly A, and the removal, through splicing, of intervening sequences, or introns.

## KEY TERMS

adaptor molecule
alternative splicing
α-amanitin
amber mutation
ambiguous code
aminoacyladenylic acid
aminoacyl site (A site)
aminoacyl tRNA
   synthetase
anticodon
beta-globin gene
CCAAT box
cell-free protein-
   synthesizing system
chain elongation
charging (tRNA)
cloverleaf model of tRNA
codon
colinear relationship
commaless code
consensus sequence
degenerate code
elongation
elongation factor (EF)
enhancer
exon

formylmethionine
frameshift mutation
gene amplification
Goldberg–Hogness
   (TATA) box
GTP-dependent release
   factor
heteroduplex
heterogeneous nuclear
   RNA (hnRNA)
hypoxanthine
information flow
initiation complex
initiation factor (IF)
initiator codon (AUG)
inosinic acid
intervening sequence
intron
messenger RNA (mRNA)
7-methylguanosine cap
monosome
MS2 phage
N-formylmethionine
   (fmet)
nonoverlapping code
nonsense codon
nonsense mutation

ochre mutation
opal mutation
ordered code
ovalbumin gene
overlapping genes
peptidyl site (P site)
peptidyl transferase
poly A
polynucleotide
   phosphorylase
polyribosome
   (polysomes)
posttranscriptional
   modification
pre-mRNA
pre-tRNA
proflavin
promoter
pseudouridine
punctuation signal
rDNA
ribosomal proteins
ribosome
ribothymidylic acid
ribozyme
RNA heteropolymer
RNA homopolymer

RNA polymerase
second genetic code
self-excision process
Shine-Dalgarno sequence
σ(sigma) subunit
small nuclear
   ribonucleoprotein
   (snRNP)
spacer DNA
spliceosome
splicing
split genes
suppression
tandem repeats
TATA factor (TFIID)
template binding
termination factors (e.g.,
   rho)
termination codons
transcription
transfer RNA (tRNA)
translation
triplet code
triplet binding assay
unambiguous code
universal code
wobble hypothesis

# INSIGHTS AND SOLUTIONS

1. Had evolution seized on 6 bases (three complementary base pairs) rather than 4 bases within the structure of DNA, calculate how many triplet codons would be possible. Would 6 bases accommodate a two-letter code, assuming 20 amino acids and start and stop codons?

   **SOLUTION:** Six things taken three at a time will produce $(6)^3$ or 216 triplet codes. If the code was a doublet, there would be $(6)^2$ or 36 two-letter codes, more than enough to accommodate 20 amino acids and start–stop punctuation.

2. In a heteropolymer experiment using 1/2C:1/4A:1/4G, how many different triplets will occur in the synthetic RNA molecule? How frequently will the most frequent triplet occur?

   **SOLUTION:** There will be $(3)^3$ or 27 triplets produced. The most frequent will be CCC, present $(1/2)^3$ or 1/8 of the time.

3. In a regular copolymer experiment, where UUAC is repeated over and over, how many different triplets will occur in the synthetic RNA, and how many amino acids will occur in the polypeptide when this RNA is translated? Be sure to consult Figure 14.6.

   **SOLUTION:** The synthetic RNA will repeat four triplets—UUA, UAC, ACU, and CUU—over and over.

   Since both UUA and CUU encode leucine, while ACU and UAC encode threonine and tyrosine, respectively, polypeptides synthesized under the directions of such an RNA contain three amino acids in the repeating sequence leu-leu-thr-tyr.

4. Actinomycin D inhibits DNA-dependent RNA synthesis. This antibiotic is added to a bacterial culture where a specific protein is being monitored. Compared to a control culture, translation of the protein declines over a period of 20 minutes, until no further protein is made. Explain these results.

   **SOLUTION:** The mRNA, which is the basis for the translation of the protein, has a lifetime of about 20 minutes. When actinomycin D is added, transcription is inhibited and no new mRNAs are made. Those already present support the translation of the protein for up to 20 minutes.

5. DNA and RNA base compositions were analyzed from a hypothetical bacterial species with the following results:

|  | (A + G)/(T + C) | (A + T)/(C + G) | (A + G)/(U + C) | (A + U)/(C + G) |
|---|---|---|---|---|
| DNA | 1.0 | 1.2 | | |
| RNA | | | 1.3 | 1.2 |

   On the basis of these data, what can you conclude about the DNA and RNA of the organism? Are the data consistent with the Watson–Crick model of DNA? Is the RNA single-stranded or double-stranded, or can't we tell? If we assume that the entire length of DNA has been transcribed, do the data suggest that RNA has been derived from the transcription of one or both DNA strands, or can't we tell from these data?

**SOLUTION:**   This problem is a theoretical exercise designed to get you to look at the consequences of base complementarity as its affects base composition of DNA and RNA. The base composition of DNA is consistent with the Watson–Crick double helix. In a double helix, we expect A + G to equal T + C (the number of purines should equal the number of pyrimidines). In this case, there is a preponderance of A/T base pairs (120 A and T pairs to every 100 G and C pairs).

Given what we know about RNA, there is no reason to expect the RNA to be double-stranded, but if it were double-stranded, then we would expect that A = U and C = G. If so, then (A + G)/(U + C) = 1. Since it doesn't equal unity, we can conclude that the RNA is not double-stranded.

If all of the DNA is transcribed, from either one or both strands, the ratio of (A + U)/(C + G) in RNA should be 1.2, and as predicted, it is. Note that this ratio will not change, whether only one or both of the strands are transcribed. This is the case because for every A = T pair in DNA, for example, transcription of RNA will yield one A *and* one U if both strands are transcribed. If just one strand is transcribed, transcription will yield one A *or* one U. In either case, the (A + U)/(C + G) ratio in RNA will reflect the (A + T)/(C + G) ratio in the DNA from which it was transcribed. To prove this to yourself, draw out a DNA molecule with 12 A = T pairs and 10 C ≡ G pairs and transcribe *both* strands and then transcribe *either* strand. Count the bases in the RNAs produced in both cases and calculate the ratios. Thus, we cannot determine whether just one or both strands are transcribed from the (A + U)/(C + G) ratio.

However, if both strands are transcribed, then the ratio of (A + G)/(U + C) should equal 1.0, and it doesn't. It equals 1.3. To verify this conclusion, examine the theoretical data you drew out on paper as you were asked to do above. One explanation for the observed ratio of 1.3 is that only one of the two strands is transcribed. If this is the case, then the (A + G)/(U + C) will reflect the proportion of A/T pairs that are A and the proportion of the G/C pairs that are G *on the DNA strand that is transcribed*. Another explanation is that transcription occurs only on one strand at any given point (e.g., for one gene), but on the other strand at other points (for other genes).

---

**PROBLEMS AND DISCUSSION QUESTIONS**

1. Early proposals regarding the genetic code considered the possibility that DNA served directly as the template for polypeptide synthesis (see the Gamow reference in Selected Readings). In eukaryotes, what difficulties would such a system pose? What observations and theoretical considerations argue against such a proposal?

2. Crick, Barnett, Brenner, and Watts-Tobin, in their studies of frameshift mutations, found that either 3 (+)s or 3 (−)s restored the correct reading frame. If the code were a sextuplet (consisting of six nucleotides), would the reading frame be restored by either of the above combinations?

3. In a mixed copolymer experiment using polynucleotide phosphorylase, 3/4G:1/4C was added to form the synthetic message. The resulting amino acid composition of the ensuing protein was determined:

|          |               |
|----------|---------------|
| Glycine  | 36/64 (56%)   |
| Alanine  | 12/64 (19%)   |
| Arginine | 12/64 (19%)   |
| Proline  | 4/64 ( 7%)    |

From this information:
(a) Indicate the percentage (or fraction) of the time each possible triplet will occur in the message.
(b) Determine one consistent base composition assignment for the amino acids present.
(c) Considering the wobble hypothesis, predict as many specific triplet assignments as possible.

4.  In a mixed copolymer experiment, messengers were created with either 4/5C:1/5A or 4/5A:1/5C. These messages yielded proteins with the following amino acid compositions. Using these data, predict the most specific coding composition for each amino acid.

| 4/5C:1/5A | | | 4/5A:1/5C | | |
|---|---|---|---|---|---|
| Proline | 63.0% | | Proline | 3.5% | |
| Histidine | 13.0% | | Histidine | 3.0% | |
| Threonine | 16.0% | | Threonine | 16.0% | |
| Glutamine | 3.0% | | Glutamine | 13.0% | |
| Asparagine | 3.0% | | Asparagine | 13.0% | |
| Lysine | 0.5% | | Lysine | 50.0% | |
| | 98.5% | | | 98.5% | |

5.  When repeating copolymers are used to form synthetic mRNAs, dinucleotides produce a single type of polypeptide that contains only two different amino acids. On the other hand, using a trinucleotide sequence produces three different polypeptides, each consisting of only a single amino acid. Why? What will be produced when a repeating tetranucleotide is used?

6.  The mRNA formed from the repeating tetranucleotide UUAC incorporates only three amino acids, but the use of UAUC incorporates four amino acids. Why?

7.  In studies using repeating copolymers, ACAC. . . . . incorporates threonine and histidine, and CAACAA. . . . . incorporates glutamine, asparagine, and threonine. What triplet code can definitely be assigned to threonine?

8.  In a coding experiment using repeating copolymers (as shown in Table 14.3), the following data were obtained:

| Copolymer | Codons Produced | Amino Acids in Polypeptide |
|---|---|---|
| AG | AGA, GAG | Arg, Glu |
| AAG | AGA, AAG, GAA | Lys, Arg, Glu |

AGG is known to code for arginine. Taking into account the wobble hypothesis, assign each of the four remaining different triplet codes to its correct amino acid.

9.  In the triplet binding technique, radioactivity remains on the filter when the amino acid corresponding to the triplet is labeled. Explain the basis of this technique.

10.  When the amino acid sequences of insulin isolated from different organisms were determined, some differences were noted. For example, alanine was substituted for threonine, serine was substituted for glycine, and valine was substituted for isoleucine at corresponding positions in the protein. List the single-base changes that could occur in triplets of the genetic code to produce these amino acid changes.

11.  In studies of the amino acid sequence of wild-type and mutant forms of tryptophan synthetase in *E. coli*, the following changes have been observed:

```
                    thr
                   ↗
            arg → ser
          ↗       ↘
        ↗           ileu
   gly
        ↘
          ↘       ↗ val
            glu ↗
                ↘ ala
```

Determine a set of triplet codes in which only a single nucleotide change produces each amino acid change.

**12.** Why doesn't polynucleotide phosphorylase synthesize RNA *in vivo*?

**13.** Refer to Table 14.1. Can you hypothesize why Poly U + Poly A does not stimulate incorporation of $^{14}$C-phenylalanine into protein?

**14.** Predict the amino acid sequence produced during translation by the following short theoretical mRNA sequences. Note that the second sequence was formed from the first by a deletion of only one nucleotide.

Sequence 1:   AUGCCGGAUUAUAGUUGA
Sequence 2:   AUGCCGGAUUAAGUUGA

What type of mutation gave rise to Sequence 2?

**15.** In 1962, F. Chapeville and others reported an experiment where they isolated radioactive $^{14}$C-cysteinyl-tRNA$^{cys}$ (charged tRNA$^{cys}$ + cysteine). They then removed the sulfur group from the cysteine, creating alanyl-tRNA$^{cys}$ (charged tRNA$^{cys}$ + alanine). When alanyl-tRNA$^{cys}$ was added to a synthetic mRNA calling for cysteine but not alanine, a polypeptide chain was synthesized containing alanine. What can you conclude from this experiment?

**16.** A short RNA molecule was isolated that demonstrated a hyperchrome shift indicating secondary structure. Its sequence was determined to be AGGCGCCGAC-UCUACU.
(a) Propose a two-dimensional model for this molecule.
(b) What DNA sequence would give rise to this RNA molecule through transcription?
(c) If the molecule were a tRNA fragment containing a CGA anticodon, what would the corresponding codon be?
(d) If the molecule were an internal part of a message, what amino acid sequence would result from it following translation? (Refer to the code chart in Figure 14.6.)

**17.** A glycine residue exists at position 210 of the tryptophan synthetase enzyme of wild-type *E. coli*. If the codon specifying glycine is GGA, how many single base substitutions will result in an amino acid substitution at position 210? What are they? How many will result if the wild-type codon is GGU?

**18.** (a)  Shown below is a theoretical viral mRNA sequence. Assuming that it could arise from overlapping genes, how many different polypeptide sequences can be produced? Using Figure 14.6, what are the sequences?

5'-AUGCAUACCUAUGAGACCCUUGGGA-3'

(b)  A mutation at one position in the DNA giving rise to the above sequence occurred, eliminating the synthesis of all but one polypeptide. The base substitution of the mutant mRNA is shown below. Using Figure 14.6, determine why.

5'-AUGCAUACCUAUGUGACCCUUGGA-3'

**19.** Most proteins have more leucine than histidine residues, but more histidine than tryptophan residues. Correlate the number of codons for these three amino acids with the above information.

**20.** Define and differentiate between transcription and translation. Where do these processes fit into the central dogma of molecular genetics?

**21.** What was the initial evidence for the existence of mRNA?

**22.** Describe the structure of RNA polymerase. What is the core enzyme? What is the role of the sigma subunit?

23. In a written paragraph, describe the abbreviated chemical reactions shown on pages 455 and 456 that summarize RNA polymerase-directed transcription. What differences exist between the "visualization of transcription" studies of *E. coli* and *Notophthalmus*? Why?

24. List all of the molecular constituents present in a functional polyribosome.

25. Contrast the roles of tRNA and mRNA during translation and list all enzymes that participate in the transcription and translation process.

26. Francis Crick proposed the "adaptor hypothesis" for the function of tRNA. Why did he choose that description?

27. What molecule bears the codon? The anticodon?

28. The $\alpha$ chain of eukaryotic hemoglobin is composed of 141 amino acids. What is the minimum number of nucleotides in an mRNA coding for this protein chain? Assuming that each nucleotide is 0.34 nm long in the mRNA, how many triplet codes can at one time occupy space in a ribosome that is 20 nm in diameter?

29. Summarize the steps involved in charging tRNAs with their appropriate amino acids.

30. In order to carry out its role, each transfer RNA requires at least four specific recognition sites that must be inherent in its tertiary structure. What are they?

31. Messenger RNA molecules are very difficult to isolate in prokaryotes because they are rather quickly degraded in the cell. Can you suggest a reason why this occurs? Eukaryotic mRNAs are more stable and exist longer in the cell than do prokaryotic mRNAs. Is this an advantage or disadvantage for a pancreatic cell making large quantities of insulin?

32. The following represent deoxyribonucleotide sequences derived from the template strand of DNA:

    Sequence 1:   CTTTTTTGCCAT
    Sequence 2:   ACATCAATAACT
    Sequence 3:   TACAAGGGTTCT

    (a) For each strand, determine the mRNA sequence that would be derived from transcription.
    (b) Using Figure 14.6, determine the amino acid sequence that would result from translation of these mRNAs.
    (c) If we assume that each sequence has been derived from the same gene, which represents the initial part? The middle region? The terminal portion?
    (d) For Sequence 1, what is the sequence of the partner strand?
    (e) For Sequence 3, draw the structural formula of the polypeptide sequence resulting from translation (see Chapter 15).

**SELECTED READINGS**

ALBERTS, B., et al. 1994. *Molecular biology of the cell*. 3rd ed. New York: Garland.

BARALLE, F. E. 1983. The functional significance of leader and trailer sequences in eukaryotic mRNAs. *Int. Rev. Cytol.* 81:71–106.

BARRELL, B. G., AIR, G., and HUTCHINSON, C. 1976. Overlapping genes in bacteriophage $\phi$X174. *Nature* 264:34–40.

BARRELL, B. G., BANKER, A. T., and DROUIN, J. 1979. A different genetic code in human mitochondria. *Nature* 282:189–94.

BIRNSTIEL, M., BUSSLINGER, M., and STRUB, K. 1985. Transcription termination and 3′ processing: The end is in site. *Cell* 41:349–59.

BONITZ, S. G., et al. 1980. Codon recognition rules in yeast mitochondria. *Proc. Natl. Acad. Sci.* 77:3167–70.

BREITBART, R. E., and NADAL-GINARD, B. 1987. Developmentally induced, muscle-specific *trans* factors control the differential splicing of alternative and constitutive troponin T-exons. *Cell* 49:793–803.

BRENNER, S. 1989. *Molecular Biology: A selection of papers.* Orlando: Academic Press.

BRENNER, S., JACOB, F., and MESELSON, M. 1961. An unstable intermediate carrying information from genes to ribosomes for protein synthesis. *Nature* 190:575–80.

BRENNER, S., STRETTON, A. O. W., and KAPLAN, D. 1965. Genetic code: The nonsense triplets for chain termination and their suppression. *Nature* 206:994–98.

CECH, T. R. 1986. RNA as an enzyme. *Scient. Amer.* (Nov.) 255, 5:64–75.

———. T. R. 1987. The chemistry of self-splicing RNA and RNA enzymes. *Science* 236:1532–39.

CECH, T. R., ZAUG, A. J., and GRABOWSKI, P. J. 1981. *In vitro* splicing of the ribosomal RNA precursor of *Tetrahymena.* Involvement of a guanosine nucleotide in the excision of the intervening sequence. *Cell* 27:487–96.

CHAMBON, P. 1975. Eucaryotic nuclear RNA polymerases. *Ann. Rev. Biochem.* 44:613–38.

———. 1981. Split genes. *Scient. Amer.* (May) 244:60–71.

CHAPEVILLE, F., et al. 1962. On the role of soluble ribonucleic acid in coding for amino acids. *Proc. Natl. Acad. Sci.* 48:1086–93.

CIGAN, A. M., FENG, L., DONAHUE, T. F. 1988. tRNA$^{met}$ functions in directing the scanning ribosome to the start site of translation. *Science* 242:93–98.

COLD SPRING HARBOR LABORATORY. 1966. The genetic code. *Cold Spr. Harb. Symp.,* Vol. 31.

CRICK, F. H. C. 1962. The genetic code. *Scient. Amer.* (Oct.) 207:66–77.

———. 1966a. The genetic code: III. *Scient. Amer.* (Oct.) 215:55–63.

———. 1966b. Codon-anticodon pairing: The wobble hypothesis. *J. Mol. Biol.* 19:548–55.

———. 1979. Split genes and RNA splicing. *Science* 204:264–71.

CRICK, F. H. C., BARNETT, L., BRENNER, S., and WATTS-TOBIN, R. J. 1961. General nature of the genetic code for proteins. *Nature* 192:1227–32.

DAHLBERG, A. E. 1989. The functional role of ribosomal RNA in protein synthesis. *Cell* 57:525–29.

DARNELL, J. E. 1983. The processing of RNA. *Scient. Amer.* (Oct.) 249:90–100.

———. 1985. RNA. *Scient. Amer.* (Oct.) 253:68–87.

DICKERSON, R. E. 1983. The DNA helix and how it is read. *Scient. Amer.* (Dec.) 249:94–111.

DUGAICZK, A., et al. 1978. The natural ovalbumin gene contains seven intervening sequences. *Nature* 274:328–33.

FIERS, W., et al. 1976. Complete nucleotide sequence of bacteriophage MS2 RNA: Primary and secondary structure of the replicase gene. *Nature* 260:500–507.

GAMOW, G. 1954. Possible relation between DNA and protein structures. *Nature* 173:318.

GUTHRIE, C., and PATTERSON, B. 1988. Spliceosomal snRNAs. *Ann. Rev. Genet.* 22:387–419.

HALL, B. D., and SPIEGELMAN, S. 1961. Sequence complementarity of T2-DNA and T2-specific RNA. *Proc. Natl. Acad. Sci.* 47:137–46.

HAMKALO, B. 1985. Visualizing transcription in chromosomes. *Trends in Genet.* 1:255–60.

HELMAN, J. D., and CHAMBERLIN, M. J. 1988. Structure and function of bacterial sigma factors. *Ann. Rev. Biochem.* 57:839–72.

HOLLEY, R. W., et al. 1965. Structure of a ribonucleic acid. *Science* 147:1462–65.

HUMPHREY, T., and PROUDFOOT, N. J. 1988. A beginning to the biochemistry of polyadenylation. *Trends in Genet.* 4:243–45.

JUDSON, H. F. 1979. *The eighth day of creation.* New York: Simon and Schuster.

JUKES, T. H. 1963. The genetic code. *Amer. Scient.* 51:227–45.

KHORANA, H. G. 1967. Polynucleotide synthesis and the genetic code. *Harvey Lectures* 62:79–105.

LAKE, J. A. 1981. The ribosome. *Scient. Amer.* (Aug.) 245:84–97.

LEHNINGER, A. H., NELSON, D. C., and COX, M. M. 1993. *Principles of Biochemistry.* New York: Worth Publishers.

MANIATIS, T., AND REED, R. 1987. The role of small nuclear ribonucleoprotein particles in pre-mRNA splicing. *Nature* 325:673–78.

MILLER, O. L., and BEATTY, B. R. 1969. Portrait of a gene. *J. Cell Physiol.* 74 (Supplement 1): 225–32.

MILLER, O. L., HAMKALO, B., and THOMAS, C. 1970. Visualization of bacterial genes in action. *Science* 169:392–95.

MIN JOU, W., HAGEMAN, G., YSEBART, M., and FIERS, W. 1972. Nucleotide sequence of the gene coding for bacteriophage MS2 coat protein. *Nature* 237:82–88.

MOORE, P. B. 1988. The ribosome returns. *Nature* 331:223–27.

NIRENBERG, M. W. 1963. The genetic code: II. *Scient. Amer.* (March) 190:80–94.

NIRENBERG, M. W., and LEDER, P. 1964. RNA codewords and protein synthesis. *Science* 145: 1399–1407.

NIRENBERG, M. W., and MATTHAEI, H. 1961. The dependence of cell-free protein synthesis in *E. coli* upon naturally occurring or synthetic polyribosomes. *Proc. Natl. Acad. Sci.* 47: 1588–1602.

NOLLER, H. F. 1973. Assembly of bacterial ribosomes. *Science* 179:864–73.

———. 1984. Structure of ribosomal RNA. *Ann. Rev. Biochem.* 53:119–62.

NOMURA, M. 1984. The control of ribosome synthesis. *Scient. Amer.* (Jan.) 250:102–14.

O'MALLEY, B., et al. 1979. A comparison of the sequence organization of the chicken ovalbumin and ovomucoid genes. In *Eucaryotic gene regulation*, Axel, R., et al., 281–99. Orlando, Fla: Academic Press.

PADGETT, R. A., GRABOWSKI, P. J., KONARSKA, M. M., SEILER, S., and SHARP, P. A. 1986. Splicing of messenger RNA precursors. *Ann. Rev. Biochem.* 55:1119–50.

REED, R., and MANIATIS, T. 1985. Intron sequences involved in lariat formation during pre-mRNA splicing. *Cell* 41:95–105.

RICH, A., and HOUKIM, S. 1978. The three-dimensional structure of transfer RNA. *Scient. Amer.* (Jan.) 238:52–62.

RICH, A., WARNER, J. R., and GOODMAN, H. M. 1963. The structure and function of polyribosomes. *Cold Spr. Harb. Symp.* 28:269–85.

ROSS, J. 1989. The turnover of mRNA. *Scient. Amer.* (April) 260:48–55.

ROULD, M. A., et al. 1989. Structure of *E. coli* glutaminyl-tRNA synthetase complexed with tRNA$^{gln}$ and ATP at 2.8 Å resolution. *Science* 246:1135–42.

SHARP, P. 1987. Splicing of messenger RNA precursors. *Science* 235:766–71.

SHARP, P. A., and EISENBERG, D. 1987. The evolution of catalytic function. *Science* 238:729–30.

SHATKIN, A. J. 1985. mRNA cap binding proteins: Essential factors for initiating translation. *Cell* 40:223–24.

SMITH, J. D. 1972. Genetics of tRNA. *Ann. Rev. Genet.* 6:235–56.

STEITZ, J. A. 1988. Snurps. *Scient. Amer.* (June) 258(6):56–63.

VOLKIN, E., and ASTRACHAN, L. 1956. Phosphorus incorporation in *E. coli* ribonucleic acids after infection with bacteriophage T2. *Virology* 2:149–61.

VOLKIN, E., ASTRACHAN, L., and COUNTRYMAN, J. L. 1958. Metabolism of RNA phosphorus in *E. coli* infected with bacteriophage T7. *Virology* 6:545–55.

WARNER, J., and RICH, A. 1964. The number of soluble RNA molecules on reticulocyte polyribosomes. *Proc. Natl. Acad. Sci.* 51:1134–41.

WATSON, J. D. 1963. Involvement of RNA in the synthesis of proteins. *Science* 140:17–26.

WATSON, J. D., HOPKINS, N. H., ROBERTS, J. W., STEITZ, J. A., and WEINER, A. M. 1987. *Molecular biology of the gene*. 4th ed. Menlo Park, CA: Benjamin-Cummings.

WITTMAN, H. G. 1983. Architecture of prokaryotic ribosomes. *Ann. Rev. Biochem.* 52:35–65.

ZUBAY, G. L., and MARMUR, J. 1973. *Papers in biochemical genetics*. 2nd ed. New York: Holt, Rinehart and Winston.

# 15

# PROTEINS: THE END PRODUCTS OF GENES

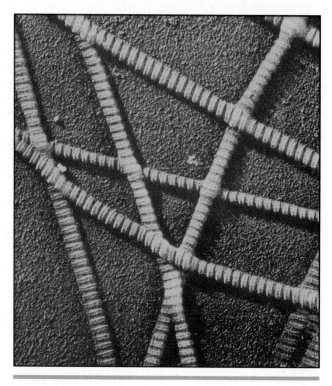

Fibers of collagen, the most abundant protein in vertebrate organisms.

*The end products of most genes are polypeptide chains. They achieve a three-dimensional conformation based on their primary amino acid sequences, often interacting with other such chains to create functional protein molecules. The function of any protein is closely tied to its structure, which can be disrupted by mutation, leading to a distinctive phenotypic effect.*

In Chapter 14 we established that there is a genetic code that stores information in the form of triplet nucleotides in DNA, and that this information can be expressed through the orderly processes of transcription and translation. The final product of gene expression, in almost all instances, is a polypeptide chain consisting of a linear series of amino acids, whose sequence has been prescribed by the genetic code. In this chapter, we will review the evidence that confirmed that proteins are the end products of genes. Then we will discuss briefly the various levels of protein structure, diversity, and function. This information provides an important foundation in our understanding of how mutations, which arise in DNA, can result in the variety of phenotypic effects observed in organisms.

## GARROD AND BATESON: INBORN ERRORS OF METABOLISM

The first insight into the role of proteins in genetic processes was provided by observations made by Sir Archibald Garrod and William Bateson early in this century. Garrod was born into an English family of medical scientists. His father was a physician with a strong interest in the chemical basis of rheumatoid arthritis, and his eldest brother was a leading zoologist in London. Thus, it is not surprising that, as a practicing physician, Garrod became interested in several human disorders that seemed to be inherited. Although he also studied albinism and cystinuria, we will describe his investigation of the disorder **alkaptonuria**. Individuals afflicted with this disorder cannot metabolize the alkapton 2,5-dihydroxyphenylacetic acid, also known as **homogentisic acid**. As a result, an important metabolic pathway (Figure 15.1) is blocked. Homogentisic acid accumulates in cells and tissues and is excreted in the urine. The molecule's oxidation products are black and thus are easily detectable in the diapers of newborns and the urine of individuals. The products tend to accumulate in cartilaginous areas, causing a darkening of the ears and nose. In joints, this deposition leads to a benign arthritic condition. This rare disease is not serious, but it persists throughout an individual's life.

Garrod studied alkaptonuria by increasing dietary protein or adding to the diet the amino acids phenylalanine or tyrosine, which are chemically related to homogentisic acid. Under such conditions, homogentisic acid levels increase in the urine of alkaptonurics but not in unaffected individuals. He concluded that normal individuals can break down, or catabolize, this alkapton but that afflicted individuals cannot. By studying the patterns of inheritance of the disorder, Garrod further concluded that alkaptonuria was inherited as a simple recessive trait.

On the basis of these conclusions, Garrod hypothesized that the hereditary information controls chemical reactions in the body and that the inherited disorders he studied are the result of alternative modes of metabolism. While the terms *genes* and *enzymes* were not familiar during Garrod's time, he used the corresponding concepts of *unit factors* and *ferments*. Garrod published his initial observations in 1902.

Only a few geneticists, including Bateson, were familiar with or referred to Garrod's work. His ideas fit nicely with Bateson's belief that inherited conditions were caused by the lack of some critical substance. In 1909, Bateson published *Mendel's Principles of Heredity*, in which he linked ferments with heredity. However, for almost thirty years, most geneticists failed to see the relationship between genes and enzymes. Garrod and Bateson, like Mendel, were ahead of their time.

### Phenylketonuria

Described first in 1934, **phenylketonuria (PKU)** may result in mental retardation and is inherited as an autosomal recessive disease. Afflicted individuals have still another step blocked in the metabolic pathway just dis-

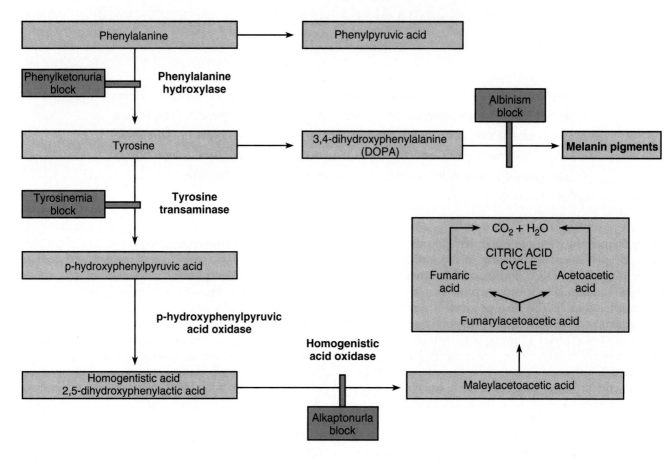

**FIGURE 15.1**    Metabolic pathway involving phenylalanine and tyrosine. Various metabolic blocks resulting from mutations lead to the disorders phenylketonuria, alkaptonuria, and albinism.

cussed. They are unable to convert the amino acid phenylalanine to the amino acid tyrosine (see Figure 15.1). These molecules differ by only a single hydroxl group (—OH), present in tyrosine, but absent in phenylalanine. The reaction is catalyzed by the enzyme **phenylalanine hydroxylase**, which is inactive in affected individuals and active at about a 30 percent level in heterozygotes. The enzyme functions in the liver. While the normal blood level of phenylalanine is about 1 mg/100 ml, phenylketonurics show a level as high as 50 mg/100 ml.

As phenylalanine accumulates, it may be converted to phenylpyruvic acid and, subsequently, other derivatives. These are less efficiently resorbed by the kidney and tend to spill into the urine more quickly than phenylalanine. Both phenylalanine and derivatives enter the cerebrospinal fluid, resulting in elevated levels in the brain. The presence of these substances during early development is thought to be responsible for retardation.

Retardation can be prevented by PKU screening of newborns. When the condition is detected in the analysis of an infant's blood, a strict dietary regimen is instituted. A low phenylalanine diet can reduce such byproducts as phenylpyruvic acid, and abnormalities characterizing the disease can be diminished. Screening of newborns occurs routinely in almost every state in this country. Phenylketonuria occurs in approximately 1 in 11,000 births.

Knowledge of inherited metabolic disorders such as alkaptonuria and phenylketonuria has caused a revolution in medical thinking and practice. No longer is human disease attributed solely to the action of invading microorganisms, viruses, or parasites. We know now that literally thousands of abnormal physiological conditions are caused by errors in metabolism that are the result of mutant genes. These human biochemical disorders are far-ranging and include all classes of organic biomolecules.

# THE ONE-GENE: ONE-ENZYME HYPOTHESIS

In two separate investigations beginning in 1933, George Beadle was to provide the first convincing experimental evidence that genes are directly responsible for the synthesis of enzymes. The first investigation, conducted in collaboration with Boris Ephrussi, involved *Drosophila* eye pigments. Encouraged by these findings, Beadle then joined with Edward Tatum to investigate nutritional mutations in the pink bread mold *Neurospora crassa*. The latter investigation led to the **one-gene: one-enzyme hypothesis**.

## Beadle and Ephrussi: *Drosophila* Eye Pigments

The studies of Beadle and Ephrussi involved **imaginal disks** in *Drosophila*. These embryonic cells found in the larvae of insects, upon metamorphosis, differentiate into a variety of adult structures. Among others, adult eyes, legs, antennae, and genital structures are derived from these groups of cells. Beadle and Ephrussi found that if an imaginal disk were removed from one larva and transplanted into the abdomen of another, that disk would differentiate during metamorphosis and its corresponding adult structure could be recovered from the abdomen of the adult fly. For example, if an eye disk were transplanted, it would develop into an eyelike structure that could be recovered and analyzed.

These investigators wondered whether a mutant eye disk, transplanted into a wild-type larva, would be altered by its new environment. Would it appear normally pigmented, or would it develop its characteristic mutant color? The first mutant used was *vermilion*, a sex-linked, bright red eye color mutation. Mutant development was reversed! The *vermilion* eye disk was altered to produce a normally pigmented wild-type eye. The normal wild-type color is brick red, produced as a combination of brown and bright red pigments. The *vermilion* mutant makes the bright red pigment, but lacks the brown pigment, indicating that its synthesis is inhibited. However, in the wild-type abdomen, synthesis of brown pigment also occurred normally in the mutant disk, producing wild-type color.

Beadle and Ephrussi then proceeded to perform similar experimentation with 25 other mutants. Only one, *cinnabar*, another bright red eye mutation missing the brown pigment, behaved like *vermilion* when transplanted; it developed a normal eye color. The remaining 24 behaved "autonomously" by exhibiting their respective mutant color when differentiated.

They concluded that the flies with either *vermilion* or *cinnabar* phenotypes are mutant due to the absence of some *diffusible substance* in their disks during development. Because this substance was diffusible, it was present in the wild-type abdomen and thus accessible to the transplanted *v* or *cn* disks. Thus, it enabled full pigment production in the transplant. However, in the case of the 24 mutations that developed autonomously, the substance that might have permitted normal pigment production was not diffusible but confined to the host cells forming the eye. Thus, no "correction" was observed.

Beadle and Ephrussi then asked whether the same substance was lacking in both the *vermilion* and *cinnabar* mutants. To answer this question, they transplanted *vermilion* eye disks into *cinnabar* larvae, and vice versa. What they found was intriguing. The *vermilion* disks were "cured" or converted to wild-type color in *cn* abdomens, but *cinnabar* disks developed autonomously in *vermilion* hosts and retained their mutant phenotype. These experiments are diagrammed in Figure 15.2.

To explain such results, they proposed that two sequential biochemical reactions are involved in normal synthesis of the brown pigment. One is inactive in flies with the *vermilion* mutation, and the other is inactive in flies with the *cinnabar* mutation. Each reaction produces a product that is diffusible in the tissues of wild type. As shown in Figure 15.2, the substance produced under the direction of the wild-type allele of the *vermilion* locus (substance Y) occurs first in the pathway. It is then converted to a second substance in the pathway. This conversion is controlled by the wild-type allele of the *cinnabar* locus (substance Z).

The order of this pathway explains the results of the reciprocal transplantations. Consider first the *vermilion* (*v*) disk in the *cinnabar* (*cn*) host. The *vermilion* disk lacks the $v^+$ enzyme but has the $cn^+$ enzyme. As the host, the *cinnabar* tissue can make substance Y because it has the wild-type allele of the *vermilion* gene ($v^+$). Substance Y is diffusible and enters the *vermilion* disk where the $cn^+$ enzyme converts it to substance Z. Substance Z is then converted to the brown pigment, restoring the wild-type eye color.

Now consider the *cinnabar* disk in the *vermilion* host. The host tissue, lacking the $v^+$ gene, cannot make the $v^+$ enzyme, and so cannot make substance Y. Even though the $cn^+$ enzyme is present, because no Y is available to be converted to Z, no Z is produced either. Thus, the transplanted *cn* disk is still blocked. It can make substance Y but cannot complete the transition through Z to the brown pigment. Thus, it remains bright red!

Within a few years, other workers had investigated the biochemical pathway leading to the brown xanthommatin pigment in *Drosophila*. As shown in Figure 15.3,

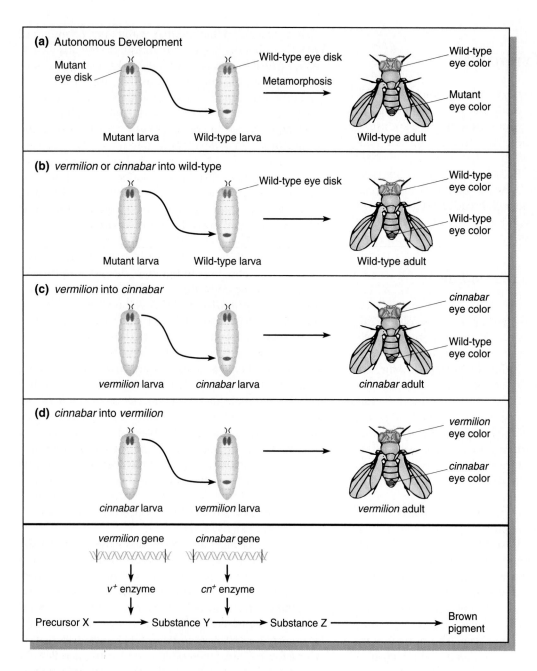

**FIGURE 15.2**   Beadle and Ephrussi's transplantation experiments involving *Drosophila* imaginal disks that develop into eyes. A mutant disk, transplanted into a wild-type larval abdomen, either develops autonomously, resulting in a mutant eye color, or the mutation is "cured" and normal eye color results, as illustrated in (a) and (b), respectively. When a *vermilion* disk is transplanted into a *cinnabar* host (c), normal wild-type eye color develops in the transplanted disk. The reciprocal experiment (d) results in autonomous development of the mutant eye color in the transplanted disk. On the basis of these results, they concluded that the biosynthetic step controlled by the $v^+$ gene product occurs prior to that controlled by the $cn^+$ gene product within the same pathway.

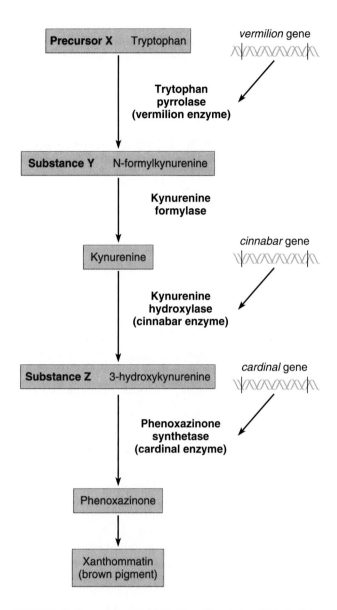

**FIGURE 15.3**    The biosynthetic pathway leading to the conversion of tryptophan to the brown eye pigment xan-thommatin. Three different mutant blocks are shown, each interrupting synthesis at a different place in the pathway. In the absence of xanthommatin, all three mutations, when homozygous, result in the identical phenotype, bright red eyes.

the brown pigment is a derivative of the amino acid tryptophan. The reactions controlled by the specific enzymes altered by the *vermilion* and *cinnabar* mutations have been identified. Other autosomal recessive mutations, including *scarlet* and *cardinal*, have also been studied and shown to affect enzymes in this pathway. Most relevant to our discussion was the confirmation in the 1940s that mutant genes controlling distinctive morphological phenotypes were linked directly to biochemi-

cal "errors," most likely due to a lack of enzyme function.

## Beadle and Tatum: *Neurospora* Mutants

In the early 1940s, Beadle and Tatum chose to work with the fungus *Neurospora crassa* because much was known about its biochemistry, and mutations could be induced and isolated with relative ease. By inducing mutations, they produced strains that had genetic blocks of reactions essential to the growth of the organism.

Beadle and Tatum knew that *Neurospora* could manufacture nearly everything necessary for normal development. For example, using rudimentary carbon and nitrogen sources, this organism can synthesize 9 water-soluble vitamins, 20 amino acids, numerous carotenoid pigments, and purines and pyrimidines. Beadle and Tatum irradiated asexual spores with X rays to increase the frequency of mutations and allowed them to be grown on "complete" medium containing all necessary growth factors (vitamins, amino acids, etc.). Under such growth conditions, a mutant strain that would be unable to grow on minimal medium would be able to grow in the enriched complete medium by virtue of supplements present. All cultures were then transferred to minimal medium. If growth occurred on minimal medium, the organisms were able to synthesize all necessary growth factors themselves, and it was concluded that the culture did not contain a mutation. If no growth occurred, then it contained a nutritional mutation, and the only task remaining was to determine its type. Both cases are illustrated in Figure 15.4(a).

Many thousands of individual spores derived by this procedure were isolated and grown on complete medium. In subsequent tests on minimal medium, many cultures failed to grow, indicating that a nutritional mutation had been induced. To identify the mutant type, the mutant strains were tested on a series of different minimal media [Figure 15.4(b) and (c)], each containing groups of supplements, and subsequently single vitamins, amino acids, purines, or pyrimidines until one specific supplement that permitted growth was found. Beadle and Tatum reasoned that the supplement that restores growth is the molecule that the mutant strain could not synthesize.

The first mutant strain isolated required vitamin B-6 (pyridoxin) in the medium, and the second one required vitamin B-1 (thiamine). Using the same procedure, Beadle and Tatum eventually isolated and studied hundreds of mutants deficient in the ability to synthesize other vitamins, amino acids, etc.

The findings derived from testing over 80,000 spores convinced Beadle and Tatum that genetics and biochemistry have much in common. It seemed likely that each

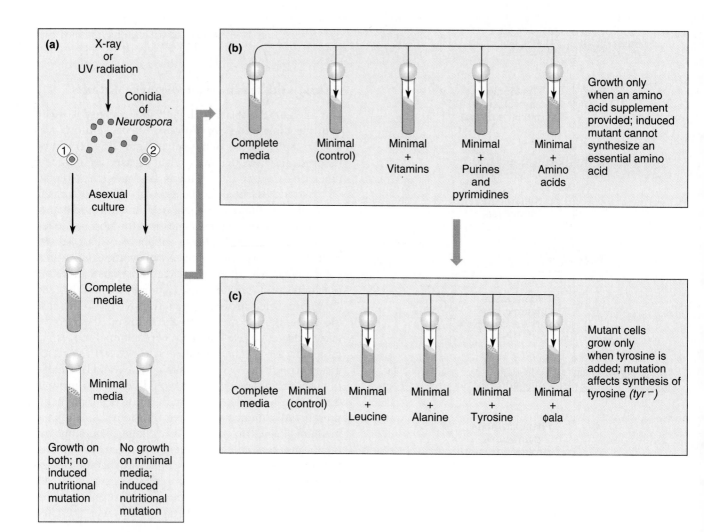

**FIGURE 15.4**    Induction, isolation, and characterization of a nutritional auxotrophic mutation in *Neurospora*. In (a), conidium 1 is not affected, but conidium 2 contains such a mutation. In (b) and (c), the precise nature of the mutation is determined to involve the biosynthesis of tyrosine.

nutritional mutation caused the loss of the enzymatic activity that facilitates an essential reaction in wild-type organisms. It also appeared that a mutation could be found for nearly any enzymatically controlled reaction. Beadle and Tatum had thus provided sound experimental evidence for the hypothesis that **one gene specifies one enzyme**, an idea alluded to over thirty years earlier by Garrod and Bateson. With modifications, this concept was to become a major principle of genetics.

## Genes and Enzymes: Analysis of Biochemical Pathways

The one-gene:one-enzyme concept and its attendant methods have been used over the years to work out many details of metabolism in *Neurospora, Escherichia coli*, and a number of other microorganisms. One of the first metabolic pathways to be investigated in detail was that leading to the synthesis of the amino acid arginine in *Neurospora*. By studying seven mutant strains, each requiring arginine for growth ($arg^-$), Adrian Srb and Norman Horowitz were able to ascertain a partial biochemical pathway leading to the synthesis of this molecule. The rationale followed in their work illustrates how genetic analysis can be used to establish biochemical information.

Srb and Horowitz tested each mutant strain's ability to reestablish growth if either **citrulline** or **ornithine**, two compounds with close chemical similarity to arginine, was used as a supplement to minimal medium. If either was able to substitute for arginine, they reasoned that it must be involved in the biosynthetic pathway of arginine. They found that both molecules could be substituted in one or more strains.

Of the seven mutant strains, four of them (*arg 1–4*) grew if supplied with either citrulline, ornithine, or arginine. Two of them (*arg 5* and *6*) grew if supplied with citrulline or arginine. One strain (*arg 7*) would grow only if arginine were supplied. Neither citrulline nor ornithine could substitute for it. From these experimental observations, the following pathway and metabolic blocks for each mutation were deduced:

```
        arg 1-4          arg 5-6          arg 7
Precursor ——|—→ Ornithine ——|—→ Citrulline ——|—→ Arginine
        Enzyme A         Enzyme B         Enzyme C
```

The reasoning supporting these conclusions is based on the following logic. If mutants *arg 1* through *4* can grow regardless of which of the three molecules is supplied as a supplement to minimal medium, the mutations preventing growth must cause a metabolic block that occurs *prior to* the involvement of ornithine, citrulline, or arginine in the pathway. When any of these three molecules is added, its presence bypasses the block. As a result, it can be concluded that both citrulline and ornithine are involved in the biosynthesis of arginine. How-

ever, the sequence of their participation in the pathway cannot be determined on the basis of these data.

On the other hand, the *arg 5* and *6* mutations grow if supplied citrulline but not ornithine. Therefore, ornithine must occur in the pathway *prior to* the block. Its presence will not overcome the block. Citrulline, however, does overcome the block, so it must be involved beyond the point of blockage. Therefore, the conversion of ornithine to citrulline represents the correct sequence in the pathway.

Finally, it can be concluded that *arg 7* represents a mutation preventing the conversion of citrulline to arginine. Neither ornithine nor citrulline can overcome the metabolic block because both participate earlier in the pathway.

Taken together, these reasons support the sequence of biosynthesis outlined here. Since Srb and Horowitz's work in 1944, the detailed pathway has been worked out and the enzymes controlling each step characterized. The chemical pathway is shown in Figure 15.5.

The concept of one-gene:one-enzyme developed in the early 1940s was not accepted immediately by all geneticists. This is not surprising, because it was not yet

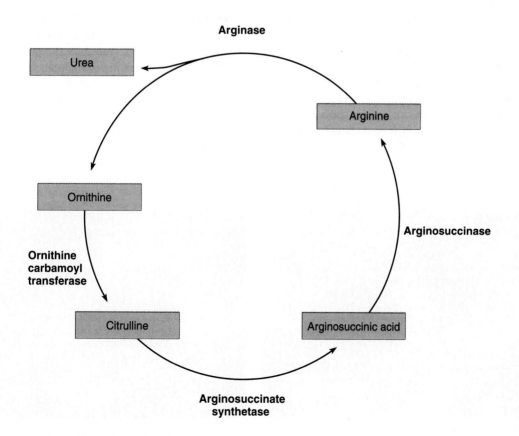

**FIGURE 15.5**     Arginine biosynthesis in *Neurospora*.

clear how mutant enzymes could cause variation in many phenotypic traits. For example, *Drosophila* mutants demonstrated altered eye size, wing shape, wing vein pattern, and so on. Plants exhibited mutant varieties of seed texture, height, and fruit size. How an inactive, mutant enzyme could result in such phenotypes was puzzling to many geneticists. Another reason for their reluctance to accept this concept was the paucity of information then available in molecular genetics. It was not until 1944 that Avery, MacLeod, and McCarty showed DNA to be the transforming factor and not until the early 1950s that most geneticists believed that DNA serves as the genetic material (see Chapter 8). However, by this time, the evidence in support of the concept of enzymes as gene products was overwhelming, and the concept was accepted as valid. Nevertheless, the question of how DNA specifies the structure of enzymes remained unanswered.

# ONE-GENE:ONE-PROTEIN/ ONE-GENE:ONE-POLYPEPTIDE

Two factors soon modified the one-gene:one-enzyme hypothesis. First, while nearly all enzymes are proteins, not all proteins are enzymes. As the study of biochemical genetics proceeded, it became clear that all proteins are specified by the information stored in genes, leading to the more accurate phraseology **one-gene:one protein**. Second, proteins were often shown to have a subunit structure consisting of two or more **polypeptide chains**. This is the basis of the **quaternary structure** of proteins, which we will discuss later in this chapter. Because each distinct polypeptide chain is encoded by a separate gene, a more modern statement of Beadle and Tatum's basic principle is **one-gene:one-polypeptide chain**. These modifications of the original hypothesis became apparent during the analysis of hemoglobin structure in individuals afflicted with sickle-cell anemia.

## Sickle-Cell Anemia

The first direct evidence that genes specify proteins other than enzymes came from the work on mutant hemoglobin molecules derived from humans afflicted with the disorder **sickle-cell anemia**. Affected individuals contain erythrocytes, which, under low oxygen tension, become elongated and curved because of the polymerization of hemoglobin. The "sickle" shape of these erythrocytes is in contrast to the biconcave disc shape characteristic of normal individuals (Figure 15.6). Individuals with the disease suffer attacks when red blood cells aggregate in the venous side of capillary systems, where oxygen tension is very low. As a result, a variety of tissues may be deprived of oxygen and suffer severe damage. When this occurs, an individual is said to experience a sickle-cell crisis. If untreated, a crisis may be fatal. The kidneys, muscles, joints, brain, gastrointestinal tract, and lungs may be affected.

In addition to suffering crises, these individuals are anemic because their erythrocytes are destroyed more rapidly than normal red blood cells. Compensatory physiological mechanisms include increased red cell production by bone marrow and accentuated heart action. These mechanisms lead to abnormal bone size and shape as well as dilation of the heart.

In 1949, James Neel and E. A. Beet demonstrated that the disease is inherited as a Mendelian trait. Pedigree

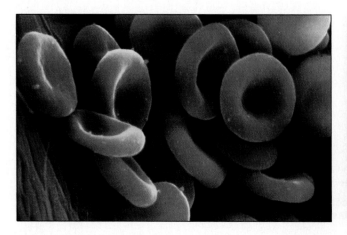
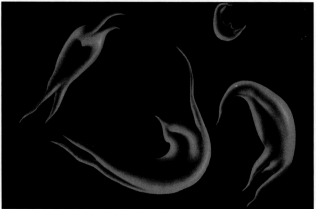

**FIGURE 15.6**    A comparison of erythrocytes derived from HbA (left) and HbS (right).

analysis revealed three genotypes and phenotypes controlled by a single pair of alleles, *A* and *S*. Normal and affected individuals result from the homozygous genotypes *AA* and *SS*, respectively. The red blood cells of the *AS* heterozygote, which exhibits the **sickle-cell trait** but not the disease, undergo much less sickling since over half of their hemoglobin is normal. Although largely unaffected, such persons are carriers of the disorder.

In the same year, Linus Pauling and his coworkers provided the first insight into the molecular basis of the disease. They showed that hemoglobins isolated from diseased and normal individuals differ in their rates of electrophoretic migration. In this technique, charged molecules migrate in an electric field. If the net charge of two molecules is different, the rate of migration will vary.

On this basis, Pauling and his colleagues concluded that a chemical difference exists between the two types of hemoglobin. The two molecules are now designated **HbA** and **HbS**.

Figure 15.7(a) illustrates the migration pattern of hemoglobin derived from individuals of all three possible genotypes when subjected to **starch gel electrophoresis**. The gel provides the supporting medium for the molecules during migration. In this experiment, samples are placed at a point of origin between the cathode (−) and the anode (+), and an electric field is applied. The migration pattern reveals that all molecules move toward the anode, indicating a net negative charge. However, HbA migrates farther than HbS, suggesting that its net negative charge is greater. The electrophoretic pattern of

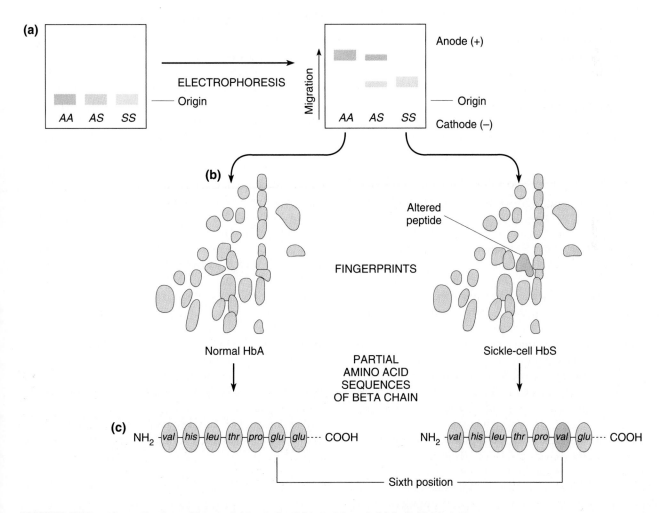

**FIGURE 15.7**    Investigation of hemoglobin derived from *AA* and *SS* individuals using electrophoresis, fingerprinting, and amino acid analysis. Hemoglobin from individuals with sickle-cell anemia (*SS*): (a) migrates differently in an electrophoretic field, (b) contains an altered peptide in fingerprint analysis, and (c) contains an altered amino acid, valine, at the sixth position in the β chain. During electrophoresis, heterozygotes (*AS*) reveal both forms of hemoglobin.

hemoglobin derived from carriers reveals the presence of both HbA and HbS, confirming their heterozygous genotype.

Pauling's findings suggested two possibilities. It was known that hemoglobin consists of four nonproteinaceous, iron-containing **heme groups** and a **globin portion** that contains four polypeptide chains. The alteration in net charge in HbS could be due, theoretically, to a chemical change in either component.

Work carried out between 1954 and 1957 by Vernon Ingram resolved this question. He demonstrated that the chemical change occurs in the primary structure of the globin portion of the hemoglobin molecule. Using the **fingerprinting technique**, Ingram showed that HbS differs in amino acid composition compared to HbA. Human adult hemoglobin contains two identical alpha ($\alpha$) chains of 141 amino acids and two identical beta ($\beta$) chains of 146 amino acids in its quaternary structure.

The fingerprinting technique involves enzymatic digestion of the protein into peptide fragments. The mixture is then placed on absorbent paper and exposed to an electric field, where migration occurs according to net charge. The paper is then turned at a right angle and placed in a solvent, where chromatographic action causes migration of the peptides in the second direction. The end result is a two-dimensional separation of the peptide fragments into a distinctive pattern of spots or "fingerprints." Ingram's work revealed that HbS and HbA differed by only a single peptide fragment [Figure 15.7(b)]. Further analysis then revealed a single amino acid change: Valine was substituted for glutamic acid at the sixth position of the $\beta$ chain, accounting for the peptide difference [Figure 15.7(c)].

The significance of this discovery has been multifaceted. It clearly establishes that a single gene provides the genetic information for a single polypeptide chain. Studies of HbS also demonstrate that a mutation can affect the phenotype by directing a single amino acid substitution. Also, by providing the explanation for sickle-cell anemia, the concept of **molecular disease** was firmly established. Finally, this work led to a thorough study of human hemoglobins, which has provided valuable genetic insights.

In the United States, sickle-cell anemia is found almost exclusively in the black population. Extensive research is now under way to determine modes of treatment for those with the disease. It affects about one in every 625 black infants born in this country. Currently, about 50,000 to 75,000 individuals are afflicted. In about one of every 145 black married couples, both partners are heterozygous carriers, where each of their children has a 25 percent chance of having the disease.

**Table 15.1**   Chain Compositions of Human Hemoglobins from Conception to Adulthood

| Hemoglobin Type | Chain Composition |
|---|---|
| Embryonic-Gower 1 | $\zeta_2\epsilon_2$ |
| Fetal-HbF | $\alpha_2{}^A\gamma_2$ $\alpha_2{}^G\gamma_2$ |
| Adult-HbA | $\alpha_2\beta_2$ |
| Minor adult-HbA$_2$ | $\alpha_2\delta_2$ |

## Human Hemoglobins

Molecular analysis reveals that a variety of hemoglobin molecules are produced in humans. All are tetramers consisting of numerous combinations of seven distinct polypeptide chains, each encoded by a separate gene. In our discussion of sickle-cell anemia, we learned that **HbA** contains two **alpha ($\alpha$)** and two **beta ($\beta$) chains**. HbA represents about 98 percent of all hemoglobin found in an individual's erythrocytes after the age of six months. The remaining 2 percent consists of **HbA$_2$**, a minor adult component. This molecule contains two alpha and two **delta ($\delta$) chains**. The delta chain is very similar to the beta chain, consisting of 146 amino acids.

During embryonic and fetal development, quite a different set of hemoglobins is found. The earliest set to develop is called **Gower 1** containing two **zeta ($\zeta$) chains**, which are alphalike, and two **epsilon ($\epsilon$) chains**, which are betalike. By eight weeks of gestation, this embryonic form is gradually replaced by still another hemoglobin molecule with still different chains. This molecule is called **HbF**, or **fetal hemoglobin**, and consists of two alpha chains and two **gamma ($\gamma$) chains**. There are two types of gamma chains designated $^G\gamma$ and $^A\gamma$. Both are betalike and differ from each other by only a single amino acid.

The nomenclature and sequence of appearance of the five tetramers described so far are summarized in Table 15.1.

## COLINEARITY

Once it was established that genes specify the synthesis of polypeptide chains, the next logical question was

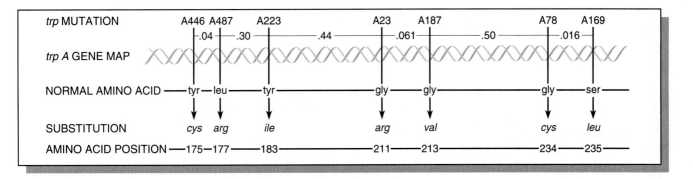

**FIGURE 15.8**    Demonstration of colinearity between the genetic map of various *trp* A mutations in *E. coli* and the affected amino acids in the protein product. The values shown between mutations represent linkage distances.

how genetic information contained in the nucleotide sequence of a gene can be transferred to the amino acid sequence of a polypeptide chain. It seemed most likely that a **colinear relationship** would exist between the two molecules. That is, the order of nucleotides in the DNA of a gene would correlate directly with the order of amino acids in the corresponding polypeptide.

While we have introduced this concept earlier in Chapter 14, it is useful to examine experimental evidence in its support. In studies of the A subunit of the enzyme **tryptophan synthetase** in *E. coli*, Charles Yanofsky sought to demonstrate colinearity. He isolated many independent mutants that had lost activity of the enzyme. He was able to map these mutations and establish their location with respect to one another within the gene. Then, he determined where the amino acid substitution had occurred in each mutant protein. When the two sets of data were compared, the colinear relationship was apparent. The location of each mutation in the *trp*A gene correlates with the position of the altered amino acid in the A polypeptide of tryptophan synthetase. This comparison is illustrated in Figure 15.8. Recall that we have already discussed the details of how information is transferred from DNA to protein (transcription and translation) in Chapter 14.

# PROTEIN STRUCTURE AND FUNCTION

Having established that the genetic information is stored in DNA and influences cellular activities through the proteins it encodes, we turn now to a discussion of protein structure and function. How is it that these molecules play such a critical role in determining the complexity of cellular activities? As we will see, the structure of proteins is intimately related to the functional diversity of these molecules.

## Protein Structure

First, we should differentiate between the terms **polypeptides** and **proteins**. Both describe molecules composed of amino acids. The molecules differ, however, in their state of assembly and functional capacity.

Polypeptides are the precursors of proteins. As assembled on the ribosome during translation, the molecule is called a polypeptide. When released from the ribosome following translation, a polypeptide folds up and assumes a higher order of structure. When this occurs, a three-dimensional conformation in space is produced. In many cases, several polypeptides interact to produce this conformation. Whether one or more than one polypeptide is involved, the three-dimensional conformation is essential to the function of the molecule. When the functional state is achieved, the molecule is appropriately called a protein.

The polypeptide chains of proteins, like nucleic acids, are linear nonbranched polymers. There are 20 amino acids that serve as the building blocks or subunits of proteins. Each amino acid has a **carboxyl group**, an **amino group**, and an **R (radical) group** (or side chain) bound covalently to a central carbon atom. The R group gives each amino acid its chemical identity. Figure 15.9 illustrates the 20 different R groups, which show a variety of configurations and may be divided into four main classes: (1) **nonpolar** or **hydrophobic**; (2) **polar** or **hydrophilic**; (3) **negatively charged**; and (4) **positively charged**. Because polypeptides are often long polymers, and because each position may be occupied by any one of 20 amino acids with unique chemical

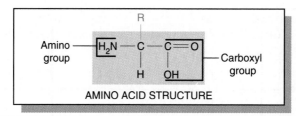

AMINO ACID STRUCTURE

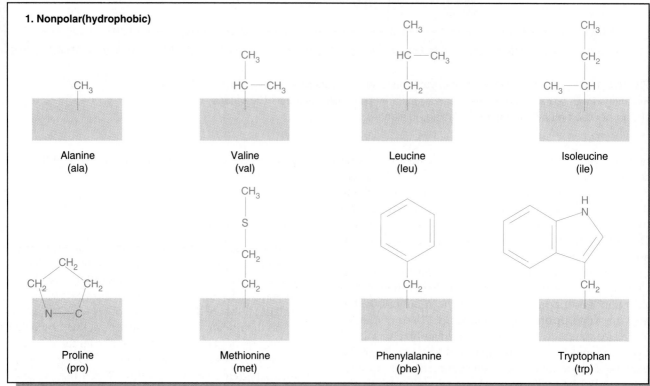

**1. Nonpolar(hydrophobic)**

Alanine
(ala)

Valine
(val)

Leucine
(leu)

Isoleucine
(ile)

Proline
(pro)

Methionine
(met)

Phenylalanine
(phe)

Tryptophan
(trp)

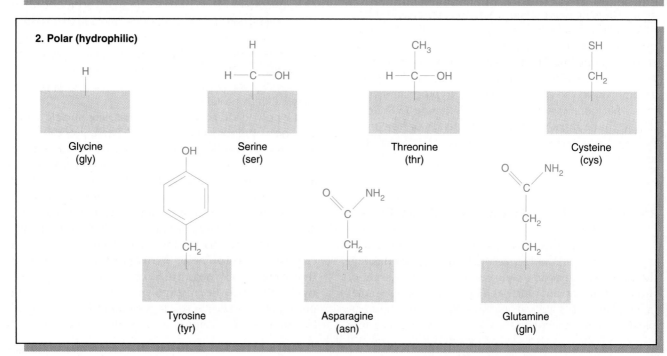

**2. Polar (hydrophilic)**

Glycine
(gly)

Serine
(ser)

Threonine
(thr)

Cysteine
(cys)

Tyrosine
(tyr)

Asparagine
(asn)

Glutamine
(gln)

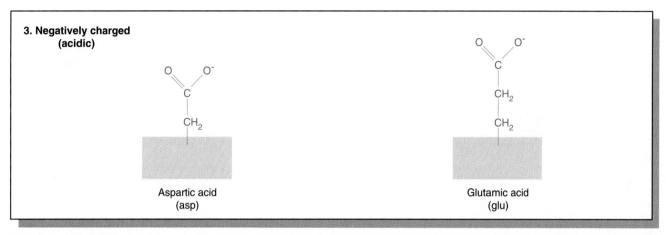

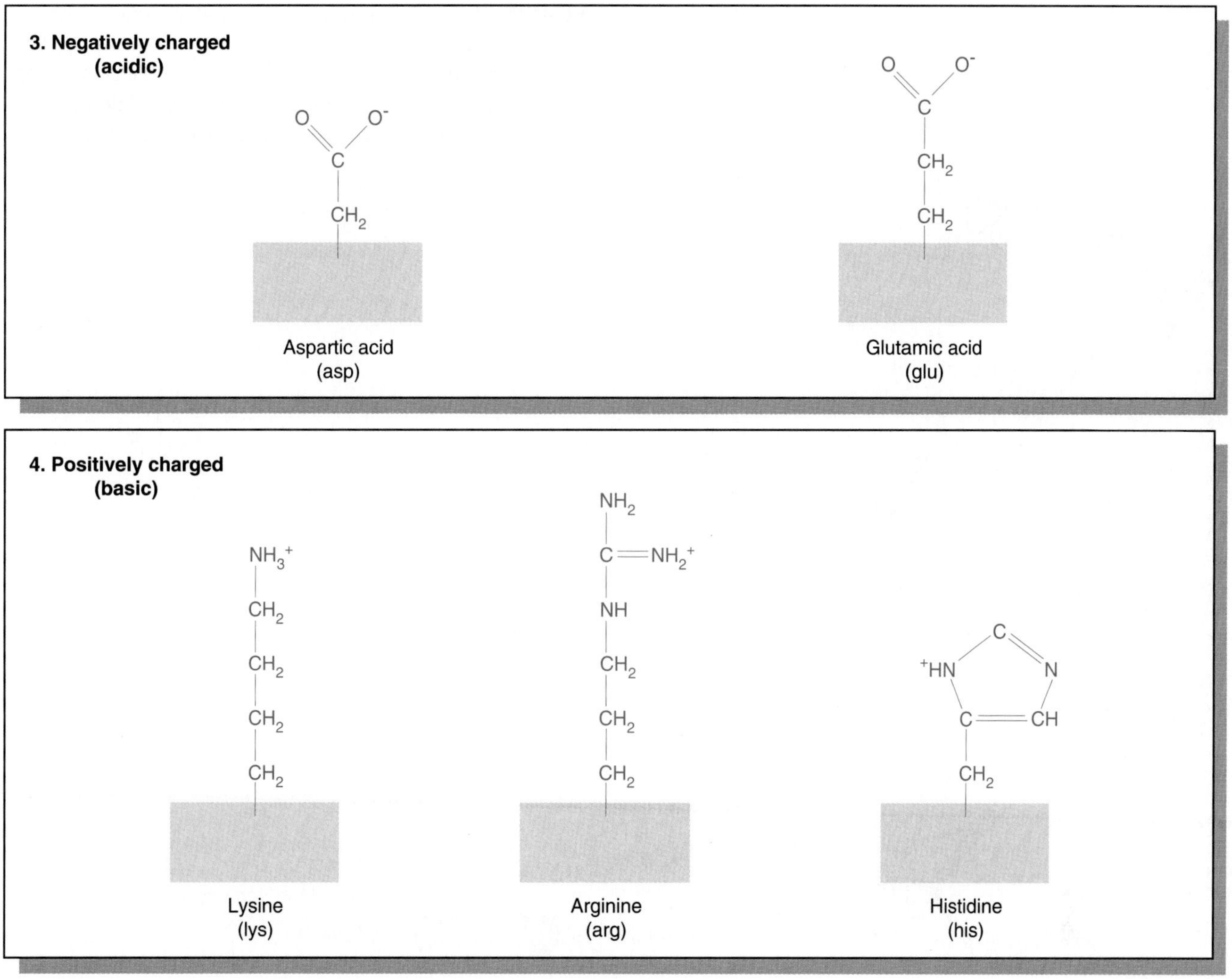

**FIGURE 15.9**    Chemical structures of the 20 amino acids found in living organisms, divided into four major categories.

properties, an enormous variation in chemical activity is possible. For example, if an average polypeptide is composed of 200 amino acids (molecular weight of about 20,000 daltons), $20^{200}$ different molecules, each with a unique sequence, can be created using 20 different building blocks.

Around 1900, a German chemist Emil Fischer determined the manner in which the amino acids are bonded together. He showed that the amino group of one amino acid can react with the carboxyl group of another amino acid during a dehydration reaction, releasing a molecule of $H_2O$. The resulting covalent bond is known as a **peptide bond** (Figure 15.10). Two amino acids linked to-

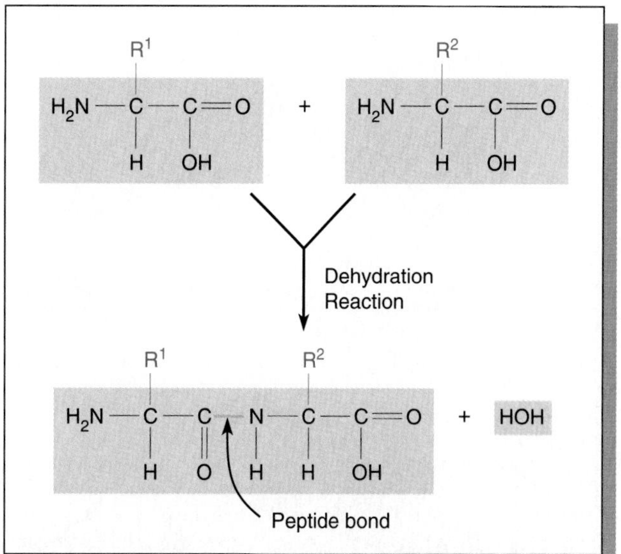

**FIGURE 15.10**    Peptide bond formation between two ▶ amino acids, resulting from a dehydration reaction.

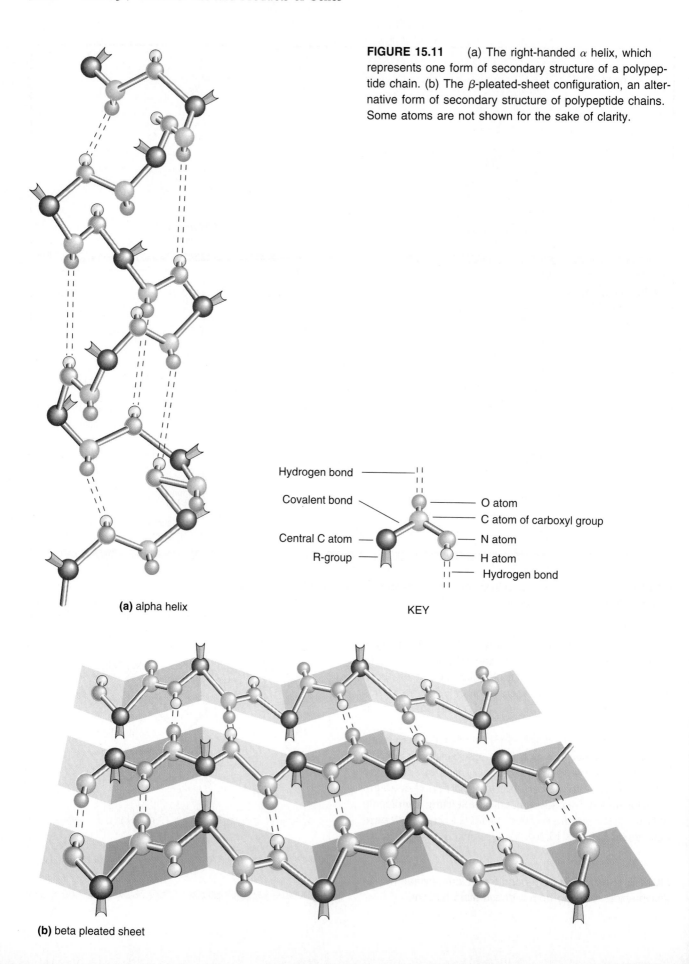

**FIGURE 15.11** (a) The right-handed $\alpha$ helix, which represents one form of secondary structure of a polypeptide chain. (b) The $\beta$-pleated-sheet configuration, an alternative form of secondary structure of polypeptide chains. Some atoms are not shown for the sake of clarity.

Hydrogen bond

Covalent bond

Central C atom

R-group

O atom

C atom of carboxyl group

N atom

H atom

Hydrogen bond

**(a)** alpha helix

KEY

**(b)** beta pleated sheet

gether constitute a **dipeptide**, three a **tripeptide**, etc. When more than ten amino acids are linked by peptide bonds, the chain is referred to as a **polypeptide**. Generally, no matter how long a polypeptide is, it will contain a free amino group at one end (the **N-terminus**) and a free carboxyl group at the other end (the **C-terminus**).

Four levels of protein structure are recognized: **primary (I°)**; **secondary (II°)**; **tertiary (III°)**; and **quaternary (IV°)**. The sequence of amino acids in the linear backbone of the polypeptide constitutes its **primary structure**. This sequence is specified by the sequence of deoxyribonucleotides in DNA via an mRNA intermediate. The primary structure of a polypeptide determines the specific characteristics of the higher orders of organization as a protein is formed.

The **secondary structure** refers to a regular or repeating configuration in space assumed by amino acids aligned closely to one another in the polypeptide chain. In 1951, Linus Pauling and Robert Corey predicted, on theoretical grounds, an **α (alpha) helix** as one type of secondary structure. The α-helix model [Figure 15.11(a)] has since been confirmed by X-ray crystallographic studies. It is rodlike and has the greatest possible theoretical stability. The helix is composed of a spiral chain of amino acids stabilized by hydrogen bonds.

The side chains of amino acids extend outward from the helix, and each amino acid residue occupies a vertical distance of 1.5 Å in the helix. There are 3.6 residues per turn. While left-handed helices are theoretically possible, all proteins demonstrating an α helix are right-handed.

Also in 1951, Pauling and Corey proposed a second structure, the **β-pleated-sheet** configuration. In this model, a single polypeptide chain folds back on itself, or several chains run in either parallel or antiparallel fashion next to one another. Each such structure is stabilized by hydrogen bonds formed between atoms present on adjacent chains [Figure 15.11(b)]. A single zigzagging plane is formed in space with adjacent amino acids 3.5 Å apart.

As a general rule, most proteins demonstrate a mixture of α and β structure. Globular proteins, most of which are round in shape and water soluble, usually contain a core of β-pleated-sheet structure as well as many areas demonstrating α helical structure. The more rigid structural proteins, many of which are water insoluble, rely on more extensive β-pleated-sheet regions for their rigidity. For example, **fibroin**, the protein made by the silk moth, depends extensively on this form of secondary structure.

While the secondary structure describes the arrangement of amino acids within certain areas of a polypeptide chain, **tertiary protein structure** defines the three-dimensional conformation of the entire chain in space. The molecule twists and turns and loops around itself in a very specific fashion, characteristic of the specific protein. Three aspects of this III° structure are most important in determining this conformation and in stabilizing the molecule.

1. Covalent disulfide bonds form between closely aligned cysteine residues to form the unique amino acid cystine.

2. Nearly all of the polar, hydrophilic R groups are located on the surface, where they may interact with water.

3. The nonpolar, hydrophobic R groups are usually located on the inside of the molecule, where they interact with one another, avoiding interaction with water.

It is important to emphasize that the three-dimensional conformation achieved by any protein is the direct result of the primary (I°) structure of the polypeptide. Thus, the genetic code need only specify the sequence of amino acids in order to encode information leading to the complete structure of proteins. The three stabilizing factors listed above depend on the location of each amino acid relative to all others in the chain. As folding occurs, the most thermodynamically stable conformation possible results.

The three-dimensional tertiary structure of the enzyme RNase T1 is shown in Figure 5.12. This level of organization is extremely important because the specific function of any protein is the direct result of its three-dimensional conformation.

The **quaternary level** of organization is characteristic of proteins composed of more than one polypeptide chain. The IV° structure indicates the conformation of the various chains in relation to one another. This type of protein is called **oligomeric**, and each chain is called a **protomer**, or less formally, a **subunit**. The individual protomers have native conformations that fit together in a specific complementary fashion. Hemoglobin, an

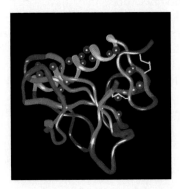

**FIGURE 15.12**    A ribbon drawing illustrating the tertiary (III°) level of protein structure in the enzyme RNase T1. Spheres depict hydrogen bonds formed early in the folding pathway.

**FIGURE 15.13**    Schematic drawing illustrating the quaternary level (IV°) of protein structure as seen in hemoglobin. Four chains (2 alpha and 2 beta) interact with four heme groups to form the functional molecule.

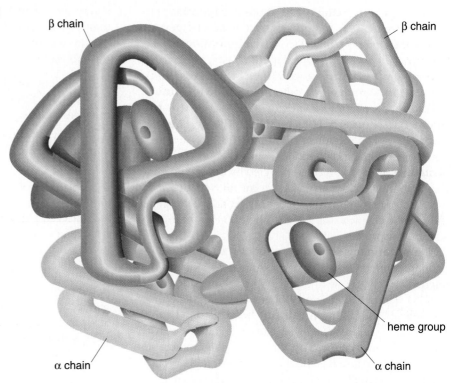

β chain

β chain

heme group

α chain

α chain

oligomeric protein consisting of four polypeptide chains, has been studied in great detail. Its IV° structure is shown in Figure 15.13. Most enzymes, including DNA and RNA polymerase, demonstrate IV° structure.

# POSTTRANSLATIONAL MODIFICATION AND PROTEIN TARGETING

Before turning to a discussion of protein function, it is important to point out that polypeptide chains, like RNA, are often modified once they have been synthesized. This concept is broadly described as **posttranslational modification**. While many of these alterations are detailed biochemical transformations and beyond the scope of this discussion, you should be aware that they occur, and that they are critical to the functional capability of the final protein product. Furthermore, many such modifications are involved in the regulation of protein activity.

Several examples are presented below, the last of which will be discussed in slightly more detail.

1. **The N-terminus and C-terminus amino acids are usually removed or modified**. For example, the initial N-terminal formymethionine residue in bacterial polypeptides is usually removed enzymatically. Often, the amino group of the initial methionine residue is removed, and the amino group of the N-terminal residue is acetylated in eukaryotic polypeptide chains.

2. **Individual amino acid residues are sometimes modified**. For example, phosphates may be added to the hydroxyl groups of certain amino acids such as tyrosine. Modifications such as this create negatively charged residues that may ionically bond with other molecules. The process of phosphorylation is extremely important in regulating many cellular activities as a result of the action of enzymes called **kinases**. In other proteins, methyl groups may be added enzymatically.

3. **Carbohydrate side chains are sometimes attached**. These are added covalently, producing **glycoproteins**, an important category of molecules that includes many antigenic determinants.

4. **Polypeptide chains may be trimmed**. For example, insulin is first translated into a longer molecule that is enzymatically trimmed to its final 51 amino acid form.

5. **Signal sequences are removed**. At the N-terminal end of some proteins is found a sequence of up to 30 amino acids that plays an important role in directing the protein to the location in the cell where it functions. Therefore, this is called a **signal sequence**, and it determines the final destination of a

protein in the cell. This process is called **protein targeting**.

For example, proteins, whose fate involves secretion or which are to become part of the plasma membrane, are dependent on specific sequences for their initial transport into the lumen of the endoplasmic reticulum. While the signal sequence of various proteins with a common destination might differ in their primary amino acid sequence, they share many chemical properties. For example, those destined for secretion all contain a string of up to 15 hydrophobic amino acids preceded by a positively charged amino acid at the N-terminus of the signal sequence. Once transported, but prior to achieving their functional status as proteins, the signal sequence is enzymatically removed from these polypeptides.

6.  **The function of some proteins is dependent on their affinity to complex with metals.** Hemoglobin, containing four iron atoms along with four polypeptide chains, is a good example.

These various types of posttranslational modifications are no doubt important in the conversion of newly translated polypeptide chains into their final three-dimensional conformation in space. Therefore, such modifications are critical in achieving the functional status specific to proteins.

## Protein Function

Proteins are the most abundant macromolecules found in cells. As the end products of genes, they play many diverse roles. For example, the respiratory pigments **hemoglobin** and **myoglobin** transport oxygen, which is essential for cellular metabolism. **Collagen** and **keratin** are examples of structural proteins associated with the skin, connective tissue, and hair of organisms. **Actin** and **myosin** are contractile proteins, found in abundance in muscle tissue. Still other examples are the **immunoglobins**, which function in the immune system of vertebrates; **transport proteins**, involved in movement of molecules across membranes; some of the **hormones** and their **receptors**, which regulate various types of chemical activity; and **histones**, which bind to DNA in eukaryotic organisms.

The largest group of proteins with a related function are the **enzymes**. These molecules specialize in catalyzing biological reactions. Enzymes increase the rate at which a chemical reaction reaches equilibrium, but they do not alter the end point of the chemical equilibrium. Their remarkable, highly specific catalytic properties largely determine the biomolecular nature of any cell type. The specific functions of many enzymes involved in the genetic and cellular processes of cells are described throughout the text.

Catalysis is a process whereby the **energy of activation** for a given reaction is lowered (Figure 15.14). The energy of activation is the increased kinetic energy state that molecules must usually reach before they react with one another. While this state can be attained as a result of elevated temperatures, enzymes allow biological reactions to occur at lower physiological temperatures. In this way, enzymes make possible life as we know it.

The catalytic properties and specificity of an enzyme are determined by the chemical configuration of the

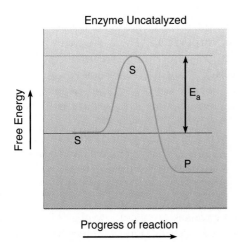
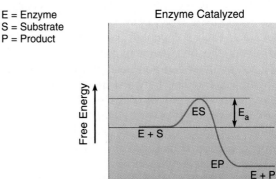

E = Enzyme
S = Substrate
P = Product

**FIGURE 15.14**    Energy requirements of an uncatalyzed versus an enzymatically catalyzed chemical reaction. The energy of activation ($E_a$) necessary to initiate the reaction is substantially lower as a result of catalysis.

molecule's **active site**. This site is associated with a crevice, a cleft, or a pit on the surface of the enzyme, which binds the reactants, or substrates, facilitating their interaction. Enzymatically catalyzed reactions control metabolic activities in the cell. Each reaction is either **catabolic** or **anabolic**. Catabolism is the degradation of large molecules into smaller, simpler ones with the release of chemical energy. Anabolism is the synthetic phase of metabolism, yielding nucleic acids, proteins, lipids, and carbohydrates. Metabolic pathways that serve the dual function of anabolism and catabolism are **amphibolic**.

## Protein Structure and Function: The Collagen Fiber

In order to provide a more in-depth example of the relationship between protein structure and function, we shall consider one of the most interesting proteins found in vertebrates: **collagen**. In mammals, it is the most abundant protein, constituting up to 25 percent of the total in an individual. It has been extensively studied and clearly illustrates this relationship.

Collagen is found in many places in the body, including tendons, ligaments, bone, connective tissue, skin, blood vessels, teeth, and the lens and cornea of the eye. In each case, the presence of collagen increases the tensile strength of, and provides support to, the specific tissue of which it is a part. An inspection of the above list makes it evident that collagen must be a versatile protein, providing a variable degree of flexibility and support to these tissues.

While there are several types of collagen produced in vertebrates, we will restrict our discussion to just the major one, called *type I*. Its main subunit or building block is called **tropocollagen**. This consists of three polypeptide chains wrapped around one another in a triple helix. This unit is extremely large, being 15 Å in diameter and 3000 Å long. Each polypeptide contains about 1000 amino acids, and the three interacting chains of the helix are stabilized by hydrogen bonds between them. There is no hydrogen bonding between amino acids within a single chain. Thus, no $\alpha$-helical or $\beta$-pleated-sheet secondary structure exists in collagen.

The maturation of the polypeptides into tropocollagen and the subsequent condensation of these helical structures into the more densely coiled collagen fibers is a fascinating tale. The polypeptides are first synthesized by fibroblasts as even longer units called procollagen. These are secreted into extracellular spaces, where they are cleaved and shortened at both the N-terminus and C-terminus by specific enzymes called **procollagen peptidases** [Figure 15.15(a)].

Once tropocollagen has been formed, these triple-helical rods associate spontaneously to form dense collagen fibers. The association follows an orderly pattern, where rows of end-to-end molecules line up in a staggered fashion next to one another [Figure 15.15(b)]. In each row a gap of approximately 400 Å exists between each tropocollagen unit. This complex structural arrangement creates protein fibers that strengthen and support a variety of tissues.

Although we have concentrated our discussion on type I collagen, there are about 20 different collagen genes, each encoding a slightly different primary chain. At least ten different combinations of chains have been discovered, each constituting a variant collagen molecule. It is likely that each imparts slightly different properties and characteristics to the different forms of collagen.

## The Genetics of Collagen

We began this chapter with a discussion of an inherited biochemical disorder first studied early in this century by Garrod. It is appropriate that we conclude this chapter with a short discussion of more modern discoveries of inherited biochemical disorders, this time related to collagen.

At least two inherited disorders known to involve defects in collagen synthesis and assembly are fairly well characterized genetically. For example, the inability to adequately transform procollagen to collagen leads to the human connective tissue disorder **Ehlers-Danlos syndrome**. Individuals exhibiting this disorder have fragile, stretchable skin and are loose-jointed, providing hypermobility. Some forms of the syndrome render individuals susceptible to arterial and colon ruptures as well as periodontal problems. They exhibit elevated levels of procollagen and decreased activity of the enzyme procollagen peptidase.

Our second example involves the lethal human disorder **osteogenesis imperfecta**. There are many variations, but all affect long bone formation and result from some defect in collagen structure. In its most severe form (*type II*), widespread bone defects occur in the fetus and newborn; fractures occur spontaneously and death often occurs soon after birth.

Type I, in comparison, is fairly mild, with an onset sometimes as late as age 35 to 45. However, at that point, declining health is associated with the degeneration of collagen-rich tissues: Blood vessels weaken, bones become fragile, and they fracture frequently. A stroke or heart attack usually causes premature death.

In one instance, the specific genetic defect has been uncovered and found to involve but a single glycine residue in the primary procollagen chain. At position 988, a glycine residue has been altered to the amino acid cys-

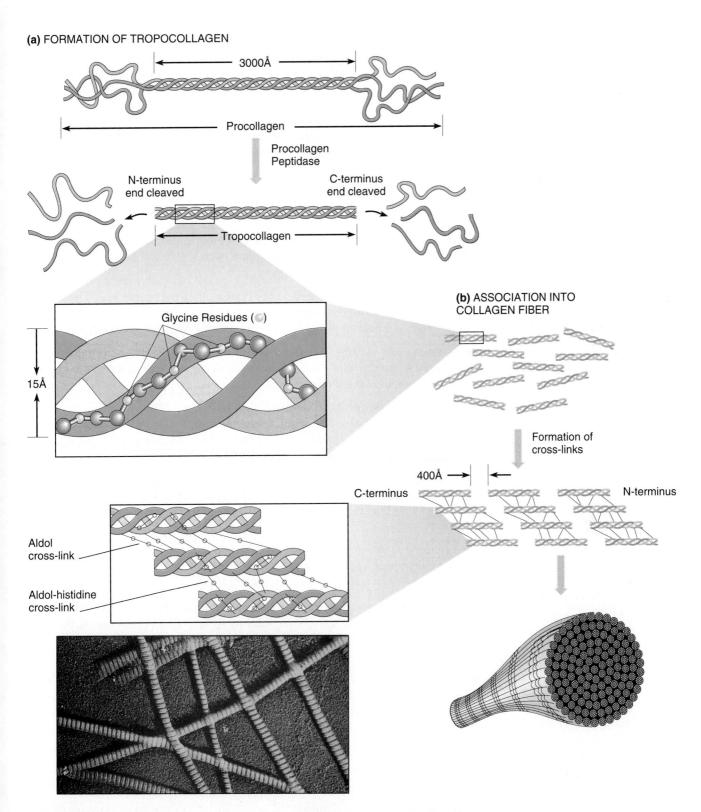

**(a)** FORMATION OF TROPOCOLLAGEN

3000Å

Procollagen

Procollagen
Peptidase

N-terminus
end cleaved

C-terminus
end cleaved

Tropocollagen

Glycine Residues (○)

15Å

**(b)** ASSOCIATION INTO
COLLAGEN FIBER

Formation of
cross-links

400Å

C-terminus

N-terminus

Aldol
cross-link

Aldol-histidine
cross-link

**FIGURE 15.15**     Various steps and intermediate structures involved in the formation
of collagen. (a) The process of posttranslational modification of polypeptides during the
formation of tropocollagen, the precursor of collagen. (b) The formation of cross-links
leading to the assembly of the collagen fiber. An electron micrograph illustrates such
fibers.

teine by a mutation of a single base change in the gene. This alteration of the primary structure inhibits the formation of highly ordered fibers that make up the bundles of collagen and leads to all of the aberrant phenotypic effects associated with the disease.

Collagen disorders are quite prevalent in the human population. It is estimated that over 100,000 individuals worldwide suffer from osteogenesis imperfecta. While Ehlers-Danlos syndrome is much rarer, there are about 20,000 individuals who suffer from still another collagen-

based disorder called **Marfan syndrome**. Long, thin arms, legs, and digits of the hands and feet characterize the appearance of affected individuals. Aortic heart disease most frequently is the cause of premature death. Abraham Lincoln is suspected to have suffered from this disorder. Given the complexity of the biochemistry of collagen and its wide distribution throughout the body, it is not surprising that these and other genetic disorders exist.

## CHAPTER SUMMARY

1.  Basic to Garrod's studies in the early 1900s of inborn, human metabolic disorders such as cystinuria, albinism, and alkaptonuria, was the concept that genes control the synthesis of specific metabolic products.

2.  The investigation of eye pigments in *Drosophila* and nutritional requirements in *Neurospora* by Beadle and colleagues made it clear that mutations cause the loss of enzyme activity. Their work led to the concept of one gene:one enzyme.

3.  The one gene:one enzyme hypothesis was later revised. Pauling and Ingram's investigations of hemoglobins from patients with sickle-cell anemia led to the discovery that one gene directs the synthesis of only one polypeptide chain.

4.  Thorough investigations have revealed the existence of several major types of human hemoglobin molecules, found in the embryo, fetus, and the adult. Specific genes control each polypeptide chain constituting these various hemoglobin molecules.

5.  The proposal suggesting that a gene's nucleotide sequence specifies in a colinear way the sequence of amino acids in a polypeptide chain was confirmed by Yanofsky's experiments involving mutations in the tryptophan synthetase gene in *E. coli.*

6.  Proteins, the end products of genes, demonstrate four levels of structural organization that together provide the chemical basis for their three-dimensional conformation.

7.  Of the myriad functions performed by proteins, the most influential role is assumed by enzymes. These highly specific, cellular catalysts play a central role in the production of all classes of molecules in living systems.

8.  Collagen is an abundant, specialized protein in vertebrates, serving a structural role in a variety of tissues. Ehlers-Danlos syndrome, osteogenesis imperfecta, and Marfan syndrome are examples of inherited human disorders resulting from mutations that alter collagen structure and its function.

## KEY TERMS

| | | | |
|---|---|---|---|
| actin | amphibolic reaction | citrulline | energy of activation |
| active site | anabolic reaction | colinear relationship | enzyme |
| alkaptonuria | beta chain | collagen | epsilon chain |
| alpha chain | β-pleated-sheet | C-terminus | fetal hemoglobin |
| α (alpha) helix | configuration | delta chain | fibroin |
| amino acid | carboxyl group | dipeptide | fingerprinting technique |
| amino group | catabolic reaction | Ehlers-Danlos syndrome | gamma chain |

globin chains
glycoprotein
Gower 1 hemoglobin
HbA
HbA$_2$
HbF
HbS
heme groups
hemoglobin
histone
homogentisic acid
hormone
hydrophilic group
hydrophobic group
imaginal disk

immunoglobin
keratin
Marfan syndrome
molecular disease
myoglobin
myosin
N-terminus
oligomeric protein
one-gene:one-enzyme
    hypothesis
one gene:one
    polypeptide chain
one gene:one protein
osteogenesis imperfecta
peptide bond

phenylalanine hydroxlase
phenylketonuria (PKU)
polypeptide chain
posttranslational
    modification
primary (I°) protein
    structure
procollagen peptidase
protein
protein targeting
protomer
quaternary (IV°) protein
    structure
R (radical) group

secondary (II°) protein
    structure
sickle-cell anemia
sickle-cell trait
signal sequence
starch gel electrophoresis
subunit
tertiary (III°) protein
    structure
transport proteins
tripeptide
tropocollagen
tryptophan synthetase
zeta chain

# INSIGHTS AND SOLUTIONS

1. In Beadle and Ephrussi's studies involving imaginal disk transplants, most eye color mutants behaved autonomously (i.e., they were not cured, or converted to wild-type color) when allowed to develop in a wild-type abdominal environment. One concludes that in such cases, the product influenced by the gene under study is *not* diffusible. The *vermilion* (*v*) and *cinnabar* (*cn*) eye disks were the exception, developing into wild-type eyes, because the products controlled by these genes *are* diffusible and supplied by the host. Recall that *cn* disks developing in *v* hosts remain as mutant *cn* eyes, but that *v* disks developing in *cn* hosts are cured and develop wild-type color. This is the basis of determining that the $v^+$ gene product acts at a point in the biosynthetic pathway prior to the $cn^+$ gene product.

   Suppose that a new bright red mutation is discovered and the product controlled in this gene is not diffusible. However, the gene product acts prior to the steps controlled by both the $v^+$ and $cn^+$ gene products in the biosynthesis of the brown xanthommatin pigment. Predict the outcome of a transplant experiment of the new mutant eye disk into a wild-type larva.

   **SOLUTION:** Since the block in the new mutation precedes either the $cn^+$ or the $v^+$ steps, the wild-type host will make both of these diffusible products. As a result, the mutant disk will be cured and develop wild-type color. This happens because these diffusible products can enter the disk and allow the block in the new mutation to be bypassed.

2. The following growth responses were obtained using four mutant strains of *Neurospora* and the related compounds A, B, C, and D. None of the mutations grow on minimal medium. Draw all possible conclusions.

|          | Growth Product | | | |
|----------|:-:|:-:|:-:|:-:|
| Mutation | C | D | B | A |
| 1        | – | – | – | – |
| 2        | – | + | + | + |
| 3        | – | – | + | + |
| 4        | – | – | + | – |

**SOLUTION:** First, nothing can be concluded about mutation 1, except that it is lacking some essential growth factor, perhaps even unrelated to the biochemical pathway represented by mutations 2–4; nor can anything be concluded about compound C. If it is involved in the pathway, it is a product synthesized prior to the synthesis of A, B, and D.

We must now analyze these three compounds and the control of their synthesis by the enzymes encoded by genes 2, 3, and 4. Since product B allows growth in all three cases, it may be considered the "end product." It bypasses the block in all three instances. Using similar reasoning, product A precedes B in the pathway since it allows a bypass in two or the three steps. Product D precedes B, yielding the more complete solution:

$$C(?) \longrightarrow D \longrightarrow A \longrightarrow B$$

Now determine which mutations control which steps. Since mutation 2 can be alleviated by products D, B, and A, it must control a step prior to all three products, perhaps the direct conversion to D, although we can't be certain. Mutation 3 is alleviated by B and A, so its effect must precede them in the pathway. Thus, we will assign it as controlling the conversion of D to A. Likewise, we can assign mutation 4 to the conversion of A to B, leading to the more complete solution:

$$C(?) \xrightarrow{2(?)} D \xrightarrow{3} A \xrightarrow{4} B$$

1. Discuss the potential difficulties involved in designing a diet to alleviate the symptoms of phenylketonuria.

2. Phenylketonurics cannot convert phenylalanine to tyrosine. Why don't these individuals exhibit a deficiency of tyrosine?

3. Phenylketonurics are often more lightly pigmented than normal individuals. Can you suggest a reason why this is so?

4. Consider the following hypothetical circumstances involving reciprocal transplantations of three mutant imaginal eye disks. The three mutations, *a*, *b*, and *c*, are responsible for the enzymatic pathway shown below. However, it is not known which mutant gene controls which step in the reactions.

$$w \longrightarrow x \longrightarrow y \longrightarrow z$$

The results of the transplantations are shown below. Assuming that the *w*, *x*, *y*, and *z* substances are diffusible, determine which reactions are controlled by which genes.

| Donor | Host | Transplant Eye Pigment | Development |
|---|---|---|---|
| *a* | *c* | Mutant | Autonomous |
| *a* | *b* | Mutant | Autonomous |
| *c* | *b* | Mutant | Autonomous |
| *c* | *a* | Wild type | Nonautonomous |
| *b* | *c* | Wild type | Nonautonomous |
| *b* | *a* | Wild type | Nonautonomous |

**5.** What $F_1$ and $F_2$ ratios will occur in a cross between *vermilion* females and *cinnabar* males? Recall that $v$ is sex-linked and $cn$ is autosomal, and that these mutants produce identical bright red phenotypes. What ratios will occur in the reciprocal cross ($v$ males and $cn$ females)?

**6.** The synthesis of flower pigments is known to be dependent upon enzymatically controlled biosynthetic pathways. In the crosses shown below, postulate the role of mutant genes and their products in producing the observed phenotypes.

  (a) $P_1$: white strain A × white strain B
   $F_1$: all purple
   $F_2$: 9/16 purple: 7/16 white

  (b) $P_1$: white × pink
   $F_1$: all purple
   $F_2$: 9/16 purple: 3/16 pink: 4/16 white

**7.** A series of mutations in the bacterium *Salmonella typhimurium* result in the requirement of either tryptophan or some related molecule in order for growth to occur. From the data shown below, suggest a biosynthetic pathway for tryptophan.

| Mutation | Minimal Medium | Anthranilic Acid | Indole Glycerol Phosphate | Indole | Tryptophan |
|---|---|---|---|---|---|
| *trp*-8 | − | + | + | + | + |
| *trp*-2 | − | − | + | + | + |
| *trp*-3 | − | − | − | + | + |
| *trp*-1 | − | − | − | − | + |

Supplement-Growth Response (header spanning the middle columns)

**8.** The study of biochemical mutants in organisms such as *Neurospora* has demonstrated that some pathways are branched. The data shown below illustrate the branched nature of the pathway resulting in the synthesis of thiamine. Why don't the data support a linear pathway? Can you postulate a pathway for the synthesis of thiamine in *Neurospora*?

| Mutation | Minimal Medium | Pyrimidine | Thiazole | Thiamine |
|---|---|---|---|---|
| *thi*-1 | − | − | + | + |
| *thi*-2 | − | + | − | + |
| *thi*-3 | − | − | − | + |

Growth Response Supplement (header spanning columns)

**9.** Explain why the one-gene:one-enzyme concept is not considered accurate today.

**10.** Why is an alteration of electrophoretic mobility interpreted as a change in the primary structure of the protein under study?

**11.** Contrast the polypeptide chain components of each of the hemoglobin molecules found in humans.

**12.** Using sickle-cell anemia as a basis, describe what is meant by a molecular or genetic disease. What are the similarities and dissimilarities between this type of a disorder and a disease caused by an invading microorganism?

13. Contrast the contributions of Pauling and Ingram to our understanding of the genetic basis for sickle-cell anemia.

14. Hemoglobins from two individuals are compared by electrophoresis and by fingerprinting. Electrophoresis reveals no difference in migration, but fingerprinting shows an amino acid difference. How is this possible?

15. Describe what colinearity means.

16. Certain mutations called *amber* in bacteria and viruses result in premature termination of polypeptide chains during translation. Many *amber* mutations found at different points along the gene coding for a head protein in phage T4 have been detected. How might this system be further investigated to demonstrate and support the concept of colinearity?

17. Define and compare the four levels of protein organization.

18. List as many different protein functions as you can, with an example of each.

19. How does an enzyme function? Why are enzymes essential for living organisms on earth?

20. Discuss the role of the following in the production of the collagen fiber: (a) glycine residues in the primary chain, (b) procollagen peptidase, (c) prolyl hydroxylase, (d) aldol cross-links, (e) hydrogen bonds.

21. Why does Fiers' work with phage MS2, discussed in Chapter 14, constitute more direct evidence in support of colinearity than Yanofsky's work with the *trp* A locus in *E. coli* (discussed in this chapter)?

22. Shown below are several amino acid substitutions in the $\alpha$ and $\beta$ chains of human hemoglobin. Using the code table (Figure 14.6), determine how many of them can occur as a result of a single nucleotide change.

| Hb Type | Normal Amino Acid | Substituted Amino Acid |
|---|---|---|
| HbJ Toronto | ala | asp ($\alpha$-5) |
| HbJ Oxford | gly | asp ($\alpha$-15) |
| Hb Mexico | gln | glu ($\alpha$-54) |
| Hb Bethesda | tyr | his ($\beta$-145) |
| Hb Sydney | val | ala ($\beta$-67) |
| HbM Saskatoon | his | tyr ($\beta$-63) |

23. Three independently assorting genes are known to control the following biochemical pathway that provides the basis for flower color in a hypothetical plant. Homozygous recessive mutations, which interrupt each step, are known.

$$\text{Colorless} \xrightarrow{\text{A-}} \text{yellow} \xrightarrow{\text{B-}} \text{green} \xrightarrow{\text{C-}} \text{speckled}$$

Determine the phenotypic results in the $F_1$ and $F_2$ generation resulting from the following $P_1$ crosses involving true breeding plants.

(a) speckled (*AABBCC*) × yellow (*AAbbCC*)
(b) yellow (*AAbbCC*) × green (*AABBcc*)
(c) colorless (*aaBBCC*) × green (*AABBcc*)

24. How would the results vary in cross (a) of Problem 24 if genes A and B were linked with no crossing over between them? How would cross (a) vary if genes A and B were linked and 20 map units apart?

## SELECTED READINGS

BARTHOLOME, K. 1979. Genetics and biochemistry of phenylketonuria—Present state. *Hum. Genet.* 51:241–45.

BATESON, W. 1909. *Mendel's principles of heredity.* Cambridge, England: Cambridge University Press.

BEADLE, G. W. 1945. Genetics and metabolism in *Neurospora. Physiol. Rev.* 25:643.

BEADLE, G. W., and EPHRUSSI, B. 1937. Development of eye colors in *Drosophila*: Diffusible substances and their interrelations. *Genetics* 22:76–86.

BEADLE, G. W., and TATUM, E. L. 1941. Genetic control of biochemical reactions in *Neurospora. Proc. Natl. Acad. Sci.* 27:499–506.

BRENNER, S. 1955. Tryptophan biosynthesis in *Salmonella typhimurium. Proc. Natl. Acad. Sci.* 41:862–63.

BYERS, P. H. 1989. Inherited disorders of collagen gene structure and expression. *Am. J. Med. Genet.* 34:72–80.

COHN, D. H., BYERS, P. H., STEINMANN, B., and GELINAS, R. E. 1986. Lethal osteogenesis imperfecta resulting from a single nucleotide change in one human pro-$\alpha$ 1(I) collagen allele. *Proc. Natl. Acad. Sci.* 83:6045–47.

DICKERSON, R. E., and GEIS, I. 1983. *Hemoglobin: Structure, function, evolution, and pathology.* Menlo Park, Calif.: Benjamin/Cummings.

DOOLITTLE, R. F. 1985. Proteins. *Scient. Amer.* (Oct.) 253:88–99.

EPHRUSSI, B. 1942. Chemistry of eye color hormones of *Drosophila. Quart. Rev. Biol.* 17:327–38.

GARROD, A. E. 1902. The incidence of alkaptonuria: A study in chemical individuality. *Lancet* 2:1616–20.

———. 1909. *Inborn errors of metabolism.* London: Oxford University Press. (Reprinted 1963, Oxford University Press, London.)

GARROD, S. C. 1989. Family influences on A. E. Garrod's thinking. *J. Inher. Metab. Dis.* 12:2–8.

INGRAM, V. M. 1957. Gene mutations in human hemoglobin: The chemical difference between normal and sickle cell hemoglobin. *Nature* 180:326–28.

———. 1963. *The hemoglobins in genetics and evolution.* New York: Columbia University Press.

KOSHLAND, D. E. 1973. Protein shape and control. *Scient. Amer.* (Oct.) 229:52–64.

LADU, B. N., ZANNONI, V. G., LASTER, L., and SEEGMILLER, J. E. 1958. The nature of the defect in tyrosine metabolism in alkaptonuria. *J. Biol. Chem.* 230:251.

LEHNINGER, A. L., NELSON, D. L., and COX, M. M. 1993. *Principles of Biochemistry.* 2nd ed. New York: Worth Publishers.

MECHANIC, G. 1972. Cross-linking of collagen in a heritable disorder of connective tissue: Ehlers-Danlos syndrome. *Biochem. Biophys. Res. Commun.* 47:267–72.

NEEL, J. V. 1949. The inheritance of sickle-cell anemia. *Science* 110:64–66.

PAULING, L., ITANO, H. A., SINGER, S. J., and WELLS, I. C. 1949. Sickle cell anemia, a molecular disease. *Science* 110:543–48.

PFEFFER, S. R., and ROTHMAN, J. E. 1987. Biosynthetic protein transport and sorting by the endoplasmic reticulum and golgi. *Ann. Rev. Biochem.* 56:829–52.

RICHARDS, F. M. 1991. The protein folding problem. *Scient. Amer.* (Jan.) 264:54–63.

SCOTT-MONCRIEFF, R. 1936. A biochemical survey of some Mendelian factors for flower colour. *J. Genet.* 32:117–70.

SCRIVER, C. R., and CLOW, C. L. 1980. Phenylketonuria and other phenylalanine hydroxylation mutants in man. *Ann. Rev. Genet.* 14:179–202.

SRB, A. M., and HOROWITZ, N. H. 1944. The ornithine cycle in *Neurospora* and its genetic control. *J. Biol. Chem.* 154:129–39.

SYKES, B. 1985. The molecular genetics of collagen. *BioEssays* 3:112–17.

WAGNER, R. P., and MITCHELL, H. K. 1964. *Genetics and metabolism.* 2nd ed. New York: Wiley.

YANOFSKY, C., DRAPEAU, G., GUEST, J., and CARLTON, B. 1967. The complete amino acid sequence of the tryptophan synthetase A protein and its colinear relationship with the genetic map of the *A* gene. *Proc. Natl. Acad. Sci.* 57:296–98.

ZIEGLER, I. 1961. Genetic aspects of ommochrome and pterin pigments. *Adv. in Genet.* 10:349–403.

# 16

# GENETIC REGULATION IN BACTERIA AND BACTERIOPHAGES

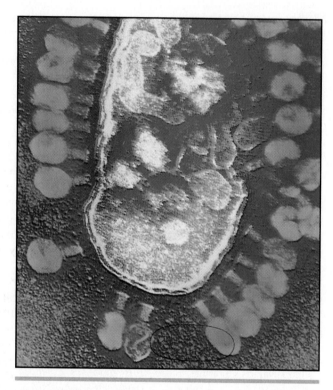

Transmission electron micrograph of many bacteriophage T$_4$ viruses infecting *E. coli*.

*Efficient expression of genetic information is dependent on regulatory mechanisms that either activate or repress gene activity. In bacteria, mechanisms exist that regulate genes in ways that meet metabolic needs of the cell. Gene regulation in bacteriophages determines the mode of existence of the virus within the bacterial host. In bacteria and their phages, these mechanisms are responses to the prevailing cellular and extracellular conditions.*

In earlier chapters, we established how DNA is organized into genes, how genes store genetic information, and how this information is expressed. We now consider one of the most fundamental issues in molecular genetics: **How is genetic expression regulated?** The evidence in support of the idea that genes can be turned on and off is very convincing. Detailed analysis of proteins in *Escherichia coli* has shown that for 4000 or so polypeptide chains encoded by the genome, there is a vast range of concentration of gene products. Some proteins may be present in as few as 5 to 10 molecules per cell, whereas others, such as ribosomal proteins and the many proteins involved in the glycolytic pathway, are present in as many as 100,000 copies per cell. While a basal level of most gene products exists, it is clear that this level can be increased and subsequently decreased in prokaryotes. Thus, fundamental regulatory mechanisms to control the expression of the genetic information must exist.

In this chapter, we will explore what is known about the regulation of genetic expression in bacteria and bacteriophages. Highly detailed information has now been obtained. The following chapter will focus on the regulation of gene expression in eukaryotes.

## BACTERIAL MUTATION AND GROWTH

The analysis of genetic regulation in bacteria depends upon our ability to isolate and study mutations in these organisms. Although it was known well before 1943 that pure cultures of bacteria could give rise to small numbers of cells exhibiting heritable variation, particularly with respect to survival under different environmental conditions, the source of the variation was hotly debated. The majority of bacteriologists believed that environmental factors induced changes in certain bacteria that led to their survival or adaptation to the new condi-

tions. For example, strains of *E. coli* are known to be *sensitive* to infection by the bacteriophage T1. Infection by the bacteriophage leads to the reproduction of the virus at the expense of the bacterial cell, which is lysed or destroyed (see Figure 8.5). If a plate of *E. coli* is homogeneously sprayed with T1, almost all cells are lysed. Rare *E. coli* cells, however, survive infection and are not lysed. If these cells are isolated and established in pure culture, all descendants are *resistant* to T1 infection. The **adaptation hypothesis**, put forth to explain this type of observation, implied that the interaction of the phage and bacterium is essential to the acquisition of immunity. In other words, the phage had "induced" resistance in the bacteria.

The occurrence of **spontaneous mutations** provided an alternative model to explain the origin of T1 resistance in *E. coli*. In 1943, Salvador Luria and Max Delbruck presented the first convincing evidence that bacteria, like eukaryotic organisms, are capable of spontaneous mutation. This experiment (called the **fluctuation test**) marked the initiation of modern bacterial genetic study. With only minor debate, spontaneous mutation is considered to be the major source of genetic variation in bacteria.

Mutant cells that arise spontaneously in an otherwise pure culture can be isolated and established independently from the parent strain by using efficient selection techniques. As a result, mutations for almost any desired characteristic can now be induced and isolated. Since bacteria and viruses are haploid, all mutations are expressed directly in the descendants of mutant cells, adding to the ease with which these microorganisms can be studied.

Bacteria are grown in either a liquid culture medium or in a Petri dish on a semisolid agar surface. If the nutrient components of the growth medium are very simple and consist only of an organic carbon source (such as a glucose or lactose) and a variety of ions, including $Na^+$, $K^+$, $Mg^{++}$, $Ca^{++}$, and $NH_3^+$ present as inorganic salts, it

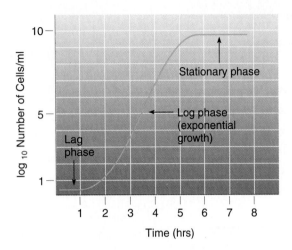

**FIGURE 16.1**   A typical bacterial population growth curve illustrating the initial lag phase, the subsequent log phase, where exponential growth occurs, and the stationary phase that results as nutrients are exhausted.

is called **minimal medium**. In order to grow on such a medium, a bacterium must be able to synthesize all essential organic compounds (e.g., amino acids, purines, pyrimidines, sugars, vitamins, fatty acids). A bacterium that can accomplish this remarkable biosynthetic feat—one that we ourselves cannot duplicate—is termed a **prototroph**. It is said to be wild type for all growth requirements. On the other hand, if a bacterium loses, through mutation, the ability to synthesize one or more organic components, it is said to be an **auxotroph**. For example, if it loses the ability to make histidine, then this amino acid must be added as a supplement to the minimal medium in order for growth to occur. The resulting bacterium is designated as a *his⁻* auxotroph, as opposed to its prototrophic *his⁺* counterpart.

In order to study mutant bacteria in a quantitative fashion, an inoculum of bacteria is placed in liquid culture medium. A characteristic growth pattern is exhibited, as illustrated in Figure 16.1. Initially, during the **lag phase**, growth is slow. Then, a period of rapid growth ensues called the **log phase**. During this phase, cells divide many times with a fixed time interval between cell divisions, resulting in logarithmic growth. When a cell density of about $10^9$ cells per milliliter is reached, nutrients and oxygen become limiting and cells enter the **stationary phase**. Since the doubling time during the log phase may be as short as 20 minutes, an initial inoculum of a few thousand cells can easily achieve a maximum cell density in an overnight culture.

Cells grown in liquid medium may be quantitated by transferring them to semisolid medium in a Petri dish. Following incubation and many divisions, each cell gives rise to a visible colony on the surface of the medium. If the number of colonies is too great to count, then serial dilutions of the original liquid culture can be made and plated, until the colony number is reduced to the point where it can be counted (Figure 16.2). This technique allows one to calculate the number of bacteria present in the original culture. Such calculations are useful in a variety of studies.

For example, in the three plates shown in the photograph in Figure 16.2, let's assume that a liquid culture of bacteria has been sampled. An initial milliliter (ml) is withdrawn, and it is our desire to determine how many cells are present per milliliter. If the initial milliliter is subjected to a series of serial dilutions, this can be accomplished easily.

Assume that the three petri dishes in Figure 16.2 represent dilutions of $10^{-3}$, $10^{-4}$, and $10^{-5}$, respectively (left to right). We need to select only the dish where the number of colonies can be accurately counted. Since each colony presumably arose from a single bacterium, that number times the dilution factor represents the number of bacteria in the initial milliliter. In our case, the dish farthest to the right contains 15 colonies. Since it represents a dilution of $10^{-5}$, the initial number of bacterium is estimated to be $15 \times 10^5$ per milliliter.

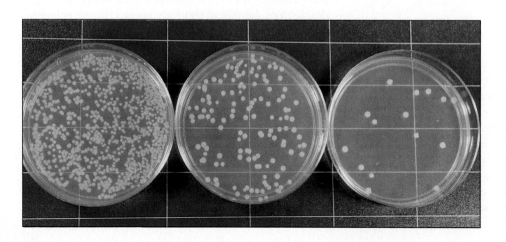

**FIGURE 16.2**   Results of the serial dilution technique and subsequent culture of bacteria. Each dilution varies by a factor of 10. Each colony was derived from a single bacterial cell.

# GENETIC REGULATION IN PROKARYOTES: AN OVERVIEW

The regulation of gene expression has been extensively studied in prokaryotes, particularly in *E. coli*. What has been found is that highly efficient genetic mechanisms have evolved that turn genes on and off, depending on the cell's metabolic need for the respective gene products. The activity of enzymes may also be regulated once they have been synthesized in the cell, but we will focus primarily on what is known about regulation at the level of gene transcription. It is important to remember that it is the resulting proteins ultimately present or absent that are critical to efficient cell function under varying environmental conditions.

It is not a particularly new concept that microorganisms regulate the synthesis of gene products. As early as 1900 it was shown that when the galactose-glucose-containing disaccharide **lactose** is present in the growth medium of yeast, enzymes specific to lactose metabolism are produced. When this substrate is absent, the enzymes are not manufactured. It was soon shown that bacteria "adapt" to their chemical environment, producing certain enzymes only when specific substrates are present. Such enzymes were thus referred to as **adaptive**. In contrast, those produced continuously regardless of the chemical makeup of the environment were called **constitutive**. Since then, the term *adaptive* in descriptive enzymology has been replaced with a more accurate term. We now call them **inducible** enzymes, reflecting the role of the substrate, or **inducer**, in their production.

More recent investigation has revealed other cases where the presence of a specific molecule causes inhibition of genetic expression. This is usually true for molecules that are end products of biosynthetic pathways. For example, an amino acid such as tryptophan can be synthesized by bacterial cells. If an exogenous supply of this amino acid is present in the environment or culture medium, it is energetically inefficient to synthesize the enzymes necessary for tryptophan production. As a result, a mechanism exists whereby tryptophan plays a role in repressing transcription of RNA essential to the production of the appropriate biosynthetic enzymes. In contrast to the inducible system controlling lactose metabolism, that governing tryptophan is said to be **repressible**.

As we will soon see, instances of regulation, whether inducible or repressible, may be under either **negative** or **positive control**. Under negative control, genetic expression occurs unless it is shut off by some form of regulator. This is in contrast to positive control, where transcription does not occur unless a regulator molecule directly stimulates RNA production. In theory, each type of control can govern inducible or repressible systems. Examples discussed in the ensuing sections of this chapter will help distinguish between these possible mechanisms.

# LACTOSE METABOLISM IN *E. COLI*: AN INDUCIBLE GENE SYSTEM

The most extensively studied system of gene regulation has involved the metabolism of lactose in *E. coli*. Beginning in 1946 with the studies of Jacques Monod and continuing through the next decade with significant contributions by Joshua Lederberg, François Jacob, and Andre L'woff, genetic and biochemical evidence was amassed. This research provided clear insights into the way in which the genes responsible for lactose metabolism are turned off, or repressed, when lactose is absent but activated or induced when it is available. In the presence of lactose, the concentration of the enzymes responsible for its metabolism increases rapidly from 5 to 10 molecules to thousands per cell. Thus, the enzymes are **inducible**, and lactose serves as the **inducer**.

Paramount to the understanding of genetic regulation in this system was the discovery of one gene and one chromosome region that serve strictly in a regulatory capacity. Neither encodes enzymes necessary for lactose metabolism. Three other genes are responsible for the production of enzymes involved in lactose metabolism. Together, these five genetic units function in an integrated fashion and provide a rapid response to the presence or absence of lactose.

## Structural Genes

Genes coding for the primary structure of the enzymes are called **structural genes**. The so-called *lac* Z gene specifies the amino acid sequence of the **β-galactosidase enzyme**, which converts lactose to glucose and galactose (Figure 16.3). This conversion is essential if lactose is to serve as the primary energy source in glycolysis. The second gene, *lac* Y, specifies the primary structure of **β-galactoside permease**, which facilitates the entry of lactose into the bacterial cell. The third gene, *lac* A, codes for a **transacetylase** enzyme whose physiological role is unrelated to our discussion.

Studies of the genes coding for these three enzymes relied on the isolation of numerous mutations that eliminated the function of one or the other enzyme. Such *lac⁻* mutants were isolated and studied by Lederberg. Mutant cells fail to produce either active β-galactosidase or permease molecules and so are unable to utilize lactose as an energy source. Mutations also were found in

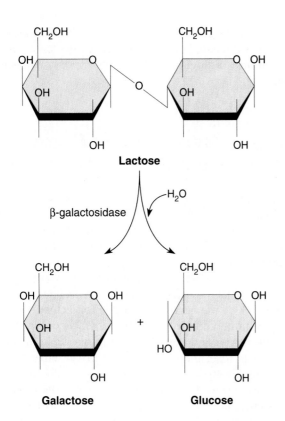

**FIGURE 16.3**    The catabolic conversion of the disaccharide lactose into its monosaccharide units, galactose and glucose.

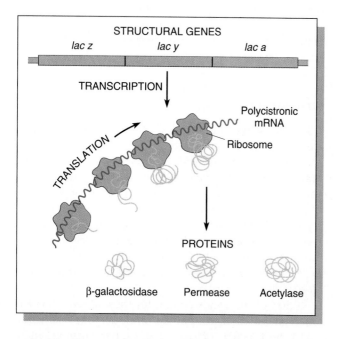

**FIGURE 16.4**    The structural genes of the *lac* operon in *E. coli*. The three genes are transcribed into a single polycistronic mRNA, which is sequentially translated into the three enzymes encoded by the operon.

the transacetylase gene. Mapping studies by Lederberg established that all three genes are closely linked or contiguous to one another in the order Z-Y-A (Figure 16.4).

Two other observations are relevant to what became known about the structural genes. First, knowledge of their close linkage led to the discovery that all three genes are transcribed together, resulting in a single **polycistronic message** or **mRNA** (Figure 16.4). Additionally, it has been shown that upon induction by lactose, the rapid appearance of the enzymes results from the *de novo* synthesis of this mRNA. Although this finding might seem obvious or expected, it is in contrast to the proposal that induced enzyme activity may result from the activation of existing but inactive forms of the enzymes, which does not require gene activation.

## The Discovery of Regulatory Mutations

How, then, can lactose activate structural genes and induce the synthesis of the related enzymes? The discovery and study of **gratuitous inducers** ruled out one possibility. These molecules, which are chemical analogues of lactose, serve as inducers but not as substrates for the enzymatic reaction. One such gratuitous inducer is the sulfur analogue **isopropylthiogalactoside (IPTG)**, shown in Figure 16.5. The discovery of such molecules is strong evidence that the primary induction event is not the result of the interaction between the inducer and the enzyme. What, then, is the role of lactose in induction?

The answer to this question required the study of a second class of mutation, the **constitutive mutants**. In this type of mutant, the enzymes are produced regardless of the presence or absence of lactose. These mutations served as the basis of studies that defined the regulatory scheme for lactose metabolism.

Maps of the first type of constitutive mutation, *lac* I⁻, showed that it is located at a site on the DNA close to, but distinct from, the structural genes. As we will soon see, the I gene is appropriately called a **repressor gene**. A second set of mutations produced identical effects but was found in a region immediately adjacent to the struc-

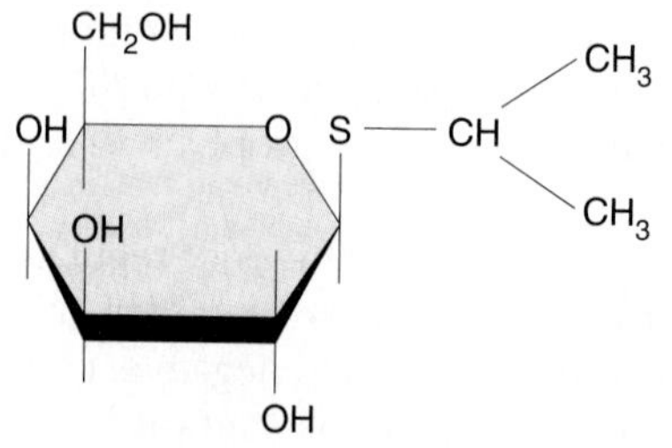

**FIGURE 16.5**    The gratuitous inducer isopropylthiogalactoside (IPTG).

tural genes. This class is designated *lac* O$^c$ and represents the **operator region**. Because the enzymes are continually produced for both types of constitutive mutation, they clearly represent regulatory units.

## The Operon Model: Negative Control

In 1961, Jacob and Monod proposed a scheme of negative control of regulation, which they called the **operon model**. In this model, the **operon** consists of the structural genes as well as the adjacent region of DNA represented by the *lac* O$^c$ mutation [Figure 16.6(a)]. They

proposed that the *lac* I gene regulates the transcription of the structural genes by producing a **repressor molecule**. The repressor was hypothesized to be **allosteric**. This concept is applied particularly to proteins that reversibly interact with another smaller molecule, causing a conformational change in three-dimensional shape.

Jacob and Monod suggested that the repressor normally interacts with the DNA sequence of the operator region. When it does so, it inhibits the action of RNA polymerase, effectively repressing the transcription of the structural genes [Figure 16.6(b)]. However, in the presence of lactose, this disaccharide binds to the repres-

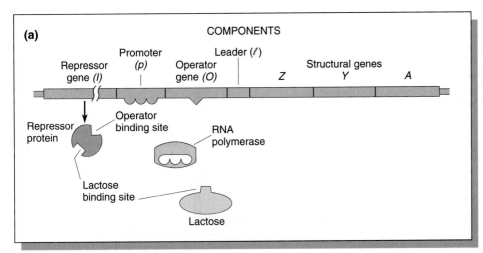

**(a)**

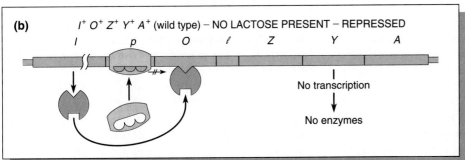

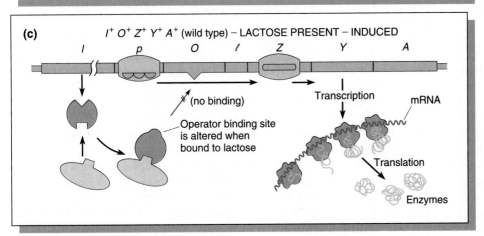

FIGURE 16.6    The components involved in the regulation of the *lac* operon and their interaction under various genotypic conditions, as described in the text.

**FIGURE 16.6** (Continued)

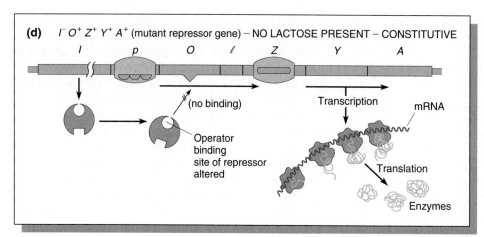

**(d)** $I^- O^+ Z^+ Y^+ A^+$ (mutant repressor gene) – NO LACTOSE PRESENT – CONSTITUTIVE

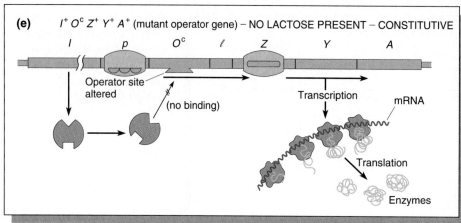

**(e)** $I^+ O^c Z^+ Y^+ A^+$ (mutant operator gene) – NO LACTOSE PRESENT – CONSTITUTIVE

sor, causing a conformational change. This change alters the binding site of the repressor, rendering it incapable of interacting with operator DNA [Figure 16.6(c)]. In the absence of the repressor–operator interaction, RNA polymerase transcribes the structural genes, and the enzymes necessary for lactose metabolism are translated. *Since transcription occurs only in the absence of the repressor,* **negative control** is exerted.

The operon model uses these potential molecular interactions to explain the efficient regulation of the structural genes. In the absence of lactose, the enzymes encoded by the genes are not needed and so are repressed. When lactose is present, it indirectly induces the activation of the genes by binding with the repressor.* If all lactose is metabolized, none is available to bind to the repressor, which is again free to bind to operator DNA and repress transcription.

Both the $I^-$ and $O^c$ constitutive mutations interfere with these molecular interactions, allowing continuous transcription of the structural genes. In the case of the $I^-$

mutant, the repressor product is altered and cannot bind to the operator region, so the structural genes are always turned on. In the case of the $O^c$ mutant, the nucleotide sequence of the operator DNA is altered and will not bind with a normal repressor molecule. The result is the same: Structural genes are always transcribed. Both types of constitutive mutations are illustrated diagrammatically in Figure 16.6(d) and (e).

### Genetic Proof of the Operon Model

The operon model is a good one because there are predictions derived from it that can be tested to determine its validity. The major assumptions to be tested are (1) the I gene produces a diffusible cellular product; (2) the O region is involved in regulation but does not produce a product; and (3) the O region must be adjacent to the structural genes in order to regulate transcription.

The construction of partial diploid bacteria (see Chapter 19) allows an assessment of these assumptions. For example, certain mating schemes make it possible to construct genotypes where an $I^+$ gene has been introduced into an $I^-$ host, or where an $O^+$ region is added to

---

*Technically, allolactose, an isomer of lactose, is the inducer. Allolactose is produced during the metabolism of lactose.

an O$^c$ host. In such cases, the entire host chromosome is present plus one or a few genes of choice. They are inserted as part of a plasmid called the **F factor**, designated F′ (see Chapter 19). The Jacob-Monod operon model predicts that adding an I$^+$ gene to an I$^-$ cell should restore inducibility, because a normal repressor would again be produced. Adding an O$^+$ region to an O$^c$ cell should have no effect on constitutive enzyme production, since regulation depends on an O$^+$ region immediately adjacent to the structural genes.

The results of these experiments are shown in Table 16.1, where Z represents the structural genes. The inserted genes are listed after the designation F′. In both cases described above, the Jacob-Monod model is upheld (part B of Table 16.1). Part C shows the reverse experiments, where either an I$^-$ gene or an O$^c$ region is added to cells of normal inducible genotypes. The model predicts that inducibility will be maintained in these partial diploids, and it is.

Another prediction of the operon model is that certain mutations in the I gene should have the opposite effect of I$^-$. That is, instead of being constitutive by failing to interact with the operator, mutant repressor molecules should be produced, which cannot interact with the inducer, lactose. As a result, the repressor would always bind to the operator sequence, and the structural genes would be permanently repressed. If this were the case,

the presence of an additional I$^+$ gene would have little or no effect on repression.

In fact, as shown in Table 16.1D, such a mutation, I$^s$, was discovered where the operon is "superrepressed." An additional I$^+$ gene does not effectively relieve repression of gene activity. These observations again support the operon model for gene regulation.

## Isolation of the Repressor

Although the operon theory of Jacob and Monod succeeded in explaining many aspects of genetic regulation in prokaryotes, the nature of the repressor molecule was not known when their landmark paper was published in 1961. While they had assumed that the allosteric repressor was a protein, RNA was also a candidate, since activity of the molecule required the ability to bind to DNA. Despite many attempts to isolate and characterize the hypothetical repressor molecule, no direct chemical evidence was immediately forthcoming. A single *E. coli* cell contains no more than 10 or so copies of the *lac* operon repressor. Direct chemical identification of 10 molecules in a population of millions of proteins and RNAs in a single cell presented a tremendous challenge.

In 1966, Walter Gilbert and Benno Müller-Hill reported the isolation of the *lac* repressor in partially purified form. To achieve the isolation, they used a *regulator quantity* (I$^q$) mutant strain that contains about 10 times as much repressor as wild-type *E. coli* cells. Also instrumental in their success were the use of the gratuitous inducer, IPTG (Figure 16.5), which binds to the repressor, and the technique of **equilibrium dialysis**. Extracts of I$^q$ cells were placed in a dialysis bag and allowed to attain equilibrium with an external solution of radioactive IPTG, which is small enough to diffuse freely in and out of the bag. At equilibrium, the concentration of IPTG was higher inside the bag than in the external solution, indicating that an IPTG-binding material was present in the cell extract and that it was too large to diffuse across the wall of the bag. Ultimately the binding material was purified; it was shown to be heat labile and to have other characteristics of a protein as well. Extracts of I$^-$ constitutive cells having no *lac* repressor activity did not exhibit IPTG-binding activity, strongly suggesting that the isolated protein was the repressor molecule.

To demonstrate this further, Gilbert and Müller-Hill grew *E. coli* cells on a medium containing radioactive sulfur and isolated the IPTG-binding protein, which was labeled in its sulfur-containing amino acids. This protein was mixed with DNA from a strain of phage lambda (λ) which carries the *lac* O$^+$ gene. The DNA sediments at 40S, while the IPTG-binding protein sediments at 7S. The DNA and protein were mixed and sedimented in a glycerol gradient in an ultracentrifuge. If the radioactive protein binds to the DNA, then it should sediment at a much

**Table 16.1** A Comparison of Gene Activity (+ or −) in the Presence or Absence of Lactose for Various *E. coli* Genotypes

| Genotype | Presence of β-galactosidase Activity | |
|---|---|---|
| | Lactose Present | Lactose Absent |
| A. I$^+$O$^+$Z$^+$ | + | − |
| I$^+$O$^+$Z$^-$ | − | − |
| I$^-$O$^+$Z$^+$ | + | + |
| I$^+$O$^c$Z$^+$ | + | + |
| B. I$^-$O$^+$Z$^+$/F′I$^+$ | + | − |
| I$^+$O$^c$ Z$^+$/F′O$^+$ | + | + |
| C. I$^+$O$^+$Z$^+$/F′I$^-$ | + | − |
| I$^+$O$^+$Z$^+$/F′O$^c$ | + | − |
| D. I$^s$O$^+$Z$^+$ | − | − |
| I$^s$O$^+$Z$^+$/F′I$^+$ | − | − |

NOTE: In parts B to D, many genotypes are partially diploid, containing an F factor plus attached genes (F′).

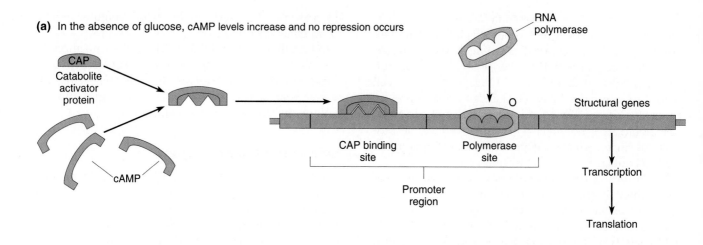

**(a)** In the absence of glucose, cAMP levels increase and no repression occurs

**(b)** In the presence of glucose, cAMP levels decrease and catabolite repression occurs

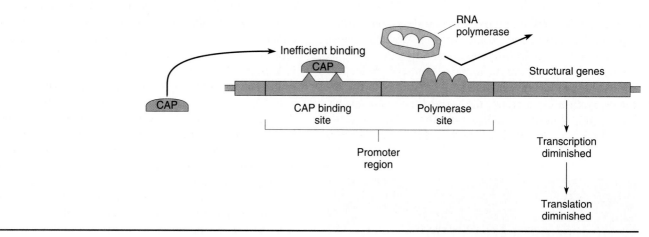

**(c)** Formation of cAMP from ATP

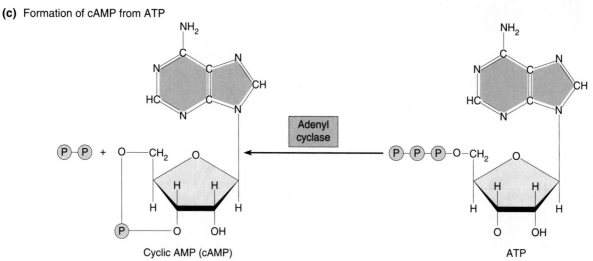

**FIGURE 16.7**    Catabolite repression. (a) In the absence of glucose, cAMP levels increase, resulting in cAMP—CAP binding and the ensuing stimulation of transcription. (b) In the presence of glucose, cAMP levels decrease, cAMP—CAP binding diminishes, and CAP fails to stimulate transcription. In (c) the formation of cAMP from ATP, catalyzed by adenyl cyclase, is diagrammed.

faster rate, moving in a band with the DNA. This was found to be the case. Further experiments showed that the IPTG-binding or repressor protein binds only to DNA containing the *lac* region. It was also shown not to bind to *lac* DNA containing an operator constitutive ($O^c$) mutation.

Subsequent study of the repressor protein has revealed that it consists of four identical subunits of 38,000 daltons. Each can bind to one molecule of IPTG. The repressor protein has a low binding affinity to DNA in general, but has an affinity about 1000-fold greater for the operator region, which consists of about 30 nucleotide pairs.

### The Catabolite Activating Protein (CAP): Positive Control of the *lac* Operon

When the *lac* repressor is bound to the inducer, the *lac* operon is activated and transcription of the structural genes proceeds under the direction of RNA polymerase. As we have learned in earlier chapters, this process is initiated as a result of the binding that occurs between this enzyme and a nucleotide sequence constituting the **promoter region**, found upstream (5′) from the initial coding sequences. Within the *lac* operon, the promoter is found between the I gene and the operator ($O^c$) region. Careful examination has revealed that, in this case, polymerase binding is not very efficient unless another protein is present to facilitate the process. This protein is called the **catabolite activating protein (CAP)**.

The discovery of CAP and the region within the promoter where it binds (the **CAP binding site**) occurred as a result of investigation of a most interesting phenomenon. Even if cellular conditions are inducible (in the presence of lactose), *transcription of the operon is inhibited in the presence of glucose*. Since glucose is a catabolic product of the *lac* enzymes, this inhibition is called **catabolite repression**.

The regulation of the *lac* operon by CAP and the role of glucose in catabolite repression is a fascinating story, summarized in Figure 16.7. In the absence of glucose, CAP exerts **positive control**, and when bound to the CAP binding site, transcription is facilitated. CAP binding has been shown to be dependent on still another molecule, **cyclic adenosine monophosphate (cAMP)**. For binding to occur, cAMP must be linked to CAP. The level of cAMP is itself dependent on an enzyme **adenyl cyclase**, which catalyzes the conversion of ATP to cAMP.

The role of glucose in catabolite repression is now clear. It inhibits the activity of adenyl cyclase, causing a decline in the level of cAMP in the cell. Under this condition, CAP cannot form the CAP–cAMP complex essential to the positive control of transcription.

Regulation of the *lac* operon by catabolite repression results in efficient energy utilization, because the presence of glucose will override the need for the metabolism of lactose, should it also be available to the cell. Catabolite repression involving CAP has also been observed for other inducible operons, including those controlling the metabolism of galactose and arabinose.

### THE *ARA* REGULATOR PROTEIN— POSITIVE AND NEGATIVE CONTROL

Before turning to a consideration of repressible systems, we want to make mention of a rather unique inducible operon in *E. coli* in which the same regulatory protein exerts both positive and negative control (Figure 16.8).

In *E. coli*, the metabolism of the sugar arabinose is under the direction of the enzymatic products of the *ara* B, A, and D genes. These genes are regulated by the protein encoded by the *ara* C gene. In the absence of arabinose, the protein behaves as a repressor, binding to regulatory sites within the *ara* O and *ara* I regions of the operon. Control is negative, and transcription is inhibited.

If arabinose *is* present, the protein binds to it and this complex becomes an activator. Attaching to only the *ara* I site on the operon stimulates the transcription of the B, A, and D genes. Under this condition, control is positive. A product must bind to induce gene expression.

A further complication is of interest in this system. Like the *lac* operon, it is sensitive to glucose and under the positive control of the CAP protein. When glucose is present and the level of cyclic AMP is low, the CAP binding site is empty. A repressive condition is created, preventing the *ara* C–arabinose complex from activating the BAD gene promoter. Thus, both cyclic AMP and arabinose must be present for the expression of the BAD genes in this operon.

### TRYPTOPHAN METABOLISM IN *E. COLI*: A REPRESSIBLE GENE SYSTEM

Although induction had been known for some time, it was not until 1953 that Monod and his coworkers discovered **enzyme repression**. Wild-type *E. coli* are capable of producing the necessary enzymes essential to the biosynthesis of amino acids as well as other essential macromolecules. Monod focused his studies on the amino acid tryptophan and the enzyme **tryptophan synthetase**.

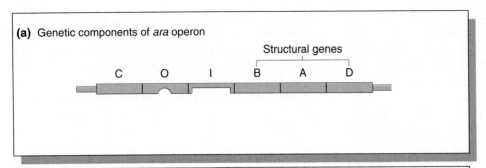

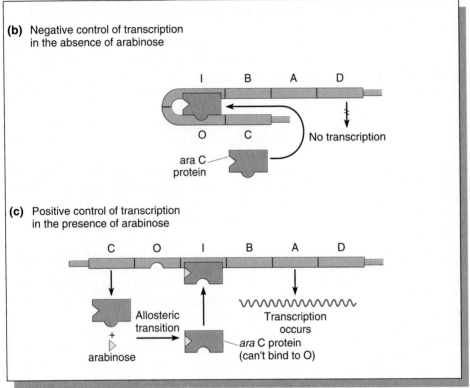

FIGURE 16.8    Gene regulation of the *ara* operon.

He discovered that if tryptophan is present in sufficient quantity in the growth medium, the enzymes necessary for its synthesis are **repressed**. Energetically, such enzyme repression is highly economical to the cell, because synthesis is unnecessary in the presence of exogenous tryptophan.

Further investigation showed that a series of enzymes encoded by five contiguous genes on the *E. coli* chromosome is involved in tryptophan synthesis. These genes are part of an operon, and in the presence of tryptophan, all of these genes are coordinately repressed. As a result, none of the enzymes are produced. Because of the great similarity between this repression and the induction of enzymes for lactose metabolism, Jacob and Monod proposed a model of gene regulation analogous to the *lac* system (Figure 16.9).

To account for repression, they suggested that an *inactive repressor is normally made* that alone cannot interact with the operator region of the operon. However, it is allosteric, and can interact with tryptophan, if present. The complex of repressor and tryptophan is activated, and it may now bind to the operator, repressing transcription. Thus, when tryptophan, the end product of this anabolic biosynthesis pathway, is present, enzymes are not made. Since the regulatory complex inhibits transcription of the operon, this **repressible system** is under **negative control**. Because tryptophan participates in repression, it is referred to as a **co-repressor** in this regulatory scheme.

## Evidence for and Concerning the *trp* Operon

Support for the concept of a repressible operon was soon forthcoming. Primarily, two distinct categories of constitutive mutations were isolated. The first class, *trp*

R⁻, maps at a considerable distance from the structural genes. This locus represents the gene coding for the repressor. Presumably, the R⁻ mutation either inhibits the interaction of a mutant repressor with tryptophan or inhibits repressor formation entirely. Thus, no repression ever occurs. The additional presence of an R⁺ gene restores repressibility.

The second constitutive mutant is analogous to that of the operator of the lactose operon because it maps immediately adjacent to the structural genes. Furthermore, the addition of a wild-type operator gene into mutant cells does not restore enzyme repression. This is predictable if the mutant operator can no longer interact with the repressor–tryptophan complex.

The entire *trp* operon has now been well defined, as shown in Figure 16.9. Five contiguous structural genes (*trp* E, D, C, B, and A) are transcribed as a polycistronic message that directs translation of the polypeptide components. These components catalyze the biosynthesis of tryptophan. As in the *lac* operon, a promotor region (*trp* P), representing the binding site for RNA polymerase, and an operator region (*trp* C), which binds the repressor, have been demonstrated. In the absence of binding, transcription is initiated within the overlapping *trp* P-*trp* O region and proceeds along a **leader sequence**. Transcription is initiated 162 nucleotides prior to the first structural gene (E). Within this leader sequence, still another regulatory site has been demonstrated. This component, called an attenuator, has been investigated extensively by Charles Yanofsky and his colleagues. As we shall see, this regulatory unit is an integral part of the control mechanism of this operon.

## The Attenuator

Yanofsky observed that, whether tryptophan is pres-

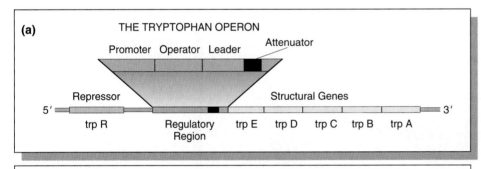

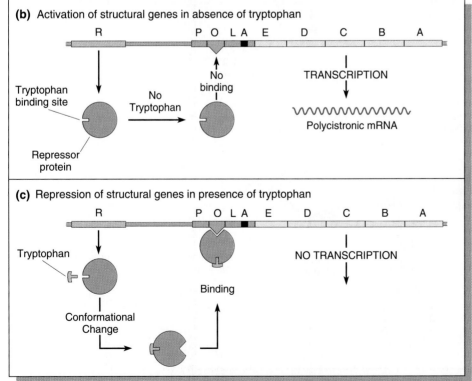

**FIGURE 16.9** (a) The components involved in the regulation of the tryptophan operon. (b) Regulatory conditions involving either activation or (c) repression of the structural genes. In the absence of tryptophan, an inactive repressor is made that cannot bind to the operator (*O*), thus allowing transcription to proceed. In the presence of tryptophan, it binds to the repressor, causing an allosteric transition to occur. This complex binds to the operator region, resulting in repression of the operon.

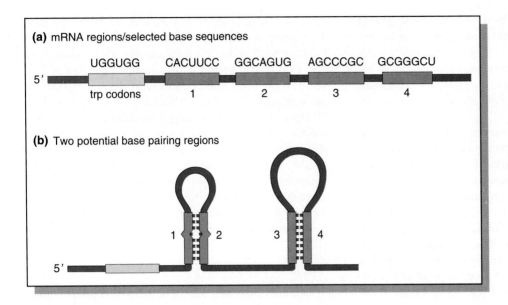

**(a)** mRNA regions/selected base sequences

UGGUGG    CACUUCC    GGCAGUG    AGCCCGC    GCGGGCU

5′

trp codons    1    2    3    4

**(b)** Two potential base pairing regions

1    2    3    4

5′

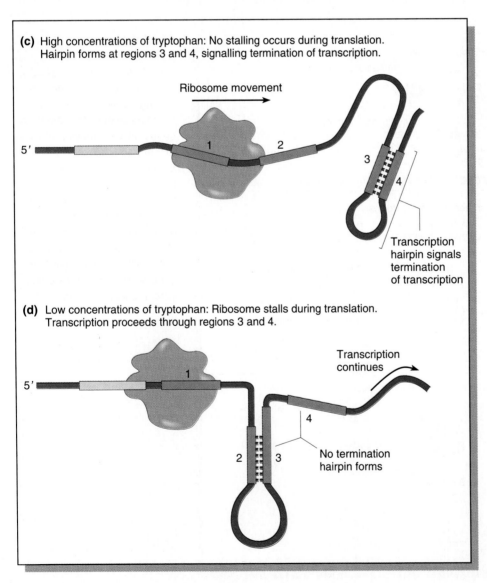

**(c)** High concentrations of tryptophan: No stalling occurs during translation. Hairpin forms at regions 3 and 4, signalling termination of transcription.

Ribosome movement

5′    1    2    3    4

Transcription hairpin signals termination of transcription

**(d)** Low concentrations of tryptophan: Ribosome stalls during translation. Transcription proceeds through regions 3 and 4.

Transcription continues

5′    1    4    2    3

No termination hairpin forms

**FIGURE 16.10**    Diagram of the role of the leader sequence of the mRNA transcript of the *trp* operon of *E. coli* during attenuation. In (a) and (b), the critical nucleotide sequences and two potential base pairing regions are shown. In (c) where tryptophan is plentiful, no ribosome stalling occurs during translation of the trp codons. As a result, the ribosome proceeds through regions 1 and 2 and a "termination hairpin" forms in regions 3 and 4, leading to attenuation of transcription. In (d) where tryptophan is scarce, ribosome stalling occurs, the termination hairpin does not form, and transcription continues unabated. The model assumes simultaneous transcription and translation, as is known to occur in bacteria.

ent or not, initiation of transcription of this leader sequence usually occurs. Thus, the activated repressor, even when bound to the operator region, does not strongly inhibit expression of the operon. As transcription of the leader sequence proceeds, if tryptophan is present in high concentration, mRNA synthesis is terminated at a point about 140 nucleotides along the transcript. It is this region that is called the **attenuator**. If, however, tryptophan is absent or present in low concentrations, this process of **attenuation** is overcome, and transcription of the entire operon occurs, with the subsequent production of the enzymes essential to the biosynthesis of tryptophan.

The identification of the site of the attenuator was made possible by the isolation of deletion mutations in the region 123 to 150 nucleotides into the leader sequence. Such mutations abolish attenuation. The phenomenon of attenuation appears to be common to other bacterial operons, including those regulating the biosynthesis of histidine and leucine.

An explanation of how attenuation occurs and how it is overcome has been put forward by Yanofsky and is summarized in Figure 16.10. Present in the region of the attenuator is a DNA sequence that, when transcribed, gives rise to an RNA molecule that immediately folds back on itself and forms a **hairpin loop**. This is then followed by eight uridine residues. Such a structure is typical of that found at the 3' end of RNA transcripts, leading to the termination phase of their transcription. Thus, even when the operon should be repressed, transcription begins, but the process is immediately terminated.

The question then remains as to how the absence (or a low concentration) of tryptophan allows attenuation somehow to be bypassed. Yanofsky's proposal is based on the discovery that the leader transcript contains an AUG initiator codon closely followed by two triplets encoding tryptophan. Even though this leader sequence is not part of the structural genes in the operon, following its transcription, the AUG sequence prompts the initiation of its translation by ribosomes. Recall that in bacteria, translation of an mRNA is initiated well before transcription is completed.

If cells are starved of tryptophan (which is a rather rare amino acid in proteins anyway), the ribosome "stalls" during the translation of triplets calling for charged tRNA$^{trp}$. This event influences the secondary structure of the transcript such that the termination hairpin does not form [Figure 16.10(b)], attenuation is bypassed, and simultaneous transcription and translation of the entire set of structural genes then proceeds. However, when adequate tryptophan is present, charged tRNA$^{trp}$ is present, no stalling occurs during translation, and the termination hairpin is formed [Figure 16.10(a)].

The details of the secondary structures of the transcript have now been worked out and described. They satisfy conditions leading to either continued transcription or termination (attenuation). Additionally, various mutations in the leader sequence predicted to alter the secondary structure of the transcript and its impact on attenuation have been isolated. In each case, the predicted result has been upheld. The model, while complex, has significantly extended our knowledge of genetic regulation. Furthermore, the supporting evidence represents an integrated approach involving genetic and biochemical analysis, which is becoming commonplace in molecular genetic research.

## GENETIC REGULATION IN PHAGE LAMBDA: LYSOGENY OR LYSIS?

Our understanding of genetic regulation at the transcriptional level has benefited from studies of bacteriophage lambda as well as from studies of operons in bacteria. Lambda DNA contains about 45,000 base pairs, enough bioinformation for about 35 to 40 average-sized genes. After injection of lambda DNA into *E. coli*, either the **lysogenic** or **lytic** pathway may be followed (see Chapter 19). In the first pathway, the phage DNA becomes integrated into the host genome and is almost totally repressed; in the second, the phage DNA is expressed and viral reproduction ensues.

If the lysogenic pathway is followed, the genes responsible for phage reproduction and lysis are turned off by the **λ repressor protein**. This protein is produced by one of the virus's own genes, *cI*. If the lytic pathway is followed, the expression of the *cI* gene is repressed by a second protein, **Cro**. This protein is produced by still another viral gene of the same name. Both the cI repressor (sometimes called the λ repressor) and the Cro protein have been isolated and characterized. Their interaction with λ DNA has also been determined. The various sites of the regulatory system are diagrammed in Figure 16.11. The development of this information, which enhances our understanding of gene regulation, is a fascinating story.

The λ repressor was isolated and characterized by Mark Ptashne in 1967. It is a protein consisting of 236 amino acids. The Cro protein, isolated more recently, consists of 66 amino acids. When the product of the *cI* gene is produced, it recognizes two different operator regions in the λ DNA, $O_L$ and $O_R$, found on either side (left or right) of the *cI* gene. The $O_R$ region consists of 80 DNA base pairs and is located between the *cI* gene and the *cro* gene. There are three binding sites in $O_R$, each with a distinctive nucleotide sequence. The three regions

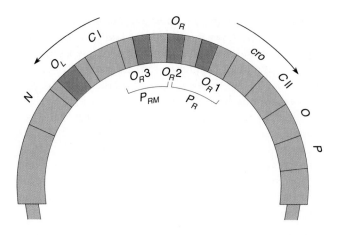

**FIGURE 16.11**    The regulatory sites controlling lysogeny and lysis in phage λ.

are labeled $O_R1$, $O_R2$, and $O_R3$, and each consists of 17 nucleotides. Overlapping this region are the promoters for the *cI* and *cro* genes, called $P_{RM}$ and $P_R$, respectively. Transcription of these genes occurs in opposite directions, using opposite strands of the helix to achieve 5'-to-3' RNA synthesis.

When the repressor is bound to both the $O_R$ and $O_L$ regions, two sets of so-called early genes are repressed, causing the remainder of the λ genes to be turned off as well. In this case, the repressor behaves as a negative control element, and the lysogenic pathway is achieved.

It is now clear that during this same binding, the λ repressor also stimulates transcription of its own *cI* gene. As shown in Figure 16.12, *cI* is self-regulating. There are three $O_R$ binding sites. At low levels of the *cI* product, $O_R1$ and $O_R2$ are bound, stimulating *cI* transcription and inhibiting *cro*. As concentrations of the λ repressor increase, $O_R3$ also binds and transcription is inhibited. As the levels decline, the cycle is repeated.

When the repressor binding to $O_L$ and $O_R$ is absent, initiation at the respective promoters ($P_{RM}$ and $P_R$) proceeds, resulting in the production of two proteins, N and Cro. The N protein functions as an antiterminator, allowing completion of transcription of genes essential to reproduction and lysis. The Cro protein, as stated earlier, serves as a repressor of *cI* gene transcription. Mutational analysis supports this model. Mutations in either $O_L$ or $O_R$ prevent repressor binding and abolish the potential for lysogeny. Mutations in the *cro* gene abolish the potential for lysis.

The greatest surprise was that both the λ repressor and Cro protein demonstrate a binding affinity to the three sites of $O_R$. However, the binding of the two proteins to the three sites varies, depending on the concentration of the proteins. In addition, when either protein is bound to these sites, the other cannot bind. The λ repressor binds

initially to $O_R1$ and $O_R2$ coordinately and then to $O_R3$, but only when present at very high concentrations. The Cro protein behaves in the opposite fashion. It initially binds $O_R3$ and subsequently binds $O_R1$ and $O_R2$.

Following initial infection, the determination of lysogeny versus lysis appears to depend on which operator region is bound first. If $O_R1$ and $O_R2$ are bound, $P_R$ is inhibited and the *cro* gene is not transcribed. $P_{RM}$, on the other hand, is available to RNA polymerase and the *cI* gene is transcribed, resulting in increased amounts of the repressor. Lysogeny is now established. However, if $O_R3$ is bound by the Cro protein, $P_{RM}$ is inhibited and the *cI* gene is repressed. This allows initiation at the $P_R$ site, and transcription essential to lysis proceeds.

While this description explains how the determination is made, it does not address why one pathway is favored over the other. Still another gene and its product are critical to this decision: *cII*. Both the cII protein and the Cro protein accumulate during early infection. The product of the *cII* gene affects $P_{RE}$, which activates the *cI* gene, causing the production of the repressor. This favors lysogenesis. Apparently, the activity of the protein expressed by *cII* changes under certain cellular conditions. For example, if glucose is plentiful, the activity is reduced, thus favoring the lytic cycle. The cII protein, which is unstable, may serve to monitor growth conditions. When it fails to prevail, the Cro protein successfully represses the *cI* gene and viral reproduction commences.

To pursue this question further, Ptashne and others have found it useful to examine the molecular events that occur when an induction event transforms the lysogenic pathway to the lytic sequence. Discovered in 1946 by L'woff, treatment of lysogenized *E. coli* by various agents (including UV light and mutagens) induces a lytic response. While such an event occurs spontaneously about once in a million bacterial cell divisions, virtually all bacteria may be induced to enter the lytic cycle.

Induction events have one common property: They all involve interaction with DNA. In so doing, they stimulate what has been called SOS response (see Chapter 11). During this response, the protein produced by the *rec*A gene, normally essential to recombination, is synthesized. Under inductive conditions, it behaves as a proteolytic enzyme and plays an essential role in the induction of lysis.

The λ repressor protein contains two structural domains and may exist in monomer or dimer form. It is the dimer form that binds strongly to $O_R$. The RecA protein cleaves the monomers at a sensitive region between the two domains. Cleavage of monomers prevents their dimerization; as a result, the repressor protein can no longer bind to $O_R$, stimulating the expression of the *cro* gene. The Cro protein, also active in dimer form, then

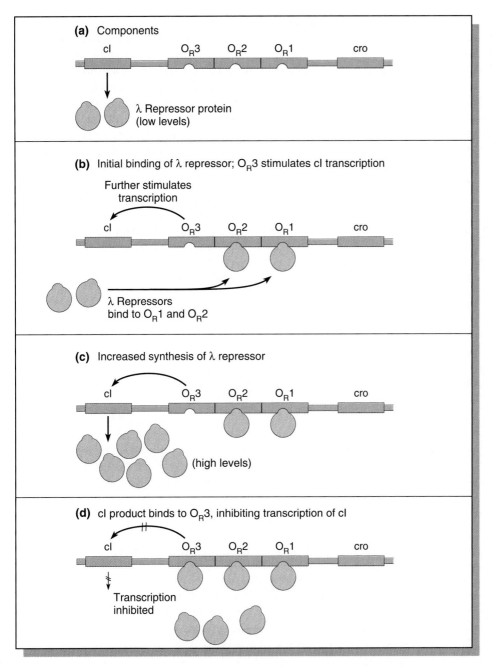

**FIGURE 16.12** Self-regulation of the *cI* repressor gene, based on the concentration of the cI repressor.

**(a)** Components

cI    $O_R3$    $O_R2$    $O_R1$    cro

$\lambda$ Repressor protein
(low levels)

**(b)** Initial binding of $\lambda$ repressor; $O_R3$ stimulates cI transcription

Further stimulates
transcription

cI    $O_R3$    $O_R2$    $O_R1$    cro

$\lambda$ Repressors
bind to $O_R1$ and $O_R2$

**(c)** Increased synthesis of $\lambda$ repressor

cI    $O_R3$    $O_R2$    $O_R1$    cro

(high levels)

**(d)** cI product binds to $O_R3$, inhibiting transcription of cI

cI    $O_R3$    $O_R2$    $O_R1$    cro

Transcription
inhibited

binds to $O_R$, inhibiting further transcription of the *cI* gene. Genetic events essential to viral reproduction then proceed, new viruses are constructed, and lysis of the bacterial cell occurs.

The research leading to the information just described is remarkable. It illustrates the depth of analysis attainable with regard to what at first may appear as a simple question: How is lysogeny regulated? Important insights have also been gained from this research regarding protein/nucleic acid interactions during genetic regulation. Molecular interactions such as these undoubtedly play major roles in many other genetic phenomena.

## PHAGE TRANSCRIPTION DURING LYSIS

When a bacteriophage invades its host bacterium and initiates the lytic cycle, it faces a novel problem. How can its genes be transcribed without the presence of virus-specific RNA polymerase, itself a product of transcription? The strategy that has evolved is an interesting example of parasite–host interaction.

The study of phage T7, which invades *E. coli*, has provided some answers to this question. This phage has evolved DNA sequences that are recognized as promoter sites by the *E. coli* RNA polymerase. As a result, upon the phage's entry to the host cell, a set of so-called **early genes** is transcribed into mRNAs, which are translated on bacterial ribosomes. One of these gene products is a viral-specific RNA polymerase that recognizes the promoters for the remaining T7 genes. A second early gene product, a protein kinase, inhibits bacterial transcription. The site of inhibitory interaction is presumably the bacterial RNA polymerase.

With the host cell's genetic apparatus secured, the remaining viral genes are expressed. The major products are DNA-replicating enzymes, head and tail components, enzymes involved in DNA packaging, and, finally, a lytic enzyme to break open the host cell.

Studies of the phage **SPO1**, which invades *Bacillus subtilis*, have revealed an even finer control or regulation

of transcription during lysis (Figure 16.13). As in T7, **early genes** are read by the host RNA polymerase. One of the early gene products is a protein subunit (a type of sigma subunit) that associates with the host RNA polymerase. This complex can now interact with the promoter regions of the **middle genes** of the phage. This association between the early gene subunit and the bacterial polymerase curbs transcription of the early genes, and the middle genes associated with DNA replication are expressed. Two of the middle genes produce additional polypeptides that can interact with the host polymerase, again altering its specificity. As a result, the polymerase now directs the expression of the **late genes** responsible for producing components that package the new phage. Thus, sequential gene action results, providing a finer mechanism for the regulation of gene expression during the lytic cycle.

SPO1 PHAGE GENE CONTROL

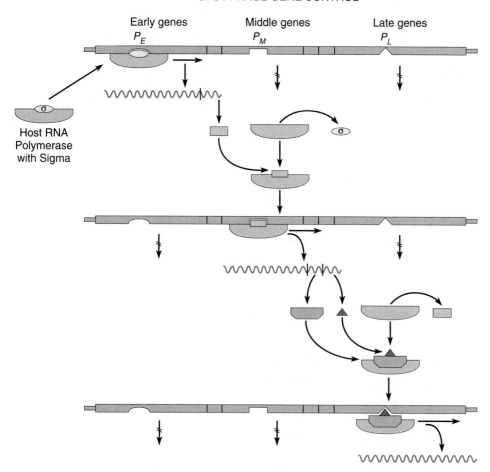

**FIGURE 16.13**    Initiation and regulation of phage SPO1 gene expression during lysis, as described in the text.

**CHAPTER SUMMARY**

1. Genetic analysis of bacteria and viruses has been pursued successfully since the 1940s. The ease of obtaining large quantities of pure cultures of bacteria and mutant strains of viruses as experimental material, and the efficient selection and isolation of naturally occurring and induced mutations have made these the organisms of choice in numerous types of genetic studies.

2. A system of genetic regulation must exist if the complete genome is not to be continuously active in transcription throughout the life of every cell of all species. Highly refined mechanisms have evolved that regulate transcription, maintaining genetic efficiency.

3. Both inducible and repressible operons, illustrated by the *lac* and *trp* gene complexes, respectively, have been documented and studied in bacteria. They involve genes of a regulatory nature in addition to structural genes that code for the enzymes of the system. Both operons are under the negative control of a repressor molecule.

4. The catabolite activating protein (CAP) facilitates the binding of RNA polymerase to the promoter in the *lac* and other operons. Catabolite repression, a mechanism that represses the operon when glucose is present, has evolved since glucose can be utilized more efficiently than lactose.

5. An additional regulatory step, referred to as attenuation, has been studied in the *trp* operon, whereby transcription begins, but is interrupted under appropriate cellular conditions.

6. The regulator protein of the *ara* gene exerts both positive and negative control over expression of the genes specifying the enzymes that metabolize the sugar arabinose.

7. Repression of transcription in phage lambda has been shown to be due to the product of one of its own genes, *cI*. Another gene product, Cro, acts to inhibit the transcription of *cI* and promotes the lytic pathway when the activity of cII protein is reduced by the presence of glucose.

8. During lysis, bacteriophages must rely on the host RNA polymerase to initiate transcription of their genome. Certain viral promoter sequences are recognized by the host polymerase. One of the early gene products in phage T7 is the RNA polymerase that transcribes the remainder of the genome.

## KEY TERMS

adaptation hypothesis
adaptive enzyme
adenyl cyclase
allosteric repressor
arabinose operon
attenuation
auxotroph
β-galactosidase enzyme
β-galactoside permease
CAP binding site
catabolite activating
    protein (CAP)
catabolite repression
constitutive enzyme
constitutive mutation

cyclic adenosine
    monophosphate
    (cAMP)
co-repressor
cro gene
early genes (in phage)
enzyme repression
equilibrium dialysis
F factor
fluctuation test
gratuitous inducers
hairpin loop
inducer
inducible enzyme
isopropylthiogalactoside
    (IPTG)

lactose
lag phase
late genes (in phage)
λ phage
λ repressor protein
leader sequence
log phase
lysogenic phage
lysis
lytic cycle
middle genes (in phage)
minimal medium
mRNA
negative control
operator region
operon

operon model
polycistronic message
positive control
promoter region
prototroph
repressible enzymes
repressible system
repressor gene
repressor molecule
spontaneous mutations
SPO1 phage
stationary phase
structural gene
transacetylase
tryptophan synthetase

# INSIGHTS AND SOLUTIONS

1.  A theoretical operon (theo) in E. coli contains several structural genes encoding enzymes that are involved sequentially in the biosynthesis of an amino acid. Unlike the lac operon, where the repressor gene is separate from the operon, the gene encoding the regulator molecule is contained within the theo operon. When the end product (the amino acid) is present, it combines with the regulator molecule, and this complex binds to the operator, repressing the operon. In the absence of the amino acid, the regulatory molecule fails to bind to the operator, and transcription proceeds.

    Characterize this operon; then consider the following mutations as well as the situation in which the wild-type gene is present along with the mutant gene in partially diploid cells (F'). In each case, will the operon be active or inactive in transcription, assuming that the mutation affects the regulation of the theo operon? Compare each response to the equivalent situation of the lac operon.
    (a) Mutation in the operator gene.
    (b) Mutation in the promoter region.
    (c) Mutation in the regulator gene.

    **SOLUTION:** The theo operon is repressible and under negative control. When there is no amino acid present in the medium (or the environment), the product of the regulatory gene cannot bind to the operator region, and transcription proceeds under the direction of RNA polymerase. The enzymes necessary for the synthesis of the amino acid are made, as is the regulator molecule. If the amino acid is present, or after sufficient synthesis occurs, the amino acid binds to the regulator, forming a complex that interacts with the operator region, causing repression of transcription of the genes within the operon.

    The theo operon is similar to the tryptophan system with the exception that the regulator gene is within the operon rather than being located separate from it. Therefore, in the theo operon, the regulator gene is itself regulated by the presence or absence of the amino acid.

(a) As in the *lac* operon, a mutation in the *theo* operator gene inhibits binding with the repressor complex, and transcription occurs constitutively. The presence of an F′ plasmid bearing the wild-type allele would have no effect since it is not adjacent to the structural genes.

(b) A mutation in the *theo* promoter region would no doubt inhibit binding to RNA polymerase and therefore inhibit transcription. This would also happen in the *lac* operon. A wild-type allele present in an F′ plasmid would have no effect.

(c) A mutation in the *theo* regulator gene, as in the *lac* system, may either inhibit its binding to the repressor or its binding to the operator gene. In both cases, transcription will be constitutive because the *theo* system is repressible. Both cases result in the failure of the regulator to bind to the operator, allowing transcription to proceed. In the *lac* system, failure to bind the co-repressor lactose would permanently repress the system. The addition of a wild-type allele would restore repressibility, provided that this gene was transcribed constitutively.

2. Analysis of the lysis and lysogeny of *E. coli* by phage λ involved many mutations that disrupted the normal capabilities of the virus. Consider the following mutations.

(a) Phage λ bearing a temperature-sensitive mutation in the *cI* gene is unable to lysogenize *E. coli* (it cannot integrate its chromosome into the host genome). If such a mutant phage is allowed to infect a host bacterial cell that is already lysogenized by a wild-type phage, predict the outcome of infection and explain this outcome.

**SOLUTION:** The mutant phage will be unable to lyse the cell because the wild-type phage (integrated into the bacterial chromosome) is expressing the normal product of the *cI* gene that represses the genetic system leading to lysis. This repressor is effective for all incoming phages as well as repressing its own genome.

(b) Other phage λ mutants have been isolated that, like that in part (a), cannot independently lysogenize host bacteria. However, they *can* lyse bacteria that are lysogenized by wild-type phages. Can you offer an explanation for this observation?

**SOLUTION:** The mutation is not in the *cI* gene. Instead, it must be in the "receptor" gene(s) that normally responds to the repressor. The mutation renders it (them) insensitive to the repressor.

---

**PROBLEMS AND DISCUSSION QUESTIONS**

1. Contrast the need for the enzymes involved in the metabolism of lactose and tryptophan in bacteria in the presence and absence of lactose and tryptophan, respectively.

2. Contrast positive and negative control systems.

3. Contrast the role of the repressor in an inducible system and in a repressible system.

4. Describe how the *lac* repressor was isolated. What properties demonstrate it to be a protein? Describe the evidence that it indeed serves as a repressor within the operon scheme.

5. Even though the *lac* Z, Y, and A structural genes are transcribed as a single polycistronic mRNA, each gene contains the appropriate initiation and termination signals essential for translation. Predict what will happen when a cell growing in the presence of lactose contains a deletion of one nucleotide early in the Z gene. Early in the A gene?

6. For the following *lac* genotypes, predict whether the structural genes (Z) are constitutive, permanently repressed, or inducible in the presence of lactose.

| Genotype | Constitutive | Repressed | Inducible |
|---|---|---|---|
| $I^+O^+Z^+$ | | | × |
| $I^-O^+Z^+$ | | | |
| $I^+O^cZ^+$ | | | |
| $I^-O^+Z^+/F'I^+$ | | | |
| $I^+O^cZ^+/F'O^+$ | | | |
| $I^sO^+Z^+$ | | | |
| $I^sO^+Z^+/F'I^+$ | | | |

7. For the following genotypes and condition (lactose present or absent), predict whether functional enzymes are made, nonfunctional enzymes are made, or no enzymes are made.

| Genotype | Condition | Functional Enzyme Made | Nonfunctional Enzyme Made | No Enzyme Made |
|---|---|---|---|---|
| $I^+O^+Z^+$ | No lactose | — | — | × |
| $I^+O^c Z^-$ | Lactose | | | |
| $I^-O^+Z^-$ | No lactose | | | |
| $I^-O^+Z^-$ | Lactose | | | |
| $I^-O^+Z^+/F'I^+$ | No lactose | | | |
| $I^+O^c Z^+/F'O^+$ | Lactose | | | |
| $I^+O^+Z^-/F'I^+O^+Z^+$ | Lactose | | | |
| $I^-O^+Z^-/F'I^+O^+Z^+$ | No lactose | | | |
| $I^s O^+Z^+/F'O^+$ | No lactose | | | |
| $I^+O^c Z^-/F'O^+Z^+$ | Lactose | | | |

8. In a theoretical operon, genes A, B, C, and D represent the repressor gene, the promoter sequence, the operator gene, and the structural gene, *but not necessarily in that order*. This operon is concerned with the metabolism of a theoretic molecule (tm). From the data given below, first decide if the operon is inducible or repressible. Then assign A, B, C, and D to the four parts of the operon. (AE = active enzyme; IE = inactive enzyme; NE = no enzyme)

| Genotype | tm Present | tm Absent |
|---|---|---|
| $A^+B^+C^+D^+$ | AE | NE |
| $A^-B^+C^+D^+$ | AE | AE |
| $A^+B^-C^+D^+$ | NE | NE |
| $A^+B^+C^-D^+$ | IE | NE |
| $A^+B^+C^+D^-$ | AE | AE |
| $A^-B^+C^+D^+/F'A^+B^+C^+D^+$ | AE | AE |
| $A^+B^-C^+D^+/F'A^+B^+C^+D^+$ | AE | NE |
| $A^+B^+C^-D^+/F'A^+B^+C^+D^+$ | AE + IE | NE |
| $A^+B^+C^+D^-/F'A^+B^+C^+D^+$ | AE | NE |

9. If the *cI* gene of phage $\lambda$ contained a mutation with effects similar to the $I^-$ mutation in *E. coli*, what would be the result?

10. Describe the level of genetic activity of the *lac* operon as well as the status of the *lac* repressor and the CAP protein under the following cellular conditions:

|  | lactose | glucose |
|---|---|---|
| (a) | − | − |
| (b) | + | − |
| (c) | − | + |
| (d) | + | + |

11. A bacterial operon is responsible for the production of the biosynthetic enzymes needed to make a theoretical amino acid tisophane (tis). The operon is regulated by a separate gene, R. Deletion of the R gene causes the loss of synthesis of the enzymes. In the wild-type condition, when tis is present, no enzymes are made. In the absence of tis, the enzymes are made. Mutations in the operator gene ($O^-$) result in repression regardless of the presence of tis.

    Is the operon under positive or negative control? Propose a model for (1) repression of the genes in the presence of tis in wild-type cells; and (2) the $O^-$ mutations.

12. A culture of an auxotrophic *leu*⁻ strain of bacteria was irradiated and incubated until it reached the stationary phase. A control culture, which was not irradiated, was also studied. These cultures were then serially diluted, and 0.1 ml of various dilutions was plated on minimal medium plus leucine and on minimal medium. The results, shown below, were used to determine the spontaneous and X-ray-induced mutation rate of *leu*⁻ to *leu*⁺.

| Culture Condition | Culture Medium | Dilution | Number of Colonies |
|---|---|---|---|
| Irradiated | (1) Minimal medium plus leucine | $10^{-9}$ | 24 |
|  | (2) Minimal medium | $10^{-2}$ | 12 |
| Control | (3) Minimal medium plus leucine | $10^{-9}$ | 12 |
|  | (4) Minimal medium | $10^{-1}$ | 3 |

(a) Describe what is represented by each value obtained. Which values should be approximately equal? Are they?

(b) Determine the induced and spontaneous mutation rate leading to prototrophic growth ($leu^-$ to $leu^+$).

**SELECTED READINGS**

ANDERSON, J. E., PTASHNE, M., and HARRISON, S. C. 1985. A phage repressor-operator complex at 7Å resolution. *Nature* 316:596–601.

BECKWITH, J., and ROSSOW, P. 1974. Analysis of genetic regulatory mechanisms. *Ann. Rev. Genet.* 8:1–13.

BECKWITH, J. R., and ZIPSER, D., eds. 1970. *The lactose operon.* Cold Spring Harbor, NY: Cold Spring Harbor Laboratory.

BERTRAND, K., et al. 1975. New features of the regulation of the tryptophan operon. *Science* 189:22–26.

ENGLESBERG, E., and WILCOX, G. 1974. Regulation: Positive control. *Ann. Rev. Genet.* 8:219–42.

GILBERT, W., and MÜLLER-HILL, B. 1966. Isolation of the *lac* repressor. *Proc. Natl. Acad. Sci.* 56:1891–98.

————. 1967. The *lac* operator is DNA. *Proc. Natl. Acad. Sci.* 58:2415–21.

GILBERT, W., and PTASHNE, M. 1970. Genetic repressors. *Scient. Amer.* (June) 222:36–44.

GOLDBERGER, R. F., DEELEY, R. G., and MULLINIX, K. P. 1976. Regulation of gene expression in prokaryotic organisms. *Adv. in Genet.* 18:1–67.

GOTTESMAN, S. 1984. Bacterial regulation: Global regulatory networks. *Ann. Rev. Genet.* 18:415–41.

HERSHEY, A. D., ed. 1971. *The bacteriophage lambda.* Cold Spring Harbor, NY: Cold Spring Harbor Laboratory.

JACOB, F., and MONOD, J. 1961. Genetic regulatory mechanisms in the synthesis of proteins. *J. Mol. Biol.* 3:318–56.

JOHNSON, A. D., et al. 1981. λ repressor and *cro*—Components of an efficient molecular switch. *Nature* 294:217–23.

KELLER, E. B., and CALVO, J. M. 1979. Alternative secondary structures of leader RNAs and the regulation of the *trp, phe, his, thr,* and *leu* operons. *Proc. Natl. Acad. Sci.* 76:6186–90.

LEWIN, B. 1974. Interaction of regulator proteins with recognition sequences of DNA. *Cell* 2:1–7.

LURIA, S. E., and DELBRUCK, M. 1943. Mutations of bacteria from virus sensitivity to virus resistance. *Genetics* 28:491–511.

MANIATIS, T., and PTASHNE, M. 1976. A DNA operator-repressor system. *Scient. Amer.* (Jan.) 234:64–76.

MILLER, J. H., and REZNIKOFF, W. S. 1978. *The operon.* Cold Spring Harbor, NY: Cold Spring Harbor Laboratory.

PETRUSEK, R. H., DUFFY, J., and GEIDUSCHEK, E. 1976. Control of gene action in phage SPO1 development: Phage-specific modifications of RNA polymerase and a mechanism of positive control. In *RNA polymerase,* eds. R. Losick and M. Chamberlin, pp. 567–85. Cold Spring Harbor, NY: Cold Spring Harbor Laboratory.

PLATT, T. 1981. Termination of transcription and its regulation in the tryptophan operon of *E. coli. Cell* 24:10–23.

PTASHNE, M. 1986. *A genetic switch*. Palo Alto: Blackwell Scientific Publ.

PTASHNE, M., et al. 1980. How the λ repressor and *cro* work. *Cell* 19:1–11.

PTASHNE, M., and GANN, A. 1990. Activators and targets. *Nature* 346:329–31.

PTASHNE, M., JOHNSON, A. D., and PABO, C. O. 1982. A genetic switch in a bacterial virus. *Scient. Amer.* (Nov.) 247:128–40.

RAIBAUD, O., and SCHWARTZ, M. 1984. Positive control of transcription initiation in bacteria. *Ann. Rev. Genet.* 18:173–206.

REZNIKOFF, W. S. 1972. The operon revisited. *Ann. Rev. Genet.* 6:133–56.

STROYNOWSKI, I., and YANOFSKY, C. 1982. Transcript secondary structures regulate transcription termination at the attenuator of *S. marcescens* tryptophan operon. *Nature* 298:34–38.

STUDIER, F. W. 1972. Bacteriophage T7. *Science* 176:367–76.

UMBARGER, H. E. 1978. Amino acid biosynthesis and its regulation. *Ann. Rev. Biochem.* 47:533–606.

YANOFSKY, C. 1981. Attenuation in the control of expression of bacterial operons. *Nature* 289:751–58.

YANOFSKY, C., and KOLTER, R. 1982. Attenuation in amino acid biosynthetic operons. *Ann. Rev. Genet.* 16:113–134.

# 17

# GENETIC REGULATION IN EUKARYOTES

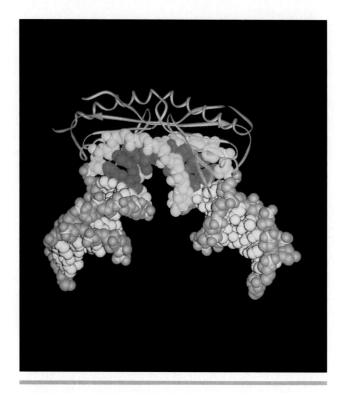

Molecular complex formed between the TATA-box binding protein (green) and the 8 base-pair TATA box of DNA (red), itself part of DNA (yellow).

*Genetic regulation in eukaryotes occurs at the level of transcription and during posttranscriptional events. Transcription is modulated by the interaction of various regulatory molecules with short DNA sequences most often located upstream from affected genes. Posttranscriptional mechanisms involve the selection of alternative products from a single transcript and the control of mRNA stability.*

# EUKARYOTIC GENE REGULATION: AN OVERVIEW

The discovery of the operon in *E. coli* by Jacob and Monod in 1960 was the first step in unraveling the mechanisms that govern gene expression. In the period that followed this breakthrough, operons were invoked to explain everything from bacterial gene expression to the developmental expression of hemoglobin synthesis in humans. While some of the basic principles learned from the *lac* operon are applicable in eukaryotes, it is now clear that eukaryotes regulate gene action in fundamentally different and more complex ways. The discovery of introns and the splicing of eukaryotic messages in 1977 confirmed and accelerated the reconsideration of how gene regulation is accomplished in eukaryotes.

Consideration of the fundamental differences in the evolutionary history of prokaryotes and eukaryotes provides us with important clues as to why operons do not constitute the basic mechanism for the regulation of eukaryotic genes. Eukaryotic cells contain a much greater amount of genetic information, and this DNA is complexed with histones and other proteins to form chromatin, while in prokaryotic organisms, the DNA does not form into chromatin. The genetic information in eukaryotes is carried on many chromosomes, rather than one, and these chromosomes are enclosed within a nuclear membrane. In eukaryotes, the process of transcription is spatially and temporally separated from that of translation. The transcripts of eukaryotic genes are processed, cleaved, and realigned before transport to the cytoplasm, and many transcribed sequences never leave the nucleus. In addition, eukaryotes for the most part are multicellular. One consequence of this is that highly differentiated cells often synthesize large amounts of a restricted number of gene products even though they contain a complete set of genetic information. In addition, during the development of multicellular organisms, changes in gene expression are often directed by external signals such as inducers and hormones.

Without belaboring the point, it seems evident that these properties of eukaryotes are likely to require more complex mechanisms for the regulation of genetic expression than those found in bacteria. Regulation of gene expression can potentially occur at many levels, including transcription, processing, transport, and translation (Figure 17.1). Although the details of gene regulation in eukaryotes have not yet been described for many genes, enough is known to allow some general features to be presented.

# REGULATION OF TRANSCRIPTION

As in prokaryotes, transcriptional control in eukaryotes involves the interaction of DNA sequences adjacent to the regulated genes and DNA-binding proteins known as **transcription factors**. As outlined in Chapter 10, the regulatory sequences adjacent to a wide range of genes have been identified, mapped, and sequenced. The development of DNA protection and gel retardation assays have allowed the detection of DNA-binding proteins in cell extracts, and new affinity chromatography techniques have been used to isolate transcriptional factors present in very low concentrations. The result has been a virtual explosion of information about eukaryotic transcription factors. In general, two classes of eukaryotic transcription factors are recognized: those that assemble at **promoter** regions adjacent to the site of transcription, and those that bind at more distant regions called **enhancers**. In this section, we will review some of this new information, including the upstream organization of eukaryotic genes, the characteristics of transcription factors, their interaction with DNA sequences, other regulatory factors, and the initiation of transcription by RNA polymerase.

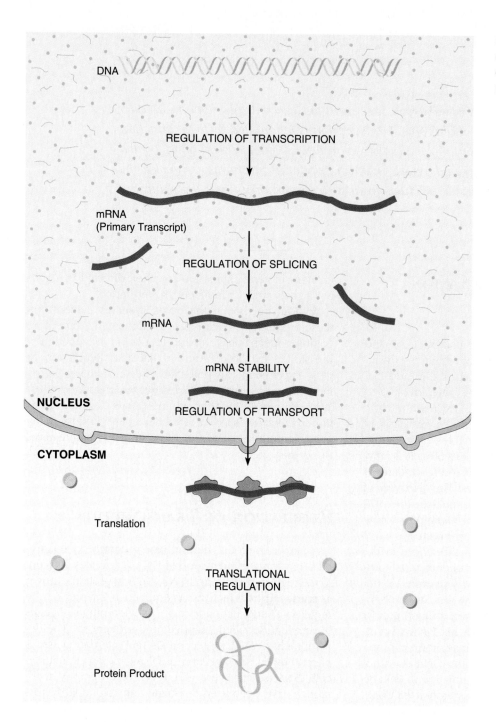

**FIGURE 17.1**    Various levels of regulation that are possible during the expression of the genetic material.

DNA

REGULATION OF TRANSCRIPTION

mRNA
(Primary Transcript)

REGULATION OF SPLICING

mRNA

mRNA STABILITY

REGULATION OF TRANSPORT

**NUCLEUS**

**CYTOPLASM**

Translation

TRANSLATIONAL
REGULATION

Protein Product

## Promoters in Eukaryotes

Eukaryotic genes have two major types of regulatory regions. The first consists of nucleotide sequences that serve as the recognition point for RNA polymerase binding. Such regions are called **promoters**, as they are in prokaryotes. The promoter represents the region necessary to *initiate* transcription. Promoters might be thought of as establishing the "potential" for transcription to occur.

In spite of having a similar function in both eukaryotes and prokaryotes, major differences exist. Foremost is the fact that, in prokaryotes, regulatory regions are usually very close to the site where transcription is initiated, whereas in eukaryotes, the upstream region is more extensive and may be farther away from the gene that is being regulated. A second major difference is the variation observed in the promoter regions in eukaryotes. This is undoubtedly related to the fact that there are three different forms of **RNA polymerase** that transcribe

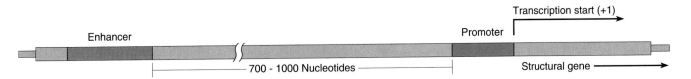

**FIGURE 17.2**    Organization of typical eukaryotic gene and transcriptional control regions. The promoter region consists of modular DNA sequences, usually located within 100 to 110 bp of the site of transcription initiation. Promoter elements can consist of sequences that include the TATA box (−25 to −30 bp), the CAAT box (−70 to −80), and the GC box (at about −110).

three different groups of genes. These are classified as type I (ribosomal RNAs), type II (mRNAs and snRNAs), and type III (tRNAs, 5 SRNA, and several other small cellular RNAs). Each category requires different recognition sites within the promoter.

Promoters that are recognized by RNA polymerase II are composed of short modular sequences of DNA usually located within 100 bp upstream (in the 5′ direction) of the gene. In the promoter of most genes is a sequence called the **TATA box** (Figure 17.2). Located some 25 to 30 bases upstream from the initial point of transcription (designated as −25 to −30), it consists of an 8-base-pair consensus sequence (a sequence conserved in most or all genes studied) composed only of T = A pairs, often flanked on either side by G ≡ C-rich regions. Mutations in the TATA box severely restrict transcription, and deletions often alter the initiation point of transcription (Figure 17.3).

A second modular component of many promoters is called the **CAAT box**. Its consensus sequence is CAAT or CCAAT, it frequently appears in the region −70 to −80 bp from the start site, and it can function in either orientation. Mutational analysis suggests that the CAAT box is critical to the promoter's ability to facilitate transcription. A third modular element of some promoters is called the **GC box**, which has the consensus sequence

GGGCGG and is often found at about position −110. This module is found in either orientation and often occurs in multiple copies.

The variation in the organization of the upstream regulatory elements in several eukaryotic genes is shown in Figure 17.4. Note that in different genes there are great differences in the number and orientation of promoter elements and the distance between them. Just how a functional promoter can be composed of different elements scattered along a DNA molecule is not clear. One model suggests that the DNA molecule may fold or loop to bring distant upstream sites into juxtaposition (Figure 17.5), implying that the spacing between elements is more important than the identity of the promoter modules.

## Transcription Factors That Bind to Promoters

Proteins that are not part of the RNA polymerase molecule itself, but are needed for the initiation of transcription, are called **transcription factors**. These proteins have at least two functional domains: one that binds to DNA sequences present in promoters and enhancers, and another that activates transcription via protein–protein interaction by binding to RNA polymerase or to

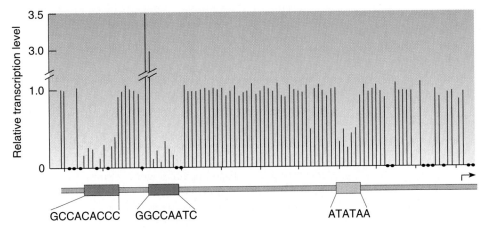

**FIGURE 17.3**    Effect of point mutations on transcription of the beta globin gene. Each line represents the level of transcription produced by a single nucleotide mutation (relative to wild type) in the promoter region. Dots represent nucleotides where no mutation was obtained. Note that mutations within promoter elements have the greatest effect on the level of transcription.

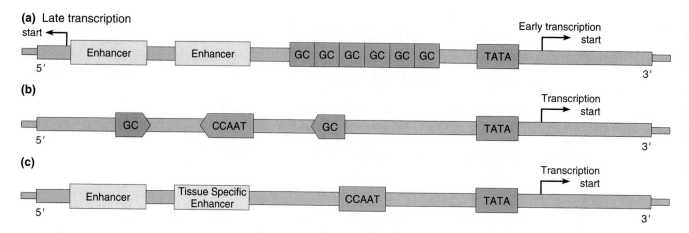

**FIGURE 17.4**    Upstream organization of several eukaryotic genes, illustrating the variable nature, number, and arrangement of controlling elements. (a) SV40 control region, with early genes to right, late genes to left. (b) Thymidine kinase. (c) Insulin gene.

other transcription factors. These domains of eukaryotic transcription factors show several structural adaptations to their functional requirements.

While there has been a variety of such factors isolated and studied, no clear picture has yet emerged as to how these "activators" of transcription work. However, it is believed that the various binding interactions with DNA and with one another provide a functional link between the **transcriptional apparatus** (i.e., the promoter–RNA polymerase complex) and the **regulatory apparatus** (i.e., the various DNA units essential for stimulating transcription to occur).

Some insights into the transcriptional apparatus of type II genes are now available and involve a series of transcriptional factors (e.g., TFIIA, TFIIB, TFIIC, TFIID, and TFIIE). It appears that in order to initiate formation of the apparatus, TFIID must first bind to the TATA box of class II promoters. About 20 base pairs of DNA are involved. This has been described as the "*commitment*" stage (Figure 17.6). Then, other TFII factors and RNA polymerase II can bind. Unlike the situation in prokaryotes where the polymerase binds directly to the DNA promoter region, eukaryotic polymerases bind to transcription factor proteins that are in turn bound to the DNA promoter region. At this point, transcription of the DNA downstream may ensue at a minimal "*basal*" level. The final stage involves the achievement of the "*induced*" state, where transcription is stimulated above the basal level. This state is not yet well defined, but involves other areas of the promoter region, enhancers, and numerous transcription factors.

## Enhancers in Eukaryotes

In addition to the promoter region, transcription of most, if not all, eukaryotic genes is regulated by addi-

tional DNA sequences called **enhancers**. They are not involved in the direct binding of RNA polymerase, but instead interact with regulatory protein molecules. Their role is to regulate whether or not, and the rate at which, transcription *actually* occurs. Thus, there is some analogy between enhancers and operator regions in prokaryotes. However, enhancers appear to be much more complex in both structure and function. Enhancers greatly stimulate the transcriptional activity of a promoter and can be distinguished from promoters by several characteristics:

1. The position of the enhancer need not be fixed; it can be placed upstream, downstream, or within the gene it regulates.

2. Its orientation can be inverted without significant effect on its action.

3. Enhancers are not restricted to specific genes. If an enhancer is moved to another location in the genome, or if an unrelated gene is placed near an enhancer, transcription of the adjacent gene is enhanced.

Most eukaryotic genes are under the control of enhancers. In the immunoglobulin heavy chain genes, an enhancer is located in an intron between two coding regions. In this case, the enhancer is located *within* the gene it regulates. It is also active only in the cells in which the immunoglobulin genes are being expressed, indicating that tissue-specific gene expression can be modulated through enhancers.

Internal enhancers have been discovered in other eukaryotic genes, including the immunoglobulin light chain gene. Downstream enhancers are found in the

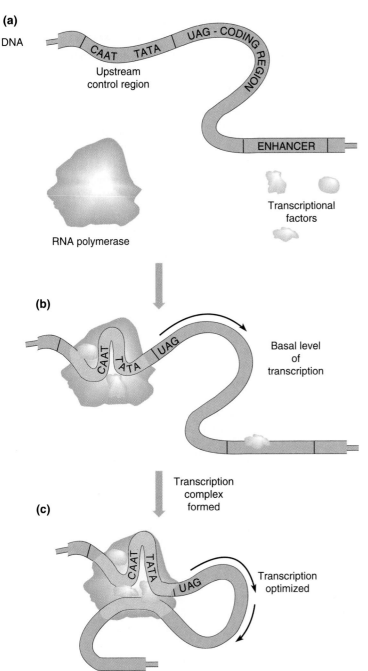

**(a)**

DNA

CAAT   TATA   UAG - CODING REGION

Upstream
control region

ENHANCER

RNA polymerase

Transcriptional
factors

**(b)**

CAAT   TATA   UAG

Basal level
of
transcription

Transcription
complex
formed

**(c)**

CAAT   TATA   UAG

Transcription
optimized

**FIGURE 17.5**    A model depicting the components [part (a)] and interactions [parts (b) and (c)] of transcription factors with different DNA control regions of a gene. A basal level of transcription is achieved initially as a result of interactions involving the CAAT and TATA boxes, producing a conformation that attracts RNA polymerase to the promoter region. Further interaction involving the enhancer region optimizes transcription.

human beta globin gene and the chicken thymidine kinase gene. In chickens, an enhancer is located between the beta globin and the epsilon globin genes and apparently works in one direction to control the epsilon gene during embryonic development and in the opposite direction to regulate the beta gene during adult life. In yeast, regulatory sequences similar to enhancers, called **upstream activator sequences (UAS)**, can function upstream at variable distances and in either orientation. They differ from enhancers in that they cannot function downstream of the transcription start point.

Like promoters, enhancers contain modules composed of short DNA sequences. The enhancer of the SV40 virus consists of two tandemly repeated 72-bp sequences located some 200 bp upstream from a transcriptional start point (Figure 17.7). If one or the other of these sequences is deleted, there is no effect on transcription, but if both are deleted, *in vivo* transcription is greatly reduced. This and other enhancers contains a copy of an octamer sequence (ATGCAAATNA) that is found in both promoters and enhancers, but so far, it has proven difficult to determine which sequences constitute the basic units of enhancer structure. These postulated units, called **enhansons**, are probably short sequences

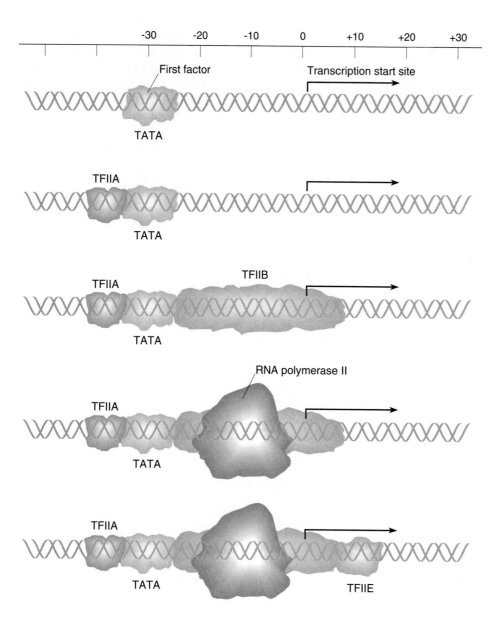

FIGURE 17.6    The assembly of transcription factors required for the initiation of transcription by RNA polymerase II. The first factor binds to the TATA box, and then TFIIA binds, forming a complex with the TATA factor and the DNA. The next factor, TFIIB, binds before RNA polymerase II. The polymerase binds to the transcription factors through protein–protein interactions. Before transcription begins, factor IIE binds, and transcription then begins from the initiation site at +1.

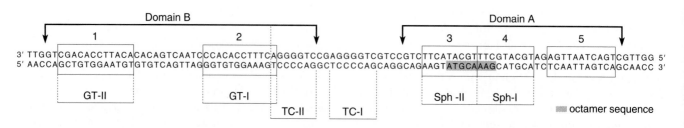

**FIGURE 17.7**    DNA sequence for the SV40 enhancer. Below the sequence are the various control sequences contained within this region.

that represent the binding sites for transcription factors, and may have sequence-specific properties for the binding of different factors.

Promoter sequences are essential for transcription, while enhancers are not. Enhancers are able to stimulate transcription at a distance, while this is not necessarily true of promoters. In addition, many enhancers are tissue-specific in their mode of action. As a group, promoters contain some modules (the TATA box) that must be located at a fixed site upstream of the transcription start point, while enhancers do not contain these modules. Although enhancers and promoters share some modules, the modules in enhancers are most often contiguous, while in promoters, they are spaced apart. On the other hand, both enhancers and promoters have some striking similarities: they both have modular organization, with some modular sequences in common, and they both bind transcription factors.

The most intriguing question about enhancers is how they are able to exert control over transcription at a great distance from either promoters or the transcriptional start site. While there is no clear answer yet, the consensus of opinion is that transcription factors bind to enhancers and alter the configuration of chromatin by bending or looping the DNA to bring distant enhancers and promoters into direct contact to more readily form complexes of promoter transcription factors and polymerases. In the new configuration, transcription is stimulated to a higher level, increasing the overall rate of RNA synthesis.

While the general features of enhancer activity are known, there are several important issues about the function of enhancers that are unresolved. It is not known what fraction of all genes require the action of an enhancer, nor do we understand what role enhancers might play in tissue-specific gene expression, nor how enhancers work at the molecular level. It is clear that such regulatory sequences work by binding protein transcription factors, and that these factors, in turn, control transcription.

## Transcription Factors That Bind to Enhancers

The domains of eukaryotic transcription factors take on several forms. The **DNA-binding domains** or **structural motifs** of eukaryotic transcription factors analyzed to date have been classified into three major types: **zinc fingers**, **helix-turn-helix (HTH) motifs**, and **leucine zippers** (Figure 17.8). This classification is not exhaustive, and other new groups will undoubtedly be established as new factors are characterized.

Zinc fingers are one of the major structural families of eukaryotic transcription factors, and they are involved in many aspects of gene regulation. Zinc fingers were originally discovered in the *Xenopus* transcription factor TFIIIA. This structural motif has now been identified in proto-oncogenes (Chapter 18), genes that regulate development in *Drosophila* (Chapter 21), in proteins whose synthesis is induced by growth factors and differentiation signals, and in transcription factors.

Zinc fingers consist of amino acid sequences containing two cysteine and two histidine residues at repeating intervals. The consensus amino acid repeat is $Cys-N_{2-4}-Cys-N_{12-14}-His-N_3-His$. The interspersed cysteine and histidine residues covalently bind zinc atoms, folding the amino acids into loops known as zinc fingers (Figure 17.8). Each finger consists of approximately 23 amino acids, with a loop of 12 to 14 amino acids between the Cys and His residues, and a linker between loops consisting of 7 or 8 amino acids. The amino acids in the loop interact with and bind to specific DNA sequences. Two-dimensional nuclear magnetic resonance (NMR) studies have shown that zinc fingers bind in the major groove of the DNA helix and wrap at least part way around the DNA. Within the major groove, the zinc finger makes contact with a set of DNA bases and may form hydrogen bonds with the bases, especially in G-rich strands.

The number of fingers in a zinc finger transcription factor varies from 2 to 13, as does the length of the DNA-binding sequence. Because of these variations, it seems plausible that there may not be a single mode of zinc-finger binding. Understanding the three-dimensional structure of a number of such transcription factors may reveal factors that bind to AT-rich sequences and establish some general rules for zinc-finger binding to DNA.

The second DNA-binding domain is called the helix-turn-helix (HTH) motif and was the first DNA-binding motif to be identified. In prokaryotes, HTH motifs have been identified in the *cro* repressor, the DNA-binding domain of the repressor, the *lac* repressor, *trp* repressor, and other proteins. Studies indicate that the HTH motif is present in many prokaryotic DNA-binding proteins. This motif is characterized by its geometric conformation rather than a distinctive amino acid sequence. Two adjacent $\alpha$ helices separated by a "turn" of several amino acids provide the binding potential of the protein to DNA (and after which the motif is named). In the *cro* repressor, one of the helices binds within a major groove containing six bases, while the other helix stabilizes the interaction by binding to the DNA backbone. Unlike several of the other DNA-binding motifs, the HTH pattern cannot fold or function alone, but is always part of a larger DNA-binding domain. Amino acid residues outside the HTH motif are important in regulating DNA recognition and binding.

The potential for forming helix-turn-helix geometry has been recognized in distinct regions of a large num-

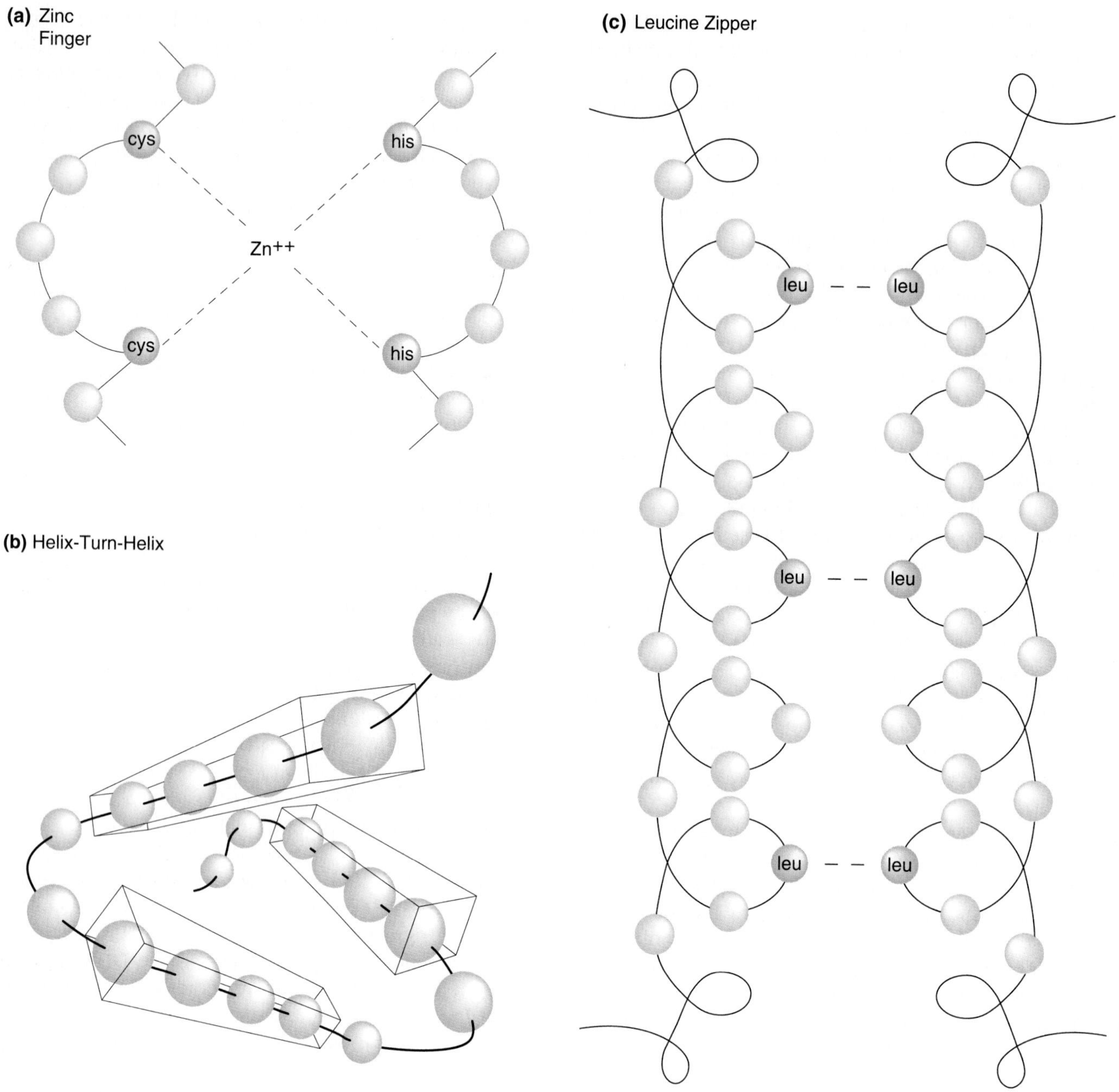

**(a)** Zinc Finger

**(b)** Helix-Turn-Helix

**(c)** Leucine Zipper

**FIGURE 17.8**   Three structural motifs of DNA-binding proteins. (a) A *zinc finger*, where cysteine and histidine residues bind to a $Zn^{++}$ atom; (b) a *helix-turn-helix* or *homeodomain*, where three different planes of the $\alpha$ helix are established; and (c) a *leucine zipper*, where dimers result from leucine residues at every other turn of the $\alpha$ helix in facing stretches of one or two polypeptide chains. Each motif provides a unique way to bind to DNA.

ber of eukaryotic genes known to regulate developmental processes. Present almost universally in eukaryotic organisms and called the **homeobox**, a stretch of 180 bp specifies a 60-amino-acid **homeodomain** sequence that can form a helix-turn-helix structure. Of the 60 amino acids, many are basic (arginine and lysine), and a conserved sequence is found among these many genes.

We shall return to homeobox genes in Chapter 21 because of their significance to developmental processes. Important here is the recognition of a particular binding motif that appears to be characteristic of a large class of genes, called **homeotic genes**, the regulation of which plays a critical role during the differentiation of organisms.

The third type of domain is represented by the so-called leucine zipper. First seen as a stretch of 35 amino acids in a nuclear protein in rat liver, four leucine residues are spaced seven amino acids apart, and flanked by basic amino acids. The leucine-rich regions form a helix with leucine residues protruding at every other turn. When two such molecules dimerize (Figure 17.8), the leucine residues "zip" together. This is possible because if the alpha helix of the protein is viewed as a wheel, the leucine residues all protrude on the same side of the helix at every other turn. Leucine zippers almost exclusively involve binding of protein dimers to DNA. The DNA-binding region of these proteins involves a helix-loop-helix region.

In addition to domains that bind DNA, transcription factors contain domains that activate transcription. These regions can occupy from 30 to 100 amino acids and are distinct from the DNA-binding domains. Activator domains were first identified in the yeast transcription factors *GAL4* and *GCN4* as short stretches of acidic (negatively charged) amino acids that can form alpha helices, and they have been identified in other eukaryotic transcription factors. These stretches of acidic amino acids are proposed to function by interacting with other transcription factors (such as those that bind to the TATA sequence) or directly with the RNA polymerase. Two other domains, one glutamine-rich and the other proline-rich, have been identified in transcription factors from insect and mammalian systems. In all cases, these domains are likely to function by contacting or interacting with other proteins. The nature of the interaction is still unknown, but since RNA polymerase II requires four or five transcription factors for activation, and most transcription factors contain multiple subunits, the existence of a number of different protein–protein interaction mechanisms seems likely.

Overall, the picture of transcriptional regulation in eukaryotes is somewhat complex, but a number of gen-

eralizations can be drawn. The structural organization of chromatin and alterations in chromatin structure to allow binding of transcription factors is considered to be the primary level of regulation. The regulation of transcription by protein factors is largely positive. The binding of one or more factors at promoter regions is a prerequisite to transcriptional activation of a locus. Promoter and enhancer sequences are recognized and bound by transcription factors, rather than by RNA polymerase. Because of the modular nature of recognition sequences, different transcription factors may compete for binding to a given DNA sequence, or the same site may bind different factors in a cell-specific fashion. For example, sequences in the TATA box and CAAT box can bind to transcription factors that recognize these modules. Finally, because of the variability in the location and modular nature of promoters and enhancers, and the variety of transcription factors, an initiation complex (consisting of transcription factors and RNA polymerase II) can be constructed in a number of ways, all involving protein–protein interaction. Because of this variability, it is impossible to provide a general transcriptional activation scheme for all genes. However, to illustrate how eukaryotic transcription factors operate, we will consider the details of transcriptional regulation in a yeast gene, and the related phenomenon of steroid hormone activation of transcription.

## Galactose Metabolism in Yeast

We shall begin by examining the genetic control mechanisms involving the enzymes that participate in the metabolism of galactose in the yeast *Saccharomyces cerevisiae*. The activities of four enzymes are required for galactose metabolism. As we shall see, these are several similarities between this regulatory model and bacterial operons. However, as with many genes in more advanced eukaryotes, regulation involves a positive system characterized by **transcriptional activation** rather than repression.

The activities of a cluster of three genes, all linked on chromosome II, are involved in the metabolism of galactose. A fourth gene (*GAL2*) located on chromosome XII is involved in galactose transport. The genes on chromosome II, *GAL1*, *GAL7*, and *GAL10*, encode a kinase, a transferase, and an epimerase, respectively. All three enzymes play a role in the conversion of galactose to glucose-6-phosphate. The latter molecule serves as an entry point into glycolysis, thus allowing galactose to serve as an energy source (Figure 17.9).

The chromatin conformation of *GAL* genes is altered to permit transcription at any time, so in this case, regulation is controlled by the presence or absence of galac-

# TOOLS AND TECHNIQUES

## Enhancer Trapping

Development is a complex process that involves a large number of genes organized into circuits with highly regulated patterns of expression; that is, expression of a given gene typically occurs only at certain times and in certain cells and tissues. Some of this regulation is controlled by *cis*-acting DNA regulatory sequences called enhancers. Enhancers characteristically increase transcription of genes, and their action is usually tissue-specific. However, enhancers are not promoter-specific and can increase transcription from any nearby promoter. To hunt for enhancers that regulate events in development, a reporter gene lacking an enhancer is constructed. In *Drosophila*, this generally consists of a modified P element containing a bacterial reporter gene, beta galactosidase (*lacZ*), arranged so that the *lacZ* gene is under the control of a promoter in the P element. This P element promoter is able to be controlled by many *Drosophila* enhancers. This construct is then used to transform *Drosophila*, becoming inserted into many places in the genome. Because there are many enhancers scattered throughout the genome, insertion often occurs near an enhancer, resulting in a tissue-specific pattern of expression that can be visualized by staining for beta galactosidase activity. In such cases, the P element carrying the reporter gene is said to have "trapped" an enhancer.

Many strains of *Drosophila* that carry single insertion copies of P-*lacZ* fusions in random locations have been studied. Analysis of embryos from a particular strain show a reproducible pattern of expression; cells transcribing the *lacZ* gene stain blue, while those where the gene is inactive are unstained. Different insertions display different staining patterns, and the staining patterns change as time and development proceed. From these patterns, it is apparent that each P-*lacZ* element is responding to regulation from a different nearby enhancer.

The chromosomal location of the P-*lacZ* fusion element in each strain can be determined by *in situ* hybridization of a labeled P element probe to polytene chromosomes. In cases where there is independent evidence about the expression pattern of nearby genes, it has been shown that this pattern correlates with the pattern observed by the beta-galactosidase staining. This confirms that the system does report on the operation of enhancers residing close to the site of the P-*lacZ* insertion.

To facilitate identification of genes regulated by enhancers discovered by trapping, a more complicated P-element-derived vector has been developed. This allows for the rapid isolation and cloning of DNA adjacent to the P-element insertion site by a process known as plasmid rescue. This P element consists of a plasmid vector with several convenient restriction enzyme cutting sites. DNA from a *Drosophila* strain carrying such an insert is cut with one of these restriction enzymes, followed by *in vitro* ligation to yield circular DNA molecules, some of which contain the plasmid vector and segments of the *Drosophila* genome adjacent to the insertion site. These circular molecules are able to replicate and can be used to transform bacterial hosts and generate clones for further study. The genomic DNA isolated by plasmid rescue can be analyzed to isolate and characterize structural genes near the trapped enhancers, or to determine whether other enhancer elements are present.

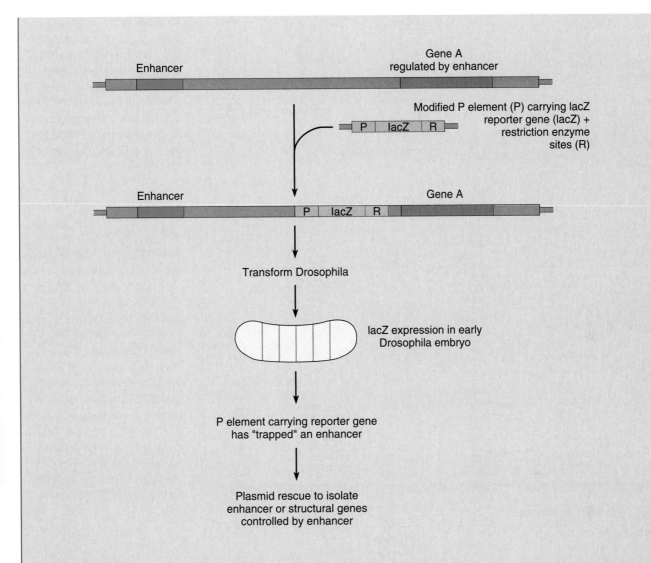

Once structural genes have been isolated by plasmid rescue, they can be used as probes for *in situ* hybridization to detect cells that are transcribing the complementary mRNA. When the cellular pattern of hybridization matches that of beta-galactosidase expression, it is likely that the transcribed gene is regulated by the enhancer initially caught in the enhancer trap.

This method has become an important tool to identify genes involved in development and to study their regulation. In particular, it allows for identification of genes whose functions are so central to development that they cannot be disrupted without causing catastrophic (usually lethal) effects. Conversely, it also allows the identification of genes whose effects are so subtle that their mutants produce little effect upon the phenotype.

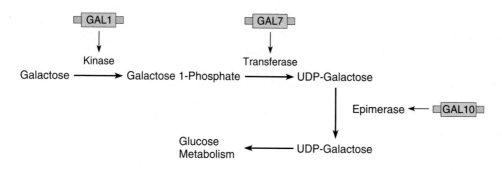

FIGURE 17.9    Metabolic pathway for galactose utilization. The genes for the first three enzymes are tightly linked in yeast, and activity of all three genes is tightly regulated at the transcriptional level by two gene products (from the *GAL4* gene and the *GAL80* gene).

**(a)** Components

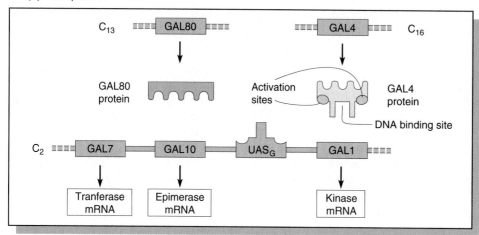

**(b)** Repression: no galactose

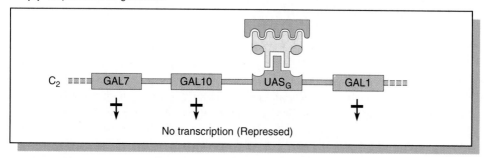

**(c)** Induction: galactose present

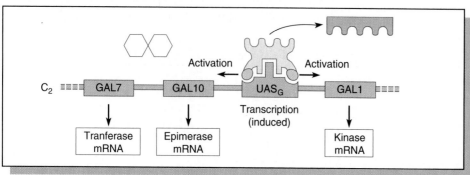

FIGURE 17.10    A representation of the interactions of the products of the *GAL80* and *GAL4* genes during the regulation of the galactose operon in yeast. Part (a) depicts the various genes and their products. Parts (b) and (c) contrast the interaction of the *GAL80* and *GAL4* products with the upstream activating sequence (UAS$_G$) in the absence and in the presence of galactose. Activation of transcription of the operon is under positive control of the *GAL4* protein.

tose. In the absence of galactose, these three genes are not transcribed. If galactose is added to the growth medium, immediate transcription of the three genes commences. Constitutive mutations in one or more genes at other sites have been isolated, indicating the existence of genetic regulation of transcription. As a result of further study, two other genes, *GAL4* (on chromosome XVI) and *GAL80* (on chromosome VII), have been implicated in such regulation, leading to the model described below and diagrammed in Figure 17.10.

For the purpose of our discussion, we shall consider *GAL1*, 7, and 10 as a unit of structural genes, since they are all activated together. However, it is important to note their arrangement on chromosome II. *GAL1* and 10 are closely linked, separated only by a critical 600-base-pair sequence called **UAS$_G$** (upstream activating sequence). Their polarities of transcription, [Figure 17.10(a)] are in opposite directions. *GAL7* is not as closely linked and is distal (in the 5′ direction) from *GAL10*.

Analysis of regulatory mutations has led to support for the following model, depicted in Figure 17.10(b). The product of the *GAL4* gene is a transcription activator that functions by binding to the UAS$_G$ region. If, however, no galactose is present, the *GAL80* gene product binds to the *GAL4* activator, rendering it ineffective in stimulating transcription of the structural genes. If galactose *is* present, this sugar binds to the *GAL80* product, preventing it

from binding to the *GAL4* activator, and transcription is initiated.

In these cases, we must assume that different combinations of binding induce conformational changes that provide alternative conditions of recognition. This assumption is supported by the findings that different portions of the regulatory molecules play different roles. The *GAL4* activator protein has now been studied extensively by Mark Ptashne and his colleagues.

The *GAL4* gene product is a protein 881 amino acids long, with three functions: binding to DNA sequences in the UAS$_G$, interacting with the *GAL80* protein, and activating transcription. By cloning truncated *GAL4* genes and assaying their ability to bind DNA and activate transcription, a region at the N-terminus of the gene product consisting of the first 98 or the first 147 amino acids was found to be the location of the DNA-binding site (Figure 17.11). This derived protein binds to DNA in the UAS$_G$, but fails to activate transcription and cannot bind the *GAL80* protein. Analysis of the amino acid composition of this region has identified a single zinc finger responsible for the DNA-binding function. Similar experiments have identified two regions of the protein involved in activation of transcription: region I, consisting of amino acids 149–196, and region II, encompassing amino acids 768–881 (Figure 17.11). Constructs that contain the DNA-binding domain and region I (amino acids 1–96 and 149–196) or the DNA-binding domain and region II

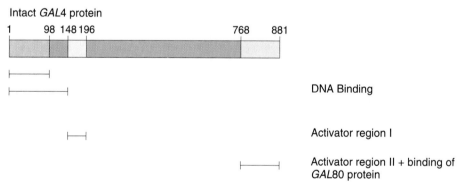

**FIGURE 17.11**    The *GAL4* protein has three domains that participate in the activation of the *GAL1* and *GAL10* loci. Amino acids 1–98 or 1–147 bind to DNA recognition sites in UAS$_G$. Amino acids 148–196 and 768–881 are required for transcriptional activation. *GAL80* binding activity resides in the amino acid region 851–881.

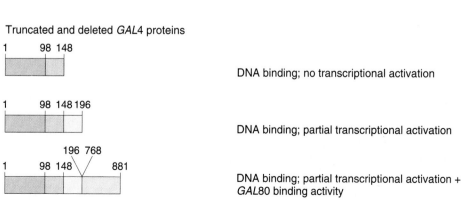

(amino acids 1–96 and 768–881) have reduced transcriptional activity. Deletions of the amino acids 852–881 destroy the ability to bind *GAL*80 and cause the truncated protein to constitutively activate the *GAL*1 and *GAL*10 loci.

The nature of the activating regions was explored by starting with a protein that contains the DNA-binding region and activating region I (*GAL4* 1–238) and isolating point mutations in region I that result in increased activation. Most single mutations that increase activation represent amino acid substitutions that increase the negative charge of region I, and most exert a threefold increase on transcription. A multiple mutant that contains four more negative charges than the starting protein has a ninefold increase in transcriptional activation, and has almost 80 percent of the activity of the complete *GAL4* protein. However, negative charge alone is not responsible for the activating function, since some mutants with decreased activity do not have fewer negative charges than the parental *GAL4* 1–238 molecule.

These results strongly suggest that transcriptional activation involves interaction between the activating domain of the transcription factor and other proteins. These other proteins must normally be components of transcription complexes, and might include the polymerase molecule itself, or any of the four or five accessory proteins required for transcription (Figure 17.6). An attractive candidate for the target factor is TFIID, which binds to the TATA sequence and is required for transcription. TFIID is a multiple-subunit factor consisting of TBP (TATA-binding protein) and multiple TAFs (TATA-associated factors). This idea is supported by the observation that some *GAL4* constructs stimulate transcription in mammalian cell extracts, and change the conformation of DNA-bound TFIID.

A model of eukaryotic transcriptional regulation based on the regulation of galactose in yeast proposes that many transcriptional activators work through an interaction with TFIID, either directly as in the case of *GAL4*, or indirectly, through a second factor, like OCT-1 or VP16 (Figure 17.12). Further experiments on TATA-binding proteins should reveal the nature and organization of their functional domains and shed more light on the molecular mechanisms involved in transcriptional activation.

## Transcriptional Regulation by Steroid Hormones

Evidence that gene expression in eukaryotes is regulated by effector molecules originating outside the cell

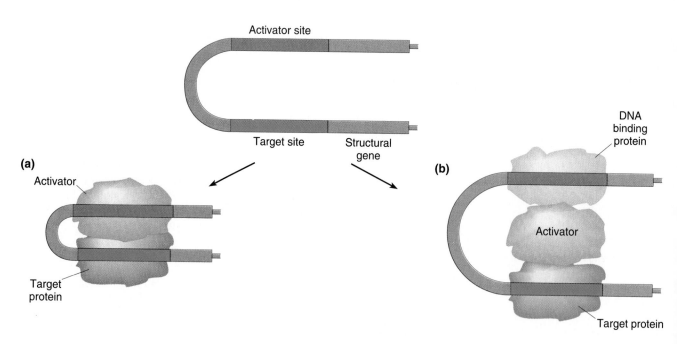

**FIGURE 17.12**    A model for gene activation by two classes of proteins. In (a), the activator contains a DNA-binding site and an acidic activating region that interacts with a target protein, activating the transcriptional complex. A second class of activators (b) contains only a single functional site, either DNA binding or target activating, requiring the action of two factors to activate the target protein and stimulate transcription.

was elegantly provided by Ulrich Clever and Peter Karlson with their discovery that the steroid hormone **ecdysone** induces specific changes in the puffing pattern in polytene chromosomes (recall from Chapter 2 that puffs represent the transcription of specific genes). In eukaryotes, steroid hormones are used to regulate growth and development and to maintain homeostasis. The major sex hormones are all steroids, as is vitamin D. The homeostatic adrenal hormones (more than 30 different types) that regulate glucose metabolism and mineral utilization are also steroids (Figure 17.13).

Steroid hormones enter the cell by passing through the plasma membrane and binding to a specific receptor protein in the cytoplasm. The receptor–hormone complex is translocated to the nucleus and activates transcription of one or more specific genes. Hormone receptors and the DNA sequences to which they bind have been intensively studied over the last decade. The structural genes encoding most every known hormone receptor protein have been cloned, and enough is now known to provide a general picture of how these receptors work. All receptors analyzed to date have three functional domains: a variable N-terminus region; a short, highly conserved central region; and a fairly well-

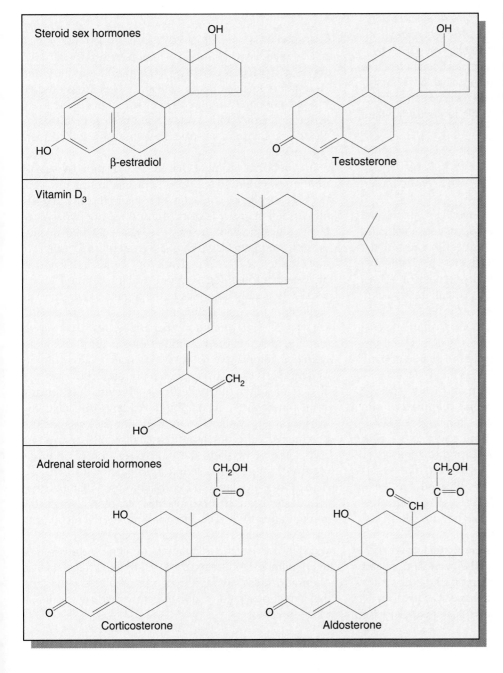

**FIGURE 17.13**    Structure of several classes of steroid hormones that activate transcription by binding to cytoplasmic receptor proteins that are then transported to the nucleus where they bind to specific DNA sequences.

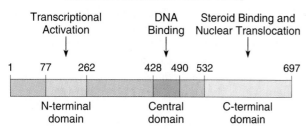

**FIGURE 17.14** Steroid hormone receptors have three functional domains, as exemplified by the glucocorticoid receptor: the N-terminal domain (amino acids 77–262 in the glucocorticoid receptor), the central domain (amino acids 428–490), and the C-terminal domain (amino acids 532–697). The central domain binds DNA sequences known as the glucocorticoid response element (GRE) by means of two zinc fingers. The C-terminal domain binds to the steroid hormone and may be responsible for allowing the hormone–receptor complex to be translocated to the nucleus. This region is also responsible for regulating the activity of the receptor.

**Table 17.1** DNA Receptor Sequences for Hormone Binding Proteins

| Hormone | Consensus Sequence | Element Name |
|---|---|---|
| Glucocorticoid | GGTACANNNTGTTCT | GRE |
| Estrogen | AGGTCANNNTGACCT | SRE |
| Thyroid | AGGTCA . . . TGACCT | TRE |

conserved C-terminus (Figure 17.14). The central region contains two zinc fingers.

In addition to activating the regulated gene, the C-terminal domain of steroid receptors interacts with and binds to the steroid hormone. (It has been suggested that in the absence of bound hormone, the receptor is unable to recognize the hormone receptor element of DNA). The C-terminal region may also be partially responsible for controlling the translocation of the hormone–receptor complex from the cytoplasm to the nucleus. The N-terminal region of the receptor is not as well understood, but it apparently modulates the degree of gene activation.

The zinc fingers in hormone receptors bind to specific DNA sequences known as **hormone-responsive elements** or **HREs**. In steroid receptors, the finger closer to the N-terminus binds in a base-specific way to the HRE and determines the specificity of binding to a particular target gene. The C-terminus finger contacts the sugar-phosphate residues adjacent to the HRE and may have other functions. In general, HREs share some characteristics with promoters and enhancers. They are composed of short consensus sequences that are related but not always identical (Table 17.1). HREs are often located several hundred bases upstream from the transcription start site, and may be present in multiple copies. Often, HREs are present within promoter or enhancer sequences.

Binding of the receptor to the HRE is necessary for activation of a specific gene (Figure 17.15), but it may not be sufficient. Binding of the receptor to the HRE serves to place regions of both the C-terminus and N-terminus in position to activate a specific promoter, but activation may require the participation of other transcription factors, and, as described below, involves the alteration of chromatin structure in the HRE and adjacent regions.

It is important to note that gene activation by steroid hormones requires alteration of chromatin structure near the regulated gene. Genes in chromatin that are actively transcribed or are able to be transcribed are hypersensitive to digestion with the enzyme DNase I, while inactive or repressed genes are relatively resistant to DNase I. *In vitro* studies of steroid activation of genes in chromatin indicate that following hormone administration, DNase I sensitivity greatly increases in the regulated gene and its flanking regions, including the hormone receptor binding site. The interpretation of these results is consistent with the idea that the binding of one or more receptor–hormone complexes (receptor may need to act as dimers or tetramers to activate genes) to the HRE causes a change or shift in the nucleosome structure of chromatin at the binding site, making other binding sites, including the promoter, available to transcription factors and RNA polymerase II, resulting in the initiation of transcription.

In some ways, then, the biology of steroid hormone regulation of gene transcription in eukaryotes is similar to the *lac* operon system in prokaryotes in that an external effector binds to a cytoplasmic receptor, changes the configuration of the receptor, and moves the receptor to the DNA, where it acts as a transcription factor. In other ways, however, the regulation of gene expression is quite different, in that the DNA-binding sequence is often at a great distance from the regulated gene, and that alterations in chromatin structure mediate the action of the receptor.

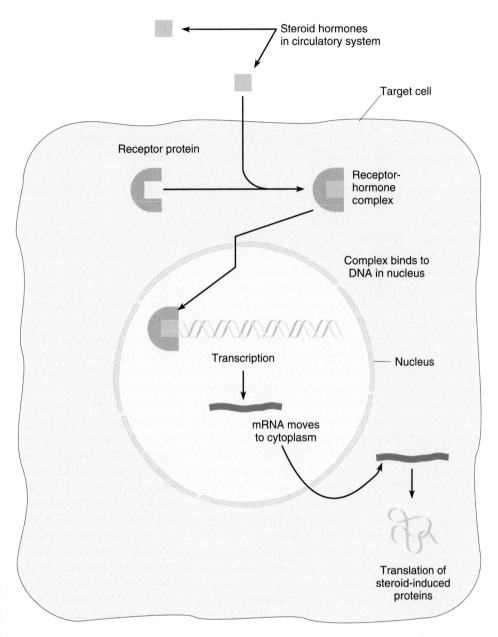

Steroid hormones
in circulatory system

Target cell

Receptor protein

Receptor-
hormone
complex

Complex binds to
DNA in nucleus

Transcription

Nucleus

mRNA moves
to cytoplasm

Translation of
steroid-induced
proteins

**FIGURE 17.15**    Stages of steroid hormone effects on gene expression. Steroids in the circulating system pass through the plasma membranes of target cells and bind to cytoplasmic receptor proteins. The steroid–receptor complex moves to the nucleus and binds to DNA at hormone receptor elements, stimulating transcription and translation of steroid-induced genes.

## GENOMIC ALTERATIONS AND GENE EXPRESSION

The concept that chromatin conformation may be involved in the regulation of transcription initiation was introduced in the preceding section. This reversible alteration in the genome is one of several ways in which gene expression can be regulated by physical changes in DNA. In this section we will consider two additional alterations that play a role in gene regulation. One is a chemical modification of DNA that involves adding or removing methyl groups to the bases in DNA, and the other is changes in the number of copies of a DNA sequence brought about by amplifying a specific gene or set of genes in order to increase the level of expression.

### DNA Methylation

The DNA of most eukaryotic organisms is modified after replication by the enzyme-mediated addition of methyl groups to bases and sugars. Base methylation most often involves cytosine, and approximately 5 per-

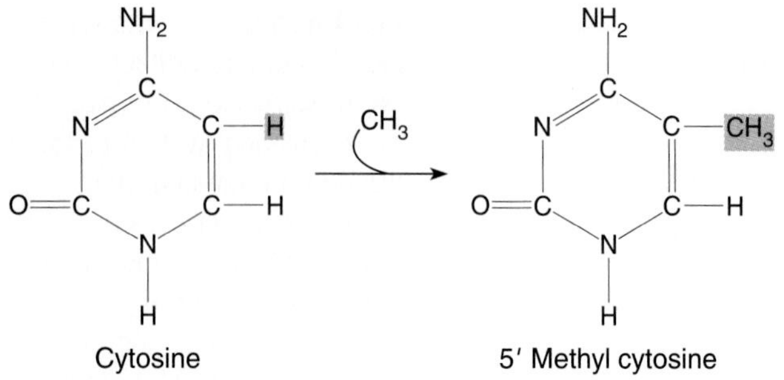

**FIGURE 17.16** Methylation of cytosine to yield 5'-methyl cytosine. In mammals, methylation of DNA almost exclusively involves the attachment of a methyl group to the 5' carbon of cytosine. Often, the methylated cytosines in DNA are followed on the 3' side by a nucleotide containing guanine; these are written as CpG dinucleotides. Up to 75 percent of the CpG dinucleotides in a mammalian genome are usually methylated.

cent of the cytosine residues are methylated in the genome of any given eukaryotic species, although the extent of methylation can be tissue-specific and varies from less than 2 percent to over 7 percent.

The ability of base methylation to alter gene expression is known from studies on the *lac* operon in *E. coli.* Methylation of DNA in the operator region, even at a single cytosine residue, can cause a marked change in the affinity of the repressor for the operator. Methylation of cytosine occurs at the 5' position (Figure 17.16), causing the methyl group to protrude into the major groove of the DNA helix where it can alter the binding of proteins to the DNA.

Methylation occurs most often in the cytosine of CG doublets in DNA, usually in both strands:

$$5'-{}^{m}CpG\ -3'$$
$$3'-\ GpC^{m}\text{-}5'$$

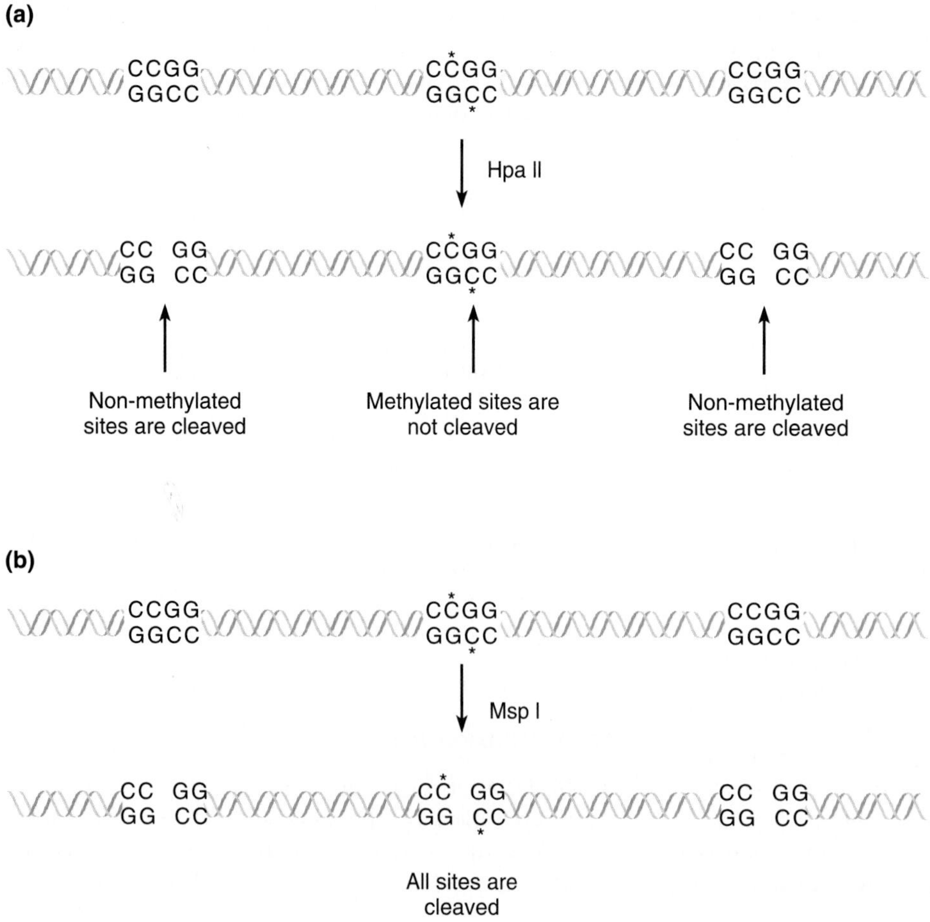

**(a)**

**FIGURE 17.17** The restriction enzymes *Hpa*II and *Msp*I recognize and cut at CCGG sequences. (a) If the second cytosine is methylated (indicated by an asterisk), *Hsp*II will not cut. (b) The enzyme *Msp*I cuts at all CCGG sites, whether or not the second cytosine is methylated. Thus the state of methylation of a given gene in a given tissue can be determined by cutting DNA extracted from that tissue with *Hpa*I and by *Msp*II.

**(b)**

The state of DNA methylation can be determined by restriction enzyme analysis. The enzyme *Hpa*II cleaves at the recognition sequence CCGG; however, if the second cytosine is methylated, the enzyme will not cut the DNA. The enzyme *Msp*I cuts at the same CCGG site whether or not the second cytosine is methylated. If a segment of DNA is unmethylated, both enzymes produce the same restriction pattern of bands. As shown in Figure 17.17, if one site is methylated, digestion with *Hpa*II produces an altered pattern of fragments. Using this method to analyze the state of methylation for a given gene in different tissues shows that a gene is not methylated in tissues in which it is expressed, and that some copies of the gene are methylated in tissues in which the gene is not expressed, indicating that the connection between methylation and transcriptional activity is not absolute.

The evidence for the role of methylation as a factor in the regulation of enkaryotic gene expression is somewhat indirect and is based on a number of observations. First, there is an inverse relationship between the degree of methylation and the degree of expression. That is, low amounts of methylation are associated with high levels of gene expression, and high levels of methylation are associated with low levels of gene expression. In mammalian females the inactivated X chromosome, which is almost totally inactive in gene expression, has a higher level of methylation than the active X chromosome. Within the inactive X, those regions that escape inactivation have much lower levels of methylation than those seen in adjacent, inactive regions.

Second, methylation patterns are tissue-specific and, once established, are heritable for all cells of that tissue. Perhaps the strongest evidence for the role of methylation in gene expression comes from studies using base analogs. The nucleotide 5'-azacytidine is incorporated into DNA in place of cytidine and cannot be methylated because the 5' position is occupied (Figure 17.18), causing undermethylation of sites where it is incorporated. Incorporation of 5'-azacytidine causes changes in the pattern of gene expression and can stimulate expression of alleles on inactivated X chromosomes.

Recently, 5'-azacytidine has been used in clinical trials for the treatment of sickle-cell anemia. In this autosomal recessive condition, a mutant hemoglobin protein with a defect in the beta subunit causes changes in shape of red blood cells that, in turn, generate a cascade of clinical symptoms. During embryogenesis, the epsilon (ε) and gamma (γ) globin genes are expressed, but normally become transcriptionally inactive after birth when beta globin synthesis begins. Treatment of affected individuals with 5'-azacytidine causes a reduction in the amount of methylation in the epsilon (ε) and gamma (γ) genes, and initiates re-expression of these embryonic and fetal

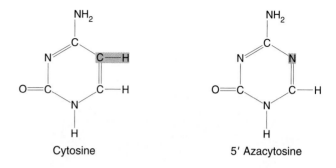

**FIGURE 17.18**    The base 5'-azacytosine has a nitrogen at the 5' position. When incorporated into a nucleotide, 5'-azacytidine, this base can be incorporated into DNA in place of cytidine. The 5'-azacytidine cannot be methylated, causing undermethylation of CpG dinucleotides wherever it has been incorporated.

genes. The epsilon and gamma proteins replace beta globin in hemoglobin molecules, bringing about a reduction in the amount of sickling in the red blood cells.

As stated above, the evidence available indicates that the absence of methyl groups in DNA is related to increases in gene expression. Methylation cannot, however, be regarded as a general mechanism for gene regulation because methylation is not a general phenomenon in eukaryotes. In *Drosophila*, for example, there is no methylation of DNA. Thus, methylation may represent only one of a number of ways in which gene expression can be regulated by genomic changes, and it seems likely that similar mechanisms remain to be discovered.

## Gene Amplification

One way to regulate the amount of a given gene product synthesized over a limited time period is to increase the number of copies of the structural gene that is being transcribed. This process, termed **gene amplification**, plays a role in the developmentally regulated expression of some gene sets, including the ribosomal RNA genes of *Xenopus* (described in Chapter 10) and the chorion genes in *Drosophila*. This amplification results in a high rate of gene expression over a limited period of time. During oogenesis in *Xenopus*, the developing oocyte, containing amplified ribosomal RNA genes, accumulates $10^{12}$ ribosomes in about 60 to 80 days, at an average rate of 300,000 per second.

In the somatic cells of *Xenopus*, the tandemly repeated ribosomal RNA genes (rDNA) are present in about 1000 copies per diploid genome. Without amplification, it would require over a thousand days to accumulate this

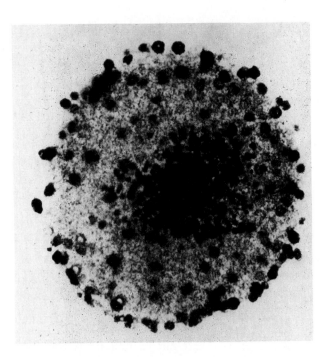

**FIGURE 17.19** Photomicrograph of a mature oocyte of *Xenopus laevis.* The dark circles in the interior and at the periphery are amplified nucleoli, each containing thousands of copies of the ribosomal genes. These amplified genes make it possible to synthesize billions of ribosomes over a relatively short time period.

number of ribosomes. In the developing *Xenopus* oocyte, the nucleus contains hundreds of nucleoli (Figure 17.19), each containing several thousand copies of the ribosomal gene cluster. Simultaneous transcription of these amplified loci allows ribosome accumulation to be completed in 60 to 80 days.

In *Drosophila*, the amplification of specific protein-encoding genes is a feature of oocyte development. One set of genes encoding eggshell proteins is clustered on chromosome 3. At the time of expression, these genes are amplified up to 60 times in the cells that synthesize the proteinaceous shell of the *Drosophila* egg (Figure 17.20). This amplification extends as a gradient centered on the chorion genes and covers a distance of some 50 kb on either side of this gene set. Sequential amplification of several protein-encoding loci occurs during development in other insects, including the fungal gnat, *Sciara*, where salivary gland chromosome puffs contain amplified DNA sequences (Figure 17.21), increasing the copy number up to 16-fold.

In mammals, gene amplification does not appear to play a developmental role, but is often associated with the appearance of drug resistance in malignant tumors during chemotherapy, and in generating the multiple copies of tumor-associated genes known as oncogenes (see Chapter 18) that are associated with the progressive changes that occur during tumor growth. Amplified DNA sequences are often visible in such mammalian cells as chromosome structures known as **extended chromosome regions (ECRs)** and **homogeneously staining regions (HSRs)**, or as extrachromosomal elements known as **double minute (DM) chromosomes**.

Several mechanisms have been proposed to explain gene amplification, and they have been grouped into two types: mechanisms associated with DNA replication and those associated with recombination and segregation (Table 17.2).

One of the replication-based mechanisms, the onion-skin model, associated with the amplification of genes in *Drosophila* and other insects, is shown in Figure 17.20. In this model, amplification is the result of multiple reinitiations of DNA synthesis within a single cell cycle. This causes the growth of nested replication complexes resembling the layers of an onion. Coupled with recombination events that occur during replication, this model predicts a wide range of outcomes, including the origin of extrachromosomal DNA that might grow into DM structures after further rounds of replication.

Perhaps the most straightforward example of recombination- and segregation-driven models of gene amplification is the sister chromatid exchange model (Figure 17.22). If two sister chromatids engage in an unequal exchange, segments of DNA will be duplicated on one chromatid, and lost from the other. If this process is repeated for several cell divisions, the outcome will be a cell that contains one chromosome containing an ESR or HSR with many amplified copies of a gene, and another homologue with a single copy of the gene. Cytogenetic analysis of cultured cells that have undergone amplification in response to drug selection confirm this expectation. When drug selection is relaxed, there is often a reduction in the copy number of the amplified gene, and a loss or reduction of the modified chromosomal region. Thus the sister chromatid model provides for a reversible means of altering the copy number of structural genes. It is hoped that future work on amplification in tumors and cultured cells will provide information about the organization of chromosomes and the chromosomal loci that serve as origins of DNA replication.

## POSTTRANSCRIPTIONAL REGULATION OF GENE EXPRESSION

As outlined above, there are many opportunities along the pathway from DNA to protein where regulation of genetic expression can occur. Although transcriptional

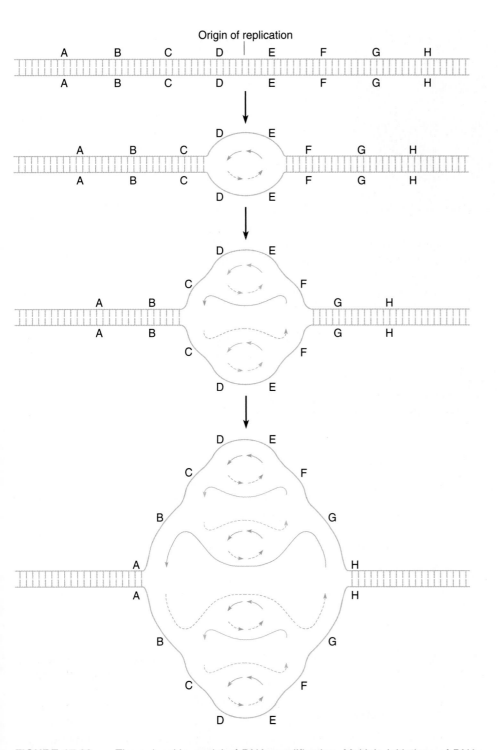

**FIGURE 17.20**    The onionskin model of DNA amplification. Multiple initiations of DNA replication at an origin of replication (marked by small arrows) gives rise to a nested complex of replicating DNA molecules. Multiple recombination events within this complex can generate an intrachromosomal linear array, while recombination events within the same duplex can generate circular, extrachromosomal DNA.

**FIGURE 17.21**   DNA puffs in *Sciara coprophila.* A polytene chromosome from the larval salivary gland of *Sciara* showing two prominent DNA puffs. In these puffs, DNA segments are amplified thousands of times, permitting an increase in the amount of transcription of selected genes over a limited time period.

**Table 17.2** MOLECULAR MECHANISMS FOR GENE AMPLIFICATION

| Replication-Based Models | Segregation and Recombination-Based Models |
| --- | --- |
| Onionskin model | Deletion plus episome model |
| Extrachromosomal double rolling circle model | Sister chromatid exchange model |
| Chromosomal spiral model | |

control is perhaps the most obvious and widely used mode of regulation in eukaryotes, posttranscriptional modes of regulation are also employed in many organisms. For example, nuclear RNA transcripts are modified prior to translation, noncoding **introns** are removed, the remaining **exons** are precisely spliced together, and the mRNA is modified by the addition of a poly A tail at the 3′ end. These processing steps offer several possibilities for regulation.

## Alternative Processing Pathways for mRNA

In discussing gene regulation through processing of transcripts, we will first consider the case in which a single gene is transcribed, but following processing, two mRNAs, differing in their untranslated leader sequences, are produced. In mice, the enzyme alpha-amylase is produced in both salivary glands and liver. The alpha-amylase gene contains two promoters, one used in salivary glands, and the other in liver (Figure 17.23). The salivary gland promoter is about 2850 bp upstream from the liver promoter, and two different exons are adjacent to the promoters. The use of the two different promoters creates two different pre-mRNAs that are differentially spliced to yield tissue-specific mRNAs (see Figure 17.23). It is important to emphasize that the protein product in both tissues is identical. The differences lie in the exons at the 5′ end of the mRNA, a region that constitutes part of the 5′ untranslated leader sequence.

Splicing variations can also generate different mRNAs responsible for directing the synthesis of different polypeptides. Studies have now documented that splicing may include or exclude an entire exon, leading to distinctive mRNAs that specify different polypeptides.

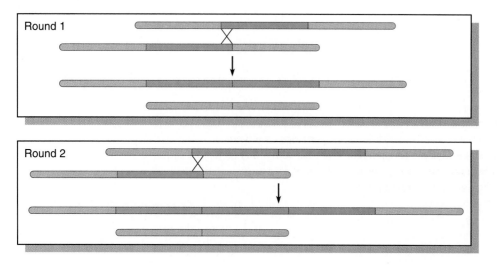

**FIGURE 17.22**   Sister chromatid recombination model of gene amplification. Misalignment and recombination between sister chromatids results in one chromatid with a duplicated segment (blue portion) and a sister chromatid with a deleted segment. Further rounds of misalignment and recombination involving the chromosome carrying the duplication will result in a linear amplification of the shaded sequence.

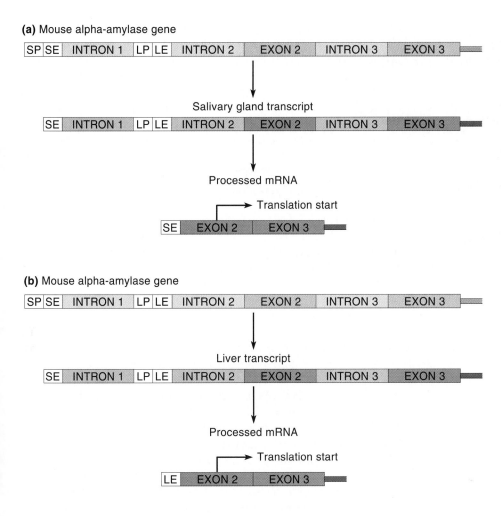

**(a)** Mouse alpha-amylase gene

**(b)** Mouse alpha-amylase gene

**FIGURE 17.23**   Organization of the mouse alpha-amylase gene. The salivary promoter (SP) and its adjacent exon (SE) are located approximately 3000 bp upstream from the liver promoter (LP) and the liver exon (LE). The use of alternative promoters results in the transcription of different pre-mRNAs in liver and salivary glands. Processing of these transcripts results in mRNAs that start with different exons in the 5′ untranslated region.

Figure 17.24 illustrates an example where the polypeptide products derived from a single type of pre-mRNA are clearly distinct from one another. Shown is a diagram of the initial bovine pre-mRNA transcript that is processed into one or two **preprotachykinin mRNAs (PPT mRNAs)**. The precursor mRNA molecule potentially includes the genetic information specifying two neuropeptides called **P** and **K**. These two peptides are members of the family of sensory neurotransmitters referred to as **tachykinins**, and are believed to play different physiological roles. While the P neuropeptide is largely restricted to tissues of the nervous system, the K neuropeptide is found more predominantly in the intestine and thyroid.

The RNA sequences for both neuropeptides are derived from the same gene. However, processing of the initial RNA transcript can occur in two different ways. In one case, exclusion of the K-exon during processing results in the α-PPT mRNA, which upon translation yields neuropeptide P but not K. On the other hand, processing that includes both the P and K exons yields β-PPT

mRNA, which upon translation results in the synthesis of both the P and K neuropeptides. Analysis of the relative levels of the two types of RNA has demonstrated striking differences between tissues. In nervous system tissues, α-PPT mRNA predominates by as much as a threefold factor, while β-PPT mRNA is the predominant type in the thyroid and intestine.

Given the existence of alternative splicing, how many different polypeptides can be derived from the same pre-mRNA? Work on alpha-tropomyosin has provided a partial answer to this question. In rats, the alpha-tropomyosin gene contains a total of 14 exons, 6 of which make up three pairs that are alternatively spliced. Only one member of each pair ends up in the finished mRNA, never both. Alternative splicing of this pre-mRNA results in 10 different forms of alpha-tropomyosin, many of which are expressed in a tissue-specific manner. Another gene, for the muscle form of troponin T, a protein that regulates calcium needed for contraction, produces 64 known forms of the protein from a single pre-mRNA by alternative splicing.

**(a)**

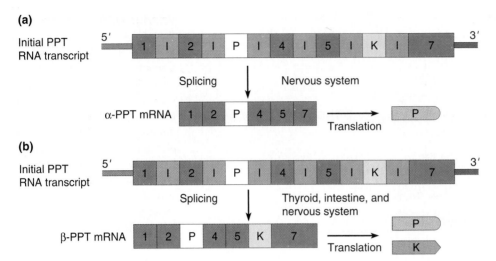

**(b)**

FIGURE 17.24    Diagrammatic representation of the alternative splicing of the initial RNA transcript of the preprotachykinin gene (PPT). Introns are unshaded and are labeled I. Exons are either numbered or designated by the letters P or K. Inclusion of the P and K exons leads to β-PPT mRNA, which upon translation yields both the P and K tachykinin neuropeptides. When the K exon is excluded, α-PPT mRNA is produced, and only the P neuropeptide is synthesized.

Alternative splicing not only has the potential to produce many different forms of a protein from a single gene, but can affect the outcome of developmental processes. Perhaps the most striking example of this involves sex determination in *Drosophila*, where splicing involving a single exon eventually determines whether the embryo will develop as a male or a female. This topic will be discussed in more detail in Chapter 21.

## Controlling mRNA Stability

After mRNA precursors are processed and transported, they enter the population of cytoplasmic mRNA molecules, from which messages are recruited for translation. Although regulation at the level of translation might be unexpected and seem inefficient, a growing body of evidence suggests that, in fact, the stability of mRNA engaged in translation may be a significant control point for gene regulation in eukaryotes. One of the best-studied examples of what is termed **autoregulation** is the synthesis of alpha and beta tubulins, the subunit components of eukaryotic microtubules. Treatment of a cell with the drug colchicine leads to a rapid disassembly of microtubules, and an increase in the concentration of the alpha and beta subunits. Under these conditions, the synthesis of alpha and beta tubulins drops dramatically. However, when cells are treated with vinblastine, a drug which also causes microtubule disassembly, the synthesis of tubulins is increased. The difference between the two drugs is that in addition to causing microtubule disassembly, vinblastine precipitates the subunits, lowering the concentration of free alpha and beta subunits. At low concentrations, synthesis of tubulins is stimulated, while at higher concentrations, synthesis is inhibited.

Work by Don Cleveland and his colleagues has described the conditions necessary for tubulin mRNA to be regulated at the level of translation (Figure 17.25). Fusion of tubulin gene fragments to a cloned thymidine kinase gene shows that the first 13 nucleotides at the 5' end of the messenger RNA following the transcription start site (encoding the amino terminal region of the protein) are necessary and sufficient to cause the hybrid mRNA to be regulated as efficiently as intact tubulin mRNA. These nucleotides encode the amino acids Met-Arg-Glu-Lys (abbreviated as MREI). Deletions, translocations, or point mutations in this 13-nucleotide segment abolish regulation. Examination of the cytoplasmic distribution of mRNAs demonstrate that only tubulin mRNAs in polysomes were depleted by drug treatment. Copies of tubulin mRNAs not bound to polysomes were protected from degradation. Third, translation must proceed to at least codon 41 of the mRNA in order for regulation to occur.

A model, shown in Figure 17.25, has been proposed to explain these observations. In this model, regulation occurs after the process of translation has begun. The first four amino acids (Met-Arg-Glu-Lys, abbreviated as MREI) of the tubulin gene product (in this case, beta-tubulin) constitute a recognition element to which the regulatory factors bind. The concentration of alpha and beta tubulin subunits in the cytoplasm may serve this regulatory function. This protein–protein interaction invokes the action of an RNase that may be a ribosomal component or a nonspecific cytoplasmic RNase. Action by the RNase degrades the tubulin mRNA in the act of translation, shutting down tubulin biosynthesis. As an alternative, binding of the regulatory element may cause the ribosome to stall in its translocation along the mRNA,

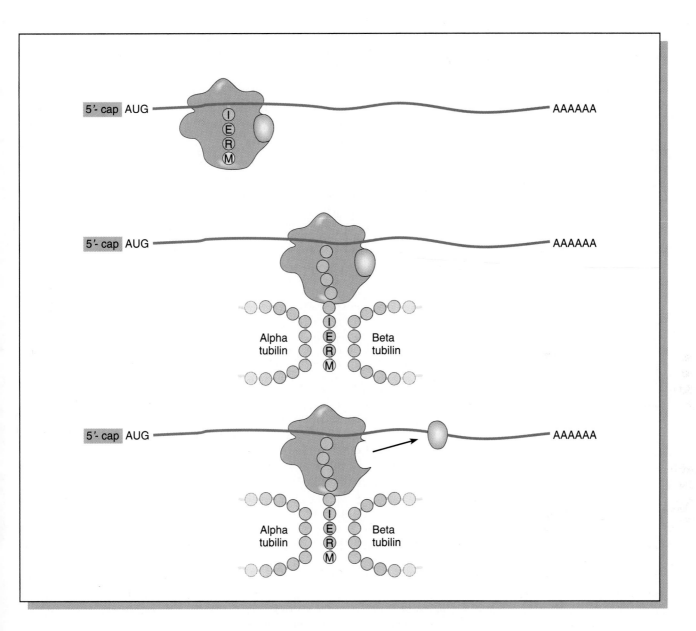

**FIGURE 17.25**    Model of posttranslational regulation of tubulin synthesis. Tubulin subunits (or another effector) bind to the amino terminal tubulin tetrapeptide as it emerges from the large ribosomal subunit. Binding activates an RNase (in this case, a ribosomal RNase) that breaks down the polysomal tubulin mRNA.

leaving it exposed to the action of RNase. Translation must proceed to at least codon 41, because the first 30 to 40 amino acids translated occupy a tunnel within the large ribosomal subunit, and only the translation of this number of amino acids will make the MREI recognition sequence accessible for binding.

RNA instability has been proposed as a regulatory mechanism for other genes, including histones, some transcription factors, lymphokines, and cytokines. This suggests the existence of a regulatory mechanism that acts on mRNAs in the act of translation, probably by protein–protein interaction between a regulatory element (most likely the gene product) and the nascent polypeptide chain, resulting in the activation of RNases.

## CHAPTER SUMMARY

1. Mechanisms controlling gene regulation in eukaryotes are governed by the properties of multicellularity, expanded genome size, and the spatial and temporal separation between transcription and translation.

2. The roles played by DNA organization and structure, especially chromatin alterations, transcriptional control, and signal transduction, as well as the expression of specific gene sets such as hormone receptors and oncogenes, are studied as models for gene regulation in higher organisms.

3. Transcription in eukaryotes is controlled by regulatory DNA sequences known as promoters and enhancers. Regulatory sequences, including the TATA box and the CCAAT box, are elements of promoters found near the transcriptional starting point. Enhancer elements, which appear to control the degree of transcription, can be located before, after, or within the gene expressed.

4. Transcription factors are proteins that bind to DNA recognition sequences within the promoters and enhancers and activate transcription through protein–protein interaction.

5. Gene expression in eukaryotes can be regulated by effector molecules originating outside the cell. Such signals are transduced by proteins that shuttle from the cytoplasm to the nucleus and regulate patterns of transcription.

6. Gene amplification, a variant of DNA organization, maximizes the accumulation of gene product per unit of time by increasing the number of gene copies necessary during times of cell growth and differentiation.

7. Several types of posttranscriptional control of gene expression are possible in eukaryotes. One such mechanism is alternative processing of a single class of pre-mRNAs to generate different mRNA species. Another mechanism provides a feedback system to mRNA regulation in response to cytoplasmic concentration of the gene product, controlling mRNA stability.

## KEY WORDS

autoregulation
CAAT box
DNA-binding domain
double minute (DM)
    chromosome
ecdysone
enhancer
enhanson
extended chromosome
    region (ECR)

GC box
gene amplification
helix-turn-helix motifs
homeobox
homeodomain
homeotic gene
homogeneously staining
    region (HSR)
hormone-responsive
    element (HRE)

intron
K neuropeptide
leucine zippers
P neuropeptide
preprotachykinin mRNA
    (PPT mRNA)
promoter
regulatory apparatus
RNA polymerase

structural motif
tachykinin
TATA box
transcriptional activation
transcriptional apparatus
transcription factor
$UAS_G$
zinc fingers

# INSIGHTS AND SOLUTIONS

1. Regulatory sites for eukaryotic genes are usually located within a few hundred nucleotides of the transcriptional starting site, but can be located up to several kilobases away. DNA sequence-specific binding assays have been used to detect and isolate protein factors present at low concentrations in nuclear extracts. In these experiments, DNA-binding sequences are bound to material on a column, and nuclear extracts are passed over the column. The idea is that if proteins that specifically bind to the sequence represented on the column are present, they will bind to the DNA, and can be recovered from the column after all other non-binding material has been washed away. Once a DNA-binding protein has been identified and isolated, the problem is to devise a general method for screening cloned libraries for genes encoding these and other DNA-binding factors. Determining the amino acid sequence of the protein and constructing synthetic oligo-nucleotide probes is time-consuming and useful only for one factor at a time. Knowing of the strong affinity for binding between the protein and its DNA recognition sequence, how would you screen for genes encoding binding factors?

   **SOLUTION:** Several general strategies have been developed, and one of the most promising has been devised by McKnight's laboratory. cDNA isolated from cells expressing the binding factor is cloned into the lambda vector, gt11. Plaques of this library containing proteins derived from expression of cDNA inserts is adsorbed onto nitrocellulose filters and probed with double-stranded, radioactive DNA corresponding to the binding site. If a fusion protein corresponding to the binding factor is present, it will bind DNA probe. After washing off unbound probe, the filter is subjected to autoradiography, and plaques corresponding to the DNA-binding signals can be identified. An added advantage of this strategy is recycling, by washing the bound DNA from the filters for reuse. This ingenious procedure is similar to the colony hybridization and plaque hybridization procedures described for library screening in Chapter 12, and provides a general method for isolating genes encoding DNA-binding factors.

## PROBLEMS AND DISCUSSION QUESTIONS

1. Why is gene regulation assumed to be more complex in a multicellular eukaryote than in a prokaryote? Why is the study of this phenomenon in eukaryotes more difficult?

2. List and define the levels of gene regulation discussed in this chapter.

3. What is the evidence from heterokaryons that suggests that cytoplasmic factors play an important role in regulating genetic activity or inactivity of the nucleus of a cell?

4. Distinguish between the regulatory elements referred to as promoters and enhancers.

5. Is the binding of a transcription factor to its DNA recognition sequence necessary and sufficient for an initiation of transcription at a regulated gene? What else plays a role in this process?

6.  In the autoregulation of tubulin synthesis, two models for the mechanism were proposed: first, that the tubulin subunits bind to the mRNA, or that the subunits interact with the nascent tubulin polypeptides. To distinguish between these two models, Cleveland and colleagues introduced mutations into the 13-base regulatory element of the beta-tubulin gene. Some of the mutations resulted in amino acid substitution, while others did not. In addition, they shifted the reading frame of the intact 13-base sequence. Below are results from a mutagenesis study of the mRNA. Which of the two models is supported by the results? What experiments would you do to confirm this?

|  |  | met | arg | glu | lys | Autoregulation |
|---|---|---|---|---|---|---|
| **Wild Type** | | **AUG** | **AGG** | **GAA** | **ATC** | **+** |
| | | | UGG | | | − |
| Second | | | GGG | | | − |
| codon | | | CGG | | | + |
| mutations | | | AGA | | | + |
| | | | AGC | | | − |
| Third | | | | GAC | | + |
| codon | | | | AAC | | − |
| mutations | | | | UAU | | − |
| | | | | UAC | | − |

7.  Assuming your answer to Problem 6 is correct, predict the results of the following experiment. Site-directed mutagenesis is used to insert four bases just in front of the AUG-initiating codon of the 13-base sequence, keeping the sequence intact, but shifting its position in the mRNA by four bases, and shifting the reading frame to an alternative sequence different from the normal *met-arg-glu-lys* sequence. Will tubulin subunits regulate translation of this mRNA? Why or why not?

## SELECTED READINGS

BAKER, B. S. 1989. Sex in flies: The stuff of life. *Nature* 340:521–24.

BEATO, M. 1989. Gene regulation by steroid hormones. *Cell* 56:335–44.

BRAWERMAN, G. 1989. mRNA decay: Finding the right targets. *Cell* 57:9–10.

BRENNAN, R. 1993. The winged-helix DNA-binding motif: Another helix-turn-helix takeoff. *Cell* 74:773–776.

BURCH, J. B. E., EVANS, M. I., FRIEDMAN, T. M., and O'MALLEY, P. J. 1988. Two functional estrogen response elements are located upstream of the major chicken vitellogen gene. *Mol. Cell. Biol.* 9:1123–31.

BUSCH, S. J., and SASSONE-CORSI, P. 1990. Dimers, leucine zippers and DNA-binding domains. *Trends in Genet.* 6:36–40.

CLEVELAND, D. W. 1988. Autoregulated instability of tubulin mRNAs: A novel eukaryotic regulatory mechanism. *Trends in Biochem. Sci.* 13:339–43.

CONAWAY, R., and CONAWAY, J. 1993. General initiation factors for RNA polymerase II. *Ann. Rev. Biochem.* 62:161–190.

DRAPKIN, R., MERINO, A., REINBERG, D. 1993. Regulation of RNA polymerase II transcription. *Curr. Opin. Cell Biol.* 5:469–476.

DYNAN, W. S. 1988. Modularity in promoters and enhancers. *Cell* 58:1–4.

JOHNSON, P. F., and McKNIGHT, S. L. 1989. Eukaryotic transcriptional regulatory proteins. *Ann. Rev. Biochem.* 58:799–839.

KAKIDANI, H., and PTASHNE, M. 1988. GAL4 activates gene expression in mammalian cells. *Cell* 52:161–67.

KARIN, M., CASTRILLO, J-L., and THEILL, L. E. 1990. Growth hormone regulation: A paradigm for cell type-specific gene activation. *Trends in Genet.* 6:92–96.

LEIDEN, J. 1993. Transcriptional regulation of T cell receptor development. *Ann. Rev. Immunol.* 11:539–570.

LEVINE, M., and MANLEY, J. L. 1989. Transcriptional repression of eukaryotic promoters. *Cell* 59:405–8.

LINDAHL, T. 1993. Instability and decay of the primary structure of DNA. *Nature* 362:709–715.

MANIATIS, T., GOODBOURN, S., and FISCHER, J. A. 1987. Regulation of inducible and tissue-specific expression. *Science* 236:1237–45.

MEEHAN, R., LEWIS, J., CROSS, S., NAN, X., JEPPESEN, P., and BIRD, A. 1992. Transcriptional repression by methylation of CpG. *J. Cell Sci. Suppl.* 16:9–14.

MILBURN, M. V., TONG, L., deVOS, A. M., BRUNGER, A., YAMAIZUMI, Z., NISHIMURA, S., and KIM, S. H. 1990. Molecular switch for signal transduction: Structural differences between active and inactive forms of proto-oncogenic *ras* proteins. *Science* 247:939–45.

MITCHELL, P. J., and TJIAN, R. 1989. Transcriptional regulation in mammalian cells by sequence-specific DNA binding proteins. *Science* 245:371–78.

MIZEJEWSKI, G. 1993. An apparent dimerization motif in the third domain of alpha-fetoprotein: Molecular mimicry of the steroid/thyroid nuclear receptor superfamily. *Bioessays* 15:427–432.

O'HALLORAN, T. 1993. Transition metals in control of gene expression. *Science* 261:715–725.

PIATORGORSKY, J., and WISTOW, G. J. 1989. Enzyme/crystallins: Gene sharing as an evolutionary strategy. *Cell* 57:197–99.

PIELER, T., and THEUNISSEN, O. 1993. TFIIIA: Nine fingers—Three hands? *Trends Biochem. Sci.* 18:226–230.

PTASHNE, M. 1988. How eukaryotic transcriptional activators work. *Nature* 335:683–89.

PTASHNE, M., and GANN, A. A. F. 1990. Activators and targets. *Nature* 346:329–31.

ROBINS, D., and SAMUELSON, L. 1992. Retrotransposons and the evolution of mammalian gene expression. *Genetica* 86:191–201.

SCHIMKE, R. T. 1989. The discovery of gene amplification in mammalian cells: To be in the right place at the right time. *Bioessays* 11:69–73.

SIBLEY, E., KASTELIC, T., KELLY, T. J., and LANE, M. D. 1989. Characterization of the mouse insulin receptor gene promoter. *Proc. Natl. Acad. Sci.* 86:9732–36.

STARK, G. R., DEBATISSE, M., GIULOTTO, E., and WAHL, G. 1989. Recent progress in understanding mechanisms of gene amplification. *Cell* 57:901–8.

STRINGER, K. F., INGLES, C. J., and GREENBLATT, J. 1990. Direct and selective binding of an acidic transcriptional activation domain to the TATA-box factor TFIID. *Nature* 345:783–86.

STRUHL, K. 1993. Yeast transcription factors. *Curr. Opin. Cell Biol.* 5:513–520.

TATE, P., and BIRD, A. 1993. Effects of DNA methylation on DNA-binding proteins and gene expression. *Curr. Opin. Genet. Dev.* 3:226–231.

YEN, T. J., GAY, D. A., PACHTER, J. S., and CLEVELAND, D. W. 1988. Autoregulated changes in stability of polyribosome-bound beta-tubulin mRNAs are specified by the first 13 translated nucleotides. *Mol. Cell. Biol.* 8:1224–35.

# 18

# THE GENETICS
# OF CANCER

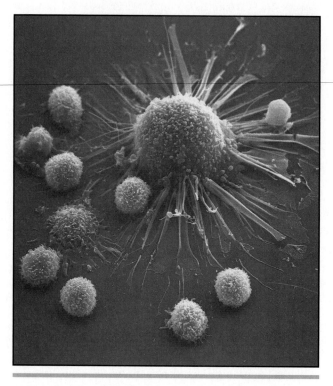

A large cancer cell, with extended cytoplasmic processes,
surrounded by an array of immunological killer cells.

*Cancer is now recognized as a genetic disorder at the cellular level that involves the mutation of a small number of genes. Many of these genes normally act to suppress or stimulate progression through the cell cycle, and loss or inactivation of these genes causes uncontrolled cell division and tumor formation. Environmental factors including viruses also play a role in the genetic alterations that are necessary for the transition of normal cells into cancerous cells.*

Although often viewed as a single disease, cancer is actually a group of complex diseases that affects a wide range of cells and tissues. A genetic link to cancer was first proposed early in this century, and this idea has served as one of the foundations of cancer research. Mutations that result in an alteration of the genome or an altered expression of gene products are now regarded as a common feature of all cancers. In some cases, such mutations are part of the germ line and are inherited. More often, mutations arise in somatic cells and are not passed on to future generations through the germ cells. Sometimes, the inherited mutation must be accompanied by a somatic mutation at the homologous locus, creating homozygosity. Whichever the case, *cancer is now considered a genetic disorder at the cellular level.*

Genomic alterations associated with cancer can involve changes as small as a single nucleotide substitution or large-scale events including chromosome rearrangement, chromosome gain or loss, or even the integration of viral genomes into chromosomal sites. Large-scale genomic alterations are a common feature of cancer; indeed the majority of human tumors are characterized by visible chromosomal changes. Some of these chromosomal changes, particularly in leukemia, are so specific they can be used to diagnose and classify the disorder and to make accurate predictions about the severity and course of the disease.

Familial forms of cancer have been known for over two hundred years. In many of these cases, no well-defined pattern of inheritance can be established. In a small number of cases, however, a Mendelian pattern of dominant or recessive inheritance can be established, indicating the hereditary nature of the cancer. Perhaps the best-known example is **familial adenomatous polyposis (FAP)**, a dominant condition that, without treatment, develops into cancer of the colon.

If mutation is the underlying cause of cancer, there will always be a baseline rate of cancer, just as there is a background rate of spontaneous mutation. In addition, environmental agents that promote mutation should also play a role in the development of cancer. There is, in fact, clear evidence that environmental factors such as ionizing radiation, chemicals, and viruses are cancer-causing agents, and almost all of these act by generating mutations. Given that mutations play a role in cancer, it is reasonable to ask a series of questions about how mutations initiate the changes that convert normal cells into malignant tumors, which mutant genes are most likely to result in cancer, and how many mutations are required to cause cancer.

One approach to answering these questions is to consider what properties cancer cells possess that distinguish them from normal cells and to ask what genes may control these properties. Cancer cells have two properties in common: (1) uncontrolled growth and (2) the ability to spread or metastasize from their original site to other locations in the body. Cell proliferation is the result of cells traversing the cell cycle and is regulated in a cell-specific fashion. In cancer cells, control over the cell cycle is lost, and cells rapidly proliferate. Investigations into the genetic control of the cell cycle are beginning to provide clues concerning the origins of cancer.

The metastasis of cancer cells is apparently controlled by gene products that become localized on the cell surface, and the genetics of metastasis is related to an understanding of how cells interact with the extracellular matrix and with other cells through cell surface molecules. Though this field is less well-developed than the study of the cell cycle, it is beginning to provide some insights into the secondary events in tumor progression.

In this chapter, we will consider the relationship between genes and cancer, with emphasis on the relationship between the cell cycle and genetic disorders associated with cancer. We will also examine the relationship between mutation and cancer, the identification of genes that in mutant form initiate the transformation of cells, and an estimate of the number of mutations involved in tumor formation. We will also discuss the relationship between chromosomal changes and cancer, and the role of environmental agents in the genesis of cancer.

# THE CELL CYCLE AND CANCER

As described in Chapter 2, the cell cycle represents the sequence of events that occurs between mitotic divisions in a eukaryotic cell. In years past, work on control of the cell cycle has been conducted mainly by two groups: (1) geneticists working with yeasts, especially *Saccharomyces cerevisiae* and *Schizosaccharomyces pombe*; and (2) developmental biologists, studying the newly fertilized eggs of organisms such as frogs, sea urchins, and newts. In the last few years, as these groups have succeeded in identifying and characterizing genes involved in the cell cycle, it has become apparent that their work is converging and overlapping with important areas of cancer biology, particularly studies on growth factors and the genes that suppress or promote tumor formation. This consolidation of fields has had a synergistic effect, resulting in new insights into the processes that control cell division and how regulation of the cell cycle is coupled to the transcription of selected genes. Because of these recent developments, it is necessary to spend some time describing what is now known about the events in the cell cycle and the genes that regulate progression through the cycle before undertaking a discussion of the genetics of cancer.

## The Cell Cycle

Stripped to its essentials, the cell cycle progresses from a period of chromosomal DNA replication (S phase) to the segregation of chromosomes into two nuclei during mitosis (M phase). Interspersed between these are two gaps, called $G_1$ and $G_2$. Together, $G_1$, S, and $G_2$ make up the interphase portion of the cell cycle (Figure 18.1). The $G_1$ stage begins after mitosis; synthesis of many cytoplasmic elements including ribosomes, enzymes, and membrane-derived organelles occurs at this time. In the S phase, DNA replication takes place, producing a duplicate copy of each chromosome. Then a second period of growth and synthesis known as $G_2$ takes place as a prelude to mitosis.

Because mitosis occurs rapidly, usually lasting less than an hour, cells spend most of the cell cycle in interphase. However, the duration of the cell cycle (the time

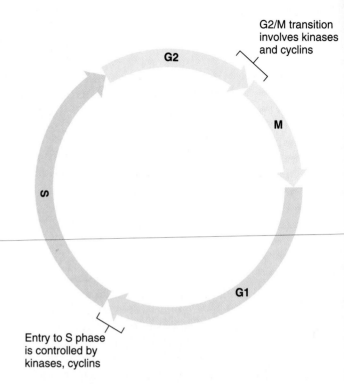

**FIGURE 18.1**    The cell cycle is controlled at two and possibly more checkpoints, one at the $G_2$/M transition, and another in late $G_1$ phase before entry into S phase. These checkpoints involve interaction between transitory proteins, called cyclins, and kinases that add phosphate groups to proteins. Phosphorylation of target proteins triggers a cascade of events allowing progress through the cell cycle.

between mitotic divisions) can vary widely in the life cycle of an organism and between different cell types in the same organism. For example, animal cells exhibit *in vivo* cycles ranging from a few minutes to several months. Most of this variation can be traced to the time spent in $G_1$, while the time necessary to complete S and $G_2$ remains relatively constant in most cell types.

While some cells such as meristematic cells in plants and dermal cells in human skin cycle continuously, other cell types, including many nerve cells, withdraw from the $G_1$ phase, and permanently enter a nondividing state known as $G_0$. Still other cell types such as white blood cells can be recruited to return from $G_0$ and re-enter the cell cycle.

Taken together, these observations suggest that the cell cycle is tightly regulated and is dependent on the life history and differentiated state of a given cell. A great deal of information about the regulation of the cell cycle has become available in the last few years. A summary of what is currently known about the genetic regulation of the cell cycle will serve as a prelude to an overview of the genetics of cancer.

## Checkpoints and Control of the Cell Cycle

As work on the cell cycle in several organisms has converged, a universal model of the cell cycle and its regulation is beginning to emerge. While the details of all molecular events or even the exact number and sequence of steps are not yet known, all eukaryotic cells probably employ a common series of biochemical pathways to regulate events in the cell cycle. If the universal aspect of the current model of the cell cycle is upheld, results gathered from the study of yeasts or clam eggs can be used to understand and predict events in normal human cells and in mutated cells that have become cancerous. As a result, discoveries in cell cycle research will probably continue to be fast-moving and spectacular.

The emerging picture indicates that the cell cycle is regulated at the $G_2/M$ transition and at a point within $G_1$. The events at the $G_2/M$ transition have been well documented, but the regulatory point within late $G_1$, several hours before the initiation of S phase, is not yet well characterized.

At each of these points, a decision is made to proceed or halt progression through the cell cycle. This decision is controlled through the interaction of two classes of proteins. One is a group of enzymes called **protein kinases** that selectively phosphorylate target proteins. Although there are a large number of different protein kinases in the cell, only a few are involved in regulation of the cell cycle. The second group of proteins involved in controlling progression through the cell cycle is called **cyclins**. These proteins, first identified in the embryos of developing invertebrates, are synthesized and degraded in a synchronous pattern throughout the cell cycle (Figure 18.2). Several cyclins, including C, D1, D2, and E, accumulate during the $G_1$ phase, with only D1 persisting after the S phase. Cyclin A accumulates during late S and persists until the $G_2/M$ transition; cyclin B accumulates during $G_2$ and persists until the end of M. Coupling of a kinase and a cyclin produces a regulatory molecule that controls the movement of the cell from $G_2$ into M and from $G_1$ into S.

The onset of M in most eukaryotic cells is controlled by a kinase called *CDK1* (cyclin-dependent kinase), which was biochemically characterized in maturing amphibian eggs and genetically identified in yeast as the product of the *cdc2* gene. This protein has a highly conserved sequence in all eukaryotes examined; its function is necessary for entry into M phase.

Several events mark the entry from $G_2$ into mitosis (M), including condensation of the chromatin to form chromosomes, breakdown of the nuclear membrane, and reorganization of the cytoskeleton. Major events in this transition are regulated by the formation of an active CDK1/cyclin B complex. The CDK1 component cata-

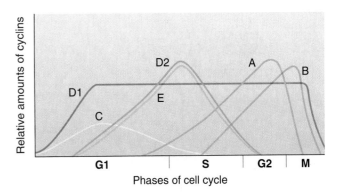

**FIGURE 18.2**    Relative expression times and amounts of cyclins during the cell cycle. D1 accumulates early in $G_1$ and is expressed at a constant level through most of the cycle. Cyclin C accumulates in $G_1$, reaches a peak, and declines by mid S phase. Cyclins D2 and E begin accumulating in the last half of $G_1$, reach a peak just after the beginning of S, and then decline by early $G_2$. Cyclin A appears in late $G_1$, accumulates through S, reaches a peak in $G_2$, and is degraded rapidly as M phase begins. Cyclin B appears in mid S phase, peaks at the $G_2/M$ transition, and is rapidly degraded.

lyzes phosphorylation, which brings about nuclear membrane breakdown and rearrangement of the cytoskeleton. It also phosphorylates histone H1, which may play a role in chromatin condensation (Figure 18.3). The function of cyclin B in this complex is not clear, but it may control the cellular localization or target molecule specificity. Although several lines of experiments suggest that cyclin A is also involved in the progression from $G_2$ to M, its functions are not clearly understood.

While the action of the B cyclin and CDK1 are responsible for passage into M, it appears that several cyclins, including D and E, can move cells from $G_1$ into S. Experiments indicate that the CDK1 kinase is also active in phosphorylation in $G_1$, but that instead of combining with cyclin B (which is not present), CDK1 combines with $G_1$ cyclins, directing the phosphorylation of $G_1$ stage-specific protein substrates.

Altogether, almost a dozen different cyclins have been identified, and a growing number of cyclin-dependent kinases are being described, indicating that multiple control points in the cell cycle exist, or that these kinases and cyclins have multiple functions.

## Cell Cycle Regulation and Cancer

Obviously, mutations that disrupt any step in cell cycle regulation are candidates for the study of cancer-causing genes. For example, mutations in genes that encode the

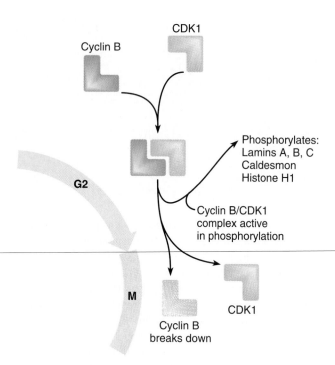

**FIGURE 18.3**   Transition to M from $G_2$ is controlled by CDK1 and cyclin B. These molecules interact to form a complex that adds phosphate groups to cellular components that break down the nuclear membrane (lamins A, B, and C), reorganize the cytoskeleton (caldesmon), and initiate chromosome condensation (histone H1). Cyclin B may specify cellular localization or target molecules. Other cyclins, especially cyclin A, is thought to be involved at this stage, but its functions are not known.

kinases and cyclins may be important in generating malignant transformation in cells. Evidence is accumulating that the $G_1$ checkpoint is aberrant in many forms of cancer, and mutant $G_1$ cyclins and kinases are thought to be the best candidates for cancer-promoting genes. Recently, a form of cyclin D called D1 has been shown to be identical to a gene product that is overexpressed in certain forms of leukemia.

In the next section, we will summarize what is known about the genetics of selected cancers, and wherever possible, relate this information to what has been discovered about the cell cycle and its regulation.

# GENES AND CANCER

Genetic studies of several different cancers have identified a small number of genes that must be mutated in order to bring about the development of cancer or maintain the growth of malignant cells. It is clear that the two

main properties of cancer, uncontrolled cell division and the ability to spread or metastasize are the result of genetic alterations. As mentioned previously, these alterations can involve large-scale genomic instability, resulting in chromosome loss, rearrangement or the insertion of foreign (often viral) DNA sequences into loci on human chromosomes. Smaller-scale alterations, such as changes in a nucleotide sequence or more subtle modifications that alter only the amount of a gene product that is present or the time over which the gene product is active may also be involved. Two interrelated questions arise from considering cancer as a genetic disorder at the cellular level: (1) are there mutant alleles that predispose an organism or specific cell types to cancer, and (2) if so, how many mutational events are necessary to cause cancer?

## Genes That Predispose Cells to Cancer

Single genes do indeed predispose cells to becoming malignant, and it is possible to identify families in which certain forms of cancer are inherited. Many studies have documented families with high frequencies of certain forms of cancer, such as breast, colon, or kidney cancers. In most cases, however, it is difficult to identify a clear, simple pattern of inheritance. One example of such a predisposition is the inheritance of **retinoblastoma (RB)**, a cancer of the retinal cells of the eye. Retinoblastoma occurs with a frequency that ranges from 1 in 14,000 to 1 in 20,000, and most often appears between the ages of 1 and 3 years. Two forms of retinoblastoma are known. One is a familial form (about 40% of all cases) inherited as an autosomal dominant trait. Because the trait is dominant, family members have a 50 percent chance of inheriting the mutant allele. Those that inherit the mutant RB allele are predisposed to develop eye tumors, and 90 percent of these individuals will develop retinal tumors, usually in both eyes. In addition, those family members that inherit the mutant allele are predisposed to developing other forms of cancer, such as osteosarcoma, a bone cancer, even if they do not develop retinoblastoma.

A second form of retinoblastoma is also known, amounting to 60 percent of all cases. In this form, retinoblastoma is not familial, and these cases apparently develop spontaneously. In such cases, retinoblastoma usually develops only in one eye, and onset occurs at a much later age than in the familial form.

Other types of dominantly inherited familial cancer are also known (Table 18.1). These include **Wilms tumor (WT)**, a cancer of the kidney inherited as an autosomal dominant condition, and **Li-Fraumeni syndrome**, a rare autosomal dominant condition that predisposes to a number of different cancers, including breast cancer.

**Table 18.1**  DOMINANTLY INHERITED
PREDISPOSITIONS TO TUMORS

| Tumor Predisposition Gene | Chromosome |
|---|---|
| Early onset familial breast cancer | 17q |
| Familial adenomatous polyposis | 5q |
| Familial melanoma | 9p |
| Gorlin syndrome | 9q |
| Hereditary nonpolyposis colon cancer | 2p |
| Li-Fraumeni syndrome | 17p |
| Multiple endocrine neoplasia, type 1 | 11q |
| Multiple endocrine neoplasia, type 2 | 22q |
| Neurofibromatous, type 1 | 17q |
| Neurofibromatous, type 2 | 22q |
| Retinoblastoma | 13q |
| von Hippel–Lindau syndrome | 3p |
| Wilms tumor | 11p |

## How Many Mutations Are Needed?

The study of genes that predispose to cancer has provided insight into the number and sequence of mutational events that are necessary to cause different forms of cancer. By studying the two different forms of retinoblastoma, Alfred Knudson and his colleagues developed a model that suggests that retinoblastoma is caused by the presence of two mutated copies of the RB gene in the same retinal cell. In the familial form of retinoblastoma, one mutant RB allele is inherited, and therefore, is carried by all cells of the body, including the retinal cells (Figure 18.4). If a second, spontaneous mutational event in the retinoblastoma gene occurs in the remaining normal RB allele in any retinal cell, formation of retinal tumors will result. Thus, those carrying an inherited single mutation of the RB gene are predisposed to develop retinoblastoma, since only one additional mutational event is required to cause retinoblastoma. This does not happen in all cases, however, since 10 percent of those that

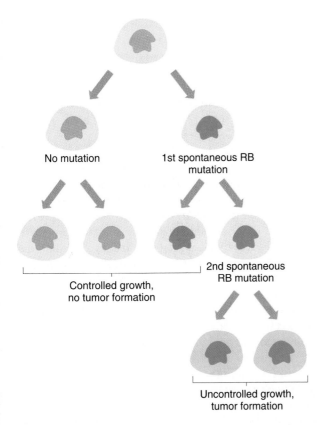

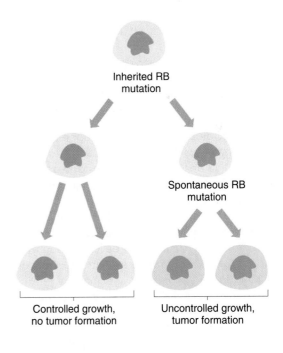

**FIGURE 18.4**    In spontaneous cases of retinoblastoma (left), two mutations in the retinoblastoma gene are acquired in a single cell, causing uncontrolled cell growth and division, resulting in tumor formation. In familial cases of retinoblastoma (at right), one mutation is inherited and present in all cells. A second mutation at the retinoblastoma locus in any retinal cell will result in uncontrolled cell growth and tumor formation.

**Table 18.2**   NUMBER OF MUTATIONS ASSOCIATED WITH SOME CANCERS

| Cancer | Chromosome Sites | Minimum Number of Mutations Required |
|---|---|---|
| Retinoblastoma | 13q | 2 |
| Wilms tumor | 11p | 2 |
| Colon cancer | 5p, 12p, 17p, 18q | 4–5 |
| Small-cell lung cancer | 3p, 11p, 13q, 17p | 10–15 |

inherit a mutant RB gene do not develop retinoblastoma. In the sporadic cases of retinoblastoma, two independently occurring somatic mutations of the RB gene must occur in the same retinal cell for a tumor to develop. As might be expected from this model, these random events are rare and occur at a later age than the familial form of retinoblastoma.

Similar studies on the distribution and inheritance of predispositions to other forms of cancer have led investigators to conclude that the number of mutations necessary for the development of cancer ranges from 2 to perhaps as many as 20 (Table 18.2).

## TUMOR SUPPRESSOR GENES

In general, the cell cycle can be regulated in two ways: (1) by genes that normally function to suppress cell division, and (2) by genes that normally function to promote cell division. The first class, called **tumor suppressor genes**, normally function to inactivate or repress passage through the cell cycle and the resulting cell division. These genes and/or their gene products must be absent or inactive for cell division to take place. If these genes become permanently inactivated or lost through mutation, control over cell division is lost, and the cell begins to proliferate in an uncontrollable fashion.

The second class of genes, called **proto-oncogenes**, normally function to promote cell division, and these genes and/or their gene products must be inactivated in order to halt cell division. If these genes become permanently switched on or overexpressed through mutation, then cell division occurs in an uncontrolled fashion, leading to tumor formation. The mutant forms of proto-oncogenes are known as **oncogenes**.

We shall consider some examples of how mutations in tumor suppressor genes can lead to a loss of control over cell division and the development of cancer. Following that discussion, we will consider proto-oncogenes and oncogenes, how the normal alleles act to regulate cell division, and how the mutant oncogenes work to promote tumor formation.

## Retinoblastoma (RB)

As described previously, retinoblastoma is a tumor of the retinal layer of the eye. The RB gene is located on chromosome 13. It encodes a protein (pRb) that is 928 amino acids long and confined to the nucleus. pRb is present in all cell and tissue types examined to date and is found in both resting ($G_0$) cells and cells active in the cell cycle. Moreover, pRb is present at all stages of the cell cycle. Because the gene is functionally active at all times, regulation occurs through reversible modifications to pRb by alternately adding or removing phosphate groups from the protein. pRb becomes phosphorylated in the S and the $G_2$/M stages of the cell cycle, but has fewer or no phosphate groups in the $G_0$ and $G_1$ stages of the cycle.

The observation that pRb alternately has phosphate groups added and removed, and that this pattern of modification occurs in synchrony with phase of the cell cycle suggested that pRb may be part of a control point in $G_1$. According to this model, when pRb is less phosphorylated (hypophosphorylated) in $G_1$, the protein is functionally active and suppresses passage through the cell cycle. The inactivation of pRb and the initiation of tumor formation in retinoblastoma provides indirect evidence for the role of this protein in suppressing cell proliferation. Conversely, just after entering S phase, pRb is modified by the addition of more phosphate groups (hyperphosphorylation) and it becomes inactive, allowing cells to pass through S and the subsequent events leading to cell division. Direct evidence for the role of pRb in controlling the cell cycle has come from several types of experiments.

Cultured osteosarcoma cells carry mutant RB genes and no functional pRb. When osteosarcoma cells are injected into a cancer-prone strain of mice, they produce tumors. If a normal RB gene is transferred to the cancer cells, pRb is produced and no tumors occur when these genetically modified cells are injected into mice. In a separate two-step experiment, a normal RB gene was transferred into osteosarcoma cells grown in culture, resulting in the production of pRb and cessation of cell division. When D or E cyclins were added to these pRb-blocked cells, cell division resumed. Analysis of pRb after the addition of cyclins indicates that the exogenous cyclins cause phosphorylation of pRb, presumably by activating CDK1 or another kinase. These results demonstrate that the presence of pRb will stop cell division, that pRb is a target of a $G_1$ cyclin/CDK complex, and that phosphorylation of pRb allows passage through the cell cycle and subsequent division, confirming the role of pRb in $G_1$ control.

At the molecular level, pRb, which is found only in the nucleus, interacts with a protein called E2F, a well-characterized transcription factor (Figure 18.5). In $G_1$, hypophosphorylated pRb binds to and inactivates E2F (recall from Chapter 17 that transcription factors bind to specific regions of DNA and regulate transcription). As the cell moves from $G_1$ to S, the CDK/cyclin complexes act to phosphorylate pRb, causing pRb to release E2F, in turn resulting in the transcription of genes required for progression through the cell cycle. This would make

pRb an important link between the cell cycle and gene transcription. In cases where pRb is missing or inactivated, as in retinoblastoma, E2F is not regulated, allowing permanent expression of genes required for progression through S and cell division, resulting in uncontrolled cell growth and tumor formation.

More recent results suggest, however, that while this explanation for the mechanism of pRb action is consistent with its involvement in tumor formation, it may not explain all the functions of pRb. Mouse embryos genetically engineered to be homozygous for mutant RB genes develop normally until day 12 or 13 of development. Since these embryos develop through many rounds of cell division, control of the cell cycle checkpoint in $G_1$ may be shared by pRb and other proteins, and/or pRb may play a role in deciding when cells leave the cell cycle and differentiate.

## Wilms Tumor

Wilms tumor (WT) or nephroblastoma is a cancer of the kidney found primarily in children. It occurs with a frequency of about 1 in 10,000 births, and like retinoblas-

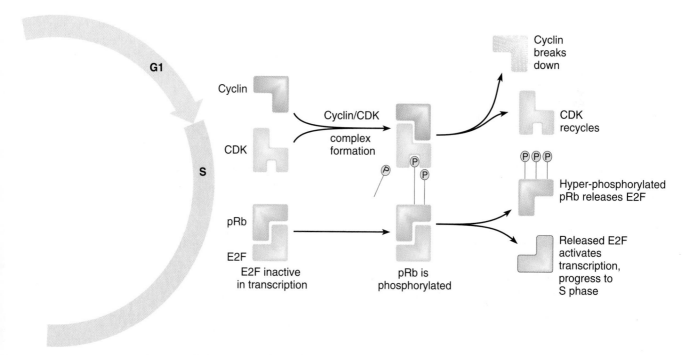

**FIGURE 18.5**    In the nucleus during $G_1$, pRb, the product of the retinoblastoma gene, interacts with and inactivates transcription factor E2F. As the cell moves from $G_1$ to S, a CDK/cyclin complex forms and adds phosphate groups to pRb. As pRb becomes hyperphosphorylated, E2F is released and becomes transcriptionally active, allowing the cell to pass through S phase. Phosphorylation of pRb is transitory; as cyclin is degraded, phosphorylation declines. The identity of the $G_1$ cyclin is uncertain.

toma, is found in two forms, a noninherited spontaneous form and a familial form conferring a predisposition to WT. This form is inherited as an autosomal dominant trait. According to the model originally developed by Knudson and colleagues, familial cases inherit one mutant allele through the germ line and develop a second mutation in the remaining normal allele in a somatic cell. Sporadic cases, on the other hand, require two independent mutations of the WT gene within the same cell and more often develop tumors that involve only one kidney. Because mutation of the gene and/or the loss of a functional gene product are associated with the development of tumors, the normal gene is regarded as a tumor suppressor.

The WT gene has been mapped to the short arm of chromosome 11. The gene product encoded by the WT gene contains four contiguous zinc finger domains. Recall from Chapter 17 that these domains are regions with interspersed cysteine and histidine residues that covalently bind zinc atoms, folding the amino acid chains into loops known as zinc fingers. Such motifs are characteristic of DNA binding proteins that regulate transcription. In addition, the amino acid sequences of the WT protein upstream from the zinc fingers are similar to those of other proteins that are transcription factors.

Unlike the RB gene, the WT gene has a very restricted pattern of expression. The WT gene is active only in a single cell type: the mesenchymal cells of the fetal kidney, and only during the brief time when the nephron (the basic filtration unit of the kidney) is being formed.

The only other cell type to express the WT gene is the tumorous nephroblastoma cell. Expression of this gene is barely detectable in cells of the adult kidney and is absent in all other adult cell and tissue types tested.

The structure of the WT gene product, its pattern and time of expression and its mutant phenotype all suggest that the WT gene encodes a nuclear protein that functions to turn off genes that sustain cell proliferation. Alternatively, the gene product may switch on genes that begin the process of differentiation of the mesenchymal cells into kidney structures. In either case, it appears that in the mutation causing Wilms tumor, the gene is switched on at the appropriate time in the appropriate cells, but the altered gene product is unable to regulate its target genes, resulting in continued proliferation of the mesenchymal cells, aberrant differentiation, and tumor formation (Figure 18.6). To completely resolve the chain of events in Wilms tumor, it will be necessary to identify the loci regulated by the WT gene and determine the nature of their gene products and modes of action.

Although both the RB and WT genes encode gene products restricted to the nucleus normally involved in the suppression of tumor formation, there are conspicuous differences in their properties and modes of action. The RB protein does not bind to DNA, but instead, interacts with other nuclear proteins that are transcription factors. The WT gene product has all the characteristics of a DNA-binding protein that regulates cell division by acting as a transcription factor. The RB protein is expressed in all dividing cells, while expression of WT is restricted

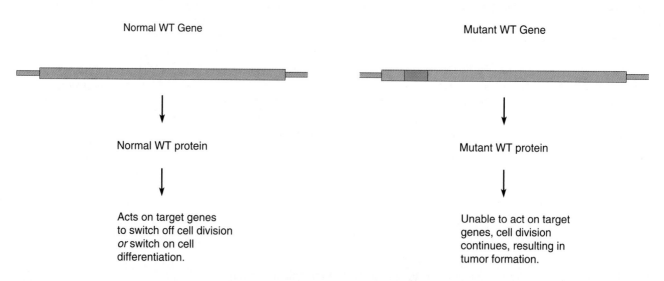

Normal WT Gene

Mutant WT Gene

Normal WT protein

Mutant WT protein

Acts on target genes
to switch off cell division
*or* switch on cell
differentiation.

Unable to act on target
genes, cell division
continues, resulting in
tumor formation.

**FIGURE 18.6**    The proposed role of the WT1 protein in regulating cell division. In normal cells (at left), the WT1 protein is produced and acts on a set of target genes to switch off cell division or switch on cell differentiation. Any mutation that causes the loss or inactivation of the WT1 protein (at right) will result in a failure to regulate the target genes, allowing cell division and tumor formation to occur.

to a single cell type during a restricted period of prenatal development. Thus, RB appears to be a general regulator of cell division, while WT is probably a cell- or tissue-specific regulator of gene activity during fetal or neonatal development. These differences emphasize that while tumor suppressor genes act to regulate cell division and/or differentiation, it is not yet clear that a general model for the action of tumor suppressor genes can be developed. For now, it is still necessary to characterize them individually and determine their mode of action.

## ONCOGENES

**Oncogenes** are genes that induce or maintain uncontrolled cellular proliferation associated with cancer. The existence of specific genes associated with the transformation of normal cells into cancerous cells was inferred by Peyton Rous in 1910–1911. Using cells from a tumor of chickens known as a **sarcoma**, Rous demonstrated that injection of cell-free extracts from these tumors could induce sarcoma formation in healthy chickens. He postulated the existence of a "filterable agent" that was responsible for transmitting the disease.

Decades later, his filterable agent was shown by other investigators to be a virus, now known as the Rous sarcoma virus (RSV). Somewhat belatedly, Rous received the Nobel prize in 1966 for his pioneering work on the relationship between viruses and cancer.

The Rous sarcoma virus belongs to a group known as **retroviruses** because the RNA genome must be transcribed by an enzyme, **reverse transcriptase**, into single-stranded DNA. The single-stranded DNA is converted to double-stranded DNA and integrated into the genome of the infected cell, where it can be transcribed to produce an infectious RNA. In the case of RSV, the tumor-forming ability results from a single gene, called the *src* gene, present in the viral genome. This single gene, responsible for the induction of tumor formation in chicken cells, is designated as an oncogene, and retroviruses that carry oncogenes are known as **acute trans-**

**forming viruses**. Other retroviruses can cause tumor formation but do not carry oncogenes. These retroviruses induce the activity of cellular genes that bring about tumor formation. They are designated as **nonacute** or **nondefective viruses**.

Oncogenes carried by acute transforming viruses are acquired from the host's genome during infection, when a portion of the viral genome is exchanged for a cellular gene (Figure 18.7), as in bacterial transduction. The oncogenes (*onc*) carried by retroviruses are called *v-onc*, and the cellular version of the gene is called a *c-onc* gene or **proto-oncogene**. Retroviruses that carry a *v-onc* gene are able to infect and transform a specific type of host cell into a tumor cell. For RSV, the oncogene captured from the chicken genome is called *v-src*, and it confers the ability to transform chicken cells into tumorous growths known as sarcomas. The cellular version of the same gene, found in the chicken genome, is called *c-src*. More than 20 oncogenes have been identified by their presence in retroviral genomes, and over 50 oncogenes have been identified. Some of these are listed in Table 18.3

Two questions about oncogenes come to mind: (1) are *v-onc* and *c-onc* versions of a gene different, and (2) what mechanisms allow oncogenes to bring about cellular transformation and tumor formation? Comparison of *v-onc* DNA sequences with the corresponding *c-onc* sequence reveals several examples (such as *v-ras, v-mos*) in which there are only minimal differences, probably generated by point mutations. In other oncogenes, segments of the *c-onc* sequence have been replaced or lost. In *v-src*, 19 amino acids of the *c-onc* sequence have been replaced with 12 different amino acids. In *v-erb*, the *v-onc* version retains only the C-terminus half of the gene (see Table 18.4). Mutations, therefore, play a role in differentiating some *v-onc* sequences from their *c-onc* counterparts. However, not all oncogenes are mutant versions of cellular genes carried by retroviruses. We must, therefore, consider both qualitative and quantitative mechanisms in oncogenetic events.

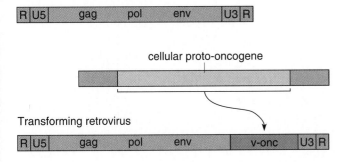

**FIGURE 18.7**    A transforming retrovirus has acquired a copy of a gene from the host genome, converting it from a *c-onc* or proto-oncogene into an oncogene that confers on the virus the ability to transform a specific type of host cell into a cancerous cell.

**Table 18.3**    ONCOGENES

| c-onc | Origin | Species | Human Chromosome Location |
|-------|--------|---------|---------------------------|
| A. Cellular Oncogenes with Viral Equivalents | | | |
| src | Rous sarcoma virus | Chicken | 20 |
| fos | FBJ osteosarcoma virus | Mouse | 2 |
| sis | Simian sarcoma virus | Monkey | 22 |
| fes | ST feline virus | Cat | 15 |
| abf | Abelson murine leukemia virus | Mouse | 9 |
| erb-B | Avian erythrobeastosis virus | Chicken | 7 |
| Ha-ras-1 | Harvey murine sarcoma virus | Mouse | 11 |
| myc | Avian MC29 myelocytomatosis virus | Chicken | 8 |
| B. Oncogenes without Viral Equivalents | | | |
| N-ras | Neuroblastoma, leukemia | Human | 1 |
| N-myc | Neuroblastoma | Human | 12 |
| neu | Neuroglioblastoma | Human, rat | 17 |
| man | Mammary carcinoma | Human, mouse | 2 |

**Table 18.4**    STRUCTURAL ALTERATIONS IN ONCOGENES

| Gene | Number of Codons | Number of Changed Amino Acids | Region Missing in v-onc Protein |
|------|------------------|-------------------------------|---------------------------------|
| H-ras | 189 | 3 | None |
| K-ras | 189 | 7 | None |
| mos | 369 | 11 | None |
| myc | 417 | 2 | None |
| erbA | 408 | 22 | Deletion at N-terminus |
| src | 533 | 16 | Deletion at C-terminus |
| erbB | 1210 | 49 | Deletion at both N-terminus and C-terminus |

**Table 18.5**    CONVERSION OF PROTO-ONCOGENES TO ONCOGENES

| Mechanism | Example |
|-----------|---------|
| Point mutation | ras |
| Translocation | abl |
| Overexpression of gene product | |
|    New promoter by viral insertion | mos, myb |
|    New enhancer by viral insertion | myc |
|    Amplification of proto-oncogene | myc |

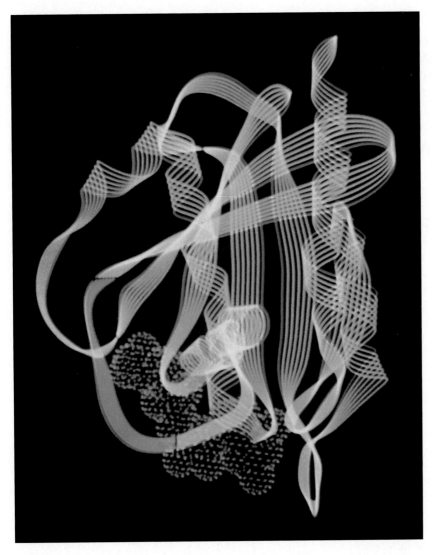

**FIGURE 18.8**    A three-dimensional computer-generated image of *ras* proteins in two different conformations. Normal *ras* proteins act as molecular switches controlling cell growth and differentiation. The switch is "on" when GTP binds to the protein, and "off" when the GTP is hydrolyzed to GDP. Switching the protein between states alters the conformation of the protein in the two regions shown in blue and yellow. Oncogenic mutations of *ras* are stuck in the "on" state, continuously signaling for cell growth.

## Oncogenes and Gene Expression

At least three general mechanisms can be invoked to explain the conversion of proto-oncogenes into oncogenes. These mechanisms include **point mutations**, **translocations**, and **overexpression** (see Table 18.5). While some of these are mediated by viruses, others operate in the absence of retroviruses, and are generated solely by intracellular events.

The *ras* gene family, encoding a protein of 189 amino acids involved with the transduction of external signals across the cell membrane, illustrates how point mutations as small as single base changes underlie the differences between the normal *c-onc* sequence and the oncogenic version of the gene. The normal *ras* protein functions as a molecular switch, alternating between an on and off state (Figure 18.8). In some mutants, the *ras* protein is stuck in the "on" position, stimulating cell

growth. In some human tumors, this event is not mediated by a retrovirus, and is presumed to be the result of a somatic mutation. Comparison of the amino acid sequence of *ras* proteins from a number of different human carcinomas reveals that *ras* mutations involve single amino acid substitutions at position 12 or 61 (Figure 18.9).

One of the well-characterized examples of oncogene activation by translocation is that of *c-abl*, an oncogene associated with chronic myelogenous leukemia (CML). In this case, described in detail in a later section, the translocation results in altered gene activity that causes tumor formation.

At least three separate mechanisms of proto-oncogene activation are associated with overexpression. First, the *c-onc* may acquire a new promoter, causing an increase in the level of transcript production, or causing a silent locus to become activated. This is the case in avian leu-

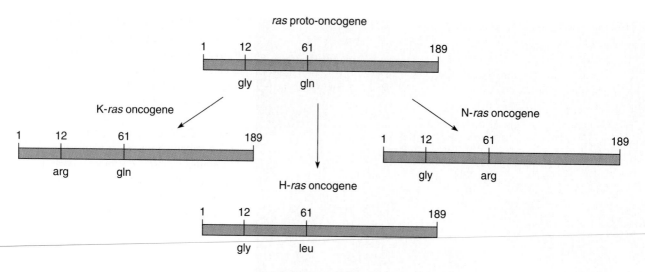

**FIGURE 18.9**   The *ras* proto-oncogene encodes a protein of 189 amino acids. In the normal protein, glycine is encoded at position 12, and glutamine at position 61. Analysis of *ras* oncogene proteins from several tumors shows a single amino acid substitution at one of these positions. A single amino acid substitution resulting from a single base change can convert a proto-oncogene into a tumor-promoting oncogene.

kosis, where strong viral promoters are inserted adjacent to a proto-oncogene, causing an increase in mRNA production and an increase in the amount of the gene product. The second mechanism of overexpression involves the acquisition of new upstream regulatory sequences including enhancers. The third mechanism involves the amplification of the proto-oncogene. In human tumors, members of the *myc* family of oncogenes are frequently amplified. The *c-myc* proto-oncogene is found amplified up to several hundred copies in some human tumors.

The protein products of proto-oncogenes are found in and associated with the plasma membrane, cytoplasm, and nucleus (Table 18.6). While their specific functions vary widely, all products characterized to date alter gene expression in a direct or an indirect manner.

**Table 18.6**   CELLULAR LOCATION OF *c-onc* AND *v-onc* PROTEINS

| Gene | Location of c-onc Protein | Location of v-onc Protein |
|------|---------------------------|---------------------------|
| *src* | Membranes | Membranes |
| *ras* | Membranes | Membranes |
| *myc* | Nucleus | Nucleus |
| *fps* | Cytoplasm | Cytoplasm and membranes |
| *abl* | Nucleus | Cytoplasm |
| *erbB* | Plasma membrane | Plasma membrane and Golgi |

# Colon Cancer: A Genetic Model for Cancer

Even in the small number of cases where it has been studied in detail, it is clear that cancer is a multi-step process resulting from a number of specific genetic alterations. Although the study of tumors such as retinoblastoma and Wilms tumor have been useful in establishing that a limited number of steps are required to transform a normal cell into a malignant one, these tumors are of limited value in establishing the molecular nature and order of the genetic events that lead to tumor formation and metastasis. For these questions, the study of colon cancer offers several advantages. First, malignant tumors of the colon and rectum develop from preexisting benign tumors, and cells from tumors at all stages of development are available for study. Moreover, two forms of colon cancer are known: one in which a predisposition is inherited in an autosomal dominant fashion (known as **familial adenomatous polyposis** or **FAP**), and one that is completely spontaneous, making it possible to study the interaction of genetic and environmental factors in the genesis of tumors.

Through an analysis of mutations in tumors at various stages, ranging from small benign growths known as **adenomas**, through intermediate stages, to **malignant tumors** and **metastatic tumors**, it has been possible to define the number and nature of the genetic and molecular steps involved in changing normal intestinal epithelial cells into tumor cells and to develop a genetic model for colon cancer. This model is shown in Figure 18.10. The first feature of this model is that multiple mutations are required. Mutations in four or five specific genes are needed to bring about malignant growth. If fewer changes are present, benign growths or intermediate stages of tumor formation result. Second, based on the analysis of many tumors, the order of mutations usually follow a *preferred* sequence (as indicated in the figure). Ultimately, however, it is the accumulation of a critical number of specific mutations that is more important than the order in which they occur.

The first mutation in the preferred sequence converts normal epithelium into a rapidly proliferating epithelial region and results in the formation of one or more benign tumors. In cases of FAP, this mutation is inherited and results in the development of dozens or hundreds of benign adenomas in the colon and rectum. In spontaneous cases, the evidence suggests that the initial mutational event takes place in a single cell and that the resulting adenoma consists of a clone of cells, all of which carry the mutation. This first mutation takes place in a gene called **APC** located on the long arm of chromosome 5. The loss of the corresponding allele on the homologous copy of chromosome 5 is *not* necessary for proliferation and adenoma formation, suggesting that the normal allele of the APC gene can be classified as a tumor suppressor gene. The relative order of subsequent

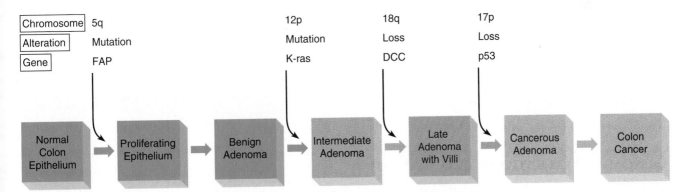

**FIGURE 18.10** A model for the step-wise production of colon cancer. The first step is the loss or inactivation of both alleles of the FAP gene on chromosome 5. In familial cases, one mutation of the FAP gene is inherited. Loss of both alleles leads to the formation of benign adenomas. Subsequent mutations involving genes on chromosomes 12, 18, and 17 in cells of the benign adenomas can lead to a malignant transformation resulting in colon cancer. Although the mutations in chromosomes 12, 17, and 18 usually occur at a later stage than those involving chromosome 5, the sum of the changes is more important than the order in which they occur.

mutations is shown in Figure 18.10. Mutations in the *ras* oncogene may precede or follow the loss of a segment of chromosome 18p. In either case, the accumulation of these two mutations in adenoma cells with a preexisting mutation on chromosome 5 causes the adenoma to grow larger and develop a number of finger-like villous outgrowths. Finally, a mutation on 17p, involving the loss or inactivation of a gene called p53, causes the transition to a cancerous cell. Mutations in the p53 gene are pivotal to the development of a number of cancers, including lung, brain, and breast cancers. The normal allele of p53 is a tumor suppressor gene that normally functions to regulate the passage of cells from late $G_1$ to S phase, perhaps by regulating transcription of a set of genes required at this stage. The p53 gene product has DNA-binding properties, and mutations in p53 may alter DNA binding to confer a new function on the gene product.

The process of metastasis occurs after the formation of cancerous cells and involves an unknown number of mutational steps, although work on other tumors suggests that one or two mutations may be sufficient to allow tumor cells to detach and spread to secondary sites.

In sum, the genetic model for colon cancer involves mutations in oncogenes, tumor suppressor genes, and disruption of the cell cycle at a specific transition point, although the nature and function of normal and mutant gene products of the p53 gene have not yet been identified with certainty. In cases of predisposition to colon cancer, the first mutation is inherited, and the rest occur as a result of the action of environmental agents, pointing up the role of the environment in the development of cancer. This multi-step model appears to have applications to other forms of cancer, but many questions remain unanswered, including the normal functions of the genes involved and the molecular mechanisms by which the mutated genes and/or their gene products bring about tumor formation.

## GENOMIC ALTERATIONS AND CANCER

Alterations in chromosome structure and/or number are associated with many forms of cancer. In many cases, the relationship between changes in chromosome number or arrangement and the development of cancer is not clear. For example, individuals with Down syndrome carry an extra copy of chromosome 21. This quantitative alteration in genetic content is associated with a 20-fold increased risk of leukemia as compared to the general population.

## Chromosome Rearrangements and Cancer

In a limited number of cases where specific chromosome rearrangements are associated with cancer, more direct information is available to indicate how the rearrangement is associated with the development and/or maintenance of the cancerous state.

The relationship between chromosome aberrations and cancer is most clearly seen in leukemias, where the presence of specific chromosome changes is well defined and often diagnostic (Table 18.7). In solid tumors, the task of identifying the primary cytogenetic alteration is often obscured by subsequent chromosomal changes. One of the best-studied examples of an association between a chromosome rearrangement and the development of cancer is the translocation between chromosomes 9 and 22 that is associated with **chronic myelocytic leukemia (CML)** (Figure 18.11). Originally, this translocation was described as an abnormal chromosome 21 and called the **Philadelphia chromosome**. Later, it was shown by Janet Rowley that the Philadelphia chromosome results from the exchange of genetic material between chromosomes 9 and 22. This translocation is never seen in a normal cell and is found only in the white blood cells involved in CML. These observations indicate that the translocation is a primary and causal event in the generation of CML, and that this form of cancer may originate from a single cell bearing this translocated chromosome.

By examining a large number of cases involving this translocation, it was possible to establish the exact location of the break points on chromosomes 9 and 22 (Fig-

**Table 18.7**    SPECIFIC CHROMOSOME ABERRATION AND CANCER

| Cancer | Chromosome Alteration |
| --- | --- |
| Chronic myelogenous leukemia | t(9;22) |
| Acute promyelocytic leukemia | t(15;17) |
| Acute lymphocytic leukemia | t(4;11) |
| Acute myelogenous leukemia | t(8;21) |
| Prostate cancer | del(10q) |
| Synovial sarcoma | t(X;18) |
| Testicular cancer | i(12p) |
| Retinoblastoma | del(13q) |
| Wilms tumor | del(11p) |

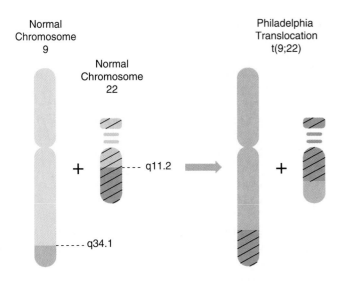

**FIGURE 18.11**    A reciprocal translocation involving the long arms of chromosomes 9 and 22 result in the production of a characteristic chromosome, the Philadelphia chromosome that is associated with cases of chronic myelogenous leukemia (CML).

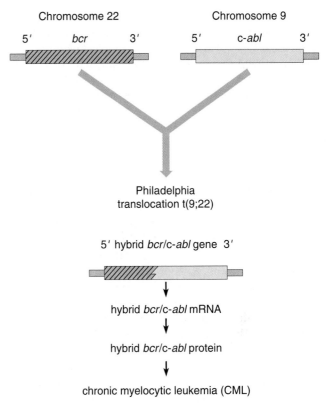

**FIGURE 18.12**    The t(9;22) translocation associated with chronic myelogenous leukemia (CML) results in the fusion of the *c-abl* oncogene on chromosome 9 with the *bcr* gene on chromosome 22. The normal *c-abl* protein is a kinase and the normal *bcr* protein activates a phosphorylation reaction. The fusion protein is a powerful hybrid that allows cells to escape control of the cell cycle, resulting in leukemia.

ure 18.12). Genetic mapping studies using recombinant DNA techniques established that a proto-oncogene *c-abl* maps to the breakpoint region on chromosome 9, and that the gene *bcr* maps near the breakpoint on chromosome 22. In the translocation event, all or most of the *c-abl* gene is translocated to a region within the *bcr* gene. This generates a hybrid *bcr/c-abl* gene that is transcriptionally active and produces a hybrid 200-kd protein product (Figure 18.12) that has been implicated in the generation of CML.

A combination of cytogenetic and molecular approaches has been applied to the cytogenetic events in lymphomas involving translocation breakpoints on chromosome 8 [these include t(8;14), t(8;22), and t(2;8)]. The breakpoint on chromosome 8 in all these cases is the same, and the proto-oncogene *c-myc* has been mapped to this locus. The loci at the breakpoint on the other chromosomes involved in this series of translocations all have various immunoglobulin genes at the breakpoints. The movement of the *c-myc* gene to a position near these immunoglobulin genes and activation of the *c-myc* gene leads to an alteration in the expression of the immunoglobulin genes, resulting in the transformation of the lymphoid cells into leukemic cells.

Other genes at translocation breakpoints have also been isolated and characterized. In these cases and in CML, the translocation results in the formation of an abnormal gene product that causes the cell to undergo a malignant transformation even though a second and presumably normal copy of the gene is present and active in synthesizing a normal gene product. If the abnormal gene products at other translocation loci associated with leukemia can be identified, it is hoped that therapeutic strategies can be developed to administer high levels of the normal gene product, or that the abnormal gene product can be inactivated.

## Genomic Instability and Cancer

The term *genomic instability* is used to describe the events that result in genomic alterations that are characteristic of cancer cells. At least three classes of genetic defects can lead to genomic instability: defects in DNA repair and replication, aberrant chromosome segregation, and defects in cell cycle control. Cultured cells from individuals with Fanconi anemia, ataxia telangiectasia (AT), and Werner syndrome all show an increase in chromosome breaks and translocations, indicating that genomic instability is an heritable trait. The main defect in these syndromes appears to be in DNA repair and/or faulty control of DNA replication.

The model for the development of colon cancer in familial adenomatous polyposis (FAP) illustrates the role of genetic instability in cancer, as evidenced by a number of mutations distributed throughout the genome. A more dramatic relationship between genomic instability and colon cancer has been recently discovered for another form of colon cancer, not associated with polyp formation. This form, called hereditary nonpolyposis colon cancer, accounts for up to 15 percent of all cases of colon cancer. The gene responsible for susceptibility to this form of colon cancer, called FCC, has been mapped to chromosome 2.

Somewhat surprisingly, malignant cells from affected individuals do not show alterations at a site on chromosome 2, but instead show changes in short, repetitive DNA sequences (called microsatellite DNA or variable nucleotide tandem repeats; see Chapter 10) scattered throughout the genome. These large-scale genomic alterations, representing perhaps thousands of alterations in the genome, indicate that the FCC gene may affect the accuracy of DNA replication, and when mutant, causes widespread genomic instability. The gene has recently been identified and cloned, and it is estimated that the mutant gene may be carried by 1 in 200 individuals in the Western world, making it one of the most common genetic disorders known.

Results from studies on tumor development *in vivo* have established that in many cases, genomic instability precedes the formation of tumors. Individuals with a chronic regurgitation of stomach acid into the esophagus have a high risk of esophageal cancer. Cytogenetic studies on esophageal cells of those affected by the reflux of stomach acid show that the first detectable change is a progressive alteration in cell cycle control in small clones of cells (descended from a single ancestral cell), followed by the development of aneuploidy, and finally by the appearance of cancer. In addition, preliminary studies on the FCC gene indicate that microsatellite instability precedes the formation of colon tumors and may be an early event in cancer development.

The relationship between cell cycle control and genomic instability can be illustrated by considering the consequences of mutation in a cell cycle control gene known as p53. As might be expected, mutations in this gene are characteristic of many types of cancer. In mammalian cells, the p53 gene functions at the $G_1$/S check-

point in the cell cycle. As cells enter the $G_1$ to S transition, normally there is a pause in the cell cycle while any broken DNA strands are identified and repaired. In cells where such repairs are not made, replication of DNA during the S phase converts single-strand breaks into double-strand breaks, producing chromosome fragments, which in turn lead to a cascade of further events in the form of translocations and the generation of other chromosome breaks. The p53 gene encodes a transcription factor that is activated at the $G_1$/S checkpoint, leading directly or indirectly to the expression of genes that repair DNA strand breaks. In mutant p53 cells, the $G_1$/S checkpoint is eliminated, and the cells are genomically unstable. This instability plays a role in the final stages of transformation from a normal cell to a malignant cell, once again emphasizing the role of cell cycle regulation in the development of cancer.

## CANCER AND THE ENVIRONMENT

The relationship between environmental agents and the genesis of cancer is often elusive, and in the early stages of investigation, this relationship is based on indirect evidence. Such studies often begin with an epidemiological survey, comparing the cancer death rates among different geographic populations. When differences are found in the death rate for a given type of cancer, or a given age group, or a cluster of related occupations such as chemical workers, further work is necessary to identify one or more environmental factors that correlate with these cancer deaths. These correlations are not conclusive, but serve to identify factors that upon additional investigation may turn out to be directly related to the development of cancer. Finally, extensive laboratory investigations may establish the mechanism by which an environmental agent generates cancer. In other cases, such as certain viruses, the relationship between cancer and the environment is more straightforward.

### Viruses and Cancer

Epidemiological surveys have shown that individuals who develop a form of liver cancer known as **hepatocellular carcinoma (HCC)** are infected with the **hepatitis B virus (HBV)** (Figure 18.13). In fact, for those carrying HBV, the risk of cancer is increased by a factor of 100. Aside from the risk of cancer, HBV infection is a public health risk that affects some 300 million persons worldwide, and is most prevalent in Asia and tropical regions of Africa. Infection produces a wide range of

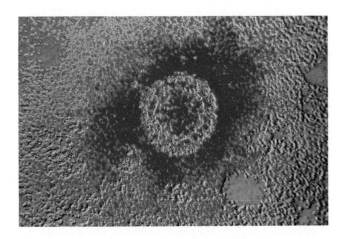

**FIGURE 18.13**    A false color transmission electron micrograph of a hepatitis B virus. Infection with this virus often results in cancer of the liver. Worldwide, viral infections of all kinds are thought to be responsible for about 15 percent of the total number of cancers.

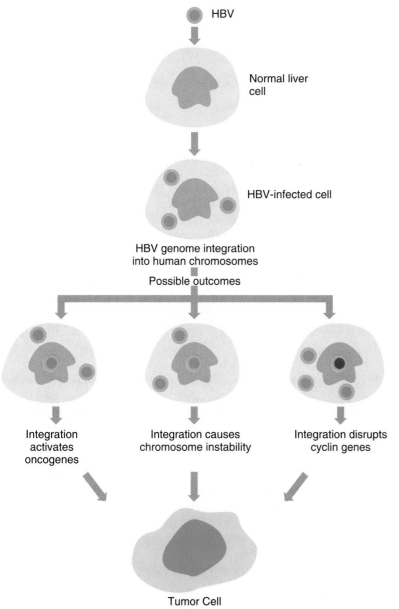

**FIGURE 18.14**    The possible outcomes following cellular infection with hepatitis B virus (HBV). Following integration of the viral chromosome into the human genome, oncogenes can be activated, resulting in tumor formation. Integration of the HBV genome can also cause chromosome instability, including the production of translocations and deletions that trigger a malignant cyclin A gene, causing abnormal regulation of the cell cycle and cellular proliferation resulting in cancer formation.

responses, from a chronic, self-limiting infection with few symptoms, to active hepatitis and cirrhosis, to fatal liver disorders and hepatocellular carcinoma. In the last few years, efforts have centered on understanding the mechanism by which HBV replicates and its role in causing liver cancer. The genome of HBV is a mostly double-stranded DNA molecule of 3200 nucleotides. After cellular infection, the DNA moves to the nucleus where it inserts into a chromosome. The viral DNA is transcribed into an RNA molecule and packaged into a viral capsid. The capsid containing the RNA pregenome moves to the cytoplasm where it is reverse-transcribed into a DNA strand, which in turn is made double stranded. The copied DNA genome is repackaged into a new capsid for release from the cell or returns to the nucleus for another round of replication.

The key to the role of HBV in carcinogenesis apparently resides in its entry into the nucleus. While in the nucleus, the HBV genome can insert itself at many different sites into human chromosomes (Figure 18.14). Insertion often results in cytogenetic alterations of the host genome that include translocations, deletions, or amplification of adjacent regions. As outlined earlier, most forms of cancer are associated with chromosomal rearrangements of these types, and it is possible that such rearrangements trigger the development of a cancerous transformation. Alternatively, the HBV genome has been shown to integrate into the cyclin A gene, and abnormal expression of this gene can disrupt the cell cycle and normal control of proliferation. Similarly, HBV integration has resulted in activation of proto-oncogene loci, and altered expression of such loci has been clearly implicated in the development of malignant cell growth. Thus, there is a strong link between the HBV virus as an environmental agent and the development of cancer in infected individuals.

## Other Environmental Agents

Many surveys of cancer incidence have pointed to the role of the physical environment and personal behavior as factors that contribute to the development of cancer. The precise role of the environment in the genesis of cancer can be difficult to ascertain. Obviously, induction of cancer can involve an interaction between the genotype and environmental agents (Table 18.8). The differential expression of a specific genotype in various environments is often neglected in making calculations about the role of the environment in cancer, but it has been estimated that at least 50 percent of all cancers are environmentally induced. Environmental agents responsible for cancer include background levels of radiation,

**Table 18.8**  EPIDEMIOLOGY OF CANCER

| Country | Death Rate per 100,000 | |
| --- | --- | --- |
| | Male | Female |
| Australia | 212 | 125 |
| Canada | 214 | 136 |
| Dominican Republic | 54 | 48 |
| Egypt | 39 | 18 |
| England | 248 | 156 |
| Greece | 188 | 103 |
| Israel | 174 | 145 |
| Japan | 190 | 109 |
| Nicaragua | 22 | 35 |
| Portugal | 180 | 108 |
| Singapore | 249 | 130 |
| United States | 216 | 137 |
| Venezuela | 135 | 128 |

occupational exposure to physical and chemical agents, exposure to sunlight, and personal behavior such as diet and the use of tobacco. To show how environmental factors are identified, let us consider a recent study on the risk of colon cancer.

Epidemiological surveys reveal that cancer of the colon occurs with a much higher frequency in North America and Western Europe than in Asia, Africa, or other parts of the developing world. In addition, when people migrate from low-risk areas to high-risk areas, their rate of colon cancer increases to match that of the native residents. These and other observations suggest that environmental factors, especially diet, may play a role in the incidence of colon cancer. In a long-range health study of more than 88,000 registered nurses across the United States, dietary habits have been monitored by questionnaires since 1980. In that population, 150 cases of colon cancer were observed through 1986. Detailed analysis of dietary intake adjusted for diet composition indicated that the risk of colon cancer was positively associated with the intake of animal fat. Nurses who had daily meals of pork, beef, or lamb had a 2.5-fold higher risk of colon cancer than those who ate such meals less than once a month. This correlation strongly suggests that animal fat in the diet is an environmental risk factor for colon cancer.

Because there was a negative correlation between eating skinned chicken meat and the incidence of colon cancer, what remains to be sorted out from this study is the role of red meat and/or fat alone as risk factors. Finally, if animal fat is identified as an important risk factor for colon cancer, laboratory studies to identify the mechanisms by which fat brings about a cancerous transformation of the intestinal epithelium will be undertaken. As described in an earlier section, these transformations are associated with mutations involving five or six different genes. However, even in the absence of information about the mechanism of action, it would seem prudent to reduce intake of animal fat in order to reduce the risk of colon cancer.

The accumulated results of research on the role of external factors as a cause of cancer indicate that diet, tobacco, and other agents, including medical and dental X rays, viruses, drugs, and ultraviolet light, play significant roles and are responsible for up to 50 percent of all cases of cancer. This means that it is not the environment in general and pollution in particular that is responsible for a large fraction of cancer cases. Instead, personal choices (e.g., the composition of food in the diet, tobacco use, suntanning) are responsible for a large proportion of all cancers. Education and more judicious choices on the part of individuals could lead to the prevention of a high percentage of all human cancers.

## CHAPTER SUMMARY

1. Cancer can be defined as a genetic disorder at the cellular level that can result from the mutation of a given subset of genes, or from alterations in the timing and amount of gene expression. Although some forms of cancer show familial patterns of inheritance, few show clear evidence for Mendelian inheritance.

2. Cancer results from the uncontrolled proliferation of cells and from the ability of such cells to metastasize or migrate to other sites and form secondary tumors. Mutant forms of genes involved in the regulation of the cell cycle are obvious candidates for cancer-causing genes. The cell cycle is apparently regulated at two sites, with two types of gene products primarily involved: cyclins and kinases. The action of these genes and the genes they control are important in regulating cell division, and links between these genes and the process of tumor formation are being discovered.

3. Although Mendelian patterns of cancer development are not clear-cut, there are genes that predispose to cancer. Study of these genes, such as retinoblastoma, has provided insight into the number and sequence of mutations that result in the development of tumors.

4. Tumor suppressor genes normally act to suppress cell division. When these genes are mutated or have altered expression, control over cell division is lost. Molecular analysis of two such genes, retinoblastoma (RB) and Wilms tumor (WT), indicates that these genes work to control gene expression in different ways, either by affecting the activity of transcription factors or by themselves acting as transcription factors.

5. Oncogenes are genes that normally function to maintain cell division, and these genes must be mutated or inactivated to halt cell division. If these genes escape control and become permanently switched on, cell division occurs in an uncontrolled fashion.

6. The cells of most tumors have visible chromosomal alterations, and the study of these aberrations has provided some insight into the steps involved in the development of cancer. This relationship between cancer and chromosome alterations has been best studied in leukemias, where the formation of hybrid genes or substitution of regulatory sequences is associated with the transformation of normal cells into malignant tumors.

7.  Although cancer is the result of genetic alterations, it appears that the environment is an important factor in cancer induction. Environmental agents include occupational exposure to physical or chemical substances, viruses, diet, and other personal choices such as the use of tobacco.

## KEY TERMS

acute transforming viruses
adenoma
APC
chronic myelocytic leukemia (CML)
cyclins

familial adenomatous polyposis (FAP)
hepatitis B virus (HBV)
hepatocellular carcinoma (HCC)
Li-Fraumeni syndrome
malignant tumor

metastatic tumor
nonacute (nondefective) viruses
oncogenes
Philadelphia chromosome
point mutations
protein kinases

proto-oncogene
retinoblastoma (RB)
retrovirus
reverse transcriptase
sarcoma
translocations
tumor suppressor
Wilms tumor (WT)

## INSIGHTS AND SOLUTIONS

1.  In disorders such as retinoblastoma, a mutation in one allele of the retinoblastoma (RB) gene can be inherited from the germline, causing an autosomal dominant predisposition to the development of eye tumors. In order to develop tumors, a somatic mutation in the second copy of the RB gene is necessary. In sporadic cases, two independent mutational events, involving both RB alleles, are necessary for tumor formation. Given that the first mutation can be inherited, what are the ways in which a second mutational event can occur?

    **SOLUTION:**   In considering how this second mutation can arise, several levels of mutational events need to be considered, including changes in nucleotide sequence as well as events that involve whole chromosomes or chromosome parts. Retinoblastoma results when both copies of the RB locus have been lost or inactivated. With this in mind, perhaps the best way to proceed is to prepare a list of the phenomena that can result in a mutational loss or inactivation of a gene.

    The first and most obvious way for the second RB mutation to occur is by a nucleotide alteration that converts the remaining normal RB allele to a mutation. This alteration may occur through a base substitution or by a frameshift mutation caused by the insertion or deletion of one or more nucleotides. A second method involves loss of the chromosome carrying the normal allele. Since the retinal cells are active in cell division, this event would take place during mitosis, resulting in chromosome 13 monosomy, with the mutant copy of the gene left as the only RB allele. This mechanism does not necessarily have to involve loss of the entire chromosome; deletion of the long arm (RB is on 13q) or an interstitial deletion involving the RB locus and some surrounding material would have the same result.

As an alternative, a chromosome aberration involving loss of the normal copy of the RB gene might be followed by a duplication of the chromosome carrying the mutant allele, restoring two copies of chromosome 13 to the cell, but no normal allele of the RB gene would be present. Lastly, a recombination event followed by chromosome segregation could produce a homozygous combination of mutant RB alleles.

Having made such a list, the next step is to analyze the cells from tumors using a combination of cytogenetic and molecular techniques (such as RFLP analysis and hybridizations to look for deletions) to see which of these mechanisms are actually found in tumors and to what extent they play a role in generating the second mutation. While such analysis is still in the preliminary stages, all mechanisms proposed have been found in tumors, indicating that a variety of spontaneous events can bring about the second mutation that triggers retinoblastoma.

## PROBLEMS AND DISCUSSION QUESTIONS

1. As a genetic counselor, you are asked to assess the risk for a couple who plans to have children, but where there is a family history of retinoblastoma. In this case, both the husband and wife are phenotypically normal, but the husband has a sister with bilateral, familial retinoblastoma. What is the probability that this couple will have a child with retinoblastoma? Are there any tests you can think of that you could recommend to help in this assessment?

2. Review the stages of the cell cycle. What events occur in each of these stages? Which stage is most variable in length?

3. Where are the major regulatory points in the cell cycle?

4. Progression through the cell cycle depends on the interaction of two types of regulatory proteins: kinases and cyclins. List the functions of each and describe how they interact with each other to cause cells to move through the cell cycle.

5. What is the difference between saying that cancer is inherited and saying that predisposition to cancer is inherited?

6. Define tumor suppressor genes. Why are most tumor suppressor genes expected to be recessive?

7. Review the differences between the transcriptional activity, tissue distribution, and function of the tumor suppressor genes associated with retinoblastoma and Wilms tumor? What properties do they have in common? Which are most different?

8. Distinguish between oncogenes and proto-oncogenes. In what ways can proto-oncogenes be converted to oncogenes?

9. How do translocations such as the Philadelphia chromosome lead to oncogenesis?

10. Given that 50 percent of all cancers are environmentally induced, and most of these are caused by actions such as smoking, suntanning, and diet choices, how much of the money spent on cancer research do you think should be devoted to research and education on the prevention of cancer rather than on finding a cure?

## Selected Readings

Ames, B., Magraw, R., and Gold, L. 1990. Ranking possible cancer hazards. *Science* 236:71–80.

Ames, B., Profet, M., and Gold, L. 1990. Dietary pesticides (99.99% all natural). *Proc. Nat. Acad. Sci.* 87:7777–81.

Benedict, W., Xu, H., Hu, S., and Takahashi, R. 1990. Role of the retinoblastoma gene in the initiation and progression of a human cancer. *J. Clin. Invest.* 85:988–93.

Birrer, M., and Minna, J. 1989. Genetic changes in the pathogenesis of lung cancer. *Ann. Rev. Med.* 40:305–17.

Bishop, J. 1987. The molecular genetics of cancer. *Science* 235:306–11.

Cairns, J. 1985. The treatment of diseases and the war against cancer. *Scient. Amer.* (November) 253:51–59.

Cotter, F. 1993. Molecular pathology of lymphomas. *Cancer Surv.* 16:157–174.

Croce, C., and Klein, G. 1985. Chromosome translocations and human cancer. *Scient. Amer.* (March) 252:54–60.

Damm, K. 1993. ErbA: Tumor suppressor turned oncogene? *FASEB J.* 7:904–909.

Easton, D., Bishop, D., Ford, D., Crockford, G., and the Breast Cancer Linkage Consortium. 1993. Genetic linkage analysis in familial breast and ovarian cancer: Results from 214 families. *Am. J. Hum. Genet.* 52:678–701.

Feldman, M., and Eisenbach, L. 1988. What makes a tumor cell metastatic? *Scient. Amer.* (November) 259:60–85.

Grignani, F., Fagioli, M., Ferrucci, P., Alcalay, M., and Pelicci, P. 1993. The molecular genetics of acute promyelocytic leukemia. *Blood Rev.* 7:87–93.

Haber, D., and Buckler, A. 1992. WT1: A novel tumor suppressor gene inactivated in Wilms tumor. *New Biol.* 4:97–106.

Hastie, N. 1993. Wilms' tumour gene and function. *Curr. Opin. Genet. Dev.* 3:408–413.

Huan, B. and Siddique, A. 1993. Regulation of hepatitis B virus gene expression. *J. Hepatol.* 17:Suppl. 3:S20–23.

Lasko, D., Cavanee, W., and Nordenskjold, M. 1991. Loss of constitutional heterozygosity in human cancer. *Ann. Rev. Genet.* 25:281–314.

Leach, F., Nicolaides, N., Papadopoulos, N., Liu, B., Jen, J., Parsons, R., Peltomäki, P., Sistonen, P., *et al.* 1993. Mutations of a *mut*S homolog in hereditary nonpolyposis colorectal cancer. *Cell* 75:1215–1225.

Murray, A., and Kirschner, M. 1991. What controls the cell cycle. *Scient. Amer.* (March) 264:56–63.

Norby, C., and Nurse, P. 1992. Animal cell cycles and their control. *Ann. Rev. Biochem.* 61:441–70.

Peltomäki, P., Aaltonen, L., Sistonen, P., Pylkkänen, L., Mecklin, J. P., Järvinen, H., Green, J., Jass, J., Weber, J., Leach, F., Petersen, G., Hamilton, S., de la Chapelle, A., and Vogelstein, B. 1993. Genetic mapping of a locus predisposing to human colorectal cancer. *Science* 260:810–12.

Rustgi, A., and Podolsky, D. 1992. The molecular basis of colon cancer. *Ann. Rev. Med.* 43:61–68.

Sager, R. 1989. Tumor suppressor genes: The puzzle and the promise. *Science* 246:1406–1412.

Solomon, E., Voss, R., Hall, V., Bodmer, W., Jars, J., Jeffreys, A., Lucibello, F., Patel, I., and Rider, S. 1987. Chromosome 5 allele loss in human colorectal carcinomas. *Nature* 328:616–29.

STANBRIDGE, E. 1992. Functional evidence for human tumor suppressor genes: Chromosome and molecular genetic studies. *Cancer Sun.* 12:43–57.

SUGIMURA, T. 1990. Cancer prevention: Underlying principles. *Basic Life Sci.* 52:225–32.

VON LINDERN, M., FORNEROD, M., SOEKARMAN, N., VAN BAAL, S., JAEGLE, M., HAGEMEIJER, A., BOOTSMA, D., and GROSVELD, G. 1992. Translocation t(6;9) in acute non-lymphocytic leukaemia results in the formation of a DEK-CAN fusion gene. *Baillieres Clin. Haematol.* 5:857–879.

WELNBERG, R. 1985. The action of oncogenes in the cytoplasm and nucleus. *Science* 203:770–776.

WHITE, R. 1992. Inherited cancer genes. *Curr. Opin. Genet. Dev.* 2:53–57.

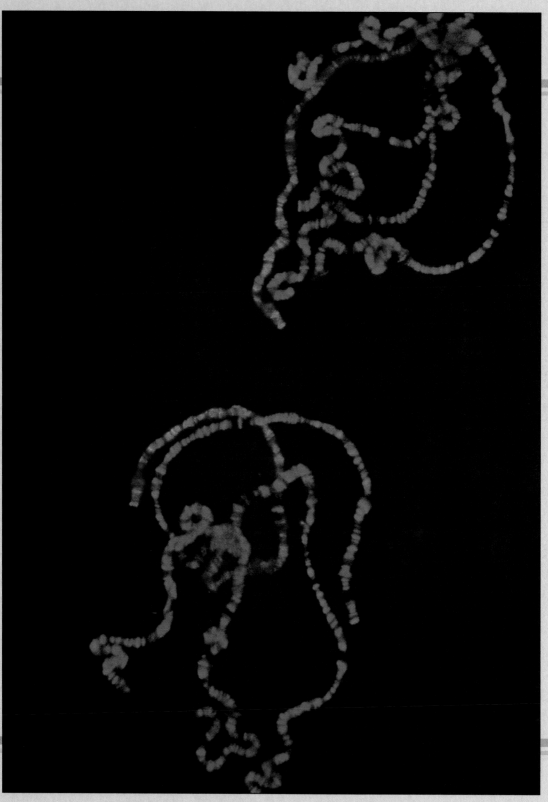

Analysis of male polytene nuclei of *Drosophila*, where the normal protein product of the *maleless* (*mle*) gene (pink) is localized to the X chromosome, whereas the autosomes (blue) do not show immunohistochemical binding. This protein, and others (*msl* proteins) have been shown to play a role in dosage compensation.

# PART 4

# GENETIC ANALYSIS

# 19

# MUTATION AND GENETIC RECOMBINATION IN BACTERIA AND BACTERIOPHAGES

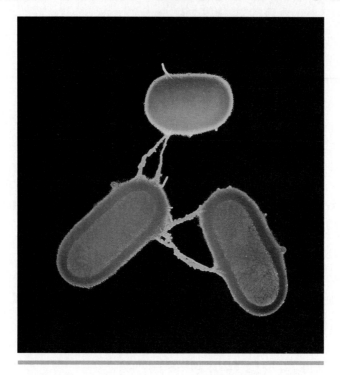

Transmission electron micrograph of conjugating *E. coli*, clearly demonstrating sex pili.

*Bacteria and viruses that have bacteria as their hosts (bacteriophages) have been the subject of extensive genetic analysis and demonstrate mechanisms by which genetic exchange occurs. These processes serve as the basis for genetic mapping. Bacteria may contain extrachromosomal DNA in the form of plasmids, which, in addition to bacteriophage DNA, can be integrated into the bacterial chromosome.*

The use of bacteria and bacteriophages has been essential to the accumulation of knowledge in many areas of genetic study. For example, much of what is known about molecular genetics, recombinational phenomena, and gene structure has been initially derived from experimental work with these organisms. Their successful use in genetic studies is due to numerous factors. Both bacteria and their viruses have extremely short reproductive cycles. Literally hundreds of generations, giving rise to billions of organisms, can be produced in short periods of time. Furthermore, they can be studied in pure cultures. That is, a single species or mutant strain of bacteria or one type of virus can be investigated independently of other similar organisms. Pure culture techniques for the study of microorganisms were developed in the latter part of the nineteenth century.

It was not until the 1940s and thereafter that genetic studies with microorganisms became productive. In 1943, Salvador Luria and Max Delbruck established that bacteria, like eukaryotic organisms, undergo spontaneous mutation. These mutations were demonstrated to be the source of variation observed in bacteria, and this finding paved the way for genetic investigation. Mutant cells that arise in an otherwise pure culture can be isolated and established independently from the parent strain by using efficient selection techniques. As a result, mutations for almost any desired characteristic can now be induced and isolated in pure culture. Since bacteria and viruses are haploid, all mutations are expressed directly in the descendants of mutant cells, adding to the ease with which these microorganisms are studied.

In this chapter, we will review a number of the historical developments leading to the extensive use of bacteria and viruses in genetics. We will begin by discussing an old and a new controversy involving the origin of mutation in bacteria. As will become evident, however, our major emphasis will be on genetic recombination that occurs in both bacteria and bacteriophages. Detailed processes have evolved that facilitate genetic recombination within populations of these microorganisms and viruses.

## MUTATION IN BACTERIA

While it was known well before 1943 that pure cultures of bacteria could give rise to small numbers of cells exhibiting heritable variation, particularly with respect to survival under different environmental conditions, the source of the variation was hotly debated. The majority of bacteriologists believed that environmental factors induced changes in certain bacteria that led to their survival or adaptation to the new conditions. For example, strains of *E. coli* are known to be *sensitive* to infection by the bacteriophage T1. Infection by the bacteriophage leads to the reproduction of the virus at the expense of the bacterial cell, which is lysed or destroyed (see Figure 8.5). If a plate of *E. coli* is homogeneously sprayed with T1, almost all cells are lysed. Rare *E. coli* cells, however, survive infection and are not lysed. If these cells are isolated and established in pure culture, all descendants are *resistant* to T1 infection. The **adaptation hypothesis**, put forth to explain this type of observation, implied that the interaction of the phage and bacterium was essential to the acquisition of immunity. In other words, the phage had somehow "induced" resistance in the bacteria.

The occurrence of **spontaneous mutations** provided an alternative model to explain the origin of T1 resistance in *E. coli*. In 1943, Luria and Delbruck presented the first direct evidence that bacteria, like eukaryotic organisms, are capable of spontaneous mutation. This experiment marked the initiation of modern bacterial genetic study.

## The Luria–Delbruck Fluctuation Test

In a beautiful example of analytical and theoretical work, Luria and Delbruck carried out experiments to differentiate between the adaptation and spontaneous mutation hypotheses. In their work, they used the *E. coli*/T1 systems just described. Many small individual liquid cultures of phage-sensitive *E. coli* were grown up. Then, numerous aliquots from each culture were added to Petri dishes containing agar medium to which T1 bacteriophages had been previously added. Following incubation, each plate was scored for the number of phage-resistant bacterial colonies. This was easy to ascertain because only mutant cells were not lysed, and they survived to be counted. The precise number of total bacteria added to each plate prior to incubation was also determined so that quantitative data could be obtained.

The experimental rationale for distinguishing between the two hypotheses was as follows:

1. **Adaptation**: Every bacterium has a small but constant probability of acquiring resistance as a result of contact with the phages on the Petri dish. Therefore, the number of resistant cells will depend only on the number of bacteria and phages added to each plate. The final results should be independent of all other experimental conditions.

   The adaptation hypothesis predicts, therefore, that if a constant number of bacteria and phages is used for each culture and if incubation time is constant, then little fluctuation in the number of resistant cells should be noted on different plates and from experiment to experiment.

2. **Spontaneous Mutation**: On the other hand, if resistance is acquired as a result of spontaneous mutation, resistance will occur randomly at a low rate during the incubation in liquid medium prior to plating. If a greater number of mutations occur *early* during incubation, which they may do randomly, subsequent reproduction of the mutant bacteria will produce large numbers of resistant cells. If most mutations occur relatively *late* during incubation, far fewer resistant cells will be present.

   The mutation hypothesis predicts, therefore, that significant fluctuation in the number of resistant cells will be observed from experiment to experiment, depending on the time when most spontaneous mutations occurred while in liquid culture.

Table 19.1 shows a representative set of data from the Luria–Delbruck experiments. The middle column shows the number of mutants recovered from a series of aliquots derived from one large individual liquid culture. In a large culture, experimental differences are evened out

**Table 19.1**  THE LURIA–DELBRUCK EXPERIMENT DEMONSTRATING THAT SPONTANEOUS MUTATIONS ARE THE SOURCE OF PHAGE-RESISTANT BACTERIA

| Sample No. | Number of T1-Resistant Bacteria | |
| --- | --- | --- |
| | Same Culture (Control) | Different Cultures |
| 1 | 14 | 6 |
| 2 | 15 | 5 |
| 3 | 13 | 10 |
| 4 | 21 | 8 |
| 5 | 15 | 24 |
| 6 | 14 | 13 |
| 7 | 26 | 165 |
| 8 | 16 | 15 |
| 9 | 20 | 6 |
| 10 | 13 | 10 |
| Mean | 16.7 | 26.2 |
| Variance | 15.0 | 2178.0 |

SOURCE: After Luria and Delbruck, 1943.

because the culture is constantly mixed. As a result, these data serve as a control because the number in each aliquot should be nearly identical. As predicted, little fluctuation is observed. The right-hand column, however, shows the number of resistant mutants recovered from a single aliquot from each of ten independent liquid cultures. The amount of fluctuation in the data will support only one of the two alternative hypotheses. For this reason, the experiment has been designated the **fluctuation test**. A great fluctuation *is* observed between cultures, thus supporting the spontaneous mutation theory. Fluctuation is measured by the amount of variance, a statistical calculation.

The conclusion reached in the Luria–Delbruck experiment that spontaneous-mutation accounts for inherited variation in bacteria received further support from other experimentation. Nevertheless, until the 1950s, staunch supporters of the adaptation theory were not convinced.

## Adaptive Mutation in Bacteria

While the concept of spontaneous mutation in viruses, bacteria, and even higher organisms is no longer disputed, the possibility that organisms such as bacteria might also be capable of "selecting" a specific set of mutations occurring as a result of environmental pressures has long intrigued geneticists. Two independent investigations, published in 1988 by John Cairns and Barry Hall and their colleagues, have provided preliminary evidence that this might be the case. While the findings presented below are still controversial, this general topic is the subject of a current investigation of potentially great significance.

Cairns devised an experimental protocol that improves on that used in the fluctuation test of Luria and Delbruck, discussed above. The procedures were designed to detect *nonrandom mutations* arising in response to factors in the environment in which bacteria are cultured. The results of this work suggest that some bacteria may select mutations that are "adaptive" to the environment.

Instead of using the characteristic of phage T1 resistance, where the only surviving cells are those which have mutated, the Cairns work involved a strain of bacteria that contained a lactose mutation ($lac^-$). Such bacteria cannot use lactose as a carbon source. (See the discussion of the *lac* operon in Chapter 16.) Cells first were grown in a rich liquid medium that provided an adequate carbon source other than lactose. Thus, the $lac^-$ cells as well as any spontaneous $lac^+$ mutants were able to grow and reproduce quite well. This gene was subsequently followed by plating aliquots of cells on Petri plates containing minimal medium to which lactose had been added. The $lac^-$ cells, present in the vast majority, will survive on the plates, but since they lack a carbon source that they can metabolize, they cannot proliferate and form colonies. On the other hand, cells that have mutated to $lac^+$, whether in the original liquid culture or as they sit on the plate, will form colonies and be detected. Those cells that have not mutated to $lac^+$ in liquid medium have the opportunity to do so after they are plated, and they may also be detected, since they will then form colonies. This is the major experimental difference between this approach and that of Luria and Delbruck some 45 years earlier.

Using a slightly more sophisticated mathematical analysis than Luria and Delbruck, Cairns attempted to identify the same sort of "fluctuation," or the lack of it, as an indication of either spontaneous or adaptive mutations, respectively. Cairns found both types of data! The composite distribution was calculated for the case in which both spontaneous mutations occurred in the liquid cultures at random times (creating "fluctuation") and directed mutations arose as a response to the presence of lactose on the Petri plates. Indeed, such a composite distribution was observed.

But, how were they to prove that, in fact, the elevated frequency of mutations occurred on the plates in response to lactose? To answer this question, the Cairns group asked whether another mutation, unrelated to the metabolism of lactose, had also occurred on the plates. They chose to assay for the production of $Val^R$ muta-

tions, which confer resistance to high concentrations of the amino acid valine. The assay was easily accomplished by overlaying the plates with agar containing valine and glucose. Only $Val^R$ mutants will grow. Cairns reasoned that if the $lac^+$ mutations were not necessarily in response to the lactose, then $Val^R$ mutations should arise with a similar distribution as earlier observed for the *lac* mutations. The result was that plates accumulating $lac^+$ mutants were not, at the same time, accumulating $Val^R$ mutations. They concluded that the increased frequency of mutations affecting lactose metabolism was the result of the presence of lactose in the medium. This is, of course, an indirect inference.

Cairns' work suggests that the bacterial cell's genetic machinery can respond to its environment by producing adaptive mutations. This conclusion is contrary to the current thinking that mutations are completely spontaneous events, many of which occur as random errors during replication of DNA. As Cairns has pointed out, these findings, at first glance, seem to support the romantic notion that life is mysterious and that certain observations cannot be explained in simple, straightforward terms based on the laws of physics. Whatever the case may be, Cairns' observations are not isolated examples.

Barry Hall's work demonstrated a similar adaptive response by *E. coli* to growth on salicin. In this case, interestingly, a two-step genetic change was required. The first occurred in response to salicin even though that mutation offered no selective advantage without the second genetic alteration, the removal, or deletion, of a segment of nucleotides called an insertion sequence (IS). Further, the changes must occur sequentially. They do so at a much higher frequency than predicted, provided salicin is present in the growth medium. The observed frequency is several orders of magnitude higher!

What possible explanation can account for observations such as these? As Hall has pointed out, our failure to adequately explain such phenomena may be attributable to our ignorance of fundamental mechanisms and rates of mutations *in nongrowing cells*. Such information and knowledge is extremely important because this physiological condition is much more akin to a natural environment as opposed to that found with cells grown in rich media under laboratory conditions. One suggestion is that under stressful nutritional conditions (starvation), bacteria may be capable of activating mechanisms that create a hypermutable state in genes that will enhance survival.

In the context of a textbook such as this, an important idea to be realized from this discussion is that knowledge may often be derived from observations that initially cannot be explained. This, and the ensuing debates and surrounding controversy, are what constantly maintain intrigue and interest in science!

# MECHANISMS OF DNA TRANSFER AND RECOMBINATION IN BACTERIA

The development of isolation techniques and the information derived from the subsequent study of bacterial mutations led to detailed investigations of the arrangement of genes on the bacterial chromosome. These studies showed that bacteria undergo **conjugation**, a parasexual process in which the genetic information from one bacterium is transferred to and recombined with that of another bacterium. Like meiotic crossing over in eukaryotes, genetic recombination in bacteria led to methodology for chromosome mapping.

Two other phenomena, **transformation** and **transduction**, also result in the transfer of genetic information from one bacterium to another. These processes also have served as a basis for determining the arrangement of genes on the bacterial chromosome. Transformation, as it occurs in nature (see Chapter 8), involves the entry and integration of a piece of DNA from one bacterium into the chromosome of another intact organism. Transduction, on the other hand, is a phage-mediated transfer of small pieces of DNA from one bacterial cell to another.

In the ensuing sections, we shall discuss each of these processes in bacteria. We will then turn to a consideration of what is known about genetic recombination in bacteriophages.

# GENETIC RECOMBINATION IN BACTERIA: CONJUGATION

In 1946, Joshua Lederberg and Edward Tatum initiated studies showing that bacteria undergo **conjugation**, a parasexual process in which the genetic information from one bacterium is transferred to and recombined with that of another bacterium. Like meiotic crossing over in eukaryotes, genetic recombination in bacteria led to methodology for chromosome mapping.

Lederberg and Tatum's initial experiments were performed with two multiple auxotroph strains (nutritional mutants) of *E. coli* K12. Strain A required methionine and biotin in order to grow, while strain B required threonine, leucine, and thiamine (Figure 19.1). Neither strain would grow on minimal medium. The two strains were first grown separately in supplemented media, and then cells from both were mixed and grown together for several more generations. They were then plated on minimal medium. Any bacterial cells that grew on minimal medium would be prototrophs (wild-type bacteria). It

was highly improbable that any of the cells that contained two or three mutant genes would undergo spontaneous mutation simultaneously at two or three locations. Therefore, any prototrophs recovered must have arisen as a result of some form of genetic exchange and recombination.

In this experiment, prototrophs were recovered at a rate of $1/10^7$ ($10^{-7}$) cells plated. The controls for this experiment involved separate plating of cells from strains A and B on minimal medium. No prototrophs were recovered. Lederberg and Tatum therefore concluded that genetic exchange had occurred!

## $F^+$ and $F^-$ Bacteria

There soon followed numerous experiments designed to elucidate the genetic basis of conjugation. It became evident quickly that different strains of bacteria were involved in a unidirectional transfer of genetic material. Cells of one strain could serve as donors of parts of their chromosomes and are designated as $F^+$ **cells** (F for "fertility"). Recipient bacteria, which undergo genetic recombination by receiving donor DNA and recombining it with part of their own, are designated as $F^-$ **cells**.

It was established subsequently that cell contact is essential to chromosome transfer. Support for this concept was provided by Bernard Davis, who designed a U-tube in which to grow $F^+$ and $F^-$ cells (Figure 19.2). At the base of the tube was a sintered glass filter with a pore size that allowed the passage of the liquid medium, but was too small to allow the passage of bacteria. $F^+$ cells were placed on one side and $F^-$ cells on the other side of the filter. The medium was moved back and forth across the filter so that the bacterial cells essentially shared a common medium during incubation. Samples from both sides of the tube were then plated independently on minimal medium, but no prototrophs were found. Davis concluded that *physical contact is essential to genetic recombination.* This physical interaction is the initial stage of the process of conjugation and is mediated through a structure called the **F** or **sex pilus**. Bacteria often have many pili, which are microscopic extensions of the cell. Different types of pili perform different cellular functions, but all pili are involved in some way with adhesion. After contact has been initiated between mating pairs (Figure 19.3), transfer of DNA begins.

Evidence was provided subsequently that $F^+$ cells contained a **fertility factor**, conferring their ability to donate part of their chromosome during conjugation. In experiments by Joshua and Esther Lederberg and by William Hayes and Luca Cavalli-Sforza, it was shown that certain conditions could eliminate donor ability in otherwise fertile cells. However, if these cells were then grown with fertile donor cells, fertility was reestablished.

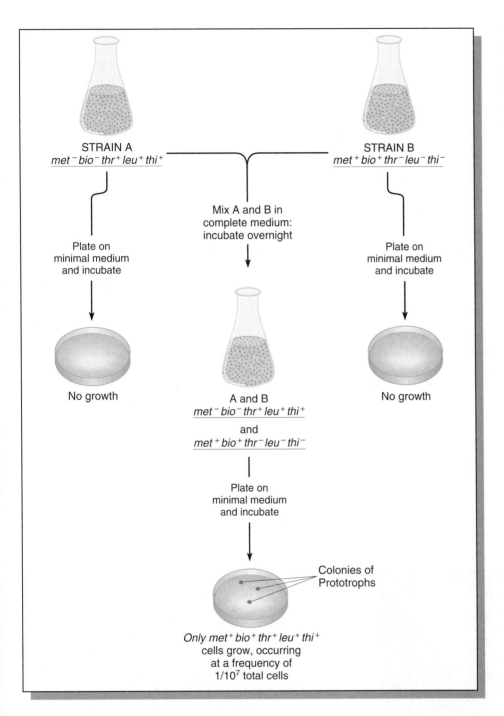

STRAIN A
*met⁻ bio⁻ thr⁺ leu⁺ thi⁺*

STRAIN B
*met⁺ bio⁺ thr⁻ leu⁻ thi⁻*

Mix A and B in
complete medium:
incubate overnight

Plate on
minimal medium
and incubate

Plate on
minimal medium
and incubate

No growth

A and B
*met⁻ bio⁻ thr⁺ leu⁺ thi⁺*
and
*met⁺ bio⁺ thr⁻ leu⁻ thi⁻*

No growth

Plate on
minimal medium
and incubate

Colonies of
Prototrophs

*Only met⁺ bio⁺ thr⁺ leu⁺ thi⁺*
cells grow, occurring
at a frequency of
1/10⁷ total cells

The conclusion was drawn that an **F factor**\* exists that cells can lose and regain.

This conclusion was further supported by the observation that following conjugation and genetic recombination, recipient cells always become F⁺. Thus, in addition to the *rare* case of gene transfer, the F factor is passed to *all* recipient cells. On this basis, the initial crosses of Lederberg and Tatum (Figure 19.1) may be designated:

STRAIN A     STRAIN B
F⁺     ×     F⁻
DONOR     RECIPIENT

Confirmation of these conclusions is based on the isolation of the F factor. It has been shown to consist of a

\* As we soon shall see, the F factor is in reality a plasmid (see page 607). However, in our historical coverage of its discovery, we will continue to refer to it as a "factor."

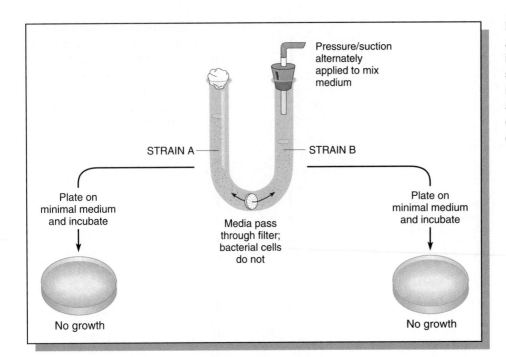

STRAIN A

STRAIN B

Pressure/suction
alternately
applied to mix
medium

Plate on
minimal medium
and incubate

Plate on
minimal medium
and incubate

Media pass
through filter;
bacterial cells
do not

No growth

No growth

**FIGURE 19.2**    When strain A and B auxotrophs are grown in a common medium but are separated by a filter, no genetic recombination occurs and no prototrophs are produced. This apparatus is called a Davis U-tube.

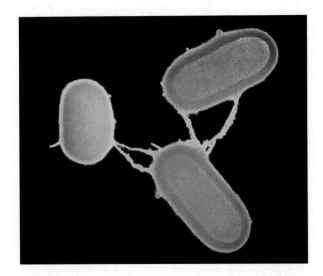

**FIGURE 19.3**    An electron micrograph of conjugation between an F$^+$ E. coli cell and two F$^-$ cells. The sex pili are visible.

circular, double-stranded DNA molecule that is distinct from the bacterial chromosome and contains about 100,000 nucleotide base pairs (100 kb). This amount of DNA is equivalent to about 2 percent of that making up the bacterial chromosome. Contained in the F factor DNA, among others, are 19 genes involved in the transfer of genetic information (*tra* genes) including those essential to the formation of the sex pilus.

It is believed that the transfer of the F factor during conjugation involves strand separation of the F factor DNA and the movement of one of the two strands into the recipient, while the other remains in the donor cell. Both of these parental strands serve as templates for semiconservative DNA replication, resulting in two intact F factors, one in each of the two F$^+$ cells. This process is diagrammed in Figure 19.4.

To summarize, E. coli cells may or may not contain the F factor. When it is present, the cell is able to form a sex pilus and potentially serve as a donor of genetic information. During conjugation, a copy of the F factor is always

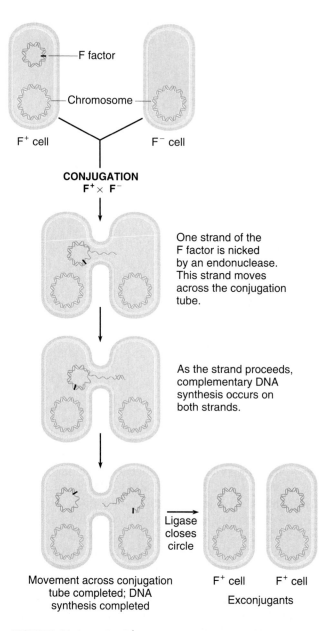

**FIGURE 19.4**    An F⁺ × F⁻ mating demonstrating how the recipient F⁻ cell is converted to F⁺. During conjugation, the DNA of the F factor is replicated with one new copy entering the recipient cell, converting it to F⁺.

One strand of the F factor is nicked by an endonuclease. This strand moves across the conjugation tube.

As the strand proceeds, complementary DNA synthesis occurs on both strands.

Ligase closes circle

Movement across conjugation tube completed; DNA synthesis completed

F⁺ cell    F⁺ cell

Exconjugants

transferred from the F⁺ cell to the F⁻ recipient, converting it to the F⁺ state. The question remains as to exactly how a very low number of F⁻ cells undergo genetic recombination. As we shall see, the answer awaited further experimentation.

## Hfr Bacteria and Chromosome Mapping

Subsequent discoveries not only clarified how genetic recombination occurs, but defined a mechanism by which the E. coli chromosome could be mapped. We shall first address chromosome mapping.

In 1950, Cavalli-Sforza treated an F⁺ strain of E. coli K12 with nitrogen mustard, a potent mutagen. From these treated cells he recovered a strain of donor bacteria that underwent recombination at a rate of $1/10^4$ $(10^{-4})$, 1000 times more frequently than the original F⁺ strains. In 1953, Hayes isolated another strain demonstrating a similar elevated frequency. Both strains were designated **Hfr**, or **high-frequency recombination**. Since Hfr cells behave as chromosome donors, they are a special class of F⁺ cells.

Another important difference was noted between Hfr strains and the original F⁺ strains. If the donor is an Hfr strain, recipient cells, while sometimes displaying genetic recombination, never become Hfr; that is, they remain F⁻. In comparison, then,

$$F^- \times F^- \rightarrow F^- \text{ (low rate of recombination)}$$
$$\text{Hfr} \times F^- \rightarrow F^- \text{ (higher rate of recombination)}$$

Perhaps the most significant characteristic of Hfr strains is the nature of recombination. In any given strain, certain genes are more frequently recombined than others, and some not at all. This *nonrandom pattern of gene transfer* was shown to vary from Hfr strain to Hfr strain. While these results were puzzling, Hayes interpreted them to mean that some physiological alteration of the F factor had occurred, resulting in the production of Hfr strains of E. coli.

In the mid-1950s, experimentation by Ellie Wollman and François Jacob helped explain why cells are Hfr compared to F⁺ and showed how these strains allow genetic mapping of the E. coli chromosome. In their experiments, Hfr and F⁻ strains with suitable marker genes were mixed and recombination of specific genes assayed at different times. To accomplish this, a culture containing a mixture of an Hfr and an F⁻ strain was first incubated and samples removed at various intervals and placed in a blender. The shear forces created in the blender separated conjugating bacteria so that the transfer of the chromosome was terminated effectively. The cells were then assayed for genetic recombination.

This process, called the **interrupted mating technique**, demonstrated that specific genes of a given Hfr strain were transferred and recombined sooner than others. Figure 19.5 illustrates this point. During the first 8 minutes after the two strains were initially mixed, no genetic recombination could be detected. At about 10 minutes, recombination of the *azi⁺* gene could be detected, but no transfer of the *tonˢ*, *lac⁺*, or *gal⁺* genes was noted. By 15 minutes, 70 percent of the recombinants were *azi⁺*; 30 percent were now also *tonˢ*; but none was *lac⁺* or *gal⁺*. By 20 minutes, the *lac⁺* gene was found among the recombinants; and by 30 minutes,

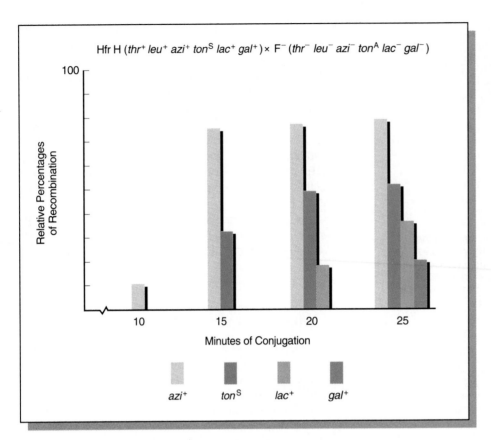

Hfr H (*thr⁺ leu⁺ azi⁺ ton^S lac⁺ gal⁺*) × F⁻ (*thr⁻ leu⁻ azi⁻ ton^A lac⁻ gal⁻*)

**FIGURE 19.5**   The progressive transfer during conjugation of various genes from a specific Hfr strain of *E. coli* to an F⁻ strain. Certain genes (*azi* and *ton*) are transferred sooner than others and recombine more frequently. Others (*lac* and *gal*) take longer to be transferred and recombine with a lower frequency.

*gal⁺* was also being transferred. Therefore, Wollman and Jacob had demonstrated an *oriented transfer of genes* that was correlated with the length of time conjugation was allowed to proceed.

It appeared that the chromosome of the Hfr bacterium was transferred linearly and that the gene order and distance between them, as measured in minutes, could be predicted from such experiments (Figure 19.6). This information served as the basis for the first genetic map of the *E. coli* chromosome. Apparently, conjugation does not usually last long enough to allow the entire chromosome to pass across to the recipient cell.

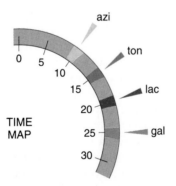

**FIGURE 19.6**   A time map of the genes studied in the experiment depicted in Figure 19.5.

Wollman and Jacob then repeated the same type of experimentation with other Hfr strains, obtaining similar results with one important difference. While genes were always transferred linearly with time, as in their original experiment, which genes entered first and which followed later seemed to vary from Hfr strain to Hfr strain [Figure 19.7(a)]. When they reexamined the rate of entry of genes, and thus the different genetic maps for each strain, a definite pattern emerged. The major differences between all strains were simply the point of the origin (*O*) and the direction in which entry proceeded from that point [Figure 19.7(b)].

In order to explain these results, Wollman and Jacob postulated that the *E. coli* chromosome is circular. If the point of origin (*O*) varied from strain to strain, a different sequence of genes would be transferred in each case. But what determines *O*? They proposed that in various Hfr strains, the F factor integrates into the chromosome at different points. Its position determines the *O* site. One such case of integration is shown in Figure 19.8. During subsequent conjugation between this Hfr and an F⁻ cell, the position of the F factor determines the initial point of transfer. Those genes adjacent to *O* are transferred first. *The F factor becomes the last part to be transferred.* Apparently, conjugation rarely, if ever, lasts long enough to allow the entire chromosome to pass across the conjugation tube. This proposal explains why recipient cells, when mated with Hfr cells, remain F⁻.

**(a)**

| Hfr Strain | Order of Transfer $\longrightarrow$ | | | | | | | |
|:---:|:---:|:---:|:---:|:---:|:---:|:---:|:---:|:---:|
| H | thr – | leu – | azi – | ton – | pro – | lac – | gal – | thi |
| 1 | leu – | thr – | thi – | gal – | lac – | pro – | ton – | azi |
| 2 | pro – | ton – | azi – | leu – | thr – | thi – | gal – | lac |
| 7 | ton – | azi – | leu – | thr – | thi – | gal – | lac – | pro |

**(b)**

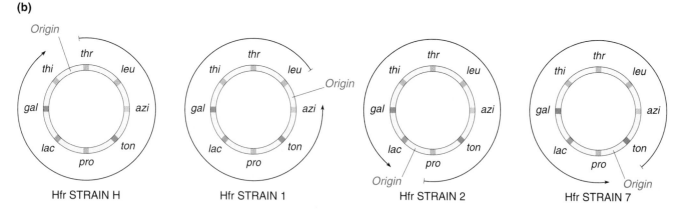

Hfr STRAIN H        Hfr STRAIN 1        Hfr STRAIN 2        Hfr STRAIN 7

**FIGURE 19.7**    (a) The order of gene transfer in four Hfr strains, suggesting that the *E. coli* chromosome is circular. (b) The position of the origin of transfer is identified in each strain. Note that transfer can proceed in either direction, depending on the strain. The origin is determined by the point of integration into the chromosome of the F factor, and the direction of transfer is determined by the orientation of the F factor as it integrates.

Figure 19.8 also depicts the way in which the double helix unwinds during transfer, allowing for the entry of single-stranded DNA into the recipient. Following replication of the entering DNA, it now has the potential to recombine with its homologous region of the host chromosome. The DNA strand that remains in the donor also undergoes replication.

The use of the interrupted mating technique with different Hfr strains has provided the basis for mapping the entire *E. coli* chromosome. Mapped in time units, strain K12 (or *E. coli* K12) is 60 minutes long. Over 900 genes have now been placed on the map. In most instances, only a single copy of each gene exists.

## Recombination in F⁺ × F⁻ Matings: A Reexamination

The above model has helped geneticists to understand better how genetic recombination occurs during the F⁺ × F⁻ matings. Recall that recombination occurs much less frequently in them than in Hfr × F⁻ matings, and that random gene transfer is involved. The current belief is that when F⁺ and F⁻ cells are mixed, conjugation occurs readily and that each F⁻ cell involved in conjugation receives a copy of the F factor, *but that no genetic recombination occurs*. However, at a very low frequency in a population of F⁺ cells, the F factor integrates spontaneously at a random point into the bacterial chromosome. This integration converts each such F⁺ cell to the Hfr state. Therefore, in F⁺ × F⁻ crosses, the very low frequency of genetic recombination ($10^{-7}$) is attributed to the rare, newly formed Hfr cells, which then also undergo conjugation with F⁻ cells. Since the point of integration is random, a nonspecific gene transfer ensues, leading to the low-frequency, random genetic recombination observed in the F⁺ × F⁻ experiment.

## The F′ State and Merozygotes

In 1959, during experiments with Hfr strains of *E. coli*, Edward Adelberg discovered that the F factor could lose its integrated status, causing reversion to the F⁺ state (Figure 19.9). When this occurs, the F factor frequently carries several adjacent bacterial genes along with it. He

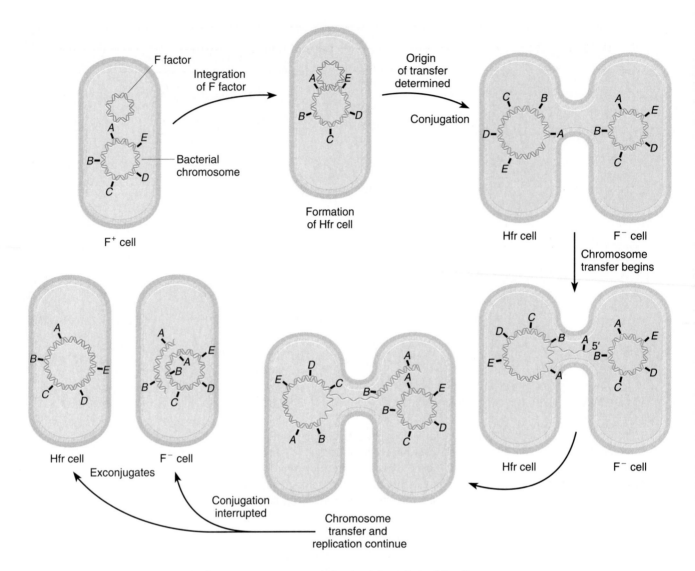

**FIGURE 19.8**   Conversion of $F^+$ to an Hfr state occurs by the integration of the F factor into the bacterial chromosome. During conjugation, the F factor is nicked by an enzyme, creating the origin of transfer of the chromosome. The major portion of the F factor is now on the end of the chromosome adjacent to the origin and is the last part to be transferred. Conjugation is usually interrupted prior to complete transfer. Above, only the A and B genes are transferred.

labeled this condition **F′** to distinguish it from $F^+$ and Hfr. F′ is thus a special case of $F^+$.

The presence of bacterial genes within a cytoplasmic F factor creates an interesting situation. An F′ bacterium behaves as an $F^+$ cell, initiating conjugation with $F^-$ cells. When this occurs, the F factor, along with the chromosomal genes, is transferred to the $F^-$ cell. As a result, whatever chromosomal genes are part of the F factor are now present in duplicate in the recipient cell because the recipient still has a complete chromosome. This creates a partially diploid cell called a **merozygote**. Pure cultures of F′ merozygotes may be established. They have been

extremely useful in the study of bacterial genetics, particularly in genetic regulation.

## THE *REC* GENES AND BACTERIAL RECOMBINATION

Having established that a unidirectional transfer of DNA occurs between bacteria, determining how the actual recombination event occurs in the recipient cell has been of interest. Just how does the genetic exchange

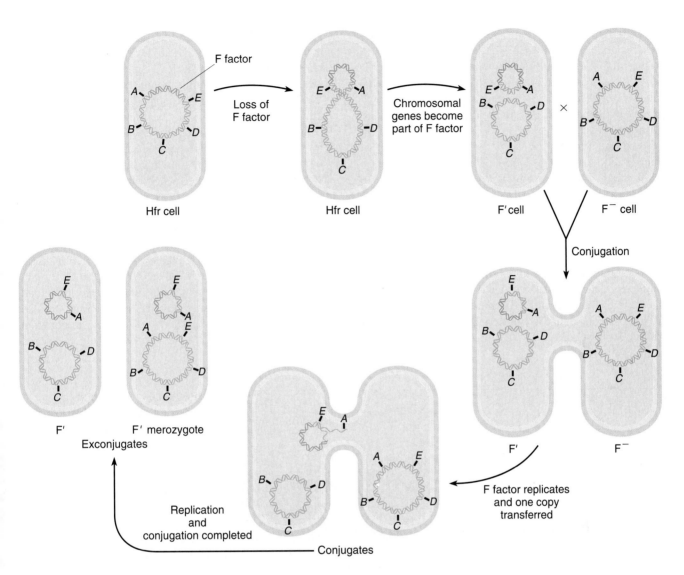

**FIGURE 19.9**    Conversion of an Hfr bacterium to F′ and its subsequent mating with an F⁻ cell. The conversion occurs when the F factor loses its integrated status and carries with it one or more chromosomal genes (A and E). Following conjugation with such an F′ cell, an F⁻ recipient becomes partially diploid and is called a merozygote. It also becomes F′.

occur? As with many systems, the mechanism by which a biochemical event occurs is often deciphered as a result of genetic studies. Such an approach has been particularly fruitful in the case of the phenomenon of recombination between DNA molecules in bacteria. Great insights were gained as a result of the isolation of a group of mutations involving the *rec* genes. This discovery illustrates quite well the value of isolating mutations, establishing their phenotypes, and determining the biological role of the normal, wild-type gene as a result of subsequent investigation.

The first relevant observation in this case involved a series of mutant genes labeled *rec*A, *rec*B, *rec*C, and *rec*D. The first mutant gene, *rec*A, was found to diminish genetic recombination in bacteria 1000-fold, nearly eliminating it altogether. The other *rec* mutations reduced recombination by about 100 times! Clearly, the normal wild-type products of these genes play some essential role in the process.

By looking for a functional gene product present in normal cells, but missing in mutant cells, two *rec* gene products were subsequently isolated and studied. They have been found to play a direct role in the process of DNA recombination in bacterial cells. The first is a protein called the RecA protein. The second is a more complex protein called the RecBCD product, consisting of polypeptide subunits derived from the B, C, and D genes. The roles of these proteins have now been elucidated. As a result of this genetic research, our knowledge of the process of recombination has been extended considerably.

The RecA protein has a strong affinity *in vitro* for single-stranded DNA (ssDNA). The binding creates a DNA–protein complex which invades and probes an independent duplex of DNA. The complex migrates along until the ssDNA locates its homologous region within the duplex. When the homologous region is encountered, the invading ssDNA, which is complementary to one of the two strands, base pairs with it, replacing its counterpart in the initial duplex. Once the hybrid duplex is formed, the protein encoded by the *rec*A gene is released. The displaced region of DNA juts out of the duplex, forming what is called a **D-loop**. This general scheme is illustrated in Figure 19.10.

The RecBCD protein, on the other hand, has an affinity *in vitro* for double-stranded DNA (dsDNA). It binds to and migrates along the duplex, separating the helix as it goes. The separated strands reassociate, but more slowly than they separated. This leaves behind loops of single-stranded DNA (ssDNA). Since mutations in any of the *rec*BCD genes diminish recombination, we can speculate that separation of strands, as described above, may be an essential step in allowing the invading RecA–ssDNA complex to locate its homologous region within the duplex.

Together, the products of *rec* genes appear to have the effect of positioning one of the donor DNA strands from the $F^+$ cells at the appropriate location within the $F^-$ chromosome in order to facilitate genetic recombination. What remains to be accomplished, of course, is the actual insertion of the $F^+$ DNA. This may be accomplished by the RecBCD protein, which has been shown to also be able to make cuts in DNA. DNA ligase might then complete the task of recombination by enzymatically

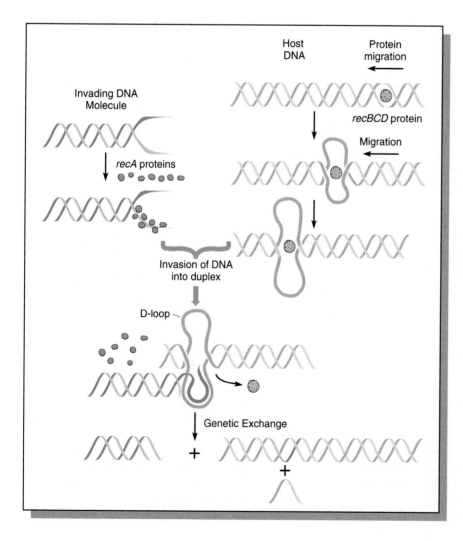

**FIGURE 19.10**    A theoretical model illustrating the possible role of the RecA and RecBCD proteins in the process of genetic recombination between bacterial DNA molecules. The RecA protein has an affinity to bind to single-stranded regions of DNA and to facilitate their invasion of double-stranded DNA. The RecBCD protein has an affinity for double-stranded DNA. This protein migrates along the helix, causing the complementary strands to dissociate. Once the invasive region finds its homologous DNA sequence, the RecA proteins are shed, and a hybrid complement is formed. The displaced strand forms a D-loop, and genetic exchange occurs.

sealing the ends of the invading DNA with those of the DNA of the recipient cell. This picture will undoubtedly be clarified as further research is conducted. Figure 19.10 illustrates the general role of the RecBCD protein as well as a way in which both gene products might work in an integrated fashion.

## PLASMIDS

In the above sections, we introduced and discussed extensively the extrachromosomal heredity unit called the F factor. When existing autonomously in the bacterial cytoplasm, this unit takes the form of a double-stranded closed circle of DNA. These characteristics place the F factor in the more general category of the genetic structure called a **plasmid**. These structures contain one or more genes and often, quite a few. Their replication depends on the host cell's enzymes and they are distributed to daughter cells along with the host chromosome during cell division.

Plasmids are generally classified according to the genetic information specified by their DNA. The F factor confers fertility and contains genes essential for sex pilus formation. Other examples include the **R** and the **Col plasmids**.

Most R plasmids consist of two components: the **RTF** (**resistance transfer factor**) and one or more **r-determinants**. The RTF contains information essential to transfer of the plasmid between bacteria, and r-determinants are genes conferring antibiotic resistance. While the DNA of RTFs is quite similar in a variety of plasmids from different bacteria species, there is a wide variation in r-determinants. Each is specific for resistance to one class of antibiotic. If a bacterial cell contains r-determinant plasmids but no RTF is present, the cell is resistant but cannot transfer genetic material to recipient cells. Therefore, resistance cannot be transferred. However, the most commonly studied plasmids consist of the RTF as well as one or more r-determinants. Resistances to tetracycline, streptomycin, ampicillin, sulfonamide, kanamycin, and chloramphenicol are most frequently encountered.

As we shall see in the ensuing section, still other DNA sequences may associate with an R plasmid. As a result, r-determinants from different R plasmids may combine into single units, conferring multiple resistance to several antibiotics (Figure 19.11).

The ColE1 plasmid, derived from *E. coli*, encodes one or more proteins which are highly toxic to bacterial strains that do not harbor the same plasmid. These proteins, called **colicins**, may kill neighboring bacteria. Those that carry the plasmid are said to be colicinogenic.

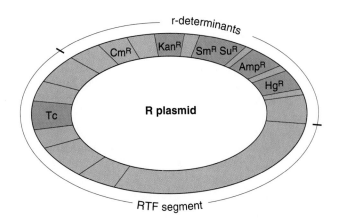

**FIGURE 19.11**    R plasmid containing resistance transfer factors (RTFs) and multiple r-determinants (Tc—tetracycline; Km—kanamycin; Sm—streptomycin; Su—sulfonamide; Ap—ampicillin; Hg—mercury).

Present in 10 to 20 copies per cell, the plasmid also contains a gene encoding an immunity protein which protects the host cell from the toxin. Unlike the F and R plasmids, the Col plasmid is not transmissible to other cells unless one of the former structures is present to confer the ability to engage in conjugation.

Interest in plasmids has increased dramatically because of their role in the genetic technology referred to as recombinant DNA research. Specific genes from any source may be inserted into a plasmid, which may be inserted into a bacterial cell. As the cell replicates its DNA and undergoes division, the foreign gene is treated like one of the cell's own. This was the major topic of Chapters 12 and 13.

## BACTERIAL TRANSFORMATION

In Chapter 8 we described how the study of transformation provided direct evidence that DNA is the genetic material. We have also discussed artificial transformation in Chapters 12 and 13 in our coverage of recombinant DNA technology. Natural transformation, like conjugation, provides a mechanism for the recombination of genetic information in some bacteria. In those bacterial species where it occurs, transformation can also be used to map bacterial genes, although in a more limited way.

This process (Figure 19.12) consists of numerous steps that can be divided into two main categories: (1) entry of exogenous DNA into a recipient cell, and (2) recombination of the donor DNA with its homologous region in the recipient chromosome. In a population of cells, only those in a particular physiological state, referred to as **competence**, take up DNA. Entry apparently occurs at a limited number of receptor sites on the surface of the bacterial cell. The efficient length of transforming DNA is

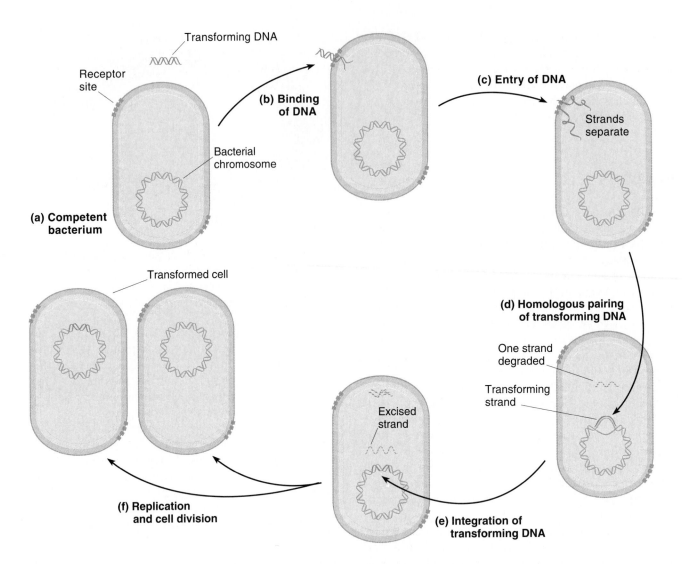

**FIGURE 19.12**   Proposed steps leading to transformation of a bacterial cell by exogenous DNA. Only one of the two strands of the entering DNA is involved in the transformation event, which is completed following cell division.

about 10,000 to 20,000 base pairs, an amount equal to about 1/200 of the *E. coli* chromosome. Passage across the cell wall and membrane is an active process requiring energy and specific transport molecules. This concept is supported by the fact that substances that inhibit energy production or protein synthesis in the recipient cell also inhibit the transformation process.

During the process of entry, one of the two strands of the double helix is digested by nucleases, leaving only a single strand to participate in transformation. This DNA segment aligns with its complementary region of the bacterial chromosome. In a process involving several enzymes, an even number of exchanges (crossovers) results in the segment replacing its counterpart in the chromosome, which is excised and degraded. Any DNA can be taken into the cell, but only homologous DNA will undergo recombination leading to transformation.

In order for recombination to be detected, the transforming DNA must be derived from a genetically different strain of bacteria. Once it is integrated into the chromosome, that region contains one strand originally present and one mutant strand. Since these strands are not genetically identical, this helical region is referred to as a **heteroduplex**. Following one round of semiconservative replication, one chromosome is identical to that of the original recipient, and the other contains the mutant gene. Following cell division, one nonmutant and one mutant cell are produced.

## Transformation and Linked Genes

Because transforming DNA is usually about 10,000 to 20,000 base pairs long, it contains a sufficient number of nucleotides to encode several genes. Genes that are adjacent or very close to one another on the bacterial chromosome may be carried on a single piece of DNA of this size. Because of this fact, a single event may result in the **cotransformation** of several genes simultaneously. Genes that are close enough to each other to be cotransformed are said to be linked. Note that here *linkage* refers to the proximity of genes, in contrast to the use of this term in eukaryotes to indicate all genes on a single chromosome.

If two genes are not linked, simultanous transformation can occur only as a result of two independent events involving two distinct segments of DNA. As in double crossing over in eukaryotes, the probability of two independent events occurring simultaneously is equal to the product of the individual probabilities. Thus, the frequency of two unlinked genes being transformed simultaneously is much lower than if they are linked.

Linked genes were first demonstrated in 1954 during studies of *Pneumococcus* by Rollin Hotchkiss and Julius Marmur. They were examining transformation at the *streptomycin* and *mannitol* loci. Recipient cells were $str^s$ and $mtl^-$, meaning that they were sensitive to streptomycin and could not ferment mannitol. Cells that were $str^r mtl^+$, which are resistant to streptomycin and able to ferment mannitol, were used to derive the transforming DNA. If these genes are unlinked, simultaneous transformation for both genes would occur with such a minimal probability as to be nearly undetectable. However, as shown in Table 19.2, a low but detectable rate (0.17%) of double transformation did occur.

In order to confirm that the genes were indeed linked, a second experiment was performed. Instead of using $str^r mtl^+$ DNA, a mixture of $str^r mtl^-$ and $str^s mtl^+$ DNA simultaneously served as the donor DNA. Thus, both the $str^r$ and $mtl^+$ genes were available to recipient cells, but on separate segments of DNA. If the observed rate of double transformants in the previous experiment had been due to two independent events and the genes unlinked, then a similar rate would also be observed under these conditions. Instead, the number of double transformants was 25 times fewer, thus confirming linkage (Table 19.2). Careful examination of the data in this table not only confirms the linkage of these loci, but raises other interesting questions. Why, for example, is the *str* locus always transformed more frequently than the *mtl* locus? This observation suggests that entry into the cell is not totally random.

Subsequent studies of transformation have shown that a variety of bacteria readily undergo this process of recombination (*e.g., Bacillus subtilis* and *Shigella paradysenteriae*). In other cases, culture conditions can be adjusted to "induce" artificial transformation. Under certain conditions, studies have shown that relative distances between linked genes can be determined from the recombination data provided by transformation experiments. While analysis is more complex, such data are interpreted in a manner analogous to chromosome mapping in eukaryotes.

There is still a third mode involving genetic transfer between bacteria, called **transduction**. We shall delay our discussion temporarily until we have considered the genetics of bacteriophages and other viruses, since transduction is a process mediated by bacteriophages. As we discussed in Chapter 16, and as we shall see in the next section, bacterial viruses are capable of either lysing the bacterial host or inserting their DNA into the host chro-

**Table 19.2** The Results of Several Transformation Experiments, Which Establish Linkage Between the *STR* and *MTL* Loci in *Pneumococcus*

| Donor DNA | Recipient Cell Genotype | Percentage of Transformed Genotypes | | |
|---|---|---|---|---|
| | | $str^r mtl^-$ | $str^s mtl^+$ | $str^r mtl^+$ |
| $str^r mtl^+$ | $str^s mtl^-$ | 4.3 | 0.40 | 0.17 |
| $str^r mtl^-$ and $str^s mtl^+$ | | 2.8 | 0.85 | 0.0066 |

Source: Data from Hotchkiss and Marmur, 1954, p. 55.

mosome where it may remain genetically quiescent but is replicated along with the host DNA and passed to all daughter cells. In some cases, it is when this integrated DNA is extricated from the host chromosome that transduction becomes possible.

# The Genetic Study of Bacteriophages

So far, we have considered recombination involving bacterial genes. Recombination has also been observed in bacterial viruses, but only when genetically distinguishable strains have simultaneously infected the same host cell. In this section, we will first review briefly the life cycle of a typical T-even bacteriophage. Then, we will discuss how these phages are studied during their infection of bacteria. Finally, we will contrast the two possible modes of behavior once the phage DNA is injected into the bacterial host. This information will serve as background for our discussion of transduction. We will then entertain more esoteric topics involving viral genetics.

## Phage T4 Infection and Reproduction

In Chapter 8, we discussed the Hershey–Chase experiment, which utilized bacteriophage T2 in providing support for the concept that DNA is the genetic material. Infection and reproduction in phage T4 is very similar and typical of the family of T-even phages. It contains double-stranded DNA in a quantity sufficient to encode more than 150 average-sized genes. The genetic material is enclosed by an icosahedral protein coat (a polyhedron with 20 faces) making up the head of the virus (see Figure 19.14). This is connected to a tail that contains a collar and a contractile sheath that surrounds a central core. Six tail fibers protrude from the tail; their tips contain binding sites that specifically recognize unique areas of the outer membrane forming the cell surface of the *E. coli* cell wall.

Binding of tail fibers to these unique areas of the cell wall is the first step of infection. Then, an ATP-driven contraction of the tail sheath causes the central core to penetrate the cell wall. The DNA in the head is extruded where it then moves across the cell membrane into the bacterial cytoplasm. Within minutes, all bacterial DNA, RNA, and protein synthesis is inhibited, and synthesis of viral macromolecules begins. This initiates the **latent period**, in which the production of viral components occurs prior to the assembly and appearance of mature virus particles.

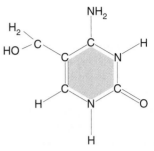

5-hydroxymethylcytosine

**FIGURE 19.13**    The chemical structure of 5-hydroxymethylcytosine, which is present in some bacteriophage DNA. This modified nitrogenous base protects the viral DNA from the enzymatic degradation that affects the bacterial host DNA during infection.

RNA synthesis is initiated by the transcription of viral genes by the host cell RNA polymerase. The viral genes are divided into three groups: (1) **immediate early genes**, (2) **delayed early genes**, and (3) **late genes**. The products of the first two groups are made prior to the initiation of T4 DNA synthesis. One of the earliest gene products is a DNase enzyme that degrades the *E. coli* chromosome. The T4 DNA molecule is protected from digestion by this enzyme because its cytosine residues are modified to form 5-hydroxymethylcytosine (Figure 19.13); as a result, this DNase is unable to use T4 DNA as a substrate.

A plethora of early and late gene products are subsequently synthesized using the host cell ribosomes. For example, about 20 different proteins are part of the protein capsid of the head. Many others are structural components of the tail and its fibers. Additionally, numerous enzymes participate in assembly of mature bacteriophages, but are not themselves structural components. One late gene product, **lysozyme**, digests the bacterial cell wall, leading to its rupture and release of mature phages once they are assembled.

Prior to the synthesis of most viral proteins, semiconservative DNA replication begins, leading to a pool of viral DNA molecules that are available to be packaged into phage heads. It is at this time that recombination may occur between DNA molecules.

Assembly of mature viruses is a complex process that has been well studied by William Wood, Robert Edgar, and others. Three major independent pathways occur, leading to: (1) DNA packaged into heads; (2) construction of tails; and (3) synthesis and assembly of tail fibers. A mutation that interrupts any one of the three pathways does not inhibit the other two. As shown in Figure 19.14,

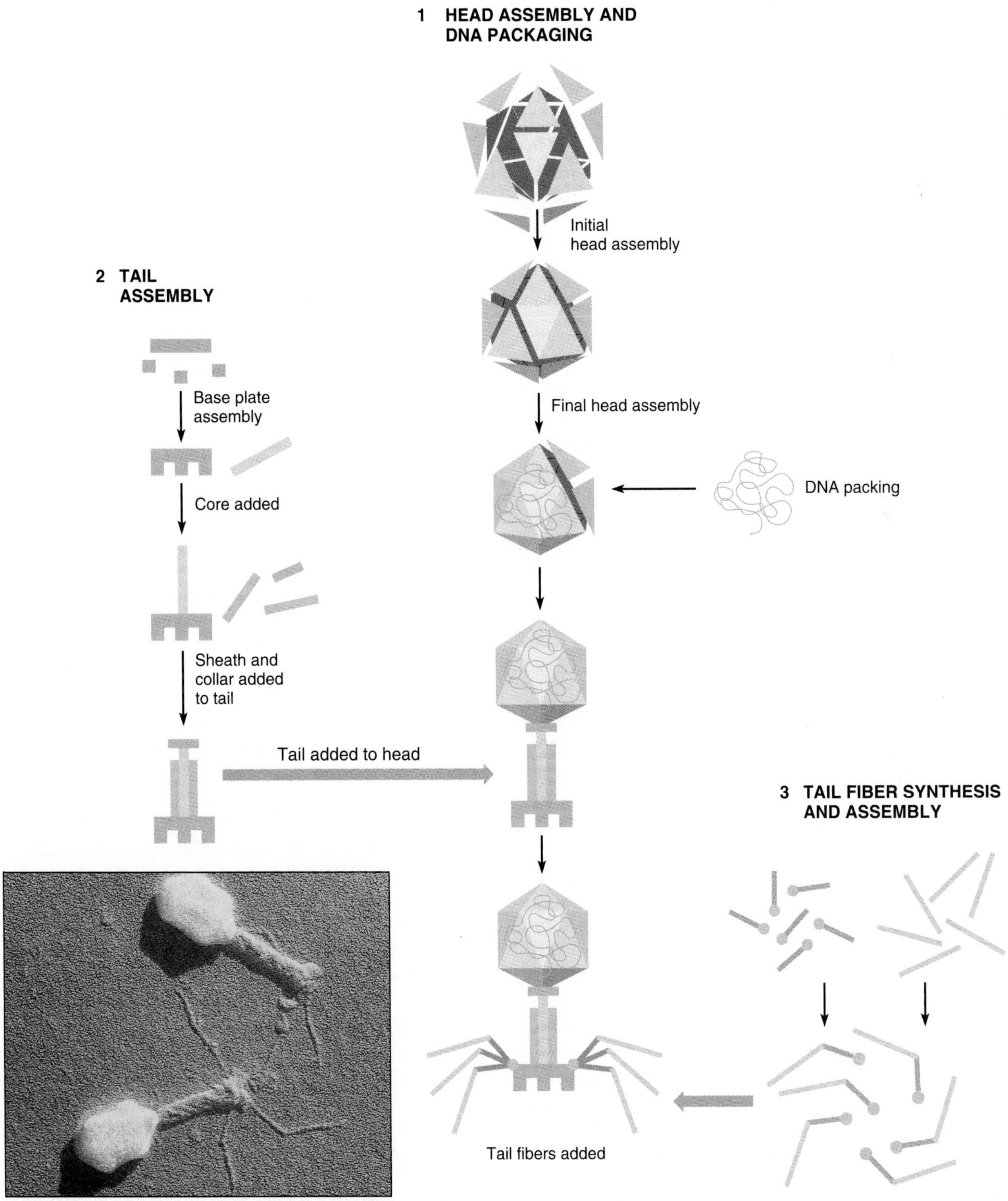

**FIGURE 19.14**    The structure and assembly of bacteriophage T4 following infection of *E. coli*. Three separate pathways lead to the formation of an icosahedral head filled with DNA, a tail containing a central core, and tail fibers. The tail is added to the head and then tail fibers are added. Each step of each pathway is influenced by one or more phage genes.

the head is assembled and DNA is packaged into it. This complex then combines with the tail, and only then are the tail fibers added. Total construction is a combination of self-assembly and enzyme-directed processes. When approximately 200 viruses are assembled, the bacterial cell is ruptured by the action of lysozyme and the mature phages are released. If this occurs on a lawn of bacteria, the 200 phages will infect other bacterial cells and the process will be repeated over and over again. For each isolated phage on the lawn, this leads to a clear area where all bacteria have been lysed. This area is called a **plaque** and represents multiple clones of the single-infecting T4 bacteriophage.

## The Plaque Assay

The experimental study of bacteriophages and other viruses has played a critical role in our understanding of molecular genetics. During infection of bacteria, enormous quantities of bacteriophages may be obtained for investigation. Often, over $10^{10}$ viruses per milliliter of culture medium are produced. Many genetic studies have relied on the ability to quantitate the number of phages produced following infection under specific culture conditions. The technique utilized is called the **plaque assay**.

This assay is illustrated in Figure 19.15, where actual plaque morphologies are also shown. A serial dilution of the original virally infected bacterial culture is first performed. Then, a 0.1-ml aliquot from one or more dilutions is added to a small volume of melted nutrient agar (about 3 ml) to which a few drops of a healthy bacterial culture have been mixed. The solution is then poured evenly over a base of solid nutrient agar in a Petri dish and allowed to solidify prior to incubation. As described in the above section, viral plaques occur at each place where a single virus has initially infected one bacterium in the lawn that has grown up during incubation. If the dilution factor is too low, the plaques are plentiful and will fuse, lysing the entire lawn. This has occurred in the $10^{-3}$ dilution in Figure 19.15. On the other hand, if the dilution factor is increased, fewer phages are present in a given aliquot and plaques can be counted. From such data, the density of viruses in the initial culture can be estimated. The calculation is the same type as that used for determining bacterial density by counting colonies following serial dilution of an initial culture:

$$(\text{plaque number/ml}) \times (\text{dilution factor})$$

Using the results shown in Figure 19.15, it is observed that there are 23 phage plaques derived from the 0.1-ml aliquot of the $10^{-5}$ dilution. Therefore, we can estimate that there are 230 phages per milliliter at this dilution. This initial viral density in the undiluted sample, where

23 plaques are observed from 0.1 ml, is calculated as

$$(230/\text{ml}) \times (10^5) = 230 \times 10^5/\text{ml}$$

Since this figure is derived from the $10^{-5}$ dilution, we can estimate that there will be only 0.23 phage per 0.1 ml in the $10^{-7}$ dilution. As a result, when 0.1 ml from this tube is assayed, it is expected that no phage particles will be present. This possibility is borne out in Figure 19.15, where an intact lawn of bacteria exists. The dilution factor is simply too great.

The use of the plaque assay has been invaluable in mutational and recombinational studies of bacteriophages. We will apply the above technique more directly later in this chapter, when Seymour Benzer's elegant genetic analysis of a single gene in phage T4 is discussed.

## Lysis and Lysogeny

As we first discussed in Chapter 16, the relationship between virus and bacterium does not always result in viral reproduction and lysis. As early as the 1920s it was known that some bacteriophages could enter a bacterial cell and establish a symbiotic relationship with it. The precise molecular basis of this symbiosis is now well understood. Upon entry, the viral DNA, instead of replicating in the bacterial cytoplasm, is integrated into the bacterial chromosome, a step that characterizes the **lysogenic pathway**. Subsequently, each time the bacterial chromosome is replicated, the viral DNA is also repli-

**FIGURE 19.15**    Diagrammatic illustration of the plaque assay for bacteriophage analysis. Serial dilutions of a bacterial culture infected with bacteriophages are first made. Then three of the dilutions ($10^{-3}$, $10^{-5}$, and $10^{-7}$) are analyzed using the plaque assay technique. In each case, 0.1 ml of the bacterial–viral culture is mixed with a few drops of a healthy bacterial culture in nutrient agar. This mixture is spread over and allowed to solidify on a base of agar. Each clear area that arises is a plaque and represents the initial infection of one bacterial cell by one bacteriophage. Each generation of viral reproduction involves more bacteria, creating a visible plaque. In the $10^{-3}$ dilution, so many phages are present that all bacteria are lysed. In the $10^{-5}$ dilution, 23 plaques are produced. In the $10^{-7}$ dilution, the dilution factor is so great that no phages are present in the 0.1-ml sample, and thus no plaques form. From the 0.1-ml aliquot of the $10^{-5}$ dilution, the original bacteriophage density can be calculated as $23 \times 10 \times 10^5$ phages/ml ($23 \times 10^5$ or $230 \times 10^5$), as described in the text. The photo illustrates serial dilutions of phage T2 on *E. coli*.

▶

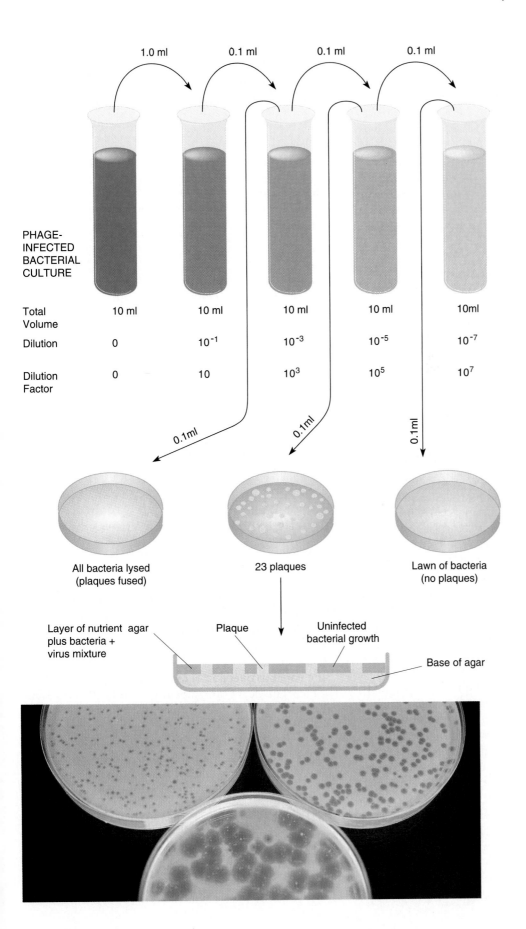

PHAGE-
INFECTED
BACTERIAL
CULTURE

|  | 1.0 ml | 0.1 ml | 0.1 ml | 0.1 ml |
|---|---|---|---|---|

| | | | | | |
|---|---|---|---|---|---|
| Total Volume | 10 ml | 10 ml | 10 ml | 10 ml | 10ml |
| Dilution | 0 | $10^{-1}$ | $10^{-3}$ | $10^{-5}$ | $10^{-7}$ |
| Dilution Factor | 0 | 10 | $10^3$ | $10^5$ | $10^7$ |

0.1ml    0.1ml    0.1ml

All bacteria lysed
(plaques fused)

23 plaques

Lawn of bacteria
(no plaques)

Layer of nutrient agar
plus bacteria +
virus mixture

Plaque

Uninfected
bacterial growth

Base of agar

cated and passed to daughter bacterial cells following division. No new viruses are produced and no lysis of the bacterial cell occurs. However, under certain stimuli, such as chemical or ultraviolet light treatment, the viral DNA may lose its integrated status and initiate replication; phage reproduction, and lysis of the bacterium.

Several terms are used to describe this relationship. The viral DNA, integrated in the bacterial chromosome, is called a **prophage**. Viruses that can either lyse the cell or behave as a prophage are called **temperate**. Those that can only lyse the cell are referred to as **virulent**. A bacterium harboring a prophage is said to be **lysogenic**; that is, it is capable of being lysed as a result of induced viral reproduction. The viral DNA, which can replicate either in the bacterial cytoplasm or as part of the bacterial chromosome, may be classified as an **episome**.

The relationship of a prophage and its host cell is symbiotic because lysogenic bacteria are immune to viral attacks by the same phage whose genetic information it harbors. This immunity is due to the synthesis of a prophage repressor molecule that regulates the expression of the viral DNA (Chapter 16).

# MUTATION AND RECOMBINATION IN VIRUSES

Much of what is known concerning viral genetics has been derived from studies of bacteriophages. Phage mutations often affect the morphology of the plaques formed following lysis of the bacterial cells. For example, in 1946 Alfred Hershey observed unusual T2 plaques on plates of *E. coli* strain B. While the normal T2 plaques are small and have a clear center surrounded by a turbid, diffuse halo, the unusual plaques were larger and possessed a sharp outer perimeter. When the viruses were isolated from these plaques and replated on *E. coli* B cells, an identical plaque appearance was noted. Thus, the plaque phenotype was an inherited trait resulting from the reproduction of mutant phages. Hershey named the mutant *rapid lysis* (*r*) because the plaques were larger, apparently resulting from a more rapid or more efficient life cycle of the phage. It is now known that wild-type phages undergo an inhibition of reproduction once a particular-sized plaque has been formed. The *r* mutant T2 phages are able to overcome this inhibition, producing larger plaques with a sharp perimeter.

Another bacteriophage mutation, *host range* (*h*), was discovered by Luria. This mutation extends the range of bacterial hosts that the phage can infect. Although wild-type T2 phages can infect *E. coli* B, they cannot normally attach or be absorbed to the surface of *E. coli* B-2. The *h*

**Table 19.3**   SOME MUTANT TYPES OF T-EVEN PHAGES

| Name | Description |
|---|---|
| minute | Small plaques |
| turbid | Turbid plaques on *E. coli* B |
| star | Irregular plaques |
| uv-sensitive | Alters UV sensitivity |
| acriflavin-resistant | Forms plaques on acriflavin agar |
| osmotic shock | Withstands rapid dilution into distilled water |
| lysozyme | Does not produce lysozyme |
| amber | Grows in *E. coli* K12, but not B |
| temperature-sensitive | Grows at 25°C, but not at 42°C |

mutation, however, provides the basis for adsorption and subsequent infection.

Table 19.3 lists other representative types of mutations that have been isolated and studied in the T-even series of bacteriophages (T2, T4, T6, etc.). These mutations are important to the study of genetic phenomena in bacteriophages. Conditional, temperature-sensitive mutations have been particularly valuable in the study of essential genes where mutations eliminate the ability of the phage to reproduce.

## Genetic Exchange Between Viruses

Around 1947, several research teams demonstrated that recombination occurs between viruses. This discovery of genetic exchange was made during experiments in which two mutant strains of bacteriophages were allowed to simultaneously infect the same bacterial culture. These **mixed infection experiments** were designed so that the number of viral particles sufficiently exceeded the number of bacterial cells so as to ensure simultaneous infection of most cells by both viral strains.

For example, in one study using the T2/*E. coli* system, the viruses were of either the $h^+r$ or $hr^+$ genotype. If no recombination occurred, these two parental genotypes (wild-type host range restriction, rapid lysis; and extended host range, normal lysis) would be the only expected phage progeny. However, the recombinants $h^+r^+$ and $hr$ were detected in addition to the parental genotypes (Figure 19.16). As with eukaryotes, the percentage of recombinant plaques divided by the total number of plaques reflects the relative distance between the genes:

$$\frac{(h^+r^+) + (hr)}{\text{total plaques}} \times 100 = \text{recombinational frequency}$$

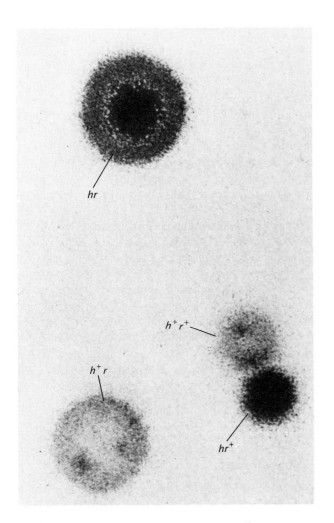

**FIGURE 19.16** Actual plaque phenotypes observed following simultaneous infection of *E. coli* by two strains of phage T2, *h⁺r* and *hr⁺*. In addition to the parental genotypes, recombinant plaques *hr* and *h⁺r⁺* were also recovered.

**Table 19.4** RESULTS OF A CROSS INVOLVING THE *H* AND *R* GENES IN PHAGE T2

| Genotype | Plaques | Designation |
|---|---|---|
| *hr⁺* | 42 | Parental Progeny 76% |
| *h⁺r* | 34 | |
| *h⁺r⁺* | 12 | Recombinants 24% |
| *hr* | 12 | |

SOURCE: Data derived from Hershey and Rotman, 1949.

Sample data for the *h* and *r* loci are shown in Table 19.4.

Similar recombinational studies have been performed with large numbers of mutant genes in a variety of bacteriophages. Data are analyzed in much the same way as they are in eukaryotic mapping experiments. Two- and three-point mapping crosses are possible, and the percentage of recombinants in the total number of phage progeny is calculated. This value is proportional to the relative distance between two genes along the DNA molecule constituting the chromosome.

An interesting observation in phage crosses is **negative interference**. Recall that in eukaryotic mapping crosses, positive interference is the rule. Fewer than expected double-crossover events are observed. In phage crosses, often just the reverse occurs. In three-point analysis, a greater than expected frequency of double exchanges is observed. Negative interference is explained on the basis of the dynamics of the conditions leading to recombination within the bacterial cell. The available evidence supports the concept that recombination between phage chromosomes involves a breakage and reunion process similar to that of eukaryotic crossing over. The process is facilitated by nucleases that nick and reseal DNA strands. A fairly clear picture of the dynamics of viral recombination has emerged.

Following the early phase of infection, the chromosomes of the phages begin replication. As this stage progresses, a pool of chromosomes accumulates in the bacterial cytoplasm. If double infection by phages of two genotypes has occurred, then the pool of chromosomes initially consists of the two parental types. Genetic exchange between these two types will occur before, during, and after replication, producing recombinant chromosomes.

In the case of the *h⁺r* and *hr⁺* example just discussed, *h⁺r⁺* and *hr* chromosomes are produced. Each of these recombinant chromosomes may undergo replication and is also free to undergo new exchange events with the other and with the parental types. Furthermore, recombination is not restricted to exchange between two chromosomes—three or more may be involved. As phage development progresses, chromosomes are randomly removed from the pool and packed into the phage head, forming mature phage particles. Thus, parental and recombinant genotypes are produced.

With powerful selection systems, it is possible to detect *intragenic* recombination in viruses, where exchanges occur at points within a single gene. Such studies have led to the fine-structure analysis of the gene, as discussed in an ensuing section of this chapter. Before we turn to that topic, however, we will first discuss the third type of bacterial recombination, transduction, which is mediated by bacteriophages.

# TRANSDUCTION: VIRUS-MEDIATED BACTERIAL DNA TRANSFER

In 1952, Joshua Lederberg and Norton Zinder were investigating possible recombination in the bacterium *Salmonella typhimurium.* Although they recovered prototrophs from mixed cultures of two different auxotrophic strains, subsequent investigations revealed that recombination was occurring in a manner different from that attributable to the presence of an F factor, as in *E. coli.* What they were to discover was a process of bacterial recombination mediated by bacteriophages and now called **transduction**.

## The Lederberg–Zinder Experiment

Lederberg and Zinder mixed the *Salmonella* auxotrophic strains LA-22 and LA-2 together and recovered prototroph cells when the mixture was plated on minimal medium. LA-22 was unable to synthesize the amino acids phenylalanine and tryptophan (*phe⁻ trp⁻*), and LA-2 could not synthesize the amino acids methionine and histidine (*met⁻ his⁻*). Prototrophs (*phe⁺ trp⁺ met⁺ his⁺*) were recovered at a rate of about $1/10^5$ ($10^{-5}$) cells.

Although these observations at first appeared to suggest that the type of recombination involved was the kind observed earlier in conjugative strains of *E. coli*, experiments using the Davis U-tube soon showed otherwise (Figure 19.17). When the two auxotrophic strains were separated by a glass-sintered filter, thus preventing cell contact but allowing growth to occur in a common medium, a startling observation was made. When samples were removed from both sides of the filter and plated independently on minimal medium, prototrophs were recovered, but only from one side of the tube. Recall that if conjugation were responsible, the conditions in the Davis U-tube would have prevented recombination altogether.

Prototrophs were recovered only when cells from the side of the tube containing LA-22 bacteria were plated. Obviously, the presence of LA-2 cells on the other side of the tube was essential for recombination, since LA-2 cells were the source of the new genetic information. Because the genetic information responsible for recombination had somehow to pass across the filter, but in an unknown form, it was initially designated simply as a **filterable agent** (**FA**).

Three subsequent observations made it clear that this recombination was quite distinct from any other form of recombination.

1. The FA would not pass across filters with a pore diameter of less than 100 nm, a size that normally allows passage of small DNA molecules.

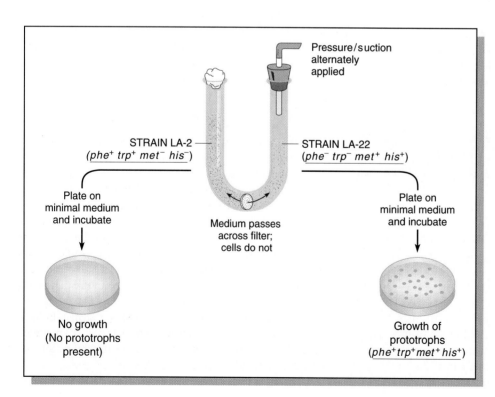

**FIGURE 19.17** The Lederberg–Zinder experiment using *Salmonella.* After mixing two auxotrophic strains in a Davis U-tube, Lederberg and Zinder recovered prototrophs from the side containing LA-22 but not from the side containing LA-2. These initial observations led to the discovery of the phenomenon called transduction.

Pressure/suction alternately applied

STRAIN LA-2
(*phe⁺ trp⁺ met⁻ his⁻*)

STRAIN LA-22
(*phe⁻ trp⁻ met⁺ his⁺*)

Plate on minimal medium and incubate

Medium passes across filter; cells do not

Plate on minimal medium and incubate

No growth
(No prototrophs present)

Growth of prototrophs
(*phe⁺trp⁺met⁺his⁺*)

**2.** Testing the FA in the presence of DNase, which will enzymatically digest DNA, showed that the FA was not destroyed. These first two observations demonstrate that FA is not naked DNA.

**3.** The third observation was particularly important. It was observed that FA was produced by the LA-2 cells only when they were grown in association with LA-22 cells. If LA-2 cells were grown independently and that culture medium was then added to LA-22 cells, recombination was not observed. Therefore, LA-22 cells play some role in the production of FA by LA-2 cells and do so only when sharing common growth medium. This was a key observation in determining the basis of genetic recombination.

These observations were explained by the presence of a prophage (P22) present in the LA-22 *Salmonella* cells. Rarely, P22 prophages entered the vegetative or lytic phase, reproduced, and lysed some of the LA-22 cells. This phage, being much smaller than a bacterium, was then able to cross the filter and lyse some of the LA-2 cells, because this strain was not immune to attack. In the process of lysis, the P22 phages produced in LA-2 often acquired a region of the LA-2 chromosome along with their own genetic material. If this region contained the *phe*$^+$, and *trp*$^+$ genes present in LA-2, and if the phages subsequently passed back across the filter and reinfected LA-22 cells, prototrophs were produced. The exact nature of how this occurred was not immediately clear.

## The Nature of Transduction

Further studies revealed the existence of transducing phages in other species of bacteria. For example, *E. coli*, *Bacillus subtilis*, and *Pseudomonas aeruginosa* can be transduced by the phages P1, SP10, and F116, respectively. The precise mode of transfer of DNA during transduction has also been established. The process most often begins when a prophage enters the lytic cycle and progeny viruses subsequently infect and lyse other bacteria. During infection, the bacterial DNA is degraded into small fragments and the viral DNA is replicated. As packaging of the viral chromosomes in the protein head of the phage occurs, errors are sometimes made. Rarely, instead of viral DNA, segments of bacterial DNA are packaged.

Even though the initial discovery of transduction involved lysogenic bacteria, the same process can occur during the normal lytic cycle. Following infection, the bacterial chromosome is degraded into small pieces. If, during bacteriophage assembly, a small piece of bacterial DNA is packaged along with the viral chromosome, subsequent transduction is possible.

Sometimes, *only* bacterial DNA is packaged! Regions as large as 1 percent of the bacterial chromosome may become enclosed randomly in the viral head. Following lysis, these aberrant phages, lacking their own genetic material, are released in the culture medium. Because the ability to infect is a property of the protein coat, they can initiate infection of other unlysed bacteria. When this occurs, bacterial rather than viral DNA is injected into the bacterium and can either remain in the cytoplasm or recombine with its homologous region of the bacterial chromosome. If the bacterial DNA remains in the cytoplasm, it does not replicate but may remain in one of the progeny cells following each division. When this happens, only a single cell, partially diploid for the transduced genes, is produced—a phenomenon called **abortive transduction**. If the bacterial DNA recombines with its homologous region of the bacterial chromosome, the transduced genes are replicated as part of the chromosome and passed to all daughter cells. As in conjugation and transformation, an even number of crossovers is necessary for recombination to occur. This process is called **complete transduction**. Both abortive and complete transduction are subclasses of the broader category of **generalized transduction**. As described above, transduction is characterized by the random nature of DNA fragments and genes transduced. Each fragment has a finite but small chance of being packaged in the phage head. Most cases of generalized transduction are of the abortive type; some data suggest that complete transduction occurs 10 to 20 times less frequently. This finding may be related to the fact that double-stranded DNA is involved. In comparison with single-stranded DNA, which is integrated during transformation, it may be much more difficult for double-stranded DNA to become integrated.

## Mapping and Specialized Transduction

As with transformation, transduction has been used in linkage studies and mapping of the bacterial chromosome. Cotransduced genes must be aligned closely to one another along the chromosome. By concentrating on two or three linked genes, transduction studies can also determine the precise order of genes. Such an analysis is predicated on the same rationale underlying other mapping techniques, where outcomes resulting from single events occur much more frequently than those relying on two or more independent events occurring simultaneously.

Another aspect of virally mediated bacterial recombination involves **specialized transduction**. Compared with generalized transduction, where all genes have an equal probability of being transduced, specialized transduction is restricted to certain genes. One of the best

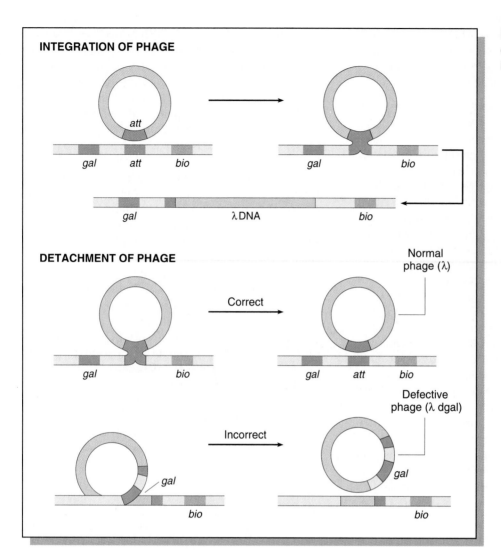

**FIGURE 19.18** Production of a defective phage λ, leading to specialized transduction.

examples involves transduction of *E. coli* by the pro-phage λ. In this case, only the *gal* (galactose) or *bio* (biotin) genes are transduced.

The reason why specialized transduction occurs became clear when it was learned how the λ DNA integrates into the *E. coli* chromosome during lysogeny. A region of λ DNA, designated *att*, is some 15 nucleotides long and is responsible for integration. A precisely homologous region exists on the *E. coli* chromosome, which is flanked by the *gal* and *bio* loci on either end. Therefore, λ DNA always integrates at a location in the chromosome between these genes (Figure 19.18).

Phage λ DNA can be induced to excise (detach) from the chromosome and cause lysis. Sometimes the excision process occurs incorrectly and carries either the *gal* or *bio E. coli* genes in place of part of the viral DNA. This happens when the recombinational event leading to detachment occurs incorrectly, outside the *att* region of the λ chromosome. The resulting chromosome is defective

because it has lost some of its genetic information, but it is nevertheless replicated and packaged during the formation of mature phage particles. Once the previously lysogenized cell is lysed, the virus can inject the defective chromosome into another bacterial cell.

In a process involving lysogeny by a nondefective chromosome, the defective viral chromosome is integrated into the bacterial chromosome and is replicated along with it. Such cells are diploid for the *gal* or *bio* genes. If the recipient cells are *gal⁻* and are unable to use galactose as a carbon source, the presence of the transducing *gal⁺* DNA will cause them to revert to a *gal⁺* phenotype, where they can use this carbohydrate. In a similar way, *bio⁻* cells can be transduced to a *bio⁺* phenotype.

As is evident from this discussion, specialized transduction occurs in quite a different way from generalized transduction. Because it is limited to certain genes, it is not as useful in linkage or mapping studies.

# INTRAGENIC RECOMBINATION IN PHAGE T4

We conclude this chapter with an account of an ingenious example of **genetic analysis**. In the early 1950s Seymour Benzer undertook a detailed examination of a single locus, *rII*, in phage T4. Benzer successfully designed experiments to recover the extremely rare genetic recombinants arising as a result of **intragenic exchange**. He demonstrated that such recombination indeed occurs between the DNA of individual phages during simultaneous infection of the host bacterium *E. coli*. In effect, such recombination is equivalent to eukaryotic crossing over, but in this case, within the gene rather than between two genes.

The end result of Benzer's work was the production of a detailed map of the *rII* locus. As we shall see, the locus consists of two functional parts, leading to two separate gene products. Because of the extremely detailed information provided from his analysis, Benzer's work is often described as the **fine-structure analysis of the gene**. Because these experiments occurred decades before DNA sequencing techniques were developed, the insights concerning the internal structure of the gene took on added significance.

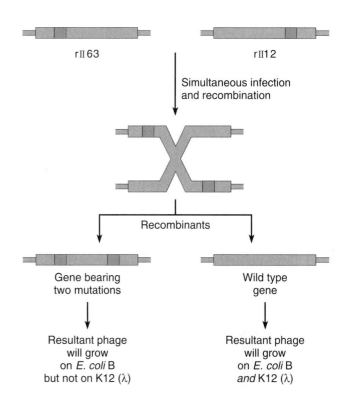

**FIGURE 19.19**    Production of recombinant wild-type phages following genetic exchange within the *rII* locus of phage T4.

## The *rII* Locus of Phage T4

The primary requirement in genetic analysis is the isolation of a large number of mutations in the gene being investigated. Mutants at the *rII* locus produce distinctive plaques when plated on *E. coli* strain B, allowing their easy identification (see Figure 19.16). Benzer's approach was to isolate many independent *rII* mutants—he eventually obtained about 20,000—and to attempt to perform recombinational studies in order to produce a genetic map of this locus.

*The key to his analysis* was that *rII* mutant phages, while capable of infecting and lysing *E. coli* B, could not successfully lyse a second related strain, *E. coli* K12 ($\lambda$). However, wild-type phages could lyse both the B and the K12 strains. Benzer realized that these conditions provided the potential for a highly sensitive screening system. He reasoned that if phages from any two independent mutant strains were allowed to simultaneously infect *E. coli* B, any exchanges between the two mutant sites within the locus would produce rare wild-type recombinants (Figure 19.19). If the phage population, which contained over 99.9 percent *rII* phages and less than 0.1 percent wild-type phages, were then allowed to infect strain K12, only those wild-type recombinants would be successful in reproducing and producing wild-

type plaques. This is the critical step in recovering and quantifying rare recombinants.

By using serial dilution techniques, it is possible to determine the total number of mutant *rII* phages produced on *E. coli* B and the total number of recombinant wild-type phages that would lyse *E. coli* K12 ($\lambda$). These data provide the basis for calculating the frequency of recombination. This value is considered to be proportional to the distance (within the gene) between the two mutations being studied. When information from many such experiments is combined, a detailed map of the locus is possible.

As we will quickly see, this experimental design was extraordinarily sensitive. It was possible for Benzer to detect as few as one recombinant wild-type phage among 100 million mutant phages. Such resolution is truly remarkable!

## Complementation by *rII* Mutations

When Benzer initiated his studies, he soon realized that many pairs of mutations demonstrated **complementation** if plated directly on *E. coli* K12 ($\lambda$). This was unexpected because only wild-type phages reproduce on this strain. If indeed the *rII* mutations represent a sin-

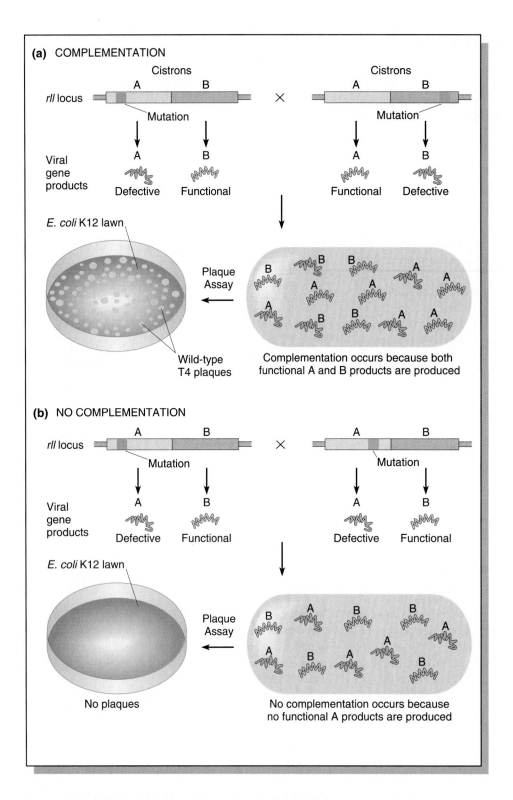

**FIGURE 19.20** Comparison of two *rII* mutations that either complement one another (a), or do not (b). Complementation occurs when each mutation is in a separate cistron and results in lysis of the cell. Failure to complement occurs if the two mutations are in the same cistron and results in no lysis.

gle gene producing a single product, this should have been impossible. However, to explain this observation, Benzer suggested that more than one functional product was encoded by this locus. Upon testing, all mutations fell into one of two possible complementation groups, A or B. Benzer coined the term **cistron** to describe each group, which he defined as the smallest functional genetic unit.

We have discussed complementation before in the text (see Chapter 11). The mechanism by which the phenomenon results in the lysis of *E. coli* K12 (λ) (during simultaneous infection by pairs of mutant phage strains) is diagrammed in Figure 19.20(a). Provided that at least some of the normal gene products from both the A and B cistrons are produced following infection, wild-type phages will be produced. As shown in Figure 19.20(b), if two independent mutations are in the same cistron, this is not possible and no complementation occurs.

As a result of these initial studies, Benzer was able to place all *rII* mutations in either the A or the B cistron. He was now poised to return to his recombination studies, testing mutations in the A cistron against each other and mutations in the B cistron against each other.

## Recombinational Analysis

Of the approximately 20,000 *rII* mutations, roughly half fell into each cistron. Benzer set about to map the mutations within each one. For example, if two *rII*A mutants were first allowed to infect *E. coli* B in a liquid culture, and if a recombination event occurred between the mutational sites in the A cistron, then wild-type progeny viruses would be produced at low frequency. If samples of the progeny viruses from such an experiment were then plated on *E. coli* K12, only the recombinants would lyse the bacteria and produce plaques. The total number of nonrecombinant progeny viruses were also determined by plating samples on *E. coli* B.

This experimental protocol is illustrated in Figure 19.21. The percentage of recombinants can be determined by counting the plaques at the appropriate dilution in each case. As in eukaryotic mapping experiments, the frequency of recombination can be taken as an estimate of the distance between the two mutations within the cistron.

For example, if the number of recombinants is equal to $4 \times 10^3$/ml, and the total number of progeny is $8 \times 10^9$/ml, then the frequency of recombination between the two muants is

$$2\left(\frac{4 \times 10^3}{8 \times 10^9}\right) = 2(0.5 \times 10^{-6})$$
$$= 10^{-6}$$
$$= 0.000001$$

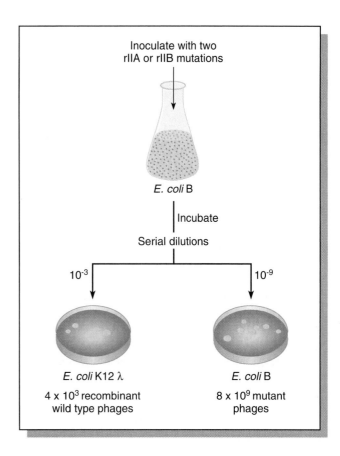

**FIGURE 19.21**    The experimental protocol for recombinational studies between pairs of *rII*A or *rII*B mutations.

Multiplying by 2 is necessary because each recombinant event yields two reciprocal products. Only one of them— the wild type—is detected (see Figure 19.19).

## Deletion Testing of the *rII* Locus

While the selective system of recombination just described was available to map mutations within each cistron, testing 1000 mutants two at a time in all combinations would have required millions of experiments. Fortunately, Benzer was able to overcome this obstacle when he discovered that some of the *rII* mutations were in reality **deletions** of small parts of each cistron. That is, the genetic changes giving rise to the *rII* properties were not point mutations involving a single nucleotide change, but instead were due to the loss or deletion of a variable number of nucleotides within the cistron. These deletions could be identified by their failure to revert to wild type. Doing so is a characteristic of point mutations. Furthermore, *and this is the key point,* it was observed that a deletion, when tested during simultaneous infection with a point mutation located in the deleted part of the same cistron, *never* yielded any wild-type recombi-

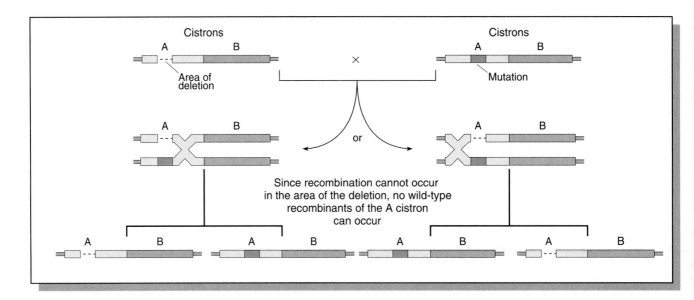

**FIGURE 19.22**    Demonstration that recombination between a phage chromosome with a deletion in the A cistron and another with a mutation overlapped by that deletion cannot yield a chromosome with wild-type A and B cistrons.

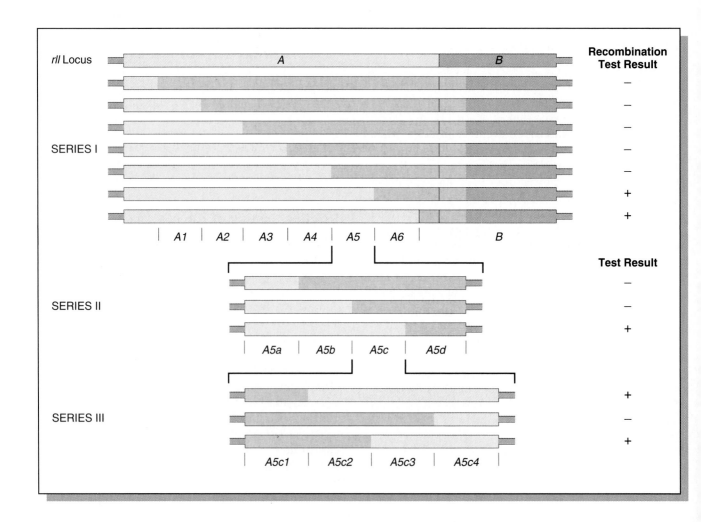

nants. The basis for the failure to do so is illustrated in Figure 19.22. Thus, a method was available that could roughly but quickly localize any mutation, provided it was contained within a region covered by a deletion.

As shown in Figure 19.23, this second property served as the basis for localization of each mutation. Seven overlapping deletions (shown in color) covered variable portions of the A and B regions and were used for initial screening of the point mutations. Each point mutation was ultimately assigned to an area of the cistron corresponding to one specific deletion. Then, further deletions within each of the seven areas were used to localize or map each *rII* point mutation more specifically. Remember that in each case, a point mutation is localized in the area of a deletion when it fails to give rise to any wild-type recombinants.

## The *rII* Gene Map

After several years of work, Benzer produced a genetic map of the two cistrons composing the *rII* locus of phage T4 (Figure 19.24). Of the 20,000 mutations analyzed, 307 distinct sites within this gene had been mapped in relation to one another. Many mutations fell into the same site. Areas containing many mutations were designated as **hot spots**. It appeared that such areas were more susceptible to mutation than areas where only one or a few mutations are found. Additionally, Benzer found areas within the cistrons where no mutations were localized. He estimated that as many as 200 recombinational units had not been localized by his studies.

The significance of Benzer's work is his application of genetic analysis to what had previously been considered an abstract unit—the gene. Benzer demonstrated that a gene is not an indivisible particle but instead consists of mutational and recombinational units. His research was performed shortly after the publication of Watson and Crick's work on DNA and some time before the genetic code was unraveled in the early 1960s. Thus, his analysis is considered one of the classic examples of genetic experimentation.

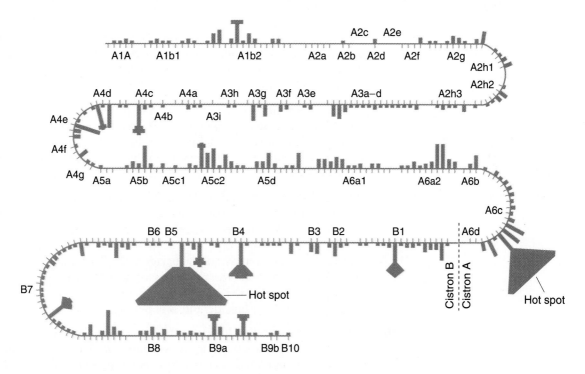

**FIGURE 19.23**    Three series of overlapping deletions in the *rII* locus used to localize the position of an unknown *r* mutation. For example, if a mutant strain tested against each deletion (shaded in gray) in Series 1 for the production of recombinant wild-type progeny shows the results at the right (+ or −), the mutation must be in segment A5. In Series II, the mutation is further narrowed to segment A5c, and in Series III to segment A5c3.

**FIGURE 19.24**    A partial map of mutations in the A and B cistrons of the *rII* locus of phage T4. Each square represents an independently isolated mutation. Note the two areas where the largest number of mutations are present, referred to as "hot spots" (A6cd and B5).

**CHAPTER SUMMARY**

1. Although the concept of spontaneous mutation is supported experimentally and is widely upheld, recent evidence suggests that some mutations that are adaptive for the survival of the organism may arise in response to selective pressure from the environment.

2. Genetic recombination in bacteria may be accomplished as a result of three different genetic modes: conjugation, transformation, and transduction.

3. Conjugation is initiated by a bacterium housing a plasmid called the F factor. If the F factor is in the cytoplasm ($F^+$), one strand is transferred to and replicated in the recipient cell, converting it to the $F^+$ status.

4. If the F factor is integrated into the donor cell chromosome (Hfr), recombination is initiated with the recipient cell, with genetic information flowing unidirectionally. Time mapping of the bacterial chromosome is based on the orientation and position of the F factor in the donor chromosome.

5. The products of a group designated as the *rec* genes have been found to be directly involved in the process of recombination between the invading DNA and the recipient bacterial chromosome.

6. Plasmids, such as the F factor, are autonomously replicating DNA molecules found in the bacterial cytoplasm. Some plasmids may contain genes such as those conferring antibiotic resistance, as well as those necessary for their transfer during conjugation.

7. The phenomenon of natural transformation, which does not require cell contact, involves the entry of single-stranded exogenous DNA into the host chromosome of a recipient bacterial cell. Linkage mapping of closely aligned genes may be performed using this process.

8. Bacteriophages may be studied using the plaque assay. The discovery of mutations in plaques, such as those causing variation in plaque morphology and the alteration of the host range of infectivity, have allowed the analysis of recombination between viruses. Such genetic exchange occurs during simultaneous (or mixed) infection of bacterial cells by two distinct mutant viral strains.

9. Transduction, or virus-mediated bacterial DNA recombination, requires the formation of a symbiotic relationship between bacteria and bacteriophages that infect them. When a lysogenized bacterium enters the lytic cycle, it may serve as the vehicle for the transfer of host (bacterial) DNA. In the process of generalized transduction, a random part of the bacterial chromosome is transferred. In specialized transduction, only specific genes adjacent to the point of insertion of the prophage into the bacterial chromosome are transferred. Transduction may also be used for bacterial linkage and mapping studies.

10. Genetic analysis of the *rII* locus in bacteriophage T4 allowed Seymour Benzer to study intragenic recombination. By isolating *rII* mutants, performing recombinational studies, and relying on deletion mapping, Benzer was able to locate 20,000 mutations at 307 distinct sites within the two cistrons of the *rII* locus.

## KEY TERMS

abortive transduction
adaptation hypothesis
bacteriophage
cistron
colicin
Col plasmid
competence
complementation
complete transduction
conjugation
cotransformation
delayed early genes
deletions
D-loop
eclipse period

episome
F⁻ cells
F⁺ cells
F′ cells
fertility factor (F factor)
filterable agent (FA)
fine-structure analysis of
   the gene
fluctuation test
F pilus
generalized transduction
genetic analysis
heteroduplex
Hfr (high-frequency
   recombination)

hot spots
immediate early genes
interrupted mating
   technique
intragenic exchange
late genes
latent period
lysogenic pathway
lysogeny
lysozyme
merozygote
mixed infection
   experiments
negative interference
plaque

plaque assay
plasmid
prophage
r-determinant
*rec* genes
resistance transfer factor
   (RTF)
R plasmid
sex pilus
specialized transduction
spontaneous mutation
temperate phage
transduction
transformation

# INSIGHTS AND SOLUTIONS

1.  Time mapping was performed in a cross involving the genes *his, leu, mal*, and *xyl*. After 25 minutes, mating was interrupted with the following results in recipient cells.

    90% were *xyl*
    80% were *mal*
    20% were *his*
    none were *leu*

    What are the positions of these genes relative to the origin (*O*) of the F factor and to one another?

    **SOLUTION:**  Since the *xyl* gene was transferred most frequently, it is closest to *O* (very close). The *mal* gene is next and reasonably close to *xyl*, followed by the *his* gene. The *leu* gene is well beyond these three, since no recombinants are recovered that include it.

    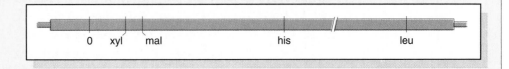

2.  Three strains of bacteria, each bearing a separate mutation, $a^-$, $b^-$, or $c^-$, were used as the sources of donor DNA in a transformation experiment involving recipient cells that were wild type for those genes, but which expressed the mutant $d^-$.

    (a) Based on the following data, and assuming that the location of the *d* gene precedes the *a*, *b*, and *c* genes, propose a linkage map for the four genes:

| Donor DNA | Recipient | Transformants | Frequency of ++ Transformants |
|-----------|-----------|---------------|-------------------------------|
| $a^-\ d^+$ | $a^+\ d^-$ | $a^+\ d^+$ | 0.21 |
| $b^-\ d^+$ | $b^+\ d^-$ | $b^+\ d^+$ | 0.18 |
| $c^-\ d^+$ | $c^+\ d^-$ | $c^+\ d^+$ | 0.63 |

**SOLUTION:** These data reflect the relative distances between each of the *a*, *b*, *c* genes and the *d* gene. The *a* and *b* genes are about the same distance away from the *d* gene and are thus tightly linked to one another. The *c* gene is more distant. Assuming that the *d* gene precedes the others, the map looks like this:

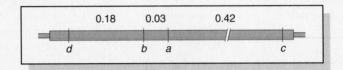

(b) If the donor DNA was wild type and recipient cells were either $a^-b^-$, $a^-c^-$, or $b^-c^-$, in which case would wild-type transformants be expected most frequently?

**SOLUTION:** Since the *a* and *b* genes are closely linked, they may be co-transformed in a single event. Thus, recipient cells of $a^-b^-$ are most likely to be converted to wild type.

3. In four Hfr strains of bacteria, all derived from an original $F^+$ culture grown over several months, a group of hypothetical genes were studied and shown to be transferred in the following orders:

| Hfr Strain | Order of Transfer |
|------------|-------------------|
| 1 | E R I U M B |
| 2 | U M B A C T |
| 3 | C T E R I U |
| 4 | R E T C A B |

Assuming that B is the first gene along the chromosome, determine the sequence of all genes shown. One strain creates an apparent dilemma. Which one is it? Explain why the dilemma is only apparent and not real.

**SOLUTION:** The solution is arrived at by overlapping the genes in each strain in the sequence in which they were transferred:

```
        2  U M B A C T
Strain: 3          C T E R I U
        1              E R I U M B
```

Starting with B, the sequence of genes is: BACTERIUM!

Strain 4 creates an apparent dilemma, which is resolved by realizing that the F factor is integrated in the opposite orientation; thus the genes enter in the opposite sequence; starting with gene R.

MUIRETCAB
$\longrightarrow$

**4.** In Benzer's fine-structure analysis of the *rII* locus in phage T4, he was able to perform complementation testing of any pair of mutations once it was clear that the locus contained two cistrons. Complementation was assayed by simultaneously infecting *E. coli* K12 with two phage strains, each with an independent mutation, neither of which could alone lyse K12. From the following data, determine which mutations are in which cistron, assuming that mutation 1 (M-1) is in the A cistron and mutation 2 (M-2) is in the B cistron. Are there any cases where the mutation cannot be properly assigned?

| Test Pair | Results |
|-----------|---------|
| 1, 2 | + |
| 1, 3 | − |
| 1, 4 | − |
| 1, 5 | + |
| 2, 3 | − |
| 2, 4 | + |
| 2, 5 | − |

**SOLUTION:**   M-1 and M-5 complement one another and are therefore not in the same cistron. Thus, M-5 must be in the B cistron. M-2 and M-4 complement one another. Using the same reasoning, M-4 is not with M-2 and therefore is in the A cistron.

M-3 fails to complement either M-1 or M-2 and would not seem to be in either cistron. One explanation is that the physical basis of M-3 somehow overlaps both the A and B cistron. It might be a double mutation with one in each cistron. It might also be a deletion that overlaps both cistrons, making it impossible for it to complement either M-1 or M-2.

**5.** Another mutation, M-6, was tested with the results shown below. Draw all possible conclusions about M-6.

| Test Pair | Results |
|-----------|---------|
| 1, 6 | + |
| 2, 6 | − |
| 3, 6 | − |
| 4, 6 | + |
| 5, 6 | − |

**SOLUTION:**   These results are consistent with assigning M-6 to the B cistron.

**6.** Recombination testing was then performed for M-2, M-5, and M-6 in order to map the B cistron. Recombination analysis using both *E. coli* B and K12 showed that recombination occurred between M-2 and M-5 and between M-5 and M-6, but not between M-2 and M-6. Why not?

**SOLUTION:**   Either M-2 and M-6 represent identical mutations, or one of them may be a deletion that overlaps the other, but not M-5. Furthermore, the data cannot rule out the possibility that both are deletions.

**7.** In recombination studies, what is the significance of the value determined by calculating growth on the K12 versus the B strains of *E. coli* following simultaneous infection in *E. coli*? Which value is always greater?

**SOLUTION:**   By performing plaque analysis on *E. coli* B, where wild-type and mutant phages are both lytic, the total number of phages per milliliter can be determined. Since almost all cells are *rII* mutants of one type or another, this value is much larger. To avoid total lysis of the plate, extensive dilutions are necessary. On K12, *rII* mutations will not grow, but wild-type phages will. Since wild-type phages are the rare recombinants, there are relatively few of them and extensive dilution is not required.

---

**PROBLEMS AND
DISCUSSION
QUESTIONS**

1.  Contrast the two major hypotheses explaining the origin of heritable variation in bacteria.

2.  Discuss the experiments performed by Luria and Delbruck that clarified the origin of heritable variation in bacteria. What was the rationale for their conclusions? Why are these experiments called the "fluctuation tests"?

3.  Distinguish between the three modes of recombination in bacteria.

4.  With respect to $F^+$ and $F^-$ bacterial matings, answer the following questions:
    (a) How was it established that physical contact was necessary?
    (b) How was it established that chromosome transfer was unidirectional?
    (c) What is the physical–chemical basis of a bacterium being $F^+$?

5.  List all major differences (a) between the $F^+ \times F^-$ and the Hfr $\times F^-$ bacterial crosses; and (b) between $F^+$, $F^-$, Hfr, and $F'$ bacteria.

6.  Describe the basis for chromosome mapping in the Hfr $\times F^-$ crosses.

7.  Why are the recombinants produced from an Hfr $\times F^-$ cross almost never $F^+$?

8.  Describe the origin of $F'$ bacteria and merozygotes.

9.  Describe what is known about the mechanisms of the transformation process.

10. In a transformation experiment involving a recipient bacterial strain of genotype $a^- b^-$, the following results were obtained:

| Transforming DNA | % Transformants | | |
|---|---|---|---|
| | $a^+ b^-$ | $a^- b^+$ | $a^+ b^+$ |
| $a^+ b^+$ | 3.1 | 1.2 | 0.04 |
| $a^+ b^-$ and $a^- b^+$ | 2.4 | 1.4 | 0.03 |

What can you conclude about the location of the *a* and *b* genes relative to each other?

11. In a transformation experiment, donor DNA was obtained from a prototroph bacterial strain $(a^+ b^+ c^+)$, and the recipient was a triple auxotroph $(a^- b^- c^-)$. The following transformant classes were recovered:

$$
\begin{array}{ll}
a^+ b^- c^- & 180 \\
a^- b^+ c^- & 150 \\
a^+ b^+ c^- & 210 \\
a^- b^- c^+ & 179 \\
a^+ b^- c^+ & 2 \\
a^- b^+ c^+ & 1 \\
a^+ b^+ c^+ & 3 \\
\end{array}
$$

What general conclusions can you draw about the linkage relationships among the three genes?

12. Explain the observations that led Zinder and Lederberg to conclude that the protorophs recovered in their transduction experiments were not the result of F-mediated conjugation.

13. Define plaque, lysogeny, and prophage.

14. Differentiate between generalized and restricted transduction. Which can be used in mapping and why?

15. Two theoretical genetic strains of a virus ($a^- b^- c^-$ and $a^- b^+ c^+$) are used to simultaneously infect a culture of host bacteria. Of 10,000 plaques scored, the following genotypes were observed:

| | | | |
|---|---|---|---|
| $a^+ b^+ c^+$ | 4100 | $a^- b^+ c^-$ | 160 |
| $a^- b^- c^-$ | 3990 | $a^+ b^- c^+$ | 140 |
| $a^+ b^- c^-$ | 740 | $a^- b^- c^+$ | 90 |
| $a^- b^+ c^+$ | 670 | $a^+ b^+ c^-$ | 110 |

Determine the genetic map of these three genes on the viral chromosome. Determine whether interference was positive or negative.

16. Describe the conditions under which genetic recombination may occur in bacteriophages.

17. If a single bacteriophage infects one *E. coli* cell present on a lawn of bacteria, and upon lysis yields 200 viable viruses, how many phages will exist in a single plaque if only three more lytic cycles occur?

18. A phage-infected bacterial culture was subjected to a series of dilutions and a plaque assay was performed in each case with the results shown below. What conclusion can be drawn in the case of each dilution?

| | Dilution Factor | Assay Results |
|---|---|---|
| (a) | $10^4$ | All bacteria lysed |
| (b) | $10^5$ | 14 plaques |
| (c) | $10^6$ | 0 plaques |

19. In complementation studies of the *rII* locus of phage T4, three groups of three different mutations were tested. For each group, only two combinations were tested. On the basis of each set of data, predict the results of the third experiments.

| Group A | Group B | Group C |
|---|---|---|
| $d \times e$—lysis | $g \times h$—no lysis | $j \times k$—lysis |
| $d \times f$—no lysis | $g \times i$—no lysis | $j \times l$—lysis |
| $e \times f$—? | $h \times i$—? | $k \times l$—? |

20. In an analysis of other *rII* mutants, complementation testing yielded the following results:

| Mutants | Results (lysis) |
|---|---|
| 1, 2 | + |
| 1, 3 | + |
| 1, 4 | − |
| 1, 5 | − |

(a) Predict the results of testing 2 and 3, 2 and 4, and 3 and 4 together.

(b) If further testing yielded the following results, what would you conclude about mutant 5?

| Mutants | Results |
|---------|---------|
| 2, 5 | − |
| 3, 5 | − |
| 4, 5 | − |

(c) Following mixed infection of mutants 2 and 3 on *E. coli* B, progeny viruses were plated in a series of dilutions on both *E. coli* B and K12 with the following results. What is the recombination frequency between the two mutants?

| Strain plated | Dilution | Colonies |
|---------------|----------|----------|
| *E. coli* B | $10^{-5}$ | 2 |
| *E. coli* K12 | $10^{-1}$ | 5 |

(d) Another mutation, 6, was then tested in relation to mutations 1 through 5. In initial testing, mutant 6 complemented mutants 2 and 3. In recombination testing with 1, 4, and 5, mutant 6 yielded recombinants with 1 and 5, but not with 4. What can you conclude about mutation 6?

21.  During the analysis of seven *rII* mutations in phage T4, mutants 1, 2, and 6 were in cistron A, while mutants 3, 4, and 5 were in cistron B. Of these, mutant 4 was a deletion overlapping mutant 5. The remainder were point mutations. Nothing was known about mutant 7.
(a) Predict the results of complementation (+ or −): 1 and 2; 1 and 3; 2 and 4; and 4 and 5.
(b) In recombination studies between 1 and 2, the following results were obtained. Calculate the recombination frequency.

| Strain | Dilution | Colonies | Phenotypes |
|--------|----------|----------|------------|
| *E. coli* B | $10^{-7}$ | 4 | r |
| *E. coli* K12 | $10^{-2}$ | 8 | + |

(c) When mutant 6 was tested for recombination with mutant 1, the data were the same for strain B as shown above, but not for K12. The researcher lost the K12 data but remembered that recombination was 10 times more frequent than when mutants 1 and 2 were tested. What were the lost values (dilution and colony numbers)?
(d) Mutant 7 failed to complement any of the other mutants (1–6). Define the nature of mutant 7.

**SELECTED READINGS**

ADELBERG, E. A. 1960. *Papers on bacterial genetics*. Boston: Little, Brown.

ADELBERG, E. A., and PITTARD, J. 1965. Chromosome transfer in bacterial conjugation. *Bacteriol. Rev.* 29:161–72.

BENZER, S. 1955. Fine structure of a genetic region in bacteriophage. *Proc. Natl. Acad. Sci.* 42:344–54.

————. 1961. On the topography of the genetic fine structure. *Proc. Natl. Acad. Sci.* 47:403–15.

————. 1962. The fine structure of the gene. *Scient. Amer.* (Jan.) 206:70–87.

BIRGE, E. A. 1988. *Bacterial and bacteriophage genetics—An introduction*. New York: Springer-Verlag.

BRODA, P. 1979. *Plasmids*. New York: W. H. Freeman.

CAIRNS, J., OVERBAUGH, J., and MILLER, S. 1988. The origin of mutants. *Nature* 335:142–45.

CAIRNS, J., STENT, G. S., and WATSON, J. D., eds. 1966. *Phage and the origins of molecular biology*. Cold Spring Harbor, NY: Cold Spring Harbor Laboratory.

CAMPBELL, A. M. 1976. How viruses insert their DNA into the DNA of the host cell. *Scient. Amer.* (Dec.) 235:102–13.

CAVALLI-SFORZA, L. L., and LEDERBERG, J. 1956. Isolation of pre-adaptive mutants in bacteria by sib selection. *Genetics* 41:367–81.

FOX, M. S. 1966. On the mechanism of integration of transforming deoxyribonucleate. *J. Gen. Physiol.* 49:183–96.

HALL, B. G. 1988. Adapative evolution that requires multiple spontaneous mutations. I. Mutations involving an insertion sequence. *Genetics* 120:887–97.

———. 1990. Spontaneous point mutations that occur more often when advantageous than when neutral. *Genetics* 126:5–16.

HAYES, W. 1953. The mechanisms of genetic recombination in *Escherichia coli. Cold Spr. Harb. Symp.* 18:75–93.

———. 1968. *The genetics of bacteria and their viruses.* 2nd ed. New York: Wiley.

HERSHEY, A. D. 1946. Spontaneous mutations in a bacterial virus. *Cold Spr. Harb. Symp.* 11: 67–76.

HERSHEY, A. D., and CHASE, M. 1951. Genetic recombination and heterozygosis in bacteriophage. *Cold Spr. Harb. Symp.* 16:471–79.

HERSHEY, A. D., and ROTMAN, R. 1949. Genetic recombination between host range and plaque-type mutants of bacteriophage in single cells. *Genetics* 34:44–71.

HOTCHKISS, R. D., and MARMUR, J. 1954. Double marker transformations as evidence of linked factors in deoxyribonucleate transforming agents. *Proc. Natl. Acad. Sci.* 40:55–60.

JACOB, F., and WOLLMAN, E. L. 1961a. *Sexuality and the genetics of bacteria.* Orlando: Academic Press.

———. 1961b. Viruses and genes. *Scient. Amer.* (June) 204:92–106.

LANDY, A., and ROSS, W. 1977. Viral integration and excision: Structure of the lambda *att* sites. *Science* 197:1147–60.

LEDERBERG, J. 1986. Forty years of genetic recombination in bacteria: A fortieth anniversary reminiscence. *Nature* 324:627–28.

———. 1989. Replica plating and indirect selection of bacterial mutants: Isolation of preadaptive mutants in bacteria by sib selection. *Genetics* 121:395–99.

LOW, B., and PORTER, R. 1978. Modes of genetic transfer and recombination in bacteria. *Ann. Rev. Genet.* 12:249–87.

LURIA, S. E., and DELBRUCK, M. 1943. Mutations of bacteria from virus sensitivity to virus resistance. *Genetics* 28:491–511.

LWOFF, A. 1953. Lysogeny. *Bacteriol. Rev.* 17:269–337.

MESELSON, M., and WEIGLE, J. J. 1961. Chromosome breakage accompanying genetic recombination in bacteriophage. *Proc. Natl. Acad. Sci.* 47:857–68.

MORSE, M. L., LEDERBERG, E. M., and LEDERBERG, J. 1956. Transduction in *Escherichia coli* K12. *Genetics* 41:141–56.

NOVICK, R. P. 1980. Plasmids. *Scient. Amer.* (Dec.) 243:102–27.

OZEKI, H., and IKEDA, H. 1968. Transduction mechanisms. *Ann. Rev. Genet.* 2:245–78.

SMITH, H. O., DANNER, D. B., and DEICH, R. A. 1981. Genetic transformation. *Ann. Rev. Biochem.* 50:41–68.

SMITH-KEARY, P. F. 1989. *Molecular genetics of Escherichia coli.* New York: Guilford Press.

STAHL, F. W. 1979. *Genetic recombination: Thinking about it in phage and fungi.* New York: W. H. Freeman.

———. 1987. Genetic recombination. *Scient. Amer.* (Nov.) 256:91–101.

STENT, G. S. 1963. *Molecular biology of bacterial viruses.* New York: W. H. Freeman.

———. 1966. *Papers on bacterial viruses.* 2nd ed. Boston: Little, Brown.

TESSMAN, I. 1965. Genetic ultrafine structure in the T4 *rII* region. *Genetics* 51:63–75.

VISCONTI, N., and DELBRUCK, M. 1953. The mechanism of genetic recombination in phage. *Genetics* 38:5–33.

WOLLMAN, E. L., JACOB, F., and HAYES, W. 1956. Conjugation and genetic recombination in *Escherichia coli* K-12. *Cold Spr. Harb. Symp.* 21:141–62.

ZINDER, N. D. 1953. Infective heredity in bacteria. *Cold. Spr. Harb. Symp.* 18:261–69.

———. 1958. Transduction in bacteria. *Scient. Amer.* (Nov.) 199:38.

ZINDER, N. D., and LEDERBERG, J. 1952. Genetic exchange in *Salmonella J. Bacteriol.* 64:679–99.

# 20

# EXTRACHROMOSOMAL INHERITANCE

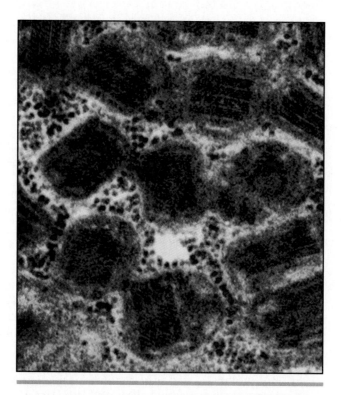

Transmission electron micrograph of abnormal mitochondria derived from cells of an individual afflicted with the maternally inherited myoclonic epilepsy and ragged red fiber disease (MERRF).

*Some traits in eukaryotes do not adhere to expected patterns associated with biparental inheritance. In most cases, these traits are controlled by genes located outside the nucleus. Transmission is usually through the female parent, whose gametes either contain maternal gene products that influence development, or are the sole source of chloroplasts and mitochondria that affect the offspring's phenotype.*

Throughout the history of genetics, occasional reports have challenged the basic tenets of transmission genetics—the production of the phenotype through the transmission of genes located on chromosomes of both parents. Instead, genetic results have been observed that do not reflect either Mendelian or neo-Mendelian principles. These reports have indicated an apparent extranuclear or extrachromosomal influence on the phenotype. Such observations were often regarded with skepticism. However, with the increasing knowledge of molecular genetics and the discovery of DNA in mitochondria and chloroplasts, **extrachromosomal inheritance** is now recognized as an important aspect of genetics.

There are many diverse examples of these unusual modes of inheritance. In this chapter we will focus on three general types of extrachromosomal genetic phenomena: (1) maternal influence resulting from the effect of stored products of nuclear genes of the female parent during early development; (2) organelle heredity resulting from the expression of DNA contained in mitochondria and chloroplasts; and (3) infectious heredity resulting from the symbiotic or parasitic association of microorganisms with eukaryotic cells. Each has the effect of producing inheritance patterns that vary from those predicted by the concepts of Mendelian and neo-Mendelian genetics.

# MATERNAL EFFECT

**Maternal effect**, also referred to as maternal influence, implies that an offspring's phenotype for a particular trait is strongly influenced by the nuclear genotype of the maternal parent. This is in contrast to most cases, where inheritance of traits is biparental. In cases of maternal effect, the genetic information of the female gamete is transcribed, and these genetic products (either proteins or yet untranslated mRNAs) are present in the egg cytoplasm. Following fertilization, these products influence patterns or traits established during early development. Two examples will illustrate the influence of the maternal genome on particular traits.

## *Ephestia* Pigmentation

In the Mediterranean meal moth, *Ephestia kuehniella*, the wild-type larva has a pigmented skin and brown eyes as a result of the dominant gene *A*. The pigment is derived from a precursor molecule, kynurenine, which is in turn a derivative of the amino acid tryptophan. A mutation, *a*, results in red eyes and little pigmentation in larvae when homozygous. As illustrated in Figure 20.1, different results are obtained in the cross $Aa \times aa$, depending on which parent carries the dominant gene. When the male is the heterozygous parent, a 1:1 brown/red-eyed ratio is observed in larvae as predicted by Mendelian segregation. However, when the female is heterozygous for the *A* gene, all larvae are pigmented and have brown eyes. As these larvae develop into adults, one-half of them gradually develop red eyes, reestablishing the 1:1 ratio.

One explanation of these results is that the *Aa* oocytes synthesize kynurenine or an enzyme necessary for its synthesis and accumulate it in the ooplasm prior to the completion of meiosis. Even in *aa* progeny (whose mothers were *Aa*), this pigment is distributed in the cytoplasm of the cells of the developing larvae—thus, they develop pigmentation and brown eyes. Eventually, the pigment is diluted among many cells and used up, resulting in the conversion to red eyes as adults. The *Ephestia* example demonstrates the maternal effect in which a cytoplasmically stored nuclear gene product influences the larval phenotype, and at least temporarily, overrides the genotype of the progeny.

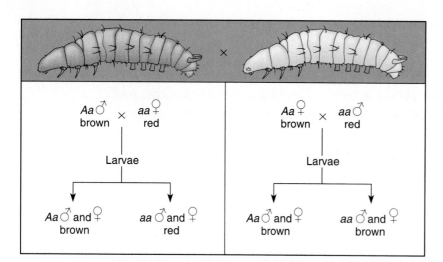

**FIGURE 20.1**    Illustration of maternal influence in the inheritance of eye pigment in the meal moth *Ephestia kuehniella*. Multiple light receptor structures (eyes) are present on each side of the anterior portion of larvae.

## *Limnaea* Coiling

Shell coiling in the snail *Limnaea peregra* represents a permanent rather than a transitory maternal effect. Some strains of this snail have left-handed or sinistrally coiled shells (*dd*), while others have right-handed or dextrally coiled shells (*DD* or *Dd*). These snails are hermaphroditic and may undergo either cross- or self-fertilization, thus providing a variety of types of matings.

Figure 20.2 illustrates the results of reciprocal crosses between true-breeding snails. As can be seen, these crosses yielded different outcomes, even though both crosses were between sinistral and dextral organisms. In the F₁, it seems apparent that *phenotypes* of the progeny depend on the *genotypes* of the female parents. If we adopt that conclusion as a working hypothesis, we can test it by examining the offspring in subsequent generations of self-fertilization events. In each case, the hypothesis is upheld. *The coiling pattern of the progeny snails is determined by the genotype of the parent producing the egg, regardless of the phenotype of that parent.*

Investigation of the developmental events in these snails reveals that the orientation of the spindle in the first cleavage division after fertilization apparently determines the direction of coiling. The orientation of the spindle, in turn, influences cell divisions following fertilization and establishes the permanent adult coiling pattern.

The dextral allele (*D*) produces an active gene product that causes right-handed coiling. If ooplasm from dextral eggs is injected into uncleaved sinistral eggs, they cleave in a dextral pattern. However, in the converse experiment, sinistral ooplasm has no effect when injected into dextral eggs. Apparently, the sinistral allele is the result of a classical recessive mutation inactivating the gene product.

## ORGANELLE HEREDITY: MATERNAL INHERITANCE

In this section we will examine examples of inheritance patterns of phenotypes related to chloroplast and mitochondrial function. Prior to the discovery of DNA in these organelles and extensive characterization of the genetic processes occurring within them, these patterns were grouped under the category of **cytoplasmic inheritance**. That is, certain mutant phenotypes seemed to be inherited through the cytoplasm rather than through the genetic information of chromosomes. Transmission was most often from the maternal parent through the ooplasm, and thus the results of reciprocal crosses varied. Such patterns are now more appropriately considered examples of **maternal inheritance**.

While such results are similar to the examples of maternal effect presented in the preceding section, a major difference exists. Maternal effects (due to the presence of nuclear gene products) are transitory in the sense that they are not necessarily heritable. With maternal inheritance, however, the phenotype is stable and is continually passed to future generations.

Analysis of the hereditary transmission of mutant alleles of chloroplast and mitochondrial DNA has been difficult. First, the function of these organelles is dependent upon gene products of both nuclear and organelle DNA. Second, the number of organelles contributed to each progeny often exceeds one. If many chloroplasts and/or mitochondria are contributed and only one or a few of them contain a mutant gene, the corresponding mutant phenotype may not be revealed. Analysis is thus much more involved than for Mendelian characters.

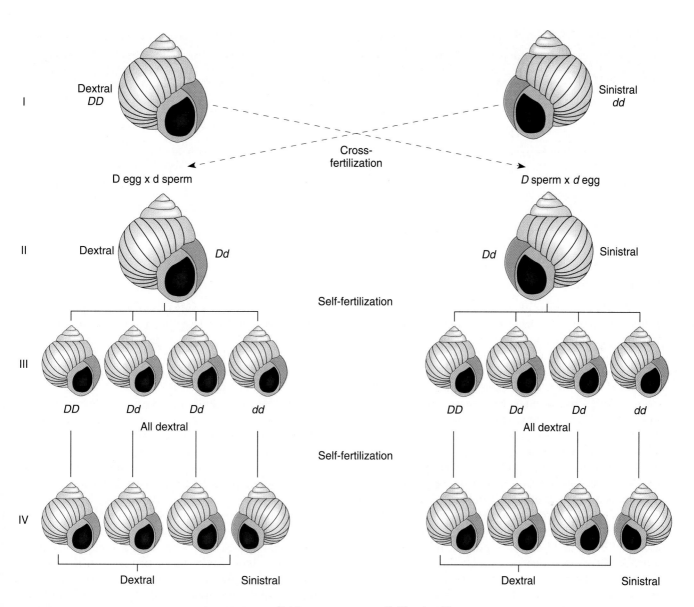

**FIGURE 20.2**    Inheritance of coiling in the snail *Limnaea peregra*. Coiling is either dextral (right-handed) or sinistral (left-handed). A maternal effect is evident in generations II and III, where the genotype of the maternal parent controls the phenotype of the offspring rather than the offspring's own genotype.

In this section, we shall discuss examples of inheritance patterns related to these organelles. We have provided a fairly detailed analysis of the molecular genetics of them in Chapter 10. You should review and relate this information to the discussion that follows.

## Chloroplasts: Variegation in Four O'Clock Plants

In 1908, Carl Correns (one of the rediscoverers of Mendel's work) provided the earliest example of inheritance linked to chloroplast transmission. Correns discovered a variety of the four o'clock plant, *Mirabilis jalapa*, which had branches with either white, green, or variegated leaves. As shown in Table 20.1, inheritance in all

possible combinations of crosses is strictly determined by the *phenotype* of the ovule source. For example, if the seeds (representing the progeny) were derived from ovules on branches with green leaves, all progeny plants bore only green leaves regardless of the phenotype of the source of pollen.

Correns concluded that inheritance was through the cytoplasm of the maternal parent because the pollen, which contributes little or no cytoplasm to the zygote, had no apparent influence on the progeny phenotypes. Since the leaf coloration involves the chloroplast, either genetic information in that organelle, or in the cytoplasm and influencing the chloroplast, could be responsible for this inheritance pattern.

**Table 20.1**    Crosses Between Flowers from Various Branches of Variegated Four O'clock Plants

| Source of Pollen | Source of Ovule | | |
|---|---|---|---|
| | White Branch | Green Branch | Variegated Branch |
| White branch | White | Green | White, green, or variegated |
| Green branch | White | Green | White, green, or variegated |
| Variegated branch | White | Green | White, green, or variegated |

## *Iojap* in Maize

A phenotype similar to *Mirabilis* but with a different pattern of inheritance has been analyzed in maize by Marcus M. Rhoades. In this case, green, colorless, or green-and-colorless striped leaves are under the influence not only of the cytoplasm, but, in addition, of a nuclear gene located on the seventh linkage group. This locus is called ***iojap***, after the recessive mutation *ij* located there. The wild-type allele is designated *Ij*. Plants homozygous for the mutation (*ij/ij*) may have green-and-white striped leaves. However, when reciprocal crosses are made between plants with striped leaves and green leaves, the results are seen to vary, depending upon which parent is mutant (Figure 20.3). If the female is striped (*ij/ij*) and the male is green (*Ij/Ij*), plants with colorless, striped, and green leaves are observed as progeny. If the male parent is striped and the female is green, only green plants are produced! In both types of crosses, all offspring have identical genotypes (*Ij/ij*). We may conclude that although a nuclear gene is somehow involved, the inheritance pattern is influenced maternally.

We can understand this pattern better by examining the offspring resulting from self-fertilization of heterozygous plants (Figure 20.3). The striped plant gives rise to progeny with colorless, striped, and green leaves, regardless of their genotype. The green plants give rise to both green and striped progeny in a 3:1 ratio. The results of these self-fertilizations substantiate that the mutant phenotype is controlled solely through the female cytoplasm, regardless of the plant's nuclear genotype.

The role of the nucleus in the origin of the mutant phenotype is unclear. However, the colorless areas of the leaf are due to the lack of the green chlorophyll pigment in chloroplasts. Once acquired, colorless chloroplasts are transmitted through the egg cytoplasm, establishing the phenotypes of leaves of progeny plants.

## *Chlamydomonas* Mutations

The unicellular green alga *Chlamydomonas reinhardi* has provided an excellent system for the investigation of maternal inheritance. The organism is eukaryotic and contains a single large chloroplast as well as numerous mitochondria. Matings are followed by meiosis and the various stages of the life cycle are easily studied in culture in the laboratory. The first cytoplasmic mutant, *streptomycin resistance* (*sr*), was reported in 1954 by Ruth Sager. Although *Chlamydomonas'* two mating types—$mt^+$ and $mt^-$—appear to make equal cytoplasmic contributions to the zygote, Sager determined that the *sr* phenotype is transmitted only through the $mt^+$ parent.

Since this discovery, a number of other *Chlamydomonas* mutations (including an acetate requirement as well as resistance to or dependence on a variety of bacterial antibiotics) have been discovered that show a similar maternal inheritance pattern. These mutations have been linked to the transmission of the chloroplast, and their study has extended our knowledge of chloroplast inheritance.

Following fertilization, the single chloroplasts of the two mating types fuse. After the resulting zygote has undergone meiosis, it is apparent that the genetic information of the chloroplasts of progeny cells is derived only from the $mt^+$ parent. The $mt^-$ chloroplast has been destroyed!

Further, studies suggest that these chloroplast mutations, representing numerous gene sites, form a single circular linkage group—the first such extrachromosomal unit to be established in a eukaryotic organism. It is also apparent that the linkage groups from the two mating types are capable of undergoing recombination following zygote formation. All available evidence supports the hypothesis that this linkage group consists of DNA residing in the chloroplast.

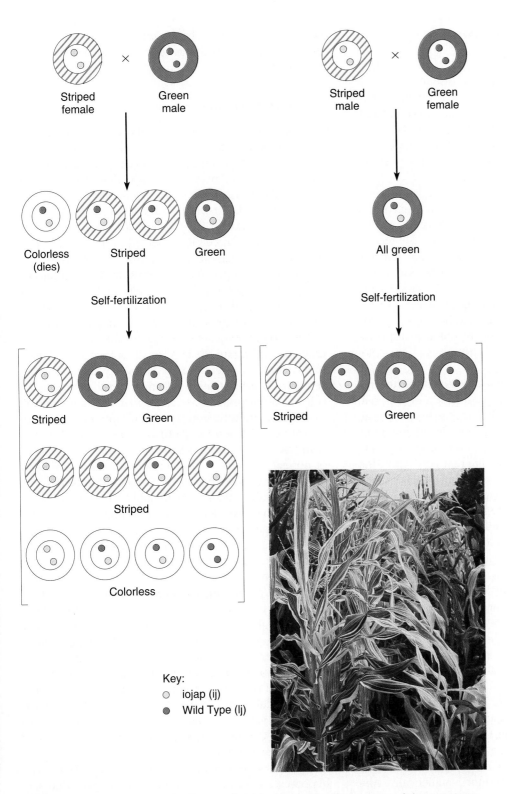

**FIGURE 20.3**    Maternal inheritance of striping in maize. Regardless of the genotype of the *iojap* genes, offspring reflect the maternal phenotypes in their appearance. If the maternal parent is striped (containing green *and* colorless areas), progeny are either colorless, green, or striped. If the maternal parent has green leaves, so do all progeny plants. Since color is due to chloroplasts in the leaves, inheritance is controlled by these organelles as they are passed through the maternal cytoplasm.

## Mitochondria: *poky* in *Neurospora*

Mitochondria, like chloroplasts, play a critical role in cellular bioenergetics and contain a distinctive genetic system. Mutants affecting mitochondrial function have been discovered and studied. As with chloroplast mutants, these are transmitted through the cytoplasm and result in non-Mendelian inheritance patterns. Additionally, the mitochondrial genetic system has now been extensively characterized.

In 1952, Mary B. and Hershel K. Mitchell discovered a slow-growing mutant strain of the mold *Neurospora crassa* and called it *poky*. (It is also designated *mi-1* for maternal inheritance.) Studies have shown slow growth to be associated with impaired mitochondrial function specifically related to certain cytochromes essential to electron transport. The results of genetic crosses between wild-type and *poky* strains suggest that the trait is maternally inherited. If the female parent is *poky* and the male parent is wild type, all progeny colonies are *poky*. The reciprocal cross produces normal wild-type colonies.

Studies with *poky* mutants illustrate the use of **heterokaryon** formation during the investigation of maternal inheritance in fungi. Occasionally, hyphae from separate mycelia fuse with one another, giving rise to structures containing two or more nuclei in a common cytoplasm. If the hyphae contain nuclei of different genotypes, the structure is called a heterokaryon. The cytoplasm will contain mitochondria derived from both initial mycelia. A heterokaryon may give rise to haploid spores, or **conidia**, that produce new mycelia. The phenotypes of these structures may be determined.

Heterokaryons produced by the fusion of *poky* and wild-type hyphae initially show normal rates of growth and respiration. However, mycelia produced through conidia formation become progressively more abnormal until they show the *poky* phenotype. This occurs in spite of the presumed presence of both wild-type and *poky* mitochondria in the cytoplasm of the hyphae.

To explain the initial growth and respiration pattern, it is assumed that the wild-type mitochondria support the respiratory needs of the hyphae. The subsequent expression of the *poky* phenotype suggests that the presence of the *poky* mitochondria may somehow prevent or depress the function of these wild-type mitochondria. Perhaps the *poky* mitochondria replicate more rapidly and "wash out" or dilute wild-type mitochondria numerically. Another possibility is that *poky* mitochondria produce a substance that inactivates the wild-type organelle or interferes with the replication of its DNA (mtDNA). As a result of this type of interaction, *poky* is an example of a broader category referred to as **suppressive muta-**

tions. This general phenomenon is characteristic of many other suspected mitochondrial mutations of *Neurospora* and yeast.

### *Petite* in *Saccharomyces*

Another extensive study of mitochondrial mutations has been performed with the yeast *Saccharomyces cerevisiae*. The first such mutation, ***petite***, was described by Boris Ephrussi and his coworkers in 1956. The mutant is so named because of the small size of the yeast colonies. Many independent *petite* mutations have since been discovered and studied. They all have a common characteristic: deficiency in cellular respiration involving abnormal electron transport. Fortunately, this organism is a facultative anaerobe and can grow by fermenting glucose through glycolysis. Thus, although colonies are small, the organism may survive the loss of mitochondrial function by generating energy anaerobically.

The complex genetics of *petite* mutations is diagrammed in Figure 20.4. A small proportion of these mutants exhibit Mendelian inheritance and are called ***segregational petites***, indicating that they are the result of nuclear mutations. The remainder demonstrates cytoplasmic transmission, producing one of two effects in matings. The ***neutral petites***, when crossed to wild type, produce ascospores that give rise only to wild-type or normal colonies. The same pattern continues if progeny of this cross are back-crossed to *neutral petites*. This is because the majority of neutrals lack mtDNA completely or have lost a substantial portion of it. Thus, the wild-type gamete is the effective source of normal mitochondria capable of reproduction. Neutral petites have now been shown to lack most, if not all, mitochondrial DNA (mtDNA).

A third type, the ***suppressive petites***, behaves similarly to *poky* in *Neurospora*. Crosses between mutant and wild type give rise to mutant diploid zygotes, which, upon undergoing meiosis, immediately yield all mutant cells. Under these conditions, the *petite* mutation behaves "dominantly" and seems to suppress the function of the wild-type mitochondria. *Suppressive petites* also represent deletions of mtDNA, but not nearly to the extent of the *neutral petites*. Buoyant density studies show a lower G-C content in suppressive mtDNA compared with normal mtDNA.

Suppressiveness remains unexplained. Two major hypotheses have been advanced. One explanation suggests that the mutant (or deleted) mtDNA replicates more rapidly, and thus mutant mitochondria "take over" or dominate the phenotype by numbers alone. The second explanation suggests that recombination occurs between the mutant and wild-type mtDNA, introducing er-

**FIGURE 20.4** The outcome of crosses involving the three types of *petite* mutations affecting mitochondrial function in the yeast *Saccharomyces cerevisiae.* The photograph shows normal yeast colonies.

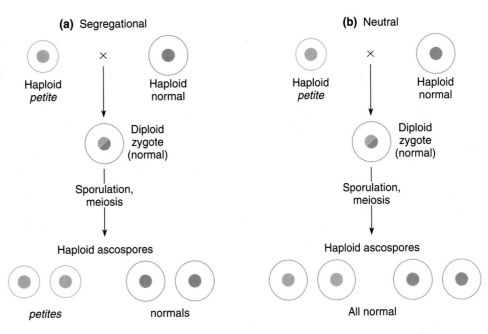

**(a)** Segregational

Haploid *petite* × Haploid normal

Diploid zygote (normal)

Sporulation, meiosis

Haploid ascospores

*petites*    normals

**(b)** Neutral

Haploid *petite* × Haploid normal

Diploid zygote (normal)

Sporulation, meiosis

Haploid ascospores

All normal

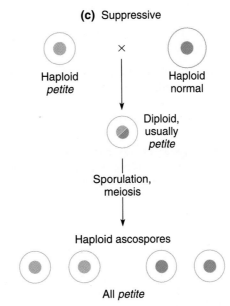

**(c)** Suppressive

Haploid *petite* × Haploid normal

Diploid, usually *petite*

Sporulation, meiosis

Haploid ascospores

All *petite*

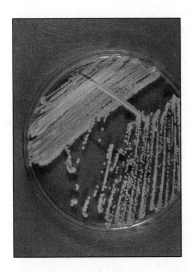

rors into or disrupting the normal mtDNA. It is not yet clear which, if either, of these explanations is correct.

# MITOCHONDRIAL DNA AND HUMAN DISEASES

As you may recall from Chapter 10, the DNA found in mitochondria **(mtDNA)** has been extensively characterized in a variety of organisms, including humans. Human mtDNA, which is circular, contains 16,569 base pairs and

is strictly inherited maternally. The mitochondrial gene products encoded include:

13 proteins, required for oxidative phosphorylation (*ox-phos*)
22 tRNAs, required for translation
2 rRNAs, required for translation

Unlike the nuclear genome, mitochondria contain little extraneous DNA that fails to code for gene products. Thus, mitochondrial function is particularly vulnerable to mutations in mtDNA, since such genetic alterations will potentially disrupt translation of all gene products *or* result in mutant *oxphos* proteins. In the former case, diminution or complete loss of translation capability within the organelle would very likely have a lethal effect.

On the other hand, a zygote receives a large number of organelles through the egg, so if only one of them contains a mutation, its impact is severely diluted because there will be many more mitochondria that will function normally. During early development, cell division disperses the initial population of mitochondria present in the zygote, and in newly formed cells, these organelles reproduce autonomously. Therefore, adults will exhibit cells with a variable mixture of normal and abnormal organelles, should a deleterious mutation arise or already be present in the initial population of organelles. This is a condition called **heteroplasmy**.

In order for a human disorder to be attributable to genetically altered mitochondria, several criteria must be met:

1. Inheritance must exhibit a maternal rather than a Mendelian pattern.

2. The disorder must reflect a deficiency in the bioenergetic function of the organelle.

3. A specific genetic mutation in one of the mitochondrial genes must be documented.

Thus far, several such cases are known that demonstrate these characteristics.

**Myoclonic epilepsy and ragged red fiber disease (MERRF)** demonstrates a pedigree consistent with maternal inheritance (Figure 20.5). Individuals with this rare disorder express deafness and dementia in addition to seizures. Both muscle fibers and mitochondria are abnormal in appearance. Such aberrant mitochondria are evident in Figure 20.5. Upon analysis of mtDNA, a single nucleotide substitution has been found in the transfer RNA specific for lysine (tRNA$^{lys}$). This change apparently interferes with translation within the organelle. Presumably, the deficiency in efficient mitochondrial function is related to the various manifestations of the disorder.

A second disorder, **Leber's hereditary optic neuropathy (LHON)** also exhibits maternal inheritance as well as mtDNA lesions. The disorder is characterized by sudden bilateral blindness. The average age of vision loss is 27, but onset is quite variable. Four different mutations have been identified, all of which disrupt normal *ox-phos* function. Over 50 percent of cases are due to a missense mutation involving a G-to-A transition at a specific position in the mitochondrial gene encoding a subunit of NADH dehydrogenase. The arginine residue at position 340 is converted to histidine. In large families, this mutation is transmitted to all maternal relatives. It is also of interest to note that in many instances of LHON, there is no family history. It appears that a significant number of cases may result from "new" mutations.

In a third disorder, **Kearns-Sayre syndrome (KSS)**, severely affected individuals lose their vision, undergo hearing loss, and display heart conditions. The genetic basis of KSS involves deletions at various positions within mtDNA. Many KSS patients are sympton-free as children but display progressive symptoms as adults.

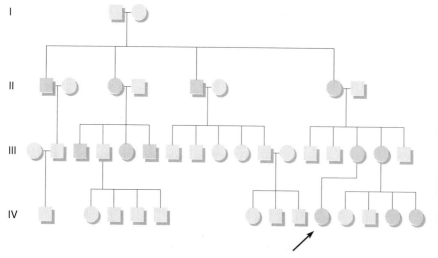
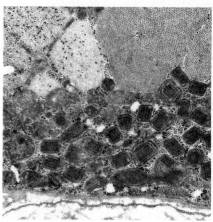

**FIGURE 20.5**    A partial pedigree illustrating maternal inheritance of myoclonic epilepsy. Some of the individuals were not available for analysis, including the great-grandparents (I) of the proband (arrow). The photograph demonstrates abnormal mitochondria derived from cells of an afflicted individual. Note the highly abnormal cristae.

Analysis has revealed that the proportion of mtDNAs that reveal deletions increases as the severity of symptoms increases.

The study of hereditary, mitochondrial-based disorders provides insights into the importance and genetic basis of this organelle during normal development as well as the relationship between mitochondrial function and neuromuscular disorders. Furthermore, such study has suggested an hypothesis for aging, based on the progressive accumulation of mtDNA mutations and the accompanying loss of *ox-phos* capacity.

# INFECTIOUS HEREDITY

There are numerous examples of cytoplasmically transmitted phenotypes in eukaryotes that are due to an invading microorganism or particle. The foreign invader coexists in a symbiotic relationship, is usually passed through the maternal ooplasm to progeny cells or organ-

isms, and confers a specific phenotype that may be studied. We shall consider several examples illustrating this phenomenon.

## Kappa in *Paramecium*

First described by Tracy Sonneborn, certain strains of *Paramecium aurelia* are called **Killers** because they release a cytoplasmic substance called **paramecin** that is toxic and sometimes lethal to sensitive strains. This substance is produced by particles called **kappa** that replicate in the Killer cytoplasm, contain DNA and protein, and depend for their maintenance on a dominant nuclear gene *K*. One cell may contain 100 to 200 such particles. Paramecia are diploid protozoans that can undergo sexual exchange of genetic information through the process of **conjugation**. In some instances, cytoplasmic exchange also occurs. Thus, there is a variety of ways in which the *K* gene and kappa can be transmitted.

The genetic events occurring during conjugation in *Paramecium aurelia* are shown in Figure 20.6. Parame-

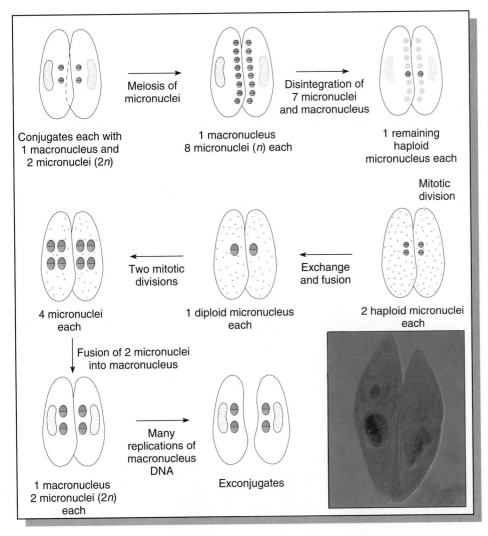

**FIGURE 20.6** Genetic events occurring during conjugation in *Paramecium*. The photographic insert shows a pair of organisms undergoing conjugation.

cia contain two diploid micronuclei. Early in conjugation, both micronuclei in each mating pair undergo meiosis, resulting in eight haploid nuclei. However, seven of these degenerate, and the remaining one undergoes a single mitotic division. Each cell then donates one of the two haploid nuclei to the other, recreating the diploid condition in both cells. As a result, exconjugates are of identical genotypes.

In a similar process involving only a single cell, **autogamy** occurs. Following meiosis of both micronuclei, seven products degenerate and one survives. This nucleus divides, and the resulting nuclei fuse to recreate the diploid condition. If the original cell was heterozygous, autogamy results in homozygosity because the newly formed diploid nucleus was derived solely from a single haploid meiotic product. In a population of cells that were originally heterozygous, half of the new cells express one allele and half express the other allele.

Figure 20.7 illustrates the results of crosses between *KK* and *kk* cells, without and with cytoplasmic exchange. When no cytoplasmic exchange occurs, even though the resultant cells may be *Kk* (or *KK* following autogamy), they remain sensitive if no kappa particles are transmitted. When exchange occurs, the cells become Killers provided the kappa particles are supported by at least one dominant *K* allele.

Kappa particles are bacterialike and may contain temperate bacteriophages. One theory holds that these viruses of kappa may become vegetative; during this multiplication, they produce the toxic products that are released and kill sensitive strains.

## Infective Particles in *Drosophila*

Two examples of similar phenomena are know in *Drosophila*: **$CO_2$ sensitivity** and **sex-ratio**. In the former, flies that would normally recover from carbon dioxide anesthetization instead become permanently paralyzed and are killed by $CO_2$. Sensitive mothers pass this trait to all offspring. Furthermore, extracts of sensitive flies induce the trait when injected into resistant flies. Phillip L'Heritier has postulated that sensitivity is due to the presence of a virus, **sigma**. The particle has been visualized and is smaller than kappa. Attempts to transfer the virus to other insects have been unsuccessful, demonstrating that specific nuclear genes support the presence of sigma in *Drosophila*.

A second example of infective particles comes from the study of *Drosophila bifasciata*. A small number of these flies were found to produce predominantly female offspring if reared at 21°C or lower. This condition, designated sex-ratio, was shown to be transmitted to daughters but not to the low percentage of males produced. This phenomenon was subsequently investigated in *Drosophila willistoni*. In these flies, the injection of ooplasm from sex-ratio females into normal females induced the condition. This observation suggests that an extrachromosomal element is responsible for the sex-ratio phenotype. The agent has now been isolated and shown to be a protozoan. While the protozoan has been found in both males and females, it is lethal primarily to developing male larvae. There is now some evidence that a virus harbored by the protozoan may be responsible for producing a male-lethal toxin.

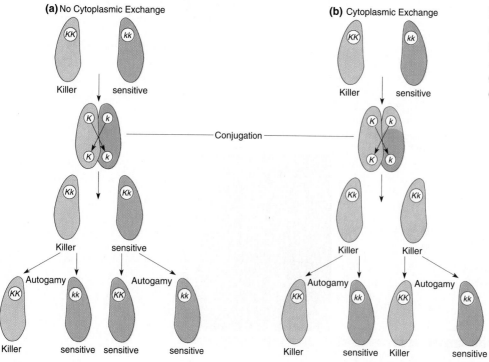

**FIGURE 20.7**    Results of crosses between *Killer* (*KK*) and *sensitive* (*kk*) strains of *Paramecium*, with and without cytoplasmic exchange during conjugation.

## CHAPTER SUMMARY

1. Patterns of inheritance sometimes vary from that expected of nuclear genes. In such instances, phenotypes most often appear to result from genetic information transmitted through the egg.

2. Maternal-effect patterns result when nuclear gene products controlled by the maternal genotype of the egg influence early development. *Ephestia* pigmentation and coiling in snails are examples.

3. Maternal inheritance patterns result from the genotypes of chloroplast and mitochondrial DNA as these organelles are transmitted through the egg. These are cases of true extrachromosomal inheritance.

4. Chloroplast mutations affect the photosynthetic capabilities of plants, while mitochondrial mutations affect cells highly dependent on ATP generated through cellular respiration. The resulting mutants display phenotypes related to the loss of function of these organelles.

5. Another form of extrachromosomal inheritance is due to the transmission of infectious microorganisms, which establish symbiotic relationships with their host cells. Kappa particles and $CO_2$-sensitivity and sex-ratio determinants are examples.

## KEY TERMS

autogamy
$CO_2$ sensitivity
conjugation
cytoplasmic inheritance
extrachromosomal
    inheritance
heterokaryon

heteroplasmy
*iojap* locus (*ij*)
kappa
Kearns-Sayre syndrome
Killer strains
Leber's hereditary optic
    neuropathy (LHON)

maternal effect
maternal inheritance
mtDNA
myoclonic epilepsy and
    ragged red fiber dis-
    ease (MERRF)
*neutral petites*

paramecin
*petite* mutation
*poky* mutation
*segregational petites*
sex-ratio
sigma virus
*suppressive petites*

## INSIGHTS AND SOLUTIONS

1. Analyze the following theoretical pedigree and determine the most consistent interpretation of how the trait is inherited and any inconsistencies.

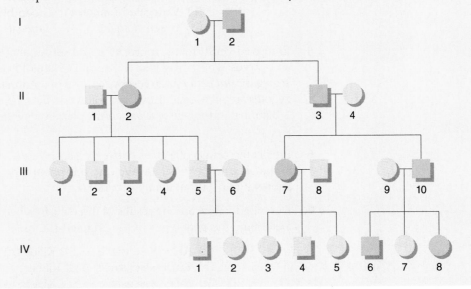

**SOLUTION:** The trait is passed from all male parents to all but one offspring but *never* passed maternally. Individual IV-7 (a female) is the only exception.

2. Can the above explanation be attributed to a gene on the Y chromosome? Defend your answer.

**SOLUTION:** No, since male parents pass the trait to their daughters as well as to their sons.

3. Is the above case an example of a paternal effect or of paternal inheritance?

**SOLUTION:** It has all the earmarks of paternal inheritance since males pass the trait to all of their offspring. To assess whether the trait is due to a paternal effect (resulting from a nuclear gene in the male gamete), analysis of further matings would be needed.

## PROBLEMS AND DISCUSSION QUESTIONS

1. What genetic criteria distinguish a case of extrachromosomal inheritance from a case of Mendelian autosomal inheritance? From a case of sex-linked inheritance?

2. In *Limnaea*, what results would be expected in a cross between a *Dd* dextrally coiled and a *Dd* sinistrally coiled snail, assuming cross-fertilization occurs as shown in Figure 20.2? What results would occur if the *Dd* dextral produced only eggs and the *Dd* sinistral produced only sperm?

3. Streptomycin resistance in *Chlamydomonas* may result from a mutation in a chloroplast gene or in a nuclear gene. What phenotypic results would occur in a cross between a member of an $mt^+$ strain resistant in both genes and a member of an $mt^-$ strain sensitive to the antibiotic? What results would occur in the reciprocal cross?

4. A plant may have green, white, or green and white (variegated) leaves on its branches owing to a mutation in the chloroplast (which produces the white leaves). Predict the results of the following crosses.

| Ovule Source | | Pollen Source |
|---|---|---|
| (a) Green branch | × | White branch |
| (b) White branch | × | Green branch |
| (c) Variegated branch | × | Green branch |
| (d) Green branch | × | Variegated branch |

5. In diploid yeast strains, sporulation and subsequent meiosis can produce haploid ascospores. These may fuse to reestablish diploid cells. When ascospores from a *segregational petite* strain fuse with those of a normal wild-type strain, the diploid zygotes are all normal. However, following meiosis, ascospores are 1/2 *petite* and 1/2 normal. Is the *segregational petite* phenotype inherited as a dominant or recessive trait?

6. Predict the results of a cross between ascospores from a *segregational petite* strain and a *neutral petite* strain. Indicate the phenotype of the zygote and the ascospores it may subsequently produce.

7. Described below are the results of three crosses between strains of *Paramecium*. Determine the genotypes of the parental strains.

(a) Killer × sensitive → 1/2 Killer:1/2 sensitive
(b) Killer × sensitive → all Killer
(c) Killer × sensitive → 3/4 Killer:1/4 sensitive

**8.** *Chlamydomonas*, a eukaryotic green alga, is sensitive to the antibiotic erythromycin that inhibits protein synthesis in prokaryotes.
(a) Explain why.
(b) There are two mating types in this alga, $mt^+$ and $mt^-$. If an $mt^+$ cell sensitive to the antibiotic is crossed with an $mt^-$ cell that is resistant, all progeny cells are sensitive. The reciprocal cross ($mt^+$ resistant and $mt^-$ sensitive) yields all resistant progeny cells. Assuming that the mutation for resistance is in chloroplast DNA, what can be concluded?

**9.** In *Limnaea*, a cross where the snail contributing the eggs was dextral but of unknown genotype mated with another snail of unknown genotype and phenotype. All $F_1$ offspring exhibited dextral coiling. Ten of the $F_1$ snails were allowed to undergo self-fertilization. One-half produced only dextrally coiled offspring, while the other half produced only sinistrally coiled offspring. What were the genotypes of the original parents?

**10.** In *Drosophila subobscura*, the presence of a recessive gene called *grandchildless* (*gs*) causes the offspring of homozygous females, but not homozygous males, to be sterile. Can you offer an explanation as to why females but not males are affected by the mutant gene?

## SELECTED READINGS

BOGORAD, L. 1981. Chloroplasts. *J. Cell Biol.* 91:256s–70s.

COHEN, S. 1973. Mitochondria and chloroplasts revisited. *Amer. Sci.* 61:437–45.

FREEMAN, G., and LUNDELIUS, J. W. 1982. The developmental genetics of dextrality and sinistrality in the gastropod *Lymnaea peregra*. *Wilhelm Roux Arch.* 191:69–83.

GILLHAM, N. W. 1978. *Organelle hereditary*. New York: Raven Press.

GOODENOUGH, U., and LEVINE, R. P. 1970. The genetic activity of mitochondria and chloroplasts. *Scient. Amer.* (Nov.) 223:22–29.

GRIVELL, L. A. 1983. Mitochondrial DNA. *Scient. Amer.* (March) 248:78–89.

LANDER, E. S., et al. 1990. Mitochondrial diseases: Gene mapping and gene therapy. *Cell* 61:925–26.

LEVINE, R. P., and GOODENOUGH, U. 1970. The genetics of photosynthesis and of the chloroplast in *Chlamydomonas reinhardi*. *Ann. Rev. Genet.* 4:397–408.

MARGULIS, L. 1970. *Origin of eukaryotic cells*. New Haven, Conn.: Yale University Press.

MITCHELL, M. B., and MITCHELL, H. K. 1952. A case of maternal inheritance in *Neurospora crassa*. *Proc. Natl. Acad. Sci.* 38:442–49.

PREER, J. R. 1971. Extrachromosomal inheritance: Hereditary symbionts, mitochondria, chloroplasts. *Ann. Rev. Genet.* 5:361–406.

ROSING, H. S., et al. 1985. Maternally inherited mitochondrial myopathy and myoclonic epilepsy. *Ann. Neurol.* 17:228–37.

SAGER, R. 1965. Genes outside the chromosomes. *Scient. Amer.* (Jan.) 212:70–79.

———. 1985. Chloroplast genetics. *BioEssays* 3:180–84.

SCHWARTZ, R. M., and DAYHOFF, M. O. 1978. Origins of prokaryotes, eukaryotes, mitochondria and chloroplasts. *Science* 199:395–403.

SLONIMSKI, P. 1982. *Mitochondrial genes*. Cold Spring Harbor, N.Y.: Cold Spring Harbor Laboratory.

SONNEBORN, T. M. 1959. Kappa and related particles in *Paramecium. Adv. in Virus Res.* 6:229–356.

STRATHERN, J. N., et al., eds. 1982. *The molecular biology of the yeast* Saacharomyces: *Life cycle and inheritance*. Cold Spring Habor, N.Y.: Cold Spring Harbor Laboratory.

STURTEVANT, A. H. 1923. Inheritance of the direction of coiling in *Limnaea. Science* 58:269–70.

TZAGOLOFF, A. 1982. *Mitochondria*. New York: Plenum Press.

WALLACE, D. C. 1992. Mitochondrial genetics: A paradigm for aging and degenerative diseases. *Science* 256:628–32.

WALLACE, D. C., et al. 1988. Familial mitochondrial encephalomyopathy (MERRF): Genetic, pathophysiological and biochemical characterization of a mitochondrial DNA disease. *Cell* 55:601–10.

# 21

# GENETIC CONTROL OF DEVELOPMENT

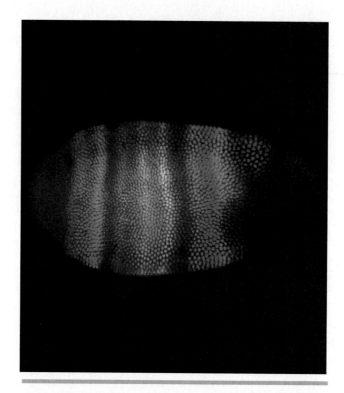

Regional differentiation of the developing *Drosophila* embryo based on differential expression of three genes. Each of the three colors (red, blue, and yellow) is the result of an immunofluorescent reaction specific to the presence of a unique protein produced during the early segmentation stage.

*The genetic basis of development is differential gene action. Gene actions vary over time in the same tissue, and between different tissues present at the same developmental stage. Gene action is required to bring about and maintain adult structures. Processes such as regeneration involve a reiteration of the embryonic program of gene action.*

In multicellular plants and animals, a fertilized egg, without further stimulus, begins a cycle of developmental events that ultimately give rise to an adult member of the species from which the egg and sperm were derived. Thousands, millions, or even billions of cells are organized into a cohesive and coordinated unit that we perceive as a living organism. The heterogeneous series of events whereby organisms attain their final adult form is studied by developmental biologists. This area of study is perhaps the most intriguing in biology because comprehension of developmental processes requires knowledge of many biological disciplines.

In the past one hundred years, investigations in embryology, genetics, biochemistry, molecular biology, cell physiology, and biophysics have contributed to the study of development. Largely, the findings have pointed out the tremendous complexity of developmental processes. Unfortunately, this description of what actually happens does not answer the "why" and "how" of development. Over the last two decades, the use of genetic analysis and molecular biology has allowed the identification of the genes that regulate developmental processes. We are now beginning to understand how the action and interaction of these genes control basic developmental processes in a wide range of prokaryotic and eukaryotic organisms.

In this chapter the primary emphasis will be on the role of gene action in regulating development. Because genetic information directs cellular function and determines the cellular phenotype, the genetics of development is being actively studied using a wide range of experimental organisms and techniques.

## DEVELOPMENTAL CONCEPTS

The current thrust in developmental genetics is to provide a molecular explanation of developmental processes including temporal changes in gene expression and cellular differentiation in order to establish a causal relationship between the presence or absence of molecules, receptors, transcriptional events, cell and tissue interactions, and the observable morphological events that accompany the process of development. To explain the sequence of developmental events that take place in any organism, geneticists rely on certain heuristic concepts. These include the theory of variable gene activity, and the principles of differential transcription and the stability of the differentiated state. These concepts have been derived from the study of a wide range of organisms used as model systems, and these ideas form the theoretical framework that is used to explain how the developmental potential present in a single cell becomes transformed into a recognizable yet individual form with a certain behavioral repertoire we identify as an adult organism.

## THE VARIABLE GENE ACTIVITY THEORY

From a genetic perspective, development may be described as the attainment of a differentiated state. For example, an erythrocyte active in hemoglobin synthesis is differentiated, but a cell in a blastula-stage embryo is undifferentiated. In order to accomplish this specialization, certain genes are actively transcribed but many other genes are not. Because most eukaryotic organisms are composed of a large number of cell types, differential transcription patterns characterize functionally diverse cells.

The concept of differential transcription has led to the **variable gene activity hypothesis** of differentiation. This theory, first entertained by Thomas H. Morgan in 1934, later proposed by Edgar and Ellen Stedman as well as Alfred Mirsky in the 1950s, is articulated in modern form by Eric Davidson in his book, *Gene Activity in Early Development*. The theory holds that the differentiated state assumed by any specific cell type is qualitatively determined by those genes that are actively transcribed in that cell. Its underlying assumption is that each cell contains an entire genome and that differential transcription of selected genes controls the development and differentiation of that cell.

The variable gene activity theory is a very useful model system for experimental design. The remainder of this chapter examines the validity of this theory and its premises and provides examples that offer experimental support from both prokaryote and eukaryote systems. It is important to remember, however, that this type of approach initially tells us what happens. We are only beginning to understand why certain genes are active and others inactive and why qualitatively different gene activity occurs within different cells.

## DIFFERENTIAL TRANSCRIPTION DURING DEVELOPMENT IN PROKARYOTES

Studies of viruses and bacteria have yielded information pertinent to development. These organisms are used because they are often more amenable to experimental approaches than higher organisms. In contrast to complex multicellular eukaryotes, their biochemical and developmental responses are less dependent on the interactions of many differentiated cell and tissue types. Thus, before describing more complex organisms, we shall briefly discuss a model system that has been studied in some detail. Through this discussion we will also find that developmental events in apparently simple systems are in reality complex mechanisms requiring the coordinated action of many gene sets.

### Bacterial Sporulation

Certain gram-positive bacilli such as *Bacillus subtilis* have developed a mechanism of survival under adverse conditions—the formation of spores. These structures are highly resistant to heat, desiccation, and toxic conditions and are induced to form in nutritionally depleted environments. The development of the spore takes place over a period of several hours in a structure known as a **sporangium**, that consists of two chambers, the mother cell and the forespore (Figure 21.1). The two compartments contain identical copies of the chromosome but undertake different programs of transcription. Recent work has demonstrated that there are also interactions between the two compartments during development, that coordinate the program of gene expression in both the forespore and the mother cell.

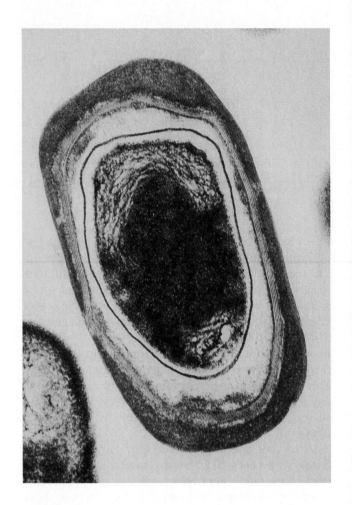

**FIGURE 21.1**    Transmission electron micrograph of sporulating *B. subtilis* cell. The sporangium consists of two chambers, the larger mother cell surrounding the smaller forespore.

Normally, cells of *B. subtilis* grow by doubling in length and divide by the formation of a central septum to produce two identical daughter cells. This stage is designated as Stage 0 in the sporulation cycle (Figure 21.2). Under conditions of starvation, the cell divides in an asymmetric fashion, and a septum forms between the two compartments (Stage II). In Stage III, the smaller prespore is engulfed by the larger mother cell. The engulfed prespore, now called the **forespore**, becomes enclosed in a modified cell wall (Stage IV). A proteinaceous coat is deposited on the outside of the spore in Stage V. During Stage VI, there is little change in the morphology of the spore, but during this stage of maturation the spore acquires the properties of resistance, dormancy, and the ability to germinate. In Stage VII, the mature spore is released, accompanied by lysis of the mother cell (Figure 21.2).

The genes controlling this developmental process have been identified by mutagenesis followed by ge-

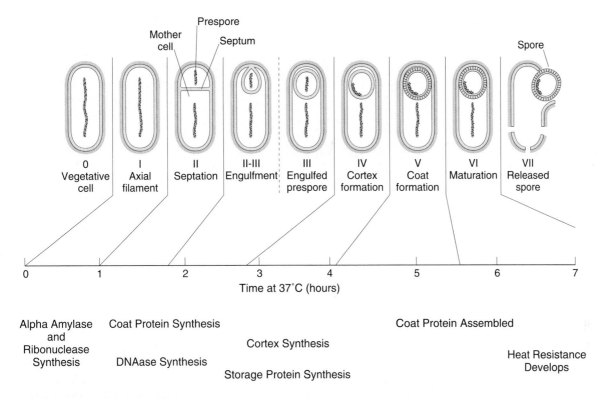

**FIGURE 21.2**    Morphological stages and some of the biochemical events during sporulation in species of the bacterium *Bacillus.*

netic screens to recover mutants of sporulation. To date, more than 80 genes involved in sporulation have been identified and grouped into four classes. The largest class (over 50 genes), required specifically for spore formation, are called *spo* genes. The *ger* genes are important in the process of spore germination; the *ssp* genes and *cot* genes encode soluble proteins and coat proteins, respectively.

Most of the approximately 50 *spo* genes have been cloned and sequenced, and information about the functions of the gene products and their temporal expression and cellular localization is available. These studies indicate that *spo* genes are primarily regulated at the level of transcription; there is little evidence of posttranscriptional or translational regulation. Analysis of mutants that affect the sequential pattern of gene expression reveals that the regulation of many genes is dependent on the appearance of specific transcription factors that activate new sets of genes. These transcription factors are **σ-factors** that bind to RNA polymerase and confer specificity for a set of promoters. σ-factors are DNA-binding proteins that recognize DNA sequences in clusters upstream from promoters. Different σ-factors bind to different DNA sequences in the promoter region. When combined with the core RNA polymerase to form the holoenzyme, the σ-factors confer promoter specificity on the enzyme.

In the preseptal cell, and for the first 1 to 2 hours after septum formation, transcription is controlled by two σ-factors, $\sigma^E$ and $\sigma^F$. After septum formation, transcription in the two compartments becomes differentially regulated by the appearance of $\sigma^G$ in the forespore and $\sigma^K$ in the mother cell [Figure 21.3(a) and (b)]. $\sigma^G$ and $\sigma^K$ each control the transcription of other genes, including compartment-specific σ-factors that activate still other gene sets during later stages of development. Thus, the genes encoding these two σ-factors can be regarded as major regulatory loci that determine cell type (forespore or mother cell) in a manner analogous to selector genes in eukaryotic embryos (discussed below).

Understanding the regulation of $\sigma^G$ and $\sigma^K$ is essential to learning how differential gene expression in the forespore and mother cell are regulated. $\sigma^G$ is encoded by the gene *spo*IIIG and is regulated in two ways (Figure 21.4). In early stages, before septum formation, transcription of *spo*IIIG is activated at a low level, perhaps by $\sigma^F$. After compartmentalization, the production of more $\sigma^G$ is autoregulated, with $\sigma^G$ stimulating the production of more $\sigma^G$ (Figure 21.4). How transcription of *spo*IIIG by $\sigma^F$ becomes compartment-specific is not yet understood.

Regulation of $\sigma^K$ expression in the mother cell is somewhat more complex and involves several separate

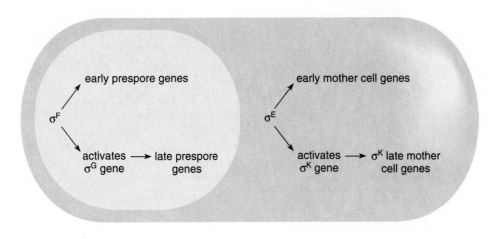

**FIGURE 21.3**   Function of $\sigma^E$ and $\sigma^F$ factors early in sporulation. After septal formation, each factor controls synthesis of specific gene sets in either the mother cell ($\sigma^E$) or prespore cell ($\sigma^F$). In the mother cell, $\sigma^E$ activates transcription of $\sigma^K$, which controls transcription of mother cell genes expressed later in development. In the prespore, $\sigma^F$ activates transcription of $\sigma^G$, which controls transcription of forespore-specific genes.

**(a)**

**(b)**

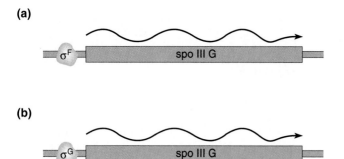

**FIGURE 21.4**   Regulation of *spo*IIIG transcription. $\sigma^G$ is encoded by the *spo*IIIG gene. (a) Before and perhaps just after septum formation, transcription of *spo*IIIG is controlled by $\sigma^F$. This early mode of transcription allows production of enough $\sigma^G$ that it can regulate its own production. (b) Autoregulation of the *spo*IIIG gene by $\sigma^G$ occurs only after septum formation and is restricted to the prespore compartment.

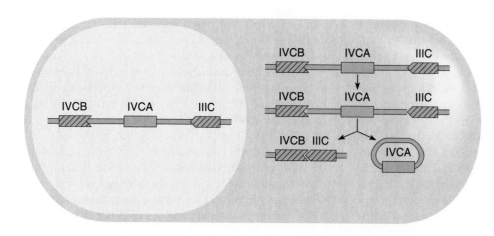

**FIGURE 21.5**   Regulation of $\sigma^K$ expression. In vegetative cells, the inactive $\sigma^K$ is encoded by two loci, IVCB and IIIC. After septation, a DNA rearrangement in the chromosome of the mother cell produces a composite $\sigma^K$ gene. The recombinase enzyme required for this rearrangement is encoded in IVCA, the excised intervening sequence. No such rearrangement takes place in the prespore compartment, restricting $\sigma^K$ transcription to the mother cell compartment.

mechanisms. In vegetative cells, $\sigma^K$ is encoded by two truncated genes, *spo*IVCB and *spo*IIIC. During sporulation, site-specific recombination in the mother cell involving short, repeated DNA sequences located within each of these genes (Figure 21.5) generates an intact $\sigma^K$ gene and an excised circle consisting of the intervening 4 kb of DNA. The recombinase enzyme that catalyzes this rearrangement is encoded by the *spo*IVCA gene located in the excised 4-kb sequence. The rearrangement and transcription of the *spo*IVCB/*spo*IIIC composite gene in mother cells accounts for the compartmentalized expression of $\sigma^K$. The recombination event and mother cell–specific expression are in turn under the control of a gene known as *spo*IIID. Unfortunately, little is known about regulation of the mother cell *spo*IIID gene.

Gene expression in the mother cell is coordinated with developmental events in the forespore by the action of the *spo*IVF gene (Figure 21.6). $\sigma^K$ is produced as an inactive precursor and is activated by an enzyme encoded by the gene *spo*IVC. Developmental events in the forespore activate a signal that crosses the membrane into the mother cell, which activates the *spo*IVC gene, initiating the processing of the $\sigma^K$ precursor. The result is a coordinated program of gene expression in the two cell types which is analogous to forms of cell–cell interaction that are important in multicellular embryonic development in eukaryotes (see below).

As a model system for the study of development, sporulation in gram-positive bacteria such as *B. subtilis* offers several advantages. Only two cell types are involved. The developmental events occur over a restricted time period (8–10 hours) and these events can be studied by genetic analysis and the techniques of molecular biology. Even a relatively simple developmental event such as sporulation turns out to be a genetically complex event, involving more than 80 genes and multiple levels of control. The primary level of regulation is transcriptional, involving cell-specific transcription factors that control gene expression. This small set of regulatory genes encoding transcription factors are themselves subject to multiple levels of control. In addition, interaction between genes and gene products and among gene products contributes to the regulation of gene expression.

Many of the features we have described above are also found in developmental processes in multicellular eukaryotes. Of particular interest are the mechanisms by which cells with identical genomes begin patterns of transcription that lead to different outcomes in cell structure and function. In the following sections, we will explore how these events are regulated in eukaryotic organisms.

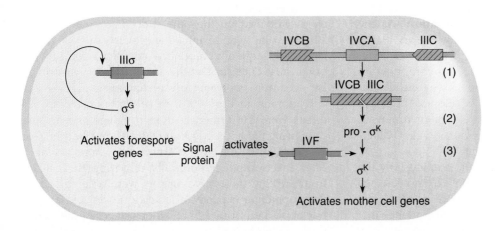

**FIGURE 21.6** Levels of gene regulation during sporulation in *B. subtilis*. Differential gene expression in the two compartments is governed by compartment-specific transcription factors, $\sigma^G$ and $\sigma^K$. In the forespore activation of $\sigma^G$ transcription allows the $\sigma^G$ protein to autoregulate expression of the $\sigma^G$ (*spo*IIIG) gene. In the mother cell, recombination (1) produces an intact $\sigma^K$ gene that is transcribed to produce an inactive precursor protein (2), called pro-$\sigma^K$. Conversion of the precursor to an active transcription factor occurs in response to events in the forespore, where a signal is generated that crosses the septum and activates the *spo*IVF gene in the mother cell. The gene product of the *spo*IVF gene converts inactive pro-$\sigma^K$ protein into the active $\sigma^K$-factor (3), which then initiates transcription of mother cell–specific genes.

## DIFFERENTIAL TRANSCRIPTION DURING DEVELOPMENT IN EUKARYOTES

Higher eukaryotic organisms develop from a **zygote**, a cell formed by the fusion of sperm and oocyte. The oocyte is a single cell with a complex cytoplasm that is heterogeneous and nonuniform in distribution within the cell. Following fertilization and early cell division, the nuclei of progeny cells will therefore find themselves in different environments as the maternal cytoplasm is distributed into the new cells. Evidence suggests that the cytoplasm exerts variable influences on the genetic material of different cells, causing differential transcription at specific points during development. Although such cells show no immediate evidence of structural or functional specialization, their position in the developing embryo seems to determine the ultimate form they will assume. It is as if their fate has been programmed prior to the actual events leading to specialization.

Early gene products synthesized by the zygote further alter the cytoplasm of each cell, producing a still different cellular environment that may in turn lead to the activation of other genes, and so on. In this way, cells embark on pathways of **differentiation** as development proceeds. Differentiation is the process by which cells attain their adult form and function. When a specific developmental fate for cells becomes fixed prior to the actual events of differentiation or specialization, the cells are said to have undergone **determination**. As the number of cells increases, they influence one another in a process of **cell–cell interaction**. The total environment acting on the genetic material, therefore, now includes the cell's individual cytoplasm as well as the influence of other cells. Thus, as the developing organism becomes more complex, so does the total environment. In this way, different forms of determination and differentiation occur during development.

In the following sections, we will examine how the processes of determination, differentiation, and cell–cell interaction relate to patterns of gene expression in development. We will begin by examining the evidence that cells of multicellular organisms each contain an intact genome and that development in eukaryotes involves transcription, translation, morphogenesis, and supracellular assembly in different embryonic cells. We will next examine how the cytoplasm influences the program of gene expression in the early stages of embryonic development in *Drosophila*, how activation of key switch genes is accomplished by transcription factors, and how processing of extracellular signals and alternative splicing of pre-mRNA begin the process of determination in early embryos. Later, we will discuss how development is accompanied by progressive restriction of the developmental options available to cells, and consider the role of cell–cell interactions in the development of multicellular organisms.

### Genomic Equivalence

The variable gene activity hypothesis is based on the premise that the somatic cells of multicellular organisms each contain a complete set of genetic information. While biochemical and cytophotometric analysis has shown the quantity of DNA in each cell within an organism to be equivalent and equal to the diploid content in cells of most species studied, other investigations have approached the question differently. Is it possible to show that differentiated cells are genetically equivalent? In other words, is a complete set of genes present in each cell?

One approach is to show that a differentiated nucleus is **totipotent** and thus capable of giving rise to a complete organism under the proper conditions. In the early part of this century, Hans Spemann demonstrated that a nucleus from a cell of the 16-cell stage of a newt embryo was capable of supporting development of a complete organism. He first constricted a newly fertilized egg prior to the first cell division, forcing the zygote nucleus into one half of the cell. The half containing the nucleus proceeded to undergo division and produce a cluster of 16 cells. He then loosened the constriction, and one of the 16 nuclei was allowed to pass back into the nonnucleated cytoplasm on the other side. Both halves of the embryo, one with 15 nuclei and one with a single nucleus, subsequently produced complete but separate organisms. Therefore, at the 16-cell stage in this organism, genomic equivalence was demonstrated.

Beginning in the 1950s, more sophisticated experiments were performed by Robert Briggs and Thomas King using the grass frog *Rana pipiens,* and by John Gurdon in the 1960s using the African frog, *Xenopus laevis.* After inactivating or surgically removing the nucleus from an egg, these investigators were able to test the developmental capacity of somatic nuclei by injecting nuclei isolated from cells at various stages of differentiation. Nuclei derived from the blastula stage of development were capable of supporting the development of complete and normal adults when transplanted into enucleated oocytes. In *Rana* (Figure 21.7), transplanted nuclei derived from later stages such as the gastrula and neurula usually allowed only partial development. In *Xenopus,* however, Gurdon's experiments showed that epithelial gut nuclei of tadpoles were able to support the development of an adult frog when transplanted into enucleated oocytes. In both cases, it is clear that under normal developmental conditions, progressive restric-

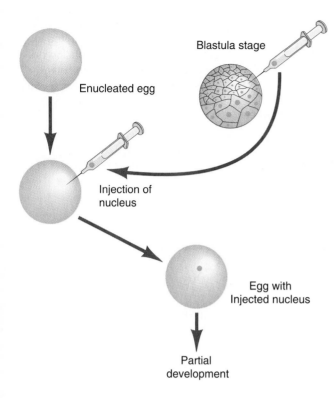

Blastula stage

Enucleated egg

Injection of
nucleus

Egg with
Injected nucleus

Partial
development

**FIGURE 21.7**     The process of nuclear transplantation in
the frog. The unfertilized egg is activated by needle punc-
ture, and the nucleus is removed with a micropipette. A
nucleus from a blastula embryo is injected.

tions that prevent the expression of totipotency are
placed on differentiating nuclei.

To test whether the nucleus of a highly differentiated
adult cell is irreversibly specialized or can support the
development of a normal embryo, Gurdon used nuclei
from adult frog skin cells in serial transplant experiments
(Figure 21.8). In these experiments, a donor nucleus was
transplanted into an enucleated egg, and the recipient
was allowed to develop for a short time, say to the blastula stage. The blastula cells were then dissociated and a
nucleus removed from one of them. This nucleus was
transplanted into still another enucleated egg and development was allowed to occur. Such serial transfers were
repeated a number of times. Subsequently, the blastula
was not dissociated, but instead was allowed to continue
development as far as it would go. Gurdon found that 30
percent of the serial nuclear transplants resulted in the
formation of advanced tadpoles (Table 21.1). Because
nuclei from fully differentiated adult epidermal cells can
eventually direct the synthesis of gene products such as
myosin, hemoglobin, and crystallin and promote the
organization of cells and tissues into a tadpole, we infer
that such cells contain a complete copy of the genome.

These experiments show that under the proper circumstances, genes not expressed in more specialized cells
can become reactivated.

In plants, the totipotency of terminally differentiated,
nonmeristematic cells has been demonstrated using the
carrot. Fredrick Steward observed that when individual
phloem cells are explanted into a liquid culture medium,
each cell divides and eventually forms a mass called a
callus. Under appropriate conditions, this cell mass will
differentiate into a mature plant.

These studies all convincingly argue that differentiated
adult cells have not lost any of the genetic information
present in the zygote. Instead, the majority of genes in
any given cell type fail to be activated or are repressed,
but can be reprogrammed to direct normal development. We are only beginning to understand the nature of
the molecular processes controlling nuclear differentiation during cellular specialization, but the nuclear transplant experiments indicate that the cytoplasm plays a
role in controlling gene expression.

## Transcriptional Control and Determination: *myo*D and *Notch*

As part of the progressive restriction of transcriptional
capacity that accompanies development, certain genes
act as switch points, decreasing the number of alternative developmental pathways. Each decision point is
usually binary, that is, there are two alternatives present,
and the action of a switch gene programs the cell to
follow one of these pathways. We will briefly describe
two such switch genes: (1) *myo*D-related proteins,
which control muscle cell determination and differentiation in both invertebrates and vertebrates; and (2) *Notch*,
a gene that controls cell determination in the *Drosophila*
embryo.

One of the best-characterized classes of switch genes
includes genes that control muscle cell formation in invertebrates and vertebrates. These genes are the **myoblast-determining** or *myo***D** genes. This family of four
genes (Table 21.2) is highly conserved in the vertebrates
but is more distantly related to the *myo*D-related genes
of invertebrates. The *myo*D genes are major regulatory
genes that normally initiate events leading to skeletal
muscle cell differentiation in embryonic cells known as
myoblasts. Experimentally induced expression of *myo*D
genes in nonmuscle-forming cell types, including fibroblasts (connective tissue cells of the skin), adipocytes
(fat cells), neurons (nerve cells), and melanocytes (pigment cells), results in the activation of muscle-specific
genes and, in some cases, causes the formation of completely differentiated muscle cells that contain contractile
fibers. The ability to switch the phenotype of differentiated cells by the activation of a single gene provides

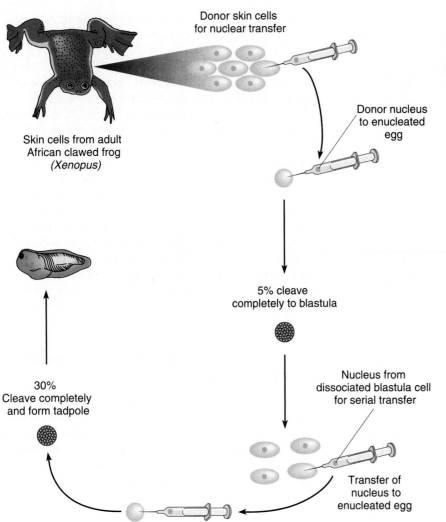

**FIGURE 21.8**    Serial transplantation experiments using adult frog skin cell nuclei. Serial transplantation dramatically increases the percentage (from 5 to 30%) of recipient eggs that undergo complete cleavage.

**Table 21.1**    First Transfer versus Serial Transplantation in *Xenopus laevis*

| Donor Cells | Percentage Reaching Tadpole Stage with Differentiated Cell Types | |
| --- | --- | --- |
| | First Transfer | Serial Transfer |
| Intestinal epithelial cells from tadpoles | 1.5 | 7 |
| Cells from adult skin | 0.037 | 8 |
| Blastula or gastrula | 36 | 57 |

Source: From Gurdon, 1974, p. 24.

**Table 21.2**    The *myo*D Gene Family

| Xenopus | Rodent | Human |
| --- | --- | --- |
| *XMyo*D | *myo*D | *myf*3 |
| *XMyogenin* | *myogenin* | *myf*4 |
| *XMyf*5 | *myf*5 | *myf*5 |
| *XMRF*4 | *MRF*4 | *myf*6 |

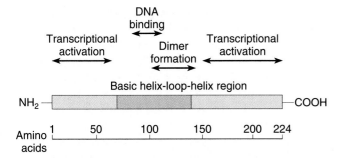

**FIGURE 21.9** Structure of myogenin, a 224-amino-acid protein belonging to the *myo*D family. This protein has a characteristic basic region and helix-loop-helix domain near the center. An adjacent region controls dimer formation necessary for activity. Regions near either end control transcriptional activation.

strong evidence for the central role of *myo*D genes in the regulation of muscle cell formation.

Members of the *myo*D gene family encode DNA-binding proteins with a helix-loop-helix (HLH) motif (see Chapter 17 for a review of eukaryotic transcription factors). Members of the *myo*D class of proteins contain an 80-amino-acid basic region (Figure 21.9) that mediates DNA binding and confers specificity for transcriptional activation. These proteins form heterodimers with proteins known as E proteins and recognize the sequence CANNTG (where N is any nucleotide) present in the enhancers and promoters of most skeletal muscle genes.

Results from cell culture studies on myoblasts indicate that *myo*D proteins themselves are under control of cell growth factors. Myoblasts cultured in the presence of high concentrations of growth factors undergo growth and division, and expression of *myo*D and *myf*5 are re-

pressed, or are expressed only at very low levels. When growth factors are withdrawn from the culture medium, the myoblasts exit the cell cycle, enter a $G_0$ state, and *myo*D and/or *myf*5 are derepressed. The action of these transcription factors activates expression of **myogenin**, which in turn triggers expression of a cascade of contractile protein genes, causing the myoblast to begin differentiation as a muscle cell. Differences in the timing of expression for the *myo*D-related proteins suggest that they each have separate and distinct functions. According to this model: (1) *myf*5 is the earliest factor to be expressed and is the switch gene involved in determination, (2) *myo*D works with *myf*5 to maintain the myoblast identity, (3) all four factors coordinately initiate the process of terminal differentiation into muscle, and (4) *myf*4 works to maintain the differentiated adult phenotype.

A similar pattern of a single gene controlling determination can be seen in the *Notch* gene of *Drosophila*. During embryogenesis, cells on the dorsal surface (ectodermal cells) of the embryo will form epidermis, while the ectodermal cells on the ventral surface will form both epidermis and neuroblasts (nerve cell precursors that will form the ventral nerve cord). On the ventral surface, the decision to form either epidermis or neuroblast is not regionally controlled, but is made at the level of the individual cell. The *Notch* gene, normally transcribed early in embryogenesis, controls this switch point. In the absence of *Notch* transcription, all the ventral cells develop into neural precursors, but the fate of the dorsal cells is largely unaffected. Thus, in *Notch* mutants, embryos die lacking epidermal cells on ventral and head surfaces and having a large excess of neuroblasts.

The *Notch* gene encodes a transmembrane protein with a structure similar to that of proteins involved in the reception of signals originating outside the cell (Figure

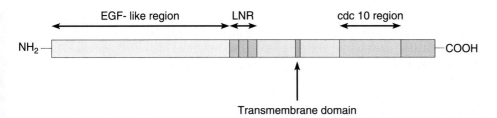

**FIGURE 21.10** Structure of the *Notch* protein. At the amino terminal, the protein resembles the epidermal growth factor (EGF) of vertebrates. This is followed by three repeats of a short stretch of amino acids found in other transmembrane proteins, including the *lin*-12 protein of *C. elegans*. The transmembrane domain anchors the protein in the plasma membrane of the cell. The cytoplasmic side of the protein contains a region similar to that found in the *cdc*-10 protein in yeast, which processes signals for cell division.

21.10), and it is likely that the *Notch* protein functions to receive and process extracellular signals that stimulate or inhibit the selection of a developmental pathway. Although the mechanism of action is not yet known, the *Notch* protein is important in determining cell fate and, perhaps of more fundamental significance, in generating a spatially ordered pattern of cell determination, a topic we will discuss in a later section.

## Developmental Pathways and mRNA Processing: Sex Determination in *Drosophila*

While the *myo*D and *Notch* genes determine cell fate in a region of the developing embryo, other mechanisms of determination can direct selection of a developmental pathway that affects the whole body. One such example is the role of alternative splicing of pre-mRNA in sex determination in *Drosophila*. As outlined in Chapter 6, sex in *Drosophila* is determined by the ratio of X chromosomes to autosomes (X:A). When the ratio is 0.5 (1X:2A), males are produced, even when no Y chromosome is present; when the ratio is 1.0 (2X:2A), females are produced. At intermediate ratios (2X:3A), intersexes are produced. The chromosomal ratios are interpreted by a small number of genes that initiate a cascade of developmental events resulting in the production of male or female somatic cells and the corresponding male or female phenotypes. The genes in this pathway are *Sex-lethal* (*Sxl*), *transformer* (*tra*), *transformer-2* (*tra-2*), *doublesex* (*dsx*), and *intersex* (*ix*).

In females, interpretation of the chromosomal ratio activates transcription of the *Sxl* gene and the production of *Sxl* protein. This activation involves expression of several intermediate genes, including *sisterless* and *daughterless,* and perhaps a small number of other genes. In males, the chromosomal ratio does not activate production of *Sxl* protein (Figure 21.11). When the *Sxl* protein is present, it binds to and controls the splicing of the pre-mRNA for the next gene in the pathway, *transformer* (*tra*). The pre-mRNA of *tra* includes a termination codon in exon 2. The female-specific *Sxl* protein binds to the *tra* pre-mRNA and directs the splicing to remove exon 2 from the mature mRNA. In males, where no *Sxl* protein is present, splicing proceeds along the ground state or default pathway, and the stop codon is incorporated into the mature mRNA (Figure 21.12). In males, translation of *tra* mRNA is prematurely terminated by the stop codon, resulting in an inactive gene product, while in females, translation produces a functional gene product. The result is a female-specific *tra* protein that acts in conjunc-

tion with the *tra-2* gene to produce the female phenotype (Figure 21.11). The *tra-2* gene is active in both males and females and contains an RNA-binding domain that binds to the pre-mRNA of the *dsx* gene, but only in the presence of the *tra* protein.

The *dsx* gene is a critical control point in the development of sexual phenotype, since it produces a functional mRNA and protein in both females and males. However, the pre-mRNA is processed in a sex-specific manner to yield different transcripts. In females, the *tra* protein and the *tra-2* protein bind to the *dsx* pre-mRNA and direct its processing in an alternative, female-specific manner. In males, where only the *tra-2* protein is present, default splicing of the *dsx* pre-mRNA results in a male-specific mRNA and protein product. The female *dsx* protein causes female sexual differentiation in somatic tissues by repressing male sexual development through coordinated action with the *ix* gene product (Figure 21.11). The male *dsx* protein brings about male sexual development by suppressing the female pathway. In the absence of *dsx* function, flies develop into phenotypic intersexes.

The *Sxl* gene acts as a switch gene in selecting the pathway of sexual development by eventually controlling the processing of the *dsx* transcript in a female-specific fashion. *Sxl* exerts this control by encoding a splicing factor that causes transcripts from the *tra* gene to be processed in a female-specific fashion. The *tra* protein interacts with the *tra-2* protein to control the splicing of the *dsx* transcript in a female-specific fashion. The *Sxl* gene is activated only in embryos with an X:A ratio of 1. Failure to control the splicing of the *dsx* transcript in a female mode (failure to activate the *Sxl* gene) results in default splicing of the transcript in a male mode, leading to production of a male phenotype.

These examples should serve to reinforce the idea that determination can be accomplished in a number of ways, including cytoplasmic gradients, transcription factors, receptors that process external signals, and splicing factors that control alternative splicing of pre-mRNA. However, determination is not a one-step process that exists in isolation, but is a multistep process that restricts the developmental potential of a cell over time. This continual process of developmental restriction is then followed by differentiation of the terminal phenotype programmed by determination. In the following section, we will examine the progressive nature of determination and the increasingly restrictive developmental outcomes that result from these decisions. At the molecular level, we will emphasize the role of genes encoding transcription factors in the establishment of cell fates during development.

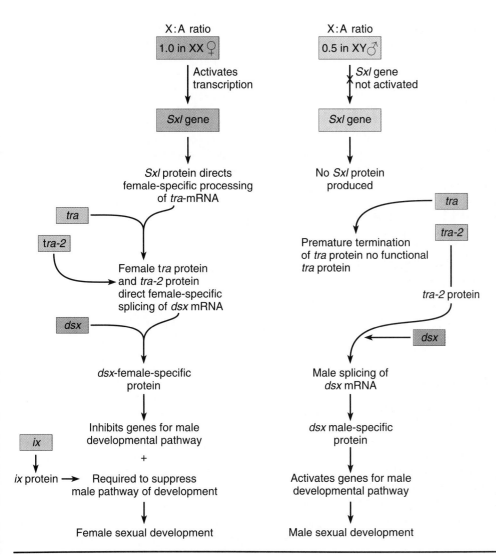

**FIGURE 21.11** Hierarchy of gene regulation for sex determination in *Drosophila*. In females, the X:A ratio activates transcription of the S*xl* gene. The product of this gene binds to pre-mRNA of the *tra* gene and directs its splicing in a female-specific fashion. The female-specific *tra* protein, in combination with the *tra*-2 protein directs female-specific splicing of *dsx* pre-mRNA, resulting in a female-specific *dsx* protein. This protein, in combination with *ix* protein, suppresses the pathway of male sexual development and activates the female pathway. In males, the X:A ratio does not activate S*xl*. The result is a male-specific processing of *tra* pre-mRNA, resulting in no functional *tra* protein. This in turn leads to male-specific processing of *dsx* transcripts, resulting in male-specific *dsx* protein that activates the pathway for male sexual development.

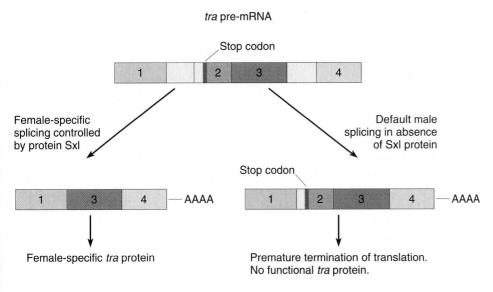

**FIGURE 21.12** Sex-specific splicing of *tra* gene transcripts. In females, the S*xl* protein binds to the pre-mRNA and directs splicing to eliminate exon 2, which contains a stop codon. This allows translation of a female-specific protein that directs female-specific splicing of *dsx* transcripts. In males, no S*xl* protein is present, and the *tra* pre-mRNA is spliced to include the stop codon in exon 2. The result is a nonfunctional *tra* protein fragment that is unable to direct processing of *dsx* transcripts.

# PROGRESSIVE RESTRICTION OF CELL FATE

We have now come to the central issue of developmental biology. In the preceding discussion, we examined factors responsible for making cells different from one another (i.e., differential transcription). However, why one cell's fate is different from another's during development is a broader and more critical issue. This question is distinct from that discussed in Chapter 17, where we asked what actually turns any given gene on or off at the cellular level. We consider here the question of how cells eventually acquire one differentiated identity or another.

As alluded to earlier, there is no simple answer to this question. However, information derived from the study of many different organisms provides a starting point. We shall examine one of these examples, embryonic development in *Drosophila*, which reflects the direction and scope of our knowledge in this area.

## Overview of *Drosophila* Development

The formation of the body plan of the larva and adult of *Drosophila melanogaster* is an elegant example of the interaction between cytoplasm and nucleus. The body of *Drosophila* is composed of head, thoracic, and abdominal segments. After fertilization, several rounds of nuclear divisions followed by enclosure in membranes result in the formation of a single layer of cells, the **blastoderm**, over the embryo surface (Figure 21.13). Each segment of the adult body is formed from the descendants of cells set aside at the blastoderm stage into discrete structures called **imaginal disks** (Figure 21.14). These disks, formed from hollow sacs of cells, are determined to form specific parts of the adult body. There are 12 bilaterally paired disks and one genital disk. For example, there are eye-antennal disks, leg disks, wing

**FIGURE 21.13**    Early stages of embryonic development in *Drosophila*. (a) Fertilized egg with zygotic nucleus, about 30 minutes after fertilization. (b) Nuclear divisions occur about every 10 minutes, producing a multinucleate cell. (c) After approximately nine divisions (512 nuclei), the nuclei migrate to the outer surface or cortex of the egg. (d) At the surface, four additional rounds of nuclear division occur. A small cluster of cells, the pole cells, form at the posterior pole about $2\frac{1}{2}$ hours after fertilization. These cells will form the germ cells of the adult. (e) About 3 hours after fertilization, the nuclei become enclosed in membranes, forming a single layer of cells over the embryo surface. This stage is known as the blastoderm.

**(a)**

Diploid zygote nucleus produced by fusion of parental gamete nuclei.

**(b)**

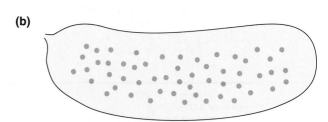

Nine rounds of nuclear divisions produce multinucleated syncytium.

**(c)**

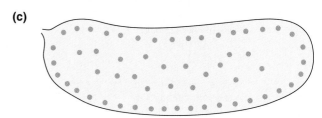

Nuclei migrate to outer surface.

**(d)**

Pole cells form at posterior pole (precursors to germ cells).

Approximately four further divisions take place at surface.

**(e)**

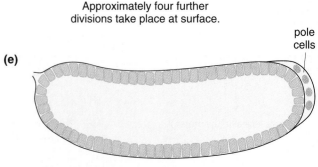

pole cells

Nuclei become enclosed in membranes, forming a single layer of cells over embryo surface.

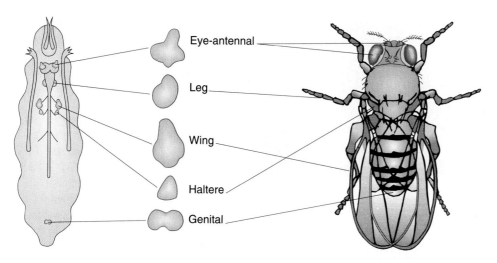

**FIGURE 21.14** Imaginal disks of the *Drosophila* larva and the adult structures that are derived from them.

disks, and so forth. Other cells in the blastoderm will form the internal and external structures of the larva. During metamorphosis, most of the larval body parts histolyze, or break down, and the imaginal disks undergo mitosis and differentiate into adult structures of the head, thorax, and abdomen.

Studies using a combination of physical and genetic techniques have revealed that the major patterns of the larval and adult body plans are specified by the blastoderm stage of embryonic development in *Drosophila*. The two-dimensional projections of the external adult body parts on the surface of the blastoderm are known as **fate maps** (Figure 21.15).

One of the most striking features of fate maps is that the embryo is organized along an anterior-posterior axis that reflects the body axis of the larva, pupa, and adult (Figure 21.15). The existence of a fate map implies that developmental fate is governed by the location to which

nuclei migrate during blastoderm formation, and that positional information in the form of an anterior-posterior axis and a dorsal-ventral axis direct the nuclei to adopt developmental programs that result in the morphologically distinct segments of the adult body.

At a slightly later stage of development, the precursors of the adult body structures are present as distinct morphological segments in the embryo (Figure 21.16). Even at this stage of development, cells in each segment are determined to form either an anterior or a posterior portion or **compartment** of the segment. Progressive restrictions continue to occur throughout development so that cells in each compartment are committed to forming a limited number of adult structures.

The positional information laid down by the gradients along the anterior-posterior axis of the embryo is interpreted by two gene sets: (1) **segmentation genes**, which divide the embryo into a series of stripes or seg-

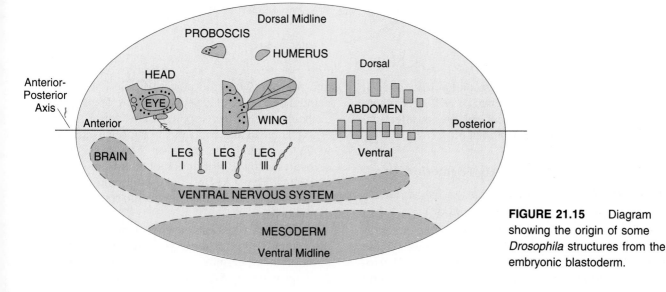

**FIGURE 21.15** Diagram showing the origin of some *Drosophila* structures from the embryonic blastoderm.

**(a)**

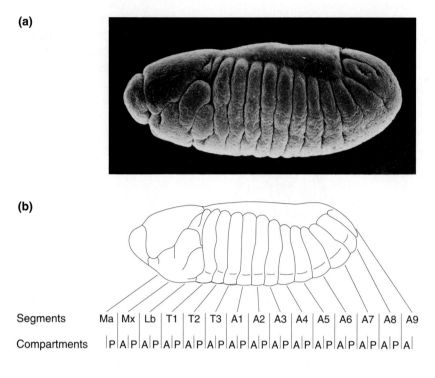

**(b)**

**FIGURE 21.16** Segmentation in *Drosophila*. (a) Scanning electron micrograph of *Drosophila* embryo at about 10 hours after fertilization. By this stage, the segmentation pattern of the body is clearly established. (b) The segments, compartments, and parasegments of the *Drosophila* embryo. Ma, Mx, and Lb represent the segments that will form head structures. T1–T3 are thoracic segments and A1–A9 are abdominal segments. Each segment is divided into anterior (A) and posterior (P) compartments. Parasegments represent an early pattern specification that is later refined into the segmental plan of the body. Note that all the parasegments are shifted forward by one compartment. (c) The segmented embryo and the adult structures that will form each segment.

| Segments | Ma | Mx | Lb | T1 | T2 | T3 | A1 | A2 | A3 | A4 | A5 | A6 | A7 | A8 | A9 |
|---|---|---|---|---|---|---|---|---|---|---|---|---|---|---|---|
| Compartments |P|A|P|A|P|A|P|A|P|A|P|A|P|A|P|A|P|A|P|A|P|A|P|A|P|A|P|A|P|A|

**(c)**

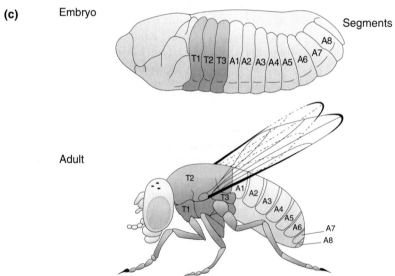

ments and define the number, size, and polarity of each segment; and (2) **selector genes**, which specify the identity or fate of each segment but do not affect the number or size of the segments (Figure 21.17).

## Cytoplasmic Gradients: Axis Formation in the *Drosophila* Embryo

The shape of the *Drosophila* egg provides a set of morphological landmarks that delineate the anterior-posterior and dorsal-ventral axes of the egg (Figure 21.18). The anterior pole of the egg contains the micropyle, a specialized structure for the entrance of sperm

into the egg, while the posterior pole is rounded and marked by a series of aeropyles that are openings for gas exchange during development. The dorsal side of the egg is flattened and contains the chorionic appendages, while the ventral side is curved. The shape of the egg correlates with the distribution of internal molecular gradients that play a key role in establishing the developmental fates of nuclei that migrate into specific regions of the embryo.

During embryogenesis, a series of nuclear divisions in the central region of the egg occurs immediately after fertilization. When nine divisions have occurred, most of the approximately 512 nuclei migrate to the egg's outer

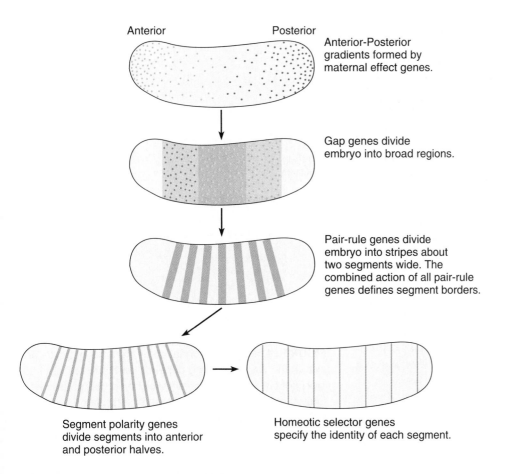

**Anterior**    **Posterior**

Anterior-Posterior gradients formed by maternal effect genes.

Gap genes divide embryo into broad regions.

Pair-rule genes divide embryo into stripes about two segments wide. The combined action of all pair-rule genes defines segment borders.

Segment polarity genes divide segments into anterior and posterior halves.

Homeotic selector genes specify the identity of each segment.

**FIGURE 21.17**    Overview of progressive restriction of cell fate during development in *Drosophila*. The process begins with the anterior-posterior axis laid down in oogenesis and activated immediately after fertilization. This axis restricts cells to form either anterior or posterior structures. Gene products from this gradient activate transcription of the gap genes that divide the embryo into a limited number of broad regions. Gap proteins are transcription factors that activate pair-rule genes whose products divide the embryo into regions about two segments wide. The combined action of all gap genes defines segment borders. The pair-rule genes in turn activate the segment polarity genes that divide the segments into anterior and posterior compartments. The collective action of the maternal genes that form the anterior-posterior axis and the segmentation genes define fields of action for the homeotic selector genes that specify the identity of each segment.

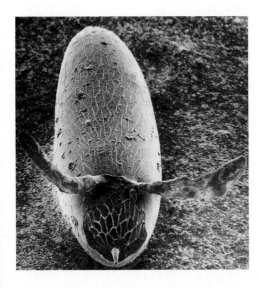

**FIGURE 21.18**    Scanning electron micrograph of the anterior pole of a newly oviposited *Drosophila* egg.

surface or cortex, where further divisions take place. The nuclei later become enclosed in membranes, forming a single layer of cells (cellular blastoderm) over the embryo surface (Figure 21.13). As nuclei migrate into different regions of the cytoplasm, maternally derived transcripts and gene products localized in these regions initiate a program of gene expression in the nuclei that results in the formation of the anterior-posterior axis of the embryo. Table 21.3 lists some of the maternal genes that are required for formation of the anterior-posterior and dorsal-ventral embryonic axes. Maternal genes are transcribed by the maternal genotype during oogenesis and transported as mRNA or as translated protein into the oocyte for storage. These transcripts and gene products are utilized during early stages of development.

Of the genes required to form the anterior portion of the embryo, the most important is *bicoid* (Table 21.3). Embryos of *bicoid* mutants lack head and thoracic structures and also have abnormalities of the first four segments of the abdomen. During oogenesis and the first hours of embryonic development, *bicoid* mRNA is localized to a small region at the anterior pole of the egg [Figure 21.19(a)]. How does *bicoid* mRNA come to be localized within the anterior region of the egg, and what keeps it in place? The actions of at least two other genes, *swallow* and *exuperantia,* are involved in localizing the *bicoid* mRNA by docking it to anterior cytoskeletal components, but their mechanisms of action are not yet known.

During early cleavages, *bicoid* mRNA is translated, and the *bicoid* protein becomes distributed in a gradient along the anterior-posterior axis, with very low levels in

**(a)** Localization of *bicoid* mRNA in oocyte

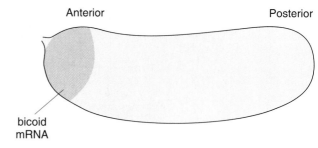

**(b)** Localization of bicoid protein

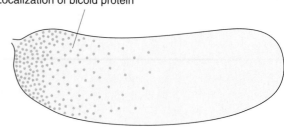

**FIGURE 21.19**    The formation of the anterior-posterior axis of the *Drosophila* egg depends on the action of three independent systems. The anterior portion of the axis is formed by a gradient of *bicoid* protein translated from mRNA beginning at fertilization.

the posterior one-third of the egg [Figure 21.19(b)]. This gradient of concentration plays a critical role in determining the developmental pattern of the egg. The *bicoid* protein contains a homeodomain (see Chapter 17 for a review of these domains), indicating that it is a transcription control factor. In fact, the *bicoid* protein binds to the upstream regulatory region of *hunchback,* one of the first zygotic genes to be activated early in development, which is expressed in the anterior half of the embryo. The upstream promoter region of *hunchback* contains five sites, each with the consensus sequence 5'-TCTAATCCC-3', that bind to the *bicoid* protein. The presence of the consensus binding sequence causes the gene to respond to binding of the *bicoid* transcription factor, and the number of sites occupied by the bicoid protein determines the degree of response. In regions of the gradient, where there is a higher concentration of *bicoid* protein, there will be a higher response by the *hunchback* gene, leading to the formation of a gradient of *hunchback* protein along the anterior-posterior axis (Figure 21.20).

The gradient of *bicoid* protein causes transcription of *hunchback* in the anterior half of the embryo, but not in the posterior half. The *hunchback* protein represses activity of genes that form the abdomen and stimulates transcription of genes that form the head and thorax.

The posterior of the embryo is formed through the

**Table 21.3**    Maternally Transcribed and Zygotically Transcribed Genes That Control the Anterior-Posterior Axis of the *Drosophila* Embryo

|  | Anterior | Posterior | Terminal |
|---|---|---|---|
| Maternal |  |  |  |
|  | *bicoid* | *staufen* | *trunk* |
|  | *exuperantia* | *oskar* | *fs(1)nasrat* |
|  | *swallow* | *vasa* | *fs(1)polehole* |
|  | *staufen* | *valois* | *torso* |
|  |  | *tudor* | *l(1)polehole* |
|  |  | *mago nashi* |  |
|  |  | *nanos* |  |
|  |  | *pumilio* |  |
| Zygotic |  |  |  |
|  | *hunchback* | *knirps* | *tailless* |
|  |  | *giant* | *huckebein* |

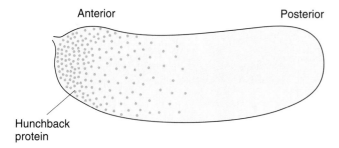

Anterior                                    Posterior

Hunchback
protein

**FIGURE 21.20**    The gradient of *bicoid* protein triggers expression of the *hunchback* gene, forming a gradient of *hunchback* protein along the anterior-posterior axis.

action of eight genes (Table 21.3). Mutants of these genes all have a similar phenotype: the posterior portion of the body is dwarfish and abdominal segments are lacking. The names of these mutants reflect this phenotype: *oskar* (the dwarf in the Günter Grass novel, *The Tin Drum*) and *nanos,* derived from the Spanish word for dwarf. The *nanos* gene is transcribed by the maternal genome during oogenesis, and its mRNA is stored at the posterior pole of the egg (Figure 21.21). The *nanos* protein acts to control formation of posterior structures in the developing embryo, but does so in a different way than the *bicoid* protein. In contrast to the action of anterior genes, the posterior genes work by a negative effect on a maternal gene transcript. The *nanos* protein acts to prevent the translation of any *hunchback* mRNA that might be present in the posterior region of the embryo, ensuring the expression of abdomen-forming genes (Figure 21.22). The other genes involved in posterior for-

mation ensure that the *nanos* mRNA is packaged and stored at the posterior pole (*oskar, staufen, valois, vasa*) and is distributed properly (*pumilio*).

The terminal genes represent a third set of genes that controls the formation of the anterior-posterior axis (Table 21.3). Mutations in these genes result in the loss of the unsegmented structures that develop from the posterior pole of the embryo. Cloning of two of the terminal genes, *torso* and *fs*(1)*polehole,* suggests that this group contains genes that encode transmembrane signal receptors associated with the reception and transduction of an extracellular signal.

Together, these three groups of maternal genes establish the anterior-posterior axis of the embryo as a spatial pattern of gene products. The anterior of the embryo is organized by a gradient of the *bicoid* protein that acts to activate anterior-specific genes and to repress posterior-specific genes. The posterior of the embryo is formed through the action of the *nanos* protein, which after migration from the posterior pole, suppresses the formation of the *hunchback* protein, relieving the *hunchback*-mediated inhibition of abdominal-specific gene expression. The critical gene in formation of the terminal region of the embryo is the *torso* gene product, which allows expression of the genes that form the unsegmented terminal structures. Two of these systems function by sequestering a specific mRNA at the poles of the egg (*bicoid* at the anterior, *nanos* at the posterior) and by the formation of a gradient of protein. The third system probably works via a signal transduction system coupled to the anterior-posterior axis system during oogenesis. The dorsal-ventral axis of the embryo is specified by a family of maternal genes that depends on the formation of a single protein gradient (dorsal) along the dorsal-ventral axis.

These three groups of maternal genes (anterior, posterior, and terminal establish a spatial pattern of gradients during oogenesis, leading to the formation of the developmental axes in the *Drosophila* embryo. Following fertilization and the appearance of embryonic nuclei, these genes regulate transcriptional activities of cells along the anterior-posterior and dorsal-ventral axes and serve to illustrate the critical role of cytoplasmic gradients in determining the fates of cells in different embryonic regions. In the next section, we will describe how components of the anterior-posterior axes control transcription of gene sets that regulate further development of the embryo.

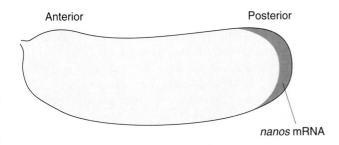

Anterior                                    Posterior

*nanos* mRNA

**FIGURE 21.21**    The posterior end of the anterior-posterior axis is formed by action of the *nanos* gene. mRNA from *nanos* transcribed during oogenesis is stored at the posterior pole of the oocyte. After fertilization, the mRNA is distributed throughout the posterior region of the embryo by action of the *pumpilo* gene. The *nanos* protein acts to suppress the action of *hunchback* protein, ensuring the formation of abdominal structures in the posterior region.

## Segmentation Genes

The segmentation genes are expressed early in embryonic development, soon after the *bicoid* gradient has

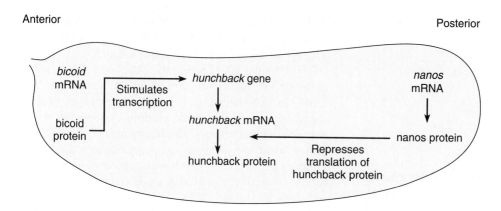

**FIGURE 21.22**    Coordinated action of *bicoid* and *nanos* proteins produces the anterior-posterior axis of the early embryo. Throughout the anterior end, *bicoid* protein activates the *hunchback* gene. The *hunchback* protein activates transcription of head and thorax genes. In the posterior region, the *nanos* protein inhibits production of *hunchback* protein, allowing the expression of abdominal genes.

been established. Over 20 segmentation loci have been identified (Table 21.4). They are classified on the basis of their mutant phenotypes: (1) the gap genes (*Krüppel*, *knirps*) delete a group of adjacent segments; (2) the pair-rule genes (*even-skipped*, *fushi-tarazu*) affect every other segment and eliminate a specific part of each affected segment; and (3) the segment polarity genes (*hedgehog*, *gooseberry*), which when mutant cause defects in homologous portions of each segment.

Transcription of the gap genes is activated or inactivated by gene products previously expressed along the anterior-posterior axis from *bicoid* and the other genes of the maternal gradient systems. Transcription of the gap genes divides the embryo into a series of broad regions within which different combinations of gene activity will eventually specify both the type of segment that will form and the proper order of segments in the body

of the larva, pupa, and adult. Gap genes such as *Krüppel* and *knirps* encode transcription factors with zinc finger motifs and establish patterns of gene activity by directly controlling transcription of pair-rule genes (Figure 21.23).

The **pair-rule genes** function to divide the broad regions established by the gap genes into regions about one segment wide. Mutations in pair-rule genes eliminate segment-size regions at every other segment. The pair-rule genes are expressed in narrow bands or stripes of nuclei that extend circumferentially around the embryo. Expression of members of this gene set first establishes the boundaries of segments and compartments. Their action then establishes the developmental fate of the cells within each segment by controlling the segment polarity genes. At least eight pair-rule genes act to divide the embryo into a series of stripes. However, the boundaries of these stripes overlap, meaning that cells in the stripes express different combinations of pair-rule genes in an overlapping fashion (Figure 21.23).

Many pair-rule genes encode transcription factors containing helix-turn-helix domains, also called homeodomains. Transcription of the pair-rule genes is mediated by the action of gap gene products, but the pattern is resolved into highly delineated stripes by the interaction among the gene products of the pair-rule genes themselves (Figure 21.24).

Transcription of **segment polarity** genes is controlled by transcription factors encoded by the pair-rule genes. Each segment polarity gene becomes active in a single band of cells within each segment that extends around the embryo. This concerted action divides the embryo into 14 segments and controls the cellular identity within each segment. Some of the segment polarity

| **Table 21.4**   SEGMENTATION GENES IN DROSOPHILA | | |
| --- | --- | --- |
| **Gap Genes** | **Pair-Rule Genes** | **Segment Polarity Genes** |
| *Krüppel* | *hairy* | *engrailed* |
| *knirps* | *even-skipped* | *wingless* |
| *hunchback* | *runt* | *cubitis interruptus*[D] |
| *giant* | *fushi-tarazu* | *hedgehog* |
| *tailless* | *odd-paired* | *fused* |
| *huckebein* | *odd-skipped* | *armadillo* |
| | *sloppy-paired* | *patched* |
| | *paired* | *gooseberry* |

**(a)**

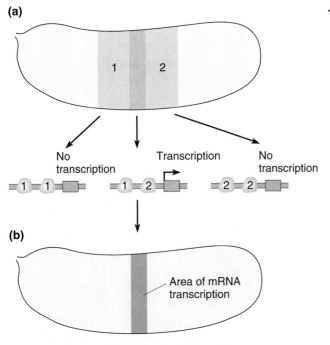

◀ **FIGURE 21.23**    New patterns of gene expression can be generated by overlapping regions containing gene products. (a) Transcription factors 1 and 2 are present in an overlapping region of expression. If both transcription factors must bind to a promoter in order to trigger expression of a target gene, the gene will be active only in cells containing both factors (most likely in the zone of overlap). (b) Expression of the target gene in the restricted region of the embryo.

**(b)**

Area of mRNA transcription

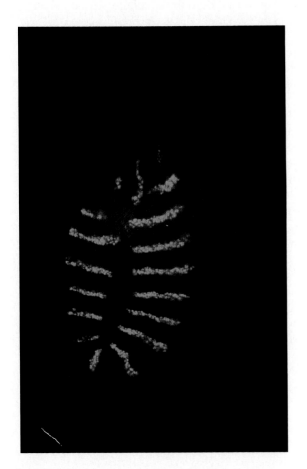

**FIGURE 21.24**    Stripe pattern of gene expression in *Drosophila* embryo. In this embryo, expression of a gene is limited to a group of cells within a segment.

genes, including *engrailed,* encode transcription factors. Rather than activating transcription, however, the *engrailed* protein competitively inhibits activation by other homeodomain proteins, resulting in the establishment of a segment border.

The genes that control development in *Drosophila* act in a temporally and spatially ordered cascade, beginning with the genes that establish the anterior-posterior and dorsal-ventral axes of the egg and early embryo. The gradients of mRNAs and proteins along the anterior-posterior axis activate the gap genes, which subdivide the embryo into broad bands. The gap genes in turn activate the pair-rule genes, which divide the embryo into segments. Finally, the segment polarity genes divide each segment into anterior and posterior regions arranged linearly along the anterior-posterior axis. This progressive restriction of the developmental potential of the cells in the *Drosophila* embryo, all of which occurs during the first one-third of embryogenesis, involves transcriptional control of gene sets and is accomplished by regulating the activity of proteins that bind to promoter and enhancer regions of genes and act as transcription factors.

## Selector Genes

As segment boundaries are being established by the action of the segmentation genes, the **selector genes** are activated. Expression of these genes determines the structures to be formed by each segment, including the antennae, mouth parts, legs, wings, thorax, and abdomen. Mutants of these genes are known as **homeotic**

# TOOLS AND TECHNIQUES

## Knockout Mice

"Knockout mice" are genetically altered so that a particular gene (the target gene) has been disrupted and made nonfunctional. The technique is complex and uses a combination of recombinant DNA technology, cell culture, and embryo manipulations. It takes about a year to construct a knockout mouse, and analyzing the effects of the knockout takes up to four or five years. The method was first developed in 1989 by several groups, including Mario Cappechi at the University of Utah, Oliver Smithies at the University of North Carolina, and Elizabeth Robertson at Harvard Medical School; to date, over 100 different genes have been knocked out in mice.

The technique begins with a cloned target gene from a normal mouse. A marker gene for antibiotic resistance is inserted into this gene, disrupting its function, and providing a marker to follow the gene through subsequent procedures. The altered gene is transferred into embryonic stem cells (ES). In some of these cells, the altered gene will replace the normal gene via a recombination event. Cells carrying one copy of the recombined gene and one normal allele are selected for and grown in culture. These genetically altered cells are introduced into a mouse embryo where they will participate in the formation of adult tissues and organs. The chimeric embryo is transferred to a foster mother to complete development. By using coat color genes as markers, it is possible to select chimeric mice that carry germ cells derived from the genetically altered ES cells. The heterozygous mice can be used to breed mice homozygous for the disrupted gene; these are called "knockout mice" because a single gene has been mutated or knocked out.

Using this method, it is possible to study the developmental effects of knocking out a single gene and to create mouse models for human genetic disorders. For example, knockout mice have been created that lack RAG-1 and RAG-2, two genes shown to be important in V(D)J recombination in the immune system, and the basis for generating antibody diversity. These mice are being studied in germ-free and normal environments to provide a clearer understanding of how these genes work during the maturation and operation of the immune system.

This technique has also been important in research directed toward the understanding and treatment of human genetic disorders. For example, a mouse model has been created for Gaucher's disease (a progressive and fatal autosomal recessive disorder of lysosomal enzymes). Somewhat surprisingly, the homozygous recessive phenotype is more severe in the mice and leads to death hours after birth. This finding led to the realization that some affected humans also die as newborns, providing new insight into this disorder. A mouse model of cystic fibrosis has been created by the knockout technique, and will prove valuable for testing gene therapy methods before they are applied to humans.

However, some surprises are also resulting from knockout studies. A knockout mouse lacking the HGPRT enzyme responsible for Lesch–Nyhan syndrome (an X-linked fatal disorder in humans) is per-

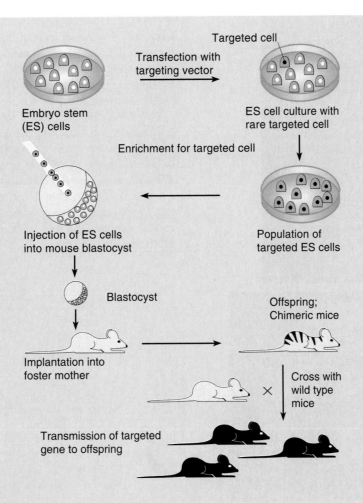

Targeted cell

Transfection with
targeting vector

Embryo stem
(ES) cells

ES cell culture with
rare targeted cell

Enrichment for targeted cell

Injection of ES cells
into mouse blastocyst

Population of
targeted ES cells

Blastocyst

Offspring;
Chimeric mice

Implantation into
foster mother

Cross with
wild type
mice

Transmission of targeted
gene to offspring

fectly normal and has no detectable defects. Knockouts for two genes thought to be essential for muscle development, *myo*D and *myf*5, undergo normal prenatal development, and as adults the *myo*D knockout mice are indistinguishable from normal mice. In the case of the *myf*5 knockout mice, muscles are normal, but ribs are absent. The frequency with which such results are turning up suggests that there are alternative biochemical and developmental pathways to compensate for gene loss, indicating that genetic control of development may be more complex than previously thought.

Now under development are a second generation of knockout mice carrying multiple gene knockouts, involving several genes in a developmental pathway, and mice that will have genes knocked out in selective tissues, or will carry disabled instead of inactivated genes. For example, a new knockout mouse for cystic fibrosis is being constructed that will carry the most common mutation (delta 508) found in the human population.

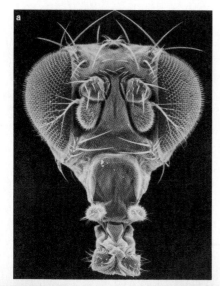

mutants. For example, the wild-type allele of the *Antennapedia* (*Antp*) gene is required to specify structures in the second thoracic segment (which carries a leg). Dominant gain of function *Antp* mutations lead to expression of the gene in the head and thorax, and the antennae of the fly are transformed into legs (Figure 21.25).

The *Drosophila* homeotic selector genes are found in two clusters on chromosome 3. The *Antp* complex contains five genes (Table 21.5) required for specification of structures in the head and first two thoracic segments. The *bithorax* complex contains three genes required for specification of structures formed by the posterior portion of the second thoracic segment, the entire third thoracic segment, and the abdominal segments (Figures 21.26 and 21.27).

**FIGURE 21.25** *Antennapedia* (*Antp*) mutation in *Drosophila*. (a) Head from wild type *Drosophila* showing the structure of the antenna and other head parts. (b) Head from an *Antp* mutant, showing the replacement of normal antenna structures with a leg. This is caused by activation of the *Antp* gene in the head region.

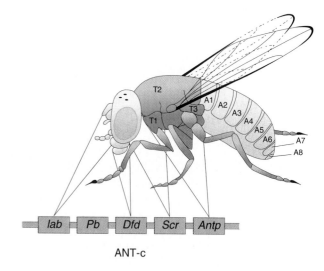

ANT-c

**FIGURE 21.26** Genes of the *Antennapedia* complex and the adult structures they specify. The *labial* (*lab*) and *Deformed* (*Dfd*) genes control the formation of head segments. The *Sex comb reduced* (*Scr*) and *Antennapedia* (*Ant*) genes specify the identity of the first two thoracic segments. The remaining gene in the complex, *proboscipedia,* may not act during embryogenesis, but may be required to maintain the differentiated state in adults. In mutants, the labial palps are transformed into legs.

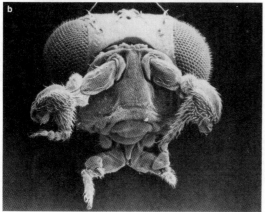

**Table 21.5** HOMEOTIC SELECTOR GENES OF *DROSOPHILA*

| *Antennapedia* Complex | *Bithorax* Complex |
| --- | --- |
| labial | Ultrabithorax |
| Antennapedia | abdominal A |
| Sex comb reduced | Abdominal B |
| Deformed | |
| proboscipedia | |

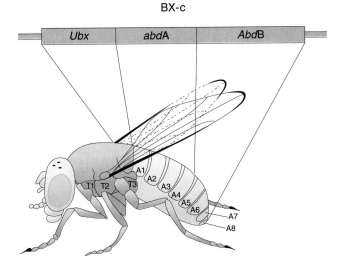

**FIGURE 21.27**    Genes of the *bithorax* complex and the adult structures they specify. *Ultrabithorax* (*Ubx*) controls structures in the posterior compartment of T2 and structures in T3. The two other genes, *abdominal A* (*abdA*) and *Abdominal B* (*AbdB*), specify the segmental identities of the eight abdominal segments (A1–A8).

Activation of the selector genes is under the control of the gap genes and the pair-rule genes. Although selector gene expression is confined to certain domains in the embryo, the timing and patterns of expression are rather complex and involve interactions between segmentation and selector genes as well as interactions among the various selector genes. For example, the *Antennapedia* gene has two promoters (P1 and P2) and can be transcribed to produce two different pre-mRNAs (Figure 21.28). Transcription from P1 is stimulated by *Krüppel* protein (a gap gene product) and repressed by *Ultrabithorax* protein (a selector gene product). Transcription from P2 is activated by *hunchback* and *fushitarazu* proteins (maternal gradient and gap gene products) and inhibited by *oskar* protein (a maternal gradient gene product). Paradoxically, the protein produced from these two pre-mRNAs is identical.

Regulation of expression of selector genes is also accomplished by alternative splicing of pre-mRNAs to yield different transcripts (Figure 21.29). The *Ultrabithorax* gene transcripts produced during early embryogenesis can be processed at varying 5′ sites in exon 1 and will include two "micro-exons," producing a number of related proteins. Transcripts from the *Ultrabithorax* gene later in development (during formation of the nervous system) include one or neither of the micro-exons.

A number of genes that control the expression of selector genes have been identified, including *extra sex combs* (*esc*), *Polycomb* (*Pc*), *super sex combs* (*sxc*), and *trithorax* (*trx*). In the second chromosomal mutant *extra sex combs* (*esc*), some of the head and all of the thoracic and abdominal segments develop as posterior abdominal segments (Figure 21.30), indicating that this gene normally controls the expression of *BX-C* genes in all body segments. The mutation does not affect either the number or the polarity of segments. It does affect their developmental fate, indicating that the *esc*⁺ gene product that is synthesized and stored in the egg by the maternal genome may be required for the correct interpretation of the information gradient in the egg cortex.

Lewis has proposed that genes in the *bithorax* complex, and perhaps other genes involved in segmentation, arose from a common ancestral gene by tandem duplication and subsequent divergence of structure and function. In fact, each of the homeotic selector genes listed in Table 21.5 encodes a transcription factor that includes a DNA-binding domain encoded by a 180-bp sequence known as a **homeobox**. The homeobox encodes a 60-amino-acid sequence known as a **homeodomain**. Similar sequences have been found in the genomes of other eukaryotes with segmented body plans, including *Xenopus,* chicken, mice, and humans. In mammals, the complexes of homeotic selector genes are called ***Hox* gene clusters**. Homeodomains from all organisms examined to date are very similar in amino acid sequence and encode a protein associated with the transcriptional regulation of a specific gene set. This suggests that the metameric or segmented body plan may have evolved only once.

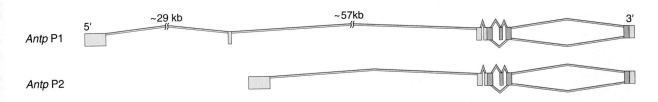

**FIGURE 21.28**    Transcription and processing of mRNA from the *Antp* gene. Transcription from P1 or P2 in the *Antp* gene results in two different pre-mRNAs.

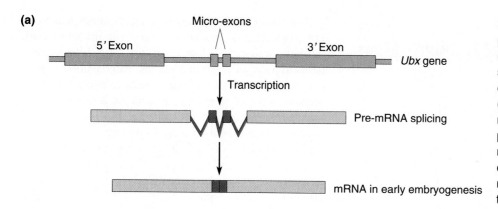

**(a)**

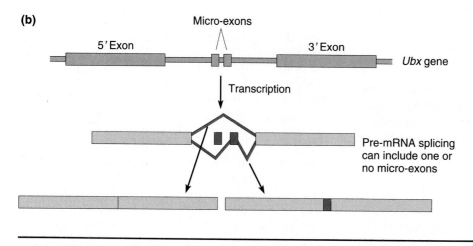

**FIGURE 21.29** Alternative processing of *Ubx* pre-mRNA. (a) In early embryogenesis, splicing includes both micro-exons in the mature mRNA. (b) During development of the nervous system, splicing of pre-mRNA includes either no micro-exons or one micro-exon, producing different but related proteins upon translation.

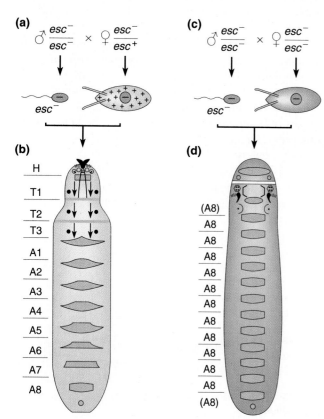

**FIGURE 21.30** Action of the *extra sex combs* mutation in *Drosophila*. (a) Heterozygous females form wild-type *esc* gene product and store it in the oocyte. (b) Fertilization of *esc⁻* egg formed at meiosis by the heterozygous female by *esc⁻* sperm still produces wild-type larva with normal segmentation pattern (maternal rescue). (c) Homozygous *esc⁻* females produce defective eggs which, when fertilized by *esc⁻* sperm, (d) produce a larva in which most of the segments of the head, thorax, and abdomen are transformed into the eighth abdominal segment. Maternal rescue demonstrates that the *esc⁻* gene product is produced by the maternal genome and is stored in the oocyte for use in the embryo. (H, head; T1–3, thoracic segments; A1–8, abdominal segments). Borderlines between the head and thorax and between thorax and abdomen are marked with arrows.

# CELL–CELL INTERACTIONS

During development in multicellular organisms, cells influence the transcriptional patterns and developmental fate of neighboring cells; these interactions involve the generation and reception of signal molecules. **Cell–cell interaction** is an important process in the development of most eukaryotic organisms, including *Drosophila* as well as vertebrates such as *Xenopus,* mice, and humans. To study the role of individual genes in developmental processes, the role of higher-level processes such as cell–cell interactions, and the genesis of behavior, Sidney Brenner began working with the soil nematode *Caenorhabditis elegans.*

## Overview of *C. elegans* Development

The adult worm is about 1 mm long and matures from a fertilized egg in about two days (Figure 21.31). The life cycle consists of an embryonic stage (about 16 hours), four larval stages (L1 through L4), and the adult stage. The diploid chromosome number is 12 (two X chromosomes and five pairs of autosomes), and it is estimated that the haploid genome consists of about 3000 genes. Adults are XX self-fertilizing hermaphrodites and can make both eggs and sperm (occasional XO males are also produced). Self-crossing mutagen-treated hermaphrodites quickly results in homozygous stocks of mutant strains, and hundreds of mutants have been generated, cataloged, and mapped.

The adult hermaphrodite consists of 959 cells, whose exact cell lineage from fertilized egg to adult has been mapped (Figure 21.32) and is invariant from individual to individual. With knowledge of the lineage of each cell, it is easy to follow the events that result from mutational alterations in cell fate or the killing of cells using laser microbeams or ultraviolet irradiation.

In *C. elegans* the fate of cells in the development of the female reproductive system is determined by cell–cell interactions and provides some insight into how gene expression and cell–cell interactions are linked in the specification of developmental pathways.

## Vulva Formation in *C. elegans*

Most adult *C. elegans* are hermaphroditic, that is, they make both male and female gametes. The gonad is connected to the outside by an opening called the vulva, located about midbody (Figure 21.33). The vulva is formed in stages during larval development and involves several rounds of cell–cell interactions.

During development in *C. elegans,* two neighboring cells, Z1.ppp and Z4.aaa, normally have different developmental fates. These cells interact with each other so that one becomes the gonadal anchor cell and the other becomes a precursor to the ventral uterus (Figure 21.34). This decision is made during the second larval stage (L2). In recessive *lin*-12(0) mutants (*lin* mutants are defective in cell lineage) both cells become anchor cells. The dominant mutant *lin*-12(d) causes both to become uterine precursors. Thus it appears that the *lin*-12 gene causes the selection of the uterine pathway, since in the absence of the *lin*-12 protein, both cells become anchor cells. However, both cells normally begin to synthesize and secrete a chemical signal (as yet unknown) for uterine differentiation and synthesize the *lin*-12 protein, which is a receptor for the signal. At a critical time in L2, the cell that by chance is secreting more of the signal molecule causes its neighboring cell to increase transcription of the *lin*-12 gene, thus increasing production of the receptor. It is thought that the cell with more *lin*-12 receptors thus becomes the ventral uterine precursor, and the other cell becomes the anchor cell. The critical factor in this first round of cell–cell interaction and determination is the *lin*-12 gene. Interestingly, the *lin*-12 protein has structural features that resemble the *Notch* protein (Figure 21.10), and they may work by similar mechanisms.

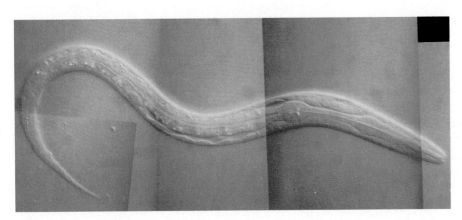

**FIGURE 21.31**    An adult *Caenorhabditis elegans*. This nematode, about 1 mm in length, consists of 959 cells and has been used to study many aspects of the genetic control of development.

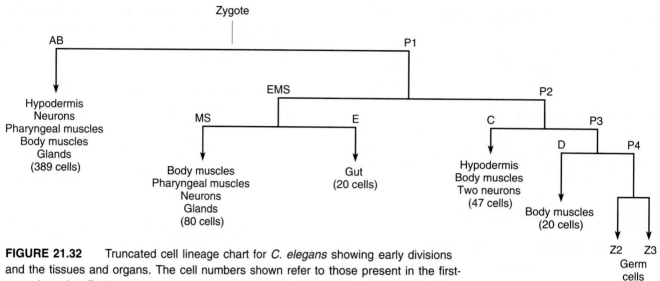

**FIGURE 21.32**     Truncated cell lineage chart for *C. elegans* showing early divisions and the tissues and organs. The cell numbers shown refer to those present in the first-stage larva L1. During subsequent larval stages, further cell divisions will produce the 959 somatic cells of the adult worm.

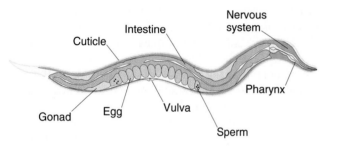

**FIGURE 21.33**     Organization of internal organs in *C. elegans*. The vulva opens near midbody.

**(a)**

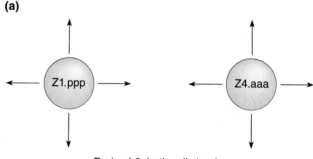

During L2, both cells begin
secreting signal for uterine differentiation

**(b)**

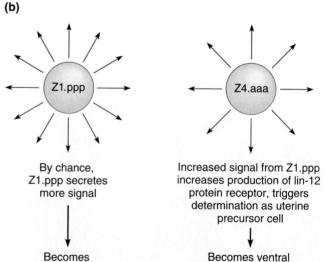

By chance,
Z1.ppp secretes
more signal

↓

Becomes
anchor cell

Increased signal from Z1.ppp
increases production of lin-12
protein receptor, triggers
determination as uterine
precursor cell

↓

Becomes ventral
uterine precursor cell

**FIGURE 21.34**     Cell–cell interaction in anchor cell determination. (a) During L2, two neighboring cells begin secretion of chemical signals for induction of uterine differentiation. (b) By chance, cell Z1.ppp secretes more of these signals, causing cell Z4.aaa to increase production of the receptor for signals. Action of increased signals causes Z4.aaa to become the ventral uterine precursor cell and allows Z1.ppp to become the anchor cell.

In L3, the third stage of larval development, a second round of cell–cell interactions involves the anchor cell, located in the gonad, and the cells that are precursors to the vulva which are located in the skin (hypodermis) adjacent to the gonad. There are six precursor prevulval (PV) cells, and at the beginning of L3, and each has three possible developmental fates: the primary vulval cell, secondary vulval precursors, or a tertiary fate as hypodermis. The fate of each of the PV cells is specified by their position relative to the anchor cell and by gene action that originates in the anchor cell (Figure 21.35). Sometime in L3, the *lin-3* gene in the anchor cell is activated and secretes a protein structurally related to the vertebrate **epidermal growth factor (EGF)**. All six of the PV cells express a receptor, most likely the product of the *let-23* gene, which is similar to the vertebrate EGF receptor. Binding of the *lin-3* protein to the *let-23* recep-

tor triggers an intracellular cascade of events that causes the PV cells to become determined to form either vulval precursor cells or secondary vulvar cells. This signal transduction cascade is the same as that seen in some oncogenes (see Chapter 18).

The PV cell closest to the anchor cell receives the strongest signal and usually becomes the primary vulval cell. The two neighboring cells receive a lower amount of signal and become secondary vulvar cells (Figure 21.35). To reinforce this determination, the primary vulvar cell secretes a signal that activates the *lin-12* gene in the neighboring secondary cells. This lateral inhibition signal prevents the secondary cells from adopting the primary fate. In other words, cells with an active *lin-12* gene cannot become primary vulvar cells. In the three remaining PV cells that receive little or no signal from the anchor cell, the hypodermal developmental pathway is

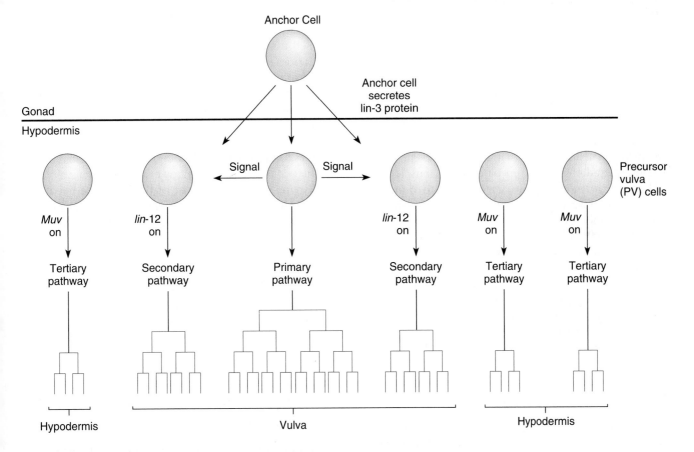

**FIGURE 21.35** Cell lineage determination in *C. elegans* vulva formation. A signal from the anchor cell in the form of *lin*-3 protein is received by three PV cells. The cells closest to the anchor cell become primary vulval precursor cells, and adjacent cells become secondary precursor cells. Primary cells secrete a signal that activates the *lin*-12 gene in secondary cells, preventing them from becoming primary cells. Flanking PV cells that receive no signal from the anchor cell activity of the *Muv* gene causes these cells to become hypodermis cells instead of vulval cells.

specified by action of the *Multivulva* (*Muv*) gene, which suppresses vulval development.

Thus, three levels of cell–cell interaction are required to specify the developmental pathway leading to vulva formation in *C. elegans*. First, two neighboring cells interact during L2 to establish the identity of the anchor cell. Second, in L3 the anchor cell interacts with three PV cells to establish the primary vulvar precursor cell. Third, the primary vulvar cell interacts with its immediate neighbors to suppress their selection of the primary vulvar pathway. Each of these interactions is accompanied by the secretion of molecular signals, and the reception and processing of these signals by neighboring cells.

This example of cell–cell interactions acting in a spatial and temporal cascade to specify the developmental fates of individual cells is a developmental theme is repeated over and over in developing organisms, ranging from prokaryotic prespore cells in *Bacillus* to higher vertebrates, including mice and probably humans.

# THE STABILITY OF THE DIFFERENTIATED STATE

Once a cell has developed specific structural and functional capacities, under normal circumstances in an adult organism the cell maintains that differentiated state. That is, kidney epithelial cells, red blood cells, muscle cells, and so on are not normally converted into other cell types. Such observations have led to the question of whether or not differentiation can be reversed under experimental conditions. Differentiation is a two-step process, beginning with determination, and followed by the biochemical and morphological changes that lead to the differentiated state. Thus the question must be addressed in two steps: Can we reverse determination, and can we reverse differentiation?

Recall that we have already established that nuclear differentiation in specialized cell types is not necessarily stable. This conclusion was drawn from nuclear transplantation experiments discussed earlier in this chapter. As we will see in the following examples, there are certain cases where determined cells and differentiated cells, not just nuclei, may alter their developmental status.

## Transdetermination

Ernst Hadorn demonstrated that if *Drosophila* imaginal disks are explanted from a larva and implanted into the abdomen of an adult, these primordia will continue to proliferate by cell division but will not differentiate. The disk subsequently may be recovered from the

adult, cut in half, and reimplanted—part to another adult abdomen, and part back into a larva (Figure 21.36). The half implanted into a second adult continues to proliferate and can be serially propagated in this way. The part placed into the larva will undergo metamorphosis and differentiate to yield adult structures.

Hadorn showed that during early transfers, the disk primordium always maintained its determined state (i.e., a leg disk always differentiated into leg structures, etc.) following experimental manipulation. After several passages through adult abdomens, atypical structures were occasionally discovered when disks were transferred to larvae; that is, an antennal disk sometimes produced wing structures, a genital disk gave rise to leg structures, and so forth. This shift in determination is referred to as **transdetermination**.

The major forms of transdetermination are summarized in Figure 21.37. All disks can be transdetermined into one or another disk type, and certain sequences of transdetermination occur; a disk can give rise only to a limited number of other disks.

Hadorn's experiments showed that developmental programming or determination can be altered by serially "subculturing" disk cells in adult abdomens. This transdetermination involves complete cells, not just the nucleus (as in nuclear transplantation experiments). Thus, it may be concluded that irreversible changes do not occur during preadult development of imaginal disks. Homeotic mutants show many of the properties of transdetermination; in fact, most of the transdeterminations are also known as homeotic mutations, emphasizing that these developmental programs are under genetic control.

## Regeneration

There are still other examples in which differentiated cells show a lack of developmental stability. Regeneration of amputated limbs occurs in the tadpoles of many frogs and toads, and in some larval and adult newts and salamanders. Since replacement of the missing structure often follows the normal developmental process, regeneration has been used to study the stability of the differentiated state as well as normal developmental pathways.

Following limb amputation, a sequence of morphogenetic events follows:

1. The cut end of the stump is covered by a distal migration of epidermal cells, forming a thin, transparent sheet over the wound.

2. Degenerative changes occur in the cells at the cut tip, and phagocytes remove cellular debris.

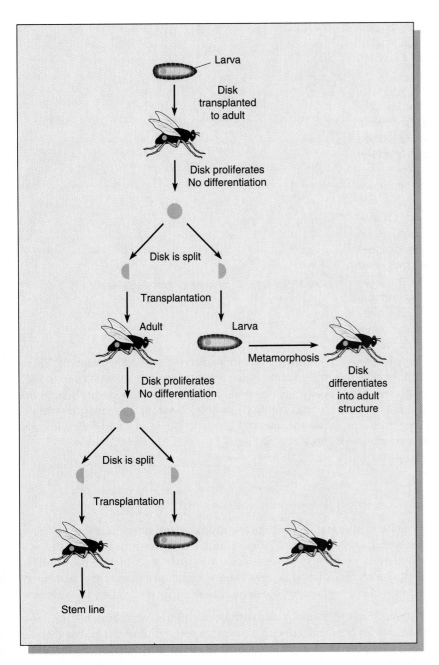

FIGURE 21.36 *In vitro* culturing of imaginal disk fragments over four generations. In each generation, a disk fragment is implanted into a larva to test its developmental potential and fate.

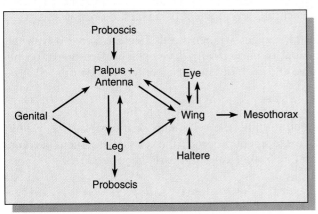

**FIGURE 21.37** Observed sequence of transdeterminations from each imaginal disk.

3. Cells from all or most tissues at the cut tip undergo a dedifferentiation and accumulate under the epidermal covering to form a **blastema**, from which all regenerating tissues will be derived.

The blastema grows by cell division into a cone projecting from the stump and then flattens into a paddle or palette (Figure 21.38). Blastemal mesenchyme cells condense into ridges, forming digits that elongate from the palette. Muscle and nerve differentiation confers movement on the regenerated limb, and growth continues until it reaches full size.

Experiments involving transplantation of $^3$H-thymidine-labeled or triploid blastema tissue onto unlabeled or diploid-amputated stumps have demonstrated that the undifferentiated blastema cells (mesenchyme) accumulate from most of the internal tissues of the adjacent stump. Cells from the cartilage, muscle, and connective tissues all contribute to the formation of the blastema. While all of the cells in the blastema have undergone dedifferentiation, some apparently retain their determined state and can redifferentiate only in one direction—a process known as **modulation**. Other cells can transform into several tissue types during differentiation. Cartilage cells, for example, can dedifferentiate into blastemal cells, but undergo modulation and regenerate into cartilage. Muscle cells, on the other hand, can redif-

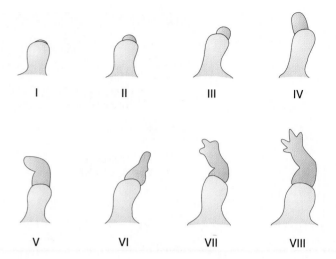

**FIGURE 21.38** Stages of limb regeneration in the newt.

ferentiate into other cell types, including cartilage. These observations again support the concept that even though cells have attained a highly specialized state where differential transcription is occurring, the potential still exists for inactive genes to be reactivated following cell proliferation.

---

**CHAPTER SUMMARY**

1. The role of genetic information during development and differentiation has been studied extensively using a simple model system: bacterial sporulation. In this system, environmental conditions induce specific gene activity, which in turn leads to altered development. By isolating developmental mutations, researchers have identified a number of the regulatory elements involved in controlling this process.

2. The variable gene activity theory, which applies to higher organisms, assumes that all somatic cells in an organism contain equivalent genetic information, but that it is differentially expressed. Studies conducted on genomic equivalence have demonstrated that somatic cells undergoing differentiation retain the entire gene set, which can be reprogrammed to direct the development of the entire organism.

3. Determination is the regulatory event whereby cells become progressively restricted in developmental capacity during early development. Determination becomes more specific with time and precedes the actual differentiation or specialization of distinctive cell types.

4. During embryogenesis, specific gene activity appears to be affected by the total environment of the cell, which includes the cell's individual cytoplasm as well as the maternal cytoplasm. As development proceeds, the cell's environment becomes further altered by the presence of other cells and early gene products.

5.  In *Drosophila,* genetic and molecular studies have confirmed that the egg contains information that specifies the body plan of the larva and adult, and that interactions of embryonic nuclei with the maternal cytoplasm initiates transcriptional programs characteristic of specific developmental pathways.

6.  The differentiated state of a cell may be stable, as in eukaryotes, or reversed, as demonstrated by transdetermination in *Drosophila* and regeneration in amphibians.

## KEY TERMS

blastema
blastoderm
cell–cell interaction
compartment
determination
differentiation

epidermal growth factor
  (EGF)
fate map
forespore
homeobox
homeodomain
*Hox* gene

imaginal disk
myoblast-determining
  (*myo*D) gene
myogenin
pair-rule gene
selector gene

$\sigma$-factor
sporangium
totipotent
transdetermination
variable gene activity
  hypothesis

## INSIGHTS AND SOLUTIONS

1.  In the slime mold, *Dictyostelium,* experimental evidence suggests that cyclic AMP (cAMP) plays a central role in the developmental program leading to spore formation. The genes encoding the cAMP cell surface receptor have been cloned and the amino acid sequence of the protein components is known. To form reproductive structures, free-living individual cells aggregate together and then differentiate into one of two cell types: prespore cells or prestalk cells. Aggregating cells secrete waves or oscillations of cAMP to foster aggregation of cells, and then continuously secrete cAMP to activate genes in the aggregated cells at later stages of development. It has been proposed that cAMP plays a central role in cell–cell interaction and gene expression in this developmental program. It is important to test this hypothesis by using several different experimental techniques. What different approaches can you devise to test this hypothesis, and what specific experimental systems would you employ to test them?

    **SOLUTION:**   Two of the most powerful forms of analysis in biology involve the use of biochemical analogs or inhibitors to block gene transcription or the action of gene products in a predictable way, and the use of mutations to alter the gene and its products. These two approaches can be used to study the role of cAMP in the developmental program of *Dictyostelium*. First, compounds chemically related to cAMP, such as GTP and GDP, can be used to test whether they have any effect on the processes controlled by cAMP. In fact, both GTP and GDP lower the affinity of cell surface receptors for cAMP, effectively blocking the action of cAMP. To inhibit the synthesis of the cAMP receptor, it is possible to construct a vector that contains a DNA sequence that transcribes an antisense RNA (a molecule that has a base sequence complementary to the mRNA). Antisense RNA forms a double-stranded structure with the mRNA, preventing it from being transcribed. If normal cells are transformed with a vector that expresses antisense RNA, no cAMP receptors will be produced. It is possible to predict that such cells will fail to respond to a gradient of cAMP and consequently will not migrate to an aggregation center. In fact that is what happens. Such cells remain dispersed and nonmigratory in the presence of cAMP. Similarly, it is possible to determine whether this response to cAMP is necessary to trigger changes in the transcriptional program by assaying for the expression of developmentally regulated genes in cells expressing this antisense RNA.

Mutational analysis can be used to dissect components of the cAMP receptor system. One way to approach this is to use transformation with wild-type genes to restore mutant function. Similarly, since the genes for the receptor proteins have been cloned, it is possible to construct mutants with known alterations in the component proteins and transform them into cells to assess their effects.

**PROBLEMS AND DISCUSSION QUESTIONS**

1. Carefully distinguish between the terms *differentiation* and *determination*. Which phenomenon occurs initially during development?

2. The *Drosophila* mutant *spineless aristapedia* ($ss^a$) results in the formation of a miniature tarsal structure (normally part of the leg) on the end of the antenna. Such a mutation is referred to as *homeotic*. From your knowledge of imaginal disks and Hadorn's transdetermination studies, what insight is provided by $ss^a$ concerning the role of genes during determination?

3. In the sea urchin, early development up to gastrulation may occur even in the presence of actinomycin D, which inhibits RNA synthesis. However, if actinomycin D is removed at the end of blastula formation, gastrulation does not proceed. In fact, if actinomycin D is present only between the 6th and 11th hours of development, gastrulation (normally occurring at the 15th hour) is arrested. What conclusions can be drawn concerning the role of gene transcription between hours 6 and 15?

4. How can you determine whether a particular gene is being transcribed in different cell types?

5. Observing that a particular gene is being transcribed during development, how can you tell whether expression of this gene is under transcriptional or translational control?

6. In studying gene action during development, it is desirable to be able to position genes in a hierarchy or pathway of action, to establish which genes are primary and in what order genes act. There are several ways of doing this. One way is to make double mutants and study the outcome. The gene *fushi-tarazu* (*ftz*) is expressed in early embryos at the seven-strip stage (see Figure 21.17). All genes involved in forming the anterior-posterior pattern affect the expression of this

gene, as do the gap genes. However, expression of segment polarity genes are affected by *ftz*. What is the location of *ftz* in this hierarchy?

7. Both the *ftz* gene and the *engrailed* gene encode homeobox transcription factors and are capable of eliciting the expression of other genes. Both genes work at about the same time during development and in the same region to specify cell fate in body segments. The question is: Does *ftz* regulate the expression of *engrailed*, or does *engrailed* regulate *ftz*? Or are they both regulated by another gene? To answer these questions, mutant analysis is performed. In *ftz*⁻ embryos (*ftz/ftz*), *engrailed* protein is absent; in *engrailed*⁻ embryos (*eng/eng*), *ftz* expression is normal. What does this tell you about the regulation of these two genes? Does the *engrailed* gene regulate *ftz*? Does the *ftz* gene regulate *engrailed*?

8. In *Bacillus*, expression of mother cell genes depends on recombination of the truncated $\sigma^K$ gene by action of the *spo*IVCA gene (see Figure 21.5). If a cloned copy of the rearranged $\sigma^K$ gene is introduced into the chromosome of a *spo*IVCA⁻ cell and restores the capacity of the mutant cell to sporulate, what does this tell you about the essential functions of the *spo*IVCA gene?

9. Nuclei from almost any source may be injected into *Xenopus* oocytes. Studies have shown that these nuclei remain active in transcription and translation. How can such an experimental system be useful in developmental genetic studies?

10. The concept of epigenesis indicates that an organism develops by forming cells that acquire new structures and functions, which become greater in number and complexity as development proceeds. This theory is in contrast to the preformationist doctrine that miniature adult entities are contained in the egg that must merely unfold and grow to give rise to a mature organism. What sorts of isolated evidence presented in this chapter might have led to the preformation doctrine? Why is the epigenetic theory held as correct today?

## SELECTED READINGS

BEACHY, P. A., HELFAND, S. L., and HOGNESS, D. S. 1985. Segmental distribution of bithorax complex proteins during *Drosophila* development. *Nature* 313:545–51.

BENDER, W., AKAM, M., KARCH, F., BEACHY, P., PEIFER, M., SPIERER, P., LEWIS, E. B., and HOGNESS, D. 1983. Molecular genetics of the *bithorax* complex in *Drosophila melanogaster*. *Science* 221:23–29.

BRIGGS, R., and KING, T. 1952. Transplantation of living nuclei from blastula cells into enucleated frog eggs. *Proc. Natl. Acad. Sci.* 38:455–63.

BROWER, D. L. 1985. The sequential compartmentalization of *Drosophila* segments revisited. *Cell* 41:361–64.

DESCHAMPS, J., and MEIJLINK, F. 1992. Mammalian homeobox genes in normal development and neoplasia. *Crit. Rev. Oncog.* 3:117–173.

EDMONDSON, D., and OLSON, E. 1993. Helix-loop-helix proteins as regulators of muscle-specific transcription. *J. Biol. Chem.* 268:755–758.

ERRINGTON, J. *Bacillus subtilis* sporulation: Regulation of gene expression and control of morphogenesis. *Microbiol. Rev.* 57:1–33.

FOSTER, S. J., and JOHNSTONE, K. 1990. Pulling the trigger: The mechanism of bacterial spore germination. *Mol. Microbiol.* 4:137–41.

GAUNT, S. 1991. Expression patterns of mouse *Hox* genes: Clues to an understanding of development and evolutionary strategies. *Bioessays* 13:505–513.

GEHRING, W. 1968. The stability of the differentiated state in cultures of imaginal disks in *Drosophila*. In *The stability of the differentiated state,* ed. H. Unsprung. New York: Springer-Verlag.

GILBERT, S. 1991. *Developmental Biology*. 3rd ed. Sunderland, MA: Sinauer Associates, Inc.

GROSSMAN, A. 1991. Integration of developmental signals and the initiation of sporulatio in *B. subtilis*. Cell 65:5–8.

GURDON, J. B. 1968. Transplanted nuclei and cell differentiation. *Scient. Amer.* (Dec.) 219: 24–35.

———. 1974. *The control of gene expression in animal development*. Cambridge, MA: Harvard University Press.

GURDON, J. B., LASKEY, R. A., and REEVES, O. R. 1975. The developmental capacity of nuclei transplanted from keratinized skin cells of adult frogs. *J. Embryol. Exp. Morphol.* 34:93–112.

HADORN, E. 1968. Transdetermination in cells. *Scient. Amer.* (Nov.) 219:110–20.

HOTTA, Y., and BENZER, S. 1973. Mapping of behavior in *Drosophila* mosaics. In *Genetic mechanisms of development,* ed. F. H. Ruddle, pp. 129–67. Orlando: Academic Press.

HUNT, P. and KRUMLAUF, R. 1992. *Hox* codes and positional specification in vertebrate embryonic axes. *Ann. Rev. Cell Biol.* 8:227–256.

JÄCKLE, H., HOCH, M., PANKRATZ, M., GERWIN, N., SAUER, F., and BRÖNNER, G. 1992. Transcriptional control by *Drosophila* gap genes. *J. Cell Sci. Suppl.* 16:39–51.

KAPPEN, C., SCHUGHART, K., and RUDDLE, F. H. 1989. Organization and expression of homeobox genes in mouse and man. *Ann. N. Y. Acad. Sci.* 567:243–52.

KAUFMANN, T., SEEGER, M., and OLSON, G. 1990. Molecular organization of the *Antennapedia* gene cluster of *Drosophila melanogaster*. *Adv. Genet.* 27:309–362.

KENNISON, J., and TAMKUN, J. 1992. Trans-regulation of homeotic genes in *Drosophila. New Biol.* 4:91–96.

KING, T. J. 1966. Nuclear transplantation in amphibia. *Methods in Cell Physiology,* ed. D. Prescott, vol. 2. Orlando: Academic Press.

KRUMLAUFF, R., and GOULD, A. 1992. Homeobox cooperativity. *Trends Genet.* 8:297–300.

KYRAICOU, C. 1992. Sex variations. *Trends Genet.* 8:261–267.

LAWRENCE, P. 1992. *The making of a fly: The genetics of animal design.* Oxford: Blackwell Scientific Publications.

LEVINE, M., RUBIN, G., and TJIAN, R. 1984. Human DNA sequences homologous to a protein coding region conserved between homoeotic genes of *Drosophila*. *Cell* 38:667–73.

LEWIS, E. B. 1976. A gene complex controlling segmentation in *Drosophila*. *Nature* 276: 565–70.

MANSEAU, L. J., and SCHÜPBACH, T. 1989. The egg came first, of course! Anterior-posterior pattern formation in *Drosophila* embryogenesis and oogenesis. *Trends Genet.* 5:400–405.

McGINNIS, W., HART, C. P., GEHRING, W. J., and RUDDLE, F. H. 1984. Molecular cloning and chromosome mapping of a mouse DNA sequence homologous to homoeotic genes of *Drosophila*. *Cell* 38:675–80.

McGINNIS, W., and KRUMLAUF R. 1992. Homeobox genes and axial patterning. *Cell* 68:283–302.

OLSON, E. 1990. *Myo*D family: A paradigm for development? *Genes Dev.* 4:1454–1461.

RUDDLE, F. H., HART, C. P., and McGINNIS, W. 1985. Structural and functional aspects of the mammalian homeobox sequences. *Trends Genet.* 1:46–50.

SANCHEZ-HERRERO, E., and AKAM, M. 1989. Spatially ordered transcription of regulatory DNA in the *bithorax* complex of *Drosophila*. *Development* 107:321–29.

SINGER, M. 1952. Influence of the nerve in regeneration of the amphibian extremity. *Quart. Rev. Biol.* 27:169–200.

SMALL, S., and LEVINE, M. 1991. The initiation of pair-rule stripes in the *Drosophila* blastoderm. *Curr. Opin. Genet. Dev.* 1:255–260.

ST. JOHNSTON, D., and NUSSLEIN-VOLHARD, C. 1992. The origin of pattern and polarity in the *Drosophila* embryo. *Cell* 68:201–219.

STRUHL, G. 1981. A gene product required for correct initiation of segmental determination in *Drosophila*. *Nature* 293:36–41.

STRUHL, G., and BROWER, D. 1982. Early role of the $esc^+$ gene product in the determination of segments in *Drosophila*. *Cell* 31:285–92.

TSONIS, P. A. 1990. Molecular approaches in limb development and regeneration. *Trends in Biochem. Sci.* 15:82–83.

WEINTRAUB, J., DAVIS, R., TAPSCOTT, S., THAYER, M., KRAUSE, M., BENEZRA, R., BLACKWELL, T., TURNER, *et al.* 1991. The *myo*D gene family: Nodal point during specification of the muscle cell lineage. *Science* 251:761–766.

WESTPHAL, H., and GRUSS, P. 1989. Molecular genetics of development studied in the transgenic mouse. *Ann. Rev. Cell Biol.* 5:181–96.

WRIGHT, W. 1992. Muscle basic helix-loop-helix proteins and the regulation of myogenesis. *Curr. Opin. Genet. Dev.* 2:243–248.

# 22

# GENETICS OF IMMUNITY

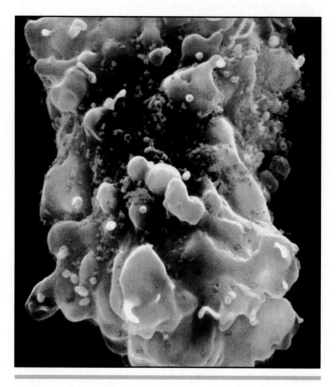

Scanning electron micrograph of human T lymphocyte infected by HIV viruses (blue dots).

*The immune system of vertebrates is a genetically controlled mechanism that defends the body against disease-causing organisms. Extensive genetic recombination in somatic cells reshuffles a limited number of genes into new combinations that produce literally millions of different antibodies and cell surface receptors. Cell surface antigens play a role in determining blood types and in the success of transfusions and organ transplants. The proliferation, differentiation, and function of cells in the immune system is under genetic control, and mutations in these genes result in disorders of the immune system.*

As vertebrates evolved, this group developed a unique and complex genetic mechanism that is crucial to the survival of such organisms. This mechanism, called the **immune system**, provides a series of defensive responses to the entry of foreign substances or the invasion of viruses and other microorganisms into the body. This response is highly specific and involves two phases: a *primary response* to the initial exposure, and a *secondary response* to subsequent exposures to the same agent.

From a genetic standpoint, the immune system has two fundamental characteristics. First, in every individual, it must recognize "self" so that an organism's cells and tissues are not attacked by its own immune system. This recognition involves a set of genes, which in humans, is located on chromosome 6. Second, the immune system must be able to produce specific molecules that will neutralize and subsequently destroy "nonself" agents. This response involves the production of *antibodies* against foreign substances or agents called *antigens*. By definition, an antigen elicits the response leading to the production of specific antibodies.

Antibody production against any and every potential antigen is one of the most remarkable genetic accomplishments of advanced organisms. From a collection of less than 300 genes, recombination events within developing antibody-producing cells create a vast immune potential. As a result of shuffling this limited number of genes, tens of millions of different antibodies can be produced.

In this chapter, we will examine the cells that make up the immune system and how these cells and their gene products are mobilized to mount an immune response. We will also examine the role of the immune system in determining blood groups and mother–fetus incompatibility, how cell-surface markers are used in transplants, and how these markers can be used in a predictive way to determine risk factors for a wide range of diseases. Finally, we will consider a number of genetic disorders of the immune system, including mutations in single genes that inactivate the entire immune system. We will also describe how acquired immmunodeficiency syndrome (AIDS) associated with infection by human immunodeficiency virus (HIV) acts to cripple the immune response of infected individuals. Taken together, work on the immune system constitutes one of the most exciting areas of genetics; one where new information and new breakthroughs occur with increasing frequency.

## COMPONENTS OF THE IMMUNE SYSTEM

The immune system provides an effective barrier against the successful invasion of potentially harmful foreign substances. The immune system represents a state of resistance whereby invading viruses, bacteria, fungi, and parasites are recognized as nonself and subsequently sequestered, inactivated, and destroyed. This **immune response** involves **antibodies** specific to foreign substances. Agents that elicit antibody production are termed **antigens**. All antibodies are proteins and are produced and secreted by specific cells of the immune system. Many different molecules can act as antigens, including proteins, nucleic acids, and polysaccharides. Usually, a distinctive structural feature of an antigen, called an **epitope**, stimulates antibody production. These invading antigens may be free molecules, or they may be part of the surface of a cell or microorganism.

Whatever the case, organisms with an immune system have the capacity to make antibodies against any antigen they encounter.

There are two main branches of the immune system: (1) **humoral immunity**, mediated by antibodies and associated with two types of blood cells, the antigen-recognizing **T cells** and the antibody-producing **B cells**; and (2) **cell-mediated immunity** involving a class of T cells known as **cytotoxic** or **killer T cells**. Other aspects of the immune response are also controlled by white blood cells; discussion of genetic mechanisms of the immune system requires knowledge of the origin, distribution, and function of these cell types.

## Cells of the Immune System

One of the key cells in the immune system is the **phagocyte**, a cell that engulfs and destroys foreign molecules and microorganisms (Figure 22.1). One type of phagocyte called a **macrophage** plays a critical role in both antibody-mediated and cell-mediated immunity by signaling other cells of the immune system that foreign antigens are present. In this signaling role, macrophages are known as *antigen-presenting cells*. All phagocytes arise from stem cells, which are mitotically active cells

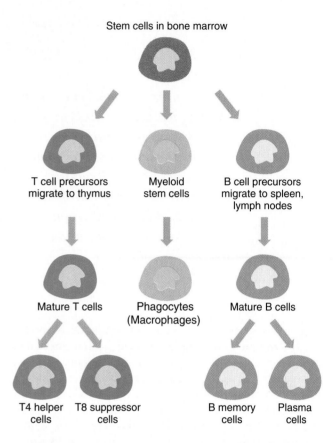

**FIGURE 22.2** Stem cells in bone marrow give rise to the precursors to the T cells, macrophages and B cells of the immune system.

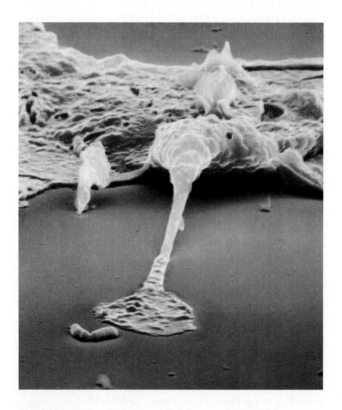

**FIGURE 22.1** A phagocyte with a cell process extended to engulf a rod-shaped bacterium in the foreground.

found in bone marrow (Figure 22.2). Mature phagocytes have the ability to leave the circulatory system by squeezing through capillary walls to enter inflamed or damaged tissues to identify, engulf, and destroy antigens.

A second cell type, the T cell, is also produced by stem cells in bone marrow (Figure 22.2). While still immature, T cells migrate to the thymus gland where they develop into several different subtypes, including *helper cells, suppressor cells,* and *killer cells*. During this period of maturation in the thymus, T cells become programmed to produce a unique type of cell receptor. Each T cell produces one and only one type of receptor, called an **antigen receptor**, that in turn will bind to only one type of antigen. There are literally tens or hundreds of millions of potential antigens, and there are a similar number of differently programmed T cells. The array of T cell receptors is produced in maturing T cells from a small number of genes by **genetic recombination** events. Mature T cells can divide, and all the offspring of a single T cell form a family group or clone. From the thymus, mature T cells become distributed throughout the circulatory and lymph systems, and also become sequestered in the lymph nodes and spleen. There are two general

## Table 22.1  Cells of the Immune System

| Cell Type | Function |
|---|---|
| T4 Cells | Also called helper T cells. Role in both cell-mediated and antibody-mediated immune responses. Master switch of the immune system. |
| T8 Cells | Suppressor cells that participate in cell- and antibody-mediated immune response. Slows down and/or turns off the immune response. |
| B Cells | Precursors to plasma cells, B cells recognize antigens, and are activated by T4 cells. |
| Plasma cells | Synthesize large quantities of a single antibody. Produced mitotically from an activated B cell. |
| Memory cells | Derived from either T cells or B cells by mitosis, these cells remain in circulation. Allow a rapid response to a second encounter with a specific antigen. |
| Killer cells | A specialized T cell that can detect and destroy virus-infected cells. |

classes of T cells: T4 helper/inducer cells and T8 cytotoxic/suppressor cells. The T4 helper cells are the master switch for the immune system; they turn on the immune response. The T8 suppressor cells are the "off" switch of the immune system; they stop or slow down the immune response (Table 22.1).

A third component of the immune system, the B cells, originate in bone marrow (Figure 22.2) and migrate to the spleen and lymph nodes. As they mature, these cells become genetically programmed to produce antibodies; each B cell produces only one type of antibody. Antibody production is triggered in B cells that mature to become **plasma cells** by the action of T4 helper cells. Memory cells represent a subpopulation of B cells that participate in subsequent encounters with a specific antigen.

# THE IMMUNE RESPONSE

The immune response involves four stages: (1) the detection of foreign antigens, (2) activation of cells of the immune system, (3) inactivation of the antigen, and (4) shutdown of the immune response. Each of these stages is directed by a specific cell type. After exposure to an antigen and the generation of an immune response, subsequent challenges to the immune system by the same antigen are directed by a *secondary immune response,* also called immunological memory. This secondary immune response is also directed by specific cell types. We will first examine antibody- and cell-mediated immune responses and then consider how immunological memory protects the body against reexposure to an antigen.

## Antibody-Mediated Immunity

A type of phagocyte known as a **macrophage**, wanders through the circulatory system and inflamed tissues and is capable of responding to foreign molecules, viruses, or microorganisms. When an antigen is encountered, it is engulfed and ingested by the macrophage (Figure 22.3). After ingestion, epitope fragments of the antigen consisting of approximately 20 amino acids are incorporated into the plasma membrane of the macrophage and displayed on its outer surface, along with the cell surface markers that identify the cell as a macrophage. As this antigen-presenting macrophage moves through the circulatory system or intercellular spaces, it may encounter a helper T4 cell with a complementary antigen receptor. The receptors on the surface of the T4 cell react with the antigen, causing activation of the helper T cell. In turn, the activated T cell identifies and activates B cells that manufacture an antibody against the antigen presented to the T cell. The stimulated B cells begin to divide and differentiate to form two types of daughter cells. One is the **memory cell**, and the other is the **plasma cell**, which synthesizes and secretes 2000–20,000 antibody molecules per *second* into the bloodstream.

Antibodies (Figure 22.4) are effective against extracellular antigens such as bacterial cells, free virus particles, and protozoan parasites in the bloodstream, or virus-infected cells that display viral antigens on their surface. Antibodies work in several ways to inactivate antigens. They can interact with the antigen to form antigen–antibody complexes. Clumping of the antigen–antibody complex tags the antigens for destruction, and they are ingested and destroyed by phagocytes. The combination of certain antibodies with antigens activates **complement**, a proteolytic system present in serum that brings about lysis of invading bacterial cells. Antibodies can also interact with antigens to destroy their ability to function. For example, some viruses have proteins on their surface that help attach the virus to the cell surface as a prelude to viral infection. Antibodies can react with this attachment protein and inactivate it, leaving the virus

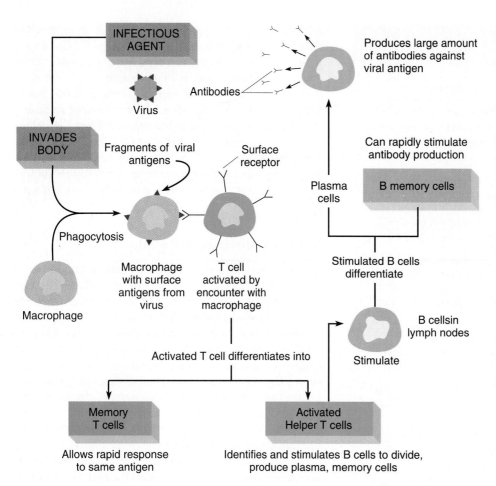

**FIGURE 22.3**    The pathway of responses in the humoral immune response. When an infectious agent such as a virus encounters a macrophage, it is ingested and destroyed. Some of the viral antigens are displayed on the surface of the macrophage and activate T4 cells. Activated T cells can differentiate into T memory cells or helper T cells. The helper T cells stimulate B cells in the lymph nodes that can produce an antibody against the viral antigen to become mitotically active. The progeny of the activated B cells can differentiate into B memory cells and plasma cells, which produce and secrete large amounts of a single antibody. The secreted antibody binds to the viral antigen, marking the viral particle for destruction by phagocytes.

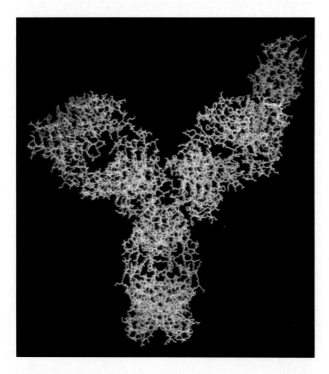

**FIGURE 22.4**    Computer graphic representation of the structure of an antibody. The backbone of the antibody molecule is represented in green, and the antigen-combining site is shown in blue.

unable to infect cells and exposed to ingestion and destruction by phagocytes.

The memory cells produced by activated B cells also produce antibodies, but instead of a life span of several days, as is the case for plasma cells, memory cells have an extended life span of several months to years. These cells play an important role in the secondary immune response.

The entire immune response is monitored by T8 suppressor cells, and when detectable antigens have been inactivated or destroyed, the T8 cells stop proliferation of B cells and turn off antibody production. Thus, T8 cells act as the "off" switch for the immune system.

## Cell-Mediated Immunity

In contrast to the action of antibodies, which tag antigens for destruction by phagocytes, direct cell–cell interactions called **cell-mediated immunity**, carried out by a class of T cells, can also result in antigen inactivation. The T cells that participate in this form of the immune reaction are known as *killer* T cells or *cytotoxic* T cells (Figure 22.5). Killer T cells originate in bone marrow and mature in the thymus gland. While the antibody-mediated immune response is effective against antigenic mol-

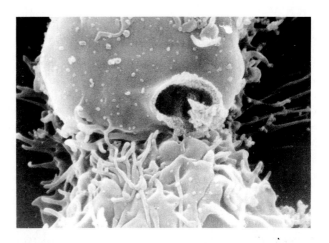

**FIGURE 22.6**    A killer cell (at bottom), has attacked a cancer cell and punctured its membrane. The contents of the tumor cell will leak out through this hole, and the cell will die. The remaining debris will be scavenged by phagocytes.

ecules and bacteria, cytotoxic T cells and cell-mediated immunity protect the body against viral infections and also kill some types of proliferating cancer cells. In cases where a cell is infected by a virus, partially lysed viral protein fragments become incorporated into the cell's plasma membrane and are displayed as a new set of antigens on the surface of the infected cell. This array of new antigens allows killer T cells to identify and kill virus-infected cells. The process of identification involves the viral antigen and a set of cell surface markers known as **histocompatibility antigens**. These antigens are encoded by the human leukocyte antigen (HLA) complex on the short arm of chromosome 6. This set of markers also plays a crucial role in grafts and transplants. Their role in cell-mediated immunity helps ensure that only virally infected cells are attacked by killer cells.

Once identified, killer cells attach to their target cells (Figure 22.6) and secrete a protein called *perforin,* which inserts into the plasma membrane of the virus-infected cell. Perforin molecules link together to form cylinders that function as pores in the plasma membrane. The cytoplasmic contents of the target cell leak out through these pores, and the virus-infected cell dies. Once detached, the killer cell is capable of detecting and killing other infected cells. In addition to killing virus-infected cells, cytotoxic T cells and a second type of killer cell known as a natural killer cell attack and kill cancer cells, fungi, and some types of parasites. Cytotoxic T cells will also attack and kill cells introduced into the body during tissue or organ transplants if the transplanted cells are recognized as foreign.

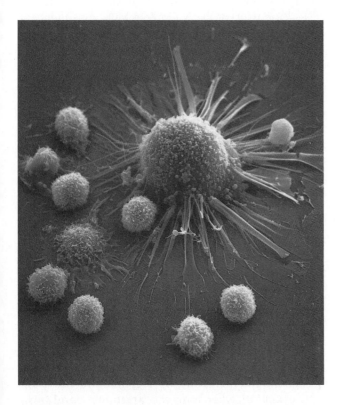

**FIGURE 22.5**    A large cancer cell, with cellular processes extended, has been detected and surrounded by an array of killer cells.

## Immunological Memory and Immunization

Even in ancient times it was known that exposure to certain diseases confers immunity to subsequent exposures to the same disease. For example, those who have had measles (an infectious disease caused by a virus) will not get measles again, even upon repeated exposure to infected individuals. This resistance to second infections is controlled by the *secondary immune response* and is the result of the production of B and T memory cells during the first exposure to the antigen. A second exposure to a molecular or microorganismic antigen results in an immediate, large-scale production of antibodies and killer T cells. Because of the presence of the memory cells, this reaction is more rapid, on a larger scale, and lasts longer than the primary immune response.

The existence of the secondary immune response is the basis for vaccination against a number of infectious diseases. A vaccine is a preparation designed to stimulate the production of memory cells against a pathogenic agent. In this procedure, the antigen is administered orally or by injection and provokes a primary immune response, including the production of memory cells. Often, a second or booster dose is administered to elicit a secondary response that raises or boosts the number of memory cells. Vaccines can be prepared from killed pathogens or weakened strains that are able to stimulate the immune system, but unable to produce symptoms of the disease. Modern recombinant DNA techniques have also been used in devising vaccines. For example, purified fragments of the protein coat of the hepatitis B virus have been prepared in this way for use as a vaccine.

# GENETIC DIVERSITY IN THE IMMUNE SYSTEM

Among vertebrates, each individual can produce millions of different types of antibodies, each responding to a different antigen. Coincidently, there are millions of different antigen recognition sites on the surface of T cells. It is now clear that the basis for this molecular diversity lies in the amino acid sequences composing the backbones of these proteins. Each protein is encoded by a family of genes, and the rearrangement of these genes in the development of B and T cells is the ultimate basis of their diversity.

The histocompatibility antigens, a third group of proteins involved in the immune response, are also encoded by a family of genes in the HLA complex, but here diversity is achieved by a large number of alleles at each locus in the family, rather than by gene rearrangement. In the

following sections, we will examine the molecular basis for diversity in antibodies and TCRs and consider diversity in the HLA complex in a later section.

## Antibodies

In humans, antibodies are produced by B cells differentiated as plasma cells. Based on their structure, five classes of antibodies or **immunoglobulins (Ig)** are recognized: **IgG**, **IgA**, **IgM**, **IgD**, and **IgE**. The first class, IgG, represents about 80 percent of the antibodies found in the blood and is the most extensively characterized class of antibodies. IgG is associated with immunological memory. The IgA class of immunoglobulins can be secreted across plasma membranes and is associated with immunological resistance to infections in the respiratory and digestive tracts. IgM antibodies are usually the first secreted by a plasma cell in response to an antigen, and as such, are associated with the early stages of the immune response. At this time, little is known about the role of the IgD class of immunoglobulins. IgE antibodies are associated with allergic responses.

A typical IgG molecule (Figure 22.7) consists of two different polypeptide chains, each present in two copies. The larger or **heavy chain (H)** consists of approximately 440 amino acids. The first 110 amino acids (about 25% of the total) at the N-terminus vary in sequence in different heavy chains and are thus known as the **variable region ($V_H$)**. The remaining C-terminal amino acids are essentially invariable within each class of H chains and are known as the **constant region ($C_H$)**. Within the constant region, there are three recognizable domains ($C_H1$, $C_H2$, and $C_H3$) that have sectors of homology, which may have arisen through duplication from a common ancestral gene. The linear region between $C_H1$ and $C_H2$ is called the **hinge region**. In humans, the H chains are encoded by genes on the long arm of chromosome 14.

The **light chain (L)** contains about 220 amino acids, with the first 110 amino acids of this chain (about 50% of the total) making up the variable region ($V_L$) (Figure 21.7) and the remaining amino acids at the C-terminus making up the constant region ($V_H$). There are two classes of L chains: the **kappa chains**, encoded by genes on human chromosome 2, and the **lambda chains**, encoded by genes on the long arm of chromosome 22.

The two heavy and two light chains that make up a functional IgG molecule are held together by disulfide bonds. The N-terminal variable regions of the heavy and light chains together form the **antibody combining site** that combines with and binds to an antigen. Within the variable region of each chain are amino acid sequences that vary more than others. There are three such

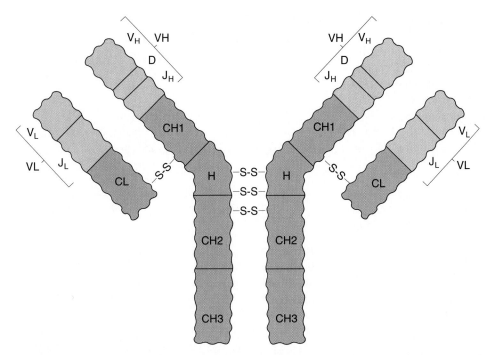

**FIGURE 22.7**    A typical immunoglobulin molecule. The longer arms are heavy chains, and the shorter arms are the light chains. For each heavy chain, the constant region is labeled CH1, CH2, and CH3, and the hinge region is H. The constant region of the light chain is labeled CL. The variable region of the heavy chain is within the bracketed region labeled VH. Within this region the subexons are indicated as $J_H$, D, and $V_H$. The variable region of the light chain is within the bracketed region labeled VL. The subexons within this region are $J_L$ and $V_L$. See the text explanation of the subexon structure and the recombination events that produce these genes. The chains are joined together by disulfide bonds indicated as —S–S—.

**hypervariable regions** in both light chains and heavy chains (Figure 22.7). These regions are important in determining the three-dimensional configuration of the antibody-combining site and are an important part of antibody diversity.

## Theories of Antibody Formation

One of the dominant features of the immune system is the continual generation of new cells that contain different combinations of antibodies and receptors. Because there are billions of such combinations, it is improbable that each combination is directly coded in the genome. One of the long-standing questions in immunogenetics is how this vast array of molecular variability in antibodies and antigen receptors *is* encoded in the genome. Over the years, three major hypotheses have been put forward and supported to one degree or another: the **germline theory**, the **somatic mutation theory**, and the **recombination theory**. The latter hypothesis has received the greatest amount of experimental support and will be considered in some detail. This theory pro-

poses that there are a limited number of genes that encode each of the various portions of the different types of H and L chains. As antibody-forming B cells mature, a mechanism invoking a form of DNA recombination rearranges these genes so that each mature B lymphocyte comes to encode, synthesize, and secrete only one specific type of antibody. In the presence of a specific antigen, the corresponding lymphocyte is stimulated to divide and differentiate. As a result, a population of differentiated plasma cells are produced, all of which synthesize one type of antibody that interacts with the antigen.

To fully understand how the diversity of antibodies is generated, we must first delve into the structure of immunoglobulins and the genes that encode the different regions of their polypeptide chains. Recall that antibody molecules are composed of H and L chains, each containing variable and constant regions ($V_L$, $C_L$, $V_H$, and $C_H$). In addition, there are five classes of immunoglobulins (IgG, IgA, IgM, IgD, and IgE). Each of these classes is characterized by its own specific heavy chain, and each of these can combine with either of two light chains, kappa or lambda. Thus, to begin with, there are ten basic classes of immunoglobulins (Table 22.2).

To this diversity, the recombination theory adds the variations found in the V regions of both H and L chains. In L chains, the V region appears to be encoded by two gene sets, designated V and J. **V genes** specify the N-terminal portion of the L chain, including the first two hypervariable regions and part of the third. The **J genes** specify the remainder of the V region, including the last part of the third hypervariable region. Because of the number of different V genes and J genes, many unique combinations of these genes can be formed when they are brought together by recombination. Given a relatively small number of different V and J genes carried in the genome, a much larger number of L chain genes can be formed via recombination in the maturing B cells.

Clear evidence for the recombination theory was first provided by Susumu Tonegawa and his colleagues in the mid-1970s. Using techniques of recombinant DNA, they demonstrated that DNA segments coding for the variable region of L chains are distant from one another in embryos, but are adjacent to one another in chromosomes isolated from antibody-producing cells. In their experiments, they isolated and characterized a cloned DNA fragment specifying a lambda $V_L$ chain derived from embryonic germline DNA. The organization of this cloned fragment was compared with the equivalent region isolated from an antibody-producing cell. In the embryo DNA, the region encoding the first 95 amino acids of the $V_L$ region was separated by 4.5 kilobases (kb) of DNA from the region encoding the remaining 13 amino acids (the J region). However, in the DNA isolated from the antibody-producing cell, the V and J genes were joined together to form a single transcription unit encoding a specific $V_L$ chain. This observation strongly supports the hypothesis that the diverse array of immunoglobulin genes found in antibody-producing cells are formed through a process of somatic recombination (Figure 22.8).

(a) Germ line DNA

(b) DNA from antibody-producing cell

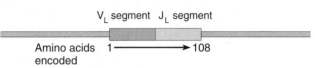

**FIGURE 22.8** (a) In DNA isolated and cloned from germline DNA, the V region of an L gene encoding the first 95 amino acids of an L chain are located 4.5 kb away from the DNA encoding the 13 amino acids of the J region. (b) In DNA isolated from an antibody-producing cell, the V and J regions were contiguous, forming a single transcription unit. This evidence provided support for the hypothesis that recombination is involved in forming antibody genes in mature B cells.

## Organization of the Immunoglobulin Genes

The work of Tonegawa and his colleagues inspired a large-scale research effort by many laboratories to study the organization and rearrangement of the immunoglobulin genes in mice and humans. From this beginning, a general picture of the origin of antibody diversity is emerging. The coding sequences for kappa and lambda genes have now been isolated from germline and B cell DNA in a variety of higher organisms. Comparative studies indicate that the basic organizational plan and mechanism of formation is similar in most mammals. In humans, the kappa L ($L_K$) gene (Figure 22.9) contains DNA segments identified as $V_K$, $J_K$, and $C_K$. The $V_K$ (variable) segment encodes the first 95 amino acids of the $L_K$ chain, the $J_K$ (joining) region encodes the middle 12 amino acids, and the $C_K$ (constant) region codes for the remaining amino acids at the C-terminus. There are 70–300 V segments per L gene, each with a different base sequence. The J region contains six different segments, and there is one C segment. The distance from the V region to the J region is about 23 kb in germline DNA.

During maturation in a B cell, one of the 300 $V_K$ regions is randomly joined by a recombination event to one of the six $J_K$ genes and the one $C_K$ segment to form an immunoglobulin gene (Figure 22.9). The remaining gene segments are lost. This newly created gene is transcribed and translated to form kappa L chains that becomes part of an antibody molecule. This rearranged

**Table 22.2** CATEGORIES AND COMPONENTS OF IMMUNOGLOBINS

| Ig Class | Light Chain | Heavy Chain | Tetramers | |
|---|---|---|---|---|
| IgA | | $\alpha$ | $\kappa_2\alpha_2$ | $\lambda_2\alpha_2$ |
| IgD | | $\delta$ | $\kappa_2\delta_2$ | $\lambda_2\delta_2$ |
| IgE | | $\epsilon$ | $\kappa_2\epsilon_2$ | $\lambda_2\epsilon_2$ |
| IgG | $\kappa$ or $\lambda$ | $\gamma$ | $\kappa_2\gamma_2$ | $\lambda_2\gamma_2$ |
| IgM | | $\mu$ | $\kappa_2\mu_2$ | $\lambda_2\mu_2$ |

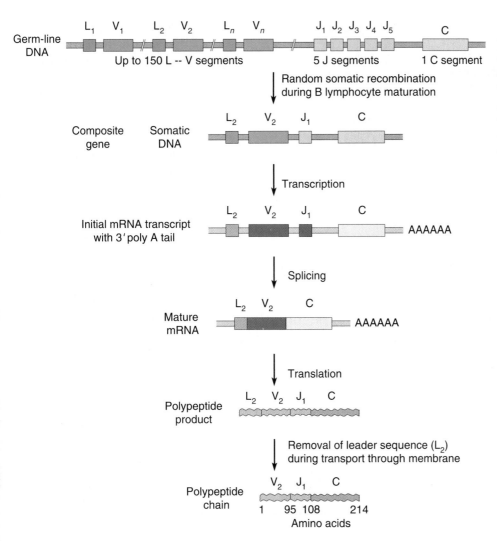

**FIGURE 22.9** The formation of the DNA segments encoding a human κ light chain and the subsequent transcription, mRNA splicing, and translation leading to the final polypeptide chain. In germline DNA, up to 300 different L–V (Leader–Variable) segments are present. These are separated from the J regions by a long noncoding sequence. The J regions are separated from a single C gene by an intervening sequence (intron) that must be spliced out of the initial mRNA transcript. Following translation, the amino acid sequence derived from leader RNA is cleaved off as the mature polypeptide chain passes across the cell membrane.

gene is stable, and is passed on to all progeny of the B cell.

Diversity in L chain production is the result of combining any of the 300 V genes with any of the six J genes. In addition, there are two other mechanisms that contribute to diversity. First, the DNA cleavage that begins the recombination reaction is imprecise, and variation in the precise cutting site generates additional diversity. Second, once formed, the $V_L$ region undergoes frequent mutation during subsequent proliferation of the B cell in which it was formed, further contributing to diversity. The joining of one of 300 V genes with any of the five J segments generates about 1500 different kappa L chain combinations. Diversity introduced by imprecise cutting and joining expands the number of different kappa chains to about $1.5 \times 10^4$.

The organization of the lambda L chain genes is similar to that of the kappa genes, except that each of the six J regions has an adjacent C segment. As is the case with the kappa chain, recombination events generate a lambda L chain immunoglobulin gene that is transcribed and translated to form a unique L chain. Diversity created by V–J joining and imprecise cutting generate about $1.5 \times 10^4$ different lambda genes.

The heavy chain genes in humans extend over a large region of DNA and include three different types of segments: $V_H$, $D_H$, and $J_H$. There are approximately 300 different $V_H$ genes, 10–50 different $D_H$ genes, and four different $J_H$ genes. In addition, there are five different $C_H$ genes (Figure 22.10). During B cell maturation, recombination randomly joins a $V_H$ region with one of the $D_H$ sequences and one of the $J_H$ sequences. This reaction occurs in two steps; first, one of the $D_H$ regions is joined to one of the $J_H$ regions, and then, the combined $D_H J_H$ segment is linked to a $V_H$ coding region (Figure 22.10). As with the L genes, the cleavage and rejoining reactions are somewhat imprecise, and deletions or insertions from 7 to 10 nucleotides at each joint contribute to additional diversity in the assembled H chain gene.

The $V_H D_H J_H$ composite assembled by recombination

Germ-line DNA (300 V segments, 10-50 D segments, 4 J segments, 5 C segments)

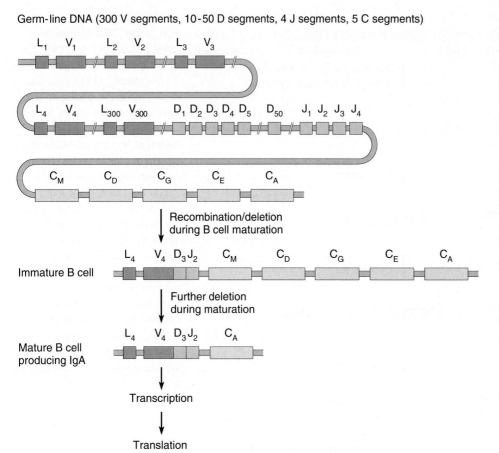

FIGURE 22.10    The variable region of the H chain is assembled by joining three different DNA segments together: $V_H$, $D_H$, and $J_H$. In embryonic DNA, these segments occur in clusters on chromosomes separated by long intervals and are adjacent to the $C_H$ region. In a maturing B cell, a random combination of one of the hundreds of $V_H$ regions with one of the 20 $D_H$ regions and one of the four $J_H$ regions produces an H chain gene that is transcribed and translated in the B cell. In other B cells, different combinations of H chain segments are joined together, producing a large number of different H chains.

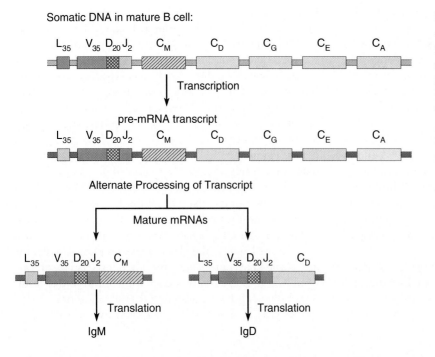

FIGURE 22.11    An alternative method of producing heavy chains with the same variable regions but different constant regions involves alternative splicing of heavy chain transcripts during B cell maturation. In the somatic cell DNA, the rearranged VDJ segment is adjacent to all five heavy chain constant regions. Transcription and alternate processing of the transcripts leads to the production of IgM and IgD containing different constant regions but the same variable region. This phenomenon is called class switching.

lies adjacent to the $C_H$ region, which encodes the constant regions of the five classes of heavy chains. A particular $V_H D_H J_H$ segment can be joined to any of the $C_H$ segments, producing an IgM, IgD, IgG, IgE, or IgA antibody, respectively. This joining can occur in several ways. One is by transcription of a long pre-mRNA molecule that begins at the 5' end of the V region and terminates at the 3' end of the $C_A$ segment. Cutting and splicing of this pre-mRNA yields mRNA molecules that can encode the same $V_H$ sequence in combination with different $C_H$ sequences. An alternative method involves a second round of recombination events that place one of the $C_H$ segments adjacent to the joined $V_H D_H J_H$ segment, and eliminate the other $C_H$ regions (Figure 22.11).

## Organization of T Cell Receptors

T cells play a role in both humoral and cell-mediated immunity. In humoral immunity, T4 cells and T8 cells turn the immune response on and off, respectively. In cell-mediated immunity, killer T cells attach to and kill virally infected cells and tumor cells. The ability of T cells to participate in immune responses depends on the presence of T cell receptors (TCRs) that interact with cell-associated antigens. The receptors are membrane-bound proteins (Figure 22.12) made up of two chains, alpha and beta, or gamma and delta. Each TCR chain is composed of an N-terminal variable portion that protrudes from the cell surface, and a C-terminal constant region that is attached to the plasma membrane. The N-terminal variable regions of the two chains form the antigen-binding site. The generation of diversity in TCRs is accomplished through somatic recombination events that

occur during T cell maturation, in a manner similar to the generation of antibody diversity in B cell maturation. The V region of the beta gene (Figure 22.13) is composed of about 25 different V genes ($V_\beta$), one D gene ($D_\beta$), and approximately six J genes ($J_\beta$) associated with each of the two C genes ($C_\beta$). The organization of the variable regions of the alpha, gamma, and beta regions are similar to those of the beta gene.

The diversity of T cell antigen receptors derives from the recombination of a V gene with a D and J gene and the association of the VDJ segments with one of the C regions. As with immunoglobulin genes, there is some imprecision in the cutting and joining of segments, contributing to receptor diversity. There is no evidence, however, of somatic mutation in the V region, so the overall diversity of TCRs is lower than that of immunoglobulin genes.

## Recombination in the Immune System

The process of **V(D)J recombination** in the formation of Ig and TCR genes is complex and involves a number of participating proteins. Of these, two closely linked loci, named **recombination-activating genes RAG-1** and **RAG-2**, have been identified in the human genome. These adjacent loci on chromosome 11 are cotranscribed, contain no introns, and are both required for recombination activity. These genes are normally expressed only in lymphocytes and only in those stages when V(D)J recombination is taking place. Transfer of these genes to fibroblasts or other cell lines results in the expression of recombination in cells that do not normally exhibit the assembly of V, D, and J segments.

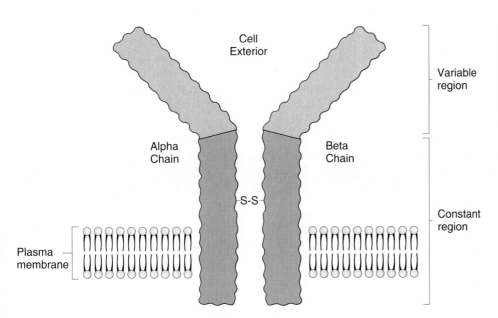

**FIGURE 22.12**    T cells produce surface receptors that recognize antigens. This T cell receptor is composed of two chains, an alpha chain and a beta chain. Each chain has a constant region and a variable region, and are encoded by separate genes containing L–V, D, J, and C segments.

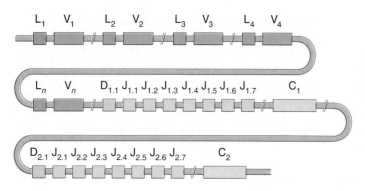

**FIGURE 22.13**    Organization of the T cell beta receptor gene. There are about 25 different L–V gene segments, one D gene, and 7 J genes associated with each of the two C genes. During T cell differentiation, recombination combines one L–V segment, one D segment, and one J segment adjacent to each C gene.

RAG-1 or RAG-2 deficient mice have no mature T or B cells, and maturation of these cells is arrested at the stage during which V(D)J recombination takes place. This and other evidence indicates that RAG-1 and RAG-2 directly participate in the somatic recombination events in maturing cells of the immune system. It seems likely that these genes work in combination with other factors and regulatory genes to coordinate recombination in the immune system, but their mechanism of action is not yet known.

# BLOOD GROUPS

Thus far, we have concentrated on the genetic basis of antibody production and cell-mediated immunity as components of an immunological defense system. We now turn our attention to the genetic basis of antigens that underlie the immunological concept of self. As we shall see, this is an equally important, closely related topic. Knowledge of the genetic determination of molecules found on the surface of blood cells is needed to carry out blood transfusions and organ transplantations. To date, more than thirty such genes have been identified, and each (along with its alleles) constitutes a blood group or blood type. To be safely transfused, the blood types of the donor and recipient must be matched. If the transfused red blood cells have a foreign antigen on their

surface, the body of the recipient will produce antibodies against this antigen, causing the transfused red cells to clump together, blocking circulation in capillaries and other blood vessels, with severe and often fatal consequences. Although there are a large number of blood groups, only two are of major immunological significance: (1) the ABO system, and (2) the Rh-blood group.

## ABO System

The genetics of the ABO system were described in Chapter 4 as an example of a gene with multiple alleles. To briefly review this system, recall that the gene $I$ (for isoagglutinin) encodes a cell surface glycoprotein and has three alleles: $I^A$, $I^B$, and $I^O$. The $I^A$ and $I^B$ alleles are codominant, each producing a slightly different version of the gene product, while $I^O$ is a recessive null allele, producing no effective antigen on the cell surface. The gene products produced by the A and B alleles behave as antigens and may stimulate the production of antibodies. For example, individuals with type A blood (genotypes AA and AO) carry the A antigen on red blood cells, so they do not have antibodies against this cell surface marker, but can make antibodies against the antigen encoded by the B allele (Table 22.3). Those with type B blood carry the B antigen on their red cells and will make antibodies against the A antigen. Individuals with AB blood carry both antigens and do not make either

**Table 22.3**    ANTIGENS, ANTIBODIES OF ABO SYSTEM

| Blood Type | Antigens on Red Blood Cells | Antibodies in Blood | Safe Transfusions To | From |
|---|---|---|---|---|
| A | A | B | A, AB | A, O |
| B | B | A | B, AB | B, O |
| AB | A, B | — | AB | A, B, AB |
| O | — | A, B | A, B, AB, O | O |

antibody, while those with type O blood have neither antigen and will make antibodies against both the A and B antigen. Thus, in transfusions, AB individuals do not make serum antibodies against A or B and can receive blood of any type, while type O individuals carry neither red cell antigen and thus are universal donors.

## Rh Incompatibility

The Rh blood group (named for the rhesus monkey in which it was discovered) consists of at least three genes (C, D, and E), each with two alleles (C, c; D, d; and E, e), making a large number of genotypic combinations possible. However, because the antigens encoded by the C and E genes are weak, they invoke only a minimal immune response, and for practical purposes, the Rh system can be considered as a simple one-gene, two-allele system involving gene D. Those of genotypes DD or Dd are said to be *Rh positive* (Rh+), and those of genotype dd are said to be *Rh negative* (Rh−).

Although the Rh blood group can play a role in transfusions, it is of major concern in immunological incompatibility between mother and fetus, a condition known as *hemolytic disease of the newborn* (*HDN*). This condition occurs when the mother is Rh− (genotype dd) and the fetus is Rh+ (genotypes DD or Dd). If the mother is Rh−, and Rh+ blood from the fetus enters the maternal circulation, antibodies against the Rh antigen will be made (Figure 22.14). The most common way fetal blood enters the maternal circulation is during the process of birth, so that the first Rh+ child is not affected. However, the maternal circulation now contains antibodies against the antigen, and a subsequent Rh+ fetus is affected because these antibodies cross the placenta and destroy the red blood cells of the fetus. The anemia that results can kill the fetus before birth, or the hemoglobin released from the lysed red blood cells can be converted into a degradation product that builds up in the brain causing neonatal death or, in survivors, severe mental retardation.

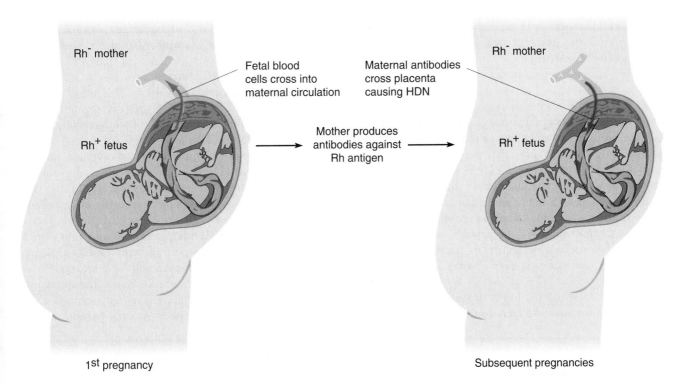

**FIGURE 22.14**    The development of hemolytic disease of the newborn (HDN). (a) When the mother is Rh− and the fetus is Rh+, any fetal blood that enters the maternal circulation will cause the production of antibodies against Rh+ red blood cells. Usually, this happens during birth, so the first Rh+ child escapes HDN. (b) In a subsequent pregnancy involving an Rh+ child, antibodies against Rh+ cells cross the placenta from the maternal circulation and destroy the red blood cells of the fetus. This causes HDN, which is characterized by anemia, jaundice, and, in extreme cases, can result in mental retardation and/or death.

To circumvent this problem, Rh− mothers are given an Rh-antibody preparation before the birth of the first Rh+ child, and in all subsequent Rh+ pregnancies. These antibodies move through the maternal circulatory system and destroy any fetal cells before the maternal immune system has a chance to make its own antibodies against the Rh antigen. The antibody preparation is administered to Rh− mothers before the first birth because she can also be sensitized by a miscarriage, abortion, or blood transfusions with Rh+ blood.

# THE HLA SYSTEM

Tissue transplants and skin grafts between unrelated individuals are usually rejected within a few weeks. Second attempts are rejected in a matter of days. The speed of the second rejection indicates that the immune system is involved in accepting or rejecting grafts and transplants. The interaction of genetically encoded cell-surface antigens of the donor with the immune system of the recipient determines whether grafts will be accepted or rejected. In laboratory mice, these antigens, known as *histocompatibility antigens,* are encoded by 20–40 genes, many of which have a large number of alleles, producing an enormous number of genotypic combinations.

Inbred strains of mice have been widely used to study the genetics of histocompatibility. Some strains are so highly inbred that all members of the strain can donate or receive grafts from any other member of the same strain. Crosses with other strains have allowed the identification and isolation of the histocompatibility genes. Results of such experiments indicate that one group of histocompatibility genes, known as the *major histocompatibility complex (MHC)* is the primary set of genes responsible for success or failure in transplants. Other minor genes are scattered over several loci in the genome.

In humans, a group of closely linked genes on chromosome 6, known as the HLA complex (similar to the mouse MHC), plays a critical role in histocompatible transplants.

## HLA Genes

The HLA complex in humans consists of four closely linked, major genes: HLA-A, HLA-B, HLA-C, and HLA-D (Figure 22.15). These genes fall into two classes: Class I antigens found on the surface of all cells in the body (HLA-A, HLA-B, and HLA-C), and Class II antigens represented by HLA-D, which encodes antigens found only on cells of the immune system, such as B lymphocytes, activated T lymphocytes, and macrophages. HLA-D is subdivided into the HLA-DR, HLA-DQ, and HLA-DP genes. Class I antigens are glycoproteins; Class II antigens are composed of two polypeptide chains that enable cells of the immune system to identify each other.

A large number of alleles have been identified in each of the HLA genes; it is one of the most highly polymorphic gene systems in the human genome. There are at least 23 alleles in HLA-A, 47 in HLA-B, 8 in HLA-C, 14 alleles in HLA-DR, 3 in DQ, and 6 in DP, making literally millions of genotypic combinations possible. Each of these alleles encodes a distinct antigen identified by a letter and number. For example, A6 is allele number 6 at the HLA-A locus, B2 is allele 2 at the HLA-B locus, etc.

The genes of the HLA system are closely linked and inherited in a codominant fashion. This close linkage means that recombination in this region is a rare event, and that the allelic combination on a single chromosome tends to be inherited as a unit. The array of HLA alleles on a given copy of chromosome 6 is known as a **haplotype**. Since humans have two copies of chromosome 6, we each have two HLA haplotypes (Figure 22.16). The codominant pattern of inheritance means that each haplotype is fully and completely expressed.

The inheritance of HLA haplotypes is shown in Figure 22.16. Because of the large number of alleles that are possible, it is rare that anyone will be homozygous at any of these loci. In the example shown, the unrelated parents have completely different haplotypes, and each child receives one haplotype from each parent, carried

**FIGURE 22.15**    The HLA region on human chromosome 6, showing the organization of the Class I and Class II regions.

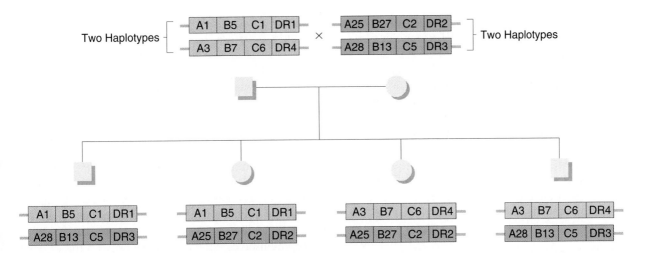

**FIGURE 22.16**    The transmission of HLA haplotypes. The cluster of HLA alleles on a single chromosome 6 is called a haplotype, each containing four major loci, encoding a different cell surface antigen. The arrangement of HLA haplotypes and their distribution to the offspring result in new combinations of haplotypes in each generation.

by the copy of chromosome 6, which is incorporated into the parental sperm or egg. The result is four new haplotype combinations possible in the offspring. Siblings have a one-in-four chance of sharing the same combination of haplotypes.

## Organ and Tissue Transplantation

Successful transplantation of organs and tissues depends to a large extent on matching HLA haplotypes between donor and recipient. Because of the large number of HLA alleles, the best chance for a match is between siblings and close relatives. Identical twins always have a perfect match. Parents have only one haplotype in common with a child, while siblings have a 1/4 chance of a perfect match. Thus, the order of preference for organ and tissue donors among relatives is: identical twin > sibling > parent > unrelated donor. Among unrelated donors and recipients, the chances for a successful match are between 1 in 100,000 and 1 in 200,000. Because HLA allele frequency differs widely between racial and ethnic groups (for example, B27 is found in 8% of American whites, but only 4% of American blacks), matches between different groups is often difficult.

When HLA types are matched, the survival of transplanted organs is improved dramatically. Figure 22.17 shows the 4-year survival rates for matched and unmatched kidney transplants. In HLA-matched transplants, over 90 percent of the transplanted kidneys survived the 4-year period, but in unmatched transplants, less than 50 percent of the kidneys were functional after 4 years. The major causes of rejection are mismatching

the HLA and/or ABO alleles. Other causes are more subtle and often difficult to define. For example, if the recipient has had blood transfusions, memory cells against HLA antigens might be present. If these antibodies act against HLA antigens on the grafted tissue, rejection is more likely.

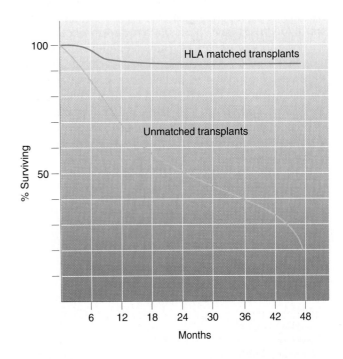

**FIGURE 22.17**    The survival of kidney transplants, when they are HLA matched surpasses that of unmatched transplants.

More recently, drugs have been used to improve the survival of transplants even when HLA matching is somewhat imperfect. The most widely used drug is cyclosporin, first isolated from a soil fungus. Its mechanism of action is not yet known with certainty, but it selectively inactivates the T cell subpopulation (the killer T cells) that is most active in tissue rejection, while not impairing other cells of the immune system. Other drugs to suppress the immune rejection of transplants are now under development, and if successful, may reduce the need for precise HLA matching for transplants, making it possible for more individuals to receive needed organs without prolonged waits for a proper donor.

## HLA and Disease

In a number of diseases, certain HLA haplotypes are found at a much higher frequency than would be expected by chance alone. For example, although allele B27 is found in 8 percent of the U.S. white population, more than 90 percent of those afflicted with a chronic inflammatory condition known as *ankylosing spondylitis* carry the B27 allele. Similarly, 70 percent of those with rheumatoid arthritis carry allele DR4, while only 28 percent of those in the general population carry this allele. A number of diseases and their HLA associations are shown in Table 22.4.

The nature of the relationship between HLA alleles and certain diseases is unknown, but a correlation exists and can often be used in diagnosing the condition. Several hypotheses have been put forward to explain this relationship, but none have proven entirely satisfactory. One hypothesis suggests that there is a direct relationship between specific HLA alleles and disease. According to this model, those carrying such an allele have a genetic susceptibility to a given disease and are therefore at much higher risk for the disease. A second hypothesis is based on the *linkage disequilibrium* that is observed in the HLA system.

Linkage disequilibrium is the nonrandom association of alleles at different loci. If, for example, the frequency of the HLA-A1 allele in a population is 0.17 and that of the HLA-B8 allele is 0.11, then if the alleles were associated at random, the frequency of their being found together in the population would be the product of their independent frequencies: $0.17 \times 0.11 = 0.019$ (1.9%). In fact, the observed frequency of this combination is much higher (about 0.09 or 9%), implying that selection is favoring this combination of alleles. If, however, another nearby locus outside the HLA complex carried an allele-conferring susceptibility to a disease, it would be inherited along with the HLA-A1/HLA-B8 combination, showing up as a disease association, even though the HLA alleles themselves have nothing to do with the disease.

A third hypothesis has suggested that the association between HLA alleles and disease may result from an interaction between certain alleles and environmental factors such as infection.

# DISORDERS OF THE IMMUNE SYSTEM

Genetically determined immunodeficiency disorders are caused by mutations that inactivate or destroy some component of the immune system. Most often, these disorders cause deficiencies in the production and functioning of one or more of the cell types in the immune system (Table 22.5). Much has been learned about how the immune system works by studying disorders that remove or disable single components. Unfortunately, not all the genetic disorders of the immune system have been described in enough detail to allow insights into the underlying molecular mechanisms. We will consider three disorders, one that affects B cells, one that affects T cells, and one that affects both B and T cells.

Other disorders of the immune system are due to factors including infections, cancer chemotherapy, and developmental errors. Two of these, developmental errors and infection are considered below.

## The Genetics of Immunodeficiency

A genetic disorder in which B cells are missing is **X-linked agammaglobulinemia** or Bruton disease. Affected individuals are almost always male, with an age of onset somewhere between 6 and 12 months. B cells and plasma cells are completely absent or, alternatively, immature B cells are present, but never become functional. The result is a complete lack of circulating antibodies and an inability to make antibodies. As a result, affected individuals have no antibody-mediated immunity and are highly susceptible to infections from micro-

**Table 22.4** HLA ALLELES AND DISEASE

| Disease | HLA Allele | Risk Factor |
|---------|-----------|-------------|
| Ankylosing spondylitis | B27 | >100 |
| Systemic lupus erythrematosus | DR3 | 3 |
| Psoriasis | B17 | 6 |
| Rheumatoid arthritis | DR4 | 6 |
| Reiter's syndrome | B27 | 50 |
| Multiple sclerosis | A3 | 3 |
| Chronic active hepatitis | B8 | 6 |

**Table 22.5**    CELL TYPES AFFECTED IN IMMUNE DISORDERS

| Immune Disorder | Affected Cell Type |
|---|---|
| X-linked agammaglobulinemia | B cells missing |
| Nucleoside phosphorylase deficiency | T cells reduced after birth |
| Severe combined immunodeficiency | T and/or B cells missing or nonfunctional |
| DiGeorge syndrome | T cells missing |
| Acquired immunodeficiency syndrome | T cells decline after infection |

organisms such as bacteria and fungi. However, these same individuals have normal levels of T cells and retain an intact cell-mediated immune system that confers immunity to most viral infections. Currently, the most effective method of treatment is a bone marrow transplant to provide a stem cell population that can give rise to functional B cells. Once these are established, they confer antibody-mediated immunity, which permanently corrects the condition.

A genetically controlled form of **T cell immunodeficiency** has been described, although the number of reported cases is very small. Affected individuals have a deficiency of the enzyme nucleoside phosphorylase, part of a biochemical pathway that salvages nucleotides. The number of B and T cells in these individuals is normal at birth, but thereafter, there is a gradual decrease in the number of T cells and cell-mediated immunity. The number of B cells is not affected, and antibody-mediated immunity is maintained. Without cell-mediated immunity, there is an increased frequency of viral infections and a high risk of certain forms of cancer. It has been suggested that the decline in T cells is caused by the build-up of purines as a result of the enzyme deficiency and that the resulting toxicity kills T cells but does not affect B cells. The structural gene for nucleoside phosphorylase maps to the long arm of human chromosome 14, and the condition is inherited as an autosomal dominant trait.

In **severe combined immunodeficiency syndrome (SCID)**, T- and B-cell populations are either absent or nonfunctional, and affected individuals have neither antibody-mediated nor cell-mediated immunity. As a result, they are susceptible to recurring and severe bacterial, viral, and fungal infections. There are autosomal and X-linked forms of SCID indicating that this disorder is genetically heterogenous. About 50 percent of the cases of SCID in the United States are X linked, 35 percent are autosomal recessive with an unknown cause, and 15 percent are autosomal recessive cases associated with a deficiency of the enzyme adenosine deaminase. In the X-linked form (XSCID), there are persistent infections beginning about 6 months after birth, with no T cells present. There are normal or even elevated levels of B cells present, but there is no antibody production by these cells. Experimental results indicate that the unknown XSCID gene product is necessary in both T and B cells and is required for B cell maturation. The longest-surviving individual with this form of SCID was a boy named David, who was born in Texas and isolated from the outside world by being placed in germ-free isolation (Figure 22.18). David died at age 12 of complications following a bone marrow transplant.

In cases of autosomally inherited SCID associated with a deficiency of the enzyme *adenosine deaminase (ADA)*, there are no T cells present, and B cells (if present) are not functional. Again, the result is a complete lack of cell-mediated and antibody-mediated immunity, and severe, recurring infections lead to death, usually by the age of 7 months. Some children affected with ADA-deficient SCID are currently receiving gene therapy to provide them with a normal copy of the gene. In this procedure, samples of their white blood cells are removed from the body and transfected with a retrovirus contain-

**FIGURE 22.18**    David, a boy born with severe combined immunodeficiency (SCID), survived by being isolated in a germ-free environment. He died at the age of 12 years following a bone marrow transplant undertaken to provide him with a functional immune system.

ing a copy of the normal gene for ADA. The cells are maintained in culture to ensure that the gene is active, and then the genetically modified white blood cells are transferred to the circulatory system with the hope that ADA expression will stimulate the development of functional T and B cells and at least partially restore a functional immune system. Although these treatments are still in their early stages, results to date have been encouraging. This represents the first attempt at gene therapy in humans; a technique that will undoubtedly become widely used as a treatment for certain genetic disorders in the future. The recombinant DNA techniques used in gene therapy are reviewed in Chapter 11.

## Acquired Immunodeficiences: DiGeorge Syndrome and AIDS

A nongenetic form of T cell deficiency called **DiGeorge syndrome** is the result of a mishap early in prenatal development. In this condition, the thymus and parathyroid glands fail to develop. The result is a complete lack of T cells that depend on the cellular environment of the thymus in order to mature. As predicted, no cell-mediated immunity results.

**Acquired immunodeficiency syndrome (AIDS)** is a collection of disorders that develop as a result of infection with a retrovirus known as the human immunodeficiency virus (HIV). The HIV virus consists of a protein coat, enclosing an RNA molecule that serves as the genetic material and an enzyme, reverse transcriptase. The entire viral particle is enclosed in a lipid coat derived from the plasma membrane of a T cell (Figure 22.19). The virus selectively infects the T4 helper cells of the immune system. Inside the cell, the RNA is transcribed

into a DNA molecule by reverse transcriptase, and the viral DNA is inserted into a human chromosome, where it can remain for months or years.

At a later time, when the infected T cell is called upon to participate in an immune response, the cellular and viral DNAs are transcribed. The viral RNA transcript is translated into viral proteins, and new viral particles are formed (Figure 22.20). These bud off the surface of the T cell, rupturing and killing the cell and setting off a new round of T cell infection. Gradually, over the course of HIV infection, there is a decrease in the number of helper T4 cells. Recall that these cells act as the master "on" switch for the immune system. As the T4 cell population falls, there is a decrease in the ability to mount an immune response. The results are increased susceptibility to infection and increased risk of certain forms of cancer. Eventually, the outcome is premature death brought about by any of a number of diseases that overwhelm the body and its compromised immune system. The relationship between the loss of T4 cells and the progression of HIV infection is strong and can be used to monitor the status of infected individuals (Figure 22.21).

HIV is transmitted by the transfer of body fluids from infected individuals to noninfected individuals; these fluids include blood, semen, vaginal secretions, and breast milk. The virus is not viable for more than 1 or 2 hours outside the body and cannot be transmitted by food, water, or casual contact. At this time, treatment options are rather limited and include the use of drugs that limit the reproduction of the virus.

## Autoimmunity

As the immune system matures, a state of **immune tolerance** develops between cells of the body and those of the immune system. This distinction between self and nonself prevents the immune system from attacking and destroying body tissues. Occasionally, immune tolerance breaks down, causing an autoimmune response in which the immune system turns against cells, tissues, and/or organs in the body. This breakdown of immune tolerance can take place in two ways: (1) a new antigen, previously unknown to the immune system can appear on cells, or (2) a new antibody produced against a foreign antigen ends up also attacking a preexisting antigen on cells of the body. For example, normally the eye is an immunologically privileged site, meaning that many proteins in the eye are never exposed to the immune system when immunological tolerance develops, and thus are not recognized by the immune system as self. In a traumatic injury, however, antigens from the eye can be exposed to the immune system, and since these are not recognized as self, antibodies against eye proteins are

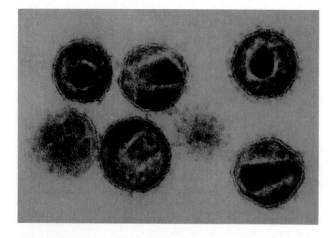

**FIGURE 22.19**   Transmission electron micrograph of the human immunodeficiency virus (HIV).

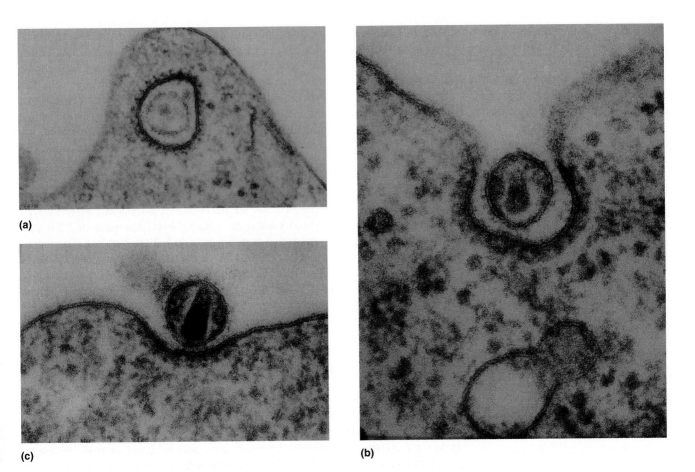

**(a)**

**(c)**

**(b)**

**FIGURE 22.20**    Release of new HIV particles from an infected cell. (a) The virus approaches the cell surface. (b) The virus buds off from the cell surface. (c) The virus, free from the cell surface, can now infect another T cell. Each round of replication releases several dozen new virus particles, gradually decreasing the number of T4 cells, reducing the body's ability to mount an immune response.

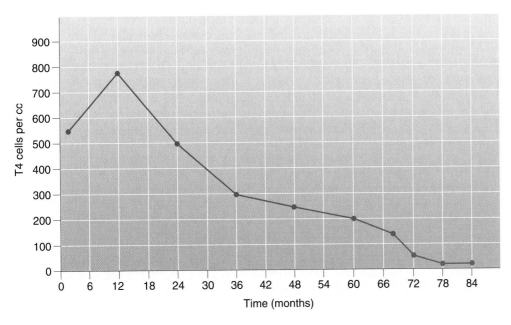

**FIGURE 22.21**    T4 cell levels during the course of an HIV infection. Following HIV infection, the number of T4 cells rise, then begin a slow decline. When levels fall below 200, the clinical symptoms of AIDS usually appear.

made. These antibodies may attack and cause blindness even in the uninjured normal eye; a condition that is known as *sympathetic ophthalmia.*

In other cases, such as infection with certain strains of streptococcus, the antibodies produced against the bacterial antigens are capable of attacking and destroying body cells that carry antigens similar to those of streptococcus. In rheumatic fever, an infection with streptococcus causes antibodies to be produced that kill the invading bacteria, but these antibodies also attack cells in the valves of the heart, causing the heart problems typically associated with rheumatic fever. Some autoimmune disorders are systemic, affecting many organs, while others are specific to individual tissues and organs. There appears to be a large number of autoimmune disorders that affect connective tissues, and these may form the underlying basis for many forms of arthritis and other connective tissue diseases.

One autoimmune disease with a specific target is insulin-dependent diabetes (IDDM), or juvenile diabetes. As the name implies, the disease starts in childhood and requires daily injections of insulin as a therapy. Insulin is a hormone produced by clusters of cells, called *islets,* embedded in the pancreas. IDDM develops when the immune system attacks and destroys the islet cells that produce insulin. Without insulin, there is no control over sugar levels in the blood. Initial symptoms include thirst and high levels of blood sugar. Without treatment, the disease progresses to kidney failure, blindness, heart disease, and premature death. More than 50 percent of affected individuals die within 40 years of onset.

The HLA alleles DR-3 and DR-4 are associated with an increased risk for IDDM. Each allele is associated with a 15-fold greater risk, and the combination of DR-3 and DR-4 is associated with a 30-fold greater risk. The mechanism by which the immune system attacks is not yet clear, and other factors are involved, since carrying a DR-3 or DR-4 allele alone is not sufficient to cause IDDM. At least one more gene, and possibly two, outside the HLA locus may play a role in stimulating the immune system, to destroy the islet cells.

## CHAPTER SUMMARY

1. The immune system is composed of two branches; one directs cell-mediated immunity and T cells, the other involves B cells and antibody-mediated immunity. The cells of the immune system originate from stem cells in bone marrow, and after maturation, move through the circulatory and lymph system as components of the immune system.

2. One of the most complex families of genes specifying proteins are those participating in the immune response. Human antibodies or immunoglobulins are characterized by amino acid sequence diversity. DNA shuffling during maturation of B cells produces the genetic variation required to match millions of different antigens with corresponding antibodies.

3. The immune response depends on the functional interrelationships of macrophages, which present foreign antigens to helper T cells, which, in turn, activate appropriate B cells to divide and produce large quantities of antibodies. The progress of the immune response is monitored by T8 suppressor cells that modulate the response and shut it down when it is no longer needed. T and B memory cells remain in circulation after the primary immune response and serve as the basis for a rapid and massive response if the immune system is challenged by the same antigen again. The formation of the memory cells is the basis for vaccination against pathogenic organisms.

4. The genetic basis of blood groups is the presence of cell surface antigens and the ability to make antibodies against foreign antigens. The success or failure of blood transfusions depends on matching blood types. Rh incompatibility between mother and fetus can result from antibodies produced in the maternal circulation against antigens on the blood cells of the fetus.

5. A class of cell surface antigens is produced by a complex of genes on human chromosome 6 called the HLA complex. With a large number of alleles, the combination carried by an individual as haplotypes constitutes a genetic signature for each individual. Successful tissue and organ transplantation depends on matching

the HLA haplotypes of the donor and recipient. Certain HLA alleles are associated with specific diseases and can be used to diagnose and predict risk factors for such diseases.

6. Mutations that disrupt the immune system work by altering the development and/or function of the component cells that participate in the immune response. These mutations can affect one cell type, a subsystem, or the entire immune system. In addition to genetic forms of disruption, the immune system can be inactivated by external factors, most notably by infection with the human immunodeficiency virus (HIV). This virus selectively infects and destroys the T cells that act as the "on" switch for the immune system, producing a gradual loss in the ability of an infected individual to mount an immune response. As a result, infectious diseases that would ordinarily be suppressed by the immune system can pose a serious risk to the life of HIV-infected individuals.

# KEY TERMS

| | | | |
|---|---|---|---|
| AIDS | epitope | immune tolerance | RAG-1, RAG-2 |
| antibodies | genetic recombination | immunoglobulins: IgG, | recombination theory |
| antibody combining site | germline theory | IgA, IgM, IgD, IgE | SCID |
| antigen receptor | haplotype | J genes | somatic mutation theory |
| antigens | heavy chain (H) | kappa chains | T cell immunodeficiency |
| B cells | hinge region | lambda chains | T cells |
| cell-mediated immunity | histocompatibility | light chain (L) | variable region |
| complement | antigens | macrophage | V(D)J recombination |
| constant region | humoral immunity | memory cell | V genes |
| cytotoxic (killer) T cells | hypervariable region | phagocyte | X-linked |
| DiGeorge syndrome | immune response | plasma cell | agammaglobulinemia |

# INSIGHTS AND SOLUTIONS

1. An antibody molecule contains two identical H chains and two identical L chains. This results in antibody specificity, with the antibody binding to a specific antigen. Recall that there are five classes of H genes and two classes of L genes. Because of the variability in the genes encoding the H and L chains, it is possible that an antibody-producing cell contains two different alleles of the H gene and two different alleles of the L gene. Yet the antibodies produced by the plasma cell contain only a single type of H chain and a single type of L chain. How can you account for this, based on the number of classes of H and L genes and the possibility of heterozygosity?

**SOLUTION:** While an antibody-producing cell contains different classes of H and L genes, and while it is entirely possible and even likely that a given antibody-producing cell will contain different alleles for the H gene and for the L gene, a phenomenon known as *allelic exclusion* allows expression of only one H allele and one L allele in any given plasma cell at a given time. That is not to say that the same antibody is produced over the life span of the antibody-producing cell, however. At first, many plasma cells produce antibodies of the IgM class, and at later times, may produce antibodies of a different class (e.g.,

IgG). Even in switching between different H genes, allelic exclusion is maintained, with only one allele of an H gene (or L gene) being expressed at a given time.

**2.** In the mouse $\kappa$ L gene, there are 250 V regions, 4 J regions, and 3 D regions. H chains have 250 V, 10 D, and 4 J regions. What are the number of combinations of these immunoglobulin components that are possible?

**SOLUTION:** First, calculate the number of different L combinations, the number of different H combinations, then the number of H and L combinations. For the L genes: 250 V $\times$ 4 J $\times$ 3 D = 3000. For the H genes: 250 V $\times$ 10 D $\times$ 4 J = 10,000. The combination of 3000 L genes with 10,000 H genes gives a total of $3 \times 10^6$ combinations of immunoglobulins. This actually is a great underestimate, since it does not take into account several other sources of diversity, including: a phenomenon called junctional diversity that produces imprecise joining between VJ or VDJ segments; the insertion of extra nucleotides into the VD or VJ segments of H chains (N regions), lambda L genes, the 5 classes of H chains, or the high level of point mutations that arise in splice antibody genes (somatic hypermutation). When these are added in, there is probably a 10,000-fold increase in diversity for about $3 \times 10^{11}$ different combinations of immunoglobulins.

---

## PROBLEMS AND DISCUSSION QUESTIONS

**1.** What do the following symbols represent in immunoglobulin structure: $V_L$, $C_H$, IgG, J, and D? What is the most acceptable theory for the generation of antibody diversity?

**2.** Figure 22.10 shows 9 unique gene sequences formed by recombination. How many other unique sequences can be formed?

**3.** If germline DNA contains 10 V, 30 D, 50 J, and 3 C genes, how many unique DNA sequences can be formed by recombination?

**4.** If there are 5 V, 10 D, and 20 J genes available to form a heavy-chain gene, and 10 V and 100 J genes available to form a light chain, how many unique antibodies can be formed?

**5.** Distinguish between an antigen and an antibody.

**6.** What are the functions of helper T cells and suppressor T cells?

**7.** If a woman has a child with hemolytic disease of the newborn (HDN), what are the possible genotypes of the mother, father, and child? After discovering that her child has HDN, the mother requests treatment with antibodies in the belief that it will prevent HDN in subsequent children. Is she correct? Why?

**8.** Why is type AB blood regarded as a universal recipient, and type O blood regarded as the universal donor? Which is more important in blood transfusion matching: the antibodies of the donor and recipient, or the antigens of the donor or recipient? Why?

**9.** How many polypeptide chains are present in an IgG antibody molecule? How many different polypeptide chains are represented in this molecule? How many gene segments from a germline cell were combined to produce the polypeptide chains in this molecule?

10. What is an HLA haplotype? How many of these haplotypes do you carry? How are these haplotypes inherited?

11. Describe the immunological processes involved in the rejection of an organ transplant.

## SELECTED READINGS

Arnett, F. 1986. HLA genes and predisposition to rheumatic diseases. *Hosp. Pract.* 20:89–100.

Chen, J., and Alt, F. 1993. Gene rearrangements and B-cell development. *Curr. Opin. Immunol.* 5:194–200.

French, D., Laskov, R., and Scharff, M. 1989. The role of somatic hypermutation in the generation of antibody diversity. *Science* 244:1152–57.

Golde, D. 1991. The stem cell. *Scient. Amer.* (December) 265:86–93.

Gonda, M. 1986. The natural history of AIDS. *Natural History* 95:78–81.

Huber, B. T. 1992. Mls genes and self-superantigens. *Trends in Genetics* 8:399–402.

Hunkapiller, T., and Hood, L. 1986. The growing immunoglobulin gene superfamily. *Nature* 323:15–16.

King, L., and Ashwell, J. 1993. Signalling for the death of lymphoid cells. *Curr. Opin. Immunol.* 5:368–373.

Leder, P. 1982. The genetics of antibody diversity. *Scient. Amer.* (May) 246:102–15.

Lieber, M. R. 1991. Site-specific recombination in the immune system. *FASEB J.* 5:2934–44.

———. 1992. The mechanism of V(D)J recombination: A balance of diversity, specificity, and stability. *Cell* 70:873–76.

Leiden, J. 1992. Transcriptional regulation during T-cell development: The alpha TCR gene as a molecular model. *Immunol. Today* 12:22–30.

Masucci, M., Gavioli, R., de Campos-Lima, P., Ahzng, Q., Trivedi, P., and Dolcetti, R. 1993. Transformation-associated Epstein-Barr virus antigens as targets for immune attack. *Ann. N.Y. Acad. Sci.* 690:86–100.

McDivitt, H. 1986. The molecular basis of antoimmunity. *Clin. Res.* 34:163–75.

Schatz, D. G., Oettinger, M. A., and Schlissel, M. S. 1992. V(D)J recombination: Molecular biology and regulation. *Ann. Rev. Immunol.* 10:359–83.

Thomson, G. 1988. HLA disease associations: Models for insulin-dependent diabetes mellitus and the study of complex human genetic disorders. *Ann. Rev. Genet.* 22:31–50.

Tonegawa, S. 1985. The molecules of the immune system. *Scient. Amer.* (October) 253:122–31.

Van Boehmer, H., and Kisielow, P. 1991. How the immune system learns about itself. *Scient. Amer.* (October) 265:74–86.

Yamamoto, F., Clausen, H., White, T., Marken, J., and Hakamori, S. 1990. Molecular genetic basis of the histo-blood group ABO system. *Nature* 345:229–33.

Yancopoulos, G., and Alt, F. 1988. Reconstruction of an immune system. *Science* 241:1581–83.

Young, J., and Cohn, Z. 1988. How killer cells kill. *Scient. Amer.* 258:38–44.

# 23

# GENETICS OF BEHAVIOR

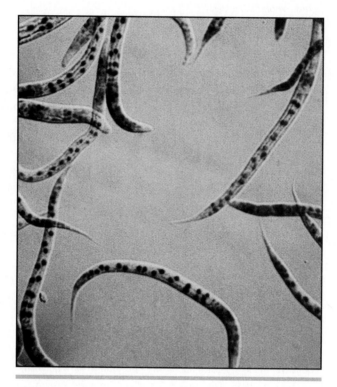

Genetically engineered *Caenohabditis elegans* roundworms that have turned blue in response to environmental stress, such as toxins or heat.

*Behavior in many organisms (both prokaryotic and eukaryotic) is controlled by both single genes and polygenic systems. At the biochemical level, these genes work by controlling cascades of metabolic reactions such as phosphorylation, methylation, and ion flux.*

Behavior is defined in general terms as a reaction to stimuli or environment. In broad terms, every action, reaction, and response represents a type of behavior. Animals run, remain still, or counterattack in the presence of a predator; birds build complex and distinctive nests; fruit flies execute intricate courtship rituals; plants bend toward light; and humans reflexively avoid painful stimuli as well as "behave" in a variety of ways as guided by their intellect, emotions, and culture.

Even though clearcut cases of genetic influence on behavior were known in the early 1900s, the study of behavior was of greater interest to psychologists, who were concerned with learning and conditioning. While some traits were recognized as innate or instinctive, behavior that could be modified by prior experience received the most attention. Such traits or patterns of behavior were thought to reflect the previous environmental setting to the exclusion of the organism's genotype. This philosophy served as the basis of the **behaviorist school**.

Such thinking provided a somewhat distorted view of the nature of behavioral patterns. It is logical that a genotype can be expressed within a series of environmental levels (e.g., cell, tissue, organ, organism, population, surrounding environment) and that a behavioral pattern must rely on the expression of the individual genotype for its execution. Nevertheless, the so-called **nature–nurture controversy** flourished well into the 1950s. By that time it became clear that while certain behavioral patterns, particularly in less complex animals, seemed to be innate, others were the result of environmental modifications limited by genetic influences. The latter condition is particularly true in organisms with more complicated nervous systems.

Since about 1950, studies of the genetic component of behavioral patterns have intensified, and support for the importance of genetics in understanding behavior has increased. The prevailing view is that all behavior patterns are influenced both genetically and environmentally. The genotype provides the physical basis and/or mental ability essential to execute the behavior and further determines the limitations of environmental influences.

Behavior genetics has blossomed into a distinct specialty within the larger field of genetics as more and more behaviors have been found to be under genetic control. This chapter provides an overview of the role of genes in behavior. More extensive treatment can be found in the list of selected readings at the end of the chapter. There seems to be no question that this topic will be one of the most exciting areas of genetics in the years to come.

## THE STUDY OF BEHAVIOR GENETICS: METHODOLOGY

Historically, three major approaches have been used in studying behavior genetics. The first involves the determination of behavioral differences between genetic strains of the same species or between closely related species. If such closely related organisms exist in similar environments and their survival needs are identical, any observed behavioral differences may be correlated with genetic differences. In the second approach, a modified behavioral trait is selected from an outbred population. If such a modified trait can be transferred to a new strain by genetic crosses, the positive influence of the genotype is established. The first two approaches identify behavioral patterns as being under the control of genes. Genetic investigation through controlled breeding experiments can then be performed in an attempt to establish inheritance patterns. In most cases where these approaches have been used successfully, it has only been shown that the inheritance is not due to simple Mendelian patterns. Instead, many behavioral traits have been attributed to polygenic inheritance (i.e., many genes controlling one trait).

The third approach, used extensively today, is to study the effects of single genes on behavior. Within inbred strains of organisms with relatively constant genotypes, spontaneous mutants with behavioral deviations may arise. Even more productive is the induction and isolation through selection of mutations altering behavior.

While a single complex behavioral trait may be controlled by many genes, disruption of the trait by a single mutation may occur and be analyzed. This approach is the most attractive because it offers the most objective information concerning the role of genes in behavior. In selected organisms, defining a gene controlling behavior is a prelude to the isolation, cloning, and molecular characterization of such genes. As we proceed through this chapter, all three approaches will be illustrated.

## COMPARATIVE APPROACHES IN STUDYING BEHAVIOR GENETICS

Before turning to specific examples of the effects of single genes on behavior, we shall present several case studies to illustrate two of the approaches used in behavior genetics. In the first approach, behavior is compared in closely related strains. Alcohol preference in mice and open-field behavior in mice are case studies representative of this method. The second approach involves selection within heterozygous populations for a modification of behavior. To illustrate this mode of investigation, we will describe studies of learning in rats and geotaxis in *Drosophila*.

### Alcohol Preference in Mice

Many studies of **alcohol preference** in mice have been reported. Different inbred strains have been compared for preference or aversion to ethanol, with all studies indicating genetic control of this response.

For example, D. A. Rogers and Gerald E. McClearn compared alcohol consumption in four strains of inbred mice over a period of three weeks. Each strain was presented with seven vessels containing either pure water or alcohol varying in strength from 2.5 to 15.0 percent. Daily consumption was measured. Table 23.1 shows the proportion of absolute alcohol to total liquid consumed on a weekly basis.

Examination of these data shows clearly that the C57BL and C3H/2 strains exhibit a preference for alcohol, while BALB/c and A/3 demonstrate an aversion toward it. Since the environment in raising the mice over many generations has been constant, the preference differences are attributed to the genotypes of each strain. Presumably, these strains vary only by the fixation of different alleles at these loci.

It has been suggested that differences in alcohol preference, metabolism, and severity of withdrawal symptoms are related to differences in the major enzymes of alcohol metabolism, particularly **alcohol dehydrogen-**

**Table 23.1**   ALCOHOL CONSUMPTION IN MICE

| Strain | Week | Proportion of Absolute Alcohol to Total Liquids | $\bar{x}$ |
|--------|------|------------------------------------------------|-----------|
| C57BL | 1 | 0.085 | |
|  | 2 | 0.093 | 9.4% |
|  | 3 | 0.104 | alcohol |
| C3H/2 | 1 | 0.065 | |
|  | 2 | 0.066 | 6.9% |
|  | 3 | 0.075 | alcohol |
| BALB/c | 1 | 0.024 | |
|  | 2 | 0.019 | 2.0% |
|  | 3 | 0.018 | alcohol |
| A/3 | 1 | 0.021 | |
|  | 2 | 0.016 | 1.7% |
|  | 3 | 0.015 | alcohol |

SOURCE: Modified from Rogers and McClearn, 1962. Reprinted by permission from *Quarterly Journal of Studies on Alcohol*, Vol. 23, pp. 26–33, 1962. Copyrighted by Journal of Studies on Alcohol, Inc. New Brunswick, NJ 08903.

**ase (ADH)** and **acetaldehyde dehydrogenase (AHD)**. The major enzymes and their isozymes have been isolated and characterized with respect to biochemical and kinetic properties, and their distribution and activities in different mouse strains have been catalogued. Genetic variants of these enzymes present in different strains of inbred mice have been used to try to establish an association between a given form of the enzyme and a form of alcohol-related behavior. Unfortunately, to date, no correlation between alcohol preference or metabolism and these biochemical markers has been established. It may be that other genetic markers, perhaps controlling the neuropharmacological effects of alcohol, will be needed to establish a link between specific genes and alcohol-related behavior in mice.

### Open-Field Behavior in Mice

First used in 1934, an **open-field test** was designed to study exploratory and emotional behavior in mice. When placed in a new environment, mice normally explore the surroundings, but they are a bit cautious or "nervous" about the new setting. The latter response is evidenced by their elevated rate of defecation and urination. To study their behaviors in the laboratory, an enclosed, brightly illuminated box was devised with the floor marked into squares. Exploration is measured by counting the number of movements into different sections, and emotion is measured by counting the number of defecations. These measurements thus attempt to quantify the two behavioral patterns.

As in the study of alcohol preference, different inbred strains of mice have been shown to vary significantly in their response to the open-field setting. Of greatest interest is the experimentation of John C. DeFries and his associates. DeFries concentrated his work on two strains, BALB/cJ and C57BL/6j. The BALB strain is homozygous for a coat color allele, $c$, and is albino, while the C57 strain has normal pigmentation ($CC$). BALB demonstrates low exploratory activity and is very emotional, while C57 is field-active and nonemotional.

DeFries proceeded to cross the two strains and then to interbreed each generation, creating $F_2$, $F_3$, $F_4$, etc., generations. Each generation beyond the $F_1$ contained albino and nonalbino mice, and these were tested. Regardless of generation, pigmented mice behaved as strain C57, while albino mice behaved as BALB. The general conclusion may be drawn that the $c$ allele behaves **pleiotropically**, affecting both coat color and behavior.

Detailed heritability and variance analysis has been performed to assess the degree of input of the $c$ gene to these behavioral patterns. Such analysis has shown that this locus accounts for 12 percent of the additive genetic variance in open-field activity and 26 percent of defecation-related emotion. Thus, these behaviors are under polygenic control, presumably in an additive way.

What might be the relation between albinism and behavior? DeFries sought an answer by testing albinos and nonalbinos under both white and red light. Red light provides little visual stimulation to mice. The behavioral differences between mice with the two types of coat pigmentation disappeared under red light. Thus, albino open-field responses are visually mediated. This is not surprising, since albino mice are photophobic, and lack of pigmentation in albinos extends to the iris as well as to the coat.

## Selection: Maze Learning in Rats

The second major approach in behavior genetics—selection for modified behavior from an outbred population and subsequent inbreeding—leads to the production of strains with differences in behavior. The end result is similar to the previous approach of initially comparing inbred lines. The classical illustration of the second approach involves measurement of **maze learning** in rats.

The first experiment of this kind was reported by E. C. Tolman in 1924. He began with 82 white rats of heterozygous ancestry and measured their ability to "learn" to

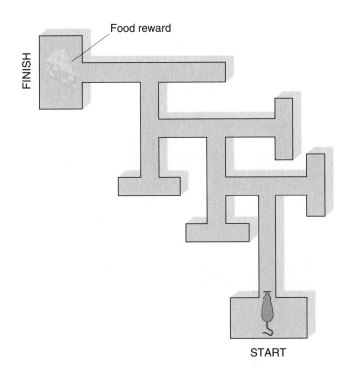

**FIGURE 23.1**    A multiple T-maze used in learning studies with rats.

obtain food at the end of a multiple T-maze (Figure 23.1) by recording the number of errors and trials. When first exposed to the maze, a rat explores all alleys and eventually arrives at the end, to be rewarded with food. In succeeding trials, fewer and fewer mistakes are made as the rat learns the correct route. Eventually, a hungry rat may proceed to the food with no errors.

From the initial 82 rats, nine pairs of each of the "brightest" and "dullest" rats were selected and mated to produce two lines. In each generation, selection was continued. Even in the first generation, Tolman demonstrated that he could select and breed rats whose offspring performed more efficiently in the maze. Subsequently, his approach was pursued by others, notably R. C. Tryon, who in 1942 published results of 18 generations of selection.

As shown in Figure 23.2, two lines, one clearly superior and one clearly inferior in maze learning, were established. There is some variation around the mean, but by the eighth generation there was no overlap between lines even in this variation. That is, the dullest of the bright rats were then superior to the brightest of the dull rats.

In other studies, bright and dull rats were tested for other related traits with varying results. Brights were found to be better in solving hunger-motivation problems but inferior in escape-from-water tests. Brights were also found to be more emotional in open-field ex-

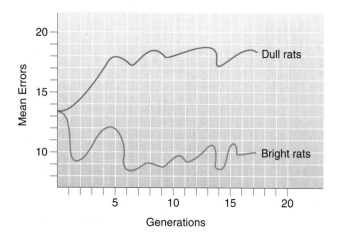

**FIGURE 23.2** Selection for the ability and inability of rats to learn to negotiate a maze.

periments. Thus, it was concluded that selection of genetic strains superior in certain traits is possible, but care must be taken not to generalize such studies to overall intelligence, which is composed of many learning parameters. Tryon seems to have selected for specific capacities rather than for general intelligence.

## Geotaxis in *Drosophila*

A **taxis** is the movement of a free-living organism toward or away from the source of an external stimulation. The response may be positive or negative, and the sources of stimulation may include chemicals, gravity, light, and so on. The investigation of **geotaxis** (response to gravity) in *Drosophila* illustrates this general behavior response as well as the selection technique. In addition, this investigation used a novel approach that determines the genetic influences of specific chromosomes on geotaxis.

Jerry Hirsch and his colleagues designed a mass-screening device that allows about 200 flies to be tested per trial, as shown in Figure 23.3. The maze is placed

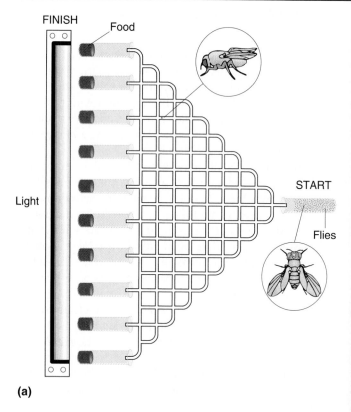

**(a)**

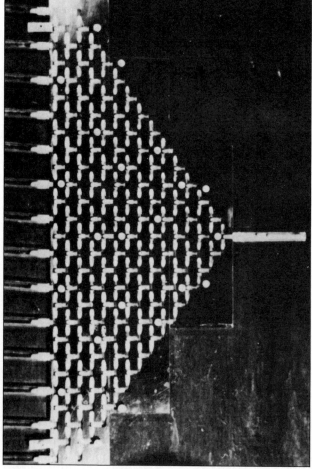

**(b)**

**FIGURE 23.3** (a) Schematic drawing and (b) photograph of a maze used to study geotaxis in *Drosophila*.

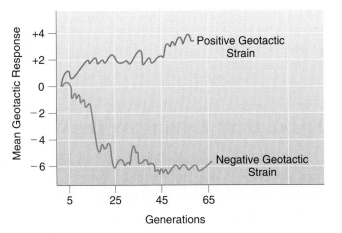

**FIGURE 23.4**    Selection for positive and negative geotaxis in *Drosophila* over many generations.

found that flies could be selected for both positive and negative geotropism, establishing the genetic influence on this behavioral response.

As shown in Figure 23.4, mean scores may vary from +4.0 to −6.0, corresponding to the number of T-junctions the fly encounters going up or down. These data show that the extreme of negative geotropism is stronger than the extreme of positive geotropism. The two lines have now undergone selection for almost 30 years, encompassing over 500 generations, and the testing of more than 80,000 flies. Throughout the experiment, clearcut but fluctuating differences were observed. Such results indicate the additive effects of polygenic inheritance.

Hirsch and his colleagues also performed an analysis of the importance of genes located on different chromosomes to geotaxis. They were able to differentiate among genes located on chromosomes 2 and 3 and the X in a rather ingenious way. They carried out a set of crosses that produced flies that were either heterozygous or homozygous for a given chromosome.

Figure 23.5 shows how this is accomplished. From a selected line, a male is crossed to a "tester" female whose chromosomes are each suitably marked with a

vertically and flies are added. Those that continue to turn up at each junction will finally arrive at the top; those that always turn down will arrive at the bottom; and those making both "up" and "down" decisions along the way will ultimately reside somewhere in between. It was

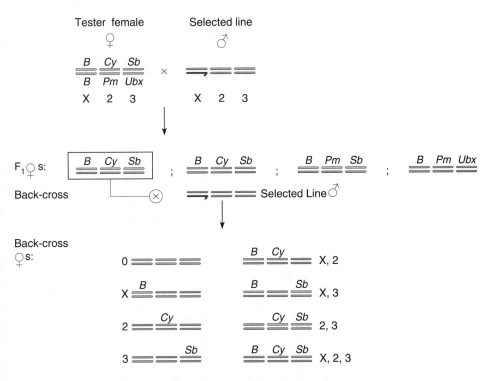

**FIGURE 23.5**    In this mating scheme in *Drosophila*, the effect of genes located on specific chromosomes that contribute to geotaxis can be assessed. The progeny produced by back-crossing the F₁ female contain all combinations of chromosomes. Examination of the phenotype of these flies make it possible to determine which chromosomes from the selected strain are present. Subsequent testing for geotaxis is then performed (*B*–Bar eyes; *Cy*–Curly wing; *Sb*–Stubble bristles). The designations alongside each genotype (X, 2, 3, etc.) indicate which chromosomes are heterozygous.

dominant mutation. Each marked chromosome carries an inversion to suppress the recovery of any crossover products. As shown, one of the $F_1$ females heterozygous for each chromosome is back-crossed to a male from the original line. The resulting female offspring contain all combinations of chromosomes. The dominant mutations make it possible to recognize which chromosomes from the selected lines are present in homozygous or heterozygous configurations. For example, combinations O and X (Figure 23.5) differ by being homozygous and heterozygous, respectively, for the X chromosome. Similar combinations exist for chromosome 2 (0 and 2) and chromosome 3 (0 and 3). Thus, by subjecting these flies to the geotaxic maze, it is possible to assess the influence of genes on any given chromosome on the behavioral response.

Many flies of each genotype have been tested, and the chromosomal effects have been compiled individually and in all combinations. For the negatively geotactic lines (flies that go up in the maze), genes on the second chromosome are the most important, followed by loci on chromosome 3 and the X (2 > 3 > X). For the positive line (flies that go down in the maze), the reverse arrangement (X > 3 > 2) indicates the relative effects of chromosomes in controlling this trait. Thus, the overall results indicate that geotaxis is under polygenic control, and that the responsible loci are distributed on all three major chromosomes of *Drosophila*.

Further genetic testing in which each chromosome from a selected line has been isolated in homozygous form in an unselected background has been used to estimate the number of genes that control the geotaxic response in *Drosophila*. While this work has not yet produced definitive results, it does indicate that a small number of genes, perhaps two to four loci, are responsible for this behavior.

## SINGLE-GENE EFFECTS ON BEHAVIOR

By far the most definitive information about the genetic influence on behavior has come from the study of the effects of single genes on behavior. In this approach, either spontaneous or induced mutations are analyzed in order to infer general principles about how normal behavior is created and regulated.

There are obvious advantages to this approach. By and large, in these laboratory studies the environmental effect on the behavioral response is minimized or eliminated. Thus, the genetic influence is more straightforward and easier to define than in studies where the environment is a major factor. As a result, it is theoretically possible to dissect a behavioral pattern into its components.

In this section we shall look at a number of single-gene influences. However, because of the large amount of information resulting from such studies, we shall be selective in our discussion.

### Nest-Cleaning Behavior in Honeybees

Honeybee nests are frequently infected with *Bacillus larvae*, a bacterium causing foulbrood disease. The dis-

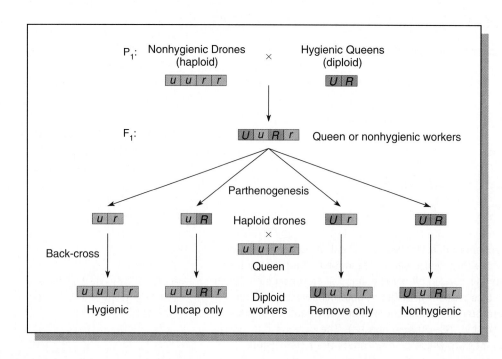

**FIGURE 23.6** Results of a honeybee cross between hygienic diploid females and nonhygienic haploid males.

ease may be counteracted by what is called **hygienic behavior** on the part of worker bees. In this behavior, worker bees open the cells containing infected larvae and the diseased organisms are removed from the hive. Hygienic hives are resistant to infection, while hives containing strains that do not display removal behavior are susceptible to the disease.

In 1964, Walter Rothenbuhler published results of his cross between a hygienic (Brown) line with a nonhygienic (Van Scoy) line. This work strongly favors the hypotheses that either two recessive independently assorting genes ($u$ and $r$) or a gene complex are responsible for hygienic behavior.

The $F_1$ hybrids were all nonhygienic. However, when $F_1$ drones were back-crossed to hygienic queens, four phenotypes were produced in roughly equal proportions, as shown in Figure 23.6. While one group was hygienic and one group nonhygienic, the other two groups were most interesting. One could uncap cells but not remove infected larvae. The fourth group, which at first appeared nonhygienic, was shown to be able to remove larvae if the cells were artificially uncapped. Thus, they were not able to uncap.

It appears that one gene pair ($u/u$) or a linked complex of genes determines uncapping behavior, and a second gene pair ($r/r$) or gene complex determines removal ability. We have begun with this example to illustrate the way a genetic study has allowed components of a more complex behavior to be dissected.

## Taxes in Bacteria: A Molecular Motor

Behavioral responses exist even in single-celled organisms such as bacteria, which lack an organized nervous system. These responses are in the form of taxes (plural of *taxis*) and are mediated by flagella or cilia. Bacteria exhibit **chemotaxis** and are attracted to or repelled by a variety of stimuli. These responses have now been carefully analyzed, and mutants have been isolated that disrupt normal behavior.

Motile bacteria such as *E. coli* and *Salmonella* are capable of responding to gradients of chemicals and moving along the gradient by controlling flagellar action (Figure 23.7). As they move, cells exhibit periods of smooth swimming called *runs* generated by counterclockwise (CCW) rotation of flagella, which act much like propellers, driving the bacteria through the medium. During runs, the flagella form a cohesive bundle, leading to movement in a single direction. During runs, which have an average duration of 1 second, the bacteria swim about 10 to 20 body lengths. Runs alternate with events called *tumbles* during which the flagella switch direction and rotate clockwise (CW). In clockwise rotation, the flagella become dispersed, and each acts independently. As a result, during tumbles, the cell covers little or no distance (Figure 23.8). Tumbles, which last only about 0.1 second, serve to change the direction of the cell. In the presence of a chemical attractant, runs that carry the cell up the gradient are extended, while those that move the cell down the gradient are not (Figure 23.9). In the presence of a chemical repellent, the parameters are inverted.

The molecular details of the chemotactic response are becoming known, with the major components of the system identified and the role of several gene products described. The response of bacteria to chemical stimuli serves as a model for the molecular mechanisms by which cells process and transduce sensory signals. In this case, chemical signals are converted into the kinetic action of flagella, moving the cell toward or away from a stimulus. *E. coli* has at least four different types of membrane-spanning receptors that sense chemicals in the environment and initiate an intracellular response. These receptor proteins are members of a gene family called **transducers** (Figure 23.10). Each protein in the family has a different receptor domain that extends into the periplasmic space and a conserved domain that extends into the cytoplasm. The cytoplasmic surface of the chemoreceptor protein forms a complex with two cytoplasmic proteins, *Che*A and *Che*W. *Che*A and the chemore-

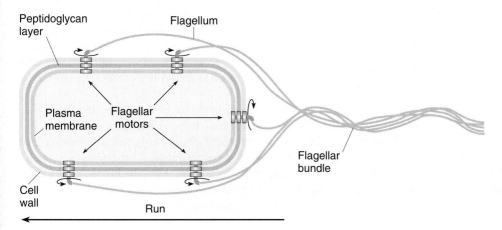

**FIGURE 23.7**    When the flagellar motors rotate in a counterclockwise (CCW) direction (as shown), the flagella form a bundle that works to push the cell along a mostly linear path called a run. When the flagellar motors rotate in a clockwise direction (CW) (not shown), the flagella operate independently, causing the cell to tumble.

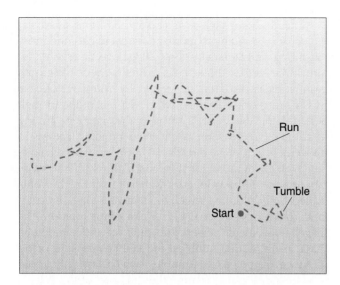

**FIGURE 23.8**    Runs and tumbles executed by an *E. coli* cell in a medium where no chemical gradient is present. Tracking of the cell began at the large dot at lower right. The cell moves via a series of runs interspersed with shorter periods of tumbling.

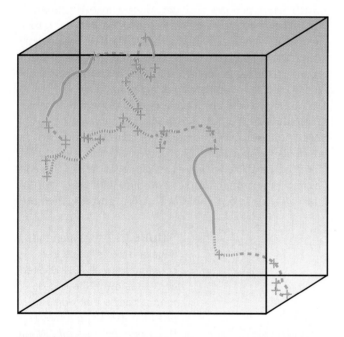

**FIGURE 23.9**    Runs and tumbles executed by a single *E. coli* cell in an exponential gradient of attractant (highest concentration at top). Movement up the gradient occurred in 35 runs, 34 tumbles. Two extended runs (indicated by the solid lines) represent most of the vertical distance traveled.

ceptor interact directly, and *Che*W apparently enhances this association. *Che*A integrates the response of the cell to chemical stimuli, as *Che*A mutants are nonchemotactic.

A change in the conformation of the chemoreceptor caused by the binding or release of a sensory molecule initiates the transmission of an intracellular signal. This signal is transmitted by phosphorylation changes in a group of cytoplasmic proteins which acts as a molecular switch to determine the direction (clockwise or counterclockwise) of flagellar rotation. In the presence of a repellent molecule, the change in chemoreceptor conformation causes an increase in the rate of *Che*A phosphorylation (Figure 23.10). Phosphorylated *Che*A then phosphorylates *Che*Y. When *Che*Y is phosphorylated, it binds directly to the base of the flagellar motor and favors clockwise rotation of the flagellum. Recall that clockwise rotation causes the cell to tumble, during which the cell switches direction. Another protein, *Che*Z turns off the action of *Che*Y by accelerating the dephosphorylation of *Che*Y.

In the presence of an attractant, the cycle of phosphorylation is reversed, decreasing the rate of *Che*A phosphorylation, which in turn slows the phosphorylation of *Che*Y. Dephosphorylated *Che*Y favors counterclockwise rotation of the flagella, moving the cell on a run toward the attractant. The goal of this behavior is to detect chemical gradients and alter cell movement accordingly, moving cells toward attractants (food) and away from repellents.

The chemoreceptors are aggregated into clusters, located at the poles of the cell, and are not clustered near the base of the flagellar motors. As expected, *Che*A and *Che*W proteins are also localized in the cytoplasm at the poles, with the *Che*A protein associated with the inner membrane of the cell. In mutants lacking all four receptors, *Che*A and *Che*W proteins are randomly distributed throughout the cytoplasm, indicating that the polar localization of these gene products depends on the presence of receptors. Studies of *Che*W mutants indicate that this protein is also required for aggregation and polar localization of the receptors. In *Che*W deletion mutants, the *Che*A receptor complexes are randomly distributed. These results suggest that signal transduction is not initiated by the formation of the complex, but through conformational changes in the complex brought about by sensory molecules.

The chemotactic behavior of *E. coli* involves a number of gene products that receive and process signals and generate a response through a metabolic network that involves protein phosphorylation. At the present time, chemotaxis in bacteria represents the best-known example of the relationship between a behavior and its underlying molecular mechanism.

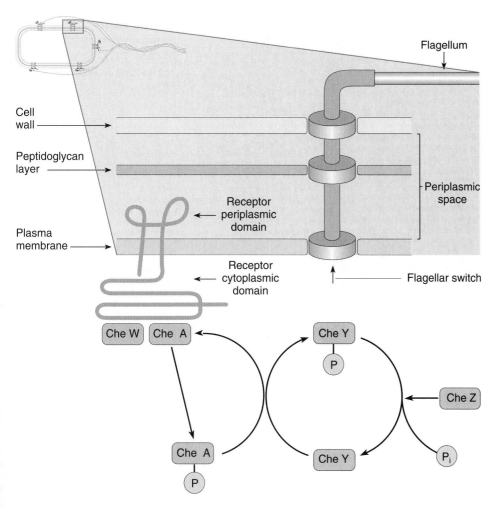

**FIGURE 23.10** Sensory transduction during bacterial chemotaxis. Stimulation of the transmembrane receptor leads to the phosphorylation of *CheA* with the aid of *CheW*. Activated *CheA* transfers a phosphate group to *CheY*. Phosphorylated *CheY* interacts with the flagellar switch to alter direction of flagellar rotation. *CheZ* inactivates *CheY* by removing the phosphate group.

## Behavior Genetics of a Nematode

In 1968, in one of the boldest attempts to define the genetic basis of behaviors in a single organism, Sidney Brenner began an investigation of the nematode *Caenorhabditis elegans*. Brenner had previously made valuable contributions in the field of molecular genetics, including studies of DNA replication, F factors, mRNA, and the genetic code. When he turned to the study of behavior, he hoped that it would be possible to dissect genetically the nervous system of *C. elegans* using techniques previously applied successfully to other organisms.

He chose this nematode because it was possible to determine the complete structure of the nervous system. Adult worms are about 1 mm long and are composed of only 959 cells, about 350 of which are neurons. As a result, it is possible to cut serial sections of an embedded organism and reconstruct the entire organism and its nervous system in the form of a three-dimensional model. Brenner then hoped to induce large numbers of behavioral mutations and to correlate aberrant behavior with structural and biochemical alterations in the ner-

vous system. Some progress has been made toward these goals, particularly in isolating large numbers of mutations. In early experiments, three general types of behavioral mutants were characterized. Worms are positively chemotactic to a variety of stimuli (cyclic AMP and GMP; anions such as $Cl^-$, $Br^-$, and $I^-$; and cations such as $Na^+$, $Li^+$, $K^+$, and $Mg^{++}$). As shown in Figure 23.11, positive attraction can be tracked in gradients on agar plates. The study of chemotactic mutants has shown that sensory receptors in the head alone mediate the orientation responses to attractants.

A second class of behavior studied involves *thermotaxis*. Cryophilic mutants move toward cooler temperatures, and thermophilic mutants move toward warmer temperatures. However, this behavior has not yet been correlated with the responsible component of the nervous system.

A third class of behavior involves generalized movement on the surface of an agar plate. Of 300 induced mutations, 77 affected the movement of the animal. While wild-type worms move with a smooth, sinuous pattern, mutants are either **uncoordinated** (*unc*) or

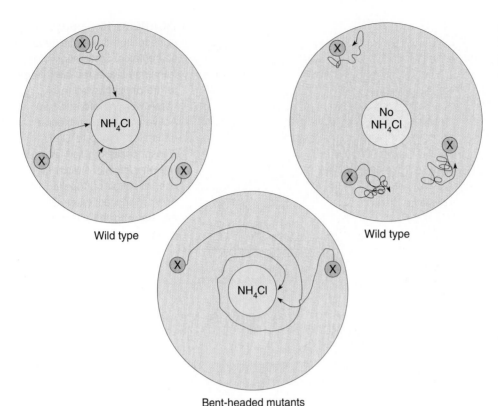

**FIGURE 23.11**    Chemotactic response to ammonium chloride (NH₄Cl) of wild-type and mutant *Caenorhabditis*.

**rollers** (*rol*). Those that are uncoordinated vary from the display of partial paralysis to small aberrations of movement, including twitching. Rollers move by rotating along their long axis, creating circular tracks on an agar surface. Many of these mutants have been correlated with defects in the dorsal or ventral nerve cord or in the body musculature.

Linkage mapping has also begun. Mutants are distributed on six linkage groups, corresponding to the haploid number of chromosomes characteristic of this organism. Numerous *unc* mutants are found on each of the six chromosomes, indicating extensive genetic control of nervous system development.

As the genetics of *C. elegans* has developed, analyses of the links among genes, the nervous system, and behavior have become more refined, beginning with a complex behavioral repertoire and isolating genes that control this behavior. As an example, feeding behavior in the worm can be used as a system to dissect genetic components of behavior. In *C. elegans*, the pharynx is a self-contained feeding pump composed of three parts: (1) the corpus, which ingests bacteria, and (2) the isthmus, which conducts bacteria to (3) the terminal bulb, where bacteria are ground up and passed into the intestine (Figure 23.12). The pharynx is surrounded by a basement membrane and contains a total of 80 cells (80 out of 959 in the organism), 20 of which are neurons making up the pharyngeal nervous system.

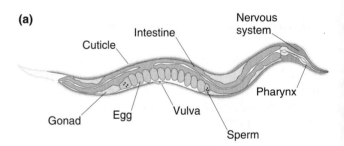

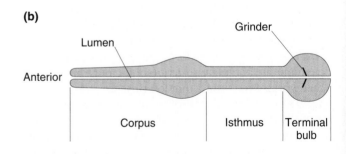

**FIGURE 23.12**    The pharynx of *C. elegans*. (a) Body plan of intact animal, showing location and arrangement of internal organs, including pharynx. (b) Pharynx, showing three regional divisions and associated structures.

The pharyngeal nervous system is somewhat self-contained, with only one pair of nerves connecting it to the rest of the nervous system; this connection can be severed without impairing the action of the pharyngeal nerves. In fact, the pharynx can be dissected from the body and will function *in vitro*. Only one of the 20 neurons, M4, which innervates the posterior isthmus, is essential for life. When M4 is missing, the isthmus remains closed and bacteria are not transported for grinding and digestion, and starvation ensues. Worms with an intact M4 neuron, but missing the 19 others, are viable, although these neurons are necessary for normal patterns of feeding.

In an ongoing study of the genetics of feeding, Leon Avery has set out to isolate and characterize the genes that control the presence or absence, the developmental fate, patterns of innervation, and function of the 20 neurons responsible for feeding behavior. Almost 38,000 progeny of mutagenized worms were screened for feeding-defective mutants. By linkage and mapping studies, 52 mutations were assigned to 35 genes located on all six chromosomes. Based on the number of mutants recovered and the number of genes involved, it is estimated that at least 60 genes are involved in feeding behavior in *C. elegans,* meaning that roughly half the genes remain to be identified.

The 52 mutations recovered to date fall into three broad phenotypic classes. The *eat* mutants affect the motion of the pharyngeal muscles; in addition, some *eat* mutants also affect the function of body wall muscles. The *pha* mutants have misshapen pharynxes that prevent normal feeding behavior. The third class of mutants, *phm* mutants, all have weak and irregular pharyngeal muscle contractions. The distribution of mutants is shown in Table 23.2. Preliminary work indicates that many of the *eat* mutants may affect function of the nervous system, or muscle functions that control contraction. In addition, *phm* mutants may have defective nervous control of muscle function. The *pha* mutants may

represent defects in morphogenesis or specification of cell fate.

Work is now proceeding on the isolation, cloning, and characterization of genes affecting muscle excitability and neuron function. Results from this study will provide insight into molecular events through which behavior is mediated by the nervous system and the muscular system.

## Single-Gene Effects in Mice

Surely the oldest recorded behavior mutant is the *waltzer* mutation in the mouse, recorded in 80 B.C. in China and described as a mouse "found dancing with its tail in its mouth." Waltzing mice run in a tight circle and demonstrate both horizontal and vertical head shaking as well as hyperirritability. Some waltzers are also deaf.

Genetic crosses between mutant and normal house mice (*Mus musculus*) reveal a simple recessive inheritance pattern characteristic of Mendel's monohybrid matings. Investigation of the inner ear of mutants has revealed degeneration of both the cochlea and semicircular canals, accounting for the deafness and circling behavior. This is an example of a mutation causing a structural anomaly which, in turn, alters behavior.

Many other single-gene behavior effects are known in the mouse, an organism particularly well characterized genetically. Of over 300 mutants discovered representing about 250 loci, over 90 are neurological in nature and alter behavior. These are classified into three groups of syndromes, as shown in Table 23.3. The neurological basis of many of these has now been determined.

The incoordination mutants *quaking* and *jumping* are due to faulty myelination of nervous tissue. *Quaking* is an autosomal recessive mutation, and *jumping* is inherited as a sex-linked recessive. The former cannot synthe-

**Table 23.2** DISTRIBUTION OF PHARYNX MUTANTS IN *C. ELEGANS* BY PHENOTYPIC CLASS

| Mutations/Gene | eat | pha | phm |
|---|---|---|---|
| 1 | 16 | 2 | 5 |
| 2 | 7 | 0 | 2 |
| 3 | 1 | 0 | 0 |
| 4 | 2 | 0 | 0 |
| Total genes: | 26 | 2 | 7 |
| Total mutations: | 41 | 2 | 9 |

SOURCE: Modified from Avery, 1993.

**Table 23.3** INHERITED NEUROLOGICAL DEFECTS IN MICE

| Group | Name |
|---|---|
| Waltzer-shaker | Waltzer, shaker, pirouette, jerker, fidget, twirler, zig-zag |
| Convulsive | Trembler, tottering, spastic |
| Incoordination | Jumping, quaking, reeler, staggerer, leaner, Purkinje cell degeneration, cerebral degeneration |

size adequate myelin, while the latter demonstrates a degenerative process involving myelin. *Cerebral degeneration* is a mutant exhibiting progressive deterioration of behavior. As its name implies, part of the brain deteriorates.

The study of such mutants clearly establishes the genetic basis of normal development. Additionally, and more germane to this chapter, abnormal development is linked directly to altered behavior by these findings.

# *DROSOPHILA* **BEHAVIOR GENETICS**

## Genes and Mating Behavior in *Drosophila*

We have already discussed geotaxis in *Drosophila* as an example of the selection methodology, but much more information on the behavior genetics of this organism is available. This is not at all surprising because of our extensive knowledge of its genetics and the ease with which it can be manipulated experimentally. Advances made beginning in the 1970s, particularly by Seymour Benzer and his colleagues, represent significant strides in the field of behavior genetics.

As early as 1915, Alfred Sturtevant observed that the sex-linked recessive gene *yellow* affects mating preference, besides conferring the more obvious pigmentation difference. Recall that such a situation is an example of **pleiotropic** gene expression (see Chapter 4). Sturtevant found that both wild-type and *yellow* females, when given the choice of wild-type or *yellow* males, prefer to mate with wild type. Wild-type and *yellow* males prefer to mate with *yellow* females. These conclusions were based on quantitative measurements of success in mating between all combinations of *yellow* and wild-type (gray-bodied) males and females.

In 1956, Margaret Bastock extended these observations by investigating which, if any, component of courtship behavior was affected by the *yellow* mutant gene.

Courtship in wild-type *Drosophila* is a complex ritual. The male first undergoes **orientation**, where he follows the female, perhaps circles her, and then orients usually at right angles and taps her on the abdomen. Once he has her attention, male wing display or **vibration** occurs. The wing closest to the female is raised and vibrates rapidly for several seconds. He then moves behind her, and contact is made with the female genitalia. Following this phase, if she has signaled acceptance by remaining in place, he mounts her and copulation occurs.

Bastock compared wild-type and *yellow* males for courtship rituals. What she observed was that *yellow* males prolong orientation but spend much less time in the vibrating and genital contact phases. Courtship patterns displayed by wild-type and *yellow* males are represented in Figure 23.13.

It appears that the *yellow* mutation has disrupted the intricate behavioral sequence of male courtship. Differences in the finer aspects of courtship have also been noted between related species of *Drosophila*. In these instances, the behavioral differences are thought to serve as possible isolating mechanisms during evolution.

## The Genetic Dissection of Behavior in *Drosophila*

In 1967, Seymour Benzer and his colleagues initiated a study of behavior genetics in *Drosophila*. Benzer's approach is an excellent example of the use of genetic techniques to "dissect" a complex biological phenomenon into its simpler components. Furthermore, the use of such techniques allows geneticists to study the underlying basis of the phenomenon—in this case, behavior. Benzer's goals in this research illustrate the "genetic dissection" approach:

1. To identify the genetic components of behavioral responses by the isolation of mutations that disrupt normal behavior.

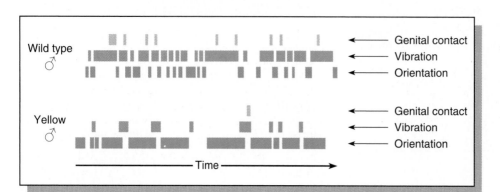

**FIGURE 23.13** Courtship patterns of wild-type and *yellow* males in *Drosophila*. The duration of the behavioral patterns of genital contact, vibration, and orientation are recorded over time.

2. To identify the mutant genes by chromosome localization and mapping.

3. To determine the actual site within the organism at which the gene expression influences the behavioral response.

4. To learn, if possible, how the particular gene expression influences behavior.

All four steps are illustrated in a discussion of **phototaxis**, one of the first behaviors studied by Benzer. Normal flies are positively phototactic; that is, they move toward a light source. Mutations are induced by feeding male flies sugar water containing **EMS** (ethylmethanesulfonate, a potent mutagen) and mating them to attached-X virgin females. As shown in Figure 23.14, the F₁ males receive their X chromosome from their fathers. Because they are hemizygous, any induced sexlinked recessive mutations are expressed.

The F₁ males are tested for their response to light, and those with abnormal behavior are isolated. Benzer found *runner* mutants, which move quickly to and from light; *negatively phototactic* mutants, which move away from light; and *nonphototactic* mutants, which show no preference for light or darkness. He established that the behavior changes were due to mutations by mating these F₁ males to attached-X virgin females. Male progeny of this cross also showed the abnormal phototactic responses, confirming them as products of sex-linked recessive mutations.

We shall now describe further work concerning just one group, the *nonphototactic* mutants. Such flies behave in the light as normal flies do in the dark. They can walk normally, but respond to light as if they were blind. Benzer and Yoshiki Hotta tested the electrical activity at the surface of mutant eyes in response to a flash of light. The pattern of electrical activity was recorded as an **electroretinogram**. Various types of abnormal responses were detected, and in none of the mutants was a normal pattern observed. When these mutations were mapped, they were not all allelic; instead, they were shown to occupy several loci on the X chromosome. Thus, it can be concluded that several gene products contribute to the formation of the behavioral response to light.

Where, within the fly, must adequate gene expression occur to yield a normal pattern? In an ingenious approach aimed at answering this question, Benzer turned to the use of mosaics. In mosaic flies, some tissues are mutant and others are wild type. If it can be ascertained which part must be mutant in order to yield the abnormal behavior, the **primary focus** of the genetic alteration can be determined.

To facilitate the production of mosaic flies, Benzer used a *Drosophila* strain that has one of its X chromo-

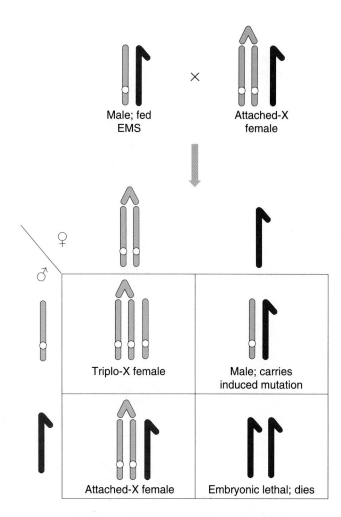

**FIGURE 23.14**    Genetic cross in *Drosophila* that facilitates the recovery of X-linked induced mutations. The female parent contains two X chromosomes that are attached, in addition to a Y chromosome. In a cross between this female and a normal male that has been fed the mutagen ethylmethanesulfonate, all surviving males receive their X chromosomes from their father and express all mutations induced on that chromosome.

somes in an unstable ring shape. When present in a zygote undergoing cell division, the ring-X is frequently lost by nondisjunction. If the zygote is female and has two X chromosomes (one normal and one ring-X), loss of the ring-X at the first mitotic division will result in two cells—one with a single X (normal X) and one with two X chromosomes (one normal and one ring-X). The former cell goes on to produce male tissue (XO) and expresses all alleles on the remaining X, while the latter produces female tissue and does not express heterozygous recessive X-linked genes. Such an occurrence is illustrated in Figure 23.15. One can see that loss of the ring-X will produce mosaic flies with male and female

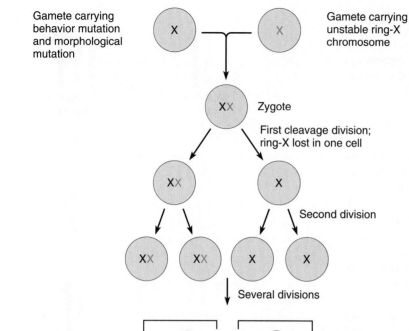

Gamete carrying behavior mutation and morphological mutation

Gamete carrying unstable ring-X chromosome

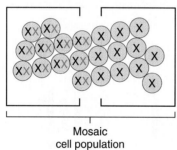

Zygote

First cleavage division; ring-X lost in one cell

Second division

Several divisions

XX female tissue; no heterozygous recessive mutations expressed

XO male tissue, expressing behavior and morphological mutation

Mosaic cell population

**FIGURE 23.15**   Production of a mosaic fruit fly as a result of fertilization by a gamete carrying an unstable ring-X chromosome (shown in color). If this chromosome is lost in one of the two cells following the first mitotic division, the body of the fly will consist of one part which is male (XO) and the other part which is female (XX). The male side will express all mutations contained on the X chromosome.

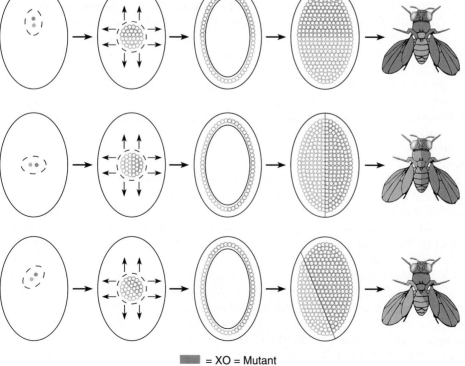

**FIGURE 23.16**   The effect of spindle orientation on the production of mosaic flies as shown in Figure 23.15.

■ = XO = Mutant
■ = XX = Wild type

parts that express or do not express sex-linked recessive genes, respectively.

In the embryo, when and where the ring-X is lost determines the pattern of mosaicism. The loss usually occurs early in development, before the cells migrate to the surface of the blastula to form the **blastoderm**. As shown in Figure 23.16, depending on the orientation of the spindle when the loss occurs, different types of mosaics will be created. If the stable X chromosome contains the behavior mutation and an obvious mutant gene (*yellow,* for example), the pattern of mosaicism will be readily apparent. For example, the fly may consist of a mutant head on a wild-type body, a normal head on a mutant body, one normal and one mutant eye on a normal or mutant body, and so on.

When such mosaics for *nonphototactic* mutants were studied, it was found that the focus of the genetic defect was in the eye itself. In mosaics where every part of the fly except the eye was normal, abnormal behavior was still detected. Even if only one eye was mutant, a modi-

fied abnormal behavior was observed. Instead of crawling straight up toward light as the normal fly does, the single mutant-eyed fly crawls upward to light in a spiral pattern. In the dark, such a fly will move in a straight line. Thus, mosaic studies have established that the focus of *nonphototactic* gene expression is in the eye itself. Additionally, the abnormal behavior is due to altered electrical conductance of the cells of the eye.

Using a combination of genetics, physiology, and biochemistry, Benzer and other workers in the field have identified a large number of genes affecting behavior in *Drosophila.* As shown in Table 23.4, mutants have been isolated that affect locomotion, response to stress, circadian rhythm, sexual behavior, visual behavior, and even learning. Some of these mutations have received very descriptive and often humorous names.

Many mutations have been analyzed with the mosaic technique in order to localize the focus of gene expression. While it was easy to predict that the focus of the *nonphototactic* mutant would be in the eye, other mu-

**Table 23.4**    SOME BEHAVIORAL MUTANTS OF *DROSOPHILA*

| Class | Name | Characteristics |
|---|---|---|
| Locomotor | *sluggish* | Moves slowly |
| | *hyperkinetic* | High consumption of $O_2$; shaking of legs; early death |
| | *wings up* | Wings perpendicular to body |
| | *flightless* | Does not fly well, although wings are well developed |
| | *uncoordinated* | Lacks coordinated movements |
| | *nonclimbing* | Fails to climb |
| Response to stress | *easily shocked* | Mechanical stock induces "coma" |
| | *stoned* | Stagger induced by mechanical shock |
| | *shaker* | Vibrates all legs while etherized |
| | *freaked out* | Grotesque, random gyrations under the influence of ether |
| | *paralyzed* | Collapses above a critical temperature |
| | *parched* | Dies quickly in low humidity conditions |
| | *tko* | Epileptic-like response |
| | *comatose* | Paralyzed by cool temperature |
| | *out-cold* | Similar to comatose |
| Circadian rhythm | *period$^o$* | Eclosion at any time; locomotor activity spread randomly over the day |
| | *period$^s$* | 19-hour cycle rather than 24-hour cycle |
| | *period$^l$* | 28-hour cycle rather than 24-hour cycle |
| Sexual | *savior-faire* | Males unsuccessful in courtship |
| | *fruitless* | Males pursue each other |
| | *stuck* | Male is often unable to withdraw after copulation |
| | *coitus interruptus* | Males disengage in about half the normal time |
| Visual | *nonphototactic* | Blind |
| | *negatively phototactic* | Moves away from light |
| Learning | *dunce* | Fails to learn conditioned response |

tants are not so predictable. For example, the focus of mutants affecting circadian rhythms has been located in the head, presumably in the brain. The *wings up* mutant might have a defect in the wings, articulation with the thorax, the thorax musculature, or the nervous system. Mosaic studies have pinpointed the indirect flight muscles of the thorax as the focus. Cytological studies have confirmed this finding, showing a complete lack of myofibrils in these muscles. Temperature-sensitive *paralytic* mutants are paralyzed at a raised temperature (29°C) but recover rapidly if the temperature is lowered. Mosaic studies have revealed that both the brain and thoracic ganglia represent the focus causing this abnormal behavior. More recent work has shown that mutant flies have defective sodium channels, and that the *paralytic* locus encodes a protein that controls the movement of sodium across the membrane of nerve cells.

The *drop-dead* mutant is most interesting. Such mutant flies appear completely normal for the first few days of adult life. Then they begin to stagger, fall over, and die. The cause of death could be related to any vital function anywhere in the body. However, mosaic study revealed that the defect is in the head. Most flies with mutant heads and normal bodies drop dead, while most with normal heads and mutant bodies do not. As shown in Figure 23.17, the brain of the *drop-dead* mutant appears to be full of holes. Examination of the brain of mutants before death shows them to be normal.

The mosaic technique has also been used to determine which regions of the brain are associated with sex-specific aspects of courtship and mating behavior. Jeffrey Hall and his associates have shown that mosaics with male cells in the most dorsal region of the brain, the protocerebrum, exhibit the initial stages of male courtship toward females. Later stages of male sexual behavior including wing vibrations and attempted copulations require male cells in the thoracic ganglion. Similar studies of female-specific sexual behavior have shown that the ability of a mosaic to induce courtship by a male depends on female cells in the posterior thorax or abdominal region. A region of the brain within the protocerebrum must be female for receptivity to copulation. Anatomical studies have confirmed that there are fine structural differences in the brains of male and female *Drosophila*, indicating that some forms of behavior may be dependent on the development and maturation of specific parts of the nervous system.

## Learning in *Drosophila*

To study the genetics of behavior such as learning, it would be advantageous to use an organism like *Drosophila* in which methods of genetic analysis are highly advanced. However, the first question is: Can *Drosophila* learn? Recent work from a number of laboratories indicates that organisms such as *Drosophila* are, in fact, capable of learning. Using a simple apparatus, flies are presented with a pair of olfactory cues, one of which is associated with an electrical shock. Flies quickly learn to avoid the odor associated with the shock. That this response is learned is indicated by a number of factors: first, performance is associated with the pairing of a stimulus/response with a reinforcement; second, the response is reversible. Flies can be trained to select an odor that they previously avoided, and flies exhibit short-term memory for the training they have received.

Other evidence suggests that more innate forms of behavior, such as courtship, are also associated with learning. Richard Siegel and his colleagues have found that following unsuccessful courtship of sexually unreceptive females, males show a reduction of courtship behavior for about 3 hours, even in the presence of sexually receptive females, Memory-deficient mutants, such as *amnesiac*, resume active courtship in about 1 hour, presumably because the previous experience has been forgotten.

The demonstration that *Drosophila* can learn opens the way to selecting mutants that are defective in learning and memory. To accomplish this, males from an inbred wild-type strain are mutagenized and mated to females from the same strain. Their progeny are recovered

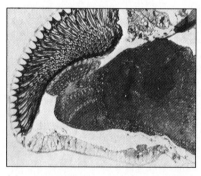

**(a)**

**(b)**

**FIGURE 23.17**    Photomicrographs of sections cut through the brains of (a) wild-type and (b) *drop-dead* mutant of *Drosophila melanogaster*.

and mated to produce populations of flies, each of which carries a mutagenized X chromosome. Mutants that affect learning are selected by testing the population for response in the olfactory/shock apparatus. A number of learning-deficient mutants including *dunce*, *turnip*, *rutabaga*, and *cabbage* have been recovered. In addition, a memory-deficient mutant, *amnesiac* (mentioned previously), that learns normally but forgets four times faster than normal has been recovered. Each of these mutations represents a single gene defect that affects a specific form of behavior. Because of the method used to recover them, all the mutants found so far are X-linked genes. Presumably similar genes controlling behavior are also located on autosomes.

## Molecular Biology of Behavior

Since in many cases, mutation results in the alteration or abolition of a single protein, biochemical study of the nervous system mutants described above can provide a link between behavior and molecular biology.

One group of *Drosophila* learning mutants has provided a link between cyclic nucleotides as a second messenger system and learning.

In cells of the nervous system as in many other cell types, cyclic AMP (cAMP) is produced from ATP by adenylyl cyclase. When produced, cyclic AMP activates target molecules, usually protein kinases that in turn alter the phosphorylation of proteins and initiate a cascade of metabolic effects that control gene transcription, memory, and learning. Behavioral mutants of *Drosophila* were among the first to show the link between cyclic AMP and learning. The *rutabaga* locus encodes one form of adenylyl cyclase, the enzyme that synthesizes cyclic AMP from ATP. In the *rutabaga* mutation a missense mutation destroys the catalytic activity of the protein and is associated with a learning deficiency in homozygous flies. The *dunce* locus encodes the structural gene for the enzyme cyclic AMP phosphodiesterase, which degrades cAMP. The *turnip* mutation occurs in a gene encoding a G protein, a class of molecules that bind GTP and, in turn, activate adenylyl cyclase. The clustering of these independently derived mutations strongly supports the idea that cyclic nucleotides play a role in learning and provide the basis for studies on the molecular basis of learning and memory.

A second group of mutations in *Drosophila* has provided insights into the mechanisms of transmission of nerve impulses and synaptic transmission. Some of these mutations, their phenotypes, and loci are shown in Table 23.5. When nerve impulses are transmitted along an axon, there is an accompanying change in concentration of sodium and potassium ions. These ions move into and out of cells through proteins that serve as channels through the cell membrane. The gene for *Shaker*, a mutation originally described over 40 years ago as a behavioral mutant, has been cloned and found to encode the protein that serves as the potassium channel. Using this clone, potassium-channel genes from a number of other species have been isolated. Studies on the *Shaker* gene and related genes from other organisms have served to elucidate how nerve impulses are transmitted, not only in *Drosophila* but in vertebrates as well. A second *Drosophila* mutation, *paralytic*, has been cloned and found to encode the sodium channel protein. Even from this brief description, it should be clear that studies on behavioral mutants in *Drosophila* have had a major impact on our understanding of nervous system function

**Table 23.5**  BEHAVIORAL MUTANTS OF *DROSOPHILA* AFFECTING NERVE IMPULSE TRANSMISSION

| Mutation | Map Location | Ion Channel Affected | Phenotype |
|---|---|---|---|
| *nap$^{ts}$* | 2–56.2 | sodium | Adults, larvae paralyzed at 37.5° C. Reversible at 25° C. |
| *para$^{ts}$* | 1–53.9 | sodium | Adults paralyzed at 29° C, larvae at 37° C. Reversible at 25° C. |
| *tip-E* | 3–13.5 | sodium | Adults, larvae paralyzed at 39–40° C. Reversible at 25° C. |
| *sei$^{ts}$* | 2–106 | sodium | Adults paralyzed at 38° C, larvae unaffected. Adults recover at 25° C. |
| *Sh* | 1–57.7 | potassium | Aberrant leg shaking in adults exposed to ether. |
| *eag* | 1–50.0 | potassium | Aberrant leg shaking in adults exposed to ether. |
| *Hk* | 1–30.0 | potassium | Ether-induced leg shaking. |
| *sio* | 3–85.0 | potassium | At 22° C, adults are weak fliers; at 38° C, adults are weak, uncoordinated. |

at the molecular level. Certainly, work in this field over the next few years will be exciting and important in clarifying the relationship between genetics and behavior.

# HUMAN BEHAVIOR GENETICS

The genetic input to behavior in humans is more difficult to characterize than that of other organisms. Not only are humans unavailable as experimental subjects in genetic investigations, but the types of responses considered to be interesting behavior are extremely difficult to study. The most popular forms of studied behavior all include some aspects of **intelligence**, **language**, **personality**, or **emotion**. There are two problems in examining such traits. First, all are difficult to define objectively and to measure quantitatively. Second, they are the traits most affected by environmental factors. In each case, while there is undoubtedly a genetic basis, it is a complex one. Furthermore, the environment is extremely important in shaping, limiting, or facilitating the final phenotype for each trait.

The study of human behavior genetics has also been hampered by two other factors. Many studies of human behavior have been performed by psychologists without adequate input form the biologist or geneticist. Second, traits involving intelligence, personality, and emotion have the greatest social and political significance. As such, these traits are more likely to be the subject of sensationalism when reported to the lay public. Because their study comes closest to infringing upon individual liberties such as the right to privacy, these traits are the basis of the most controversial investigations.

In lamenting the gulf between psychology and genetics in explaining human behavior genetics, C. C. Darlington in 1963 wrote, "Human behavior has thus become a happy hunting ground for literary amateurs. And the reason is that psychology and genetics, whose business it is to explain behavior, have failed to face the task together." Since 1963, some progress has been made in bridging this gap, but the genetics of human behavior remains a controversial area.

## Mental Disorders with a Clearcut Genetic Basis

Many genetic disorders in humans result in some behavioral abnormality. One of the most prominent examples is **Huntington disease (HD)**. Inherited as an autosomal dominant disorder, it affects the nervous system, including the brain. Symptoms of HD usually appear in the fifth decade of life with a gradual loss of motor function and coordination. Degeneration of the nervous system is progressive, and personality changes occur. The affected individual soon is unable to care for himself. Most victims die within 10 to 15 years after onset of the disease. Since onset is usually after a family has been started, all children of an affected person must live with the knowledge that they face a 50 percent probability of developing the disorder. The phenotype of HD is associated with elevated brain levels of quinolinic acid, a naturally occurring neurotoxin. The HD gene has been identified and encodes a large protein (348 kd) that is unrelated to any known gene product. The mutant form of the gene is associated with the presence of extra CAG trinucleotide repeats near the 5' end. Normal individuals have 11 to 24 repeats, but those affected by HD carry 42 to 86 CAG repeats. The expansion of a trinucleotide repeat in HD is similar to expansions reported as the basis of mutation in several other disorders that affect the brain and nervous system, indicating that such "stutter" mutations may be a common form of mutation in behavioral traits.

The **Lesch–Nyhan syndrome** is inherited as a sex-linked recessive disorder. Onset is within the first year, and the disease is most often fatal early in childhood. The disorder is metabolic, involving purine biosynthesis. Affected individuals lack hypoxanthine-guanine phosphoribosyltransferase (HGPRT), and accumulate high levels of uric acid. Mental and physical retardation occurs, and these individuals demonstrate uncontrolled self-mutilation. They also strike out at individuals attempting to care for them.

Other metabolic disorders are also known to affect mental function. For example, **Tay-Sachs disease**, an autosomal recessive disorder, involves severe mental retardation. The disease is apparent soon after birth and is fatal. **Porphyria**, under the control of an autosomal dominant allele, usually has an adult onset and is marked by recurring periods of dementia. The autosomal recessive disease **phenylketonuria**, unless detected and treated early, results in mental retardation. All of these disorders alter the normal biochemistry of the affected individuals and are inherited in a Mendelian fashion.

Chromosome abnormalities also produce syndromes with behavioral components. **Down syndrome** (trisomy 21) results in mental retardation. While there is a wide range of variability, the mean IQ of affected individuals is estimated to be 25 to 50. The onset of walking and talking is often delayed until four to five years of age. Both Klinefelter syndrome (XXY) and Turner syndrome (XO) may also result in diminished mental capacity.

## Human Behavior Traits with Less-Defined Genetic Bases

Other aspects of human behavior, notably **schizophrenia** and **manic–depressive illness**, have been the subject of extensive investigations. Studies have sought to relate the development of the mental disorder or the display of intelligence to the closeness of family relationships or twin studies. In all cases, it has been concluded that a genetic component influences the trait, but that environment also plays a substantial role.

A discussion of schizophrenia may serve to illustrate the methodology used. This mental disorder is characterized by withdrawn, bizarre, and sometimes delusional behavior. Those affected by the disease are unable to lead organized lives and are periodically disabled by the condition. It is clearly a familial disorder, with relatives of schizophrenics having a much higher incidence of this disorder than the general population. Furthermore, the closer the relationship to the index case or proband, the greater is the probability of the disorder occurring.

The concordance of schizophrenia in monozygotic and dizygotic twins has been the subject of many studies. In almost every investigation, concordance has been higher in monozygotic twins than in dizygotic twins reared together. Although these results suggest that a genetic component exists, they do not reveal the precise genetic basis of schizophrenia. Simple monohybrid and dihybrid inheritance as well as multiple gene control have been proposed for schizophrenia. However, it seems unlikely that only one or two loci are involved, nor is it likely that the control is strictly quantitative, as in polygenic inheritance. In both schizophrenia and manic–depressive illness, it is most sound to conclude that each individual is endowed with a genetic predisposition for normal or abnormal behavior and that environmental factors can serve to alter the final phenotype.

There is a long-standing controversy about genetic differences in intelligence between races. While IQ testing has established intelligence differences in populations of different races, there is currently no strong evidence to support the conclusion that this is due to a genetic component. With regard to intelligence, genetic factors may provide an upper and lower potential range, but an individual's environment may modify the development of intelligence. The environment undoubtedly has a profound effect on the type of intelligence measured by the various forms of IQ tests.

## CHAPTER SUMMARY

1. Behavioral genetics has emerged as an important specialty within the field of genetics because both genotype and environment have been found to have an impact in determining an organism's behavioral response.

2. Since both genotype and environment play a role in the expression of behavioral traits, the production of purely objective data in this area is particularly difficult.

3. Three areas of genetically influenced behavior research are being pursued: the behavior of closely related organisms from similar environments whose survival needs appear to be identical; the laboratory modification of behavioral traits for evidence of heritability; and the specific effects of a single gene on behavior patterns.

4. The studies of alcohol preference and open-field behavior in mice illustrate behaviors strongly influenced by the genotype.

5. Studies of maze learning in rats and geotaxis in *Drosophila* have successfully established both bright and dull lines of rats and either positively or negatively geotaxic *Drosophila*. These results have led researchers to ascribe the relative contributions of genes on a specific chromosome to these behaviors.

6. By isolating mutations that cause deviations from normal behavior, the role of a corresponding wild-type allele in the respective response is established. Numerous examples have been examined, including honeybee hygenic behavior; chemotaxis, thermotaxis, and general movement in nematodes; neurological mutations in mice; and a variety of behaviors in *Drosophila*.

7. Any aspect of human behavior is difficult to study because the individual's environment makes an important contribution toward trait development. In humans, studies using twins have shown that while a family may have a predisposition to schizophrenia, the expression of this disorder may be modified by the environment. General intelligence is also a product of both genetic and environmental influences.

---

## KEY TERMS

| | | | |
|---|---|---|---|
| alcohol preference | emotion | language | open-field test |
| behavior | EMS | Lesch–Nyhan syndrome | orientation |
| behaviorist school | (ethylmethanesulfonate) | manic–depressive illness | personality |
| chemotaxis | geotaxis | maze learning | phenylketonuria |
| Down syndrome | Huntington disease (HD) | nature–nurture | phototaxis |
| electroretinogram | intelligence | controversy | schizophrenia |
| | | | Tay-Sachs disease |

## INSIGHTS AND SOLUTIONS

1.  Manic depression is an affective disorder associated with recurring mood changes. It is estimated that one in four individuals will suffer from some form of affective disorder at least once in his lifetime. Genetic studies indicate that manic depression is familial, and that single genes may play a major role in controlling this behavioral disorder. In 1987 two separate studies using RFLP analysis and other genetic markers reported linkage between manic depression and markers on the X chromosome and to the short arm of chromosome 11. At the time, these reports were hailed as landmark discoveries, opening the way to the isolation and characterization of genes that control specific forms of behavior, and to the development of therapeutic strategies based on knowledge of the nature of the gene product and its action. However, further work on the same populations (reported in 1989 and 1990) demonstrated that the original results were invalid, and concluded that no linkage to markers on the X chromosome and to markers on chromosome 11 could be established. These findings do not exclude the role of major genes on the X chromosome and autosomes in manic depression, but do exclude linkage to the markers used in the original reports.

    This setback not only caused embarrassment and confusion, but it also forced a reexamination of the validity of the methods used in mapping human genes, and the analysis of data from linkage studies involving complex behavioral traits. Several factors have been proposed to explain the flawed conclusions reported in the original studies. What do you suppose some of these factors to be?

    **SOLUTION:** While some of the criticisms were directed at the choice of markers, most of the factors at work in this situation appear to be related to the phenotype of manic depression. At least three confounding elements have been identified. One is age of onset. There is a positive correlation between age and the appearance of manic depression. Therefore, at the time of a pedigree study, younger individuals who will be affected later in life may not show any signs of manic depression. Another confusing factor is that in the populations studied, there may in fact be more than one major sex-linked gene and more than one major autosomal gene that cause manic depression. A third factor relates to the diagnosis of manic depression itself. The phenotype is complex and not as easily quantified as height or weight. In addition, swings in mood are a universal part of everyday life, and it is not always easy to distinguish transient mood alterations and the role of environmental factors from affective disorders having a biological and/or genetic basis.

These factors point up the difficulty in researching the genetic basis of complex behavioral traits. Individually or in combinations, the factors described above might skew the results enough so that guidelines for proof that are adequate for other traits are not stringent enough for behavioral traits with complex phenotypes and complex underlying causes.

**PROBLEMS AND DISCUSSION QUESTIONS**

1. Contrast the methodologies used in studying behavior genetics. What are the advantages of each method with respect to the type of information gained?

2. Contrast the advantages of using *Drosophila* versus *Caenorhabditis* for studying behavior genetics.

3. Theoretical data concerning the genetic effect of *Drosophila* chromosomes on geotaxis are shown below. What conclusions can you draw?

| | Chromosome | | |
|---|---|---|---|
| | X | 2 | 3 |
| Positive geotaxis | +0.2 | +0.1 | +3.0 |
| Unselected | +0.1 | −0.2 | +1.0 |
| Negative geotaxis | −0.1 | −2.6 | +0.1 |

4. If a haploid honeybee drone of the genotype *uR* is mated to a queen of genotype *UuRR*, what ratio of behavioral phenotypes will be observed in the hive in the offspring with respect to American foulbrood disease? For *Ur* × *UURr*? For *uR* × *Uurr*?

5. Assume that you discovered a fruit fly that walked with a limp and was continually off balance as it moved. Describe how you would determine if this behavior was due to an injury (induced in the environment) or was an inherited trait. Assuming that it is inherited, what are the various possibilities for the focus of gene expression causing the imbalance? Describe how you would locate the focus experimentally if it were X-linked.

6. In humans, the chemical phenylthiocarbamide (PTC) is either tasted or not. When the offspring of various combinations of taster and nontaster parents are examined, the following data are obtained:

| PARENTS | Both tasters | Both tasters | Both tasters | One taster One nontaster | One taster One nontaster | Both nontasters |
|---|---|---|---|---|---|---|
| OFFSPRING | All tasters | ½ tasters ½ nontasters | ¾ tasters ¼ nontasters | All tasters | ½ tasters ½ nontasters | All nontasters |

Based on these data, how is PTC tasting behavior inherited?

7. Discuss why the study of human behavior genetics has lagged behind that of other organisms.

8. J. P. Scott and J. L. Fuller studied 50 traits in five pure breeds of dogs. Almost all traits varied significantly in the five breeds, but very few bred true in crosses. What can you conclude with respect to the genetic control of these behavioral traits?

**SELECTED READINGS**

AVERY, L. 1993. The genetics of feeding in *Caenorhabditis elegans*. *Genetics* 133:897–917.

AVERY, L., and HORVITZ, H. R. 1987. A cell that dies during wild-type *C. elegans* development can function as a neuron in a *ced-3* mutant. *Cell* 51:1071–78.

BASTOCK, M. 1956. A gene which changes a behavior pattern. *Evolution* 10:421–39.

———. 1967. *Courtship—An ethological study.* Chicago: Aldine.

BENZER, S. 1973. Genetic dissection of behavior. *Scient. Amer.* (Dec.) 229:24–37.

BERG, H. C. 1988. A physicist looks at bacterial chemotaxis. *Cold Spring Harb. Symp. Quant. Biol.* 53: (Part 1) 1–9.

BODMER, W. F., and CAVALLI-SFORZA, L. L. 1976. *Genetics, evolution and man.* New York: W. H. Freeman.

BRENNER, S. 1974. The genetics of *Caenorhabditis elegans*. *Genetics* 77:71–94.

BYERS, D., DAVID, R. L., and KIGER, J. A., Jr. 1981. Defect in the cyclic AMP phosphodiesterase due to the *dunce* mutation of learning in *Drosophila*. *Nature* 289:79–81.

DEVOR, E. J., and CLONINGER, C. R. 1989. Genetics of alcoholism. *Ann. Rev. Genet.* 23:19–36.

EHRMAN, L., and PARSONS, P. A. 1981. *Behavior genetics and evolution.* New York: McGraw-Hill.

FARBER, S. L. 1980. *Identical twins reared apart.* New York: Basic Books.

FULLER, J. C., and THOMPSON, W. R. 1978. *Foundations of behavior genetics.* St. Louis: C. V. Mosby.

FULLER, J. L. 1960. Behavior genetics. *Ann. Rev. Psychol.* 11:41–63.

GAILEY, D. A., HALL, J. C., and SIEGEL, R. W. 1985. Reduced reproductive success for a conditioning mutant in experimental populations of *Drosophila melanogaster*. *Genetics* 111:795–804.

GANTEZKY, B. 1989. There's a whole lot of shaking going on. *Genetics* 121:201–4.

GANTEZKY, B., and WU, C. F. 1986. Neurogenetics of membrane excitability in *Drosophila*. *Ann. Rev. Genet.* 20:13–44.

GOODMAN, R. M., ed. 1970. *Genetic disorders of man.* Boston: Little, Brown.

GOTTESMAN, I. I., and SHIELDS, J. 1972. *Schizophrenia and genetics: A twin study vantage point.* Orlando: Academic Press.

HALL, J. C., GREENSPAN, R. J., and HARRIS, W. A. 1982. *Genetic neurobiology.* Cambridge, MA: The MIT Press.

HARRIS, W. A. 1985. Genetics and development of the nervous system. *J. Neurogenet.* 2:179–96.

HAY, D. A. 1985. *Essentials of behaviour genetics.* Palo Alto, CA: Blackwell Scientific Publ.

HESTON, L. L. 1970. The genetics of schizophrenia and schizoid disease. *Science* 167:249–56.

HIRSCH, J., ed. 1967. *Behavior-genetic analysis.* New York: McGraw Hill.

HOTTA, Y., and BENZER, S. 1972. Mapping behavior of *Drosophila* mosaics. *Nature* 240:527–35.

Huntington's Disease Collaborative Group. 1993. A novel gene containing a trinucleotide repeat that is expanded and unstable on Huntington's disease chromosomes. *Cell* 72:971–83.

KAPLAN, A. R. 1976. *Human behavior genetics.* Springfield, IL: Charles C. Thomas.

KIDD, K. K., and CAVALLI-SFORZA, L. L. 1973. An analysis of the genetics of schizophrenia. *Social Biol.* 20:254–65.

LEWONTIN, R. C. 1975. Genetic aspects of intelligence. *Ann. Rev. Genet.* 9:387–405.

LINDZEY, G., LOEHLIN, J., MANOSEVITZ, M., and THIESSEN, D. 1971. Behavioral genetics. *Ann. Rev. Psychol.* 22:39–94.

LINDZEY, G., and THIESSEN, D. D. 1970. *Contributions to behavior-genetic analysis: The mouse as a prototype.* New York: Appleton-Century-Crofts.

LIVINGSTONE, K. S. 1985. Genetic dissection of *Drosophila* adenylate cyclase. *Proc. Natl. Acad. Sci.* 82:5795–99.

LOUGHNEY, K., KREBER, R., and GANTEZKY, B. 1989. Molecular analysis of the *para* locus, a sodium channel gene in *Drosophila*. *Cell* 58:1143–54.

MADDOCK, J. R., and SHAPIRO, L. 1993. Polar location of the chemoreceptor complex in the *E. coli* cell. *Science* 259:1717–23.

McCLEARN, G. E. 1970. Behavioral genetics. *Ann. Rev. Genet.* 4:437–68.

MERRELL, D. J. 1965. Methodology in behavior genetics. *J. Hered.* 56:263–66.

PARKINSON, J. S., and BLAIR, D. 1993. Does *E. coli* have a nose? *Science* 259:1701–02.

PLOMIN, R., DeFRIES, J. C., and McCLEARN, G. E. 1990. *Behavioral genetics: A primer.* 2nd ed. New York: W. H. Freeman.

QUINN, W. B., and GREENSPAN, R. J. 1984. Learning and courtship in *Drosophila*: Two stories with mutants. *Ann. Rev. Neurosci.* 7:67–93.

QUINN, W. G., and GOULD, J. L. 1979. Nerves and genes. *Nature* 278:19–23.

RICKER, J. P., and HIRSCH, J. 1988(a). Genetic changes occurring over 500 generations in lines of *Drosophila melanogaster* selected divergently for geotaxis. *Behavior Genet.* 18:13–24.

————. 1988(b). Reversal of genetic homeostasis in laboratory populations of *Drosophila melanogaster* under long-term selection for geotaxis and estimates of gene correlates: Evolution of behavior-genetic systems. *J. Comp. Psychol.* 102:203–14.

RIDDLE, D. L. 1978. The genetics of development and behavior in *Caenorhabditis elegans. J. Nematol.* 10:1–15.

ROGERS, D. A., and McCLEARN, G. E. 1962. Mouse strain differences in preference for various concentrations of alcohol. *Quart. J. Stud. Alc.* 23:26–33.

ROTHENBUHLER, W. C., KULINCEVIC, J. M., and KERR, W. E. 1968. Bee genetics. *Ann. Rev. Genet.* 2:413–38.

SCOTT, J. P., and FULLER, J. L., eds. 1974. *Dog behavior: The genetic basis.* Chicago: The University of Chicago Press.

SIEGEL, R. W., HALL, J. C., GAILEY, D. A., and KYRIACOU, C. P. 1984. Genetic elements of courtship in *Drosophila*: Mosaics and learning mutants. *Behav. Genet.* 14:383–410.

SPRINGER, M. S., GOY, M. F., and ADLER, J. 1979. Protein methylation in behavioral control mechanisms and in signal transduction. *Nature* 208:279–84.

TRYON, R. C. 1942. Individual differences. In *Comparative psychology,* 2nd ed., ed. F. A. Moss. Englewood Cliffs, NJ: Prentice-Hall.

TULLY, T. 1984. *Drosophila* learning: Behavior and biochemistry. *Behav. Genet.* 14:527–57.

VALE, J. R. 1980. *Genes, environment, and behavior: An interactionist approach.* New York: Harper & Row.

WARD, S. 1973. Chemotaxis by the nematode, *Caenorhabditis elegans*: Identification of attractants and analysis of the response by use of mutants. *Proc. Natl. Acad. Sci.* 70:817–21.

# 24

# POPULATION GENETICS

A population of king penguins, one of 17 existing species, gathered during the breeding season in the South Atlantic.

*Individuals can carry only two different alleles of a given gene. A group of individuals can carry a larger number of different alleles, giving rise to a reservoir of genetic diversity. The diversity contained in the population can be measured by the Hardy–Weinberg law. Mutation is the ultimate source of genetic variation, and other factors such as drift, migration, and selection can alter the amount of genetic variation in populations.*

When Charles Darwin published *The Origin of Species* in 1859 following work coauthored with Alfred Russel Wallace on the likely mechanism of natural selection, the foundation for the modern interpretation of evolution was established. Although organisms are capable of reproducing in a geometric fashion, Wallace and Darwin observed that this growth potential of species is not realized. Instead, population numbers remain relatively constant in nature. Both Wallace and Darwin deduced that some form of competitive struggle for survival must therefore occur. Darwin observed that there is a natural variation between individuals within a species; on this observation, he based his theory of natural selection: " . . . that any being, if it vary however slightly in any manner profitable to itself . . . will have a better chance of surviving." Darwin included in his concept of survival " . . . not only the life of the individual, but success in leaving progeny."

While Mendel was familiar with Darwin's work, Darwin was unaware of any underlying mechanism to account for the morphological variation he had observed. However, with the development of the concept of genes and alleles, the genetic basis of inherited variation was established.

As others pursued the study of evolution, it became apparent that the population rather than the individual was the unit of study in this process. In order to study the role of genetics in the process of evolution, therefore, it was necessary to consider allele frequencies in populations rather than offspring from individual matings. Thus arose the discipline of **population genetics**.

Early in the twentieth century various workers, including Gudny Yule, William Castle, Godfrey Hardy, and Wilhelm Weinberg, formulated the basic principles of this field. In the early years of population genetics, the emphasis was on theory and the development of mathematical models to describe the genetic structure of populations. Only in the past thirty years have experimentalists and field workers had the biochemical and molecular techniques to directly measure variation at the protein and DNA levels in order to test these theories and models. Allele frequencies and the forces that alter these frequencies, such as mutation, migration, selection, and random genetic drift, have been and are being examined. In this chapter, we shall consider some general aspects of population genetics and also discuss other areas of genetics relating to evolution.

## POPULATIONS, GENE POOLS, AND ALLELE FREQUENCIES

Members of a species are often distributed over a wide geographic range. A **population**, however, is a local group belonging to a single species, within which mating is actually or potentially occurring. The set of genetic information carried by all interbreeding members of a population is called the **gene pool**. For a given gene, this pool includes all the alleles of that gene present in the population. In population genetics the focus is on groups rather than on individuals, and on the measurement of allele and genotype frequencies from generation to generation rather than on the distribution of genotypes resulting from a single mating. The term **allele frequency** will be used throughout the chapter and will represent the frequency of alleles in contrast to genotype frequencies. Gametes produced by one generation represent withdrawals from the gene pool to form the zygotes of the next generation. This new generation has a reconstituted gene pool that may differ from that of the preceding generation.

Populations are dynamic; they may grow and expand or diminish and contract through changes in birth or death rates, by migration, or by merging with other populations. This has important consequences and, over time, can lead to changes in the genetic structure of the population.

One approach used to study a population's genetic structure is by measuring the frequency of a given allele

## Table 24.1  MN Blood Groups

| Genotype | Blood Type | Reaction with Antibodies | |
|---|---|---|---|
| | | Anti-M | Anti-N |
| $L^M L^M$ | M | + | − |
| $L^M L^N$ | MN | + | + |
| $L^N L^N$ | N | − | + |

## Table 24.2  Methods of Determining Allele Frequencies for Codominant Alleles

**A. Counting Alleles**

| Genotype/Phenotype | MM | MN | NN | Total |
|---|---|---|---|---|
| No. of individuals | 36 | 48 | 16 | 100 |
| No. of $M$ alleles | 72 | 48 | 0 | 120 |
| No. of $N$ alleles | 0 | 48 | 32 | 80 |
| Total no. of alleles | 72 | 96 | 32 | 200 |

Frequency of $M$ in population: $\dfrac{120}{200} = 0.6 = 60\%$

Frequency of $N$ in population: $\dfrac{80}{200} = 0.4 = 40\%$

**B. From Genotypes**

| Genotype/Phenotype | MM | MN | NN | Total |
|---|---|---|---|---|
| No. of individuals | 36 | 48 | 16 | 100 |
| Genotype frequency | 36/100 = 0.36 | 48/100 = 0.48 | 16/100 = 0.16 | 1.00 |

Frequency of $M$ in population: $36 + (1/2)48 = 0.36 + 0.24 = 0.60 = 60\%$
Frequency of $N$ in population: $16 + (1/2)48 = 0.16 + 0.24 = 0.40 = 40\%$

## Table 24.3  Frequencies of $M$ and $N$ Alleles in Various Populations

| Population | Genotype Frequency (%) | | | Allele Frequency | |
|---|---|---|---|---|---|
| | MM | MN | NN | M | N |
| Eskimos (Greenland) | 83.48 | 15.64 | 0.88 | 0.913 | 0.087 |
| U.S. Indians | 60.00 | 35.12 | 4.88 | 0.776 | 0.224 |
| U.S. whites | 29.16 | 49.38 | 21.26 | 0.540 | 0.460 |
| U.S. blacks | 28.42 | 49.64 | 21.94 | 0.532 | 0.468 |
| Ainus (Japan) | 17.86 | 50.20 | 31.94 | 0.430 | 0.570 |
| Aborigines (Australia) | 3.00 | 29.60 | 67.40 | 0.178 | 0.822 |

controlling a known trait. This is possible once the mode of inheritance and the number of different alleles present in the population have been established. Allele frequencies cannot always be determined directly because in many cases only phenotypes, and not genotypes, can be observed. However, if alleles expressed in a codominant fashion are considered, there is a 1:1 relationship between phenotypes and genotypes. Such is the case with the autosomally inherited **MN blood group** in humans.

In this case, the gene $L$ on chromosome 2 has two alleles, $L^M$ and $L^N$ (often referred to as $M$ and $N$, respectively).* Each allele controls the production of a distinct antigen on the surface of red blood cells. Thus, an individual may be type M ($L^M L^M$), N ($L^N L^N$), or MN ($L^M L^N$).

The genotypes, phenotypes, and immunological reactions of the MN blood group are shown in Table 24.1. Because they are codominant, the frequency of $M$ and $N$ in a population can be determined simply by counting the number of alleles for each phenotype. For example, consider a population of 100 individuals of which 36 are $MM$, 48 are $MN$, and 16 are $NN$. The 36 $MM$ individuals represent 72 $M$ alleles, and the 48 $MN$ heterozygotes represent 48 additional $M$ alleles. There are $72 + 48$ or 120 $M$ alleles out of a total of 200 alleles in the population (100 individuals, each with two alleles at this locus). Therefore, the frequency of the $M$ allele in this population is 0.6 ($120/200 = 0.6 = 60\%$). The frequency of the $N$ allele can be estimated in a similar fashion ($80/200 = 0.4 = 40\%$) and is 0.4. Table 24.2 illustrates two methods for computing the frequency of $M$ and $N$ alleles in a hypothetical population of 100 individuals, and Table 24.3 lists the frequencies of $M$ and $N$ alleles actually measured in several human populations.

## THE HARDY–WEINBERG LAW

In the example of the MN blood group, $M$ and $N$ are codominant. If, on the other hand, one allele had been recessive, the heterozygotes would have been phenotypically identical to the homozygous dominant individuals, and the frequency of the alleles could not have been directly determined. However, a mathematical model developed independently by the British mathematician Godfrey H. Hardy and the German physician, Wilhelm Weinberg, can be used to calculate allele frequencies in this case. The **Hardy–Weinberg law (HWL)** is one of the fundamental concepts in population genetics. In the Hardy–Weinberg law, the following conditions are presumed:

1. The population is infinitely large, or large enough that sampling error is negligible.

2. Mating within the population occurs at random.

3. There is no selective advantage for any genotype; that is, all genotypes produced by random mating are equally viable and fertile.

4. There is an absence of other factors such as mutation, migration, and random genetic drift.

In an ideal population, suppose that a locus has two alleles, $A$ and $a$. The frequency of the dominant allele $A$ in both eggs and sperm is represented by $p$, and the frequency of the recessive allele in gametes is represented by $q$. Because the sum of $p$ and $q$ represents 100 percent of the alleles for that gene in the population, then $p + q = 1$. A Punnett square can be used to represent the random combination of gametes containing these alleles and the resulting phenotypes:

|          | Sperm     |         |
|          | $A(p)$    | $a(q)$  |
|----------|-----------|---------|
| $A(p)$   | $AA$ ($p^2$) | $Aa$ ($pq$) |
| $a(q)$   | $Aa$ ($pq$) | $aa$ ($q^2$) |

Eggs

In the random combination of gametes in the population, the probability that sperm and egg both contain the $A$ allele is $p \times p = p^2$. Similarly, the chance that gametes will carry unlike alleles is $(p \times q) + (p \times q) = 2pq$, and the chance that a homozygous recessive individual will result is $q \times q = q^2$. Note that these terms also describe an important aspect of the HWL: allelic frequencies determine genotype frequencies. In other words, while the value $p^2$ is the probability that both gametes in a fertilization event will carry the $A$ allele, it is also a measure of the frequency of $AA$ homozygotes in the ensuing generation. In a similar way, $2pq$ describes the frequency of $Aa$ heterozygotes, and $q^2$ is a measure of the frequency of homozygous recessive ($aa$) zygotes. Thus, the distribution of genotypes in the next generation can be expressed as

$$p^2 + 2pq + q^2 = 1 \tag{24.1}$$

Next, consider a population in which 70 percent of the alleles for a given gene are $A$, and 30 percent are $a$. Thus, $p = 0.7$ and $q = 0.3$, and $p(0.7) + q(0.3) = 1$. The distribution of genotypes produced by random mating is shown in the following Punnett square.

*L stands for Karl Landsteiner, the geneticist for whom the locus is named.

Sperm

|  | $A(p = 0.7)$ | $a(q = 0.3)$ |
|---|---|---|
| $A$<br>$(p = 0.7)$ | $AA$<br>$p^2$<br>$(0.49)$ | $Aa$<br>$pq$<br>$(0.21)$ |
| Eggs<br><br>$a$<br>$(q = 0.3)$ | $Aa$<br>$pq$<br>$(0.21)$ | $aa$<br>$q^2$<br>$(0.09)$ |

In this new generation, 49 percent ($p^2$) of the individuals will be homozygous dominant, 42 percent ($2pq$) will be heterozygous, and 9 percent ($q^2$) will be homozygous recessive. By inspection of the Punnett square, the frequency of the $A$ allele in the new generation can be calculated as

$$p^2 + \tfrac{1}{2}(2pq) \qquad (24.2)$$
$$0.49 + \tfrac{1}{2}(0.42)$$
$$0.49 + 0.21 = 0.70$$

For $a$, the frequency is

$$q^2 + \tfrac{1}{2}(2pq) \qquad (24.3)$$
$$0.09 + \tfrac{1}{2}(0.42)$$
$$0.09 + 0.21 = 0.30$$

Since $p + q = 1$, the value for $a$ could have been calculated as

$$q = 1 - p \qquad (24.4)$$
$$q = 1 - 0.70 = 0.30$$

The frequency of $A$ and $a$ in the new generation are the same as in the previous generation.

A population in which the frequency of a given allele remains constant from generation to generation is said to be in a state of **genetic equilibrium**. In this case, the frequencies of $A$ and $a$ remain constant, and it appears that the conditions presumed for the Hardy–Weinberg law hold true in this population.

Several important points are relevant to the preceding example. First, while the hypothetical alleles we have considered were at equilibrium, not all alleles in a population are. This is particularly true when the assumptions made in the Hardy–Weinberg law do not hold. Second, the examples illustrate why dominant traits do not tend to increase in frequency as new generations are produced. Finally, the examples demonstrate that genetic equilibrium can maintain a state of **genetic variability** in a population. Once fixed in a population, allelic frequencies remain unchanged during equilibrium—a factor important to the evolutionary process.

## Sex-Linked Genes

In considering genotype and allele frequencies for autosomal recessive traits using the Hardy–Weinberg equation, we assumed that the frequency of $A$ is the same in both sperm and eggs. What about sex-linked genes? In species that have two X chromosomes, such as *Drosophila*, females carry two copies of all genes on the X chromosome, while males, on the other hand, carry only one copy of all X-linked genes. Genes on the X chromosome are therefore distributed unequally in the population, with females carrying two-thirds and males one-third of the total number. Can the Hardy–Weinberg law be applied to calculate allelic frequencies and genotypic frequencies in such circumstances?

It is easy to determine frequencies for X-linked genes in males. Because they have only one copy of all genes on this chromosome, the phenotype of both dominant and recessive alleles is expressed. Therefore, in males, the frequency of an X-linked allele is the same as the phenotypic frequency. In Western Europe, a form of sex-linked color blindness occurs with a frequency of 8 percent in males ($q = 0.08$). Since females have two doses of all genes on the X chromosome, color-blind females exhibit the phenotypic frequencies expected for complete dominance, and the genotype and allele frequencies can be calculated by using the standard Hardy–Weinberg equation. For example, since color blindness in males has a frequency of 0.08, at equilibrium, the expected frequency in females is $q^2$, or 0.0064. This means that 800 out of 10,000 males surveyed would be expected to be color-blind, but only 64 of 10,000 females would show this trait. Expected values of the frequency of X-linked traits in males and females are compared in Table 24.4.

## Linkage Disequilibrium

If the frequency of an X-linked allele differs between males and females, then the population is not in equilib-

**Table 24.4**  EXPECTED RELATIVE FREQUENCY OF X-LINKED TRAITS IN MALES AND FEMALES

| Frequency of<br>Males with Trait | Expected Frequency<br>in Females |
|---|---|
| 90/100 | 81/100 |
| 50/100 | 25/100 |
| 10/100 | 1/100 |
| 1/100 | 1/10,000 |
| 1/1000 | 1/1,000,000 |
| 1/10,000 | 1/100,000,000 |

rium. In contrast to alleles of autosomal recessive genes, equilibrium in each group will not be reached in a single generation, but will be approached over a series of succeeding generations. Because males inherit their X chromosome maternally, the allelic frequency in females will determine the frequency in males of the next generation. Daughters will inherit both a maternal and a paternal X, and their allelic frequency is the average of that found in the parents. Even though the population's overall allele frequency remains constant, there is an oscillation in frequencies for the two sexes in each generation, with the differences being halved in each succeeding generation until an equilibrium is reached. This concept is illus-

trated in Figure 24.1 for the simple case where an allele has an initial frequency of 1.0 in females and 0.0 in males.

## Multiple Alleles

In addition to autosomal recessive and sex-linked genes, it is common to find several alleles of a single locus in a population. The ABO blood group in humans is such an example. The locus $I$ (isoagglutinin) has three alleles ($I^A$, $I^B$, and $I^O$), yielding six possible genotypic combinations ($I^AI^A$, $I^BI^B$, $I^OI^O$, $I^AI^B$, $I^AI^O$, $I^BI^O$). Recall that in this case A and B are codominant alleles, and both of these are dominant to O. The result is that homozygous $AA$ and heterozygous $AO$ individuals are phenotypically identical, as are $BB$ and $BO$ individuals, so we can distinguish only four phenotypic combinations.

By adding another variable to the Hardy–Weinberg equation, we can calculate both genotype and allele frequencies for the situation involving three alleles. In an equilibrium population, the frequency of the three alleles can be described by

$$p(A) + q(B) + r(O) = 1 \qquad (24.5)$$

and the distribution of genotypes will be given by

$$(p + q + r)^2 \qquad (24.6)$$

In our hypothetical population, the genotypes $AA$, $AB$, $AO$, $BB$, $BO$, and $OO$ will be found in the ratio

$$p^2(AA) + 2pq(AB) + 2pr(AO) + q^2(BB) \\ + 2qr(BO) + r^2(OO) = 1 \quad (24.7)$$

If we know the frequencies of blood types for a population, we can then estimate the frequencies for the three alleles of the ABO system. For example, in one population sampled, the following blood types were observed: A = 0.53, B = 0.13, AB = 0.08, and O = 0.26. Since the $O$ allele is recessive, the frequency of type O blood in the population is equal to the frequency of the recessive genotype $r^2$. Thus,

$$r^2 = 0.26 \\ r = \sqrt{0.26} \\ = 0.51$$

By using the value estimated for $r$, we can estimate the allele frequencies for the $A$ ($p$) and $B$ ($q$) alleles. The $A$ allele is present in two genotypes, $AA$ and $AO$. The frequency of the $AA$ genotype is represented by $p^2$, and the $AO$ genotype by $2pr$. Therefore,

$$\text{Frequency of } A + O = p^2 + 2pr + r^2$$

where $A$ and $O$ represent the phenotypic frequencies. This can be rearranged to give

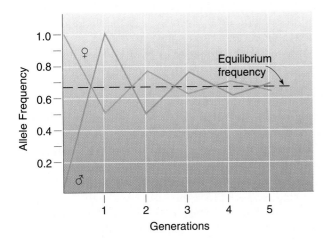

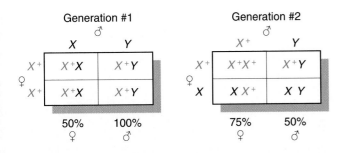

**FIGURE 24.1**    Approach to equilibrium for an X-linked trait with an initial frequency of 1.0. The fluctuations in frequency in the first two generations are shown in the Punnett square. In the parents of generation 1, the X-linked trait ($x^+$) is present in 100 percent of the females and 0 percent of the males. In the progeny, the trait is present in 50 percent of all female X chromosomes and in 100 percent of all male X chromosomes. Thus, as the graph shows, the frequency in females drops from 1.0 to 0.5 in a generation, while the frequency in males rises from 0.0 to 1.0. In generation 2, the frequency in females is 0.75, and in males 0.5. This oscillation continues until eventually an equilibrium is reached.

$$p = \sqrt{\text{Frequency of } A + O} - r$$
$$= \sqrt{0.53 + 0.26} - 0.51$$
$$= 0.89 - 0.51 = 0.38$$

Having estimated the frequencies for $A$ ($p$) and $O$ ($r$), the frequency for the $B$ allele can be estimated from the following relationship:

$$p + q + r = 1$$
$$q = 1 - (p + r)$$
$$= 1 - (0.38 + 0.51)$$
$$= 1 - 0.89$$
$$= 0.11$$

The phenotypic frequencies and allele frequencies for this population are summarized in Table 24.5.

**Table 24.5** CALCULATING GENOTYPIC AND PHENOTYPIC FREQUENCIES FOR MULTIPLE ALLELES WHERE THE FREQUENCY OF $A$($p$) Is 0.38, $B$($q$) Is 0.11, AND $O$($r$) Is 0.51

| Genotype | Genotypic Frequency | Phenotype | Phenotypic Frequency |
|---|---|---|---|
| AA | $p^2 = (0.38)^2 = 0.14$ | | |
| AO | $2pr = 2(0.38 \times 0.51) = 0.39$ | A | 0.53 |
| BB | $q^2 = (0.11)^2 = 0.01$ | | |
| BO | $2qr = 2(0.11 \times 0.51) = 0.11$ | B | 0.12 |
| AB | $2pq = 2(0.38 \times 0.11) = 0.082$ | AB | 0.08 |
| OO | $r^2 = (0.51)^2 = 0.26$ | O | 0.26 |

## Heterozygote Frequency

One of the practical applications of the Hardy–Weinberg law is the estimation of heterozygote frequency in a population. The frequency of a recessive phenotype usually can be determined by counting such individuals in a sample of the population. This information and the Hardy–Weinberg equation can be used to calculate the allele and genotype frequencies.

Albinism, an autosomal recessive trait, has an incidence of about 1/10,000 (0.0001) in some populations. Albinos are easily distinguished from the population at large by a lack of pigment in skin, hair, and iris. Since this is a recessive trait, albino individuals must be homozygous. Their frequency in a population is represented by $q^2$, provided that mating has been at random and all Hardy–Weinberg conditions have been met in the previous generation.

The frequency of the recessive allele therefore is

$$\sqrt{q^2} = \sqrt{0.0001} \qquad (24.8)$$
$$q = 0.01, \text{ or } 1/100$$

Since $p + q = 1$, then the frequency of $p$ is

$$p = 1 - q \qquad (24.9)$$
$$= 1 - 0.01$$
$$= 0.99, \text{ or } 99/100$$

In the Hardy–Weinberg equation, the frequency of heterozygotes is given as $2pq$. Therefore, we have

$$\text{Heterozygote frequency} = 2pq \qquad (24.10)$$
$$= 2[(0.99)(0.01)]$$
$$= 0.02, \text{ or } 2\%, \text{ or } 1/50$$

Thus, heterozygotes for albinism are rather common in the population (2%), even though the incidence of homozygous recessives is only 1/10,000.

In general, the frequencies of all three genotypes can be estimated once the frequency of either allele is known and Hardy–Weinberg conditions are assumed. The relationship between genotype frequency and allele frequency is shown in Figure 24.2. It is important to note how fast heterozygotes increase in a population as the values of $p$ and $q$ move away from zero. This observation confirms our conclusion that when a recessive trait like albinism is rare, the majority of those carrying the allele are heterozygotes. In some cases, where the frequencies of $p$ and $q$ are between 0.33 and 0.67, heterozygotes actually constitute the major class in the population.

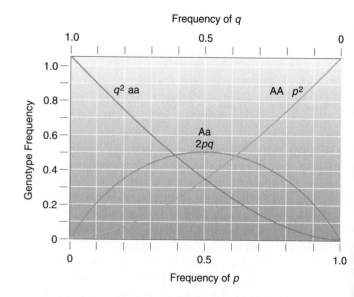

**FIGURE 24.2**    Relationship between genotype frequency and allele frequencies derived from the Hardy–Weinberg equation.

## Demonstrating Equilibrium

The Hardy–Weinberg law can also be used to determine whether a given population is in equilibrium. In order to do this, one must be able to phenotypically identify heterozygotes. If this is possible, then it must be ascertained whether the existing population fits a $p^2 + 2pq + q^2 = 1$ relationship. If not, some factor, presumably selection, mutation, migration, or some other force, is causing allele frequencies to shift with each successive generation.

The MN blood group in one human population, the Australian aborigines, is a good example of this application of the Hardy–Weinberg law. From the values of allele frequencies in Table 24.3 (0.178 for $M$ and 0.822 for $N$), the expected frequencies of blood types M, MN, and N can be calculated to determine if the population is in equilibrium:

$$\text{Expected frequency of type M} = p^2 \qquad (24.11)$$
$$= (0.178)^2 = 0.032 = 3.2\%$$

$$\text{Expected frequency of type MN} = 2pq$$
$$= 2(0.178)(0.822) = 0.292 = 29.2\%$$

$$\text{Expected frequency of type N} = q^2$$
$$= (0.822)^2 = 0.676 = 67.6\%$$

These **expected frequencies** are nearly identical to the **observed frequencies** shown in Table 24.3, confirming that the population is in equilibrium. If there were a question as to whether the observed frequencies varied significantly from the expected frequencies, a chi-square analysis could be performed (see Chapter 3). In this example, the values are so close that we shall accept them as being in agreement.

On the other hand, if the Hardy–Weinberg test demonstrates that the population is not in equilibrium, one or more of these necessary conditions are not being met. Using the values listed for $M$ and $N$ frequencies in Table 24.3, we can construct a hypothetical mixed population of 500 Australian aborigines and 500 American Indians. The frequencies for $M$ and $N$ for this hypothetical population are shown in Table 24.6. In such a mixed population, the frequency of $M$ ($p$) would be

$$0.315 + 1/2(0.324) = 0.477 \qquad (24.12)$$

and that of $N$ ($q$) would be

$$0.361 + 1/2(0.324) = 0.523 \qquad (24.13)$$

If these allelic frequencies were derived by random mating, then the proportions of phenotypes we should expect are MM ($p^2$), 22.8 percent; MN ($2pq$), 49.8 percent; and NN ($q^2$), 27.4 percent. Since the expected genotype frequencies do not fit those observed (and a chi-square test would confirm this), we would conclude that the population is in a state of nonequilibrium brought about, obviously, by a lack of random mating in this hypothetical population.

What would happen, however, if such a mixed population were to be established on an island with no previous inhabitants and mating did take place at random? How long would it take for equilibrium to be established? Applying the Hardy–Weinberg equation, we can predict that after one generation the observed and expected genotype frequencies would converge. Using the $p$ and $q$ values just calculated, and remembering that the allele frequencies represent those of the total gene pool of the first generation mixed population, we can construct the following Punnett square:

| ♀ \ ♂ | $M(p = 0.477)$ | $N(q = 0.523)$ |
|---|---|---|
| $M$<br>($p = 0.477$) | MM<br>$p^2$<br>0.228 | MN<br>$pq$<br>0.249 |
| $N$<br>($q = 0.523$) | MN<br>$pq$<br>0.249 | NN<br>$q^2$<br>0.274 |

We can see that after one generation, the observed genotype frequencies of 22.8 percent for $MM$ ($p^2$); 49.8 percent for $MN$ ($2pq$); and 27.4 percent for $NN$ ($q^2$) are those expected for an equilibrium population. The allele frequencies of $M = 0.477$ and $N = 0.523$ are also those expected for an equilibrium population. However, to rigorously demonstrate HWL equilibrium, we would need to calculate the frequency of matings.

**Table 24.6** FREQUENCIES OF $M$ AND $N$ ALLELES AND GENOTYPES IN NATURAL AND ARTIFICIAL POPULATIONS

| Population | Genotype Frequencies | | | Allele Frequencies | |
|---|---|---|---|---|---|
| | MM | MN | NN | p | q |
| Australian aborigines | 0.03 | 0.296 | 0.674 | 0.178 | 0.822 |
| American Indians | 0.60 | 0.351 | 0.049 | 0.776 | 0.224 |
| Mixed population (500 + 500): | | | | | |
| Observed | 0.315 | 0.324 | 0.361 | 0.477 | 0.523 |
| Expected | 0.228 | 0.498 | 0.274 | | |

# FACTORS THAT ALTER ALLELE FREQUENCIES

The Hardy–Weinberg law allows us to compare allele frequencies and genotype frequencies in populations in which our initial assumptions about random mating, absence of selection and mutation, and equal viability and fecundity hold true. Obviously, it is difficult—if not impossible—to find natural populations in which all these conditions are met. Populations in nature are dynamic, and changes in size and structure are part of their life cycles. This does not mean that the Hardy–Weinberg relationship is invalid; rather, it provides an opportunity to study those forces that introduce and maintain genetic diversity in populations. In this section, we will discuss factors that prevent populations from reaching equilibrium, or drive populations toward a different equilibrium, and the relative contribution of these factors to evolutionary change.

## Mutation

Within a population, the gene pool is reshuffled each generation to produce new combinations in the genotypes of the offspring. Because the number of possible genotypic combinations is so large, the members of a population alive at any given time represent only a fraction of all possible genotypes. Although an enormous genetic reserve is present in the gene pool, new genotypic combinations are produced by Mendelian assortment and recombination, but these processes do not produce any new alleles. **Mutation** alone acts to create new alleles. But in the absence of other forces, mainly selection, mutation is a negligible force in changing allele frequencies.

For our purposes, we need only consider that mutational events occur at random—that is, without regard for any possible benefit or disadvantage to the organism. Here we will discuss only the generation of mutant alleles and in a later section will consider the spread and distribution of new alleles through the population.

To know whether mutation is a significant force in changing allele frequencies, we must measure the rate at which mutations are produced. Since most mutations are recessive, it is difficult to observe mutation rates directly in diploid organisms. Indirect methods using probability and statistics or large-scale screening programs must be employed. For certain dominant mutations, however, a direct method of measurement can be used. To ensure accuracy, several conditions must be met:

1. The trait must produce a distinctive phenotype that can be distinguished from similar traits produced by recessive alleles.

2. The trait must be fully expressed or completely penetrant so that mutant individuals can be identified.

3. An identical phenotype must never be produced by nongenetic agents such as drugs or chemicals.

Mutation rates can be expressed as the number of new mutant alleles per given number of gametes. Suppose for a given gene that undergoes mutation to a dominant allele, 2 out of 100,000 births exhibit a mutant phenotype. In these two cases, the parents are phenotypically normal. Because the zygotes that produced these births each carry two copies of the gene, we have actually surveyed 200,000 copies of the gene (or 200,000 gametes). If we assume that the affected births are each heterozygous, we have uncovered 2 mutant alleles out of 200,000. Thus the mutation rate is 2/200,000 or 1/100,000, which in scientific notation would be written as $1 \times 10^{-5}$.

In humans, a dominant form of dwarfism known as **achondroplasia** fulfills the requirements outlined above for the measurement of mutation rates. Individuals with this skeletal disorder have an enlarged skull, short arms and legs, and can be diagnosed by radiologic examination at birth. In a survey of almost 250,000 births, the mutation rate ($\mu$) for achondroplasia has been calculated as

$$1.4 \times 10^{-5} \pm 0.5 \times 10^{-5} \qquad (24.14)$$

Knowing the rate of mutation, we can then employ the Hardy–Weinberg law to estimate the change in frequency for each generation. If initially only the normal $d$ allele exists (i.e., all individuals are homozygous $dd$), the frequency of $d$ ($q_0$) is 1.0 and the frequency of $D$ (dwarf) is $p_0 = 0$. If the rate of mutation from $d$ to $D$ is $\mu$, then in the next generation,

$$\text{Frequency of } D = p_1 = q_0\mu \qquad (24.15)$$

The new frequency for $d$, which will be lowered by the rate of mutation, can be expressed as

$$\text{Frequency of } d = q_1 = 1 - p_0\mu \qquad (24.16)$$

Although the rate of mutation is constant from generation to generation, the rate of change in the mutant allele is initially high, but declines to zero at mutational equilibrium.

As a more general example, let us assume that the mutation rate for genes in the human genome is $1.0 \times 10^{-5}$. With such a low mutation rate, changes in allele frequency brought about by mutation alone are very small. In fact, Figure 24.3 shows that if $A$ is the only allele of the $a$ locus in the population ($p = 1$), with a mutation rate of $1.0 \times 10^{-5}$ for $A$ to $a$, it would require about

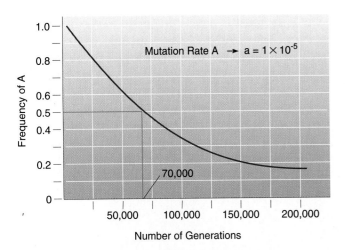

**FIGURE 24.3**    Rate of replacement of an allele by mutation alone, assuming an average mutation rate of $1.0 \times 10^{-5}$.

70,000 generations to reduce the frequency of $A$ to 0.5. Even if the rate of mutation were to increase through exposure to higher levels of radioactivity or chemical mutagens, the impact of mutation on allele frequencies would be extremely weak. This example once again emphasizes that mutation is a major force in generating genetic variability, but by itself has an insignificant role in changing allele frequencies.

## Migration

Frequently, a species of plant or animal becomes divided into subpopulations that to some extent are geographically separated. Differences in mutation rate and selective pressures can establish different allele frequencies in the subpopulations. **Migration** occurs when individuals move between these populations. Consider a single pair of alleles, $A$ ($p$) and $a$ ($q$). The change in the frequency of $A$ in one generation can be expressed as

$$\Delta p = m(p_m - p) \tag{24.17}$$

where

$p$ = frequency of $A$ in existing population
$p_m$ = frequency of $A$ in immigrants
$\Delta p$ = change in one generation
$m$ = coefficient of migration (proportion of migrant genes entering the population per generation)

For example, assume that $p = 0.4$ and $p_m = 0.6$, and that ten percent of the parents giving rise to the next generation are immigrants ($m = 0.1$). Then, the change in the frequency of $A$ in one generation is

$$\begin{aligned} \Delta p &= m(p_m - p) \\ &= 0.1(0.6 - 0.4) \\ &= 0.1(0.2) \\ &= 0.02 \end{aligned}$$

In the next generation, the frequency of $A$ ($p_1$) will increase as follows:

$$\begin{aligned} p_1 &= p + \Delta p \tag{24.18} \\ &= 0.40 + 0.02 \\ &= 0.42 \end{aligned}$$

If either $m$ is large or if $p$ is very different from $p_m$, then a rather large change in the frequency of $A$ will occur in a single generation. All other factors being equal (including migration), an equilibrium will be attained when $p = p_m$. These calculations reveal that the change in allele frequency attributable to migration is proportional to the differences in allele frequency between the donor and recipient populations and to the rate of migration. Since the migration coefficient can have a wide range of values, the effect of migration can substantially alter allele frequencies in populations. Although migration can be somewhat difficult to quantify, it can often be estimated.

Migration can also be regarded as the flow of genes between two populations that were once but are no longer geographically isolated. For example, most of the blacks in the United States are descended from ancestors in West Africa. In Africa, the frequency of the Duffy blood group allele $Fy^0$ is almost 100 percent. In Europe, the source of most U.S. whites, the frequency is close to zero percent, and almost all individuals are $Fy^a$ or $Fy^b$. By measuring the frequency of $Fy^a$ or $Fy^b$ among U.S. blacks, we can estimate the amount of gene flow into the black population. Figure 24.4 shows the frequency of the $Fy^a$ allele among blacks in three African nations and in several regions of the United States. Using the average frequency and assuming an equal rate of gene flow in each generation, we can estimate the migration of the $Fy^a$ allele at about 5 percent per generation.

## Natural Selection

Mutation and migration introduce new alleles into populations. **Natural selection**, on the other hand, is the principal force that shifts allele frequencies within large populations and is one of the most important factors in evolutionary change.

Darwin's main contribution to the study of evolution was his recognition that natural selection is the mechanism that leads to the divergence and eventual separation of populations into distinct species. In any population at a given time, there are individuals with different genotypes. Because of these inherent genetic differences, some of the individuals in this population will be

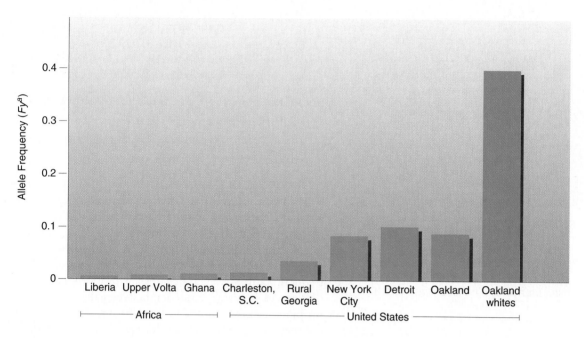

**FIGURE 24.4**    Frequencies of the $Fy^a$ allele in African and U.S. black populations.

better adapted to the environment than others, leading to the differential survival and reproduction of some genotypes over others. Natural selection is thus the differential reproduction of genotypes. Therefore, allelic frequencies will change over time. Natural selection, then, is a major force in changing allelic frequencies and therefore represents a departure from the Hardy–Weinberg assumption that all genotypes have equal viability and fertility.

When a particular genotype/phenotype combination confers an advantage to organisms in competition with others harboring an alternate combination, selection occurs. The relative strength of selection varies with the amount of advantage provided. The probability that a particular phenotype will survive and leave offspring is a measure of its **fitness**. Thus, fitness refers to a total reproductive potential or efficiency. The concept of fitness is usually expressed in relative terms by comparing a particular genotypic/phenotypic combination with one regarded as optimal. Fitness is a relative concept because as environmental conditions change, so does the advantage conferred by a particular genotype.

Mathematically, the difference between the fitness of a given genotype and another regarded as optimal is called the **selection coefficient (s)**. For a phenotype conferred by the genotype $aa$, when only 99 of every 100 organisms successfully reproduce, $s = 0.01$. If the $aa$ genotype is a homozygous lethal and $AA$ and $Aa$ have equal fitness, then $s = 1.0$, and the $a$ allele is propagated only in the heterozygous carrier state. Starting with any

original frequencies of $p_0$ and $q_0$ in a population, when $s = 1.0$, the effect of selection on any successive generation can be calculated using the formula

$$q_n = \frac{q_0}{1 + nq_0} \qquad (24.19)$$

where $n$ is the number of generations elapsing since $p_0$ and $q_0$.

Figure 24.5 demonstrates the change in allele frequencies when $A(p_0) = a(q_0) = 0.5$. Initially, because of the high percentage of $aa$ genotypes, the frequency of the $a$ allele is reduced rapidly. The frequency of $a$ is halved (0.25) in only two generations. By the sixth generation, the frequency of $a$ is again reduced twofold (0.12). By now, however, the majority of $a$ alleles are carried by heterozygotes. Since selection operates on phenotypes and not on components of the genotype, this means that heterozygotes are not selected against. Subsequent reductions in frequency occur very slowly in successive generations and depend on the rate at which heterozygotes are removed from the population. For this reason, it is difficult to eliminate a recessive allele from the population by selection as long as heterozygotes continue to mate.

When $s$ is less than 1.0, it is possible to calculate the effects of selection on each successive generation with any $s$, $p_0$, and $q_0$ values using the formula

$$q_1 = \frac{q_0 - sq_0^2}{1 - sq_0^2} \qquad (24.20)$$

| Generation | p | q | p² | 2pq | q² |
|---|---|---|---|---|---|
| 0 | 0.50 | 0.50 | 0.25 | 0.50 | 0.25 |
| 1 | 0.67 | 0.33 | 0.44 | 0.44 | 0.12 |
| 2 | 0.75 | 0.25 | 0.56 | 0.38 | 0.06 |
| 3 | 0.80 | 0.20 | 0.64 | 0.32 | 0.04 |
| 4 | 0.83 | 0.17 | 0.69 | 0.28 | 0.03 |
| 5 | 0.86 | 0.14 | 0.73 | 0.25 | 0.02 |
| 6 | 0.88 | 0.12 | 0.77 | 0.21 | 0.01 |
| 10 | 0.91 | 0.09 | 0.84 | 0.15 | 0.01 |
| 20 | 0.95 | 0.05 | 0.91 | 0.09 | <0.01 |
| 40 | 0.98 | 0.02 | 0.95 | 0.05 | <0.01 |
| 70 | 0.99 | 0.01 | 0.98 | 0.02 | <0.01 |
| 100 | 0.99 | 0.01 | 0.98 | 0.02 | <0.01 |

**FIGURE 24.5**    Changes in allele frequency under selection where $s = 1.0$. The frequency of $a$ is halved in two generations, and halved again by the sixth generation. Subsequent reductions occur slowly because the majority of $a$ alleles are carried by heterozygotes.

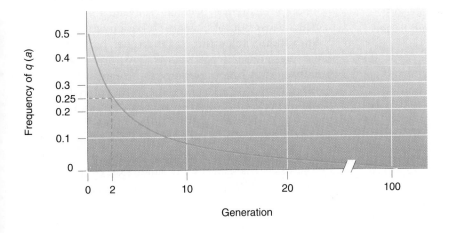

Figure 24.6 demonstrates a variety of initial conditions and the change of the frequency of $q(a)$ through a large number of generations. If $q_0$ is initially 0.5 and $s = 0.5$, five generations must elapse before $q$ is halved. Then, another nine generations must elapse to cut $q$ in half again. Figure 24.6 shows the relative trends as $q_0$ and $s$ change.

There has been much study of selection on both laboratory and natural populations. A classic example of selection in natural populations is that of the peppered moth, *Biston betularia*, in England. Before 1850, 99 percent of the moth population was light colored, allowing the moths to fit well into their surroundings. As toxic gases produced by industry killed the lichens and mosses growing on trees and buildings, and soot deposits darkened the landscape, the light-colored moths became easy targets for their predators. The rare dark-colored moths suddenly gained a great selective advantage because of their natural camouflage. A rapid shift in frequency of this phenotype occurred, probably in 50 generations or less.

Laboratory experiments have demonstrated that the dark form is due to a single dominant allele, $C$. Figure 24.7 demonstrates the protective advantage from predators provided by the dark phenotype. Following the adoption of laws to restrict environmental pollution in 1964, the frequency of nonmelanic forms of the moth has started to increase around the Manchester area, showing the close relationship between the environment and selection.

Because selection acts on the organism's genotypic/phenotypic combination, it also acts on polygenic or

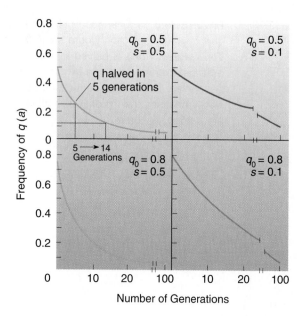

**FIGURE 24.6**    Changes in allele frequencies under different selection coefficients and different initial gene frequencies. When $q_0 = 0.5$ and $s = 0.5$, the frequency of $q$ is halved in 5 generations. When $q_0 = 0.8$ and $s = 0.5$, the frequency of $q$ is halved in 5 generations and halved again in an additional 5 generations.

quantitative traits—those that are controlled by a number of genes and may be susceptible to environmental influences. Such quantitative traits, including adult body height and weight in humans, often demonstrate a continuously varying distribution resembling a bell-shaped curve. Selection for such traits can be classified as (1) directional, (2) stabilizing, or (3) disruptive.

In **directional selection**, important to plant and animal breeders, desirable traits, often representing phenotypic extremes, are selected. In fact, if the trait is polygenic, the most extreme phenotypes that the genotype can express will appear in the population only after prolonged selection. An example is the selection for oil content in domestic corn kernels. In upward selection, C. M. Woodworth and others were able to raise the oil content of corn more than threefold, from about 4 percent to just over 15 percent in 50 generations, with no sign of a plateau being reached (Figure 24.8). Similarly, downward selection reduced the content to about 1 percent. Directional selection favors an extreme phenotype and tends to produce genetic uniformity in the population. In nature, directional selection can occur when one of the phenotypic extremes becomes selected for or against, usually as a result of changes in the environment.

**Stabilizing selection**, on the other hand, tends to favor intermediate types, with both extreme phenotypes being selected against. One of the clearest demonstrations of stabilizing selection is the data of Mary Karn and Sheldon Penrose on human birth weight and survival. Figure 24.9 shows the distribution of birth weight for 13,730 children born over an 11-year period and the percentage of mortality at four weeks of age. Infant mortality increases dramatically on either side of the optimal birth weight of 7.5 pounds. At the genetic level, stabiliz-

**FIGURE 24.7**    The speckled and melanic forms of the peppered moth, *Biston,* seen on a normal, lichen-covered tree trunk (left) and on a darkened, soot-covered trunk (right).

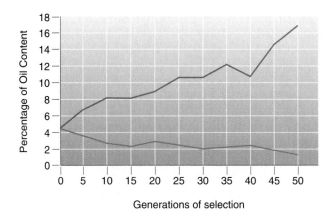

**FIGURE 24.8**    Long-term selection to alter the oil content of corn kernels.

The types and effects of selection are summarized in Figure 24.10.

## Genetic Drift

In laboratory crosses, one condition essential to realizing theoretical genetic ratios (i.e., $3:1$, $1:2:1$, $9:3:3:1$) is a fairly large sample size. This condition is equally important to the study of population genetics as allele and genotype frequencies are examined or predicted. For example, if a population consists of 1000 randomly mating heterozygotes (*Aa*), the next generation will consist of approximately 25 percent *AA*, 50 percent *Aa*, and 25 percent *aa* genotypes. Provided that the initial and sub-

ing selection represents a situation where a population is well adapted to its environment.

**Disruptive selection** is selection against intermediates and for both phenotypic extremes. It may be viewed as the opposite of stabilizing selection. In one set of experiments, John Thoday applied disruptive selection to progeny of *Drosophila* strains with high and low bristle number. Matings were carried out using females from one strain and males from the other. Progeny were selected for high and low bristle number, and matings were carried out for a number of generations, and the two lines diverged rapidly. In natural populations, such a situation may exist for a population in a heterogeneous environment.

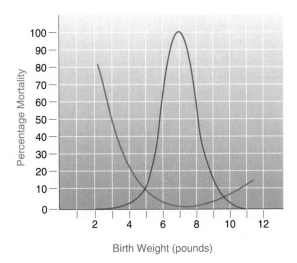

**FIGURE 24.9**    Relationship between birth weight and mortality in humans.

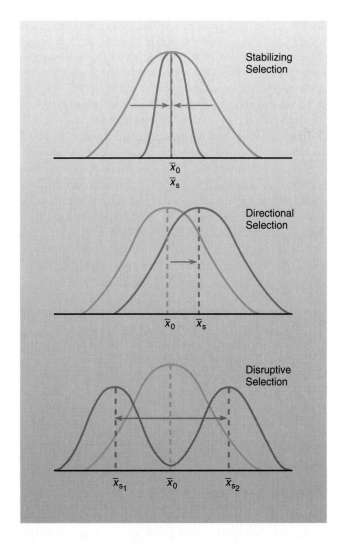

**FIGURE 24.10**    Comparison of the impact of stabilizing, directional, and disruptive selection. In each case, $\bar{x}_0$ is the mean of an original population, and $\bar{x}_s$ is the mean of the population following selection.

sequent populations are large, there will be at most only minor deviations from this mathematical ratio, and the frequency of $A$ and $a$ will remain about equal (0.50).

However, if a population is formed from only one set of heterozygous parents and they produce only two offspring, allele frequency can change drastically. In this case, the probability of the occurrence of genotypes and allele frequencies in the subsequent generation can be predicted as shown in Table 24.7. As shown, in 10 of 16 times such a cross is made, the new allele frequencies would be altered. In 2 of 16 times, either the $A$ or $a$ allele would be eliminated in a single generation.

The conditions used to generate these calculations are extreme, but they illustrate why large interbreeding populations are essential to the Hardy–Weinberg equilibrium. In small populations, significant random fluctuations in allelic frequencies are possible strictly on the basis of chance deviation. The degree of fluctuation will increase as the population size decreases. Such changes illustrate the concept of **genetic drift**. In the extreme case, genetic drift may lead to the chance fixation of one allele to the exclusion of another allele.

To study genetic drift in laboratory populations of *Drosophila melanogaster*, Warwick Kerr and Sewall Wright set up over 100 lines, with four males and four females as the parents for each line. Within each line, the frequency of the sex-linked bristle mutant *forked* ( $f$ ) and its wild-type allele ( $f^+$ ) was 0.5. In each generation, four males and four females were chosen at random to be parents of the next generation. After 16 generations, fixation had occurred in 70 lines—29 in which only the *forked* allele was present and 41 in which the wild-type allele had become fixed. The rest of the lines were still

segregating the two alleles or had been lost. If fixation had occurred randomly, then an equal number of lines should have become fixed for each allele. In fact, the experimental results do not differ statistically from the expected ratio of 35:35, demonstrating that alleles can spread through a population and eliminate other alleles by chance alone.

How are small populations created in nature? In one instance, a large group of organisms may be split by some natural event, creating a small, isolated subpopulation. A natural disaster such as an epidemic might also occur, leaving a small number of survivors to constitute the breeding population. Third, a small group might emigrate from the larger population, as founders, to a new environment, such as a volcanically created island.

Allelic frequencies in certain human isolates best support the role of drift as an evolutionary force in natural populations. The Pingelap atoll in the western Pacific Ocean (lat. 6° N, long. 160° E) has in the past been devastated by typhoons and famine, and in 1775 there were only 30 survivors. Today there are fewer than 2000 inhabitants, and their ancestry can be traced to the typhoon survivors. Approximately 5 percent of the current population is affected by the recessively determined disorder **achromatopsia**, which causes ocular disturbances and a form of color blindness. This disorder is extremely rare in the human population as a whole. However, the responsible allele is present at a relatively high frequency in the Pingelap population. From genealogical reconstruction, it was found that one of the original survivors was a chief who was heterozygous for the condition. If we assume that he was the only carrier in the founding population, the initial gene frequency was 1/60, or 0.016. Since some 5 percent of the current population is affected, they represent homozygous recessives, and the frequency of the gene in the present population is calculated as 0.23.

Another example of genetic drift involves the Dunkers, a small religious isolate living in Pennsylvania who emigrated from the German Rhineland. Since their religious beliefs do not permit marriage with outsiders, the population has grown only through marriage within the group. When the frequencies of the ABO and MN blood group alleles are compared, significant differences are found among the Dunkers, the German population, and the U.S. population. The frequency of blood group A in the Dunkers is about 60 percent, in contrast to 45 percent in the U.S. and German populations. The $I^B$ allele is nearly absent in the Dunkers. Type M blood is found in about 45 percent of the Dunkers, compared with about 30 percent in the U.S. and German populations. Since there is no evidence for a selective advantage of these alleles, it is apparent that their observed frequencies are the result of chance events in a relatively small isolated population.

**Table 24.7**    ALL POSSIBLE PAIRS OF OFFSPRING PRODUCED BY TWO HETEROZYGOUS PARENTS ($Aa \times Aa$)

| Possible Genotypes of Two Offspring | Probability | Allele Frequency | |
|---|---|---|---|
| | | $A$ | $a$ |
| $AA$ and $AA$ | $(1/4)(1/4) = 1/16$ | 1.00 | 0.00 |
| $aa$ and $aa$ | $(1/4)(1/4) = 1/16$ | 0.00 | 1.00 |
| $AA$ and $aa$ | $2(1/4)(1/4) = 2/16$ | 0.50 | 0.50 |
| $Aa$ and $Aa$ | $(2/4)(2/4) = 4/16$ | 0.50 | 0.50 |
| $AA$ and $Aa$ | $2(2/4)(1/4) = 4/16$ | 0.75 | 0.25 |
| $aa$ and $Aa$ | $2(1/4)(2/4) = 4/16$ | 0.25 | 0.75 |

# Inbreeding

One of the assumptions of the Hardy–Weinberg equilibrium is random mating within the population. This is an ideal condition and as such does not often occur in nature. Within a widespread population, individuals tend to mate with those in physical proximity rather than with those separated by some distance. The result is the restriction of gene flow within that population.

Nonrandom mating occurs in many species, including humans. One form of nonrandom mating is called **assortative**. In humans, pair bonds may be established by religious practices, physical characteristics, professional interests, and so forth on a nonrandom basis. In nature, phenotypic similarity plays the same role.

Another form of nonrandom mating is **inbreeding**, where mating occurs between relatives. Inbreeding results in increased homozygosity. To illustrate this concept, we shall consider the most extreme form of inbreeding, **self-fertilization**.

Figure 24.11 shows the results of four generations of self-fertilization starting with a single individual heterozygous for one pair of alleles. By the fourth generation, only about 6 percent of the individuals are still heterozygous. Note that the frequencies of $A$ and $a$ still remain at 50 percent.

In humans, inbreeding (called **consanguineous marriages**) is related to population size, mobility, and social customs governing marriages among relatives. To determine the probability that two alleles at the same locus in an individual are derived from a common ancestor, Sewall Wright devised the **coefficient of inbreeding**. Expressed as $F$, the coefficient can be defined as the probability that two alleles of a given gene in an individual are derived from a common allele in an ancestor. If $F = 1$, all genotypes are homozygous and derived from a common ancestor. If $F = 0$, no alleles present are derived from a common ancestor. Shown in Figure 24.12 are pedigrees of first- and second-cousin marriages. A lethal allele, $a$, is assumed to be present in one heterozygous parent. For each relative, the probability of the presence of this lethal allele is shown. The pedigrees then culminate in the final probability, $F$, that the lethal allele will become homozygous in the offspring of first or second cousins. Since every population carries as its genetic burden a number of heterozygous recessive alleles that are lethal or deleterious in homozygotes, mortality will rise in matings between relatives.

Thus, from an evolutionary standpoint, if a population is split into smaller subpopulations, the effects of inbreeding will become evident. **Inbreeding depression**, an expression of reduced fitness in a population caused by inbreeding, may occur; and because an increase in homozygous individuals results, genetic variability will decrease. Both favorable and unfavorable alleles will become more frequently homozygous. From the standpoint of Darwinian fitness, the adaptive nature of the population will decrease as well.

This knowledge has been applied in the breeding of domesticated plant and animal species. First, an inbreeding program is initiated. As homozygosity increases, some populations will become fixed for favorable alleles and others for unfavorable alleles. By selecting the more viable and vigorous plants or animals, the proportion of individuals carrying desirable traits can be increased.

If members of two favorable inbred lines are mated, hybrid offspring are often more vigorous in desirable traits than is either of the parental lines. This phenomenon has been called **hybrid vigor**. Such an approach was used in the breeding programs established for corn, and crop yields increased tremendously. Unfortunately, the hybrid vigor extends only through the first generation. Many hybrid lines are sterile, and those that are fertile show subsequent declines in yield. Consequently, the hybrids must be regenerated each time by crossing the inbred parental lines.

Hybrid vigor has been explained in two ways. The first hypothesis, the **dominance hypothesis**, incorporates the obvious reversal of inbreeding depression, which inevitably must occur in out-crossing. Consider a cross between two strains of corn with the following genotypes:

$P_1$:  Self-fertilization $Aa$

| | AA | Aa | aa |
|---|---|---|---|
| $F_1$: | 0.250 | 0.500 | 0.250 |
| $F_2$: | 0.375 | 0.250 | 0.375 |
| $F_3$: | 0.437 | 0.125 | 0.437 |
| $F_4$: | 0.468 | 0.063 | 0.468 |
| $F_n$: | $\dfrac{1-\frac{1}{2}n}{2}$ | $\dfrac{1}{2^n}$ | $\dfrac{1-\frac{1}{2}n}{2}$ |

**FIGURE 24.11**    Reduction in heterozygote frequency brought about by self-fertilization. After $n$ generations, the frequencies of the genotypes can be calculated according to the formulas in the bottom row.

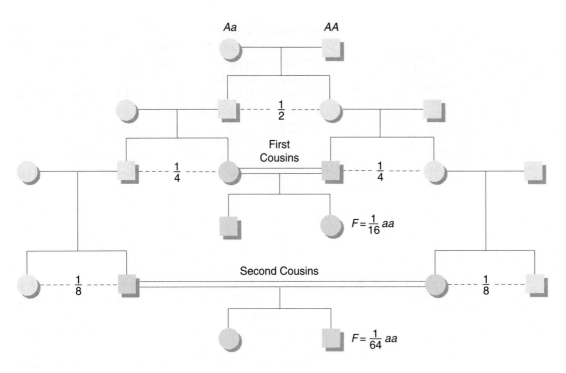

**FIGURE 24.12**    Changes in the coefficient of inbreeding (*F*) brought about by consanguineous matings. Numbers indicate the probability of carrying the recessive allele *a*.

Strain A         Strain B
$$aaBBCCddee \times AAbbccDDEE$$
$$\downarrow$$
$$AaBbCcDdEe$$

The $F_1$ hybrids are heterozygotes at all loci shown. Deleterious recessive alleles present in the homozygous form in the parents will be masked by the more favorable dominant alleles in the hybrids. Such masking is thought to cause hybrid vigor.

The second theory, **overdominance**, holds that in many cases the heterozygote is superior to either homozygote. This may be related to the fact that in the heterozygote, there may be two forms of a gene product present, providing a form of biochemical diversity. Thus, the cumulative effect of heterozygosity at many loci accounts for the hybrid vigor. Most likely, such vigor results from a combination of phenomena explained by both hypotheses.

## CHAPTER SUMMARY

1. The idea that the population rather than the individual is the effective unit in the process of evolution has led to the study of allele frequencies and to the emergence of the discipline of population genetics.

2. The Hardy–Weinberg law established the conditions for genetic equilibrium within a population. These conditions include a large, randomly breeding population, and the absence of selection, mutation, and migration. If any of these conditions are not met, allele or genotype frequencies will change from generation to generation.

3. The Hardy–Weinberg formula is also useful in demonstrating whether or not a population is in equilibrium and in calculating the degree of heterozygosity within a population for any particular locus.

4. The process of evolution requires conditions leading to change, such as changing allele frequencies that provide genetic variation. The mathematical derivation of the Hardy–Weinberg equilibrium has served to assess the forces of evolution: selection, mutation, migration, genetic drift, and inbreeding.

5. Mutation and migration introduce new alleles into a population, but the retention or loss of these alleles depends on the degree of fitness conferred and the action of selection.

6. Natural selection is the most powerful force altering a population's genetic constitution. In small populations, however, inbreeding and genetic drift may be important factors.

## KEY TERMS

achondroplasia
achromatopsia
allele frequency
assortive mating
coefficient of inbreeding
consanguineous
    marriages
directional selection

disruptive selection
dominance hypothesis
fitness
gene pool
genetic drift
genetic equilibrium
genetic variability

Hardy–Weinberg law
  (HWL)
hybrid vigor
inbreeding
inbreeding depression
migration
MN blood group
mutation

natural selection
overdominance
population
population genetics
selection coefficient
self-fertilization
stabilizing selection

## INSIGHTS AND SOLUTIONS

1. Color blindness is caused by a recessive allele on the X chromosome. In a survey, 16 women out of 20,000 were found to be color blind. What is the expected frequency of color-blind males in this population?

   **SOLUTION:**   Remember that for X-linked traits, females exhibit the phenotypic frequencies expected for complete dominance. This means that the frequency of color blindness in women ($16/20{,}000 = 0.0008$) is equal to $q^2$. If $0.0008 = q^2$, then $q = 0.0282$, the value for color blindness in males. The expected frequency in males would be 2.82 percent or about 1 in 34.

2. Eugenics is the term employed for the selective breeding of humans to bring about improvements in the species. As a eugenic measure, it has been suggested (sometimes by force of law) that individuals suffering from serious genetic disorders should be prevented from reproducing (by sterilization, if necessary) in order to reduce the frequency of the disorder in future generations. Suppose that such a trait were present in the population at a frequency of 1 in 40,000, and that affected individuals did not reproduce. In 100 generations, or about 2,500 years, what would be the frequency of the condition? What are your assumptions about this condition? Are the eugenic measures effective in this case?

**SOLUTION:**    Let us assume that the trait is recessive, and that homozygous recessive individuals are prevented from reproducing. In this case, selection against the homozygous recessive phenotype is equal to 1. Using formula 24.19, we can calculate the frequency after 100 generations as follows:

$$q_n = \frac{q_0}{1 + nq_0}$$

$$= \frac{0.000025}{1 + 100(0.000025)}$$

$$= \frac{0.000025}{1.0025}$$

where

$$q_n = 1/41{,}666$$
$$s = 1$$
$$n = 100$$
$$q_0 = 0.000025$$

The frequency of the condition has been reduced from 1 in 40,000 to 1 in 41,666 in 100 generations, indicating that this eugenic measure is ineffective.

---

1. The ability to taste the compound PTC is controlled by a dominant allele $T$, while individuals homozygous for the recessive allele $t$ are unable to taste this compound. In a genetics class of 125 students, 88 were able to taste PTC, 37 could not. Calculate the frequency of the $T$ and $t$ alleles in this population, and the frequency of the genotypes.

2. Calculate the frequencies of the $AA$, $Aa$, and $aa$ genotypes after one generation if the initial population consists of 0.2 $AA$, 0.6 $Aa$, and 0.2 $aa$ genotypes. What frequencies will occur after a second generation?

3. Consider rare disorders in a population caused by an autosomal recessive mutation. From the following frequencies of the disorder in the population, calculate the percentage of heterozygous carriers:
   (a) 0.0064
   (b) 0.000081
   (c) 0.09
   (d) 0.01
   (e) 0.10

4. What must be assumed in order to validate the answers in Problem 3?

5. In a population where only the total number of individuals with the dominant phenotype is known, how can you calculate the percentage of carriers and homozygous recessives?

6. For the following two sets of data, determine whether they represent populations that are in equilibrium. Use chi-square analysis if necessary.
   (a) MN blood groups: *MM*, 60%; *MN*, 35.1%; *NN*, 4.9%
   (b) Sickle-cell hemoglobin: *AA*, 75.6%; *AS*, 24.2%; *SS*, 0.2%

7. If 4 percent of the population in equilibrium expresses a recessive trait, what is the probability that the offspring of two individuals who do not express the trait will express it?

8. Consider the allele frequencies $p = 0.7$ and $q = 0.3$ where selection occurs against the homozygous recessive genotype. What will be the allele frequencies after one generation if
   (a) $s = 1.0$?
   (b) $s = 0.5$?
   (c) $s = 0.1$?
   (d) $s = 0.01$?

9. If initial allele frequencies are $p = 0.5$ and $q = 0.5$, and $s = 1.0$, what will be the frequencies after 1, 5, 10, 25, 100, and 1000 generations?

10. Determine the frequency of allele $A$ after one generation under the following conditions of migration:
    (a) $p = 0.6$; $p_m = 0.1$; $m = 0.2$
    (b) $p = 0.2$; $p_m = 0.7$; $m = 0.3$
    (c) $p = 0.1$; $p_m = 0.2$; $m = 0.1$

11. Assume that a recessive autosomal disorder occurs in one of 10,000 individuals (0.0001) in the general population. Assume that in this population about 2 percent (0.02) of the individuals are carriers for the disorder. Estimate the probability of this disorder occurring in the offspring of a marriage between first cousins and between second cousins. Compare these probabilities to the population at large.

12. What is the basis of inbreeding depression?

13. Does inbreeding cause an increase in frequency of recessive alleles within a population?

14. Describe how inbreeding may be used in the domestication of plants and animals. Discuss the theories underlying these techniques.

15. Evaluate the following statement: inbreeding increases the frequency of recessive alleles in a population.

16. In a breeding program to improve crop plants, which of the following mating systems should be employed to produce a homozygous line in the shortest possible time?
    (a) self-fertilization
    (b) brother–sister matings
    (c) first-cousin matings
    (d) random matings
    Illustrate your choice with pedigree diagrams.

**17.** If the recessive trait albinism ($a$) is present in 1/10,000 individuals, calculate the frequency of
(a) The recessive mutant allele.
(b) The normal dominant allele.
(c) Heterozygotes in the population.
(d) Matings between heterozygotes.

**18.** One of the first Mendelian traits identified in man was a dominant condition known as *brachydactyly*. This gene causes an abnormal shortening of the fingers and/or toes. At the time, it was thought by some that this dominant trait would spread until 75 percent of the population would be affected (since the phenotypic ratio of dominant to recessive is 3:1). Show how this line of reasoning is incorrect.

**19.** Achondroplasia is a dominant trait that causes a characteristic form of dwarfism. In a survey of 50,000 births, five infants with achondroplasia were identified. Three of the affected infants had affected parents, while the rest had normal parents. Calculate the mutation rate for achondroplasia and express the rate as the number of mutant genes per given number of gametes.

**SELECTED READINGS**

BARTON, N. H., and TURELLI, M. 1989. Evolutionary quantitative genetics: How little do we know? *Ann. Rev. Genet.* 23: 337–70.

BODMER, W. F., and CAVALLI-SFORZA, L. L. 1976. *Genetics, evolution and man.* New York: W. H. Freeman.

CAVALLI-SFORZA, L. L., and BODMER, W. F. 1971. *The genetics of human populations.* New York: W. H. Freeman.

COOK, L. M., ASKEW, R. R., and BISHOP, J. A. 1970. Increasing frequency of the typical form of the peppered moth in Manchester. *Nature* 227: 1155.

CROW, J. F. 1986. *Basic concepts in population, quantitative and evolutionary genetics.* New York: W. H. Freeman.

CROW, J. F., and KIMURA, M. 1970. *An introduction to population genetic theory.* New York: Harper and Row.

DAVID, J. R., and CAPY, P. 1988. Genetic variation of *Drosophila melanogaster* natural populations. *Trends Genet.* 4: 106–11.

DOBZHANSKY, T. 1955. A review of some fundamental concepts and problems of population genetics. *Cold Spr. Harb. Symp.* 20: 1–15.

EDWARDS, J. H. 1989. Familarity, recessivity and germline mosaicism. *Ann. Hum. Genet.* 53: 33–47.

FISHER, R. A. 1930. *The genetical theory of natural selection.* Oxford, England: Clarendon Press. (Reprinted in 1958 by Dover Press.)

FREIRE-MAIA, N. 1990. Five landmarks in inbreeding studies. *Am. J. Med. Genet.* 35: 118–20.

GAYLE, J. S. 1990. *Theoretical population genetics.* Boston: Unwin–Hyman.

HARTL, D. L. 1988. *A primer of population genetics.* 2nd ed. Sunderland, MA: Sinauer Associates.

JONES, J. S. 1981. How different are human races? *Nature* 293: 188–90.

KARN, M. N., and PENROSE, L. S. 1951. Birth weight and gestation time in relation to maternal age, parity and infant survival. *Ann. Eugen.* 16: 147–64.

KERR, W. E., and WRIGHT, S. 1954. Experimental studies of the distribution of gene frequencies in very small populations of *Drosophila melanogaster*. I. *Forked. Evolution* 8: 172–77.

KETTLEWELL, H. B. D. 1961. The phenomenon of industrial melanism in *Lepidoptera. Ann. Rev. Entomol.* 6: 245–62.

———. 1973. *The evolution of melanism: The study of a recurring necessity, with special reference to industrial melanism in the Lepidoptera.* London: Oxford University Press.

KHOURY, M. J., and FLANDERS, W. D. 1989. On the measurement of susceptibility to genetic factors. *Genet. Epidemiol.* 6: 699–701.

LI, C. C. 1977. *First course in population genetics.* Pacific Groves, CA: Boxwood Press.

METTLER, L. E., GREGG, T., and SCHAFFER, H. E. 1988. *Population genetics and evolution.* 2nd ed. Englewood Cliffs, NJ: Prentice-Hall.

NEEL, J. V., and ROTHMAN, E. 1981. Is there a difference among human populations in the rate with which mutation produces electrophoretic variants? *Proc. Natl. Acad. Sci.* 78: 3108–12.

REED, T. E. 1969. Caucasian genes in American Negroes. *Science* 165: 762–68.

ROBERTS, D. F. 1988. Migration and genetic change. *Hum. Biol.* 60: 521–39.

SPIESS, E. B. 1989. *Genes in populations.* 2nd ed. New York: Wiley.

WALLACE, B. 1981. *Basic population genetics.* New York: Columbia University Press.

———. 1989. One selectionist's perspective. *Quart. Rev. Biol.* 64: 127–45.

WOODWORTH, C. M., LENG, E. R., and JUGENHEIMER, R. W. 1952. Fifty generations of selection for protein and oil in corn. *Agron. J.* 44: 60–66.

# 25

# EVOLUTIONARY GENETICS

Lady-bird beetles capping a daisy in the Chiracahua Mountains in Arizona.

*Evolution is concerned with the effects of allele frequency changes within the gene pool of a population. Changes in frequency lead to changes in genetic diversity and the ability to undergo evolutionary divergence. The process of speciation divides an originally homogeneous gene pool into two or more reproductively separate gene pools. This division may be accompanied by changes in morphology, physiology, and adaptation to the environment.*

In Chapter 24, we described populations in terms of allele frequencies and genetic equilibria and outlined the forces that alter the genetic constitution of populations. Mutation, migration, selection, and drift are some of the forces that individually and collectively alter gene frequencies and lead to evolutionary divergence and species formation. This process depends not only upon genetic divergence but also upon environmental or ecological diversity. If a population is spread over a geographic range encompassing a number of subenvironments or **niches**, the subpopulations occupying these niches will adapt and become genetically differentiated. This process may lead to the formation of **races** or subspecies. Because the formation and maintenance of these races depend ultimately on the interaction of the genotype and environment, races are dynamic entities that may remain in existence, become extinct, merge with the parental population, or continue to diverge from the parental population until they become reproductively isolated and form new species.

It is difficult to define the exact moment of transition between a subspecies and species. A **species** can be defined as one or more groups of interbreeding or potentially interbreeding organisms that are reproductively isolated in nature from all other organisms. The process of **speciation** depends on changing allele frequencies and involves the division of a homogeneous population (a single gene pool) into two or more reproductively isolated subunits (separate gene pools). Changes in morphology, physiology, and adaptation to an ecological niche may also occur, but are not necessary components of the speciation event. Speciation can take place gradually over a long time period or within a generation or two. Throughout this process, the degree of genetic diversity is the key factor. Individuals of a species are members of a common gene pool that distinguishes them from members of other species. Genetic divergence should be the starting point for a discussion of speciation and evolution.

In this chapter, we will focus on three central topics: first, the extent of genetic diversity in closely related organisms and its potential contribution to speciation; second, analysis of the evolution of diverse organisms at the molecular level; and third, the role of mutation in evolution, a matter of current controversy.

## GENETIC DIVERSITY

If a population of organisms constituting a species is well suited to inhabit and reproduce successfully in its environment, how much genetic diversity does it possess? We might assume that members of a population are fit because the most favorable allele at each locus has become fixed. Certainly, an examination of most populations of plants and animals reveals them to be phenotypically similar. However, current evidence supports the opinion that a high degree of heterozygosity is maintained within the genome of diploid individuals in a population. This built-in diversity is concealed, so to speak, because it is not necessarily apparent in the phenotype. Such diversity may better adapt a population to the inevitable changes in the environment and may promote the survival of the species.

The detection of this concealed genetic variation is not a simple task. Nevertheless, such investigation has been successful using several techniques.

### Inbreeding Depression

One method of measuring genetic diversity is to determine what fraction of the alleles carried by individuals of a population are deleterious. Deleterious alleles in a population can be detected by monitoring increases in homozygosity that result from inbreeding. Because inbreeding involves mating between relatives, mates have genotypic similarities that extend across all segregating

**Table 25.1**  PERCENTAGE OF HOMOZYGOUS CHROMOSOMES TESTED THAT REVEAL RECESSIVE MODIFIERS OF VIABILITY AND FERTILITY

| Drosophila Species | Chromosome | Lethals and Sublethals | Subvitals | Sterility | Female Sterility | Male Sterility |
|---|---|---|---|---|---|---|
| D. pseudoobscura | 2 | 33.0 | 62.6 | <0.1 | 10.6 | 8.3 |
| (Mather, Calif.) | 3 | 25.0 | 58.7 | <0.1 | 13.6 | 10.5 |
|  | 4 | 22.7 | 51.8 | <0.1 | 4.3 | 11.8 |
| D. persimilis | 2 | 25.5 | 49.8 | 0.2 | 18.3 | 13.2 |
| (Mather, Calif.) | 3 | 22.7 | 61.7 | 2.1 | 14.3 | 15.7 |
|  | 4 | 28.1 | 70.7 | 0.3 | 18.3 | 8.4 |
| D. prosaltans | 2 | 32.6 | 33.4 | <0.1 | 9.2 | 11.0 |
| (Brazil) | 3 | 9.5 | 14.5 | 3.0 | 6.6 | 4.2 |
| D. willistoni | 2 | 38.8 | 57.5 | <0.1 | 40.5 | 64.8 |
| (Venezuela, British Guiana, and Trinidad) | 3 | 34.7 | 47.1 | 1.0 | 40.5 | 66.7 |

NOTE: Lethal alleles cause premature death, often before sexual maturity and reproduction; subvital mutations kill less than 50 percent of carriers before maturation; supervital mutations increase viability above the wild-type level.
SOURCE: From Spiess, 1977, p. 288.

loci. As a result of inbreeding, recessive alleles previously present in the heterozygous condition become homozygous, and if these alleles are deleterious, a reduction in viability called **inbreeding depression** is evident in successive generations.

In using inbreeding depression to study genetic diversity, Theodosius Dobzhansky and his colleagues succeeded in making groups of alleles and sometimes entire chromosomes of *Drosophila* species completely homozygous. When this is accomplished, viability is depressed. As shown in Table 25.1, such studies reveal that lethal, semilethal, subvital, and male and female sterility alleles are prominent in natural populations of different species. In *Drosophila pseudoobscura,* almost 100 percent of chromosomes 2 and 3 from a sample of the population cause a depression in viability when made homozygous. In *Drosophila willistoni,* almost 70 percent of chromosomes 2 and 3 contain male sterile alleles when made homozygous.

Although these and other studies reveal the substantial genetic burden carried by organisms, we discuss them here as examples of genetic variation maintained in natural populations. If detrimental alleles are maintained in a heterozygous state, certainly alleles that are effectively neutral or potentially advantageous must also be retained in the heterozygous condition. As recessive alleles, they represent concealed genetic variation within the "normal" heterozygous genotypes characterizing the population.

## Protein Polymorphisms

The use of Mendelian genetics to estimate genetic diversity is both laborious and slow. Gel electrophoresis (see Appendix A), on the other hand, is a simpler technique that can be used to separate protein molecules on the basis of differences in electrical charge (Figure 25.1). If variation in a structural gene results in the substitution of a charged amino acid such as glutamic acid for an uncharged amino acid such as glycine, the net charge on the protein will be altered. This change in charge can be detected by electrophoresis. In the mid-1960s, John Hubby and Richard Lewontin introduced the use of gel electrophoresis to measure protein variation in natural populations of *Drosophila*. Since then, genetic variation has been studied using gel electrophoresis in a wide range of organisms.

When two alleles from one locus produce slightly different electrophoretic forms, yet perform identical functions, the resultant proteins are designated as **allozymes**. As shown in Table 25.2, a surprisingly large percentage of loci examined from diverse species produce allozymes. Of the populations shown in Table 25.2, approximately 30 loci per species were examined, and about 30 percent of the loci were polymorphic (i.e., they revealed allozymes). An approximate average of 10 percent heterozygosity per diploid genome was revealed.

These values apply only to genetic variation detectable by altered protein migration in an electric field.

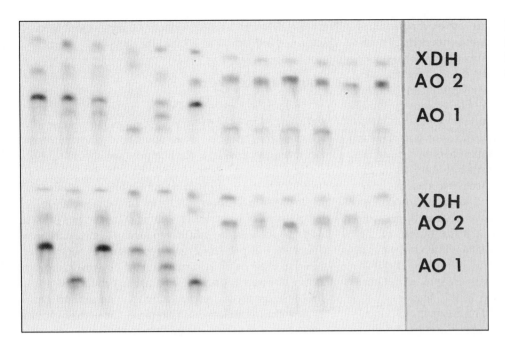

**FIGURE 25.1**   Photograph of gel electrophoretic separation of isozymes of the enzyme alcohol dehydrogenase.

Electrophoresis probably detects only about 30 percent of the actual variation that is due to amino acid substitutions because many substitutions do not change the net electric charge on the molecule. Richard Lewontin has thus estimated that about two-thirds of all loci are polymorphic in a population, and that in any individual within that group about one-third of the loci exhibit genetic variation in the form of heterozygosity.

There is some controversy over the significance of genetic variation as detected by electrophoresis. Some argue that the allozymes are biochemically equivalent and therefore do not play any role in evolution. We shall address this argument later in this chapter.

## Chromosomal Polymorphisms

While the chromosome number is usually invariant for a given species, the arrangement of chromosomal material is often polymorphic as a result of chromosomal inversions and translocations.

**Inversions** are produced by two breaks in a chromosome followed by rejoining of the broken fragment rotated 180° (see Chapter 6). The presence of an inversion does not imply that any new genes are present, nor does the presence of a variety of specific inversions prove that allele frequencies vary between any two arrangements. However, we learned in Chapter 6 that inversion hetero-

| **Table 25.2** | Heterozygosity at the Molecular Level | | | |
|---|---|---|---|---|
| Species | Number of Populations Studied | Number of Loci Examined | Loci per Population | Heterozygosity per Locus |
| *Homo sapiens* (humans) | 1 | 71 | 28 | 6.7 |
| *Mus musculus* (house mouse) | 4 | 41 | 29 | 9.1 |
| *Drosophila pseudoobscura* (fruit fly) | 10 | 24 | 43 | 12.8 |
| *Limulus polyphemus* (horseshoe crab) | 4 | 25 | 25 | 6.1 |

Source: From Lewontin, 1974, p. 117.

zygotes only rarely produce viable crossover products involving genes contained within the inversion. Therefore, an inversion suppresses the recovery of crossover gametes. In so doing, inversions tend to preserve specific allele arrangements along the inverted stretch of the chromosome involved. One effect of inversions in populations, therefore, can be a reduction in genetic variability. In addition, inversions reduce heterozygote fitness because of the production of inviable crossover gametes. Therefore, if inversions are present in a species, it implies a causal relationship with selective forces that favor inversion heterozygotes. In other words, perhaps the inversion has preserved a favorable gene arrangement, which at least partially accounts for the fitness of the population in a specific habitat. Linkage then becomes a possible criterion for selection. The role of inversions in natural populations has been intensively studied in *Drosophila* and will be discussed later in this section.

**Translocations**, in which chromosome segments move to other, nonhomologous chromosomes, are also found in populations. In simple translocations, a segment of one chromosome becomes joined to a nonhomologous chromosome. In reciprocal translocations, chromosome parts are exchanged between two nonhomologous chromosomes. Translocations can also reduce genetic variability in populations because translocation heterozygotes are usually subfertile, producing 50 percent chromosomally abnormal gametes. Although somewhat rare in animal populations, translocation heterozygotes are found in many plants. In a population of the flowering plant *Clarkia elegans,* Harlan Lewis found that 13 percent of the individuals carried a translocation. In the extreme case, the angiosperm *Rhoeo discolor* has all of its chromosomes in translocation heterozygotes.

### DNA Sequence Polymorphisms

The most direct way to estimate genetic variation is by examination of the actual nucleotide sequence diversity between individuals of a population. With the development of techniques for cloning and sequencing DNA, nucleotide sequence variations have been cataloged for an increasing number of gene systems. Using restriction endonucleases to detect polymorphisms (see Chapter 12 for the methodology), Alec Jeffrey surveyed 60 unrelated individuals to estimate the total number of DNA sequence variants in humans. His results show that within the genes of the beta-globin cluster, 1 in 100 base pairs shows polymorphic variation. If this region is representative of the genome, this indicates that at least $3 \times 10^7$ nucleotide variants per genome are possible. Data from other organisms such as *Drosophila,* the rat, and the mouse have produced similar estimates of nucleotide diversity, indicating that there is an enormous reservoir of genetic variability within a population, and that at the level of DNA, most and perhaps all genes exhibit diversity from individual to individual.

### The Adaptive Norm

A population of interbreeding organisms is more or less adapted to its surrounding environment. Those phenotypes that enhance adaptation to a given environment, termed adaptive phenotypes, are a consequence of the array of genotypes possessed by the individuals constituting the group. These genotypes are present as a result of the evolutionary history of the population.

Dobzhansky has called the array of genotypes present in the population the **adaptive norm**. Ideally, each individual should possess a genotype and phenotype best suited to the immediate environment. However, the preceding discussion on genetic variability in natural populations suggests that there is a wide variation in genotypes within populations. Therefore, the adaptive norm tends to be one of balanced heterozygosity. Because numerous recessive alleles are concealed and do not alter the phenotype, it follows that a variety of genotypes can be well adapted to the same environment. As Dobzhansky has pointed out, the variety making up the adaptive norm constitutes a set of genotypes that must be able to meet environmental diversity and stresses over space and time.

Alleles that at one point in time do not seem to be of major significance to individual fitness may be of great value to the population in future generations, as shown by the melanic forms of the peppered moth, discussed in Chapter 24. Under changing environmental conditions, previously insignificant or even detrimental alleles may become essential to the maintenance of fitness. The concept of **preadaptation** describes the so-called hidden or concealed genetic variation as a storehouse of genetic information available to enhance survival under new environmental conditions.

## SPECIATION

In the classical sense, the process of splitting a genetically homogeneous population into two or more populations that undergo genetic differentiation and eventual reproductive isolation is called **speciation**. According to Ernst Mayr, species originate in two predominant ways. In the first mode, often called **phyletic evolution** or **anagenesis**, species A over a long period of time becomes, through genetic change, transformed into species B. In the second method, one species gives rise to one or more derived species, bringing about multiplication of species or **cladogenesis**. This second process can occur over a long period of time or rarely, in a gener-

## Table 25.3   Modes of Speciation

I. Transformation of Species (phyletic evolution)
  1. Autogenous speciation
II. Reduction in Number of Species (fusion of two species)
III. Multiplication of Species (true species)
  A. Instant speciation (through individuals)
    1. Genetic
      (a) Single mutation in asexual species
      (b) Macrogenesis
    2. Cytological
      (a) Chromosomal mutation (translocations, etc.)
      (b) Autopolyploidy
      (c) Amphidiploidy
  B. Gradual speciation (through populations)
    1. Sympatric speciation
    2. Semigeographic speciation
    3. Geographic speciation (allopatric)

Source: From Mayr, 1963, Table 15.1.

ation or two. Table 25.3 summarizes the principal methods of speciation.

The most developed classical model of speciation is **geographic** or **allopatric speciation**, first proposed by Moritz Wagner in 1868. According to Wagner, physical isolation of populations by geographic features such as lakes, rivers, or mountains that act as barriers to gene flow is the first step toward species formation. In a second step, these isolated populations undergo independent genetic changes and may diverge to produce two distinct species.

Twentieth-century workers such as Mayr and Dobzhansky have refined and updated this model but have retained the general features of Wagner's hypothesis. In the refined model, the first step requires that populations become separated; that is, gene flow must be interrupted. The absence or interruption of gene flow is a prerequisite for the development of genetic differences brought about by adaptation to local conditions. Genetic diversity arising by natural selection, or by random genetic drift, will be reflected in the presence of new alleles, changes in allele frequency, or the presence of new chromosomal arrangements. Eventually, a point will be reached when the populations have enough genetic differences that they can be identified as distinct races or semispecies. This process may continue uninterrupted until two or more species are present.

If at any time during the process of genetic divergence, the conditions that prevent gene flow between the populations are removed, two outcomes are possible: (1) the two populations may fuse into a single gene pool, because hybridization does not necessarily reduce fertility or viability; or (2) the gene pools of the populations may have diverged to the point where biological isolating mechanisms may have arisen.

The various biological and behavioral properties of organisms that act to prevent or reduce interbreeding are called **reproductive isolating mechanisms**. These mechanisms are classified in Table 25.4. For example, genetic divergence may have reached the stage where the viability or fertility of hybrids is reduced. Hybrid zygotes may be formed, but all or most may be inviable. Alternatively, the hybrids may be viable but have reduced fertility or be sterile. In another possibility, the hybrids themselves may be fertile, but their progeny may have lowered viability or fertility. These mechanisms, called **postzygotic**, act at or beyond the level of the zygote and are a byproduct of genetic divergence. Such

## Table 25.4   Reproductive Isolating Mechanisms

**Prezygotic Mechanisms** (prevent fertilization and zygote formation)
  1. Geographic or ecological: The populations live in the same regions but occupy different habitats.
  2. Seasonal or temporal: The populations live in the same regions but are sexually mature at different times.
  3. Behavioral (only in animals): The populations are isolated by different and incompatible behavior before mating.
  4. Mechanical: Cross-fertilization is prevented or restricted by differences in reproductive structures (genitalia in animals, flowers in plants).
  5. Physiological: Gametes fail to survive in alien reproductive tracts.

**Postzygotic Mechanisms** (fertilization takes place and hybrid zygotes are formed, but these are nonviable or give rise to weak or sterile hybrids)
  1. Hybrid nonviability or weakness.
  2. Developmental hybrid sterility: Hybrids are sterile because gonads develop abnormally or meiosis breaks down before completion.
  3. Segregational hybrid sterility: Hybrids are sterile because of abnormal segregation to the gametes of whole chromosomes, chromosome segments, or combinations of genes.
  4. $F_2$ breakdown: $F_1$ hybrids are normal, vigorous, and fertile, but $F_2$ contains many weak or sterile individuals.

Source: From G. Ledyard Stebbins, *Processes of Organic Evolution*, 3rd edition, © 1977, p. 143. Reprinted by permission of Prentice-Hall, Inc., Englewood Cliffs, N.J.

isolating mechanisms waste gametes and zygotes and lower the reproductive fitness of hybrid survivors. Selection will therefore favor the spread of alleles that will reduce the formation of hybrids, leading to the development of **prezygotic isolating mechanisms**. However, speciation and the development of isolating mechanisms can also occur in the absence of selection, as when populations remain permanently isolated on islands. Selection accelerates speciation in cases where some reproductive isolation has resulted from genetic diversity, but not all isolating mechanisms are represented in each speciation event. During speciation, usually at least two isolating mechanisms are developed by natural selection drawing on the genetic variability present in the evolving populations.

## Formation of Races and Species

Although the theory of geographic speciation is well developed, it has been more difficult to observe directly the processes involved. Diversification of isolated populations occurs gradually over thousands or hundreds of thousands of years. In addition, the geographic changes that paralleled this divergence may be complex, or even completely unknown. In most cases, therefore, the formation of species is an historical event, and biologists studying this process must rely on the present-day distribution of races, subspecies, and sibling species to reconstruct stages in the evolutionary process.

To study speciation, therefore, biologists must first find examples in nature where all or most of the stages of race formation and speciation can be documented. The intensive studies carried out on natural populations of

*Drosophila* provide good examples of the stages involved in **geographic** or **allopatric** speciation.

To illustrate the first step, the formation of races, we shall consider studies on *Drosophila pseudoobscura* conducted by Dobzhansky and his colleagues. This species is found over a wide range of environmental habitats, including the western and southwestern United States. Although the flies throughout this range are morphologically similar, Dobzhansky discovered that populations from different locations vary substantially in the arrangement of genes on chromosome 3. He found a variety of different inversions in this chromosome detected by loop formations in the salivary chromosomes. Each particular inversion sequence was named after the locale in which it was first discovered (e.g., AR = Arrowhead, British Columbia; CH = Chiricahua Mountains, etc.). The inversion sequences were compared with one standard sequence, designated ST.

Figure 25.2 shows a comparison of three arrangements detected in populations found at three different elevations in the Sierra Nevada in California's Yosemite region. The ST arrangement is most common at low elevations but declines in frequency as elevation increases. At 8000 feet, AR is the most common and ST least common. In these populations, the frequency of the CH arrangement gradually increases with elevation. The gradual change in inversion frequencies is probably the result of natural selection and thus parallels the gradual environmental changes occurring at ascending elevations. Since the populations along this gradient show a continuous and gradual change in inversion frequencies, it is difficult to classify a fly as a member of one racial group.

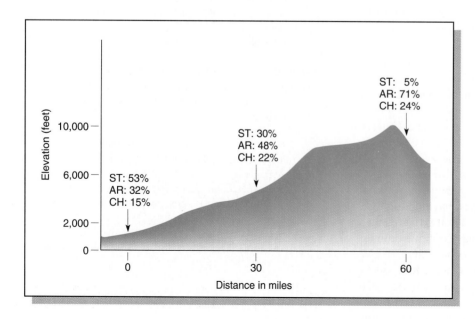

**FIGURE 25.2** Inversions in chromosome 3 of *D. pseudoobscura* found at different elevations in the Sierra Nevada near Yosemite National Park.

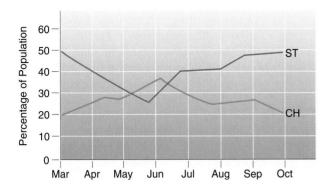

FIGURE 25.3    Changes in the ST and CH arrangements in *D. pseudoobscura* throughout the year.

Dobzhansky also found that if populations were collected at a single site throughout the year, inversion frequencies also changed. During the different seasons, cyclic variation in chromosome arrangements occurred, as shown in Figure 25.3. Such variation was consistently observed over a period of several years. The frequency of ST always declined during the spring, with a concomitant increase in CH during the same period.

To test the hypothesis that this cyclic change is a response to natural selection, Dobzhansky devised the following laboratory experiment: he constructed a large population cage from which samples could be periodically removed and studied. He began with a population of a known inversion frequency, 88 percent CH and 12 percent ST. He reared it at 25°C and sampled it over a one-year period. As shown in Figure 25.4, the frequency of ST increased gradually until it was present at a level of 70 percent. At this point, an equilibrium between ST and

CH was reached. When the same experiment was performed at 16°C, no change in inversion frequency occurred. It can be concluded that the equilibrium reached at 25°C was in response to the elevated temperature, the only variable in the experiment.

The results of Dobzhansky's experiment are strong evidence that the presence of a balanced condition of the two inversions and their respective gene arrangements in a population is superior to either inversion by itself. The equilibrium attained presumably represents the greatest degree of fitness in the population under controlled laboratory conditions. This interpretation of the experiment suggests that natural selection is the driving force toward equilibrium.

In a more extensive study, Dobzhansky and his colleagues sampled populations over a much broader geographic range. Twenty-two different gene arrangements were found in populations from 12 locations. In Figure 25.5, the frequencies of five inversions are shown according to geographic location. The differences are largely quantitative, with most populations differing only in relative frequencies of inversions. However, in some cases, there are qualitative differences. For example, PP is absent in all four California locales where ST is pre-

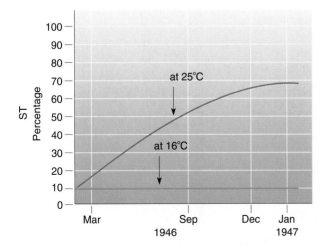

FIGURE 25.4    Increase in the ST arrangement of *D. pseudoobscura* in population cages under laboratory conditions.

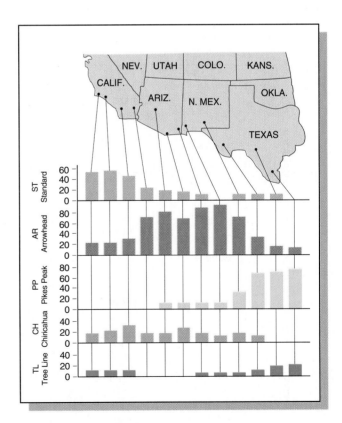

FIGURE 25.5    Frequencies of five chromosomal inversions in *D. pseudoobscura* in different geographic regions.

dominant. In the Texas locations, ST is absent or present in very low frequency and PP is predominant. Some general trends are also apparent. ST increases in frequency from east to west, while just the opposite is true of PP. AR is least common in the east-west extremes, yet predominant in Arizona and New Mexico.

Because these locations represent varied environments and because inversions preserve different gene combinations, we can conclude that numerous races of *D. pseudoobscura* have been formed. Not only do the studies of *D. pseudoobscura* illustrate the principles of the initial step in speciation, but they also strongly support the concept that the adaptive norm consists of balanced genotypes within a population.

For speciation to occur, the development of races must be followed by a second step, reproductive isolation. We might then ask whether the evolution of *D. pseudoobscura* has gone beyond the formation of races. Dobzhansky investigated this question by examining the chromosome structure of other closely related species called **sibling species**. Sibling species are reproductively isolated from one another, but remain very similar morphologically.

One sibling species, *Drosophila persimilis,* has provided very interesting information. *D. persimilis* and *pseudoobscura* each have five pairs of chromosomes

and carry inversions within chromosome 3. When Dobzhansky examined the chromosome 3 inversions carefully, 11 such arrangements were found. One, ST, is also found in *D. pseudoobscura*. Based on the common arrangement, it was possible to construct a phylogenetic sequence for all arrangements found in both species.

As shown in Figure 25.6, ST is shared by both species. Only one hypothetical arrangement is necessary to complete the continuity of the tree. It appears that an ancestral population with the ST arrangement gave rise to many different inversions. Some were incorporated into races as members of the *pseudoobscura* species, and others gave rise to the *persimilis* species.

Today, even when the geographic distributions of these sibling species overlap, several isolating mechanisms keep them from interbreeding. They are isolated by prezygotic mechanisms such as habitat selection, with *persimilis* preferring high elevations and cooler temperatures. Differences in courtship rituals allow females to distinguish between males of the two species and choose only males of their own species for mating. Even the time of day when courtship and mating occur is different in the species, with *persimilis* tending to court in the morning and *pseudoobscura* more active in the evening. Postzygotic mechanisms also maintain reproductive isolation in these species. When cross-fertilized

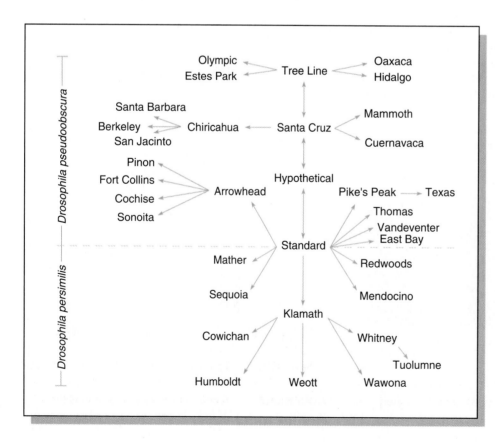

**FIGURE 25.6**    Inversion phylogeny for arrangements of chromosome 3 in *D. pseudoobscura* and *D. persimilis.* The ST arrangement is shared by both species.

in the laboratory, the species produce sterile $F_1$ hybrid males, with male sterility being associated with interactions between the X chromosome and chromosome 2. Back-crosses between $F_1$ females and parental males exhibit hybrid breakdown through lowered viability of the offspring.

## Quantum Speciation

At the heart of the classical theory of speciation is the concept of **gradualism**. According to this concept, speciation is a microevolutionary event resulting from the accumulation of many minute gene differences over a long period of time under the influence of natural selection. Recently, however, more **stochastic** or **catastrophic models of speciation** have been proposed. These models emphasize the role of evolutionary events that occur suddenly and intermittently and are therefore referred to as **quantum speciation**. We will briefly discuss one model derived from the fossil record, and two examples that depend on chromosome rearrangements as isolating mechanisms. Finally, we will consider a mechanism that has long been known to exist in angiosperms—speciation through polyploidy.

Drawing on evidence from the fossil record, Niles Eldredge and Stephen Jay Gould have proposed a mechanism of species formation known as **punctuated equilibrium**. According to this model, evolutionary changes are not gradual and continuous, but occur intermittently and rapidly as events that punctuate or interrupt long periods of evolutionary equilibrium during which little or no change occurs. In paleontological terms, *rapidly* means within thousands or even hundreds of thousands of years. The results of such evolutionary alterations are thought to cause abrupt changes in the fossil record, since relatively few fossils will be formed during such changes, compared to the large numbers of fossils formed during long periods (extending into millions of years) of evolutionary equilibrium.

The evolutionary spurts that punctuate periods of stasis are thought to be related to catastrophic events that produce mass extinctions, or events that take place in small populations located at the fringes of the species range. In these locales, environmental stresses disrupt genetic stability, eventually causing the emergence of new species. Proponents of this hypothesis point out that such changes leading to speciation take place through natural selection of individual variations, and do not invoke any unusual genetic mechanisms. On the other hand, some population geneticists argue that the mechanism of punctuated equilibria unnecessarily separates the process of species formation from other evolutionary events that generate diversity between members of a species.

A second model of quantum speciation, proposed by Hampton Carson and based on his study of Hawaiian *Drosophila,* is called the **founder-flush theory**. According to this model, populations and even species can be started by a single individual. The founder principle is not new, as this idea was advanced much earlier by Mayr. In Carson's proposal, however, gradualism is not a necessary component of speciation, and more importantly, reproductive isolation *precedes* adaptation rather than being a consequence of genetic diversification. In other words, Carson has changed the order of steps in the classical model of speciation. According to the founder-flush theory, a single fertilized female can colonize an isolated territory previously unoccupied by members of this species. If conditions are favorable, the population founded by this individual will undergo a **flush**, or rapid expansion. After several generations, it is likely that the population growth will outstrip the environment's carrying capacity, causing a **population crash**. The crash causes the death or dispersal of almost the entire population in a random fashion. A single survivor, or at most a few survivors, may rebuild the population, which eventually undergoes several flush–crash cycles before coming to equilibrium with the environment. This cycle is diagrammed in Figure 25.7. Through

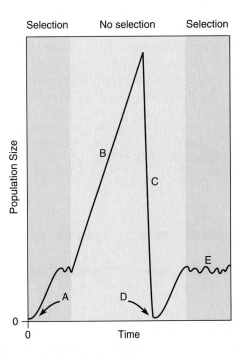

**FIGURE 25.7**    The flush–crash cycle. Small populations descended from a single founder (A) undergo a population flush (B), followed by a crash (C). A single survivor in the form of a fertilized female (D) builds a new population (E).

the genetic revolution it has undergone, the colony will, in all probability, have acquired adaptations making it unable to interbreed with the parental population.

According to Carson's theory, genetic changes are brought about in two ways. First, the founding of the population by one or a very few individuals can establish allele frequencies different from those in the ancestral population. Second and most important, selection is relaxed during a population flush. If descendants of the founder can invade a new niche, they expand their numbers in a flush. Carson proposes that normally some blocks of genes on chromosomes remain tightly linked or "closed" to recombination because of some selective advantage conferred by their configuration. Any new genotypes produced by recombination within these closed areas have reduced fitness. In fact, the advantage conferred by balanced polymorphisms for inversion heterozygotes may derive from the protection of such closed gene complexes. A diagram of open and closed chromosomal regions is shown in Figure 25.8.

During the flush period, genotypes produced by recombination in closed regions of the genome may survive under the relaxed conditions of selection. These variants may often be incompatible with the normal closed system of the species, but they survive because of relaxed selection. After a crash, the survivor's reshuffled genome is acted upon by selection to produce a new combination of open and closed gene groups adapted to the environment. Several such cycles can produce enough genetic differences so that crosses between the progenitor and colony populations produce hybrids with lowered fitness commonly displayed by interspecific hybrids. Carson views cycles of disorganization and reorganization of the genomes as the essence of speciation rather than as the gradual genetic divergence of isolated populations over long periods of time.

The evolution of *Drosophila* species in the Hawaiian

Islands appears to have followed such a founder–flush cycle. Geologic evidence indicates that the northwesternmost islands are the oldest, and that the southeastern island of Hawaii (produced by volcanic action) is the youngest at about 700,000 years old. The relationships among species of *Drosophila* can be traced by mapping the location and frequency of inversions in the banded polytene chromosomes of larval salivary glands. One group, the *planitibia* complex, has three species with the same basic set of chromosome inversions: *D. planitibia, D. heteroneura,* and *D. silvestris. D. planitibia* is found on the older island of Maui, and the other two are found on Hawaii. From this evidence, it is postulated that an immigrant fertilized female belonging to an ancestral stock on Maui, chromosomally related to the present-day *D. planitibia,* crossed the Alenuihaha Channel to Hawaii. Subsequent flush–crash cycles led to the reconstruction of the colony's genotype, giving rise to the two species found on Hawaii— *D. heteroneura* and *D. silvestris* (Figure 25.9). An alternative hypothesis proposes that one of the species on Hawaii could have arisen as the result of a second colonization from Maui. Similar evidence indicates that two other groups on Maui have given rise to a total of five species on Hawaii.

In laboratory tests of this theory, Jeffrey Powell has found that 15 generations after several flush–crash cycles, laboratory strains of *Drosophila* species showed significant behavioral (prezygotic) isolation from other

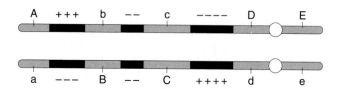

**FIGURE 25.8**    Model for open and closed regions of chromosomes. Products of crossing over anywhere within the open system (blue) result in offspring with high fitness, while crossovers in the closed regions (black) produce zygotes with low fitness. The letters represent genes or polygenes, and the pluses and minuses represent internally balanced gene complexes. Recombination within these blocks produces unfit gene combinations.

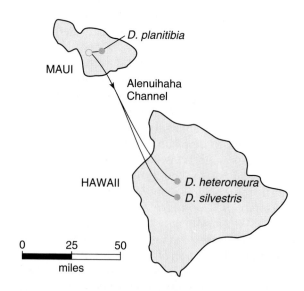

**FIGURE 25.9**    Proposed pathway of colonization of Hawaii by members of the *D. planitibia* species complex. Open circles represent a population ancestral to the three present-day species.

strains. Subsequent testing several months later showed that these differences were stable. These experimental results are important in establishing that the first stages of speciation can occur rapidly under certain circumstances.

The third model of quantum speciation that we will discuss, Michael White's **statispatric speciation**, was originally derived to account for the evolution of flightless grasshoppers in Australia. In this model, a chromosomal aberration such as a translocation arises by chance in a small population. If heterozygotes for the translocation have a slightly reduced fitness, perhaps caused by abnormal meiosis, selection will favor either homokaryotype (two copies of the translocation or two normal chromosomes). A translocation homokaryotype, with a chromosome number different from the original, may arise, spread, and partially displace the ancestral population. Semispecies would then be reproductively isolated because hybrids would have an unbalanced chromosome complement and a lowered fitness. Thus, in White's as well as Carson's model, reproductive isolation precedes the development of genetic variability. The statispatric model has also been applied to explain the origin of closely related species of mole rats differing in chromosome number, the *Spalax ehrenbergi* complex.

The final example of quantum speciation involves polyploidy in plants. The formation of species by polyploidy in animals is rare, but has been an important factor in the evolution of plants. It is estimated that one-half of all flowering plants have evolved by polyploidy. One such form of polyploidy is **allopolyploidy** (see Chapter 6), produced by doubling the chromosome number in an interspecific hybrid. If two species of related plants have the genetic constitution $SS$ and $TT$, where $S$ and $T$ represent the haploid set of chromosomes in each species, then the $F_1$ hybrid would have the chromosome constitution $ST$. Normally such a plant would be sterile because there are few or no homologous chromosome pairs, and aberrations would arise during meiosis. If the hybrid undergoes a spontaneous doubling of chromosome number, however, a tetraploid $SSTT$ would be produced. This might occur in somatic tissue, giving rise to a partially tetraploid plant that would produce some tetraploid flowers. Alternatively, aberrant meiotic events may produce $ST$ gametes, which when fertilized would yield $SSTT$ zygotes. The $SSTT$ plants would be fertile because they would possess homologous chromosomes producing viable $ST$ gametes. This new, true-breeding tetraploid would have a combination of characters derived from the parental species, and would be reproductively isolated from them because $F_1$ hybrids would be triploids and consequently sterile. The tobacco plant *Nicotiana tabacum* ($2n = 48$) is the result of the dou-

bling of the chromosome number in the hybrid between *N. otophora* ($2n = 24$) and *N. silvestris* ($2n = 24$).

# MOLECULAR EVOLUTION

The diversity of life forms inhabiting the earth is overwhelming. Nearly two million species of plants and animals have been described, and surely there are many yet to be classified. Nevertheless, all organisms share the same chemical features: they are composed principally of carbon, hydrogen, nitrogen, and oxygen atoms; they use nucleic acids to store and transfer chemical information; and proteins are, for them, the indispensable products of the stored genetic information.

The recent development of techniques to analyze and sequence proteins and nucleic acids has allowed biologists to infer relatedness of organisms and to construct molecular clocks and molecular phylogenetic trees. In this section we shall review some of the findings in this area of evolutionary study and examine the genetic variability demonstrated at the molecular level within populations.

## Amino Acid Sequences and the Molecular Clock

Evolutionary relatedness or divergence can be measured by comparing the amino acid sequences of proteins common to various organisms. The first protein to be sequenced was **insulin**, composed of only 51 amino acids. In the early 1950s, the amino acid sequence of insulin was examined in a variety of mammals, including cattle, pigs, horses, sperm whales, and sheep. With the exception of a stretch of three amino acids, the protein was shown to contain an identical sequence in each of these mammals.

**Cytochrome c** is another commonly investigated protein. It is a respiratory pigment found in the mitochondria of eukaryotes. The molecule consists of 104 amino acids in many vertebrates and a slightly higher number in most other organisms. Cytochrome c has changed very slowly during evolution. For example, the amino acid sequence in humans and chimpanzees is identical; between humans and rhesus monkeys only one amino acid is different. This is remarkable considering that lines leading to humans and monkeys diverged from a common ancestral form approximately 20 million years ago.

Table 25.5 shows the number of amino acid differences in cytochrome c among a variety of organisms. Even as distantly removed from humans as yeasts are,

**Table 25.5**  A Comparison of the Number of Amino Acid Differences and the Minimal Mutational Distance in Cytochrome c

| Organism | Number of Amino Acid Differences | Minimal Mutational Distance |
|---|---|---|
| Human | 0 | 0 |
| Chimpanzee | 0 | 0 |
| Rhesus monkey | 1 | 1 |
| Rabbit | 9 | 12 |
| Pig | 10 | 13 |
| Dog | 10 | 13 |
| Horse | 12 | 17 |
| Penguin | 11 | 18 |
| Moth | 24 | 36 |
| Yeast | 38 | 56 |

SOURCE: From W. M. Fitch and E. Margoliash, "Construction of phylogenetic trees," *Science* 155:279–84, 20 January 1967. © 1967 by the American Association for the Advancement of Science.

only 38 amino acids are different. In a comparison of a large number of organisms, more than 15 percent of the sequences remain unchanged.

In addition to the number of amino acid differences, it is possible to assess the minimum number of nucleotide changes that must have occurred during the evolution of a protein. This assessment requires knowledge of the genetic code and is a more refined analysis. For example, more than one nucleotide change may have been necessary to establish any given amino acid change found between two organisms. Thus, over evolutionary time, two or more independent mutations may have been essential in order to produce the observed change. When all nucleotide changes necessary for all amino acid differences are totaled, the **minimal mutational distance** between any two species is established. Table 24.5 shows such an analysis of the genes coding for cytochrome c in ten organisms. One can see that, as expected, these values are larger than the corresponding number of amino acids separating humans from the other nine organisms.

The data on amino acid substitutions and mutational distance can be combined with paleontology to construct a **molecular clock**. The information on amino acid differences provides a way of measuring the number of mutational events that have accumulated since any two organisms shared a common ancestor. The fossil record provides information about the time that has elapsed since the two organisms shared a common ancestor. Assuming that the rate of amino acid replacements occurred at a regular rate proportional to absolute time, the differences in amino acid content can be used

as a molecular clock, measuring the time since the two species diverged from a common ancestor.

## Phylogenetic Trees

On the basis of information provided by a molecular clock, it is possible to construct **divergence dendrograms** or **phylogenetic trees** based on the analysis of amino acid sequences of a single protein from diverse organisms. This is perhaps the most fascinating application of this information to evolutionary study.

The underlying assumption of this analysis is that all present-day sequences from different species represent gene products that diverged from common ancestral sequences at various points in evolutionary time. By determining minimal mutational distances among all species under analysis and in which specific amino acids have changed, the most likely relationships among the species can be determined. This analysis can also establish the point in these relationships at which a now extinct ancestral sequence must have existed in order to lead to evolutionary divergence.

This information may be summarized in the form of a phylogenetic tree. The phylogenetic tree shown in Figure 25.10 is based on the sequences used to derive the

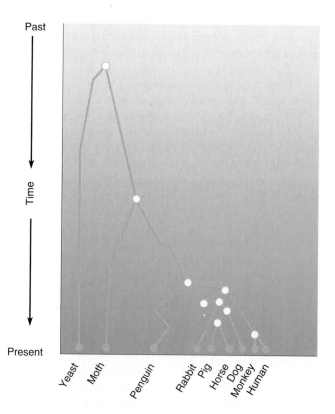

**FIGURE 25.10**    Phylogenetic sequence constructed by comparison of homologies in cytochrome c amino acid sequences.

data of Table 25.5. The tree is plotted so that the ordinate represents proportional amounts of distance. If a constant mutation rate is assumed, the ordinate represents a relative estimate of geologic time.

When phylogenetic trees are constructed in this way, they agree remarkably well with trees constructed using more conventional approaches such as morphological and paleontological evidence. Once a number of proteins from a variety of organisms are sequenced and analyzed simultaneously, even more accurate trees may be constructed.

When closely related species are to be examined in this way, proteins that have evolved more rapidly than cytochrome c are more useful. The 115-amino-acid protein carbonic anhydrase has been used to analyze more accurately the relationship between humans and several other primates (Figure 25.11).

This phylogenetic approach may also be applied to related molecules that have arisen during evolution through gene duplication. The various hemoglobin chains and myoglobin have been analyzed in this way. As in the comparison of species, each protein chain may be studied using comparative sequences. As shown in Figure 25.12, the oxygen-carrying myoglobin molecule and all hemoglobin chains are proposed to have arisen from a common ancestral sequence. Furthermore, all four hemoglobin chains have a common origin.

Single chains such as the alpha and beta globins can also be compared between species. The differences between humans and their close relatives are noteworthy.

Human and chimpanzee alpha and beta chains are identical in sequence, but the gorilla differs from the human in one amino acid in each chain. Table 25.6 shows the number of amino acid differences between humans and a number of species.

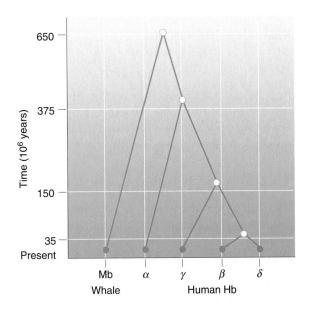

**FIGURE 25.12** Phylogenetic sequence of myoglobin and hemoglobin proteins. These genes arose by duplication and subsequently diverged in amino acid sequence.

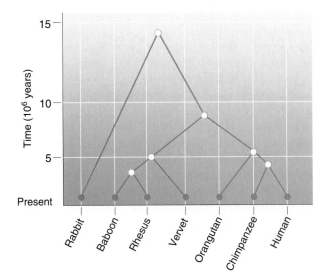

**FIGURE 25.11** Phylogenetic sequence of carbonic anhydrase amino acid substitutions.

**Table 25.6** A COMPARISON OF THE ALPHA AND BETA HEMOGLOBIN CHAINS BETWEEN HUMANS AND OTHER ORGANISMS

| Organism | Amino Acid Differences Between Humans and Various Organisms | |
| --- | --- | --- |
| | $\alpha$ Chains | $\beta$ Chains |
| Chimpanzee | 0 | 0 |
| Gorilla | 1 | 1 |
| Macaque | 5 | 10 |
| Mouse | 19 | 31 |
| Sheep | 26 | 33 |
| Pig | 20 | 28 |
| Horse | 22 | 30 |
| Rabbit | 28 | 16 |
| Chicken | 45 | — |
| Kangaroo | — | 54 |
| Carp | 93 | — |
| Lamprey | 113 | — |

SOURCE: Data from various sources.

All in all, amino acid sequence data have been extremely useful in evolutionary studies. These independent analyses complement other types of evolutionary evidence. Perhaps their greatest value is that these studies are performed directly at the level of translated genetic information, which is an underlying resource supporting evolutionary change.

## Nucleotide Sequence Phylogeny

The molecular hybridization of DNA from different sources has been a valuable technique in evolutionary studies. Although it has been discussed in Chapter 8 and is described in Appendix A, we shall briefly review it here.

Hybridization of nucleic acids is based on nucleotide sequence similarity. Molecules are heated until they dissociate (melt) into single strands and are then allowed to reanneal by slow cooling. In reannealing, complementary sequences join together in a stable double-stranded form at the lower temperature. Molecular hybridization will occur between mixtures of DNA strands (DNA:DNA) or between mixtures of DNA and RNA strands (DNA:RNA).

In studies of sequence diversity between species, radioactive DNA from one species is prepared and those sequences present once in the haploid genome (single copy fraction) are isolated. This single copy DNA is melted into individual strands and can be reassociated with melted single copy DNA from the same species (homologous reaction) or from other species (heterologous reaction) to form a double-stranded duplex. By constructing a melting profile (review melting profiles in Chapter 8 and see Figure 25.15), a melting temperature ($T_m$) can be determined for these newly created homologous and heterologous DNA. The difference in thermal stability ($\Delta T_m$) between homologous and heterologous duplex molecules is a measure of the nucleotide sequence divergence between the two species. For example, a $\Delta T_m$ of 1°C roughly corresponds to 1 percent mismatching in nucleotide sequences. The results of experiments with two species of sea urchin are shown in Figure 25.13. From differences in thermal stability, it can be calculated that there is a sequence divergence between *Strongylocentrotus purpuratus* and *S. franciscanus* of about 19 percent. A similar experiment shows about 7 percent sequence diversity between *S. purpuratus* and another species, *S. drobachiensis*. From the data on nucleotide diversity and what evidence is available in the fossil record, it has been proposed that *S. purpuratus* diverged from *S. franciscanus* some 15 to 20 million years ago.

Where the nucleotide sequence divergence can be

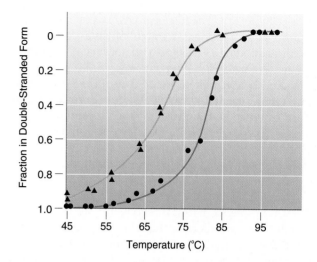

**FIGURE 25.13**     Measurement of nucleotide sequence diversity in the sea urchin. The thermal stability of *S. purpuratus* single copy DNA reassociated with *S. purpuratus* (circles) and *S. franciscanus* DNA is a measurement of nucleotide sequence divergence.

compared with other indicators of genetic variability, such as protein polymorphisms and chromosomal rearrangements, the evidence indicates that nucleotide sequence diversity is a more sensitive indicator of evolutionary divergence than amino acid replacements. *Drosophila heteroneura* and *D. silvestris,* found only on the island of Hawaii, are estimated to have diverged only about 300,000 years ago, but it is difficult to demonstrate significant differences between these species in chromosomal inversion patterns or protein polymorphisms (Figure 25.14). Figure 25.15 shows the results obtained when labeled single copy DNA from *D. silvestris* is hybridized with itself and with DNA from *D. heteroneura* and *D. picticornis.* (*D. picticornis* is another member of the *planitibia* group and is found only on Kauai, a much older island.) The sequence diversity between the two species from Hawaii is about 0.55 percent, but *D. silvestris* and *D. picticornis* show a 2.1 percent difference in nucleotide sequences. Thus, nucleotide sequence diversity may precede the development of protein or chromosomal polymorphisms.

The fact that *D. heteroneura* and *D. sylvestris* share identical chromosome arrangements, have almost no detectable protein differences, are 99 percent homologous in DNA sequence, and yet are classified as separate species may seem paradoxical. However, the two species are clearly separated from each other by different and incompatible courtship and mating behaviors (a prezygotic isolating mechanism), by morphology, and by pigmentation of the body and wings (Figure 25.16). The available evidence suggests that these differences are controlled by a relatively small number of genes. In a

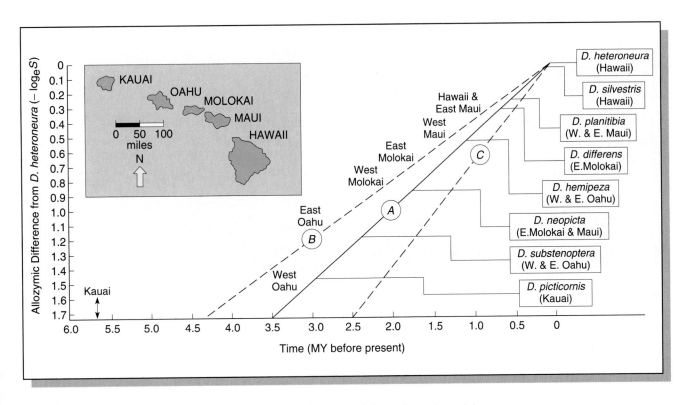

FIGURE 25.14    Detectable allozyme differences and proposed time of species origin in some Hawaiian *Drosophila* (line A). Line B assumes a slower accumulation of genetic differences and C is a faster accumulation. In any case, genetic differences based on enzymes in the most recently evolved species *(planitibia, sylvestris, heteroneura)* are difficult to quantify.

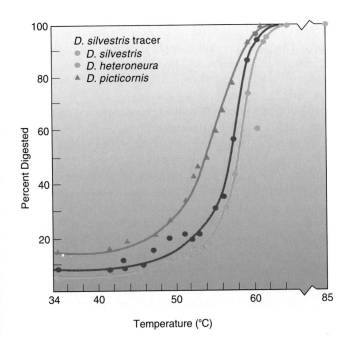

FIGURE 25.15    Nucleotide sequence diversity in the *Drosophila planitibia* species complex. The degree of shift to the left by the heterologous hybrids is an indication of the degree of nucleotide sequence divergence.

(a)                                                        (b)

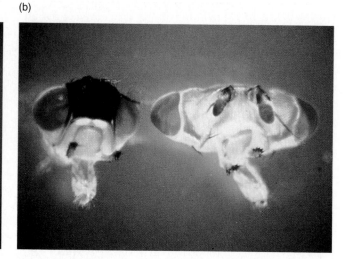

**FIGURE 25.16**    (a) Differences in pigmentation patterns in *D. sylvestris* (left) and *D. heteroneura* (right). (b) Head morphology in *D. sylvestris* (left) and *D. heteroneura* (right).

genetic analysis of the differences in head shape and pigmentation, F. C. Val has estimated that only 15 to 19 loci are responsible for the morphological differences between these species, demonstrating that the process of speciation need only involve a small number of genes.

## Molecular Studies on Human Evolution

Problems similar to those encountered in measuring genetic differences in the Hawaiian *Drosophila* are also encountered in studies of higher primates. The hominoid primates include the chimpanzees, gorilla, orangutan, gibbon, and human. Past chromosomal studies and data on protein and isozyme differences have failed to accurately resolve the taxonomic relationships among the chimpanzee, gorilla, and humans, because they are so closely related. Using hybridization of single copy DNA sequences from a large number of individuals and using calibrations derived from the fossil record, Charles Sibley and Jon Ahlquist have clarified the evolutionary branching pattern in the hominoids and have estimated the times at which divergence occurred (Figure 25.17). According to their data, humans and chimpanzees are more closely related than either is to the gorilla. The measure of relatedness is called the $\Delta T_{50}H$ and is related to the $\Delta T_m$ discussed earlier. Between humans and chimpanzees, the $\Delta T_{50}H$ is 1.6, somewhat less than the value of 2.1 to 2.3 for the distance between the gorilla line and the chimpanzee/human line. Their calculation of divergence from the chimpanzee line is 8 to 10 million years (MY) for the gorilla, 6.3 to 7.7 MY for humans, and 2.4 to 3.0 MY for the pygmy chimpanzee. The high degree of nucleotide homology, similarity in chro-

mosome patterns, and protein homology are reminiscent of the situation in *D. heteroneura* and *D. sylvestris* and lend support to the notion that distinct species are not necessarily separated by large numbers of gene differences.

The close evolutionary relationship among the two species of chimpanzees and the human species revealed by DNA hybridization studies poses an interesting taxonomic problem. In other areas of taxonomy, differences of less than 2 in the $\Delta T_{50}H$ for the DNA of two species are usually found within the same genus and perhaps the same family. Yet humans are currently placed in different families than gorillas and chimpanzees (humans in Hominidae, gorillas and chimps in Pongidae) and in different genera: humans in the genus *Homo,* and chimpanzees in the genus *Pan.* The problems and challenges posed by this anthropomorphic inconsistency and the evolutionary relationship between humans and chimpanzees is explored in the book *The Third Chimpanzee* by Jared Diamond.

The question of where the human species originated has spawned several theories. Africa, Asia, Europe, and the Americas have been suggested as the site of human origin, and another theory proposes that modern humans arose simultaneously in several locations. The available fossil record indicates that *Homo sapiens* was present in Southern Africa more than 100,000 years ago, and was found in Asia at least 50,000 years ago, and it appears that *Homo sapiens* replaced Neanderthals in Europe about 30,000 to 40,000 years ago. The bulk of evidence from fossils supports the origin of humans in Africa, probably by phyletic evolution from *Homo erectus,* with a rapid spread to non-African regions. Over the

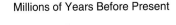

Millions of Years Before Present

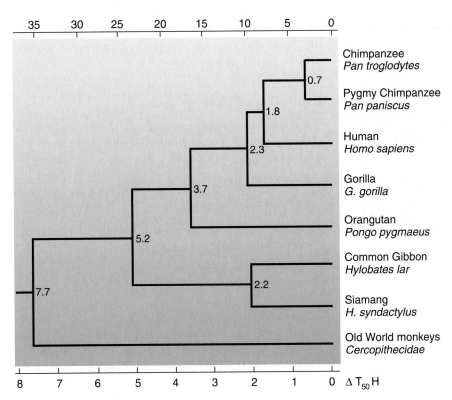

**FIGURE 25.17** Phylogenetic sequence of the hominoid primates and the Old World monkeys as estimated by DNA hybridization. The numbers at the branch points are $\Delta T_{50}H$ measurements. The evolutionary branch points are dated by reference to the fossil record and nucleotide divergence.

past decade, molecular techniques, including RFLP, analysis, have been used to provide additional evidence for the origin and spread of the human species across the globe.

In Chapter 13 we discussed the use of restriction fragment length polymorphisms (RFLPs) to map genes to specific chromosomes and/or chromosomal regions. RFLPs can be generated by single nucleotide base changes in a DNA sequence that is recognized by a restriction enzyme. Since restriction enzymes make double-stranded cuts in DNA, alterations in recognition sequences result in changes in the pattern of cuts made in the DNA. These detectable changes or polymorphisms are inherited in a codominant Mendelian fashion. RFLP analysis has been used to construct both a molecular clock and phyletic trees in order to reconstruct events in human evolution.

RFLP analysis has been used on both nuclear genes and mitochondrial genes to study the evolutionary relationships among various human populations. Using five different restriction sites clustered within the beta globin complex as markers, J. S. Wainscoat and colleagues studied the distribution of these sites in eight population groups. In all, 32 combinations of these five restriction sites are possible, and 14 of these were actually observed in the populations surveyed. Three combinations of re-

striction sites (known as haplotypes) are common in non-African populations, and are rare in African populations. These haplotypes are represented as $+ - - - -$, $- + + + +$, and $- + + - +$. The African populations studied have high frequencies of two haplotypes: $- - - - +$ and $- + - - +$, both of which are rare or absent in all other populations studied.

Using fossil evidence and the RFLP data, a dendrogram was constructed for the eight populations studied (Figure 25.18). The model is consistent with the suggestion that *Homo sapiens* originated in Africa, and that a small founder population migrated from Africa and later gave rise to all non-African populations. Further RFLP screening of other nuclear genes will help to resolve the issues raised in this study, and to answer other questions about evolutionary events, such as the size of the founding population, and the time scale over which these evolutionary events may have transpired.

A study of restriction polymorphisms in mitochondrial DNA (which is maternally inherited) by Cavalli-Sforza and his colleagues first suggested that African populations are ancestral to others.

The use of RFLPs in mitochondrial DNA as a molecular clock to construct a phyletic tree to study human origins by Allan Wilson and his colleagues has led to a controversy over the appearance of modern humans as mem-

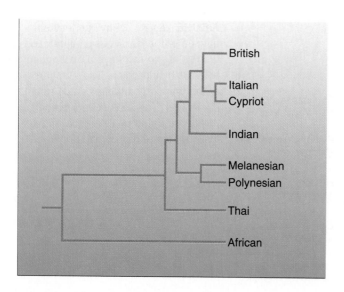

**FIGURE 25.18** Evolutionary dendrogram of human populations constructed from RFLP analysis of alleles adjacent to the beta globin gene.

bers of the species *Homo sapiens.* In this case, there is little dispute over the evidence that the earlier ancestors of the human species arose in Africa, and that an ancestral species, *Homo erectus* spread from Africa to populate Europe and Asia beginning about 1 million years ago. What is in dispute is how and where *Homo sapiens* originated. Two hypotheses are currently the subject of debate. One, advanced by those using trees constructed from mitochondrial DNA, argues that the modern human lineage arose in Africa around 200,000 years ago. According to this hypothesis, the maternal lineage of humans (in the form of mitochondrial DNA) has an ultimate common female ancestor, and the modern human species, *Homo sapiens,* arose in Africa. From here, modern *Homo sapiens* spread throughout the Old World, replacing the several human lineages descended from *Homo erectus.*

The opposing hypothesis agrees with the idea that humans originated in Africa and spread throughout Europe and Asia as *Homo erectus.* However, this hypothesis holds that the bulk of genetic evidence based on nuclear genes and the fossil record support the idea that modern humans in the form of *Homo sapiens* arose not in Africa around 200,000 years ago, but in multiple regions throughout the Old World as part of an interbreeding network of human lineages descended from *Homo erectus.* This debate is often contentious and bitterly contested. Most recently, the accuracy of the molecular clock as measured by nucleotide substitutions and the method used to construct the mitochondrial tree have

been called into question, raising doubts about the first hypothesis. Still others argue that data yet to be gathered from the larger and more complex nuclear genome will ultimately resolve this question; alternatively, it may be that molecular genetics in its present form is not the best tool to settle this issue.

## Evolution of Genome Size

An early study on the relationship between cellular DNA content and evolution by Alfred Mirsky and Hans Ris in 1951 concluded that:

1. Nuclear DNA content increases from the lower to higher invertebrates.

2. Closely related organisms usually have similar amounts of DNA.

3. The development of terrestrial vertebrates has been accompanied by reductions in nuclear DNA content.

DNA content measurements from a wide range of species are now available, and organisms can be classified into four somewhat overlapping groups, as shown in Figure 25.19. The lowest DNA content in free-living organisms is found in bacteria (Group 1), with a range between 0.003 and 0.01 picogram (pg) per cell. Fungi (Group 2) average less than 0.1 pg. Group 8 includes most animals and plants, including sea urchins, reptiles, birds, and mammals. Group 4 is composed of organisms whose genomes are larger than 10 pg, including sala-

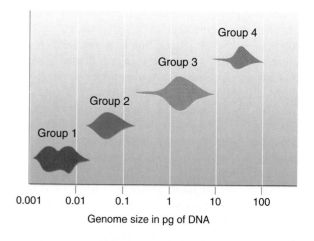

**FIGURE 25.19** Classes of organisms grouped according to their DNA content. Group 1, bacteria; Group 2, fungi; Group 3, most animals and some plants; Group 4, many plants, salamanders, and some fish.

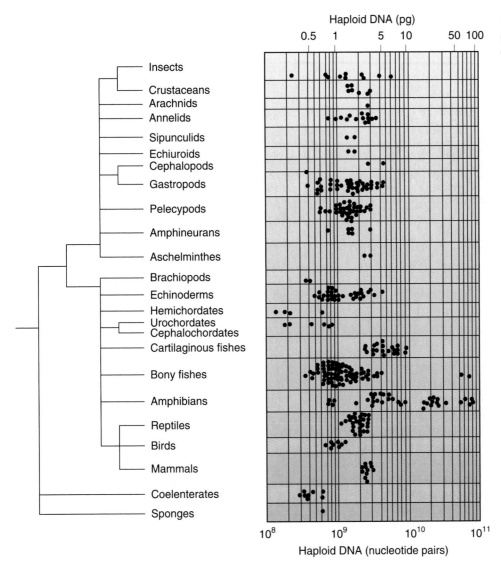

**FIGURE 25.20**   Haploid genome size clusters in higher organisms.

manders, some fish, and many plant species. The clustering of haploid DNA content in higher organisms and the clustering of related organisms are shown in Figure 25.20.

Although the relationship between DNA content and evolution is somewhat complex, several general statements can be made. A large increase in DNA content occurred during evolution from bacteria to higher plants. Few higher eukaryotes have genomes smaller than 0.1 pg, suggesting that a minimum DNA content is necessary to support this level of organization. Interestingly, while there is no necessary correlation between DNA content and chromosome number, there is a direct relationship between DNA content and nuclear and cellular size. In fact, a rough estimate of DNA content can be made from measurements of nuclear size.

Several explanations have been offered for the trend toward increases in DNA content per haploid genome during evolution, including the development of control system redundancy, generation of repetitive DNA sequences, and slower cell division and development. Unfortunately, we are not sure whether increases in DNA content are the raw material for selection or are produced as a byproduct. An examination of genome size in closely related organisms suggests that changes in DNA content have occurred mainly by duplication of small segments of DNA rather than by saltational increases associated with polyploidy. Figure 25.21 shows the distribution of DNA content in teleost fishes. If changes had arisen through polyploidy, a distribution into discrete classes reflecting multiples of a basic genome size would be expected. However, differences in DNA content are the result of small incremental increases. Such studies confirm the role of gene duplication in evolution.

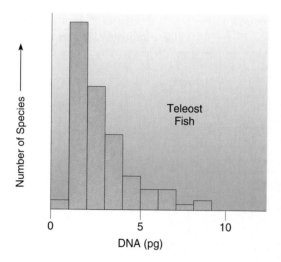

**FIGURE 25.21**    Haploid genome size in teleost fish.

## Evolution of Gene Families by Duplication

We have already mentioned the origin of the globin genes from an ancestral oxygen-carrying molecule. Cloning and nucleotide sequencing have produced a great deal of information about the evolution of this gene family. By examining functionally distinct but sequence-related genes that arose by duplication, we can study the evolution of gene function and the development of regulatory systems. Figure 25.22 summarizes what is known about the organization of the human

alpha- and beta-globin gene families. Several features of these gene clusters are worth noting:

1.  Both families consist of several genes packed together over a relatively short distance, all of which are oriented in the same direction with 5′ ends to the left and 3′ ends to the right.

2.  In addition to sequences that code for all known globin polypeptides, several related gene sequences called **pseudogenes** ($\psi_{\alpha_1}$, $\psi_{\beta_1}$, $\psi_{\beta_2}$) are present. Pseudogenes are stretches of DNA that are homologous with protein-coding regions, but are themselves inactive.

3.  The genes in each family are arranged in the order in which they are expressed in development.

In the beta-globin cluster, the embryonic epsilon form ($\epsilon$) is first, followed by the fetal gamma genes ($G_\gamma$, $A_\gamma$), and the adult beta ($\beta$) and delta ($\delta$) genes. It is not known whether this order is related to mechanisms regulating gene expression or is a reflection of evolutionary history.

The complete DNA sequences for all these genes are now available. A phylogenetic tree based on these sequences is shown in Figure 25.23. According to this tree, which was constructed from differences in nucleotide composition, the original duplication that gave rise to the alpha and beta gene clusters occurred some 500 million years (MY) ago. The adult sequences diverged from the prenatal forms about 200 MY ago, sometime during the evolution of reptiles, which are ancestral to mammals. Separate embryonic and fetal sequences diverged at the beginning of the mammalian radiation some 100 MY ago, and this event may be linked to the development of placental mammals with their unique problems of oxygen transport. The adult beta and delta genes

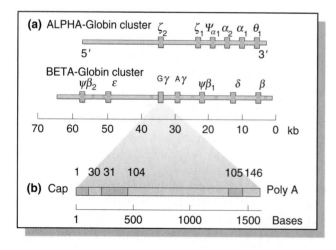

**FIGURE 25.22**    (a) Organization of the alpha- and beta-globin genes in humans. The genes are represented as rectangles, and repeated DNA sequences within the beta-globin gene cluster are shown as circles. (b) The coding (solid) and noncoding regions of the $^G\gamma$-globin gene. The top numbers are amino acid residues.

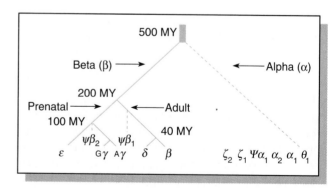

**FIGURE 25.23**    Phylogenetic sequence of globin genes based on nucleotide sequence of coding regions. Dotted line indicates relationships not yet clarified.

diverged some 40 MY ago, before the separation of higher primates into two lines—one leading to the New World monkeys and the other to the Old World monkeys, including great apes and humans.

Comparison of the nucleotide sequences in the coding and noncoding regions of the genes indicates that these regions have continued to diverge by base substitutions as well as by duplication and deletion of short regions. An unexpected outcome of this analysis has been the detection of pseudogenes in each gene family. Pseudogene sequences can arise as a result of increased crossing over between homologous chromosome regions, or by reverse transcription from mRNA. These extra copies—free from selection because they have no phenotypic effect—can accumulate mutations that would cause the gene to become inactive, but retain many of the features associated with active genes such as promoter sites, splicing sequences, and poly-A addition sites. In the human globin gene families, pseudogenes separate the embryonic and adult alpha genes, and the prenatal and adult beta genes. The significance and function of such pseudogenes are matters of intense speculation. Oliver Smithies has proposed that because of their location in human globin families, pseudogenes modulate the expression of the protein-producing alpha and beta genes. Philip Leder and his colleagues, on the other hand, have shown that pseudogenes of the alpha-globin complex in the mouse are located on different chromosomes from the active genes. Dispersed single pseudogenes (called **orphons**) derived from tandemly repeated families, including histone and ribosomal genes, have been found in a wide range of organisms. Drawing on these results, Leder suggests that the genomes of higher eukaryotes are mosaics of dispersed pseudogenes, which serve as a reservoir of sequences that can evolve new functions. This interpretation implies that gene duplication leading to the formation and dispersal of pseudogenes has been a critical factor in the evolution of higher organisms.

# THE ROLE OF MUTATION IN EVOLUTION

## Selectionists Versus Neutralists

When we discussed the concept of the adaptive norm in an earlier section, it was pointed out there is a controversy over the importance of extensive genetic variation within a population. This controversy centers particularly on the variation detected at the molecular level. On the one hand, the **classical hypothesis** proposes that natural selection favors the fixation of the most favorable alleles at each locus. Homozygosity therefore should be the rule rather than the exception. On the other hand,

the **balance hypothesis** favors the maintenance through natural selection of a high degree of heterozygosity in populations.

The classical hypothesis is supported by the observation that induced mutations almost invariably lower fitness. Therefore, if most mutations are detrimental, the accumulation of genetic variability will result in a substantial genetic burden to the fitness of a population. According to the classical theory, this detrimental variation will be purged by natural selection.

Dobzhansky and his colleagues provided the initial support for the balance hypothesis. Having observed a high degree of heterozygosity in adapted *Drosophila* populations, they postulated that populations displaying genetic polymorphisms will have added fitness. The observation of protein polymorphisms by Lewontin in *Drosophila* and by Henry Harris in humans, and the evidence for numerous amino acid substitutions in single proteins throughout evolution, attest to a high degree of heterozygosity, particularly at the molecular level, in many diploid and polyploid populations. These findings lend support to the balance hypothesis and argue against the classical theory.

However, the existence of a high degree of genetic variation does not prove that it is the basis of evolutionary fitness. A third theory, the **neutralist hypothesis** proposed by James Crow and Motoo Kimura, argues that mutations leading to amino acid substitutions are rarely favorable. They are sometimes detrimental, but most often neutral or genetically equivalent to the allele that is replaced. Those that are favorable or detrimental are either preserved or removed from the population, respectively, by natural selection. However, neutral genetic changes will not be affected by selection and will instead become randomly fixed in the population. Their frequency will be determined by the rate of mutation and the principles of random genetic drift.

The neutralist interpretation, as most recently articulated by Kimura, is based on several observations and theoretical considerations. These ideas focus on several important aspects of genetic variation as an evolutionary phenomenon. The neutralist theory is based on the following:

1. The rate of amino acid substitution in a given protein is about the same in organisms with diverse lineages.

2. The types of substitutions do not seem to demonstrate any particular pattern, but are instead random.

With regard to these observations, the neutralists believe that if selection favored the amino acid changes even slightly, the rate would vary in different organisms exposed to different selective pressures. The substitutions would therefore not be random.

3.  The overall rate of mutation leading to amino acid substitutions is relatively high (it is about equal to the substitution of one nucleotide per genome every two years in mammals) and has remained relatively constant for a long period of time.

For example, when nucleotide substitution as estimated from amino acid changes in seven proteins from 16 pairs of mammals is extimated over the last 150 MY, the points fall close to a straight line. A constant rate of change is indicated. The neutralists argue that if natural selection were acting on all such mutations, the rate of fixation would not be constant in fluctuating environments over time.

4.  The number of amino acid substitutions is higher in the less important parts of molecules (those parts not critical to tertiary structure and the active site of proteins) and in molecules less critical to the organism.

This facet of the argument centers on **functional constraint**. That is, those regions that determine whether the molecule functions or not are less able to tolerate change. The regions of lesser importance—those determining only how well the molecule functions—can tolerate greater amounts of change. It is argued that if such changes in less-constrained regions of proteins are neutral, the rate of amino acid substitution in these regions will be substantially higher than in constrained regions. The same argument is applicable to total proteins of lesser importance to organisms. Examination of less-constrained regions of proteins and less-critical proteins does show a higher amino acid substitution rate. This is interpreted by the neutralists as support for their contentions.

5.  The rate of amino acid substitution is much too high to be accounted for by selection.

If all heterozygosity within a population represented alleles with even a slight advantage, the neutralists argue that the **cost of selection** necessary to fix and maintain them in populations would be astronomically high. The cost of selection is the genetic death of those organisms that must be replaced by those bearing the more favorable change. Mathematical calculations can be made to support this theory.

The neutralist theory runs somewhat counter to the theory of natural selection as set down by Darwin and as expanded by modern knowledge of genetics. In the 1960s there was general agreement that all biological characteristics could be interpreted as arising by adaptive evolution through natural selection. In this sense, most mutations that have survived through evolution should be at least slightly adaptive in homozygotes, heterozygotes, or both.

Those who oppose the neutralist theory have been called **selectionists**. They point out examples where enzyme or protein polymorphism is associated with adaptation to certain environmental conditions. The well-known advantage of sickle-cell anemia carriers in malarial regions is such an example.

Selectionists also stress that enzyme polymorphisms may often appear to offer no advantage, but exist in such a frequency that it is impossible to explain as a random occurrence. Thus, even though no currently available analytical technique can detect any physiological difference, it cannot be proved that some slight advantage associated with any given amino acid substitution does not exist.

On many specific points of contention, selectionists are able to offer persuasive theoretical arguments and calculations. For example, although the nucleotide substitution rate appears on the average to be constant, localized variation is sometimes as much as 2.5 times higher than expected by chance.

Even though this controversy is highly theoretical and esoteric, it is important that we not lose sight of several factors. For example, the neutralists do not discount natural selection as a guiding force in evolution. Rather, they suggest that some features of organisms' genotypes are nonadaptive, fluctuate randomly, and may have been fixed by genetic drift. On the other hand, the selectionists certainly do not deny that genetic drift is an important factor in establishing gene frequencies.

Finally, it should be pointed out that the two theories are not mutually exclusive. It is difficult to argue against the notion that some genetic variation must be neutral. The difference between the two theories is in the degree of neutrality that exists. While current data are insufficient to resolve the problem, one important point has emerged from the arguments. In considering natural selection, there are clearly two levels that must be examined: the phenotypic level, including all morphological and physiological characteristics imparted by the genotype, and the molecular level, represented by the precise nucleotide and amino acid sequence of DNA and proteins. There is no question about selection acting at the phenotypic level. The extent to which selection occurs at the molecular level, however, is in question.

## Recent Trends

One of the major premises of the classical theory of speciation is that mutations (along with recombination) are the source of all variation. In this context, new alleles are inherited in Mendelian fashion and may gradually increase in frequency if they confer greater fitness in accord with the principles of population genetics. Speciation involves a large number of allelic substitutions sequentially incorporated into the gene pool. We have already seen that the concept of gradualism is being challenged by chromosomal theories of quantum speciation. Work in molecular biology, particularly the discovery of intervening sequences, transposable elements, and pseudogenes, is causing fundamental changes in our concepts of genome structure, stability, and mutation.

The genome of higher organisms has been regarded as a largely static entity, with low rates of mutation and chromosome aberration. We now know that the genome is highly dynamic and in constant flux, with families of repeated sequences being created, fixed, deleted, and replaced. Previously, mutation has been thought of in terms of nucleotide substitution or, at most, small duplications or deletions. However, nucleotide sequence analysis of the *bithorax* gene complex in *Drosophila* has shown that most mutations in these genes involve the insertion or deletion of pieces of DNA several thousand base pairs in length. Insertion of transposable elements into genes produces mutant phenotypes, and their removal produces reversion to wild type. In addition, since many transposable elements contain start and stop signals for transcription, the dispersal and mobility of such elements in the genome can create new patterns of gene expression in a single generation.

The evolutionary role of regulatory mutations in development is also receiving increased attention. Many workers now believe that mutations that change the timing of gene expression in embryogenesis can have a significant impact on the rates of speciation. Such mutations would not change the frequency of protein polymorphisms in a population, but rather the order or duration of gene expression during embryonic or preadult development, producing instant morphological variation. Such a mutant has already been described in the *bithorax* region of *Drosophila*. By transposing a gene some 40 kb within the complex, the mutation changes the gene's time of expression during development. This gene, *Cbx,* produces a fly with a greatly altered phenotype.

Finally, pseudogenes are seen as possible reservoirs of inactive genes that have the potential to be mobilized to perform new functions. The classical theory of speciation appears ready to undergo a new and dramatic synthesis, and it is likely that this synthesis will arise from within molecular biology.

## CHAPTER SUMMARY

1. Speciation, which depends on genetic variation and forces controlling its distribution (e.g., natural selection), is initiated when a population becomes separated into smaller, reproductively semi-isolated breeding groups.

2. The partitioning of a population's gene pool allows each group to adapt to its new environment. Races form as each group acquires its own unique set of genetic variations. Races may become reproductively isolated from one another and form new species as evolution proceeds. The formation of races and speciation are illustrated by several studies of *Drosophila*.

3. Two approaches have been especially fruitful in evolutionary study at the molecular level: (1) comparison of complementary sequences present in the DNA of different organisms, and (2) comparison of amino acid substitutions in proteins common to a variety of organisms. Such comparisons provide not only a measure of evolutionary relatedness, but further allow the assessment of genetic variation during evolution.

4. The central question concerning the importance of variation at the molecular level is whether or not all nucleotide and amino acid changes preserved through evolution in DNA and proteins are adaptive to the population. Some believe that the majority of these changes are neutral or genetically equivalent, while others adhere to the theory of adaptive evolution through natural selection.

## KEY TERMS

adaptive norm
allopolyploidy
allozymes
balance hypothesis
cladogenesis
classical hypothesis
cost of selection
cytochrome c
divergence dendrogram
flounder–flush theory
functional constraint

geographic (allopatric)
  speciation
gradualism
inbreeding depression
insulin
inversions
minimal mutational
  distance
molecular clock
neutralist hypothesis
niches
orphons

phyletic evolution
  (anagenesis)
phylogenetic trees
population crash
postzygotic
preadaptation
prezygotic isolating
  mechanisms
pseudogenes
punctuated equilibrium
quantum speciation

races
reproductive isolating
  mechanisms
selectionists
sibling species
speciation
species
statispatric speciation
stochastic (catastrophic)
  models of speciation
translocations

## INSIGHTS AND SOLUTIONS

1.  Sequence analysis of DNA can be accomplished by a number of techniques (see Appendix A for a detailed description). Protein sequencing, on the other hand, is made more complex by the fact that twenty different subunits need to be unambiguously identified and enumerated, rather than the four nucleotides of DNA. Because of their unique properties, the N-terminal and C-terminal amino acids in a protein are easy to identify, but the array in between offer a difficult challenge, since many proteins contain hundreds of amino acids. How is it that protein sequencing is accomplished?

    **SOLUTION:** The strategy for protein sequencing is the same as for DNA sequencing: divide and conquer. To accomplish this, specific enzymes are used that reproducibly cleave proteins between certain amino acids. The use of different enzymes produces overlapping fragments. Each fragment is isolated and its amino acid sequence is determined by chemical means. Sequences from overlapping fragments are then assembled to give sequence for the entire protein.

2.  A single plant twice the size of others in the same population suddenly appears. Normally, plants of this species reproduce by self-fertilization and by cross-fertilization. Is this new giant plant simply a variant, or could it be a new species? How would you determine this?

    **SOLUTION:** One of the most widespread mechanisms of speciation in higher plants is polyploidy, the multiplication of entire chromosome sets. The result of polyploidy is usually a larger plant with larger flowers and seeds. There are two ways of testing this new variant to determine whether it is a new species. First, the giant plant should be crossed with a normal-sized plant to see if it produces viable, fertile offspring. If it does not, the two different types of plants would appear to be reproductively isolated. Second, the giant plant should be cytogenetically screened to examine its chromosome complement. If it has twice the number of its normal-sized neighbors, it is a tetraploid that may have arisen spontaneously. If the chromosome number differs by a factor of two, and the new plant is reproductively isolated from its normal-sized neighbors, it is a new species.

1. Discuss the rationale behind the statement that inversions in chromosome 3 of *Drosophila pseudoobscura* represent genetic variation.

2. Describe the process of race formation. What is the role of natural selection?

3. Contrast the classical and balance hypotheses as they relate to the adaptive norm.

4. What types of nucleotide substitutions will not be detected by electrophoretic studies of a gene's protein product?

5. In a sequencing experiment (using the numbers 1 through 6 to represent amino acids), the following two sets of peptide fragments were obtained in independent experiments with different proteolytic enzymes:

| Proteins | Fragments | | |
|---|---|---|---|
| Enzyme I | 624 | 24635 | 135 |
| Enzyme II | 136 | 356 | 524   24 |

Determine the sequence of fragments and amino acids in the protein.

6. Shown below are two homologous lengths of the alpha and beta chains of human hemoglobin. Consult the genetic code dictionary (Figure 11.8) and determine how many amino acid substitutions may have occurred as a result of a single nucleotide substitution. For any that cannot occur as the result of a single change, determine the minimal mutational distance.

| *Alpha:* | Ala | Val | Ala | His | Val | Asp | Asp | Met | Pro |
|---|---|---|---|---|---|---|---|---|---|
| *Beta:* | Gly | Leu | Ala | His | Leu | Asp | Asn | Leu | Lys |

7. Determine the minimal mutational distances between the following amino acid sequences of cytochrome c from various organisms. Compare the distance between humans and each organism.

| *Human:* | Lys | Glu | Glu | Arg | Ala | Asp |
|---|---|---|---|---|---|---|
| *Horse:* | Lys | Thr | Glu | Arg | Glu | Asp |
| *Pig:* | Lys | Gly | Glu | Arg | Glu | Asp |
| *Dog:* | Thr | Gly | Glu | Arg | Glu | Asp |
| *Chicken:* | Lys | Ser | Glu | Arg | Val | Asp |
| *Bullfrog:* | Lys | Gly | Glu | Arg | Glu | Asp |
| *Fungus:* | Ala | Lys | Asp | Arg | Asn | Asp |

8. The genetic difference between *D. heteroneura* and *D. sylvestris* as measured by nucleotide diversity is about 1.8 percent. The difference between chimpanzees *(Pan troglodytes)* and humans *(Homo sapiens)* is about the same, yet the latter species are classified in different genera. In your opinion, is this valid? If so, why; if not, why not?

9. As an extension of the previous question, consider the following: In sorting out the complex taxonomic relationships among birds, species with $\Delta T_{50}H$ values of 4.0 are placed in the same genus, even by traditional taxonomy based on morphology. Using the data in Figure 25.17, construct a phylogeny that obeys this rule, using any of the appropriate genus names *(Pongo, Pan, Homo),* or constructing new ones.

10. The use of nucleotide sequence data to measure genetic variability is complicated by the fact that the genes of higher eukaryotes are complex in organization and contain 5′ and 3′ flanking regions as well as introns. Slightom and colleagues have compared the nucleotide sequence of two cloned alleles of the gamma-globin gene from a single individual and found a variation of 1 percent. Those differences include 13 substitutions of one nucleotide for another, and three short DNA segments that have been inserted in one allele or deleted in the other. None of the

changes take place in the exons (coding regions) of the gene. Why do you think this is so, and should it change the concept of genetic variation?

**11.** Discuss the arguments supporting the neutral mutation theory. What counterarguments are proposed by the selectionists?

**12.** Of what value to our understanding of genetic variation and evolution is the debate concerning the neutral mutation theory?

**SELECTED READINGS**

ANDERSON W., et al. 1975. Genetics of natural populations, XLII. Three decades of genetic change in *Drosophila pseudoobscura. Evolution* 29:24–36.

AYALA, F. J. 1976. *Molecular evolution.* Sunderland, MA: Sinauer Associates.

———. 1984. Molecular polymorphism: How much is there, and why is there so much? *Dev. Genet.* 4:379–91.

BACHMAN, K. O. B., et al. 1972. Nuclear DNA amounts in vertebrates. In *Evolution of genetic systems,* ed. H. B. Smith, Brookhaven Symposium in Biology, vol. 23.

BARTON, N. H., and HEWITT, G. M. 1989. Adaptation, speciation and hybrid zones. *Nature* 341: 497–503.

BERNARDI, G., MOUCHROUD, D., GAUTIER, C., and BERNARDI, G. 1988. Compositional patterns in vertebrate genomes: Conservation and change in evolution. *J. Mol. Evol.* 28:7–18.

BRITTEN, R. J., and DAVIDSON, E. H. 1971. Repetitive and non-repetitive DNA sequences and a speculation on the origin of evolutionary novelty. *Quart. Rev. Biol.* 46:111–38.

CARSON, H. 1970. Chromosome tracers of the origin of species. *Science* 168:1414–18.

———. 1975. The genetics of speciation at the diploid level. *Amer. Natur.* 109:83–92.

COYNE, J. A. 1992. Genetics and speciation. *Nature* 355:511–15.

DAYHOFF, M. O. 1969. Computer analysis of protein evolution. *Scient. Amer.* (July) 221: 86–95.

DIAMOND, J. 1992. *The third chimpanzee: The evolution and future of the human animal.* New York: Harper–Collins.

DOBZHANSKY, T. 1947. Adaptive changes induced by natural selection in wild populations of *Drosophila. Evolution* 1:1–16.

———. 1948. Genetics of natural populations, XVI. Altitudinal and seasonal changes produced by natural selection in certain populations of *Drosophila pseudoobscura* and *Drosophila persimilis. Genetics* 33:158–76.

———. 1955. *Genetics of the evolutionary process.* New York: Columbia University Press.

DOBZHANSKY, T., et al. 1966. Genetics of natural populations, XXXVIII. Continuity and change in populations of *Drosophila pseudoobscura* in western United States. *Evolution* 20:418–27.

DOVER, G. 1982. Molecular drive: A cohesive mode of species evolution. *Nature* 299:11–17.

DOVER, G. A., and FLAVELL, R. B., eds. 1982. *Genome Evolution.* Orlando: Academic Press.

ELDREDGE, N. 1985. *Time frames: The evolution of punctuated equilibria.* Princeton, NJ: Princeton University Press.

EFSTRATIADIS, A., et al. 1980. The structure and evolution of the human beta-globin gene family. *Cell* 21:653–68.

FITCH, W. M. 1973. Aspects of molecular evolution. *Ann. Rev. Genet.* 7:343–80.

FITCH, W. M., and MARGOLIASH, E. 1967. Construction of phylogenetic trees. *Science* 155:279–84.

———. 1970. The usefulness of amino acid and nucleotide sequences in evolutionary studies. *Evol. Biol.* 4:67–109.

GILLESPIE, J. H. 1992. *The causes of molecular evolution.* New York: Oxford University Press.

GOULD, S. J. 1982. Darwinism and the expansion of evolutionary theory. *Science* 216:380–87.

HALL, T., GRULA, J., DAVIDSON, E. H., and BRITTEN, R. J. 1980. Evolution of sea urchin non-repetitive DNA. *J. Mol. Evol.* 16:95–110.

HINEGARDNER, R. 1976. Evolution of genome size. In *Molecular evolution,* ed. F. J. Ayala, pp. 179–99. Sunderland, MA: Sinauer Associates.

Hunt, J., et al. 1981. Evolution distance in Hawaiian *Drosophila*. *J. Mol. Evol.* 17:361–67.

Jeffrey, A. 1979. DNA sequence variation in the $^G\gamma$-, $^A\gamma$-, $\delta$- and B-globin genes of man. *Cell* 18:1–10.

Jukes, T. H. 1966. *Molecules and evolution.* New York: Columbia University Press.

Kimura, M. 1979a. Model of effectively neutral mutations in which selective constraint is incorporated. *Proc. Natl. Acad. Sci.* 76:3440–44.

————. 1979b. The neutral theory of molecular evolution. *Scient. Ameri.* (Nov.) 241:98–126.

————. 1989. The neutral theory of molecular evolution and the world view of the neutralists. *Genome* 31:24–31.

King, M. C., and Wilson, A. C. 1975. Evolution at two levels: Molecular similarities and biological differences between humans and chimpanzees. *Science* 188:107–16.

Kohne, D. E. Chiscon, J. A., and Hoyer, B. H. 1972. Evolution of primate DNA sequences. *J. Hum. Evol.* 1:627–44.

Lande, R. 1989. Fisherian and Wrightian theories of speciation. *Genome* 31:221–27.

Leder, A., Swan, D., Ruddle, F. H., D'eustachio, P., and Leder P. 1981. Dispersion of alpha-like globin genes of the mouse to three different chromosomes. *Nature* 293:196–200.

Lewin, R. 1993. *Human evolution.* 3rd ed. Cambridge, MA: Blackwell Scientific Publications.

Lewontin, R. C., and Hubby, J. L. 1966. A molecular approach to the study of genic heterozygosity in natural populations. II. Amount of variation and degree of heterozygosity in natural populations of *Drosophila pseudoobscura*. *Genetics* 54:595–609.

Mayr, E. 1963. *Animal species and evolution.* Cambridge, MA: Harvard University Press.

Nagel, R. L., and Labie, D. 1989. DNA haplotypes and the beta s globin gene. *Prog Clin. Biol. Res.* 316B: 371–93.

Pagel, M. D., and Harvey, P. H. 1989. Comparative methods for examining adaptation depend on evolutionary models. *Folia Primatol.* 53:203–20.

Powell, J. 1978. The founder-flush speciation theory: An experimental approach. *Evolution* 32:465–74.

Ridley, M. 1993. *Evolution.* Cambridge, Blackwell Scientific Publications.

Roberts, D. F. 1988. Migration and genetic change. *Hum. Biol.* 60:521–39.

Sibley, C. and Ahlquist, J. 1984. The phylogeny of the hominoid primates, as indicated by DNA-DNA hybridization. *J. Mol. Evol.* 20:2–15.

Sibley, C. G., Comstock, J. A., and Ahlquist, J. E. 1990. DNA evidence of hominoid phylogeny: A Re-analysis of the data. *J. Mol. Evol.* 30:202–236.

Smith, J. M. 1989. *Evolutionary genetics.* New York: Oxford University Press.

Smithies, O., Blechl, A. E., Shen, S., Slightom, J. L., and Vanin, E. F. 1981. Co-evolution and control of globin genes. In *Levels of genetic control in development,* eds. S. Subtelny and U. Abbot, pp. 185–200. New York: Alan R. Liss.

Stebbins, G. L. 1977. *Processes of organic evolution.* 3rd ed. Englewood Cliffs, NJ: Prentice-Hall.

Stebbins, G. L., and Ayala, F. J. 1981. Is a new evolutionary synthesis necessary? *Science* 213:967–71.

Tashian, R. E., and Carter, N. D. 1976. Biochemical genetics of carbonic anhydrase. In *Advances in human genetics,* eds. H. Harris and K. Hirschhorn, pp. 1–56. New York: Plenum Press.

Templeton, A. R. 1985. Phylogeny of the hominoid primates: A statistical analysis of the DNA-RNA hybridization data. *Mol. Biol. Evol.* 2:420–33.

Thorne, A. G., and Wolpoff, M. H. 1992. The multiregional evolution of humans. *Scient. Amer.* (April) 266:76–83.

Val, F. C. 1977. Genetic analysis of the morphological differences between two interfertile species of Hawaiian *Drosophila*. *Evolution* 31:611–29.

White, M. J. D. 1977, *Modes of speciation.* New York: W.H. Freeman.

Wilson, A. C., and Cann R. L. 1992. The recent African genesis of humans. *Scient. Amer.* (April) 266:68–73.

Yunis, J. J., and Prakash, O. 1982. The origin of man: A chromosomal pictorial legacy. *Science* 215:1525–30.

# APPENDIX A

# EXPERIMENTAL METHODS

In addition to the techniques of genetic analysis, physical and chemical techniques for the separation and analysis of macromolecular components of the cell nucleus and cytoplasm have been instrumental in advancing our understanding of genetics at the molecular level. In this appendix we will describe the background and theoretical basis of some techniques that have been important in molecular genetics.

## ISOTOPES

**Isotopes** are forms of an element that have the same number of protons and electrons but differ in the number of neutrons contained in the atomic nucleus. For example, the most common form of carbon has an atomic number of 6 (the number of protons in the nucleus) and an atomic weight of 12 (the sum of the protons and neutrons in the nucleus). In a very small percentage of carbon atoms, a seventh neutron is present, producing an atom with an atomic weight of 13. This is an example of a so-called **heavy isotope**. Since the number of protons and electrons, and thus the net charge, has not changed, the atom has the same chemical properties as **carbon-12 ($^{12}$C)** and differs only in mass. **Carbon-13 ($^{13}$C)** is thus a stable, heavy isotope of carbon.

Although the addition of neutrons does not alter the chemical properties of an atom, it can produce instabilities in the atomic nucleus. If another neutron is added to a carbon-13 atom, the isotope **carbon-14 ($^{14}$C)** results.

However, the presence of eight neutrons and six protons is an unstable condition, and the atom undergoes a nuclear reaction in which radiation is emitted during the transition to a more stable condition. Therefore, carbon-14 is a **radioactive isotope** of carbon.

The type of radiation emitted and the rate at which these nuclear events take place are characteristic of the element. Table A.1 lists types of radioactivity. The rate at which a radioactive isotope emits radiation is expressed as its **half-life**, which is the time required for a given amount of a radioactive substance to lose one-half of its radioactivity. Table A.2 lists some of the isotopes available for use in research.

**Table A.2** SOME ISOTOPES USED IN RESEARCH

| Element | Isotope | Half-life | Radiation |
|---------|---------|-----------|-----------|
| H | $^2$H | — | Stable |
|   | $^3$H | 12.3 years | $\beta$ |
| C | $^{13}$C | — | Stable |
|   | $^{14}$C | 5700 years | $\beta$ |
| N | $^{15}$N | — | Stable |
| O | $^{18}$O | — | Stable |
| P | $^{32}$P | 14 days | $\beta$ |
| S | $^{35}$S | 87 days | $\beta$ |
| K | $^{40}$K | $1.2 \times 10^9$ years | $\beta$, gamma |
| Fe | $^{59}$Fe | 45 days | $\beta$, gamma |
| I | $^{125}$I | 60 days | $\beta$, gamma |
|   | $^{131}$I | 8 days | $\beta$, gamma |

**Table A.1** PROPERTIES OF IONIZING RADIATION

| Type | Relative Penetration | Relative Ionization | Range in Biological Tissue |
|------|----------------------|---------------------|----------------------------|
| Alpha particle (2 protons + 2 neutrons) | 1 | 10,000 | Microns |
| Beta particle (electron) | 100 | 100 | Microns–mm |
| Gamma ray | >1000 | <1 | $\infty$ |

# DETECTION OF ISOTOPES

The choice of which isotope to use as a tracer in biological experiments depends on a combination of its physical and chemical properties, which enable the investigator to quantitate the amount of radioactivity or measure the ratio of heavy to light isotopes. For the detection of heavy isotopes, two methods are commonly employed: **mass spectrometry** and **equilibrium density gradient centrifugation** (discussed in the following section). Although the use of heavy isotopes has been more restricted than that of radioisotopes, they have been instrumental in several basic advances in molecular biology [e.g., demonstrating the semiconservative nature of DNA replication and the existence of messenger RNA (mRNA)].

There are a number of methods to detect radioisotopes, the foremost being **liquid scintillation spectrometry**, which provides quantitative information about the amount of radioactive isotope present in a sample, and **autoradiography**, which is used to demonstrate the cytological distribution and localization of radioactively labeled molecules.

In recording radioactivity by liquid scintillation counting, a small sample of the material to be counted is solubilized and immersed in a solution containing a **phosphor**, an organic compound that emits a flash of light after it absorbs energy released by decay of the radioactive compound. The counting chamber of the liquid scintillation spectrometer is equipped with very sensitive photomultiplier tubes that record the light flashes emitted by the phosphor. The data are recorded as counts of radioactivity per unit time and are displayed on a printout or can be fed into a computer for storage.

In autoradiography, a gel, chromatogram, or plant or animal part is placed against a sheet of photographic film. Radioactive decay from the incorporated isotope behaves just as light energy does and reduces silver grains in the emulsion. After exposure, the sheet or film is developed and fixed, revealing a deposit of silver grains corresponding to the location of the radioactive substance (Figure A.1).

Alternatively, to record the subcellular localization of an incorporated labeled isotope, cells or chromosomal preparations that have been incubated with radioactively labeled compounds are affixed to microscope slides and covered with a thin layer of liquid photographic emulsion. After they are exposed, the slides are processed to develop and fix the reduced silver grains in the emulsion. After staining, microscopic examination reveals the location and extent of labeled isotope incorporation (Figure A.2).

# CENTRIFUGATION TECHNIQUES

The centrifugation of biological macromolecules is widely employed to provide information about their physical characteristics (e.g., size, shape, density, and molecular weight) and to purify and concentrate cells, organelles, and their molecular components.

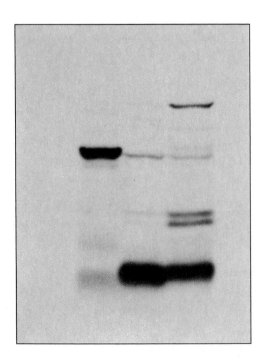

**FIGURE A.1**    Autoradiogram of radioactive bacterial proteins synthesized in a minicell system.

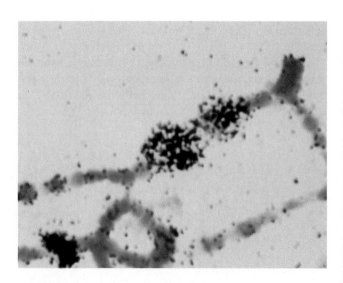

**FIGURE A.2**    Autoradiogram of RNA synthesis in salivary glands of *Drosophila* larva. Silver grains are deposited over sites of RNA synthesis at chromosome puffs.

**Differential centrifugation** is commonly used to separate materials such as cell homogenates according to size. Initially, the homogenate is distributed uniformly in the centrifuge tube. After a period of centrifugation, the pellet obtained is enriched for the largest and most dense particles, such as nuclei, in the homogenate. After each step, the supernatant can be recentrifuged at higher speeds to pellet the next heavier component. A typical fractionation scheme for cell homogenates is shown in Figure A.3. Further fractionation using density gradient techniques can be used to purify any of the fractions obtained by differential centrifugation.

**Rate zonal centrifugation** is used to separate particles on the basis of differences in their sedimentation rates. It may be used to separate mixtures of macromolecules such as proteins or nucleic acids and cellular organelles such as mitochondria. In this technique, which employs a medium of increasing density, the rate at which particles sediment depends on size, shape, density, and the frictional resistance of the solvent.

In addition to the preparation and purification of macromolecules and cellular components, rate zonal centrifugation can be used to determine the **sedimentation coefficients** and **molecular weights** of biological macromolecules. If a purified molecule such as a protein is spun in a centrifugal field, the molecule will eventually sediment toward the bottom at a constant velocity. At this point, the molecular weight ($M$) can be calculated as

$$M = f \times v/\omega^2 r$$

where $f$ is the frictional coefficient of the solvent system (which has been calculated from other measurements)

and $v/\omega^2 r$ is the rate of sedimentation per unit applied centrifugal field. The latter value is given the symbol $S$, or sedimentation coefficient. The $S$ value for most proteins is between $1 \times 10^{-13}$ sec and $2 \times 10^{-11}$ sec. A sedimentation coefficient of $1 \times 10^{-13}$ is defined as one **Svedberg unit ($S$)**; this unit is named for The Svedberg, a pioneer in the field of centrifugation. Thus, a protein with a sedimentation value of $2 \times 10^{-11}$ sec would have a value of 200$S$. Figure A.4 shows the $S$ values of selected molecules and particles.

**Isopycnic centrifugation**, or **equilibrium density gradient centrifugation**, is one of the most widely used techniques in genetics and molecular biology. In this technique, the solvent varies in density from one end of the tube to the other. A mixture of molecules layered on top and centrifuged through this gradient will migrate toward the bottom of the tube until each particle reaches its isopycnic point—that is, the place in the gradient where the density of the solvent equals the buoyant density of the particle. When each molecular species in the mixture migrates to its own characteristic isopycnic point, it no longer moves and is at equilibrium no matter how much longer the centrifugal field is applied. After separation by this method, components may be recovered by puncturing the bottom of the tube and collecting fractions. Two methods of forming density gradients are commonly employed (Figure A.5), one using preformed gradients of sucrose, or soluble salts of heavy metals such as cesium chloride or cesium sulfate. In the second method, the gradient is formed by the action of the centrifugal field on the salt solution.

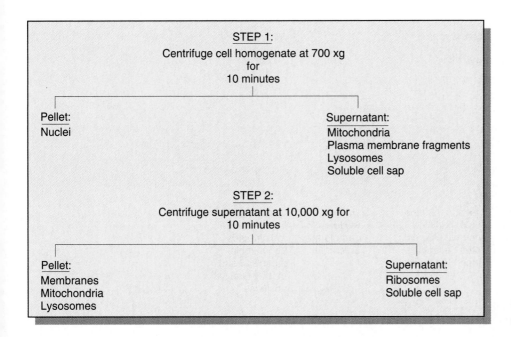

**FIGURE A.3**    Fractionation scheme for cell homogenates.

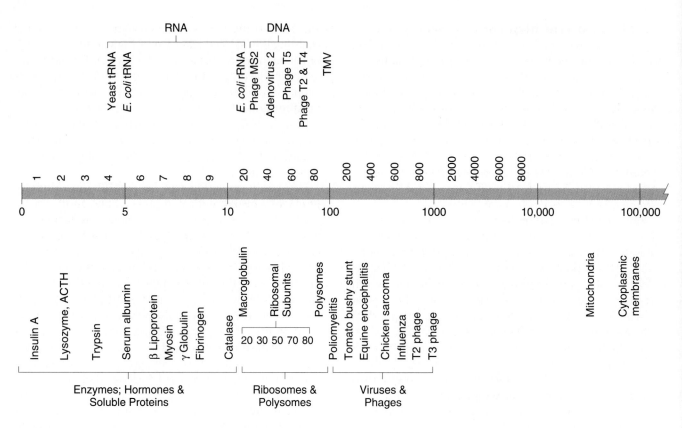

**FIGURE A.4**    *S* values of some common biological molecules and particles.

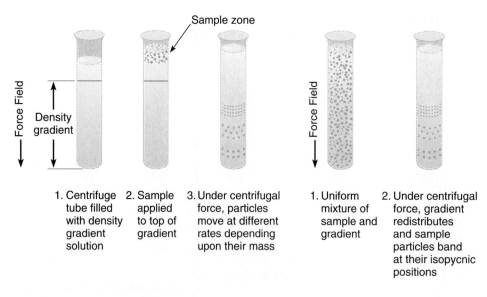

**FIGURE A.5**    Two methods of forming density gradients.

# RENATURATION AND HYBRIDIZATION OF NUCLEIC ACIDS

The ability of separated complementary strands of nucleic acids to unite and form stable, double-stranded molecules has been used to measure the relatedness of nucleic acids from different parts of the cell, different organs, and even different species. If both strands are DNA, the process is known as **renaturation** or **reassociation**. If one strand is RNA and the other DNA, it is known as **hybridization**. Renaturation and hybridization involve two steps: (1) a rate-limiting step, in which collision or nucleation between two homologous strands initiates base pairing, and (2) a rapid pairing of complementary bases, or "zippering" of the strands, to form a double-stranded molecule. For DNA renaturation, the formation of double-stranded molecules can be assayed at any time during an experiment. A sample is passed over a column of **hydroxyapatite**, which selectively binds double-stranded DNA but allows single-stranded molecules to pass through. The double-stranded molecules can be released from the column by raising the temperature or salt concentration.

The process of renaturation follows second-order kinetics according to the equation

$$\frac{C}{C_0} = \frac{1}{1 + kC_0 t}$$

where $C$ is the single-stranded DNA concentration at time $t$, $C_0$ is the total DNA concentration, and $k$ is a second-order rate constant. It is usually convenient to express the data from renaturation experiments as the fraction of single-stranded DNA at any time $t$ versus the product of total DNA concentration and time, as shown in Figure A.6.

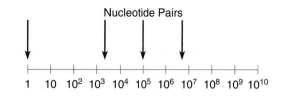

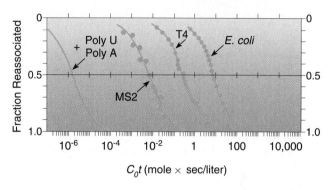

**FIGURE A.7**    Changes in $C_0 t$ value as genome size increases.

In the process of renaturation, the DNA is initially single-stranded and is renatured to double-stranded structures in the final state. The time necessary to reassociate half of the DNA in a sample at a given concentration should be proportional to the number of different pieces of DNA present. Consequently, the half reassociation time should be proportional to the DNA content of the genome, with smaller genomes having shorter half-renaturation rates. Figure A.7 confirms this expectation and shows increases in $C_0 t$ values as the size of the genome increases. This proportional relationship between $C_0 t$ and genome size is valid only in cases where repetitive DNA sequences are absent from the DNA being studied. DNA from calf thymus (and many other eukaryotic sources) exhibits a complex pattern of reassociation, indicating that bovine DNA contains some sequences (in this case, 40% of the total DNA) that reassociate rapidly and others that reassociate more slowly. The rapidly reassociating fraction must therefore contain sequences that are present in many copies. The more slowly renaturing DNA, however, contains sequences present in only one copy per genome. The *E. coli* DNA shown in Figure A.8 renatures with a pattern close to that of an ideal second-order reaction, indicating the absence of significant amounts of repeated DNA sequences.

In the case of hybridization between DNA and RNA, two approaches can be used—either the RNA or the DNA can be in excess. RNA-excess hybridization is usually preferred because most double-stranded molecules

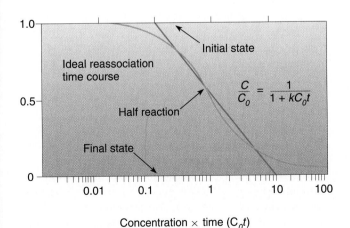

**FIGURE A.6**    Idealized $C_0 t$ curve.

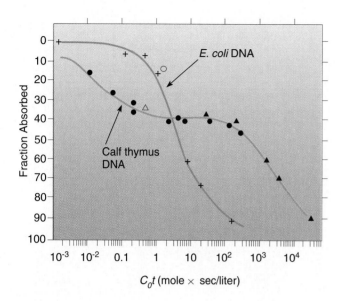

**FIGURE A.8**    Renaturation curve for *E. coli* DNA, containing no repetitive DNA sequences, and for calf thymus DNA, containing several families of repetitive DNA.

that are formed are RNA:DNA hybrids. Since DNA is present in low concentration, and since single-stranded RNA molecules lack complementary RNA strands in the mixture, the number of RNA:RNA and DNA:DNA hybrids is negligible.

The extent of hybridization can be followed by using hydroxyapatite columns or by using radioactively labeled RNA or DNA. In practice, one of the components of the hybridization reaction is usually immobilized on a substrate such as nitrocellulose paper. For example, DNA may be sheared to a uniform size, denatured to single strands, immobilized on nitrocellulose filters, and hybridized with an excess of labeled RNA. After hybridization, the unbound RNA is removed by washing, and the hybrids assayed by liquid scintillation counting. DNA:RNA hybridization can also be performed using cytological preparations, a technique called *in situ* **hybridization**. DNA, which is a part of an intact chromosome preparation fixed to a slide, can be denatured and hybridized to radioactive RNA. Hybrid formation is detected using autoradiography (Figure A.9).

## ELECTROPHORESIS

**Electrophoresis** is a technique that measures the rate of migration of charged molecules in a liquid-containing medium when an electrical field is applied to the liquid. Negatively charged molecules (anions) will migrate toward the positive electrode (anode) and positively charged molecules (cations) will migrate toward the negative electrode (cathode). Several factors affect the rate of migration, including the strength of the electrical field and the molecular sieving action of the medium (paper, starch, or gel) in which migration takes place. Since proteins and nucleic acids are electrically charged, electrophoresis has been used extensively to provide information about the size, conformation, and net charge of these macromolecules. On a larger scale, electrophoresis provides a method of fractionation that can be used to isolate individual components in mixtures of proteins or nucleic acids.

More recently, the analytical separation of proteins has been enhanced with the development of two-dimensional electrophoresis techniques, in which separation in the first dimension is by net charge, and in the second dimension by size and molecular weight.

In practice, electrophoresis employs a buffer system; a medium (paper, cellulose acetate, starch gel, agarose gel, or polyacrylamide gel); and a source of direct current. Samples are applied, and current is passed through the system for an appropriate time. Following migration

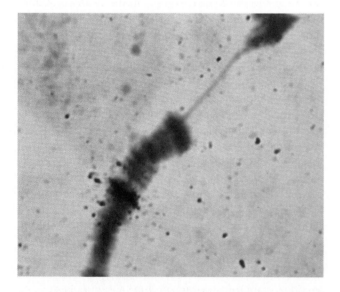

**FIGURE A.9**    Light micrograph of *in situ* hybridization of radioactive 5*S* RNA to a single band of polytene chromosome in the salivary gland of a *Drosophila* larva.

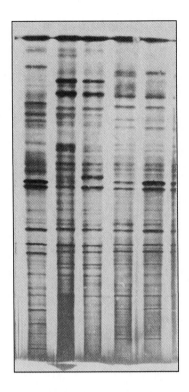

**FIGURE A.10**    Coomassie blue-stained protein slab gel.

of the molecules, the gel or paper may be treated with selective stains to reveal the location of the separated components (Figure A.10).

# DNA SEQUENCING

The ability to determine directly the sequence of DNA segments has added immensely to our understanding of gene structure and the mechanisms of gene regulation. Although techniques were available in the 1940s to determine the base composition of DNA, it was not until the 1960s that methods providing direct chemical analysis of nucleotide sequence were developed and used. These first methods were based on those used in determining the amino acid sequence of proteins, and were time-consuming and laborious. For example, in 1965 Robert Holley determined the sequence of a tRNA molecule consisting of 74 nucleotides, a task that required almost a year of concentrated effort.

In the 1970s more efficient and direct methods of analysis of nucleotide sequencing were developed. These methods developed in parallel with recombinant DNA techniques that allow the isolation of large quantities of purified DNA segments from any organism. A chemical method, developed by Allan Maxam and Walter Gilbert, cleaves DNA at specific bases. A second method, developed by Fred Sanger and colleagues, synthesizes a stretch of DNA that terminates at a given base. In both methods, the DNA to be sequenced is subjected to four individual reactions (one for each base). The products of the four reactions are a series of DNA fragments that differ in length by only one nucleotide. These reaction products are electrophoretically separated in four adjacent lanes on a gel. Each band on the gel corresponds to a base, and the sequence of the DNA segment can be read from the bands on the gel.

In the Maxam–Gilbert method, the complementary strands of the DNA fragment to be sequenced are separated and recovered. The strand to be sequenced is labeled at its 5' end with radioactive $^{32}P$ using the enzyme polynucleotide kinase. This provides a means of identifying specific DNA fragments after gel electrophoresis. In the next step, aliquots of the DNA are subjected to each of four chemical treatments that cleave the strand at a specific nucleotide. The reaction is carried out for a limited time so that any given molecule is cleaved at only a small number of the target nucleotides. The result is a collection of fragments, all labeled at the 5' end, but differing in length, depending on the point of cleavage.

Four different reactions cleave DNA at guanine (G > A), adenine (A > G), cytosine alone (C), or cytosine and thymine (C + T). In these reactions, purines are cleaved by using dimethylsulfate. This reagent methylates guanine far more efficiently than adenine, and when heat is applied, the strand is broken at the methylated site, producing a DNA fragment most frequently cleaved at a G residue (G > A). This can be reversed by cleaving the strand in acid, producing fragments most often cleaved at an A residue (A > G). The reaction for pyrimidines uses hydrazine, which cleaves both cytosine and thymine; however, in high salt (2M NaCl), only cytosine reacts. Thus, in the two reactions, one represents only cytosine (C), and the other represents cytosine and thymine (C + T).

For each set of reactions, the fragments produced are then subjected to gel electrophoresis in adjacent lanes on a gel. Each reaction contains a series of fragments in which strand breakage has occurred at one of the target bases, resulting in a series of fragments that differ in length. Under the influence of the electric field during electrophoresis, the different-sized fragments separate from each other, with the smallest fragments migrating the farthest. Since the DNA fragments contain a radioactive 5' end, the gel is subjected to autoradiography by placing a sheet of X-ray film over the gel for an appropriate exposure time. The DNA sequence is then analyzed by reading the bands on the gel from the bottom up, reading across all four lanes. In the example shown in Figure A.11, the first bases on the gel read GCGG.

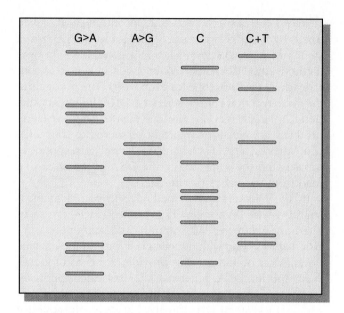

**FIGURE A.11**    Nucleotide sequence derived by the Maxam–Gilbert method.

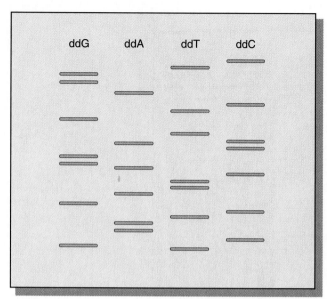

**FIGURE A.12**    Nucleotide sequence derived by the Sanger method.

In the Sanger method for DNA sequencing, which relies on an initial enzymatic treatment, the DNA strand to be sequenced is used as a template for the synthesis of a new DNA strand catalyzed by DNA polymerase I. This method employs a modified dideoxynucleoside triphosphate to generate a series of DNA fragments. The dideoxynucleoside triphosphates lack a 3′ OH group. This allows them to be added to a DNA strand undergoing synthesis, but since they lack a 3′ OH group, no nucleotide can be added to them, causing termination of strand synthesis, producing a DNA fragment. In use, four separate reaction mixtures are set up, each with a template DNA strand, a primer, all four radioactive nucleoside triphosphates, and a small amount of a single dideoxyribonucleoside triphosphate. Each of the four reactions contains a different dideoxynucleoside triphosphate that acts as a chain terminator. Because only a small amount of the modified nucleoside is used in each reaction, the newly synthesized strands are randomly terminated, producing a collection of fragments.

After synthesis, the radioactive fragments are electrophoresed in four adjacent lanes, one corresponding to each of the reactions. The fragments are visualized by autoradiography, and the gel is analyzed by reading the sequence from the bottom, as in the Maxam–Gilbert reaction. In the example shown in Figure A.12, the first band is in the lane with ddT, so it is a T residue, and the next few bands have the sequence GCAATCG.

DNA sequencing methods have generated a great deal of information about the structure and organization of many genes in a wide range of organisms. In the case of some viruses, the sequence of the entire genome is known, and large portions of the genome of other organisms, including *E. coli*, have been sequenced. To date, only a very small portion of the human genome has been sequenced, but the U.S. Department of Energy and the National Institutes of Health are coordinating efforts to develop the technology to sequence the more than 3 billion bases that constitute the haploid human genome.

## THE POLYMERASE CHAIN REACTION

The polymerase chain reaction (PCR) technique, described in Chapter 12, relies on the fact that the enzyme DNA polymerase requires a double-stranded primer section of DNA in order to initiate DNA synthesis. In the PCR technique, that primer is supplied in the form of a synthetic oligonucleotide, allowing the replication of a specific region of DNA. In other words, DNA polymerase can be directed to replicate only one specific region from an entire genome. Because these same primer sites are copied onto the newly synthesized strands of DNA, after each round of replication they can be used as primers. By repeating the replication process, a region of DNA can be amplified millions or billions of times, producing enough copies of the DNA for experimental purposes without the need for cloning (Figure A.13).

Instead of a single technique, the PCR reaction has become a versatile tool used in combination with other methods and has become the basis of a technical revolu-

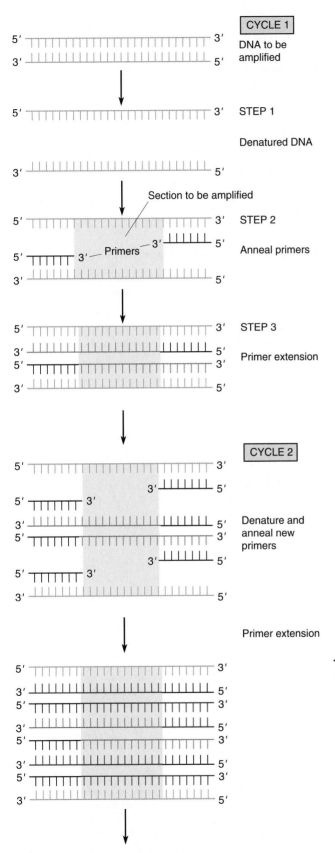

CYCLE 1

DNA to be amplified

STEP 1

Denatured DNA

Section to be amplified

STEP 2

Anneal primers

Primers

STEP 3

Primer extension

CYCLE 2

Denature and anneal new primers

Primer extension

Cycles 3-25 for greater than a $10^6$-fold increase in DNA

tion in molecular genetics. The PCR method has been used to detect the presence of mutations, produce *in vitro* mutations, diagnose genetic disorders, prepare DNA for sequencing, identify viruses and bacteria in infectious diseases, amplify DNA from fossils, analyze genetic defects in gametes and single cells from human embryos, and a host of other applications.

## Mutation Detection

A PCR-based method called single-strand conformation polymorphism (SSCP) has been used to screen for mutations caused by single base substitutions. Under certain conditions, single-stranded DNA fragments fold into nucleotide-sequence-dependent conformations. These conformations affect the mobility of the fragment during electrophoresis. Single base substitutions change the conformation of the DNA, altering its electrophoretic mobility. The method is fast, simple, and does not require DNA sequencing to detect single base changes.

To detect mutations using SSCP, the DNA of interest is first amplified by PCR. The double-stranded amplified DNA is then converted to a single-stranded form by boiling and is loaded onto a gel for electrophoresis. DNA from a normal gene is similarly amplified, denatured, and loaded in adjacent lanes. After electrophoresis, any variations in base sequence in the DNA being tested can be detected as a shifted band. The method gives best results when the DNA fragments being tested are around 200 base pairs, so screening an entire gene requires the use of several to many fragments, each about 200 base pairs in length.

In the development of new PCR-based techniques, one technique is often piggy-backed onto another. While SSCP is a rapid and simple technique, its accuracy is somewhat limited. DNA sequencing is a more accurate technique for detecting single base changes, but it is rel-

◄ **FIGURE A.13**    PCR amplification. In the polymerase chain reaction (PCR) method, the DNA to be amplified is first denatured into single strands, and then each strand is annealed to a different primer. The primers are synthetic oligonucleotides complementary to sequences flanking the region to be amplified. DNA polymerase is added along with nucleotides to extend the primers in the 3′ direction, resulting in a double-stranded DNA molecule with the primers incorporated into the newly synthesized strand. In a second PCR cycle, the products of the first cycle are denatured into single strands, primers are added, and they are annealed and extended by DNA polymerase. Repeated cycles can amplify the original DNA sequence by more than a million times.

atively slow and laborious. Recently, these two techniques have been combined into a method called dideoxy fingerprinting (ddF). The technique begins by amplifying a DNA segment by PCR. The amplified segment is used as a template for one of the Sanger dideoxy sequencing reactions, and the gel is run under conditions where the position of the band reflects both size (like DNA sequencing) and conformation (like SSCP). This combined technique is three times faster than sequencing alone and although more complex than SSCP, has a degree of accuracy approaching 100 percent.

## PCR Cycle Sequencing

Conventional DNA sequencing requires subcloning of fragments into vectors such as a plasmid, followed by growth of a host colony and extraction and purification of the cloned DNA. This DNA is then used as a template in the four reactions for DNA sequencing, followed by gel electrophoresis. The technique of cycle sequencing incorporates parts of the PCR reaction with parts of the standard reaction for the dideoxy method of DNA sequencing. The double-stranded DNA segment to be sequenced is first heated to form single strands that serve as templates. Primers are added, and DNA polymerase then begins the process of DNA replication. Extension of the primer continues until a labeled dideoxynucleotide is incorporated. The labeled strand is separated from the template (by heating) and is then used in another round of replication. The amplified extension products are then analyzed by gel electrophoresis. This adaptation of DNA sequencing is faster than conventional sequencing, requires only a very small amount of template, and can be used to directly sequence plasmid or phage clones. To further simplify the process, commercial kits are available that provide all materials and solutions for this procedure.

## Versatility: The Advantage of PCR

The examples above illustrate the versatility of the polymerase chain reaction and some of its applications in molecular genetics. This remarkable technique has so many adaptations and variations that several journals are devoted to reports on PCR. As evidence of its impact on molecular genetics, between 1991 and 1993 there were over 16,000 research reports on PCR.

# GLOSSARY

**A-DNA** An alternative form of the right-handed double-helical structure of DNA in which the helix is more tightly coiled, with 11 base pairs per full turn of the helix. In this form, the bases in the helix are displaced laterally and tilted in relation to the longitudinal axis. It is not yet clear whether this form has biological significance.

**abortive transduction** An event in which transducing DNA fails to be incorporated into the recipient chromosome. (See *transduction*.)

**acentric chromosome** Chromosome or chromosome fragment with no centromere.

**acquired immunodeficiency syndrome (AIDS)** An infectious disease caused by a retrovirus designated as human immunodeficiency virus (HIV). The disease is characterized by a gradual depletion of T lymphocytes, recurring fever, weight loss, multiple opportunistic infections, and rare forms of pneumonia and cancer associated with collapse of the immune system.

**acridine dyes** A class of organic compounds that bind to DNA and intercalate into the double-stranded structure, producing local disruptions of base pairing. These disruptions result in additions or deletions in the next round of replication.

**acrocentric chromosome** Chromosome with the centromere located very close to one end. Human chromosomes 13, 14, 15, 21, and 22 are acrocentric.

**active immunity** Immunity gained by direct exposure to antigens followed by antibody production.

**active site** That portion of a protein, usually an enzyme, whose structural integrity is required for function (e.g., the substrate binding site of an enzyme).

**adaptation** A heritable component of the phenotype that confers an advantage in survival and reproductive success. The process by which organisms adapt to the current environmental conditions.

**additive genes** See *polygenic inheritance*.

**additive variance** Genetic variance that is attributed to the substitution of one allele for another at a given locus. This variance can be used to predict the rate of response to phenotypic selection in quantitative traits.

**albinism** A condition caused by the lack of melanin production in the iris, hair, and skin. In humans, most often inherited as an autosomal recessive.

**aleurone layer** In seeds, the outer layer of the endosperm.

**alkaptonuria** An autosomal recessive condition in humans caused by the lack of an enzyme, homogentisic acid oxidase. Urine of homozygous individuals turns dark upon standing due to oxidation of excreted homogentisic acid. The cartilage of homozygous adults blackens from deposition of a pigment derived from homogentisic acid. Such individuals often develop arthritic conditions.

**allele** One of the possible mutational states of a gene, distinguished from other alleles by phenotypic effects.

**allele frequency** Measurement of the proportion of individuals in a population carrying a particular allele.

**allele-specific nucleotide (ASO)** Synthetic nucleotides, usually 15 to 20 bp in length that by hybridization under carefully controlled conditions will hybridize only to a complementary sequence with a perfect match. Under these same conditions, ASOs with a one-nucleotide mismatch will not hybridize.

**allelic exclusion** In plasma cell heterozygous for an immunoglobulin gene, the selective action of only one allele.

**allelism test** See *complementation test*.

**allopatric speciation** Process of speciation associated with geographic isolation.

**allopolyploid** Polyploid condition formed by the union of two or more distinct chromosome sets with a subsequent doubling of chromosome number.

**allosteric effect** Conformational change in the active site of a protein brought about by interaction with an effector molecule.

**allotetraploid** Diploid for two genomes derived from different species.

**allozyme** An allelic form of a protein that can be distinguished from other forms by electrophoresis.

**alpha fetoprotein (AFP)** A 70-kd glycoprotein synthesized in embryonic development by the yolk sac. High levels of this protein in the amniotic fluid are associated with neural tube defects such as spina bifida. Lower than normal levels may be associated with Down syndrome.

***Alu* sequence** An interspersed DNA sequence of approximately 300 bp found in the genome of primates that is cleaved by the restriction enzyme *Alu* I. *Alu* sequences are composed of a head-to-tail dimer, with the first monomer approximately 140 bp and the second approximately 170 bp. In humans, they are dispersed throughout the genome and are present in 300,000 to 600,000 copies, constituting some 3 to 6 percent of the genome. See *SINEs*.

**amber codon** The codon UAG, which does not code for an amino acid but for chain termination.

**Ames test** An assay developed by Bruce Ames to detect mutagenic and carcinogenic compounds using reversion to histidine independence in the bacterium *Salmonella typhimurium*.

**amino acid** Any of the subunit building blocks that are covalently linked to form proteins.

**aminoacyl tRNA**   Covalently linked combination of an amino acid and a tRNA molecule.

**amniocentesis**   A procedure used to test for fetal defects in which fluid and fetal cells are withdrawn from the amniotic layer surrounding the fetus.

**amphidiploid**   See *allotetraploid*.

**anabolism**   The metabolic synthesis of complex molecules from less complex precursors.

**analogue**   A chemical compound structurally similar to another, but differing by a single functional group (e.g., 5-bromodeoxyuridine is an analogue of thymidine).

**anaphase**   Stage of cell division in which chromosomes begin moving to opposite poles of the cell.

**aneuploidy**   A condition in which the chromosome number is not an exact multiple of the haploid set.

**angstrom**   Unit of length equal to $10^{-10}$ meter. Abbreviated Å.

**antibody**   Protein (immunoglobulin) produced in response to an antigenic stimulus with the capacity to bind specifically to the antigen.

**anticipation**   A phenomenon first observed in myotonic dystrophy, where the severity of the symptoms increases from generation to generation and the age of onset decreases from generation to generation. This phenomenon is caused by the expansion of trinucleotide repeats within or near a gene.

**anticodon**   The nucleotide triplet in a tRNA molecule that is complementary to, and binds to, the codon triplet in an mRNA molecule.

**antigen**   A molecule, often a cell surface protein, that is capable of eliciting the formation of antibodies.

**antiparallel**   Describing molecules in parallel alignment, but running in opposite directions. Most commonly used to describe the opposite orientations of the two strands of a DNA molecule.

**apoenzyme**   The protein portion of an enzyme that requires a cofactor or prosthetic group to be functional.

**ascospore**   A meiotic spore produced in certain fungi.

**ascus**   In fungi, the sac enclosing the four or eight ascospores.

**asexual reproduction**   Production of offspring in the absence of any sexual process.

**assortative mating**   Nonrandom mating between males and females of a species. Selection of mates with the same genotype is positive; selection of mates with opposite genotypes is negative.

**ATP**   Adenosine triphosphate.

**attached-X chromosome**   Two conjoined X chromosomes that share a single centromere.

**attenuator**   A nucleotide sequence between the promoter and the structural gene of some operons that can act to regulate the transit of RNA polymerase and thus control transcription of the structural gene.

**autogamy**   A process of self-fertilization resulting in homozygosis.

**autoimmune disease**   The production of antibodies that results from an immune response to one's own molecules, cells, or tissues. Such a response results from the inability of the immune system to distinguish self from nonself. Diseases such as arthritis, scleroderma, systemic lupus erythematosus, and perhaps diabetes are considered to be autoimmune diseases.

**autopolyploidy**   Polyploid condition resulting from the replication of one diploid set of chromosomes.

**autoradiography**   Production of a photographic image by radioactive decay. Used to localize radioactively labeled compounds within cells and tissues.

**autosomes**   Chromosomes other than the sex chromosomes. In humans, there are 22 pairs of autosomes.

**autotetraploid**   An autopolyploid condition composed of four similar genomes. In this situation, genes with two alleles (*A* and *a*) can have five genotypic classes: *AAAA* (quadraplex), *AAAa* (triplex), *AAaa* (duplex), *Aaaa* (simplex), and *aaaa* (nulliplex).

**auxotroph**   A mutant microorganism or cell line which requires a substance for growth that can be synthesized by wild-type strains.

**B-DNA**   See *double helix*.

**back-cross**   A cross involving an $F_1$ heterozygote and one of the $P_1$ parents (or an organism with a genotype identical to one of the parents).

**bacteriophage**   A virus that infects bacteria (synonym is *phage*).

**bacteriophage mu**   A group of phages whose genetic material behaves as an insertion sequence that can cause inactivation of host genes and rearrangement of host chromosomes.

**bacteriostatic**   A compound that inhibits the growth of bacteria, but does not kill them.

**balanced lethals**   Recessive, nonallelic lethal genes, each carried on different homologous chromosomes. When organisms carrying balanced lethal genes are interbred, only organisms with genotypes identical to the parents (heterozygotes) survive.

**balanced polymorphism**   Genetic polymorphism maintained in a population by natural selection.

**Barr body**   Densely staining nuclear mass seen in the somatic nuclei of mammalian females. Discovered by Murray Barr, this body is thought to represent an inactivated X chromosome.

**base analogue**   See *analogue*.

**base substitution**   A single base change in a DNA molecule that produces a mutation. There are two types of substitutions: *transitions,* in which a purine is substituted for a purine or a pyrimidine for a pyrimidine; and *transversions,* in which a purine is substituted for a pyrimidine, or vice versa.

**bidirectional replication**   A mechanism of DNA replication in which two replication forks move in opposite directions from a common origin of replication.

**biometry**   The application of statistics and statistical methods to biological problems.

**biotechnology**   Commercial and/or industrial processes that utilize biological organisms or products.

**bivalents**   Synapsed homologous chromosomes in the first prophase of meiosis.

**Bombay phenotype**   A rare variant of the ABO system in which affected individuals do not have A or B antigens, and thus appear as blood type O, even though their genotype may carry unexpressed alleles for the A and/or B antigens.

**BrdU (5-bromodeoxyuridine)**  A mutagenically active analogue of thymidine in which the methyl group at the 5' position in thymine is replaced by bromine.

**buoyant density**  A property of particles (and molecules) that depends upon their actual density, as determined by partial specific volume and degree of hydration. Provides the basis for density gradient separation of molecules or particles.

**CAAT box**  A highly conserved DNA sequence found about 75 base pairs 5' to the site of transcription in eukaryotic genes.

**cAMP**  Cyclic adenosine monophosphate. An important regulatory molecule in both prokaryotic and eukaryotic organisms.

**canonical sequence**  See *consensus sequence*.

**CAP**  Catabolite activator protein. A protein that binds cAMP and regulates the activation of inducible operons.

**carcinogen**  A physical or chemical agent that causes cancer.

**carrier**  An individual heterozygous for a recessive trait.

**cassette model**  First proposed to explain mating type interconversion in yeast, this model proposes that both genes for mating types, *a* and *alpha* are present as silent or unexpressed genes in transposable DNA segments (cassettes) that are activated (played) by transposition to the mating type locus.

**catabolism**  A metabolic reaction in which complex molecules are broken down into simpler forms, often accompanied by the release of energy.

**catabolite activator protein**  See *CAP*.

**catabolite repression**  The selective inactivation of an operon by a metabolic product of the enzymes encoded by the operon.

***cdc* mutation**  A class of mutations in yeast that affect the timing and progression through the cell cycle.

**cDNA**  DNA synthesized from an RNA template by the enzyme reverse transcriptase.

**cell cycle**  Sum of the phases of growth of an individual cell type; divided into $G_1$ (gap 1), S (DNA synthesis), $G_2$ (gap 2), and M (mitosis).

**cell-free extract**  A preparation of the soluble fraction of cells, made by lysing cells and removing the particulate matter, such as nuclei, membranes, and organelles. Often used to carry out the synthesis of proteins by the addition of specific, exogenous mRNA molecules.

**CEN**  In yeast, fragments of chromosomal DNA, about 120 bp in length, that when inserted into plasmids confer the ability to segregate during mitosis. These segments contain at least three types of sequence elements associated with centromere function.

**centimeter**  A unit of length equal to $10^{-2}$ meter. Abbreviated cm.

**centimorgan**  A unit of distance between genes on chromosomes. One centimorgan represents a value of 1 percent crossing over between two genes.

**central dogma**  The concept that information flow progresses from DNA to RNA to proteins. Although exceptions are known, this idea is central to an understanding of gene function.

**centric fusion**  See *Robertsonian translocation*.

**centriole**  A cytoplasmic organelle composed of nine groups of microtubules, generally arranged in triplets. Centrioles function in the generation of cilia and flagella and serve as foci for the spindles in cell division.

**centromere**  Specialized region of a chromosome to which the spindle fibers attach during cell division. Location of the centromere determines the shape of the chromosome during the anaphase portion of cell division. Also known as the primary constriction.

**centrosome**  Region of the cytoplasm containing the centriole.

**character**  An observable phenotypic attribute of an organism.

**charon phages**  A group of genetically modified lambda phages designed to be used as vectors for cloning foreign DNA. Named after the ferryman in Greek mythology who carried the souls of the dead across the River Styx.

**chemotaxis**  Negative or positive response to a chemical gradient.

**chiasma** (pl., **chiasmata**)  The crossed strands of nonsister chromatids seen in diplotene of the first meiotic division. Regarded as the cytological evidence for exchange of chromosomal material, or crossing over.

**chiasmatype theory**  The theory that crossing over between nonsister chromatids is the cause of chiasma formation.

**chi-square ($\chi^2$) analysis**  Statistical test to determine if an observed set of data fits a theoretical expectation.

**chloroplast**  A cytoplasmic self-replicating organelle containing chlorophyll. The site of photosynthesis.

**chorionic villus sampling (CVS)**  A technique of prenatal diagnosis that intravaginally retrieves fetal cells from the chorion and uses them to detect cytogenetic and biochemical defects in the embryo.

**chromatid**  One of the longitudinal subunits of a replicated chromosome, joined to its sister chromatid at the centromere.

**chromatin**  Term used to describe the complex of DNA, RNA, histones, and nonhistone proteins that make up chromosomes.

**chromatography**  Technique for the separation of a mixture of solubilized molecules by their differential migration over a substrate.

**chromocenter**  An aggregation of centromeres and heterochromatic elements of polytene chromosomes.

**chromomere**  A coiled, beadlike region of a chromosome most easily visible during cell division. The aligned chromomeres of polytene chromosomes are responsible for their distinctive banding pattern.

**chromosomal aberration**  Any change resulting in the duplication, deletion, or rearrangement of chromosomal material.

**chromosomal mutation**  See *chromosomal aberration*.

**chromosomal polymorphism**  Alternative structures or arrangements of a chromosome that are carried by members of a population.

**chromosome**    In prokaryotes, an intact DNA molecule containing the genome; in eukaryotes, a DNA molecule complexed with RNA and proteins to form a threadlike structure containing genetic information arranged in a linear sequence.

**chromosome banding**    Technique for the differential staining of mitotic or meiotic chromosomes to produce a characteristic banding pattern or selective staining of certain chromosomal regions such as centromeres, the nucleolus organizer regions, and GC- or AT-rich regions. Not to be confused with the banding pattern present in unstained polytene chromosomes, which is produced by the alignment of chromomeres.

**chromosome map**    A diagram showing the location of genes on chromosomes.

**chromosome puff**    A localized uncoiling and swelling in a polytene chromosome, usually regarded as a sign of active transcription.

**chromosome theory of inheritance**    The idea put forward by Walter Sutton and Theodore Boveri that chromosomes are the carriers of genes and the basis for the Mendelian mechanisms of segregation and independent assortment.

***cis* configuration**    The arrangement of two mutant sites within a gene on the same homologue, such as

$$\frac{a^1 \ a^2}{+ \ +}$$

Contrasts with a *trans* arrangement, where the mutant alleles are located on opposite homologues.

***cis* dominance**    The ability of a gene to affect the expression of other genes adjacent to it on the chromosome.

***cis-trans* test**    A genetic test to determine whether two mutations are located within the same cistron.

**cistron**    That portion of a DNA molecule that codes for a single polypeptide chain; defined by a genetic test as a region within which two mutations cannot complement each other.

**cline**    A gradient of genotype or phenotype distributed over a geographic range.

**clonal selection**    Theory of the immune system that proposes that antibody diversity precedes exposure to the antigen, and that the antigen functions to select the cells containing its specific antibody to undergo proliferation.

**clone**    Genetically identical cells or organisms all derived from a single ancestor by asexual or parasexual methods. For example, a DNA segment that has been enzymatically inserted into a plasmid or chromosome of a phage or a bacterium and replicated to form many copies.

**cloned library**    A collection of cloned DNA molecules representing all or part of an individual's genome.

**code**    See *genetic code*.

**codominance**    Condition in which the phenotypic effects of a gene's alleles are fully and simultaneously expressed in the heterozygote.

**codon**    A triplet of bases in a DNA or RNA molecule that specifies or encodes the information for a single amino acid.

**coefficient of coincidence**    A ratio of the observed number of double-crossovers divided by the expected number of such crossovers.

**coefficient of inbreeding**    The probability that two alleles present in a zygote are descended from a common ancestor.

**coefficient of selection**    A measurement of the reproductive disadvantage of a given genotype in a population. If for genotype *aa*, only 99 of 100 individuals reproduce, then the selection coefficient (*s*) is 0.1.

**colchicine**    An alkaloid compound that inhibits spindle formation during cell division. Used in the preparation of karyotypes to collect a large population of cells inhibited at the metaphase stage of mitosis.

**colicin**    A bacteriocidal protein produced by certain strains of *E. coli* and other closely related bacterial species.

**colinearity**    The linear relationship between the nucleotide sequence in a gene (or the RNA transcribed from it) and the order of amino acids in the polypeptide chain specified by the gene.

**competence**    In bacteria, the transient state or condition during which the cell can bind and internalize exogenous DNA molecules, making transformation possible.

**complementarity**    Chemical affinity between nitrogenous bases as a result of hydrogen bonding. Responsible for the base pairing between the strands of the DNA double helix.

**complementation test**    A genetic test to determine whether two mutations occur within the same gene. If two mutations are introduced into a cell simultaneously and produce a wild-type phenotype (i.e., they complement each other), they are often nonallelic. If a mutant phenotype is produced, the mutations are noncomplementing and are often allelic.

**complete linkage**    A condition in which two genes are located so close to each other that no recombination occurs between them.

**complexity**    The total number of nucleotides or nucleotide pairs in a population of nucleic acid molecules as determined by reassociation kinetics.

**complex locus**    A gene within which a set of functionally related pseudoalleles can be identified by recombinational analysis (e.g., the *bithorax* locus in *Drosophila*).

**concatemer**    A chain or linear series of subunits linked together. The process of forming a concatemer is called concatenation (e.g., multiple units of a phage genome produced during replication).

**concordance**    Pairs or groups of individuals identical in their phenotype. In twin studies, a condition in which both twins exhibit or fail to exhibit a trait under investigation.

**conditional mutation**    A mutation that expresses a wild-type phenotype under certain (permissive) conditions and a mutant phenotype under other (restrictive) conditions.

**conjugation**    Temporary fusion of two single-celled organisms for the sexual exchange of genetic material.

**consanguine**    Related by a common ancestor within the previous few generations.

**consensus sequence**    A nucleotide sequence most often found in a defined segment of DNA.

**continuous variation**    Phenotype variation exhibited by quantitative traits distributed from one phenotypic extreme to another in an overlapping or continuous fashion.

**cosmid**    A vector designed to allow cloning of large segments of foreign DNA. Cosmids are hybrids composed of the cos sites of lambda inserted into a plasmid. In cloning, the recombinant DNA molecules are packaged into phage protein coats, and after infection of bacterial cells, the recombinant molecule replicates and can be maintained as a plasmid.

**coupling conformation**   See *cis configuration.*

**covalent bond**   A nonionic chemical bond formed by the sharing of electrons.

**cri-du-chat syndrome**   A clinical syndrome in humans produced by a deletion of a portion of the short arm of chromosome 5. Afflicted infants have a distinctive cry which sounds like that of a cat.

**crossing over**   The exchange of chromosomal material (parts of chromosomal arms) between homologous chromosomes by breakage and reunion. The exchange of material between nonsister chromatids during meiosis is the basis of genetic recombination.

**cross-reacting material (CRM)**   Nonfunctional form of an enzyme, produced by a mutant gene, which is recognized by antibodies made against the normal enzyme.

**C-terminal amino acid**   The terminal amino acid in a peptide chain which carries a free carboxyl group.

**C terminus**   The end of a polypeptide that carries a free carboxyl group of the last amino acid. By convention, the structural formula of polypeptides is written with the C terminus at the right.

**C value**   The haploid amount of DNA present in a genome.

**C value paradox**   The apparent paradox that there is no relationship between the size of the genome and the evolutionary complexity of species. For example, the C value (haploid genome size) of amphibians varies by a factor of 100.

**cyclins**   A class of proteins found in eukaryotic cells that are synthesized and degraded in synchrony with the cell cycle, and regulate passage through stages of the cycle.

**cytogenetics**   A branch of biology in which the techniques of both cytology and genetics are used to study heredity.

**cytokinesis**   The division or separation of the cytoplasm during mitosis or meiosis.

**cytological map**   A diagram showing the location of genes at particular chromosomal sites.

**cytoplasmic inheritance**   Non-Mendelian form of inheritance involving genetic information transmitted by self-replicating cytoplasmic organelles such as mitochondria, chloroplasts, etc.

**cytoskeleton**   An internal array of microtubules, microfilaments, and intermediate filaments that confers shape and the ability to move on a eukaryotic cell.

**dalton**   A unit of mass equal to that of the hydrogen atom, which is $1.67 \times 10^{-24}$ gram. A unit used in designating molecular weights.

**Darwinian fitness**   See *fitness.*

**deficiency (deletion)**   A chromosomal mutation involving the loss or deletion of chromosomal material.

**degenerate code**   Term used to describe the genetic code, in which a given amino acid may be represented by more than one codon.

**deletion**   See *deficiency.*

**deme**   A local interbreeding population.

**denatured DNA**   DNA molecules that have been separated into single strands.

**de novo**   Newly arising; synthesized from less complex precursors rather than having been produced by modification of an existing molecule.

**density gradient centrifugation**   A method of separating macromolecular mixtures by the use of centrifugal force and solvents of varying density. In sedimentation velocity centrifugation, macromolecules are separated by the velocity of sedimentation through a preformed gradient such as sucrose. In density gradient equilibrium centrifugation, macromolecules in a cesium salt solution are centrifuged until the cesium solution establishes a gradient under the influence of the centrifugal field, and the macromolecules sediment until the density of the solvent equals their own.

**deoxyribonuclease**   A class of enzymes that breaks down DNA into oligonucleotide fragments by introducing single-stranded breaks into the double helix.

**deoxyribonucleic acid (DNA)**   A macromolecule usually consisting of antiparallel polynucleotide chains held together by hydrogen bonds, in which the sugar residues are deoxyribose. The primary carrier of genetic information.

**deoxyribose**   The five-carbon sugar associated with the deoxyribonucleotides found in DNA.

**dermatoglyphics**   The study of the surface ridges of the skin, especially of the hands and feet.

**determination**   A regulatory event that establishes a specific pattern of gene activity and developmental fate for a given cell.

**diakinesis**   The final stage of meiotic prophase I in which the chromosomes become tightly coiled and compacted and move toward the periphery of the nucleus.

**dicentric chromosome**   A chromosome having two centromeres.

**differentiation**   The process of complex changes by which cells and tissues attain their adult structure and functional capacity.

**dihybrid cross**   A genetic cross involving two characters in which the parents possess different forms of each character (e.g., tall, round × short, wrinkled peas).

**diploid**   A condition in which each chromosome exists in pairs; having two of each chromosome.

**diplotene**   A stage of meiotic prophase I immediately after pachytene. In diplotene, one pair of sister chromatids begins separating from the other, and chiasmata become visible. These overlaps move laterally toward the ends of the chromatids (terminalization).

**directional selection**   A selective force that changes the frequency of an allele in a given direction, either toward fixation or toward elimination.

**discontinuous replication of DNA**   The synthesis of DNA in discontinuous fragments on the lagging strand of the replication fork. The fragments, known as Okazaki fragments, are joined by DNA ligase to form a continuous strand.

**discontinuous variation**   Phenotypic data that fall into two or more distinct classes that do not overlap.

**discordance**   In twin studies, a situation where one twin shows a trait but the other does not.

**disjunction**   The separation of chromosomes at the anaphase stage of cell division.

**disruptive selection**   Simultaneous selection for phenotypic extremes in a population, usually resulting in the production of two discontinuous strains.

**dizygotic twins**   Twins produced from separate fertilization events; two ova fertilized independently. Also known as fraternal twins.

**DNA**   See *deoxyribonucleic acid*.

**DNA footprinting**   See *footprinting*.

**DNA gyrase**   One of the DNA topoisomerases that functions during DNA replication to reduce molecular tension caused by supercoiling. DNA gyrase produces, then seals, double-stranded breaks.

**DNA ligase**   An enzyme that forms a covalent bond between the 5′ end of one polynucleotide chain and the 3′ end of another polynucleotide chain. Also called polynucleotide-joining enzyme.

**DNA polymerase**   An enzyme that catalyzes the synthesis of DNA from deoxyribonucleotides and a template DNA molecule.

**DNase**   Deoxyribonucleosidase, an enzyme that degrades or breaks down DNA into fragments or constitutive nucleotides.

**dominance**   The expression of a trait in the heterozygous condition.

**dosage compensation**   A genetic mechanism that regulates the levels of gene products at certain autosomal loci; this results in homozygous dominants and heterozygotes having the same amount of a gene product. In mammals, random inactivation of one X chromosome in females leads to equal levels of X chromosome-coded gene products in males and females.

**double-crossover**   Two separate events of chromosome breakage and exchange occurring within the same tetrad.

**double helix**   The model for DNA structure proposed by James Watson and Francis Crick, involving two antiparallel, hydrogen-bonded polynucleotide chains wound into a right-handed helical configuration, with 10 base pairs per full turn of the double helix. Often called B-DNA.

**duplication**   A chromosomal aberration in which a segment of the chromosome is repeated.

**dyad**   The products of tetrad separation or disjunction at the first meiotic prophase. Consists of two sister chromatids joined at the centromere.

**effector molecule**   Small, biologically active molecule that acts to regulate the activity of a protein by binding to a specific receptor site on the protein.

**electrophoresis**   A technique used to separate a mixture of molecules by their differential migration through a stationary phase in an electrical field.

**endocytosis**   The uptake by a cell of fluids, macromolecules, or particles by pinocytosis, phagocytosis, or receptor-mediated endocytosis.

**endogenote**   The segment of the chromosome in a partially diploid bacterial cell (merozygote) that is homologous to the chromosome transmitted by the donor cell.

**endomitosis**   Chromosomal replication that is not accompanied by either nuclear or cytoplasmic division.

**endonuclease**   An enzyme that hydrolyzes internal phosphodiester bonds in a polynucleotide chain or nucleic acid molecule.

**endoplasmic reticulum**   A membraneous organelle system in the cytoplasm of eukaryotic cells. The outer surface of the membranes may be ribosome-studded (rough ER) or smooth ER.

**endopolyploidy**   The increase in chromosome sets that results from endomitotic replication within somatic nuclei.

**endosymbiont theory**   The proposal that self-replicating cellular organelles such as mitochondria and chloroplasts were originally free-living organisms that entered into a symbiotic relationship with nucleated cells.

**enhancer**   Originally, a 72-bp sequence in the genome of the virus, SV40, that increases the transcriptional activity of nearby structural genes. Similar sequences that enhance transcription have been identified in the genomes of eukaryotic cells. Enhancers can act over a distance of thousands of base pairs and can be located 5′ or 3′ to the gene they affect, and thus are different from promoters.

**enhanson**   The DNA sequence that represents the core sequence of an enhancer.

**environment**   The complex of geographic, climatic, and biotic factors within which an organism lives.

**enzyme**   A protein or complex of proteins that catalyzes a specific biochemical reaction.

**epigenesis**   The idea that an organism develops by the appearance and growth of new structures. Opposed to preformationism, which holds that development is the growth of structures already present in the egg.

**episome**   A circular genetic element in bacterial cells that can replicate independently of the bacterial chromosome or integrate and replicate as part of the chromosome.

**epistasis**   Nonreciprocal interaction between genes such that one gene interferes with or prevents the expression of another gene. For example, in *Drosophila,* the recessive gene *eyeless,* when homozygous, prevents the expression of eye color genes.

**epitope**   That portion of a macromolecule or cell that acts to elicit an antibody response; an antigenic determinant. A complex molecule or cell can contain several such sites.

**equational division**   A division of each chromosome into longitudinal halves that are distributed into two daughter nuclei. Chromosome division in mitosis is an example of equational division.

**equatorial plate**   See *metaphase plate*.

**euchromatin**   Chromatin or chromosomal regions that are lightly staining and are relatively uncoiled during the interphase portion of the cell cycle. The region of the chromosomes thought to contain most of the structural genes.

**eugenics**   The improvement of the human species by selective breeding. Positive eugenics refers to the promotion of breeding of those with favorable genes, and negative eugenics refers to the discouragement of breeding among those with undesirable traits.

**eukaryotes**   Those organisms having true nuclei and membranous organelles and whose cells demonstrate mitosis and meiosis.

**euploid** Polyploid with a chromosome number that is an exact multiple of a basic chromosome set.

**evolution** The origin of plants and animals from preexisting types. Descent with modifications.

**excision repair** Repair of DNA lesions by removal of a poly-nucleotide segment and its replacement with a newly syn-thesized, corrected segment.

**exogenote** In merozygotes, the segment of the bacterial chromosome contributed by the donor cell.

**exon (extron)** The DNA segment(s) of a gene that are tran-scribed and translated into protein.

**exonuclease** An enzyme that breaks down nucleic acid mol-ecules by breaking the phosphodiester bonds at the 3′ or 5′ terminal nucleotides.

**expression vector** Plasmids or phages carrying promoter regions designed to cause expression of cloned DNA se-quences.

**expressivity** The degree or range in which a phenotype for a given trait is expressed.

**extranuclear inheritance** Transmission of traits by genetic information contained in cytoplasmic organelles such as mi-tochondria and chloroplasts.

**F$^+$ cell** A bacterial cell having a fertility (F) factor. Acts as a donor in bacterial conjugation.

**F$^-$ cell** A bacterial cell that does not contain a fertility (F) factor. Acts as a recipient in bacterial conjugation.

**F factor** An episome in bacterial cells that confers the ability to act as a donor in conjugation.

**F′ factor** A fertility (F) factor that contains a portion of the bacterial chromosome.

**F$_1$ generation** First filial generation; the progeny resulting from the first cross in a series.

**F$_2$ generation** Second filial generation; the progeny result-ing from a cross of the F$_1$ generation.

**F pilus** See *pilus.*

**facultative heterochromatin** Chromatin that may alternate in form between euchromatic and heterochromatic. The Y chromosome of many species contains facultative hetero-chromatin.

**familial trait** A trait transmitted through and expressed by members of a family.

**fate map** A diagram or "map" of an embryo showing the location of cells whose development fate is known.

**fertility (F) factor** See *F factor.*

**filial generations** See F$_1$, F$_2$ *generations.*

**fingerprint** The pattern of ridges and whorls on the tip of a finger. The pattern obtained by enzymatically cleaving a pro-tein or nucleic acid and subjecting the digest to two-dimen-sional chromatography or electrophoresis.

**FISH** See *fluorescent in situ hybridization.*

**fitness** A measure of the relative survival and reproductive success of a given individual or genotype.

**fixation** In population genetics, a condition in which all members of a population are homozygous for a given allele.

**fixity of species** The idea that members of a species can give rise only to other members of the species, thus implying that all species are independently created.

**fluctuation test** A statistical test developed by Salvadore Luria and Max Delbruck to determine whether bacterial mu-tations arise spontaneously or are produced in response to selective agents.

**fluorescent *in situ* hybridization (FISH)** A method of *in situ* hybridization that utilizes probes labeled with a fluores-cent tag, causing the site of hybridization to fluoresce when viewed under the microscope.

**fMet** See *formylmethionine.*

**footprinting** A technique for identifying a DNA sequence that binds to a particular protein, based on the idea that the phosphodiester bonds in the region covered by the protein are protected from digestion by deoxyribonucleases.

**formylmethionine (fMet)** A molecule derived from the amino acid methionine by attachment of a formyl group to its terminal amino group. This is the first amino acid inserted in all bacterial polypeptides. Also known as N-formyl methi-onine.

**founder effect** A form of genetic drift. The establishment of a population by a small number of individuals whose geno-types carry only a fraction of the different kinds of alleles in the parental population.

**fragile site** A heritable gap or nonstaining region of a chro-mosome that can be induced to generate chromosome breaks.

**fragile X syndrome** A genetic disorder caused by the ex-pansion of a CGG trinucleotide repeat and a fragile site at Xq27.3, within the FMR-1 gene.

**frameshift mutation** A mutational event leading to the in-sertion of one or more base pairs in a gene, shifting the codon reading frame in all codons following the mutational site.

**fraternal twins** See *dizygotic twins.*

**gamete** A specialized reproductive cell with a haploid num-ber of chromosomes.

**gap genes** Genes expressed in contiguous domains along the anterior–posterior axis of the *Drosophila* embryo, which regulate the process of segmentation in each domain.

**gene** The fundamental physical unit of heredity whose exis-tence can be confirmed by allelic variants and which occu-pies a specific chromosomal locus. A DNA sequence coding for a single polypeptide.

**gene amplification** The process by which gene sequences are selected for differential replication either extrachromo-somally or intrachromosomally.

**gene conversion** A directional process that changes one al-lele into another specific allele. In fungi, this process in-volves meiosis and a recombinational step, but in other organisms, such as trypanosomes, meiosis may not be required.

**gene duplication** An event in replication leading to the pro-duction of a tandem repeat of a gene sequence.

**gene flow** The gradual exchange of genes between two populations, brought about by the dispersal of gametes or the migration of individuals.

**gene frequency**   The percentage of alleles of a given type in a population.

**gene interaction**   Production of novel phenotypes by the interaction of alleles of different genes.

**gene mutation**   See *point mutation.*

**gene pool**   The total of all genes possessed by reproductive members of a population.

**generalized transduction**   The transduction of any gene in the bacterial genome by a phage.

**genetic burden**   Average number of recessive lethal genes carried in the heterozygous condition by an individual in a population. Also called genetic load.

**genetic code**   The nucleotide triplets that code for the 20 amino acids or for chain initiation or termination.

**genetic counseling**   Analysis of risk for genetic defects in a family and the presentation of options available to avoid or ameliorate possible risks.

**genetic drift**   Random variation in gene frequency from generation to generation. Most often observed in small populations.

**genetic engineering**   The technique of altering the genetic constitution of cells or individuals by the selective removal, insertion, or modification of individual genes or gene sets.

**genetic equilibrium**   Maintenance of allele frequencies at the same value in successive generations. A condition in which allele frequencies are neither increasing nor decreasing.

**genetic fine structure**   Intragenic recombinational analysis that provides mapping information at the level of individual nucleotides.

**genetic load**   See *genetic burden.*

**genetic polymorphism**   The stable coexistence of two or more discontinuous genotypes in a population. When the frequencies of two alleles are carried to an equilibrium, the condition is called balanced polymorphism.

**genetics**   The branch of biology that deals with heredity and the expression of inherited traits.

**genome**   The array of genes carried by an individual.

**genomic imprinting**   A condition where the expression of a trait depends on whether the trait has been inherited from a male or a female parent.

**genotype**   The specific allelic or genetic constitution of an organism; often, the allelic composition of one or a limited number of genes under investigation.

**germplasm**   Hereditary material transmitted from generation to generation.

**Goldberg–Hogness box**   A short nucleotide sequence 20 to 30 bp 5' to the initiation site of eukaryotic genes to which RNA polymerase II binds. The consensus sequence is TATAAAA. Also known as TATA box.

**graft versus host disease (GVHD)**   In transplants, reaction by immunologically competent cells of the donor against the antigens present on the cells of the host. In human bone marrow transplants, often a fatal condition.

**gynandromorph**   An individual composed of cells with both male and female genotypes.

**gyrase**   One of a class of enzymes known as topoisomerases. Gyrase converts closed circular DNA to a negatively super-coiled form prior to replication, transcription, or recombination.

**H substance**   The carbohydrate group present on the surface of red blood cells. When unmodified, it results in blood type O; when modified by the addition of monosaccharides, it results in type A, B, and AB.

**haploid**   A cell or organism having a single set of unpaired chromosomes. The gametic chromosome number.

**haplotype**   The set of alleles from closely linked loci carried by an individual and usually inherited as a unit.

**Hardy–Weinberg law**   The principle that both gene and genotype frequencies will remain in equilibrium in an infinitely large population in the absence of mutation, migration, selection, and nonrandom mating.

**heat shock**   A transient response following exposure of cells or organisms to elevated temperatures. The response involves activation of a small number of loci, inactivation of previously active loci, and selective translation of heat shock mRNA. Appears to be a nearly universal phenomenon observed in organisms ranging from bacteria to humans.

**helicase**   An enzyme that participates in DNA replication by unwinding the double helix near the replication fork.

**helix-turn-helix motif**   The structure of a region of DNA-binding proteins in which a turn of four amino acids holds two alpha helices at right angles to each other.

**hemizygous**   Conditions where a gene is present in a single dose. Usually applied to genes on the X chromosome in heterogametic males.

**hemoglobin (Hb)**   An iron-containing, conjugated respiratory protein occurring chiefly in the red blood cells of vertebrates.

**hemophilia**   A sex-linked trait in humans associated with defective blood-clotting mechanisms.

**heredity**   Transmission of traits from one generation to another.

**heritability**   A measure of the degree to which observed phenotypic differences for a trait are genetic.

**heterochromatin**   The heavily staining, late replicating regions of chromosomes that are condensed in interphase. Thought to be devoid of structural genes.

**heteroduplex**   A double-stranded nucleic acid molecule in which each polynucleotide chain has a different origin. These structures may be produced as intermediates in a recombinational event, or by the *in vitro* reannealing of single-stranded, complementary molecules.

**heterogametic sex**   The sex that produces gametes containing unlike sex chromosomes.

**heterogeneous nuclear RNA (hnRNA)**   The collection of RNA transcripts in the nucleus, representing precursors and processing intermediates to rRNA, mRNA, and tRNA. Also represents RNA transcripts that will not be transported to the cytoplasm, such as snRNA (small nuclear RNA).

**heterogenote**   A bacterial merozygote in which the donor (exogenote) chromosome segment carries different alleles than does the chromosome of the recipient (endogenote). A heterozygous merozygote.

**heterokaryon** A somatic cell containing nuclei from two different sources.

**heterosis** The superiority of a heterozygote over either homozygote for a given trait.

**heterozygote** An individual with different alleles at one or more loci. Such individuals will produce unlike gametes and therefore will not breed true.

**Hfr** A strain of bacteria exhibiting a high frequency of recombination. These strains have a chromosomally integrated F factor that is able to mobilize and transfer all or part of the chromosome to a recipient $F^-$ cell.

**histocompatibility antigens** See *HLA*.

**histones** Proteins complexed with DNA in the nucleus. They are rich in the basic amino acids arginine and lysine and function in the coiling of DNA to form nucleosomes.

**HLA** Cell surface proteins, produced by histocompatibility loci, which are involved in the acceptance or rejection of tissue and organ grafts and transplants.

**hnRNA** See *heterogeneous nuclear RNA*.

**holandric** A trait transmitted from males to males. In humans, genes on the Y chromosome are holandric.

**Holliday structure** An intermediate in bidirectional DNA replication seen in the transmission electron microscope as an X-shaped structure showing four single-stranded DNA regions.

**homeobox** A sequence of about 180 nucleotides that encodes a 60-amino-acid sequence called a homeodomain, which is a DNA-binding protein that acts as a transcription factor.

**homeotic mutation** A mutation that causes a tissue normally determined to form a specific organ or body part to alter its differentiation and form another structure. Alternatively spelled: homoeotic.

**homogametic sex** The sex that produces gametes that do not differ with respect to sex chromosome content; in mammals, the female is homogametic.

**homogeneously staining regions (hsr)** Segments of mammalian chromosomes that stain lightly with Giemsa following exposure of cells to a selective agent. These regions arise in conjunction with gene amplification and are regarded as the structural locus for the amplified gene.

**homogenote** A bacterial merozygote in which the donor (exogenote) chromosome carries the same alleles as the chromosome of the recipient (endogenote). A homozygous merozygote.

**homologous chromosomes** Chromosomes that synapse or pair during meiosis. Chromosomes that are identical with respect to their genetic loci and centromere placement.

**homozygote** An individual with identical alleles at one or more loci. Such individuals will produce identical gametes and will therefore breed true.

**homunculus** The miniature individual imagined by preformationists to be contained within the sperm or egg.

**human immunodeficiency virus (HIV)** A human retrovirus associated with the onset and progression of acquired immunodeficiency syndrome (AIDS).

**hybrid** An individual produced by crossing two parents of different genotypes.

**hybridoma** A somatic cell hybrid produced by the fusion of an antibody-producing cell and a cancer cell, specifically, a myeloma. The cancer cell contributes the ability to divide indefinitely, and the antibody cell confers the ability to synthesize large amounts of a single antibody.

**hybrid vigor** See *heterosis*.

**hydrogen bond** An electrostatic attraction between a hydrogen atom bonded to a strongly electronegative atom such as oxygen or nitrogen and another atom that is electronegative or contains an unshared electron pair.

**identical twins** See *monozygotic twins*.

**Ig** See *immunoglobulin*.

**imaginal disk** Discrete groups of cells set aside during embryogenesis in holometabolous insects which are determined to form the external body parts of the adult.

**immunoglobulin** The class of serum proteins having the properties of antibodies.

**inborn error of metabolism** A biochemical disorder that is genetically controlled; usually an enzyme defect that produces a clinical syndrome.

**inbreeding** Mating between closely related organisms.

**inbreeding depression** A loss of reduction in fitness that usually accompanies inbreeding of organisms.

**incomplete dominance** Expression of heterozygous phenotype which is distinct from, and often intermediate to, that of either parent.

**incomplete linkage** Occasional separation of two genes on the same chromosome by a recombinational event.

**independent assortment** The independent behavior of each pair of homologous chromosomes during their segregation in meiosis I. The random distribution of genes on different chromosomes into gametes.

**inducer** An effector molecule that activates transcription.

**inducible enzyme system** An enzyme system under the control of a regulatory molecule, or inducer, which acts to block a repressor and allow transcription.

**initiation codon** The triplet of nucleotides (AUG) in an mRNA molecule that codes for the insertion of the amino acid methionine as the first amino acid in a polypeptide chain.

**insertion sequence** See *IS element*.

***in situ* hybridization** A technique for the cytological localization of DNA sequences complementary to a given nucleic acid or polynucleotide.

**intercalary deletion** A form of chromosome deletion where material is lost from within the chromosome. Deletions that involve the end of the chromosome are called terminal deletions.

**intercalating agent** A compound that inserts between bases in a DNA molecule, disrupting the alignments and pairing of bases in the complementary strands (e.g., acridine dyes).

**interference** A measure of the degree to which one crossover affects the incidence of another crossover in an adjacent region of the same chromatid. Positive interference increases the chances of another crossover; negative interference reduces the probability of a second crossover event.

**interferon**   One of a family of proteins that acts to inhibit viral replication in higher organisms. Some interferons may have anticancer properties.

**interphase**   That portion of the cell cycle between divisions.

**intervening sequence**   See *intron.*

**intron**   A portion of DNA between coding regions in a gene which is transcribed, but which does not appear in the mRNA product.

**inversion**   A chromosomal aberration in which the order of a chromosomal segment has been reversed.

**inversion loop**   The chromosomal configuration resulting from the synapsis of homologous chromosomes, one of which carries an inversion.

***in vitro***   Literally, in glass; outside the living organism; occurring in an artificial environment.

***in vivo***   Literally, in the living; occurring within the living body of an organism.

**IS element**   A mobile DNA segment that is transposable to any of a number of sites in the genome.

**isoagglutinogen**   An antigenic factor or substance present on the surface of cells that is capable of inducing the formation of an antibody.

**isochromosome**   An aberrant chromosome with two identical arms and homologous loci.

**isolating mechanism**   Any barrier to the exchange of genes between different populations of a group of organisms. In general, isolation can be classified as spatial, environmental, or reproductive.

**isotopes**   Forms of a chemical element that have the same number of protons and electrons, but differ in the number of neutrons contained in the atomic nucleus. Unstable isotopes undergo a transition to a more stable form with the release of radioactivity.

**isozyme**   Any of two or more distinct forms of an enzyme that have identical or nearly identical chemical properties, but differ in some property such as net electrical charge, pH optima, number and type of subunits, or substrate concentration.

**kappa particles**   DNA-containing cytoplasmic particles found in certain strains of *Paramecium aurelia.* When these self-reproducing particles are transferred into the growth medium, they release a toxin, paramecin, which kills other sensitive strains. A nuclear gene, *K*, is responsible for the maintenance of kappa particles in the cytoplasm.

**karyokinesis**   The process of nuclear division.

**karyotype**   The chromosome complement of a cell or an individual. Often used to refer to the arrangement of metaphase chromosomes in a sequence according to length and position of the centromere.

**kilobase**   A unit of length consisting of 1000 nucleotides. Abbreviated kb.

**kinetochore**   A fibrous structure with a size of about 400 nm, located within the centromere. It appears to be the location to which microtubules attach.

**Klenow fragment**   A part of bacterial DNA polymerase that lacks exonuclease activity, but retains polymerase activity. It is produced by enzymatic digestion of the intact enzyme.

**Klinefelter syndrome**   A genetic disease in human males caused by the presence of an extra X chromosome. Klinefelter males are XXY instead of XY. This syndrome is associated with enlarged breasts, small testes, sterility, and, occasionally, mental retardation.

**lagging strand**   In DNA replication, the strand synthesized in a discontinuous fashion, 5' to 3' away from the replication fork. Each short piece of DNA synthesized in this fashion is called an Okazaki fragment.

**lampbrush chromosomes**   Meiotic chromosomes characterized by extended lateral loops, which reach maximum extension during diplotene. Although most intensively studied in amphibians, these structures occur in meiotic cells of organisms ranging from insects through humans.

**leader sequence**   That portion of an mRNA molecule from the 5' end to the beginning codon; may contain regulatory or ribosome binding sites.

**leading strand**   During DNA replication, the strand synthesized continuously 5' to 3' toward the replication fork.

**lectins**   Carbohydrate-binding proteins found in the seeds of leguminous plants such as soybeans. Although their physiological role in plants is unclear, they are useful as probes for cell surface carbohydrates and glycoproteins in animal cells. Proteins with similar properties have also been isolated from organisms such as snails and eels.

**leptotene**   The initial stage of meiotic prophase I, during which the chromosomes become visible and are often arranged in a bouquet configuration, with one or both ends of the chromosomes gathered at one spot on the inner nuclear membrane.

**lethal gene**   A gene whose expression results in death.

**leucine zipper**   A structural motif in a DNA-binding protein that is characterized by a stretch of leucine residues spaced at every seventh amino acid residue, with adjacent regions of positively charged amino acids. Leucine zippers on two polypeptides may interact to form a dimer that binds to DNA.

**LINEs**   Long interspersed elements are repetitive sequences found in the genomes of higher organisms, such as the 6-kb Kpnl sequences found in primate genomes.

**linkage**   Condition in which two or more nonallelic genes tend to be inherited together. Linked genes have their loci along the same chromosome, do not assort independently, but can be separated by crossing over.

**linkage group**   A group of genes that have their loci on the same chromosome.

**linking number**   The number of times that two strands of a closed, circular DNA duplex cross over each other.

**locus**   The site or place on a chromosome where a particular gene is located.

**lod score**   A statistical method used to determine whether two loci are linked or unlinked. A lod (log of the odds) score of +3 by convention indicates linkage.

**long period interspersion**   Pattern of genome organization in which long stretches of single copy DNA are interspersed with long segments of repetitive DNA. This pattern of genome organization is found in *Drosophila* and the honeybee.

**long terminal repeat (LTR)**   Sequence of several hundred base pairs found at the ends of retroviral DNAs.

**Lutheran blood group**   One of a number of blood group systems inherited independently of the ABO, MN, and Rh systems. Alleles of this group determine the presence or absence of antigens on the surface of red blood cells. Gene is on human chromosome 19.

**Lyon hypothesis**   The idea proposed by Mary Lyon that random inactivation of one X chromosome in the somatic cells of mammalian females is responsible for dosage compensation and mosaicism.

**lysis**   The disintegration of a cell brought about by the rupture of its membrane.

**lysogenic bacterium**   A bacterial cell carrying a temperate bacteriophage integrated into its chromosome.

**lysogeny**   The process by which the DNA of an infecting phage becomes repressed and integrated into the chromosome of the bacterial cell it infects.

**lytic phase**   The condition in which a temperate bacteriophage loses its integrated status in the host chromosome (becomes induced), replicates, and lyses the bacterial cell.

**major histocompatibility loci**   See *MHC*.

**map unit**   A measure of the genetic distance between two genes, corresponding to a recombination frequency of 1 percent. See *centimorgan*.

**maternal effect**   Phenotypic effects on the offspring produced by the maternal genome. Factors transmitted through the egg cytoplasm that produce a phenotypic effect in the progeny.

**maternal influence**   See *maternal effect*.

**maternal inheritance**   The transmission of traits via cytoplasmic genetic factors such as mitochondria or chloroplasts.

**mean**   The arithmetic average.

**median**   The value in a group of numbers below and above which there is an equal number of data points or measurements.

**meiosis**   The process in gametogenesis or sporogenesis during which one replication of the chromosomes is followed by two nuclear divisions to produce four haploid cells.

**melting profile**   See $T_m$.

**merozygote**   A partially diploid bacterial cell containing, in addition to its own chromosome, a chromosome fragment introduced into the cell by transformation, transduction, or conjugation.

**messenger RNA**   See *mRNA*.

**metabolism**   The sum of chemical changes in living organisms by which energy is generated and used.

**metacentric chromosome**   A chromosome with a centrally located centromere, producing chromosome arms of equal lengths.

**metafemale**   In *Drosophila,* a poorly developed female of low viability in which the ratio of X chromosomes to sets of autosomes exceeds 1.0. Previously called a superfemale.

**metamale**   In *Drosophila,* a poorly developed male of low viability in which the ratio of X chromosomes to sets of autosomes is less than 0.5. Previously called a supermale.

**metaphase**   The stage of cell division in which the condensed chromosomes lie in a central plane between the two poles of the cell, and in which the chromosomes become attached to the spindle fibers.

**metaphase plate**   The arrangement of mitotic or meiotic chromosomes at the equator of the cell during metaphase.

**MHC**   Major histocompatibility loci. In humans, the HLA complex; and in mice, the H2 complex.

**micrometer**   A unit of length equal to $1 \times 10^{-6}$ meter. Previously called a micron. Abbreviated $\mu$m.

**micron**   See *micrometer*.

**migration coefficient**   An expression of the proportion of migrant genes entering the population per generation.

**millimeter**   A unit of length equal to $1 \times 10^{-3}$ meter. Abbreviated mm.

**minimal medium**   A medium containing only those nutrients that will support the growth and reproduction of wild-type strains of an organism.

**mismatch repair**   DNA repair by a mechanism known as cut and patch. In this repair, the defective single-stranded region is excised, followed by the synthesis of a new segment, using the complementary strand as a template.

**missense mutation**   A mutation that alters a codon to that of another amino acid, causing an altered translation product to be made.

**mitochondrion**   Found in the cells of eukaryotes, a cytoplasmic, self-reproducing organelle that is the site of ATP synthesis.

**mitogen**   A substance that stimulates mitosis in nondividing cells (e.g., phytohemagglutinin).

**mitosis**   A form of cell division resulting in the production of two cells, each with the same chromosome and genetic complement as the parent cell.

**mode**   In a set of data, the value occurring in the greatest frequency.

**monohybrid cross**   A genetic cross between two individuals involving only one character (e.g., $AA \times aa$).

**monosomic**   An aneuploid condition in which one member of a chromosome pair is missing; having a chromosome number of $2n - 1$.

**monozygotic twins**   Twins produced from a single fertilization event; the first division of the zygote produces two cells, each of which develops into an embryo. Also known as identical twins.

**mRNA**   An RNA molecule transcribed from DNA and translated into the amino acid sequence of a polypeptide.

**mtDNA**   Mitochondrial DNA.

**multigene family**   A gene set descended from a common ancestor by duplication and subsequent divergence from a common ancestor. The globin genes represent a multigene family.

**multiple alleles**   Three or more alleles of the same gene.

**multiple-factor inheritance**   See *polygenic inheritance*.

**multiple infection**    Simultaneous infection of a bacterial cell by more than one bacteriophage, often of different genotypes.

**mu phage**    A phage group in which the genetic material behaves like an insertion sequence, capable of insertion, excision, transposition, inactivation of host genes, and induction of chromosomal rearrangements.

**mutagen**    Any agent that causes an increase in the rate of mutation.

**mutant**    A cell or organism carrying an altered or mutant gene.

**mutation**    The process that produces an alteration in DNA or chromosome structure; the source of most alleles.

**mutation rate**    The frequency with which mutations take place at a given locus or in a population.

**muton**    The smallest unit of mutation in a gene, corresponding to a single base change.

**nanometer**    A unit of length equal to $1 \times 10^{-9}$ meter. Abbreviated nm.

**natural selection**    Differential reproduction of some members of a species resulting from variable fitness conferred by genotypic differences.

**nearest-neighbor analysis**    A molecular technique used to determine the frequency with which nucleotides are adjacent to each other in polynucleotide chains.

**neutral mutation**    A mutation with no immediate adaptive significance or phenotypic effect.

**noncrossover gamete**    A gamete that contains no chromosomes that have undergone genetic recombination.

**nondisjunction**    An accident of cell division in which the homologous chromosomes (in meiosis) or the sister chromatids (in mitosis) fail to separate and migrate to opposite poles; responsible for defects such as monosomy and trisomy.

**nonsense codon**    The nucleotide triplet in an mRNA molecule that signals the termination of translation. Three such codons are known: UGA, UAG, and UAA.

**nonsense mutation**    A mutation that alters a codon to one which encodes no amino acid, that is, UAG (amber codon), UAA (ochre codon), or UGA (opal codon). Leads to premature termination during the translation of mRNA.

**NOR**    See *nucleolar organizer region*.

**normal distribution**    A probability function that approximates the distribution of random variables. The normal curve, also known as a Gaussian or bell-shaped curve, is the graphic display of the normal distribution.

**np**    See *nucleotide pair*.

**N-terminal amino acid**    The terminal amino acid in a peptide chain that carries a free amino group.

**N terminus**    The end of a polypeptide that carries a free amino group of the first amino acid. By convention, the structural formula of polypeptides is written with the N terminus at the left.

**nu body**    See *nucleosome*.

**nuclease**    An enzyme that breaks bonds in nucleic acid molecules.

**nucleoid**    The DNA-containing region within the cytoplasm in prokaryotic cells.

**nucleolar organizer region (NOR)**    A chromosomal region containing the genes for rRNA; most often found in physical association with the nucleolus.

**nucleolus**    A nuclear organelle that is the site of ribosome biosynthesis; usually associated with or formed in association with the NOR.

**nucleoside**    A purine or pyrimidine base covalently linked to a ribose or deoxyribose sugar molecule.

**nucleosome**    A complex of four histone molecules, each present in duplicate, wrapped by two turns of a DNA molecule. One of the basic units of eukaryotic chromosome structure. Also known as a nu body.

**nucleotide**    A nucleoside covalently linked to a phosphate group. Nucleotides are the basic building blocks of nucleic acids. The nucleotides commonly found in DNA are deoxyadenylic acid, deoxycytidylic acid, deoxyguanylic acid, and deoxythymidylic acid. The nucleotides in RNA are adenylic acid, cytidylic acid, guanylic acid, and uridylic acid.

**nucleotide pair**    The pair of nucleotides (A and T, or G and C) in opposite strands of the DNA molecule that are hydrogen-bonded to each other.

**nucleus**    The membrane-bounded cytoplasmic organelle of eukaryotic cells that contains the chromosomes and nucleolus.

**nullisomic**    Describes an individual with a chromosomal aberration in which both members of a chromosome pair are missing.

**ochre codon**    A codon that does not code for the insertion of an amino acid into a polypeptide chain, but signals chain termination. The ochre codon is UAA.

**Okazaki fragment**    The small, discontinuous strands of DNA produced during DNA synthesis.

**oncogene**    A gene whose activity promotes uncontrolled proliferation in eukaryotic cells.

**open reading frame (ORF)**    The interval between the start and stop codon that encodes amino acids for insertion into polypeptide chains.

**operator region**    A region of a DNA molecule that interacts with a specific repressor protein to control the expression of an adjacent gene or gene set.

**operon**    A genetic unit that consists of one or more structural genes (that code for polypeptides) and an adjacent operator gene that controls the transcriptional activity of the structural gene or genes.

**orphon**    Single copies of tandemly repeated genes found dispersed in the genome. For example, histone genes are members of a multigene family present in several hundred copies clustered in a tandem array. A single copy of a histone gene found elsewhere in the genome is said to have lost its family and is regarded as an orphon.

**overdominance**    The phenomenon where heterozygotes have a phenotype that is more extreme than either homozygous genotype.

**overlapping code**    A genetic code first proposed by George

Gamow in which any given nucleotide is shared by two adjacent codons.

**pachytene**   The stage in meiotic prophase I when the synapsed homologous chromosomes split longitudinally (except at the centromere), producing a group of four chromatids called a tetrad.

**pair rule genes**   Genes expressed as stripes around the blastoderm embryo during development of the *Drosophila* embryo.

**palindrome**   A word, number, verse, or sentence that reads the same backward or forward (e.g., *able was I ere I saw elba.*). In nucleic acids, a sequence in which the base pairs read the same on complementary strands ($5' \rightarrow 3'$). For example: $5'$GAATTC$3'$, $3'$CTTAAG$5'$. These often occur as sites for restriction endonuclease recognition and cutting.

**pangenesis**   A discarded theory of development that postulated the existence of pangenes, small particles from all parts of the body that concentrated in the gametes, passing traits from generation to generation, blending the traits of the parents in the offspring.

**paracentric inversion**   A chromosomal inversion that does not include the centromere.

**parasexual**   Condition describing recombination of genes from different individuals that does not involve meiosis, gamete formation, or zygote production. The formation of somatic cell hybrids is an example.

**parental gamete**   See *noncrossover gamete.*

**parthenogenesis**   Development of an egg without fertilization.

**partial diploids**   See *merozygote.*

**partial dominance**   See *incomplete dominance.*

**passive immunity**   A form of immunity produced by receipt of antibodies synthesized by another individual.

**patroclinous inheritance**   A form of genetic transmission in which the offspring have the phenotype of the father.

**pedigree**   In human genetics, a diagram showing the ancestral relationships and transmission of genetic traits over several generations in a family.

**P element**   Transposable DNA elements found in *Drosophila* that are responsible for hybrid dysgenesis.

**penetrance**   The frequency (expressed as a percentage) with which individuals of a given genotype manifest at least some degree of a specific mutant phenotype associated with a trait.

**peptide bond**   The covalent bond between the amino group of one amino acid and the carboxyl group of another amino acid.

**pericentric inversion**   A chromosomal inversion that involves both arms of the chromosome and thus involves the centromere.

**permissive condition**   Environmental conditions under which a conditional mutation (such as temperature-sensitive mutant) expresses the wild-type phenotype.

**phage**   See *bacteriophage.*

**phenocopy**   An environmentally induced phenotype (nonheritable) that closely resembles the phenotype produced by a known gene.

**phenotype**   The observable properties of an organism that are genetically controlled.

**phenylketonuria (PKU)**   A hereditary condition in humans associated with the inability to metabolize the amino acid phenylalanine. The most common form is caused by the lack of the liver enzyme phenylalanine hydroxylase.

**phosphodiester bond**   In nucleic acids, the covalent bond between a phosphate group and adjacent nucleotides, extending from the $5'$ carbon of one pentose (ribose or deoxyribose) to the $3'$ carbon of the pentose in the neighboring nucleotide. Phosphodiester bonds form the backbone of nucleic acid molecules.

**photoreactivation enzyme (PRE)**   An exonuclease that catalyzes the light-activated excision of ultraviolet-induced thymine dimers from DNA.

**photoreactivation repair**   Light-induced repair of damage caused by exposure to ultraviolet light. Associated with an intracellular enzyme system.

**phyletic evolution**   The gradual transformation of one species into another over time; vertical evolution.

**pilus**   A filamentlike projection from the surface of a bacterial cell. Often associated with cells possessing F factors.

**plaque**   A clear area on an otherwise opaque bacterial lawn caused by the growth and reproduction of phages.

**plasmid**   An extrachromosomal, circular DNA molecule (often carrying genetic information) that replicates independently of the host chromosome.

**platysome**   Term originally used in electron and X-ray diffraction studies to describe the flattened appearance of the DNA in the nucleosome core.

**pleiotropy**   Condition in which a single mutation simultaneously affects several characters.

**ploidy**   Term referring to the basic chromosome set or to multiples of that set.

**point mutation**   A mutation that can be mapped to a single locus. At the molecular level, a mutation that results in the substitution of one nucleotide for another.

**polar body**   A cell produced at either the first or second meiotic division in females that contains almost no cytoplasm as a result of an unequal cytokinesis.

**polycistronic mRNA**   A messenger RNA molecule that encodes the amino acid sequence of two or more polypeptide chains in adjacent structural genes.

**polygenic inheritance**   The transmission of a phenotypic trait whose expression depends on the additive effect of a number of genes.

**polylinker**   A segment of DNA that has been engineered to contain multiple sites for restriction enzyme digestion. Polylinkers are usually found in engineered vectors such as plasmids.

**polymerase chain reaction (PCR)**   A method for amplifying DNA segments that uses cycles of denaturation, annealing to primers, and DNA polymerase–directed DNA synthesis.

**polymerases**   The enzymes that catalyze the formation of

DNA and RNA from deoxynucleotides and ribonucleotides, respectively.

**polymorphism**    The existence of two or more discontinuous, segregating phenotypes in a population.

**polypeptide**    A molecule made up of amino acids joined by covalent peptide bonds. This term is used to denote the amino acid chain before it assumes its functional three-dimensional configuration.

**polyploid**    A cell or individual having more than two sets of chromosomes.

**polyribosome**    See *polysome.*

**polysome**    A structure composed of two or more ribosomes associated with mRNA, engaged in translation. Formerly called polyribosome.

**polytene chromosome**    A chromosome that has undergone several rounds of DNA replication without separation of the replicated chromosomes, forming a giant, thick chromosome with aligned chromomeres producing a characteristic banding pattern.

**population**    A local group of individuals belonging to the same species, which are actually or potentially interbreeding.

**position effect**    Change in expression of a gene associated with a change in the gene's location within the genome.

**postzygotic isolation mechanism**    Factors that prevent or reduce inbreeding by acting after fertilization to produce nonviable, sterile hybrids or hybrids of lowered fitness.

**preadaptive mutation**    A mutational event that later becomes of adaptive significance.

**preformationism**    The discredited idea that an organism develops by growth of structures already present in the egg or sperm.

**prezygotic isolation mechanism**    Factors that reduce inbreeding by preventing courtship, mating, or fertilization.

**Pribnow box**    A 6-bp sequence 5' to the beginning of transcription in prokaryotic genes, to which the sigma subunit of RNA polymerase binds. The consensus sequence for this box is TATAAT.

**primary protein structure**    Refers to the sequence of amino acids in a polypeptide chain.

**primary sex ratio**    Ratio of males to females at fertilization.

**primer**    In nucleic acids, a short length of RNA or single-stranded DNA that is necessary for the functioning of polymerases.

**prion**    An infectious pathogenic agent devoid of nucleic acid and composed mainly of a protein, PrP, with a molecular weight of 27,000 to 30,000 daltons. Prions are known to cause scrapie, a degenerative neurological disease in sheep, and are thought to cause similar diseases in humans, such as kuru and Creutzfeldt-Jakob disease.

**probability**    Ratio of the frequency of a given event to the frequency of all possible events.

**proband**    See *propositus.*

**probe**    A macromolecule such as DNA or RNA that has been labeled and can be detected by an assay such as autoradiography or fluorescence microscopy. Probes are used to identify target molecules, genes, or gene products.

**product law**    The law that holds that the probability of two independent events occurring simultaneously is the product of their independent probabilities.

**proflavin**    An acridine dye that acts as a mutagen. See *acridine dyes.*

**progeny**    The offspring produced from a mating.

**prokaryotes**    Organisms lacking nuclear membranes, meiosis, and mitosis. Bacteria and blue-green algae are examples of prokaryotic organisms.

**promoter**    Region having a regulatory function and to which RNA polymerase binds prior to the initiation of transcription.

**proofreading**    A molecular mechanism for correcting errors in replication, transcription, or translation. Also known as editing.

**prophage**    A phage genome integrated into a bacterial chromosome. Bacterial cells carrying prophages are said to be lysogenic.

**propositus** (female, **proposita**)    An individual in whom a genetically determined trait of interest is first detected. Also known as a proband.

**protein**    A molecule composed of one or more polypeptides, each composed of amino acids covalently linked together.

**proto-oncogene**    A cellular gene that normally functions to control cellular proliferation. Proto-oncogenes can be converted to oncogenes by alterations in structure or expression.

**protoplast**    A bacterial or plant cell with the cell wall removed. Sometimes called a spheroplast.

**prototroph**    A strain (usually microorganisms) that is capable of growth on a defined, minimal medium. Wild-type strains are usually regarded as prototrophs.

**pseudoalleles**    Genes that behave as alleles to one another by complementation, but that can be separated from one another by recombination.

**pseudodominance**    The expression of a recessive allele on one homologue caused by the deletion of the dominant allele on the other homologue.

**pseudogene**    A nonfunctional gene with sequence homology to a known structural gene present elsewhere in the genome. They differ from their functional relatives by insertions or deletions and by the presence of flanking direct repeat sequences of 10 to 20 nucleotides.

**puff**    See *chromosome puff.*

**punctuated equilibrium**    A pattern in the fossil record of brief periods of species divergence punctuated with long periods of species stability.

**quantitative inheritance**    See *polygenic inheritance.*

**quantitative trait loci (QTL)**    Two or more genes that act on a single polygenic trait.

**quantum speciation**    Formation of a new species within a single or a few generations by a combination of selection and drift.

**quaternary protein structure**    Types and modes of interaction between two or more polypeptide chains within a protein molecule.

**R point**    The point (also known as the restriction point) during the $G_1$ stage of the cell cycle when either a commitment is made to DNA synthesis and another cell cycle, or the cell withdraws from the cycle and becomes quiescent.

**race** A phenotypically or geographically distinct subgroup within a species.

**rad** A unit of absorbed dose of radiation with an energy equal to 100 ergs per gram of irradiated tissue.

**radioactive isotope** One of the forms of an element, differing in atomic weight and possessing an unstable nucleus that emits ionizing radiation.

**random mating** Mating between individuals without regard to genotype.

**reading frame** Linear sequence of codons (groups of three nucleotides) in a nucleic acid.

**reannealing** Formation of double-stranded DNA molecules from dissociated single strands.

**recessive** Term describing an allele that is not expressed in the heterozygous condition.

**reciprocal cross** A paired cross in which the genotype of the female in the first cross is present as the genotype of the male in the second cross, and vice versa.

**reciprocal translocation** A chromosomal aberration in which nonhomologous chromosomes exchange parts.

**recombinant DNA** A DNA molecule formed by the joining of two heterologous molecules. Usually applied to DNA molecules produced by *in vitro* ligation of DNA from two different organisms.

**recombinant gamete** A gamete containing a new combination of genes produced by crossing over during meiosis.

**recombination** The process that leads to the formation of new gene combinations on chromosomes.

**recon** A term coined by Seymour Benzer to denote the smallest genetic units between which recombination can occur.

**reductional division** The chromosome division that halves the diploid chromosome number. The first division of meiosis is a reductional division.

**redundant genes** Gene sequences present in more than one copy per haploid genome (e.g., ribosomal genes).

**regulatory site** A DNA sequence that is involved in the control of expression of other genes, usually involving an interaction with another molecule.

**rem** Radiation equivalent in man; the dosage of radiation that will cause the same biological effect as one roentgen of X rays.

**renaturation** The process by which a denatured protein or nucleic acid returns to its normal three-dimensional structure.

**repetitive DNA sequences** DNA sequences present in many copies in the haploid genome.

**replicating form (RF)** Double-stranded nucleic acid molecules present as an intermediate during the reproduction of certain viruses.

**replication** The process of DNA synthesis.

**replication fork** The Y-shaped region of a chromosome associated with the site of replication.

**replicon** A chromosomal region or free genetic element containing the DNA sequences necessary for the initiation of DNA replication.

**replisome** The term used to describe the complex of proteins, including DNA polymerase, that assembles at the bacterial replication fork to synthesize DNA.

**repressible enzyme system** An enzyme or group of enzymes whose synthesis is regulated by the intracellular concentration of certain metabolites.

**repressor** A protein that binds to a regulatory sequence adjacent to a gene and blocks transcription of the gene.

**reproductive isolation** Absence of interbreeding between populations, subspecies, or species. Reproductive isolation can be brought about by extrinsic factors, such as behavior, and intrinsic barriers, such as hybrid inviability.

**resistance transfer factor (RTF)** A component of R plasmids that confers the ability for cell-to-cell transfer of the R plasmid by conjugation.

**resolution** In an optical system, the shortest distance between two points or lines at which they can be perceived to be two points or lines.

**restriction endonuclease** Nuclease that recognizes specific nucleotide sequences in a DNA molecule, and cleaves or nicks the DNA at that site. Derived from a variety of microorganisms, those enzymes that cleave both strands of the DNA are used in the construction of recombinant DNA molecules.

**restriction fragment length polymorphism (RFLP)** Variation in the length of DNA fragments generated by a restriction endonuclease. These variations are caused by mutations that create or abolish cutting sites for restriction enzymes. RFLPs are inherited in a codominant fashion and can be used as genetic markers.

**restrictive condition** Environmental conditions under which a conditional mutation (such as a temperature-sensitive mutant) expresses the mutant phenotype.

**restrictive transduction** See *specialized transduction*.

**retrovirus** Viruses with RNA as genetic material that utilize the enzyme reverse transcriptase during their life cycle.

**reverse transcriptase** A polymerase that uses RNA as a template to transcribe a single-stranded DNA molecule as a product.

**reversion** A mutation that restores the wild-type phenotype.

**R factor (R plasmid)** Bacterial plasmids that carry antibiotic resistance genes. Most R plasmids have two components: an r-determinant, which carriers the antibiotic resistance genes, and the resistance transfer factor (RTF).

**RFLP** See *restriction fragment length polymorphism*.

**Rh factor** An antigenic system first described in the rhesus monkey. Recessive *r/r* individuals produce no Rh antigens and are Rh negative, while *R/R* and *R/r* individuals have Rh antigens on the surface of their red blood cells and are classified as Rh positive.

**ribonucleic acid** A nucleic acid characterized by the sugar ribose and the pyrimidine uracil, usually a single-stranded polynucleotide. Several forms are recognized, including ribosomal RNA, messenger RNA, transfer RNA, and heterogeneous nuclear RNA.

**ribose** The five-carbon sugar associated with the ribonucleotides found in RNA.

**ribosomal RNA** See *rRNA*.

**ribosome** A ribonucleoprotein organelle consisting of two subunits, each containing RNA and protein. Ribosomes are the site of translation of mRNA codons into the amino acid sequence of a polypeptide chain.

**RNA**    See *ribonucleic acid.*

**RNA polymerase**    An enzyme that catalyzes the formation of an RNA polynucleotide strand using the base sequence of a DNA molecule as a template.

**RNase**    A class of enzymes that hydrolyze RNA molecules.

**Robertsonian translocation**    A form of chromosomal aberration that involves the fusion of long arms of acrocentric chromosomes at the centromere.

**roentgen**    A unit of measure of the amount of radiation corresponding to the generation of $2.083 \times 10^9$ ion pairs in one cubic centimeter of air at 0°C at an atmospheric pressure of 760 mm of mercury. Abbreviated R.

**rolling circle model**    A model of DNA replication in which the growing point or replication fork rolls around a circular template strand; in each pass around the circle, the newly synthesized strand displaces the strand from the previous replication, producing a series of contiguous copies of the template strand.

**rRNA**    The RNA molecules that are the structural components of the ribosomal subunits. In prokaryotes, these are the 16$S$, 23$S$, and 5$S$ molecules; and in eukaryotes, they are the 18$S$, 28$S$, and 5$S$ molecules.

**RTF**    See *resistance transfer factor.*

**S₁ nuclease**    A deoxyribonuclease that cuts and degrades single-stranded molecules of DNA.

**satellite DNA**    DNA that forms a minor band when genomic DNA is centrifuged in a cesium salt gradient. This DNA usually consists of short sequences repeated many times in the genome.

**SCE**    See *sister chromatid exchange.*

**secondary protein structure**    The alpha helical or pleated-sheet form of a protein molecule brought about by the formation of hydrogen bonds between amino acids.

**secondary sex ratio**    The ratio of males to females at birth.

**secretor**    An individual having soluble forms of the blood group antigens A and/or B present in saliva and other body fluids. This condition is caused by a dominant, autosomal gene unlinked to the *ABO* locus (*I* locus).

**sedimentation coefficient**    See *Svedberg coefficient unit.*

**segment polarity genes**    Genes that regulate the spatial pattern of differentiation within each segment of the developing *Drosophila* embryo.

**segregation**    The separation of homologous chromosomes into different gametes during meiosis.

**selection**    The force that brings about changes in the frequency of alleles and genotypes in populations through differential reproduction.

**selection coefficient (s)**    A quantitative measure of the relative fitness of one genotype compared with another.

**selfing**    In plant genetics, the fertilization of ovules of a plant by pollen produced by the same plant. Reproduction by self-fertilization.

**semiconservative replication**    A model of DNA replication in which a double-stranded molecule replicates in such a way that the daughter molecules are composed of one parental (old) and one newly synthesized strand.

**semisterility**    A condition in which a proportion of all zygotes are inviable.

**sex chromatin body**    See *Barr body.*

**sex chromosome**    A chromosome, such as the X or Y in humans, which is involved in sex determination.

**sexduction**    Transmission of chromosomal genes from a donor bacterium to a recipient cell by the F factor.

**sex-influenced inheritance**    Phenotypic expression that is conditioned by the sex of the individual. A heterozygote may express one phenotype in one sex and the alternate phenotype in the other sex.

**sex-limited inheritance**    A trait that is expressed in only one sex even though the trait may not be X-linked.

**sex linkage**    The pattern of inheritance resulting from genes located on the X chromosome.

**sex ratio**    See *primary* and *secondary sex ratio.*

**sexual reproduction**    Reproduction through the fusion of gametes, which are the haploid products of meiosis.

**Shine–Dalgarno sequence**    The nucleotides AGGAGG present in the leader sequence of prokaryotic genes that serve as a ribosome binding site. The 16$S$ RNA of the small ribosomal subunit contains a complementary sequence to which the mRNA binds.

**short period interspersion**    Pattern of genome organization in which stretches of single-copy DNA (about 1000 bp) are interspersed with short segments of repetitive DNA (300 bp). This pattern is found in *Xenopus,* humans, and the majority of organisms examined to date.

**shotgun experiment**    The cloning of random fragments of genomic DNA into a vehicle such as a plasmid or phage, usually to produce a bank or library of clones from which clones of specific interest will be selected.

**sibling species**    Species that are morphologically almost identical, but which are reproductively isolated from one another.

**sickle-cell anemia**    A genetic disease in humans caused by an autosomal recessive gene, usually fatal in the homozygous condition. Caused by an alteration in the amino acid sequence of the beta chain of globin.

**sickle-cell trait**    The phenotype exhibited by individuals heterozygous for the sickle-cell gene.

**sigma factor**    A polypeptide subunit of the RNA polymerase that recognizes the binding site for the initiation of transcription.

**SINEs**    Short interspersed elements are repetitive sequences found in the genomes of higher organisms, such as the 300-bp *Alu* sequence.

**sister chromatid exchange (SCE)**    A crossing over event that can occur in meiotic and mitotic cells; involves the reciprocal exchange of chromosomal material between sister chromatids (joined by a common centromere). Such exchanges can be detected cytologically after BrdU incorporation into the replicating chromosomes.

**site-directed mutagenesis**    A process that uses a synthetic oligonucleotide containing a mutant base or sequence as a primer for inducing a mutation at a specific site in a cloned gene.

**small nuclear RNA (snRNA)**  Species of RNA molecules ranging in size from 90 to 400 nucleotides. The abundant snRNAs are present in $1 \times 10^4$ to $1 \times 10^6$ copies per cell. snRNAs are associated with proteins and form RNP particles known as snRNPs or snurps. Six uridine-rich snRNAs known as U1–U6 are located in the nucleoplasm, and the complete nucleotide sequence of these is known. snRNAs have been implicated in the processing of pre-mRNA and may have a range of cleavage and ligation functions.

**snurps**  See *small nuclear RNA (snRNA)*.

**solenoid structure**  A level of eukaryotic chromosome structure generated by the supercoiling of nucleosomes.

**somatic cell genetics**  The use of cultured somatic cells to investigate genetic phenomena by parasexual techniques.

**somatic cells**  All cells other than the germ cells or gametes in an organism.

**somatic mutation**  A mutational event occurring in a somatic cell. In other words, such mutations are not heritable.

**somatic pairing**  The pairing of homologous chromosomes in somatic cells.

**SOS response**  The induction of enzymes to repair damaged DNA in *E. coli*. The response involves activation of an enzyme that cleaves a repressor, activating a series of genes involved in DNA repair.

**spacer DNA**  DNA sequences found between genes, usually repetitive DNA segments.

**special creation**  An idea that each species originated through a special act of creation by a divine force.

**specialized transduction**  Genetic transfer of only specific host genes by transducing phages.

**speciation**  The process by which new species of plants and animals arise.

**species**  A group of actually or potentially interbreeding individuals that is reproductively isolated from other such groups.

**spheroplast**  See *protoplast*.

**spindle fibers**  Cytoplasmic fibrils formed during cell division that are involved with the separation of chromatids at anaphase and their movement toward opposite poles in the cell.

**spliceosome**  The nuclear macromolecule complex within which splicing reactions occur to remove introns from pre-mRNAs.

**spontaneous generation**  The origin of living systems from nonliving matter.

**spontaneous mutation**  A mutation that is not induced by a mutagenic agent.

**spore**  A unicellular body or cell encased in a protective coat that is produced by some bacteria, plants, and invertebrates; is capable of survival in unfavorable environmental conditions; and can give rise to a new individual upon germination. In plants, spores are the haploid products of meiosis.

**SRY**  A gene (sex-determining region of the Y) found near the pseudoautosomal boundary of the Y chromosome. Accumulated evidence indicates that this gene is the testis-determining factor (TDF).

**stabilizing selection**  Preferential reproduction of those individuals having genotypes close to the mean for the population. A selective elimination of genotypes at both extremes.

**standard deviation**  A quantitative measure of the amount of variation in a sample of measurements from a population.

**standard error**  A quantitative measure of the amount of variation in a sample of measurements from a population.

**sterility**  The condition of being unable to reproduce; free from contaminating microorganisms.

**strain**  A group with common ancestry that has physiological or morphological characteristics of interest for genetic study or domestication.

**structural gene**  A gene that encodes the amino acid sequence of a polypeptide chain.

**sublethal gene**  A mutation causing lowered viability, with death before maturity in less than 50 percent of the individuals carrying the gene.

**submetacentric chromosome**  A chromosome with the centromere placed so that one arm of the chromosome is slightly longer than the other.

**subspecies**  A morphologically or geographically distinct interbreeding population of a species.

**sum law**  The law that holds that the probability of one or the other of two mutually exclusive events occurring is the sum of their individual probabilities.

**supercoiled DNA**  A form of DNA structure in which the helix is coiled upon itself. Such structures can exist in stable forms only when the ends of the DNA are not free, as in a covalently closed circular DNA molecule.

**superfemale**  See *metafemale*.

**supermale**  See *metamale*.

**suppressor mutation**  A mutation that acts to restore (completely or partially) the function lost by a previous mutation at another site.

**Svedberg coefficient unit**  A unit of measure for the rate at which particles (molecules) sediment in a centrifugal field. This unit is a function of several physico-chemical properties, including size and shape. A sedimentation value of $1 \times 10^{-13}$ sec is defined as one Svedberg coefficient ($S$) unit.

**symbiont**  An organism coexisting in a mutually beneficial relationship with another organism.

**sympatric speciation**  Process of speciation involving populations that inhabit, at least in part, the same geographic range.

**synapsis**  The pairing of homologous chromosomes at meiosis.

**synaptonemal complex (SC)**  An organelle consisting of a tripartite nucleoprotein ribbon that forms between the paired homologous chromosomes in the pachytene stage of the first meiotic division.

**syndrome**  A group of signs or symptoms that occur together and characterize a disease or abnormality.

**synkaryon**  The nucleus of a zygote that results from the fusion of two gametic nuclei. Also used in somatic cell genetics to describe the product of nuclear fusion.

**syntenic test**  In somatic cell genetics, a method for determining whether or not two genes are on the same chromosome.

$T_m$ The temperature at which a population of double-stranded nucleic acid molecules is half-dissociated into single strands. This is taken to be the melting temperature for that species of nucleic acid.

**target theory** In radiation biology, a theory stating that damage and death from radiation is caused by the inactivation of specific targets within the organism.

**TATA box** See *Goldberg–Hogness box.*

**tautomeric shift** A reversible isomerization in a molecule brought about by a shift in the localization of a hydrogen atom. In nucleic acids, tautomeric shifts in the bases of nucleotides can cause changes in other bases at replication and are a source of mutations.

**TDF (testis-determining factor)** A gene on the Y chromosome that controls the developmental switch point for the development of the indifferent gonad into a testis. See *SRY.*

**telocentric chromosome** A chromosome in which the centromere is located at the end of the chromosome.

**telomerase** The enzyme that adds short, tandemly repeated DNA sequences to the ends of eukaryotic chromosomes.

**telomere** The terminal chromomere of a chromosome.

**telophase** The stage of cell division in which the daughter chromosomes reach the opposite poles of the cell and re-form nuclei. Telophase ends with the completion of cytokinesis.

**temperate phage** A bacteriophage that can become a prophage and confer lysogeny upon the host bacterial cell.

**temperature-sensitive mutation** A conditional mutation that produces a mutant phenotype at one temperature range and a wild-type phenotype at another temperature range.

**template** The single-stranded DNA or RNA molecule that specifies the nucleotide sequence of a strand synthesized by a polymerase molecule.

**teratocarcinoma** Embryonal tumors that arise in the yolk sac or gonads and are able to undergo differentiation into a wide variety of cell types. These tumors are used to investigate the regulatory mechanisms underlying development.

**terminalization** The movement of chiasmata toward the ends of chromosomes during the diplotene stage of the first meiotic division.

**tertiary protein structure** The three-dimensional structure of a polypeptide chain brought about by folding upon itself.

**test cross** A cross between an individual whose genotype at one or more loci may be unknown and an individual who is homozygous recessive for the genes in question.

**tetrad** The four chromatids that make up paired homologues in the prophase of the first meiotic division. The four haploid cells produced by a single meiotic division.

**tetrad analysis** Method for the analysis of gene linkage and recombination using the four haploid cells produced in a single meiotic division.

**tetranucleotide theory** An early theory of DNA structure proposing that the molecule was composed of repeating units, each consisting of the four nucleotides adenosine, thymidine, cytosine, and guanine.

**tetraparental mouse** A mouse produced from an embryo that was derived by the fusion of two separate blastulas.

**theta structure** An intermediate in the bidirectional replication of circular DNA molecules. At about midway through the cycle of replication, the intermediate resembles the Greek letter theta.

**thymine dimer** A pair of adjacent thymine bases in a single polynucleotide strand between which chemical bonds have formed. This lesion, usually the result of damage caused by exposure to ultraviolet light, inhibits DNA replication unless repaired by the appropriate enzyme system.

**topoisomerase** A class of enzymes that convert DNA from one topological form to another. During DNA replication, these enzymes facilitate the unwinding of the double-helical structure of DNA.

**totipotent** The ability of a cell or embryo part to give rise to all adult structures. This capacity is usually progressively restricted during development.

**trailer sequence** A transcribed but nontranslated region of a gene or its mRNA that follows the termination signal.

**trait** Any detectable phenotypic variation of a particular inherited character.

***trans* configuration** The arrangement of two mutant sites on opposite homologues, such as

$$\frac{a^1 \ +}{+ \ a^2}$$

Contrasts with a *cis* arrangement, where they are located on the same homologue.

**transcription** Transfer of genetic information from DNA by the synthesis of an RNA molecule copied from a DNA template.

**transdetermination** Change in developmental fate of a cell or group of cells.

**transduction** Virally mediated genes transfer from one bacterium to another, or the transfer of eukaryotic genes mediated by retrovirus.

**transfer RNA** See *tRNA.*

**transformation** Heritable change in a cell or an organism brought about by exogenous DNA.

**transgenic organism** An organism whose genome has been modified by the introduction of external DNA sequences into the germline.

**transition** A mutational event in which one purine is replaced by another, or one pyrimidine is replaced by another.

**translation** The derivation of the amino acid sequence of a polypeptide from the base sequence of an mRNA molecule in association with a ribosome.

**translocation** A chromosomal mutation associated with the transfer of a chromosomal segment from one chromosome to another. Also used to denote the movement of mRNA through the ribosome during translation.

**transmission genetics** The field of genetics concerned with the mechanisms by which genes are transferred from parent to offspring.

**transposable element** A defined length of DNA that translocates to other sites in the genome, essentially independent of sequence homology. Usually such elements are flanked by short, inverted repeats of 20 to 40 base pairs at each end. Insertion into a structural gene can produce a mutant pheno-

Glossary  **B-19**

type. Insertion and excision of transposable elements depends on two enzymes, transposase and resolvase. Such elements have been identified in both prokaryotes and eukaryotes.

**transversion**  A mutational event in which a purine is replaced by a pyrimidine, or a pyrimidine is replaced by a purine.

**triploidy**  The condition in which a cell or organism possesses three haploid sets of chromosomes.

**trisomy**  The condition in which a cell or organism possesses two copies of each chromosome, except for one, which is present in three copies. The general form for trisomy is therefore $2n + 1$.

**tritium ($^3$H)**  A radioactive isotope of hydrogen, with a half-life of 12.46 years.

**trivalent**  An association between three homologous chromosomes.

**tRNA**  Transfer RNA; a small ribonucleic acid molecule that contains a three-base segment (anticodon) that recognizes a codon in mRNA, a binding site for a specific amino acid, and recognition sites for interaction with the ribosomes and the enzyme that links it to its specific amino acid.

**tumor suppressor gene**  A gene that encodes a gene product that normally functions to suppress cell division. Mutations in tumor suppressor genes result in the activation of cell division and tumor formation.

**Turner syndrome**  A genetic condition in human females caused by a 45,X genotype (XO). Such individuals are phenotypically female but are sterile because of undeveloped ovaries.

**unequal crossing over**  A crossover between two improperly aligned homologues, producing one homologue with three copies of a region and the other with one copy of that region.

**unique DNA**  DNA sequences that are present only once per genome. Single copy DNA.

**universal code**  The assumption that the genetic code is used by all life forms. In general, this is true, but with some exceptions found in mitochondria, ciliates, and mycoplasmas.

**unwinding proteins**  Nuclear proteins that act during DNA replication to destabilize and unwind the DNA helix ahead of the replicating fork.

**up promoter**  A promoter sequence, often mutant, that increases the rate of transcription initiation. Also known as strong promoter.

**variable number tandem repeats (VNTRs)**  Short DNA sequences (2–20 nucleotides) present as tandem repeats between two restriction enzyme sites. Variations in the number of repeats creates DNA fragments of differing lengths following restriction enzyme digestion.

**variable region**  Portion of an immunoglobulin molecule that exhibits many amino acid sequence differences between antibodies of differing specificities.

**variance**  A statistical measure of the variation of values from a central value, calculated as the square of the standard deviation.

**variegation**  Patches of differing phenotypes, such as color, in a tissue.

**vector**  In recombinant DNA, an agent such as a phage or plasmid into which a foreign DNA segment will be inserted.

**viability**  The measure of the number of individuals in a given phenotypic class that survive, relative to another class (usually wild type).

**virulent phage**  A bacteriophage that infects and lyses the host bacterial cell.

**VNTR**  See *variable number tandem repeats.*

**W, Z chromosomes**  Sex chromosomes in species where the female is the heterogametic sex (WZ).

**western blot**  A technique in which proteins are separated by gel electrophoresis and transferred by capillary action to a nylon membrane or nitrocellulose sheet. A specific protein can be identified through hybridization to a labeled antibody.

**wild type**  The most commonly observed phenotype or genotype, designated as the norm or standard.

**wobble hypothesis**  An idea proposed by Francis Crick stating that the third base in an anticodon can align in several ways to allow it to recognize more than one base in the codons of mRNA.

**writhing number**  The number of times that the axis of a DNA duplex crosses itself by supercoiling.

**X inactivation**  In mammalian females, the random cessation of transcriptional activity of one X chromosome. This event, which occurs early in development, is a mechanism of dosage compensation. Molecular basis of inactivation is unknown, but loci on the tip of the short arm of the X can escape inactivation. See *Barr body, Lyon hypothesis.*

**X linkage**  See *sex linkage.*

**X-ray crystallography**  A technique to determine the three-dimensional structure of molecules through diffraction patterns produced by X-ray scattering by crystals of the molecule under study.

**Y chromosome**  Sex chromosome in species where the male is heterogametic (XY).

**Y linkage**  Mode of inheritance shown by genes located on the Y chromosome.

**Z-DNA**  An alternative structure of DNA in which the two antiparallel polynucleotide chains form a left-handed double helix. Z-DNA has been shown to be present along with B-DNA in chromosomes and may have a role in regulation of gene expression.

**zein**  Principal storage protein of corn endosperm, consisting of two major proteins, with molecular weights of 19,000 and 21,000 daltons.

**zinc finger** A DNA-binding domain of a protein that has a characteristic pattern of cysteine and histidine residues that complex with zinc ions, throwing intermediate amino acid residues into a series of loops or fingers.

**zygote** The diploid cell produced by the fusion of haploid gametic nuclei.

**zygotene** A stage of meiotic prophase I in which the homologous chromosomes synapse and pair along their entire length, forming bivalents. The synaptonemal complex forms at this stage.

# APPENDIX C

## Solutions to Selected Even-Numbered Problems and Discussion Questions

### CHAPTER 1 An Introduction to Genetics

**2.** *Epigenesis* is the theory that organisms are derived from the assembly and reorganization of substances in the egg. *Preformationism* is a 17th century theory which states that the sex cells (eggs or sperm) contain miniature adults called homunculi.

**4.** Darwin proposed that physical units (gemmules) represented each body part and gathered in the semen. He also proposed that such gemmules could adapt to an individual's environment.

**6.** *Transmission* genetics, *cytological investigations, molecular and biochemical* analysis including *recombinant DNA technology, and population* genetics.

**8.** Borlaug applied Mendelian principles of hybridization and trait selection to the development of superior varieties of wheat to change in worldwide agricultural food production.

**10.** Lysenko directed Soviet genetics based on the idea of inheritance of acquired characteristics; that plant productivity could be improved in an inherited fashion by changes in the environment.

### CHAPTER 2 Cell Division and Chromosomes

**2.** Chromosomes that are homologous share many properties including: *overall length; position of the centromere; banding patterns; type and location of genes; autoradiographic pattern. Diploidy* means that both members of a homologous pair of chromosomes are present. *Haploidy* means that each cell contains *one chromosome of each homologous pair of chromosomes*.

**6.** Metacentric, submetacentric, acrocentric, telocentric

**12.** Each spermatogonium and primary spermatocyte produces four spermatids, whereas each oogonium and primary oocyte produces one ootid. Unequal distribution of cytoplasm in oogenesis evolved to provide sufficient information and nutrients to support development until the transcriptional activities of the zygotic nucleus begin to provide products. Polar bodies probably represent nonfunctional by-products of such evolution.

**14.** Four unique combinations are possible.

**16.** **(a)** Same; **(b)** eight combinations with homologues paired; **(c)** eight different meiotic (haploid) products.

**18.** **(a)** *Crossing over* is known to have occurred by pachynema; **(b)** *synapsis* begins at zygonema with the homology search while intimate pairing (synapsis) is completed during pachynema; **(c)** leptonema; **(d)** diplonema.

**20.** Folded-fiber model

**22.** Numerous DNA replications, pairing of homologues, and absence of strand separation or cytoplasmic division

**24.** Meiotic genetic processes (vertebrate oocytes)

### CHAPTER 3 Mendelian Genetics

**2.** **(a)** Parents = *Aa*, nonalbino children = *AA* or *Aa*, albino child = *aa*; **(b)** male = *AA* or *Aa*, female = *aa*, children = *Aa*; **(c)** male = *Aa*, female = *aa*, nonalbino children = *Aa*, albino children = *aa*.

**6.** *P* = checkered; *p* = plain

|  |  | F₁ Progeny | |
| --- | --- | --- | --- |
|  | $P_1$ Cross | Checkered | Plain |
| (a) | $PP \times PP$ | *PP* | |
| (b) | $PP \times pp$ | *Pp* | |
| (c) | $pp \times pp$ | | *pp* |
| (d) | $PP \times pp$ | *Pp* | |
| (e) | $Pp \times pp$ | *Pp* | *pp* |
| (f) | $Pp \times Pp$ | *PP,Pp* | *pp* |
| (g) | $PP \times Pp$ | *PP,Pp* | |

**8.** Symbolism as before:
$w$ = wrinkled seeds  $g$ = green cotyledons
$W$ = round seeds  $G$ = yellow cotyledons
**(a)** $WwGG \times WwGG$ or $WwGG \times WwGg$
**(b)** $wwGg \times WwGg$
**(c)** $WwGg \times WwGg$
**(d)** $WwGg \times wwgg$

**10.** Independent assortment

**12.** Overall length, position of the centromere, banding patterns, type and location of genes, autoradiographic pattern

**14.**

| | Phenotypic Ratio | Genotypic Ratio |
|---|---|---|
| **(a)** | 9:3:3:1 | 1:2:2:4:1:2:1:2:1 |
| **(b)** | 9:3:3:1 | 1:2:2:4:1:2:1:2:1 |
| **(c)** | 3:1 | 1:2:1 |

**16.**

| | Genotypic Ratio | Phenotypic Ratio |
|---|---|---|
| **(a)** | 1:1:1:1:2:2:2:2:1:1:1:1 | 3:1 |
| **(b)** | 1:2:1:1:2:1 | 3:1:3:1 |
| **(c)** | 27 classes | 27:9:9:9:3:3:3:1 |

**18.** 1/2 black and 1/2 white

**20.** No gene is recessive; incomplete dominance exists.

**22.** **(a)** $\chi^2 = 0.5$ does not deviate significantly from expected; **(b)** $\chi^2 = 0.35$ does not deviate significantly from the expected; **(c)** $\chi^2 = 0.01$ does not deviate significantly from the expected.

**24.** The stringency of failing to reject the null hypothesis is increased.

**26.** Autosomal recessive; I-1 ($AA$ or $Aa$), I-2 ($aa$), I-3 ($Aa$), I-4 ($Aa$); II-1 ($Aa$), II-2 ($Aa$), II-3 ($Aa$), II-4 ($Aa$), II-5 ($aa$), II-6 ($AA$ or $Aa$), II-7 ($AA$ or $Aa$); III-1 ($AA$ or $Aa$), III-2 ($aa$), III-3 ($AA$ or $Aa$).

**28.** 1/8

**30.** 3/4

**32.** 4/9

**34.** **(a)** = 1/32
**(b)** = 5/16
**(c)** = 5/16
**(d)** 1/32 + 1/32 = 1/16

**36.** Incomplete dominance or codominance

# CHAPTER 4 Modification of Mendelian Ratios

**4.** $S$ = short, $s$ = long
Cross 1: $Ss \times ss \rightarrow$ 1/2 $Ss$, 1/2 $ss$
Cross 2: $Ss \times Ss \rightarrow$ 1/4 $SS$, 2/4 $Ss$, 1/4 $ss$

**6.** $I^A I^B$ (AB), $I^A I^O$ (A), $I^B I^O$ (B), $I^O I^O$ (O):   1:1:1:1

**8.** $S$ = secretor  $s$ = nonsecretor
Cross 1: $SS \times SS \rightarrow SS$
Cross 2: $ss \times ss \rightarrow ss$
Cross 3: $SS \times ss \rightarrow Ss$
Cross 4: $Ss \times Ss \rightarrow$ 1/4 $SS$; 2/4 $Ss$; 1/4 $ss$

**10.** **(a)** $c^a c^a \times c^{cb} c^a \rightarrow$ 1/2 chinchilla; 1/2 albino
**(b)** $c^a c^a \times Cc^a \rightarrow$ 1/2 full color; 1/2 albino
**(c)** $c^h c^a \times c^h c^a \rightarrow$ 3/4 Himalayan; 1/4 albino

**12.** $RRPPDD \times rrppdd \rightarrow RrPpDd$ (pink, personate, tall); 18/64

**14.** **(a)** Incomplete dominance; $C^{cb} C^{cb}$ = chestnut, $C^c C^c$ = cremello, $C^{cb} C^c$ = palomino; **(b)** $F_1$ = all palomino, $F_2$ = 1:2:1

**16.** **(a)** $F_1$ = $AaCc$ (gray), $F_2$ = 9:3:4; **(b)** (1) $AACc$, (2) $AaCC$, (3) $AaCc$

**18.** **(a)** Gray; **(b)** gray; **(c)** 16:9:3:3:1; **(d)** 9:3:4; **(e)** 3:1:4

**20.** Epistasis; 3/16 type A, 6/16 type AB, 3/16 type B, 4/16 type O

**22.** **(a)** $A\_B\_$ = 9/16 (tall), $A\_bb$ = 3/16 (dwarf), $aaB\_$ = 3/16 (dwarf), $aabb$ = 1/16 (dwarf); **(b)** 3/7, $AAbb$, $aaBB$, and $aabb$

**24.** Cross 1 = (c), Cross 2 = (d), Cross 3 = (b), Cross 4 = (e), Cross 5 = (a). Pattern of a 9:3:3:1 ratio because of homozygosity for the $A$ locus and heterozygosity for both the $B$ and $R$ loci.

**26.** **(a)** $AAB- \times aaBB$ (other configurations possible)
**(b)** $AaB- \times aaBB$ (other configurations are possible)
**(c)** $AABb \times aaBb$  **(d)** $AABB \times aabb$
**(e)** $AaBb \times Aabb$  **(f)** $AaBb \times aabb$
**(g)** $aaBb \times aaBb$  **(h)** $AaBb \times AaBb$
True breeding = $AABB$, all genotypes that are $bb$, and $aaBB$.

**28.** 1/8; $AabbCc$

**30.** 4 inches = $aabbccdd$; 8 inches = $AAbbccdd$ (and others); 12 inches = $aaBBCCdd$ (and others); 16 inches = $AABBCCdd$ (and others); 20 inches = $AABBCCDD$

**32.** **(a)** The genotypes of the parents would be combinations of alleles that would produce a 6-cm ($aabbcc$) tail and a 30-cm ($AABBCC$) tail while the 18-cm offspring would have a genotype of $AaBbCc$; **(b)** variety of possibilities.

**34.** **(a)** 1/4; **(b)** 1/2; **(c)** 1/4; **(d)** zero

**36.** **(a)** $F_1$: 1/2 $X^+ X^{sd}$ (female normal), 1/2 $X^{sd}/Y$ (male, scalloped); $F_2$: 1/4 $X^+ X^{sd}$ (female, normal), 1/4 $X^{sd} X^{sd}$ (female, scalloped), 1/4 $X^+/Y$ (male, normal), 1/4 $X^{sd}/Y$ (male, scalloped). **(b)** $F_1$: 1/2 $X^+ X^{sd}$ (female, normal), 1/2 $X^+/Y$ (male, normal); $F_2$: 1/4 $X^+ X^+$ (female, normal), 1/4 $X^+ X^{sd}$ (female, normal), 1/4 $X^+/Y$ (male, normal), 1/4 $X^{sd}/Y$ (male, scalloped). If the *scalloped* gene were not X-linked, then all of the $F_1$ offspring would be wild (phenotypically) and a 3:1 ratio of normal to scalloped would occur in the $F_2$.

**38.** $F_1$: 1/2 $X^+ X^v$; $su-v^+/su-v$ (female, normal)
1/2 $X^+/Y$; $su-v^+/su-v$ (male, normal)
$F_2$: 8/16 wild-type females, 5/16 wild-type males, 3/16 vermilion males

**40.** **(a)** $F_1$: $w^+/w$; $se^+/se$ = wild females
$w/Y$; $se^+/se$ = white-eyed males
$F_2$: 3/16 males wild, 4/16 males white, 1/16 males sepia, 3/16 females wild, 4/16 females white, 1/16 females sepia
**(b)** $F_1$: $w^+/w$; $se^+/se$ = wild females
$w^+/Y$; $se^+/se$ = wild males
$F_2$: 3/16 males wild, 4/16 males white, 1/16 males sepia, 6/16 females wild, 2/16 females sepia

**42.** Sex-influenced inheritance; $RR$ = red, $Rr$ = red in females, $Rr$ = mahogany in males, $rr$ = mahogany. $P_1$: female: $RR$ (red) × male: $rr$ (mahogany). $F_1$: $Rr$ = females red; males mahogany, 1/2 females (red), 1/2 males (mahogany)

| $F_2$: | 1/2 females | 1/2 males |
|---|---|---|
| 1/4 $RR$ | 1/8 red | 1/8 red |
| 2/4 $Rr$ | 2/8 red | 2/8 mahogany |
| 1/4 $rr$ | 1/8 mahogany | 1/8 mahogany |

# CHAPTER 5 Linkage, Crossing Over, and Chromosome Mapping

**2.** Proximity through synapsis, chiasmata

**4.** Because crossing over occurs at the four-strand stage of the cell cycle (that is, after S phase), notice that each single crossover involves only two of the four chromatids.

**6.** Physical constraints

**8.**
```
dp — cl ———————————— ap
   3 mu        39 mu
```

**10.** $RY/ry$ × $ry/ry$; 10 map units between the $R$ and $Y$ loci.

**12.** $PZ/pz$ × $pz/pz$; 14 map units between the two genes.

**14.**

| | Female A | Female B | Frequency |
|---|---|---|---|
| NCO | 3, 4 | 7, 8 | First |
| SCO | 1, 2 | 3, 4 | Second |
| SCO | 7, 8 | 5, 6 | Third |
| DCO | 6, 5 | 1, 2 | Fourth |

**16. (a)** $y\,w\,+/+\,+\,ct \times y\,w\,+/Y$
**(b)**
```
y ————— w ———————————————— ct
0.0     1.5                  20.0
```
**(c)** $0.185 \times 0.015 \times 1000 = 2.775$ double crossovers expected. **(d)** No

**18. (a)** 0.20 wild type, 0.05 ebony, 0.05 pink, 0.20 pink, ebony, 0.2 dumpy, 0.05 dumpy, ebony, 0.05 dumpy, pink, 0.20 dumpy, pink, ebony; **(b)** 0.25 wild type, 0.25 pink, ebony, 0.25 dumpy, 0.25 dumpy, pink, ebony

**20. (a,b)** $+\,b\,c/a\,+\,+$; $a - b = 7$ map units, $b - c = 2$ map units; **(c)** $+\,+\,c$ and $a\,b\,+$

**22.** It would take two crossovers to "isolate" the *singed* gene for a spot, and one exchange for a twin spot. Therefore the likelihood of a twin spot is greater than the likelihood of a singed spot. The least frequent event would probably be forked spots because it would take two crossovers to "isolate" the *forked* gene.

**24.** The synaptonemal complex is probably required for crossing over.

# CHAPTER 6 Chromosome Variation and Sex Determination

**2.** Nondisjunction in which fertilization, by a Y-bearing sperm cell, of those female gametes with two X chromosomes would produce the XXY Klinefelter syndrome. Fertilization of the "no-X" female gamete with a normal X-bearing sperm will produce Turner syndrome.

**8.** Such phenotypic mosaicism is dependent on the heterozygous condition of genes on the two X chromosomes.

**12.** Temperature influences sex determination in alligators and most turtles; however, in most lizards, temperature is not influential.

**16.** A significant maternal age effect

**18.** At least 20 percent of all conceptions are terminated in natural abortion.

**22.** An interspecific hybridization occurred followed by chromosome doubling.

**24.** Crossover chromatids that end up being abnormal in genetic content fail to produce viable (or competitive) gametes or lead to zygotic or embryonic death.

**26.** Genes $a$, $b$, $c$, $d$ are included in an inversion. The minimum distance between loci $d$ and $e$ can be estimated as 10 map units; however, this is actually the distance from the $e$ locus to the breakpoint that includes the inversion.

**28.** Gene duplication has been essential in the origin of new genes.

**30.** Mother

**32.** Autotetraploids; $3n$ individuals are more likely to be sterile because there are trivalents at meiosis I.

**36.** $WWWW \times wwww$ (assuming that chromosomes pair as bivalents at meiosis)

**38. (a)** Crossing over in the inversion loop of an inversion (in the heterozygous state). **(b)** A significant proportion (perhaps 50%) of the children of the man will be similarly influenced by the inversion. **(c)** Since the karyotypic abnormality is observable, it may be possible to detect some of the abnormal chromosomes of the fetus by amniocentesis or CVS. However, depending on the type of inversion and the ability to detect minor changes in banding patterns, all abnormal chromosomes may not be detected.

# CHAPTER 7 Advanced Topics in Transmission Genetics

**2.** Height, general body structure, skin color, and, perhaps most common, behavioral traits including some components of intelligence.

**6. (a)** 140 cm; **(b)** 374.18; **(c)** 19.34; **(d)** 0.70

**8.** 0.53

**10.** 10 map units

**12.** Cross 1 = 50 map units, Cross 2 = 12 map units

**14. (a)** 1 = NP, 2 = T, 3 = P, 4 = NP, 5 = T, 6 = P, 7 = T. **(b)** The genes are linked. **(c)** Centromere to $c$ distance = 7.9 map units; centromere to $d$ distance = 16.4 map units. **(d)** 20 map units. **(e)** The discrepancy between the two mapping systems is caused by the manner in which first and second division segregation products are scored.

## CHAPTER 8 DNA: Structure and Analysis

**2.** Proteins are composed of as many as twenty different subunits (amino acids), thereby providing ample structural and functional variation for the multiple tasks that must be accomplished by the genetic material. Nucleic acids seemed to have insufficient variability to account for the diverse roles of the genetic material.

**4.** DNase eliminates DNA and transformation; therefore, DNA must be the transforming principle.

**6.** The T2 phage, in its mature state, contains very little if any RNA; therefore, DNA would be interpreted as being the genetic material in T2 phages.

**10.** Linkages among the three components require the removal of water ($H_2O$).

**12.** Guanine: 2-amino-6-oxypurine; cytosine: 2-oxy-4-amino-pyrimidine; thymine: 2,4-dioxy-5-methylpyrimidine; uracil: 2,4-dioxyprimidine

**16.** A lack of pairing of these bases would favor a single-stranded structure or some other nonhydrogen-bonded structure or A=G and T=C, which would require purines to pair with purines and pyrimidines to pair with pyrimidines. This would have contradicted the data from Franklin and Watkins that called for a constant diameter for the double-stranded structure.

**18.** (1) Uracil in RNA replaces thymine in DNA; (2) ribose in RNA replaces deoxyribose in DNA; and (3) RNA often occurs as both single- and double-stranded forms, whereas DNA most often occurs in a double-stranded form.

**20.** Nitrogenous bases, UV absorption is greater in single-stranded molecules (hyperchromic shift) as compared to double-stranded structures, and A−T rich DNA denatures more readily than G−C rich DNA; therefore, one can estimate base content by denaturation kinetics.

**22.** G≡C base pairs are more dense than A=T pairs.

**24.** Curve *A* in the problem; there is evidence for a rapidly renaturing species (repetitive) and a slowly renaturing species (unique). Fraction *B* contains mostly unique, relatively complex DNA.

**26.** It takes more energy (higher temperature) to separate G≡C pairs than A=T pairs.

**30.** MS-2 = 200 base pairs, *E. coli* = $2 \times 10^6$ base pairs

**32.** (a) = left; (b) = right

## CHAPTER 9 DNA: Replication, Synthesis and Recombination

**4.** **(a)** Under a conservative scheme all of the newly labeled DNA will go to one sister chromatid, while the other sister chromatid will remain unlabeled. **(b)** Under a dispersive scheme all of the newly labeled DNA will be interspersed with unlabeled DNA.

**6.** A DNA template, a divalent cation ($Mg^{++}$), and all four of the deoxyribonucleoside triphosphates: dATP, dCTP, dTTP, and dGTP.

**8.** Comparison of base composition, and by *nearest neighbor frequencies*

**10.** DNA polymerase I is slow, capable of degrading as well as synthesizing DNA, and a strain of *E. coli* (*polA1*) was discovered that still replicated its DNA but was deficient in DNA polymerase I activity.

**12.** The DNA is capable of supporting typical metabolic activities of the cell or organism and is capable of faithful reproduction.

**18.** *Helicase, dnaA,* and *single-stranded DNA-binding* proteins initially unwind, open, and stabilize DNA at the initiation point. *DNA gyrase* relieves supercoiling generated by helix unwinding.

**22.** **(a)** No; **(b)** yes; **(c)** no

**24.** **(a)** About 4,000,000 bp; **(b)** $1.36 \times 10^6$ nm

## CHAPTER 10 DNA: Organization in Chromosomes and Genes

**6.** *Heterochromatin* is chromosomal material that stains deeply, remains condensed, replicates late in S phase, and is relatively genetically inactive. Telomeres and the areas adjacent to centromeres are composed of heterochromatin.

**8.** Volume of DNA = $1.57 \times 10^8 \, Å^3$; volume of capsid = $2.67 \times 10^8 \, Å^3$. The DNA will fit into the capsid.

**10.** *Exon shuffling* is where exons of genes, each coding for functional domains of proteins, can be "shuffled" or rearranged to facilitate the evolution of a new protein.

## CHAPTER 11 DNA: Mutation, Repair, and Transposable Elements

**2.** Mutations are the "windows" through which geneticists look at the normal function of genes, cells, and organisms.

**4.** All mutations may not be deleterious. Those few, rare variations that are beneficial will provide a basis for possible differential propagation of the variation.

**6.** A diploid organism possesses at least two copies of each gene (except for "hemizygous" genes) and in most cases, the amount of product from one gene of each pair is sufficient for production of a normal phenotype.

**12.** Tautomeric forms, caused by single proton shifts, could result in mutations by allowing hydrogen bonding of normally noncomplementary bases.

**14.** Yes

**16.** In contrast to UV light, X rays penetrate surface layers of cells and break chromosomes. Ions and free radicals are formed in the paths of X rays and these interact with components of DNA to cause mutations. UV light generates pyrimidine dimers, primarily thymine, which distort the normal conformation of DNA and inhibit normal function.

**18.** X rays are known to be mutagenic. This subject is presently under considerable debate.

**22.** Group 1: *XP1, XP2, XP3*; Group 2: *XP4*; Group 3: *XP5, XP6, XP7*

**26.** Reverse transcriptase, in making DNA, provides a DNA segment that is capable of integrating into the yeast chromosome.

## CHAPTER 12 DNA: Cloning and Manipulation

**2.** Eukaryotic mRNAs typically have a 3′ polyA tail. The poly dT segment provides a double-stranded section that serves to prime the production of the complementary strand.

**4.** $N = 1.38 \times 10^5$

**6.** Two of the plasmids can join to form a dimer that migrates high in the gel.

**8.** Size of the foreign DNA; selective qualities of the vector

**10.** Assuming a random distribution of all four bases, the four-base sequence would occur (on average) every 256 base pairs ($4^4$), the six-base sequence every 4096 base pairs ($4^6$), and the eight-base sequence every 65,536 base pairs ($4^8$). One might use an eight-base restriction enzyme to produce a relatively few large fragments.

**12.**

```
              II   I
       200  |    |        950
       ―――――――――――――――――――――――――
                150
```

**14.** Reverse transcriptase may not completely synthesize the DNA from the RNA template. The 3′ end of the copied DNA tends to fold back on itself, thus providing a primer for the DNA polymerase. Additional preparation of the cDNA requires some digestion at the folded region. Since this folded region corresponds to the 5′ end of the mRNA, some of the message is often lost.

## CHAPTER 13 DNA: Applications of Recombinant Technology

**4.** **(a,b)** (1) Integration into the host must be cell-specific so as not to damage nontarget cells. (2) Retroviral integration into host cell genomes only occurs if the host cell is replicating. (3) Insertion of the viral genome might influence nontarget but essential genes. (4) Retroviral genomes have a low cloning capacity and can not carry large inserted sequences as are many human genes. (5) There is a possibility that recombination with host viruses will produce an infectious virus that may do harm.

**8.** Negative results can occur even though the person carries the gene for CF. The specific probes (or allele-specific oligonucleotides) that have been developed will not necessarily be useful for screening all mutant genes.

**10.** The child in question is a carrier of the deletion in the beta globin gene, just as the parents are carriers. Its genotype is therefore $\beta^A\beta^S$.

**12.** The amino acid sequence of the protein can be used to produce the gene synthetically. Or, since the introns are spliced out of the hnRNA in the production of mRNA, if mRNA can be obtained, it can be used to make DNA through the use of reverse transcriptase.

## CHAPTER 14 Storage and Expression of Genetic Information

**2.** No

**4.** Proline:      $C_3$, and one of the $C_2A$ triplets
Histidine:      one of the $C_2A$ triplets
Threonine:      one $C_2A$ triplet, and one $A_2C$ triplet
Glutamine:      one of the $A_2C$ triplets
Asparagine:      one of the $A_2C$ triplets
Lysine:      $A_3$

**6.** Degeneracy is not revealed and all the codons produce unique amino acids.

**8.** Given that AGG = arg, then information from the AG copolymer indicates that AGA also codes for arg and GAG must therefore code for glu. Coupling this information with that of the AAG copolymer, GAA must also code for glu, and AAG must code for lys.

**10.**

| Original | | Substitutions |
|---|---|---|
| *threonine* | → | *alanine* |
| $\underline{AC}$(U, C, A, or G) | | $\underline{GC}$(U, C, A, or G) |
| *glycine* | → | *serine* |
| $\underline{GG}$(U or C) | | $\underline{AG}$(U or C) |
| *isoleucine* | → | *valine* |
| $\underline{AU}$(U, C or A) | | $\underline{GU}$(U, C or A) |

**12.** *In vivo*, the concentration of ribonucleoside diphosphates is low and the degradative process is favored.

**14.** Sequence 1: met-pro-asp-tyr-ser-(term)
Sequence 2: met-pro-asp-(term)

**16.** **(b)** TCCGCGGCTGAGATGA (use complementary bases, substituting T for U); **(c)** GCU; **(d)** assuming that the AGG . . . is the 5′ end of the mRNA:*arg-arg-arg-leu-tyr.*

**18.** **(a)** met-his-thr-tyr-glu-thr-leu-gly; met-arg-pro-leu-gly. **(b)** In the shorter of the two reading sequences a UGA triplet was introduced.

**20.** DNA produces, through transcription, RNA, which is "decoded" (during translation) to produce proteins.

**24.** mRNA, charged tRNA, large and small ribosomal subunits, elongation and perhaps initiation factors, peptidyl transferase, GTP, $Mg^{++}$, nascent proteins, possibly GTP-dependent release factors.

**28.** 423 code letters (nucleotides), 426 including a termination codon; the approximate number of triplet codes is 20.

**30.** Attachment of the specific amino acid, interaction with the aminoacyl tRNA synthetase, interaction with the ribosome, and interaction with the codon (anticodon).

**32.** **(a)** Sequence 1: GAAAAAACGGUA
Sequence 2: UGUAGUUAUUGA
Sequence 3: AUGUUCCCAAGA
**(b)** Sequence 1: *glu-lys-thr-val*
Sequence 2: *cys-ser-tyr*
Sequence 3: *met-phe-pro-arg*
**(c)** Sequence 1 = middle, Sequence 2 = terminal portion, Sequence 3 = initial portion; **(d)** GAAAAAACGGTA.

## CHAPTER 15 Proteins: The End Products of Genes

**2.** Both phenylalanine and tyrosine can be obtained from the diet.

**4.**

**6.** (a)

(b)

**8.**

**10.** The electrophoretic mobility of a protein is based on a variety of factors, primarily the net charge of the protein and to some extent, the conformation in the electrophoretic environment.

**14.** If the amino acid is substituted with an amino acid of like charge and similar structure, there is a chance that factors which influence electrophoretic mobility (primarily net charge) will not be altered.

**16.** The nearer the mutations are to the 5′ end of the mRNA, the shorter will be the polypeptide product.

**20.** **(a)** R-group on glycines facilitate the formation of the triple helical structure. **(b)** Procollagen peptidases shorten the procollagen fibers. **(c)** Prolyl hydroxylase adds hydroxyl groups to proline. **(d)** Cross-links may form to yield a highly stable structure. **(e)** Stability.

**22.** All involve one base change.

**24.** No difference

## CHAPTER 16 Genetic Regulation in Bacteria and Bacteriophages

**2.** Under *negative* control the regulatory molecule interferes with transcription, while in *positive* control the regulatory molecule stimulates transcription.

**6.** $I^+O^+Z^+$ = inducible, $I^-O^+Z^+$ = constitutive, $I^+O^cZ^+$ = constitutive, $I^-O^+Z^+/F'I^+$ = inducible, $I^+O^cZ^+/F'O^+$ = constitutive, $I^sO^+Z^+$ = repressed, $I^sO^+Z^+/F'I^+$ = repressed

**8.** $c$ = structural gene, $b$ = promoter, $a$ = operator, $d$ = repressor

**10.** **(a)** Operon is off; **(b)** operon is transcribing; **(c)** operon is off; **(d)** operon is off.

**12.** **(b)** Spontaneous mutation rate = $0.25 \times 10^{-8}$; induced mutation rate = $0.5 \times 10^{-7}$

## CHAPTER 17 Genetic Regulation in Eukaryotes

**2.** Organization of DNA, gene amplification, transcription, processing and transport, translation

**4.** *Promoters* influence transcription from the "upstream" side (5′) of mRNA coding genes. *Enhancers* are *cis*-acting and may be upstream, downstream, or within the gene being regulated. The orientation may be inverted without significantly influencing its action. Enhancers are not gene-specific.

**6.** It is the sequence of the amino acids, not the mRNA that is critical in the process of autoregulation. The model that depicts binding of factors to the nascent polypeptide chain is supported.

## CHAPTER 18 The Genetics of Cancer

**2.** The $G_1$ stage begins after mitosis and is involved in the synthesis of many cytoplasmic elements. In the S phase DNA synthesis occurs. $G_2$ is a period of growth and preparation for mitosis. Most cell cycle time variation is caused by changes in the duration of $G_1$. $G_0$ is the nondividing state.

**4.** Kinases regulate other proteins by adding phosphate groups. Cyclins bind to the kinases, switching them on and off. Several cyclins, including D and E, can move cells from $G_1$ to S. At the $G_2$/mitosis border a CDK1 (cyclin dependent kinase) combines with another cyclin (cyclin B). Phosphorylation occurs, bringing about a series of changes in the nuclear membrane, cytoskeleton, and histone 1.

**6.** A tumor suppressor gene is a gene that normally functions to suppress cell division. Natural selection would favor the evolution of recessive. If a tumor suppressor gene makes a product that regulates the cell cycle favorably, cellular conditions have evolved in such a way that sufficient quantities of this gene product are made from just one gene (of the two present in each diploid individual) to provide normal function.

**8.** Oncogenes are mutant forms of proto-oncogenes that normally function to regulate cell division.

## CHAPTER 19 Mutation and Genetic Recombination in Bacteria and Bacteriophages

**4.** **(a)** By placing a filter in a U-tube. **(b)** By treating cells with streptomycin, an antibiotic, it was shown that recombination would not occur if one of the two bacterial strains was inactivated. **(c)** An $F^+$ bacterium contains a circular, double-stranded, structurally independent, DNA molecule that can direct recombination.

**8.** An $F^+$ element can enter the host bacterial chromosome and upon returning to its independent state, it may pick up a piece of a bacterial chromosome, producing a partial diploid, or merozygote.

**10.** No linkage

**12.** A filter was placed between the two auxotrophic strains that would not allow contact.

**14.** In *generalized transduction* virtually any genetic element from a host strain may be included in the phage coat and thereby be transduced. In *specialized transduction* only those genetic elements of the host that are closely linked to the insertion point of the phage can be transduced. Because only certain genetic elements are involved in specialized transduction, it is not useful in determining linkage relationships.

**16.** When there is a sufficiently high number of infecting viruses so that there is a high likelihood that more than one type of phage will infect a given bacterium.

**18.** **(a)** The concentration of phages is greater than $10^4$. **(b)** The concentration of phages is around $1.4 \times 10^6$. **(c)** The concentration of phages is less than $10^6$.

**20.** **(a)** $2 \times 3 =$ no lysis; $2 \times 4 =$ lysis; $3 \times 4 =$ lysis. **(b)** Major alteration in the gene such that both cistrons are altered. **(c)** $5 \times 10^{-4}$. **(d)** Because mutant 6 complemented mutations 2 and 3, it is likely to be in the cistron with mutants 1, 4, and 5. Mutant 6 is probably a deletion that overlaps mutation 4.

## CHAPTER 20 Extrachromosomal Inheritance

**2.** All of the offspring must have the phenotype of the mother's genotype, which is dextral.

**4.** **(a)** green; **(b)** white; **(c)** variegated (patches of white and green); **(d)** green

**6.** 1:1 ratio of normals to petites

**8.** **(a)** There are many similarities among mitochondrial, chloroplast, and prokaryotic molecular systems. **(b)** The $mt^+$ strain is the donor of the cpDNA.

**10.** The female provides so much vital material and information to the egg, including the cytoplasm necessary for germline

determination, it is not surprising that such maternal effect genes exist.

## CHAPTER 21 Genetic Control of Development

**2.** One "selector" gene distinguishes aristal from tarsal structures. In transdetermination notice that a "one-step" change is involved in the interchange of leg and antennal structures.

**4.** Protein products, chromosomal puffs in some cases, labeled probes (*in situ* hybridization)

**6.** That segment polarity genes are influenced by *ftz* indicates that *ftz* functions earlier, thus placing *ftz* in the pair-rule group of genes.

**8.** The essential function of the *spoIVCA* gene appears to be the production of a rearranged *sigK* gene.

## CHAPTER 22 Genetics of Immunity

**2.** 18 more combinations, 27 total possible

**4.** $1000 \times 1000 = 10^6$

**6.** *Helper T cells* "turn on" the immune response. *Suppressor T cells* are the "off" switch of the immune system.

**8.** Individuals with type AB blood do not produce antibodies against the A or B antigens. Since type O blood does not have A or B antigens, it can be given to all other blood types (assuming other antigens or factors are compatible).

**10.** The *HLA haplotype* is an array of HLA alleles on a given copy of chromosome 6 in humans.

## CHAPTER 23 Genetics of Behavior

**2.** *Drosophila* = immense repertoire of chromosomal and single-gene alterations, and a fairly elaborate set of various behaviors (reproductive, locomotor, taxic, etc.).

**4.** 1/2 *UuRR* = nonhygienic, 1/2 *uuRR* = uncap only. For *Ur × UURr*: 1/2 *UURr* = nonhygienic, 1/2 *UUrr* = remove only. For *uR × Uurr*: 1/2 *UuRr* = nonhygienic, 1/2 *uuRr* = uncap only.

**6.** Dominant gene

**8.** Conditioning or complex genetic factors may have influences. There may also be different interactions with environmental stimuli from subtle genetic variations.

## CHAPTER 24 Population Genetics

**2.** $AA = 0.25$ or 25%; $Aa = 0.5$ or 50%; $aa = 0.25$ or 25%

**4.** The population must be in equilibrium.

**6.** **(a)** $MM = 0.6014$ or 60.14%; $MN = 0.3482$ or 34.82%; $NN = 0.0504$ or 5.04% (in equilibrium). **(b)** $AA = 0.7691$ or

76.91%; $AS = 0.2157$ or 21.57%; $SS = 0.0151$ or 1.51%; $\chi^2 = 1.47$ (in equilibrium).

8. **(a)** $q_1 = 0.23$, $p_1 = 0.77$; **(b)** $q_1 = 0.267$, $p_1 = 0.733$; **(c)** $q_1 = 0.293$, $p_1 = 0.707$; **(d)** $q_1 = 0.299$, $p_1 = 0.701$

10. **(a)** $p_1 = 0.5$; **(b)** $p_1 = 0.35$; **(c)** $p_1 = 0.11$

12. Reduction in fitness, because of increased homozygosity, observed in populations that are inbred.

16. Self-fertilization or brother-sister matings

18. The frequency of a gene is determined by a number of factors including the fitness it confers, mutation rate, and input from migration.

## CHAPTER 25 Evolutionary Genetics

4. Because of degeneracy in the code, there are some nucleotide substitutions, especially in the third base, that do not change amino acids. In addition, if there is no change in the overall charge of the protein, it is likely that electrophoresis will not separate the variants.

6. All of the amino acid substitutions (Ala → Gly, Val → Leu, Asp → Asn, Met → Leu) require only one nucleotide change. The last change from Pro (CC−) → Lys (AAA, G) requires two changes (the minimal mutational distance).

8. DNA sequence divergence is not always directly proportional to morphological, behavioral, or ecological divergence. While the genus classifications provided in this problem seem to be invalid, other factors, well beyond simple DNA sequence comparison, must be considered in classification practices.

12. Debate and controversy are natural components of scientific understanding. Some genes (like histones) will not tolerate nucleotide substitutions to a significant degree and the neutral mutation theory will not hold. However, there are other genes that produce quite variable products and provide support for the neutral mutation theory.

# INDEX

genetic material in, 6, 313, 318
herbicide resistance and, 433
Holmes ribgrass (HR) virus, 251
mutations, 347
recombination in, 614–616
SV40 virus, 314
temperature effect on, 211
tobacco mosaic virus (TMV), 446
transduction, *see* Viral transduction
types of, *see specific types of viruses*
Vitamin D, 553
Volkin, Elliot, 456
*v-onc* gene, 577, 580
Von Tschermak, Eric, 5, 62, 383
von Winiwarter, H., 166
*v-src* gene, 577

W

Wagner, Moritz, 757
Wallace, Alfred Russel, 62, 731
Walzer, Stanley, 168–169
Wang, Andrew, 264
Wang, James, 312
Watermelons, seedless, 183
Watson, James:
    DNA studies, 241
    DNA structure model, 257–261, 281–282, 350, 623
    genetic material study, 255–256

Watts-Tobin, R. J., 447
Weinberg, Wilhelm, 731, 733
Weismann, August, 4–5
Weiss, Samuel, 457
Werner syndrome, 584
Western blot, 399
Wheat:
    advances in, 12–13
    amino acid composition, 346
    quantitative characteristics in, 219
    and rye hybridization, 184–185
White, Michael, 763
White, Ray, 419
Whitehouse, Harold L. K., 302
Wiener, Alexander, 89
Wild type allele, 83
Wilkins, Maurice:
    DNA x-ray diffraction analysis, 256–257
    genetic material study, 255
Wilms Tumor (WT), 572, 574–577
Wilms Tumor gene, 575–577
Wilson, Allan, 769
Wobble hypothesis, 453, 474
Woff, Casper, 3
Wollman, Ellie, 601–602
Wood, William, 610
Woods, Philip, 284–286
Wright, Sewall, 744–745

X

X chromosome:
    in *Drosophila*, *see Drosophila*, X chromosome in
    genes on, 103–105
    in humans, 166
    meiosis and, 25
    *sgl* gene on, 175
    structure of, 24
X-inactivation center (XIC), 173
X-linked agammaglobulinemia, 698–700
X rays:
    diffraction analysis, DNA structure study, 256–257, 262
    impact of, 149, 364–365
*Xenopus laevis*:
    clone selection, 395
    differentiation study, 652
    gene amplification in, 557
    mtDNA in, 316
    oogenesis in, 557–558
    ribosomes in, 189
    RNA transcription, 471
    tandem repeat families in, 334
    transplantation and, 654
    zinc fingers in, 545
Xerodoma pigmentosum (XP), ultraviolet radiation and, 362–364
XIST (X-inactive specific transcript), 173

XYY individuals, frequency of, 168

Y

Y chromosome:
    discovery of, 166
    in humans, 166
    male development, 169–170
    meiosis and, 25
    structure of, 24
Yanofsky, Charles, 499, 525, 527
Yeast, *see Saccharomyces cerevisiae*
Yeast artificial chromosome (YAC), gene transfer, 403–405
Young, Michael, 372
Yule, Gudny, 100, 731

Z

Z-DNA, 262
*Zea mays*, *see* Corn
Zinc fingers:
    binding of, 554
    defined, 545
*ZIP1*, 40
Zygote(s):
    defined, 25
    diploid, 223
    production of, 36
Zygotene stage of meiosis, *see* Meiosis, zygotene stage of

# CREDITS

Region of Chromosome 17." Science 236:1100-1102, Fig. 1. © 1987 by the American Association for the Advancement of Science.

**p. 430** David Parker/SPL/Photo Researchers

**p. 432** Courtesy of Bayer AG, Germany

**p. 434** Charles J. Arntzen, Ph. D. Director, Plant Bio-technology Program at the Institute of Biosciences and Technology, Houston, Texas.

**p. 434** Dr. Edwin Westbrook, Argonne National Laboratory, Argonne, Ill.

PART III

**p. 440** M. Wurtz/Biozentrum, Univ. of Basel/SPL/Science Source, Photo Researchers

CHAPTER 14

**p. 444** Loren Williams, MIT/MAC DUPE

**p. 462** from Miller, Hankalo, and Thomas, 1970. © 1970 by the American Association for the Advancement of Science.

**p. 463** from Miller & Beatty, 1969, by permission of Alan R. Liss, Inc.

**p. 466** Courtesy Bert W. O'Malley

**p. 478** from Rich et al., 1963. Reproduced by permission of the Cold Springs Harbor Laboratory.

**p. 478** Courtesy of E.V. Kiselva

CHAPTER 15

**p. 486** J. Gross, Biozentrum/SPL/Science Source/Photo Researchers

**p. 494** Dennis Kunkel/CNRI/Phototake

**p. 494** Francis Leroy, Biocosmos/Photo Researchers

**p. 501** Frank M. Raushel/Dept. of Chemistry, Texas A&M University

**p. 505** J. Gross, Biozentrum/SPL/Photo Researchers

CHAPTER 16

**p. 514** Lee D. Simon/Science Source, Photo Researchers

**p. 516** Visuals Unlimited

CHAPTER 17

**p. 538** From Paul B. Sigler et al. Nature vol. 365, 10/7/93, p. 487. Photo courtesy of Paul B. Sigler, Dept. of Molecular Biophysics and Biochemistry and the Howard Hughes Med. Institute, Yale University

**p. 558** Dr. Donald D. Brown, Carnegie Institution of Washington

**p. 560** Dr. Suzanne McCutcheon/sent to us by Dr. Michael R. Cummings

CHAPTER 18

**p. 568** Copyright Boehringer Ingelheim International GmbH, Photo Lennart Nilsson, "THE BODY VICTORIOUS," Dell Publishing Company.

**p. 579** Prof. Sung-Hou Kim/Department of Chemistry, University of California, Berkeley

**p. 585** Science Photo Library/Photo Researchers

PART 4: Genetic Analysis

**p. 592** From M. Kuroda, M. Kernan, R. Kreber, B. Ganetzky, and B. Baker, Cell 66, 1991, 95; cover of Sept. 6, 1991 issue.

CHAPTER 19

**p. 594** Dr. L. Caro/SPL/Science Source, Photo Researchers

**p. 600** Dr. L. Caro/SPL/Photo Researchers

**p. 611** W. Wurtz, Biozentrum, University of Basel/SPL/Photo Researchers

**p. 613** Bruce Iverson

**p. 615** from Hershey & Chase, 1951.

CHAPTER 20

**p. 632** Courtesy of Dr. Ed Coe, Curtis Hall, University of Missouri, Columbia, Mo.

**p. 637** Courtesy of Dr. Ed Coe, Curtis Hall, University of Missouri

**p. 639** Raymond B. Otero/Visuals Unlimited

**p. 640** Eduardo Bonilla, M.D., College of Physicians & surgeons of Columbia University

**p. 641** A.M. Seigelman/Visuals Unlimited

CHAPTER 21

**p. 646** Courtesy of James Langeland, Stephen Paddock, and Sean Carrol, HHMI, University of Wisconsin

**p. 648** Dr. Tony Brain/SPL/Photo Researchers

**p. 660** F.R. Turner Ph.D. Department of Biology, Indiana University

**p. 661** Dr. William S. Klug/Biology Dept., Trenton State College, Trenton, N.J.

**p. 665** Courtsey of James Langeland, Stephen Paddock & Sean Carroll, HHMI, University of Wisconsin. (MAC DUPE)

**p. 668** from Kaufmann, T. et al. 1990. Advanced Genetics 27: 309-362, Department of Biology, Indiana University.

**p. 671** Erik Jorgensen, Ph.D./Laboratory of H. Robert Horvitz, M.I.T. MAC DUPE

CHAPTER 22

**p. 682** Lennart Nilsson/Courtesy Beohringer Ingelheim International GmbH.

**p. 684** Lennart Nilsson/Courtesy Beohringer Ingelheim International GmbH.

**p. 686** Institute Pasteur/Phototake

**p. 687** Lennart Nilsson/Courtesy Beohringer Ingelheim International GmbH.

**p. 687** Gilla Kaplan/Rockefeller University

**p. 699** Baylor Collection of Medical Photography/Peter Arnold, Inc.

**p. 700** Hans Gelderblom/Visuals Unlimited

**p. 701** Hans Gelderblom/Visuals Unlimited

**p. 701** Hans Gelderblom/Visuals Unlimited
**p. 701** Hans Gelderblom/Visuals Unlimited

CHAPTER 23

**p. 706** P. Candido and Eve G. Stringham, The Journal of Experimental Zoology 266:227-233 (1993). Fig. 2, part 6, p. 229
**p. 710** Jerry Hirsch, Dept. of Psychology University of Illinois at UrbanaChampaign.
**p. 722** A.M. Siegelman/Visuals Unlimited
**p. 722** A.M. Siegelman/Visuals Unlimited

CHAPTER 24

**p. 730** George Holton/Photo Researchers
**p. 742** Breck P. Kent
**p. 742** Breck P. Kent

CHAPTER 25

**p. 752** Edward S. Ross
**p. 755** no credit
**p. 768** Kenneth Kaneshiro/University of Hawaii, Dept. of Entomology
**p. 768** Kenneth Kaneshiro/University of Hawaii, Dept. of Entomology

Appendix A

**p. 220**
**p. A-2**  Ralph M. Sinibaldi/MRC
**p. A-9**  Ralph M. Sinibaldi/MRC
**p. A-10** Courtesy of Suzanne McCutcheon

# WATSON-CRICK MODEL OF DNA

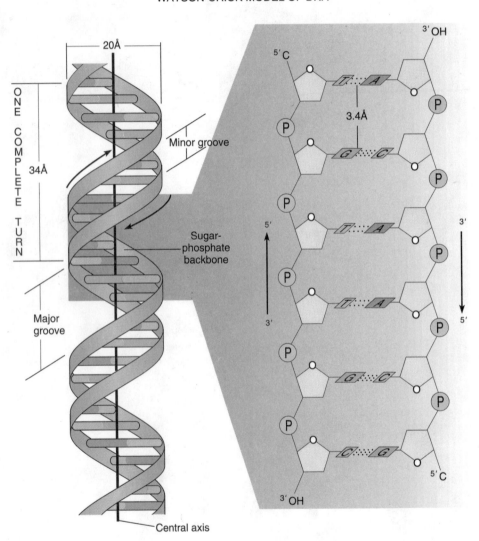